FEATURED REVIEWS

in Mathematical Reviews

1997–1999

Featured Reviews in
Mathematical Reviews

FEATURED REVIEWS

in Mathematical Reviews

1997–1999

With Selected Reviews
of Classic Books and Papers
from 1940–1969

Edited by
Donald G. Babbitt and Jane E. Kister

American Mathematical Society
Providence, Rhode Island

2000 *Mathematics Subject Classification.* Primary 00–XX.

ISBN 0-8218-2166-0
ISSN 1094-3706

∞ The paper used in this book is acid-free and falls within the guidelines
established to ensure permanence and durability.
Visit the AMS home page at URL: http://www.ams.org/

10 9 8 7 6 5 4 3 2 1 05 04 03 02 01 00

CONTENTS

PREFACE

This second collection of reprints of Featured Reviews published in *Mathematical Reviews* makes widely available informed reviews of some of the best mathematics published recently. This volume contains those reviews published in 1997–99. (The first volume contained the Featured Reviews from 1995 and 1996.)

Featured Reviews were introduced in MR in part to provide some guidance to the literature. With the exponential growth of publications in mathematical research in the first half-century of MR, it had become essentially impossible for the user of MR to identify in the monthly MR issue (or in MathSciNet or MathSciDisc) the most noteworthy books and papers. Active researchers are often familiar with much of what is going on within their own field but even they may have difficulties finding out about outstanding recent publications across the areas covered by MR. By identifying some of the "best" new publications, books and papers that are expected to have a significant impact on the area of pure or applied mathematics with which they are concerned and which contain interesting new ideas or applications or a deep synthesis of existing ideas or a combination of these, it is intended to lead the user to important new research across all fields covered by MR.

The MR editors, sometimes using suggestions from distinguished outside mathematicians, are responsible for the initial selection of potential Featured Review items. For each such item, a leading expert in the field is consulted for advice before a final decision is made. Frequently, the expert goes on to write the Featured Review. The reviewer is asked to set the item in context, perhaps by giving some historical background, state the main results, outline (in not too technical a fashion) the main new ideas in the item and include an evaluation of the item. The review should be accessible to more than just the experts in the area.

We would like to take this opportunity to acknowledge that many papers worthy of Featured Review treatment are not given a Featured Review. The MR editors cannot, even with outside advice, identify correctly **the** best papers among the more than 5,000 they consider each month. Although some papers are recognized at the preprint stage as being important contributions to mathematics, there are others, often the most innovative, whose worth is not appreciated until long after publication. In some cases, even when a paper has been selected for Featured Review treatment, it may be impossible to acquire a review of the right caliber within a reasonable time frame. The importance of timely reviews may sometimes require that then a "regular" review be published.

The main part of this volume is divided into six sections, A–F, with each section covering an area of mathematics represented by several related sections of the 1991 Mathematics Subject Classification (1991 MSC), the classification used in MR in 1997–99. Each section opens with a list of the Featured Reviews in that part. The reviews within each section are printed in the order used in MR (except that in section A reviews in computer science follow immediately after those in logic and precede those in number theory); that is, the reviews are in classification order and within a given 5-digit

MSC classification, alphabetically by first author. (Further information on the 1991 MSC can be found in the 1998 MR Annual Index or on e-MATH (www.ams.org).)

This volume also contains a supplementary section, G, which contains reprints of selected reviews published in 1940-69. (These reviews are ordered first by their original 2-digit classifications and within each classification in MR number order.) To be selected for inclusion in this section, the item under review was required to be of considerable mathematical significance; in addition, the review had to be of the caliber for a Featured Review. (Many seminal items are not included because the review was not appropriate for this volume.) The MR editors gave valuable advice on which reviews to include but the final choice was ours. We hope that readers will find that this compilation of reviews of some of the most important mathematics during this 30-year period provides an informative and enjoyable overview of what was in many areas of mathematics a "golden age". (A similar compilation of reviews, published in 1970–94, is planned for the next volume of Featured Reviews.)

The references within the reviews have been updated whenever possible. Also, a reference to another review reprinted in this volume includes the paging on which the review appears. Author and reviewer indexes appear at the end of the volume.

We would like to take this opportunity to thank the many mathematicians who have supported Featured Reviews in MR by suggesting papers to be given a Featured Review and by advising the MR editors on items the MR editors have selected for consideration for Featured Review treatment. Featured Reviews would not appear without the continuing efforts of the MR staff, and most particularly the editors.

The authors of this collection, the distinguished mathematicians who wrote the Featured Reviews published in 1997–99 and the earlier reviews selected for inclusion here, deserve our particular thanks on behalf of the mathematical community for their considerable volunteer efforts. The reviews are informed pieces of scholarship that taken together provide both a glimpse of the mathematics of the mid twentieth century and a wide-ranging view of some important mathematical advances of the last several years.

Donald G. Babbitt
Jane E. Kister

A. REVIEWS IN LOGIC, COMPUTER SCIENCE AND NUMBER THEORY

98m:03002 03-02 03E05 03E35 03E40 03E45 03E50
Shelah, Saharon (Jerusalem)
★**Proper and improper forcing. (English summary)**
Second edition.
Perspectives in Mathematical Logic.
Springer-Verlag, Berlin, 1998. xlviii+1020 pp. $189.00. ISBN 3-540-51700-6
Saharon Shelah's proper forcing has become a staple part of the methods of modern set theory, with its applications powerful and wide-ranging and the development of its theory a fount of continuing research. Shelah came to proper forcing in the late 1970's, and in a timely tract (the first edition of the present text) [*Proper forcing*, Lecture Notes in Math., 940, Springer, Berlin, 1982; MR 84h:03002] communicated the subject to an excited set theory community. The tract in fact went far beyond its title to bring together much of Shelah's work in set theory to that time, and the wealth of information and techniques therein greatly stimulated set-theoretic research in the ensuing years.

Time passed, and with Shelah's inimitable English and expositional gaps having become legion and a steady stream of new results having been established, it became apparent that a mature, magisterial presentation of proper forcing in all of its aspects was needed. The book under review is the result, far larger in scope than even the original tract and most welcome as the authoritative exposition, the more so after several years of being held up because of word processing difficulties. The appearance, finally, of the book serves as an occasion not only to describe Shelah's work on proper forcing, but also to reflect on Shelah's standing in and approach to set theory.

Saharon Shelah is a phenomenal mathematician, preeminent in both model theory and set theory. His work, beginning in the early 1970's, has tremendously advanced both subjects, and even now, in his fifties, he is going from strength to strength, continuing to produce results at a furious pace. He has over 600 items in his bibliography, the majority of them long or substantial papers, and although many are jointly authored, it is often the case that he established the lion's share of the results presented. (Shelah's recent papers in LaTeX files together with input files can be retrieved by anonymous ftp at math.rutgers.edu in the directory pub/shelah/all. Lists of his papers are also at math.rutgers.edu in the directory pub/shelah. On the WWW, the above together with a list of papers in hypertext format can be found at http://math.rutgers.edu/~shelah/.)

In set theory Shelah is initially stimulated by specific problems. He typically makes a direct, frontal attack, bringing to bear extraordinary powers of concentration, a remarkable ability for sustained effort, an enormous arsenal of accumulated techniques, and a fine, quick memory. When he is successful on the larger problems, it is often as if a resilient, broad-based edifice has been erected, the traditional serial constraints loosened in favor of a wide, fluid flow of ideas and the final result almost incidental to the larger structure. What has been achieved is more than just a succinctly stated theorem but rather the establishment of a whole network of robust constructions and arguments. A telling point is that when some local flaw is pointed out to Shelah, he is usually able to come up quickly with another idea for crossing that bridge.

Shelah's written accounts have acquired a certain notoriety that in large part has to do with his insistence that his edifices be regarded as autonomous mental constructions. Their life is to be captured in the most general forms, and this entails the introduction of many parameters. Often, the network of arguments is articulated by complicated combinatorial principles and transient hypotheses, and the forward directions of the flow are rendered as elaborate transfinite inductions carrying along many side conditions. The ostensible goal of the construction, the succinctly stated result that is to encapsulate it in memory, is often lost in a swirl of conclusions. This can make for difficult and frustrating reading, with the usual problem of presenting a mathematical argument in linear form exacerbated by the emphasis on the primacy of the construction itself and its overarching generality.

Further difficulties ensue from the nature of the enterprise. Shelah regards the written word as necessary and central for capturing and fixing a construction, and so for him getting everything down on paper is of crucial importance. The tensions among the robustness of the construction, the variability of its possible renditions, and the need to convey it all in print are inevitably complicated by the speed with which he is able to establish new results. The papers have to be written quickly, previous constructions are newly refreshed and modified, and so a labyrinthian network may result over a series of related papers. In mathematics one often aspires to the most elegant or definitive treatment; in contrast, Shelah's work features a continuing, dynamic self-dialogue, one that pushes to the limits of exposition.

Many may consider Shelah's work to be "technical", but as T. S. Eliot has written "We cannot say at what point 'technique' begins or where it ends" [*The Sacred Wood*]. While there is a particular drive to solve specific problems, Shelah with his generalizing approach is able to draw out larger, recurring patterns that lead to new techniques that soon get elevated to methods. One primary instance is the whole complex of approaches and results he developed under the general rubric of proper forcing.

Shelah started out in model theory, developing an abstract classification theory for models which is a continuing research program for him and model theorists to this day. In the mid-1970's, in his first major body of results in set theory, Shelah resolved a long-standing problem in abelian group theory, Whitehead's problem, by establishing both the consistency and the independence of the corresponding proposition. It is through these beginnings, motivated by the set-theoretic problems that arose, that Shelah started to develop a general theory of iterated forcing for the continuum.

After Cohen invented forcing in 1963, a plethora of relative consistency results were established by this new method of extending models of set theory by introducing new sets in a carefully controlled manner. It was quickly found that iterated forcing held the key to establishing the consistency of many propositions of the form $\forall x \exists y$ by iteratively creating for each x a corresponding y, taking care of every x that could arise during the iteration. For the iteration to be efficacious some of the underlying structure would have to be preserved, e.g. that the cardinals of the original model remain cardinals in the final extension.

Analyzing in particular Jensen's argument for the consistency of the continuum hypothesis plus Suslin's hypothesis, Shelah hit upon properness as a general property that preserves cardinals and is moreover preserved through iterations, particularly countable support iterations. A succinct, though not very useful, formulation can be given: A notion of forcing is proper iff it preserves the stationary subsets of $\mathscr{P}_{\omega_1}\lambda$ for every uncountable cardinal λ. (Here, $\mathscr{P}_{\omega_1}\lambda$ denotes the collection of countable subsets of λ, and there is a well-known concept of stationariness of subsets of $\mathcal{P}_{\omega_1}\lambda$.) The useful formulation is in the model-theoretic terms of countable elementary substructures of structures $\langle H(\mu), \in, <^*, \cdots\rangle$, where $H(\mu)$ denotes the collection of sets hereditarily of cardinality less than μ and $<^*$ is a well-ordering of H_μ. In fact, the elaboration of forcing in terms of model-theoretic approximations is the elegant basis of Shelah's conceptualization, and his approach of taking elementary substructures of $H(\mu)$'s has become a convenient and familiar strategem in set-theoretic research.

The book tries to be self-contained beyond a beginning course in post-Cohen set theory, and though rich and thick in content is pleasing to the eye. There is a guiding annotated table of contents, and each of its 18 chapters has a helpful introduction that establishes the historical and mathematical context with citations and accreditations of the published literature. The reviewer cannot hope to provide a substantively informative description of its one thousand pages, but the following description paints it in very broad strokes and emphasizes the expansions beyond the 1982 tract.

The first two chapters provide the basics of forcing, and Shelah takes the opportunity to include some of his more recent consistency results that are presentable at this level. Chapter III introduces proper forcing and establishes the first relative consistency results with it. Chapter IV describes oracle-chain-condition forcing, rather set apart from proper forcing, and its main application to establishing the consistency of "$\mathcal{P}(\omega)$/finite has no (nontrivial) automorphisms". Chapter V is a first incursion into variants of properness and the particular problem of preserving the property of "not adding any new reals" through an iteration. Chapter VI comes to grips with general preservation theorems for preserving properness plus φ for various properties φ of reals. One pivotal property is "not adding any new function $f:\omega \to \omega$ that eventually dominates every old function". This is the first substantial new development beyond the 1982 tract, and, in connection with it, new sections have been added that present some of Shelah's more recent results about various consistencies in the ordering of cardinal invariants of the continuum. Chapters VII and VIII give axiomatic presentations of variants of proper forcing in the style of Martin's axiom, pursuing the idea of having applicable black boxes for establishing consistency results and illustrating this approach with applications. Chapter IX introduces the general notion of a free limit for iteration, and provides important applications to special and non-special Aronszajn trees. Chapter X introduces semi-proper forcing, an important local version of proper forcing that allows for changing the cofinality of cardinals $\geq \aleph_2$ to ω. The relevant form of iteration is revised countable support. Shelah establishes the preservation theorems and the early consistency results relative to large cardinals. Chapter XI then sharpens the work to equi-consistency results, and Chapter XII makes a remarkable connection between the semi-properness of Namba forcing and Chang's conjecture.

From Chapter XIII on, Shelah shifts the focus from results in his 1982 tract to results established in more recent papers. Semi-proper forcing achieved a new notoriety with its use by M. Foreman, M. Magidor and Shelah [Ann. of Math. (2) **127** (1988), no. 1, 1–47; MR 89f:03043; correction; MR 90d:03110]. That paper not only established a maximal form of Martin's axiom called Martin's Maximum, but also featured new consistency results relative to large cardinals and overturned a long-held view about how large cardinal properties would have to be scaled down. Hugh Woodin, stimulated by this work, then established his definitive relative consistency results about the axiom of determinacy. Chapter XIII, a reworking of a published paper, establishes results about ideals over ω_1 that interweave through these developments. Chapter XIV, a reworking of another published paper, explores possible generalizations of revised countable support to uncountable supports. There are only weak analogues of properness that are preserved, and this is connected with a basic problem: consistency results with proper forcing generally result in the size of the continuum being $\aleph_2$, and so the consistency of various propositions is unknown for a large continuum. Chapter XV provides a means of generalizing the work of Chapter XI to incorporate the negation of the continuum hypothesis. Chapter XVI sharpens the work of Chapter XIII; while the earlier chapter had liberally applied the consistency strength of supercompact cardinals, the new chapter works at the level of what are now called Woodin cardinals. Chapter XVII is drawn from yet another published paper, and sorts out interconnections growing out of the Foreman-Magidor-Shelah paper among Martin's Maximum and forcing axioms for semi-proper forcing. Lastly, Chapter XVIII is the repository for how far Shelah can go in the direction of preserving the property "not adding any new reals" through iteration limit stages.

The great value of this book lies in its authoritative synthesis of the many aspects of properness. Given the time since the appearance of the 1982 tract, the book provides the definitive mature and comprehensive presentation of proper forcing, retiring several of Shelah's recent papers. For those who had struggled through the former tract, the book can present a satisfying, even serene, retrospective. In the end, however, we are still left to marvel, like Caliban peering at Prospero, at how all of this could have been accomplished, and, unable to take it all in, to gaze silent. *A. Kanamori* (Boston, MA)

99k:03043 03E45 03E10 03E55 04A10 04A15

Steel, John R. (Berkeley, CA)

★The core model iterability problem.

Lecture Notes in Logic, 8.

Springer-Verlag, Berlin, 1996. iv+112 pp. $49.95. ISBN 3-540-61938-0

This monograph is an essential building-block in what has been somewhat loosely described as the "inner model program", but which perhaps more informatively could be stated as the systematic investigation of inner models of the universe V of all sets under various axioms of strong infinity. The theory of such models has in the last two decades been the impetus for much of the current work at the forefront of set theory (by which we mean the theory of sets as axiomatised by Zermelo and Fraenkel, ZF, set theory, usually in these contexts with the addition of the axiom of choice—AC—yielding the theory ZFC). The reader is urged to refer to A. Kanamori's review of the previous monograph in this series, jointly authored by the present author and W. J. Mitchell [*Fine structure and iteration trees*, Springer, Berlin, 1994; MR 95m:03099], for a historical and motivational survey of what this program tries to achieve. We only briefly gloss this here. Gödel's universe of constructible sets L (with which he proved the consistency of the axiom of choice—as well as the generalised continuum hypothesis—that for every cardinal number κ, the size of the power set of κ was precisely the next possible cardinal number κ^+, together with the other ZF axioms) is the paradigm of an inner model: it has an extremely orderly structure, with a definable well-order of its reals, and a wealth of ordinal combinatorial structure as exemplified by its "fine structure" discovered by R. B. Jensen in the late 1960s. Jensen's landmark theorem, the "covering lemma" for L, showed how much of its cardinal structure was a true reflection of the cardinal structure of the whole universe V itself. The necessary condition for this theorem to be true was that L should admit no nontrivial elementary embeddings into itself. The next query was apparent: if this requirement failed, could one find a model that would generalise L and serve as a substitute "core" of the universe V? The first such core model, K, stepping out from L, was invented by Dodd and Jensen in the mid-1970s. This K possessed the same kind of fine structure as L and enjoyed a covering lemma, again under the assumption of its non-embeddability. The latter is equivalent to stating that a certain axiom of strong infinity (in this case that of a measurable cardinal) was not itself exemplified in some inner model of the ZFC axioms. Since this formative step, core models were constructed by Mitchell which accommodated the possibility that the universe V did in fact harbour inner models with many measurable cardinals, indeed up to the maximum number of possible measures on any one cardinal number. The goal of constructing a core model was seen to be a task undertaken assuming some "anti-large cardinal axiom" was true, constructing the model somehow, showing that it had the requisite pleasant properties, and proving the truth of some form of covering lemma for the model concerned. The failure of this lemma would then show that the "large cardinal axiom" indeed had to be true in some inner model.

A key feature of the construction of such models was the need for comparison of the putative components. The original Dodd-Jensen core model could be conceived of as an agglomeration of such components, albeit comparable by the method of iterated ultrapowers of Kunen. The construction of later models would involve an inductive process that required a method of comparing such models as an intrinsic part of the defining procedure. One can see that, roughly speaking, if one is to adjoin a normal measure F, say, to a set-sized initial segment M of a model to be, and if comparability is to be achieved using Kunen's theory of iterated ultrapowers, then we should require that M have well-founded ultrapowers by F and its images. An additional requirement is to choose the measures so that this well-founded iterability will last throughout the inductive defining process of further enlargements of the F sequence. As long as this involved a linear progression of inductively defining the next step or next measure (or its generalisation—the next "extender"), all was seen (not necessarily easily) to be well.

This well-being would last only as long as the models contained only large cardinals that were amenable to linear comparisons, but this would inevitably have to break down. Steel (1985) invented a comparison process that involved "trees" of models, which was a necessary step if large cardinals at the level of "Woodin" cardinals were to be involved. Such cardinals had been seen to be crucial in the developing theory of $L(\mathbf{R})$ through the work of Woodin, Foreman-Magidor-Shelah and Shelah (which we do not have the space to go into here).

To coin a slogan, the monograph previously cited, *Fine structure and iteration trees*, married this iteration process to a method of defining set models with a fine structure. The problem left open in that volume, and that alluded to in the title of the current volume, and here solved, is that of establishing conditions for the iterability of the desired full final model, and thus its very existence.

In one sense this is a solution to a technicality, but the monograph also provides a description of the core model "below" a Woodin cardinal that will itself be a model for further core models to come—as indeed it already has.

There are further applications to consistency results outlined in the book's introduction in §7. We first outline some of the fruits of this work—the applications, and then very briefly sketch the contents of the main body of the book. Shelah has shown that given a Woodin cardinal, there is a forcing extension of V which adds no reals but causes there to be a presaturated ideal on ω_1. Steel shows that the existence of such an ideal, coupled with that of a measurable cardinal Ω, implies the existence of an $M \subseteq V_\Omega$ with $M \models$ "there is a Woodin cardinal". The existence of such an Ω is an artefact of the solution to the iterability problem: some kind of assumption appears necessary, although the measurability of Ω may be weakened. The conclusion that there is a transitive set model of ZFC + there exists a Woodin cardinal may also be drawn if (a) Martin's Maximum holds; (b) the proper forcing axiom holds (via work of Schimmerling and Todorcevic); (c) every set of reals definable over $L_{\omega_1}(\mathbf{R})$ is weakly homogeneous; (d) there is a strongly compact cardinal; and (e) all reals have sharps and $\Sigma^1_3(x)$-separation holds for all reals x.

Subsequent work of Mitchell-Schimmerling-Steel has yielded a covering lemma for the K constructed here; work of Zeman and Schimmerling has shown that the rich combinatorial principle $\square_\kappa$ holds here also.

To briefly describe the volume's structure: the proof uses as template the construction by Mitchell of his model for measure sequences. A "background" model K^c is constructed first (this uses the measurable cardinal Ω), and the ultimate K will be a Skolem hull of this one. The inductive definition of K^c involves choosing partial extenders from V to put on the sequence being built which have sufficient strength to ensure the final iterability of K^c. This sufficiency has to be "certified" in a certain way by objects in V. (In the previous volume a model containing a Woodin cardinal was built, albeit assuming that there were sufficient full extenders in V for certification—the crucial distinction here that is the root of the "problem" requiring solution.) The extra assumption of the measurable Ω is needed to show that K^c is rich enough. The iterability results are outlined in §2, although the full proof is deferred to §9. Some preliminary work on the requisite hull and definability properties of Mitchell takes up §§3, 4 and the construction itself takes up §5. In §6 it is shown how to give an inductive definition of K which (under the requisite assumptions) yields the same model as the Skolem hull definition. This definition shows that $K \cap \mathrm{HC}$ is $\Sigma_1(L_{\omega_1}(\mathbf{R}))$ (known to be optimal by work of Woodin). §8 gives further information about K and its embeddability properties; §7 the applications already mentioned.

It is inevitable that the readership for this volume will be specialists or motivated would-be specialists in inner model theory. Throughout, reference is made to notation, concepts, and whole proofs from the previously cited monograph. Nevertheless the construction is a template for future work in the field, and is a landmark that belies its short 100-odd-page length. *Philip D. Welch* (Kobe)

99d:68077a 68Q15 03B35 03D15 68Q25
Arora, Sanjeev (Princeton, NJ); **Safra, Shmuel** (Tel Aviv)
Probabilistic checking of proofs: a new characterization of NP. (English summary)
J. ACM **45** (1998), *no.* 1, 70–122.

99d:68077b 68Q15 03B35 03D15 68Q25
Arora, Sanjeev (Princeton, NJ); **Lund, Carsten** (Murray Hill, NJ);
Motwani, Rajeev (Stanford, CA); **Sudan, Madhu** (Cambridge, MA);
Szegedy, Mario (Murray Hill, NJ)
Proof verification and the hardness of approximation problems. (English summary)
J. ACM **45** (1998), *no.* 3, 501–555.

These two papers make important contributions to two areas of complexity theory: the study of approximation algorithms and the study of probabilistic proof systems.

Complexity theory is concerned with the attempt to understand the nature of efficient algorithms, procedures and systems. This attempt is at the very heart of theoretical computer science. Arguably, the most important types of algorithms are those designed to solve optimization problems. Fundamental works by Cook, Karp and Levin, done in the early 1970's, have provided strong systematic evidence for the intractability of many important optimization problems by showing them to be NP-hard [see M. R. Garey and D. S. Johnson, *Computers and intractability*, Freeman, San Francisco, Calif., 1979; MR 80g:68056]. Approximation algorithms are supposed to bypass the difficulty of finding the best solution to an optimization problem by finding a solution which is "nearly" the best one. The main question to be investigated in this context concerns the trade-off between the running-time of an algorithm and the quality of the approximation that it can achieve. However, very little progress in resolving this question was made in the 1970's and 1980's. The connections established by U. Feige et al. [J. ACM **43** (1996), no. 2, 268–292; MR 97h:68037] (and further developed in the second paper under review) between probabilistically checkable proof (PCP) systems and the difficulty of approximating several central optimization problems, and the results concerning such proof systems, have provided a breakthrough in the investigation of approximation algorithms.

PCP systems are proof systems which allow super-fast verification by probing the proof at very few random locations. Thus, the alleged proof is in redundant form, and the verification procedure is efficient and probabilistic. For every valid statement there exists a valid proof which is always accepted by the verification procedure, whereas for a non-valid statement each false proof is rejected with probability at least $1/2$ (taken over the coin tosses of the verification procedure). Furthermore, proofs relative to any automatic verification procedure (known as "NP-proofs") can be efficiently transformed into proofs of the form above. Also, by going over all possible random choices, one may regain absolute certainty in the validity of a probabilistically checkable proof. The most important parameters of a PCP system are the number of coins tossed by the verification procedure and the number of locations (in the alleged proof) probed by the procedure.

The main result of the first paper is that any NP-proof system can be transformed into a PCP system in which both parameters mentioned above are logarithmic in the length of the original proof (and/or the claimed statement). In the second paper the number of probes is reduced to an absolute constant (independent of the statement to be proven). Subsequent works have reduced this constant to 5 (cf. the two papers under review as well as the book by the reviewer [*Modern cryptography, probabilistic proofs and pseudorandomness*, Springer, Berlin, 1999] for a survey of subsequent improvements and a wider perspective).

Using the connection established by Feige et al., it is shown that, for some constant $c > 0$, the problem of approximating the size of a maximum clique in an N-vertex graph up to a factor of N^c is NP-hard (and thus probably infeasible). Subsequent work has shown that c may be any constant smaller than 1. The second paper also presents a new connection between PCP and approximation. Via this connection a large number of optimization problems (in the class known as MaxSNP) are shown to be NP-hard to approximate (to within some problem-specific constants). Again, subsequent work has improved on these constants, reaching tight results for some natural optimization problems (e.g., Max3SAT).

Thus the works mentioned above are milestones in the study of probabilistically checkable proof systems and their relation to the difficulty of approximating several central optimization problems, which in turn constitutes the most important development in complexity theory in the current decade. Both papers are also very interesting from a technical point of view. The first paper introduces the paradigm of "proof composition" as a tool in the design of PCP systems. Indeed this paradigm has played a key role in all subsequent developments. Both papers present important improvements in the analysis of "low-degree tests" (i.e., algorithms which, by probing a given function at a few random inputs, test whether it is a low-degree polynomial or far from any such polynomial). Finally, the second paper presents two PCP systems which are especially amenable to "proof composition" and are aimed at achieving very probe-efficient PCP systems. *Oded Goldreich* (Rehovot)

99a:68088 68Q25 94C10
Håstad, Johan (Stockholm)
The shrinkage exponent of De Morgan formulas is 2. (English summary)
SIAM J. Comput. **27** (1998), *no.* 1, 48–64 (electronic).

Proving nontrivial lower bounds on the computational complexity of specific computing problems in the framework of various computational models is of fundamental value to the understanding of computation as well as to the classification of computing tasks according to their difficulty. Proving a lower bound on the amount of computing resources needed to solve a concrete computing problem corresponds to proving the nonexistence of efficient computations solving the given problem. As usual in the natural sciences (and especially in mathematics), the proofs of nonexistence are some of the hardest ones. So, it is not surprising that the fundamental problems (like P versus NP) remain open.

This paper is devoted to the development of lower bound proof methods. The complexity measure considered is the length (size) of a De Morgan formula. A De Morgan formula can be viewed as a binary tree in which each leaf is labeled by a literal from the set $\{x_1, \cdots, x_n, \overline{x}_1, \overline{x}_2, \cdots, \overline{x}_n\}$ of Boolean variables and their negations. Each internal node v is labeled by an operation which is either $\wedge$ or $\vee$. The size of such a formula is defined as the number of leaves. For a long time the highest lower bound for the formula size of a specific function was the $\Omega(n^2)$ bound obtained by V. M. Khrapchenko in 1971 [Mat. Zametki **9** (1971), 35–40; MR **44** #8051; Mat. Zametki **10** (1971), 83–92; MR **44** #5183]. In 1987 A. E. Andreev [Vestnik Moskov. Univ. Ser. I Mat. Mekh. **1987**, no. 1, 70–73, 103; MR 88c:68033] observed that the old method found in a paper by B. A. Subbotovskaya [Soviet Math. Dokl. **2** (1961), 110–112; MR **25** #1088] could provide much more than the lower bounds proved in that paper, and as a consequence the new record lower bound $\Omega(n^{2.5-o(1)})$ was established [A. E. Andreev, op. cit.]. The crucial idea of this lower bound method is based on the creation of a random restriction. One, independently for each variable, keeps it as a variable with a probability p and otherwise assigns it the constant value 0 or 1 with the same probability $(1-p)/2$. The question is what the expected formula size of the induced function will be after the application of a random restriction according to a probability p. Any improvement of the estimation on the size of the formulas obtained results in a higher lower bound. Motivated by Andreev [op. cit.], a few further papers achieved some improvements of lower bounds in this way.

It seems that in this paper Johan Håstad closes the effort in this direction. He gives a precise analysis of the probability of reducing the size of the formula. As a consequence the lower bound $\Omega(n^{3-o(1)})$ on the formula size of the specific function considered by Andreev is established. This is probably the best possible lower bound that could be provided by this proof method. The paper is very well written and I strongly recommend it to everyone interested in proving lower bounds on complexity.

Juraj Hromkovič (Aachen)

98k:68150 68T15 03-04 03B35 03G05 06E05

McCune, William (Argonne, IL)

Solution of the Robbins problem. (English summary)

J. Automat. Reason. **19** (1997), *no.* 3, 263–276.

With the help of his automated theorem proving program EQP for equational logic, the author has solved affirmatively Problem 1.1 on page 245 of [L. Henkin, J. D. Monk and A. Tarski, *Cylindric algebras. Part I*, North-Holland, Amsterdam, 1971; MR **47** #3171]. This problem asks whether an algebra $\langle A,+,-\rangle$, where $+$ is a binary commutative associative operation on A, and $-$ is a unary operation on A satisfying the "Robbins equation"

$$-[-(x+y)+-(x+-y)]=x,$$

is necessarily a Boolean algebra. This question was first asked by Herbert Robbins in about 1933. It is related to an old result of Huntington, which states that the commutative and associative laws and a postulate very similar to the Robbins equation do characterize Boolean algebras. Extra conditions that were previously known to be sufficient for the algebra to be Boolean include (1) A contains elements c and d such that $c+d=d$, and (2) A contains elements c and d such that $-(c+d)=-d$. These conditions were proved to be sufficient by S. Winker [J. Algebra **153** (1992), no. 2, 414–423; MR 93j:06015]. Winker obtained his proofs about 1980 with assistance from the automated reasoning program AURA [cf. L. T. Wos and S. Winker, in *Automated theorem proving (Denver, Col., 1983)*, 73–88, Contemp. Math., 29, Amer. Math. Soc., Providence, RI, 1984; see MR 85d:68005]. A more recent article by Winker [J. Automat. Reason. **6** (1990), no. 4, 465–489; MR 91i:68136] described in detail how automated reasoning assisted him in finding his proofs.

EQP's proof uses paramodulation with built-in associative-commutative (AC) unification, and simplification with built-in AC matching, to derive the equation $-[-(3x)-x]+2x=2x$ (where $2x$ is $x+x$ and $3x$ is $2x+x$) from the Robbins equation. This establishes Winker's sufficient condition (1). The first successful search for a proof established Winker's sufficient condition (2); it took almost 8 days of computer time, and used about 30 megabytes of memory. Subsequently the author found the proof of Winker's condition (1) that is presented in the paper; this proof comprises 12 steps, which are explained in detail in the paper, and took about 10 days of computer time. A proof that Winker's condition (1) is sufficient for the algebra to be Boolean, found automatically by EQP, is also included in the paper. EQP has also found direct proofs of Huntington's postulate from the Robbins equation, but these proofs are not included in the paper because of their length. The proofs in this paper were found automatically by EQP, unlike Winker's proofs, which were almost entirely done by hand.

Highlights of previous work in which automated solutions have been obtained to open questions in mathematics and logic are listed on pages 6 and 7 of McCune and R. Padmanabhan's *Automated deduction in equational logic and cubic curves* [Lecture Notes in Comput. Sci., 1095, Springer, Berlin, 1996; MR 98m:68238]. In comparison with these previously open questions, the question answered in the present paper has been longer outstanding, was more widely known, and seems to have been intrinsically more difficult.

This paper shows how a suitably programmed computer may be able to assist in solving a problem in equational logic that has been too difficult for unaided humans. McCune's solution has been described as "the first real breakthrough in automated theorem proving" (Stanley Burris), and as "the most remarkable result in automated theorem proving in 30 years" and "clearly a form of computer thinking" (Robert Boyer). Whether McCune's work may have philosophical implications remains controversial, but it does seem likely to have profound implications for theorem proving in equational logic. *J. A. Kalman* (Auckland)

99k:11004 11-06 11D41 11F80 11Fxx 11G05

★**Modular forms and Fermat's last theorem.**

Papers from the Instructional Conference on Number Theory and Arithmetic Geometry held at Boston University, Boston, MA, August 9–18, 1995.
Edited by Gary Cornell, Joseph H. Silverman and Glenn Stevens.
Springer-Verlag, New York, 1997. xx+582 pp. $49.95. ISBN 0-387-94609-8

Contents: Glenn Stevens, An overview of the proof of Fermat's last theorem (1–16); Joseph H. Silverman, A survey of the arithmetic theory of elliptic curves (17–40); David E. Rohrlich, Modular curves, Hecke correspondence, and L-functions (41–100); Lawrence C. Washington, Galois cohomology (101–120); John Tate, Finite flat group schemes (121–154); Stephen Gelbart, Three lectures on the modularity of $\overline{\rho}_{E,3}$ and the Langlands reciprocity conjecture (155–207); Bas Edixhoven, Serre's conjecture (209–242); Barry Mazur, An introduction to the deformation theory of Galois representations (243–311); Bart de Smit and Hendrik W. Lenstra, Jr., Explicit construction of universal deformation rings (313–326); Jacques Tilouine, Hecke algebras and the Gorenstein property (327–342); Bart de Smit, Karl Rubin and René Schoof, Criteria for complete intersections (343–356); Fred Diamond and Kenneth A. Ribet, l-adic modular deformations and Wiles's "main conjecture" (357–371); Brian Conrad, The flat deformation functor (373–420); Ehud de Shalit, Hecke rings and universal deformation rings (421–445); Alice Silverberg, Explicit families of elliptic curves with prescribed mod N representations (447–461); Karl Rubin, Modularity of mod 5 representations (463–474); Fred Diamond, An extension of Wiles' results (475–489); Fred Diamond and Kenneth Kramer, Appendix: classification of $\overline{\rho}_{E,l}$ by the j-invariant of E (491–498); H. W. Lenstra, Jr. and P. Stevenhagen, Class field theory and the first case of Fermat's last theorem (499–503); Michael Rosen, Remarks on the history of Fermat's last theorem 1844 to 1984 (505–525); Gerhard Frey, On ternary equations of Fermat type and relations with elliptic curves (527–548); Henri Darmon, Wiles' theorem and the arithmetic of elliptic curves (549–569).

The story of Fermat's last theorem (FLT) and its resolution is now well known. It is now common knowledge that Frey had the original idea linking the modularity of elliptic curves and FLT, that Serre refined this intuition by formulating precise conjectures, that Ribet proved a part of Serre's conjectures, which enabled him to establish that modularity of semistable elliptic curves implies FLT, and that finally Wiles proved the modularity of semistable elliptic curves.

The purpose of the book under review is to highlight and amplify these developments. As such, the book is indispensable to any student wanting to learn the finer details of the proof or any researcher wanting to extend the subject in a higher direction. Indeed, the subject is already expanding with the recent researches of Conrad, Darmon, Diamond, Skinner and others.

The book has 21 chapters. Chapters 18 and 19 are historical. Chapter 18 by Lenstra and Stevenhagen discusses class field theory and the first case of Fermat's last theorem. The early history (1844–1984) is given in Chapter 19 by Rosen, and it may be a good idea to begin the reading of this book with these chapters. The new ideas of the last two decades begin with Chapter 1 by Stevens, where an overview of the proof is given. The story begins with a possible solution of Fermat: $a^p + b^p = c^p$, and the construction of the corresponding Frey curve: $y^2 = x(x-a^p)(x+b^p)$, whose minimal discriminant is $\Delta = 2^{-8}(abc)^{2p}$ and conductor is $N = \prod_{l|abc} l$. Since N is squarefree, this Frey curve is semistable.

What is the meaning of all these terms? Chapter 2 by Silverman covers the basic terminology and gives a brief survey of elliptic curves. Following a suggestion of Taniyama in 1957, Shimura and Weil, in the period between 1957 and 1967, formulated the modularity conjecture more precisely. It says the following: Given any elliptic curve E over $\mathbf{Q}$, define $a_p(E)$ by setting it equal to $p+1-\#E(\mathbf{F}_p)$ where $\#E(\mathbf{F}_p)$ is the number of points on the curve mod p. (Technically, this is well-defined for all p coprime to N.) Then there exists a modular form $f(z)$ with Fourier expansion

$$f(z) = \sum_{n=1}^{\infty} a_n(f) e^{2\pi i n z}$$

such that $a_p(f) = a_p(E)$. Phrased another way, this says that the L-function of E coincides with the L-function of a modular form. This is to be viewed as a (nonabelian) reciprocity law. The relevant details about L-functions are found in Chapter 3 by Rohrlich.

Given a cusp form f of weight 2 and level N which is a normalized (i.e. $a_1(f) = 1$) eigenform of the Hecke operators, with $a_p(f)$ rational integers, Eichler and Shimura have described how one can construct an elliptic curve E_f such that we have the equality of the L-series: $L(E_f, s) = L(f, s)$. More precisely, fix $z_0 \in \mathbf{C}$ and consider the set

$$L_f = \left\{ \int_{z_0}^{\gamma z_0} f(z)dz \colon \gamma \in \Gamma_0(N) \right\}.$$

It is not difficult to see that L_f is independent of the choice of $z_0 \in \mathbf{C}$. Shimura showed that L_f is a lattice in $\mathbf{C}$, the elliptic curve $E_f = \mathbf{C}/L_f$ is defined over $\mathbf{Q}$ and its L-function is equal to $L(s, f)$. The curves E_f are called modular elliptic curves and the modularity conjecture says that every elliptic curve over $\mathbf{Q}$ arises in this way.

The proof of FLT breaks up naturally into two parts: Ribet's theorem that the modularity conjecture for semistable elliptic curves implies FLT, and Wiles' theorem that every semistable elliptic curve over $\mathbf{Q}$ (and in particular, the above Frey curve) satisfies the modularity conjecture. To understand Ribet's theorem, one needs to understand Serre's conjectures, which are explained in Chapter 7 by Edixhoven. In a nutshell, Serre's conjectures predict that every continuous, irreducible, odd representation

$$\rho \colon \mathrm{Gal}(\overline{\mathbf{Q}}/\mathbf{Q}) \to \mathrm{GL}_2(\overline{\mathbf{F}}_l)$$

"comes from a cusp form f of weight $k(\rho)$ and level $N(\rho)$" whenever l is a prime > 2. What this means is that

$$\mathrm{tr}\,\rho(\mathrm{Frob}_p) \equiv a_p(f) \pmod{l}$$

where Frob_p is the Frobenius automorphism. Serre makes a precise conjecture about what $k(\rho)$ and $N(\rho)$ should be. This is an amazing conjecture in itself and opens the way for a "Langlands program" mod p. Given an elliptic curve E over $\mathbf{Q}$, one can consider the action of $G_{\mathbf{Q}} = \mathrm{Gal}(\overline{\mathbf{Q}}/\mathbf{Q})$ on the l-division points $E[l]$ of E. Since $E[l]$ is a rank-2 $\mathbf{Z}/l\mathbf{Z}$ module, this gives us a representation $\rho_l \colon G_{\mathbf{Q}} \to \mathrm{GL}_2(\mathbf{F}_l)$. By using Mazur's theorem on the classification of torsion of E over $\mathbf{Q}$, one can show that ρ_l is irreducible for $l \geq 5$. By a theorem of Serre, this forces ρ_l to be absolutely irreducible for $l \geq 5$. For $l = 3$, one has to modify this slightly, as Serre did when he formulated his conjectures, and this is important in the case of Wiles since $l = 3$ is the case he deals with initially.

Serre's prediction implies that the ρ_l that arises from the action of $G_{\mathbf{Q}}$ on the l-division points of the Frey curve constructed out of a possible solution of Fermat must come from a cusp form of level 2 and weight 2. Since there are no such forms, there is no nontrivial solution to FLT.

Serre's conjecture also implies the modularity conjecture. Indeed, given an elliptic curve E of conductor N, Serre's conjectures imply that for each prime l there is an eigenform f_l such that

$$a_p(E) = \mathrm{Tr}(\rho_l(\mathrm{Frob}_p)) \equiv a_p(f_l) \pmod{l},$$

for all primes $l \geq 5$. Since there are only finitely many possibilities for f_l, we must have, for some f, $f_l = f$ for infinitely many l. Thus, for infinitely many l, we have $l \mid (a_p(f) - a_p(E))$, which can only happen if $a_p(f) = a_p(E)$ for all primes p: this is the modularity conjecture.

Ribet proved a fundamental theorem about the level $N(\rho)$ in Serre's conjecture which validates the argument of the penultimate paragraph. This is explained in some detail in Chapter 7. The subject of finite flat group schemes arises in both Ribet's work and Wiles' work, so this is expounded in Chapter 5 by Tate.

According to theorems which are due to Serre and Faltings, two elliptic curves are isogenous if and only if their l-adic representations arising from the action of $G_{\mathbf{Q}}$ on the Tate module $V_l(E) = \varprojlim E[l^n] \otimes \mathbf{Q}_l$ are isomorphic for some prime l. This is the same as saying that the corresponding L-functions are equal. Thus, to show that E is modular, it "suffices" to find a modular elliptic curve E' such that the corresponding l-adic representations are isomorphic for some prime l. Wiles chooses $l = 3$ and begins by considering the action of $G_{\mathbf{Q}}$ on $E[3]$. This gives a representation of $G_{\mathbf{Q}}$ into $\mathrm{PGL}_2(\mathbf{F}_3)$ which is isomorphic to S_4. By a celebrated theorem of Langlands and Tunnell, there is a modular form g of weight one such that $\operatorname{Tr} \rho_3(\mathrm{Frob}_p) = a_p(g)$, which means that $a_p(E) \equiv a_p(g) \pmod 3$. Since the weight is "wrong" Wiles multiplies g by an Eisenstein series E_1 of weight one (with Nebentypus) such that $E_1 \equiv 1 \pmod 3$. This gives a form f of weight 2 which can be lifted to an eigenform by a theorem of Deligne. This at least proves the modularity conjecture mod 3! For this, one needs to use one of the central results in the Langlands program. The object of Chapter 6 by Gelbart is to describe the proof of the important Langlands-Tunnell theorem.

If one can extend the congruence modulo higher powers of 3, then the modularity conjecture follows. This is where Mazur's universal deformation ring enters the picture. Let G_S be the Galois group of the maximal extension of $\mathbf{Q}$ unramified outside S. Let $\overline{\rho}\colon G_S \to \mathrm{GL}_2(\mathbf{F}_l)$ be given. Suppose $\overline{\rho}$ is absolutely irreducible. Suppose further that there is a complete local Noetherian ring A whose residue field is $\mathbf{F}_l$ and such that ρ lifts to $\rho_A\colon G_S \to \mathrm{GL}_2(A)$ and $\rho = \rho_A \circ \mathrm{pr}$ where pr is the natural projection of A onto $\mathbf{F}_l$. Among all such A, Mazur showed that there is a "universal one" in the following sense. There is a complete Noetherian local ring R with residue field $\mathbf{F}_l$ which lifts $\overline{\rho}$ to $\rho_R\colon G_S \to \mathrm{GL}_2(R)$, which commutes with the natural projection of $\mathrm{GL}_2(R) \to \mathrm{GL}_2(\mathbf{F}_l)$ and which satisfies the following universal property: For every representation $\rho_A\colon G_S \to \mathrm{GL}_2(A)$ there is a unique homomorphism $\pi\colon R \to A$ such that $\rho_A = \rho_R \cdot \pi$. R is called the universal deformation ring. In Wiles' proof, one needs to impose additional conditions on R. These technical details are discussed in Chapter 8 by Mazur and Chapter 9 by de Smit and Lenstra.

Some of these lifts ρ_R arise from modular forms (as is suggested by the Eichler-Shimura construction, for example). These modular lifts of the given $\overline{\rho}$ give a representation into $\mathrm{GL}_2(T)$, where T is the Hecke ring. By the universal property, T is a quotient of R. The main theorem of Wiles shows that $R = T$, that is, that all lifts are modular. To this end, he needs to use many deep properties about the Hecke ring T, the main one being that T is Gorenstein. One of the features of Wiles' proof is his simple numerical criterion for $R = T$ in terms of Galois cohomology. The preliminary discussion on Galois cohomology is done in Chapter 4 by Washington. The discussion of the Gorenstein property is done in Chapter 10 by Tilouine, Chapter 11 by de Smit, Rubin and Schoof, Chapter 12 by Diamond and Ribet, Chapter 13 by Conrad and Chapter 14 by de Shalit. These chapters form the major portion of the text.

All of this discussion applies if ρ_3 is irreducible. If it is reducible, Wiles considers the action of $G_{\mathbf{Q}}$ on the 5-division points. By a clever use of the Chebotarev density theorem and Hilbert's irreducibility theorem Wiles shows that there is a semistable curve E' such that $\rho_{5,E} \simeq \rho_{5,E'}$ and $\rho_{3,E'}$ is irreducible. Thus E' is modular. But then, $\rho_{5,E} \simeq \rho_{5,E'}$ implies that E and E' are isogenous by the Serre-Faltings theorem.

These ideas, as well as the modularity of $\mathbf{F}_4$ and $\mathbf{F}_5$ representations, are discussed in Chapter 15 by Silverberg and Chapter 16 by Rubin.

In Chapter 17, Diamond extends the work of Wiles to prove the modularity of elliptic curves which are semistable at 3 and 5. Very recently, Conrad, Diamond, Taylor and Christophe Breuil have proved that *all* elliptic curves over the rationals are modular. Their methods are further developments of the approach outlined in the articles of this volume. In Chapter 20, Frey discusses the generalised Fermat equation. Finally, in the last chapter, Darmon discusses the connection of Wiles' theorem to the Birch and Swinnerton-Dyer conjectures about the order of zeros at the central critical point.

The book has a nicely prepared index and with its numerous references is a welcome addition to the library of every student and researcher in the area. The student may also consult the related volume [*Seminar on Fermat's Last Theorem (Toronto, ON, 1993–1994)*, Amer. Math. Soc., Providence, RI, 1995; MR 96f:11004], which appeared several years earlier and attempted to achieve the same goal of giving a detailed overview of Wiles' proof.

FLT deserves a special place in the history of civilization. Because of its simplicity, it has tantalized amateurs and professionals alike, and its remarkable fecundity has led to the development of large areas of mathematics such as, in the last century, algebraic number theory, ring theory, algebraic geometry, and in this century, the theory of elliptic curves, representation theory, Iwasawa theory, formal groups, finite flat group schemes and deformation theory of Galois representations, to mention a few. It is as if some supermind planned it all and over the centuries had been developing diverse streams of thought only to have them fuse in a spectacular synthesis to resolve FLT. No single brain can claim expertise in all of the ideas that have gone into this "marvelous proof". In this age of specialization, where "each one of us knows more and more about less and less", it is vital for us to have an overview of the masterpiece such as the one provided by this book.

M. Ram Murty (Kingston, ON)

97m:11023 11B65 11L07 11N25

Granville, Andrew (Athens, GA); **Ramaré, Olivier** (Vandoeuvre-les-Nancy)

Explicit bounds on exponential sums and the scarcity of squarefree binomial coefficients.

Mathematika **43** (1996), *no.* 1, 73–107.

Ever since Chebyshev there has been an interest in the study of the prime divisors of binomial coefficients. It is readily observed that unless k is fairly close to one of the ends of the range $0 \le k \le n$ then the coefficient $\binom{n}{k}$ contains many prime factors and many of these occur with high multiplicity. Erdős made a number of conjectures quantifying these observations, one well-known conjecture being that the middle coefficient $\binom{2n}{n}$ is not squarefree for any $n > 4$. A. Sárközy [J. Number Theory **20** (1985), no. 1, 70–80; MR 86c:11002] proved that this is indeed true for all sufficiently large n.

In the current paper the authors make a number of substantial contributions to this area. In their first theorem they prove the above conjecture of Erdős. In their second theorem they sharpen quite considerably an already nice theorem of J. W. Sander [Bull. London Math. Soc. **24** (1992), no. 2, 140–142; MR 93g:11019] by proving that if $\binom{n}{k}$ is squarefree then either k or $n-k$ is $n^{o(1)}$ (in fact this is only a weak consequence of their result). They also prove results which set a limit to bounds of the above type; for example Theorem 4 states that there are infinitely many n such that $\binom{n}{k}$ is squarefree for all $k \leq \frac{1}{5}\log n$. A nice further result is Corollary 1, which gives a precise version of the statement that "on average, there are approximately ten-and-two-thirds squarefree entries in a row of Pascal's triangle".

The above questions are brought into contact with the methods of analytic number theory by means of an observation of Kummer that the exponent of a prime p dividing $\binom{n}{m}$ is given by the number of integers $j \geq 0$ for which $\{m/p^j\} > \{n/p^j\}$; here $\{\ \}$ is the fractional part. Via Fourier techniques this latter inequality may be reduced to the study of the exponential sums $\sum_n \Lambda(n)e(x/n)$ as n runs over an interval.

There are a number of estimates known for the above sums. Perhaps the first is due to M. Jutila [J. Indian Math. Soc. (N.S.) **38** (1974), no. 1, 2, 3, 4, 125–130 (1975); MR **53** #2864], building on ideas of van der Corput and Vinogradov. Previous bounds for this exponential sum have however been given with unspecified constants. The authors go to considerable effort to give such bounds in a completely explicit form, as they are required to do for their applications. Because these exponential sums have been applied elsewhere the authors' explicit bounds will be useful in such other cases.

The reviewer has not stated all the results contained in this very fine piece of work. The reader is encouraged to look at the paper. *John B. Friedlander* (West Hill, ON)

97c:11029 11B73 11M06 11M41

Hauss, Michael (Aachen)

★**Verallgemeinerte Stirling, Bernoulli und Euler Zahlen, deren Anwendungen und schnell konvergente Reihen für Zeta Funktionen. (German) [Generalized Stirling, Bernoulli and Euler numbers and their applications and fast convergent series for zeta functions]**

Dissertation, RWTH Aachen, Aachen, 1995.

Berichte aus der Mathematik. [Reports from Mathematics]

Verlag Shaker, Aachen, 1995. iv+209 pp. ISBN 3-8265-0863-7

In this very valuable thesis, the author generalizes several well-known sequences of numbers and employs them in generating interesting, rapidly convergent series representations for several functions, including the Riemann zeta function $\zeta(s)$, the Hurwitz zeta function, the Lerch zeta function, polylogarithms, and the Dirichlet L-function with primitive, nonprincipal character of modulus 4.

The first number sequences that the author examines are generalized Stirling numbers of the first and second kinds. For example, the generalized Stirling numbers of the second kind, $S_t(n,k)$, are generated by $z^n = \sum_{k=0}^{n} S_t(n,k)(z|t)_k$, $n \in \mathbf{N}_0$, $z \in \mathbf{C}$, where, for $k \in \mathbf{N}$, $(z|t)_k := \prod_{\nu=0}^{k-1}(z-t_\nu)$, where $t_0, t_1, \cdots, t_{k-1}$ are arbitrary complex numbers. In Chapters 3 and 4 the author provides a clear exposition of the theory of generalized Stirling numbers of the first and second kinds.

In Chapter 5 the author examines fractional p-factorial numbers and functions of the first kind. First define a certain generalization of Stirling numbers of the first kind, $S(\alpha,k)$, by

$$[z]_\alpha := \frac{\Gamma(z+1)}{\Gamma(z-\alpha+1)} = \sum_{k=0}^{\infty} S(\alpha,k)z^k, \quad \alpha\in\mathbf{C},\ |z|<1.$$

The fractional p-factorial numbers of the first kind are defined by

$$[z,\alpha;p] := [z]_{\alpha/p}\prod_{\nu=1}^{p-1}[-z\omega_p^\nu]^{\alpha/p}.$$

Here, $\alpha,z\in\mathbf{C}$, $p\in\mathbf{N}$, $z\notin\mathbf{Z}^-$, $z+\alpha\notin\mathbf{Z}_0^-$, $\alpha/p-z\omega_p^\nu\notin\mathbf{Z}_0^-$, and $\omega_p=\exp(2\pi i/p)$. Lastly, the p-factorial functions of the first kind, $a_p(\alpha,k)$, are defined by $[z,\alpha;p]=\sum_{k=0}^\infty a_p(\alpha,k)z^k$, $|z|<\rho$, where ρ is a certain number with $0<\rho<1$. The author develops the theory of these numbers.

Chapter 6 contains a development of generalized Bernoulli numbers and polynomials.

The highlights of this thesis are Chapters 7 and 8, providing systematic developments of various series and integral representations of $\zeta(s)$ and allied functions. One of the author's main tools is a new, very general series transformation for $\sum_{n=N_0+1}^\infty f(n)/s_n^{m+1}$, where f is any function defined on $\{N_0+1,N_0+2,\cdots\}$ and s_n are arbitrary nonzero complex numbers. Of course, it is assumed that the series converges. The transformation involves generalized Stirling numbers of the first kind.

Portions of the author's theoretical developments and special cases of his particular formulas had been previously derived over a period of more than two centuries by a variety of authors. One of the strengths of this thesis is that it organizes, unifies, and generalizes a large body of material. The author has been very careful to discuss and cite all previous work done on the topics at hand. His bibliography of 149 items is extremely useful and contains several items previously unknown to the reviewer.

Bruce C. Berndt (Urbana, IL)

99a:11043 11E04 11E57 22E40 58F11

Eskin, Alex (Chicago, IL);
Margulis, Gregory [Margulis, Grigorii A.] (New Haven, CT);
Mozes, Shahar (Jerusalem)

Upper bounds and asymptotics in a quantitative version of the Oppenheim conjecture.

Ann. of Math. (2) **147** (1998), *no.* 1, 93–141.

Let Q be an indefinite nondegenerate quadratic form in n variables. It is a classical problem of number theory to study the distribution of the set $Q(\mathbf{Z}^n)$ of values of Q at integral points. If $Q(x_1,x_2)=x_1^2-\alpha x_2^2$ then the structure of the set $Q(\mathbf{Z}^2)$ is determined by Diophantine properties of α. For instance, if $\alpha=\sqrt{2}$ then the set $Q(\mathbf{Z}^2)$ has 0 as an isolated point.

The Oppenheim conjecture formulated in 1929 states that if $n\geq 3$ and Q is irrational (not proportional to a form with rational coefficients) then the set $Q(\mathbf{Z}^n)$ is dense in $\mathbf{R}$. One can easily show that the validity of the conjecture for some $m\geq 3$ implies its validity for all $n\geq m$. Hence it suffices to prove it for $n=3$.

By methods of analytic number theory the Oppenheim conjecture was proved in the late 50s for $n\geq 21$; see the paper by Margulis [in *Fields Medallists' lectures*,

272–327, World Sci. Publishing, River Edge, NJ, 1997; MR 99e:11046] for a detailed survey. In full scope the conjecture was proved by Margulis [C. R. Acad. Sci. Paris Sér. I Math. **304** (1987), no. 10, 249–253; MR 88f:11027; in *Number theory, trace formulas and discrete groups (Oslo, 1987)*, 377–398, Academic Press, Boston, MA, 1989; MR 90k:22013a]. The proof used a connection noticed by Raghunathan between the conjecture and orbit closures of certain homogeneous actions. It turns out that (a slightly weaker form of) the Oppenheim conjecture would follow if one proves that any relatively compact orbit of the left action of $\mathrm{SO}(2,1)$ on the homogeneous space $\mathrm{SL}(3,\mathbf{R})/\mathrm{SL}(3,\mathbf{Z})$ is compact; precisely this was done by Margulis.

The statement on orbit closures follows easily from the so-called Raghunathan conjecture, which states that given a homogeneous space G/Γ of finite volume (i.e. a space admitting a finite G-invariant measure) and a unipotent subgroup $U \subset G$ (i.e. such that all the adjoint operators Ad_u, $u \in U$, are unipotent on the Lie algebra of G), for every $x \in G/\Gamma$ there exists a subgroup $F \subset G$ such that $\overline{Ux} = Fx$ and Fx has a finite F-invariant measure. More generally, Margulis conjectured that the same holds for any subgroup $U \subset G$ generated by unipotent elements (example: $U = \mathrm{SO}(2,1) \subset G = \mathrm{SL}(3,\mathbf{R})$). The Raghunathan conjecture (and that of Margulis for connected U) was proved by M. Ratner [Duke Math. J. **63** (1991), no. 1, 235–280; MR 93f:22012]. The proof is based on another fundamental result: Ratner's measure classification theorem [Ann. of Math. (2) **134** (1991), no. 3, 545–607; MR 93a:22009].

An alternative approach for the deduction of the Raghunathan conjecture from Ratner's measure theorem was proposed by S. G. Dani and Margulis [in *I. M. Gel'fand Seminar*, 91–137, Amer. Math. Soc., Providence, RI, 1993; MR 95b:22024]. That paper also contained additional information about the structure of the set $Q(\mathbf{Z}^n)$. Namely, let $D_T = \{x \in \mathbf{R}^n\colon |x| < T\}$ and

$$V_{(a,b)}(T) = \{x \in D_T\colon a < Q(x) < b\}.$$

One can show that $\mathrm{Vol}(V_{(a,b)}(T)) \sim \lambda(b-a)T^{n-2}$ as $T \to \infty$ for some constant $\lambda = \lambda(Q)$. It is of importance to know the asymptotics of the number $N_{(a,b)}(T)$ of integral points in $V_{(a,b)}(T)$. One might expect that $V_{(a,b)}(T) \sim \mathrm{Vol}(V_{(a,b)}(T))$ as $T \to \infty$ for any indefinite nondegenerate irrational form Q of $n \geq 3$ variables, which can be interpreted as "uniform distribution" of the sets $Q(\mathbf{Z}^n \cap D_T) \subset \mathbf{R}$. Dani and Margulis supported this conjecture by proving the exact lower bound

$$\liminf_{T\to\infty} \frac{N_{(a,b)}(T)}{\mathrm{Vol}(V_{(a,b)}(T))} \geq 1.$$

In addition, this estimate holds if one varies Q in a compact family of indefinite nondegenerate irrational forms. For a single Q, the fact that the lower limit is positive was obtained earlier by Dani and Mozes and also by Ratner (unpublished).

The method of the above-mentioned paper of Dani and Margulis is not applicable for finding upper bounds of $N_{(a,b)}(T)$. This is done completely in the paper under review. It turns out that the upper bound may not coincide with the lower one. In fact, it was noticed by Sarnak that given a quadratic form $Q_\alpha(x) = x_1^2 + x_2^2 - \alpha^2 x_3^2$ the number $N_{(a,b)}(T)$ may grow faster than T^{n-2} if α is too rapidly approximated by rationals. The authors prove that the same holds for quadratic forms

$$Q_\alpha(x) = x_1^2 + x_2^2 - \alpha^2(x_3^2 + x_4^2).$$

Moreover, for every $\varepsilon > 0$ there exist an irrational form Q of signature $(2,q)$ with $q = 1,2$, a sequence $T_j \to \infty$ and a constant $c > 0$ such that

$$N_{(a,b)}(T_j) > cT_j^q(\log T_j)^{1-\varepsilon}.$$

On the other hand, for any form of signature $(2,q)$ with $q = 1,2$ one can find a constant $c > 0$ such that $N_{(a,b)}(T) < cT^q \log T$, $T > 2$. Moreover, for almost all such forms one has $N_{(a,b)}(T) \sim \lambda(b-a)T^q$.

As for irrational forms with signatures (p,q), $p \geq 3$, $q \geq 1$, there are no "counterexamples", i.e. the lower bound is always the upper one:

$$N_{(a,b)}(T) \sim \lambda(b-a)T^{p+q-2}, \quad T \to \infty.$$

In addition, the authors obtain uniform upper bounds over some compact families of indefinite forms of the same signature (earlier the lower bound was found by Dani and Margulis).

As before, all the results are obtained by using the theory of homogeneous actions. Actually, the study of $N_{(a,b)}(T)$ is related to the estimation of certain integrals over orbits of semisimple subgroups on $\mathrm{SL}(n,\mathbf{R})/\mathrm{SL}(n,\mathbf{Z})$. The authors were able to handle complications arising from the structure at infinity of the space $\mathrm{SL}(n,\mathbf{R})/\mathrm{SL}(n,\mathbf{Z})$. Since the proof uses different facts from ergodic theory, the bounds are not effective. For instance, as is mentioned in the paper, it is not clear how to determine T such that $N_{(a,b)}(T) \neq 0$ (this should be done in terms of Diophantine properties of the coefficients of Q). *Alexander Starkov* (Moscow)

99b:11040 11F22 11F46 11F55 14J28 17B67 81R10

Gritsenko, Valeri A. [Gritsenko, V. A.] (St. Petersburg);
Nikulin, Viacheslav V. [Nikulin, Vyacheslav Valentinovich] (Moscow)
Automorphic forms and Lorentzian Kac-Moody algebras. I. (English summary)
Internat. J. Math. **9** (1998), *no.* 2, 153–199.

This paper continues a long series of papers by the authors devoted to generalized Kac-Moody algebras associated to certain families of algebraic $K3$ surfaces. The construction depends on the existence of a special automorphic form Ψ on the period space Ω_T of the family $\mathbf{M}_T$ of $K3$ surfaces whose lattice of transcendental cycles admits a primitive embedding in a given even lattice T of signature $(2,t)$. The period space is equal to the Grassmannian $\mathrm{Gr}(2,\mathbf{R}^{2,t})_+$ of positive definite oriented planes in the inner product space $\mathbf{R}^{2,t} = T_{\mathbf{R}}$. The function Ψ must be automorphic with respect to a subgroup G of finite index in $\mathrm{O}(T)^+$, where $\mathrm{O}(T)^+$ is the subgroup of the orthogonal group of T which acts identically on the discriminant group T^*/T. It must be holomorphic, with the set of zeroes contained in the union of quadratic divisors $H_\delta = \{v \in \mathrm{Gr}(2,\mathbf{R}^{2,t})_+ \colon v \perp \delta\}$, where δ belongs to a set $\Delta(T)$ of primitive vectors in T with negative norm which define a reflection $s_\delta \in \mathrm{O}(T)$. The authors call a lattice T reflective if Ω_T admits such an automorphic form Ψ, called a reflective automorphic form. The union of quadratic divisors H_δ with $(\delta,\delta) = -2$ is the discriminant of $\mathbf{M}_T$. Its points parametrize the periods of $K3$ surfaces from $\mathbf{M}_T$ which do not contain an ample divisor class orthogonal to T. The first examples of automorphic forms Ψ on Ω_T were constructed by R. E. Borcherds [Invent. Math. **120** (1995), no. 1, 161–213; MR 96j:11067]. Their Fourier expansion at a cusp of Ω_T has a particular property allowing one to expand it in a certain infinite product. This infinite product is used to define a certain generalized Kac-Moody Lie superalgebra for which the infinite product is equal to the denominator in the generalized Kac-Weyl character formula. The lattice T for which such Ψ exists with zeroes of order one only is called a Lie reflective lattice. The authors conjecture that the number of such lattices is finite up to multiplication of its quadratic form by a positive integer and address the problem of their classification. The reason behind the conjecture is the following. A cusp on the domain Ω_T is determined by a choice of an isotropic primitive vector $f \in T$. The lattice $S = (T \cap f^{\perp})/\mathbf{Z}f$ is of signature $(1, t-1)$ and is reflective in the sense that the semi-direct product $W \cdot A(\mathcal{M})$ of its reflection group W and the group of symmetries $A(\mathcal{M})$ of a fundamental polyhedron $\mathcal{M}$ with respect to W is of finite index in its orthogonal group $\mathrm{O}(S)$. Earlier results of the second author and È. Vinberg on reflection groups in Lobachevskiĭ spaces can be used to show that the set of such lattices of rank ≥ 3 is finite [see V. V. Nikulin, Izv. Ross. Akad. Nauk Ser. Mat. **60** (1996), no. 2, 73–106; MR 97f:17026]. Let us fix a set of vectors α in S perpendicular to the faces of $\mathcal{M}$ which span S such that each α defines a reflection s_α of S (that is, $2(\alpha,\alpha) \mid (\alpha, S)$). If one additionally assumes the existence of a Weyl vector $\rho \in S \otimes \mathbf{Q}$ satisfying the property that $(\rho, \alpha) = -(\alpha,\alpha)/2$ for any $\alpha \in P(\mathcal{M})$, then one can assign to $(S, W, P(\mathcal{M}))$ a hyperbolic (or Lorentzian) Kac-Moody algebra $\mathfrak{g}(S, W, P(\mathcal{M}))$ with the root lattice S, the set of simple real roots $\Pi = P(\mathcal{M})$, the Weyl group W, the set of real roots $\Delta^{\mathrm{re}} = W(\Pi)$, the set of imaginary roots $\Delta^{\mathrm{im}} = W(V^+(S) \cap S)$ (where $V^+(S)$ is the Tits cone), the light cone in the Lobachevskiĭ space $S_{\mathbf{R}} = S \otimes \mathbf{R}$. As usual, the set of roots $\Delta = \Delta^{\mathrm{re}} \cup \Delta^{\mathrm{im}}$ is divided into the disjoint sum of positive and negative roots $\Delta_+ \coprod \Delta_-$. The Kac-Weyl denominator formula for $\mathfrak{g}(S, W, P(\mathcal{M}))$ is

$$\sum_{w \in W} \det(w) e^{-2\pi i (w(\rho), z)} = \prod_{\alpha \in \Delta_+} (1 - e^{2\pi i (\alpha, z)})^{\mathrm{mult}(\alpha)}.$$

The function $\Psi'(z)$ equal to one of the sides in this formula is a holomorphic function on the tube domain $D_S = S_{\mathbf{R}} + iV^+(S) \subset S \otimes \mathbf{C}$ which is anti-invariant with respect to the group $W \cdot A(\mathcal{M})$ (i.e. $\Psi'(w \cdot a(z)) = \det(w)\Psi'(z)$ for any $w \in W,\ a \in A(\mathcal{M})$). The group $W \cdot A(\mathcal{M})$ can be identified with a parabolic subgroup of $\mathrm{O}(T)^+$ and is not an automorphic form for the latter groups. The reflective automorphic form $\Psi(z)$ should be thought of as an extension of the infinite product for $\Psi'(z)$ when the set of simple roots $P(\mathcal{M})$ is extended to $P(\mathcal{M}) \cup H$. The additional simple roots from H are imaginary, i.e. satisfy $(\alpha, \alpha) > 0$. The corresponding infinite-dimensional Lie algebra is not a Kac-Moody algebra anymore but rather a generalized Kac-Moody algebra, or a Borcherds algebra [see R. E. Borcherds, J. Algebra **115** (1988), no. 2, 501–512; MR 89g:17004]. The product of the additional factors in the expansion of $\widetilde{\Psi}$ is called the automorphic correction for the hyperbolic Kac-Moody algebra $\mathfrak{g}(S, W, P(\mathcal{M}))$. Thus the classification of Lie reflective lattices T is more or less equivalent to the classification of reflective lattices S of signature $(1, t-1)$ such that $T = S \perp U(m)$ is Lie reflective. Here $U(m)$ is the lattice generated by two isotropic vectors f, g with $(f, g) = m$. Note that the lattice S defines a family of $K3$ surfaces whose Picard lattice contains S as a primitive sublattice. This is the "mirror family" of the family $\mathbf{M}_T$ as defined by the reviewer [J. Math. Sci. **81** (1996), no. 3, 2599–2630; MR 97i:14024]. Thus the denominator formula for the Borcherds algebra associated to a reflective lattice S with a Weyl vector is a holomorphic automorphic function on the period space of the mirror family with zeroes on discriminant hypersurfaces.

This is the mirror symmetry for $K3$ surfaces as defined by the authors [Math. Res. Lett. **3** (1996), no. 2, 211–229; MR 97e:14050]. The present paper discusses this mirror symmetry and also presents a complete classification of hyperbolic reflective lattices S of rank 3 satisfying the following properties: S admits a Weyl vector ρ, the polyhedron $\mathcal{M}$ is not compact but of finite volume, $P(\mathcal{M})$ can be chosen such that each $\alpha \in P(\mathcal{M})$ satisfies $\alpha = \lambda(\alpha)\delta(\alpha)$, where $(\delta,\delta) = -2$ and $\lambda(\alpha)$ is a positive integer. The classification is given in terms of the coefficients $\lambda(\alpha)$ and the matrix $A = (\delta(\alpha), \delta(\alpha'))$. There are 60 lattices S as above. A partial list of 12 of them (satisfying the additional property that $\lambda(\alpha) = 1$) was previously announced by the authors [Mat. Sb. **187** (1996), no. 11, 27–66; MR 98a:11059].

{For Part II see the following review [MR 99b:11041, see p. A23 in this volume].}

I. Dolgachev (Ann Arbor, MI)

99b:11041 11F22 11F46 11F55 14J28 17B67 81R10

Gritsenko, Valeri A. [Gritsenko, V. A.] (St. Petersburg);

Nikulin, Viacheslav V. [Nikulin, Vyacheslav Valentinovich] (Moscow)

Automorphic forms and Lorentzian Kac-Moody algebras. II. (English summary)

Internat. J. Math. **9** (1998), *no.* 2, 201–275.

In this second part of their work [cf. Part I, Internat. J. Math. **9** (1998), no. 2, 153–199; MR 99b:11040, see p. A21 in this volume; see the preceding review], the authors classify all reflective lattices T of signature $(2,3)$ which admit a reflective automorphic form Ψ with zeroes in the union of quadratic divisors H_δ with $(\delta,\delta) = -2$. In this case the period space Ω_T is naturally isomorphic to the Siegel half-plane space $\mathcal{Z}_2$ of symmetric complex (2×2)-matrices with positive imaginary part. The list consists of 14 lattices $T = U^2 \perp \langle 2k\rangle$, $k = 2,3,4$ and $T = U(k) \perp U(k) \perp \langle 2\rangle$, $1 \le k \le 8$, $k = 10, 12, 16$. Here $\langle n\rangle$ is the rank-1 lattice generated by a vector e with $(e,e) = n$, and $U(k)$ is the rank-2 lattice generated by vectors f,g with $(f,f) = (g,g) = 0$, $(f,g) = k$. For example, the reflective automorphic form corresponding to the lattice $U(4) \perp U(4) \perp \langle 2\rangle$ [resp. $U(8) \perp U(8) \perp \langle 2\rangle$] is the Igusa modular form Δ_5 [resp. Δ_2]. This was shown in the authors' earlier paper [Amer. J. Math. **119** (1997), no. 1, 181–224; MR 98g:11056]. The lattice $U \perp U \perp \langle 2\rangle$ corresponds to the Igusa modular form Δ_{35}, as was shown by the authors in [Mat. Sb. **187** (1996), no. 11, 27–66; MR 98a:11059]. They also prove the reflectiveness of some other lattices T of rank 5.

The method of constructing a reflective automorphic form Ψ is based on the first author's version of Maass lifting of modular forms of half-integral weight [H. Maaß, Invent. Math. **60** (1980), no. 1, 85–104; MR 83e:10036]. It was proposed in his earlier papers [see, e.g., in *Number theory (Paris, 1992–1993)*, 103–126, Cambridge Univ. Press, Cambridge, 1995; MR 96d:11049] and is called an arithmetic lifting. It starts with a holomorphic Jacobi form $\varphi_{k,t}(\tau,z)$ of weight k of integral or half-integral index t with some character and defining

$$\mathrm{Lift}_\mu(\varphi) = \sum_{m\equiv\mu \bmod Q,\, m>0} m^{2-k}(\widetilde{\varphi}|_k T^{(Q)}(m))(Z).$$

Here $\mu \in (\mathbf{Z}/Q\mathbf{Z})^*$, $\widetilde{\varphi} = \varphi_{k,t}(\tau,z)e^{2\pi i t w}$, $Z = \left(\begin{smallmatrix} \tau & z \\ z & w\end{smallmatrix}\right) \in \mathcal{Z}_2$, and $\cdot|_k T^{(Q)}(m)$ are some Hecke operators. For example, the reflective automorphic form $\Delta_1(Z)$ corresponding to the lattice $T = U(12) \perp U(12) \perp \langle 2\rangle$ is equal to $\mathrm{Lift}_1(\eta(\tau)\theta(\tau,z))$, where $\eta(\tau)$ is the Dedekind eta function and $\theta(\tau, z)$ is the Riemann theta function. Notice that another, more general approach to constructing reflective automorphic forms which contains the arithmetic liftings as a special case was proposed by R. E. Borcherds [Invent. Math. **132** (1998), no. 3, 491–562; MR 99c:11049, see p. A25 in this volume].

The infinite product for the Igusa modular form $\Delta_5(Z)$ has the form $p^{1/2}q^{1/2}y^{1/2}\prod_{(k,l,m)>0}(1-p^kq^ly^r)^{f(mn,l)}$, where $p=e^{2\pi i w}$, $q=e^{2\pi i\tau}$, $y=e^{2\pi i z}$, $(k,l,m)>0$ means that either $k>0$, $l\geq 0$, or $k\geq 0$, $l>0$, or $k=l=0$, $r<0$, and $\varphi_{0,1}(\tau,z)=\sum_{n\geq 0, l\in\mathbf{Z}} f(n,l)e^{2\pi i(n\tau+lz)}$ is the Fourier expansion of the Jacobi form $\varphi_{0,1}$. In the recent work of the physicists R. Dijkgraaf, G. Moore, E. Verlinde and H. Verlinde [Comm. Math. Phys. **185** (1997), no. 1, 197–209; MR 98g:81191], the inverse of $\prod_{(k,l,m)>0}(1-p^kq^ly^r)^{f(mn,l)}$ was identified with the generating series $\sum_{m=0}^{\infty} p^m Z[M^{(m)}](\tau,z)$, where $Z[M^{(m)}]$ is the orbifold elliptic genus of the mth symmetric product of a $K3$ surface M. This is in striking analogy with the formula of L. Göttsche [Math. Ann. **286** (1990), no. 1-3, 193–207; MR 91h:14007] for the generating function of the orbifold Euler characteristic of $M^{(m)}$,

$$\sum_{m=0}^{\infty} q^m e[M^{(m)}](\tau)=\prod_{n>0}\frac{1}{(1-q^n)^{24}}=\frac{q}{\eta(\tau)^{24}}.$$

There are other connections of the work under review with physics. For example, the Gromov-Witten prepotential for the $K3$-fibred Calabi-Yau 3-fold $V=X(30,16,12,1,1)$ when restricted to the subspace of the cohomology group $H^2(V,\mathbf{C})$ coming from the monodromy invariant cohomology of a general fibre (which belongs to the $K3$-family mirror to the family $\mathbf{M}_T$, where $T=U\perp U\perp\langle 2\rangle$) is expressed in terms of the Igusa modular forms $\Delta_5(Z)$ and $\Delta_{35}(Z)$ [T. Kawai, Phys. Lett. B **372** (1996), no. 1-2, 59–64; MR 97b:81107; Phys. Lett. B **397** (1997), no. 1-2, 51–62; MR 98b:81200]. The generalized Kac-Moody algebras associated to the $K3$ fibres of a $K3$-fibred Calabi-Yau 3-fold via the product formula for the reflective automorphic form on the space of periods of the mirror family are presumably related to the algebras of BPS states attached to a Calabi-Yau 3-fold in the works of the physicists J. A. Harvey and Moore [Nuclear Phys. B **463** (1996), no. 2-3, 315–368; MR 97h:81163, see p. F31 in this volume].

I. Dolgachev (Ann Arbor, MI)

98b:11047 11F33 11F67 11F85 11G07 14E30

Coleman, Robert F. (Berkeley, CA)

***p*-adic Banach spaces and families of modular forms.**

Invent. Math. **127** (1997), *no.* 3, 417–479.

Let p be a prime and $F=\sum_n a_nq^n$ a modular form on $X_1(Np)$ with coefficients a_n in a finite extension K of $\mathbf{Q}_p$, the field of p-adic numbers, contained in $\mathbf{C}_p$, the completion of an algebraic closure of $\mathbf{Q}_p$, where N is an integer coprime to p. Let U_p be the Hecke operator on $X_1(Np)$ which takes $\sum_n a_nq^n$ to $\sum_n a_{np}q^n$. A modular form F is said to have slope $\alpha\in\mathbf{Q}$ if there is a polynomial $R(T)$ over $\mathbf{C}_p$ such that the Newton polygon of $R(T)$ has only one side and its slope is $-\alpha$ (this means that if λ is a root of $R(T)$ then $v(\lambda)=\alpha$, where v is the valuation on $\mathbf{C}_p$ such that $v(p)=1$). F is said to be ordinary if $\alpha=0$. The main result of the paper concerns the existence of families of modular forms of given slope α. Such families generalize

Hida's families of ordinary forms [H. Hida, Invent. Math. **85** (1986), no. 3, 545–613; MR 87k:11049] which correspond to the case $\alpha = 0$.

Theorem. Suppose $\alpha \in \mathbf{Q}$ and $\varepsilon\colon (\mathbf{Z}/p\mathbf{Z})^\times \to \mathbf{C}_p^\times$ is a character. Then there exists a number n_0 which depends on p, N, ε and α with the following property: If $k \in \mathbf{Z}$, $k > \alpha + 1$ and there is a unique normalized cusp form F on $X_1(Np)$ of weight k, character $\varepsilon\omega^{-k}$ and slope α and if $k' > \alpha + 1$ is an integer congruent to k modulo p^{n+n_0}, for any positive integer n, then there exists a unique normalized cusp form F' on $X_1(Np)$ of weight k', character $\varepsilon\omega^{-k'}$ and slope α (ω denotes the Teichmüller character). Moreover this form $F'(q)$ satisfies the congruence $F'(q) \equiv F(q) \bmod p^{n+1}$.

The proof is based on the study of the characteristic power series $P_{N,i}(k, T)$ of the operator U_p acting on overconvergent forms of weight k on $X_1(Np)$ and of character ω^{i-k}, where $0 \le i \le p-1$, and by definition, the (infinite-dimensional) p-adic Banach space of overconvergent forms of weight k on $X_1(Np)$ consists of sections of an appropriate invertible sheaf on a rigid analytic space attached to $X_1(Np)$. It turns out that these p-adic Banach spaces together with the action of the operator U_p can be naturally included in a p-adic rigid analytic family parametrized by k (Theorem B). Then certain spectral properties of the action of U_p on the (finite-dimensional) subspaces of classical forms $M_{k,\mathrm{cl}}(\omega^{i-k})$ are described in Theorems D and C. The results of the paper are related to the theory of p-adic representations of the Galois group of $\overline{\mathbf{Q}}/\mathbf{Q}$ (a joint work in progress of the author with B. Mazur; see also a paper by Mazur [in *Galois groups over* **Q** *(Berkeley, CA, 1987)*, 385–437, Springer, New York, 1989; MR 90k:11057] where the ordinary case is treated).

Alexey A. Panchishkin (Grenoble)

99c:11049 11F37 11F22 14J28 17B67 57R57

Borcherds, Richard E. (Cambridge)

Automorphic forms with singularities on Grassmannians. (English summary)

Invent. Math. **132** (1998), *no.* 3, 491–562.

The paper completes the work of many people on the correspondence which takes modular forms of half-integral weight of level 1 on the upper half-plane $\mathscr{H}$ which are holomorphic at cusps to modular forms of integral weight on a higher-dimensional Hermitian homogeneous domain isomorphic to a connected component of the Grassmannian $\mathrm{Gr}(2, \mathbf{R}^{2,n})_+ \cong \mathrm{O}(2,n)/\mathrm{O}(2) \times \mathrm{O}(n)$ of positive definite oriented planes in the inner product space of signature $(2, n)$. It generalizes the similar constructions of G. Shimura ($n = 1$), Doi-Naganuma ($n = 2$), H. Maass ($n = 3$) and V. A. Gritsenko (any n). In his earlier work [Invent. Math. **120** (1995), no. 1, 161–213; MR 96j:11067] the author extended this correspondence to the case when the modular form is allowed to have poles at cusps. Subsequently the physicists J. A. Harvey and G. W. Moore [Nuclear Phys. B **463** (1996), no. 2-3, 315–368; MR 97h:81163, see p. F31 in this volume] used some divergent integrals to extend Howe's correspondence to modular forms with singularities at cusps. This led to simplifying Borcherds' correspondence in the case of level 1 forms with singularities. In this paper the author uses this approach to simplify the proofs and generalize his earlier results to include the case of level greater than 1. The core of the construction is the integral

$$\Phi_M(v, F) = \int_{\tau \in \mathscr{H}/\Gamma} \overline{\Theta}(\tau; v) F(\tau) y^{b^+/2} \frac{dx\,dy}{y^2}.$$

Here the integration is taken over a fundamental domain of the modular group $\Gamma = \mathrm{SL}(2,\mathbf{Z})$, $F(\tau)$ is a $\mathbf{C}[M^*/M]$-valued modular form with respect to Γ of weight $(b^+ - b^-)/2$ which is holomorphic on $\mathscr{H}$, $\overline{\Theta}(\tau, v^+)$ is the Siegel theta function of a lattice M in $\mathbf{R}^{b^+,b^-}$, $v \in \mathrm{Gr}(b^+, \mathbf{R}^{b^+,b^-})_+$, where the bar means complex conjugation. Recall that the Siegel theta function $\Theta_M(\tau, v)$ of a unimodular lattice M is defined by $\Theta_M(\tau, v) = \sum_{m\in M} e^{\pi i(\tau m_+{}^2 + \bar{\tau} m_-{}^2)}$, where m_+ is the orthogonal projection of m to the subspace $v \in \mathrm{Gr}(b^+, \mathbf{R}^{b^+,b^-})_+$ and $m_- = m - m_+$. In the non-unimodular case one has to modify this formula a little bit. There is also a generalization of this formula for not necessarily holomorphic modular forms, and also including an additional polynomial parameter p. Although the integral is often divergent, there is a way to regularize it. Also, allowing F to be vector-valued takes care of scalar-valued holomorphic forms of higher level, as well as Siegel modular forms which were dealt with by Gritsenko.

Here are the main properties of the function $\Phi_M(v, F)$: (i) $\Phi_M(v, F)$ is invariant under $\sigma \in \mathrm{Aut}(M)$ in the sense that it is equal to $\Phi_M(\sigma(v), \sigma(F))$, where σ acts on values of F via its action on the discriminant group M^*/M. (ii) $\Phi_M(v, F)$ is holomorphic on $\mathrm{Gr}(b^+, \mathbf{R}^{b^+,b^-})$ outside a locally finite set of codimension b^+ sub-Grassmannians of the form $m^\perp = \{v\colon (v, m) = 0\}$ for some $m \in M$ of negative norm (isomorphic to $\mathrm{Gr}(b^+, b^- - 1)$). (iii) There is an explicit Fourier expansion at a cusp which is defined by a choice of an isotropic vector $f \in M$. It is given in terms of a similar function Φ_K, where $K = (M \cap f^\perp)/\mathbf{Z}f$.

Now let us assume that $b^+ = 2$. Let

$$F = \sum_{\delta\in M^*/M} e_\delta\Big(\sum_{r\in\mathbf{Q}} c_\delta(r) e^{2\pi i r\tau}\Big)$$

be the Fourier expansion of F. The choice of a cusp as in (iii) allows one to identify $\mathrm{Gr}(2, \mathbf{R}^{2,b^-})$ with its tube domain realization $\Omega_f = K_{\mathbf{R}} + \sqrt{-1}K_{\mathbf{R}}^+ \subset K_{\mathbf{C}}$. Here $K_{\mathbf{R}}^+$ is the light cone in the hyperbolic space $K_{\mathbf{R}} \cong \mathbf{R}^{1,b^- -1}$. There exists a meromorphic function $\Psi_f(Z, F)$ on Ω_f with the following properties: (1)

$$\Phi_M(Z, F) = -4\log|\Psi_f(Z, F)| - 2c_0(0)\Big(\log|Y| + \frac{\Gamma'(1)}{2} + \log\sqrt{2\pi}\Big),$$

where $Z = X + iY \in \Omega_f$. (2) $\Psi_f(Z, F)$ is an automorphic form of weight $c_0(0)/2$ for a subgroup $\Gamma_M = \{\sigma \in \mathrm{Aut}(M)\colon \sigma(m^*) - m^* \in M,\ \forall m^* \in M^*\}$ with respect to its unitary character. (3) The only zeroes or poles of $\Psi_f(Z, F)$ lie on rational quadratic divisors $m^\perp$ as in (ii). (4) Let W be the subgroup of $\mathrm{Aut}(M)$ generated by reflections and C be its Weyl chamber in $K_{\mathbf{R}}^+$. The restriction of $\Psi_f(Z, F)$ to a neighborhood of $i\infty$ intersected with $K_{\mathbf{R}} + iC$ admits an infinite product representation of the form

$$e^{2\pi i(\rho, Z)} \prod_{\substack{\lambda\in K\\ (\lambda, C)>0}} \prod_{\substack{\delta\in M^*/M\\ \delta|f^\perp=\lambda}} (1 - e^{2\pi i((\lambda, Z)+(\delta, f'))})^{c_\delta((\lambda,\lambda)/2)},$$

where $f \in M^*$ satisfies $(f', f) = 1$, K' is the set of primitive vectors in K and ρ is a Weyl vector in K (dependent on the choice of C) which is determined by the property that $8\sqrt{2}(\rho, v) = \|v\|\Phi_M(v/\|v\|, F)$ for any $v \in C$.

It is an interesting problem to describe the pairs (M, f) as above for which one can construct the form $\Psi_{M,f}(Z)$ with properties 2–4. Some of these lattices occur as the lattices of transcendental lattices on a $K3$ surface, and in the case $b^- = 3$ they were essentially classified in the work of Gritsenko and V. V. Nikulin [see Internat. J. Math. **9** (1998), no. 2, 153–199; MR 99b:11040, see p. A21 in this volume; Internat. J. Math. **9** (1998), no. 2, 201–275; MR 99b:11041, see p. A23 in this volume]. The case when $M = \Lambda \perp U \perp U$, where Λ is the Leech lattice and U is the unimodular even lattice of signature $(1,1)$, was treated by the author in his above-cited earlier paper. It corresponds to $\Psi_f(Z,F)$ where $f \in U$, $F = 1/\eta^{24}(\tau)$, and $\eta(\tau)$ is the Dedekind eta-function. The infinite product is interpreted as the denominator formula for the fake Monster Lie algebra. This is also true for any M as above for which $\Phi_M(Z,F)$ can be defined. The infinite product for $\Psi_f(Z)$ is the denominator formula for a certain generalized Kac-Moody Lie superalgebra whose root lattice is K and whose Weyl group is the reflection group of K.

The lattice $M = E_{10}(2) \perp U$, where E_{10} is the unimodular even lattice of signature $(1,9)$ and (2) indicates that the inner product is multiplied by 2, defines an automorphic form on the period space of Enriques surfaces. It allowed the author to prove that the coarse moduli space of such surfaces is quasi-affine, solving a problem which was inaccessible by conventional algebraic geometrical techniques [Topology **35** (1996), no. 3, 699–710; MR 97e:14051].

Finally, there is a beautiful relationship between the functions $\Phi_M(v,F)$ defined on the hyperbolic space $\mathrm{Gr}(1,\mathbf{R}^{1,b^-})_+$ and Donaldson polynomials. Both of these functions are piecewise polynomial and have similar wall crossing formulas. In fact, when $M = E_{10}$ the wall crossing formula for $\Phi_M(v,F)$ coincides with the wall crossing formula for Donaldson polynomials defined on the cohomology of simply connected algebraic elliptic surfaces $X(p,q)$, hence the difference of the two functions is a polynomial diffeomorphism invariant independent of the choice of chamber. *I. Dolgachev* (Ann Arbor, MI)

99j:11047 11F46 11F30 11G18 14G40

Kudla, Stephen S. (College Park, MD)

Central derivatives of Eisenstein series and height pairings.

Ann. of Math. (2) **146** (1997), *no.* 3, 545–646.

The paper under review provides evidence that, in many cases, the central derivative of a Siegel Eisenstein series (i.e., the derivative evaluated at the center of symmetry) with odd functional equation has important relations with objects from arithmetical algebraic geometry.

There is a rich history leading to this paper. The immediate origins of the work lie in joint research of the author with B. Gross, D. Zagier and M. Harris on the L-function attached to a triple of modular forms of weight 2 by P. Garrett. However, we should go back another century in order to understand better the importance of the questions addressed in the paper. We begin with Adolf Hurwitz's paper [Math. Ann. **25** (1885), 157–196; JFM 17.0154.02], which used modular equations of elliptic functions to prove anew Kronecker's class number relation for binary quadratic forms of negative determinant.

For each positive integer D one defines the Hurwitz class number $\mathbf{H}(D)$ to be the number of $\mathrm{SL}_2(\mathbf{Z})$-equivalence classes of positive definite binary quadratic forms over $\mathbf{Z}$ with determinant D, except that the forms equivalent to $e(x_1^2+x_2^2)$ [resp. $e(x_1^2+x_1x_2+x_2^2)$] are counted with multiplicity $\frac{1}{2}$ [resp. $\frac{1}{3}$]. If $m \geq 1$ is *not* a perfect square, define

$$G(m) = \sum_{t\in\mathbf{Z}} \mathbf{H}(4m - t^2) \quad (t^2 \le 4m).$$

Hurwitz showed that

$$G(m) = \sum_{dd'=m} \max(d, d'),$$

which is the Kronecker-Hurwitz relation

$$\sum_{dd'=m} \max(d, d') = \sum_{t\in\mathbf{Z}} \mathbf{H}(4m - t^2) \quad (t^2 \le 4m).$$

It will be instructive to review Hurwitz's approach to the computation of $G(m)$. Let $j(\tau)$ denote the elliptic modular function on the upper half-plane $\mathfrak{h}$, and let j, j' be indeterminates. Let m be an integer. Kronecker's modular polynomial $\varphi_m \in \mathbf{C}[j, j']$ is characterized by the relation

$$\varphi_m(j(\tau), j(\tau')) = \prod_A (j(\tau) - j(A\tau')),$$

a finite product, where A runs over 2×2 integral matrices of determinant m taken modulo the action of $\mathrm{SL}_2(\mathbf{Z})$ by right multiplication.

Kronecker already knew that $\varphi_m(j, j') \in \mathbf{Z}[j, j']$, and that $\varphi_m(j, j) \in \mathbf{Z}[j]$ is nonzero provided that m is not a perfect square. Moreover, he knew the complete factorization of $F_m(x) = \varphi_m(x, x)$: namely, for each negative discriminant d and each integer f such that $f^2 \mid d$, let $w(d/f^2)$ be the number of roots of unity in the quadratic order of discriminant d/f^2 in $\mathbf{Q}(\sqrt{d})$ and let

$$f_d(x) = \prod_{f^2\mid d} \prod_{[\tau]} (x - j(\tau))^{2/w(d/f^2)},$$

with $[\tau]$ running over equivalence classes of integral elements τ in $\mathbf{Q}(d)$ taken modulo the right action of $\Gamma = \mathrm{SL}_2(\mathbf{Z})$; then

$$F_m(x) = \pm \prod_{t\in\mathbf{Z}} f_{t^2-4m}(x) \quad (t^2 < 4m).$$

The Kronecker-Hurwitz relation results from computing the degrees of both sides of the above equation.

In their striking paper [Invent. Math. **112** (1993), no. 2, 225–245; MR 94h:11046], Gross and K. Keating cast a new light on this. The numbers that Hurwitz had calculated (in the first part of his paper) were now interpreted as intersection numbers of divisors T_m defined by $\varphi_m = 0$ on the complex surface $S_{\mathbf{C}} = \mathrm{Spec}(\mathbf{C}[j, j'])$. Namely, in the Gross-Keating optic, Hurwitz's formula

$$\operatorname{degree} \varphi_m(x, x) = \sum_{dd'=m} \max(d, d') = \sum_{t\in\mathbf{Z}} \mathbf{H}(4m - t^2) \quad (t^2 \le 4m)$$

represents the computation

$$(T_m \cdot T_1)_{S_{\mathbf{C}}} = \sum_{t\in\mathbf{Z}} \mathbf{H}(4m - t^2) \quad (t^2 \le 4m).$$

More generally, Hurwitz computed $(T_m \cdot T_{m'})_{S_{\mathbf{C}}}$ for pairs m, m' of relatively prime integers. Gross and Keating observed that

$$(T_m \cdot T_{m'})_{S_{\mathbf{C}}} = \dim \mathbf{C}[j, j']/(\varphi_m, \varphi_{m'}).$$

They went on in their paper to compute the "arithmetic intersection number"

$$(T_m \cdot T_{m'} \cdot T_{m''})_S = \log \#\{\mathbf{Z}[j, j']/(\varphi_m, \varphi_{m'}, \varphi_{m''})\}$$

on the 3-dimensional scheme $S = \operatorname{Spec} \mathbf{Z}[j, j']$ in all cases where the ring $\mathbf{Z}[j, j']/(\varphi_m, \varphi_{m'}, \varphi_{m''})$ is finite.

Their calculations revealed a startling phenomenon: the intersection numbers, without exception, are related to the Fourier coefficients of weight-2 Eisenstein series on various Siegel half-spaces. To give examples, recall that for an element $Z = X + iY$ of Siegel's upper half-space of genus g, and for a complex number s of sufficiently large real part, one sets

$$E(Z, s) = \sum \det(CZ + D)^{-2} \frac{\det(Y)^s}{|\det(CZ + D)|^{2s}},$$

where the sum is taken over representatives $\left(\begin{smallmatrix} * & * \\ C & D \end{smallmatrix}\right)$ for the left cosets of the Siegel parabolic $\left(\begin{smallmatrix} * & * \\ 0 & * \end{smallmatrix}\right)$ in $\mathrm{Sp}_{2g}(\mathbf{Z})$. The series converges uniformly on compact subsets to a function $E(Z, s)$ that has a Fourier expansion

$$E(Z, s) = \sum_M c(M, Y, s) e^{\pi i \operatorname{tr}(MZ)},$$

where the sum is taken over even integral $g \times g$ symmetric matrices (and the meromorphic continuation in s follows from this expansion).

Specializing to (genus) $g = 2$ Eisenstein series, one has

$$(T_{m_1} \cdot T_{m_2})_{S_{\mathbf{C}}} = c' \cdot \sum_{M > 0} c(M) \quad (\operatorname{diag}(M) = (m_1, m_2)),$$

where c' is a rational constant independent of m_1 and m_2, and $c(M) = c(M, Y, 0)$ is independent of Y.

For $g = 3$, an interesting phenomenon appears. The function $E(Z, s)$ vanishes identically at $s = 0$, but $(\partial E / \partial s)(Z, 0)$ is a non-holomorphic modular form of weight 2 for the genus-3 Siegel modular group $\mathrm{Sp}_6(\mathbf{Z})$. This partial derivative is invariant under translation by symmetric integral matrices. It has a Fourier expansion with coefficients

$$c'(M) = \frac{\partial c}{\partial s}(M, Y, 0)$$

that depend on the symmetric half-integral matrix $M > 0$ but are independent of Y.

Calculations of Kudla and Zagier suggested that

$$(T_{m_1} \cdot T_{m_2} \cdot T_{m_3})_S = c_3 \sum_m c'(M)$$

with the sum taken over integral $M > 0$ with $\operatorname{diag} M = (m_1, m_2, m_3)$. This formula was later explained conceptually by Kudla [Duke Math. J. **86** (1997), no. 1, 39–78; MR 98e:11058].

The paper under review is part of a program that goes back to the author's work in the late 1980s with M. Harris, with Zagier and with Gross. For example, in their joint paper [Compositio Math. **81** (1992), no. 2, 143–209; MR 93g:11047], Gross and Kudla studied an L-function $L(f \otimes g \otimes h, s)$ associated to a triple (f, g, h) of new cusp forms of weight 2. In case f, g and h are all of the same square-free level, N, using a version (over the adèles) of Paul Garrett's integral representation of the L-function in terms of the genus-3 Eisenstein series $E(z, s-2)$, they obtained precise analytic information about $L(F, s)$ for $F = f \otimes g \otimes h$, namely, the holomorphy and the functional equation of $L(F, s)$ when s is replaced by $4-s$. In case the L-function has even functional equation, they obtained an expression for the central critical value $L(F, 2)$. Namely, if t is the number of prime factors of N, then $L(F, 2)$ factors as a period, $(F, F)/2\pi N \cdot 2^t$, times a real algebraic number that arises as an algebraic height pairing of certain homologically trivial cycles on a 3-fold product of a curve over $\mathbf{Q}$ with itself. Moreover, in those cases where $L(F, s)$ has *odd* functional equation, so that the central critical value $L(F, 2) = 0$, they presented an analogous conjectural formula for the value $L'(F, 2)$ of the derivative.

The paper under review provides evidence, at least in the case of genus 2, for an extension of the conjecture just mentioned. Take a Siegel Eisenstein series $E(Z, s-2)$ whose functional equation is odd, and consider its Fourier expansion. Such an "incoherent" Eisenstein series, as these are called, vanishes at the center of symmetry for its functional equation and, hence, so do all of its Fourier coefficients $c(M, Y, s)$. Consider instead the central value of the derivative $(\partial c/\partial s)(M, Y, 0)$ of a non-singular Fourier coefficient. Its value should have a close connection with arithmetical algebraic geometry, and in particular, with height pairings of certain algebraic cycles on Shimura varieties. More concretely, pulling the central derivatives back to suitable subgroups should give rise to sums that express the global height pairings of such algebraic cycles in terms of local height pairings. This conjecture is the motivation for the paper.

The first part of the paper establishes formulas for the derivatives of nonsingular Fourier coefficients for incoherent Siegel Eisenstein series and for their pullbacks to certain subgroups.

The second part of the paper specializes to the case of genus 2 and weight $\frac{3}{2}$. Let $W = W_1 + W_2$, where W_1 and W_2 are copies of the nondegenerate 2-dimensional symplectic space W', let $G_{\mathbf{A}} = \mathrm{Mp}(W_{\mathbf{A}})$ and let $G'_{\mathbf{A}} = \mathrm{Mp}(W'_{\mathbf{A}})$. There is an embedding $\iota: G'_{\mathbf{A}} \times G'_{\mathbf{A}} \to G_{\mathbf{A}}$. To each quadratic character χ is associated a global degenerate principal series representation $I(s, \chi)$ for $G_{\mathbf{A}}$. Standard sections Φ of $I(s, \chi)$ give rise to Eisenstein series $E(g, s, \Phi)$ on $G_{\mathbf{A}}$.

To construct incoherent Eisenstein series, one fixes an indefinite quaternion algebra B over $\mathbf{Q}$. Let V be the kernel of trace on B. For suitable quadratic characters χ, V is given a quadratic space structure. There is a natural map

$$\lambda: S(V(\mathbf{A}_f)^2) \to \Pi_2(V)_f \subset I_2(0, \chi)_f,$$

where $\Pi_2(V)$ is an irreducible constituent of $I_2(0, \chi)$. For a standard section Φ of weight $\frac{3}{2}$ such that $\Phi_f(0) = \lambda(\varphi)$ with $\varphi \in S(V(\mathbf{A}_f)^2)$, the Eisenstein series $E(g, s, \Phi)$ is incoherent.

Let $F(g_1, g_2, \Phi)$ be the pullback of the central derivative $E'(g, 0, \Phi)$ via ι. The Fourier coefficients $F_{d_1,d_2}(g_1, g_2, \Phi)$ of F are indexed by pairs of rational numbers d_1, d_2. Each coefficient has a nonsingular part which decomposes as a sum, over all primes p (including ∞), of terms $F_{d_1,d_2}(g_1, g_2, \Phi)_p$. It is these terms that are to be identified with local height pairings.

The last part of the paper is mainly concerned with weighted 0-cycles on the family of Shimura curves attached to the indefinite quaternion algebra B, now assumed to be a division algebra. Very roughly, the main result is that pulling the central derivative back via $\imath$ gives an expression for the Gillet-Soulé global height pairing for the weighted 0-cycles.

We now sketch some details. Let $H = B^{\times}$. For a compact open subgroup $K \subset H(\mathbf{A}_f)$ let X_K be the associated Shimura curve. On these the author constructs a family of weighted 0-cycles $Z(d,\varphi;K)$ on X_K parameterized by rational $d>0$ and K-invariant Schwartz functions φ on $V(\mathbf{A}_f)$. Each such cycle is a finite linear combination of special points, and the family is compatible with projections between the X_K's. For $g \in G'_{\mathbf{R}}$, the cycles are multiplied by the value $w(g)$ of a confluent hypergeometric function and are equipped with a Green function to define elements $\widehat{\mathcal{Z}}(g,d,\varphi,K)$ of the arithmetic Chow group $\widehat{\mathrm{CH}^1(X_K)}$ in the sense of Gillet and Soulé.

Suppose that $\widehat{\mathcal{Z}}(g_1,d_1,\varphi_1,K)$ and $\widehat{\mathcal{Z}}(g_2,d_2,\varphi_2,K)$ are two such elements. That is, for $i=1,2$, $g_i \in G'_{\mathbf{R}}$, $d_i \in \mathbf{Q}$, and $\varphi_i \in S(V(\mathbf{A}_f))^K$. To $\varphi_1 \otimes \varphi_2$ one associates, via the map λ, a standard section Φ of weight $\frac{3}{2}$. One part of the main result is that

$$F_{d_1,d_2}(g_1,g_2,\Phi)_\infty = 2\pi^2 \mathrm{vol}(K)\cdot\langle\widehat{\mathcal{Z}}(g_1,d_1,\varphi_1,K),\widehat{\mathcal{Z}}(g_2,d_2,\varphi_2,K)\rangle_\infty^{\mathrm{ns}},$$

where the superscript ns indicates the nonsingular part of the Archimedean height pairing. In case $d_1 d_2$ is not a square, and with some restrictions on K, a similar result holds for finite primes p in place of ∞, aside from a finite set of primes depending on χ and B.

Kudla's paper carries us further along a strong and beautiful current of mathematical ideas. *Martin L. Karel* (Wilmington, DE)

97j:11023 11F70 11F66 11M41 22E50 22E55

Bump, Daniel (Stanford, CA); **Ginzburg, David** (Tel Aviv); **Hoffstein, Jeffrey** (Providence, RI)

The symmetric cube.

Invent. Math. **125** (1996), *no.* 3, 413–449.

Let π be a cuspidal automorphic representation of $\mathrm{GL}(2,\mathbf{A})$, where $\mathbf{A}$ is the adèle ring of a global field F. Suppose further that F contains the cube roots of unity and that the central character χ of π is unitary. In this paper, the authors construct and analyze a Rankin-Selberg integral for the symmetric cube L-function $L(s,\pi,\vee^3)$. The meromorphic continuation of this L-function was first obtained by Langlands and Shahidi [cf. F. Shahidi, Compositio Math. **70** (1989), no. 3, 245–273; MR 90m:11081], by the study of Eisenstein series on the exceptional group G_2. In particular, Shahidi, making use of his theory of local coefficients, was able to establish a functional equation with L-factors at every place, and to prove that this L-function has no poles for $\mathrm{Re}(s)\geq 1$, except when π is a monomial representation attached to a Hecke character of a quadratic extension and $s=1$. The Rankin-Selberg integral exhibited here allows the authors to improve this: Suppose that for every Archimedean place v of F, χ_v is trivial, and let S be a finite set of places of F containing all Archimedean places and all places where π_v is not a spherical principal series representation. Then the authors show that the partial L-function $L_S(s,\pi,\vee^3)$ has no pole with $\mathrm{Re}(s)\geq\frac{3}{4}$ except possibly at $s=1$.

This improvement has an important application: the standard L-function $L(s,\pi)$ does not have any Siegel zeroes. This follows from the result cited together with the work of Hoffstein-Ramakrishnan, which in turn builds on previous work of Gelbart-Jacquet, Bump-Ginzburg, Hoffstein-Lockhart, and Goldfeld-Hoffstein-Lieman.

The Rankin-Selberg integral exhibited is one case of a general, still mostly undeveloped, principle originating in the work of Bump and Hoffstein [Duke Math. J. **53** (1986), no. 4, 1047–1072; MR 88d:11044]. To describe this, suppose for the moment that the base field contains the nth roots of unity, and let θ_n be the theta function on the n-fold cover of $\mathrm{GL}(n)$. This object is the residue of an Eisenstein series on the n-fold cover of $\mathrm{GL}(n)$, does have a Whittaker model, and may be described as a representation of small Gel′fand-Kirillov dimension [cf. D. A. Kazhdan and S. Patterson, Inst. Hautes Études Sci. Publ. Math. No. 59 (1984), 35–142; MR 85g:22033; correction; MR 87h:22024]. Then the Bump-Hoffstein principle is that convolving (in the Rankin-Selberg sense) an automorphic object with θ_n is always Eulerian! For $n = 2$, this is already a theme in the work of Shimura; the new idea here is the case $n > 2$. This generalization is surprising, since θ_2 may be constructed using the Weil representation, while for $n > 2$ no such construction is known. Though the theory of Rankin-Selberg convolutions for automorphic forms on various algebraic groups is somewhat developed, most of these Rankin-Selberg integrals do not carry over to the metaplectic context. Thus, carrying out the convolution envisioned by Bump-Hoffstein, even in low-rank cases, typically requires new ideas.

In the paper under review, the authors achieve the convolution of π with θ_3. To do so, they introduce and study a new Rankin-Selberg integral (not the standard one for $\mathrm{GL}(2)\times\mathrm{GL}(3)$). Namely, they induce θ_3 up to a metaplectic Eisenstein series E_s on the cubic cover of $\mathrm{GSp}(6)$, via the Siegel parabolic. They then integrate over a subgroup of the form $\mathrm{GL}(2)U$, where $\mathrm{GL}(2)$ is embedded diagonally and U is a unipotent subgroup. The integrand is E_s times a function in the space of π and a character of U.

The authors analyze this convolution by unfolding, and carry out the unramified computation. In this, they make use of the Whittaker values of θ_3, computed by Bump-Hoffstein and by Proskurin. The local factor at a good place is shown to be the quotient of $L(s,\pi_v,\vee^3)$ by the product of two $\mathrm{GL}(1)$ L-functions. They next analyze the local integrals at the bad places, showing for the non-Archimedean places that they have analytic continuation and that the local data may be chosen so as to make them nonzero at a given point. At the Archimedean places they prove absolute convergence in a certain right half-plane. When this is combined with a list of the possible poles of the Eisenstein series E_s, the theorem follows. Further study of the Archimedean integral could lead to further improvements in the study of the poles.

The authors note that "except for the previously mentioned examples of Bump and Hoffstein, the present paper is the first known example of an Eulerian Rankin-Selberg integral employing an n-fold metaplectic cover with $n > 2$". This was the case when the core calculations described here were carried out, but other such examples are now known. T. Goetze [Trans. Amer. Math. Soc. **350** (1998), no. 3, 975–1011; MR 98e:11061] gives such an integral by convolving a generic automorphic form on the cubic cover of $\mathrm{GSp}(4)$ with θ_3. T. Suzuki [Duke Math. J. **90** (1997), no. 3, 577–630; MR 99c:11053] has convolved automorphic forms on the n-fold cover of $\mathrm{GL}(r)$ with θ_n for any n to obtain Euler products. And Bump and the reviewer [in *Automorphic forms, automorphic representations, and arithmetic (Fort Worth, TX,*

1996), 1–17, Proc. Sympos. Pure Math., Part 2, Amer. Math. Soc., Providence, RI, 1999] have analyzed such integrals by the study of non-unique models in certain cases. *Solomon Friedberg* (Chestnut Hill, MA)

97k:11080 11F70 11F41 11R39 22E50 22E55
Bump, Daniel (Stanford, CA)
★**Automorphic forms and representations.**
Cambridge Studies in Advanced Mathematics, 55.
Cambridge University Press, Cambridge, 1997. xiv+574 pp. $79.95.
ISBN 0-521-55098-X

This outstanding book covers, from a variety of points of view, the theory of automorphic forms on GL(2). Along the way the reader will encounter a host of information and techniques which go beyond GL(2). It comprises a solid foundation for the serious student or the researcher interested in automorphic forms, and contains a wealth of information which is not readily accessible in textbook form elsewhere. The book also features numerous and well-thought-out exercises, at a variety of levels.

The book is actually four courses in one: each of its four chapters is a complete course, which may be studied on its own; however, the four chapters are complementary. The first chapter is written in a classical language (functions but, except in §1.8, neither representations nor the adèles), but it is far more comprehensive and sophisticated than most introductions to the theory. It begins with a treatment of Dirichlet L-functions. This is followed by a discussion of the modular group $\mathrm{SL}(2,\mathbf{Z})$ and of holomorphic modular forms. Hecke operators are defined and used to obtain the Euler product associated to such a form. Weil's converse theorem is stated and proved. Next, the non-holomorphic Eisenstein series are introduced, and the Rankin-Selberg method for $\mathrm{GL}(2)\times\mathrm{GL}(2)$ is explained. After a brief review of relevant algebraic number theory, Hecke characters are defined, and the continuation of their L-functions is proved following Hecke. Hilbert modular forms are defined, and the base change theorem of Doi and Naganuma is stated. This is the result that given a Hecke eigenform f for $\mathrm{SL}(2,\mathbf{Z})$ and a real quadratic field K, there is a Hilbert modular form F whose L-series is the product of $L(s,f)$ and $L(s,f,\chi_K)$, where χ_K is the quadratic character associated to $K/\mathbf{Q}$. Artin L-functions are defined, and the Langlands conjectures stated in rough form. In particular, it is shown how the quadratic base change result is a special case of the functoriality conjecture. The strategy exploited by Doi and Naganuma is explained heuristically. Maass forms are introduced, and the Maass forms associated to a real quadratic extension are constructed. Finally, the quadratic base change map is produced following Doi and Naganuma: by taking the Rankin-Selberg convolution of the given modular form f with the Maass forms associated to K, one obtains the continuation of the desired product and its twists, and by Weil's converse theorem, the result is thus the L-function of a Hilbert modular form.

The second chapter systematically develops the connection between the representation theory of $\mathrm{GL}(2,\mathbf{R})$ and automorphic forms on the Poincaré upper half plane. The focus is on the compact quotient case. The chapter begins by posing the question as to why there are precisely two types of automorphic forms on the Poincaré upper half plane, classical holomorphic modular forms and Maass forms, and no others.

The theory of the Laplacian is developed, and the spectral problem is formulated. Basic Lie theory is introduced, and the Laplacian recast in terms of the center of the universal enveloping algebra. Gel′fand's result that the spherical Hecke algebra for $\mathrm{GL}(n)$ is commutative is proved. Hilbert-Schmidt theory is introduced, and the discreteness of the spectrum is proved. Then the basic vocabulary of the representation theory of $\mathrm{GL}(n, \mathbf{R})$ is given: smooth vector, K-finite vector, admissible representation, $(\mathfrak{g}, K)$-module, infinitesimal equivalence. The irreducible $(\mathfrak{g}, K)$-modules for $\mathrm{GL}(2, \mathbf{R})$ are classified, and their unitarizability is studied. Finally, the representation theory developed is used to refine the spectral problem and to answer the question posed at the beginning of the chapter. The chapter concludes with brief sections on Whittaker models and on a result of Harish-Chandra concerning convolutions.

Chapter 3 provides an introduction to the modern theory of automorphic forms. The chapter begins with the adèles and Tate's thesis. The spectral theory is revisited, this time for the non-compact quotient case, and the fundamental result of Gel′fand, Graev, and Piatetski-Shapiro is used to decompose the space of cuspidal square integrable functions into discrete irreducible subspaces of finite multiplicity. Automorphic forms on $\mathrm{GL}(n)$ are defined, and the tensor product theorem of Flath is proved in full detail. Whittaker models are introduced, and the existence and uniqueness of global Whittaker models for automorphic representations of $\mathrm{GL}(2)$ are proved (with some local results stated here which are then proved in Chapter 4). The functional equation for the standard L-function on $\mathrm{GL}(2)$ is also proved (once again with some local results quoted from Chapter 4). The adelization of classical automorphic forms is covered. Next the author describes the $\mathrm{GL}(2)$ Eisenstein series, and proves their analytic continuation and functional equation from an examination of their Fourier expansions. The constant term in the Fourier expansion leads to the intertwining integrals, which are developed further in Chapter 4. The Rankin-Selberg L-function for $\mathrm{GL}(2) \times \mathrm{GL}(2)$ is defined and studied. The Langlands functoriality conjecture and the L-group are described. The chapter concludes with a brief discussion of the triple-product L-function.

The last chapter concerns the representation theory of $\mathrm{GL}(2)$ over a local field. The emphasis is on techniques which generalize to $\mathrm{GL}(n)$ or beyond, and as in the previous chapters, some results are stated for $\mathrm{GL}(n)$ but proved for $\mathrm{GL}(2)$. The chapter begins with a brief discussion of $\mathrm{GL}(2)$ over a finite field, including the Weil representation; exercises include such important topics as the Stone-von Neumann theorem and Harish-Chandra's philosophy of cusp forms. Turning to a local field, the author introduces smooth and admissible representations and studies their basic properties. The tools of distributions and sheaves are introduced, following Bernstein and Zelevinsky, in order to ultimately extend Mackey theory to locally compact groups. The uniqueness of the local Whittaker model is proved, and conversely it is shown that an irreducible admissible representation of $\mathrm{GL}(2)$ which has no Whittaker functional must be one-dimensional. In proving this, Jacquet functors are introduced, and proved to be exact. Next the principal series representations are constructed by induction from a Borel subgroup. The spherical representations are studied, and the Macdonald formula for the spherical functions and the explicit formula for the spherical Whittaker function are proved by the method of Casselman-Shalika. The unitarizability of the local principal series is discussed. The local functional equation is proved; this was needed in Chapter 3 above. Supercuspidal representations are constructed from the Weil representation, following Jacquet-

Langlands. The compatibility of the local gamma factors with parabolic induction (up to GL(2)) is also proved, using the Weil representation. The last section describes the local Langlands correspondence, and how the construction of supercuspidals achieved above fits into this.

In summary, this book covers a large amount of material concerning automorphic forms on GL(2), and the material is presented from a vantage far broader. The careful reader will have taken a significant step on the path to contemporary research in this important and beautiful area.

{Reviewer's remarks: The author is maintaining a web page on the book, with links to a virtual study group and to a list of errata, at http://math.stanford.edu/~bump/book.html.}

Solomon Friedberg (Chestnut Hill, MA)

98c:11045a 11F70 11F72 11G09 14G35 22E55
Laumon, Gérard (Orsay)
★**Cohomology of Drinfeld modular varieties. Part I.**
Geometry, counting of points and local harmonic analysis.
Cambridge Studies in Advanced Mathematics, 41.
Cambridge University Press, Cambridge, 1996. xiv+344 pp. $64.95.
ISBN 0-521-47060-9

98c:11045b 11F70 11F72 11G09 14G35 22E55
Laumon, Gérard (Orsay)
★**Cohomology of Drinfeld modular varieties. Part II. (English summary)**
Automorphic forms, trace formulas and Langlands correspondence.
With an appendix by Jean-Loup Waldspurger.
Cambridge Studies in Advanced Mathematics, 56.
Cambridge University Press, Cambridge, 1997. xii+366 pp. $64.95.
ISBN 0-521-47061-7

Eichler-Shimura theory describes the action of the Galois group on the l-adic cohomology of a modular curve in terms of Hecke correspondences and thus provides a fundamental link between the arithmetic of modular curves and modular forms. A great deal of progress in the theory of automorphic forms over the past thirty years has come from attempts to formulate and prove natural generalizations of Eichler-Shimura theory. Langlands and later Kottwitz developed a beautiful, conjectural formulation of Eichler-Shimura theory for higher-dimensional Shimura varieties using sophisticated notions from representation theory (theory of endoscopy, the Langlands group, Arthur's extended L-packets, etc.). The extension to automorphic forms over function fields was pioneered by Drinfel′d in his seminal work on Drinfel′d modules and their applications to the Langlands conjectures for GL(2) over a function field.

The two volumes under review take up the project of extending Drinfel′d's theory from GL(2) to GL(n). This is a formidable task which brings together heavy amounts of local and global harmonic analysis on GL(n) with a careful study of the cohomology of the moduli spaces for Drinfel′d modules. Fortunately for those wishing to learn this beautiful field, the author has produced an outstanding work which is equally a mathematical and an expository achievement.

Whether one is dealing with Shimura varieties attached to a reductive group G or Drinfel′d modular varieties, the basic goal is always the same: to decompose the appropriate l-adic cohomology (e.g., cohomology with compact supports, intersection cohomology, etc.) relative to the commuting actions of the Galois group and the ring of Hecke operators. In fact, after passing to the limit, the Hecke operators give way to a representation of the adelic group $G(\mathbf{A}^\infty)$. Here $\mathbf{A}^\infty$ denotes the ring of adèles away from ∞ where ∞ is a set of places of the global field F. If F is a number field, ∞ is the set of Archimedean places, and in the function field case it is the singleton consisting of the arbitrary place of F that must be fixed to define Drinfel′d modules. The joint action of Galois and Hecke yields a correspondence $\pi \to \sigma(\pi)$ between the finite parts $\pi_{\mathrm{f}} = \bigotimes_{v \neq \infty} \pi_v$ of certain cuspidal representations of $G(\mathbf{A})$ and l-adic representations of the Galois group $\mathrm{Gal}(\overline{F}/F)$. It is then of interest to describe $\sigma(\pi)$ in terms of π.

This goal is achieved in the case at hand, namely for $\mathrm{GL}(n)$ over a function field, in Chapter 12 of Volume II. The cuspidal representations π intervening in the description of the cohomology (with compact supports) of the Drinfel′d modular varieties are precisely those whose local component at the place ∞ are Steinberg representations. Theorem 12.4.1 asserts that, as expected, the correspondence $\pi \to \sigma(\pi)$ is the Langlands correspondence, i.e., it is characterized by the condition that at almost all places v, the characteristic polynomial of the conjugacy class of a Frobenius element coincides, under a suitable normalization, with the characteristic polynomial of the Langlands class attached to the local component π_v. The Ramanujan-Petersson conjecture for π is also derived using work of Jacquet-Shalika, Grothendieck and Deligne. We might add that it is precisely this sort of easily stated relation between π and $\sigma(\pi)$ which does *not* hold when the group is not $\mathrm{GL}(n)$. The correct relation can be formulated only after the introduction of endoscopic groups, and this is what substantially complicates the conjectural statements of Langlands-Kottwitz for general Shimura varieties. The hypothesis that π_∞ is a Steinberg representation is intrinsic to the approach via Drinfel′d modules taken by Laumon. To treat arbitrary cuspidal representations, it is necessary to use Drinfel′d's theory of shtukas. This was carried out by Drinfel′d himself for $\mathrm{GL}(2)$ and is currently being pursued for $\mathrm{GL}(n)$ in work of Laurent Lafforgue.

Although Eichler-Shimura theory rests on the congruence relation for Hecke correspondences, most higher-dimensional versions, including the present one, use a different approach originating in work of Ihara and Langlands, namely, a comparison of the Arthur-Selberg trace formula and the Grothendieck-Lefschetz fixed point formula. This method of proof involves two main stages, the first being concerned with preparations for the trace formula and the second with the trace formula itself. The material is correspondingly divided between the two volumes.

The first three chapters of Volume I introduce the Drinfel′d modular varieties and derive a formula, familiar from the case of Shimura varieties, for the Lefschetz numbers of an operator of the form $\mathrm{Frob}_o^r \times T$, where Frob_o is a power of the Frobenius at an (unramified) place o and T is a Hecke operator away from o and ∞. Laumon first writes the formula as a sum over an appropriate set of elliptic elements of certain terms, each of which is a product of volume factors, a twisted orbital integral at o, and an orbital away from o and ∞. The final formula only involves ordinary orbital integrals, but the idea of introducing the twisted orbital integral at this point is due to Kottwitz (in the Shimura variety case), who realized that a suitable case of the Fundamental Lemma (as formulated by Langlands) would then show that the twisted orbital integral is equal to an ordinary orbital integral. The case of the Fundamental Lemma needed here is proved in Chapter 4. The rest of Volume I is devoted to the construction of a particular function f_∞ on $\mathrm{GL}_n(F_\infty)$ whose orbital integrals give the ∞-adic contribution to the volume factors. Laumon introduces the notion of a very cuspidal function and proves that the f_∞ he has constructed is very cuspidal and is also a pseudo-coefficient for the Steinberg representation. The end result is that the Lefschetz numbers are expressed as global orbital integrals of a certain function $f_{\mathbf{A}}$ on $\mathrm{GL}_n(\mathbf{A})$. Volume I also contains an excellent discussion of the unramified principal series and four useful appendices devoted to simple algebras, Dieudonné modules, some combinatorics, and representation theory.

Most of Volume II is devoted to computing what happens when you plug $f_{\mathbf{A}}$ into Arthur's non-invariant trace formula. This trace formula asserts the equality of two complicated expressions called the geometric and spectral sides of the trace formula. Although the non-invariant trace formula depends on the choice of an arbitrary truncation parameter, Laumon develops the trick of Kazhdan, which asserts that when the power r of Frobenius is sufficiently large, this dependence disappears and the geometric side of the trace formula reduces precisely to the Lefschetz number computed in Volume I. It then remains to compute the terms on the spectral side to achieve the desired goal of relating the Lefschetz numbers to the trace of Hecke operators. This is done in (the difficult) Chapter 11 of Volume II, at least for the terms attached to a so-called regular cuspidal datum. The result for general cuspidal data, due to Lafforgue, is stated without proof and is used to derive the main theorems in Chapter 12. The last chapter of Volume II discusses some conjectures related to the intersection cohomology of Drinfel′d modular varieties. This volume also contains four very useful appendices on aspects of global harmonic analysis. For some reason, they are labelled D, E, F, G even though the fourth appendix of Volume I is also D.

As described above, these two volumes contain many results that are new and important. However, they are also the best source available for learning about the approach to zeta functions via the theory of automorphic representations. They contain a wealth of information, theorems, and calculations, laid before the reader in Laumon's superb expository style. A wide range of topics from arithmetic geometry and local or global harmonic analysis are handled with ease and completeness. In short, these two volumes are a welcome addition to the literature on automorphic representations and are highly recommended.

Jonathan David Rogawski (Los Angeles, CA)

98g:11058 11F72 11F12 11M36 22E40 57M50 57S30

Elstrodt, J. (Münster); **Grunewald, F.** (Düsseldorf); **Mennicke, J.** (Bielefeld)

★**Groups acting on hyperbolic space. (English summary)**

Harmonic analysis and number theory.

Springer Monographs in Mathematics.

Springer-Verlag, Berlin, 1998. xvi+524 pp. $89.95. ISBN 3-540-62745-6

The book under review is a very welcome contribution to harmonic analysis of hyperbolic three-manifolds and the related number theory. It is the first time that most of the relevant results in this area have been brought together in such a coherent way. The book is written very carefully, with much expertise, and the authors have taken every effort to provide as clear an exposition as possible. All chapters, and many sections, begin with a description of the purpose they serve. Chapters are concluded with helpful remarks on omissions and history. Remarks of this kind are also to be found throughout the text. The authors treat the case of dimension three only, and there truly is enough to say about this. The mathematical language is a classical one, and representation theory is not used. In this respect, the book is a complement to much of the recent literature on automorphic forms. The ten chapters, which we divide into three parts, are as follows.

Chapter 1 introduces the standard models of hyperbolic space in dimension 3. The metric, geodesics, and Laplace operator Δ are described in each case and isometries between the models are provided, as well as the exceptional isomorphisms between the matrix groups realizing the isometry group in each case. The rest of the book mostly deals with the upper half-space model $\mathbf{H}$ and its isometry group $\mathrm{PSL}(2, \mathbf{C})$. Let $\Gamma \subset \mathrm{PSL}(2, \mathbf{C})$ be a discrete group. In Chapter 2 a fundamental domain for Γ is constructed. The geometry of a certain fundamental domain for Γ is related to a presentation of Γ. We furthermore find Shimizu's lemma and Jørgensen's inequality, which give information on entries of (some pairs of) matrices in Γ. This implies lower bounds for covolumes of discrete cofinite groups acting on $\mathbf{H}$. Thurston's results on the hierarchy of covolumes of cofinite groups are described. Mostow's rigidity and Margulis' lemma are stated and a hyperbolic lattice point problem is discussed. Selberg's lemma is proved.

The next group consists of Chapters 3 to 6 and contains the most difficult parts of the book. Chapter 3 deals with the general theory of automorphic functions, which is analogous to the Maass-Selberg theory of nonholomorphic modular functions. The central objects are Poincaré and Eisenstein series (certain families of Γ-invariant functions on $\mathbf{H}$), their Fourier expansion, and the Maass-Selberg relations. A strong version of Godement's theorem is proved and the Selberg transform, linking the eigenvalues of Δ and of invariant integral operators, is explained. Chapter 4 proves the main facts concerning the action of $-\Delta$ on $L^2(\Gamma\backslash\mathbf{H})$: it is essentially self-adjoint and positive semidefinite, and the Maass-Selberg series from Chapter 3 is its resolvent kernel and has analytic continuation. This kernel is of Hilbert-Schmidt type in the cocompact case. In the cofinite case, the continuous spectrum of $-\Delta$ is calculated and the cuspidal eigenvalues are shown to be square-summable. In Chapter 5, a version of Selberg's trace formula for cocompact Γ, connecting the structure of $\Gamma \subset \mathrm{PSL}(2, \mathbf{C})$ and the spectral theory of Δ, is deduced. We also find Selberg zeta functions, Weyl's asymptotic law on the distribution of $-\Delta$-eigenvalues, and analogues of the Riemann conjecture and of some theorems of analytic number theory, providing geometric information: namely, the prime geodesic theorem. The case of cofinite but not cocompact groups is dealt with in Chapter 6. Selberg's trace formula is deduced in full generality. The language of eigenpackets is introduced and again the number of Δ-eigenvalues less than a given bound is estimated, this time not implying Weyl's law, which is not known to hold for general cofinite Γ.

The last four chapters deal with examples and to a large extent are self-contained. Chapters 7 to 9 deal with $\Gamma = \mathrm{PSL}(2, \mathcal{O})$, where $\mathcal{O}$ is the ring of integers in an imaginary quadratic number field. Explicit fundamental domains are provided and the group structure of Γ is explained geometrically. Some hints on the (conjectural) number-theoretic significance of the (co-)homology of Γ are given. $\frac{3}{4}$ is shown to be a lower bound for the first positive eigenvalue of $-\Delta$. Selberg's conjecture says that 1 should be a lower bound. Chapter 8 re-proves some facts concerning Eisenstein series for Γ by using number theory. This then leads to the validity of Weyl's asymptotic formula for congruence subgroups. The covolume of Γ is computed in two ways. Chapter 9 investigates binary Hermitian forms and their reduction theory, representation numbers, a mass formula and a third computation of the covolume. In Chapter 10 quaternion algebras are used to construct cocompact groups. Unit groups of quadratic forms enter via the Kleinian model of hyperbolic space and arithmeticity of discrete groups is discussed. We also find a detailed treatment of all discrete groups which possess a tetrahedral fundamental domain.

To conclude, the book under review is a brilliant introduction for non-experts to its subject. For the expert it will be a major source of inspiration and indispensable for reference. It should be part of every mathematical library, though the reviewer strongly recommends not leaving it there. *Stefan Kühnlein* (Karlsruhe)

97d:11090a 11F85 14G20
Coleman, Robert F. (Berkeley, CA)
Classical and overconvergent modular forms.
Invent. Math. **124** (1996), *no.* 1-3, 215–241.

97d:11090b 11F85 14G20
Coleman, Robert F. (Berkeley, CA)
Classical and overconvergent modular forms.
Les Dix-huitièmes Journées Arithmétiques (Bordeaux, 1993).
J. Théor. Nombres Bordeaux **7** (1995), *no.* 1, 333–365.

The theory of p-adic modular forms was created by B. Dwork, J.-P. Serre, and N. M. Katz, who, in the 1970s, showed how to define objects which were recognizably modular forms but which reflected the p-adic topology in crucial ways. Serre [in *Modular functions of one variable, III (Antwerp, 1972)*, 191–268, Lecture Notes in Math., 350, Springer, Berlin, 1973; MR **53** #7949a; correction; MR **53** #7949b] defined p-adic modular forms as limits (with respect to the p-adic topology) of Fourier expansions of classical modular forms, and showed that such objects have "weights" which are not necessarily integers, but rather belong to the "p-adic weight space" $\mathbf{Z}_p \times \mathbf{Z}/(p-1)\mathbf{Z}$. Katz [in *Modular functions of one variable, III (Antwerp, 1972)*, 69–190, Lecture Notes in Mathematics, 350, Springer, Berlin, 1973; MR **56** #5434] and Dwork [in *Modular functions of one variable, III (Antwerp, 1972)*, 57–67, Lecture Notes in Math., 350, Springer, Berlin, 1973; MR **48** #10985] developed an equivalent definition of p-adic modular forms as sections of line bundles over spaces obtained from classical modular curves over $\mathbf{Z}_p$ by deleting the "supersingular disks", i.e., the set of points corresponding to curves with supersingular reduction. They also pointed out that it is interesting to consider the result of restricting one's attention to those sections that can be extended to a slightly bigger space, including some points corresponding to supersingular curves. This amounts to imposing conditions on the growth of the coefficients of the Laurent expansions of these forms around the supersingular disks; such forms are called "overconvergent". Classical modular forms are, of course, overconvergent, but the space of overconvergent modular forms of a given weight and level is an infinite-dimensional p-adic Banach space.

In the 1980s, H. Hida called attention to another subspace of p-adic modular forms, the space of "ordinary" modular forms [see Ann. Sci. École Norm. Sup. (4) **19** (1986), no. 2, 231–273; MR 88i:11023; Sūgaku **39** (1987), no. 2, 124–139; MR 89a:11061]. This subspace is topologically generated by generalized eigenforms for the Atkin-Lehner U operator whose eigenvalues are p-adic units. Hida showed that the space of ordinary p-adic modular forms is particularly easy to control, and proved two crucial theorems. First, he showed that ordinary forms fall into p-adic-analytic families of varying weight. Second, he showed that any ordinary p-adic modular form of integral weight $k \geq 2$ is in fact a classical modular form.

One sees quickly that Hida's results do not extend to the full space of p-adic modular forms. For example, for any $\alpha \neq 0$ the p-adic modular forms of any weight k which are generalized eigenforms for the U operator with eigenvalues having p-adic valuation α span an infinite-dimensional space (and therefore cannot all be classical). On the other hand, one can show that any ordinary form is overconvergent. (Both results appeared in a monograph of Gouvêa [*Arithmetic of p-adic modular forms*, Lecture Notes in Math., 1304, Springer, Berlin, 1988; MR 91e:11056].) Furthermore, the Atkin-Lehner U operator is a completely continuous operator on the space of overconvergent p-adic modular forms of a given weight, which means that it has a spectral theory, and that the generalized eigenspaces for the U operator corresponding to nonzero eigenvalues will be finite-dimensional. These considerations, plus extensive numerical evidence, led Gouvêa and B. C. Mazur to specific conjectures that extended Hida's result to overconvergent forms corresponding to nonzero eigenvalues of U (the conjectures were stated in classical terms in a paper by Gouvêa and Mazur [Math. Comp. **58** (1992), no. 198, 793–805; MR 93d:11049], and their p-adic version was given in a paper by Gouvêa [in *Elliptic curves and related topics*, 85–99, Amer. Math. Soc., Providence, RI, 1994; MR 95a:11039]). The article under review settles one of those conjectures, proving a "control theorem" that locates the classical forms within the space of all overconvergent forms.

Fix a prime number $p \geq 5$, let N be an integer prime to p, and let k be an integer. We write $M_k(N)$ for the space of overconvergent p-adic modular forms of weight k on $\Gamma_1(N)$, which contains the space $M_{k,\mathrm{cl}}(N)$ of classical modular forms of weight k on $\Gamma_1(N)$ (since we want to consider the subspaces of classical forms, we restrict to the case when k is an integer). $M_k(N)$ also contains the space $M_{k,\mathrm{cl}}(N;p)$ of classical forms of weight k on $\Gamma_1(N) \cap \Gamma_0(p)$. One says $f \in M_k(N)$ is of slope $\alpha \in \mathbf{Q}$ if f belongs to the sum of the generalized eigenspaces for U corresponding to eigenvalues with p-adic valuation α. If $f \in M_{k,\mathrm{cl}}(N;p)$ has slope α, then $0 \leq \alpha \leq k-1$.

In this language, Hida's control theorem says that if $f \in M_k(N)$ has slope 0 and $k \geq 2$, then f is classical, i.e., $f \in M_{k,\mathrm{cl}}(N;p)$. Generalizing this, Gouvêa and Mazur conjectured that if $f \in M_k(N)$ had slope α and $k \geq 2\alpha + 2$, then f was classical. Coleman, in the paper under review, proves a still stronger statement: If $k \in M_k(N)$ is overconvergent of slope $\alpha < k-1$, then f is classical. This almost completely settles the issue of locating the classical forms, since no classical form can have slope greater than $k-1$. The only ambiguous case is the case of "critical slope", i.e., when $\alpha = k-1$. Coleman also obtains several results that begin to clarify what happens in this case.

The main technique in the paper is to recast the whole theory in terms of rigid analytic geometry and then to relate overconvergent modular forms to the de Rham cohomology of a coherent sheaf with a connection on an algebraic curve. Modular forms are interpreted as sections of invertible sheaves over the rigid-analytic space that is obtained by starting from a classical modular curve over $\mathbf{Q}_p$ and deleting some of the points that correspond to elliptic curves with supersingular reduction. The author considers the cohomology of these sheaves, defines the U operator in this context, and uses the Gauss-Manin connection to introduce an operator Θ, which, at the level of modular forms, is $q\,d/dq$ on q-expansions. Θ does not preserve overconvergence in general [see R. F. Coleman, F. Q. Gouvêa and N. Jochnowitz, Internat. Math. Res. Notices **1995**, no. 1, 23–41 (electronic); MR 96d:11047], but the author shows that for any $k \geq 0$ the operator Θ^{k+1} maps $M_{-k}(N)$ to $M_{k+2}(N)$. The cokernel of Θ^{k+1} can be related to a de Rham cohomology group, and this serves as the crucial tool in the proof of the theorem. An additional result gives information about the case of critical slope: any Hecke eigenform of weight $k+2>1$ and slope $k+1$ is classical if it does not equal $\Theta^{k+1}g$ for some $g \in M_{-k}(N)$. (The author gives an example to show that there *are* classical eigenforms which are in the image of Θ^{k+1}, so this does not completely resolve the issue.)

Finally, the author uses his techniques to define "overconvergent forms on $\Gamma_1(Np)$", and generalizes his results to this case. This allows him to work in a larger "weight space" similar to the one considered by Serre. He notes, also, that the methods introduced here can be used to prove a major portion of the other half of the Gouvêa-Mazur conjectures, having to do with p-adic analytic families of modular forms. These results will appear in a forthcoming paper by Coleman [Invent. Math. **127** (1997), no. 3, 417–479; MR 98b:11047, see p. A24 in this volume], and they have already led to other significant advances. For example, one can use Coleman's results to prove, in a special case, that the full deformation space of a modular residual Galois representation can be described in modular terms [see F. Q. Gouvêa and B. Mazur, in *Computational perspectives in number theory (Chicago, IL, 1995)*, 127–142, Amer. Math. Soc., Providence, RI, 1998; MR 99a:11056]. This important paper is sure to continue to have a significant impact on the field.

{The two papers listed in the heading are identical.}

{For the entire collection containing the second paper see MR 97d:11003.}

Fernando Q. Gouvêa (Waterville, ME)

97i:11062 11G09 11L05 11R58

Goss, David (Columbus, OH)

★**Basic structures of function field arithmetic.**

Ergebnisse der Mathematik und ihrer Grenzgebiete (3) [Results in Mathematics and Related Areas (3)], 35.

Springer-Verlag, Berlin, 1996. xiv+422 pp. \$124.50. ISBN 3-540-61087-1

The book under review is a thorough and very readable introduction to the arithmetic of function fields of one variable over a finite field, by an author who has made fundamental contributions to the field. It serves as a definitive reference volume, as well as offering graduate students with a solid understanding of algebraic number theory the opportunity to quickly reach the frontiers of knowledge in an important area of mathematics.

The theory presented in the book had its first beginning in the 1930s, when Leonard Carlitz discovered an "exponential" function $\mathbf{e}$ associated to the field $\mathbf{k} =$

$\mathbf{F}_q(T)$ of rational functions in one variable over a finite field. This function was given by a power series with coefficients in $\mathbf{k}$, and in modern terms it should be viewed as an analytic function from $\mathbf{C}_\infty$ to itself, where $\mathbf{C}_\infty$ is the $1/T$-adic completion of an algebraic closure of the Laurent series field $k_\infty = \mathbf{F}_q((1/T))$. In Carlitz's theory, the ring $A = \mathbf{F}_q[T]$ and the fields $\mathbf{k}$, $\mathbf{k}_\infty$, and $\mathbf{C}_\infty$ played roles analogous to those played by the integers, the rational numbers, the real numbers, and the complex numbers respectively in the theory of the classical exponential. Over a period of years, Carlitz and his students expanded this analogy in a variety of ways. For example, the function $\mathbf{e}$ is periodic, with periods consisting of the set of multiples $a\xi$ for $a \in A$ and ξ a fixed number in $\mathbf{C}_\infty$; this ξ is a "characteristic p" version of $2\pi i$, and Carlitz's student Wade proved it transcendental in the 1940s. Hayes (in the early 1970s) used Carlitz's ideas to construct the maximal abelian extension of $\mathbf{k}$, giving an explicit version of class field theory for $\mathbf{k}$ which beautifully mirrors the cyclotomic theory for the field of rational numbers.

In 1974, Drinfel′d published his spectacular paper on elliptic modules. In that paper, Drinfel′d, apparently unaware of Carlitz's and Hayes's earlier work, introduced his notion of elliptic (now called Drinfel′d) module of rank d, which vastly generalized Carlitz's analogy. In some sense, Drinfel′d's theory of elliptic modules is to Carlitz's work as the theory of general elliptic and abelian functions is to the exponential function. In particular, the analytic functions arising from Drinfel′d modules of higher rank are multi-periodic, and the theory of these multi-periodic functions leads to a theory of modular varieties analogous to the modular curves and Shimura varieties of classical number theory. Drinfel′d's goal in introducing this theory was to advance the Langlands program for function fields, where his methods have proved extremely successful [see, e.g., G. Laumon, M. Rapoport and U. Stuhler, Invent. Math. **113** (1993), no. 2, 217–338; MR 96e:11077].

The theory of Drinfel′d modules has proved to have ramifications far beyond Drinfel′d's original interest in the Langlands program. One of these ramifications is the study of the Drinfel′d modular varieties, which show fascinating similarities with (and differences from) the classical modular curves. A second is the study of the arithmetic of function fields in the tradition of Carlitz and Hayes. Some of the more striking developments in this field over the past twenty years have been Goss's invention of "characteristic p-valued" L-functions, Hayes's explicit class field theory over general function fields, Anderson's generalization of Drinfel′d modules to T-motives and his work with Thakur relating division points of such motives to special values of Goss's L-functions, Yu's results on transcendence of zeta-values, Thakur's gamma functions, and the recent work of Taguchi and Wan on meromorphic continuation of Goss's L-functions. Many of these developments are described in detail in Goss's book, and Goss provides pointers to those topics not so treated.

Chapters I and II are introductory material, including a review of non-Archimedean analysis, which applies to all of the analytic functions (exponentials, gamma functions, L-series, and so on) which arise in the theory. Chapter III is a review of Carlitz's work, which serves as a motivating example for the more advanced topics in the remainder of the book.

Chapters IV, V, and VI treat respectively Drinfel′d modules, T-motives, and shtukas. These three types of objects (all closely related) are the key arithmetic tools in the theory of function fields. The treatment of Drinfel′d modules brings out their similarities with elliptic curves and abelian varieties, and discusses their analytic properties and the structure of their rings of endomorphisms. T-motives and shtukas are generalizations of Drinfel′d modules due to Anderson and Drinfel′d. They play important roles in different parts of the theory; for example, T-motives give a concrete construction of the tensor products of Drinfel′d modules, while shtukas are important for the construction of Drinfel′d modules over function fields other than $\mathbf{k}$.

Chapter VII applies the theory developed in the earlier chapters to the class field theory of function fields. Many of the results are originally due to Hayes, who showed how to construct abelian extensions of function fields by adjoining torsion points of appropriate Drinfel′d modules. The crucial concept is that of "sign-normalized rank-one Drinfel′d modules". This chapter gives a complete description of the significance of these modules for the class field theory of a finite extension $\mathbf{L}$ of $\mathbf{k}$. The end of the chapter includes recent work on topics related to class field theory, including a summary of the role shtukas play in the construction of Drinfel′d modules over $\mathbf{L}$.

Chapters VIII and IX treat "analytic" number theory: L-functions and gamma functions. The L-functions of this theory, introduced by Goss, are not the L-functions of Artin, Hasse, and Weil, which arise from the action of Frobenius on étale cohomology and are functions of a complex variable; rather, they are (non-Archimedean) analytic functions on the space $S_\infty = \mathbf{C}_\infty^* \times \mathbf{Z}_p$. These functions do not satisfy a functional equation, and as a result they have two important classes of special values (those where $s \in S_\infty$ is a (suitably interpreted) "positive integer", and those where s is a "negative integer"). The special values for "negative integers", which Goss calls the arithmetic special values, are related to class groups in the tower of constant field extensions of the base field $\mathbf{k}$, while those for positive integers, which Goss calls the geometric special values, are related to the class groups in the tower of extensions arising from the torsion points of a Drinfel′d module. In addition to an extended discussion of these relationships, Chapter VIII presents very recent work and conjectures on the zeroes of Goss's L-functions, and a generalization of these functions to higher-dimensional fields due to Kapranov.

Chapter IX, on gamma functions, is an exposition of work primarily due to Thakur and Anderson on gamma functions and the related notions of (analogues of) Gauss and Jacobi sums. There are (at least) three different sorts of gamma functions which arise in this theory, each of which captures a different property of the classical gamma function. Much of the exposition here concentrates on the arithmetic significance of special values of gamma functions and relationships with Thakur's Gauss and Jacobi sums. The chapter includes a discussion of Thakur's work applying shtukas to obtain the factorization of Jacobi sums.

Chapter X surveys some of the current areas of research not treated in detail in the text. Notable among these areas is the work of Anderson on "solitons" and Sinha's application of this to the function field version of Deligne reciprocity, and a discussion of the work of Taguchi on the Tate conjecture for Drinfel′d modules and the work of Taguchi and Wan on meromorphic continuation of Goss's L-functions.

The seventeen-page bibliography is a comprehensive overview of the subject.

The arithmetic of function fields is a universe filled with beautiful surprises, in which familiar objects from classical number theory reappear in new guises, and in which entirely new objects play important roles. Goss's clear exposition and lively style make this book an excellent introduction to this fascinating field.

Jeremy T. Teitelbaum (Chicago, IL)

98e:11072 11G09 11R58
Sinha, Samarendra K. (Princeton, NJ)
Periods of t-motives and transcendence.
Duke Math. J. **88** (1997), *no.* 3, 465–535.

This is a very interesting paper which contains a number of important results in the arithmetic of function fields over finite fields. The paper uses the theory of solitons in characteristic p to study the transcendency properties of special values of gamma functions associated to the Carlitz module. So this paper is an outgrowth of the techniques originated by G. W. Anderson [in *The arithmetic of function fields (Columbus, OH, 1991)*, 51–73, de Gruyter, Berlin, 1992; MR 94a:11183; Duke Math. J. **73** (1994), no. 3, 491–542; MR 95b:11059].

The results in the paper under review are motivated by classical results in transcendency theory and, in particular, by a beautiful result of Wüstholz which we now recall. Theorem: Let Y be a smooth curve over the algebraic closure of $\mathbf{Q}$. Let ω be a nonexact meromorphic differential and let D be the polar divisor of ω. Let λ be a closed path representing a homology class in $H_1(Y \setminus D, \mathbf{Z})$. Then $\int_\lambda \omega$ is either 0 or transcendental. In particular, by applying the theorem to the periods of some quotients of the Fermat curves, one can deduce the transcendency of certain special values of the classical Euler beta-function.

To begin the function field theory, we need to describe the setup of the "two T's". In the arithmetic of $\mathbf{F}_q[T]$, the element T really plays two distinct roles: in extensions of $k := \mathbf{F}_q(T)$ it is a scalar, whereas via a Drinfel′d module, or T-module, it is an operator. In function field theory, it is possible to actually separate these two distinct roles of T by using rings of the form $\mathscr{A} := \mathbf{F}_q[T] \otimes_{\mathbf{F}_q} \mathbf{F}_q[T]$, etc. Here the first copy of $\mathbf{F}_q[T]$ represents the scalars while the second represents the operators. In order to have a simplicity of notation, the author lets $T := T \otimes 1 =$ "T as scalar" and $t := 1 \otimes T =$ "T as operator". For example, let C be the Carlitz module given by $C_T = T\tau^0 + \tau$ (where $\tau(x) = x^r$). This can then be written in two-T's form $t \cdot x = Tx + x^r$, so that $t - T$ just equals τ. The two-T's approach precisely leads to a very useful reformulation of much of the theory of Drinfel′d modules in the category of $\mathscr{A}$-modules, etc.

Next we discuss the particular Γ-function that is being studied in this paper; for more details we refer the reader to Section 9.9 of *Basic structures of function field arithmetic* [D. Goss, Springer, Berlin, 1996; MR 97i:11062, see p. A41 in this volume]. The Γ-function will live in the universe of scalars. It is defined as

$$\Gamma(z) := \frac{1}{z} \prod_{n \in \mathbf{F}_q[T] \text{ monic}} \left(1 + \frac{z}{n}\right)^{-1};$$

it clearly converges for all $z \neq n$ where n is a monic polynomial or 0. It is called the "geometric gamma function" and is easily related to the exponential of the Carlitz module.

It has been known for over two decades that it is possible, using rings like $\mathscr{A}$, to describe Drinfel′d modules in terms of locally free sheaves in a manner following

Krichever's work on KdV [see, e.g., D. Mumford, in *Proceedings of the International Symposium on Algebraic Geometry (Kyoto Univ., Kyoto, 1977)*, 115–153, Kinokuniya Book Store, Tokyo, 1978; MR 83j:14041]. Let f be a monic polynomial in t and let ζ_f be a generator of the $\mathbf{F}_q[t]$ module of f-division points under the Carlitz module. Let X_f be a smooth projective model of the field $k_f := \mathbf{F}_q(T, \zeta_f)$. One knows that k_f/k is a geometric extension (i.e., contains no finite extensions of $\mathbf{F}_q$) and that $\mathrm{Gal}(k_f/k) \cong \mathbf{F}_q[t]/(f)^*$. For each $a \in \mathbf{F}_q[t]/(f)^*$ we let $[a]$ be the unique isomorphism of $X_f/\mathbf{P}^1$ such that $[a]\Theta = \sigma_a \circ \Theta$, where Θ is a generic point of X_f realizing the above description of its function field. Following the analogy with KdV, Anderson was able to construct certain functions defined on an open subset of $X_f \times_{\mathrm{Spec}(\mathbf{F}_q)} X_f$ which he called "solitons". A little more precisely, Anderson showed that there is a meromorphic function φ on $X_f \times X_f$ such that for all a, as above, and all positive integers N, we have

$$1 - \varphi(\mathrm{Frob}^N \Theta, [a]\Theta) = \prod_{\substack{n \in \mathbf{F}_q[T] \text{ monic} \\ \deg(n) = N-1}} \left(1 + \frac{a}{fn}\right), \tag{1}$$

where "Frob" is the Frobenius morphism. Note the similarity between the above product and the definition of the geometric Γ-function.

The calculation of the divisor of the soliton by Anderson and Sinha established that it satisfied an equation quite similar to that of the "shtuka function" associated to sign-normalized rank-one Drinfel′d modules [D. Goss, op cit. (Subsection 7.11)]. This then allows the author to construct certain "geometric" t-modules which play a role similar to that of the Jacobians of Fermat curves. These t-modules naturally come equipped with complex multiplication, which makes them behave like rank-one modules. On the other hand, these t-modules are not 1-dimensional and so it is not automatic that they are uniformizable (i.e., that their exponential functions are surjective). The key to the author's uniformizability results is the long diagram in Subsection 1.3.3. This diagram, due to Anderson and D. S. Thakur [Ann. of Math. (2) **132** (1990), no. 1, 159–191; MR 91h:11046], reinterprets the canonical long exact sequence of the exponential of the Carlitz module in terms of homological algebra. The author then writes down a similar sequence for his geometric t-modules. The uniformizability is thus established through the use of a "Green's function" which plays a role very similar to the one it plays in classical arithmetic [see B. H. Gross, in *Arithmetic geometry (Storrs, Conn., 1984)*, 327–339, Springer, New York, 1986; see MR 89b:14029].

Equation 1 implies that there should be some relationship between these geometric Γ-functions and the periods of the geometric t-modules of the author, and indeed such results are established by the author. Transcendency results are then deduced via results of Gel′fond-Schneider-Lang type due to J. Yu [e.g., Duke Math. J. **58** (1989), no. 3, 559–575; MR 90j:11056].

The reader may well wonder about an analogue of the "two T's" for classical arithmetic. However, as $\mathbf{Z} \otimes_{\mathbf{Z}} \mathbf{Z}$ *is* $\mathbf{Z}$ such a separation is impossible (and one needs to check which role $\mathbf{Z}$ is playing in any given situation). By abuse of language and science, one can view this as a sort of "wave-particle duality" in number theory....

It is my pleasure to thank Dinesh Thakur for kindly sharing his very valuable insights into the paper under review. *David Goss* (Columbus, OH)

98f:11062 11G09 11G05 11G10

Pink, Richard (Mannheim)

The Mumford-Tate conjecture for Drinfeld-modules. (English summary)

Publ. Res. Inst. Math. Sci. **33** (1997), *no.* 3, 393–425.

This paper establishes results of fundamental importance in the arithmetic of Drinfel′d modules. It is concerned with Galois-theoretic properties of Drinfel′d modules and is very readable; the interested reader is highly encouraged to also consult it directly. Our notation here will be different than the author's and well suited for the study of L-series (mentioned at the end of this review). Let $\mathscr{X}$ be a smooth, projective, geometrically irreducible curve over the finite field $\mathbf{F}$ of characteristic $p>0$. Let $\infty\in\mathscr{X}$ be a fixed closed point and let $\mathbf{A}$ be the ring of those rational functions on $\mathscr{X}$ which are regular outside ∞. It is well known that $\mathbf{A}$ is a Dedekind domain with a finite class group and unit group $\mathbf{F}^*$. Let $\mathbf{k}$ be the quotient field of $\mathbf{A}=$ the field of rational functions on $\mathscr{X}$, and let $\mathbf{K}=\mathbf{k}_\infty$. There are basic analogies $\mathbf{A}\sim\mathbf{Z}$, $\mathbf{k}\sim\mathbf{Q}$, and $\mathbf{K}\sim\mathbf{R}$—for instance, $\mathbf{A}$ is discrete and co-compact in $\mathbf{K}$ as $\mathbf{Z}$ is in $\mathbf{R}$. The rings $\mathbf{A}$, $\mathbf{k}$, etc., will be the "rings of operators". We now take another copy of these rings, which will be denoted by non-bold characters; these are the "rings of scalars". Let θ be the canonical isomorphism from bold to non-bold rings. The basic objects in the theory are then defined over rings of scalars and possess actions by the rings of operators. Thus, let $\mathscr{F}$ be a finitely generated field extension of k. Over $\mathscr{F}$ we have the ring of algebraic endomorphisms of $\mathbf{G}_a$ (= the non-commutative ring of polynomials in the pth power mapping), which we denote by $\mathrm{End}_{\mathscr{F}}(\mathbf{G}_a)$. A Drinfel′d module, then, is an $\mathbf{F}_r$-algebra map $\varphi\colon\mathbf{A}\to\mathrm{End}_{\mathscr{F}}(\mathbf{G}_a)$ such that for $a\in\mathbf{A}$ we have (1) $\varphi_a(x)=\theta(a)x+\{\text{higher terms involving } x^{p^i}\}$, and (2) for some $a\in\mathbf{A}$, $\varphi_a(x)\neq\theta(a)x$. Once one has this basic concept, one easily defines analogues of standard notions from elliptic curves such as Tate modules, heights and so on. These Tate modules are free modules over the completions $\mathbf{A}_\lambda$, $\lambda\in\mathrm{Spec}(\mathbf{A})$, of constant rank d (called the "rank of φ").

Such elementary algebraic analogies between Drinfel′d modules and elliptic curves (or abelian varieties) lead one to expect deeper analogies as well. The ones of interest here are precisely those involving Galois representations on Tate modules as follows. Let $\overline{\mathscr{F}}$ be a fixed algebraic closure of $\mathscr{F}$ and let $\mathscr{F}^{\mathrm{sep}}\subset\overline{\mathscr{F}}$ be the separable closure. Let λ and $T_\lambda(\varphi)$ be as above. Since the canonical Galois action commutes with the $\mathbf{A}$-action via φ, we deduce the existence of a representation $\rho_\lambda\colon\mathrm{Gal}(\mathscr{F}^{\mathrm{sep}}/\mathscr{F})\to\mathrm{Aut}_{\mathbf{A}_\lambda}(T_\lambda(\varphi))$. Upon tensoring with the completion $\mathbf{k}_\lambda$, we obtain the rational Tate module $V_\lambda(\varphi)$ equipped with the obvious Galois action.

Before stating the results established by the author, it is worthwhile to recall some basic results about these rational Tate modules that are needed by the author. Theorem 1: $V_\lambda(\varphi)$ is a semisimple $\mathbf{k}_\lambda[\mathrm{Gal}(\mathscr{F}^{\mathrm{sep}}/\mathscr{F})]$-module. This result was originally proved by Y. Taguchi [J. Number Theory **44** (1993), no. 3, 292–314; MR 94k:11064] when $\mathscr{F}$ is finite over $\mathbf{k}$. The author extends Taguchi's result to arbitrary finitely generated $\mathscr{F}$. Theorem 2 ("Tate conjecture for Drinfel′d modules"): The natural map

$$\mathrm{End}_{\mathbf{k}}(\varphi)\otimes_{\mathbf{A}}\mathbf{k}_\lambda\to\mathrm{End}_{\mathbf{k}_\lambda[\mathrm{Gal}(\mathscr{F}^{\mathrm{sep}}/\mathscr{F})]}(V_\lambda(\varphi))$$

is an isomorphism. This is due to Taguchi [Proc. Amer. Math. Soc. **123** (1995), no. 11, 3285–3287; MR 96a:11052; see also J. Number Theory **60** (1996), no. 1, 124–141; MR

97e:11067] and simultaneously to A. Tamagawa [Sūrikaisekikenkyūsho Kōkyūroku No. 925 (1995), 89–94; see MR 97f:11002].

We now turn to the results of the author. These are motivated by classical results and techniques of J.-P. Serre on elliptic curves and abelian varieties [see, e.g., *Abelian l-adic representations and elliptic curves*, W. A. Benjamin, Inc., New York-Amsterdam, 1968; MR **41** #8422; Invent. Math. **15** (1972), no. 4, 259–331; MR **52** #8126]. Note that $\mathrm{End}_{\mathscr{F}}(\varphi)$ acts on $V_\lambda(\varphi)$ and commutes with the Galois action.

Theorem 3: Suppose that $\mathrm{End}_{\mathscr{F}}(\varphi) = \mathrm{End}_{\overline{\mathscr{F}}}(\varphi)$. Let Λ be a finite set of primes of $\mathbf{A}$. Then the image of the natural map

$$\mathrm{Gal}(\mathscr{F}^{\mathrm{sep}}/\mathscr{F}) \to \prod_{\lambda\in\Lambda} \mathrm{Cent}_{\mathrm{Aut}_{\mathbf{k}_\lambda}(V_\lambda(\varphi))}(\mathrm{End}_{\mathscr{F}}(\varphi))$$

is open.

In particular, if $\mathrm{End}_{\overline{\mathscr{F}}}(\varphi) = \mathbf{A}$ then $\mathrm{Cent}_{\mathrm{Aut}_{\mathbf{k}_\lambda}(V_\lambda(\varphi))}(\mathrm{End}_{\mathscr{F}}(\varphi)) = \mathrm{Aut}_{\mathbf{k}_\lambda}(V_\lambda(\varphi))$.

Now let X be an integral scheme of finite type over $\mathbf{F}_r$ whose function field is $\mathscr{F}$. Note that as $\mathscr{F}$ is an extension of k, the ring of absolute endomorphisms of φ is an order in a finite field extension of $\mathbf{k}$.

Theorem 4: Let Δ be the finite quotient of $\mathrm{Gal}(\mathscr{F}^{\mathrm{sep}}/\mathscr{F})$ which acts faithfully on $\mathrm{End}_{\overline{\mathscr{F}}}(\varphi)$. Let p^e be the degree of the totally inseparable part of $\mathrm{End}_{\overline{\mathscr{F}}}(\varphi)$ over $\mathbf{A}$. (a) Let x be any closed point of X where φ has good reduction. Then the height of this reduction is divisible by p^e. (b) Let l be an integer ≥ 1. Then the set of closed points of X at which φ has good reduction of height $p^e l$ has Dirichlet density

$$\frac{\mathrm{card}\{\delta \in \Delta \mid \mathrm{ord}(\delta) = l\}}{\mathrm{card}(\Delta)}.$$

As a corollary, if $\mathrm{End}_{\overline{\mathscr{F}}}(\varphi) = \mathbf{A}$ then the set of closed points where φ has good ordinary reduction has Dirichlet density 1. For some results on those points with non-ordinary reduction, we refer the reader to M. L. Brown [Invent. Math. **110** (1992), no. 2, 419–439; MR 93m:11048] and C. David [J. Number Theory **56** (1996), no. 2, 366–380; MR 96j:11080].

For the proof of his results, the author first reduces to the case where φ has absolutely no complex multiplications. Then Theorems 1 and 2 are used to show that the Zariski closure of the image is $\mathrm{Aut}_{\mathbf{k}_\lambda}(V_\lambda(\varphi))$. The openness eventually follows from results of D. R. Hayes on the rank-1 case [in *Studies in algebra and number theory*, 173–217, Academic Press, New York, 1979; MR 81d:12011], and results of the author on the arithmetic of φ and algebraic groups ["Compact subgroups of linear algebraic groups", Preprint, 1996; per bibl.].

As the representation-theoretic properties of Drinfel′d modules mirror so well those of classical objects, it is reasonable to hope that this also holds for the associated L-functions [see D. Goss, *Basic structures of function field arithmetic*, Springer, Berlin, 1996; MR 97i:11062, see p. A41 in this volume (Subsections 8.6, 8.24)]. Let $f(u)$ be any entire power series over a complete non-Archimedean field L. It is easy to see that any complete field containing the zeroes must *also* contain the coefficients; so a factorization of $f(u)$ into a product $\prod f_i(u)$ of entire power series with coefficients generating a *proper* extension of L automatically implies that the zeroes of f cannot *all* be in L. Conversely, let $\mathbf{k} = \mathbf{F}_r(T)$, etc., and let φ be a Drinfel′d module defined over $\mathscr{F} = k$ with no complex multiplications over $\bar{k}$. From the results presented above, one knows that the Tate modules are absolutely irreducible Galois modules; in particular, the Brauer-Nesbitt theorem (as treated by J.-P. Serre [see D. Goss, in *The arithmetic of function fields (Columbus, OH, 1991)*, 313–402, de Gruyter, Berlin, 1992; MR 94b:11048 (§3.8)] and semi-simplicity imply that the L-series of φ determines φ up to isogeny. Moreover, from the author's results we deduce that the L-series of φ over k does *not* inherit a factorization arising from representation theory. Therefore it is reasonable to inquire as to whether the roots of this L-series generate a "small" extension of $\mathbf{K}$. This is now known for the Carlitz module where this small extension is seen to be $\mathbf{K}$ itself [see D. Q. Wan, J. Number Theory **58** (1996), no. 1, 196–212; MR 97c:11064; J. Diaz-Vargas, J. Number Theory **59** (1996), no. 2, 313–318; MR 97m:11078; J. Sheats, J. Number Theory **71** (1998), no. 1, 121–157; MR 99b:11127, see p. A96 in this volume]. The notation used here *suggests* that these fields of roots will eventually be given a natural interpretation in terms of the Drinfel′d action. Finally, it should be noted that D. Wan has also pointed out to the present author how these "RH" results at ∞ also apply (at least for a very large set of exponents) to the v-adic interpolations of $L(\varphi, s)$ at primes of degree 1. Computer calculations also show that these results will *not* extend trivially to interpolations at primes of higher degree. However, it does appear that whatever theory is ultimately behind these results should involve statements at *all* the places of k, and this is very different from what is expected, at the moment, in classical number theory. *David Goss* (Columbus, OH)

99h:11062 11G09 46F99 47S10
Yang, Zifeng (Columbus, OH)
Locally analytic functions over completions of $\mathbf{F}_r[U]$. (English summary)
J. Number Theory **73** (1998), *no.* 2, 451–458.

In his original paper on the function $\zeta(s) := \sum_{n=1}^{\infty} n^{-s}$, $\mathrm{Re}(s) > 1$, B. Riemann presented two distinct proofs of its meromorphic continuation to the whole complex plane with poles only at $s = 0$ and $s = 1$ (see "Über die Anzahl der Primzahlen unter einer gegebenen Grösse" ("On the number of primes less than a given magnitude") [Monatsber. Konigl. Preuss. Akad. Wiss. Berlin **1859**, 671–680; reprinted in *Gesammelte mathematische Werke und wissenschaftlicher Nachlass*, Art. VII, Teubner, Leipzig, 1876; see JFM 08.0231.03]). The first proof is a clever use of contour integration. The second proof uses a theta function and has been the model for much of what followed in number theory. Here, briefly, is how the proof goes (albeit in a slightly different form from the original): Let

$$\theta(\tau) := \frac{1}{2} \sum_{n=-\infty}^{\infty} e^{in^2\pi\tau} = \frac{1}{2} + \sum_{n=1}^{\infty} e^{in^2\pi\tau}$$

be the theta function, and let $\Lambda(s) := \pi^{-s}\Gamma(s)\zeta(2s)$ where, of course, $\Gamma(s)$ is Euler's gamma function. Put $d\theta := (\theta(it) - \frac{1}{2})\, dt/t$. As may be checked through a simple substitution, one has

$$\Lambda(s) = \int_0^{\infty} t^s\, d\theta = \int_0^{\infty} t^s \left(\theta(it) - \frac{1}{2}\right) \frac{dt}{t}.$$

Now $\theta(\tau)$ satisfies the well-known functional equation $\theta(-1/\tau) = (\tau/i)^{1/2}\theta(\tau)$. Thus we find

$$\Lambda(s) = \int_1^\infty (t^s + t^{1/2-s})\, d\theta - \frac{1}{2s} - \frac{1}{1-2s}.$$

This immediately gives the desired continuation as well as the functional equation of $\zeta(s)$ (in the form of invariance under $s \mapsto \frac{1}{2} - s$). The theta function is an example of an automorphic form (and one might call $d\theta$ the corresponding "automorphic measure"). In the general Langlands philosophy, L-series coming from arithmetic (via various types of Galois representations) should be among those coming from automorphic forms—this association is called "nonabelian reciprocity" when the representations are higher-dimensional. Thus, conjecturally, the original L-series can be analytically continued via integration over the associated (adelic) automorphic measure; the functional equation then follows easily. Moreover, if the automorphic measure corresponds to an irreducible cuspidal representation of GL_n, $n \geq 2$, then it is easy to see that one also *automatically* obtains the analytic continuation and functional equation—without poles—of the twists of the original L-series via abelian characters. Thus one automatically deduces the analytic continuation of all partial L-series (i.e., where one only sums the terms in the original L-series over some residue class mod m for some modulus m). Consequently the association of an automorphic form (and measure) to an L-series $L(s)$ is more powerful in general than merely giving the analytic nature of $L(s)$. This is important to keep in mind for what is mentioned below.

Now let $\mathscr{X}$ be a smooth, complete, geometrically connected curve over the finite field $\mathbf{F}_r$. Let $\mathbf{k}$ be the field of rational functions on $\mathscr{X}$; thus $\mathbf{k}$ is a global field of characteristic p. One expects, and knows in many cases, that the Langlands philosophy also holds for characteristic-0 Galois representations associated to $\mathbf{k}$. On the other hand, within the past couple of decades, L-series have been constructed which are naturally associated to *finite* characteristic Galois representations (and so these representations do not a priori fit into the Langlands philosophy). The paper under review provides results needed to formulate a good "integral calculus" approach to the analytic continuation of these L-series. To begin with, following Drinfel′d, we choose a fixed closed point on $\mathscr{X}$ which we call ∞. We then define $\mathbf{A}$ to be the affine algebra of $\mathscr{X} \setminus \infty$. The completion of $\mathbf{k}$ at ∞ is denoted $\mathbf{K}$ and the completion of a fixed algebraic closure of $\mathbf{K}$ (with the canonical absolute value) is called $\mathbf{C}_\infty$. These bold-face algebras are the rings of "operators". One then takes another non-bold copy of these rings A, k, etc. These are the rings of "scalars". The canonical map "bold to non-bold" $(a \mapsto \overline{a})$ makes the non-bold algebras into $\mathbf{A}$-algebras. Drinfel′d modules, T-modules, shtukas, etc., are now defined over the scalars and, by definition, come equipped with actions by $\mathbf{A}$, etc. So for instance, the simplest Drinfel′d module, the "Carlitz module" C, is defined over $k = \mathbf{F}_r(T)$ and is given by $C_T(x) = \overline{T}x + x^r$. While the "two T's" notation of bold and non-bold algebras may seem to be overkill in the category of Drinfel′d modules, there are actually two very good reasons for using it (following Drinfel′d and Anderson). The first is that with general T-modules (such as higher powers of the Carlitz module), the differential of the action of an element $a \in \mathbf{A}$ is *not* scalar multiplication by $\overline{a}$; rather, it differs from this action by a nontrivial nilpotent (so is "infinitesimally close" to scalar multiplication). Secondly, one can naturally describe these objects as certain modules over two-dimensional algebras such as $k \otimes_{\mathbf{F}_r} \mathbf{A}$. (It is quite instructive to realize that both reasons evaporate in classical arithmetic, as is readily seen.) In any case, let L be a finite extension of k and let φ be a Drinfel′d $\mathbf{A}$-module over L. One knows that almost all "finite" (i.e., lying over $\mathrm{Spec}(A)$) primes $\mathfrak{P}$ of L are "good" for φ in the sense that one can reduce φ to obtain a Drinfel′d module of the *same* rank over the associated finite field. As with elliptic curves, or abelian varieties, the Frobenius $F_{\mathfrak{P}}$ over this finite field gives rise to an endomorphism of the reduced Drinfel′d module, and thus an action on Tate spaces T_v where v is prime to $\mathfrak{P}$—note that by definition T_v is a module over the completed operator algebra $\mathbf{A}_v$. Again in analogy with abelian varieties, the polynomial in t, $\det(1-tF_{\mathfrak{P}}|T_v)$, has coefficients in $\mathbf{A}$ which are independent of v. Moreover, V. G. Drinfel′d [Mat. Sb. (N.S.) **102(144)** (1977), no. 2, 182–194, 325; MR **55** #12644] has established that the roots of this polynomial satisfy the "Riemann hypothesis" in terms of their absolute value at infinite primes. Very similar results are known for T-modules, etc.

Therefore, following classical theory it is only natural to define the L-series of φ as

$$L(\varphi, s) := \prod_{\mathfrak{P}\text{ good}} f_{\mathfrak{P}}(n\mathfrak{P}^{-s})^{-1}$$

where n is the ideal norm of $\mathfrak{P}$ to $\mathbf{A}$. This works, except that we need an appropriate "s". In classical arithmetic s is a complex number, of course. Here s is taken in the space $S_\infty := \mathbf{C}_\infty^* \times \mathbf{Z}_p$. Moreover, if $s=(x,y)\in S_\infty$ and I is a fractional $\mathbf{A}$-ideal, one sets $I^s := x^{\deg I}\cdot\langle I\rangle^y$ where $\langle I\rangle \in \mathbf{C}_\infty^*$ is a 1-unit defined as follows: Upon fixing a uniformizer $\pi\in\mathbf{K}^*$ one deduces the "sign" of an element $a\in\mathbf{A}$ (completely analogous to taking the coefficient of highest degree in a polynomial), and so the notion of "positive elements" (i.e., elements with sign $=1$). Now from the finiteness of the class group one deduces the existence of a positive integer v such that $I^v=(i)$ where i is positive. Then one sets $\langle I\rangle := (\pi^{\deg i} i)^{1/v}$, where the unique 1-unit root is taken—the elements $\langle I\rangle$ all belong to a finite, totally inseparable extension $\mathbf{K}_1$ of $\mathbf{K}$. The definition I^s satisfies all the usual properties and, combined with the Riemann hypothesis over finite fields, gives an L-series with a "half-plane of convergence" in S_∞ (i.e., those (x,y) where $|x|_\infty$ is sufficiently large). Very similar definitions are given v-adically and one sees that the theory at the finite and infinite primes is much closer here than in classical theory.

The paper under review presents some results related to the analytic continuation of these L-series via integral calculus. Basic to this is the concept of a non-Archimedean measure: Let Y be a compact totally disconnected space (for example, $Y=\mathbf{Z}_p, \mathbf{F}_q[\![u]\!]$, or a Galois group). Let K be any local non-Archimedean field (e.g., $\mathbf{Q}_p$ or $\mathbf{F}_q((u))$). Then a non-Archimedean measure μ is a finitely additive K-valued function on the compact open subsets of Y whose image lies in a compact subset of K (i.e., is bounded). Given a continuous function $f\colon Y\to K$, one can form the "Riemann sums" in the obvious way; a nice calculation shows that they converge to an element of K naturally denoted by $\int_Y f(t)\,d\mu(t)$. In practise, such measures are characterized in the following fashion. The continuous K-valued functions on Y form a "non-Archimedean Banach space equipped with an orthonormal basis". That is, there exists a countable collection $\{b_i(t)\}_{i=0}^\infty$ of continuous functions such that every continuous function $f(t)$ may be written (uniquely) as $f(t)=\sum_{i=0}^\infty c_i b_i(t)$ where the coefficients c_i are in K and tend to 0 as $i\to\infty$, and such that $\max_{y\in Y}|f(y)|=\max_i|c_i|$. Thus $\int f(t)d\mu(t)=\sum c_i\beta_i$ where $\beta_i := \int b_i(t)d\mu(t)$. Conversely one obtains a measure simply by choosing any bounded sequence $\{\beta_i\}$

and setting $\int f := \sum c_i \beta_i$; the sum converges precisely because the sequence $\{\beta_i\}$ is bounded. The simplest measures of all (the "Dirac measures") are defined by the sequence $\delta_y := \{b_i(\alpha)\}$ where $\alpha \in Y$; thus $\int f(t) d\delta_\alpha(t) = f(\alpha)$.

There are (at least) two canonical choices of basis functions $b_i(t)$: In the "classical case" of p-adic valued functions on $\mathbf{Z}_p$, one sets $b_i(t) = \binom{t}{i}$ and the existence of such an expansion is Mahler's theorem. In the case of $\mathbf{F}_q[\![u]\!]$, the canonical construction is due to L. Carlitz [Duke Math. J. **6** (1940), 486–504; MR **1**, 324a] and the corresponding expansion is due to C. G. Wagner [Acta Arith. **17** (1970/1971), 389–406; MR **44** #207]—Carlitz's functions also provide a good description of the $\mathbf{F}_q$-linear continuous functions. In a fundamental paper published in 1964, Y. Amice gave an abstract construction of such a basis in a great many cases generalizing Mahler's result [see Y. Amice, Bull. Soc. Math. France **92** (1964), 117–180; MR **32** #5638; see also M. Bhargava and S. Kedlaya, "Continuous functions on compact subsets of local fields", Acta Arith., to appear]. The advantage of the above concrete bases over the abstract ones is that the concrete bases also satisfy certain functional identities which precisely allow one to compute the convolution algebra of measures (where the convolution product is just as in classical measure theory). In the case of $\mathbf{Z}_p$ and measures with values in $\mathbf{Z}_p$, one obtains an isomorphism with the formal power series domain $\mathbf{Z}_p[\![X]\!]$; this may be viewed as a manifestation of the Fourier transform in p-adic theory. In the case of $\mathbf{F}_q[\![u]\!]$ and $\mathbf{F}_q[\![u]\!]$-valued measures, the Carlitz functions provide an isomorphism with the algebra of formal divided power series $\{\sum \alpha_i X^i/i!\colon \{\alpha_i\} \subset \mathbf{F}_q[\![u]\!]\}$. As of this writing, this identification remains very mysterious, although a hint may lie in viewing such divided power series as being the divided differential operators on the Tate algebra of rigid analytic functions on the closed unit disc.

Amice [op. cit.] also provided a beautiful characterization of those functions on $\mathbf{Z}_p$ or $\mathbf{F}_q[\![u]\!]$ which possess various degrees of analyticity. The main result for $\mathbf{Z}_p$ may be stated as follows: Let $f(t) = \sum c_i \binom{t}{i}$ be a continuous function from $\mathbf{Z}_p$ to $\mathbf{Q}_p$. Then the function $f(t)$ is locally analytic if and only if $\liminf_i \{v_p(c_i)/i\} > 0$. A similar result was established by Amice for the general bases that she constructed. In the paper under review, Yang shows how to deduce from Amice a completely similar result for the Carlitz polynomials by using them to obtain explicit bases for various Banach spaces of analytic functions; in fact, Yang's proof works for *any* polynomial orthonormal basis for the space of continuous functions. As a fundamental consequence, one can still "integrate" locally analytic functions against sequences $\{\beta_i\}$ which are no longer bounded as long as the sum $\sum_i c_i \beta_i$ still converges. Such sequences are said to give rise to "tempered distributions".

It is now possible to explain the connection of all of this with L-series. Let $L(s)$, $s \in S_\infty$, be some L-series arising from arithmetic which we write additively as $L(s) = \sum_I a_I I^{-s}$ where I runs over the ideals of $\mathbf{A}$. We now associate to $L(s)$ the family of measures $\mu_x = \mu_x^L$ defined by $\mu_x = \sum_I a_I x^{-\deg I} \delta_{\langle I \rangle}$. For all x large enough so that (x, y), $y \in \mathbf{Z}_p$, lies in the half-plane of convergence of the Euler product for $L(s)$, it is trivial to see that μ_x converges to a measure. By definition,

$$\int_{\widetilde{U}_1} t^{-y}\, d\mu_x = L(x, y)$$

where $\widetilde{U}_1 \subset \mathbf{K}_1^*$ is the group of 1-units. A completely similar definition is given v-adically for finite primes v. In an elegant paper [J. Number Theory **35** (1990), no. 1,

1–17; MR 91e:11139] D. S. Thakur has calculated the v-adic measures associated to the simplest L-series $\zeta_A(s) := \sum_{a \in \mathbf{F}_r[T] \text{ monic}} a^{-s}$. Remarkably, it is clear from Thakur's result that when the v-adic measure μ_x blows up (i.e., becomes unbounded), it only blows up *logarithmically* as a function of i; so μ_x remains tempered! Thus for *all* x (i.e., not just those coming from the convergence of the Euler product), one can still integrate the locally analytic function which is t^{-y} (or more precisely its v-adic cousin) on the units and 0 elsewhere. This precisely gives the v-adic analytic continuation of $\zeta_A(s)$! If one integrates t^{-y} against the measure $\chi_V(t)d\mu_x$, where χ_V is the characteristic function of some residue class V, then we also obtain the analytic continuation of partial L-functions. Thus, in one stroke we obtain a powerful analog of classical theory. Note that there are not enough characters with values in completions of $\mathbf{k}$ to obtain this result simply by twisting as in classical theory!

One expects this "logarithmic blowing up" phenomenon to work in complete generality; for *all* L-series coming from arithmetic at *all* the places of $\mathbf{k}$. To show it for abelian L-series one needs Artin reciprocity (in order to use generalized class groups, etc.). In general, it is not clear what sort of "reciprocity" may be involved. Moreover, J. T. Sheats [J. Number Theory **71** (1998), no. 1, 121–157; MR 99b:11127, see p. A96 in this volume] has presented some calculations which seem to imply that one should, at the very least, expect that the roots of *all* L-series coming from arithmetic lie in finite extensions of the various completions $\mathbf{k}_w$, w a place of $\mathbf{k}$. One can also say that the integration techniques discussed here and these "Riemann hypothesis" calculations arise because the standard estimates used in the theory [D. Goss, *Basic structures of function field arithmetic*, Springer, Berlin, 1996; MR 97i:11062, see p. A41 in this volume (§8.8)] are *not* always best possible. Rather, one can easily show, e.g., that if $W \subseteq \mathcal{F}$ is a d-dimensional $\mathbf{F}_q$-subspace of a field, then the sum $\sum_{w \in W} w^i$ vanishes for all $i < r^d - 1$. This accounts for the logarithmic blowing up in the abelian case. What might account for such blowing up in the nonabelian case cannot yet even be speculated on. It does appear that in order to handle completely general partial L-series some sort of adelic theory will ultimately be needed. In any case, past success gives good reason for optimism!

David Goss (Columbus, OH)

98a:11080 11G30 11G20 14G25 14H30

Mochizuki, Shinichi (Kyoto)

The profinite Grothendieck conjecture for closed hyperbolic curves over number fields. (English summary)

J. Math. Sci. Univ. Tokyo **3** (1996), *no.* 3, 571–627.

In this impressive paper, Mochizuki proves Grothendieck's conjecture for complete curves over number fields (Theorem A), to the effect that a smooth, geometrically connected, projective algebraic curve X of genus $g \geq 2$ over a number field K can be functorially recovered from the action of the Galois group of K on the geometric fundamental group of X.

Grothendieck's conjecture, sometimes known as "Grothendieck's conjecture on anabelian geometry", was introduced in a letter which Alexander Grothendieck sent to G. Faltings, inspired by Faltings' proof of Mordell's conjecture [G. Faltings, Invent. Math. **73** (1983), no. 3, 349–366; MR 85g:11026a; erratum; MR 85g:11026b].

A copy of this letter is now available in a book edited by L. Schneps and P. Lochak [*Geometric Galois actions. Vol. 1*, Cambridge Univ. Press, Cambridge, 1997; MR 98e:14003].

To state Grothendieck's ideas, some notation is necessary. Given a field K, denote by Γ_K the Galois group of K. If X is a variety over K, let $\pi_1(X)$ be the algebraic fundamental group of K. One has a natural continuous homomorphism $\pi_1(X) \to \Gamma_K$, which is functorially associated to X (up to inner automorphisms). The kernel of this homomorphism is called the geometric fundamental group of X.

What Grothendieck conjectured was that, for a suitable class of fields K (in particular, number fields) and a suitable class of varieties X (in particular, complete curves of genus ≥ 2 over K, and spectra of finite extensions of K), the functor which associates to X the homomorphism $\pi_1(X) \to \Gamma_K$ should be fully faithful.

Thus the main result of the paper under review solves this conjecture for the case of isomorphisms of complete curves over number fields.

The proof in the present paper is based on a result of A. Tamagawa [Compositio Math. **109** (1997), no. 2, 135–194; MR 99a:14035], who proved an analogous result for affine curves. In fact, Mochizuki uses Tamagawa's result over finite fields.

The main ideas for the proof are as follows: in order to prove Theorem A, the author shows that it suffices to consider an analogous, slightly weaker result over local fields (Theorem C). A clever trick (which should be useful in other situations) using index-p covers allows one to replace the curve by one which has stable bad reduction (even if the original curve has good reduction). The author now considers the special fiber using the theory of logarithmic structures. This allows him to reduce Theorem C to a strong version of Grothendieck's conjecture for an appropriate class of singular logarithmic curves over finite fields (Theorem B).

One notices that such a logarithmic curve is obtained by taking a finite number of irreducible affine curves, putting them together by adding nodes, and endowing the resulting curve with a suitable logarithmic structure. Tamagawa's result dealt with the irreducible affine case, and Mochizuki proceeds to show that the added features (the set of irreducible components, the nodes, the logarithmic structure, etc.) can be recovered from the Galois action. Especially remarkable is Section 1 (on recovering the set of irreducible components), which is a true mathematical gem.

The paper is long and detailed. The bulk of the paper is devoted to the proof of Theorem B. A thorough understanding of logarithmic structures [K. Kato, in *Algebraic analysis, geometry, and number theory (Baltimore, MD, 1988)*, 191–224, Johns Hopkins Univ. Press, Baltimore, MD, 1989; Zbl 776.14004; Amer. J. Math. **116** (1994), no. 5, 1073–1099; MR 95g:14056] as well as logarithmic admissible coverings [S. Mochizuki, Publ. Res. Inst. Math. Sci. **31** (1995), no. 3, 355–441; MR 96j:14017] is assumed.

It should be noted that, since the appearance of this paper, Mochizuki has made major advances in the subject. In particular, in the preprint "The local pro-p anabelian geometry of curves" [Preprint 1097, Res. Inst. Math. Sci. (RIMS), Kyoto Univ., Kyoto, 1996; per revr.], Mochizuki proves the conjecture for dominant morphisms between smooth pro-varieties and hyperbolic pro-curves over a field K, where K is any subfield of a field finitely generated over the field of p-adic numbers. This subsumes both Theorem A and Theorem C of the present paper. The proof of this new result, which uses p-adic Hodge theory, seems to be a major breakthrough in the subject. *Dan Abramovich* (Boston, MA)

98k:11087 11G30 11G10 14H30 14H40
Poonen, Bjorn (Princeton, NJ); **Schaefer, Edward F.** (Santa Clara, CA)
Explicit descent for Jacobians of cyclic covers of the projective line. (English summary)
J. Reine Angew. Math. **488** (1997), 141–188.
Fermat's notion of "infinite descent" has informed work on Diophantine equations, including Kummer's famous assaults on Fermat's Last Theorem, for centuries. But in terms of modern terminology, the seminal example of "descent" was due to Mordell, who in 1922 proved Poincaré's conjecture that the rational points on an elliptic curve over the rationals form a finitely-generated abelian group. The result was extended by Weil to all abelian varieties A over any number field K.

The proof breaks into two parts. The first—the so-called weak Mordell-Weil theorem—shows that for some integer $m>1$, $A(K)/mA(K)$ is finite. Therefore if $Q_1,\cdots,Q_l$ is a set of representatives for $A(K)/mA(K)$, for any $P\in A(K)$ we can recursively define $P_n\in A(K)$ by $P_0=P$, and $P_{n-1}=mP_n+Q(n)$, where $Q(n)=Q_i$ for the appropriate i. The second part of the proof attaches a height $h(Q)$ to a point $Q\in A(K)$, and shows that there are only finitely many $Q\in A(K)$ of bounded height, and that for any $P\in A(K)$, with P_n defined as above, there is a sufficiently large n so that $h(P_n)<B$ for some bound B. So $\{Q_i\}$ and the points of height less than B generate $A(K)$.

It is really the second part of the proof which is the "descent", but the word now also generally denotes any process by which $A(K)/mA(K)$ is found, from which the rank of the free part of $A(K)$ (the Mordell-Weil rank) can be gleaned. In practice, $A(K)/mA(K)$ is usually just bounded—the existence of a nontrivial m-part in the Tate-Shafarevich group of A over K keeps current methods from always finding the Mordell-Weil rank of $A(K)$.

There are both computational and theoretical reasons for wanting to compute the Mordell-Weil rank. Data from descents on elliptic curves helped B. J. Birch and H. P. F. Swinnerton-Dyer [J. Reine Angew. Math. **212** (1963), 7–25; MR **26** #3669; J. Reine Angew. Math. **218** (1965), 79–108; MR **31** #3419] frame their famous conjectures relating the arithmetic of the curve to its L-series. J. E. Cremona's tables of elliptic curves [*Algorithms for modular elliptic curves*, Cambridge Univ. Press, Cambridge, 1992; MR 93m:11053; second edition, 1997; MR 99e:11068] are of value to all researchers in the field, and contain ranks computed via descents. One can also sometimes derive information about the rational points on a curve from the Mordell-Weil rank of its Jacobian. If the rank is 0, all the points on the curve can be found. But even if the rank r is positive, R. F. Coleman [Duke Math. J. **52** (1985), no. 3, 765–770; MR 87f:11043] has derived a bound on the number of rational points on the curve if r is less than the genus. In some cases Coleman's techniques can be adapted to find all the rational points on the curve; see J. W. S. Cassels and E. V. Flynn [*Prolegomena to a middlebrow arithmetic of curves of genus* 2, Cambridge Univ. Press, Cambridge, 1996; MR 97i:11071] for applications to genus 2 curves, and W. G. McCallum [Math. Ann. **299** (1994), no. 3, 565–596; MR 95c:11079] for applications to Fermat curves.

The standard approach to descent comes from taking the long exact sequence in Galois cohomology attached to the exact sequence

$$0\to A[\varphi]\to A(\overline{K})\xrightarrow{\varphi}A(\overline{K})\to 0,$$

where $\overline{K}$ is an algebraic closure of K, $\varphi: A \to A$ is a nontrivial isogeny, and $A[\varphi]$ is the kernel of φ, and getting the short exact sequence

$$0 \to A(K)/\varphi A(K) \to H^1(G, A[\varphi]) \to H^1(G, A)[\varphi] \to 0,$$

where G is the Galois group of $\overline{K}$ over K. The image of $A(K)/\varphi A(K)$ is contained in the finite group $H^1(G, A[\varphi], S)$ of cocycle classes unramified outside S, where S consists of all the infinite places of K, the places of bad reduction for A, and the primes dividing the degree of φ. So the problem is to compute the kernel of $H^1(G, A[\varphi], S) \to H^1(G, A)[\varphi]$. Elements of $H^1(G, A)[\varphi]$ can be identified with certain homogeneous spaces of A, namely the twists of A for the cover $\varphi: A \to A$. A homogeneous space A' represents the trivial class precisely when it has a K-rational point. However, it is hard to test for the existence of a K-rational point! The best one usually does is to test whether A' has solutions over every localization of K. Hence the obstruction to the whole procedure is the φ-part of the Tate-Shafarevich group of A over K, which by definition is the set of homogeneous spaces in $H^1(G, A)[\varphi]$ which are everywhere locally trivial.

The technology for descents on elliptic curves is well developed [see, e.g., J. H. Silverman, *The arithmetic of elliptic curves*, Springer, New York, 1986; MR 87g:11070]. But already in doing descents for the Jacobians of curves of genus 2, the homogeneous spaces take a lot of work to compute and to compute with [see D. M. Gordon and D. R. Grant, Trans. Amer. Math. Soc. **337** (1993), no. 2, 807–824; MR 93h:11057]. In the case of Jacobians, there are ways to carry out a descent just by using functions on the curve. This is quite tractable for genus 2 curves [see J. W. S. Cassels, in *Arithmetic and geometry, Vol. I*, 27–60, Progr. Math., 35, Birkhäuser, Boston, Mass., 1983; MR 84k:14032; E. V. Flynn, Acta Arith. **66** (1994), no. 1, 23–43; MR 95g:11057]. More generally for hyperelliptic curves, see Schaefer [J. Number Theory **51** (1995), no. 2, 219–232; MR 96c:11066], and even more generally for superelliptic curves $y^p = f(x)$ with p a prime, see another paper by Schaefer [Math. Ann. **310** (1998), no. 3, 447–471; MR 99h:11063].

The major advantages of this approach are that one does not need to use the full homogeneous spaces, and that the descent can be carried out over K and not some extension field (which can make the computations less tractable). There has been one problem with this approach: the case of a superelliptic curve where p divides the degree of f, which is in some sense the generic case. Overcoming this obstacle in the case of $p = 2$ was discussed in the paper of Cassels cited above, and in [E. V. Flynn, B. Poonen and E. F. Schaefer, Duke Math. J. **90** (1997), no. 3, 435–463; MR 98j:11048], and for general prime p in the article under review.

The authors give a very fine cohomological account of the descent in this case, while still working with functions on the curve to perform the descent. They achieve a method that is satisfying and practical. They compute the Mordell-Weil rank of a non-hyperelliptic superelliptic curve of genus 8 which was concocted to be a proper test of the method. They even compute the rational points on the curve using Coleman's results.

The key idea of the paper is to bring the generalized Jacobian into the picture. Recall that pullbacks of isogenies on Jacobians give rise to unramified abelian covers of a curve (the pullbacks of covers by homogeneous spaces give heterogeneous spaces, in the parlance of K. R. Coombes and Grant's work [J. London Math. Soc. (2) **40** (1989), no. 3, 385–397; MR 91d:11069]). Doing the descent with functions on the curve requires (in some sense) working on covers of the curve, and the authors realized that when p divides the degree of f, the necessary covers of the curve are abelian but ramified, and so are twists of pullbacks of isogenies from a certain generalized Jacobian.

In brief detail, let K be a global field of characteristic not p. Assume K contains a primitive pth root of unity ζ. (There is no harm in adjoining ζ to K if need be, since one can relate the Mordell-Weil rank over $K(\zeta)$ to the rank over K using the $\mathbf{Z}[\zeta]$-action on the Jacobian.) Let $\pi\colon X \to \mathbf{P}^1$ be a cyclic cover of degree p over K, so that X has a model of the form $y^p = f(x)$, with f factoring over a separable closure K^{sep} as $c\prod_{i=1}^d (x-\alpha_i)^{n_i}$. Let J be the Jacobian of X. The authors give several criteria for when $\text{Pic}^0(X)(K) \cong J(K)$; for simplicity of exposition let us say that the isomorphism does hold. Let $m = \pi^*\infty \in \text{Div}(X)$, and let J_m be the corresponding generalized Jacobian, which parameterizes $\text{Pic}^0_m(X)$ (the divisors of degree 0 which are congruent to 1 mod m modulo principal divisors). Let $f_0 = c_0\prod_{i=1}^d (x-\alpha_i)$, for some chosen $c_0 \in K^*$. Set $L = K[T]/(f_0(T))$, and $L^{\text{sep}} = L \otimes K^{\text{sep}}$. A divisor is called good if its support is disjoint from m and from the divisor of zeros of y. For $D = \sum n_P P$ a good divisor, the authors define

$$(x-T)(D) = \prod (x(P)-T)^{n_P} \in L^{\text{sep}}.$$

Every divisor class can be represented by a good divisor, and there is an induced map

$$(x-T)\colon J(K)/\varphi J(K) \to L^*/(L^*)^p K^*;$$

the descent requires understanding the kernel and image of $(x-T)$ (as well as $J[\varphi]$, which the authors determine in terms of f).

Much of the paper is dedicated to finding a cohomological characterization of the $(x-T)$-map, which is achieved via a study of J_m. The kernel of $(x-T)$ turns out to depend on whether X has a K-rational divisor of degree 1. The image is contained in the kernel of the norm map from $L^*/(L^*)^p K^*$ to $K^*/(K^*)^p$, and, considered as a subgroup of $H^1(\mu_p(L^{\text{sep}})/\mu_p(K^{\text{sep}}))$, is unramified outside places of bad reduction of X, Archimedean places, and places of residue characteristic p.

David Grant (Boulder, CO)

98d:11073 11G40 14G10 19F27

Wildeshaus, Jörg (Münster)

On an elliptic analogue of Zagier's conjecture.

Duke Math. J. **87** (1997), *no.* 2, 355–407.

The goal of the theory of polylogarithms is to give an explicit description of motivic cohomology of algebraic varieties and of regulator maps. For example, the K-theory groups of a field F have a natural γ-filtration and the motivic cohomology of Spec(F) is equal to $H^i_{\mathcal{M}}(\text{Spec}(F), \mathbf{Q}(n)) = \text{gr}^n_\gamma K_{2n-i}(F)_{\mathbf{Q}}$. One would like to construct a natural complex representing $\mathbb{R}\text{Hom}(\mathbf{Q}(0), \mathbf{Q}(n))$ in the category of mixed (Tate) motives over F, with cohomology groups $H^i_{\mathcal{M}}(\text{Spec}(F), \mathbf{Q}(n))$.

For $n = 2$ the answer is given by the Bloch-Suslin complex (in degrees 1 and 2): $\delta: \mathbf{Q}[F^* - \{1\}]/R_2 \to \bigwedge^2 F^*_{\mathbf{Q}}$, where $\delta([x]) = x \wedge (1-x)$ and R_2 is generated by

$$[x] - [1/y] + [(1-x)/(1-xy)] - [1/(1-xy)] + [(1-y)/(1-xy)],$$

as $\operatorname{Coker}(\delta) = K_2(F)_{\mathbf{Q}}$ and $\operatorname{Ker}(\delta) = K_3^{\text{ind}}(F)_{\mathbf{Q}}$.

The relations in R_2 come from the functional equation of the dilogarithm function Li_2 ($\mathrm{Li}_k(z) = \sum_{n=1}^{\infty} z^n/n^k$) and its single-valued version $D_2(z) = \operatorname{Im}(\mathrm{Li}_2(z)) + \arg(1-z)\log|z|$. For $F = \mathbf{C}$ the map $\mathbf{Q}[\mathbf{C}^* - \{1\}] \to \mathbf{R}$ sending $[x]$ to $D_2(x)$ factors through $\mathbf{Q}[\mathbf{C}^* - \{1\}]/R_2$ and its restriction to $\operatorname{Ker}(\delta) = K_3^{\text{ind}}(\mathbf{C})_{\mathbf{Q}} \to \mathbf{R}$ coincides with the Borel regulator (up to a scalar).

The corresponding complex for $n > 2$ is conjecturally of the form

$$B_n(F) \to B_{n-1}(F) \otimes F^*_{\mathbf{Q}} \to B_{n-2}(F) \otimes \bigwedge^2 F^*_{\mathbf{Q}} \to \cdots \to \bigwedge^n F^*_{\mathbf{Q}}$$

with $B_n(F) = \mathbf{Q}[F^* - \{1\}]/R_n$, where R_n is given by functional equations of single-valued versions of Li_n. Geometrically, these complexes should reflect the structure of the category of mixed Tate motives over F. P. Deligne [in *Galois groups over* **Q** *(Berkeley, CA, 1987)*, 79–297, Springer, New York, 1989; MR 90m:14016] observed that such motives occur naturally inside $\pi_1(\mathbf{P}^1 - \{0, 1, \infty\})$; A. A. Beĭlinson, R. D. MacPherson and V. V. Schechtman [Duke Math. J. **54** (1987), no. 2, 679–710; MR 88f:14021] and Beĭlinson et al. [in *The Grothendieck Festschrift, Vol. I*, 135–172, Progr. Math., 86, Birkhäuser Boston, Boston, MA, 1990; MR 92h:19007] considered mixed Tate motives associated to configurations of hyperplanes in projective spaces.

D. B. Zagier [in *Arithmetic algebraic geometry (Texel, 1989)*, 391–430, Progr. Math., 89, Birkhäuser Boston, Boston, MA, 1991; MR 92f:11161] proposed an inductive definition of the groups $B_n(F)$ and of the arrow $B_n(F) \to B_{n-1}(F) \otimes F^*_{\mathbf{Q}}$. Coupled with A. Borel's theorem on values of the Dedekind zeta-function, this leads to a conjectural formula for $\zeta_F(n)$ involving values of Li_n (for every number field F). Beĭlinson and Deligne [in *Motives (Seattle, WA, 1991)*, 97–121, Proc. Sympos. Pure Math., 55, Part 2, Amer. Math. Soc., Providence, RI, 1994; MR 95a:19008; "Motivic polylogarithm and Zagier conjecture", Preprint, 1992; per revr.] and A. B. Goncharov [in *Motives (Seattle, WA, 1991)*, 43–96, Proc. Sympos. Pure Math., 55, Part 2, Amer. Math. Soc., Providence, RI, 1994; MR 94m:19003] provided a motivic and a K-theoretical interpretation of Zagier's groups $B_n(F)$.

The story of classical polylogarithms has an elliptic analogue, first investigated by Beĭlinson and A. Levin [in *Motives (Seattle, WA, 1991)*, 123–190, Proc. Sympos. Pure Math., 55, Part 2, Amer. Math. Soc., Providence, RI, 1994; MR 94m:11067]. One fixes an elliptic curve E over $\operatorname{Spec}(F)$ (more generally, an elliptic fibration $\mathscr{E} \to B$) and considers the category of "mixed elliptic motives", generated by mixed Tate motives and by $H_1(E)$. Such motives occur naturally inside $\pi_1(E - \{0\})$ (the author generalized this aspect of the theory to arbitrary Shimura varieties in his recent book [*Realizations of polylogarithms*, Lecture Notes in Math., 1650, Springer, Berlin, 1997]). The analogues of Li_n are the Kronecker-Eisenstein-Lerch series

$$K_{a,b,L}(z) = \sum_{\gamma \in L - \{0\}} \frac{\exp(2\pi i \operatorname{Im}(z\overline{\gamma})/\operatorname{Vol}(L))}{\gamma^a \overline{\gamma}^b} \qquad (E(\mathbf{C}) = \mathbf{C}/L).$$

The analogue of Borel's theorem in this context is a special case of Beĭlinson's conjectures [in *Current problems in mathematics, Vol. 24*, 181–238, Akad. Nauk SSSR, Vsesoyuz. Inst. Nauchn. i Tekhn. Inform., Moscow, 1984; MR 86h:11103] that involves special values of $L(\mathrm{Sym}^n(H_1(E)), s)$.

The aim of the article under review is to formulate an elliptic analogue of Zagier's map $B_n(F) \to B_{n-1}(F) \otimes F^*_{\mathbf{Q}}$. The author postulates the existence of $\mathbf{Q}$-vector spaces $\mathscr{L}_k = \mathscr{L}_k(\mathscr{E})$ ($\mathscr{L}_1 = (\mathscr{E} - \{0\})(B)_{\mathbf{Q}}$) and maps

$$d_k\colon \mathscr{L}_k \to \Big(\bigoplus_{l=1}^{k-1} \mathscr{L}_l\Big)^{\otimes 2}$$

with (among others) the following properties: (i) $\mathrm{Ker}(d_k)$ injects into motivic cohomology $H^{k-1}_{\mathscr{M}}(\mathscr{E}^{(k-2)}, \mathbf{Q}(k-1))$; (ii) for every nonzero section s of $\mathscr{E}$ there are elements $\{s\}_k \in \mathscr{L}_k$, with $d_k(\{s\}_k) = \{s\}_{k-1} \otimes \{s\}_1$. (iii) If $B = \mathrm{Spec}(\mathbf{C})$, then the value of the regulator map on $\sum_\alpha \lambda_\alpha \{s_\alpha\}_k \in \mathrm{Ker}(d_k)$ is given by a suitable combination of the series $K_{i,k-i}$, evaluated at s_α.

The author shows that such objects exist in the first nontrivial case $k = 2$. He explains that in general their existence would follow from a suitable theory of "motivic elliptic sheaves" (in a purely axiomatic setting). He also discusses the relation to Beĭlinson's conjectures and to the theory of the Eisenstein symbol [A. A. Beĭlinson, in *Applications of algebraic K-theory to algebraic geometry and number theory, Part I, II (Boulder, Colo., 1983)*, 1–34, Contemp. Math., 55, Amer. Math. Soc., Providence, RI, 1986; MR 88f:11060]. This is further elaborated on in the author's article, "On the generalized Eisenstein symbol" [Preprint, 1997; per revr.], which generalizes the construction of the Eisenstein symbol to non-torsion sections of $\mathscr{E} \to B$.

An alternative approach can be found in a recent work of Goncharov [in *Galois representations in arithmetic algebraic geometry (Durham, 1996)*, 147–221, Cambridge Univ. Press, Cambridge, 1998], which provides an elliptic analogue of the whole complex $\mathbb{R}\mathrm{Hom}(\mathbf{Q}(0), \mathbf{Q}(n))$, not just of its first two terms.

Jan Nekovář (Cambridge)

99f:11080 11G40 11G05 14G10 14G20

Bertolini, Massimo (Pavia); **Darmon, Henri** (Montreal, QC)

Heegner points, p-adic L-functions, and the Cerednik-Drinfeld uniformization.

Invent. Math. **131** (1998), *no.* 3, 453–491.

To an abelian variety E over a number field K one associates, via the Galois action on torsion points, a complex L-function, denoted $s \mapsto L(E, s)$, taking values in $\mathbf{C}$, and defined for s in $\mathbf{C}$ with $\mathrm{Re}(s) > \frac{3}{2}$. Such L-functions are conjectured to have an analytic continuation to all of $\mathbf{C}$, and to satisfy a given functional equation relating the values at s and $2-s$. Birch and Swinnerton-Dyer conjectured that the rank of the group $E(K)$ of K-rational points of E equals the order of vanishing at 1 of $L(E, \cdot)$, together with a formula for the leading coefficient, involving, among other things, the order of the (conjecturally finite) Tate-Shafarevich group of E, the K-rational torsion of E and of its dual E', and the determinant of the canonical height pairing between $E(K)$ and $E'(K)$ [see, e.g., J. T. Tate, in *Séminaire Bourbaki, Vol. 9*, Exp. No. 306, 415–440, Soc. Math. France, Paris, 1995]. Let us note that this conjecture has been vastly generalized by Deligne, Beĭlinson, Bloch and Kato [see, e.g., J.-M. Fontaine, Astérisque No. 206 (1992), Exp. No. 751, 4, 205–249; MR 94f:11055].

From now on, let E be an elliptic curve over $\mathbf{Q}$, and assume that E is modular, i.e., that $L(E,\cdot)$ is the Mellin transform of a modular form. The modularity of E implies (and is essentially equivalent to the condition) that $L(E,\cdot)$ does have the expected analytic continuation and functional equation. Nowadays, after Wiles, it is known that every elliptic curve over $\mathbf{Q}$ with conductor not divisible by 3^3 is modular [see B. Conrad, F. Diamond and R. L. Taylor, J. Amer. Math. Soc. **12** (1999), no. 2, 521–567; MR 99i:11037]. Another major advance was made by V. A. Kolyvagin [in *The Grothendieck Festschrift, Vol. II*, 435–483, Progr. Math., 87, Birkhäuser Boston, Boston, MA, 1990; MR 92g:11109], who showed, under the assumption that $L(E,\cdot)$ has a zero of order $r \leq 1$ at 1, that the rank of $E(\mathbf{Q})$ equals r, together with the finiteness of the Tate-Shafarevich group of E. Kolyvagin's proof involves an imaginary quadratic field K such that $L(E_K,\cdot)$ has order 1 at 1, and uses a result of Gross and Zagier relating the derivative $L'(E_K,1)$ to the canonical height of a so-called Heegner point α_K in $E(K)$, as expected from the Birch and Swinnerton-Dyer conjecture. More recently, Kato has re-proved Kolyvagin's result in the case where $L(E,1)\neq 0$ [see K. Rubin, in *Galois representations in arithmetic algebraic geometry (Durham, 1996)*, 351–367, Cambridge Univ. Press, Cambridge, 1998; A. J. Scholl, ibid. 379–460], and a paper by Bertolini and Darmon [Ann. of Math. (2) **146** (1997), no. 1, 111–147; MR 99f:11079] contains a proof of the fact that $E(\mathbf{Q})$ is finite if E is semistable and $L(E,1)\neq 0$.

The article under review is not concerned with complex L-functions, but with p-adic L-functions. (For a recent overview of the theory of p-adic L-functions, see P. Colmez ["Fonctions L p-adiques", Exp. 851 (Séminaire Bourbaki), Astérisque, to appear].) Such L-functions are locally analytic $\mathbf{C}_p$-valued functions (with $\mathbf{C}_p$ the p-adic completion of an algebraic closure of $\mathbf{Q}_p$) on p-adic spaces such as the group of $\mathbf{C}_p^*$-valued characters of the Galois group $\mathbf{Z}_p^*$ of the p-cyclotomic extension of $\mathbf{Q}$. They are characterized by their values at characters of finite order, where they agree, up to a simple factor, with the complex L-function, twisted by the corresponding complex characters. Mazur, Tate and Teitelbaum have formulated an analog of the Birch and Swinnerton-Dyer conjecture for the p-adic L-functions $L_p(E,\cdot)$, for p a prime of multiplicative or good and ordinary reduction of E [see B. C. Mazur, J. T. Tate and J. T. Teitelbaum, Invent. Math. **84** (1986), no. 1, 1–48; MR 87e:11076; R. Greenberg and G. Stevens, in *p-adic monodromy and the Birch and Swinnerton-Dyer conjecture (Boston, MA, 1991)*, 183–211, Contemp. Math., 165, Amer. Math. Soc., Providence, RI, 1994; MR 95j:11057]. In this case, $L_p(E,\cdot)$ is a p-adic L-function associated to cyclotomic extensions of $\mathbf{Q}$. It seems that Kato can prove that the rank of $E(\mathbf{Q})$ is at most the order of vanishing of $L_p(E,\cdot)$.

Now we come to the article under review. Let K be an imaginary quadratic field such that E has good reduction at the primes that are ramified in K, and semistable reduction at all primes that are inert in K. Moreover, assume that there is a prime p that is inert in K and at which E has multiplicative reduction, and that the number of such primes is even. Following ideas of Gross, the authors define a p-adic L-function $L_p(E_K,\cdot)$ on the p-adic space of $\mathbf{C}_p^*$-valued characters of the Galois group of the anticyclotomic p-adic tower of K. The derivative $L_p'(E_K,1)$ is naturally an element of the p-adic completion K_p of K. The main theorem of the article then says that the p-adic uniformization map $K_p^* \to E(K_p)$ maps $L_p'(E_K,1)$ to $\alpha_K \pm \overline{\alpha}_K$

with α_K an explicitly defined Heegner point in $E(K)$. More precisely, α_K is the image, under the uniformization of E by a Shimura curve X, of a divisor consisting of complex multiplication points on X. It follows that $L'_p(E_K, 1)$ is nontrivial if and only if $\alpha_K \pm \overline{\alpha}_K$ has infinite order. The fact that the derivative of the p-adic L-functions gives the Heegner point, and not only its height, was predicted by Bertolini and Darmon [Invent. Math. **126** (1996), no. 3, 413–456; MR 97k:11100]. Unlike the proof by Greenberg and Stevens of the cyclotomic conjecture of Mazur, Tate and Teitelbaum [see R. Greenberg and G. Stevens, op. cit.], which is based on Hida theory, the proof of this result uses directly the p-adic uniformization of X_{K_p}, and a computation of the Abel-Jacobi map from degree-zero divisors on the p-adic upper half-plane to the (p-adic uniformization of the) Jacobian of X_{K_p}. The article is reasonably self-contained, and a pleasure to read.

Very recently, Bertolini and Darmon have proved half of their p-adic Birch and Swinnerton-Dyer conjecture for the anti-cyclotomic p-adic L-function attached to (E, K, p), where E is a modular elliptic curve over $\mathbf{Q}$, K an imaginary quadratic field, and p an ordinary prime for E. More precisely, they show that the order of vanishing of $L_p(E_K, s)$ at 1 is at least what it is conjectured to be (Bertolini gave three talks on this subject at the Centre Emile Borel in Paris, on March 2, 3 and 4 of 1999). *Bas Edixhoven* (Rennes)

97e:11080 11J68 11J13

Evertse, Jan-Hendrik (Leiden)

An improvement of the quantitative subspace theorem.

Compositio Math. **101** (1996), *no.* 3, 225–311.

As promised in the title, the author gives an improved version of the theorem. On the one hand, the conditions on the objects occurring in the theorem are relaxed, and on the other hand earlier estimates are much improved.

Let K be a number field, M_K its set of places, and $|\cdot|_v$ for $v \in M_K$ a suitably normalized absolute value. Let S be a finite set of places, of cardinality s, and containing all the Archimedean places. For $v \in S$ let $l_{1v}, \cdots, l_{nv}$ be linearly independent linear forms in n variables with coefficients in K. For $\mathbf{x} \in K^n \setminus \{\mathbf{0}\}$ set $f_{iv}(\mathbf{x}) = |l_{iv}(\mathbf{x})|_v / |\mathbf{x}|_v$, where $|\mathbf{x}|_v = \max(|x_1|_v, \cdots, |x_n|_v)$ when $\mathbf{x} = (x_1, \cdots, x_n)$. (Similarly define $|l_{iv}|_v$ in terms of its coefficient vector.) The subspace theorem, as generalized (from the reviewer's version) by Schlickewei to number fields and sets S as above, says that given $\delta > 0$, the points $\mathbf{x} \in K^n \setminus \{\mathbf{0}\}$ with

$$F(\mathbf{x}) := \prod_{v \in S} \prod_{i=1}^{n} f_{iv}(\mathbf{x}) < H(\mathbf{x})^{-n-\delta}$$

lie in finitely many proper subspaces of K^n. Here $H(\mathbf{x})$ is the height of $\mathbf{x}$, i.e., it is $\prod_{v \in M_K} |\mathbf{x}|_v$. For quantitative versions it is convenient to introduce an explicit parameter A and to distinguish solutions with $H(\mathbf{x}) < A$ and $H(\mathbf{x}) \geq A$. Following a picture suggested by L. Szpiro, one may visualize the solutions with $H(\mathbf{x}) < A$ as lying inside the sun, and the solutions with $H(\mathbf{x}) \geq A$ as lying in finitely many subspaces seen as rays emanating from the sun. It is clear that there are only finitely many solutions with $H(\mathbf{x}) < A$, which therefore lie in finitely many subspaces.

In Schlickewei's version [H. P. Schlickewei, Compositio Math. **82** (1992), no. 3, 245–273; MR 93f:11050] of the quantitative subspace theorem one assumes K to be normal, and one takes $A=\max((n!)^{9/\delta}, H^{dns/\delta})$, where $d=\deg K$ and H is a bound for the "heights" of the linear forms l_{iv}. It is shown that when $0<\delta<1$, the solutions of

$$F(\mathbf{x}) < \Big(\prod_{v\in S}\prod_{i=1}^{n} |l_{iv}|_v\Big) H(\mathbf{x})^{-n-\delta}$$

with $H(\mathbf{x})>A$ lie in at most $\exp(2^{34nd}s^6\delta^{-2}\log(8sd))$ proper subspaces.

In the paper being reviewed, Evertse drops the assumption that K be normal, and that the l_{iv} have coefficients in K. Instead they just have coefficients in $\overline{\mathbf{Q}}$, and each absolute value $|\cdot|_v$ $(v\in M_K)$ is extended in an arbitrary but fixed way to an absolute value of $\overline{\mathbf{Q}}$. He sets $A=H$, where again H is a bound for the heights of the linear forms l_{iv}. He proves that (i) the solutions $\mathbf{x}\in K^n\setminus\{\mathbf{0}\}$ with $H(\mathbf{x})\geqq H$ of

$$(*)\qquad F(\mathbf{x}) < \Big(\prod_{v\in S} |\det(l_{1v},\cdots,l_{nv})|_v\Big) H(\mathbf{x})^{-n-\delta}$$

lie in at most $(2^{60n^2}\delta^{-7n})^s\log 4D\cdot\log\log 4D$ proper subspaces, where D is a bound for the degrees $[K(l_{iv}):K]$ $(v\in S,\ 1\le i\le n)$, and (ii) the solutions of $(*)$ with $H(\mathbf{x})<H$ lie in at most $(150n^4\delta^{-1})^{ns+1}(2+\log\log 2H)$ proper subspaces.

An essential tool is a sharpened version of Roth's lemma. This version of the lemma had earlier been deduced by Evertse [Acta Arith. **73** (1995), no. 3, 215–248; MR 96h:11068] from Faltings' product theorem. (R. Ferretti had deduced a similar version of Roth's lemma.) For the case $n=2$ (Roth's theorem), Bombieri and others had used the Dyson-Esnault-Viehweg lemma instead, but this lemma in its present state does not seem to work for $n>2$, and even for $n=2$ it appears to give a bound for $(*)$ which is at least quadratic in $\log D$, hence worse than Evertse's bound. The estimates of the paper are not likely to be best possible in any sense, except that a factor $\log\log 2H$ is known to be needed in the bound (ii).

The rather complicated arguments include the standard division of the proof of (i) into two parts, one part (Theorem B) a reduction of the problem to a somewhat simpler problem in K^N, where $N=\binom{n}{k}$ for some k, with K^N arising as the ambient space of Grassmann coordinates of k-dimensional subspaces, and another part (Theorem C) dealing with this simpler problem in K^N, the simplicity being that one assumes to have $N-1$ simultaneous solutions $\mathbf{x}_1,\cdots,\mathbf{x}_{N-1}$ of roughly the same height of $(*)$ (where n is replaced by N). Among the innovations of the author are a very strong "gap principle" using his result on S-unit equations in two variables, a replacement of the usually employed Davenport lemma by the clever new "Lemma 15", and the use of a concept of "primitive points" in K^n, which generalizes the usual primitive points in $\mathbf{Z}^n$. The presentation is very careful. But due to the paper's generality and sophistication, it is not easy to develop a good intuition on the main ideas.

The paper is a great advance. However, the final word on the subject still has not been said. For example, Schlickewei ["A parametric version of the subspace theorem", to appear] gives a version (albeit a "parametric" version, further restricted in the choice of the linear forms l_{iv}) with bounds which contain no reference to $s=\operatorname{card} S$.

Wolfgang M. Schmidt (Boulder, CO)

99j:11083 11J83 22E40

Kleinbock, D. Y. (Princeton, NJ); **Margulis, G. A.** (New Haven, CT)

Flows on homogeneous spaces and Diophantine approximation on manifolds. (English summary)

Ann. of Math. (2) **148** (1998), *no.* 1, 339–360.

This important paper settles conjectures of Baker and Sprindzhuk in the theory of Diophantine approximation on manifolds formulated in the 1970s. We adopt the following notations: for $\mathbf{x} \in \mathbf{R}^n$ let $\|\mathbf{x}\| = \max_{1 \le i \le n} |x_i|$, $\Pi(\mathbf{x}) = \prod_{i=1}^n |x_i|$. One says that a vector $\mathbf{x} \in \mathbf{R}^n$ is very well approximated (VWA) if for some $\varepsilon > 0$ there are infinitely many $q \in \mathbf{Z}$ and $\mathbf{p} \in \mathbf{Z}^n$ such that $\|q\mathbf{x} + \mathbf{p}\|^n \cdot |q| \le |q|^{-\varepsilon}$. Similarly, a vector $\mathbf{x} \in \mathbf{R}^n$ is very well multiplicatively approximated (VWMA) if for some $\varepsilon > 0$ there are infinitely many $q \in \mathbf{Z}$ and $\mathbf{p} \in \mathbf{Z}^n$ such that $\Pi(q\mathbf{x} + \mathbf{p}) \cdot |q| \le |q|^{-\varepsilon}$.

Clearly, a vector is VWMA whenever it is VWA. One can also show that almost every $\mathbf{x} \in \mathbf{R}^n$ is not VWMA. A much more difficult problem arises if one restricts one's attention to a proper submanifold $M \subset \mathbf{R}^n$. In the 1930s Mahler conjectured that almost all points on the curve $M_0 = \{(t, t^2, \cdots, t^n),\ t \in \mathbf{R}\} \subset \mathbf{R}^n$ are not VWA. This was settled by V. G. Sprindzhuk [Dokl. Akad. Nauk SSSR **155** (1964), 54–56; MR **28** #2090b]. Later, Sprindzhuk in [Uspekhi Mat. Nauk **35** (1980), no. 4(214), 3–68, 248; MR 81j:10039] conjectured the following. Let $f_1, \cdots, f_n$ be real-analytic functions in $\mathbf{x} \in U$, U being a domain in $\mathbf{R}^d$, which together with 1 are linearly independent over $\mathbf{R}$. Then almost all points of M (in the sense of the natural measure on M) are not VWA.

The case $n = 2$ of the conjecture was settled by W. M. Schmidt [Monatsh. Math. **68** (1964), 154–166; MR **30** #1980], and recently V. V. Beresnevich and V. I. Bernik [Acta Arith. **75** (1996), no. 3, 219–233; MR 97b:11095] proved it for $n = 3$.

A stronger conjecture (also formulated by Sprindzhuk) states that almost all points of M are not VWMA. It has not been proved even for the curve M_0 as above (this particular case is known as Baker's conjecture) except for the case $n = 2$ [V. G. Sprindzhuk, *Metric theory of Diophantine approximations,* Translated from the Russian and edited by Richard A. Silverman, Winston, Washington, D.C., 1979; MR 80k:10048].

In the paper under review the authors prove a very general result which settles both of the conjectures of Sprindzhuk and Baker. Let $M = \{\mathbf{f}(\mathbf{x})\colon \mathbf{x} \in U\}$, where U is open in $\mathbf{R}^d$ and $\mathbf{f} = (f_1, \cdots, f_n)$ is a C^m embedding of U into $\mathbf{R}^n$. One says that a point $\mathbf{y} = \mathbf{f}(\mathbf{x})$ of the manifold M is nondegenerate if for some $l \le m$ the space $\mathbf{R}^n$ is spanned by partial derivatives of $\mathbf{f}$ at $\mathbf{x}$ of order up to l (this is an infinitesimal version of not lying in any proper affine hyperplane). Assume now that almost every point of M is nondegenerate. Then almost all points of M are not VWMA.

The proof makes use of the ergodic theory of homogeneous actions on the space $\mathscr{L}_{n+1} = \mathrm{SL}_{n+1}(\mathbf{R})/\mathrm{SL}_{n+1}(\mathbf{Z})$ (which is just the space of unimodular lattices in $\mathbf{R}^{n+1}$). More precisely, given $\mathbf{y} \in \mathbf{R}^n$ let $\Lambda_{\mathbf{y}} = \begin{pmatrix} 1 & \mathbf{y}^{\mathrm{T}} \\ 0 & I_n \end{pmatrix} \mathbf{Z}^{n+1}$ be a unimodular lattice in $\mathbf{R}^{n+1}$. For $\mathbf{t} = (t_1, \cdots, t_n)$, $t_i \ge 0$, define $t = \sum_{i=1}^n t_i$ and $g_{\mathbf{t}} = \mathrm{diag}(e^t, e^{-t_1}, \cdots, e^{-t_n}) \in \mathrm{SL}_{n+1}(\mathbf{R})$. Finally, one introduces a function δ on $\mathscr{L}_{n+1}$ by $\delta(\Lambda) = \inf_{\mathbf{v} \in \Lambda - \{0\}} \|v\|$.

It is easy to prove that for any $\mathbf{y} \in \mathbf{R}^n$ which is VWMA there exist $\gamma > 0$ and infinitely many $\mathbf{t} \in \mathbf{Z}_+^n$ such that $\delta(g_{\mathbf{t}}\Lambda_{\mathbf{y}}) \le e^{-\gamma t}$. Thus, in view of Borel-Cantelli, it is enough to prove that for any nondegenerate point $\mathbf{y}_0 = \mathbf{f}(\mathbf{x}_0) \in M$ there is a neighbourhood B of $\mathbf{x}_0 \in U$ such that given $\gamma > 0$, one has $\sum_{\mathbf{t}\in\mathbf{Z}^n_+} |E_\mathbf{t}| < \infty$, where $E_\mathbf{t} = \{\mathbf{x} \in B\colon \delta(g_\mathbf{t}\Lambda_{\mathbf{f}(\mathbf{x})}) \le e^{-\gamma t}\}$ and $|\cdot|$ stands for the Lebesgue measure.

This is an immediate corollary of the following nontrivial estimate: (1) $|E_\mathbf{t}| \le Ce^{-\gamma t/dl}|B|$ for some ball B centered at $\mathbf{x}_0$, some $C > 0$ and all $\mathbf{t} \in \mathbf{Z}^n_+$ with large enough t (here $l \le m$ is taken from the nondegeneracy condition for $\mathbf{y}_0$). The estimate is obtained by refining proofs of earlier important results on nondivergence of unipotent flows on the space of lattices.

Namely, let $\{u_x,\ x \in \mathbf{R}\} \subset \mathrm{SL}_k(\mathbf{R})$ be a one-parameter subgroup all of whose entries are polynomials in x (such a subgroup is called unipotent). A remarkable result of Margulis [in *Lie groups and their representations (Proc. Summer School, Bolyai, János Math. Soc., Budapest, 1971)*, 365–370, Halsted, New York, 1975; MR **57** #9907] states that no orbit of the action of $u_\mathbf{R}$ on $\mathscr{L}_k$ goes to infinity. By Mahler's criterion, this is equivalent to saying that given any $\Lambda \in \mathscr{L}_k$, one has $\delta(u_x\Lambda) \not\to 0$ as $x \to \infty$. Later, S. G. Dani [Ergodic Theory Dynam. Systems **6** (1986), no. 2, 167–182; MR 88e:58052] proved more: For any $c > 0$ and any $\Lambda \in \mathscr{L}_k$ there exists $\varepsilon > 0$ such that given any unipotent subgroup $u_\mathbf{R} \subset \mathrm{SL}_k(\mathbf{R})$, one has (2) $|\{x \in [0, T]\colon \delta(u_x\Lambda) < \varepsilon\}| \le cT$. Similar estimates were found by N. A. Shah [Duke Math. J. **75** (1994), no. 3, 711–732; MR 95j:22022] for any polynomial map $\mathbf{R}^d \to \mathrm{SL}_k(\mathbf{R})$ in place of the polynomial homomorphism $u\colon \mathbf{R} \to \mathrm{SL}_k(\mathbf{R})$.

Here the authors establish a quantitative relation between c and ε in (2), thus sharpening the result of Dani: it turns out that $c = c_1(k)\varepsilon^{1/k^2}$. The proof uses the notion of (C,α)-good functions introduced by A. Eskin, S. Mozes and Shah [Geom. Funct. Anal. **7** (1997), no. 1, 48–80; MR 98d:22006], and the fact that polynomials of degree at most k are $(C_k, 1/k)$ good. In a similar way they obtain (1) using the fact that in a neighbourhood of a nondegenerate point, linear combinations of $1, f_1, \cdots, f_n$ are $(C, 1/dl)$-good.

Finally, the authors note that the methods of their paper can help in solving related problems on Diophantine approximations in the p-adic case.

Alexander Starkov (Moscow)

99k:11112 11J83 11K38 11K50 11K55
Harman, Glyn (Cardiff)
★**Metric number theory.**
London Mathematical Society Monographs. New Series, 18.
The Clarendon Press, Oxford University Press, New York, 1998. xviii+297 pp.
\$125.00. ISBN 0-19-850083-1

Metric number theory was born shortly after Lebesgue's theory of measure. Borel, in papers of 1903 and 1909, showed that almost all real numbers α are *normal* to a given base $b \ge 2$ (that is, the digits $0, 1, \cdots, b-1$ appear with frequency b^{-1} in the expansion of α to base b), and classified those functions $g(n)$ (for example, $g(n) = n(\log n)^{1+\varepsilon}$) such that the partial quotients $a_n(\alpha)$ exceed $g(n)$ only finitely often, for almost all α. Weyl, in a well-known paper of 1916 on the fundamentals of uniform distribution, proved that for any sequence $\lambda_1, \lambda_2, \cdots$ of real numbers with $\lambda_{n+1} - \lambda_n \ge c > 0$, the sequence $(\lambda_n\alpha)$ is uniformly distributed (mod 1) for almost all α. He also introduced the concept of discrepancy of a sequence $x_1, x_2, \cdots$; let

$$Z_n(\gamma) = \sum_{\substack{k \leq n \\ x_k \in [0,\gamma) \pmod 1}} 1;$$

the discrepancy is $D_n = \sup\{|Z_n(\gamma) - n\gamma| \colon \gamma \in [0,1)\}$.

In the 1920s, Khinchin showed that $(n\alpha)$ has discrepancy $O((\log n)^{1+\varepsilon})$ a.e., and for the sequence $(2^n\alpha)$ found that $n^{1/2}(\log\log n)^{1/2}$ is the correct order of size of the discrepancy a.e. (the first occurrence of the law of the iterated logarithm in the literature). Another well-known result of Khinchin from 1924 concerns a positive function $\psi(n)$ for which $n\psi(n)$ is decreasing. If $\sum_{n=1}^{\infty} \psi(n)$ diverges, the inequality (1) $|n\alpha - m| < \psi(n)$ holds for infinitely many pairs n, m, for almost all α; if the series converges, (1) has finitely many solutions a.e. Khinchin also introduced strong uniform distribution, giving conditions on f under which

$$N^{-1} \sum_{n=1}^{N} f(n\alpha) \to \int_0^1 f(t)\,dt \quad \text{a.e.} \tag{2}$$

(f has period 1). He conjectured that this holds for any integrable f, and this was only disproved in 1970, by Marstrand. However, if we replace $n\alpha$ by $2^n\alpha$, the analogue of the conjecture holds, a result of Raĭkov from 1936.

From these beginnings, fine scholars such as Cassels, Davenport, Erdős, Gallagher, Mahler, Sprindzhuk and Schmidt developed more refined results. Many questions are suggested by the desire to find out whether a sequence $(a_n\alpha)$, with $a_1, a_2, \cdots$ distinct integers, behaves a.e. like $(n\alpha)$, or $(2^n\alpha)$, or neither. The author made some essential contributions from 1984 onwards, in particular treating the variant of (1) in which n, m are restricted to sequences $\mathscr{A}, \mathscr{B}$ respectively. All of the above topics, and much else in the same vein, are treated in the author's very scholarly monograph. The book is much broader, and at the same time deeper, than V. G. Sprindzhuk's volume [*Metric theory of Diophantine approximations*, Translated from the Russian and edited by Richard A. Silverman, Winston, Washington, D.C., 1979; MR 80k:10048]. The author often gives new and succinct treatments of difficult results from the literature, always with a clarity which will appeal to non-experts. This is an ideal introduction to the subject, with modest prerequisites in analysis and number theory. It might well inspire new attempts on the stubborn Duffin-Schaeffer conjecture (discussed in Chapter 2) that (1) has infinitely many solutions a.e. with coprime m, n, for any positive sequence ψ for which $\sum_{n=1}^{\infty} \psi(n)\varphi(n)/n$ diverges. *R. C. Baker* (Provo, UT)

97m:11102 11J89 11F11 12H05 13N99

Nesterenko, Yu. V. (Moscow)

Modular functions and transcendence questions. (Russian. Russian summary)

Mat. Sb. **187** (1996), *no.* 9, 65–96; *translation in Sb. Math.* **187** (1996), *no.* 9, 1319–1348.

For $k \geq 2$, let

$$E_{2k}(q) = 1 + (-1)^k (4k/B_k) \sum_{n \geq 0} \Big(\sum_{d|n} d^{2k-1} \Big) q^n$$

be the standard Eisenstein series of weight $2k$, where B_k is the kth Bernoulli number. Thus, $P = E_2$, $Q = E_4$ and $R = E_6$, in the notations of Ramanujan, who showed that the $\mathbf{Q}$-algebra generated by these three functions is stable under the differential operator $\partial = q\,d/dq$. Note that Q and R generate the ring of modular forms for $SL_2(\mathbf{Z})$. Transcendence properties of these functions have been studied for some time, using their relation with periods of elliptic integrals. For instance, let E be an elliptic curve over a subfield K of $\mathbf{C}$, and let ω, η be the periods, over a non-trivial cycle of $E(\mathbf{C})$, of a basis of the de Rham cohomology of E/K. A fundamental result of G. V. Chudnovsky [Dokl. Akad. Nauk Ukrain. SSR Ser. A **1976**, no. 8, 698–701, 767; MR **54** #12670; and in *Proceedings of the International Congress of Mathematicians (Helsinki, 1978)*, 339–350, Acad. Sci. Fennica, Helsinki, 1980; MR 81j:10051] states that $K(\omega/\pi, \eta/\pi)$ has transcendence degree at least 2 over $\mathbf{Q}$, and this implies that for any non-zero complex number q of absolute value < 1, at least two of the numbers $P(q)$, $Q(q)$, $R(q)$ are algebraically independent (over $\mathbf{Q}$). Chudnovsky's theorem still provides the only known proof that Euler's Γ-function assumes transcendental values at $\frac{1}{3}$ and $\frac{1}{4}$.

In 1995, K. Barré-Sirieix, G. Diaz, F. Gramain and G. Philibert [Invent. Math. **124** (1996), no. 1-3, 1–9; MR 96j:11103] made a breakthrough in transcendence theory by proving the long-standing conjecture of Mahler and Manin according to which the modular invariant $J = 12^3(1 - R^2/Q^3)^{-1}$ assumes transcendental values at any non-zero complex (or p-adic) algebraic number α of absolute value < 1. The main new feature in their proof (which seems so natural in retrospect) was to give a crucial role to the cusp $q = 0$ in the construction of their auxiliary function, and only then to look at (the isogeny class of) the elliptic curve with multiplicative period $q = \alpha$. Their result can therefore be considered the first modular theorem of transcendence theory.

The present article of Nesterenko provides a fundamental advance in this direction by generalizing both the Chudnovsky and the Barré-Sirieix-Diaz-Gramain-Philibert theorems, and by giving the first "truly modular" theorem on algebraic independence, viz.: Let q be a non-zero complex (or p-adic) number of absolute value < 1; then, at least three of the numbers q, $P(q)$, $Q(q)$, $R(q)$ are algebraically independent.

As with Chudnovsky's theorem, the depth of this result is best illustrated by its striking corollaries on fields of smaller transcendence degree, for which we still have no direct proof at our disposal. To wit: (1) The numbers π and e^{π} are algebraically independent; in fact, π, e^{π} and $\Gamma(\frac{1}{4})$ are algebraically independent, and more generally, for any quadratic imaginary field with discriminant $-D$ and character ε, the numbers π, $e^{\pi\sqrt{D}}$ and $\prod_{a=1,\cdots,D-1} \Gamma(a/D)^{\varepsilon(a)}$ are algebraically independent. (2) Let α be a non-zero complex or p-adic algebraic number of absolute value < 1, and let f be the expansion at ∞ of a non-constant meromorphic modular form relative to some congruence subgroup, with algebraic Fourier coefficients (e.g. $f = J$, the discriminant Δ, or a Thetanullwerte); then α is not a pole of f, and $f(\alpha)$ is transcendental; more generally, the numbers $J(\alpha)$, $\partial J(\alpha)$ and $\partial^2 J(\alpha)$ are algebraically independent.

Nesterenko complements his theorem with a measure of algebraic independence, implying in case (2) that for any non-zero element A in $\mathbf{Z}[X_1, X_2, X_3]$ of degree $\leq t$ and logarithmic height $\leq t$, the upper bound $(*)$ $-\mathrm{Log}|A(P(\alpha), Q(\alpha), R(\alpha))| < Ct^4 \mathrm{Log}(t)^{24}$ holds with an effective constant $C = C(\alpha)$. The exponent 24 in $(*)$ has been decreased to 16 by P. Philippon [J. Reine Angew. Math. **497** (1998), 1–15; MR 99f:11091], who also has given new approaches to the qualitative theorem.

The most novel ingredient in Nesterenko's work is the following functional analogue of $(*)$: For any non-zero element A in $\mathbf{C}[z][X_1, X_2, X_3]$ of degree $\leq t$ in the X_i's and degree $\leq t'$ in z, the order at $z = 0$ of the series $A(z, P(z), Q(z), R(z))$ is $< Ct't^3$, where C is an effective universal constant. The proof of this multiplicity estimate (which takes up two-thirds of the paper) uses all the techniques of commutative algebra introduced by the author since the 1970s (see, in particular, [Izv. Akad. Nauk SSSR Ser. Mat. **41** (1977), no. 2, 253–284; MR **58** #10774] and [Acta Arith. **53** (1989), no. 1, 29–46; MR 91d:11078]), together with a thorough study of the differential ideals of the differential ring $\{\mathbf{Q}[z, P, Q, R], \partial\}$ (his previous results do not apply directly, because the corresponding vector field is nonlinear and has a singularity). Thus, this article can be considered the crowning step of the remarkable methods he developed in the theory of E-functions, even if his main result (1) on π and e^π eventually turns out to owe nothing to the exponential function!

The other ingredients in the proof are, on the one hand, the slow arithmetic growth of the coefficients of modular forms at the cusp $q = 0$, and, on the other hand, the criterion of algebraic independence of Philippon [Inst. Hautes Études Sci. Publ. Math. No. 64 (1986), 5–52; MR 88h:11048], or its sharpening by M. Ably [J. Number Theory **42** (1992), no. 2, 194–231; MR 93g:11082] for $(*)$. In fact, as noticed by Philippon in his article on K-functions [to appear, op. cit.], both the criterion and the above multiplicity estimate can be disposed of in the case where $J(q)$ is algebraic, and replaced by known measures of algebraic independence for ω/π, η/π [cf. G. Philibert, Ann. Inst. Fourier (Grenoble) **38** (1988), no. 3, 85–103; MR 90b:11071]; in the p-adic case, one can appeal to the analogous measure for $\partial J(q)$, $\partial^2 J(q)$, which was recently established by A. Ya. Yanchenko [Mat. Zametki **61** (1997), no. 3, 431–440; MR 99e:11103]). This provides a much simpler proof to corollaries such as (1). The same phenomenon had occurred with Chudnovsky's theorem, which can be deduced either from Gel′fond's criterion and a zero estimate, or from Fel′dman's transcendence measure for π, and in conclusion, one can hope that Nesterenko's deep estimate $(*)$ will similarly pave the way to the proof of the algebraic independence of four numbers. Which new number will appear remains to be seen, but a natural candidate would probably come from the theory of Fourier-Jacobi forms, thereby combining the elliptic and modular approaches described in this review. *Daniel Bertrand* (Paris)

98j:11057 11Kxx 11K06 11K38
Drmota, Michael (Vienna); **Tichy, Robert F.** (Graz)
★**Sequences, discrepancies and applications.**
Lecture Notes in Mathematics, 1651.
Springer-Verlag, Berlin, 1997. xiv+503 pp. $93.00. ISBN 3-540-62606-9

In 1916 H. Weyl published a pioneering paper [Math. Ann. **77** (1916), 313–352; JFM 46.0278.06] on the theory of uniform distribution. His limit relation

$$\lim_{N\to\infty} N^{-1}\sum_{n=1}^{N} f(x_n) = \int_0^1 f(t)dt$$

has played a decisive role in the development of the theory. Subsequent research has led to the following fundamental results:

Koksma's inequality (or Koksma-Hlawka's inequality in the multidimensional case): For any $f:[0,1]\to\mathbf{R}$ having bounded variation $V(f)$ and for any N points $x_1,\cdots,x_N$ in $[0,1]$, we have

$$\left|N^{-1}\sum_{n=1}^{N} f(x_n)-\int_0^1 f(t)dt\right|\le V(f)D_N(x_n),$$

where

$$D_N(x_n)=\sup_{0\le x<y\le 1}|N^{-1}\#\{n\le N;\ x_n\in[x,y)\}-(y-x)|$$

is called the discrepancy of x_n.

Erdős-Turán's inequality (or Erdős-Turán-Koksma's inequality in the multidimensional case): There exists an absolute constant C such that

$$D_N(x_n)\le C\left(m^{-1}+\sum_{h=1}^{m}h^{-1}\left|N^{-1}\sum_{n=1}^{N}e^{2\pi ihx_n}\right|\right)$$

for any finite sequence $x_1,\cdots,x_N$ in $[0,1]$ and any positive integer m.

Schmidt's lower bound: For any infinite sequence x_n in $[0,1]$, we have $D_N(x_n)\ge C(\log N)/N$ for infinitely many N, where $C=0.046\cdots$.

As the authors note, there are six monographs on this subject [L. Kuipers and H. Niederreiter, *Uniform distribution of sequences*, Wiley-Intersci., New York, 1974; MR **54** #7415; G. Rauzy, *Propriétés statistiques de suites arithmétiques*, Presses Univ. France, Paris, 1976; MR **53** #13152; W. M. Schmidt, *Lectures on irregularities of distribution*, Tata Inst. Fund. Res., Bombay, 1977; MR 81d:10047; E. Hlawka, *Theorie der Gleichverteilung*, Bibliographisches Inst., Mannheim, 1979; MR 80j:10057; English translation, A B Acad., Berkhamsted, 1984; MR 85f:11056; L. K. Hua and Y. Wang, *Applications of number theory to numerical analysis*, Translated from the Chinese, Springer, Berlin, 1981; MR 83g:10034; J. Beck and W. W. L. Chen, *Irregularities of distribution*, Cambridge Univ. Press, Cambridge, 1987; MR 88m:11061].

The monograph of Drmota and Tichy is a very welcome addition to the literature. It covers a huge amount of research over the past 23 years. As the authors remark in the preface, "The present book attempts to summarize special developments and methods of the theory of uniform distribution since 1974 when Kuipers and Niederreiter's book appeared. We emphasize topics which are not covered by some of the above-mentioned monographs. Every section of this book consists of two parts, a self-contained one where main results and methods are established and a Notes part where the corresponding literature is discussed. References of papers published before 1974 are only taken into account if they are necessary for the presentation and proofs of the results." The general setting of the book is the multidimensional space $\mathbf{R}^k$. The authors concentrate on a quantitative point of view and systematically use the discrepancy. We now briefly describe the contents of the book. The three chapters are divided into a total of 14 sections.

Chapter 1 (Discrepancy of sequences) deals with the classical part of the theory and begins with the notion of discrepancies (usual, star, isotropic, L^p-discrepancy, diaphony) and dispersions. Included are the multidimensional Weyl criterion, Koksma-Hlawka's inequality and an improved version of the Erdős-Turán-Koksma inequality due to P. Grabner ["Harmonische Analyse, Gleichverteilung und Ziffernentwicklungen", Dissertation, Tech. Univ. Wien, Vienna, 1989; per bibl.]. Then Roth's lower bound for the discrepancy is described. The proof follows the new lines of Beck's Fourier transform technique [Invent. Math. **74** (1983), no. 3, 477–487; MR 85g:11063]. Schmidt's improvement for $\mathbf{R}^1$ and Beck's improvement for $\mathbf{R}^3$ [Compositio Math. **72** (1989), no. 3, 269–339; MR 91f:11054] are presented. The major section of this chapter is devoted to the exhibition, via discrepancies, of important classes of sequences, namely the sequence of fractional parts of $n\alpha$, the Kronecker sequence, digital sequences, normal numbers and exponential sequences. In all cases the authors discuss an upper bound for the usual discrepancy. First an explicit formula for the discrepancy of $n\alpha$ involving continued fraction expansion of α is established. The proof follows J. Schoissengeier [Acta Arith. **44** (1984), no. 3, 241–279; MR 86c:11056]. For the Kronecker sequence $(\alpha_1 n, \cdots, \alpha_k n)$ mod 1 a sharp bound for the isotropic discrepancy is given. It is optimal for badly approximable $(\alpha_1, \cdots, \alpha_k)$. Also, the authors give a metric discrepancy bound due to Beck [Ann. of Math. (2) **140** (1994), no. 2, 449–502; MR 95g:11069b] which is an analogue to Khinchin's theorem. The main digital sequences covered here are the uniformly distributed sequences $\alpha f(n) + \beta n$, where $f(n)$ is related to the q-ary digital expansion of n and α is an irrational. Then the authors present an explicit construction of normal numbers α in the q-ary digit expansion $\alpha = 0.[f(1)][f(2)]\cdots$, where $f(x)$ is a non-constant polynomial with rational coefficients, and they give the discrepancy bound for the sequence $q^n\alpha$. Part of exponential sequences is concentrated to an effectively constructible $\lambda > 1$ such that λ^n is completely uniformly distributed mod 1. Again, the explicit bound of the discrepancy is given. Another section is devoted to distribution functions. The first part presents the distribution functions of maldistributed sequences developed by G. Myerson [in *Number theory with an emphasis on the Markoff spectrum (Provo, UT, 1991)*, 163–190, Dekker, New York, 1993; MR 94a:11112]. The remaining part is devoted to the theory of recurring sequences and follows P. Kiss and the second author [Nederl. Akad. Wetensch. Indag. Math. **48** (1986), no. 1, 79–86; MR 87e:11020]. The final section of Chapter 1 is devoted to metric theory. The authors start with the Erdős-Gál-Koksma method to establish a general metric upper bound for the discrepancy of $f_n(x)$. Then they present a law of the iterated logarithm for the discrepancy function of sequences $a_n x$ provided that a_n is a lacunary sequence [S. Takahashi, Tôhoku Math. J. (2) **24** (1972), 319–329; MR **48** #9242]. Inspired by Knuth's concept of random sequences, the authors add a result of Niederreiter and the second author [Mathematika **32** (1985), no. 1, 26–32; MR 87h:11070] that x^{a_n} is completely uniformly distributed mod 1 for almost all $x > 1$. The chapter ends with a general theorem of the iterated logarithm due to W. Philipp [*Mixing sequences of random variables and probablistic number theory*, Mem. Amer. Math. Soc., 114, Amer. Math. Soc., Providence, R. I., 1971; MR **55** #10411].

Chapter 2 (General concepts of uniform distribution) starts with geometric concepts of discrepancy with respect to a discrepancy system, including balls in a compact metric space, homothetic sets, spherical caps and spherical slices in the Euclidean space. Here, the irregularity behaviour of the discrepancy is discussed. For the discrepancy system of all halfspaces intersected with a given set, a method of J. R. Alexander [Invent. Math. **103** (1991), no. 2, 279–296; MR 92f:11106] is described. It also contains the main ideas of uniformly distributed sequences on the Sierpiński gasket. The following section is devoted to classical summation methods, namely matrix limitation methods, weighted means, the logarithmic weighted mean, and Abel's summation method. Again, the authors present analoga to the Koksma-Hlawka and Erdős-Turán-Koksma inequalities. Included here is the logarithmic discrepancy bound for the logarithmic uniformly distributed sequence $\alpha n + \beta \log n \bmod 1$ and also for a more general $f(n)$. This is followed by a detailed discussion of weighted well-distributed sequences, including a proof that for a weighted mean which tends to zero too fast almost all sequences are weighted well distributed. The section ends with the notion of Abel discrepancy, an Abel discrepancy bound for $n\alpha \bmod 1$, and regularity conditions for the weighted Abel method. This is followed by a section on continuous uniform distribution. The authors start with continuous analoga to the Weyl criterion, the discrepancy $D_T(\mathbf{x}(t))$, the Koksma-Hlawka inequality, the Erdős-Turán-Koksma inequality and the difference theorem of van der Corput. There is a special entire function $f(t)$ continuously uniformly distributed mod 1. There follows a result of R. J. Taschner [Monatsh. Math. **88** (1979), no. 4, 321–330; MR 81h:10071] that the discrepancy can tend to 0 arbitrarily fast and a discrepancy bound in terms of arclength derived by the first author [Ann. Inst. Fourier (Grenoble) **39** (1989), no. 3, 501–527; MR 91b:11086] which is optimal in the one-dimensional case. The authors conjecture that, for $k \geq 2$ and $\mathbf{x}\colon [0,\infty) \to \mathbf{R}^k$ such that the arclength $s(T)$ is finite for any $T \geq 0$, we have $\limsup_{T\to\infty} TD_T(\mathbf{x}(t)) > 0$. This section concludes with a general notion of continuous uniform distribution on compact manifolds, the discrepancy with respect to a discrepancy system, metric theorems and a law of iterated logarithm with respect to the Brownian motion measure. Many of the results given are due to the authors. The chapter ends with a section on uniform distribution in abstract and discrete spaces. Firstly, uniform distribution in compact and locally compact spaces and locally compact groups is briefly sketched. There follow distribution problems in a finite set X of order q. It is here that $\sup\inf\sup_{N\geq 1} ND_N = 1 - 1/2(q-1)$, where the supremum is taken over all probability measures on X and the infimum over all sequences $x_n \in X$ [R. Tijdeman, J. Combinatorial Theory Ser. A **15** (1973), 129–137; MR **47** #8318; H. G. Meijer, Nederl. Akad. Wetensch. Proc. Ser. A **76**=Indag. Math. **35** (1973), 9–17; MR **47** #4806]. The authors consider a special $k(N)$-distribution, which is related to the usual complete uniform distribution. There is a theorem that almost all sequences are $k(N)$-distributed for all sequences $k(N)$ satisfying $\log_q N - \log_q \log_q N - k(N) \to \infty$ due to M. Goldstern [Monatsh. Math. **116** (1993), no. 3-4, 237–243; MR 95a:28002]. Next, combinatorial concepts of discrepancy are discussed. The object is to color a set with two colors so that each set in a given family will be colored as uniformly as possible. The authors present a general upper bound for the combinatorial discrepancy; for a survey they refer to Beck and V. T. Sós [in *Handbook of combinatorics, Vol. 1, 2*, 1405–1446, Elsevier, Amsterdam, 1995; MR 96m:11060]. Then the notion of uniform distribution modulo m is generalized to Dedekind domains modulo an ideal and a new notion of weak uniform distribution is described. There are criteria for uniform distribution of linear recurring sequences of second order and inhomogeneous recurring sequences $x_{n+1} = ax_n + b$. Finally, there are criteria for weak uniform distribution of $x_{n+1} = ax_n + b$ due to the second author and G. Turnwald [Manuscripta Math. **61** (1988), no. 1, 11–22; MR 89d:11060].

Chapter 3 (Numerical integration, approximation, and mathematical finance) discusses the quasi-Monte Carlo methods. It begins with numerical integration based on the Koksma-Hlawka inequality. The main sequences covered here are the van der Corput sequence, the Halton sequence, the Hammersley sequence, good lattice points, (t, m, s)-nets and (t, s)-sequences. All such sequences are low discrepancy sequences with discrepancy $D_N = O((\log N)^k/N)$. Constructions of good lattice points are due to N. M. Korobov [Dokl. Akad. Nauk SSSR **124** (1959), 1207–1210; MR **21** #2848] and Hlawka [Monatsh. Math. **66** (1962), 140–151; MR **26** #888]. The discrepancy bounds for the (t, m, s)-nets and (t, s)-sequences follow Niederreiter [Monatsh. Math. **104** (1987), no. 4, 273–337; MR 89c:11120]. Also included are an algorithm of worst-case complexity $O(N(\log N)^k)$ which computes the L^2-discrepancy, the approximation of differentiable functions, a method for numerical integration of singular functions, and two examples from mathematical finance. The following section is devoted to spherical problems. There is a Chebyshev type quadrature formula for the ultraspherical weight function which gives a spherical design. It is followed by an application of low-dispersed spherical point sequences to the piecewise linear approximation of curves. In the next section the authors describe the quasi-Monte Carlo method for computing an approximate solution of the heat equation due to W. J. Morokoff and R. E. Caflisch [SIAM J. Numer. Anal. **30** (1993), no. 6, 1558–1573; MR 94k:65009] and the Boltzmann equation due to C. Lécot [Math. Comp. **56** (1991), no. 194, 621–644; MR 91j:65008]. There is a method of Hua and Wang [Sci. Sinica **17** (1974), no. 3, 331–348; ibid. **18** (1975), no. 2, 184–198; MR **55** #12675b] for solving a general class of initial value problems via a Fourier series expansion combined with an application of the good lattice point method. In a later section the authors review the basic concepts of randomness, A. N. Kolmogorov's complexity concept [Problemy Peredači Informacii **1** (1965), vyp. 1, 3–11; MR **32** #2273], D. E. Knuth's hierarchy of definitions [*The art of computer programming. Vol. 2*, Second edition, Addison-Wesley Publishing Co., Reading, Mass., 1981; MR 83i:68003] and R. Winkler's test functions [Math. Slovaca **43** (1993), no. 4, 493–512; MR 94g:65009]. For a detailed introduction to random number generators they refer to Niederreiter [*Random number generation and quasi-Monte Carlo methods*, SIAM, Philadelphia, PA, 1992; MR 93h:65008]. Following mainly J. C. Lagarias [in *Probability and algorithms*, 65–85, Nat. Acad. Press, Washington, DC, 1992; see MR 94c:68097], the chapter ends with several examples of number-theoretic random number generators, namely the multiplicative congruential generator $y_{n+1} = ay_n + c \bmod M$, the power generator $y_{n+1} = y_n^d \bmod M$, the quadratic generator $y_{n+1} = ay_n^2 + by_n + c \bmod M$, the discrete exponential generator $y_{n+1} = g^{y_n} \bmod M$, the inverse congruential generator $y_{n+1} = ay_n^{-1} + b \bmod M$ and generators involving a matrix and a cellular automaton.

In summary, the present authors have written a very readable and informative monograph on the theory of uniform distribution. The exposition in this book is quite complete in terms of the development of the ideas and statements of the results and often presents material at the edge of current research. The authors do provide an extensive bibliography (2000 items), and their book will be a valuable resource for researchers in the field. *Oto Strauch* (Bratislava)

97j:11037 11K31 11G20 11K45
Niederreiter, Harald (Vienna); **Xing, Chaoping** (Hefei)
Quasirandom points and global function fields. (English summary)
Finite fields and applications (*Glasgow*, 1995), 269–296, *London Math. Soc. Lecture Note Ser.*, 233, *Cambridge Univ. Press, Cambridge*, 1996.

In a series of papers by the authors [see, e.g., Acta Arith. **72** (1995), no. 3, 281–298; MR 96g:11099; Acta Arith. **73** (1995), no. 1, 87–102; MR 96g:11096; Acta Arith. **75** (1996), no. 4, 383–396; MR 97d:11177; C. R. Acad. Sci. Paris Sér. I Math. **322** (1996), no. 7, 651–654; MR 97d:11098; Finite Fields Appl. **2** (1996), no. 3, 241–273; MR 97h:11080; Acta Arith. **79** (1997), no. 1, 59–76; MR 97m:11141] a new method for the construction of extremely well distributed point sets in the s-dimensional unit cube was introduced and analyzed. This new method is based on tools from number theory and on deep methods from algebraic geometry and it provides point sets with the by far best distribution properties known so far. Quasi-Monte Carlo methods essentially are based on the use of point sets with such excellent distribution properties. So the work of Niederreiter and Xing (besides its great relevance in number theory and in algebraic geometry) is an enormous breakthrough in the theory of uniform distribution and its applications (maybe the most important one in the history of quasi-Monte Carlo methods). The present paper gives an excellent survey on the method of Niederreiter and Xing and it summarizes the work of the authors done in the papers cited above.

The starting point of the construction method is the general construction principle of so-called digital (t, s)-sequences constructed over a finite field. We give this concept here in a somewhat simplified form. In particular we restrict, for the sake of explanation, to the case of fields Z_p of prime order p.

To construct a digital (t, s)-sequence over Z_p, that is, an infinite sequence of points $\mathbf{x}_0, \mathbf{x}_1, \cdots$ in the s-dimensional unit cube $I^s := [0, 1)^s$ with a certain distribution quality parameter t, choose s infinite $\mathbf{N} \times \mathbf{N}$ matrices $C_1, \cdots, C_s$ over Z_p. To generate the ith coordinate $x_n^{(i)}$ of the point $\mathbf{x}_n$, represent the non-negative integer n in base p and denote the resulting vector of digits (identified with the corresponding elements of Z_p) in this representation by $\overrightarrow{n} = (a_0, \cdots, a_r, 0, 0, 0, \cdots)^{\mathsf{T}}$. Then multiply C_i by $\overrightarrow{n}$. The result is an infinite vector over Z_p, say $C_i \times \overrightarrow{n} := (y_1, y_2, \cdots)^{\mathsf{T}}$. Then set $x_n^{(i)} := y_1/p + y_2/p^2 + \cdots$. The quality parameter t now is defined in the following way: For an arbitrary positive integer m denote by $C_i^{(m)}$ the left upper $m \times m$ submatrix of C_i, and denote by $c_i^{(m)}(j)$ its jth row. Let $R(m)$ denote the maximal integer such that for all non-negative integers $d_1, d_2, \cdots, d_s$ with $d_1 + d_2 + \cdots + d_s = R(m)$ the system of row vectors $\{c_i^{(m)}(j) \mid j = 1, \cdots, d_i;\ i = 1, \cdots, s\}$ is linearly independent over Z_p. Then $t := \sup_m (m - R(m))$. For arbitrary finite fields the concept is quite similar.

It turns out that digital (t, s)-sequences with small parameter t (the optimal case is $t = 0$) have excellent distribution properties. So for the usual discrepancy D_N of the first N elements of a digital (t, s)-sequence constructed over the finite field $\mathbf{F}_q$ the following is known:

$$D_N \leq K(q, s) \cdot q^t \cdot (\log N)^s / N.$$

($K(q, s)$ is a constant depending only on q and on the dimension s.) According to a famous conjecture in the theory of uniform distribution this is the best possible order for the discrepancy of infinite sequences at all.

So the problem is to find s-tuples of matrices $C_1, \cdots, C_s$ over a finite field $\mathbf{F}_q$ with small parameter t. It was soon recognized that for given $\mathbf{F}_q$ the optimal parameter $t = 0$ is available only up to a certain dimension $s(q)$. So, for example, already I. M. Sobol′ [Ž. Vyčisl. Mat. i Mat. Fiz. **7** (1967), 784–802; MR **36** #2321] showed that $s(2) = 2$ and he gave special examples of digital (t, s)-sequences over $\mathbf{F}_2$ with reasonably good parameters t. H. Faure [Acta Arith. **41** (1982), no. 4, 337–351; MR 84m:10050] gave special examples for $(0, s(p))$-sequences over $\mathbf{F}_p$, p prime.

In 1987 Niederreiter [Monatsh. Math. **104** (1987), no. 4, 273–337; MR 89c:11120] introduced the general concept of digital (t, s)-sequences in full detail and later [J. Number Theory **30** (1988), no. 1, 51–70; MR 89k:11064] he gave a general construction method for s-tuples $C_1, \cdots, C_s$ providing digital (t, s)-sequences over $\mathbf{F}_q$. This method was based on working in a rational function field over a finite field and it provided parameters t which were of very high quality in the following sense. It is known that for every digital (t, s)-sequence over $\mathbf{F}_q$ we must have $t \geq K(q) \cdot s$, with a constant $K(q)$ depending only on q. The generation method of Niederreiter provided for every q and every s digital (t, s)-sequences with $t \leq K'(q) \cdot s \cdot \log s$.

For some time it was believed that the order $s \cdot \log s$ is best possible and that the lower bound on t should be improved. But then, based on an idea of Niederreiter, the two authors developed their method to construct suitable s-tuples $C_1, \cdots, C_s$, which now was based on work in global function fields with many rational places, and which now provided digital (t, s)-sequences with the best possible order of the quality parameter, that is, with $t \leq K''(q) \cdot s$.

They introduced various slightly different construction methods, each of which shows certain advantages. Just to give an idea of the method, we describe, quite briefly, the main steps for one of these constructions and for deriving the above upper bound for t. Let $K/\mathbf{F}_q$ be a global function field ($\mathbf{F}_q$ is the full constant field of the algebraic function field K), let $g = g(K/\mathbf{F}_q)$ be its genus, and let $N(K/\mathbf{F}_q)$ denote the number of rational places of $K/\mathbf{F}_q$, that is, of places of degree 1. For any nonzero k in K denote by $(k)_\infty$ the pole divisor of k. Let the dimension $s \geq 1$ be given and suppose that $N(K/\mathbf{F}_q) \geq s + 1$. Let $P_\infty, P_1, P_2, \cdots, P_s$ be $s + 1$ distinct rational places of $K/\mathbf{F}_q$. If $g = 0$ then there are $h_j^{(i)} \in K$ such that $(h_j^{(i)})_\infty = jP_i$ for $1 \leq i \leq s$ and $j \geq 1$. Now choose a local uniformizing parameter z at P_∞, consider the expansions $h_j^{(i)} = \sum_{r=0}^{\infty} b_{j,r}^{(i)} z^r$ with all $b_{j,r}^{(i)} \in \mathbf{F}_q$, and define $k_j^{(i)} := h_j^{(i)} - b_{j,0}^{(i)} \in K$ for $1 \leq i \leq s$ and $j \geq 1$. Then the matrices $C_1, C_2, \cdots, C_s$ over $\mathbf{F}_q$ are defined by $C_i = (c_{j,r}^{(i)})$, $j \in \mathbf{N}$, $r \in \mathbf{N}_0$, where $k_j^{(i)} = \sum_{r=0}^{\infty} c_{j,r}^{(i)} z^{r+1}$ for $1 \leq i \leq s$ and $j \geq 1$. The case where $g > 0$ is treated quite similarly.

Let now $V_q(s) = \min g(K/F_q)$, where the minimum is extended over all global function fields with $N(K/F_q) \geq s + 1$. Then it is shown that the above construction provides digital (t, s)-sequences with $t = 0$ if $1 \leq s \leq q$ and with $t \leq 2V_q(s) - 1$ if $s \geq q + 1$.

Denote by $N_q(g)$ the maximum number of rational places that a global function field over $\mathbf{F}_q$ of genus g can have. This quantity shows a close connection to $V_q(s)$ and was studied for example by Ihara, Serre, García and Stichtenoth, Xing and various other authors. By giving some estimates on $N_q(g)$ and by means of the class field theory of global function fields, and more specifically by the construction of certain infinite class field towers, general upper bounds for $V_q(s)$ and therefore for the quality parameter t are obtained.

The above results in many cases are existence results. So some further work had to be done and actually still will be done in forthcoming papers of the authors on the explicit construction of global function fields with many rational places.

To conclude: the work of the authors summarized in this paper presents a fascinating application of deep results in algebraic geometry and in the theory of global function fields to the construction of extremely well distributed number-theoretic point sets and consequently to quasi-Monte Carlo methods. Furthermore, the other way round, the construction method of Niederreiter-Xing motivates various new investigations in the theory of global function fields and provides the reason for various new results in this field of research.

{For the entire collection see MR 97h:11002.} *Gerhard Larcher* (Salzburg)

99d:11092 11L05 11A15 11L10 11T22 11T24
Berndt, Bruce C. (Urbana, IL); **Evans, Ronald J.** (La Jolla, CA); **Williams, Kenneth S.** (Ottawa, ON)
★**Gauss and Jacobi sums.**
Canadian Mathematical Society Series of Monographs and Advanced Texts.
A Wiley-Interscience Publication.
John Wiley & Sons, Inc., New York, 1998. xii+583 pp. $69.95. ISBN 0-471-12807-4

Gauss and Jacobi sums have been a subject of high interest in number theory ever since Gauss used the former in his *Disquisitiones arithmeticae* (1801) and Jacobi used the latter in his paper "Über die Kreistheilung und ihre Anwendung auf die Zahlentheorie" (1837). In both cases, the notions had been introduced earlier: for Gauss sums, by Lagrange in 1771 under the name of "resolvents"; for Jacobi sums, by Gauss in his unpublished manuscripts, by Jacobi himself in a letter of 1827 to Gauss and, independently, by Cauchy in 1829. In the nineteenth century, among the most notable contributors to the development and applications—to "cyclotomy", quadratic partitions of primes and reciprocity laws—of these sums were Gauss, Jacobi, Cauchy, Eisenstein, Kummer and Stickelberger. In the present century, there has been a great extension of the investigations and applications of these subjects. Among the topics of research involving Gauss and Jacobi sums are the study of cyclotomic fields, the derivation of reciprocity laws, the counting of the solutions of equations over finite fields and the study of the corresponding zeta functions, the construction of Hecke characters and the determination of their conductors, the explicit evaluation and the estimates of various related exponential sums, the relations and congruences among these sums and between them and the binomial coefficients or the p-adic gamma functions, the study of quadratic forms and that of power residue distributions. Further applications are concerned with combinatorial designs, coding theory and primality testing. The literature concerning these topics is immense and scattered in a great many papers, by numerous contributors. Some of the most cited references are by Hasse, Davenport, Weil, Mordell, Carlitz, Gross, Koblitz, Katz, Patterson, Chowla, Lehmer, and others. Until recently, the main source book for Gauss sums was H. Hasse's book [*Vorlesungen über Zahlentheorie*, Zweite neubearbeitete Auflage. Die Grundlehren der Mathematischen Wissenschaften, Springer, Berlin, 1964; MR **32** #5569], which has a chapter dealing with the subject. Then a few more books treated in some detail Gauss and Jacobi sums, such as those of K. F. Ireland and M. I. Rosen [*A classical introduction to* *modern number theory*, Second edition, Springer, New York, 1990; MR 92e:11001], S. Lang [*Cyclotomic fields*, Springer, New York, 1978; MR **58** #5578; *Vol. II*, 1980; MR 81i:12004], L. C. Washington [*Introduction to cyclotomic fields*, Springer, New York, 1982; MR 85g:11001], and R. Lidl and H. Niederreiter [*Finite fields*, Addison-Wesley, Reading, MA, 1983; MR 86c:11106]. Also, a different approach to the study of Gauss and Kloosterman sums, via sheaf-theoretic methods, was presented by N. M. Katz [*Gauss sums, Kloosterman sums, and monodromy groups*, Princeton Univ. Press, Princeton, NJ, 1988; MR 91a:11028]. However, the present book is exclusively concerned with Gauss, Jacobi and closely related sums, and gives the most extensive and systematic treatment of the subject, by a relatively elementary approach.

The basic theory of Gauss and Jacobi sums, over finite fields, is presented in a simple and efficient way in Chapters 1 and 2. There is in particular a reciprocity relation for generalized quadratic Gauss sums, proved using Cauchy's residue theorem, then applied to evaluate the classical quadratic Gauss sum. The latter is also obtained by two more elementary methods. One section deals with character sums over finite residue rings of integers. The cyclotomic numbers, which are linear combinations of Jacobi sums, are considered over a prime field $\mathbf{F}_p$ in Chapter 2 and over any finite field $\mathbf{F}_{p^r}$ in Chapter 11. A main theme of the book is the evaluation of various character sums of certain orders. A Gauss sum is (to within a well-determined root of unity factor) of the form $G_r(\chi) = \sum_{t\in\mathbf{F}_q} \chi(t)\zeta_p^{\mathrm{tr}(t)}$, where χ is a multiplicative character on the finite field $\mathbf{F}_q$, with $q = p^r$, p a prime number, $\zeta_p = e^{2\pi i/p}$ in $\mathbf{C}$ and tr is the trace map in $\mathbf{F}_q|\mathbf{F}_p$. A Jacobi sum can be written $J_r(\chi^m, \chi^n) = \sum_{t\in\mathbf{F}_q} \chi^m(t)\chi^n(1-t)$. The order k of these sums is that of the character χ. Another type of Gauss sum of order (a positive integer) k is $g_r(k) = \sum_{t\in\mathbf{F}_q} \zeta_p^{\mathrm{tr}(t^k)}$. It is related to the previous one by $g_r(k) = \sum_{j=1}^{k-1} G_r(\chi^j)$, provided that k divides $q-1$, with χ any character of order k on $\mathbf{F}_q$. In Chapter 3, the Jacobi sums over $\mathbf{F}_p$ of orders $k \leq 12$, $k \neq 9, 11$, and $k = 16, 20, 24$ are determined in terms of corresponding quadratic partitions of p. In Chapter 4, expressions of $g_1(k)$ for $k = 3, 4, 6$ are stated in terms of a Weierstrass $\wp$-function, and for $k = 8, 12$, they are given with a sign ambiguity. In Chapter 6, the Jacobsthal sums, which are related to Jacobi sums, are studied and similarly evaluated for some orders of characters. In Chapter 12, the Eisenstein sums, related to Gauss sums, are studied, and those of various orders over $\mathbf{F}_q$ are evaluated, which in turn allows the evaluation of $g_r(3)$ and $g_r(4)$. Another type of sum related to the Jacobsthal and Eisenstein sums, the Brewer sum, is studied and evaluated in Chapter 13. The Gaussian (f-nomial) periods over $\mathbf{F}_p$, closely related to the Gauss sums, are considered and applied in Chapter 5, which deals with "difference sets", and they are studied further in Chapter 10, while the periods over $\mathbf{F}_q$ are introduced in Chapter 12. The properties of Gauss sums are used, in Chapters 7 and 8, to derive various power residuacity criteria of small primes modulo a fixed prime p, in terms of quadratic partitions of p, and to obtain the cubic, biquadratic and some rational reciprocity laws. In Chapter 9, congruences $(\mathrm{mod}\, p)$ for binomial coefficients, in terms of quadratic partitions of p, are established, using Jacobi sums; some of them are extended to congruences $(\mathrm{mod}\, p^2)$ using the Gross-Koblitz formula, which relates the p-adic gamma function to Gauss sums. In Chapter 10, the number of solutions of diagonal equations over $\mathbf{F}_q$ is obtained in terms of generalized Jacobi sums (depending on more than two characters); the latter are first studied and some of the properties of the previously considered Jacobi sums are extended to them; upper bounds for the number of solutions are also deduced from those expressions. Chapter 11 contains the deepest results, namely the classical Stickelberger congruence for Gauss sums—yielding the prime ideal factorizations of Gauss and Jacobi sums—the Davenport-Hasse product formula for Gauss sums and the Davenport-Hasse theorem for lifted Gauss sums; the chapter concludes with an application to irreducible cyclic codes. Finally, Chapter 14 presents an extension of the classical Eisenstein reciprocity law to the p^nth power residue symbol, using the properties of Gauss and Jacobi sums, namely their prime ideal factorization, and based on work by A. E. Western (1907–1908).

Every chapter ends with a set of problems that complement and extend the results in the text. These are followed by invaluable notes giving references to the vast literature, outlining the history of the results, and indicating links to related subjects and some current research directions. In this respect, the book ends with an interesting list of 28 open problems. The bibliography, which covers 66 pages, is fairly complete. Relatively few errors were found, most of which are easy to correct by the reader. On page 11, formula (1.1.4) is valid provided k divides $q-1$. On page 271, in Theorem 9.2.6, t is to be replaced by a_7. On page 362, before formula (11.6.1), the reference should be to Theorem 2.1.3(a) instead of Theorem 1.1.4(a). On page 365, after (11.6.4), one should read: let γ be a generator of $\mathbf{F}_q^*$ (instead of $\mathbf{F}_q$). On page 444, in formula (13.2.6), the second of the three terms in the equality ought to be $\Omega_d(a^{n/d})$. On page 468, after (14.1.1), where the domain of Φ_a is extended "to all nonzero ideals of O_K" and where $\Phi_a(\alpha)$ is defined, one should add the condition that these ideals and elements α of O_K be relatively prime to k. On page 470, the definition of a primary element as stated is not equivalent to (14.2.2) (which is the only condition used in the sequel), since if $l>2$, one may have M odd and $\varepsilon_c(v)=(-1)^M=-1$; to correct this, one may either adopt (14.2.2) as the definition or set in the existing definition $\varepsilon_c(v)=(-1)^{M(k-1)}$.

The book is carefully written, in a clear and fluent style, which makes it accessible to a broad audience. It undoubtedly reaches its goal of being a "user-friendly" introduction to the subject, focused on "examining basic properties of Gauss and Jacobi sums, providing systematic and explicit evaluations of these sums, and providing applications". Even though it does not exhaust the immense subject of exponential sums, it should prove to be a standard reference. *Charles Helou* (Media, PA)

98m:11088 11L07 11G25

Fouvry, Étienne (Orsay); **Michel, Philippe** (Orsay)

Sur certaines sommes d'exponentielles sur les nombres premiers. (English and French summaries) [Some exponential sums over primes]

Ann. Sci. École Norm. Sup. (4) **31** (1998), *no.* 1, 93–130.

The study of exponential sums plays a central role in number theory. On the one hand there is the analytic number theorist hunting for his favourite numbers and anxious to keep a count of how many he has managed to track down. For him this counting will usually involve some form of harmonic analysis, either directly or perhaps during the course of applying a sieve, and this analysis will frequently reduce to the problem of detecting cancellation in a relevant exponential sum having the form $\sum_{m\in\mathcal{M}}\psi(f(m))$. Here $\mathcal{M}$ is a set of the "favourite numbers", f is a rational function with integer coefficients, and ψ is a nontrivial additive character (i.e. a character on the additive group) of a given field. In a most basic case f is a function of one variable, the field is either the reals $\mathbf{R}$ or a finite field $\mathbf{F}_q$ and the number theorist is searching for primes.

The exponential sum over primes $\sum_{p\leq x}\psi(f(p))$ can be re-formulated by means of a combinatorial identity which expresses the characteristic function on the prime numbers in terms of a sum of functions, more complicated in appearance, but which turn out to be easier to estimate. The first identity of this type is just the sieve of Eratosthenes. The idea of using such an identity to estimate exponential sums is due to Vinogradov. More recently, useful identities of the requisite nature have been proposed by Linnik, Gallagher, Vaughan, Heath-Brown, Harman, Friedlander-Iwaniec and others. Any of these, when inserted into the above exponential sum, leads to sums

$$S_{\mathrm{I}} = \sum_{m\leq M}\alpha_m \sum_{n\in\mathscr{I}}\psi(f(mn))$$

and

$$S_{\mathrm{II}} = \sum_{m\leq M}\alpha_m \sum_{n\in\mathscr{I}}\beta_n\psi(f(mn))$$

where α_m, β_n are sequences of (usually unwieldy) complex numbers having modulus ≤ 1, $\mathscr{I}$ is an interval containing N integers and M, N are parameters in certain ranges determined by the combinatorial identity.

In the case of the real field there is a long history of results. The sums S_{I}, S_{II} can be treated using ideas originating with Weyl, Hardy-Littlewood, van der Corput and Vinogradov and it was by using such results in conjunction with his combinatorial identities that Vinogradov produced the first results for the sum $\sum_{p\leq x}\psi(f(p))$.

The methods of Weyl and his successors are ingenious and have been extensively developed; nevertheless, in the end they reduce to the estimation of sums of elementary nature (in the case of Weyl, a geometric progression). When the field of real numbers is replaced by a finite field something very beautiful happens.

There is a second sense in which the study of exponential sums plays a central role in number theory; indeed there are many who would strongly object to the use of the adjective "second" in this regard. Sums of the type $S_q = \sum_{y\in\mathbf{F}_q}\psi(f(y))$ (and more general sums over varieties) have, since Gauss, arisen in a more algebraic setting and have been studied using techniques of arithmetic geometry by, amongst others, Hasse, Weil, Grothendieck, Bombieri, Deligne, Katz, Adolphson and Sperber. The results obtained are justly celebrated for their depth and importance. Quite apart from their own intrinsic interest they have offered and continue to offer to the analytic number theorist very powerful tools. However, because of the formidable machinery required, there are unfortunately not many analytic number theorists who are comfortable working with them. The authors of the current paper are clearly exceptional in this regard.

Especially in the case in which q is prime, it is easy to see the relevance of the above sum S_q to sums over ordinary integers. However, whereas it is the sum over the full group which occurs in a natural algebraic setting, it is the "incomplete" sum, at best over an interval such as in S_{I}, S_{II}, at worst over a more complicated subset such as the primes in that interval, which concerns the analytic number theorist.

As already mentioned, the sum over primes can be reduced via a combinatorial identity to sums like S_{I}, S_{II}. The problem then is to reduce these to the "complete" sums. The inner sum in S_{I} is over n in an interval and has nice smooth weights but the inner sum in S_{II} is corrupted by the unknown weights β_n. One may rewrite this sum as $\sum_n \beta_n(\sum_m \alpha_m \psi(f(mn)))$, then apply Cauchy's inequality followed by the trivial bound $|\beta_n| \leq 1$, obtaining a majorant sum of the form

$$\sum_{m_1}\sum_{m_2} \alpha_{m_1}\overline{\alpha}_{m_2} \sum_n \psi(f(m_1 n) - f(m_2 n)).$$

Here the inner summation is free from the coefficients β_n, so that we may now hope to apply whatever ideas we use for S_{I}.

There is a standard technique, which dates at least back to Pólya in connection with Dirichlet character sums, whereby the sum over a subinterval $\mathcal{I}$ of $[1,q]$ can be expressed in terms of complete sums albeit of slightly different form. This is achieved by writing the characteristic function χ of the integers in $\mathcal{I}$ as a finite Fourier series and interchanging the order of the double sum so obtained. This widely applicable technique leads to results for many different sums. There is however associated to it a natural stopping point for the length of the interval $\mathcal{I}$ such that for intervals below this length the method gives no nontrivial result (this does not usually happen for the corresponding sum when the field is $\mathbf{R}$).

To obtain nontrivial results for shorter intervals is important (sometimes crucial) for applications and requires considerably more ingenuity. The first success in this direction was the Burgess bound for Dirichlet character sums [D. A. Burgess, Proc. London Math. Soc. (3) **13** (1963), 524–536; MR **26** #6133] and (hence) for Dirichlet L-functions. More recently, in a series of papers [e.g., in *Proceedings of the International Congress of Mathematicians, Vol. 1, 2 (Zürich, 1994)*, 363–373, Birkhäuser, Basel, 1995; MR 97e:11108] the reviewer with W. Duke and H. Iwaniec has obtained such "sub-convexity" bounds for GL_2 automorphic L-functions.

The authors consider the quite general case of the exponential sum

$$S(f;q,x) = \sum_{p\leq x} \psi(f(p))$$

over primes p, where q is a prime, ψ is any nontrivial additive character over $\mathbf{F}_q$, $f(X) = P(X)/Q(X)$ is any rational function (other than the trivial cases of f = constant, or f = linear polynomial), with P and Q relatively prime monic polynomials in $\mathbf{Z}[X]$.

The authors point out that known methods provide, as a fairly straightforward consequence of the Weil bound, nontrivial results in the range $x \geq q$, and they consider the problem of surmounting this stumbling block and dealing with the short sums, $x < q$. By means of long, difficult, and ingenious arguments they are able to break through this barrier and prove the bound

$$S(f;q,x) \ll q^{3/16+\varepsilon}x^{25/32}$$

with any $\varepsilon > 0$, f, q and ψ as above, and $1 \leq x \leq q$. Here the implied constant depends at most on ε and on the degrees of P and Q. Note that the exponents $\frac{3}{16}$ and $\frac{25}{32}$ are independent of these degrees, a very nice fact not known to happen for the real field. The bound is nontrivial throughout the range $q^{6/7+\varepsilon} < x \leq q$. In the case of the special functions $f = X^k + wX$, where $k \neq 0, 1$ and w are in $\mathbf{Z}$, the authors are able to strengthen their results and obtain a bound nontrivial through the range $q^{3/4+\varepsilon} < x \leq q$.

As an agreeable consequence of their main result they give the following: Let $f_1, \cdots, f_{33}$ be thirty-three rational functions of the above (general) type. Then the equation

$$N = f_1(p_1) + \cdots + f_{33}(p_{33}) \pmod q$$

has, as $q \to \infty$, $(\pi(q)^{33}/q)(1 + O(q^{-1/33}))$ solutions in primes $p_i \le q$, uniformly in the integer N.

Because the argument follows the thread described above, the authors prove along the way important new bounds for the relevant sums S_{I}, S_{II} and it is herein that the main innovations are to be found. These bounds may also be used, as the authors point out, to treat more general sums (than that over the primes) which can also be decomposed by combinatorial identities, including for example the sum $\sum_{n \le x} \mu(n) \psi(f(n))$.

At the risk of repetition the reviewer wishes to express how impressive he finds both the ingenuity of the arguments and the authors' virtuosity in interweaving the analytic and the algebraic aspects of the argument. The authors generously state that they got the idea for this work while studying the paper of the reviewer and Iwaniec [Ann. of Math. (2) **121** (1985), no. 2, 319–350; MR 86i:11050], but if so, they have picked up this idea and carried it a very long way indeed.

John B. Friedlander (West Hill, ON)

99i:11074 11M26

Conrey, J. B. (Stillwater, OK); **Ghosh, A. [Ghosh, Amit[1]]** (Stillwater, OK); **Gonek, S. M.** (Rochester, NY)

Simple zeros of the Riemann zeta-function.

Proc. London Math. Soc. (3) **76** (1998), *no.* 3, 497–522.

This important paper has been well known to experts in the field for many years. The results were first announced in 1983 [J. B. Conrey, in *Topics in analytic number theory (Austin, Tex., 1982)*, 28–41, Univ. Texas Press, Austin, TX, 1985; see MR 86f:11007]. Despite the long delay in its publication, this paper still provides the best conditional result on simple zeros currently known.

Let $\rho = \beta + i\gamma$ be a complex zero of the Riemann zeta-function. The Riemann hypothesis is that $\beta = \frac{1}{2}$ for all of these zeros. In addition, it has been conjectured that all the zeros are simple. The methods for addressing the problem of simple zeros attempt to show that a large proportion of the zeros are simple. We let $N(T) = |\{\rho = \beta + i\gamma\colon\ 0 < \gamma \le T,\ \zeta(\rho) = 0\}|$, $N_s(T)$ denote the number of simple zeros in the set defining $N(T)$, and $N_d(T)$ the number of distinct zeros in that set. We then define C_s to be the infimum of the ratio $N_s(T)/N(T)$ as $T \to \infty$, and C_d is defined similarly. Thus C_s is asymptotically the smallest proportion of zeros that are simple. By a modification of Levinson's method, it is known unconditionally [J. B. Conrey, J. Reine Angew. Math. **399** (1989), 1–26; MR 90g:11120] that $C_s > 0.401$ (and these zeros are on the $\frac{1}{2}$ line as well). H. L. Montgomery [in *Analytic number theory (Proc. Sympos. Pure Math., Vol. XXIV, St. Louis Univ., St. Louis, Mo., 1972)*, 181–193, Amer. Math. Soc., Providence, R.I., 1973; MR **49** #2590; MR errata, EA **49**] showed in his well-known work on pair correlation of zeros of $\zeta(s)$ that, assuming the Riemann hypothesis, $C_s > \frac{2}{3}$, and this proportion has been improved slightly to $0.6727\cdots$. Assuming Montgomery's pair correlation conjecture one gets $C_s = 1$ so that asymptotically almost all the zeros are simple. Montgomery's method is based on bounding the number of zeros weighted by their multiplicity, and does not deal directly with simple zeros. In this paper a different conditional approach is introduced that deals directly with simple zeros. The starting point is the obvious observation that ρ is a simple zero of $\zeta(s)$ if and only if $\zeta'(\rho) \neq 0$, and thus by Cauchy's inequality

$$\Big|\sum_{0<\gamma\leq T} B\zeta'(\rho)\Big|^2 \leq N_s(T) \sum_{0<\gamma\leq T} |B\zeta'(\rho)|^2,$$

where $B\zeta' = B(s)\zeta'(s)$, and $B(s)$ is any analytic function. To minimize the loss in applying Cauchy's inequality we want to choose B so that $B\zeta'$ is close to constant, and therefore B is chosen as a Dirichlet series mollifier for ζ' given by

$$B(s) = \sum_{k\leq y} \frac{b(k)}{k^s}, \quad b(k) = \mu(k)P\left(\frac{\log y/k}{\log y}\right), \quad y = T^\theta,$$

where $P(x)$ is a polynomial with real coefficients which satisfies $P(0) = 0$ and $P(1) = 1$. The main result of the paper is the following theorem: Assume the generalized Lindelöf hypothesis for Dirichlet L-functions. Then for $0 < \theta < \frac{1}{2}$, and letting $L = \log(T/2\pi)$, we have

$$S_1 := \sum_{\gamma\leq T} B\zeta'(\rho) \sim \left(\frac{1}{2} + \theta\int_0^1 P(x)\,dx\right)\frac{TL^2}{2\pi},$$

and

$$S_2 := \sum_{\gamma\leq T} B\zeta'(\rho)B\zeta'(1-\rho) \sim \lambda\frac{TL^3}{2\pi},$$

where

$$\lambda = \frac{1}{3} + \left(\theta\int_0^1 P(x)\,dx\right)^2 + \theta\int_0^1 P(x)\,dx + \frac{1}{12\theta}\int_0^1 P'(x)^2\,dx.$$

On assuming the Riemann hypothesis, $S_2 = \sum_{\gamma\leq T}|B\zeta'(\rho)|^2$, and therefore the sums in the inequality for N_s are evaluated. The only use of the Riemann hypothesis is at this point. By the calculus of variations the choice $P(x) = -\theta x^2 + (1+\theta)x$ is optimal, and on taking $\theta = \frac{1}{2} - \varepsilon$ and $\varepsilon \to 0^+$, it follows that $S_1 \sim \frac{19}{24}N(T)L$ and $S_2 \sim \frac{57}{64}N(T)L^2$ (using also $N(T) \sim TL/2\pi$). From these results it follows that $C_s \geq \frac{19}{27}$, and a simple argument then gives $C_d \geq \frac{5}{6} + \frac{1}{81}$. The assumptions for these results are the Riemann hypothesis and the generalized Lindelöf hypothesis.

The evaluation of S_1 and S_2 is long and difficult. The starting point is the formula

$$\sum_{\gamma\leq T} F(\rho) = \frac{1}{2\pi i}\int_{\mathscr{C}} \frac{\zeta'(s)}{\zeta(s)}F(s)\,ds,$$

where $\mathscr{C}$ is a rectangle with vertices at $1-c+i$, $1-c+iT$, $c+iT$, and $c+i$, where $c = 1+L^{-1}$, and $F(s)$ is an analytic function in this rectangle. Applying this formula to S_1 and S_2 and using the functional equation $\zeta(s) = \chi(s)\zeta(1-s)$, one eventually obtains a number of main terms which occur in Levinson's method and are familiar, together with the terms for $\nu = 1$ or 2,

$$M_\nu = \sum_{k \le y} \sum_{m \le kT/2\pi} \frac{a_\nu(m) b(k)}{k} e\left(-\frac{m}{k}\right),$$

where $a_1(m)$ is the Dirichlet series coefficient of $Z(s) = \frac{\zeta'}{\zeta}(s)\zeta'(s)$, $a_2(m)$ is the Dirichlet coefficient of $ZB\zeta'(s)$, and $e(u) = e^{2\pi i u}$. The interesting feature of M_ν is the factor $e(-m/k)$ which arises from the factor $\chi(s)$ in the functional equation. As in the circle method, by breaking up the sum according to the residue class into which m falls modulo k, and assuming the terms are well distributed in the reduced residue classes, one finds that the main term in M_ν is obtained by replacing $e(-m/k)$ in the sum by $\mu(K)/\varphi(K)$, where $K = k/(m,k)$. To carry out the analysis one introduces characters, and the error terms can be handled assuming the generalized Riemann hypothesis. To avoid this hypothesis the authors employ a much more complicated approach using a Vaughan type identity together with large sieve estimates. The approach falls short of being unconditional only at one point where the conjectured sixth power moment for L-functions is needed. This follows from the generalized Lindelöf hypothesis, which is the only place where this hypothesis is used.

The work in this paper has been used by D. W. Farmer [Mathematika **40** (1993), no. 1, 71–87; MR 95a:11076] to give some interesting results and conjectures on the Riemann zeta-function. *Daniel A. Goldston* (San Jose, CA)

98a:11120 11N05 11N32 11N35 11N36
Fouvry, Etienne (Orsay); **Iwaniec, Henryk** (New Brunswick, NJ)
Gaussian primes.
Acta Arith. **79** (1997), *no.* 3, 249–287.

Fermat proved that any prime $p \equiv 1 \bmod 4$ can be written as the sum of two squares and, because of this characterization (the other odd primes obviously cannot), we know that there are infinitely many primes $p = l^2 + m^2$ and also know the asymptotic formula for their frequency.

In this important paper the authors prove that there are infinitely many primes $p = l^2 + m^2$ where l also is a prime. They also establish the asymptotic formula for the frequency of such p. In fact the authors' main theorem is much more general.

Let χ denote the non-principal character of modulus four and let $\psi(l) = \prod_{p \nmid l}(1 - \chi(p)/(p-1))$. Let λ_l denote arbitrary complex numbers with $|\lambda_l| \le 1$ and Λ the von Mangoldt function. Then, for any $A > 0$,

$$\sum_{l^2+m^2 \le x} \lambda_l \Lambda(l^2+m^2) = \sum_{l^2+m^2 \le x} \lambda_l \psi(l) + O_A\big(x(\log x)^{-A}\big).$$

Thus the set of primes l can be replaced by almost any reasonably nice set of integers which has density $\gg x(\log x)^{-B}$ for some B.

The authors prove a number of other theorems of independent interest where, in the above formula, the von Mangoldt function Λ is replaced by the generalized von Mangoldt function Λ_k, which lives on almost primes or by the number of ideals of given norm in a specified abelian number field.

All of the functions in question may be viewed as being supported on a set of integers which survive a sieve and the results are obtained by sieve methods. In the most interesting and difficult case of the main theorem, the set of survivors consists of prime numbers and because of the parity problem, the authors use a combinatorial identity (in this case, Vaughan's) which requires the consideration of both linear and bilinear forms. In the current case the authors require relatively little in the way of the bilinear forms and it is this that is to some extent responsible for the great generality of their result. The reason they have comparatively easy going on this point is that they are spectacularly successful in dealing with the linear forms (also known as the remainder term in the classical form of the sieve).

This success rests on a simple, beautiful and surprisingly strong large sieve inequality for the roots v of the congruence $v^2+1\equiv 0 \bmod d$. Such roots are, as noted by Gauss, in one-to-one correspondence with the primitive representations of d as the sum of two squares $d=r^2+s^2$ with $(r,s)=1$ and $-s<r\le s$. This correspondence satisfies $v/d\equiv r/sd-\bar{r}/s \bmod 1$, where $\bar{r}r\equiv 1 \bmod s$. This allows us to replace v/d (essentially) by $\bar{r}/s$, which has much smaller denominator (roughly $d^{1/2}$ on average), and hence to deduce that distinct fractions v/d are more evenly spaced modulo 1 than are typical fractions. This leads to Lemma 2: For any complex numbers α_n,

$$\sum_{8D<d\le 9D}\ \sum_{v^2+1\equiv 0 \bmod d}\left|\sum_{n\le N}\alpha_n\, e\left(\frac{vn}{d}\right)\right|^2\le 72(D+N)\|\alpha\|^2.$$

Here $e(u)=e^{2\pi iu}$, $\|\alpha\|$ is the l_2-norm and we might have expected to see D^2 in place of D on the right side.

The authors' success with this remainder term provided motivation for, and an essential ingredient of, the recent proof by the reviewer and H. Iwaniec [Ann. of Math. (2) **148** (1998), no. 3, 945–1040; Ann. of Math. (2) **148** (1998), no. 3, 1041–1065] of the infinitude of (and the asymptotic formula for) the primes of the form a^2+b^4. *John B. Friedlander* (West Hill, ON)

99m:11106 11N37 11N64
Ford, Kevin (Austin, TX)
The distribution of totients. (English summary)
Paul Erdős (1913–1996).
Ramanujan J. **2** (1998), *no.* 1-2, 67–151.

This excellent paper is concerned with the circle of problems related to the values taken on by Euler's totient function $\varphi(m)$. In particular the author considers the function $V(x)$ which counts the number of positive integers $n\le x$ which are in the image, in other words equal to $\varphi(m)$ for some m. He also considers $A(n)$, the multiplicity of n, that is, the number of m for which $\varphi(m)=n$, and $V_k(x)$, the number of $n\le x$ for which $A(n)=k$. There are a number of well-known and difficult problems connected with these functions. In this long and comprehensive paper the author makes significant progress with a number of these.

Theorem 1, the main result, gives the precise order of magnitude of $V(x)$. The behavior of the set of values of this very basic function is rather irregular and much work had already been done on this problem. Earlier results due to Erdős, then Erdős and Hall, then C. Pomerance, then H. Maier and Pomerance [Acta Arith. **49** (1988), no. 3, 263–275; MR 89d:11083], had given successively sharper asymptotic formulae but only for log $V(x)$ and these were not of sufficient precision to determine the order of $V(x)$ itself.

In the second theorem the author shows that if an integer k occurs as the multiplicity $A(d)$ of any integer d then it occurs as the multiplicity of a positive proportion of such integers, that is, if $V_k(x)$ is positive for some x then in fact $V_k(x) \gg V(x)$.

A famous conjecture of Carmichael states that $A(n) = 1$ does not occur, that is, if an integer occurs as the value of Euler's function then it does so at least twice. The author sharpens earlier work to show that any exception to this conjecture must exceed $10^{10^{10}}$. In addition the case $k = 1$ of the previous result shows that if Carmichael's conjecture is false then it is false a positive proportion of the time.

Another well-known conjecture due to Sierpiński states that, contrary to what is expected in case $k = 1$, for each $k \geq 2$ the multiplicity $A(n) = k$ does occur. A. Schinzel [Acta Arith. **7** (1961/1962), 1–8; MR **24** #A70] has shown that this conjecture does follow from his famous Hypothesis H. In this paper the author uses a simpler argument to show that it suffices to assume the weaker prime k-tuples conjecture.

There is a great deal more in the current paper than has been indicated in this review; fifteen theorems are given. The technicalities of proof are long and complicated so it is welcome that the author gives in the last two pages of the introduction a quite readable sketch of the basic ideas behind these proofs.

It should be mentioned that there have been significant further developments since this paper was written. First, in work by Ford and S. Konyagin [in *Number theory in progress, Vol. 2 (Zakopane-Kościelisko, 1997)*, 795–803, de Gruyter, Berlin, 1999] an unconditional proof has been given of the above Sierpiński conjecture for every even integer k and of the corresponding Sierpiński conjecture (this time for all k, including $k = 1$) for the sum of divisors function $\sigma(m)$, that is, for each $k \geq 1$ there exists an integer n such that the equation $\sigma(m) = n$ has precisely k solutions m. Then, Ford ["The number of solutions of $\varphi(x) = m$", Ann. of Math., to appear] has given a much more complex argument which yields the full Sierpiński conjecture for $\varphi(m)$ itself. *John B. Friedlander* (West Hill, ON)

98g:11110 11P05 11L07

Wooley, Trevor D. (Ann Arbor, MI)

On exponential sums over smooth numbers.

J. Reine Angew. Math. **488** (1997), 79–140.

During the last decade the theory of Waring's problem has largely benefitted from the development of the new iterative method due to R. C. Vaughan [Acta Math. **162** (1989), no. 1-2, 1–71; MR 90c:11072]. This involves exponential sums over smooth numbers

$$\mathcal{A}(P, R) = \{x \leq P\colon p \mid x \Rightarrow p \leq R\} \tag{1}$$

where typically R is a small power of P. Very roughly speaking, it turns out to be much more effective to count solutions of Waring's equation (2) $x_1^k + x_2^k + \cdots + x_s^k = n$ with all or at least most variables restricted to (1) than simply to count *all* solutions of (2). The method relies on mean value estimates for

$$\int_0^1 \Big| \sum_{x \in \mathcal{A}(P,R)} e(\alpha x^k) \Big|^{2s} d\alpha, \tag{3}$$

which can be viewed as a counting problem for a homogeneous equation. Homogeneity is a very important feature in all previous applications of the new iterative method, and the method seemingly collapses in non-homogeneous situations.

The paper under review shows that a considerable portion of the power of the new iterative method can be rescued when homogeneity is lost, and provides a systematic study of sums

$$f = \sum_{x \in \mathscr{A}(P,R)} e(\alpha_1 x^{k_1} + \cdots + \alpha_t x^{k_t}) \tag{4}$$

where $1 \leq k_t < k_{t-1} < \cdots < k_1$, and of the corresponding mean values

$$\int_0^1 \cdots \int_0^1 |f|^{2s} d\alpha_1 \cdots d\alpha_t.$$

The mean values are reminiscent of Vinogradov's mean value theorem, and indeed the approach taken here combines modern technology for sums of kth powers and the differencing approach to Vinogradov's mean value theorem [T. D. Wooley, Mathematika **39** (1992), no. 2, 379–399; MR 94d:11074].

Applications of exponential sums of the type considered here are too numerous to be listed here. We only pick some examples. There are implications for systems of additive equations, but these are not discussed in the present paper [see T. D. Wooley, J. Reine Angew. Math. **419** (1991), 141–198; MR 92e:11025b; Mathematika **45** (1998), no. 2, 319–335]. When $\alpha_1 = \cdots = \alpha_t$ in (4), then f is the generating function for Waring's problem with polynomial summands. Here the author obtains the first significant progress in almost 50 years. Other applications relate to the distribution of fractional parts of polynomials $\alpha_1 n^{k_1} + \cdots + \alpha_t n^{k_t}$. For more details we refer to the carefully written introduction of the paper. *Jörg Brüdern* (Stuttgart)

99m:11124 11R23 11G40 11R42

Perrin-Riou, Bernadette (Orsay)

Systèmes d'Euler p-adiques et théorie d'Iwasawa. (English and French summaries) [p-adic Euler systems and Iwasawa theory]

Ann. Inst. Fourier (*Grenoble*) **48** (1998), *no.* 5, 1231–1307.

The paper under review develops a general theory of cohomological cyclotomic Euler systems attached to Galois representations satisfying certain conditions. To explain what these systems are and their significance we recall some history. In very general terms an Euler system consists of elements $c(n)$, for certain integers n, satisfying some relations.

Euler systems were introduced by Kolyvagin to give an effective bound on the class number of cyclotomic fields. The system in this case is of cyclotomic units $c(n) = 1-\zeta_n$ where ζ_n is a primitive nth root of unity. The result Kolyvagin obtains in this case bounds the Galois eigenspaces of the class group of the real subfield of $\mathbf{Q}(\zeta_p)$ by that of the same eigenspace in the quotient of all units by the cyclotomic units. The review of [V. A. Kolyvagin, in *The Grothendieck Festschrift, Vol. II*, 435–483, Progr. Math., 87, Birkhäuser Boston, Boston, MA, 1990; MR 92g:11109] is an excellent summary. Kolyvagin's methods were further developed to give a proof of the one- and two-variable Iwasawa main conjecture by K. Rubin [Invent. Math. **103** (1991), no. 1, 25–68; MR 92f:11151]. For a nice introduction one can consult the appendix by Rubin to the new edition of S. Lang's book [*Cyclotomic fields I and II*, Combined second edition, Springer, New York, 1990; MR 91c:11001].

In a later work Kolyvagin used Euler systems based on Heegner points on modular curves to obtain the first general results concerning the Birch and Swinnerton-Dyer conjecture for a modular elliptic curve E and in particular was able to bound the Tate-Shafarevich group of E in terms of the index of a Heegner point in $E(\mathbf{Q})$. For a very nice introduction see [B. C. Mazur, Bull. Amer. Math. Soc. (N.S.) **29** (1993), no. 1, 14–50; MR 93m:11052].

In extending the results of Kolyvagin to Kuga-Sato varieties over modular curves, J. Nekovář [Invent. Math. **107** (1992), no. 1, 99–125; MR 93b:11076] considered an Euler system in a purely cohomological context: $c(n) \in H^1(K_n, V)$, where K_n is the ring class field of an imaginary quadratic field K and V is the Galois module attached to a modular form by Deligne.

While some of the computations in Nekovář's paper are essential to the paper under review, in particular the analysis of the key operation of Kolyvagin derivatives described in Chapter 3, no general theory of Euler systems was suggested there. The need for the development of such a general theory was enhanced by the work of K. Kato [see A. J. Scholl, in *Galois representations in arithmetic algebraic geometry (Durham, 1996)*, 379–460, Cambridge Univ. Press, Cambridge, 1998], who defined an Euler system on the representation associated to a modular form based on elements in the algebraic K-theory of the associated modular curve. Such an "Euler system machine" is developed for the first time in the paper under review. We now describe the main results.

Let ρ be a continuous $\mathrm{Gal}(\overline{\mathbf{Q}}/\mathbf{Q})$ representation on a vector space V, finite-dimensional over a finite extension of $\mathbf{Q}_p$, which is unramified except over a finite set of places S. One chooses an invariant integral submodule T in V. An Euler system in more traditional use is what the author calls here an Euler system of rank 1. It consists of cohomology classes $c(m) \in H^1(\mathbf{Q}(\mu_m), T)$ given for all square-free m prime to a finite set of primes Σ. Here $\mathbf{Q}(\mu_m)$ is the cyclotomic field generated by the set μ_m of m-roots of unity. These classes should satisfy the relation

$$\mathrm{Tr}_{\mathbf{Q}(\mu_{ml})/\mathbf{Q}(\mu_m)}(c(ml)) = P_l(V^*(1), \mathrm{Frob}_l^{-1})c(m),$$

whenever l is a prime not dividing m, where for any two fields $K \subset L$, $\mathrm{Tr}_{L/K}: H^1(L,T) \to H^1(K,T)$ denotes the corestriction map and where $P_l(V^*(1), X) := \det(1 - X\rho(\mathrm{Frob}_l)^{-1} | V^*(1)^{I_l})$ is the characteristic polynomial of Frob_l acting on the inertia invariants of $V^*(1)$.

The author defines the notion of a "p-adic Euler Iwasawa system" of rank 1 where the elements $c(m)$ are also defined for integers of the form $m = m'p^n$ where m' is square-free and prime to Σ. These classes are then required to satisfy the additional relation $\mathrm{Tr}_{\mathbf{Q}(\mu_{ml})/\mathbf{Q}(\mu_m)}(c(ml)) = c(m)$ whenever $l \mid m$.

The definition of an Euler system in the style of Kolyvagin required an additional "congruence condition" relating the restriction of the classes $c(ml)$ and $c(m)$ to the decomposition group at primes above l. It was first observed by Rubin [J. Reine Angew. Math. **425** (1992), 141–154; MR 93d:11117] that a weak congruence condition can be derived from the "Iwasawa" condition. This derivation is the main focus of Chapter 2 of the present paper.

Let $\mathbf{Q}_\infty = \bigcup \mathbf{Q}_n$ be the cyclotomic $\mathbf{Z}_p$ extension of $\mathbf{Q}$. For any number field F the author defines groups $H^i_{\infty,S}(F,T)$ to be the projective limit of $H^i(G_{S,F\mathbf{Q}_n}, T)$ under corestriction maps, where $G_{S,L}$ is the Galois group of the maximal extension of L ramified only at S. When $i=1$ this group is independent of S and one denotes it simply by $H^1_\infty(F,T)$. One defines $H^2_\infty(F,T)\subset H_{\infty,S}(F,T)$ as the kernel of restrictions to the decomposition groups above S. Both $H^i_\infty(\mathbf{Q}(\mu_m),T)$ are modules for the Iwasawa algebra $\Lambda=\mathbf{Z}_p[\![\Gamma]\!]$ with $\Gamma=\mathrm{Gal}(\mathbf{Q}_\infty/\mathbf{Q})=\mathrm{Gal}(\mathbf{Q}(\mu_{mp^\infty})/\mathbf{Q}(\mu_{mp}))$.

The author defines the notion of a "p-adic Euler Iwasawa system" c_p, where

$$c_p(n)\in\bigwedge^{d_-}H^1_\infty(\mathbf{Q}(\mu_n),T)^+\oplus\bigwedge^{d_+}H^1_\infty(\mathbf{Q}(\mu_n),T)^-$$

for m square-free and prime to $p\Sigma$ and satisfying the same relations as an Euler system of rank 1. Here $(\ )^\pm$ are the eigenspaces of complex conjugation and $d_\pm$ are the dimensions of $V^\pm$. This type of element arises naturally from the author's work on p-adic L-functions [in *Proceedings of the International Congress of Mathematicians, Vol. 1, 2 (Zürich, 1994)*, 400–410, Birkhäuser, Basel, 1995; MR 97j:11021; Astérisque No. 229 (1995), 198 pp.; MR 96e:11062].

The main result is this: Suppose a p-adic Euler Iwasawa system c_p is given. One makes certain technical assumptions regarding the image of the Galois group which are discussed in Chapter 4. Whenever $c_p(m)$ is defined the author defines elements $\mathcal{F}_m, f_m\in\mathbf{Q}_p\otimes\Lambda[\mathrm{Gal}(\mathbf{Q}(\mu_m)/\mathbf{Q})]$ measuring in the Iwasawa-theoretic sense the index of $c_p(m)$ in $\bigwedge^{d_-}H^1_\infty(\mathbf{Q}(\mu_n),T)^+\oplus\bigwedge^{d_+}H^1_\infty(\mathbf{Q}(\mu_n),T)^-$ and the size of $H^2_\infty(\mathbf{Q}(\mu_m),T)$ respectively. Then the result is that $f_m\mid\mathcal{F}_m\times$ some power of p. This result is deduced from similar consequences of the existence of a p-adic Euler Iwasawa system of rank 1.

The proof of the main theorem follows the same lines as other works on the subject except that the added generality increases the level of complexity, which is already quite high to begin with. It is recommended that those new to the theory of Euler systems be introduced to it via a less general work, e.g., Nekovář's paper mentioned above.

Both Rubin and Kato have developed similar theories.

Amnon Besser (Be'ér Sheva')

97j:11053 11R29 11R18 11R23

Kuz'min, L. V. (Moscow)

On formulas for the class number of real abelian fields. (Russian. Russian summary)

Izv. Ross. Akad. Nauk Ser. Mat. **60** (1996), *no.* 4, 43–110; *translation in Izv. Math.* **60** (1996), *no.* 4, 695–761.

Let k be a real abelian extension of the field $\mathbf{Q}$ and A the l-primary part of the class group Cl_k of k, where l is a fixed prime ($l=2$ is permitted). The order of A is known: for $k=\mathbf{Q}(\zeta_p)^+$ (p prime), Kummer proved already that the class group Cl_k has the same order as U_k/C_k, where U_k denotes the group of units in k, and C_k denotes the subgroup of circular (or cyclotomic) units. This result was generalized around 1980 to all real abelian fields by Sinnott. The formulas are more complicated, in particular if k is *not* of the form $\mathbf{Q}(\zeta_n)^+$.

Research on class groups in recent times has focused on the sharper question of determining not only the order of A but also the Galois module structure, i.e. the structure of A as a $\mathbf{Z}_l[G]$-module, where G is the Galois group of k. This is a hard problem. The author focuses on the "semi-simple part" of the question: The ring

$B = \mathbf{Z}_l[G]$ splits as a direct sum of subrings $e_\varphi B$, where φ runs over the l-adic characters of the non-l-part G_0 of G, and the main result of the paper under review determines the order of the "eigenspaces" $e_\varphi A$. Let us mention that analogous results for the so-called minus class groups are known: see the papers of B. C. Mazur and A. Wiles [Invent. Math. **76** (1984), no. 2, 179–330; MR 85m:11069], D. R. Solomon [Ann. Inst. Fourier (Grenoble) **40** (1990), no. 3, 467–492; MR 92a:11133] (this work even goes beyond the semisimple case for odd l), and the reviewer [Ann. Inst. Fourier (Grenoble) **42** (1992), no. 3, 449–499; MR 93j:11071] (the focus is on $l = 2$). All these results express orders of eigenspaces in minus class groups by means of generalized Bernoulli numbers. Generally speaking, working with plus class groups (= class groups of real fields) is harder; in the work of Mazur and Wiles, one also finds a proof of the so-called Gras conjecture (away from $l = 2$), which is just the special case "l does not divide the degree of k" of the main result in the present paper. The case where the l-part of the Galois group of k is cyclic is also known: this was done by the reviewer [op. cit. (Theorem 4.14)].

Kuz′min proves two main theorems, the former for full real cyclotomic fields $k = \mathbf{Q}(\zeta_n)^+$, the latter for general real abelian k. The first expresses $|e_\varphi A|$ as λ times $|e_\varphi(U_k/C_k)[l]|$, where λ is 1 for most l and φ, and a certain power of 2 for $l = 2$ and φ trivial. The second expresses $|e_\varphi A|$ as $c^+_{k,\varphi}$ times $|e_\varphi(U_k/C_k)[l]|$, where now $c^+_{k,\varphi}$ is substantially less pleasant, but at least calculable a priori, without looking at class groups. Both results can be interpreted nicely as sharper versions of Sinnott's theorems. They are very general, and it will be difficult to go beyond them (that is, to say much about the structure as modules over the l-part of G). In the proof of his main results, the author successfully surmounted numerous obstacles; this is witnessed by the quantity of intermediate results, and the considerable length of the paper. Everything is, in the reviewer's opinion, very well presented, and most details are given.

Let us try to give an impression of the strategy of proof. First, Kuz′min states a result (Theorem 3) which again compares two quantities: X_Γ has the same order as $\widehat{U}_S(k)/C_S(k)$, and the same for φ-eigenspaces. We explain (notation not always the same as Kuz′min's): The Iwasawa module X is the projective limit of the l-parts of the l-class groups of k_n, with $\bigcup k_n = k_\infty$ the cyclotomic $\mathbf{Z}_l$-extension, and Γ is the Galois group of k_∞/k. Then X_Γ surjects onto A, sometimes with a nontrivial kernel. On the other side, $\widehat{U}_S(k)$ is, roughly speaking, the group of l-units of k which are locally universal norms from the cyclotomic tower, and $C_S(k)$ is a cognate of C_k with an ad hoc technical definition. Note however the gain: the ugly constant $c^+_{k,\varphi}$ is absent now. One may divide the paper into two parts: first the proof of this Theorem 3 (this uses traditional Iwasawa theory and descent, but also some new ideas); and second the deduction of the main results from Theorem 3. This is a clever and highly complicated calculation.

Some words about the proof of Theorem 3: One also brings into play the group of l-units which are universal norms from k_∞ globally; call it $U_S^*(k)$. Of course this is a subgroup of $\widehat{U}_S(k)$, and a previous result of the author (re-proved here for convenience) shows that the order of the quotient $\widehat{U}_S(k)/U_S^*(k)$ equals $|X^\Gamma|$. (The l-adic Gross conjecture, which is known to be true for all abelian fields, just states that the group X^Γ of Γ-invariants is indeed finite.) By standard arguments, the problem is now reduced to identifying the characteristic power series of a φ-eigenspace of X with the characteristic series of the corresponding eigenspace of $U_S(k_\infty)/C_S(k_\infty)$;

here the main conjecture comes in. For the experts we mention that one of the main technical points is the following: the Γ-coinvariants of $U_S(k_\infty)$ are exactly $\widehat{U}_S(k)$. (The analogous statement for units instead of S-units would not be true.) There is a further technical complication which we gloss over in this review (descent from a full cyclotomic base field, giving rise to groups $\widetilde{U}_S(k_\infty)$ and others).

The rest of the paper deduces Theorems 1 and 2 from Theorem 3. This is a long argument, and I'll not go into much detail. One of the ingredients I particularly liked is an equivariant version of Hasse's conductor-discriminant formula. One takes $R=\overline{\mathcal{O}}\otimes\mathcal{O}_k$, with $\overline{\mathcal{O}}$ the ring of integers in $\overline{\mathbf{Q}}_l$. The l-part of the discriminant of k equals the square of the generalized index of R in its integral closure. One now can take φ-parts of R and its integral closure, and obtain "φ-parts of the discriminant"; Kuz′min shows that these are given by Gauss sums in a very natural fashion. (I surmise that special cases of this, let's say k tame over $\mathbf{Q}$, are known to specialists in Galois module theory. Cf. also Question 3 at the end of the paper.) The main point of the argument is now that one has to make several transitions: from X_Γ (the left-hand side in Theorem 3) to A, and from the right-hand side of Theorem 3 to something which does not mention the local universal norms $\widehat{U}_S(k)$. This transition is made possible by a cancellation phenomenon: the quantity $[H(k)\colon \widehat{U}_S(k)]$ occurs in the numerator, and in the denominator. Finally, at the end of the argument, Sinnott's terminology and reasoning are used. Some interesting observations and questions conclude the paper.

Let us mention a few minor things in passing: in (2.1), first case, I feel one should omit "isomorphic to A_φ". Proposition 4.2, second formula: add a $*$ to $\widetilde{U}_S(k)$ on the right-hand side. Proposition 4.2: replace tilde on right-hand side by hat. In the proof of Theorem 4.1, there is a somewhat abrupt change from unit groups proper to S-unit groups: the main argument uses units, but the quoted Proposition 4.1 uses S-units. This makes no difference in the end, but one has to think for a while. In the proof of 5.1, paragraph treating $l\neq 2$ after display 5.16: replace C by C_S five times. Another remark: special cases of Theorem 10.1 are known [see, e.g., J. M. Kim, S. Bae and I.-S. Lee, Israel J. Math. **75** (1991), no. 2-3, 161–165; MR 93i:11120].

This is an impressive paper, because of its extreme generality, precision, and good exposition. In spite of the technical difficulties, it is highly recommended reading for everybody interested in Iwasawa theory and class groups.

Cornelius Greither (Neubiberg)

97b:11133 11R32 11R23 11S15 14F20

Schmidt, Alexander (Heidelberg)

An arithmetic site for the rings of integers of algebraic number fields.

Invent. Math. **123** (1996), *no.* 3, 575–610.

This paper develops an earlier work of K. Wingberg, and shows us that there is a remarkable class of Galois extensions of number fields which is so large as to contain all unramified extensions and the cyclotomic $\widehat{\mathbf{Z}}$-extension, and which is expected to be just so large as to have its total Galois group analogous in structure (at least for their nilpotent quotients) to the fundamental groups of algebraic curves. The results obtained support this exciting expectation.

The first key point is more related to an earlier work of Wingberg [Invent. Math. **77** (1984), no. 3, 557–584; MR 86e:11104; see also J. Reine Angew. Math. **357** (1985),

193–204; MR 86h:11101; A. Schmidt, J. Reine Angew. Math. **458** (1995), 93–126; MR 95k:11141]. So, we shall start by reviewing some background. The point is whether there is a suitable enlargement of unramified extensions for which the total Galois group has simpler structure analogous to the fundamental groups of curves and, if so, what sort of ramifications one should allow. The answer that Wingberg gave is related in some way to the complex conjugation, and is not a type of restriction of ramifications that would arise from a standard purely local thought. The following classical observation is basic. Over the field $\mathbf{Q}(\mu_p)$ of pth roots of 1 (p a prime, $\neq 2$), there are almost as many unramified $(\mathbf{Z}/p)$-extensions (expected to be all of odd type) as the p-ramified $(\mathbf{Z}/p)$-extensions of even type, and they are complementary to each other in a natural way. Here, odd [resp. even] means that the complex conjugation acts on the Galois group as $g \to g^{-1}$ [resp. $g \to g$]; "p-ramified" means unramified outside p, and "almost" is used because $\mathbf{Q}(\mu_{p^2})$ is exceptional. The latter p-ramified extensions, being even, arise from some $(\mathbf{Z}/p)$-extension of the maximal real subfield $\mathbf{Q}(\mu_p)^+$ of $\mathbf{Q}(\mu_p)$. It is natural to take these p-ramified extensions also into account in studying unramified extensions. Now, Wingberg was the first person to carry this idea beyond the border of abelian extensions, i.e., to pro-p extensions. Here, a Galois extension is called pro-p if it is a composite of finite Galois extensions of p-power degrees. A special case of his main result in the first cited paper is as follows. Let $\mathfrak{K}^{(p)}$ be the maximal pro-p extension of $\mathbf{Q}(\mu_p)$ which is unramified outside p and "positively ramified above p", i.e., the corresponding pro-p extension of the local field $\mathbf{Q}_p(\mu_p)$ arises from a pro-p extension of $\mathbf{Q}_p(\mu_p)^+$ (by adding μ_p). Then $\mathfrak{K}^{(p)}$ contains all unramified pro-p extensions of $\mathbf{Q}(\mu_{p^\infty})$, and the Galois group of $\mathfrak{K}^{(p)}/\mathbf{Q}(\mu_{p^\infty})$ has a surprisingly simple structure; it is isomorphic to the pro-p completion of the fundamental group of a compact Riemann surface of genus $g = \lambda_p^-$, the Iwasawa λ^--invariant of $\mathbf{Q}(\mu_p)$. (Its Frattini quotient is of rank $2g$, of which one half corresponds to the odd unramified $(\mathbf{Z}/p)$-extensions of $\mathbf{Q}(\mu_{p^\infty})$, and the other half, to the even p-ramified ones.) Or, almost equivalently stated, the Galois group of $\mathfrak{K}^{(p)}/\mathbf{Q}(\mu_p)$ either is a Poincaré group of dimension 3, or is isomorphic to $\mathbf{Z}_p$—this Galois group is more analogous to the finite constant field case for curves, and this is the way in which the main results of the present paper are stated. But how general is such a phenomenon? Wingberg himself previously generalized this to the case of CM-fields containing μ_p under some Iwasawa-theoretic assumption [op. cit.; MR 86e:11104]. A CM-field is, by definition, a totally imaginary quadratic extension of a totally real number field, and has the properties that the complex conjugation commutes with every embedding into $\mathbf{C}$, and that the "odd part" of the unit group is finite. The present author gained a further insight, and made use of "complex conjugation at p", and the "p-CM fields" at each p, instead, which will now be explained.

Now for the basic local definitions of the present author (under a slightly different formulation). Let $\mathbf{Q}_p^{\mathrm{ur}}$ be the maximal unramified extension of $\mathbf{Q}_p$ and put $I_p = \mathrm{Gal}(\overline{\mathbf{Q}}_p/\mathbf{Q}_p^{\mathrm{ur}})$. Then there exists an element ρ_p in a 2-Sylow subgroup of I_p such that $\rho_p(\zeta) = \zeta^{-1}$ for all p^n-th-power roots ζ of 1 ($n \geq 1$). Let $I_p^* = I_p/N_p$ be the maximal quotient on which the image ρ_p^* of ρ_p is a central involution. Now let $p \neq 2$. Then N_p, ρ_p^* are independent of the choice of ρ_p. Call $\mathbf{C}_p$ [resp. $\mathbf{R}_p$] the fixed field of N_p [resp. $\langle N_p, \rho_p \rangle$]. Then $\mathbf{C}_p/\mathbf{R}_p$ is a quadratic extension with the Galois group $\langle \rho_p^* \rangle$. (This field $\mathbf{C}_p$ is the same as the $\mathbf{Q}_p^{\mathrm{pre}}$ of the paper. It is not $\widehat{\overline{\mathbf{Q}}}_p$.) An extension l/k

in $\overline{\mathbf{Q}}_p$ is called positively ramified if (i) its Galois closure $\widehat{l}/k$ is either unramified or purely wildly ramified, and (ii) $l \subset k \cdot \mathbf{C}_p$. This notion of "positive ramifiedness" gives the key local restriction of ramifications.

Now an essential new step set forward in this paper is as follows. Using everywhere positively ramified "coverings" of $X = \operatorname{Spec}\mathcal{O}_K$ ($\mathcal{O}_K$ is the ring of integers of a number field K), the author defines a Grothendieck topology on X, called a positively ramified (p.r.) site, and establishes a 3-dimensional Poincaré duality theorem on the p.r. cohomology groups on X. The dualizing sheaf (in fact, a system of sheaves) is closely related to the multiplicative group $\mathbf{G}_m$, with the action of ρ_p^* taken into account. Thus, one can now start from any number field K, and for $X = \operatorname{Spec}\mathcal{O}_K$, can ask what the structure of the Galois group $\pi_1^{\text{pos}}(X)$ of the maximal everywhere p.r. Galois extension of K is; or, more practically, about its maximal pro-p quotient $\pi_1^{\text{pos}}(X)(p)$, for all but finitely many bad primes p. The author predicts that the structure of $\pi_1^{\text{pos}}(X)(p)$ is more or less analogous to that of the pro-p fundamental group of a curve over a finite field of char $\neq p$. With the Poincaré duality for the p.r. site on hand, the next step is to find out whether X is "$K(\pi, 1)$ at p", i.e., the group cohomology of $\pi_1^{\text{pos}}(X)(p)$ equals the p.r. sheaf cohomology on X. The author defines totally p-real fields and p-CM fields using the ρ_p^*, and proves that if K is either totally p-real or p-CM, then X is in fact $K(\pi, 1)$ at p. Here, some earlier results of O. Neumann [Izv. Akad. Nauk SSSR Ser. Mat. **39** (1975), no. 2, 259–271, 471; MR **51** #8076] and of Wingberg [J. Reine Angew. Math. **400** (1989), 185–202; MR 91a:11055; J. Reine Angew. Math. **416** (1991), 187–194; MR 92f:11158] are used. Finally, the author proves that (i) if K is p-CM and contains μ_p, then $\pi_1^{\text{pos}}(X)(p)$ is either a free pro-p group or a Poincaré group of dimension 3, and (ii) if K is totally p-real, then $\pi_1^{\text{pos}}(X)(p)$ is either free or a duality group of dimension 2. (More precise results are given in each case.) Note that each given K is totally p-real for almost all p (e.g. those unramified in $K/\mathbf{Q}$), and that then $\pi_1^{\text{pos}}(X)(p)$ is nothing but the Galois group of the maximal pro-p extension of K unramified outside p.

The reviewer believes that the direction advanced by these theories is very important and merits special attention at this stage, and he hopes that this will be developed further in the near future. *Yasutaka Ihara* (Kyoto)

98a:11150 11R33 11R27 11R42

Ritter, Jürgen (Augsburg); **Weiss, Alfred** (Edmonton, AB)

Cohomology of units and L-values at zero.

J. Amer. Math. Soc. **10** (1997), *no.* 3, 513–552.

The article under review presents quite significant progress on Stark's conjecture in the version of Tate. A considerable amount of algebraic machinery is developed to that end.

Let us consider a fixed G-Galois extension K/k of number fields. Stark's conjecture postulates a link between the group E_S of S-units in K considered as a G-module, and the Artin L-functions $L(s, \chi)$ with χ running over the characters of G. Here S is for the moment a finite set of places containing all Archimedean places, and S is "big enough" (a technical notion). More precisely: For every χ one lets $A(\chi)$ be the quotient of the Tate regulator at χ (for χ the character of the regular representation, this is essentially the classical regulator) by the leading coefficient c_χ of $L(s, \chi)$ (with

Euler factors at primes in S omitted) at $s = 0$. Then Stark's conjecture predicts strong algebraicity properties of $A(\chi)$.

However, $A(\chi)$ is not uniquely defined: it depends on the choice of a certain G-invariant map $\varphi: \Delta S \to E_S$ with finite cokernel. Here ΔS is a quite explicit "comparison module": the kernel of the augmentation $\mathbf{Z}S \to \mathbf{Z}$. This ambiguity can be remedied as follows: there are invariants $q_\varphi(\chi)$ (with a purely algebraic yet complicated definition) such that the quotient $a(\chi) = A_\varphi(\check{\chi})/q_\varphi(\chi)$ is independent of the "isogeny" φ, as a fractional ideal of a suitable auxiliary field F. (Here $\check{\chi}$ is the contragredient character of χ.) The following conjecture (made by Chinburg) is sometimes called the strong Stark conjecture: $a(\chi)$ is always the unit ideal. This conjecture in turn implies the third Chinburg conjecture for K/k modulo the kernel group $D(\mathbf{Z}G)$: $\Omega(3, K/k) \equiv 0 \bmod D(\mathbf{Z}G)$. The paper under review proves the strong Stark conjecture for all absolutely abelian K (so $k = \mathbf{Q}$) with the cautionary condition that 2 is unramified in K.

So far, a construction of $q_\varphi(\chi)$ has been available only for "big" sets S. This presents problems in the sequel of the argument, and one major novelty in the paper under review is that the construction of $q_\varphi(\chi)$ is now extended to all S (containing S_∞). The main problem of working with big S is the following. One has to compute the Tate regulators, so one needs an explicit isogeny φ, and to get that, one needs a well-behaved explicit system of S-units of full rank. When K is abelian over the rationals, everybody would suggest to use cyclotomic units (and this will be the solution ultimately), but there is an obstacle: no matter how you define "cyclotomic S-units", you seldom get full rank unless S is quite small. About one-half of the paper is concerned with a way to surmount this obstacle. The authors redo the whole construction for sets $S \supset S_\infty$ which are not "big". This entails a lot of complicated algebra. The comparison modules ΔS are replaced by ∇, a G-module which reflects ramification of K/k outside S. The theory of Tate sequences which is instrumental for the definition of $q_\varphi(\chi)$ has to be modified and extended (this is based on previous work of the authors). Then it is proved that the resulting Stark quotient $a(\chi)$ is after all independent of the set S and all other choices (Theorem B), so one might think one does not need all this since one now might choose a big set S in the first place. But that is not the point since one wants just the opposite, that is, the flexibility to choose S very small. These developments take up §§ 1–7. §8 presents some general properties (induction, etc.) of the invariants $a(\chi)$.

The second half or so of the paper is devoted to the proof of Theorem A, which states that these invariants are all trivial in the absolutely abelian case. This is again long and difficult, but we are on slightly more familiar ground. First one chooses S to be the set of all Archimedean primes (and no others!); then one constructs φ with the help of Ramachandra units (§10). One obtains a formula for the χ-determinant of the composite $\lambda\varphi$ on ΔS, where λ is the regulator map. This gives a product of three terms: a transcendental part $c_S(\check{\chi})$ (leading term of an L-series), a second term, which also appears in Ramachandra's index formula, and a third term which is a product of generalized Bernoulli numbers. (The last term essentially comes from the "minus side" and the first two terms from the "plus side".) This, in conjunction with a previously proved formula (§8), allows one to get a formula for $a(\chi)$ (first formula in §11) which contains besides totally explicit quantities the following: generalized Bernoulli numbers; the χ-part of the units modulo the Ramachandra units; and the χ-part of the class group. (We oversimplify here—one has to go l-adic before one can state and prove this formula. The authors also do a very clever reduction to the "semisimple" case, i.e., they show that one need only consider the case where χ has order prime to l.) The showdown of the paper is now to prove that *everything cancels*. The methods and tools for this are fairly standard. In fact, for some classes of characters χ, the authors are able to quote results of Wiles (and of the reviewer to cover the somewhat singular case $l = 2$). They are left with the case in which χ is an even character. Here another intricate analysis in the spirit of the Gras conjecture is needed. All these arguments are outgrowths of the Main Conjecture in Iwasawa theory.

It is natural to ask whether the techniques in this paper might be extended to prove the third Chinburg conjecture for abelian fields *in full*. We recall that at present this conjecture is only known to be true modulo the so-called kernel group, a big subgroup of $\mathrm{cl}(\mathbf{Z}G)$, either by work of Holland and the reviewer, or as a consequence of the main result of the present paper. For known results on the full conjecture in special cases, we refer to D. J. Burns [Amer. J. Math. **117** (1995), no. 4, 875–903; MR 96d:11120] and Burns and D. Holland [Proc. London Math. Soc. (3) **74** (1997), no. 1, 29–51; MR 97g:11124]. The reviewer also has some recent results towards the full conjecture, but again only for restricted classes of abelian fields.

The paper under review is a quite important contribution to Galois module theory. It is written in a very precise and condensed way, and it certainly requires some effort of the reader. On the other hand, the authors manage well to explain what they are doing. I highly recommend the introduction and §11 for a good overview.

Cornelius Greither (Neubiberg)

97d:11172 11R39 11F80 11G05
Diamond, Fred (Cambridge)
On deformation rings and Hecke rings.
Ann. of Math. (2) **144** (1996), *no.* 1, 137–166.

Many of the recent advances on the Diophantine theory of elliptic curves over the rational numbers (for instance the very important works of B. H. Gross and D. B. Zagier [Invent. Math. **84** (1986), no. 2, 225–320; MR 87j:11057] and of V. A. Kolyvagin [Izv. Akad. Nauk SSSR Ser. Mat. **52** (1988), no. 6, 1154–1180, 1327; MR 90f:11035; Izv. Akad. Nauk SSSR Ser. Mat. **52** (1988), no. 3, 522–540, 670–671; MR 89m:11056] on the Birch-Swinnerton-Dyer conjecture) are predicated on the assumption that the elliptic curve is modular (i.e. that there is a non-constant holomorphic map from a (complete) modular curve to the elliptic curve). Indeed it is only for these elliptic curves that the L-function can be shown to extend to the entire complex plane and satisfy a functional equation. It is, however, a standard conjecture (now usually referred to as the Shimura-Taniyama conjecture) that every elliptic curve over $\mathbf{Q}$ is modular, but until recently there was no general result in this direction. In the article under review Diamond proves the extremely important partial result that every elliptic curve over $\mathbf{Q}$ with semi-stable reduction at 3 and 5 is modular.

To an elliptic curve $E/\mathbf{Q}$ and a prime number l one can associate a representation of the absolute Galois group of $\mathbf{Q}$, $\rho_{E,l}\colon G_{\mathbf{Q}} \to \mathrm{GL}_2(\mathbf{Z}_l)$, by looking at the action on the l-power division points of E. By G. Faltings' theorem [Invent. Math. **73** (1983), no. 3, 349–366; MR 85g:11026a; erratum; MR 85g:11026b], to show that E

is modular it suffices to show that for one prime l, $\rho_{E,l}$ is modular in a suitable sense. A. Wiles [Ann. of Math. (2) **141** (1995), no. 3, 443–551; MR 96d:11071] pioneered a method of showing that if $l > 2$, if $\overline{\rho}: G_{\mathbf{Q}} \to \mathrm{GL}_2(\mathbf{F}_l)$ is modular and satisfies certain additional conditions and if $\rho: G_{\mathbf{Q}} \to \mathrm{GL}_2(\mathbf{Z}_l)$ is a lifting of $\overline{\rho}$ which is "semistable of weight 2 at l" then ρ is also modular. Applying this to $\rho_{E,3}$ and $\rho_{E,5}$ and using a very important result of R. P. Langlands [*Base change for* GL(2), Ann. of Math. Stud., 96, Princeton Univ. Press, Princeton, N.J., 1980; MR 82a:10032] and J. Tunnell [Bull. Amer. Math. Soc. (N.S.) **5** (1981), no. 2, 173–175; MR 82j:12015] that $\rho_{E,3}$ mod 3 is always modular, he could deduce that a wide, but rather complicated to describe, class of elliptic curves are all modular. The condition that ρ is "semistable of weight 2 at l" is a condition on the restriction of ρ to the decomposition group at l. This condition, or some weakening of it, is known to be essential. The additional conditions on $\overline{\rho}$ were of two types. Firstly, $\overline{\rho}$ should be absolutely irreducible when restricted to the quadratic extension of $\mathbf{Q}$ with discriminant l. The second condition was on the restriction of $\overline{\rho}$ to decomposition groups at primes other than l. These two conditions are presumably unnecessary. In the paper under review Diamond completely removes conditions of the second type, and hence deduces his result on modularity of elliptic curves.

Wiles [op. cit.] reduced the general problem of showing that "semistable of weight 2 at l" liftings of a fixed modular $\overline{\rho}: G_{\mathbf{Q}} \to \mathrm{GL}_2(\mathbf{F}_l)$, satisfying his two conditions, are modular, to only showing that "minimal" lifts are modular. These are lifts which are "minimally" ramified in some precise sense. The reviewer and Wiles [Ann. of Math. (2) **141** (1995), no. 3, 553–572; MR 96d:11072] showed that such "minimal" lifts were indeed modular. Neither of these steps worked in the absence of the condition of the second type. However, in both cases the problem was clear: it was the need to give a characterisation of the space of modular forms giving rise to l-adic representations which are minimal lifts of $\overline{\rho}$, in such a way that one can prove various properties of the Hecke algebra on this space. In the presence of the second condition on $\overline{\rho}$ this was rather easy using the level and the Nebentypus character. In its absence, Diamond has to use a much more intricate argument using modular forms on Shimura curves and the Jacquet-Langlands correspondence. *Richard Taylor* (Cambridge, MA)

98m:11125 11R39 11F27 11F67 11F70 22E50 22E55

Gelbart, Stephen (Rehovot); **Rogawski, Jonathan** (Los Angeles, CA); **Soudry, David** (Tel Aviv)

Endoscopy, theta-liftings, and period integrals for the unitary group in three variables.

Ann. of Math. (2) **145** (1997), *no.* 3, 419–476.

This article describes the structure of local and global L-packets for the quasisplit unitary group U(3) from the points of view of endoscopy, theta lifting and period integrals. It provides the clearest evidence to date of the connections one should expect to find for general reductive groups among these three notions which lie at the heart of the theory of automorphic representations. It is also of interest because of its connections to the theory of algebraic cycles on Picard modular surfaces. The research represents the culmination of a series of papers by the authors.

Understanding this paper requires substantial preparation. Some of this is provided by Gelbart's survey article [in *Theta functions: from the classical to the modern*, 129–174, Amer. Math. Soc., Providence, RI, 1993; MR 94f:22023]. On the specific topics of endoscopy, theta liftings and period integrals, we also mention the following introductory articles by J. G. Arthur [in *Representation theory and automorphic forms (Edinburgh, 1996)*, 433–442, Proc. Sympos. Pure Math., 61, Amer. Math. Soc., Providence, RI, 1997; MR 98j:22029], D. Prasad [in *Theta functions: from the classical to the modern*, 105–127, Amer. Math. Soc., Providence, RI, 1993; MR 94e:11043], and H. M. Jacquet [in *Representation theory and automorphic forms (Edinburgh, 1996)*, 443–455, Proc. Sympos. Pure Math., 61, Amer. Math. Soc., Providence, RI, 1997; MR 98m:22024].

To state the results of the present paper, it is necessary to recall some basic concepts. According to Langlands' theory, one expects to attach a stable trace formula to a general connected reductive group defined over a global field. The spectral side of the trace formula involves a sum over L-packets of representations of the given group, as well as its elliptic endoscopic subgroups. In the case of interest, $G = \mathrm{U}(3)$, the L-packets are understood, as is the stable trace formula, thanks to Rogawski's work [*Automorphic representations of unitary groups in three variables*, Ann. of Math. Stud., 123, Princeton Univ. Press, Princeton, NJ, 1990; MR 91k:22037]. In this case, there is a unique elliptic endoscopic group, namely, $H = \mathrm{U}(2) \times \mathrm{U}(1)$, where $\mathrm{U}(2)$ denotes the quasisplit unitary group in two variables. All of these unitary groups are defined with respect to some fixed quadratic extension E/F of global fields. For each idèle class character γ of E which extends the class field theory character of E/F, there is an associated embedding of L-groups ${}^LH \to {}^LG$ which is used to map each L-packet ρ of H to an L-packet $\Pi(\rho)$ of G.

The L-packets of primary interest are referred to as (cuspidal) endoscopic L-packets. These are the cuspidal L-packets $\Pi(\rho)$, for which $\rho = \rho_2 \times \rho_1$ is a cuspidal L-packet (consisting of infinite-dimensional representations). It is shown that each endoscopic L-packet contains a unique generic representation. The analogous local result is also proven.

If Π is an endoscopic L-packet, then its fiber $\widehat{\Pi}$ under endoscopic transfer is a set of at most three L-packets for H. In fact, Rogawski has shown that there is a natural way to identify the set $S_\Pi = \widehat{\Pi} \cup \{1\}$ with one of the groups $\mathbf{Z}/2$ or $\mathbf{Z}/2 \oplus \mathbf{Z}/2$ so that the cuspidal multiplicity of $\pi \in \Pi$ is given by a formula $m(\pi) = |S_\Pi|^{-1} \sum_{\rho \in S_\Pi} \langle \rho, \pi \rangle$, for some character $\rho \mapsto \langle \rho, \pi \rangle$ of S_Π. This is compatible with the multiplicity formulas outlined by Arthur for general groups. The multiplicity-one property of $\mathrm{U}(3)$ implies that $m(\pi) = 0$ or 1.

The endoscopic lifting of $\rho = \rho_2 \times \rho_1$ to $\Pi(\rho)$ is characterized by an L-function identity which relates the L-function of a twist of $\pi \in \Pi(\rho)$ to the product of L-functions of corresponding twists of the standard L-functions associated to the base change lifts, with respect to E/F, of ρ_2 and ρ_1 to representations of $\mathrm{GL}(2)$ and $\mathrm{GL}(1)$, respectively.

In addition to endoscopic lifting from H to G, one may construct representations of G via theta lifting from unitary groups in two variables. If $d \in F^\times$, we define a matrix $\Phi_d = \begin{pmatrix} d & 0 \\ 0 & -1 \end{pmatrix}$ and let H_d denote the associated unitary group. Then the isomorphism class of H_d is determined by the class of d modulo norms from $E^\times$. The choice of a nontrivial character ψ of $\mathbf{A}/F$ together with the choice of γ determines a theta correspondence between cuspidal representations σ of H_d and certain representations $\Theta_\psi(\sigma, \gamma)$ of G. When σ does not correspond to a representation of G, then the theta lift is considered to be zero.

The group G lies in a tower of unitary groups $\mathrm{U}(2n+1)$ and one may consider theta lifting from H_d to any of the groups in the tower. The philosophy of towers suggests that, given σ, the first occurrence of a nonzero theta lift of σ in the tower must be cuspidal. Indeed, the authors showed in a previous paper that if the theta lift of σ to $\mathrm{U}(1)$ is zero then the theta lift to G must be cuspidal, though possibly zero. In the present paper, they show that if the theta lift of $\sigma = \bigotimes_v \sigma_v$ to $\mathrm{U}(1)$ is zero then the theta lift $\Theta_\psi(\sigma,\gamma)$ is nonzero precisely when the local Howe lift of each component σ_v is nonzero.

Much of the above discussion has an obvious adaptation to the local case, which we will not provide here. In short, if E/F is local, we are interested in L-packets of irreducible, admissible representations of the quasisplit group $G(F) = \mathrm{U}(3, E/F)$ which are obtained from the elliptic, endoscopic group H by functorial transfer. Such L-packets will be called endoscopic L-packets. There are local "multiplicity pairings" $\langle\rho,\pi\rangle$ which are related to the global pairing by a product formula. Given an irreducible, admissible representation σ of the local group $H_d(F)$, we will let $H_\psi(\sigma,\gamma)$ denote the Howe lift of σ to $G(F)$.

We now describe the connection between endoscopic lifting and Howe lifting. Suppose that $\pi = H_\psi(\sigma,\gamma)$ belongs to a local endoscopic L-packet Π. Then it is shown that there exists $\rho \in \widehat{\Pi}$ such that $\rho_1 = \gamma|E^1$, where $E^1 = \mathrm{U}(1)$. On the other hand, if π is any representation which belongs to an endoscopic L-packet Π and if $\rho \in \widehat{\Pi}$ satisfies $\rho_1 = \gamma|E^1$ then there exist d, σ, ψ such that $\pi = H_\psi(\sigma,\gamma)$. Moreover, d is unique modulo norms. This leads to the definition $\varepsilon_\rho(\pi) = 1$ if d is a norm, and $\varepsilon_\rho(\pi) = -1$ if d is not a norm. It is proven that, in fact, $\varepsilon_\rho(\pi) = \langle\rho,\pi\rangle$. A more general statement of these results which applies to twists of Howe lifts by characters of E^1 is actually proven.

The p-adic endoscopic L-packets may be parametrized in terms of Howe lifts as follows. Recall that both the functorial lifting and the theta lifting depend on the choice of the character γ. For simplicity, we are using the same choice of γ for both purposes. (In the notation of the paper, we are assuming $\mu = \gamma^{-1}$.) Fix a nonnorm d_0. Then there is an analogue of the Jacquet-Langlands correspondence between the quasisplit group H_1 and the non-quasisplit group H_{d_0}. Assume ρ_2 is an L-packet on H_1. Let ρ_2' be the corresponding L-packet on H_{d_0}, if it exists, and let $\rho_2' = \varnothing$ otherwise. The authors show that the endoscopic L-packet associated to $\rho = \rho_2 \times \rho_1$, with $\rho_1 = \gamma|E^1$, is identical to the collection of all nonzero Howe lifts $H_\psi(\sigma,\gamma)$ such that $\sigma \in \rho_2 \cup \rho_2'$. Moreover, the set of all such σ (with nonzero Howe lift) is precisely described. Once again, the actual theorem proven applies more generally to twists of Howe lifts.

Returning to the global case, assume that $\pi = \bigotimes_v \pi_v$ belongs to an endoscopic L-packet $\Pi(\rho)$ and $\gamma|E^1 = \rho_1$. Let $\varepsilon_\rho(\pi) = \prod_v \varepsilon_{\rho_v}(\pi_v)$ and assume that $\varepsilon_\rho(\pi) = 1$. Then we may choose $d \in F^\times$ which is a norm at precisely those places v at which $\varepsilon_{\rho_v}(\pi_v) = 1$. We may then form an admissible representation $\sigma = \bigotimes_v \sigma_v$ of $H_d(\mathbf{A})$ by choosing σ_v such that $\pi_v = H_{\psi_v}(\sigma_v,\gamma)$. It is shown that $m(\pi) = 1$ if and only if $m(\sigma) = 1$, where $m(\sigma)$ is the cuspidal multiplicity of σ with respect to the spectrum of H_d. On the other hand, if $\varepsilon_\rho(\pi) = -1$ then $m(\pi) = 0$.

We now turn to the relation between the theta lifts and endoscopic lifts considered above and the nonvanishing of certain period integrals. The period integrals of interest arise as period integrals on Picard modular surfaces over algebraic cycles corresponding to unitary groups in two variables. They are defined by

$$P(\varphi,\eta,d)=\int_{H_d(F)\backslash H_d(\mathbf{A})}\varphi(h)\eta^{-1}(\det h)dh,$$

where φ is an automorphic form in the space of the cuspidal automorphic representation π of G, the character η is defined on $E^1\backslash E^1_{\mathbf{A}}$, and we have fixed an embedding of H_d in G. These integrals are analogous to the period integrals used by Harder, Langlands and Rapoport to prove Tate's conjectures for Hilbert modular surfaces.

If $P(\varphi,\eta,d)$ is nonzero for some d and φ, we say π is η-distinguished. An endoscopic L-packet which contains a (cuspidal) η-distinguished element is also said to be η-distinguished. It is proven that if Π is such an L-packet, then $\widehat{\Pi}$ must contain an element ρ such that $\eta=\rho_1$. If $\pi\in\Pi$ is cuspidal, then the global condition of π being ρ_1-distinguished is shown to be equivalent to the local conditions $\langle\rho_v,\pi_v\rangle=1$, for all v. In this case, the L-function $L(s,\pi\otimes\eta^{-1})$ must have a pole at $s=1$.

Given an η-distinguished representation π, we may choose $\varphi=\bigotimes_v\varphi_v$ such that $P(\varphi,\eta,d)\neq 0$ and then vary one of the components φ_v. This yields a local "period functional" on the space of π_v, that is, a nonzero linear functional λ on the space of π_v such that $\lambda(\pi_v(h)v)=\eta_v(\det(h))\lambda(v)$, for all $h\in H_d(F_v)$. An irreducible, admissible, infinite-dimensional representation of $G(F_v)$ is said to be η_v-distinguished if it admits a period functional, for some H_d. It is shown in the p-adic case that such a representation must lie in an endoscopic L-packet $\Pi(\rho_v)$ such that $\eta_v=\rho_{v1}$ and $\langle\rho_v,\pi_v\rangle=1$, and, conversely, that these conditions characterize the η_v-distinguished representations. It is also shown that a cuspidal representation π is η-distinguished if and only if it is η_v-distinguished for all v. On the other hand, if one fixes d and only considers periods relative to this choice, then it is possible to find π such that $P(\varphi,\eta,d)\equiv 0$ even though there exist (nonzero) period functionals relative to H_d at each place of F.

Putting together some of the results mentioned above, one deduces that the cuspidal representations π with nonzero periods are those cuspidal representations which are theta lifts of cuspidal representations σ of the quasisplit group H_1. Suppose, in this case, that σ_E is the base change lift of σ to $\mathrm{GL}(2)_{/E}$. The last result of the paper is the formula

$$P(\varphi,\gamma|E^1_{\mathbf{A}},d)=L_S(1,\sigma_E\otimes\gamma^{-3})\prod_{v\in S}C_v(\varphi_v),$$

in which S is a finite set of places of F, the C_v's are period functionals, and the components of the test function $\varphi=\bigotimes_v\varphi_v$ outside of S are held fixed.

Besides the work of Harder, Langlands and Rapoport on Hilbert modular surfaces, one of the most obvious precursors of this paper is Waldspurger's work on the Shimura correspondence. Many of the connections among theta lifting, special values of L-functions and period integrals were already evident in the work of Waldspurger, as well as in many subsequent applications of Jacquet's theory of relative trace formulas on GL(2). The present paper fully integrates the theory of endoscopy into this mix and it greatly increases our understanding of the theory of automorphic representations for U(3). *Jeff Hakim* (Washington, DC)

99b:11127 11R58 11G09 11M41 11R42
Sheats, Jeffrey T. (Tucson, AZ)
The Riemann hypothesis for the Goss zeta function for $\mathbf{F}_q[T]$. (English summary)
J. Number Theory **71** (1998), *no.* 1, 121–157.

In Riemann's spectacular paper "Über die Anzahl der Primzahlen unter einer gegebenen Grösse" ("On the number of primes less than a given magnitude") [Monatsber. Königl. Preuss. Akad. Wiss. Berlin **1859**, 671–680; reprinted in *Gesammelte mathematische Werke und wissenschaftlicher Nachlass*, Art. VII, Teubner, Leipzig, 1876; see JFM 08.0231.03], the function

$$\zeta(s)=\zeta_{\mathbf{Q}}(s):=\sum_{n=1}^{\infty}n^{-s}=\prod_{p\text{ prime}}(1-p^{-s})^{-1}$$

is used to study the distribution of prime numbers. That 8-page paper provides many important facts about $\zeta(s)$, among which are: (1) its meromorphic continuation to the whole complex plane; (2) the functional equation $\xi(s)=\xi(1-s)$, where

$$\xi(s):=\Pi(s/2)(s-1)\pi^{-s/2}\zeta(s)$$

is now entire and $\Pi(s)=\Gamma(s+1)$ (with $\Gamma(s)$ being Euler's gamma function); (3) the product expansion

$$\xi(s)=\xi(0)\prod_{\rho}(1-s/\rho),$$

where ρ runs over the zeroes of $\xi(s)$ and where one must pair a given root ρ with $1-\rho$. Riemann goes on to examine the function $\Xi(t):=\xi(\frac{1}{2}+it)$, which one can now readily express as a *real* power series in t with an infinite radius of convergence, and remarks that "it is very likely that all of its roots are real". This statement, the famous "Riemann hypothesis", has profound implications for the distribution of primes and is of course not yet resolved. It has launched voyage after voyage of mathematical discovery of the greatest importance in a process that continues, unabated, till this day. It goes to the heart of our understanding of both number theory and analysis.

Once mathematicians began studying the Riemann zeta function, it was soon realized that $\zeta(s)$ was but the first of many different types of "zeta and L-functions". For instance, it was realized that one could associate a generalization of $\zeta(s)$ to any number field k; these functions $\zeta_k(s)$ possess, or are conjectured to possess, properties analogous to those listed above for $\zeta(s)$, including a (conjectured) analog of the Riemann hypothesis.

In the algebraic theory of numbers, it is known that much of the basic theory works for all "global fields", i.e., those fields which are finite extensions of $\mathbf{Q}$ (number fields) or finite extensions of $\mathbf{F}_p(T)$, for some prime p (the so-called "function fields over a finite field"). So it is natural to associate a zeta function to a function field $\mathscr{K}$ where one replaces products over the places of the number field with products over the places of $\mathscr{K}$. This theory was developed by Artin, Hasse and Weil. One sees that $\zeta_{\mathscr{K}}(s)$ has an analytic continuation and a functional equation very similar to that of Riemann's zeta function. Moreover, Hasse and Weil were actually able to prove the analog of the Riemann hypothesis for $\mathscr{K}$. In fact, it turns out that for a function field over $\mathbf{F}_p$, one can express $\zeta_{\mathscr{K}}(s)$ as $R_{\mathscr{K}}(p^{-s})$, where $R_{\mathscr{K}}(u)$ is a rational function with $\mathbf{Q}$ coefficients; moreover, this function $R_{\mathscr{K}}(u)$ may be explicitly computed via certain cohomology theories associated to $\mathscr{K}$ (in analogy with the singular cohomology associated to function fields over **C** or, equivalently, Riemann surfaces). These results on $\zeta_{\mathscr{K}}(s)$ have proved to be, and still are, a very fruitful source of ideas about zeta and L-functions of number fields [see, e.g., P. Sarnak, in *Proceedings of the International Congress of Mathematicians (Berlin, 1998), Vol. I*, Doc. Math. **1998**, Extra Vol. I, 291–303 (electronic)].

Because the function $R_{\mathscr{K}}(u)$ is a rational function, the classical zeta functions of function fields are inherently algebraic objects. This makes them simpler than zeta functions of number fields, which are known to be transcendental functions; indeed, to see this one need only consider Euler's famous computation of the "special values" $\zeta(2n)$, for $n \geq 1$, in terms of π and Bernoulli numbers which are in **Q**. Euler's computation, and its generalizations, are essential to the construction of p-adic zeta and L-functions via (p-adic) interpolation. On the other hand, no such nontrivial results on special values exist for $\zeta_{\mathscr{K}}(s)$. Recently, however, it has become possible to associate to function fields certain zeta and L-functions which have a more transcendental flavor to them, and which have properties analogous to those of the Riemann zeta function not possessed by $\zeta_{\mathscr{K}}(s)$ (such as Euler's results on special values). These functions are made possible by the theory of Drinfel′d modules and their generalizations, T-modules.

In order to understand a given mathematical object it is sometimes useful to view an analog of it "outside the box" where it lives. In terms of classical zeta functions (and, in particular, $\zeta(s)$) this was traditionally very difficult as *all* varieties/schemes "live" in the same box as the integers (i.e., all varieties are fibered over Spec(**Z**)). Moreover, function fields have no "bottom", unlike number fields, as every function field possesses a nontrivial subfield of finite codimension. Drinfel′d modules, etc., "cure" these problems in the following manner: One chooses, once and for all, a fixed place (called "∞") of $\mathscr{K}$, sets $\mathbf{F}_r$, $r = p^m$, to be the full field of constants, and sets **A** to be the affine ring of those elements of $\mathscr{K}$ which are regular outside ∞. Now let $\mathscr{L}$ be any field with a given $\mathbf{F}_r$-algebra map $\imath\colon \mathbf{A} \to \mathscr{L}$. A Drinfel′d **A**-module [V. G. Drinfel′d, Mat. Sb. (N.S.) **94(136)** (1974), 594–627, 656; MR **52** #5580] φ defined over $\mathscr{L}$ is a realization of **A** as an algebra of operators on the affine line over $\mathscr{L}$ such that the constant term for φ_a, $a \in \mathbf{A}$, is $\imath(a)$ (i.e., $\varphi_a = \imath(a)F^0 + \sum_{i=0}^{m} \alpha_i F^i$, where $\{\alpha_i\} \subseteq \mathscr{L}$ and F is the p-power map on the affine line; so, in particular, F is additive: $F(x+y) = F(x) + F(y)$). A T-module [G. W. Anderson, Duke Math. J. **53** (1986), no. 2, 457–502; MR 87j:11042] is somewhat similarly defined but there we use arbitrary affine n-space. The action of **A** on a Drinfel′d module is viewed, and studied, as an analog of the canonical action of **Z** on $\mathbf{G}_m$, or an elliptic curve. Moreover, if the field $\mathscr{L}$ is an extension of the completion $\mathscr{K}_\infty$, then it is a fundamental fact that one is also able to describe Drinfel′d modules analytically via "lattices" and "exponential functions". Thus, and importantly, in the theory of Drinfel′d modules **A** precisely plays the role of **Z**, etc., giving us indeed a theory "outside the box". With these analogies in mind, one readily constructs Tate modules of Drinfel′d modules (or T-modules) which are also Galois modules in the obvious fashion; as of this writing the analogy between these Galois modules and those coming from elliptic curves over number fields has held up to a very high degree [see, e.g., R. Pink, Publ. Res. Inst. Math. Sci. **33** (1997), no. 3, 393–425; MR 98f:11062, see p. A46 in this volume]. One uses these Tate modules to provide Euler factors at the finite primes (over Spec(**A**)) just as for elliptic curves. It is, however, essential to note that these Euler factors belong to $\mathbf{A}[u]$ and so live in characteristic p. As such the L-function of a Drinfel′d module [D. Goss, *Basic structures of function field arithmetic*, Springer, Berlin, 1996; MR 97i:11062, see p. A41 in this volume (Section 8)] must live in characteristic p also. Let $\mathbf{C}_\infty$ be the completion of a fixed algebraic closure of the local field $\mathscr{K}_\infty$ (itself the $\mathbf{A}$-analog of the real numbers $\mathbf{R} = \mathbf{Q}_\infty$). The construction of the L-function as an Euler product is easily accomplished once the element "$f^{-s} \in \mathbf{C}_\infty$" for $f \in \mathbf{A}$ (with good properties as an "exponential") is defined. By making certain auxiliary choices, the definition of f^{-s} is readily given [D. Goss, op. cit. (Def. 8.2.5)], where s is taken in the space $S_\infty := \mathbf{C}_\infty^* \times \mathbf{Z}_p$ (so S_∞ plays the role classically played by $\mathbf{C}$; it is easy to see how to discretely embed the usual integral powers into S_∞). The space S_∞ introduces non-Archimedean analysis into the theory as, by definition, f^{-s} is a 1-parameter family of $\mathbf{C}_\infty$-valued analytic functions on $\mathbf{C}_\infty^*$. As the basic example, we let $\mathbf{A} := \mathbf{F}_r[T]$ and set

$$\zeta_{\mathbf{A}}(s) := \sum_{\substack{n \in \mathbf{A} \\ n \text{ monic}}} n^{-s} = \prod_{f \text{ monic prime}} (1 - f^{-s})^{-1}.$$

After using some elementary estimates, one sees that $\zeta_{\mathbf{A}}(s)$ gives rise to a 1-parameter family of power series with coefficients in the "real numbers" $\mathscr{K}_\infty = \mathbf{F}_r((1/T)) = \mathbf{F}_r(T)_\infty$ (i.e., is an "entire function on S_∞"). Similar results can be established for all abelian L-functions.

As shown in (3) (see the first paragraph), Riemann established the factorization of $\xi(s)$ as "an infinite polynomial". In the case of L-functions of Drinfel′d modules, etc. (and so, in particular, $\zeta_{\mathbf{A}}(s)$), such a factorization follows automatically once one has established the integrality of the L-series, as in non-Archimedean analysis every entire power series has such a factorization over its zeroes (which are also automatically algebraic over the field of definition of the power series). Thus subtleties in complex analysis involving entire functions and the exponential function do not exist in non-Archimedean analysis (however, other subtleties about the nature of the zeroes arise to take their place!). Moreover, L-functions of Drinfel′d modules, etc., seem to have special analytic properties even for non-Archimedean functions. Recently, for instance, the present author and Z. Yang have used the results of M. Car [J. Number Theory **66** (1997), no. 1, 148–171; MR 98h:11149] to show that those L-functions of Drinfel′d modules, etc., which have good additive expansions (i.e., they arise from abelian objects) also possess an expansion in certain polynomials analogous to a Fourier expansion. So, as a corollary, these functions are determined by the values on the subset $\{(a^{-1}, y) |\ 0 \neq a \in \mathbf{A}\}$ of S_∞. This is not true for arbitrary entire functions on S_∞.

The simplest Drinfel′d module is called the "Carlitz module" (after L. Carlitz, who studied it in the 1930s [see Duke Math. J. **1** (1935), 137–168; Zbl 012.04904]); it is an $\mathbf{F}_r[T]$-analog of $\mathbf{G}_m$ over $\mathbf{C}$ and its exponential shares remarkably many properties with the classical exponential function. (E.g., the L-series of the Carlitz module over the rational field is easily seen to be $\zeta_{\mathbf{A}}(s-1)$.) In particular, the period lattice of the Carlitz exponential has $\mathbf{A}$-rank 1 and is generated by an element ξ which is remarkably similar to $2\pi i$. This element ξ, and the Carlitz exponential, are readily used to establish a version of Euler's formula for $\zeta_{\mathbf{A}}(s)$ (i.e. we obtain an expression for the special values of $\zeta_{\mathbf{A}}(s)$ at the positive integers $\equiv 0 \pmod{r-1}$ in terms of ξ and rational functions).

One can also obtain very strong results on values of $\zeta_{\mathbf{A}}(s)$, etc., at negative integers; via interpolation these results lead to v-adic functions for $v \in \operatorname{Spec}(\mathbf{A})$ (analogous to p-adic L-functions) which have domain $\mathbf{C}_v^* \times S_v$, with $\mathbf{C}_v$ being the v-adic analog of $\mathbf{C}_\infty$, and S_v being the Cartesian product of $\mathbf{Z}_p$ and a finite abelian group [see D. Goss, op. cit. (Subsection 8.3)]. The theory of nonabelian L-functions is ultimately expected to be similar.

Having mentioned some of the strong analogies between $\zeta_{\mathbf{A}}(s)$ and Riemann's zeta function, it is quite natural to inquire about the existence of (i) a functional equation for $\zeta_{\mathbf{A}}(s)$, and (ii) an analog of the Riemann hypothesis. We postpone discussing (i) till a bit later on. On the other hand, (ii) brings us, finally, to the very important paper by Sheats under review. We begin by discussing what is known at ∞. Recall that in order to define f^{-s} certain auxiliary choices need to be made. Thus, the zeroes of $\zeta_{\mathbf{A}}(s)$ will also depend on these choices and one is led to search for what, if anything, is independent of these choices. The answer is quite provocative (and is valid in total generality, once the integrality has been shown): For each $y \in \mathbf{Z}_p$ we adjoin to $\mathscr{K}_\infty$ all the zeroes of $\zeta_{\mathbf{A}}(x,y)$. Then the maximal separable (over $\mathscr{K}_\infty$) subfield of this extension is independent of all auxiliary choices and so only depends on y (and the original Drinfel′d module, etc., used to construct the L-series). What is this maximal separable extension? In non-Archimedean analysis there is a technique, called the "Newton polygon", which is associated to a power series and gives information about its zeroes of a given magnitude. If the Newton polygon is sufficiently simple then the zeroes are themselves simple and belong to the field of definition of the power series. In the case of $\zeta_{\mathbf{A}}(s)$, this is exactly what was found to happen when $\mathbf{A} = \mathbf{F}_p[T]$ by D. Q. Wan [J. Number Theory **58** (1996), no. 1, 196–212; MR 97c:11064], implying that the zeroes lie "on the line" $\mathscr{K}_\infty = \mathbf{F}_p((1/T))$! One only has to compare this with Riemann's remark, quoted above, on the zeroes of $\Xi(t)$ to understand why this result deserves to be called a "Riemann hypothesis". Wan's proof was a difficult ad hoc calculation but soon D. Thakur realized that a combinatorial result stated, but not proved, in an old paper by Carlitz [Duke Math. J. **15** (1948), 1001–1012; MR **10**, 283h] on power sums of polynomials would allow one to give a simpler proof, and this was carried out in the case $\mathbf{A} = \mathbf{F}_p[T]$ by J. Diaz-Vargas [J. Number Theory **59** (1996), no. 2, 313–318; MR 97m:11078]. Now Carlitz stated his result for arbitrary $\mathbf{F}_r[T]$ $(r = p^m)$, but going from $\mathbf{F}_p$ to arbitrary $\mathbf{F}_r$ proved to be quite difficult. The first break came when B. Poonen found a proof for $r = 4$. This proof inspired Sheats to overcome all the difficulties and give a complete proof in this well-written and careful paper. It is important not only as a beautiful piece of combinatorics but also because it shows that the computation of the Newton polygon is not derailed by strange r-adic behavior of binomial coefficients, etc. This "Riemann hypothesis" for $\zeta_{\mathbf{A}}(s)$ seems to imply that we should expect the fields generated by the zeroes of arbitrary L-series at ∞ to be "as small as possible". Of course one would like a good conceptual reason for this to be true in terms of function field arithmetic.

Remarkably, as Wan has pointed out to the present author, the computation at ∞ can also be used to establish the similar result v-adically for (at least) an open subset of S_v and $\deg v = 1$. However, once the degree of v is sufficiently large, computations involving the Newton polygon no longer suffice to tell us the rationality of the zeroes. So exactly what a general "v-adic Riemann hypothesis" might be is completely open. However, one still can find certain fields associated to the zeroes that are independent of auxiliary choices and all such fields should have some sort of intrinsic meaning as well as description. One possible path towards describing these fields arises from "being outside the box" as follows: Suppose we have a Drinfel′d $\mathbf{F}_r[T]$-module defined over the field $\mathbf{F}_r(T)$ (e.g., the Carlitz module). We then have a situation where T plays two roles: In $\mathbf{F}_r[T]$ it is an operator (e.g., via the Carlitz module) whereas it is obviously a scalar in $\mathbf{F}_r(T)$. To keep track of these two different roles it is traditional to denote "T as scalar" by θ. It is then possible to combine these two actions into what Anderson [see op. cit.] calls "the T-motive" of the Drinfel′d module; this is a module over the two-dimensional algebra $\mathbf{F}_r(\theta)[T] = \mathbf{F}_r(\theta) \otimes_{\mathbf{F}_r} \mathbf{F}_r[T]$. It is easy to see that no similar two-dimensional algebra can be constructed classically. One is thus able to classify the objects of function field arithmetic according to whether they arise from T or θ. For instance, the lattice associated to a Drinfel′d module lies in the universe of θ ("T as scalar") and thus so do modular forms on the Drinfel′d upper half spaces. On the other hand, the Euler L-factors arising from Drinfel′d modules live in the realm of T ("T as operator") as indeed they are defined using characteristic polynomials of the $\mathbf{A}$-action via the Drinfel′d module. This suggests how L-series zeroes should be interpreted. In fact, standard results on the extension of derivations, and divided derivations, imply that the separable closures of the completions of $\mathscr{K}$ are precisely the maximal algebras that the T-action on an arbitrary T-module can be formally extended to (as a "formal module")! As such it seems reasonable to expect that there is some interpretation of the L-series zeroes in terms of this formal action; indeed, for Drinfel′d modules, the action is all one has! (See, e.g., G. Anderson [J. Number Theory **60** (1996), no. 1, 165–209; MR 97h:11058] for such an interpretation for the special value $s = 1$.) There is also precedent for such an operator-theoretic interpretation of zeroes in the theory of Drinfel′d modules (and elliptic curves) over finite fields where it is supplied by the Frobenius morphism.

Finally, we turn to functional equations. There is no known functional equation of classical type for the L-series of Drinfel′d modules. However, recall that classically functional equations arise by expressing an L-series as integral transform of automorphic forms. Now, as above, modular forms associated to Drinfel′d modules live in the universe of T as scalar; nevertheless, one is able to construct certain non-Archimedean "measures" from them [J. T. Teitelbaum, in *The arithmetic of function fields (Columbus, OH, 1991)*, 189–207, de Gruyter, Berlin, 1992; MR 93k:11056]. Using these "Schneider-Teitelbaum measures" one can construct "L-functions" associated to modular forms with domain $\mathbf{Z}_p$ and values in characteristic p [D. Goss, in *The arithmetic of function fields (Columbus, OH, 1991)*, 227–251, de Gruyter, Berlin, 1992; MR 94c:11036]; the modular symmetries then imply that such functions do indeed possess classical style functional equations, but have no (!) known connection to the L-functions of Drinfel′d modules.

In summary, the arithmetic of function fields provides us with analogs of the L-series of classical algebraic number theory. Much is known about these function field analogs, with much more remaining to be discovered. The theory of these functions seems to require the invention of some sort of new formalism in order to understand the zeroes of these new L-series. Indeed, even if the theory reduces to calculations of Newton polygons (which seems highly unlikely), there would still have to be some techniques for finding the Newton polygons of L-series associated to nonabelian objects (as such L-series do not possess good additive expansions!). So future results in this area have the potential to be of even greater interest.

David Goss (Columbus, OH)

97g:11136 11R70 11R23 11R34 11R42 19F27
Kolster, Manfred (Hamilton, ON); **Nguyen Quang Do, Thong** (Besançon); **Fleckinger, Vincent** (Besançon)
Twisted S-units, p-adic class number formulas, and the Lichtenbaum conjectures.
Duke Math. J. **84** (1996), *no.* 3, 679–717.

This is an important paper in which delicate methods from Galois cohomology and Iwasawa theory are used, in conjunction with deep results in algebraic K-theory, to establish a version of the Lichtenbaum conjectures for absolutely abelian fields F. These conjectures are far-reaching generalizations of the analytic class formula, which will be recalled just below. Before we enter into the details of the paper under review, we will try to give a general impression of what is going on. The Lichtenbaum conjectures are concerned with predictions about the order of even-numbered K-groups of the ring $\mathcal{O}_F$ of integers in a number field F, and about regulators attached to odd-numbered K-groups. The philosophy is that even-numbered K-groups behave like ideal class groups, and the odd-numbered ones behave like unit groups. At the very least, one knows that $K_{2i}(\mathcal{O}_F)$ is finite, and $K_{2i-1}(\mathcal{O}_F)$ is finitely generated with known rank, by results of A. Borel; $K_0(\mathcal{O}_F)$ is the direct sum of the ideal class group of F and a copy of $\mathbf{Z}$, and $K_1(\mathcal{O}_F)$ is exactly the unit group of $\mathcal{O}_F$. Now the analytic class number formula provides information about the class group of F, and the regulator of F, which is a quantity attached to the unit group of $\mathcal{O}_F$. Actually, when one uses the functional equation of the Dedekind zeta function ζ_F, the more usual version of the analytic class number formula, involving the residue at $s=1$, changes into the neater statement $\zeta_F(0)^* = -h_F R_F/w_F$, where h_F and R_F are the class number and the regulator of F, w_F counts the roots of unity in F, and * means: "take the first nonvanishing coefficient in the Taylor expansion at $s=0$". This can be taken to be the Lichtenbaum conjecture for $m=1$ (of course, it is a long-standing theorem).

Let us now briefly recall the Lichtenbaum conjecture LC_m for $m \geq 2$ and explain in the process what is proved in the paper under review. Briefly, the leading coefficient of the development of $\zeta_F(s)$ at $s = 1-m$ is conjectured to be given as a quotient

$$\zeta_F(1-m)^* = \frac{|K_{2m-2}(\mathcal{O}_F)|}{w_m(F)} \cdot R_m(F) \cdot r,$$

where K_* is Quillen's K-theory, the number $w_m(F)$ is a straightforward generalization of $w_F = w_1(F)$, the number of roots of unity in F, $R_m(F)$ is a certain regulator attached to $K_{2m-1}(\mathcal{O}_F)$ (it *is* a determinant as the usual regulator, but we will not attempt to explain it), and r is another rational factor.

The main result of the paper under review proves the above equality for F abelian over Q, with the following specifications and modifications: r is a certain explicit product of Euler factors, times a (nonexplicit) power of 2. Furthermore, the Quillen K-group $K_{2m-2}(\mathcal{O}_F)$ is replaced by the group $K^g_{2m-2}(\mathcal{O}_F)$, which needs to be explained: For every prime p, there is an epimorphism $\mathrm{ch}_{2,m}\colon K_{2m-2}(\mathcal{O}_F) \otimes \mathbf{Z}_p \to K^{\text{ét}}_{2m-2}(\mathcal{O}_F)$, and its target (called étale K-theory; the notation suppresses the dependency on p) is isomorphic to the étale cohomology group $H^2(\mathcal{O}_F, \mathbf{Z}_p(m)) = H^2_{\text{ét}}(\mathcal{O}_F[1/p], \mathbf{Z}_p(m))$. This is due to Dwyer and Friedlander. (The étale cohomology groups just mentioned can also be expressed by Galois cohomology.) Now the group $K^g_{2m-2}(\mathcal{O}_F)$ is the direct sum over all p of the image of a splitting of $\mathrm{ch}_{2,m}$, which exists by work of B. Kahn. So the main result of the paper under review basically replaces K-theory by étale K-theory. Another way of looking at this is the following: The kernel of $\mathrm{ch}_{2,m}$ is finite and expected to be zero for odd p; this is the so-called Quillen-Lichtenbaum conjecture, which is known to hold for $m = 2$. One may therefore say: the Lichtenbaum conjecture is proved modulo the Quillen-Lichtenbaum conjecture. (There is a quite recent preprint of Kahn ["The Quillen-Lichtenbaum conjecture at the prime 2", Preprint; per revr.] containing a proof of the 2-primary part of the Quillen-Lichtenbaum conjecture.) We should also mention that the Chern class maps $\mathrm{ch}_{2,m}$ have analogues $\mathrm{ch}_{1,m}\colon K_{2m-1}(\mathcal{O}_F)\otimes\mathbf{Z}_p \to K^{\text{ét}}_{2m-1}(\mathcal{O}_F)$; they are again onto, with finite kernel, and the kernel is again conjectured to be zero (neglecting the 2-part). However, the regulator "does not see" the unknown but finite kernel of $\mathrm{ch}_{1,m}$.

The long and clever argument begins with the identification of étale K-theory with certain cohomology groups $H^i(\mathcal{O}_F, \mathbf{Z}_p(m))$, with $i = 1$ and $i = 2$. It has been accepted for some time that these cohomology groups should behave like "twisted" unit groups [resp. class groups]. Precise results in this direction are proved in §3. To wit, there is an exact sequence

$$0 \to \overline{U}'_\infty(m-1)_{G_\infty} \to H^1(\mathcal{O}_F, \mathbf{Z}_p(m)) \to X'_\infty(m-1)^{G_\infty} \to 0.$$

The $\overline{U}'$-term is obtained by taking a projective limit of groups of p-units in a p-cyclotomic extension, twisting $m-1$ times, and "descending back" by taking coinvariants. The H^1-term is an étale K-group, and the cokernel $X'_\infty(m-1)^{G_\infty}$ (which comes from Iwasawa theory) is finite. One might think of this as a twisted version of the result which states that K^1 of a ring of integers is just the unit group. The authors also introduce $\overline{C}'_\infty$, a subgroup of $\overline{U}'_\infty$ constructed from cyclotomic units, which will be crucial in the sequel.

The principal theorem of this paper on the algebraic side (§5) is a sharpened version of the following index formula (we disregard the character decomposition to keep things simpler):

$$[H^1(\mathcal{O}_F, \mathbf{Z}_p(m))^\pm : \overline{C}'_\infty(m-1)^\pm_{G_\infty}] = |H^2(\mathcal{O}_F, \mathbf{Z}_p(m))^\pm|,$$

with plus signs for m odd and minus signs for m even. This is an analogue of Kummer's result describing the plus class number h_l^+ as an index of cyclotomic units inside a full unit group. The proof of the index formula just mentioned combines intricate cohomological calculations with Iwasawa theory in a very nice way, and it relies on most of the preceding results. The main theorem of the paper, which has already been explained above, is subsequently deduced in §6 with the help of this theorem, using many deep results of Beĭlinson, Bloch and Soulé on regulator maps and cyclotomic elements in K-theory.

Two small criticisms: The proof of the remark on p. 692 (which is very important for §5) is incorrect as it stands. The authors are aware of the problem; a corrigendum is in preparation and should appear in due course [Duke Math. J. **90** (1997), no. 3, 641–643; MR 98k:11172]. Second point: On p. 715, a certain element $C_m^D(w)$ is called "Galois generator" of a certain group $\widetilde{C}_\infty(m-1)_{G_\infty}$. (In the fifth line, read ζ_{p^n} for ζ_{pn}.) I'm not sure whether the group in question is generated by that single element; perhaps one should allow w to vary over all Nth roots of unity. The reader is referred without a page or theorem number to P. Deligne's fundamental paper [in *Galois groups over* **Q** *(Berkeley, CA, 1987)*, 79–297, Springer, New York, 1989; MR 90m:14016], which is quite long.

The last remarks should be considered minor matters. In the reviewer's opinion, this is a paper which cleverly employs many powerful techniques and proves a major result. *Cornelius Greither* (Neubiberg)

99g:11140 11R70 11R34 11S23 19A31 19F27

Chinburg, T. (Philadelphia, PA); **Kolster, M.** (Hamilton, ON); **Pappas, G. [Pappas, Georgios]** (Princeton, NJ); **Snaith, V.** (Southampton)

Galois structure of K-groups of rings of integers. (English summary)

K-Theory **14** (1998), *no.* 4, 319–369.

The analytic class number formula for a number field N expresses the product of the class number and the regulator of K as the residue of the Dedekind zeta function of N, times some explicit and easily computable factors. (The regulator "measures" in some way the unit group of the ring of integers in N.) This classical link between class numbers and unit groups (via regulators) has been the subject of far-reaching generalizations, both established and conjectural, and this paper is an important contribution. The objects of study are the K-groups $K_{2n+1}(O_N)$ (odd-numbered) and $K_{2n}(O_N)$ (even-numbered) of the ring of integers O_N, where N/K is now a G-Galois extension of number fields (or of function fields, see Chapter 7). It is helpful to think of the odd-numbered K-groups as "twisted versions" of the unit group, and of the even-numbered ones as "twisted versions" of the class group. Indeed, $K_1(O_N)$ is precisely the unit group, and $K_0(O_N)$ is isomorphic to $\mathrm{Cl}_N \oplus \mathbf{Z}$. All these K-groups are considered as G-modules. The difference $D = (K_{2n+1}(O_N) - (K_{2n}(O_N)))$ should be something simple, viewed in a suitable Grothendieck group. In the fairly weak Grothendieck group $G_0(\mathbf{Z}[G])$, one can easily find correction terms such that one obtains a torsion element (conjecturally zero) on subtracting them from D. This corrected difference is called $\Omega_n(N/K)^{\mathrm{naive}}$. The main point is now to define a lifting of this class, $\Omega_n(N/K)$, in the finer Grothendieck group $K_0(\mathbf{Z}[G])$. For $n = 0$, this is the by now classical construction of Chinburg's "multiplicative" or "third" invariant, using Tate sequences. Snaith and Pappas (independently) did the case $n = 1$, making essential use of work of B. Kahn [K-Theory **7** (1993), no. 1, 55–100; MR 94i:11094]. The authors of the present paper previously gave a definition of Ω_n for all $n \geq 1$, assuming the existence of the so-called Lichtenbaum complexes. In the present paper, the authors give an unconditional construction of Ω_n; they then prove that it does reduce to the naive omega invariant in G_0, and they give results about this new omega invariant modulo the kernel group. The proofs of the results, which were announced in [T. Chinburg et al., C. R. Acad. Sci. Paris Sér. I Math. **320** (1995), no. 12, 1435–1440; MR 96d:19010], partly depend on open conjectures (for details, see below). Important note: D. Burns and M. Flach have constructed, by very general methods, canonical elements $\Omega(L/K, M)$ to each motive $M = h^n(X)(r)$ satisfying certain "standard conjectures". In a recent paper [Amer. J. Math. **120** (1998), no. 6, 1343–1397] they prove a precise relationship between their class for $X = \mathbf{Q}(-n)$ and the class $\Omega_n(L/K)$ constructed in the paper under review.

I will try to comment chiefly on two main points: the construction of the new omega invariants, and the calculations modulo the kernel group. (There is still more in the paper: in Chapter 7, a strong result is proved for the function field case.) The principal idea is, if not simple, at least clear-cut: $K_{2n+1}(O_N)$ should "be" a first étale cohomology group of the étale sheaf $\mathbf{Z}(n+1)$, and $K_{2n}(O_N)$ should "be" a second étale cohomology group of that same sheaf. (This is just a very loose way of talking about Dwyer and Friedlander's Chern character homomorphisms.) But the road from this idea to a precise statement and a rigorous construction is long and somewhat arduous. Many a modification or amendment is needed. Actually, the sheaf $\mathbf{Z}(n+1)$ does not even make sense: the Tate twist is only defined for $\mathbf{Z}_l$ (l a prime), not for $\mathbf{Z}$, so there is a local-global game going on. Then if one wants to relate the l-completion of $K_{2n+1}(O_N)$ to the étale cohomology of $\mathbf{Z}_l(n+1)$, the base space $X_l = \mathrm{Spec}(O_{N,S}[1/l])$ for the cohomology depends on l. There are quite serious problems for the prime $l = 2$ which are overcome by replacing cohomology by so-called positive cohomology, noted with a subscript $+$; here one has to put up with a dimension shift (2 becomes 1, 1 becomes 0), and we refer to the paper for the details of the definition. By general theory, one then obtains, for each l, 2-extensions $0 \to H_l^0(\mathbf{Z}_l(r)_+) \to A \to B \to H_l^1(\mathbf{Z}_l(r)_+) \to 0$, where the G-modules A and B are cohomologically trivial. The H^0 and H^1 terms are related to, but unfortunately not equal to, l-completions of $K_{2n+1}(O_N)$ and $K_{2n}(O_N)$, respectively. The resulting elements in Ext^2, for each l, then give rise to a global 2-extension which serves as an analog of the canonical class in Chinburg's original construction. It seems that (besides a lot of fascinating detail) the arithmetical ingredient which really makes this work is furnished by vanishing theorems for the cohomology of sheaves $\mathbf{Z}_l(r)$. In fact, the passage to positive cohomology is crucial at this point. The fourth chapter is devoted to the proof that the omega invariants are well-defined, that is, they do not depend on all the choices made along the way.

In Chapter 5, it is proved that the invariant $\Omega_n(N/K)$ does, essentially, reduce to $\Omega_n(N/K)^{\text{naive}}$ under the canonical homomorphism from $K_0(\mathbf{Z}[G])$ to $G_0(\mathbf{Z}[G])$, if $n = 1$, or if n is general and the Quillen-Lichtenbaum conjecture concerning the Chern characters mentioned above is true. The argument flows, in a way, naturally from the construction; it is not totally formal, however, since one has to show that certain explicit modules have class zero.

In the sixth chapter, it is proved that $\Omega_n(N/K)$ is congruent to the root number class $W_{N/K}$ modulo the kernel group, if N is totally real and n is odd, so both the K-groups with index $2n$ and $2n+1$ are finite. When working in the setup of the Stark conjectures, there are then no regulators, so one is only concerned with L-values and q-indices. The q-indices may be expressed in terms of étale cohomology again, and the proof of the Main Conjecture by Wiles allows one to conclude. For N not necessarily totally real, and n of any parity, the same result on the omega invariant is shown to be a consequence of a conjecture (6.12) dubbed the "modified Lichtenbaum-Gross conjecture".

Some very minor remarks: The symbol Λ should have a subscript m in the line following (2.17) on page 326, and $\mathscr{D}$ should have no subscript m in the line following (2.20) on page 327. In the paragraph just before Proposition 3.14, read "2-power" instead of "even" (cf. page 349). Right after (4.81), it would perhaps be better to invoke Lemma 4.14 directly (rather than Corollary 4.15); note that $I_w = 1$ at that point. Near the bottom of page 360, the name "Voevodsky" came out a bit scrambled.

It is perhaps visible from what has been said that this is a substantial paper containing a lot of technique. However, the exposition is very readable. In fact, it is one of the many merits of this paper that the technical details are all explained, or at least provided with a precise reference. This paper will rapidly become a standard reference in its field.

{For related work see [T. Chinburg et al., in *Algebraic K-theory (Toronto, ON, 1996)*, 1–29, Amer. Math. Soc., Providence, RI, 1997; MR 99b:11130].}

Cornelius Greither (Neubiberg)

B. REVIEWS IN ALGEBRA AND ALGEBRAIC GEOMETRY

99j:05189a 05E05 33C80 33D80
Kirillov, Anatol N. (Meguro); **Noumi, Masatoshi** (Kobe)
Affine Hecke algebras and raising operators for Macdonald polynomials.
Duke Math. J. **93** (1998), *no.* 1, 1–39.

99j:05189b 05E05 33C80 33D80
Sahi, Siddhartha (New Brunswick, NJ)
Interpolation, integrality, and a generalization of Macdonald's polynomials.
Internat. Math. Res. Notices **1996,** *no.* 10, 457–471.

99j:05189c 05E05 33C80 33D80
Knop, Friedrich (New Brunswick, NJ)
Integrality of two variable Kostka functions.
J. Reine Angew. Math. **482** (1997), 177–189.

99j:05189d 05E05 33C80 33D80
Garsia, A. M. (La Jolla, CA); **Remmel, J.** (La Jolla, CA)
Plethystic formulas and positivity for q,t-Kostka coefficients. (English summary)
Mathematical essays in honor of Gian-Carlo Rota (Cambridge, MA, 1996), 245–262, *Progr. Math.*, 161, *Birkhäuser Boston, Boston, MA*, 1998.

99j:05189e 05E05 33C80 33D80
Garsia, A. M. (La Jolla, CA); **Tesler, G. [Tesler, Glenn]** (La Jolla, CA)
Plethystic formulas for Macdonald q,t-Kostka coefficients. (English summary)
Adv. Math. **123** (1996), *no.* 2, 144–222.

Symmetric functions are indispensable to modern mathematics, providing keys to computation in representation theory, geometry, topology, enumerative combinatorics, and more. Ancient as symmetric function theory is, basic new discoveries—of facts Giambelli or even Jacobi could have appreciated—are still being made today.

The modern discoveries begin with Hall-Littlewood polynomials $P_\mu(X;t)$, symmetric in variables $X=\{x_1,x_2,\cdots\}$, and depending on an extra parameter t. The index μ ranges over integer partitions. Hall-Littlewood polynomials describe, among other things, cohomology of Springer fibers, singularities of nilpotent matrix orbit closures, irreducible characters of $\mathrm{GL}_n(q)$, and subgroup enumerations in finite abelian groups.

A dramatic expansion in symmetric function theory took place in 1988, with I. G. Macdonald's discovery of two-parameter polynomials $P_\mu(X;q,t)$ generalizing both the Hall-Littlewood polynomials and the Jack and zonal polynomials of harmonic analysis [*Symmetric functions and Hall polynomials,* Second edition, Oxford Univ. Press, New York, 1995; MR 96h:05207]. We are only beginning to glimpse the full import of Macdonald polynomials. They are connected with Hilbert schemes, with the Calogero-Sutherland model in physics, and with affine root systems. In the latter setting they arise as joint eigenfunctions of operators due to I. Cherednik [Ann. of Math. (2) **141** (1995), no. 1, 191–216; MR 96m:33010], who used them to prove Macdonald's generalized constant term identities for all root systems.

Multiplying $P_\mu(X;q,t)$ by an appropriate (q,t)-hook product $c_\mu(q,t)$ yields integral form Macdonald polynomials $J_\mu(X;q,t)$, which enjoy certain pleasant duality properties. Let $S_\lambda[X(1-t)]$ denote the image of a Schur function under the ring homomorphism sending the k-th power-sum symmetric function $p_k(X)=\sum_i x_i^k$

to $p_k[X(1-t)] = (1-t^k)p_k(X)$. We can now state Macdonald's positivity conjecture, the most striking unsolved problem in this area: The rational functions of q, t (known as Kostka-Macdonald coefficients) which arise in the expansion $J_\mu(X; q, t) = \sum_\lambda K_{\lambda\mu}(q, t)S_\lambda[X(1-t)]$ are polynomials, $K_{\lambda\mu} \in \mathbf{Z}[q, t]$, with non-negative integer coefficients.

It has recently been proven that the $K_{\lambda\mu}$ belong to $\mathbf{Z}[q, t]$. Five separate proofs appeared more-or-less simultaneously, and are presented in the five papers under review. The positivity conjecture remains open. (However, a representation-theoretic conjecture of Garsia and the reviewer would imply positivity.)

The five proofs vary considerably, although some have themes in common. The proofs by Garsia-Remmel and Garsia-Tesler are set in the context of classical symmetric function theory, making skillful use of the symmetric function operation known as plethystic substitution (or, less accurately, λ-ring notation) to construct explicit formulas in which slightly modified forms of the coefficients $K_{\lambda\mu}$ for partitions $\lambda = (k, \nu)$ appear as the plethystic substitution of a quantity depending on μ into a symmetric function $k_\nu(X; q, t)$ depending only on ν.

The proofs by Knop and Sahi are based on formulas for Macdonald's nonsymmetric polynomials. These are the joint eigenfunctions (in type A) of Cherednik's operators, whose appropriate context is the affine Hecke algebra of S_n. The usual Macdonald polynomials are symmetrizations of them. Knop's formulas are derived from Hecke algebra raising operators which produce the nonsymmetric polynomials by a stepwise inductive procedure. Sahi's formulas arise from a remarkable interpolation property: Macdonald's nonsymmetric polynomials are the highest-degree homogeneous parts of certain inhomogeneous polynomials characterized by a simple vanishing condition. The interpolation property leads Sahi to raising operators similar to Knop's but for the inhomogeneous variants.

The proof by Kirillov and Noumi is set in a kind of hybrid of the Hecke algebra and symmetric function contexts. They find raising operators which produce the symmetric Macdonald polynomials J_μ directly, adding one part at a time to the indexing partition. Though these raising operators are initially defined in terms of Cherednik's Hecke algebra operators, they are given in symmetric function terms by classical-type formulas similar to those defining the original operators of Macdonald. It is possible to study them entirely in the symmetric function setting, as has been done by Kirillov and Noumi in a preprint ["q-difference raising operators for Macdonald polynomials and the integrality of transition coefficients", Preprint No. UTMS-96-25, Univ. Tokyo, 1996; per bibl.] in which the main result of the paper under review is given a much simpler proof.

We now discuss the various proofs in somewhat more detail.

All the proofs use some form of the following lemma: It suffices that $k_{\lambda\mu} \in \mathbf{Z}[q, t]$, where $J_\mu(X; q, t) = \sum_\lambda k_{\lambda\mu}(q, t)m_\lambda(X)$. Proof: Writing m_λ in terms of $S_\lambda[X(1-t)]$ we see that the $K_{\lambda\mu}(q, t)$ are fractions over $\mathbf{Z}[q, t]$ with denominator factors of the form $(1-t^k)$. Since $K_{\lambda\mu}(t, q) = K_{\lambda'\mu'}(q, t)$, the denominator is 1.

In principle the simplest of the five proofs is that of Kirillov and Noumi, though one must consult their preprint [op. cit.] for the simpler version. They construct operators B_m such that $B_m J_\mu = J_{(1^m)+\mu}$ for μ with $\leq m$ parts. Hence $J_\mu = B_{\mu'_1} \cdots B_{\mu'_l} \cdot 1$, where μ' is the conjugate partition. From their formula for B_m it is obvious that J_μ has polynomial coefficients, and the lemma above completes the proof. The work is to prove that B_m has the desired raising property. In n variables, P_μ is an eigenfunction of Macdonald's operator $D(u)$, with eigenvalue $\prod_i(1-t^{n-i}q^{\mu_i}u)$, and $e_nP_\mu = P_{(1^n)+\mu}$, where $e_n = x_1\cdots x_n$. The J_μ are normalized so that $e_nD(t)J_\mu = J_{(1^n)+\mu}$; so $e_nD(t)$ would give the required operator B_n in n variables. For $m<n$, B_m is essentially a sum of $e_mD(t)$ over all m-subsets of the variables. The proof that this works rests on a clever device: Macdonald's inner product gives rise to an explicit Cauchy kernel $\Omega_{q,t}(X,Y)$ in two sets of variables X, Y, such that operators F, G are adjoint iff $F_X\Omega = G_Y\Omega$. By taking n X and m Y variables, the raising identity for B_m in n variables is reduced to a simpler identity involving the adjoint of B_m in m variables. The final identity involves t only and is established by setting $q=t$ and reducing to Schur functions (because $P_\mu(X;t,t)=S_\mu$).

Sahi's proof begins with the following "interpolation" theorem. Given a composition $(\lambda_1,\cdots,\lambda_n)\in\mathbf{N}^n$, define $w(\lambda)=(w_1,\cdots,w_n)\in S_n$ by $w_i<w_j$ iff $\lambda_i\geq\lambda_j$, for $i<j$. Let $\overline{\lambda}_i = q^{\lambda_i}t^{n-w_i(\lambda)}$. Then there exists a unique polynomial $G_\lambda(X)\in\mathbf{Q}(q,t)[x_1,\cdots,x_n]$ of degree $|\lambda|$ such that $G_\lambda(\overline{v})=0$ for all $v\neq\lambda$, $|v|\leq|\lambda|$, and the coefficient of x^λ in G_λ is 1. Similarly there is a unique symmetric polynomial $R_\lambda(X)$ with this interpolation property for λ, v partitions. In fact, R_λ is a sum $\sum_w T_wG_\lambda$, where the operators T_w belong to the Hecke algebra $H(t)$ of S_n, acting on polynomials by

$$T_{s_i}f = ts_if+(t-1)x_i(f-s_if)/(x_i-x_{i+1}).$$

The beautiful result (due to Sahi and Knop) is that the homogeneous component of degree $|\lambda|$ in R_λ is the Macdonald polynomial P_λ. The proof uses the interpolation property to verify that R_λ is an eigenfunction of an inhomogeneous operator, whose homogeneous part is the Macdonald operator. Sahi now exhibits inductive formulas for suitably renormalized functions $c_\lambda(q,t)G_\lambda$, showing they have coefficients in $\mathbf{Z}[q,t,t^{-1}]$. Hence so does $c_\lambda R_\lambda$, and its top degree part is J_λ. The theorem now follows from the lemma above.

Knop's proof uses the nonsymmetric polynomials $E_\lambda(X;q,t)$ ($\lambda\in\mathbf{N}^n$) defined by Macdonald [Astérisque No. 237 (1996), Exp. No. 797, 4, 189–207; MR 99f:33024] in Cherednik's affine Hecke algebra context. They exist for any root system; Knop's are for GL_n, in n variables X. Monomials in X represent formal exponentials of weights. Since affine weights are affine-linear functions on the co-root space, q appears as the exponential of a constant. The affine Hecke algebra $H_a(t)$ acts by a t-analog of the affine Weyl group action, as above. In $H_a(t)$ is a commuting family of t-translation operators Y^μ, which act triangularly on monomials and preserve Cherednik's scalar product. By definition the E_λ are their eigenfunctions. Symmetrizing over $H(t)$ collapses Cherednik's scalar product to Macdonald's, and the E_λ to the P_λ. Knop puts things in explicit polynomial terms for GL_n, and proves that suitable renormalizations $\mathscr{E}_\lambda$ of E_λ are polynomials in X, q, t. The result on J_λ follows. To prove that the $\mathscr{E}_\lambda$ are polynomials, Knop gives inductive formulas similar to those of Sahi for the G_λ. Indeed, Knop has shown elsewhere [Comment. Math. Helv. **72** (1997), no. 1, 84–100; MR 98m:05204] that E_λ is the top homogeneous component of G_λ.

The preceding proofs evolved from Cherednik's and Macdonald's work and treat P_λ in n variables as associated with the root system of GL_n and its Hecke algebra. In particular the methods of proof are specific to symmetric functions in a fixed finite number of variables.

The proofs by Garsia-Remmel and Garsia-Tesler, by contrast, use formal symmetric function theory in an infinite alphabet of variables. In this context the power sums p_k are algebraically independent and may be specialized at will. Given a formal Laurent series A the plethystic substitution $p_k[A]$ is defined by replacing each variable a in A by a^k. This extends to a unique ring homomorphism defining $f[A]$ for all f. If $X = x_1 + x_2 + \cdots$ then $f[X] = f(x_1, x_2, \cdots)$, and $f[X(1-t)]$ is as we defined it in the introduction.

For a partition μ set $B_\mu(q,t) = \sum t^i q^j$, the sum taken over the coordinates (i,j) of cells in the Ferrers diagram of μ, indexed from $(0,0)$, e.g. $B_{(2,2)} = (1+q)(1+t)$. The amazing fact is that $K_{\lambda\mu}$ depends plethystically on B_μ. Precisely, given a partition ν of d, there is a unique symmetric function k_ν, inhomogeneous of highest degree d, such that for all $k \geq \nu_1$ and all partitions μ of $d+k$, we have $\widetilde{K}_{(k,\nu)\mu}(q,t) = k_\nu[B_\mu]$, where $\widetilde{K}_{\lambda\mu} = t^{n(\mu)} K_{\lambda\mu}(q, 1/t)$ and $n(\mu) = \sum (i-1)\mu_i$. For example, the values of $\widetilde{K}_{\lambda\mu}$ for λ a hook shape, originally determined by Macdonald, take the simple form $k_{(1^r)}[B_\mu] = e_r[B_\mu - 1]$. The coefficients of k_ν are Laurent polynomials in q, t, hence so are the $K_{\lambda\mu}$. It was already known that the $K_{\lambda\mu}$ have no poles at $q = 0$ or $t = 0$, so it follows that they are polynomials.

Both proofs of the plethystic formulas involve explicit calculations with the Pieri rules to show that the Hall inner product of J_μ with $e_k f$ depends plethystically on μ, assuming the same by induction for symmetric functions f of lower degree. The proof by Garsia and Remmel is relatively simple and only treats the case $k = 1$. This suffices because the operator Δ whose eigenfunctions are J_μ with eigenvalue B_μ has an explicit plethystic expression, and by judiciously interspersing multiplications by e_1 with applications of Δ it is possible to construct coefficient formulas which span all linear functionals on symmetric functions, and are given on J_μ by plethystic evaluations. The price paid for simplicity in the Garsia-Remmel proof is that the construction introduces denominator factors of the forms $(1-q^r)$ and $(1-t^r)$ both. However, L. Lapointe and L. Vinet [Adv. Math. **130** (1997), no. 2, 261–279; MR 99e:05128] had already discovered extremely simple raising operators for J_μ which only contain denominator factors of the form $(1-t^r)$, $(q-t^r)$, q and t (this construction preceded the Kirillov-Noumi raising operators, which are more complicated but are polynomial). Together with duality these results leave only $(1-t)$, $(1-q)$, q, t as possible denominator factors of J_μ, and these are ruled out by specializations of Macdonald polynomials which were already known.

The proof by Garsia and Tesler is more difficult, as they carry out the plethystic description of the Hall inner products $\langle J_\mu, e_k f\rangle$ for all k, and the Pieri formulas involved are much more intricate than for $k = 1$. The reward for the extra work is that the operator Δ does not intervene in the computation of the coefficient formulas. As a result, the end product is an explicit raising operator formula for polynomials π_ν which give the inner products $\langle J_\mu, e_k e_\nu\rangle$ plethystically. The only denominators involve factors of the form $(1-t^r)$. This makes it possible to establish that the symmetric functions k_ν which yield $\widetilde{K}_{\lambda\mu}$ plethystically are themselves Laurent polynomials in q, t.

A notable feature of the different proofs is that at present, there are no perceptible links between the methods of Garsia, Remmel, and Tesler on the one hand, and Kirillov, Noumi, Knop, and Sahi on the other. We may expect this and much else to be clarified as this beautiful and astonishingly far-reaching subject continues to develop.

{For the entire collection containing the fourth paper see MR 99a:00055.}

Mark D. Haiman (La Jolla, CA)

97e:08002a 08B05
McKenzie, Ralph (Nashville, TN)
The residual bounds of finite algebras. (English summary)
Internat. J. Algebra Comput. **6** (1996), *no.* 1, 1–28.

97e:08002b 08B05 03C05 03D35
McKenzie, Ralph (Nashville, TN)
The residual bound of a finite algebra is not computable. (English summary)
Internat. J. Algebra Comput. **6** (1996), *no.* 1, 29–48.

97e:08002c 08B05 03C05 03D35
McKenzie, Ralph (Nashville, TN)
Tarski's finite basis problem is undecidable. (English summary)
Internat. J. Algebra Comput. **6** (1996), *no.* 1, 49–104.

These three papers settle several important, long-standing open questions in the equational theory of finite algebraic systems. Moreover, the innovative techniques introduced here have already proven useful in solving various other problems for finite algebras. Taken as a whole, these papers are a major contribution to our knowledge of general algebras.

The first two papers deal with the residual character of the variety (i.e., equational class) $\mathscr{V} = V(\mathbf{A})$ generated by a finite algebra $\mathbf{A}$. For an algebra $\mathbf{A}$, let $\kappa(\mathbf{A})$ denote the least cardinal λ such that every subdirectly irreducible algebra in $V(\mathbf{A})$ has cardinality $< \lambda$, and $\kappa(\mathbf{A}) = \infty$ if no such cardinal exists. If $\kappa(\mathbf{A}) \neq \infty$, then $\mathscr{V}$ is called residually small (RS). Since every algebra in a variety can be represented by its subdirectly irreducible homomorphic images, the invariant $\kappa(\mathbf{A})$ is of fundamental importance for understanding $\mathscr{V}$. In a 1971 paper [Algebra Universalis **1** (1971/72), 265–266; MR **45** #3295] R. W. Quackenbush posed a problem, which subsequently became known as the Quackenbush conjecture: If the variety $V(\mathbf{A})$ generated by a finite $\mathbf{A}$ has arbitrarily large finite subdirectly irreducible members, then it must have an infinite one. That is, for no finite $\mathbf{A}$ does $\kappa(\mathbf{A}) = \omega$. Work on this conjecture in the 1970's suggested an even stronger result, which was articulated in the mid 1980's as the RS conjecture: If $\mathbf{A}$ is finite with $\kappa(\mathbf{A}) \geq \omega$, then $\kappa(\mathbf{A}) = \infty$.

Among the results of the first paper is the unexpected refutation of the RS conjecture. It was known that if $\mathbf{A}$ is a nontrivial finite algebra generating a residually small variety, then $\kappa(\mathbf{A}) \in \{3, 4, \cdots, \omega, \omega_1, (2^\omega)^+\}$. For each cardinal λ in this set of potential residual bounds the author constructs a 4-element algebra $\mathbf{A}_\lambda$ with $\kappa(\mathbf{A}_\lambda) = \lambda$. In these constructions, the $\mathbf{A}_\lambda$ with finite residual bound are of finite type, that is, they have a finite number of basic operations, while the three with λ infinite have infinite type.

The final section of the first paper presents an 8-element algebra $\mathbf{C}$ that refutes the RS conjecture, even for algebras of finite type. The algebra $\mathbf{C}$ has precisely seven operations and $\kappa(\mathbf{C}) = \omega_1$.

The work of the past 25 years on the classification of subdirectly irreducible algebras in finitely generated varieties led to the belief that there would be some effective procedure to determine for a given finite algebra **A** of finite type whether or not $\kappa(\mathbf{A}) < \omega$. The second paper in this series puts this belief to rest. The author exhibits a method to associate with each Turing machine $\mathcal{T}$ a finite algebra $\mathbf{A}(\mathcal{T})$ of finite type such that $\kappa(\mathbf{A}(\mathcal{T})) < \omega$ if $\mathcal{T}$ halts, and $\kappa(\mathbf{A}(\mathcal{T})) = \omega_1$ if $\mathcal{T}$ does not halt. Thus, the problem of determining if a finite algebra generates a residually finite variety is undecidable. The algebra $\mathbf{A}(\mathcal{T})$ that the author constructs from $\mathcal{T}$ is based on the 8-element algebra **C** of the first paper. It has $20m + 8$ elements, where m is the number of states of $\mathcal{T}$.

A very rough description of how $\mathbf{A}(\mathcal{T})$ is constructed from $\mathcal{T}$ is the following. Given $\mathcal{T}$ that runs on a two-way infinite tape, an algebra $\mathbf{A} = \mathbf{A}(\mathcal{T})$ is built so that a configuration of $\mathcal{T}$ that uses an interval I of tape is coded as an element of $\mathbf{A}^I$. Here a configuration is what has been written on the tape together with the current position of the read/write head and the current internal state of $\mathcal{T}$. For a given I let **B** be the subalgebra of $\mathbf{A}^I$ generated by all such configurations contained in I that arise by starting from the initial state reading the empty tape. Among the operations of **A** are some such that if one configuration of $\mathcal{T}$ leads to another, then there is a polynomial of **A**, acting coordinatewise, that maps the corresponding elements of **B**, one to the other. Operations for **A** are also chosen so that the subdirectly irreducible algebras of interest in $V(\mathbf{A})$ are of the form $\mathbf{B}/\theta$ for θ a particular congruence in which elements of **B** that code the sequence of configurations of $\mathcal{T}$ are in separate θ classes. Yet more operations are added to **A** to force $\mathbf{B}/\theta$ to not be subdirectly irreducible if a halting configuration arises when computing within the tape interval I. Thus, if $\mathcal{T}$ halts, then only for small I will $\mathbf{B}/\theta$ be subdirectly irreducible and so $\kappa(\mathbf{A}) < \omega$. If $\mathcal{T}$ does not halt, then it is possible to find a **B** containing the coding of an infinite number of configurations, for which $\mathbf{B}/\theta$ is a countably infinite subdirectly irreducible algebra. Obviously, the selection of operations for **A** to make all this happen requires extraordinary care and insight.

The third paper settles a problem of A. Tarski dating back to the early 1960's: does there exist an algorithm which, when given an effective description of a finite algebra **A**, determines whether or not **A** has a finite basis for its equational theory? The author answers Tarski's question in the negative by providing a construction that produces for every Turing machine $\mathcal{T}$ a finite algebra $\mathbf{F}(\mathcal{T})$ such that $\mathcal{T}$ halts if and only if $\mathbf{F}(\mathcal{T})$ is finitely based. The algebra $\mathbf{F}(\mathcal{T})$ has $10m$ elements, where m is the number of states of $\mathcal{T}$, and has a finite number of at most 6-ary basic operations, the exact number of operations depending on the Turing machine $\mathcal{T}$. These algebras are, of course, very exotic. However, using a type reduction theorem that the author proved in an earlier paper [Algebra Universalis **18** (1984), no. 1, 29–69; MR 86h:08011], the following version, in a more familiar setting, is obtained. Let $\mathcal{K}$ be the set of all algebras $\langle A, \circ \rangle$ for A a finite nonvoid set of integers and $\circ$ a binary operation on A. Then the set of finitely based members of $\mathcal{K}$ is not recursive, while the set of non-finitely based and the set of inherently non-finitely based members of $\mathcal{K}$ are not recursively enumerable.

The results in these three papers alter major portions of the landscape of equational logic. In the introduction to each paper the author briefly surveys this change in our knowledge, and raises some questions that emerge or remain open. Among the many questions posed in the first paper, the most prominent is the still open, strong version of the Quackenbush problem. Does there exist a finite algebra **A** of finite type with $\kappa(\mathbf{A}) = \omega$? For $\mathscr{K}$ the set of algebras defined in connection with the third paper, the author raises the problem of determining the precise Turing degree of the set of finitely based members of $\mathscr{K}$. In particular, is this set recursively enumerable?

The new techniques and constructions developed in these three papers have already been used by the author and others to prove new results that were hitherto considered unobtainable. M. Valeriote settled in the negative the analogue of the RS conjecture for simple algebras in a variety, and thereby provided a negative solution for Problem 11 of the Hobby and McKenzie monograph [D. Hobby and R. N. McKenzie, *The structure of finite algebras*, Contemp. Math., 76, Amer. Math. Soc., Providence, RI, 1988; MR 89m:08001]. R. Willard has done the same for the analogue of the Quackenbush conjecture for simple algebras. The introduction to the second paper under review raises the question of whether "$V(\mathbf{A})$ is residually small" is a decidable problem for finite algebras **A**. The author has since answered this question; it is undecidable. In 1991 Hobby [Internat. J. Algebra Comput. **1** (1991), no. 4, 437–444; MR 93d:08008] proved that the problem in tame congruence theory of determining for a finite algebra **A** if $\mathbf{3} \in \operatorname{typ}\{V(\mathbf{A})\}$ is PSPACE-hard, and he raised the question of whether this is a decidable problem. McKenzie has shown that indeed the problem is undecidable. Recently, J. Wood has announced that the same problem for type **4** is also undecidable. An alternate proof of the undecidability of the Tarski problem has been given by Willard. He argues that the algebras $\mathbf{A}(\mathscr{T})$ constructed in McKenzie's second paper are finitely based if $\mathscr{T}$ halts, and they were already known to be non-finitely based otherwise. Willard has recently constructed, on a 9-element set A, a sequence of operations f_n for $n < \omega$ such that if $\mathbf{A}_n = \langle A, (f_i\colon\ i \leq n)\rangle$, then the algebra $\mathbf{A}_{2n}$ is finitely based while $\mathbf{A}_{2n+1}$ is inherently non-finitely based. This settles a problem of G. McNulty that dates from 1982. Willard's construction subtly plays off properties of **C**, $\mathbf{A}(\mathscr{T})$, and the 4-element algebras $\mathbf{A}_\lambda$ of the first two papers under review to obtain his oscillating sequence of algebras. *Joel Berman* (Chicago, IL)

99a:12009 12H25 14F30 14G20

Christol, G. (Paris); **Mebkhout, Z.** (Paris)

Sur le théorème de l'indice des équations différentielles p-adiques. II. [On the index theorem for p-adic differential equations. II]

Ann. of Math. (2) **146** (1997), *no.* 2, 345–410.

B. Dwork's seminal work on the zeta function of a hypersurface [Inst. Hautes Études Sci. Publ. Math. No. **12** (1962), 5–68; MR **28** #3039] and its subsequent developments illuminated the key role played by the study of p-adic differential equations for an understanding of the p-adic cohomology of a variety in characteristic p. This work also made it clear that the appropriate domain for the associated differential operators was the set of overconvergent functions [see P. Berthelot, Mém. Soc. Math. France (N.S.) No. 23 (1986), 3, 7–32; MR 88a:14020].

After Dwork's pioneering work, many further results were established, but the main point, a proof of the finiteness of p-adic cohomology, remained open. The paper under review is part of an ambitious and successful project aimed at such a proof. The proof given is in the spirit of the original insight of Dwork and Monsky [P. Monsky and G. Washnitzer, Ann. of Math. (2) **88** (1968), 181–217; MR **40** #1395; P. Monsky, Ann. of Math. (2) **93** (1971), 315–343; MR **48** #296] that a complete knowledge of differential p-adic equations and, more generally, of what can be called "p-adic coefficients", lies at the base of the study of p-adic cohomology. The complete series of three articles by the authors [Part I, Ann. Inst. Fourier (Grenoble) **43** (1993), no. 5, 1545–1574; MR 95j:12009; "Sur le théorème de l'indice des équations différentielles p-adiques. III", to appear; see also F. Loeser, Astérisque No. 245 (1997), Exp. No. 822, 3, 57–81; MR 99f:12006] is devoted to giving a profound study of the p-adic differential equations with a view to p-adic cohomology and the associated index problems. Finally, Mebkhout [Amer. J. Math. **119** (1997), no. 5, 1027–1081; MR 99a:14027, see p. B45 in this volume] applied those results to give a proof of finiteness.

In the article under review the two authors give an answer to the following question: what is the appropriate p-adic analogue of a regular singularity of a differential operator in the complex case? To understand the problem one should take as a local situation a free differential module of rank m defined over an annulus of the type $\{x \in K \mid 1-\varepsilon < |x| < 1\}$ ($\varepsilon \in \mathbf{R}$, and K is an ultrametric field). A deep study of such differential modules with regard to the finiteness of the index was initiated in the 1980s by P. Robba. Lacking a sufficiently evolved theory of p-adic monodromy, he gave a totally p-adic condition for the conjectural p-adic analogue of a regular differential module in the complex case: the maximality of the radius of convergence of the solutions at the generic disk of radius r for $r \in (1-\varepsilon', 1)$. This condition is now usually referred to as Robba's condition. However, why such a condition should have been regarded as the analogue of regularity was justified only by some examples. For all of this, see Robba and Christol [*Équations différentielles p-adiques*, Hermann, Paris, 1994; MR 97g:12005].

The article under review shows why Robba's condition is, in fact, the correct one in relation to what one might expect (in a naive way) by analogy with the classical case. For p-adic differential operators satisfying Robba's condition the authors use an approximation process based on the Frobenius antecedent to define a notion of exponents. In general the exponents are elements of $(\mathbf{Z}_p/\mathbf{Z})^m$ modulo a nontrivial equivalence relation. Because of the possibility of "arithmetic" pathologies, when the differences of the exponents are non-Liouville, they are able to associate an element $(\alpha_1, \cdots, \alpha_m) \in (\mathbf{Z}_p/\mathbf{Z})^m$ (up to permutation) which coincides with the formal exponents. Indeed they go further and show that such a module has a basis in the space of overconvergent analytic functions (i.e. ε is no longer fixed) which represents the module as an iterated extension of modules of the type $d/dx - \alpha_i/x$, $\alpha_i \in \mathbf{Z}_p$, i.e. with only one regular (in the classical sense) singularity at zero. In fact they can show even more: there is a basis in which the action of the derivation is given by A/x, where A is a constant matrix with entries in K. As a corollary one has the finiteness of the index for operators satisfying Robba's condition. As one might expect from such profound results, their theory also has other applications. For example, they are also able to establish a p-adic analogue of Riemann's existence theorem: if an overconvergent differential module defined on a curve satisfies Robba's condition on each residue class and if the differences of the exponents in each class are non-Liouville then it is algebraic with only logarithmic poles and the formal exponents coincide with the p-adic ones. They propose the category of such p-adic differential modules as the correct analogue of the category of vector bundles over a Riemann surface in the complex case, or, in the étale world, the category of smooth tamely ramified l-adic sheaves, and hope to define the p-adic analogue of monodromy.

The article is very deep and is full of very interesting results. It will be a central work for all future studies on p-adic differential equations.

Bruno Chiarellotto (Padua)

99c:13005 13A35 13D45

Lyubeznik, Gennady (Minneapolis, MN)

F-modules: applications to local cohomology and D-modules in characteristic $p > 0$.

J. Reine Angew. Math. **491** (1997), 65–130.

The structure of Grothendieck's local cohomology modules $H_I^i(M)$ $(i \geq 0)$, where I is an ideal of a commutative Noetherian ring R and M is an R-module, is still full of mystery: it is hard enough to tell whether a given local cohomology module is zero; and when it is non-zero, it is rarely finitely generated, even if M is. Therefore, finiteness properties, which somehow "bound" local cohomology modules, are interesting because they improve our understanding of the structure of such modules.

C. Huneke and the reviewer [Trans. Amer. Math. Soc. **339** (1993), no. 2, 765–779; MR 93m:13008] used the Frobenius homomorphism, very much in the spirit of C. Peskine and L. Szpiro's pioneering work [Inst. Hautes Études Sci. Publ. Math. No. 42 (1973), 47–119; MR **51** #10330], to establish the following finiteness results for the local cohomology modules $H_I^i(R)$ $(i \geq 0)$ of R itself, when R is a regular ring containing a field of characteristic $p > 0$. (i) For each maximal ideal $\mathfrak{m}$ of R, the local cohomology module $H_{\mathfrak{m}}^j(H_I^i(R))$ is injective (for all $j \geq 0$). (ii) The injective dimension of $H_I^i(R)$ satisfies $\operatorname{inj\,dim} H_I^i(R) \leq \dim H_I^i(R)$. (iii) The set of associated primes of $H_I^i(R)$ is finite. (iv) All the Bass numbers of $H_I^i(R)$ are finite.

It is natural to ask whether the analogues of (i)–(iv) hold over a regular ring containing a field of characteristic 0. Lyubeznik [Invent. Math. **113** (1993), no. 1, 41–55; MR 94e:13032] used D-modules in an imaginative way to establish those analogues. In fact, he proved more: he established the corresponding statements in which the functor H_I^i is replaced by a functor T from a larger class of functors which, although too complicated to describe in detail here, includes compositions of finitely many local cohomology functors.

This, in turn, raised the question of whether the results (i)–(iv) of Huneke and the reviewer (in characteristic $p > 0$) could be extended to this larger class of functors. In the paper under review, Lyubeznik achieves this as one of his striking applications of a new theory of F-modules.

An F-module over a regular (commutative Noetherian) ring R containing a field of characteristic $p > 0$ is an R-module M endowed with an R-isomorphism $M \cong F(M)$, where $F(M)$ is the result of applying to M the Frobenius functor F of Peskine and Szpiro. The concept of F-module is the central topic of study in the paper under review; the special class of F-finite F-modules is accorded a lot of attention, and a major result established in §3 is that, when R is a finitely generated algebra over a regular local ring, then every F-finite F-module has finite length in the category of F-modules. This result plays a crucial rôle in the applications of the theory in the final two sections.

One of these applications is to cofinite $A[f]$-modules, where A is a Noetherian local ring of prime characteristic $p>0$ and $A[f]$ is the subring of $\mathrm{Hom}_{\mathbf{Z}}(A,A)$ generated by the multiplications by elements of A and the Frobenius homomorphism $f\colon A\to A$: to say that a left $A[f]$-module is cofinite is to say that it is Artinian as an A-module. Lyubeznik uses his new theory of F-modules, in conjunction with Cohen's structure theorem for complete local rings and Matlis duality, to obtain striking results about cofinite $A[f]$-modules; these are then used, in turn, to shed new light on work of R. Hartshorne and R. Speiser [Ann. of Math. (2) **105** (1977), no. 1, 45–79; MR **56** #353].

This review only hints at the wealth of ideas in this important work. The reader is encouraged to look at the paper. *Rodney Y. Sharp* (Sheffield)

98c:14004 14C15 14C35 14C40 14F20 19E08 19E15
Saito, Takeshi (Meguro); **Terasoma, Tomohide** (Meguro)
Determinant of period integrals.
J. Amer. Math. Soc. **10** (1997), *no.* 4, 865–937.

The paper is devoted to a general formula for the determinant of period integrals. The first main result, Theorem 1 of Section 4, expresses the determinant of periods of an integrable connection on an open U of a smooth projective variety X of dimension n defined over a subfield of $\mathbf{C}$ as the product of the following data: the determinant of periods of the trivial rank-1 connection; gamma factors involving residues of the connection along the divisor $D=X\setminus U$, assumed to be with normal crossings; the result of the pairing of the determinant of the connection with the relative canonical cycle (a top Chern class) in the relative Chow group $\mathrm{CH}^n(X,D)$.

This result is really the Hodge-de Rham analogue of results obtained by Saito [J. Algebraic Geom. **3** (1994), no. 3, 411–434; MR 95f:14036] in the l-adic setting. Using Theorem 1 and the l-adic results of Saito, the authors prove (Theorem 2 of Section 5) a motivic analogue of Theorem 1, assuming the validity of Gersten's conjecture. In particular, they introduce the notion of being "determinantally motivic" and prove it is stable under taking the determinant of the cohomology. As a special case, the category of algebraic Hecke character motives is proven to be stable under taking the determinant of the cohomology. This important result gives support to a conjecture of P. Deligne asserting that a motive of rank 1 should always be associated to an algebraic Hecke character [Proc. Sympos. Pure Math., XXXIII, in *Automorphic forms, representations and L-functions (Proc. Sympos. Pure Math., Oregon State Univ., Corvallis, Ore., 1977), Part 2*, 313–346, Amer. Math. Soc., Providence, R.I., 1979; MR 81d:12009].

An important part of the work consists in defining the above data. The determinant of periods of a connection and the gamma factors at infinity are defined in Section 1. Section 2 is devoted to the relative Chow group $\mathrm{CH}^n(X,D)$, defined as $\mathrm{CH}^n(X,D)=H^n(X,\mathscr{K}_n(X,D))$, where $\mathscr{K}_n$ stands for the sheaf of Quillen's K-groups, $\mathscr{K}_n(X,D)$ is the complex $\mathscr{K}_n(X)\to\bigoplus_i\mathscr{K}_n(D_i)$, and the D_i's are the components of the normal crossing divisor D. The authors give the definition of the top Chern class of a locally free $\mathscr{O}_X$-module trivialized on D and its basic properties. This extends previous constructions by Saito [Invent. Math. **113** (1993), no. 2, 389–417; MR 94m:14025]. The main result of this section is that $\mathrm{CH}^n(X,D)$ may be obtained as the cokernel of a complex $\bigoplus_{y\in X_1}B_y\to\bigoplus_{x\in X_0}A_x$. Here A_x is an extension of $\mathbf{Z}$ by $\bigoplus_{i\in I_x}\kappa(x)^\times$

(I_x is the set of i's such that x belongs to D_i and $\kappa(x)$ is the residue field), B_y is defined in terms of K_2 and the map is defined in terms of tame symbols. The proof of this result, which is purely K-theoretic and quite technical, has been put into an appendix. Section 3 is devoted to the construction of a pairing between the group of isomorphism classes of rank-one connections on U and $\mathrm{CH}^n(X,D)$. The previous adelic description of $\mathrm{CH}^n(X,D)$ is used here in an essential way. These constructions and results might turn out to be useful for other purposes. In Section 4 the authors are finally able to prove Theorem 1. The proof is much in the spirit of the one by Saito [op. cit.; MR 94m:14025]: by using Lefschetz pencils and functoriality of the constructions one reduces to the case $X=\mathbf{P}^1$, which is essentially equivalent to a product formula for periods due to Terasoma [Math. Ann. **298** (1994), no. 4, 577–589; MR 95b:32035]. Section 5 is devoted to a motivic analogue of Theorem 1. The authors introduce and study an étale version of the tame symbol which plays an important role in the proof and the statement of the result. In the last section a refinement of Theorem 1 is proven in the case of curves. In this situation the authors are able not only to prove that two objects are isomorphic but also to construct a canonical isomorphism between them via Deligne's Riemann-Roch theorem [in *Current trends in arithmetical algebraic geometry (Arcata, Calif., 1985)*, 93–177, Contemp. Math., 67, Amer. Math. Soc., Providence, RI, 1987; MR 89b:32038].

François Loeser (Paris)

98a:14012 14C25 55N22

Totaro, Burt (Chicago, IL)

Torsion algebraic cycles and complex cobordism.

J. Amer. Math. Soc. **10** (1997), *no.* 2, 467–493.

Let X be a smooth, projective variety of dimension n over the complex numbers. The Chow group, $\mathrm{CH}^r(X)$, of codimension r algebraic cycles on X modulo rational equivalence is a fundamental invariant of X. The paper under review is concerned primarily with the relationship between $\mathrm{CH}^r(X)$ and the cohomology of X. The most exciting new result pertains to the torsion subgroup of the Chow group. The usual cycle class map to Deligne cohomology gives rise to a cycle class map $\lambda^r\colon \mathrm{CH}^r(X)_{\mathrm{tors}} \to H^{2r-1}(X,\mathbf{Q}/\mathbf{Z}(r))$. On torsion cycles which are homologous to zero this may be identified with the Abel-Jacobi map studied by Griffiths. It was known classically that λ^1 is an isomorphism. More recently, work of Roĭtman established that λ^n is an isomorphism as well. Injectivity for λ^2 has been shown by Merkur′ev and Suslin. One was tempted to speculate that λ^r would be injective for all r. The present paper shows that this is in fact false. This is done by factoring the usual cycle class map to integral cohomology through a group constructed from complex cobordism.

Complex cobordism is a tool which has rarely been used to study algebraic cycles. Consequently the paper begins with a brief summary of this and related topological theories. A central role is played by the Borel-Moore complex bordism group, $M\mathrm{U}_*^{\mathrm{BM}}(S)$, which the author defines for any sufficiently well-behaved topological space S. If S is a complex manifold, then $M\mathrm{U}_*^{\mathrm{BM}}(S)$ is Poincaré dual to the complex cobordism of S. In any case, $M\mathrm{U}_*^{\mathrm{BM}}(S)$ is a module over the complex bordism ring of a point $M\mathrm{U}_*$. When Y is an arbitrary complex algebraic variety the author defines a cycle class map $\mathrm{cl}_i\colon Z_i(Y)\to M\mathrm{U}_{2i}^{\mathrm{BM}}(Y)\otimes_{M\mathrm{U}_*}\mathbf{Z}$ and a natural map $b_{2i}\colon M\mathrm{U}_{2i}^{\mathrm{BM}}(Y)\otimes_{M\mathrm{U}_*}\mathbf{Z}\to H_{2i}^{\mathrm{BM}}(Y,\mathbf{Z})$ such that the composition is the usual cycle class map to Borel-Moore homology. This uses resolution of singularities and a fundamental theorem of Quillen about complex cobordism. An extension of Quillen's theorem is employed to show that cycles which are algebraically equivalent to zero map to zero. The compatibility of the new cycle class map with proper direct image and with intersection products (when Y is smooth) is established.

In order to find explicit examples of smooth projective varieties for which λ^3 is not injective the author searches for varieties for which the map b_* is not injective and the intermediate Jacobian is zero. Satisfying these conditions is a difficult task. In fact, b_* is an isomorphism whenever the cohomology is torsion-free and examples where b_* is not surjective seem to be easier to come by than examples where b_* is not injective. Eventually the author shows that b_{24} is not injective for the quotient of a fifteen-dimensional complete intersection in $\mathbf{P}^N$ by the free action of a certain finite group. The clever and highly sophisticated proof is based in part on earlier work of Atiyah and Hirzebruch and an analysis of the complex cobordism of the classifying spaces of various groups. To deduce that λ^3 is not injective, a two-torsion element of the Chow group whose class in the Borel-Moore complex bordism group is a nonzero element of the kernel of b_{24} is constructed from Chern classes of vector bundles with finite structure group. Both the cycle and the variety may be defined over $\mathbf{Q}$.

The paper ends with a discussion of some interesting open problems.

Chad Schoen (Durham, NC)

98b:14006 14C30 14C35 19F27 57M50 57R20

Reznikov, Alexander (Jerusalem)

Rationality of secondary classes. (English summary)

J. Differential Geom. **43** (1996), *no.* 3, 674–692.

In this short but spectacular paper, the author proves Bloch's conjecture that for any smooth complex projective variety, and any representation of its fundamental group into $\mathrm{SL}(2,\mathbf{C})$, the second Chern class in Deligne cohomology of the associated holomorphic bundle is torsion. The author later proved more generally that for all flat bundles on a smooth projective variety, all the Chern classes in Deligne cohomology except the first one are torsion [Ann. of Math. (2) **141** (1995), no. 2, 373–386; MR 96a:14012]. An even stronger conjecture of Bloch's says that the ith Chern class in the Chow ring of a flat bundle on a smooth projective variety X should be in the bottom group of the conjectural $(i+1)$-step filtration of $\mathrm{CH}^i(X)\otimes\mathbf{Q}$: thus c_1 should be 0 in rational cohomology, c_2 should be 0 in rational Deligne cohomology, "and so on" [S. Bloch, in *Proceedings of the International Symposium on Algebraic Geometry (Kyoto Univ., Kyoto, 1977)*, 103–114, Kinokuniya Book Store, Tokyo, 1978; MR 82f:14009]. A reference for the conjectural filtration on Chow groups is a paper by U. Jannsen [in *Motives (Seattle, WA, 1991)*, 245–302, Proc. Sympos. Pure Math., 55, Part 1, Amer. Math. Soc., Providence, RI, 1994; MR 95c:14006].

The Chern classes of a flat bundle in Deligne cohomology are the images of its Cheeger-Chern-Simons classes $\widehat{c}_i\in H^{2i-1}(X,\mathbf{C}/\mathbf{Z})$, which are defined for flat bundles on any space. So the main theorem follows from the stronger result that the Cheeger-Chern-Simons class $\widehat{c}_2\in H^3(X,\mathbf{C}/\mathbf{Z})$ is torsion for representations $\mu\colon\pi_1(X)\to\mathrm{SL}(2,\mathbf{C})$ when X is a smooth complex projective variety. This statement can be divided into two parts, the rationality of the "Chern-Simons" class in $H^3(X,\mathbf{R}/\mathbf{Z})$ and the vanishing of the "Borel regulator" class $\mathrm{Vol}(\mu)\in H^3(X,\mathbf{R})$.

The proof runs as follows. The Cheeger-Chern-Simons classes except for the first one do not change under continuous deformation of a representation of $\pi_1(X)$. So the given representation can be deformed without changing the Cheeger-Chern-Simons class to a representation into $\mathrm{SL}(2,\overline{\mathbf{Q}})$. This representation actually goes into $\mathrm{SL}(2,O_S)$ for some ring of S-integers in a number field F, which we can assume to have no real embeddings. Using Borel's theorem that $H^3(\mathrm{SL}(O_S),\mathbf{R})$ is generated by the pullbacks of the Borel regulator $\mathrm{Vol}\in H^3(\mathrm{SL}(\mathbf{C}),\mathbf{R})$ under the various embeddings of F into $\mathbf{C}$, the author reduces the problem of rationality of the Chern-Simons class in $H^3(X,\mathbf{R}/\mathbf{Z})$ to the problem of vanishing of the Borel regulator class $\mathrm{Vol}\in H^3(X,\mathbf{R})$, for all representations of the fundamental group of X in $\mathrm{SL}(2,\mathbf{C})$.

To prove the vanishing of the Borel regulator class Vol for such representations, these number-theoretic methods are not used at all. Instead, the author shows that either a representation can be deformed into the semidirect product of SO(2) with $\mathbf{R}^2$, in which case Vol pulls back to 0 on X, or else, by the powerful analytic results of Donaldson and Corlette (which work for any Riemannian manifold), the natural hyperbolic 3-space bundle over X associated to the representation of the fundamental group in $\mathrm{SL}(2,\mathbf{C})$ has a harmonic section. Following Sampson, the Kähler property of X implies that this harmonic map has rank at most 2 everywhere. It follows that $\mathrm{Vol}=0$ in $H^3(X,\mathbf{R})$ in this case as well, which completes the proof.

Finally, the end of the paper uses related ideas to prove the following results: (1) For any compact n-dimensional hyperbolic manifold M and any representation $\mu\colon\pi_1(M)\to\mathrm{PSO}(1,n)$, $\mathrm{Vol}(\mu)\le\mathrm{Vol}(M)$. (2) For any Seifert fibered 3-manifold M and any representation $\mu\colon\pi_1(M)\to\mathrm{SL}(2,\mathbf{C})$, the Chern-Simons invariant of μ in $\mathbf{R}/\mathbf{Z}$ is rational. (3) For any compact arithmetic hyperbolic 3-manifold M with defining representation $\mu\colon\pi_1(M)\to\mathrm{PSL}(2,\mathbf{C})$, the Chern-Simons invariant of μ in $\mathbf{R}/\mathbf{Z}$ is rational. The arithmetic hyperbolic 3-manifolds considered here are those usually called "real-arithmetic". Some more general arithmetic hyperbolic 3-manifolds are expected to have irrational Chern-Simons invariant [W. D. Neumann and J. Yang, Enseign. Math. (2) **41** (1995), no. 3-4, 281–296; MR 97h:57038].

Burt Totaro (Cambridge)

98d:14012 14C35 14F99 19E08 19E15

Gillet, H. (Chicago, IL); **Soulé, C.** (Bures-sur-Yvette)

Descent, motives and K-theory.

J. Reine Angew. Math. **478** (1996), 127–176.

This fine paper draws interesting links among the three topics in the title.

The method of descent computes invariants of a scheme X by passing to another scheme X', or a family (X_i) or a diagram $i\mapsto X_i$ of schemes. This is well known for Čech cohomology, where the sheaf cohomology on X is computed by using an open covering $(U_i)_{i\in I}$ of X. The method of cohomological descent uses a different type of covering, namely proper and surjective morphisms $X'\to X$, to compute the cohomology of X by passing to a suitable diagram of schemes, usually a simplicial scheme. This is, to date, the only way to define a mixed Hodge structure on the singular cohomology of an arbitrary complex variety X, viz., one has to use

Hironaka's resolution of singularities to "resolve" X by a simplicial variety $X_\cdot$ with smooth and proper components. It is well known that such coverings do not work for algebraic K-theory: applied to a finite Galois cover $X' \to X$, this would imply Galois descent for K-theory, which does not hold. But the first-named author has shown [in *Algebraic K-theory, number theory, geometry and analysis (Bielefeld, 1982)*, 80–103, Lecture Notes in Math., 1046, Springer, Berlin, 1984; MR 86a:14016] that algebraic K'-theory (associated to coherent sheaves) satisfies descent for so-called envelopes, i.e., proper and surjective morphisms which have the lifting property for points in arbitrary fields. Fortunately, Hironaka's theorem even implies that every variety X over a field of characteristic zero has an envelope $X' \to X$ with smooth X'.

Using this technique, the authors of the present paper associate to any variety X over a field of characteristic zero a bounded complex $W(X)$ in the additive category $\mathbf{M}$ of motives which up to homotopy is unique and contravariant for proper morphisms. In fact, everything is already obtained in the category $\mathbf{C}$ of correspondences, which is the following linearization of the category $\mathbf{V}$ of smooth and proper varieties: the objects of $\mathbf{C}$ are the same as in $\mathbf{V}$, but the morphisms are algebraic cycles on $X \times Y$ of a suitable codimension, modulo rational equivalence. The composition $\mathrm{Hom}_{\mathbf{C}}(X, Y) \times \mathrm{Hom}_{\mathbf{C}}(Y, Z) \to \mathrm{Hom}_{\mathbf{C}}(X, Z)$ is defined by pulling back to $X \times Y \times Z$, intersecting, and pushing forward to $X \times Z$. There is a functor from $\mathbf{V}$ to $\mathbf{C}$ sending a morphism $f: X \to Y$ to its graph $\Gamma_f \subset X \times Y$. Grothendieck's category of (pure, effective) motives $\mathbf{M}$ is obtained from $\mathbf{C}$ by taking the Karoubian envelope, so that objects are pairs (X, p) with X in $\mathbf{V}$ and p an idempotent in $\mathrm{End}_{\mathbf{C}}(X)$ (cf. A. J. Scholl [in *Motives (Seattle, WA, 1991)*, 163–187, Proc. Sympos. Pure Math., 55, Part 1, Amer. Math. Soc., Providence, RI, 1994; MR 95b:11060] for a recent overview on motives). All the usual cohomology theories on $\mathbf{V}$ extend to $\mathbf{M}$, and Grothendieck's philosophy is that $X \mapsto M(X) = (X, \mathrm{id}_X)$ is the universal cohomology theory on $\mathbf{V}$. In any case, every relation in $\mathbf{M}$ implies one for all so-called Weil cohomology theories, and the use of $\mathbf{M}$ allows one to define interesting summands in "the" cohomology of X; e.g., the projective bundle theorem is the isomorphism $M(\mathbf{P}_X(\mathscr{E})) = \bigoplus_{i=0}^{r} M(X) \otimes L^i$ for a vector bundle $\mathscr{E}$ of rank r on X (where L is the Lefschetz motive, defined by $M(\mathbf{P}^1) = 1 \oplus L$), and classical modular forms f give rise to two-dimensional motives $M(f)$ over $\mathbf{Q}$ of great arithmetic interest.

There is the wish, going already back to Grothendieck, to extend the theory of motives to the category $\mathbf{V}'$ of all varieties (= separated algebraic schemes over the ground field), by embedding $\mathbf{M}$ in a bigger category $\mathbf{MM}$ of mixed motives. $\mathbf{MM}$ should reflect mixed structures in the cohomology, i.e., nontrivial extensions (as for mixed Hodge structures which arise only from $\mathbf{V}'$ and not from $\mathbf{V}$). Moreover, according to ideas of Deligne and Beĭlinson, $\mathbf{MM}$ (or its derived category) should play the same role for cohomology theories on $\mathbf{V}'$ that spectra play for universal cohomology theories on topological spaces. In particular, algebraic K-groups should be related to Ext-groups in $\mathbf{MM}$. But up to now, all attempts to define such a category $\mathbf{MM}$ have failed, although there are some interesting candidates, especially for $D^b(\mathbf{MM})$. This certainly provides some interest in the complexes $M(X)$ for any X in $\mathbf{V}'$ which are defined unconditionally in the paper under review.

In particular, by passing to the Grothendieck group $K_0(\mathbf{M})$, for any X in $\mathbf{V}'$ the complex $M(X)$ gives rise to a class $[X]$ in $K_0(\mathbf{M})$ which has the property that $[X] =$

$[Y]+[X-Y]$ for a closed subscheme $Y \subseteq X$. This answers positively a question of J.-P. Serre [Astérisque No. 198-200 (1991), 11, 333–349 (1992); MR 92m:14002]. As an application of the complexes $M(X)$ themselves, one gets the surprising fact that there is a well-defined integrally defined weight filtration on the cohomology with compact support $H_c^i(X, \mathbf{Z})$ of a complex variety X which gives the weight filtration of the mixed Hodge structure after extending scalars to $\mathbf{Q}$, but cannot be deduced from it. As a particularly striking application, one can "derive" any functor Γ on $\mathbf{C}$ with values in an abelian category to a "cohomology theory" $R^i\Gamma$ on $\mathbf{V}'$ such that one has the long exact sequence $\cdots \to R^i\Gamma(X-Y) \to R^i\Gamma(X) \to R^i\Gamma(Y) \to R^{i+1}\Gamma(X-Y) \to \cdots$ for $Y \subseteq X$ closed. This was used by the reviewer in his recent proof of a Hasse principle conjectured by K. Kato, and in a preprint of J. Denef and F. Loeser ["Motivic Igusa zeta functions", Jussieu Preprint No. 86, Univ. Paris VI and VII, Paris, 1996; per revr.] on Igusa type zeta functions. In down-to-earth terms, it gives the following result on resolutions. Let U be a smooth variety of dimension d, and let $X \supseteq U$ be smooth and proper such that $Y = X - U$ is a divisor with components Y_i, crossing normally. Then the homology of the complex $\Gamma(Y^{(d)}) \to \Gamma(Y^{(d-1)}) \to \cdots \to \Gamma(Y^{(1)}) \to \Gamma(X)$ depends only on U and not on (X, Y) (here $Y^{(p)} = \coprod Y_{i_1} \cap \cdots \cap Y_{i_p}$, where the disjoint union runs over all indices $i_1 < \cdots < i_p$), and a similar result can be stated for arbitrary varieties and so-called hyperenvelopes. This gives highly interesting cohomology theories, even starting from $\Gamma(X) = \mathrm{CH}^0(X) = \mathbf{Z}^{\pi_0(X)}$, where $\pi_0(X)$ is the set of connected components of X in $\mathbf{V}$.

An important ingredient in the proof is the use of the Gersten complexes $R_{q,\bullet}(X)$. Here $R_{q,i}(X) = \bigoplus K_i(\kappa(x))$, where the sum is over all points $x \in X$ of dimension $q+i$, and $K_i(\kappa(x))$ is the ith algebraic K-group of the residue field $\kappa(x)$ of x [D. Quillen, in *Algebraic K-theory, I: Higher K-theories (Proc. Conf., Battelle Memorial Inst., Seattle, Wash., 1972)*, 85–147, Lecture Notes in Math., 341, Springer, Berlin, 1973; MR **49** #2895]. There is a spectral sequence abutting in the K-theory associated to coherent sheaves on X, and $H_0(R_{q,\bullet}(X))$ is just Fulton's homological Chow group $\mathrm{CH}_q(X)$ of cycles of dimension q [H. Gillet, Adv. in Math. **40** (1981), no. 3, 203–289; MR 83m:14013]. The well-known "Manin identity principle" for motives—which is in fact just a version of the Yoneda lemma—is generalized as follows: A complex $\cdots \xrightarrow{\alpha_2} X_1 \xrightarrow{\alpha_1} X_0$ in $\mathbf{C}$ whose differentials α_i are just a $\mathbf{Z}$-linear combination of morphisms in $\mathbf{V}$ is contractible in $\mathbf{C}$ if (and only if) the total complex associated to the double complex $R_{q,\bullet}(Y \times X_\bullet)$ is acyclic for all $q \geq 0$ and all Y in $\mathbf{V}$. This provides the link between the descent property for hyperenvelopes of K-theory mentioned above, and the theory of motives.

In Chapter 4 (which is essentially independent of the rest) the authors study the behaviour of various Chow and K-groups under blow-ups. They point out that there is an alternative construction of $W(X)$ due to F. Guillén Santos and V. Navarro Aznar ["Un critère d'extension d'un foncteur défini sur les schémas lisses", Preprint, Barcelona, 1995; per bibl.] which is based on blow-ups and cubic hyperresolutions [F. Guillén Santos et al., *Hyperrésolutions cubiques et descente cohomologique (Barcelona, 1982)*, Lecture Notes in Math., 1335, Springer, Berlin, 1988; MR 90a:14024] and also gives "cohomological" counterparts of the "homological" $W(X)$. In the final chapter the authors treat a variant of motives, replacing Chow groups by K_0 in the definition of correspondences, thus defining so-called K-groups with compact supports.

A special merit of the paper is that the authors are extremely careful in introducing the topics, stating the results, writing rigid and intelligible proofs and quoting precisely from the literature. This is in pleasant contrast to habits that unfortunately have spread in mathematics, and makes the paper readable for non-experts and a reliable reference for experts. In conclusion, in this beautiful paper the expert will find an unconditional result on arbitrary varieties and motives which has concrete and useful applications, and the non-expert will encounter some important objects and techniques in the area of motives and K-theory, safely guided through the daunting technicalities by the kind authors. *Uwe Jannsen* (Cologne)

97e:14015 14D07 14D05 14J32 14N10 32G20
Givental, Alexander B. [Givental′, A. B.] (Berkeley, CA)
Equivariant Gromov-Witten invariants.
Internat. Math. Res. Notices **1996,** *no.* 13, 613–663.

This outstanding and extremely beautiful article provides a description of the quantum periods of a complete intersection $X \subset \mathbf{P}^n$ with $K_X \leq 0$ in terms of solutions of a certain Picard-Fuchs equation. This correspondence generalizes the one predicted by mirror symmetry [cf. P. Candelas et al., Nuclear Phys. B **359** (1991), no. 1, 21–74; MR 93b:32029]: roughly speaking, mirror symmetry says that certain functions constructed from symplectic invariants of X (or from its enumerative geometry) can be computed from the variation of Hodge structure parametrized by the family of deformations of a mirror variety X^*. Givental's work not only proves mirror symmetry in its strongest form for Calabi-Yau complete intersections, but also enlarges this circle of ideas to include non-Calabi-Yau varieties.

To explain the result more precisely: Given such X, the Gromov-Witten invariants $\Phi_{d,n}(\varphi^{\otimes n})$, $\varphi \in H^*(X)$, which count the number of degree-d, n-marked rational curves meeting a given representative H_i of the homology class Poincaré dual to φ at the ith marked point, are used to construct the Gromov-Witten potential

$$\Psi_q(\varphi) = \sum_{d,n\geq 3} \frac{q^d}{n!}\Phi_{d,n}(\varphi^{\otimes n}),$$

which is a function on $H^{2*}(X)$. It is known by Y. Ruan and G. Tian's work [J. Differential Geom. **42** (1995), no. 2, 259–367; MR 96m:58033] that Ψ_q satisfies the DWVV equation, which means that the quantum product $*_\alpha$ on $H^{2*}(X)$ defined by $(\varphi *_\alpha \varphi', \varphi'') = \partial^3_{\varphi\varphi'\varphi''}\Psi_q(\alpha)$, where $(\ ,\ )$ is the intersection pairing, is associative. Equivalently, the "connection" $\nabla^h_u(v) = hd_u(v) + u * v$ defined on the tangent bundle of $H^{2*}(X)$ is integrable. Restricting this connection to $H^2(X)$, we get a flat connection on the trivial bundle $\mathscr{H}$ on $H^2(X)$ with fiber $H^{2*}(X)$. In the case where X is Calabi-Yau one can define a complex variation of Hodge structure (VHS) of weight $r = \dim X$ on this bundle (that is, a filtration satisfying Griffiths transversality with respect to the quantum connection). The $\mathscr{H}^{r,0}$ bundle for this VHS is $H^0(X) \otimes \mathscr{O}_{H^2(X)}$, which has 1 as a canonical section, and the quantum periods for this VHS are the components of 1 in a ∇^h-flat basis of $\mathscr{H}$. Even in the non-Calabi-Yau case, this last definition makes sense so that we still have the quantum periods, which are functions on $H^2(X)$.

What mirror symmetry predicts is the identification of these periods with the normalized periods at infinity of the mirror family. To state it precisely, let us assume X is a hypersurface of degree d in $\mathbf{P}^{d-1}$; the mirror family is the family of quotients

Y_λ of the hypersurfaces with equation $\sum_{0\le i\le d-1} X_i^d + \lambda X_0 \cdots X_{d-1} = 0$ by a finite group. Let $\psi = \lambda^d$, $z = 1/\psi$; there is a natural section ω_ψ of the bundle with fiber $H^{d-2,0}(Y_\psi)$, and it satisfies the following Picard-Fuchs equation (where ∇ is the Gauss-Manin connection and $D = \nabla_{z\partial/\partial z}$):

$$(*) \qquad D^{d-1}\omega = zd(dD+1)\cdots(dD+d-1)\omega.$$

The normalized periods for this family are obtained by replacing ω by a more canonical section $f(z)\omega$, and the algebraic coordinate z by a canonical coordinate q. The mirror identification should identify q and e^t, where t is the linear coordinate on $H^2(X)$ corresponding to the generator $c_1(\mathcal{O}_X(1))$.

A similar statement can be formulated in the non-Calabi-Yau case. Assume now that $X \subset \mathbf{P}^n$ is of degree $d \le n+1$; the generalization of $(*)$ is the equation

$$(**) \qquad h^{n+1-d} D^n\omega = zd(dD+1)\cdots(dD+d-1)\omega.$$

Givental relates the quantum periods to the solutions of $(**)$ as follows: if $d<n$, he proves that the quantum periods are the solutions of $(**)$, with $z=e^t$; if $d=n$, they become the solutions of $(**)$ after an explicit change of variable $T = t+g(e^t)$, $z = e^T$; finally, if $d=n+1$ (the Calabi-Yau or mirror symmetry case), they become the solutions of $(*)$ after a change of variable as above and multiplication by a function $f(e^t)$.

There are two main ingredients in the proof: the introduction of a perturbed quantum connection, using S^1-equivariant Gromov-Witten invariants of $X \times \mathbf{P}^1$, with the standard S^1-action on the second factor, and the use of the torus action on $\mathbf{P}^n$, which by the fixed point formula allows one to compute everything by reduction to the set of rational curves fixed under this torus action. This second step is borrowed from a paper of M. Kontsevich [in *The moduli space of curves (Texel Island, 1994)*, 335–368, Progr. Math., 129, Birkhäuser Boston, Boston, MA, 1995; MR 97d:14077]. The first step consists in the construction of an S^1-equivariant version of the Gromov-Witten potential, defined on $H^{2*}_{S^1}(X\times\mathbf{P}^1) \cong (H^{2*}(X)^2)\otimes \mathbf{C}[h]$. It allows one to construct a connection on $H^{2*}(X)^2$ with parameter h, which turns out to be a deformation of the direct sum of two copies of ∇^h. A simple trick then allows one to express matrix solutions of the quantum differential equation using S^1-equivariant integrals on the space of n-marked genus-zero curves in $\mathbf{P}^n \times \mathbf{P}^1$. After several transformations, involving the fixed point formula for the S^1-action, this step leads to the following conclusion, which is already striking evidence for mirror symmetry: Let M_k be the projective space of polynomials $P(z)$ of degree k with coefficients in $\mathbf{C}^{n+1}$; M_k has the S^1-action $\lambda\cdot P(z) = P(\lambda z)$, and there is a class $p \in H^2_{S^1}(M_k)$ extending $c_1(\mathcal{O}_{M_k}(1))$. Let E_k be the S^1-equivariant bundle on M_k, with fiber a twist of $H^0(\mathcal{O}_{\mathbf{P}^1}(kd))$ so that if $\sigma \in H^0(\mathcal{O}_{\mathbf{P}^n}(d))$ one can define by evaluation an equivariant section $\tilde\sigma$ of E_k by $\tilde\sigma(P) = \sigma\circ P$. Clearly, the top Chern class of E_k is naively related to the enumeration of rational curves in the hypersurface $V(\sigma)$. Consider the function

$$f(t,\tau) = \sum_k e^{k\tau}\int_{M_k}^{S_1} e^{p(t-\tau)/h} c^{S^1}(E_k).$$

It is a nice observation, already made in a previous paper by Givental [Selecta Math. (N.S.) **1** (1995), no. 2, 325–345; MR 97c:14052], that for fixed τ, $f(t,\tau)$ is a solution of $(**)$. Now, instead of M_k, consider the moduli space M_k' of two-marked genus-zero rational curves of bidegree $(k,1)$ in $\mathbf{P}^n \times \mathbf{P}^1$, such that the marked points are over 0 and ∞: S^1 acts naturally on M'_k, which carries an equivariant evaluation bundle E'_k defined in a similar way as E_k. One can show that the natural birational equivariant map μ between M'_k and M_k, the reverse of which associates to a polynomial map its graph, is everywhere defined. What Givental proves is: For fixed τ the function

$$\Phi(t,\tau)=\sum_k e^{k\tau}\int_{M'_k}^{S^1} e^{\mu^* p(t-\tau)/h} c^{S^1}(E'_k)$$

is a quantum period of the degree-d hypersurface $X\subset\mathbf{P}^n$.

So, basically, what remains to be done, and this is the most technical part of the paper, is the study of the classes $\mu_*(c^{S^1}(E'_k))$.

In the course of the paper, another beautiful application of the use of the torus action on $\mathbf{P}^n$ is given. This is the way that the localization formula, which provides a canonical basis for $H^*_{S^1}(\mathbf{P}^n)$, allows one to give formulas for the canonical coordinates of the Frobenius manifold structure on $H^{2*}(X)$ defined by the quantum product (these coordinates are obtained by integrating the idempotent vector fields of the quantum algebra); this application assumes $d\leq n$ for the quantum algebra to be semisimple at the generic point.

This paper proves in particular the spectacular predictions of the virtual number of rational curves in a general quintic threefold in every degree made in the physicists' paper mentioned above and checked before only up to degree 4. It is a major mathematical contribution to the history of mirror symmetry, and together with Batyrev's work which provides mirror candidates for Calabi-Yau complete intersection in toric Fano varieties, it essentially solves the mirror symmetry conjecture for such varieties. *Claire Voisin* (Paris)

99d:14011 14D20 14D15 14N10

Li, Jun [Li, Jun[1]] (Stanford, CA); **Tian, Gang** (Cambridge, MA)

Virtual moduli cycles and Gromov-Witten invariants of algebraic varieties.

J. Amer. Math. Soc. **11** (1998), *no.* 1, 119–174.

After the idea of mirror symmetry was born in the physical context of string theory, the computations of P. Candelas et al. [Nuclear Phys. B **359** (1991), no. 1, 21–74; MR 93b:32029] suggested that a mathematical statement relating the "number of curves" on some manifolds to certain integrals governing the variation of Hodge structures of some mirror dual manifolds should be true. To formulate this statement rigorously one has to make all the ingredients of the nonformal claim above precise. One such problem is to give a precise definition of the "number of curves". The moduli spaces of stable curves which were introduced by Kontsevich furnish the correct framework in which to address this question. These spaces are, however, quite complicated and in order to correctly define the "number of curves", one first has to construct some natural class, which plays the role of the fundamental class, such that the "number of curves" is the result of an integration of a product of certain pulled-back cycles over this class.

To accomplish this feat, the authors of the present article introduce so-called virtual fundamental classes for a certain class of moduli functors. With some additional arguments this construction can be utilized in the setting of Kontsevich's functors of stable maps which are representable by Deligne-Mumford stacks. This procedure then yields virtual fundamental classes for the coarse moduli spaces. Using the above-mentioned classes the authors are able to define Gromov-Witten invariants of algebraic varieties, which furnish the suitable extension of the definition of "number of curves", and prove that they fulfill the natural requirements which were previously formulated as axioms [cf. M. Kontsevich and Y. I. Manin, Comm. Math. Phys. **164** (1994), no. 3, 525–562; MR 95i:14049].

Inspired by the technique employed in this article, K. A. Behrend and B. Fantechi [Invent. Math. **128** (1997), no. 1, 45–88; MR 98e:14022] succeeded in formulating everything in a stack-theoretic framework. Thereby they provided a direct approach which also shows the motivic origin of the Gromov-Witten invariants.

Before this algebraic approach there were symplectic constructions of Gromov-Witten invariants using a different compactification of the space of maps by Ruan and Tian. These invariants and yet another variant of the symplectic theory by Siebert all coincide with the algebro-geometric constructions in the projective category.

To define the virtual fundamental classes, Li and Tian start from the general setting of a moduli functor which admits a so-called perfect tangent obstruction complex. To be precise, the authors' work is restricted to those moduli functors whose functor of tangent spaces is induced by a sheaf-valued functor over the fibered category of modules over schemes over the functor itself which admit a perfect tangent obstruction complex.

The condition of being perfect means roughly that for any family with affine base the two-term complex defined by the tangent and the obstruction functor can locally be realized by the first and second sheaf cohomology of a two-term complex of locally free sheaves.

In the particular case of a representable functor, these structures provide the tangent space and respectively the obstruction space at each point of the moduli space. More explicitly, let $\mathcal{M}$ be the moduli space under consideration and $w \in \mathcal{M}$ a closed point; then there exists a complex of sheaves $\mathcal{E}^\bullet = [\mathcal{E}^1 \to \mathcal{E}^2]$ such that the tangent space $T_w\mathcal{M}$ is given by $T_1 = \mathfrak{h}^1(\mathcal{E}^\bullet \otimes_{\mathcal{O}_M} k(w))$ and the obstruction space is given by $T_2 = \mathfrak{h}^2(\mathcal{E}^\bullet \otimes_{\mathcal{O}_M} k(w))$, where $\mathfrak{h}^i$ is the ith sheaf cohomology of the complex.

The idea of the authors is to use the local equations given by a Kuranishi map to define a cone whose Gysin pullback yields the desired virtual fundamental class, much as in the Fulton-MacPherson normal cone construction. The key point, however, is that the cone is defined intrinsically without any reference to some ambient space. The necessity of the construction lies in the well-known fact that the actual dimension of the moduli space under consideration might not coincide with the expected or virtual dimension.

To begin the construction, the authors start by looking at a relative version of a perfect tangent obstruction complex for flat families of the type described by the moduli functor over an affine base S (relative to a formal scheme Z over S endowed with a section). For this setting, they introduce the notion of corresponding relative Kuranishi families and show that up to the action of a transitively acting group these families correspond to the given perfect relative tangent obstruction complex.

Given a perfect tangent obstruction complex together with a complex of sheaves $\mathcal{E}^\bullet$ whose cohomology is this tangent obstruction complex, Li and Tian then construct virtual normal cones using relative Kuranishi families on affines with respect to the completion along the diagonal of the product. As schemes the cones do not depend on the choice of a relative Kuranishi family and they satisfy a characteristic property ensuring their uniqueness. It is then shown that these locally defined cones can be glued together, thus extending the construction to quasiprojective bases.

By construction, these cones are naturally embedded into a vector bundle and the authors then define the virtual fundamental class to be the pullback of the class of the cycle of the virtual normal cone inside the above-mentioned vector bundle via the Gysin homomorphism. Note that this virtual fundamental class depends on the chosen tangent obstruction complex and the complex of sheaves $\mathcal{E}^{\bullet}$ which induces it.

Concerning the functor of stable maps, the virtual fundamental class is the key ingredient in defining Gromov-Witten invariants, the main aim of this paper. To apply their machinery, the authors work out the standard tangent obstruction complex, show that it is perfect and furthermore prove that in this case the virtual fundamental class is also independent of the choice of the complex whose cohomology furnishes the tangent obstruction complex. There is one more issue, namely, that the functor under consideration is only representable as a Deligne-Mumford stack and the coarse moduli space carries no universal family. The authors argue how this can be overcome either by using operational intersection theory and intersection theory on stacks or by using an atlas and the **Q**-descent of the virtual normal cones. Following the latter option they construct the respective equivariant data.

In the remaining sections of the paper, the authors use the virtual fundamental classes to define the Gromov-Witten invariants of algebraic varieties and check that they satisfy all expected properties, the most difficult part being the verification of the composition "axiom" which is equivalent to the associativity of the quantum cohomology ring. *Ralph Martin Kaufmann* (Los Angeles, CA)

97m:14013 14D25 14L30

Kollár, János (Salt Lake City, UT)

Quotient spaces modulo algebraic groups.

Ann. of Math. (2) **145** (1997), *no.* 1, 33–79.

Let S be a scheme. Let X be an algebraic space and G an affine algebraic group over S. Assume that an action of G on X is given. Then a morphism $q\colon X \to Z$, where Z is an algebraic space (over S), is called a geometric quotient (of X by the action of G) if (i) q is locally of finite type, (ii) for every algebraically closed field K, $q(K)\colon X(K)/G(K) \to Z(K)$ is an isomorphism (of sets), (iii) q is universally submersive and (iv) $\mathcal{O}_Z = (q_*\mathcal{O}_X)^G$ (see 2.7.2). This notion, in the case where X is a scheme, was introduced by D. Mumford [*Geometric invariant theory*, Academic Press, New York, 1965; MR **35** #5451]. According to the author, the main result of the paper (Theorem 1.5) says that if S is excellent, X and G are of finite type over S and the action of G on X is proper, then in the following two cases: (a) G over S is reductive, (b) $S = \operatorname{Spec}(k)$, where k is a field of positive characteristic, a geometric quotient $q\colon X \to X/G$ exists. Results of this kind had been expected (see Conjecture 1.1 of the author) and were earlier known in some special cases (e.g. where $S = \operatorname{Spec}(\mathbf{C})$; see Section 1).

The starting point of the proof is the case where X is normal (and G over S is smooth). Then the theorem follows from C. S. Seshadri [Ann. of Math. (2) **95** (1972), 511–556; errata, ibid. (2) **96** (1972), 599; MR **46** #9044] (see Section 4 of the present paper). Next the author considers the case where X is seminormal (Section 5). The main result here says that, under some assumptions, the existence of a geometric quotient of a normalization $\widetilde{X}$ of X implies the existence of a geometric quotient of a seminormalization X^{sn}. In Section 6 the author considers case (b) of the main theorem. Using the Frobenius morphism one may assume that G is smooth over S and X can be replaced by its seminormalization (Theorem 6.7). Results of Section 8 together with the above-mentioned results (of Section 4) give the existence of a geometric quotient in the general case.

In case (a), if X is an affine scheme (of finite type over excellent S), then the theorem goes back to Seshadri [Advances in Math. **26** (1977), no. 3, 225–274; MR **57** #6035]. If X is seminormal, then the main theorem follows from the normal case and results of Section 8 (and the above-mentioned results of Section 5). In the general case one first considers a seminormalization $X^{\text{sn}} \to X$ and an affine scheme U' with a surjective étale morphism $U' \to X^{\text{sn}}/G$. This gives an étale G-morphism $V' \to X^{\text{sn}}$, where V' is affine. Since there exists a geometric quotient $V' \to V'/G$, it suffices first to prove the existence of a descent giving $V \to X$, where V is affine, and then to use one of the results established in Section 2.

The statement of the main theorem does not imply some interesting features of the quotient $q\colon X \to X/G$ proved in the paper. In fact the author proves also that apart from being a geometric quotient it is also a categorical quotient (see 2.15; example 2.18 shows that in the category of algebraic spaces a geometric quotient in general is not categorical, contrary to the case when the quotient space is a scheme). Moreover, the quotient morphism is affine (3.12 together with 3.7). This last property says that quotients of proper actions are obtained by étale descent of "classical" quotients of affine schemes. It seems that quotients with this property play a distinguished role in geometric invariant theory.

The paper is very nicely written and can be read with pleasure (the reader should be aware that the letters f and q sometimes denote the same morphism (e.g., compare 2.7 and 2.8.3) but this causes little trouble).

{See also the following review [MR 97m:14014, see p. B25 in this volume].}

Andrzej Białynicki-Birula (Warsaw)

97m:14014 14D25 14L30

Keel, Seán (Austin, TX); **Mori, Shigefumi** (Kyoto)

Quotients by groupoids.

Ann. of Math. (2) **145** (1997), *no.* 1, 193–213.

This is a remarkable paper from the point of view of both the results presented and the methods used. The main theorem (Theorem 1.1) says that (in the category of algebraic spaces of finite type over a locally Noetherian base), if $j\colon R \to X \times X$ is a flat groupoid and the stabilizer $j^{-1}(\Delta_x) \to X$ is a finite morphism, then there exists a geometric quotient $q\colon X \to X/R$ which is also a categorical quotient. Moreover, if j is finite then X/R is separated. As a corollary (Corollary 1.2) one obtains the existence of geometric (categorical) quotients with separated X/G (in the category given above) of algebraic spaces X by actions of flat group schemes G with finite stabilizer. This is a vast generalization of all previous known results [see J. Kollár, Ann. of Math. (2) **145** (1997), no. 1, 33–79; MR 97m:14013, see p. B24 in this volume; see the preceding review] and seems to close the topic.

The main idea of the proof is simple, straightforward and ingenious. First, it is sufficient to prove the existence of quotients locally in étale topology (Lemma 3.2). Next, for a given $x \in X$, one can take a slice $W \to X$ (a composition of a closed embedding and an étale map) at x so that the induced groupoid structure on W is quasi-finite and flat (Lemma 3.3). Moreover, the quotient of W by the groupoid structure (if the quotient exists) coincides with the quotient of an étale neighbourhood of x in X (Lemma 3.3). In this way the problem of existence of a geometric quotient is reduced to the case where the groupoid $j: R \to X \times X$ is quasi-finite and flat. Then the authors prove that if $j: R \to X \times X$ is quasi-finite and flat then for any $x \in X$, one may find an affine étale neighbourhood W of x such that the induced groupoid on W is split into two parts, one of them a finite, flat groupoid which contains $j^{-1}(x, x)$ (Proposition 4.2). Hence it suffices to prove the existence of a geometric quotient of affine schemes by a finite groupoid. This is done in Section 5, which completes the proof.

In the last section, the authors compare their results with geometric invariant theory (GIT). They show that when X is an algebraic variety with an action of a reductive group G and $U \subset X$ is a set of stable points, then U is contained in the maximal open subset $V \subset X$ where the stabilizer is proper, a geometric quotient (in the sense of the paper) $V \to V/G$ exists (with a general non-separated quotient space V/G) and restricted to U gives the quotient in the sense of GIT.

It would be interesting to know if, as in Kollár's paper, q is affine when G is an affine group over the base.

The paper is clearly written. There is a mistake in the formulation of Lemma 6.3. Possibly, the assumption that q is GC should be replaced by the assumption that q is G. *Andrzej Białynicki-Birula* (Warsaw)

99c:14018 14E05 13H05 14B99

Cutkosky, Steven Dale (Columbia, MO)

Local factorization of birational maps.

Adv. Math. **132** (1997), *no.* 2, 167–315.

Since Castelnuovo, Abhyankar and Zariski, we have known that an inclusion $R \subset S$ of regular local rings with a common quotient field can be factored by a unique finite product of blow-ups with smooth centers if $\dim(R) = \dim(S) = 2$ (in that case, centers are closed points).

If $\dim(R) = \dim(S) \geq 3$, this is not true, even in dimension 3; counterexamples exist [see J. Sally, Trans. Amer. Math. Soc. **171** (1972), 291–300; MR **46** #9033; erratum; MR **52** #3143; D. L. Shannon, Amer. J. Math. **95** (1973), 294–320; MR **48** #8492] where S and R are essentially of finite type over a field.

In his paper, Cutkosky proves the following theorem, which gives a positive answer to Abhyankar's conjecture [S. Abhyankar, *Ramification theoretic methods in algebraic geometry*, Ann. of Math. Stud., 43, Princeton Univ. Press, Princeton, N.J., 1959; MR **21** #4158 (p. 237)]: Theorem A: Suppose that R, S are excellent local regular rings of dimension 3, containing a field k of characteristic 0, with a common quotient field K, such that S dominates R. Let V be a valuation ring of K which dominates S. Then there exists a regular local ring T, with quotient field K, such that T dominates S, V dominates T, and the inclusions $R \to T$ and $S \to T$ can be factored by sequences of monoidal transformations with regular centers.

If you only use Hironaka's theorem of desingularization, you can find T such that either $R \to T$ or $S \to T$ can be factored by sequences of monoidal transformations with regular centers, but not both. Theorem A is a very difficult result.

Define $Z(X)$ to be the set of all valuation rings V of K, in case K is the quotient field of a proper k-variety, $Z(X)$ is a naturally ringed space obtained as the inverse limit of the systems of all projective varieties with K as function field. $Z(X)$ is called the Zariski-Riemann manifold of K. Theorem A is a local result for the topology of $Z(X)$. Cutkosky gives an answer to the problem of globalization. Theorem C: Let k be a field of characteristic 0, $\Phi: X \to Y$ a birational morphism of integral nonsingular proper excellent k-schemes of dimension 3. Then there exist a nonsingular complete k-scheme Z and birational complete morphisms $\alpha: Z \to X$ and $\beta: Z \to Y$ commuting with Φ such that α and β are locally products of monoidal transformations.

You can glue local triangles given in Theorem A and, as $Z(X)$ is quasi-compact, only a finite number of triangles are necessary to get Theorem C. An important point: α and β may be nonseparated morphisms. The global problem: "given a birational map between two complete smooth varieties of the same dimension, is it possible to decompose it in a sequence of blow-ups and blow-downs along smooth centers?" is still open (it has been solved in the case of toric varieties [see D. Abramovich, K. Matsuki and S. Rashid, "A note on the factorization theorem of toric birational maps after Morelli and its toroidal extension", Preprint, http://xxx.lanl.gov/abs/math/9803126; R. Morelli, J. Algebraic Geom. **5** (1996), no. 4, 751–782; MR 99b:14056; J. Włodarczyk, Trans. Amer. Math. Soc. **349** (1997), no. 1, 373–411; MR 97d:14021]). The gap between this last problem and Theorem A is comparable to the gap between local uniformization and desingularization. These last two results together with Theorem A lead to great hope that the answer to the global problem is yes: in [D. Abramovich, K. Matsuki and S. Rashid, op. cit.], a research program is proposed to solve it.

As noted above, the difficult result in Cutkosky's paper is Theorem A. Let us say a few words about the proof, which is 146 pages long. It is carried out case by case. The valuations $v \in Z(X)$ dominating R are classified by the triples $\mathrm{rank}(v)$, $\mathrm{rrank}(v)$, $\dim_R(v)$, which satisfy Abhyankar's inequality:

$$\mathrm{rank}(v) + \dim_R(v) \le \mathrm{rrank}(v) + \dim_R(v) \le \dim(R),$$

where $\dim_R(v)$ is the transcendence degree of $R/\mathcal{M} \to \mathcal{O}_v/\mathcal{M}_v$, $\mathrm{rank}(v) = \dim(\mathcal{O}_v)$, $\mathrm{rrank}(v) = \dim(\Gamma_v) \otimes \mathbf{Q}$, Γ_v being the group of v. In the theory of birational maps, there is a general fact that is always verified: the greater $\dim(R) - (\mathrm{rrank}(v) + \dim_R(v))$, the worse the case. The (not easy) case where v has maximal rank 3 has already been proved by C. Christensen [J. Indian Math. Soc. (N.S.) **45** (1981), no. 1-4, 21–47 (1984); MR 88a:14010a]. Here the worst case is $\dim_R(v) = 0$, with Γ_v a nondiscrete subgroup of $\mathbf{Q}$. Let us concentrate on this case. First, we suppose k algebraically closed. Lemma 2.1: Let (x, y, z) be coordinates in S, let $v(y)/v(x) = \alpha/\beta$, with $(\alpha, \beta) = 1$, α', β' nonnegative integers such that $\alpha\alpha' - \beta\beta' = 1$; then there exists a unique constant $c \in k$ such that $v(y^\beta/x^\alpha - c) > 0$. Define $x_1 = x^{\alpha'}/y^{\beta'}$, $y_1 = y^\beta/x^\alpha - c$, $z_1 = z$, S_1 the localization of $S[x_1, y_1]$ at (x_1, y_1, z_1). Then $S \to S_1$ is birational and can be factored by a sequence of monoidal transformations, and v dominates S_1.

Let us note that the monoidal transformation $S \to S_1$ is quite natural: it was used in a paper by O. Zariski [Ann. of Math. **40** (1939), 639–689; MR **1**, 26d] to make the uniformization of a surface singularity along a valuation with nondiscrete group.

Let us start at the end: many computations are done to reach the last case, where there exist regular systems of parameters (r.s.p.) (u,v,w) of R, (x,y,z) of S such that $u=\gamma x^a$, $v=y$, $w=x^c\Lambda(x,y,z)$ with γ a unit, $\Lambda(x,y,0)\neq 0=\Lambda(0,0,0)$. By Theorem 2.13, $(x,y,\Lambda(x,y,z))$ is an r.s.p. of S and $a=1$. Theorem A is then trivial. Theorem 2.12: If we have $u=\omega x^{\lambda_0}$, $v=y$, with ω a unit, then there exist $R\to R'$ and $S\to S'$ compositions of monoidal transformations of the type in Lemma 2.1 such that we have the last case for $R'\subset S'$ and ν dominates S'.

The main difficulty in the proof is to desingularize $w\in S$ without losing the assumption $v=y$. The author has to work with pairs of monoidal transformations $S\to S_1$, $R\to R_1$ such that $R_1\subset S_1$ and (with clear notations) $v_1=y_1$.

Let us note that the assumption $u=\omega x^{\lambda_0}$ in Theorem 2.12 is immediate: it is a consequence of Hironaka's desingularization and is stable under any reasonable monoidal transformation of S. The second assumption is obtained in a classical way: you do an expansion of $v\in k[\![x,y,z]\!]=\widehat{S}$:

$$v=x^r(\alpha_0+\alpha_1x+\cdots+\alpha_nx^n+\beta\Lambda(y,z)x^n+x^{n+1}F(x,y,z)),$$

where the Greek letters represent elements of k. By Lemma 2.8, you can suppose $v=x^r(\alpha+\beta\Lambda(y,z)+xF)$. By Theorem 2.10, you can suppose $\mathrm{mult}(\Lambda)=1$; then Lemma 2.9 asserts that you can get the hypotheses of Theorem 2.12.

To get some flavour of the difficulties, note that many different subcases appear: the elements of the expansion of v are formal, so you have to control an extension $\widehat{\nu}$ of ν to $\widehat{S}$ [see J. Sally, op. cit.] in order to control the restriction of $\widehat{\nu}$ to $k[\![x,y]\!]$. For example, it may happen that $\mathrm{rrank}(\widehat{\nu})>\mathrm{rrank}(\nu)$; then the valuations of some elements of $\widehat{S}$ are infinitely greater than all elements of Γ_ν (see Lemmas 2.5, 2.6).

In the case when k is not algebraically closed, the author has to re-prove the results above. Base changes do not help much: you can get k algebraically closed in K; then there is an extension $\overline{\nu}$ of ν to $\overline{K}=K\otimes_k\overline{k}$, and $\overline{T}$ is the localization of $T\otimes_k\overline{k}$ at the center of $\overline{\nu}$ for any local regular ring T with fraction field K. Unfortunately, the assumption k algebraically closed in K is not stable if you have to deal with nonrational points. *Vincent Cossart* (Versailles)

98e:14010 14E15 32S45

Bierstone, Edward (Toronto, ON); **Milman, Pierre D.** (Toronto, ON)

Canonical desingularization in characteristic zero by blowing up the maximum strata of a local invariant. (English summary)

Invent. Math. **128** (1997), *no.* 2, 207–302.

Hironaka's theorem on the existence of resolutions of singularities for any algebraic or analytic variety V over a field of characteristic zero [H. Hironaka, Ann. of Math. (2) **79** (1964), 109–203; ibid. (2) **79** (1964), 205–326; MR **33** #7333, see p. G34 in this volume; J. M. Aroca, H. Hironaka and J. L. Vicente, *Desingularization theorems*, Consejo Sup. Inv. Cient., Madrid, 1977; MR 80h:32027] is an outstanding achievement of twentieth-century mathematics, by virtue of the depth both of its proof and of its applications. For compact complex-analytic V, a resolution is a proper modification $f\colon V'\to V$ from a manifold V' onto V, where "modification" means "map inducing an isomorphism $V' - f^{-1}S \xrightarrow{\sim} V - S$ for some nowhere dense subvariety S of V" (for example, $S :=$ singular locus of V). If V is embedded in a manifold M, one also considers "embedded resolutions", where V' is embedded in a manifold M' and f is induced by a proper modification $g: M' \to M$ such that $g^{-1}V$ ($\supset V'$) has only normal crossing singularities (points where there is a local coordinate system such that each local component of $g^{-1}V$ is defined by the vanishing of a subset of the coordinate functions). Often—not always—one asks for g to be a composition of blowups with smooth centers. Hironaka also dealt with, e.g., real-analytic, non-compact, non-reduced varieties, where further technicalities come into play. Moreover, one may ask for a *canonical* realization of g—one which commutes with open immersions, and in particular with group actions, or more generally with smooth base change.

The history of the problem of existence of resolutions goes back more than a century [see, e.g., J. Lipman, in *Algebraic geometry (Proc. Sympos. Pure Math., Vol. 29, Humboldt State Univ., Arcata, Calif., 1974)*, 187–230, Amer. Math. Soc., Providence, R.I., 1975; MR **52** #10730]. Zariski brought about the transition from the classical characteristic-0 approaches—which succeeded only up to dimension 2—to the modern ones. In the 1940's he proved local uniformization (resolution "along a valuation") in any dimension, and was able to globalize in dimension 3. In positive characteristic, the 2- and 3-dimensional cases were dealt with in the 1950's and 1960's by S. S. Abhyankar [see, e.g., *Resolution of singularities of embedded algebraic surfaces*, Academic Press, New York, 1966; MR **36** #164], who, like Zariski, made heavy use of valuations, a technique which has only recently come back into vogue. And the history is ongoing: work continues on the still-open positive-characteristic case (the total effort diminished in its diversity by a premature announcement of success); and novel approaches to global desingularization have just been developed by A. J. de Jong et al. (see, e.g., his paper with D. Abramovich [J. Algebraic Geom. **6** (1997), no. 4, 789–802; MR 99b:14016], and also the paper by F. A. Bogomolov and T. G. Pantev [Math. Res. Lett. **3** (1996), no. 3, 299–307; MR 97e:14024]), leading to a new generation of short, but non-constructive, proofs.

Hironaka's proof is lengthy, difficult, and non-constructive. Influential as the proof has been, few people can have checked it through entirely, even after some subsequent enhancements of the machinery—Hilbert-Samuel stratification (B. Bennett), maximal contact (J. Giraud, ...). Simplified, more algorithmic proofs are important not only for imparting better understanding of what is really involved in this great theorem, but also for their potential value in unearthing basic features of singularities and their classification. The challenge of finding more straightforward algorithmic approaches was taken up by Zariski, Abhyankar, and others, and successfully met only in the past decade by Bierstone, Milman, and Villamayor.

The paper under review presents—with care, precision, and examples—an essentially self-contained account of such an algorithm. One of the stated goals is to lead the reader along the road by which the authors have come to understand desingularization rather than simply "know" it is true. They write in the language of analysis, not in that of modern algebraic geometry. This not only enlarges the potential readership, but leads to a unified treatment covering algebraic and analytic spaces—complex, real, p-adic—as well as some categories where desingularization theorems, and applications such as Łojasiewicz's inequalities, were not previously known—for example "quasi-Noetherian" spaces and "quasi-analytic" hypersurfaces.

Let us indicate some salient features of the paper, with reference to the cultural context in which they arose.

In the early 1980's, Abhyankar announced a "canonical desingularization" procedure [*Weighted expansions for canonical desingularization*, Springer, Berlin, 1982; MR 84m:14013]. His fundamental idea was, very roughly, to construct a desingularization as a succession of blowups, specifying at every point of each variety appearing in the process a function, the "resolution invariant", which is locally invariant in the sense of compatibility with open immersions, which depends on the preceding history of the process, which takes a finite number of values in a totally ordered set, and whose maximum value locus is a smooth subvariety, upon blowing up which one lowers the said maximum value, thereby "improving the worst singularities". Eventually, by finiteness, the resolution invariant can no longer decrease—so it is constant everywhere and then the original variety has been transformed into a smooth one. (This description is not only vague but also oversimplified, since normal-crossings conditions also play a basic role; nevertheless it may suggest what everyone since has thought a constructive resolution should be.) Abhyankar stated in the preface of those notes that by going through the 220+ pages "consisting mostly of definitions ..., the reader could get an idea of the proof"; but complete details have never appeared. Abhyankar's paper [Adv. in Math. **68** (1988), no. 2, 87–256; MR 89e:14012], giving his 1966 proof of surface desingularization by "good points", was intended to serve as "a good introduction to the higher-dimensional canonical desingularization". (It is, in principle, but U. Orbanz's earlier exposition of that proof, in a book by V. Cossart, Giraud and Orbanz [*Resolution of surface singularities*, Lecture Notes in Math., 1101, Springer, Berlin, 1984; MR 87e:14032 (pp. 1–49)], is more readable.)

A local version of constructive resolution for analytic spaces, intermediate between uniformization and global resolution, was published by Bierstone and Milman in 1989 [J. Amer. Math. Soc. **2** (1989), no. 4, 801–836; MR 91c:32033]. (An earlier version for the hypersurface case was in §4 of another paper of theirs [Inst. Hautes Études Sci. Publ. Math. No. 67 (1988), 5–42; MR 89k:32011], hereafter referred to as [BM 1988].) The first published proof of global constructive resolution was Villamayor's [O. E. Villamayor U., Ann. Sci. École Norm. Sup. (4) **22** (1989), no. 1, 1–32; MR 90b:14014], subsequently refined and clarified by Villamayor [Ann. Sci. École Norm. Sup. (4) **25** (1992), no. 6, 629–677; MR 93m:14012] and Villamayor and S. Encinas [Acta Math. **181** (1998), no. 1, 109–158; MR 99i:14020, see p. B33 in this volume]. Villamayor's papers all assume Hironaka's reduction of the general problem to the "hypersurface-like" problem of "simplification of coherent sheaves of ideals" (see below), and constructivity must be interpreted accordingly. Bierstone and Milman's globalization for hypersurfaces appeared in another paper [in *Effective methods in algebraic geometry (Castiglioncello, 1990)*, 11–30, Progr. Math., 94, Birkhäuser Boston, Boston, MA, 1991; MR 92h:32053]. The present paper treats fully the general—not necessarily hypersurface—case. Another helpful exposition by Bierstone and Milman of the hypersurface case is presented elsewhere ["Resolution of singularities", *Current developments in several complex variables*, edited by Y.-T. Siu and M. Schneider, Cambridge Univ. Press, to appear].

The Bierstone-Milman approach is "bottom-up": it begins with an explicit, elementary construction of a resolution invariant with properties as above. Carefully as the invariant is motivated and described, nevertheless a good deal of patience will be required to master it. This seems inevitable: given what it accomplishes, no resolution algorithm is likely to be all that simple. The description of the invariant, depending initially on a choice of local coordinates, involves numerous formulas, not easy to absorb, at least for the reviewer. Except for the rare reader with sufficient experience and intuition, considerable time and effort will likely be needed to understand fully what to do with, say, the surface $Z^n = X^a Y^b$—indeed, the paper spends $1\frac{2}{3}$ pages on just part of the case $Z^2 = X^2 Y^3$. (The point of that example is to illustrate the workings of the invariant, not to give the complete resolution of the surface.) This is not to say that such an effort will go unrewarded; the invariant and its behavior with respect to certain blowups computably encode the essential local features of the algorithm in a concise, efficient manner; and moreover their local properties, whose establishment in Chapter II is a basic part of the paper, are strong enough to open up an easy passage from local to global resolution—a fundamental problem in the work of Zariski, Hironaka, and Abhyankar. (This passage is negotiated in Chapter II for the hypersurface case and in Chapter IV for the general case.)

In contrast, Villamayor's approach is "top-down": he conceptualizes each step of his process, and ultimately converges to the construction of a resolution invariant—not, however, as explicitly described as Bierstone-Milman's. Thus the two approaches reinforce each other, illuminating various aspects of resolution procedures in different ways. On the other hand, the respective controlling invariants are built ultimately from the same ingredients, by similar processes. For instance, Encinas has informed me that in the above example $Z^2 = X^2 Y^3$, Villamayor's algorithm blows up the same centers described by Bierstone and Milman, though the order for the blowups in "years two and three" is reversed. (By incorporating a variant of Abhyankar's "good point" notion, the algorithm in the most recent work of Villamayor and Encinas becomes shorter in some cases, though not in the preceding example. According to Bierstone and Milman, the algorithm in the present paper sometimes becomes more efficient too if one makes use of additional information—beyond the maximum value—carried by the resolution invariant.)

Let us go lightly over some of those ingredients, keeping in mind that having raw ingredients is one thing, and coming up with a recipe for refining and combining them is something else again. Hironaka made clear that the basic resolution problems, even in characteristic > 0, can be reformulated in terms of resolving "idealistic exponents": pairs (J, b) (modulo some equivalence relation) with J a coherent ideal on a non-singular variety W and b an integer—together with a collection E of hypersurfaces in W having only normal crossings [see in *Algebraic geometry (J. J. Sylvester Sympos., Johns Hopkins Univ., Baltimore, Md., 1976)*, 52–125, Johns Hopkins Univ. Press, Baltimore, Md., 1977; MR **58** #16661]. Such pairs appear, in somewhat different guise, as "basic objects" in the work of Villamayor, and as "infinitesimal presentations" in the present paper. (There is a significant technical difference between infinitesimal presentations and idealistic exponents, having to do with what are called "exceptional blowups". Resolution invariants are defined in terms of presentations, idealistic exponents, basic objects ..., so are not a priori invariant, since the defining data depend on local coordinate systems. Proving independence from coordinates is a major issue, taken care of vis-à-vis presentations in Chapter II—this is where exceptional blowups are essential.) The singular locus $S = S(J, b)$ consists of all points where J has order $\geq b$. One associates with any blowup $W' \to W$ of a "permissible" subvariety C of S (i.e., having only normal crossings in W with E) a transformed basic object $(J', b) + E'$; and the problem is to find a sequence of such blowups at the end of which the transform has empty singular locus. Hironaka proved that this is always possible in characteristic zero. He formulated the fundamental inductive strategy: roughly speaking, if S itself is "large enough" then it is permissible, and blowing it up gives a resolution; otherwise there exists a basic object in a "maximal contact" space of dimension $< \dim W$, whose resolution is equivalent to that of (J, b).

Hironaka used an integer ω, the "residual exponent" (roughly the order of J minus that of E, all divided by b), which Villamayor calls "weighted order", to measure the progress of the resolution. (When the weighted order vanishes, one has a "monomial situation", where a simple combinatoric procedure suffices.) A similar integer is a basic component of Bierstone-Milman's invariant. A key to simplifying Hironaka's very complicated induction argument seems to have been the discovery of the right way to take history into account as well: roughly, E has to be divided into two parts, one of which consists of divisors arising from blowups which have occurred since the weighted order last improved; and only those divisors should enter into the definition of weighted order. Working this idea into an inductive scheme is actually more complicated: informally speaking, the resolution invariant is a sequence of up to $\dim V$ different pairs, one for each level of the induction on dimension, each one a weighted order together with the number of exceptional divisors which were present the last time in the resolution process that weighted order, or anything at a higher-dimensional level in the invariant, decreased.

Bierstone and Milman state in the Introduction that "Essential points [of our proof] include the way we encode the history of the resolution process, originating in [BM 1988] ...". It seems then worth noting that some striking formal similarities are shared by the [BM 1988] proof and the above-mentioned early 1980's exposition by Orbanz of Abhyankar's two-dimensional good points proof [see V. Cossart, J. Giraud and U. Orbanz, op. cit.; MR 87e:14032 (pp. 35–38)], especially in the enabling representation of a power series as $p_1 \cdots p_r f$ where $f = z^n + \alpha_2(x, y)z^{n-2} + \cdots + \alpha_n(x, y)$, $p_j = z + \beta_j(x, y)$, where the p_j represent some history, and where finally induction on dimension has been used to make the α's and β's into a set of monomials totally ordered by divisibility. Any such similarities notwithstanding, the [BM 1988] proof obviously accomplishes much more, not only in what it proves—more than local uniformization in any dimension—but in its revelation of the inductive possibilities inherent in the techniques. The authors' inductive procedure, further developed in the above-mentioned paper [J. Amer. Math. Soc. **2** (1989), no. 4, 801–836; MR 91c:32033], lies at the heart of their approach to desingularization.

Hironaka's reduction of resolution problems to those for idealistic exponents involves coming up with what he calls an idealistic space associated to the Hilbert-Samuel function of an arbitrary variety. Hironaka's construction works for the étale topology. (That topology presents no obstruction to Villamayor's approach, which is compatible with arbitrary smooth base change.) In the present paper a self-contained treatment of an analogous reduction has a prominent place (Chapter III). The reduction is to a situation which can be handled by the techniques of Chapter II, where there is a self-contained proof for simplification of coherent sheaves, in particular for desingularization of hypersurfaces. A noteworthy improvement here is that, in accordance with the goal of presenting a constructive, unified treatment of desingularization for several different categories, only "regular" functions within the initial class under consideration are needed, and not functions algebraic over them (such as would be required for étale coverings).

The authors freely acknowledge their debt to Hironaka's results and philosophy. What can be gleaned from this paper concerning the natural question of its relation to the previously published work of Villamayor? Besides the listing in the References of the first two of the above-mentioned papers of Villamayor (as [V1] and [V2]), the only direct allusion is a statement that "... the way we encode the history of the resolution process, originating in [BM 1988], [is] used in a similar way in [V1])". In [V1], it might be noted, Villamayor does much more than the local induction of [BM 1988], describing a globally valid inductive proof of desingularization by blowing up permissible centers.

Bierstone-Milman and Villamayor are certainly aware of each other's work, and have communicated about it. My own impression is that both approaches were worked out more or less independently in the late 1980's and early 1990's. All three authors deserve the gratitude of the mathematics community for making both the understanding and possible applications of resolution much more accessible than was the case before.

In conclusion, Bierstone and Milman have synthesized and lovingly exposed an approach to canonical resolution whose algorithmic basis should prove of great use in the study of both local and global properties of singular spaces in a variety of categories—for some of which no form of resolution was previously available. This is a very substantial, valuable contribution to an area of fundamental importance.

Joseph Lipman (West Lafayette, IN)

99i:14020 14E15

Encinas, S. (Valladolid); **Villamayor, O. [Villamayor U., Orlando E.]** (Madrid)

Good points and constructive resolution of singularities.

Acta Math. **181** (1998), *no.* 1, 109–158.

This is a paper which accomplishes the enormous task of resolution of singularities in characteristic zero by giving an explicit resolution algorithm. Despite its technical flavour the main ideas and properties become transparent. They will be described below.

A resolution of an algebraic variety or scheme X is a proper birational morphism $\pi\colon \widetilde{X} \to X$ from a regular scheme $\widetilde{X}$ to X such that π is an isomorphism over the regular locus of X. Geometrically, the resolution interprets X as the image of a manifold living in some higher-dimensional space under a projection down to a lower-dimensional space. For instance, the plane curve C given by $x^3 = y^2$ is the projection of the smooth space curve $\widetilde{C}$ given by $x = z^2$, $y = z^3$. Algebraically, the resolution provides a parametrization of the solution set of a system of polynomial equations by a manifold.

It is a classical and difficult problem to construct a resolution for a given X. The optimum would be a direct and explicit definition of $\pi\colon \widetilde{X} \to X$ in terms of X. This works only for curves, where the normalization of $\mathcal{O}_X$ gives a resolution of X. In higher dimensions, no concrete one-step construction is known so far.

Instead, one tries to modify X in small portions so as to improve the singularities of X gradually, hoping that this process describes after finitely many iterations a complete resolution $\pi\colon \widetilde{X} \to X$. The natural and most popular approach is therefore to look for a modification X' of X which tries to loosen the singularities of X. The map from X' to X should be birational and proper. To show that the situation has become better for X' it is appropriate to consider an invariant which measures how singular the points of X are. If it drops when passing to X', the situation has improved. If the invariant belongs moreover to a well-ordered set where the minimal value corresponds to manifolds, induction applies and shows that finitely many modifications produce a resolution.

Therefore the problem splits into two parts: the definition of the modification of X, and the search for significant invariants. Both are strongly related, and changing one will require adjustments for the other to keep both in equilibrium and to make the induction work. The constructions should fulfill some basic requirements (there exist more radical viewpoints). (a) The modification $X' \to X$ should be natural in the sense that it preserves symmetries of X (e.g. lifts group actions), is compatible with respect to field extensions, restrictions to open subschemes and passage to localizations and completions, and it should be an isomorphism over the regular locus of X. (b) The modification should be compatible with embeddings of X in a regular ambient scheme W, i.e. be induced from a birational morphism $W' \to W$. (c) The modification should be a composition of blowups of regular centers (optional). (d) The invariant should be the maximum on X of local invariants, say $i_{a,X}$, defined at all points a of X. The locus where the local invariant achieves the maximum should form a closed regular subscheme of X. These are the worst points of X. The local invariants should be upper-semicontinuous with respect to deformation and localization. (e) At each point a' of X' above a point a of X where $i_{a,X}$ is maximal, the invariant should have dropped.

There are various ways to modify singularities, e.g. normalization, Nash modifications or toric modifications. All these will not be mentioned here. The main method to modify a singularity is to blow up a regular subscheme Z of X (usually inside the singular locus so as not to change the regular points). Geometrically, it consists in approximating the singularity by the directions normal to the center and in separating these by substituting the center through the projectivized normal cone. This is a rather crude way to change the singularity (though there has not been found any better way which works in all dimensions). According to the center, the singularity may, but need not, improve. If the center is too small, the singularity may even get worse (blow up the origin of the surface $x^2 + y^3z^3 = 0$). If the center is too big, e.g. if it leaves the locus of X where the order of the defining ideal is constant, the order after blowup may increase. So how to choose the center inside the singular locus of X is a subtle matter. One possibility consists in taking the stratum of X where the local invariant achieves its maximal value. In this case, the construction of such an invariant becomes the central problem.

We give a brief sketch of the development, referring to [J. Lipman, in *Algebraic geometry (Proc. Sympos. Pure Math., 29, Arcata, Calif., 1974)*, 187–230, Amer. Math. Soc., Providence, R.I., 1975; MR **52** #10730; H. Hauser, "Resolution of singularities 1860–1899", in *Resolution of singularities*, Progr. Math., Birkhäuser, Basel, to appear] for more details and the underlying definitions. For curves, the singular points are isolated, thus they are the only reasonable centers, and blowing them up yields in finitely many steps a resolution, as was already known in the second half of the nineteenth century. For surfaces, the singular locus can consist of isolated points and of several curves. Zariski proposed to refine it by the top locus S of X of those points where the order of the defining ideal of X in some ambient space (of minimal dimension) is maximal [O. Zariski, Ann. of Math. **40** (1939), 639–689; MR **1**, 26d; Ann. of Math. (2) **45** (1944), 472–542; MR **6**, 102f]. If all components of S are regular, one may blow up one of them, showing then that the situation improves. If S has singular points, induction on the dimension appears for the first time. Blowing up these points yields a new scheme X', whose top locus S' is the transform of S plus possibly some smooth components. This compatibility property of S with point blowups is crucial, but fails in higher dimensions [H. Hauser, in *Singularities (Oberwolfach, 1996)*, 289–313, Progr. Math., 162, Birkhäuser, Basel, 1998; MR 99g:14016 (ex. 9 and 10)]. By (embedded) resolution of curves, finitely many blowups transform S into a union of isolated points and regular curves with normal crossings, allowing one to choose as center one of these components (Zariski selects the component arbitrarily and may thus lose symmetry properties; see [H. Hauser, "Excellent surfaces and their taut resolution", in *Resolution of singularities*, Progr. Math., Birkhäuser, Basel, to appear] for how to avoid this by admitting centers with normal crossings).

In characteristic zero, it is not too difficult to choose invariants which show that such a choice of blowups yields a resolution. In positive characteristic, the induction argument is already more involved. There are proposals by S. Abhyankar (studying the corresponding field extensions [Ann. of Math. (2) **63** (1956), 491–526; MR **17**, 1134d]) and H. Hironaka (local resolution using asymmetric invariants extracted from the Newton polyhedron ["Desingularization of excellent surfaces", Bowdoin College, Brunswick, ME, 1967; published as an appendix to V. Cossart, J. Giraud and U. Orbanz, *Resolution of surface singularities*, Lecture Notes in Math., 1101, Springer, Berlin, 1984; MR 87e:14032]) to establish resolution of surfaces over a perfect, respectively an algebraically closed field (see [V. Cossart, Tôhoku Math. J. (2) **33** (1981), no. 1, 25–33; MR 84k:14011] for the extension of [H. Hironaka, op. cit., "Desingularization..."] to arbitrary fields, and [H. Hauser, op. cit., "Excellent surfaces..."] for a systematic treatment and global embedded resolution based on [H. Hironaka, op. cit., "Desingularization..."]). J. Lipman [Ann. Math. (2) **107** (1978), no. 1, 151–207; MR **58** #10924; Inst. Hautes Études Sci. Publ. Math. No. **36** (1969), 195–279; MR **43** #1986] proceeded somewhat differently in reducing the surface singularities by blowups to rational ones (using cohomology theory). These can then be resolved by point blowups. The impact of his work comes from the fact that it proves resolution for arbitrary excellent two-dimensional schemes. In a similar spirit are articles of Faltings, Brodmann, and Kawasaki aiming at a Macaulayfication of the singularities [G. Faltings, Math. Ann. **238** (1978), no. 2, 175–192; MR 80c:14030; M. P. Brodmann, Comment. Math. Helv. **58** (1983), no. 3, 388–415; MR 85g:14061; T. Kawasaki, "On Macaulayfication of schemes", Preprint, 1998; per revr.]. Still for surfaces, replacing blowups by Nash modifications has turned out to be effective over the complex numbers, but seems limited to this situation.

Already for threefolds, the choice of the centers represents a serious problem. There is no immediate geometric candidate, and many of the counterexamples to conjectures occur first in this dimension [H. Hauser, op. cit.; MR 99g:14016]. Aside from the correct choice of the center, the induction on the dimension and the compatibility requirements for the locus S with respect to blowups are very delicate. In particular, as the exceptional divisor is covered by several affine charts which exhibit apparently different types of improvements of the singularities, it is not clear which invariant can best take care of all these types *simultaneously*. For this reason, people tried instead to prove an easier and weaker version of resolution of singularities, the so-called local uniformization. This notion was introduced by Zariski using valuation theory, and proven by him in all dimensions over algebraically closed fields of characteristic zero [Ann. of Math. (2) **41** (1940), 852–896; MR **2**, 124a]. In dimensions 2 and 3 he showed that it implies resolution. Uniformization is a kind of local resolution in the sense that one assumes prescribed ab initio in each exceptional divisor a unique point of interest, and one asks for a sequence of blowups such that only these points become eventually regular. Though this simplifies considerably the induction argument, local uniformization is still not established in positive characteristic. There are, however, important advances by F.-V. Kuhlmann ["Valuation-theoretic and model-theoretic aspects of local uniformization", in *Resolution of singularities*, Progr. Math., Birkhäuser, Basel, to appear; "On local uniformization in arbitrary characteristic. I", to appear], B. Teissier ["Valuations, deformations and toric geometry", in *Resolution of singularities*, Progr. Math., Birkhäuser, Basel, to appear] and T. Urabe ["Resolution of singularities of germs in characteristic positive associated with valuation rings of iterated divisor type", Preprint, 1999; per revr.].

Let us return to global resolution. In 1964 Hironaka succeeded in proving resolution of singularities in arbitrary dimension over fields of characteristic zero [H. Hironaka, Ann. of Math. (2) **79** (1964), 109–203; ibid. (2) **79** (1964), 205–326; MR **33** #7333, see p. G34 in this volume], see also [H. Hironaka, in *Algebraic geometry (Baltimore, Md., 1976)*, 52–125, Johns Hopkins Univ. Press, Baltimore, Md., 1977; MR **58** #16661]. To handle non-hypersurfaces, he introduced the concept of standard bases. They allow one to compute the strict transform of an ideal under blowup by means of a selected generator system. The orders of such generators define the induction invariants ν^* and τ. Except for the difficult Chapter IV of [H. Hironaka, op. cit.; MR **33** #7333, see p. G34 in this volume], where the actual induction argument is established, Hironaka's paper is an excellent exposition of many important features of resolution of singularities, and (despite its formidable reputation) reading it is recommended to anybody interested in the subject. The induction itself is very involved and to some extent even mysterious. In any case, it shows existence without giving an effective construction. The invariant ν^* is not as natural as one may expect, in particular, because it is not semicontinuous with respect to localization. Hironaka's student Bennett studied as a substitute the Hilbert-Samuel function H of singular points, and proved that it does not increase under permissible blowup [B. M. Bennett, Ann. of Math. (2) **91** (1970), 25–87; MR **40** #5608]. He also showed the equivalence of the constancy of ν^* and H along a subscheme (normal flatness). This was exploited by Hironaka in [op. cit.; MR **58** #16661] and later by J. Giraud [Math. Z. **137** (1974), 285–310; MR **57** #705] by introducing the concept of hypersurfaces of maximal contact in zero characteristic (obtained by Abhyankar through his notion of Tschirnhausen transformations). They allow one to embed the equimultiple locus (Hilbert-Samuel stratum) into a lower-dimensional ambient space, which is a prerequisite for induction on the dimension.

In a similar vein but in positive characteristic, Abhyankar substituted Tschirnhausen transformations by coordinate changes which maximize certain orders of ideals. Using this and partly different invariants he established resolution for arithmetic surfaces as well as for threefolds over fields of characteristic > 5 [S. S. Abhyankar, in *Arithmetical Algebraic Geometry (Proc. Conf. Purdue Univ., 1963)*, 111–

152, Harper & Row, New York, 1965; MR **34** #171; *Resolution of singularities of embedded algebraic surfaces*, Academic Press, New York, 1966; MR **36** #164; second edition, Springer, Berlin, 1998; MR 99c:14021]. His theory of good and bad points [Adv. in Math. **68** (1988), no. 2, 87–256; MR 89e:14012] enabled him to achieve a certain stability of the centers of the blowup by establishing first a certain arithmetic condition on the orders of the involved ideals (see below).

At the end of the 1980s and the beginning of the 1990s, Villamayor [Ann. Sci. École Norm. Sup. (4) **22** (1989), no. 1, 1–32; MR 90b:14014; Ann. Sci. École Norm. Sup. (4) **25** (1992), no. 6, 629–677; MR 93m:14012] came up with an explicit global resolution algorithm in characteristic zero and for schemes of arbitrary dimension. At about the same time, E. Bierstone and P. D. Milman [Inst. Hautes Études Sci. Publ. Math. No. 67 (1988), 5–42; MR 89k:32011; J. Amer. Math. Soc. **2** (1989), no. 4, 801–836; MR 91c:32033] established a simple local resolution procedure (admitting centers outside the singular locus). Encinas and Villamayor ["A course on constructive desingularization and equivariance", in *Resolution of singularities*, Progr. Math., Birkhäuser, Basel, to appear] improved and deepened the original algorithm of Villamayor. Bierstone and Milman established a global and detailed algorithm in [Invent. Math. **128** (1997), no. 2, 207–302; MR 98e:14010, see p. B28 in this volume]. Though almost opposite in the presentation, both approaches and in particular the resolution invariants are rather similar, and even coincide to a large extent. Lipman's featured review of Bierstone and Milman's paper [op. cit.; MR 98e:14010, see p. B28 in this volume] provides an excellent comparison. The paper under review goes substantially beyond the mentioned papers by taking into account Abhyankar's good points and by exhibiting certain fundamental principles exploited in [S. Encinas and O. Villamayor, "A new theorem of desingularization over fields of characteristic zero", Preprint, 1999; per revr.]. Here we shall first concentrate on the actual substance of the algorithm, and then explain how Encinas and Villamayor reinterpret good points so as to make the algorithm faster. See also G. Bodnár and J. Schicho's paper presenting the implementation of the algorithm in Maple ["A computer program for the resolution of singularities", in *Resolution of singularities*, Progr. Math., Birkhäuser, Basel, to appear].

Very recently, a new approach to resolution of singularities was opened through the work of A. J. de Jong [Inst. Hautes Études Sci. Publ. Math. No. 83 (1996), 51–93; MR 98e:14011]. Instead of modifying the scheme X, fiber it in curves by taking a suitable projection to a scheme of dimension one less. By induction on the dimension, this parameter scheme can be assumed regular. Semi-stable reduction of the curves yields a scheme with very mild singularities, which can then be resolved by toroidal methods. This allows one to prove in arbitrary characteristic and dimension (including arithmetic schemes) resolution up to a finite map, which in many applications suffices. In zero characteristic, it even allows one to establish quite quickly weak resolution, a notion which differs from the above definition by admitting centers outside the singular locus (thus modifying regular points) [D. Abramovich and A. J. de Jong, J. Algebraic Geom. **6** (1997), no. 4, 789–801; MR 99b:14016; F. A. Bogomolov and T. G. Pantev, Math. Res. Lett. **3** (1996), no. 3, 299–307; MR 97e:14024].

Let us describe the algorithm of Encinas and Villamayor in more detail, simplifying at some places to avoid technicalities. It encompasses and improves the algorithm of Villamayor [op. cit., MR 90b:14014; op. cit., MR 93m:14012] and Encinas and

Villamayor [op. cit., "A course on constructive..."] (and hence also of Bierstone and Milman [op. cit.; MR 98e:14010, see p. B28 in this volume]). Moreover it exhibits general properties one should expect from a resolution algorithm.

The starting point is a singular scheme X embedded in some regular ambient scheme W of, say, dimension n. To each point a of X one associates a local invariant $i_{a,X}$ in a well-ordered set Γ (i.e., $i_{a,X}$ is defined through the local ring of X at a). This set, essentially, only depends on the embedding dimension n. It is shown—and this is very subtle—that $i_{a,X}$ satisfies the various properties assembled in (a) and (d) from above. The invariant is then used to define the center at each step of the resolution process as the (regular) stratum where it achieves its maximal value, and it is also used inductively by showing that it drops after each blowup. This second property fails in positive characteristic, as was observed by T. T. Moh [in *Algebraic geometry and singularities (La Rábida, 1991)*, 49–93, Progr. Math., 134, Birkhäuser, Basel, 1996; MR 97d:14025] and made explicit by the reviewer.

To dispose of the invariant throughout the resolution process, and to be able to take into account the configuration of the exceptional components, one associates to X at each stage two sequences of ideals $J_n, \cdots, J_1$ and $\overline{J_n}, \cdots, \overline{J_1}$, numbers $b_n, \cdots, b_1$ and a sequence $E_i = E_i^+ \cup E_i^-$ of partitions of the set of exceptional components into young and old ones. At the beginning, all E_i are empty. For convenience, let us call the sequences $J_i, \overline{J_i}, b_i, E_i^+, E_i^-$ the setup of X. It reflects accurately the singular structure of X and relates it to the blowups which have occurred so far. The setup allows one to define the invariant $i_{a,X}$, which, in turn, determines the center of blowup.

The setup at stage k of the resolution process is defined by induction on k and the dimension n. Let us first describe the setup at stage $k=0$, i.e. before any blowup. For simplicity, we will do this locally at a given point a. The construction is sufficiently canonical to globalize and to provide an invariant which is semicontinuous. Then, in a second step, we shall indicate how the setup transforms under blowup.

It was proven by Hironaka [op. cit.; MR **58** #16661] that the Hilbert-Samuel stratum of X admits an ideal sheaf J on W and a number b which identifies it with the locus $\mathrm{Sing}(J, b)$ of points of W where J has order $\geq b$. Set $J_n = \overline{J_n} = J$ and $b_n = b$. By the theory of hypersurfaces of maximal contact in characteristic zero, there exists a hypersurface W_{n-1} in W which contains $\mathrm{Sing}(\overline{J_n}, b_n)$. Giraud [op. cit.] showed how to determine such a hypersurface by taking derivatives, thus avoiding the passage to completions (see [O. Villamayor, "On the multiplicity of embedded hypersurfaces", Preprint, 1999; per revr.] for a replacement of hypersurfaces of maximal contact by taking projections). Moreover, one can construct an ideal J_{n-1} in W_{n-1}, the coefficient ideal of $\overline{J_n}$, and a number b_{n-1} such that $\mathrm{Sing}(\overline{J_n}, b_n) = \mathrm{Sing}(J_{n-1}, b_{n-1})$, the second locus taken inside W_{n-1}. Set $\overline{J_{n-1}} = J_{n-1}$. The sequences J_i, $\overline{J_i}$ and b_i can now be defined recursively until one $\overline{J_i}$ becomes 0, in which case the construction stops. As we are at the beginning of the process, all E_i are empty.

Now assume that the setup of the strict transforms of X at stages $0, \cdots, k-1$ of the resolution process is constructed, anticipating that we know already how to choose the centers up to stage $k-1$. Suppose that the following holds (trivially satisfied for $k=0$). For each i there is a factorization $J_i = H^{\alpha_i} \cdot \overline{J_i}$, where H is a collection of well-defined exceptional components and α_i are certain exponents written as multi-indices. Moreover, E_i^+ consists of the exceptional components which have appeared since the last drop of the order of $\overline{J_i}$ in the previous blowups. Then define the setup at stage k as follows, denoting with primes the objects of the new setup: J_i' and $\overline{J_i'}$ are the weak transforms (given by factoring from the total transform the new exceptional component as many times as the order of the ideal below indicates) of J_i and $\overline{J_i}$ for all $i \geq i_0$, where i_0 is the first index for which the order of $\overline{J_i}$ has dropped after blowup. For the remaining i, set $J_i' = \overline{J_i'}$ equal to the coefficient ideal of a certain ideal constructed from J_{i+1}', $\overline{J_{i+1}'}$ and E_{i+1}^-. The b_i' are then defined suitably, and $(E_i')^+$ consists of the transforms of the components of E_i^+ plus the new component. This part—oversimplified here—is the heart of the algorithm and apparently ingenious.

To make the construction work, it has to be shown that this setup is indeed a setup for the strict transform X' of X, that the J_i' are again the coefficient ideals of the $\overline{J_{i+1}'}$, and that J_i' factorizes with respect to $\overline{J_i'}$ as before. This relies on the crucial fact that the strict transform of a hypersurface of maximal contact contains the new equimultiple locus and has again maximal contact if the order has remained constant.

Given the strict transform $X^{(k)}$ of the original scheme X at stage k of the process, a point a of $X^{(k)}$ and a setup of $X^{(k)}$ at a, the invariant is then given (up to some further components which are of purely combinatorial nature) by the vector

$$i_{a,X} = (o_n, m_n, \cdots, o_1, m_1) \in \mathbf{N}^{2n},$$

where o_i denotes the order of $\overline{J_i}$ at a (set equal ∞ if $\overline{J_i} = 0$) and m_i the number of components of E_i^- which pass through a. It is shown that $i_{a,X}$ does not depend on the choice of the hypersurface W_i of maximal contact inside W_{i+1}. The center Z of the next blowup is then given by the closed subscheme where $i_{a,X}$ achieves its maximal value (except in the combinatorial situation). Let i be minimal with $o_i < \infty$. If $o_i > 0$ the center coincides with W_i, is hence regular and lies in all equimultiple loci of the $\overline{J_i}$. Moreover, blowing up Z in W restricts perfectly to all W_i, which is needed to show the above compatibility requirements. If $o_i = 0$, the situation is combinatorial and extra components are added to the invariant.

The components m_i of the invariant, often referred to as history invariants, are necessary in the algorithm to guarantee that the centers lie inside the top loci S_i of the ideals $\overline{J_i}$ (thus implying the transversality of the exceptional components with the resolved scheme) and to define the center in the combinatorial situation when the last finite o_i equals 0. There, several transversal components could be chosen, and the invariant selects a component in a canonical way.

Once these constructions and properties are established, it follows that the inequality $i_{a',X'} \leq i_{a,X}$ holds lexicographically for all points a' above a, mainly because the order of the weak transform of an ideal does not increase under permissible blowup (i.e. blowup with center along which the ideal has constant order). It is then relatively easy to show that the invariant actually drops after each blowup (to see this, recall that if all components remained constant, the plane curve J_2 in W_2 would not improve). As it is semicontinuous and $\mathbf{N}^{2n}$ well ordered, a resolution is achieved. See also [S. Encinas and O. Villamayor, op. cit., "A course on constructive…"] for a very careful description of this part of the algorithm and the computation of many examples. In [S. Encinas, "On constructive desingularization of non-embedded schemes", Preprint, 1998; per revr.], an example is given showing that the algorithm depends on the embedding of X (i.e., on the dimension of W).

The paper under review refines and improves the above algorithm. To get an impression of the concept of good points, it suffices to consider surfaces of type $f = x^o + y^r z^s \cdot g(y,z)$ with $r+s+\operatorname{ord} g \geq o$ (the monomial $y^r z^s$ expresses the presence of exceptional components having occurred so far). For simplicity let us work with fixed coordinates. If $r+\operatorname{ord}_y g \geq o$ or $s+\operatorname{ord}_z g \geq o$ the lines $x=y=0$ or $x=z=0$ are permissible (i.e., lie in the equimultiple locus of f.) Blowing up one of them leaves g invariant and shifts r [respectively s] by $-o$. This can be iterated until both $r+\operatorname{ord}_y g$ and $s+\operatorname{ord}_z g$ are $< o$. Now, if at the beginning r and s satisfy the arithmetic condition $\overline{r}+\overline{s}+\operatorname{ord} g < o$ (where the bar denotes the remainder modulo o), it will follow that these blowups yield a drop of the order o of f. Therefore, to reduce this order, it is not necessary to blow up points, and the centers are prescribed combinatorially by the exceptional monomial in front of g . This simplifies considerably the argument and speeds up the algorithm. If the arithmetic condition is not satisfied, one can establish it by prior point blowups. In the diagram below the set of exponents (r,s) is partitioned into good and bad triangles according to $\overline{r}+\overline{s}$ smaller or bigger than o (in case $\operatorname{ord} g = 0$).

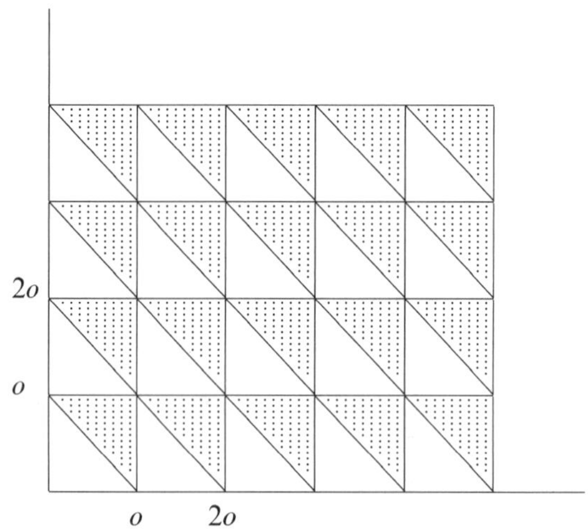

Now let the point a be fixed. It is good in the sense of Encinas and Villamayor with respect to the ideal J_i if $J_i = H^{\alpha_i} \cdot \overline{J_i}$ satisfies the arithmetic condition from above *or* if $o_i = \operatorname{ord} \overline{J_i}$ is zero (this is one of the aspects where it differs from Abhyankar's definition [op. cit.; MR 89e:14012]). Then the invariant $i_{a,X}$ is replaced by

$$i_{a,X} = (\varepsilon_n, o_n, m_n, \cdots, \varepsilon_1, o_1, m_1) \in \mathbf{N}^{3n},$$

where ε_i equals 0 if a is good for J_i, and 1 otherwise (complemented again by a vector which takes care of the combinatorial situation). This implies that for the first index i with $\varepsilon_i = 0$ combinatorial blowups suffice to reduce the order o_{i+1}, and further coefficient ideals can be neglected. The center is again chosen as the locus where the invariant is maximal.

With this effective algorithm one disposes of a precise, detailed but still complicated resolution procedure in characteristic zero. The depth of the authors' understanding of the subject, which has allowed them to overcome all intricacies and complications inherent to resolution of singularities—problems which could not be mentioned in this review—is impressive. The reviewer was informed by Encinas and Villamayor that they have since learned how to dispense with the Hilbert-Samuel function, thus using only the order of ideals for the induction argument [S. Encinas and O. Villamayor, op. cit., "A new theorem..."]. This is done by isolating key properties any reasonable resolution algorithm for setups of singular schemes should satisfy, and by showing that resolution through the passage to weak transforms also implies resolution with respect to the strict transform.

The reference to the history of the prior blowups burdens the exposition; it might have been simpler to extract instead the necessary information (which is little, namely a certain bookkeeping of exceptional components) and to add it to the actual definition of basic objects (or infinitesimal presentations in the sense of Bierstone and Milman [op. cit.; MR 98e:14010, see p. B28 in this volume]) as is done in this review via setups. As for characteristic $p>0$, it may be possible that the previous definition of the invariant can be extended with some adaptations (in particular if the concept of maximal contact is revised). However, as examples show, the invariant $i_{a,X}$ will not improve in general. This implies either that it is not appropriate for induction purposes and completely different invariants have to be chosen, or that the whole approach to resolution via induction has to be reinvented. In any case, a comprehensive understanding of the situation in characteristic zero seems obligatory for a breakthrough in positive characteristic. *Herwig Hauser* (Innsbruck)

99f:14021a 14E30 14E05 14J35
Andreatta, Marco (Trento); **Wiśniewski, Jarosław A.** (Warsaw)
On contractions of smooth varieties. (English summary)
J. Algebraic Geom. **7** (1998), *no.* 2, 253–312.

99f:14021b 14E30 14E05 14J35
Andreatta, Marco (Trento); **Wiśniewski, Jarosław A.** (Warsaw)
Contractions of smooth varieties. II. Computations and applications. (Italian summary)
Boll. Unione Mat. Ital. Sez. B Artic. Ric. Mat. (8) **1** (1998), *no.* 2, 343–360.
The basic features of the theory of algebraic surfaces were solidly established around the beginning of the 19th century but the first steps of a higher-dimensional theory appeared only in the fundamental paper of S. Mori [Ann. of Math. (2) **116** (1982), no. 1, 133–176; MR 84e:14032]. The basic question of this theory, nowadays called the minimal model program or Mori's program, is the following: Given an algebraic variety X (which we usually assume to be smooth and projective) we would like to find another variety X' which is birational to X such that the global structure of X' is as simple as possible. Moreover, we would like to find a procedure which constructs X' as the final result of a sequence of elementary steps. These steps are analogs of the inverses of blowing-ups of smooth subvarieties.

Mori [op. cit.] gave a complete list of these elementary steps in the case of smooth threefolds. Smoothness is, unfortunately, a strong restriction, since the program cannot be carried out within the category of smooth varieties. One has to allow the so-called terminal singularities introduced by M. Reid [in *Journées de Géometrie Algébrique d'Angers, Juillet 1979/Algebraic Geometry, Angers, 1979*, 273–310, Sijthoff & Noordhoff, Alphen aan den Rijn, 1980; MR 82i:14025]. It was also understood that there are two types of elementary steps in general. A contraction (or a Fano-Mori contraction) is a birational morphism $X \to Y$ such that the canonical class of

X is negative on the fibers. In some cases the resulting variety Y is very singular (the precise meaning of this is not important for the papers under review) and then we need another step, called a flip. Flips can be viewed as analogs of Dehn surgeries in the theory of topological 3-manifolds.

An approach to the study of contractions was developed by Y. Kawamata [Ann. of Math. (2) **119** (1984), no. 3, 603–633; MR 86c:14013b] and at least the general features of their behaviour seem to be well understood. The study of flips turned out to be much harder. The existence of flips in the three-dimensional case was established by Mori [J. Amer. Math. Soc. **1** (1988), no. 1, 117–253; MR 89a:14048], as the last piece of the minimal model program in dimension three. Since then most of the progress in the theory has been restricted to dimension three. Many applications of these methods were discovered and these led to the solution of numerous open problems in the theory of surfaces and threefolds. Some of these are reviewed in [J. Kollár and S. Mori, *Birational geometry of algebraic varieties*, Cambridge Univ. Press, Cambridge, 1998]. There has been very little progress on the theory of minimal models in dimensions four and higher. Recent advances of higher-dimensional algebraic geometry have followed a different path; see Kollár and Mori [op. cit. (Section 7.6)] for a quick survey.

The papers under review reach back to the beginnings of the minimal model program. They achieve in dimension four what Mori [op. cit.; MR 84e:14032] completed for threefolds. In this they constitute the first significant step of the last decade in the four-dimensional minimal model program. The results of these papers build on and partially complete earlier work of Y. Kachi [Ann. Scuola Norm. Sup. Pisa Cl. Sci. (4) **24** (1997), no. 1, 63–131; MR 98j:14016].

The main result is the theorem (see Part I, p. 256) which states the following. Let $g: X \to Z$ be a proper birational morphism of fourfolds. Assume that X is smooth, $-K_X$ is ample on the fibers of g and let F be a two-dimensional fiber such that nearby fibers have dimension at most 1. The F is either the projective plane or a quadric (which may be smooth, a cone or the union of two planes). The conclusion holds scheme theoretically and the theorem also describes the normal bundle of F. If g is not birational, there are many more cases. These are listed in Part I, Proposition 4.11.

The theorem is stated clearly and concisely, but it is not for the uninitiated. First, it is not obvious that its assumptions are reasonable, but I want to assure the reader that the case tackled by the authors is indeed the difficult one. The remaining Fano-Mori contractions of smooth fourfolds are either already understood [Y. Kawamata, Math. Ann. **284** (1989), no. 4, 595–600; MR 91e:14039] or probably not very interesting. Second, as with any long classification result, it is not clear that such a list is indeed useful. An answer to this concern is not yet clear, but I believe that applications of this result will surely come.

To gain some appreciation of the result the reader may prefer to start with Part I, Section 3. There one can find a list of carefully worked out examples of contractions. To me this was the really enjoyable part of the paper. The examples are intricate and many-faceted. It is quite amazing to see that so many do occur. Having seen so many of these, one is equally surprised that there are no more. The hard work of the paper is concentrated in Sections 4 and 5. A very careful study of the contractions using the base point freeness method and the deformation theory of curves leads to a classification.

Part II deals with some of the questions left open in the first part. It contains two main results. Section 1 gives a description of some Fano-Mori contractions of smooth fourfolds by explicit equations. These are actually quite complicated but the new point of view is likely to be useful and certainly instructive. Section 4 studies Fano-Mori contractions where a divisor is contracted to a curve. These contractions were also studied by H. Takagi [Proc. Amer. Math. Soc. **127** (1999), no. 2, 315–321; MR 99h:14016]. This case was not investigated in Part I and, as expected, there are no surprises here. *János Kollár* (Salt Lake City, UT)

97b:14019 14F10 14F30 14G20
Berthelot, Pierre (Rennes)
$\mathscr{D}$-modules arithmétiques. I. Opérateurs différentiels de niveau fini. (English summary)
[Arithmetical $\mathscr{D}$-modules. I. Differential operators of finite level]
Ann. Sci. École Norm. Sup. (4) **29** (1996), *no.* 2, 185–272.

In this article the author begins the systematic study of a theory of D-modules in mixed characteristic. The goal of this study is to establish a theory of p-adic sheaves and p-adic cohomology of algebraic varieties in characteristic p. Even though p-adic cohomology theory as provided for by crystalline cohomology [P. Berthelot, *Cohomologie cristalline des schémas de caractéristique* $p>0$, Lecture Notes in Math., 407, Springer, Berlin, 1974; MR **52** #5676] and by rigid cohomology [P. Berthelot, "Cohomologie rigide et cohomologie rigide à supports propres", in preparation] has been successful as a basic cohomology theory, there is not yet a fully developed theory of p-adic (constructible) sheaves in characteristic p. Indeed, at the moment it seems clear that the correct analogue of l-adic smooth sheaves are the overconvergent F-isocrystals [P. Berthelot, op. cit., 1974]. However, the pushforward of such an isocrystal by a closed immersion cannot be defined inside the category of isocrystals.

The structure of the theory that Berthelot has in mind is the following. For a p-adic formal scheme $\mathscr{X}$ formally smooth over $\mathbf{Z}_p$ we define a sheaf $\mathscr{D}^\dagger_{\mathscr{X},\mathbf{Q}}$ of differential operators. For any divisor $Z\subset\mathscr{X}\otimes\mathbf{F}_p$ one constructs a larger sheaf $\mathscr{D}^\dagger_{\mathscr{X}}({}^\dagger Z)_{\mathbf{Q}}$ of differential operators with overconvergent singularities along Z. Then for a closed subvariety T of $\mathscr{X}\otimes\mathbf{F}_p$ one describes overconvergent F-isocrystals on $(T\smallsetminus Z,T)$ as coherent modules over $\mathscr{D}^\dagger_{\mathscr{X}}({}^\dagger Z)_{\mathbf{Q}}$. It is after such a description that one can hope to define pushforward functors for maps $f\colon(\mathscr{X},Z,T)\to(\mathscr{X}',Z',T')$ by using techniques similar to the D-module techniques used in characteristic zero. This pushfoward will be a $\mathscr{D}^\dagger_{\mathscr{X}'}({}^\dagger Z')_{\mathbf{Q}}$-module (one hopes coherent since the original module is so) but in general will no longer correspond to an F-isocrystal on T'. Of course we will want to define regularity and holonomicity conditions for the $\mathscr{D}^\dagger_{\mathscr{X}}({}^\dagger Z)_{\mathbf{Q}}$-modules under consideration and prove that all the modules we are interested in (e.g. those coming from geometry) have these properties.

This outline does not correspond to the contents of the paper. The paper introduces the basic objects $\mathscr{D}^\dagger_{\mathscr{X},\mathbf{Q}}$ and $\mathscr{D}^\dagger_{\mathscr{X}}({}^\dagger Z)_{\mathbf{Q}}$ and studies overconvergent F-isocrystals in the case $T=\mathscr{X}\otimes\mathbf{F}_p$. There are follow-up preprints [P. Berthelot, "$\mathscr{D}$-modules arithmétiques. II, III", in preparation] which deal with inverse and direct image functors for $\mathscr{D}^\dagger$-modules and with Frobenius descent.

The method that is used is the introduction of partial divided power structures of level m for all $m\geq 0$. If I is an ideal, then such a structure on I corresponds more or less to requiring the divided powers of x^{p^m} to exist for all $x\in I$. The sheaf $\widehat{\mathscr{D}}^{(m)}_{\mathscr{X}}$

is the p-adically completed dual of the sheaf $\mathscr{P}^n_{\mathscr{X},(m)}$, which is the universal nilpotent level-m partially divided power hull of the ideal of the diagonal immersion $\mathscr{X} \to \mathscr{X} \times \mathscr{X}$. The local structure of this sheaf is simple enough so that Berthelot is able to prove coherence properties and Theorems A and B for $\widehat{\mathscr{D}}^{(m)}_{\mathscr{X}}$ and $\mathscr{D}^{\dagger}_{\mathscr{X}} = \varinjlim \mathscr{D}^{(m)}_{\mathscr{X}}$. The same is done for the sheaves $\mathscr{D}^{\dagger}_{\mathscr{X}}({}^{\dagger}Z)$. At the end of the article some of the results mentioned above on overconvergent F-isocrystals are proved.

The theory as developed in this series of papers should (eventually) lead to a good theory of p-adic constructible sheaves and so will be the long-awaited counterpart to the existing theory of l-adic étale sheaves. *Johan de Jong* (Cambridge, MA)

98m:14019 14F25 17B10 32J27

Looijenga, Eduard (Utrecht); **Lunts, Valery A.** (Bloomington, IN)

A Lie algebra attached to a projective variety. (English summary)

Invent. Math. **129** (1997), *no.* 2, 361–412.

This paper introduces a new structure on the cohomology of a projective (or compact Kähler) manifold which refines the Lefschetz decomposition of the cohomology in terms of primitive forms.

Let X be a projective manifold of dimension n; let h be the degree 0 endomorphism of $H^*(X) := H^*(X, \mathbf{C})$ which acts on $H^k(X)$ by multiplication by $k-n$. For $\kappa \in H^2(X)$, denote by e_κ the endomorphism $\alpha \mapsto \alpha\kappa$ of $H^*(X)$. If κ is an ample class, the Lefschetz theorem tells us that e^s_κ induces an isomorphism of $H^{n-s}(X)$ onto $H^{n+s}(X)$ for all s. This is equivalent to the existence of an endomorphism f_κ of $H^*(X)$, of degree -2, such that $[e_\kappa, f_\kappa] = h$; then (e_κ, h, f_κ) span a Lie algebra $\mathfrak{g}_\kappa$ isomorphic to $\mathfrak{sl}_2$. The existence of this action of $\mathfrak{sl}_2$ on $H^*(X)$ is equivalent to the Lefschetz decomposition theorem.

The basic idea of this paper is to consider simultaneously all ample classes κ. Define the Néron-Severi Lie algebra $\mathfrak{g}_{\mathrm{NS}}$ to be the Lie algebra generated by all the $\mathfrak{g}_\kappa$'s. The crucial observation is that $\mathfrak{g}_{\mathrm{NS}}$ is semisimple; this is a consequence of the positivity properties of the cup-product with respect to the Lefschetz decomposition. There are two variants of this construction for compact Kähler manifolds: one could take the Lie algebra spanned by the $\mathfrak{g}_\kappa$'s for all Kähler classes κ, or simply by all the classes $\kappa \in H^2(X)$ with the Lefschetz property. In both cases one finds semisimple Lie algebras, called the Kähler and the total Lie algebra.

In order to study these situations the authors introduce the following algebraic formalism. A Lefschetz module is a graded finite-dimensional vector space M (over $\mathbf{C}$), together with a finite-dimensional vector space $\mathfrak{a}$ and a linear map $e: \mathfrak{a} \to \mathrm{End}(M)$ such that $e(\mathfrak{a})$ consists of commuting endomorphisms of degree 2. We require moreover the following conditions (where h is the endomorphism such that $h(m) = \deg(m) \cdot m$ for any homogeneous element $m \in M$): (1) For some $a \in \mathfrak{a}$ (and therefore for all a in an open subset of $\mathfrak{a}$), e^s_a induces an isomorphism of M_{-s} onto M_s for all s; in other words, there exists an endomorphism f_a of M of degree -2 such that $[e_a, f_a] = h$; (2) the Lie algebra $\mathfrak{g}(\mathfrak{a}, M)$ spanned by the e_a's and the f_a's is semisimple.

Though the authors obtain severe restrictions on the triples $(\mathfrak{g}(\mathfrak{a}, M), h, \mathfrak{a})$, they do not arrive at a complete classification. They have to impose an additional hypothesis: (3) The endomorphisms f_a commute with each other.

In this case they call the pair $(\mathfrak{g}, h)$ a Jordan-Lefschetz pair. The graded Lie algebra $\mathfrak{g} = \mathfrak{g}(\mathfrak{a}, M)$ is of the form $\mathfrak{g} = \mathfrak{g}_{-2} \oplus \mathfrak{g}_0 \oplus \mathfrak{g}_2$, with $\mathfrak{g}_{-2}$ and $\mathfrak{g}_2$ abelian; the main theorem of the paper is the classification of Jordan-Lefschetz pairs. There are five infinite series (we give only the types of $\mathfrak{g}$ and $\mathfrak{g}_0$): (a) $(A_{2m-1}, A_{m-1} + A_{m-1})$; (b) (B_m, B_{m-1}); (c) (C_m, A_{m-1}); (d) (D_m, D_{m-1}); (d′) (D_{2m}, A_{2m-1}); and one exceptional case, (E_7, E_6).

The authors give two important classes of manifolds which give rise to Jordan-Lefschetz pairs: complex tori and hyper-Kähler manifolds. For complex tori, the total Lie algebra is of type (d′) and the Kähler Lie algebra of type (a); the algebra $\mathfrak{g}_{\mathrm{NS}}$ is of type (a), (c) or (d′). For hyper-Kähler manifolds, the total Lie algebra is of type (b) or (d); this gives an elegant way of expressing the results of M. S. Verbitsky [Geom. Funct. Anal. **6** (1996), no. 4, 601–611; MR 98a:53069].

The last section is devoted to the notion of "Frobenius-Lefschetz module", a generalisation of the Jordan-Lefschetz situation. Such modules still can be classified when the corresponding Lie algebra is simple of classical type.

The authors have clearly discovered an important tool in the study of the cohomology of projective or compact Kähler manifolds. This tool is somehow a measure of the action of $H^2(X)$ on the whole cohomology (in particular it gives something new only for $b_2 > 1$). It would definitely be of interest to find other families of manifolds with the Jordan-Lefschetz property, since in this case the classification theorem provides precise information on the cohomology. *Arnaud Beauville* (Paris)

99a:14027 14F30
Mebkhout, Z. (Paris)
Sur le théorème de finitude de la cohomologie p-adique d'une variété affine non singulière. (French summary) [On the finiteness theorem for the p-adic cohomology of a nonsingular affine variety]
Amer. J. Math. **119** (1997), *no.* 5, 1027–1081.

Given a non-singular variety X defined over a field of characteristic p, the properties of the étale cohomology with coefficients in $\mathbf{Q}_l$, $l \neq p$, and their use for calculating the zeta function of X, have long been known [P. Deligne, Inst. Hautes Études Sci. Publ. Math. No. 43 (1974), 273–307; MR **49** #5013; Inst. Hautes Études Sci. Publ. Math. No. 52 (1980), 137–252; MR 83c:14017]. But one would like to associate to X a cohomology with coefficients in a non-Archimedean field K whose residue field is of characteristic p. Such a cohomology is sometimes called crystalline, from the name of one of the first attempts to construct such a cohomology [P. Berthelot, *Cohomologie cristalline des schémas de caractéristique* $p > 0$, Lecture Notes in Math., 407, Springer, Berlin, 1974; MR **52** #5676], or sometimes, more ambiguously, p-adic cohomology, the latter name also being used for étale p-adic cohomology. Other candidates introduced over the last thirty years include Monsky-Washnitzer cohomology, and rigid cohomology. For an introduction see Berthelot [Mém. Soc. Math. France (N.S.) No. 23 (1986), 3, 7–32; MR 88a:14020]. There are at least three reasons for the importance of such a cohomology: (i) In case X is the fiber at p of a $\mathbf{Z}$-scheme $X_{\mathbf{Z}}$, then one would like to study the integral cohomology $H^{\cdot}(X_{\mathbf{Z}}, \mathbf{Z})$, and by étale theory one knows only the prime-to-p part [A. Grothendieck, in *Dix exposés sur la cohomologie des schémas*, 306–358, North-Holland, Amsterdam, 1968; MR **42** #4558]. (ii) From the fundamental article of B. Dwork [Inst. Hautes Études

Sci. Publ. Math. No. **12** (1962), 5–68; MR **28** #3039] on the rationality of the zeta function, it was clear that a correct definition of a p-adic cohomology was useful and important in order to characterize the invariants of X (Betti numbers, eigenvalues of Frobenius acting in the cohomology, etc.). (iii) Finally one would like, in the motivic world, to give a p-adic realization for the variety $X_{\mathbf{Z}}$, which should respect all the motivic compatibilities [see P. Deligne, in *Galois groups over* **Q** *(Berkeley, CA, 1987)*, 79–297, Springer, New York, 1989; MR 90m:14016].

There were several attempts at the construction of such a p-adic cohomology theory. We may mention, first (though not historically), the crystalline cohomological theory which turned out to be "good" (in Weil's sense) for proper and smooth varieties. But, on the other hand, directly following Dwork's construction, P. Monsky and G. Washnitzer, between the 1960s and the 1970s [P. Monsky and G. Washnitzer, Ann. of Math. (2) **88** (1968), 181–217; MR **40** #1395; P. Monsky, Ann. of Math. (2) **93** (1971), 315–343; MR **48** #296], constructed what is called now Monsky-Washnitzer cohomology (M-W cohomology): it applies to non-singular affine X (defined over a field of characteristic p), and has the advantage of having an explicit formulation. Starting from this definition, Monsky succeeded in proving several properties of such a cohomology, but the main point, the finiteness, remained open.

The article under review has solved this problem. We would like also to mention that all the previous p-adic cohomological theories have been linked as particular cases (for proper smooth varieties in the case of crystalline cohomology and for smooth affine varieties in the case of M-W cohomology) of rigid cohomological theory, whose finiteness (for non-singular varieties) has been proved in [P. Berthelot, Invent. Math. **128** (1997), no. 2, 329–377; MR 98j:14023]. But, although Mebkhout's proof deals with the non-singular affine case, we stress that the explicit formulation of M-W cohomology compels the author to give a totally independent proof which involves a complete and explicit study (work done together with G. Christol) of the relevant p-adic differential equations.

Let us now give a more direct account of this article. As stated above, the author proves the finiteness of the M-W cohomology. His proof is inspired by Monsky's proof of the finiteness of the algebraic de Rham cohomology (in characteristic 0) [Amer. J. Math. **94** (1972), 237–245; MR **46** #177]: Mebkhout is able to translate the problem of finiteness of cohomology of a variety in dimension n (i.e. the study of the action of the trivial differential module in dimension n) to an index problem for a differential module (an exponential module which is not, in general, "regular") in dimension one, more specifically over the affine line minus the origin. Then, thanks to deep knowledge of the behaviour of the p-adic coefficients in dimension one, he is able to conclude. (See the series of articles of the author together with Christol [Ann. Inst. Fourier (Grenoble) **43** (1993), no. 5, 1545–1574; MR 95j:12009; Ann. of Math. (2) **146** (1997), no. 2, 345–410; MR 99a:12009, see p. B11 in this volume; "Sur le théorème de l'indice des équations différentielles p-adiques. III", Preprint, 1995; per bibl.] in which a sense is given to the notions of regularity and irregularity for p-adic differential equations and finiteness of the index is proved.) Of course, to achieve such a result the author has to adapt the preliminary steps in Monsky's proof from characteristic 0 to characteristic p. Thus we find a construction of a Gysin isomorphism, a Bertini type result and a comparison lemma connecting algebraic and overconvergent indexes. The work ends with some challenging remarks on the action of the Frobenius on the Monsky-Washnitzer cohomology in connection with Weil's conjectures.

In this wonderful article we see a vindication of the philosophy inspiring the works of P. Robba and Dwork (and shared, of course, by the author): the study of the p-adic differential equations (and, in general, of the p-adic coefficients) is essential for the study of the varieties in characteristic p. *Bruno Chiarellotto* (Padua)

97e:14030 14F99 14C25 14F20 19E15
Suslin, Andrei (Cambridge, MA); **Voevodsky, Vladimir** (Cambridge, MA)
Singular homology of abstract algebraic varieties.
Invent. Math. **123** (1996), *no.* 1, 61–94.

In one of his most influential papers, A. Weil [Bull. Amer. Math. Soc. **55** (1949), 497–508; MR **10**, 592e] proved the "Riemann hypothesis for curves over functions fields", an analogue in positive characteristic algebraic geometry of the classical Riemann hypothesis. In contemplating the generalization of this theorem to higher-dimensional varieties (subsequently proved by P. Deligne [Inst. Hautes Études Sci. Publ. Math. No. 43 (1974), 273–307; MR **49** #5013] following foundational work of A. Grothendieck), Weil recognized the importance of constructing a cohomology theory with good properties. One of these properties is functoriality with respect to morphisms of varieties. J.-P. Serre showed with simple examples that no such functorial theory exists for abstract algebraic varieties which reflects the usual (singular) integral cohomology of spaces. Nevertheless, Grothendieck together with M. Artin [*Théorie des topos et cohomologie étale des schémas. Tome 1*, Lecture Notes in Math., 269, Springer, Berlin, 1972; MR **50** #7130; *Tome 2*, Lecture Notes in Math., 270, 1972; MR **50** #7131; *Tome 3*, Lecture Notes in Math., 305, 1973; MR **50** #7132] developed étale cohomology which succeeds in providing a suitable Weil cohomology theory provided one considers cohomology with finite coefficients (relatively prime to residue characteristics). This theory, presented in J. Milne's book [*Étale cohomology*, Princeton Univ. Press, Princeton, N.J., 1980; MR 81j:14002], relies on a new formulation of topology and sophisticated developments in sheaf theory.

In the present paper, the authors offer a very different solution to the problem of providing an algebraic formulation of singular cohomology with finite coefficients. Indeed, their construction is the algebraic analogue of the topological construction of singular cohomology [see, e.g., E. H. Spanier, *Algebraic topology*, McGraw-Hill, New York, 1966; MR **35** #1007], thereby being much more conceptual. Their algebraic singular cohomology with (constant) finite coefficients equals étale cohomology for varieties over an algebraically closed field. The proof of this remarkable fact involves new topologies, new techniques, and new computations reminiscent of the earlier work of Artin and Grothendieck.

To understand the authors' construction, we recall the classical theorem of A. Dold and R. Thom [Ann. of Math. (2) **67** (1958), 239–281; MR **20** #3542]. This asserts that the singular homology of a CW complex X is naturally isomorphic to the homotopy groups of the simplicial abelian group $(\mathrm{Sing}_{\bullet}(\coprod_{d\geq 0} S^d X))^+$, the group completion of the singular complex of the topological abelian monoid $\coprod_{d\geq 0} S^d X$. Now, if X is an algebraic variety, so are its symmetric products. Moreover, homotopy groups of the simplicial abelian group $(\mathrm{Sing}_{\bullet}(\coprod_{d\geq 0} S^d X))^+$ can be computed as the homology of the associated chain complex, which we denote by $(\mathrm{Sing}_{\bullet}(\coprod_{d\geq 0} S^d X))^{\sim}$. The construction of Suslin-Voevodsky, first proposed by Suslin in a talk in Luminy in 1987, is to replace the singular complex by its algebraic analogue. Algebraic singular simplices were exploited years ago by M. Karoubi and O. Villamayor [C. R. Acad. Sci. Paris Sér. A-B **269** (1969), A416–A419; MR **40** #4944] and more recently used by S. Bloch in his formulation of higher Chow groups [Adv. in Math. **61** (1986), no. 3, 267–304; MR 88f:18010].

The fundamental theorem of Suslin-Voevodsky is that if X is an algebraic scheme of finite type over an algebraically closed field k of characteristic $p \geq 0$ and if n is an integer prime to p, then the étale cohomology of X with $\mathbf{Z}/n$ coefficients can be computed as $\mathrm{Ext}^*((\mathrm{Sing}_\cdot^{\mathrm{alg}}(\coprod_{d\geq 0} S^d X))^+, \mathbf{Z}/n)$. (The published statement assumes that the ground field k is of characteristic 0; as the authors soon realized, recent work of J. de Jong giving a weak version of resolution of singularities for varieties over fields of positive characteristic enables this extension to arbitrary characteristic.) Although the formulation of this theorem is relatively elementary, its proof involves sophisticated techniques of abstract algebraic geometry as well as insights from algebraic K-theory. Indeed, the authors encountered this theorem as a part of a sweeping approach to motivic cohomology and algebraic K-theory [see, e.g., V. Voevodsky, A. Suslin and E. Friedlander, *Cycles, transfers, and motivic homology theories*, Ann. of Math. Stud., to appear]. Underlying the authors' approach to (motivic) cohomology is the utilization of algebraic cycles. Maps from a normal variety S (e.g., a standard algebraic simplex Δ^k) to a symmetric product of X correspond to cycles on $S \times X$ finite and surjective over X.

The geometric heart of the proof is the authors' determination of the relative algebraic singular homology of a relative curve in terms of the relative Picard group, just as a key first ingredient for étale cohomology is the understanding of the étale cohomology of curves. This computation leads to a general form of the rigidity theorem of O. Gabber [in *Algebraic K-theory, commutative algebra, and algebraic geometry (Santa Margherita Ligure, 1989)*, 59–70, Contemp. Math., 126, Amer. Math. Soc., Providence, RI, 1992; MR 93c:19005] and H. A. Gillet and R. W. Thomason [J. Pure Appl. Algebra **34** (1984), no. 2-3, 241–254; MR 86e:18014] which played a key role in Suslin's proof of the Quillen-Lichtenbaum conjecture for an arbitrary algebraically closed field [A. A. Suslin, J. Pure Appl. Algebra **34** (1984), no. 2-3, 301–318; MR 86d:18010]. Namely, the authors consider homotopy invariant presheaves with transfers, a basic structure which now plays a central role in their approach to motivic cohomology. The example of most interest for the present work is the "free sheaf generated by X", whose values on standard algebraic simplices determine the chain complex $(\mathrm{Sing}_\cdot^{\mathrm{alg}}(\coprod_{d\geq 0} S^d X))^\sim$. This example fits their general context of presheaf with transfers thanks to the theorem that any "qfh-sheaf" admits the structure of a presheaf with transfers.

An essential ingredient in the authors' approach to cohomology is a further generalization of the étale topology in which proper maps arising in resolutions of singularities occur as coverings. Voevodsky's "h-topology" and its quasi-finite version leading to qfh-sheaves [cf. Selecta Math. (N.S.) **2** (1996), no. 1, 111–153; MR 98c:14016] play an important role. The authors' rigidity theorem asserts the equality of various Ext-groups from sheaves associated to a homotopy invariant presheaf F with transfers to $\mathbf{Z}/n$, where these Ext-groups are computed in the étale topology and various topologies associated to the h-topology. Much of the formal effort in establishing their comparison theorems consists in analyses and manipulations of resolutions of sheaves for these topologies. *Eric M. Friedlander* (Evanston, IL)

97f:14023 14G20 11G18 14F30 14L05 14M15 20G05 20G25
Rapoport, M.; Zink, Th.
★**Period spaces for p-divisible groups.**
Annals of Mathematics Studies, 141.
Princeton University Press, Princeton, NJ, 1996. xxii+324 pp. \$59.50; \$24.95 paperbound. ISBN 0-691-02782-X; 0-691-02781-1

Let G be a connected reductive group over a field F, and let X be the variety of Borel subgroups of G. Then X is a homogeneous space for G, which acts by conjugation on X, and, if G has a Borel subgroup B defined over F, then X can be identified with G/B. In several contexts it is useful to consider certain subsets of X that are stable under the action of $G(F)$ (here we are identifying X with $X(\overline{F})$, where $\overline{F}$ denotes an algebraic closure of F).

For example, suppose that F is a finite field $\mathbf{F}_q$. Then one has a Frobenius map $\Phi: X \to X$, whose fixed point set is $X(\mathbf{F}_q)$. The group G has finitely many orbits on $X \times X$, and in fact these orbits are indexed by the elements in the Weyl group W of G. Let us denote by $O(w)$ the G-orbit indexed by $w \in W$. Then for $w \in W$ one defines a Zariski locally closed subset X_w of X as the set of points $x \in X$ such that $(x, \Phi(x)) \in O(w)$. It is clear from this definition that each subset X_w is stable under the action of the finite group $G(\mathbf{F}_q)$ and that the sets X_w partition X. Moreover, the cohomology groups with compact support of the varieties X_w provide interesting representations (Deligne-Lusztig) of $G(\mathbf{F}_q)$ (the irreducible constituents of these representations are by definition the unipotent representations). To get the remaining representations Deligne and Lusztig also introduce certain $G(\mathbf{F}_q)$-equivariant local systems on the varieties X_w and consider the cohomology with compact support of X_w with coefficients in these local systems.

Next suppose that F is the real field. Then it turns out that the group $G(\mathbf{R})$ has finitely many orbits on X, and these orbits provide the desired $G(\mathbf{R})$-stable subsets of X. Each orbit is a locally closed real-analytic submanifold of the complex manifold X. As in the case of finite fields one can construct representations of $G(\mathbf{R})$ starting from a $G(\mathbf{R})$-equivariant local system L on a $G(\mathbf{R})$-orbit Y in X. Let $L_!$ denote the extension by 0 of L to a sheaf on X, and let $\mathcal{O}$ denote the sheaf of holomorphic functions on X. Then the desired representations are the groups $\mathrm{Ext}^i(L_!, \mathcal{O})$. More generally $\mathcal{O}$ can be replaced by the sheaf of holomorphic sections of any G-equivariant line bundle on X. This construction has a long history, starting with ideas of Griffiths, Langlands and Schmid.

Now suppose that F is the p-adic field $\mathbf{Q}_p$. In this case it is harder to define the desired subsets of X. Let us follow the historical development for a moment. Consider the simplest possible case, that in which G is GL(2) and X is the projective line $\mathbf{P}^1$. Then, working out the definitions above, one sees that for both $\mathbf{F}_q$ and $\mathbf{R}$ our decomposition of $\mathbf{P}^1$ has two pieces, namely $\mathbf{P}^1(F)$ and its complement in $\mathbf{P}^1$. This suggests that for $F = \mathbf{Q}_p$ one should also decompose $\mathbf{P}^1$ into two pieces, namely $\mathbf{P}^1(\mathbf{Q}_p)$ and its complement in $\mathbf{P}^1$. The piece $\mathbf{P}^1(\mathbf{Q}_p)$ is simply a profinite set, but $\mathbf{P}^1 \setminus \mathbf{P}^1(\mathbf{Q}_p)$ is a very interesting 1-dimensional rigid-analytic space. In fact, rigid-analytic spaces of the form $\mathbf{P}^1 \setminus S$ for suitable closed subsets S of $\mathbf{P}^1(\mathbf{Q}_p)$ were used first by Tate (with $S = \{0, \infty\}$) to p-adically uniformize certain elliptic curves, and then by Mumford to uniformize certain other curves. Soon afterward Cherednik showed that certain Shimura curves have p-adic uniformizations of Mumford's type, in which $S = \mathbf{P}^1(\mathbf{Q}_p)$.

In order to understand the next step in the historical development we must first generalize our problem by allowing X to be any partial flag manifold (the quotient of G by some parabolic subgroup). For example, we can take $G = \mathrm{GL}(n)$ and $X = \mathbf{P}^{n-1}$. Generalizing what we did in case $n = 2$, we may partition $\mathbf{P}^{n-1}$ into n disjoint $G(\mathbf{Q}_p)$-stable subsets X_m $(m = 1, \cdots, n)$ as follows. Let V denote the n-dimensional vector space $\mathbf{Q}_p^n$. A point in X is a line L in V over $\overline{\mathbf{Q}}_p$, and we denote by W_L the smallest subspace of V over $\mathbf{Q}_p$ such that L is contained in $W \otimes \overline{\mathbf{Q}}_p$. Then X_m consists of all lines L such that W_L is m-dimensional. Thus X_n is the $(n-1)$-dimensional rigid-analytic space Ω^{n-1} studied by Drinfel′d; it is the complement in $\mathbf{P}^{n-1}$ of the union of all $\mathbf{Q}_p$-rational hyperplanes in $\mathbf{P}^{n-1}$. The space X_m has the following structure: it maps (via $L \mapsto W_L$) onto the profinite set of m-dimensional subspaces of V, and each fiber of the map is isomorphic to Ω^{m-1}. Drinfel′d realized Ω^d as the generic fiber of a formal scheme $\widehat{\Omega}^d$ over $\mathbf{Z}_p$, interpreted $\widehat{\Omega}^d$ as a moduli space of p-divisible groups, used this interpretation to construct covering spaces of Ω^d and conjectured that all cuspidal representations of $\mathrm{GL}_{d+1}(\mathbf{Q}_p)$ occur in the cohomology of these covering spaces.

Now we come to the book under review. One of the many things it does is to define interesting $G(\mathbf{Q}_p)$-stable partitions of partial flag manifolds for any connected reductive group G over $\mathbf{Q}_p$ (Totaro has compared this definition with a different one given by van der Put and Voskuil). The definition is quite subtle and involves in an essential way ideas of Fontaine on p-adic cohomology theories, especially the functors going back and forth between étale and de Rham cohomology.

The open pieces of these partial flag manifolds are called period domains and are generalizations of Drinfel′d's space Ω^d, though they by no means share all the properties of Ω^d. A few of the period domains are related to p-divisible groups, and for these Rapoport and Zink are able to construct a tower of rigid-analytic spaces mapping to the period domain (the maps, called period maps in analogy with the complex case studied by Griffiths, are rigid-analytic maps). In special cases such period maps had been introduced earlier by Dwork and Gross-Hopkins. One expects, as in Drinfel′d's case, that the cohomology of this tower of spaces will provide some very interesting representations of p-adic groups. For certain spaces in the tower Rapoport and Zink also construct models as formal schemes, but except for Ω^d the special fibers of these models are formal schemes rather than schemes.

In the real case certain $G(\mathbf{R})$-orbits on partial flag manifolds are Hermitian symmetric domains, and the quotients of these by arithmetic subgroups of $G(\mathbf{R})$ are Shimura varieties. In Cherednik's theorem something similar happens: for certain arithmetic subgroups Γ of $\mathrm{GL}_2(\mathbf{Q}_p)$ the quotient of a space closely related to Ω^1 by Γ is a Shimura curve. Rapoport and Zink generalize Drinfel′d's proof of Cherednik's theorem to the cases mentioned above involving p-divisible groups. However, there is one big difference. Except in cases involving Ω^d one does not get a p-adic uniformization of the whole Shimura variety. Instead one just obtains uniformizations of pieces of the Shimura variety. For example, in the case of the Shimura variety of principally polarized abelian varieties of dimension g, the most interesting piece having a p-adic uniformization is the set of points whose reductions modulo p are supersingular abelian varieties.

This book is an important advance in our understanding of p-adic groups and Shimura varieties over p-adic fields and will undoubtedly stimulate much further research. Moreover, it opens the way to a better understanding of the representation theory of p-adic groups. Thus it should interest representation theorists as well as arithmetic algebraic geometers. *Robert E. Kottwitz* (Chicago, IL)

97k:14024 14H10 14C30 17B99 20F38 57M99 57N05
Hain, Richard (Durham, NC)
Infinitesimal presentations of the Torelli groups.
J. Amer. Math. Soc. **10** (1997), *no.* 3, 597–651.

Let Σ_g be a closed oriented surface of genus g. Its mapping class group, denoted here by Γ_g, is the group of isotopy classes of orientation preserving diffeomorphisms of Σ_g. This group acts on the Teichmüller space $\mathcal{T}_g$ of Σ_g and the quotient space $\mathcal{M}_g$ is the Riemann moduli space of genus-g curves. This moduli space and the Teichmüller space are very important in both algebraic geometry and complex analysis. Although $\mathcal{M}_g$ is known to be simply connected with respect to the usual topology, it has a natural structure of a quasi-projective orbifold and Γ_g can be considered as its orbifold fundamental group.

On the other hand, in the cases where $g \geq 2$, the classifying space $\mathrm{BDiff}^+\Sigma_g$ of oriented differentiable Σ_g-bundles is an Eilenberg-Mac Lane space $K(\Gamma_g, 1)$. This is because the identity component of $\mathrm{Diff}^+\Sigma_g$ is contractible by a theorem of C. J. Earle and J. Eells [Bull. Amer. Math. Soc. **73** (1967), 557–559; MR **35** #3705]. The mapping class group also enters in the theory of 3-manifolds: if Σ_g is given as a submanifold of codimension one in an oriented 3-manifold, then opening up the manifold along Σ_g and glueing back according to an element $f \in \mathrm{Diff}^+\Sigma_g$ yields another 3-manifold whose diffeomorphism class depends only on the isotopy class of f. Thus Γ_g also plays an important rôle in topology.

The natural action of Γ_g on the first homology group $H_g := H_1(\Sigma_g; \mathbf{Z})$ of Σ_g induces a classical representation $\rho: \Gamma_g \to \mathrm{Sp}(H_g) \cong \mathrm{Sp}(2g; \mathbf{Z})$ of Γ_g onto the Siegel modular group. This representation corresponds to the natural mapping from $\mathcal{M}_g$ to $\mathcal{A}_g$, which is the moduli space of principally polarized abelian varieties, given by associating the Jacobian variety to each curve in $\mathcal{M}_g$. The Torelli group $\mathcal{I}_g$, which is the main subject of the article under review, is defined to be the kernel of this representation. Thus, one natural way of investigating the structure of Γ_g is to try to understand the structure of the Torelli group together with the analysis of the action of $\mathrm{Sp}(H_g)$ on it.

D. Johnson obtained several fundamental results concerning the structure of $\mathcal{I}_g$ in his pioneering works [Math. Ann. **249** (1980), no. 3, 225–242; MR 82a:57008; Ann. of Math. (2) **118** (1983), no. 3, 423–442; MR 85a:57005; Topology **24** (1985), no. 2, 113–126; MR 86i:57011; Topology **24** (1985), no. 2, 127–144; MR 87a:57016]. He proved that $\mathcal{I}_g$ is finitely generated for $g \geq 3$ and he determined the abelianization of $\mathcal{I}_g$ completely. This involves a natural $\mathrm{Sp}(H_g)$-equivariant homomorphism $\tau: \mathcal{I}_g \to \bigwedge^3 H_g/H_g$. He proved that this homomorphism is surjective and that its kernel maps to 2-torsion in the abelianization of $\mathcal{I}_g$. He then showed that this 2-torsion is detected by certain (Birman-Craggs) homomorphisms $\mathcal{I}_g \to \mathbf{F}_2$ that can be defined by means of the Rokhlin invariant of homology 3-spheres. The homomorphism τ is the first of a series of similar homomorphisms which are now called Johnson homomorphisms. One of the reviewers extended and generalized Johnson's works and in particular found a close connection between the structure of $\mathcal{I}_g$ and the Casson invariant [S. Morita, Topology **28** (1989), no. 3, 305–323; MR 90h:57020; Invent. Math. **111** (1993), no. 1, 197–224; MR 93j:57001].

In view of these works, it may be said that the Torelli group has a much richer and more complicated structure than expected. However, we should also point out that the structure of $\mathcal{I}_g$ still largely remains to be revealed. For example, it is not yet known whether $\mathcal{I}_g$ is finitely presentable, except for the case $g = 2$, in which case G. Mess [Topology **31** (1992), no. 4, 775–790; MR 93k:57003] proved that $\mathcal{I}_2$ is an infinitely generated free group. This contrasts with the groups $\mathrm{Sp}(2g;\mathbf{Z})$ and Γ_g because they have been known to be finitely presentable for quite some time (in fact, explicit presentations of these groups are known; see J. S. Birman [Math. Ann. **191** (1971), 59–68; MR **43** #6325] and B. Wajnryb [Israel J. Math. **45** (1983), no. 2-3, 157–174; MR 85g:57007] respectively).

The main result of the present work is what its title states, namely an explicit finite presentation of the Mal′tsev Lie algebra of the Torelli group. The Mal′tsev Lie algebra of a finitely generated discrete group Γ is defined to be the pronilpotent Lie algebra corresponding to a tower of nilpotent Lie groups made out of successive nilpotent quotients of Γ by each member of its lower central series. This is a very important step toward understanding the structure of $\mathcal{I}_g$. This kind of infinitesimal presentation of discrete groups goes back to a result of P. Deligne et al. [Invent. Math. **29** (1975), no. 3, 245–274; MR **52** #3584]. Based on Sullivan's theory of minimal models and the classical Hodge theory, they showed that the rational homotopy type of a compact Kähler manifold X is a formal consequence of its cohomology ring. For example, all its Massey products will vanish. With regard to the fundamental group, the isomorphism type of the Mal′tsev Lie algebra of the fundamental group of X can be read off from the first cohomology group $H^1(X)$ together with the cup product $\bigwedge^2 H^1(X) \to H^2(X)$: it is isomorphic to the free Lie algebra on $H_1(X)$ divided by the ideal generated by the annihilator in $\bigwedge^2 H_1(X)$ of the kernel of the above cup product. J. W. Morgan [Inst. Hautes Études Sci. Publ. Math. No. 48 (1978), 137–204; MR 80e:55020] generalized these results to the case of smooth algebraic varieties.

The author's presentation is simple enough to be outlined here. By Johnson's result mentioned above, we have an isomorphism $H_1(\mathcal{I}_g;\mathbf{Q}) \cong (\bigwedge^3 H_g/H_g)\otimes\mathbf{Q}$. It is easy to see that $U := (\bigwedge^3 H_g/H_g)\otimes\mathbf{Q}$ is an irreducible representation of the algebraic group $\mathrm{Sp}(H_g\otimes\mathbf{Q}) \cong \mathrm{Sp}(2g;\mathbf{Q})$ corresponding to the Young diagram [111] and the above isomorphism is equivariant with respect to the natural actions of $\mathrm{Sp}(H_g\otimes\mathbf{Q})$. The author determines the kernel of the cup product $\bigwedge^2 H^1(\mathcal{I}_g;\mathbf{Q}) \to H^2(\mathcal{I}_g;\mathbf{Q})$ by an explicit topological argument. It turns out that the annihilator in $\bigwedge^2 U = [111111]+[1111]+[11]+[2211]+[22]+[0]$ $(g \geq 6)$ of the above kernel is equal to $R := [111111]+[1111]+[11]+[2211]$. Then the result says that the Mal′tsev Lie algebra of $\mathcal{I}_g$ is a Lie algebra with $\mathrm{Sp}(H_g\otimes\mathbf{Q})$ action isomorphic to the free Lie algebra generated by U divided by the ideal (R) for $g \geq 6$. The author obtains similar presentations for various decorated Torelli groups of surfaces with punctures and boundaries.

Although the formulation of this result involves only group theory, the proof is deeply rooted in mixed Hodge theory. Before saying a little about the proof, it may be worth recounting what it is that makes that theory so powerful. It puts on the cohomology groups (with **Q**-coefficients) of algebro-geometric objects two filtrations, the weight filtration, defined over **Q**, and the Hodge filtration, which lives on the complexification. These are subject to certain conditions, which are abstracted by the notion of a mixed Hodge structure (abbreviated as MHS). The MHSs are the objects of a category, so that all linear maps of an algebro-geometric origin define morphisms in this category. For topologists the following two features are especially important: (1) the weight filtration often has a simple topological characterization and (2) the category of MHSs is abelian. The latter implies that passing to the graded object of the weight filtration is an exact functor (a property definitely not possessed by the category of filtered vector spaces).

The Mal′tsev Lie algebra of the fundamental group of a pointed algebraic variety X comes naturally with an MHS. Property (2) then implies that such a Lie algebra is a filtered Lie algebra isomorphic to the associated weight graded Lie algebra. In particular, weight grading commutes with taking Lie algebra cohomology, so this is a very strong property indeed. Since a presentation involves the first and second cohomology of this Lie algebra only, it is especially useful to know the weight grading of these cohomology groups.

This cannot be directly applied to the Torelli group since that group does not appear naturally as the fundamental group of an algebraic variety. To get around this, the author works on $\mathcal{M}_g$ (with base point) and with the Mal′tsev Lie algebra of its orbifold fundamental group Γ_g relative to the homomorphism $\Gamma_g \to \mathrm{Sp}(H_g \otimes \mathbf{Q})$. This is the Lie algebra of a proalgebraic group having $\mathrm{Sp}(H_g \otimes \mathbf{Q})$ as Levi quotient and whose prounipotent radical turns out to be (for $g \geq 3$) the quotient of the Mal′tsev Lie algebra of the Torelli group by a one-dimensional center. A general theory, developed by the author ["The Hodge de Rham theory of relative Malcev completion", Preprint, http://xxx.lanl.gov/abs/alg-geom/9607019], puts an MHS on the Lie algebra of the unipotent radical. But work is needed to extend that MHS naturally to the whole Mal′tsev Lie algebra of the Torelli group. Another important ingredient is an estimate of the weights appearing in the second cohomology group. Such an estimate was obtained by A. Kabanov [Compositio Math. **110** (1998), no. 2, 163–186; MR 99m:14051] with this application in mind; this eventually leads to the presentation stated above.

There are a number of applications of the main result. Here we only mention two of them. One is a proof of a conjecture made by one of the reviewers, concerning certain algebraic properties of a series of representations of Γ_g which are induced from its action on $\pi_1\Sigma_g$. Another is the consequence that the two natural filtrations on $\mathcal{I}_g$, one coming from the lower central series of $\pi_1\Sigma_g$ and the other from the lower central series of $\mathcal{I}_g$ itself, are not cofinal with respect to each other: the former is coarser and it is strictly so: the quotient is at any finite stage of dim ≥ 1 and is detected by the Casson invariant.

Summing up, the present work takes a crucial step toward a complete understanding of the structure of the Torelli group and provides a basic result concerning the structure of the mapping class group as well as the moduli space in the contexts of both algebraic geometry and topology.

Shigeyuki Morita and E. Looijenga (Tokyo and Utrecht)

98f:14024 14H50 14C20
Hartshorne, Robin (Berkeley, CA)
Families of curves in $\mathbf{P}^3$ and Zeuthen's problem. (English summary)
Mem. Amer. Math. Soc. **130** (1997), *no.* 617, viii+96 pp.
The problem which is usually referred to as Zeuthen's problem is the following one: to determine if every family of space curves—after the customary division—contains limit curves which are composed of lines. This problem was proposed in 1901 as a prize problem by the Royal Danish Academy. The interest of classical geometers in it, as it is presented in the historical notes of the introduction of this paper, was twofold: in order to give a rigorous foundation to enumerative techniques, and to have a combinatorial approach to the classification of distinct families of smooth space curves. Nodal curves whose irreducible components are lines are now called stick figures.

The prize remained without a winner; many positive results were found in the course of this century. W. Brill solved the problem for cuves of genus 0, 1 and 2; F. Gaeta proved in 1950 that every arithmetically Cohen-Macaulay curve specializes to a stick figure; the same is true for curves of genus $g \leq d-3$, for arithmetically Buchsbaum curves and for other families.

The general answer to Zeuthen's problem, however, is negative, and this is proved in Theorem 6.6 of this paper.

This paper is an important example of application of the theory of generalized divisors and it is based on a deep investigation of the behaviour of families of curves on a cubic surface: the study of the degenerations of these families (to limit curves which in most cases will lie on reducible cubic surfaces) shows that if the limit curve is a stick figure (hence it has no embedded points) then very restrictive conditions are imposed on the pair (d, g), the degree and genus of the curves of the family. By making these restrictions explicit, the author finds a first counterexample to Zeuthen's problem in degree 30 and genus 113.

As explicitly stated in the introduction, in several places of the paper the author just proves what is needed for the solution of Zeuthen's problem, suggesting to the reader many open problems. Indeed, the questions raised throughout the paper are very interesting, and the techniques developed seem to be promising for further applications. *Giorgio Bolondi* (Milan)

99g:14052 14J60 14B07 14D20 14H52 14J27 14J32 32J81 81T30
Friedman, Robert [Friedman, Robert David] (New York, NY);
Morgan, John [Morgan, John W.] (New York, NY); **Witten, Edward** (Princeton, NJ)
Vector bundles and F theory. (English summary)
Comm. Math. Phys. **187** (1997), *no.* 3, 679–743.
This is the first mathematical exposition of the theory of principal G-bundles on elliptic Calabi-Yau manifolds. The theory is motivated by a conjectural duality between heterotic string theory on an n-dimensional elliptic Calabi-Yau manifold $\pi: Z \to B$ and F-theory on an $(n+1)$-dimensional elliptic Calabi-Yau manifold $f: X \to B'$, where B' is a $\mathbf{P}^1$-bundle over B [C. Vafa, Nuclear Phys. B **469** (1996), no. 3, 403–415; MR 97g:81059]. Thus in the case $n = 1$ the duality predicts a connection between principal bundles over an elliptic curve with structure group $E_8 \times E_8$ and an elliptic $K3$ surface. Some non-mathematical expositions of subjects close

to those of this paper can be found in papers by D. R. Morrison and Vafa [Nuclear Phys. B **473** (1996), no. 1-2, 74–92; MR 97g:81060a, see p. F29 in this volume; Nuclear Phys. B **476** (1996), no. 3, 437–469; MR 97g:81060b, see p. F29 in this volume], M. Bershadsky et al. [Nuclear Phys. B **505** (1997), no. 1-2, 165–201; MR 99d:81107], and S. Katz and Vafa [Nuclear Phys. B **497** (1997), no. 1-2, 146–154; MR 98i:81209].

There is a beautiful relationship between G-bundles on an elliptic curve, del Pezzo surfaces, and versal deformations of simple elliptic singularities. This was intensively studied in the mathematical literature in the seventies. In this connection E. Looijenga [Invent. Math. **38** (1976/77), no. 1, 17–32; MR **57** #6015] and independently (based on different motivation) J. N. Bernstein and O. V. Shvartsman [Funktsional. Anal. i Prilozhen. **12** (1978), no. 4, 79–80; MR 80d:32007] showed that semistable holomorphic G-bundles over an elliptic curve with G a simply-connected semisimple algebraic group are parametrized by a weighted projective space $\mathbf{P}(1, s_1, \cdots, s_r)$, where s_i are the coefficients of the highest coroot of G. A large part of the present paper explains this result in more explicit terms for various types of groups G. The authors also give a uniform construction of such bundles as a deformation of some particular non-stable bundle. To describe stable G-bundles over an elliptic fibration $f: Z \to B$ one collects together the moduli spaces of G-bundles on each fibre to form a certain weighted projective bundle $\mathscr{W}$ over the base B. This bundle is locally trivial in the sense that $\mathscr{W}$ is equal to the projectivization of a rank $r+1$ vector bundle E which is the sum of line bundles $\mathscr{O} \oplus (\bigoplus_{i=1}^r \mathscr{L}^{-d_j})$, where d_j are the degrees of independent Casimir invariants of G. In defining the projectivization, $\mathbf{C}^*$ acts diagonally on the $\mathscr{L}^{-d_j}$ with weight s_j under appropriate matching of the weights and the degrees. In the case when the base is the upper half-plane the local triviality of $\mathscr{W}$ is due to K. Saito. For general base B and G not of type E_8 this result is due to K. Wirthmüller [Compositio Math. **82** (1992), no. 3, 293–354; MR 94f:14044]. The authors somehow dismiss the fact that in most cases the fibration has degenerate fibres so that $\mathscr{W}$ may degenerate, too. A stable G-bundle V on Z defines a section $s: B \to \mathscr{W}$ (again neglecting the fact that this section may be only a rational map). This section defines a line bundle $\mathscr{N} = s^*(\mathscr{O}(1))$ whose first Chern class is related to the second Chern class of V. In the case $G = \mathrm{SU}(r)$ the authors describe the section s by means of a spectral curve which is a certain r-sheeted cover $C \to B$. The bundle V is obtained as a push-forward of a certain line bundle on C. In this way one obtains a map from the moduli space of G-bundles to the space of sections of $\mathscr{W}$ whose fibres are the Jacobians of the corresponding spectral curves. The paper does not elaborate much on this construction. More details for a general group G can be found in a recent work of R. Y. Donagi [Asian J. Math. **1** (1997), no. 2, 214–223; MR 99d:14010]. The last sections of the paper are devoted to computations of the Chern classes of the G-bundles on Z. The main idea is to compute the Chern classes of the unstable bundle from which V arises via deformation theory. For example in the case $G = E_8$, the authors take the unstable bundle with the structure group reduced to $\mathrm{SU}(6) \times \mathrm{SU}(2) \times \mathrm{SU}(3)$. This choice is motivated by F-theory. The first nontrivial characteristic class of V is the fundamental 4-dimensional class $\lambda(V) = c_2(V)/60$. It is found to be equal to $\eta\sigma - 15\eta^2 + 135\eta c_1(\mathscr{L}) - 310 c_1(\mathscr{L})^2$, where $\mathscr{O}(\sigma) \otimes \mathscr{L}$ is the determinant of the SU(2)-summand of V, and $\eta = \pi_*(\lambda(V))$. This formula agrees with expectations from F-theory. Another expectation from this theory is that the heterotic theory on three-dimensional Calabi-Yau Z with two E_8-bundles V_1 and V_2 satisfies $\lambda(V_1) + \lambda(V_2) + h[F] = c_2(T_Z)$, where $[F]$ is the class of the fibre, T_Z is the tangent bundle of Z and h is a positive number (equal to the number of fivebranes on Z). The authors verify another prediction from the string duality $h=\chi(X)/24$, where X is the associated Calabi-Yau fourfold.

It is a well-known problem to compute the dimension of the moduli space of G-bundles on a Calabi-Yau threefold (in this dimension the terms in the corresponding Riemann-Roch formula cancel out because of Serre's duality). The last section of the paper is devoted to a Z_2-equivariant index theorem which allows one to compute the difference between the dimensions n_e-n_o of the moduli space of invariant and anti-invariant G-bundles with respect to the inversion involution of the elliptic fibration. In many examples $n_o=0$ (e.g. when the associated fourfold X has zero H^3). In the case $G=\mathrm{SU}(r)$ the formula for n_e-n_o is verified by comparing it with the actual construction of $\mathrm{SU}(r)$-bundles on Z. *I. Dolgachev* (Ann Arbor, MI)

97d:14063 14K05 14E05

Ein, Lawrence (Chicago, IL); **Lazarsfeld, Robert** (Los Angeles, CA)

Singularities of theta divisors and the birational geometry of irregular varieties.

J. Amer. Math. Soc. **10** (1997), *no.* 1, 243–258.

Let X be a smooth projective variety over $\mathbf{C}$. X is called regular if the topological cohomology group $H^1(X,\mathbf{C})=0$ and irregular otherwise. The terminology predates cohomology theory, and it derives from the circumstance that a smooth projective surface in $\mathbf{CP}^3$ is regular. Thus, as the name indicates, irregular varieties were considered to behave peculiarly. The simplest examples of irregular varieties are abelian varieties; these are smooth projective varieties which are biholomorphic to $\mathbf{C}^g/\mathbf{Z}^{2g}$. D. Mumford's book [*Abelian varieties*, Published for the Tata Institute of Fundamental Research, Bombay, 1970; MR **44** #219] is an excellent but demanding introduction.

Albanese discovered that irregular varieties have an almost canonical map to an abelian variety. In modern language this is done as follows [cf. P. Griffiths and J. Harris, *Principles of algebraic geometry*, Wiley-Intersci., New York, 1978; MR 80b:14001 (p. 331)]. One can prove that $\dim H^1(X,\mathbf{C})=2g$ is even and there are g linearly independent global holomorphic 1-forms $\omega_1,\cdots,\omega_g\in H^0(X,\Omega^1_X)$ on X. Pick a point $0\in X$ and define the Albanese map by $x\mapsto(\int_0^x\omega_1,\cdots,\int_0^x\omega_g)$. The integral depends only on the homology class of the path chosen, thus giving a well-defined map (called Albanese map) $X\to H^0(X,\Omega^1_X)^*/H_1(X,\mathbf{Z})$. The target is called the Albanese variety. A typical algebraic variety does not have any interesting maps to other varieties, thus the existence of the Albanese map is rather remarkable in itself.

Despite this, a serious study of irregular varieties started relatively late. S. Iitaka [J. Math. Soc. Japan **23** (1971), 356–373; MR **44** #2749] called attention to the study of plurigenera of algebraic varieties. The plurigenera of a variety X are the numbers $P_m(X):=\dim H^0(X,\mathcal{O}(mK_X))$, where K_X is the canonical class and $m\geq 1$. Iitaka observed that varieties with Kodaira dimension zero (that is, $P_m(X)\in\{0,1\}$ but not always 0) occupy an especially important place in the classification theory of varieties. K. Ueno [*Classification theory of algebraic varieties and compact complex spaces*, Lecture Notes in Math., 439, Springer, Berlin, 1975; MR **58** #22062] proposed the following conjecture: Let X be a smooth projective variety of Kodaira dimension zero. Then the Albanese map is surjective, $\dim H^1(X,\mathbf{C})\leq 2\dim X$, and if equality holds then X is birational to its Albanese variety. This was proved for $\dim X = 3$ by Ueno [in *Proceedings of the International Symposium on Algebraic Geometry (Kyoto Univ., Kyoto, 1977)*, 693–708, Kinokuniya Book Store, Tokyo, 1978; MR 82f:14012]. The method of Ueno is to study the sheaves $f_*(\omega^m_{X/Y})$ for an arbitrary morphism and to prove that they have certain semi-positivity properties. The whole conjecture was proved by Y. Kawamata [Compositio Math. **43** (1981), no. 2, 253–276; MR 83j:14029]. The method of Kawamata required $P_m \leq 1$ for every m. The reviewer [Ann. of Math. (2) **123** (1986), no. 1, 11–42; MR 87c:14038] gave the first version of the result which relied on knowing P_m for only finitely many m. The best results in this direction were: (1) If $P_m(X) = 1$ for some $m \geq 3$ then the Albanese map is surjective. (2) If in addition $\dim H^1(X, \mathbf{C}) = 2 \dim X$ then the Albanese map is birational [J. Kollár, *Shafarevich maps and automorphic forms*, Princeton Univ. Press, Princeton, NJ, 1995; MR 96i:14016 (17.1)].

The paper of Ein and Lazarsfeld introduces a new method to the study of irregular varieties, and gives three distinct applications. The method relies on the generic vanishing theorems proved by M. Green and Lazarsfeld [Invent. Math. **90** (1987), no. 2, 389–407; MR 89b:32025; J. Amer. Math. Soc. **4** (1991), no. 1, 87–103; MR 92i:32021].

The Kodaira vanishing theorem and its recent generalizations [see, e.g., J. Kollár, op. cit. (Chapters 9–11)] prove the vanishing of cohomologies for divisors of the form $K_X + D$, where D is close to being ample. For the generic vanishing theorems D is allowed to be numerically trivial. The set of all numerically trivial line bundles on a variety X (essentially) forms an abelian variety $\operatorname{Pic}^\circ(X)$ (the Picard variety of X). One can then study the cohomology groups $H^i(X, \mathcal{O}(K_X + D))$ as D varies in $\operatorname{Pic}^\circ(X)$. The main results of Green and Lazarsfeld [op. cit.] are: (1) Jumps in these cohomology groups occur along quotients of linear subspaces in $\operatorname{Pic}^\circ(X)$. (2) These subspaces can be computed from the action of the operators $\bigwedge \omega \colon H^0(X, \Omega^i) \to H^0(X, \Omega^{i+1})$, where ω is a 1-form.

Theorem 4 of this paper gives an astoundingly short proof of the following result: If $P_1(X) = P_2(X) = 1$ then the Albanese map is surjective. The proof goes as follows. If the Albanese map is not surjective, let z be a linear coordinate on $H^0(X, \Omega^1_X)^*$ which is tangent to the Albanese image at a general point. Then dz, viewed as a 1-form on X, has a zero at a general point of X. This unexpected behavior can be detected through its action on the $H^0(X, \Omega^i)$. This in turn implies that $H^0(X, \mathcal{O}(K_X + D))$ is nonzero along a subvariety through $0 \in \operatorname{Pic}^\circ(X)$. Let $D \in \operatorname{Pic}^\circ(X)$ be a point on this subvariety near 0. Since the cohomology jumps occur along linear spaces, $-D$ is also on the subvariety. The image of the multiplication map

$$H^0(X, \mathcal{O}(K_X + D)) \times H^0(X, \mathcal{O}(K_X - D)) \to H^0(X, \mathcal{O}(2K_X))$$

gives a new global section, hence $P_2(X) \geq 2$.

The same method is applied to correct and settle a conjecture in [J. Kollár, op. cit. (17.9.3)]. I conjectured that if the Albanese map of X is generically finite and X is of general type then $\chi(X, \mathcal{O}(K_X)) > 0$. Theorem 3 of the present article shows that this holds if the Albanese map is birational onto its image, but not in general.

While the methods have been developed to handle the above questions, the most surprising result of the present paper concerns principally polarized abelian varieties. These abelian varieties come equipped with a special divisor, called the theta divisor. The singularities of theta divisors were already studied by Riemann, with more recent contributions by A. Andreotti and A. L. Mayer [Ann. Scuola Norm. Sup. Pisa (3) **21** (1967), 189–238; MR **36** #3792] and G. Kempf [Ann. of Math. (2) **98** (1973), 178–185; MR **50** #2180]. These articles mostly studied the singularities of the theta divisors for the Albanese varieties of curves (usually called Jacobians). Very little was known about the singularities of general theta divisors. E. Arbarello and C. De Concini [Duke Math. J. **54** (1987), no. 1, 163–178; MR 88i:14025] conjectured that theta divisors are normal, unless the abelian variety is a product. The reviewer [op. cit. (17.13)] showed how the above-mentioned general methods can be applied to the study of theta divisors. In the present paper this method is combined with generic vanishing theorems to yield the following: The theta divisor of a principally polarized abelian variety is normal with only rational singularities. As a special case this settles the above-mentioned conjecture of Arbarello-De Concini. Some special cases were proved earlier by R. C. Smith and R. Varley [Duke Math. J. **82** (1996), no. 2, 319–326; MR 97c:14043]. *János Kollár* (Salt Lake City, UT)

97f:14047 14L05 14F30
de Jong, A. J. (Cambridge, MA)
Crystalline Dieudonné module theory via formal and rigid geometry.
Inst. Hautes Études Sci. Publ. Math. No. 82 (1995), 5–96 (1996).
Let $p > 0$ be a prime number. The idea of using crystalline techniques to extend the Dieudonné-Manin classification of commutative formal Lie groups over a perfect field of characteristic p to p-divisible groups over a more general base scheme S goes back to A. Grothendieck, and was one of his main motivations for introducing these techniques. In his talk at the Nice International Congress in 1970 [in *Actes du Congrès International des Mathématiciens (Nice, 1970), Tome 1*, 431–436, Gauthier-Villars, Paris, 1971; MR **58** #28211], he indicated two methods to associate to a p-divisible group G over S a Dieudonné crystal over S, i.e. a crystal $\mathbf{D}(G)$ endowed with Frobenius and Verschiebung morphisms $F : \mathbf{D}(G)^\sigma \to \mathbf{D}(G)$, $V : \mathbf{D}(G) \to \mathbf{D}(G)^\sigma$ such that $F \circ V = p$, $V \circ F = p$ (where $\mathbf{D}(G)^\sigma$ denotes the Frobenius pullback of $\mathbf{D}(G)$). The first method, based on the crystalline nature of the universal extension of a p-divisible group by a vector group, and the theory of exponential map, was developed by W. Messing [*The crystals associated to Barsotti-Tate groups: with applications to abelian schemes,* Lecture Notes in Math., 264, Springer, Berlin, 1972; MR **50** #337]. The second one, based on Grothendieck's notion of $\natural$-extension, and leading to a comparison theorem with the first de Rham cohomology group for abelian schemes, was developed by B. Mazur and Messing [*Universal extensions and one dimensional crystalline cohomology,* Lecture Notes in Math., 370, Springer, Berlin, 1974; MR **51** #10350]. A cohomological interpretation of this construction was later obtained by L. Breen, Messing and the reviewer in terms of $\mathcal{E}xt$ sheaves on the crystalline site [P. Berthelot, L. S. Breen and W. Messing, *Théorie de Dieudonné cristalline. II,* Lecture Notes in Math., 930, Springer, Berlin, 1982; MR 85k:14023; see also P. Berthelot and W. Messing, in *Journées de Géométrie Algébrique de Rennes (Rennes, 1978), Vol. I,* 17–37, Astérisque, 63, Soc. Math. France, Paris, 1979; MR 82f:14041]. The greater flexibility achieved through cohomological methods allowed them to extend the theory to finite flat p-groups, and to prove that it satisfies the expected basic properties of Dieudonné theory.

Once these Dieudonné crystals have been constructed, the main problem is to decide to what extent they classify p-divisible groups, or finite flat p-groups. Over a perfect field, it was shown by Breen, Messing and the reviewer that they can be identified with the classical contravariant Dieudonné module, as defined for example by J.-M. Fontaine in terms of Witt covectors [*Groupes p-divisibles sur les corps locaux*, Astérisque, 47-48, Soc. Math. France, Paris, 1977; MR **58** #16699]; thus the Dieudonné crystal gives an equivalence of categories in this case. This result was extended by the reviewer to the case of p-divisible groups, or finite flat p-groups, over a perfect valuation ring [Ann. Sci. École Norm. Sup. (4) **13** (1980), no. 2, 225–268; MR 82b:14026]. For further results, additional hypotheses are needed, either on the groups or on the base scheme; it was indeed pointed out by the reviewer and A. Ogus that, due to the presence of horizontal p-torsion in divided power envelopes, the crystalline Dieudonné functor for p-divisible groups is not always fully faithful, even when the base is an Artinian ring [Invent. Math. **72** (1983), no. 2, 159–199; MR 85e:14025]. For étale or multiplicative groups, Messing and the reviewer have shown that the crystalline Dieudonné functor is an equivalence of categories over any scheme on which p is locally nilpotent [announced in *Journées de Géométrie Algébrique de Rennes (Rennes, 1978), Vol. I*, 17–37, Astérisque, 63, Soc. Math. France, Paris, 1979; MR 82f:14041; with complete proofs given in *The Grothendieck Festschrift, Vol. I*, 173–247, Progr. Math., 86, Birkhäuser Boston, Boston, MA, 1990; MR 92h:14012]. The full faithfulness of the crystalline Dieudonné functor was proved by the same authors (in the above-cited papers) for p-divisible and for finite flat p-groups on a normal base scheme of characteristic p on which there exists locally a p-basis. In that case, the problem of describing the essential image of the crystalline Dieudonné functor remained open, except for the case of connected p-divisible groups when the base scheme is smooth over a perfect ring, which had been previously solved by S. J. Bloch in an unpublished preprint ["Dieudonné crystals associated to p-divisible formal groups", Preprint, 1974; per revr.].

With this article, de Jong makes decisive progress on the classification problem in two directions. On the one hand, he completes the program for p-divisible groups on smooth schemes over a field with a finite p-basis: he proves that, in this case, the crystalline Dieudonné functor is an equivalence of categories (in a somewhat more general context; see below). Similar results over a perfect field had also been obtained independently by K. Kato for $p \neq 2$ [e.g., S. J. Bloch and K. Kato, "p-divisible groups and Dieudonné theory", Preprint, 1992; per revr.] (the extension to the case $p = 2$ has been done by Messing). On the other hand, over any scheme of finite type over a field with a finite p-basis, he solves the problem of full faithfulness up to isogeny, a question raised by the reviewer and Messing [op. cit.; MR 92h:14012], but which had never been studied until now.

Let us now briefly indicate the main ideas of the proofs. In order to solve the equivalence problem, de Jong actually solves it in a more general context, by extending the construction of the Dieudonné crystal associated to a p-divisible group to the case of p-divisible groups over an adic formal scheme $\mathfrak{X}$ of characteristic p. Then his main theorem asserts that, when $\mathfrak{X}$ is formally smooth, and its reduced subscheme is of finite type over a field with a finite p-basis, the crystalline Dieudonné functor is an equivalence of categories. The proof goes by induction on the dimension of $\mathfrak{X}_{\mathrm{red}}$, the most difficult step being the case where $\dim \mathfrak{X}_{\mathrm{red}} = 0$. When $\mathfrak{X}$ is the spectrum of a field with a finite p-basis, the argument is based on a classification result for formal p-divisible groups proved in the author's thesis [Invent. Math. **114** (1993), no. 1, 89–137; MR 94j:14043], and on the study of the correspondence between extensions of étale groups by multiplicative groups, and similar extensions of their Dieudonné crystals. The study of the nonreduced case and the induction argument make use of deformation theory to extend a p-divisible group knowing a deformation of its Dieudonné crystal. In higher dimension, the proof uses the full faithfulness for truncated p-divisible groups to go from the fraction field of $\mathfrak{X}$ to an open subset of $\mathfrak{X}$ (in the reduced case), and a descent argument to glue groups defined on an open subscheme and on the formal completion along the complementary closed subscheme.

To prove the full faithfulness theorem up to isogeny, de Jong proves a more precise result: If S is a reduced scheme of finite type over a field k with a finite p-basis, and if G_1, G_2 are two p-divisible groups over S, there exists a decomposition $\mathrm{Hom}(G_1, G_2) \oplus T = \mathrm{Hom}(\mathbf{D}(G_2), \mathbf{D}(G_1))$, where T is a torsion subgroup. The proof is long and delicate. It uses rigid analytic methods, viewing Dieudonné crystals up to isogeny as convergent isocrystals (in the sense of the reviewer [Mém. Soc. Math. France (N.S.) No. 23 (1986), 3, 7–32; MR 88a:14020]). One can reduce to the case where S is affine, and use deformation theory to find liftings H_1, H_2 of G_1, G_2 over some base $\mathfrak{Y}$ over a Cohen ring of k, which are simultaneously versal. Given a morphism $\varphi : \mathbf{D}(G_2) \to \mathbf{D}(G_1)$, the general idea is to construct a closed formal subscheme $\mathfrak{Z} \subset \mathfrak{Y}$ containing S such that the wanted morphism $\psi : G_1 \to G_2$ (such that $\varphi - \mathbf{D}(\psi)$ is torsion) lifts to $\mathfrak{Z}$. Using induction on the dimension of S, the problem can be reduced to a problem of glueing two formal subschemes of $\mathfrak{Y}$ given on a regular open subscheme and on the formal completion along its complement. This is done by passing to the generic fiber of $\mathfrak{Y}$, which is a rigid analytic space, and by giving a direct construction of the generic fiber Z of $\mathfrak{Z}$. The existence of $\mathfrak{Z}$ follows then from a general result on formal schemes and rigid analytic spaces which is proved in the last section of the paper.

Let us point out another general result on rigid analytic varieties of independent interest, proved in a separate section of the article and used in the construction of Z: If X is a quasi-compact and connected rigid analytic variety, then any two points in X can be connected by a curve.

To conclude, this article makes an essential contribution to the general problem of classifying p-divisible groups through "semi-linear algebra" structures, and can be viewed as a major achievement of Grothendieck's philosophy about the relations between commutative p-groups schemes and Dieudonné crystals.

Pierre Berthelot (Rennes)

98h:14055a 14L15 20C20 20G10
Friedlander, Eric M. (Evanston, IL); **Suslin, Andrei** (Evanston, IL)
Cohomology of finite group schemes over a field.
Invent. Math. **127** (1997), *no.* 2, 209–270.

98h:14055b 14L15 20C20 20G10
Suslin, Andrei (Evanston, IL); **Friedlander, Eric M.** (Evanston, IL); **Bendel, Christopher P.** (Evanston, IL)
Infinitesimal 1-parameter subgroups and cohomology.
J. Amer. Math. Soc. **10** (1997), *no.* 3, 693–728.

98h:14055c 14L15 20C20 20G10
Suslin, Andrei (Evanston, IL); **Friedlander, Eric M.** (Evanston, IL);
Bendel, Christopher P. (Evanston, IL)
Support varieties for infinitesimal group schemes.
J. Amer. Math. Soc. **10** (1997), *no.* 3, 729–759.

Many major advances in the last two decades in modular representation theory have been made through the investigation and use of the cohomological spectrum. For finite groups, L. Evens [Trans. Amer. Math. Soc. **101** (1961), 224–239; MR **25** #1191] proved that the cohomology ring is Noetherian. Subsequent results by D. Quillen followed [Ann. of Math. (2) **94** (1971), 549–572; ibid. (2) **94** (1971), 573–602; MR **45** #7743] on the structure of the spectrum of this cohomology ring. In 1977, J. Alperin defined the important notion of the complexity of a module by looking at the rate of growth of the minimal projective resolution of the given module. Major progress in this area during this same time period can be attributed to J. F. Carlson [J. Algebra **85** (1983), no. 1, 104–143; MR 85a:20004], who defined a cohomological variety for a given module whose dimension equals the complexity of the module. These varieties are often referred to as support varieties. A stratification theorem providing some understanding about the structure of these support varieties (generalizing the prior work of Quillen) was proved by G. S. Avrunin and L. L. Scott [Invent. Math. **66** (1982), no. 2, 277–286; MR 83h:20048].

In a series of papers [see, e.g., Invent. Math. **86** (1986), no. 3, 553–562; MR 88f:17018], Friedlander and B. J. Parshall extended the theory of support varieties for restricted Lie algebras. In this setting they proved that the cohomology ring is finitely generated, and this result together with work of J. C. Jantzen [Abh. Math. Sem. Univ. Hamburg **56** (1986), 191–219; MR 88e:17019] provided a realization of these varieties as a subvariety of the given Lie algebra. For finite-dimensional graded connected cocommutative Hopf algebras, finite generation of the cohomology ring was proved by C. Wilkerson [Trans. Amer. Math. Soc. **264** (1981), no. 1, 137–150; MR 82e:16019], and in later work by A. M. Bajer and H. Sadofsky [J. Pure Appl. Algebra **94** (1994), no. 2, 115–126; MR 95g:18009]. These graded Hopf algebras arise naturally when one looks at finite-dimensional Hopf subalgebras of the Steenrod algebra. A theory of support varieties was developed within this context, similar to the one for finite groups, by Nakano and J. H. Palmieri [Math. Z. **227** (1998), no. 4, 663–684; MR 99d:55011]. Aside from these specific examples, it was not known in general whether the cohomology ring of an arbitrary finite-dimensional cocommutative Hopf algebra is finitely generated and whether a theory of supports could be developed. For many years the general question of finite generation remained as an open, difficult and elusive problem. The three papers under review provide both important and exciting breakthroughs in this direction.

In the first paper, the authors prove the finite generation of cohomology for a finite group scheme (or equivalently a finite-dimensional cocommutative Hopf algebra). The proof given by the authors involves first embedding a finite group scheme G into some general linear group GL_n and then constructing universal extension classes in certain degrees. These extensions in essence provide the generators for the cohomology ring. An interesting twist on this theme is the use of the polynomial functors developed by V. Franjou, J. Lannes and L. Schwartz in order to construct such extensions. The theory of such functors is related to the study of modules for the classical Schur algebras. Other interesting results in this paper involve computation of the cohomology of the infinite general linear group over a finite field.

With finite generation of cohomology firmly established, Friedlander, Suslin and Bendel are now armed to investigate the spectrum of the cohomology ring for infinitesimal group schemes in the second paper. The authors work in the context of schemes rather than varieties. The first step is to construct a scheme $V_r(G)$ consisting of homomorphisms from one-dimensional additive subgroups to the given group scheme G. This scheme in the case when the height equals one is the well-known scheme of p-nilpotent elements in $\mathrm{Lie}(G)$. More generally, $V_r(G)$ can be identified with r-tuples of p-nilpotent, pairwise commuting matrices of GL_n given a closed embedding of G into GL_n. This object was first defined and studied by D. Gross. The authors proceed to show that there exists a natural homomorphism of rings $\Phi\colon H^{2*}(G,k)\to k[V_r(G)]$, where $k[V_r(G)]$ is the coordinate algebra of $V_r(G)$. Much of the paper is devoted to determining characteristic classes associated to one-parameter subgroups of GL_n. This information is valuable because it enables the authors to give a concrete interpretation of the composition of $V_r(G)\to \mathrm{Spec}(H^{2*}(G,k))\to \mathrm{gl}_n^{(r)\times r}$, where the second composition is the map induced by a natural map obtained by looking at the universal classes defined in the first paper.

The third paper involves using the information obtained in the second paper to give a concrete realization of support varieties for infinitesimal group schemes. A key ingredient to the results is the use of embeddings of one-parameter subgroups into G to show that cohomology classes are detected modulo nilpotence on these subgroups. One of the main results which uses this detection principle is that the map Φ is indeed an inseparable isogeny. This allows the authors to give an explicit non-cohomological realization of the support variety of a rational G-module as a certain subscheme of $V_r(G)$. At the end of this paper several applications, including some generalizations of known results in the group algebra case, are given for infinitesimal group schemes. Explicit calculations of support varieties for simple and induced modules for Frobenius kernels of SL_2 are also included.

These three papers serve as an important contribution to the understanding of the cohomology and representation theory of finite group schemes. The authors manage to answer difficult questions in a very elegant manner. The advent of these new and beautiful ideas makes one optimistic that more interesting results in this direction will arise in the not too distant future. *Daniel K. Nakano* (Logan, UT)

99e:14061a 14N10 14H10 14J26 14M25
Caporaso, Lucia (Cambridge, MA);
Harris, Joe [Harris, Joseph Daniel] (Cambridge, MA)
Parameter spaces for curves on surfaces and enumeration of rational curves. (English summary)
Compositio Math. **113** (1998), *no.* 2, 155–208.

99e:14061b 14N10 14H10 14J26 14M25
Caporaso, Lucia (Cambridge, MA);
Harris, Joe [Harris, Joseph Daniel] (Cambridge, MA)
Enumerating rational curves: the rational fibration method. (English summary)
Compositio Math. **113** (1998), *no.* 2, 209–236.

The study of the geometry of rational curves on complex varieties, and in particular the enumerative geometry of such curves when the variety is "positive enough", has been reinvigorated and revolutionized by the introduction of the moduli space of stable maps by M. Kontsevich. In particular, there are many surprising patterns in these numbers implied by the WDVV equation (and other topological recursions, as well as Göttsche's conjecture and the fundamental Virasoro conjecture). In these two papers, an important contribution to the enumerative geometry of curves on surfaces, the authors introduce several powerful new ideas. However, they do this not by using stable maps (they use the Hilbert scheme), but by extracting important lessons from recent work in maps: (i) the deformation theory of rational curves in varieties can be quite simple if the variety is "positive enough"; (ii) one can use one-parameter families of rational curves to understand their enumerative geometry, so only "codimension-1" degenerations need to be studied; and (iii) families of rational curves are very special, allowing one to use techniques such as cross-ratio (see the first paper) or the simple topology of such families (see the second paper). (Nonetheless, the reader familiar with Gromov-Witten theory will find it enlightening to read the paper while simultaneously translating the arguments into the language of maps.)

The main ideas are as follows. Fix a surface S with a smooth divisor E. In applications, S will be $\mathbf{P}^2$ (and E a line), or the rational ruled surface $\mathbf{F}_n = \mathbf{P}_{\mathbf{P}^1}(\mathcal{O} \oplus \mathcal{O}(n))$ (with E the section of self-intersection $-n$). (Note that $\mathbf{F}_n$ is not convex, which had earlier stymied enumerative geometers.) The authors define "Severi varieties" parametrizing rational curves in a fixed divisor class, intersecting E at various unspecified points with specified multiplicities. In suitably nice situations, the components of the boundary of the Severi variety can be interpreted in terms of other Severi varieties. (The curve could break into two rational pieces, of the form described above, or it could break into the union of E and several rational pieces each of the form described above.) The authors analyze the deformation space of a high-order tacnode (the key technical lemma) and use this to understand the local geometry of the universal curve at a general point of the boundary. As a result, they have a family of nodal genus-0 curves whose geometry they understand well.

In the first paper, the authors extend Kontsevich's "cross-ratio trick" (which is what the WDVV equation boils down to in such situations). To count rational curves through n points (where n is chosen so that the number is finite), instead examine the one-parameter families of curves through $n-1$ points (with two of the points labeled p and q), with two more marked points r and s on curves L and M on S. By looking at the cross-ratio of the four points (in particular looking at the degree of the cross-ratio morphism from the one-parameter base to $\overline{M}_{0,4} \cong \mathbf{P}^1$ by counting pre-images—with multiplicity—of two boundary points), one finds a recursion for this number in terms of solutions to similar, simpler problems. Applications include a recursion counting irreducible rational curves in any divisor class on $\mathbf{F}_2$, a formula for rational curves on $\mathbf{F}_n$, and a beautiful closed form count of rational curves in a certain class on $\mathbf{F}_n$.

In the second paper, the authors use a similar one-parameter family (without marked points). If U is the total space of the family, with morphism $\pi: U \to S$, and L and M are divisor classes on S, then $(\pi^* L) \cdot (\pi^* M)$ is $L \cdot M$ times the solution to the desired enumerative problem. The Néron-Severi group of U is generated by the class of a section of the fibration and classes of components of fibers; the intersection product on these generators is not hard to calculate. The authors express $\pi^* L$ and $\pi^* M$ in terms of these generators (the coefficients involve solutions to similar problems), and then intersect these classes to produce recursions. Applications include another quick proof of Kontsevich's recursion for the plane, and general recursions counting irreducible rational curves on $\mathbf{F}_2$ and $\mathbf{F}_3$.

The methods are elegant, explicit, and easy to understand. The best introduction to the ideas of the papers is through the examples in the first author's survey [in *European Congress of Mathematics, Vol. I (Budapest, 1996)*, 136–149, Progr. Math., 168, Birkhäuser, Basel, 1998; MR 99j:14050].

The authors introduce several new ideas, which (when seen from the perspective of stable maps) have been influential in Gromov-Witten theory and elsewhere. The enumerative invariants are probably best understood as "relative Gromov-Witten invariants" (which have yet to be properly defined in algebraic geometry). The deformation theory of the high-order tacnode is the key technical result in the authors' incredible recursions [Invent. Math. **131** (1998), no. 2, 345–392; MR 99i:14064], counting curves of any degree and genus in the plane. "Nodal reduction" is essentially semistable reduction for maps in an explicit form. Finally, the ideas introduced here and in [L. Caporaso and J. Harris, op. cit.] can be extended in many ways (e.g. to count curves of any degree and genus on any $\mathbf{F}_n$ and del Pezzo surfaces of degree at least 3).

In conclusion, these two papers form a substantial contribution to the field. The methods are simple, beautiful, and concrete, yet very powerful.

Ravi D. Vakil (Cambridge, MA)

99e:14062 14N10 14D20 14J32

Lian, Bong H. (Waltham, MA); **Liu, Kefeng** (Stanford, CA); **Yau, Shing-Tung** (Cambridge, MA)

Mirror principle. I. (English summary)

Asian J. Math. **1** (1997), *no.* 4, 729–763.

The starting point for much of the current mathematical progress in enumerative geometry was the paper by the physicists P. Candelas, X. C. de la Ossa, P. S. Green and L. Parkes [Nuclear Phys. B **359** (1991), no. 1, 21–74; MR 93b:32029]. In that remarkable work the authors gave predictions for the numbers n_d of degree-d rational curves on a general quintic hypersurface X in four-dimensional complex projective space, for any d. Their main ingredient in the calculation was the so-called mirror symmetry, which asserts that a certain physical theory defined on a three-dimensional Calabi-Yau manifold is equivalent to another theory defined on its "mirror manifold". Using this, they were able to transform the numbers n_d, being ingredients of the one physical theory on the Calabi-Yau threefold X, into more accessible ones on the mirror manifold of X. Their prediction expresses the generating function for the n_d in terms of a certain hypergeometric series.

As mirror symmetry of Calabi-Yau threefolds still puzzles mathematicians, only modest progress has been made in the years since the paper of Candelas et al. to reproduce the numbers n_d by mathematically rigorous methods. By classical techniques, the cases $d = 1$ and $d = 2$ could be checked easily, but already the case $d = 3$ was quite difficult [G. Ellingsrud and S. A. Strømme, Math. Scand. **76** (1995), no. 1, 5–34; MR 96g:14045], and the case of general d was out of reach.

For general d the very first problem is already the definition of the numbers n_d— it is only conjectured but not yet known that the number of rational curves of fixed degree on a general quintic threefold is finite. Hence one has to define the n_d as suitable "virtual" numbers. This has been solved by M. Kontsevich and his moduli spaces of stable maps [in *The moduli space of curves (Texel Island, 1994)*, 335–368, Progr. Math., 129, Birkhäuser Boston, Boston, MA, 1995; MR 97d:14077]. Using torus actions on the moduli spaces of stable maps to projective four-space and the Bott residue formula, Kontsevich was in principle also able to compute the numbers n_d; however, his formula was extremely complicated and could only be checked numerically for low values of d to coincide with the physicists' prediction.

A few years later, A. B. Givental′ [Internat. Math. Res. Notices **1996**, no. 13, 613–663; MR 97e:14015, see p. B20 in this volume] pursued Kontsevich's idea of using the Bott residue formula further and applied the theory of equivariant cohomology to the problem. He derived the original hypergeometric series for the n_d conjectured by Candelas et al.; however, his proof was hard to understand and at some points incomplete.

The current paper of Lian, Liu, and Yau now gives the first complete rigorous proof of the physicists' formula for the numbers n_d. It uses essentially all the ingredients that other authors have applied to the problem before. It should be noted that the paper under review is considerably more general than stated here: it can be applied to many other problems of computing characteristic classes of vector bundles on the moduli spaces of rational stable maps to projective space.

The outline of the authors' technique, which they call the "mirror principle", is as follows. The numbers n_d are defined in terms of Euler classes of obstruction bundles on the spaces of genus-zero stable maps to $\mathbf{P}^4$. The first step of the calculation is to replace the spaces of degree-d stable maps by the so-called linear sigma model, i.e. by the spaces $N_d = \mathbf{P}^{5d+5}$ parametrizing five degree-d polynomials in the two coordinates of $\mathbf{P}^1$. Pushing forward the above Euler classes to the linear sigma model for each d, one gets a sequence Q_d of equivariant cohomology classes on N_d. These classes satisfy certain identities that can be derived from the behavior of the obstruction bundles at fixed points of the torus action. The sequence Q_d then gets "linked" to another such sequence P_d of classes on N_d satisfying the same identities, but which is easier to handle. Finally, this suffices to compute the Q_d in terms of the P_d (this is actually what the authors call the mirror transformation). Passing from equivariant to ordinary cohomology, one can finally read off the n_d from the Q_d.

The authors also announce a second paper ["Mirror principle. II", in preparation], which will include various further generalizations.

Andreas Gathmann (Cambridge, MA)

97c:14055 14Q20 14B05 14Q05 32S65 93B25

Gabrielov, Andrei [Gabrièlov, A. M.] (Ithaca, NY)

Multiplicities of zeroes of polynomials on trajectories of polynomial vector fields and bounds on degree of nonholonomy. (English summary)

Math. Res. Lett. **2** (1995), *no.* 4, 437–451.

In many important problems of analysis the following situation is encountered: to arrive at an answer, one has to perform recurrent algebraic computations (an algebraic algorithm) and check some polynomial inequality-type conditions. If these conditions are satisfied, the algorithm terminates with either a definite positive or a definite negative answer to the original problem. If the conditions are not satisfied, the situation remains indeterminate and additional iterations of the algorithm are required.

As a paradigm, consider the problem of determining the nature of a critical point of a real analytic function in one variable. In this model case the algorithm consists in computing consecutive Taylor coefficients, and the polynomial inequality-type condition consists in checking the sign of the coefficient obtained: if the sign is nonzero, then the answer can easily be found in terms of it and the parity of the step. If the last computed coefficient is zero, the next coefficient is to be considered, etc.

Analyzing various versions of this construction, appearing in connection with different problems, V. I. Arnol′d arrived in 1970 at the notion of an algebraically solvable problem of local analysis [Vestnik Moskov. Univ. Ser. I Mat. Mekh. **25** (1970), no. 2, 52–56; MR **43** #633]. According to Arnol′d, the local problem is called algebraically solvable if on the level of k-jets a positive or negative answer can be guaranteed for k-jets from certain semialgebraic sets of the jet space, and the indeterminacy occurs on a semialgebraic complement whose codimension grows to infinity with $k \to \infty$. Many problems are known to be algebraically solvable in this sense: in particular, the question whether a real analytic planar vector field with a pair of imaginary eigenvalues $\pm i\omega$, $\omega > 0$, has a center or a focus is algebraically solvable. On the other hand, the problem of Lyapunov stability was proved to be not algebraically solvable [V. I. Arnol′d, Funkcional. Anal. i Priložen. **4** (1970), no. 3, 1–9; MR **42** #7989].

The definition of algebraic solvability applies to local analytic (not necessarily algebraic) objects. In the analytic settings there obviously cannot be any upper bound for the number of iterations of the algorithm. However, if the problem itself is algebraic, i.e. all k-jet spaces eventually stabilize for $k \geq d$ (the degree), then the new phenomenon appears: in many cases one can prove that after some number of iterations there is no indeterminacy component in the corresponding N-jet space. This usually follows from the fact that the polynomial rings are Noetherian, hence all descending chains of algebraic varieties eventually stabilize.

In our simplified example of determining the type of a critical point of an analytic function in one variable, this corresponds to the following (obvious) claim: if a function is known to be polynomial of degree $\leq d$ and the above algorithm gave no answer after d iterations, then the function is constant and hence has (nonstrict) local minimum. The problem of determining N (in terms of d) becomes therefore a most important question after the algebraic solvability is established.

Probably the best known result of this type is the Dulac theorem on quadratic vector fields. Recall that a singular point of a planar analytic vector field with a pair of imaginary eigenvalues possesses the first return map (the Poincaré map), and the Taylor coefficients of this map can be recursively computed using algebraic operations. To be a center, the singular point should have the identical first return map, that is, all higher Taylor coefficients should vanish. The Dulac theorem [H. Dulac, Bull. Sci. Math. (2) **32** (1908), 230–252; JFM 39.0374.01] claims that if a planar quadratic vector field has a singular point with two imaginary eigenvalues and the first seven Taylor coefficients vanish, then the point is of the center type, i.e. the first return map is identical. Similar results are known for vector fields with homogeneous cubic nonlinearities, but the general case is totally obscure and constitutes one of the most challenging problems.

After this lengthy introduction aimed at illustrating the importance of obtaining explicit bounds on the length of solutions in algebraically solvable algebraic problems, we proceed with discussing the paper itself. It contains solutions of the following two problems.

The first of them is referred to as the Risler problem on the maximal order of tangency [J.-J. Risler, Theoret. Comput. Sci. **157** (1996), no. 1, 129–136; MR 96m:93065]. Consider a polynomial vector field in $\mathbf{R}^n$ and its nonsingular trajectory γ. Then for any algebraic hypersurface Σ the curve γ may entirely belong to Σ or have a finite order of tangency. If the hypersurface is given by the equation $P(x)=0$, where $P\in \mathbf{R}[x]$ is a polynomial, and γ is parameterized as $t\mapsto x(t)$, then the order of tangency is the order of zero of the restriction $f(t)=P(x(t))$ at a certain point $t=t_0$, and without loss of generality we may put $t_0=0$. The problem of determining this order is tantamount to computing the first nonzero Taylor coefficient of the (analytic) function $f(t)$, and it is obvious that the problem of determining whether $\gamma\subset\Sigma$ or not is algebraically solvable in the above sense. The first main result by Gabrièlov is an explicit upper bound for the maximal nontrivial order of tangency in the case when the degree of the polynomial vector field and the surface are explicitly known together with the dimension of the problem.

Theorem 1. A nonsingular trajectory of a polynomial vector field of degree $\le d_1$ in $\mathbf{R}^n$ and an algebraic hypersurface of degree $\le d_2$ may have tangency of order $\le d_1^{2^{n-2}-1}\cdot d_2^{2^{n-2}}\cdot(d_1+d_2)^{2^{n-2}}$.

It is worth noting that this upper bound is double exponential in the number of variables, though there is no claim as to the sharpness of this bound.

Besides proving this elegant and impressive result (only the planar case was previously known [A. M. Gabrièlov, J.-M. Lion and R. Moussu, C. R. Acad. Sci. Paris Sér. I Math. **319** (1994), no. 3, 219–221; MR 95g:32058]), the author gives an application of it to the problem of effective bounds of nonholonomy degree. The second main result of the paper concerns the geometric control theory but has deep connections in partial differential equations (the so-called Hörmander condition), stochastic processes and differential (sub-Riemannian) geometry. Let $\xi_1,\cdots,\xi_k$ be a collection of (germs of) analytic vector fields in a neighborhood of a point, say, the origin in $\mathbf{R}^n$, at which $\xi_1(0)\neq 0$. Consider the Lie algebra $\mathscr{L}(\xi_1,\cdots,\xi_k)$, generated by these fields, i.e. the linear space generated by the fields ξ_i and their iterated commutators of all orders. Then the values of all vector fields from $\mathscr{L}$ evaluated at the origin generate some d-dimensional subspace of the tangent space. It is important to know the maximal order of commutators necessary to generate this d-dimensional subspace.

Theorem 2. The number of iterated commutators necessary to generate the subspace of maximal dimension is no greater than some explicit expression polynomial in d and double exponential in n.

The paper under review is written in the very succinct style characteristic of the author. Thus we will use this opportunity to make several comments that we hope will highlight the proofs of the above-mentioned results. Consider first the problem of estimating the order of contact between an algebraic hypersurface $\{p_0(x)=0\}\subset \mathbf{R}^n$, $p_0\in\mathbf{R}[x]$, and a trajectory of the polynomial vector field $\dot{x}_j = v_j(x)$, $j = 1,\cdots,n$, $v_j\in\mathbf{R}[x]$.

Let R be a commutative ring with unit and $D\colon R\to R$ a derivation of this ring. Then for any element $p_0\in R$ one can form a sequence of derivatives $p_{k+1}=Dp_k$ and the ascending chain of ideals $I_k=(p_0,\cdots,p_k)$, $I_0\subseteq I_1\subseteq\cdots\subseteq I_k\subseteq\cdots\subseteq R$.

The basic and very simple observation is that this chain must be strictly ascending, i.e. if $I_k=I_{k+1}$, then from that moment on the chain of ideals stabilizes: $I_{k+1}= I_{k+2}=\cdots\stackrel{\text{def}}{=}I_\infty$. This observation is applied to the problem of order of tangency as follows.

Let $L\colon\mathbf{R}[x]\to\mathbf{R}[x]$ be the (Lie) derivation of the polynomial ring into itself along the vector field $v(x)=(v_1(x),\cdots,v_n(x))$, $Lp=\sum_{j=1}^n v_j(\partial p/\partial x_j)$. Consider the consecutive derivatives $p_{k+1}=Lp_k$, $k=0,1,2,\cdots$, the ascending chain $J_k = (p_0,p_1,\cdots,p_k)\subseteq\mathbf{R}[x]$ of polynomial ideals and the descending chain $Z_k=\{x\in \mathbf{C}^n\colon\ p_0(x)=\cdots=p_k(x)=0\}\subseteq\mathbf{C}^n$ of their respective complexified zero loci. By construction, the set Z_k consists of points of contact of order $\geq k$ between the hypersurface and the field. Therefore our problem is to determine the moment of stabilization of the chain $\{Z_k\}$, i.e. the first index ν such that $Z_\nu=Z_{\nu+1}=\cdots\stackrel{\text{def}}{=}Z_\infty$.

The upper bound for the length of the chain Z_k is obtained upon analyzing the dimension of the (complex semialgebraic) sets $Z_k\smallsetminus Z_\infty$. Let the latter locus be r-dimensional (as a complex stratified subvariety of $\mathbf{C}^n$). The vector field v can be tangent to $Z_k\smallsetminus Z_\infty$ only on a proper subset of strictly smaller dimension (otherwise the entire trajectories would lie on Z_k, which is impossible outside Z_∞). Let $a\notin Z_\infty$ be a point where v is transversal to some r-dimensional stratum of $Z_k\smallsetminus Z_\infty$. Take an $(n-r)$-dimensional analytic transversal T to Z_k, invariant by the local flow of v. The restrictions of all polynomials from J_k on this transversal span an ideal I_k in the local ring R of analytic germs $(T,a)\to\mathbf{C}$, and this ideal is of a finite codimension $\mu_k=\dim_{\mathbf{C}} R/I_k<\infty$ in R, because the intersection $T\cap Z_k=\{a\}$ is isolated. The invariance of T by v allows one to define the derivation D of R, conjugate to the original derivation L. The (sub)chain of ideals $\{J_j\}_{j=k}^\infty$ after restriction on T generates the ascending chain of ideals $\{I_j\}_{j=k}^\infty$ in the ring R of germs, spanned by consecutive iterations of D. All ideals I_j have finite codimension in R. By the basic observation, this chain has to be strictly ascending, hence the codimensions $\mu_j = \dim_{\mathbf{C}} R/I_j$ must strictly decrease with j. Thus the length l of this chain does not exceed the codimension μ_k of the starting ideal I_k, and the latter can be estimated by the Bézout inequality.

The chain I_j cannot stabilize on a nontrivial ideal $I_\infty\subsetneqq R$, since this would imply $a\in Z_\infty$. Therefore $I_{k+l}=R$, so $p_{k+l}(a)\neq 0$ and $a\notin Z_{k+l}$. In other words, the locus Z_{k+l} must be less than r-dimensional. Iterating these arguments by induction and using explicitly the Bézout inequalities, one arrives at the inequality asserted in Theorem 1.

The proof of the second main result is based on the following construction, which reduces the problem of the degree of nonholonomy to that of estimating the order of tangency between a vector field and some algebraic surface. The author constructs inductively the chain $\chi_0, \chi_k, \cdots, \chi_{d-1}$ of vector fields (recall that d is the dimension of the subspace generated by the original vector fields ξ_i and their commutators at the origin), as follows.

The chain begins with, say, ξ_1, and either each next χ_i is a linear combination of the preceding χ_j with $j<i$ and their commutators, or (in the case when all such commutators belong to the subspace spanned by $\chi_0, \cdots, \chi_{i-1}$) the next field χ_i is chosen to be one of the fields ξ_j that never occurred previously in the construction. This choice can be done in such a way that the vector fields $\chi_0, \cdots, \chi_{d-1}$ are linearly independent on an open subset of $\mathbf{R}^n$, and, moreover, the wedge product $\Omega = \chi_0 \wedge \cdots \wedge \chi_{d-1}$ is not identically zero on the integral trajectory of some linear combination $\chi = \sum_{j=0}^{d-1} \lambda_j \chi_j$ of the fields χ_j (note that this product may well be zero at the origin).

Assuming that the fields ξ_j were polynomial, one can ensure that the fields χ_j are also polynomial, and their degrees will be explicitly bounded then. Their wedge product Ω is therefore a (vector) polynomial of some known degree, and the fact that it is not identically zero on the trajectory of χ means that the order of contact between χ and the algebraic surface $\{\Omega = 0\}$ is finite. Applying Theorem 1, one may obtain an upper bound for the corresponding multiplicity ν of contact. After differentiating Ω along χ the specified number ν times, we arrive at an expression that is nonzero at the origin. On the other hand, the explicit formulas for derivation of the wedge product show that this expression is a linear combination of dth degree wedge products of the fields ξ_j and their iterated commutators of order $\leq \nu$. Thus at least one such monomial wedge product is nonzero, hence the corresponding fields and their commutators of order $\leq \nu$ span the d-dimensional subspace at the origin. In other words, in order to determine the order of nonholonomy it is sufficient to compute the iterated commutators of orders $\leq \nu$, q.e.d.

Reading the paper under review is not an easy thing, but the efforts are well rewarded by learning the fascinating results, which make a very important contribution to this interesting and rapidly progressing subject. *Sergeĭ Yu. Yakovenko* (Rehovot)

97i:16007 16G60 16E30
de la Peña, J. A. (México); **Skowroński, A. [Skowroński, Andrzej]** (Toruń)
Geometric and homological characterizations of polynomial growth strongly simply connected algebras.
Invent. Math. **126** (1996), *no.* 2, 287–296.

In the study of the module theory for finite-dimensional algebras over an algebraically closed field a lot of attention was given in the seventies to investigating the algebras of finite type, that is, those where there are only a finite number of indecomposable (finitely generated) modules up to isomorphism. The language of representations of quivers (with relations) was useful in this connection. The next level of complication for the module structure was given by the tame algebras, that is, those algebras for which in each dimension there are only a finite number of one-parameter families of indecomposable modules and only a finite number of discrete indecomposable modules. For path algebras of finite connected quivers the ones of finite type are those where the underlying graph of the quiver is a Dynkin diagram, and the tame ones (not of finite type) correspond to extended Dynkin diagrams. There are, however, not many results of a general nature on tame algebras, so a lot of effort has gone into trying to get a better understanding of special classes of tame algebras. One such class consists of the algebras of polynomial growth, where by definition the minimal number of one-parameter families grows polynomially with the dimension. For self-injective algebras, in particular for group algebras of finite groups over an algebraically closed field, there is a lot of information available on this class, by Skowroński, Erdmann and others.

In the paper under review the authors deal with tame algebras, in particular those of polynomial growth. They start out with a class of algebras called strongly simply connected, which includes for example the algebras where the graph of the associated quiver is a tree. It is hoped to reduce the study of tame algebras to the tame strongly simply connected algebras, via covering techniques.

The paper contains both geometric and homological characterizations of polynomial growth algebras. Already for the study of finite type it was useful to view the modules with a fixed finite set of composition factors as a variety, with a group acting on the variety in such a way that orbits correspond to isomorphism classes of modules. There is a natural quadratic form, called the Tits form, associated with a finite-dimensional algebra, and a geometric argument was used by Tits to show that this form is positive definite for the path algebra of a Dynkin quiver. Geometric methods and associated quadratic forms have also been important for tame algebras, and it had been shown earlier [see J. A. de la Peña, Comm. Algebra **19** (1991), no. 6, 1795–1807; MR 92i:16016] that the Tits form of a tame algebra whose quiver has no oriented cycles must be weakly nonnegative. A characterization of polynomial growth is given in terms of dimensions of associated varieties and a quadratic form called the Euler form (closely related to the Tits form). One of the main homological results in this paper is that an algebra is of polynomial growth if and only if the Tits form is weakly nonnegative and $\mathrm{Ext}^2(X,X)=0$ for each indecomposable module X. The main results fail if the assumption of strongly simply connected is dropped. The proofs use previous detailed investigations of algebras of polynomial growth, including tests for algebras being of this type, as developed by the authors and others. *Idun Reiten* (Dragvoll)

98k:16019 16G60

Skowroński, Andrzej (Toruń)

Simply connected algebras of polynomial growth. (English summary)

Compositio Math. **109** (1997), *no.* 1, 99–133.

Finite-dimensional algebras over an algebraically closed field are divided into two disjoint classes depending on the structure of their module categories. One class consists of the tame algebras, for which, in each dimension, there are only a finite number of one-parameter families of indecomposable modules and only a finite number of discrete indecomposable modules. The second class is formed by the wild algebras, for which the classification of indecomposable modules contains the well-known open problem of classification of pairs of non-commuting endomorphisms up to conjugation.

Until the beginning of the eighties much attention was paid to the simplest case of the tame situation, that is, the case of algebras accepting only finitely many indecomposable modules up to isomorphism. Techniques were developed by several authors in order to classify all indecomposable modules over those algebras. For the general class of tame algebras few general results were known. There were important examples, including the path algebras of quivers whose underlying graph is of extended-Dynkin type, the tame concealed algebras and the tubular algebras defined by C. M. Ringel [*Tame algebras and integral quadratic forms*, Lecture Notes in Math., 1099, Springer, Berlin, 1984; MR 87f:16027]. For these examples the structure of the module category was clearly understood.

The representation theory of tame algebras is slowly emerging. At the moment, an accessible kind of tame algebras consists of the polynomial growth algebras, where the minimal number of one-parameter families of indecomposable modules grows polynomially with the dimension. This kind of algebras includes the above-mentioned examples, certain group algebras of finite groups and others. Particularly interesting are the strongly simply connected algebras, which are by definition algebras A without oriented cycles in the Gabriel quiver and such that the first Hochschild cohomology $H^1(C)$ vanishes for any full subcategory C of A. It is hoped to reduce the study of any polynomial growth algebra to the polynomial growth strongly simply connected algebras via covering techniques. In the paper under review the author provides a very complete description of the structure of the module category over any strongly simply connected algebra of polynomial growth. The work is the conclusion of the research on the topic by the author of the paper and other authors. It will be a necessary reference for further research on tame algebras.

Let A be a strongly simply connected algebra. The main result of the paper shows that A is of polynomial growth if and only if A does not contain any convex subcategory which is pg-critical or hypercritical. This is a handy criterion for checking whether or not A is of polynomial growth since the strongly simply connected algebras which are pg-critical and those which are hypercritical have been already classified [see, respectively, R. Nörenberg and A. Skowroński, Colloq. Math. **73** (1997), no. 2, 301–330; MR 98f:16012; L. Unger, Bayreuth. Math. Schr. No. 31 (1990), 145–154; MR 91i:16033]. Moreover, A is of polynomial growth if and only if every cycle of nonzero morphisms between indecomposable modules is finite and this is equivalent to the vanishing of $\operatorname{rad}^\infty(X, X)$ for every indecomposable A-module X. Further, a description of the components of the Auslander-Reiten quiver of a polynomial growth algebra A is given. Namely, an appropriate generalization of the notion of tube is called a coil [I. Assem and A. Skowroński, Math. Scand. **71** (1992), no. 1, 31–61; MR 94c:16020]; then a multicoil consists of a finite number of coils glued together by some cycle-free parts. The algebra A is of polynomial growth if and only if every cycle of maps between indecomposables consists of modules on a coil of a multicoil of the Auslander-Reiten quiver of A. The main part of the proof given in the paper consists of a careful analysis of the structure of the components of the Auslander-Reiten quiver of an algebra A which does not contain pg-critical or hypercritical convex subcategories. An appropriate induction argument shows that these components are multicoils.

A number of consequences are given in the paper. There is a natural quadratic form, called the Tits form, associated with a finite-dimensional algebra. It has been shown that for tame algebras the Tits form is weakly non-negative [J. A. de la Peña, Comm. Algebra **19** (1991), no. 6, 1795–1807; MR 92i:16016]. In the paper under review it is shown that a strongly simply connected algebra A having no pg-critical convex subcategories is tame if and only if the associated Tits form is weakly non-negative. Other consequences have been given in other papers; for example, a strongly simply connected algebra A is of polynomial growth if and only if the Tits form is weakly non-negative and $\mathrm{Ext}^2(X, X) = 0$ for every indecomposable A-module X [J. A. de la Peña and A. Skowroński, Invent. Math. **126** (1996), no. 2, 287–296; MR 97i:16007, see p. B69 in this volume]. *J. Antonio de la Peña*

98c:16034a 16S34 16N20 20C07
Passman, D. S. (Madison, WI)
The Jacobson radical of group rings of locally finite groups. (English summary)
Trans. Amer. Math. Soc. **349** (1997), *no.* 12, 4693–4751.

98c:16034b 16S34 16N20 20C07
Passman, D. S. (Madison, WI)
The semiprimitivity problem for twisted group algebras of locally finite groups. (English summary)
Proc. London Math. Soc. (3) **73** (1996), *no.* 2, 323–357.

The main result of these beautiful papers, the culmination of a quarter-century of work in which the author has throughout played the leading role, is a description of the Jacobson radical of the group algebra of a locally finite group. Groups acting by linear transformations on vector spaces over a field K are, of course, ubiquitous throughout mathematics. Typically, one tries to analyse such actions in terms of the irreducible ones, these being the ones for which no proper subspace is left invariant by the group. These are the irreducible modules for the group algebra KG, and they can provide no information about those K-linear combinations of group elements which annihilate all such modules, so one is immediately led to try to identify this ideal of KG. This is the Jacobson radical $J(KG)$, whose study spans our century, starting from Maschke's theorem of 1898 stating that, for finite G, $J(KG) \neq 0$ if and only if K has characteristic $p > 0$ and G contains an element of order p.

Confronted by Maschke's result one might make the wild surmise that the same result holds for all groups G. When K has characteristic 0 it is possible that this is indeed true: In 1950 C. Rickart used the dense embedding of $\mathbf{C}G$ in $L^1(G)$ to show that $J(\mathbf{C}G) = 0$ for all groups G [C. E. Rickart, Ann. of Math. (2) **51** (1950), 615–628; MR **11**, 670d]. Following I. Kaplansky's highlighting of the problem in his seminal Ram's Head Inn problem list [*Problems in the theory of rings. Report of a conference on linear algebras, June, 1956, pp. 1–3*, Nat. Acad. Sci., Washington, Publ. 502, 1957; MR **20** #3179], S. A. Amitsur used techniques from associative algebra (and essentially no group theory) to extend Rickart's result to all characteristic-0 fields K which are not algebraic over $\mathbf{Q}$ [Michigan Math. J. **6** (1959), 251–253; MR **21** #7256]; and there, in char 0, matters rest, all the major developments of the past 40 years having concerned the case of positive characteristic.

The naive extension of Maschke's theorem to all groups is false in positive characteristic. Indeed there are semisimple group algebras KG over fields of characteristic p with G having non-trivial normal abelian p-subgroups: $G = C_p \wr C_\infty$ will do, as observed by Passman [Proc. Amer. Math. Soc. **20** (1969), 515–519; MR **38** #4576]. Stimulated in part by the first, beautiful interventions into the field by A. E. Zalesskiĭ, who determined when the group algebra of a soluble group is semisimple and in so doing popularised the notion of a controlling subgroup for an ideal, Passman formulated the conjecture that $J(KG) = J(K\Lambda^+(G))KG$, where $\Lambda^+(G) = \{x \in G\colon |x| < \infty$ and $|H\colon C_H(x)| < \infty$ for all finitely generated subgroups H of $G\}$. Since $\Lambda^+(G)$ is a characteristic locally finite subgroup of G, the truth of this conjecture would reduce the problem of describing $J(KG)$ to the case where G is locally finite. Unfortunately, very little progress has been made on this conjecture since it was stated in 1974. However, the first of the papers under review completes the second half of the 1974 programme, by describing $J(KG)$ for all fields K and locally finite groups G.

To formulate the main theorem, say that a finite subgroup A of the locally finite group G is locally subnormal in G if A is subnormal in every finite subgroup B of G which contains A. Fix a prime p and set $S^p(G)$ to be the subgroup of G generated by the set of all locally subnormal subgroups of G which are generated by their p-elements. Let K be a field of characteristic p, and assume that G has no non-trivial normal p-subgroup—that is, that $O_p(G) = 1$. Then $J(KG) = J(KS^p(G))KG$; and $J(KS^p(G)) = \bigcup\{J(KA)\colon A \in \ \}$. Thus the word "describing" at the end of the last paragraph means here "reducing to the case of finite groups".

The proof is too long and complex to be summarised here. Much of the work needed is group-theoretic; amongst the factors on which it hinges, we highlight (i) the classification of finite simple groups, which has been used crucially in several places; (ii) the return of Zalesskiĭ to the subject, resulting in his joint paper with Passman [Proc. London Math. Soc. (3) **67** (1993), no. 2, 243–276; MR 94g:16036]; (iii) the work of J. I. Hall and R. E. Phillips on finitary locally finite groups [e.g., J. I. Hall, in *Finite and locally finite groups (Istanbul, 1994)*, 147–188, Kluwer Acad. Publ., Dordrecht, 1995; MR 96j:20042; R. E. Phillips, in *Finite and locally finite groups (Istanbul, 1994)*, 111–146, Kluwer Acad. Publ., Dordrecht, 1995; MR 96m:20080]; and (iv) the need to work with twisted group algebras. The last point continues a pattern which prevailed in the 70s also—and it manifests itself now in that the proof of the main theorem of the first paper (quoted above) uses crucially the main results of the second paper, and of another of the author's papers [Israel J. Math. **96** (1996), part B, 481–509; MR 98e:16025], which together give necessary and sufficient conditions for a twisted group algebra of a locally finite group to be semiprimitive.

Finally let me endorse the words of the introduction to the second paper explaining why the latter is dedicated to the memory of Brian Hartley. I first learned about group rings (and in particular about Passman's conjectural description of $J(KG)$) in a graduate course given by Brian Hartley at Warwick in 1974. Throughout his life Brian took a keen interest in the semisimplicity problem, and was himself a leading advocate of the use of the deepest group-theoretic methods and results, allied to ring theory, as a way of advancing the subject. He would greatly have enjoyed reading the papers under review. *Kenneth A. Brown* (Glasgow)

99m:17006 17B05 17B65

Kac, Victor G. (Cambridge, MA)

Classification of infinite-dimensional simple linearly compact Lie superalgebras.

Adv. Math. **139** (1998), *no.* 1, 1–55.

A. J. Coleman [Math. Intelligencer **11** (1989), no. 3, 29–38; MR 90f:01047] describes W. Killing's 1888 paper [Math. Ann. **33** (1888), 1–48; JFM 20.0368.03] as "the greatest mathematical paper of all time". In that paper, among many other important things, Killing classified the possible (finite-dimensional) simple Lie algebras over the complex numbers. Killing's results were taken up by É. Cartan in his famous thesis. Cartan gave a clear exposition of Killing's results, corrected some errors, and filled in some details of proofs. (Then Cartan went far beyond Killing and set the stage for the development of Lie group theory in the 20th century.)

The Cartan-Killing classification is the first of four classification results that I want to discuss in this review. The remaining three are: (i) The list by Cartan [Ann. Sci. École Norm. Sup. (3) **26** (1909), 93–161; JFM 40.0193.02] of the "simple infinite transformation groups". Cartan's proof of his classification contained a gap. The complete proof was given in [V. Guillemin, D. Quillen and S. Sternberg, Proc. Nat. Acad. Sci. U.S.A. **55** (1966), 687–690; MR **34** #3594], using the "integrability of characteristics" theorem [see V. W. Guillemin, D. Quillen and S. Sternberg, Comm. Pure Appl. Math. **23** (1970), no. 1, 39–77; MR **57** #1582]. (ii) The classification of the finite-dimensional simple Lie superalgebras achieved by Kac [V. G. Kac, Advances in Math. **26** (1977), no. 1, 8–96; MR **58** #5803]; see also the exposition by M. Scheunert [*The theory of Lie superalgebras* , Springer, Berlin, 1979; MR 80i:17005]. Finally (iii) the paper under review, in which Kac classifies the infinite-dimensional simple linearly compact Lie superalgebras.

The reader might be surprised at the almost sixty-year hiatus between Cartan's work and Guillemin, Quillen and Sternberg's 1966 paper (cited above) and the shift in terminology from "group" to "pseudogroup" or "linearly compact algebra". The reasons are related. In the interim the "finite transformation groups" of Lie had been developed (by Cartan, Weyl and others) into a theory of abstract Lie groups as manifolds with a compatible group structure. One must remember that the very definition of a manifold was first introduced by Hermann Weyl in 1913 [cf. *Die Idee der Riemannschen Fläche*, Teubner, Leipzig; JFM 44.0492.01], and it was Weyl's magnificent series of papers in the 1920's [Math. Z. **23** (1925), 271–309, Math. Z. **24** (1926), 328–376, 377–395; JFM 51.0319.01; Math. Z. **24** (1926), 789–791; JFM 52.0116.02] that brought Cartan's work on "finite continuous groups" and their representations to the attention of the mathematical public. Thus it was the study of abstract Lie groups, their representations, and more generally their smooth actions on manifolds which occupied center stage. What was needed was an appropriate category for studying the "infinite transformation groups" in an abstract setting.

This was provided in [I. M. Singer and S. Sternberg, J. Analyse Math. **15** (1965), 1–114; MR **36** #911; V. W. Guillemin and S. Sternberg, Bull. Amer. Math. Soc. **70** (1964), 16–47; MR **30** #533] with the notion of a linearly compact Lie algebra L. (Linearly compact means that the topology is such that L admits a fundamental system of neighborhoods of 0 consisting of subspaces of finite codimension, and such that L is complete in this topology.) A subalgebra L^0 is called fundamental if it is proper and contains no nonzero closed ideals. Given L and L^0 we define $L^{-1} := L$ and inductively L^i for $i > 0$ by

$$L^i = \{X \in L^{i-1} | \, [X, Y] \in L^{i-1} \; \forall Y \in L^{-1}\}.$$

We will have $\bigcap L^i = \{0\}$ and the L^i provide a descending filtration giving a system of neighborhoods of 0. Associated with this filtration we get of course a **Z**-graded Lie algebra $\mathfrak{g} = \bigoplus \mathfrak{g}_i$, where $\mathfrak{g}_i := L^i/L^{i-1}$ so that the grading starts at -1. (An important refinement of this filtration and grading was introduced by B. Weisfeiler [Funkcional. Anal. i Priložen **2** (1968), no. 1, 94–95; MR **38** #1134], who chose L^{-1}

to be a subspace containing L^0 which generates L as a subalgebra, defined L^{-j} inductively by $L^{-j-1} := [L_{-1}, L_{-j}] + L_{-j}$ for $j \geq -1$ and defined L^i as above but with this new choice of L^{-1}. Then the associated filtration starts at some finite negative integer, not necessarily -1.) Guillemin and Sternberg's paper [op. cit.] is reprinted as an appendix in [S. Sternberg, *Lectures on differential geometry*, Second edition, Chelsea, New York, 1983; MR 88f:58001].

An example of all this is to take L to be Vect(n), the algebra of all formal power series vector fields in n variables, so that an element of Vect(n) is an expression of the form

$$X = X^1 \frac{\partial}{\partial x_1} + \cdots + X^n \frac{\partial}{\partial x_n}$$

where the X^r are formal power series in the (complex) variables $x_1, \cdots, x_n$ and bracket is the usual Lie bracket of vector fields; here L^i consists of vector fields whose coefficients vanish together with all their partial derivatives up to order i at the origin. The associated graded pieces $\mathfrak{g}_i$ then consist of vector fields whose components are homogeneous polynomials of degree $i+1$. The Lie algebra Vect(n) corresponds to one of Cartan's simple "groups of infinite type", the "group" of all (locally defined) transformations on an n-dimensional manifold. In [V. W. Guillemin and S. Sternberg, op. cit. (Theorem III)] it was proved that every linearly compact Lie algebra is isomorphic to a subalgebra of some Vect(n). In [R. J. Blattner, Trans. Amer. Math. Soc. **144** (1969), 457–474; MR **46** #7338a] this "realization theorem" was proved using the tools of induced and produced representations, and the analogue of this result for Lie superalgebras was of use in the classification of the finite-dimensional simple Lie superalgebras; see for example Section 4.5 of Scheunert's book [op. cit.] and also Example 1.3 in the paper under review.

Having established the class of objects under study, the next project was to translate many of the concepts and theorems of Cartan expressed in the framework of exterior algebra (and also important work by D. C. Spencer [Ann. of Math. (2) **76** (1962), 306–398; MR **27** #6287a; Ann. of Math. (2) **76** (1962), 399–445; MR **27** #6287b; Ann. of Math. (2) **81** (1965), 389–450; MR **31** #4052]) into the more familiar language of commutative algebra. This was done in [I. M. Singer and S. Sternberg, op. cit.; V. W. Guillemin and S. Sternberg, op. cit.] and especially in the important letter from Serre reproduced in the latter paper. For example, once this was done, we were able to classify the possible $\mathfrak{g}_0$ which could arise from an infinite-dimensional L, and which acted irreducibly on the corresponding $\mathfrak{g}_{-1}$ by a straightforward application of Hilbert's Nullstellensatz [see I. M. Singer and S. Sternberg, op. cit.; V. W. Guillemin, D. Quillen and S. Sternberg, J. Analyse Math. **18** (1967), 107–112; MR **36** #221].

But the key step in completing the proof of Cartan's classification of the simple transitive infinite "groups" depended on a new idea which went beyond commutative algebra. This was the "integrability of characteristics" theorem, which had to do with the relation between a filtered associative ring whose graded ring is commutative and the behavior of radicals of ideals in the graded ring under the induced Poisson structure. A statement of this theorem can be found on pages 418–419 of [S. Sternberg, op. cit.]. In [V. W. Guillemin, D. Quillen and S. Sternberg, op. cit.; MR **57** #1582] we stated this theorem in general, and proved a version of it for the case of the ring of analytic differential operators in n complex variables, our version being sufficient for our applications. Our proof involved a process of localization which used the formal analogue of pseudodifferential operators, what we might call formal "microlocalization"; and we were able to prove the integrability of characteristics at all points of the maximal-dimensional components of the characteristic variety. M. Sato, T. Kawai and M. Kashiwara [in *Hyperfunctions and pseudo-differential equations (Proc. Conf., Katata, 1971; dedicated to the memory of André Martineau)*, 265–529, Lecture Notes in Math., 287, Springer, Berlin, 1973; MR **54** #8747] extended our techniques to show that the result holds at all points of the characteristic variety. In my seminar in Tel Aviv in 1980 I posed the problem of finding a better version of noncommutative localization which could give a direct proof of the integrability of characteristics theorem, and within a few weeks Ofer Gabber came up with such a proof [see Amer. J. Math. **103** (1981), no. 3, 445–468; MR 82j:58104]. (In view of Gabber's result, Kac's comment in the paper under review that our solution of Cartan's problem made use of a "rather complicated result from analysis" seems now unwarranted.)

Using the Weisfeiler filtration mentioned above, Weisfeiler and Kac gave an alternative proof of the Cartan classification which did not make use of the integrability of characteristics theorem [see V. G. Kac, Izv. Akad. Nauk SSSR Ser. Mat. **32** (1968), 1323–1367; MR **41** #4590; B. Yu. Weisfeiler, Funkcional. Anal. i Priložen **2** (1968), no. 1, 94–95; MR **38** #1134; and related papers]. Also, Kac [op. cit.; MR **41** #4590] introduced a rather different class of algebras (perhaps more famous today than the Cartan algebras), which are currently known as the Kac-Moody algebras. On the other hand, Guillemin [J. Differential Geometry **4** (1970), 257–282; MR **42** #3132] used the characteristic variety to give a direct proof of the existence of a unique maximal fundamental subalgebra in a primitive infinite-dimensional linearly compact Lie algebra. In the paper under review Kac extends Guillemin's method of proof to the superalgebra case, introducing a "super Nullstellensatz" and a concomitant "characteristic supervariety", and Kac's extension of Guillemin's method plays a key role in his classification argument. Additionally, extensions of the Spencer cohomology and an extension of the study of $\mathbf{Z}$-graded algebras to the "super" case [cf. S.-J. Cheng and V. G. Kac, Adv. Theor. Math. Phys. **2** (1998), no. 5, 1141–1182; Transform. Groups **4** (1999), no. 2-3, 219–272] play an essential role in the classification theorem in the paper under review. This completes what I want to say about Cartan's classification of the infinite-dimensional simple Lie algebras, except for one piquant detail. In the classification of A. I. Kostrikin and I. R. Shafarevich [Izv. Akad. Nauk SSSR Ser. Mat. **33** (1969), 251–322; MR **40** #5680] of the simple p-Lie algebras, the analogues of the Cartan algebras make their appearance—things which are infinite-dimensional over characteristic zero may have finite-dimensional analogues in characteristic p (see in this connection [V. G. Kac, Uspehi Mat. Nauk **26** (1971), no. 3(159), 199–200; MR **47** #293]). More on this later.

Let me now turn to the classification of the finite-dimensional simple Lie superalgebras. The idea of "super" mathematics, i.e. having objects which are $\mathbf{Z}/2\mathbf{Z}$ graded, even and odd, was introduced by Grassmann in the 19th century. Thus the exterior algebra of a vector space is the standard example of an associative commutative superalgebra. (The physicists, in fact, use the name Grassmann algebra.) The fundamental papers by H. Cartan in 1950 [e.g. in *Colloque de topologie (espaces fibrés), Bruxelles, 1950*, 15–27, Georges Thone, Liège, 1951; MR **13**, 107e] implicitly make use of a Lie superalgebra. The Schouten bracket of polyvector fields on a manifold is a Lie superalgebra [J. A. Schouten and W. v. d. Kulk, *Pfaff's Problem and Its Generalizations*, Oxford, at the Clarendon Press, 1949; MR **11**, 179e]. (The formal version of this Lie superalgebra is one of the items on Kac's list in the paper under review.) But, to the best of my knowledge, the first place where a Lie superalgebra is clearly defined is in [A. Nijenhuis, Nederl. Akad. Wetensch. Proc. Ser. A. **58** = Indag. Math. **17** (1955), 390–397, 398–403; MR **17**, 661c]. M. Gerstenhaber made use of them in his remarkable series of papers on deformations of rings and algebras [Ann. of Math. (2) **78** (1963), 267–288; MR **28** #5102; Ann. of Math. (2) **79** (1964), 59–103; MR **30** #2034; Ann. of Math. **84** (1966), 1–19; MR **34** #7608; Ann. of Math. (2) **88** (1968), 1–34; MR **39** #1521], as did Milnor and Moore in their paper on Hopf algebras [J. W. Milnor and J. C. Moore, Ann. of Math. (2) **81** (1965), 211–264; MR **30** #4259]. The unfortunate prefix "super" had not yet entered the mathematical lingo; on the other hand the term "graded" that *was* used was unsatisfactory, as it conflicted with the term "graded" as used above for an ordinary Lie algebra with a **Z** (or other) grading.

In the Western physics literature the concept of a Lie superalgebra was introduced in one of the early papers on string theory by P. Ramond [Phys. Rev. D (3) **3** (1971), 2415–2418; MR **49** #1971], where it was used as a spectrum generating algebra. (See [P. Ramond, Phys. D **15** (1985), no. 1-2, 25–41; MR 86d:81031] for a retrospective overview.) In the Soviet Union, the idea had been introduced even earlier by G. L. Stavraki ["A non-local model of field self-interactions and the algebra of field operators (Russian)", in *High energy physics and theory of elementary particles (Russian) (Yalta, 1966)*, "Naukova Dumka", Kiev, 1967; per revr.], which led to an important 1971 paper by Kac [in *Eleventh All-Union Algebraic Colloquium (Russian)*, 140–141, Redakcionno-Izdat. Otdel Akad. Nauk Moldav. SSR, Kishinev, 1971; per revr.], which was a precursor of the classification result of [V. G. Kac, op. cit.; MR **58** #5803]. Also in 1971, Ya. Yu. Gol′fand and E. P. Likhtman [Piśma Zh. Èksper. Teoret. Fiz. **13** (1971), no. 8, 452–455] showed how to extend the Poincaré algebra so as to include fermionic "charges". With the advent of the Wess-Zumino models in physics, the subject became popular in the physics community in 1974 [see, e.g., L. Corwin, Y. Ne'eman and S. Sternberg, Rev. Modern Phys. **47** (1975), 573–603; MR **55** #11828]. The race was on to find the simple Lie superalgebras. There were four groups working on this problem: Corwin, Ne'eman, and I in Tel Aviv, Freund and Kaplansky in Chicago, Nahm, Rittenberg and Scheunert in Bonn, and Kac in Moscow. Some of us were aware of the existence of some of the other groups working on the problem, so the term "race" is appropriate. And the winner was—Victor Kac, of course. In [L. Corwin, Y. Ne'eman and S. Sternberg, op. cit.] we found many of the classes of the simple Lie superalgebras, but there is a moral to be learned from the ones that we *missed*. If one formally replaces the ring of formal power series by the exterior algebra $\bigwedge V$ of a finite-dimensional vector space, then the Cartan simple infinite Lie algebra becomes replaced by the superalgebra $W(V)$ of all superderivations of $\bigwedge V$, and this superalgebra is finite-dimensional and simple if $\dim V \geq 2$. In a similar way the other infinite simple Cartan algebras have their analogues as finite-dimensional simple Lie superalgebras; in fact Kac called them Cartan Lie superalgebras. Historians of science sometimes discuss what someone "should have known" in the light of the then current knowledge. In this case, in spite of the fact that I "should have known" about these Lie superalgebras, especially in view of Kostrikin and Shafarevich's work on p-Lie algebras of which I was well aware, these were the algebras that I missed! So much for what one "should have known".

Some twenty years later we now have the work under review which, completing the cycle, gives the classification of the linearly compact infinite-dimensional simple Lie superalgebras. There are eight series of (completed) graded superalgebras, two series of filtered deformations, and six exceptional ones. Particularly beautiful is the "largest" exceptional algebra $E(5,10)$ which is (the formal version of) the Lie superalgebra whose even part consists of divergence-free vector fields in five dimensions and whose odd part consists of closed two-forms on five-space. The vector fields act on the forms by Lie derivative. The bracket of two closed two-forms is their exterior product, a closed four-form which can be identified with a divergence-free vector field via the interior product of the vector field with the volume five-form.

A marvelous achievement. *Shlomo Sternberg* (Cambridge, MA)

98f:17007 17B10

Serganova, Vera (Berkeley, CA)

Kazhdan-Lusztig polynomials and character formula for the Lie superalgebra $\mathfrak{gl}(m|n)$. (English summary)

Selecta Math. (*N.S.*) **2** (1996), *no.* 4, 607–651.

A Lie superalgebra $\mathfrak{g}=\mathfrak{g}_0\oplus\mathfrak{g}_1$ is a Z_2-graded algebra (over some field K) equipped with a bracket satisfying two axioms (the superanalogues of antisymmetry and the Jacobi identity). Simple Lie superalgebras over **C** were classified by V. G. Kac [Advances in Math. **26** (1977), no. 1, 8–96; MR **58** #5803]. Just as $\mathfrak{gl}(n)$ is the generic example for Lie algebras, for Lie superalgebras the generic family consists of the general linear Lie superalgebras $\mathfrak{gl}(m|n)$. Much of the structure theory developed for Lie algebras has a direct analogue for Lie superalgebras. For representation theory, however, the picture is rather different. Lie superalgebras have in general finite-dimensional indecomposable representations, for example.

Part of the representation theory for Lie superalgebras was also developed by Kac. For a Lie superalgebra $\mathfrak{g}$ of basic classical type, Kac [Comm. Algebra **5** (1977), no. 8, 889–897; MR **56** #3075] proved a character formula for irreducible finite-dimensional typical representations. The problem of giving a character formula for the remaining atypical representations (the constituents of the indecomposable ones) has turned out to be very difficult, even for the Lie superalgebras $\mathfrak{g}=\mathfrak{gl}(m|n)$. Several conjectures and partial results for this problem have been proposed in the past [J. N. Bernstein and D. A. Leĭtes, C. R. Acad. Bulgare Sci. **33** (1980), no. 8, 1049–1051; MR 82j:17020a; J. Van der Jeugt et al., Comm. Algebra **18** (1990), no. 10, 3453–3480; MR 91j:17046; V. G. Kac and M. Wakimoto, in *Lie theory and geometry*, 415–456, Progr. Math., 123, Birkhäuser Boston, Boston, MA, 1994; MR 96j:11056].

In a previous paper [in *I. M. Gel′fand Seminar*, 151–165, Amer. Math. Soc., Providence, RI, 1993; MR 94k:17005], the author defined Kazhdan-Lusztig polynomials for $\mathfrak{gl}(m|n)$ in terms of a homology ring. The character of a simple module (i.e. irreducible representation) can be expressed in terms of such Kazhdan-Lusztig polynomials and known characters. In order to evaluate Kazhdan-Lusztig polynomials (and hence the characters), a conjecture regarding the semisimplicity of a functor on the category of $\mathfrak{g}$-modules was formulated. In the present paper, this earlier conjecture is proved, and it is shown how to calculate the character of an atypical $\mathfrak{gl}(m|n)$-module. At the same time, multiplicities of the composition factors in a Kac module are determined. The main results are presented in the beginning, and in the later sections the tools and (more technical) proofs are developed. The important objects are: a definition of Kazhdan-Lusztig polynomials and their properties; a counterpart of the Vogan functor U; some properties of the tensor product of modules; recurrence relations for polynomials related to U.

With this paper, the solution to an interesting problem which has been open for almost 20 years has been settled. *Joris Van der Jeugt* (Ghent)

97f:17014 17B37 16W30 18D20 81R50
Etingof, Pavel (Cambridge, MA); **Kazhdan, David** (Cambridge, MA)
Quantization of Lie bialgebras. I. (English summary)
Selecta Math. (*N.S.*) **2** (1996), *no.* 1, 1–41.

The main result of the paper is a construction of a quantization for an arbitrary Lie bialgebra over a field k of characteristic zero. In a previous paper, V. G. Drinfel′d [in *Quantum groups (Leningrad, 1990)*, 1–8, Lecture Notes in Math., 1510, Springer, Berlin, 1992; MR 94a:17006] formulated a number of problems in quantum group theory. In particular, he raised the question of the existence of a quantization for Lie bialgebras, which arose from the problem of quantization of Poisson Lie groups. The present paper gives a positive answer to a number of Drinfel′d's questions, using the methods and ideas of Kazhdan and G. Lusztig [J. Amer. Math. Soc. **7** (1994), no. 2, 335–381; MR 94g:17048].

In the first part of the paper the quantization of a finite-dimensional Lie bialgebra $(\mathfrak{a},\delta)$ is constructed, where $\mathfrak{a}$ is a Lie algebra, δ is a 1-cocycle of $\mathfrak{a}$ with coefficients in $\mathfrak{a}\otimes\mathfrak{a}$, and the map $\delta^*\colon\mathfrak{a}^*\to\mathfrak{a}^*\otimes\mathfrak{a}^*$ is a Lie bracket on $\mathfrak{a}^*$. The quantization is a topological Hopf algebra A over $k[[h]]$ such that $A/hA\simeq U(\mathfrak{a})$ and $h^{-1}(\Delta(x)-\Delta^{\mathrm{op}}(x))\equiv\delta(x_0) \bmod h$ for any $x_0\in\mathfrak{a}$, $x\in A$ and $x=x_0 \bmod h$. The vector space $\mathfrak{g}=\mathfrak{a}\oplus\mathfrak{a}^*$ has the natural structure of a Lie algebra (moreover, of a quasitriangular Lie bialgebra) and is called a double of $\mathfrak{a}$. The subspaces $\mathfrak{g}_+=\mathfrak{a}$ and $\mathfrak{g}_-=\mathfrak{a}^*$ are Lie subalgebras isotropic with respect to a nondegenerate invariant inner product identifying $\mathfrak{g}_+$ with $\mathfrak{g}_-^*$, i.e. $(\mathfrak{g},\mathfrak{g}_+,\mathfrak{g}_-)$ is a Manin triple. Following Drinfel′d the authors consider a category $\mathscr{M}$ whose objects are $\mathfrak{g}$-modules and whose morphisms $\mathrm{Hom}_{\mathscr{M}}(U,W)=\mathrm{Hom}(U,W)[[h]]$ are formal power series with coefficients in $\mathrm{Hom}(U,W)$. This category carries a braided monoidal structure defined via an associator Φ [V. G. Drinfel′d, Algebra i Analiz **1** (1989), no. 6, 114–148; MR 91b:17016; Algebra i Analiz **2** (1990), no. 4, 149–181; MR 92f:16047]. Let $\mathscr{A}$ be the category of topologically free $k[[h]]$-modules. The underlying functor $F\colon\mathscr{M}\to\mathscr{A}$ which assigns to every $\mathfrak{g}$-module M the $k[[h]]$-module $M[[h]]$ is equivalently described as $F(M)=\mathrm{Hom}(U(\mathfrak{g}),M)$. The regular module $U(\mathfrak{g})$ is identified with the tensor product of Verma modules $M_\pm=\mathrm{Ind}_{\mathfrak{g}_\pm}^{\mathfrak{g}}\mathbf{1}$. This decomposition allows the authors to define a tensor structure on the functor F. Then, according to general categorical principles, the algebra $\mathrm{End}(F)=U_h(\mathfrak{g})$ of endomorphisms of the functor F is a quasitriangular (topological) Hopf algebra with undeformed underlying algebra $U(\mathfrak{g})[[h]]$ and with comultiplication and antipode obtained from the usual ones by means of the Drinfel′d twist $J\in U(\mathfrak{g})^{\otimes 2}[[h]]$ (the expression for J involves the associator Φ). Moreover, it is proved that $U_h(\mathfrak{g})$ is a quantum double $U_h(\mathfrak{g}_+)\otimes U_h(\mathfrak{g}_-)$. The main result is that $U_h(\mathfrak{g})$ and $U_h(\mathfrak{g}_\pm)$ are quantizations of the Lie bialgebras $\mathfrak{g}$ and $\mathfrak{g}_\pm$ respectively.

As corollaries it is proved that any classical r-matrix over an associative algebra A can be quantized in the form $R = 1 + hr + O(h^2)$. A method for quantization of a quasitriangular Lie bialgebra $\mathfrak{a}$ (not necessarily finite-dimensional) is also proposed which produces a quantized universal enveloping algebra isomorphic to $U(\mathfrak{g})[\![h]\!]$. Another paper by the authors ["Quantization of Poisson algebraic groups and Poisson homogeneous spaces", Preprint, http://xxx.lanl.gov/abs/q-alg/9510020] is essentially based on the present paper. It is proved there that any Poisson formal group and any Poisson algebraic group admit a local quantization. A quantization method for a certain class of Poisson homogeneous spaces is also proposed. All these results answer some other questions of Drinfel′d. The second part of the present paper generalizes results of the first part to the case of infinite-dimensional Lie bialgebras. These considerations involve topological vector spaces instead of ordinary ones. Infinite-dimensional generalization is essential for the proof of functoriality of the main construction.

Results about functoriality are announced to appear shortly in Part II of the present study [Selecta Math. (N.S.) **4** (1998), no. 2, 213–231]. In part II it is shown that the quantization procedure of the present paper is given by universal acyclic formulas and defines an equivalence between the category of Lie bialgebras over $k[\![h]\!]$ and the category of quantized universal enveloping (QUE) algebras. Part III [P. Ètingof and D. Kazhdan, Selecta Math. (N.S.) **4** (1998), no. 2, 233–269] is devoted to Lie bialgebras of $\mathfrak{g}$-valued functions on a punctured rational or elliptic curve, where $\mathfrak{g}$ is a finite-dimensional simple Lie algebra. Of course, existence of quantization for such bialgebras follows from general considerations of the first paper. But in this case a more effective quantization procedure is given, which does not use an associator.

The paper under review and subsequent papers give complete solutions for a series of important open problems in quantum group theory. The existence of quantization of Lie bialgebras, quasitriangular Lie bialgebras, the classical r-matrix, and Poisson algebraic groups are proved under the most general assumptions. For Lie bialgebras associated to rational and elliptic curves with punctures, an explicit quantization procedure is proposed. *Yu. N. Bespalov* (Kiev)

98j:17020 17B50 17B20
Strade, H. (Hamburg)
The classification of the simple modular Lie algebras. VI. Solving the final case. (English summary)
Trans. Amer. Math. Soc. **350** (1998), *no.* 7, 2553–2628.

The problem of classifying the finite-dimensional simple Lie algebras of characteristic $p > 0$ is a long-standing one. Work on this question during the last thirty years has been directed by the Kostrikin-Shafarevich Conjecture of 1966, which states: Over an algebraically closed field of characteristic $p > 5$ a finite-dimensional restricted simple Lie algebra is classical or of Cartan type.

The classical algebras are the analogues of the finite-dimensional simple complex Lie algebras, while the Cartan-type algebras correspond to the four infinite families W, S, H, K (Witt, special, Hamiltonian, contact) of infinite-dimensional complex Lie algebras of Cartan. If the notion of Cartan-type algebra is expanded to include the simple algebras arising from Cartan-type algebras by twisting by an automorphism, then one can formulate the Generalized Kostrikin-Shafarevich Conjecture by removing the restrictedness assumption in the statement above.

The Generalized Kostrikin-Shafarevich Conjecture is now a theorem for $p > 7$. First announced by Strade and R. L. Wilson in 1991, its proof is spread over a number of papers. As the title suggests, the paper under review, which solves the final case, is the capstone of work on the classification problem.

The main contributions to the proof include A. I. Kostrikin and I. R. Shafarevich's construction and investigation of the Cartan-type algebras [Izv. Akad. Nauk SSSR Ser. Mat. **33** (1969), 251–322; MR **40** #5680]; generalizations of their construction by V. G. Kac [Uspehi Mat. Nauk **26** (1971), no. 3(159), 199–200; MR **47** #293; Izv. Akad. Nauk SSSR Ser. Mat. **38** (1974), 800–834; MR **51** #5685], Wilson [Bull. Amer. Math. Soc. **75** (1969), 987–991; MR **42** #3135; J. Algebra **40** (1976), no. 2, 418–465; MR **54** #366] and G. Y. Shen [Chinese Ann. Math. Ser. B **4** (1983), no. 3, 329–346; MR 85k:17010]; Kac's Recognition Theorem [Izv. Akad. Nauk SSSR Ser. Mat. **34** (1970), 385–408; MR **43** #2033] and its reworking by the reviewer, T. Gregory, and A. Premet (in preparation); R. E. Block's determination [Ann. of Math. (2) **90** (1969), 433–459; MR **40** #4319] of the differentiably simple Lie algebras which allowed the finite-dimensional semisimple Lie algebras of characteristic p to be described; B. Weisfeiler's work [J. Algebra **53** (1978), no. 2, 344–361; MR 80b:17011; Bull. Amer. Math. Soc. **84** (1978), no. 1, 127–130; MR 81f:14027] on graded Lie algebras, which is critical in the local analysis of simple algebras; the classification by Block and Wilson [J. Algebra **114** (1988), no. 1, 115–259; MR 89e:17014] of the restricted simple Lie algebras (thereby proving the original (restricted) Kostrikin-Shafarevich Conjecture); and Strade's monumental series of articles starting in 1989 [Part I, Ann. of Math. (2) **130** (1989), no. 3, 643–677; MR 91a:17023; Part II, J. Algebra **151** (1992), no. 2, 425–475; MR 93j:17042; Part III, Ann. of Math. (2) **133** (1991), no. 3, 577–604; MR 92g:17024; Part IV, Ann. of Math. (2) **138** (1993), no. 1, 1–59; MR 94k:17039; Part V, Abh. Math. Sem. Univ. Hamburg **64** (1994), 167–202; MR 95h:17023].

The Block-Wilson classification marked a major breakthrough in the theory in that not only did it determine the restricted simple Lie algebras, but it also provided a framework for the classification of the nonrestricted simple Lie algebras as well. It was Strade's insight that p-envelopes could be used to replace the restrictedness assumption. A p-envelope for a Lie algebra L is a restricted Lie algebra $(L_p, [p])$ and an embedding $\psi: L \to L_p$ such that the p-subalgebra generated by $\psi(L)$ is L_p. Since Cartan subalgebras and tori in characteristic-p Lie algebras need not be conjugate, and the root space decompositions may look quite different depending on which Cartan subalgebra or torus is used, a judicious choice of torus needs to be made. What seems to work best for classification purposes is a torus T of maximal dimension in the p-envelope L_p such that for each root α there exists a value $1 \leq i \leq p-1$ such that $\alpha([L_{i\alpha}, L_{-i\alpha}]) = 0$. Such a torus is termed optimal, and Strade in 1989 showed that optimal tori always exist for L simple of characteristic $p > 7$. In that same paper Strade analyzed the 1-sections $L(\alpha) = \sum_{i=0}^{p-1} L_{i\alpha}$ and 2-sections $L(\alpha, \beta) = \sum_{i,j=0}^{p-1} L_{i\alpha+j\beta}$ of L determined by roots α and β with respect to an optimal torus.

The characterization of the 1-sections enabled the reviewer, J. M. Osborn and the author [Trans. Amer. Math. Soc. **341** (1994), no. 1, 227–252; MR 94c:17035] to show that in each 1-section $L(\alpha)$ relative to an optimal torus T there is a unique subalgebra $Q(\alpha)$ satisfying the following properties: (1) The solvable radical

$\operatorname{rad} L(\alpha)$ of $L(\alpha)$ is contained in $Q(\alpha)$; (2) $Q(\alpha)$ is solvable or $Q(\alpha)/\operatorname{rad} Q(\alpha)$ is a classical simple algebra; and (3) $\dim L(\alpha)/Q(\alpha) \leq 2$. The sum $Q = \sum_\alpha Q(\alpha)$ is either L or a maximal T-invariant subalgebra. When L is restricted, then the Mills-Seligman axiomatic characterization of the classical algebras can be used to show that if $Q = L$, then L is classical. Otherwise, in the restricted case, it can be argued that Q is a maximal subalgebra of the type required to apply Kac's Recognition Theorem. The nonrestricted case does not behave so nicely. What one can say in the nonrestricted case is that the subalgebra Q equals L or it can be embedded in a maximal subalgebra. In most situations a maximal subalgebra containing Q is the appropriate choice for applying the recognition theorem. Further analysis to show that result requires knowledge of the 1-sections and 2-sections. The socle of a semisimple Lie algebra is the direct sum of ideals which are of the form $S_i \otimes A(m_i; \underline{1})$, where S_i is a simple Lie algebra and $A(m_i; \underline{1})$ is a truncated polynomial algebra in m_i indeterminates for some $m_i \in \{0, 1, 2, \cdots\}$. The sum $\sum_i S_i$ of the simple algebras is called the core. The core of a nonsolvable 1-section is by definition just the core of $L(\alpha)/\operatorname{rad} L(\alpha)$, which in this case must be a simple algebra. At this juncture in the classification it is convenient to divide considerations into four cases: (a) Every 1-section is solvable. (b) Every 1-section is solvable or has core sl(2), and a nonsolvable 1-section exists. (c) There exists a 1-section with a core which is a nonclassical simple Lie algebra, and no 2-section has the simple Hamiltonian Lie algebra $\mathrm{H}(2; \underline{1}; \Phi(\tau))^{(1)}$ as its core. (d) There exist a 1-section with a core which is a nonclassical simple Lie algebra and a 2-section with $\mathrm{H}(2; \underline{1}; \Phi(\tau))^{(1)}$ as its core.

Simple Lie algebras satisfying (b) were classified under one additional assumption by the reviewer [Comm. Algebra **18** (1990), no. 11, 3633–3638; MR 91f:17015] and in full generality by Strade (1991)—they are precisely the classical simple Lie algebras. The algebras for which (c) holds are of Cartan type (Strade, 1993), and the ones satisfying (d) are Cartan-type Lie algebras belonging to the Hamiltonian series (Strade, 1994). The Lie algebras which satisfy (a) are determined in the paper under review, and they are the last piece of the classification puzzle. For many reasons this is the most difficult case because the assumptions preclude the customary investigations using sl(2)-theory, a mainstay of classification work. The 3-sections need to be analyzed and the simple Lie algebras of toral rank 3 need to be determined. This is one of the few places where the 1-sections and 2-sections are insufficient. Strade shows in this paper, which is a real tour de force, that the simple Lie algebras satisfying (a) are the Block algebras (certain Cartan-type Lie algebras in the Hamiltonian series) or the special series algebras $\mathrm{S}(m; \underline{n}; \Phi(\tau))^{(1)}$. With these results the Generalized Kostrikin-Shafarevich Conjecture for $p > 7$ (as announced by Strade and Wilson in 1991) is now known to hold. Work on this final case was begun in collaboration with Wilson, and Strade in the article acknowledges Wilson's substantial influence, particularly on Sections 5 and 7 of the paper.

As the classification has evolved over the last thirty years, and especially over the last ten, many arguments needed to solve particular cases have required extension to handle subsequent cases. Now that the whole story is complete and the essential ingredients are understood, it is apparent that the exposition can be streamlined considerably. Work in this direction has already begun.

It is believed that the Generalized Kostrikin-Shafarevich Conjecture holds in characteristic 7. In characteristic 5, the Melikyan algebras provide the only known counterexamples, while in characteristic 3 and 2 there are a host of algebras that are neither classical nor Cartan type. One of the main obstacles in the prime characteristic theory (and why it is orders of magnitude harder than the characteristic-zero case) is the failure of Lie's theorem. This is particularly critical when it comes to saying that the derived algebra of a Cartan subalgebra of a simple Lie algebra L acts nilpotently on L. This last result was proven by Wilson [Trans. Amer. Math. Soc. **234** (1977), no. 2, 435–446; MR **58** #806] for $p > 7$ and by Premet [J. Algebra **167** (1994), no. 3, 641-703; MR 95f:17019] for $p = 7$. Premet also showed that, with one exception, the Cartan subalgebras of the Melikyan algebras have this property. Recent papers by Skryabin (1996) and Premet and Strade (1996) have provided the local analysis of the 1-sections and 2-sections in simple Lie algebras of characteristics 5 and 7 using "sandwich element" techniques. M. I. Kuznetsov [Comm. Algebra **19** (1991), no. 4, 1281–1312; MR 92d:17004] has developed a structural characterization of the Melikyan algebras; and, as mentioned earlier, Kac's Recognition Theorem, a major tool in the classification, has been reworked and now accommodates the Melikyan algebras, which were omitted in the original statement and proof. These results give reason to be very optimistic that the classification problem for $p = 5$ and 7 some day soon will be solved. Currently there is less optimism for the other small characteristics, although some classification results are beginning to emerge.

The list of the finite-dimensional simple Lie algebras of characteristic $p > 7$ consists of algebras having very natural characteristic-zero analogues, and it bears striking resemblance to the list of finite-dimensional complex simple Lie superalgebras. Reasons for this latter coincidence would be interesting to explore. Some of the methods involved in the classification—sandwich theory, for example—play an important role in Burnside-type problems in group theory. In studying the asymptotic behavior of finite p-groups and their associated Engel Lie algebras, one is led very naturally to Lie algebras of Cartan type. These relations hint at much deeper hidden interconnections between these topics, and why they exist is intriguing and still very much open to speculation. Rather than ending the story, Strade's paper may just be the beginning of an even bigger tale. *Georgia M. Benkart* (Madison, WI)

97e:17029 17B56 14M15 22E60 58F07

Kostant, Bertram (Cambridge, MA)

Flag manifold quantum cohomology, the Toda lattice, and the representation with highest weight ρ.

Selecta Math. (*N.S.*) **2** (1996), *no.* 1, 43–91.

The motivation for this work comes from the description of "quantum cohomology of flag manifolds" given by I. Ciocan-Fontanine [Internat. Math. Res. Notices **1995**, no. 6, 263–277; MR 96h:14071], putting on a rigorous basis earlier work of A. Givental and B. Kim. Let y be the $n \times n$ matrix with -1 below the diagonal, indeterminates $x_1, x_2, \cdots, x_n$ on the diagonal and indeterminates $q_1, q_2, \cdots, q_{n-1}$ above the diagonal. One may view these two sets of variables as defining an affine space Y. Then, expansion of the characteristic polynomial $\det(\lambda I - y)$ gives regular functions $F_1, F_2, \cdots, F_{n-1}$ on Y. Let $\mathcal{J}$ be the ideal they generate in the space $A(Y)$ of regular functions on Y. Then the main result of the above authors, which is the starting point of the present work, is that the quantum cohomology algebra identifies with the quotient space $A(Y)/\mathcal{J}$.

The above setup is closely related to that used in several earlier works of the author, particularly those on Lie algebra cohomology [1] [Ann. of Math. (2) **77** (1963), 72–144; MR **26** #266], on the null cone [2] [Amer. J. Math. **85** (1963), 327–404; MR **28** #1252, see p. G41 in this volume] and on the Toda lattice problem [3] [Adv. in Math. **34** (1979), no. 3, 195–338; MR 82f:58045]. In the present work the author makes an in-depth study of the subvariety Y_0 of Y defined by $\mathcal{J}$. A particular point is to describe an open dense subspace Y_0^* of Y_0 parametrized by affine coordinates $u_1, u_2, \cdots, u_{n-1}$ and to describe the variables x_i, q_j as functions of the former. A further result is a partial description of the singular locus of Y_0.

Let $\mathfrak{g}$ be a semisimple Lie algebra with Cartan subalgebra $\mathfrak{h}$. Let $S(\mathfrak{g})$ [resp. $U(\mathfrak{g})$] be the symmetric [resp. enveloping] algebra of $\mathfrak{g}$. Let e_i, $i = 1, 2, \cdots, l$, be the root vectors in $\mathfrak{g}$ corresponding to a choice of simple roots $\{\alpha_1, \alpha_2, \cdots, \alpha_l\}$, $\mathfrak{e}$ their linear span, and f the sum of the corresponding negative root vectors e_{-i}. Then $Y = f + \mathfrak{h} + \mathfrak{e}$ generalizes, from $\mathfrak{g} = \mathfrak{sl}(n)$, the affine space considered in the first paragraph. Here the quantum coordinates q_i are given through the Killing form $q_i(y) = (e_{-i}, y)$. Recall that $S(\mathfrak{g})^{\mathfrak{g}}$ is a polynomial algebra on homogeneous generators Φ_i, $i = 1, 2, \cdots, l$, and let $\mathcal{N}$ denote the zero locus of the Φ_i. Set $Y^* := \{y \in Y |\ q_i(y) \neq 0 \text{ for all } i\}$, $Y_0 = Y \cap \mathcal{N}$, $Y_0^* = Y_0 \cap Y^*$.

Let G be a simply-connected algebraic group with Lie algebra $\mathfrak{g}$ and B the Borel subgroup corresponding to the above choice of $\mathfrak{h}$ and of simple roots. Set $N = [B, B]$, $\mathfrak{b} = \operatorname{Lie} B$. Fix a representative $\kappa \in G$ of the unique longest element of the Weyl group. Let G_0^f be the identity component of the subgroup of G fixing f and G_*^f its intersection with the open Bruhat double coset $B\kappa N$. The author shows that Y_0^* is a smooth irreducible l-dimensional variety by establishing an isomorphism $\beta: g \mapsto (\operatorname{Ad} n(g))f$ of G_*^f onto Y_0^*, where $n(g)$ denotes the component of g in N. Furthermore, by applying a similar construction for each parabolic subalgebra $\mathfrak{p}_\gamma$, $\gamma \in \Gamma$, of $\mathfrak{g}$ (containing $\bar{\mathfrak{b}} := \kappa(\mathfrak{b})$) he obtains a stratification of Y_0 into non-singular subvarieties $(Y_\gamma)_0^*$ of dimension l_γ, where $l - l_\gamma$ is the rank of the semisimple Levi factor of $\mathfrak{p}_\gamma$. Hence Y_0 is irreducible of dimension l and, repeating an argument from [2], he shows that its ideal of definition $\mathcal{J}$ is generated by the $F_i := \Phi_{i|_y}$ and in particular is prime.

The author's next step is to construct a projectivization of G_0^f, which he calls the Peterson variety P_f. This is in order to recover the whole of Y_0. Since f is principal nilpotent there is a unique Borel subgroup B_{p_f}, namely $\overline{B} := \kappa B \kappa^{-1}$, corresponding to a G_0^f fixed point p_f in G/B. Set $U_{p_f} = B_{p_f} \kappa B_{p_f} / B_{p_f}$. The point $\tau_0 = \kappa p_f \in U_{p_f}$ corresponds to B and generates the unique G_0^f orbit Q in U_{p_f} stable under the "normalizer" of f. Then P_f is defined to be the closure of Q in G/B. Given $y \in Y_0$ there is a unique $n \in N$ such that $(\operatorname{Ad} n)f = y$ and from this one obtains a corresponding point, namely np_f, in P_f. The set of such points is denoted by R. This gives an isomorphism $Y_0 \cong R$, whereas $G_0^f \cong Q$ and $Q \cap R \cong Y_0^*$. In particular, $A(R)$ is the quotient $A(Y)/\langle F_i\colon\ i = 1, 2, \cdots, l\rangle$ and the author shows that the ordinary cohomology $H^*(G/B, \mathbf{C})$ on the flag variety is recovered from the "quantum cohomology" ring $A(R)$ by setting $q_i = 0$, $i = 1, 2, \cdots, l$, that is, $H^*(G/B, \mathbf{C}) = A(R)/I(R)$, where $I(R) = \langle q_i\colon\ i = 1, 2, \cdots, l\rangle$.

The space of finite-dimensional representations of G obtains an algebra structure through the Cartan product (the top term in the tensor product). A classical result asserts that this is just the algebra of homogeneous functions on G/B. Through this the author constructs the algebras of regular functions $A(Q)$, $A(R)$ on Q and R respectively. Indeed, for each dominant weight λ let V_λ be the simple $U(\mathfrak{g})$-module with highest weight λ and V'_λ its dual. Let $v_\lambda \in V_\lambda$ [resp. $v'_\lambda \in V'_\lambda$] be highest weight vectors normalized by $\langle v'_\lambda, \kappa^{-1} v_\lambda\rangle = 1$. Then $\psi_\lambda(g\tau_0) := \langle v'_\lambda, g^{-1}v_\lambda\rangle$, $g \in G_0^f$, defines a regular function on Q. If $\lambda = \sum n_i \rho_i$, $n_i \in \mathbf{N}$, is the decomposition of λ into the fundamental weights ρ_i, then Cartan multiplication implies that $(*)$ $\psi_\lambda = \prod \psi_i^{n_i}$, where $\psi_i = \psi_{\rho_i}$.

The author shows that $Q \cap R = \{p \in Q \mid \psi_i(p) \neq 0 \text{ for all } i\}$. Furthermore he extends $(*)$ to define $\psi_{-\alpha_i}$ on $Q \cap R$ and shows that $\psi_{-\alpha_i} = q_i$. Specializing to $\mathfrak{sl}(n)$, one can write an element $g \in G_0^f$ as a lower triangular matrix $M(g)$ with 1 on the diagonal and u_i on each $(k+i,k)$th entry. It follows that $q_i(u_1, u_2, \cdots, u_{n-1})$ can be expressed as an explicit rational function of certain cofactors of $M(g)$.

Let R_i^*, $i = 1, 2, \cdots, l$, be the codimension-1 subvarieties of R obtained from the stratification of $R \cong Y_0$ through the minimal parabolic subalgebras $\mathfrak{p}_i = \bar{\mathfrak{b}} + \mathbf{C}e_i$. Let R^s denote the set of simple points of R. The author shows that either $R_i^* \subset R^s$ or $R_i^* \cap R^s = \varnothing$. Moreover, the former holds exactly when $\dim(\mathfrak{g}^f \cap \operatorname{nilrad} \mathfrak{p}_i) = l - 1$ and when the coefficient of α_i in the unique highest root (for $\mathfrak{g}$ simple) equals 1. Consequently R is smooth only for $\mathfrak{sl}(2)$ and normal for just also $\mathfrak{sl}(3)$.

Given $z \in \mathfrak{g}^f$ let ξ^z be the derivation of $A(Q)$ so that $\xi^z_{g\tau_0}$ is the tangent vector to the curve $g \exp(-(tz)\tau_0)$ at $t = 0$. Identify Y with a subspace of $\bar{\mathfrak{b}}'$. A result of [3] shows that Y^* is a $\overline{B}$ coadjoint orbit with Y_0^* a Lagrangian subvariety. The latter is a completely integrable mechanical system, the Toda lattice, with the restriction of the quadratic invariant Ψ as Hamiltonian and the remaining invariants as constants of the motion. Furthermore the Hamiltonian vector field ξ_{Φ_i} on Y_0^* translates to some ξ^{z_i}, $z_i \in \mathfrak{g}^f$, on G_*^f, with ξ_Ψ becoming ξ^f. From this the author shows that for each weight ν and each $z \in \mathfrak{g}^f$ the logarithmic derivative $x_\nu^z = \xi^z \psi_\nu / \psi_\nu$ defined on $Q \cap R$ extends to a regular function on R. When λ is a dominant weight he also obtains the matrix expression

$$x_\lambda^z(g\tau_0) = \frac{\langle v'_\lambda, g^{-1} z v_\lambda\rangle}{\langle v'_\lambda, g^{-1} v_\lambda\rangle} \quad \text{for all } g \in G_*^f.$$

Set $x_\lambda = x_\lambda^f$. Specializing to $\mathfrak{sl}(n)$, one may recover $x_{\rho_i}(u_1, u_2, \cdots, u_{n-1})$ as an explicit rational function of appropriate cofactors of $M(g)$. Moreover, setting $x_{\rho_0} = x_{\rho_n} = 0$, one obtains the x_i of the first paragraph through $x_i = x_{\rho_i} - x_{\rho_{i-1}}$.

To recover a basis for the ordinary cohomology on G/B one must take polynomials in the x_{ρ_i}, and then the above expressions are not the most convenient as they require taking higher and higher tensor products. The author obtains alternative expressions just involving (more general) matrix elements of the representation with highest weight $\rho = \sum \rho_i$. First, inspired by J. B. Carrell [in *Proceedings of the Hyderabad Conference on Algebraic Groups (Hyderabad, 1989)*, 23–57, Manoj Prakashan, Madras, 1991; MR 92k:14051], the author shows that $\xi^z A(R) \subset I(R)$. Here, notably, $\xi^f x_{\rho_i} = -q_i$, which relates the vector field ξ^f for the Toda flow, the ordinary 2-cohomology class x_{ρ_i} and the quantum parameter q_i. Again, ξ^f implements the time derivative in Hamiltonian mechanics, so the author had viewed the above equation as the time derivative of momentum $\xi^f x_{\rho_i}$ being the position derivative $-q_i$ for potential energy, that is, as Newton's equation of motion. Now define ξ^u, $u \in U(\mathfrak{g}^f)$, as a differential operator on $A(Q)$ by composition of the

ξ^z, $z \in \mathfrak{g}^f$, and consider $x_\nu^u = \xi^u \psi_\nu / \psi_\nu$ as a regular function on $Q \cap R$. The author shows that x_ν^u extends to an element of $A(R)$. Now let $\mu_\nu \colon U(\mathfrak{g}^f) \to A(R)$ be the algebra homomorphism defined by $\mu_\nu(z) = x_\nu^z$ for all $z \in \mathfrak{g}^f$. Then, from the above, $(**)$ $\mu_\nu(u) = x_\nu^u \bmod I(R)$. On the other hand, x_ν^u is given as the matrix entry $\langle v_\nu', g^{-1} u v_\nu \rangle / \langle v_\nu', g^{-1} v_\nu \rangle$. Then from $(**)$ a monomial of the form $x_{\rho_1}^{n_1} x_{\rho_2}^{n_2} \cdots x_{\rho_l}^{n_l}$ can be written mod $I(R)$ as a matrix entry for the representation with highest weight ρ. As indicated above, this has the following application. Since $H^*(G/B, \mathbf{C}) \cong A(R)/I(R)$ and the latter is generated by the x_{ρ_i}, $i = 1, 2, \cdots, l$, it follows that for each $\sigma \in W$ there exists a polynomial p_σ in the x_{ρ_i}, $i = 1, 2, \cdots, l$, representing the Schubert class $[c_\sigma]$ of $\overline{Bn_\sigma B}/B$. (These polynomials, also known to the author, had been given by J. N. Bernstein, I. M. Gel′fand and S. I. Gelfand [Uspehi Mat. Nauk **28** (1973), no. 3(171), 3–26; MR **55** #2941].) Consequently the above Schubert classes (which form a basis for $H^*(G/B, \mathbf{C})$) can be expressed quite explicitly as matrix entries in the representation with highest weight ρ. This relates back to the paper [1] cited above and to forthcoming work of the author. Finally the author notes that the unique highest class is a generalization of the Vandermonde determinant and gives a non-singular affine open subset $O \subset P_f$ where the quantum parameters q_i are a coordinate system. Namely, if one sets $O = \{p \in Q \cap R \mid (dq_1 \wedge \cdots \wedge dq_l)_p \neq 0\}$ then there is a vector $w \in V_\rho$ such that $O = \{g\tau_0 \in G_*^f \tau_0 \mid \langle v_\rho', g^{-1} w \rangle \neq 0\}$.

Anthony Joseph (Rehovot)

99b:17033 17B67 14D25 16G20 17B35 53C25 58F05

Nakajima, Hiraku (Meguro)

Quiver varieties and Kac-Moody algebras.

Duke Math. J. **91** (1998), *no.* 3, 515–560.

In his earlier paper [Duke Math. J. **76** (1994), no. 2, 365–416; MR 95i:53051] the author constructed an important class of algebraic varieties of great interest in combinatorics and representation theory, the so-called quiver varieties. A quiver is just a finite oriented graph, and a representation of a quiver is a rule which assigns to each vertex i a finite-dimensional vector space $V(i)$ (over some field) and to each edge a linear operator. The category of such representations possesses many remarkable relations with root systems and representations of Lie algebras and quantum groups, which were gradually uncovered over the period of several decades by Gabriel, Bernstein, Gel′fand, Ponomarev, Ringel, Lusztig and other people. In all these works it is almost always the positive part of the root system or enveloping algebra which naturally appears.

Quiver varieties of Nakajima provide a natural framework for the whole enveloping algebra, not just the positive part. These varieties parametrize more complicated data than representations of quivers; in particular, to each edge (i, j) of a graph there are now associated not one but two maps between $V(i)$ and $V(j)$ in the opposite directions, thus taking into account all possible orientations at once. In addition, to each vertex i there corresponds another space $W(i)$. In some (but not all) examples quiver varieties parametrize some vector bundles on open algebraic surfaces and the $W(i)$ describe the behavior of the bundle at infinity.

The action of the full enveloping algebra U constructed by the author is achieved via certain "Hecke correspondences" relating quiver varieties with different numerical parameters. These are analogous to the well-known Hecke correspondences for vector bundles on curves. Positive and negative Chevalley generators corresponding to simple roots are represented by (slight adjustments of) the Hecke correspondences and their formal conjugates. The main result of [op. cit.] was a construction of representations of U in the space of cohomology of constructible functions on quiver varieties. In the present paper the author constructs U itself as a version of a convolution algebra formed by the Hecke correspondences regarded as elements of the homology of the product of quiver varieties. This is certainly a very important result.

The role of quiver varieties and Hecke correspondences is not restricted to the construction of representations of enveloping algebras. Among other things, they provide a wide class of holomorphic symplectic (in fact, hyper-Kähler) manifolds and nontrivial examples of symplectic reduction and Lagrangian subvarieties.

Michael M. Kapranov (Evanston, IL)

98h:19006 19K35 18G60 19D55 19L10 46H05 46L80 46M20

Cuntz, Joachim (Heidelberg)

Bivariante K-Theorie für lokalkonvexe Algebren und der Chern-Connes-Charakter. (English summary) [Bivariant K-theory for locally convex algebras and the Chern-Connes character]

Doc. Math. **2** (1997), 139–182 (electronic).

This is likely to be a fundamental and important paper. It continues the work begun by C. Kassel [K-Theory **3** (1989), no. 4, 367–400; MR 91f:19002], X. Wang [Canad. J. Math. **44** (1992), no. 2, 400–435; MR 93i:19005], and V. Nistor [Ann. of Math. (2) **138** (1993), no. 3, 555–590; MR 94i:19007] in building a "bivariant Connes-Chern character" from some version of bivariant K-theory to bivariant cyclic homology. The basic difficulty in this subject, which has affected the work of all the above authors, is that the bivariant K-theory of greatest analytic interest (especially for index theory) is Kasparov's KK-theory, which is only defined for C^*-algebras, whereas cyclic homology is only interesting for certain "smooth" subalgebras of C^*-algebras. The author's approach here is therefore to create a new version kk of Kasparov theory, defined on the category of m-algebras (complete locally convex topological algebras whose topology is defined by a family of submultiplicative seminorms). In this category, the author repeats all the usual constructions familiar in the category of C^*-algebras, but with continuous homotopy replaced by diffeotopy (smooth homotopy). Basic roles in the theory are played by the "J-functor" and the "Toeplitz extension", defined as follows. Let $\mathfrak{A}$ be an m-algebra, and let $T\mathfrak{A}$ be the completion of

$$\mathfrak{A} \oplus (\mathfrak{A} \otimes \mathfrak{A}) \oplus (\mathfrak{A} \otimes \mathfrak{A} \otimes \mathfrak{A}) \oplus \cdots$$

for the seminorms

$$\widehat{p} = p \oplus (p \otimes p) \oplus (p \otimes p \otimes p) \oplus \cdots,$$

as p runs over the submultiplicative seminorms defining the topology of $\mathfrak{A}$. The author observes that the algebra $T\mathfrak{A}$ is smoothly contractible. $J\mathfrak{A}$ is defined to be the kernel of the obvious multiplication homomorphism $T\mathfrak{A} \to \mathfrak{A}$. The Toeplitz algebra $\mathfrak{T}$, an extension of $\mathscr{C}^\infty(S^1)$ by the "smoothing operators" $\mathfrak{K}$, is the algebra of smooth Toeplitz operators on S^1. It gives rise to a homomorphism $\varepsilon: J^2\mathfrak{A} \to \mathfrak{K} \widehat{\otimes} \mathfrak{A}$, which plays the role of the inverse Bott element. (Here $\widehat{\otimes}$ denotes the projective tensor product in the sense of Grothendieck.) The $\mathbf{Z}/2$-graded bivariant K-theory kk_* is then defined by

$$kk_i(\mathfrak{A}, \mathfrak{B}) = \varinjlim \langle J^{2k+i}\mathfrak{A},\ \mathfrak{K}\widehat{\otimes}\mathfrak{B}\rangle,$$

where the angle brackets denote diffeotopy classes of continuous homomorphisms, and the limit is taken using the homomorphism ε. The author shows that this theory comes with Bott periodicity in both variables, and has an associative and bilinear product with the same properties as the product in Kasparov's KK-theory.

However, in some ways kk-theory is better behaved than Kasparov theory (because of the replacement of homotopy by diffeotopy). For example, excision automatically holds in both variables. Thus, the author is able to show that kk is the universal bivariant functor on the category of m-algebras having three natural properties: (E1) diffeotopy invariance, (E2) stability (i.e., the natural map $\iota\colon \mathfrak{A} \to \mathfrak{K}\widehat{\otimes}\mathfrak{A}$ induces an isomorphism), and (E3) half-exactness.

From this elegant characterization, the existence of the Connes-Chern character to bivariant periodic cyclic homology falls out effortlessly. At the end of the paper, the author shows that his Chern character coincides with that of Nistor [op. cit.] in the p-summable case, and that $kk_*(\mathbf{C}, \mathfrak{A})$ coincides with the K-theory of N. C. Phillips [Internat. J. Math. **2** (1991), no. 1, 77–129; MR 92e:46143] when $\mathfrak{A}$ is an m-algebra defined by only countably many submultiplicative seminorms.

{Reviewer's remark: The reference [CuSk], referred to in §5, seems to be missing from the bibliography. It should refer to [J. J. R. Cuntz and G. Skandalis, J. Operator Theory **15** (1986), no. 1, 163–180; MR 88b:46099].}

Jonathan M. Rosenberg (College Park, MD)

98h:20012 20C20 20G99

Ariki, Susumu

On the decomposition numbers of the Hecke algebra of $G(m, 1, n)$.

J. Math. Kyoto Univ. **36** (1996), *no.* 4, 789–808.

Let S_n be the symmetric group on n letters. Given any complex number ζ, we have a corresponding Iwahori-Hecke algebra $H_n(\zeta)$, defined abstractly by a presentation with generators and relations which may be regarded as "ζ-deformations" of those for S_n. This is done in such a way that if we set $\zeta = 1$ then $H_n(\zeta)$ actually is just the group algebra of S_n. In fact, by results of R. Dipper and G. D. James [Proc. London Math. Soc. (3) **52** (1986), no. 1, 20–52; MR 88b:20065; Proc. London Math. Soc. (3) **54** (1987), no. 1, 57–82; MR 88m:20084], not only is $H_n(\zeta)$ defined as a "ζ-deformation" of S_n, but also many combinatorial constructions (Young modules, Specht modules and so on) that are known for S_n have "ζ-deformed versions" for $H_n(\zeta)$.

By a classical result of Iwahori [see R. W. Carter, *Finite groups of Lie type*, Wiley, New York, 1985; MR 87d:20060], the above algebras arise naturally in representation theory: if $\zeta = q$ is a prime power then $H_n(\zeta)$ is the endomorphism algebra of the permutation module of $\mathrm{GL}_n(q)$ on the cosets of a Borel subgroup. This result is only a special case of a wider theory—the "Harish-Chandra philosophy"—which connects the representation theory of Iwahori-Hecke algebras with that of finite groups of Lie type. In another direction, V. F. R. Jones [Ann. of Math. (2) **126** (1987), no. 2,

335–388; MR 89c:46092] has shown that certain trace functions on Iwahori-Hecke algebras as above can be used to define invariants of knots and links.

In order to understand the structure and the representations of $H_n(\zeta)$, we distinguish two cases. If ζ is such that $H_n(\zeta)$ is semisimple, G. Lusztig [J. Algebra **71** (1981), no. 2, 490–498; MR 83a:20053] has constructed a canonical isomorphism between $H_n(\zeta)$ and the group algebra of S_n. Moreover, in this case, all Specht modules are irreducible and the representation theory of $H_n(\zeta)$ can be developed to a large extent analogously to that of S_n. In this sense, the structure of $H_n(\zeta)$ in the semisimple case is well-understood. It is known that this is the case unless $\zeta \neq 1$ is a root of unity of order $d \leq n$.

Let us now assume that $H_n(\zeta)$ is not semisimple and that ζ is a dth root of unity. Then the Specht modules of $H_n(\zeta)$ are not irreducible and we have a nontrivial decomposition matrix D_ζ^n which records the multiplicities of the simple modules of $H_n(\zeta)$ (which have a natural labelling by the d-regular partitions of n) in the various Specht modules (which are naturally indexed by all partitions of n). A major, notoriously difficult problem in modular representation theory (of groups, algebras, etc.) is the explicit computation of such decomposition matrices. In the present case, if d is a power of a prime l, the matrix D_ζ^n can be seen as an approximation to the l-modular decomposition matrix for S_n itself. On the other hand, by results of Dipper [J. Algebra **130** (1990), no. 1, 235–259; MR 91c:20015], D_ζ^n occurs as a submatrix of the l-modular decomposition matrix for $\mathrm{GL}_n(q)$, whenever l is not too small and coprime to q and q has multiplicative order d modulo l. The condition that l is not too small was made more precise by a conjecture of James [Proc. London Math. Soc. (3) **60** (1990), no. 2, 225–265; MR 91c:20024], which was proved in certain special cases (and was formulated for Iwahori-Hecke algebras of all types) by the reviewer [Comm. Algebra **20** (1992), no. 10, 2937–2973; MR 94a:20019].

A new point of view was introduced by A. Lascoux, B. Leclerc and J.-Y. Thibon [Comm. Math. Phys. **181** (1996), no. 1, 205–263; MR 97k:17019] by stating a conjecture which links the modular representation theory of $H_n(\zeta)$ with the theory of canonical bases in the sense of Kashiwara and Lusztig [see G. Lusztig, *Introduction to quantum groups*, Progr. Math., 110, Birkhäuser Boston, Boston, MA, 1993; MR 94m:17016]. In view of the connections of ideas involved in this conjecture and their relevance to the above applications, the paper under review, in which the Lascoux-Leclerc-Thibon conjecture is proved, may be regarded as a breakthrough in modular representation theory. The matrices D_ζ^n are now explicitly known, and one can even write down an efficient algorithm for their computation. Actually, Ariki also considers Iwahori-Hecke algebras of type B_n and, even more generally, the "cyclotomic Hecke algebras" associated to the wreath product of S_n with a cyclic group, introduced by himself and K. Koike [Adv. Math. **106** (1994), no. 2, 216–243; MR 95h:20006], and independently by M. Broué and G. Malle [Astérisque No. 212 (1993), 119–189; MR 94m:20095].

The conjecture states the following. We consider a vector space $\mathcal{F}$ with a basis given by all partitions (of all n). For any n and any d-regular partition μ of n, we define $P_\mu = \sum_\lambda d_{\lambda\mu}\lambda \in \mathcal{F}$, where λ runs over all partitions of n and $d_{\lambda\mu}$ are the entries in D_ζ^n which give the multiplicities of the simple module labelled by μ in the various Specht modules. Let $\mathcal{K} \subseteq \mathcal{F}$ be the subspace generated by P_μ (for all n, μ). Then there is an action of the affine Kac-Moody algebra of type $A_{d-1}^{(1)}$ on $\mathcal{F}$ so that $\mathcal{K}$ is an invariant subspace and is in fact an integrable highest-weight module with the empty partition as a highest-weight vector. The Chevalley generators of that Kac-Moody algebra act by operators "i-induction" and "i-restriction", which have already been studied earlier by Robinson from a combinatorial point of view; furthermore, by a result of Hayashi, this action even lifts to the level of quantized universal enveloping algebras; see Lascoux, Leclerc and Thibon [op. cit.] for further remarks and references concerning this action. Now Lascoux, Leclerc and Thibon conjectured that the basis $\{P_\mu\}$ of the module $\mathscr{H}$ is the canonical basis in the sense of Kashiwara and Lusztig and they described an efficient algorithm for its computation.

The proof of the conjecture involves an enormous amount of mathematics. It proceeds by a reduction to a statement (Proposition 4.3) about the canonical basis of the negative part of the universal enveloping algebra of type $A_{d-1}^{(1)}$ itself and a class of simple modules for an affine Hecke algebra (of which $H_n(\zeta)$ is a quotient). The advantage of working with affine Hecke algebras is that the simple modules can be described geometrically via the "Deligne-Langlands conjecture" and its variants [see D. A. Kazhdan and G. Lusztig, Invent. Math. **87** (1987), no. 1, 153–215; MR 88d:11121; N. Chriss and V. A. Ginzburg, *Representation theory and complex geometry*, Birkhäuser Boston, Boston, MA, 1997; MR 98i:22021]. On the other hand, Lusztig [J. Amer. Math. Soc. **4** (1991), no. 2, 365–421; MR 91m:17018] has described a geometric setting for canonical bases, which Ariki shows to be compatible with that for affine Hecke algebras. A more detailed sketch of the proof can also be found in the reviewer's talk at the "Séminaire Bourbaki" [Astérisque No. 252 (1998), Exp. No. 836, 3, 33–55].

Two final remarks: (1) Leclerc and Thibon [Internat. Math. Res. Notices **1996**, no. 9, 447–456; MR 97h:17023] have formulated an extension of the above conjecture to the Dipper-James q-Schur algebra. This would give the whole l-modular decomposition matrix for $\mathrm{GL}_n(q)$ (for all q, l satisfying the same conditions as above). But this conjecture is not yet proved. (2) I. Grojnowski also announced a proof of the conjecture of Lascoux, Leclerc and Thibon, following his note [Internat. Math. Res. Notices **1994**, no. 5, 215 ff., approx. 3 pp. (electronic); MR 95e:20054], but a published version of that proof has not yet appeared. *Meinolf Geck* (Paris)

98k:20015a 20C30 05E10
Ford, Ben (Cleveland, OH); **Kleshchev, Alexander S.** (Eugene, OR)
A proof of the Mullineux conjecture.
Math. Z. **226** (1997), *no.* 2, 267–308.

98k:20015b 20C30 05E10
Bessenrodt, C. (Magdeburg); **Olsson, J. B.** (Copenhagen)
On residue symbols and the Mullineux conjecture. (English summary)
J. Algebraic Combin. **7** (1998), *no.* 3, 227–251.

The irreducible representations D^λ of the symmetric group S_n over a field of characteristic p are parametrized in a canonical way by the p-regular partitions λ of n. The tensor product of D^λ with the sign representation is again an irreducible representation, so it equals D^{λ^P} for some p-regular partition λ^P. If p is greater than n then λ^P is the partition conjugate to λ, but it is difficult even to guess a rule for recovering λ^P from λ, in general. For example, if $p=5$ and $\lambda=(14)$ then $\lambda^P=(4^2,3^2)$ and if $p=3$ and $\lambda=(9,2,1)$ then $\lambda^P=(5,4,3)$.

In 1979, G. Mullineux [J. London Math. Soc. (2) **20** (1979), no. 1, 60–66; MR 80j:20016] gave a bijection on the set of p-regular partitions and he conjectured that this map coincides with the map $\lambda \mapsto \lambda^P$. Until recently, the only evidence in support of Mullineux's conjecture consisted of answers culled from decomposition matrices for small n, together with proofs that the Mullineux map was consistent with various combinatorial constraints; for example, the p-core of the image of λ under the Mullineux map is the conjugate of the p-core of λ. No progress was made on the problem of applying representation theory to describe the map $\lambda \mapsto \lambda^P$, combinatorially, before Kleshchev's series of papers on branching rules for p-modular representations of symmetric groups. Vital to these papers is the concept of a good removable node in the diagram for λ. Whether or not a removable node is good can be determined easily by taking account of the removable and indent nodes which have the same residue as the given node. In [J. London Math. Soc. (2) **54** (1996), no. 1, 25–38; MR 96m:20019c], Kleshchev obtained, as a corollary to his main results, a combinatorial description of the map $\lambda \mapsto \lambda^P$ in terms of good nodes.

It is unclear that Kleshchev's map coincides with Mullineux's map. The two papers under review give different proofs that these maps are, indeed, the same, and thereby they complete the proof of the Mullineux conjecture.

The proof by Ford and Kleshchev was the first to be discovered, and it uses the abacus notation to record partitions. On the other hand, Bessenrodt and Olsson's proof is shorter. They introduce a variation on the Mullineux symbol, called the residue symbol; it is straightforward to translate back and forth between these symbols. In each paper, a series of lemmas is required, in order to keep track of the way in which the removal of a good node influences the chosen way in which the partition is recorded; once these lemmas are proved, the Mullineux conjecture follows by induction on n.

Both Mullineux's and Kleshchev's algorithms for computing λ^P can be carried out by hand, even for large partitions. Each has its theoretical advantages, so it is important that their equivalence has been established. Two aspects are particularly striking. They are that Mullineux was able to devise such a remarkable conjecture to fit the facts that were available to him, and that the apparently innocuous question from representation theory has such an intricate solution. *G. D. James* (London)

97d:20018 20D20 20E34
Stellmacher, Bernd (Kiel)
A characteristic subgroup of Σ_4-free groups. (English summary)
Israel J. Math. **94** (1996), 367–379.

This paper is a remarkable advance in an area of finite group theory that began about forty years ago. In 1959, John Thompson proved his celebrated theorem that the Frobenius kernel of a Frobenius group is nilpotent. Five years later, he introduced a characteristic subgroup $J(S)$ of a p-group S and used it to give a much simpler proof, which depended partly on analyzing the structure of a special type of solvable group [J. Algebra **1** (1964), 43–46; MR **29** #4793]. Shortly afterwards, he extended this analysis to obtain a factorization of a solvable (or more generally, p-solvable) group H as a product of two subgroups depending on a Sylow p-subgroup of H [Pacific J. Math. **16** (1966), 371–372; MR **32** #5735].

Each of these results reduces to a case in which there are a prime p and a p-solvable group H such that p divides the order of each non-trivial normal subgroup of H. Let S be a Sylow p-subgroup of H and $Z(S)$ be the center of S. Thompson defined $J(S)$ to be the subgroup of S generated by the abelian subgroups of maximal rank in S. For his factorization result, he proved that $H = C_H(Z(S))N_H(J(S))$ unless $p \leq 3$ and the group $\mathrm{SL}(2, p)$ is involved in H, i.e., there exist subgroups K, L of H such that $L \lhd K$ and $K/L \cong \mathrm{SL}(2, p)$. (He also obtained a stronger "triple factorization" when $p \geq 5$.) The conclusion may be false if $\mathrm{SL}(2, p)$ is involved. For instance, for $p = 2$ and $H = \Sigma_4$ (the symmetric group of degree 4), $C_H(Z(S)) = N_H(J(S)) = S$, so that $C_H(Z(S))N_H(J(S)) < H$.

Now let p be a fixed prime and S be a fixed non-trivial p-group. For any group H, there is a unique maximal normal p-subgroup, called $O_p(H)$. By extending Thompson's methods, the reviewer showed in 1967 (for a slightly different subgroup $J(S)$) that, for p odd, $Z(J(S)) \lhd H$ in the situation above or, more generally, whenever (A) S is a Sylow p-subgroup of H, and $O_p(H)$ contains its centralizer $C_H(O_p(H))$; and (B) $\mathrm{SL}(2, p)$ is not involved in H [Canad. J. Math. **20** (1968), 1101–1135; MR **37** #6365]. Thus, there is a non-trivial characteristic subgroup of S that is normal (in fact, characteristic) in H, and independent of the choice of H. This result was applied to the study of a simple non-abelian group G in which S is a Sylow p-subgroup; then $N_G(Z(J(S)))$ contains every subgroup H of G satisfying (A) and (B). Later, I. M. Isaacs found a simpler proof by using a subgroup defined by L. Puig instead of $Z(J(S))$ (Appendix B in *Local analysis for the odd order theorem* by H. Bender and the reviewer [Cambridge Univ. Press, Cambridge, 1994; MR 96h:20036]). More recently, the author used the "amalgam method" to prove the existence of an analogous subgroup without having to specify the subgroup [Proc. Amer. Math. Soc. **109** (1990), no. 4, 925–929; MR 90k:20045; erratum; MR 92e:20012].

These results naturally raise the question of an analogue for $p = 2$, i.e., a non-trivial characteristic subgroup $W(S)$ of S such that $W(S) \lhd H$ for every H satisfying (A) and (B). (Examples show that $Z(J(S))$ does not work.) But this case is much more difficult. For odd p, hypothesis (B) yields a strong condition (p-stability) about commutators for normal subgroups. But, for $p = 2$, it appears to yield comparatively weak information. Moreover, in many ways, 2-groups are much more complicated than p-groups for odd p. Therefore, this problem stimulated considerable effort for nearly thirty years. In the present paper, the author finally solves the problem by showing that $W(S)$ exists.

A special case of this result for a family of solvable groups was proved by M. Hayashi [Math. Z. **181** (1982), no. 2, 215–227; MR 84f:20019]. A weaker consequence of (A) and (B) for $p = 2$, namely, a "triple factorization" similar to Thompson's above and inspired by a special case in Thompson's N-group work, was proved by the reviewer in 1977 [*Factorizations in local subgroups of finite groups*, Amer. Math. Soc., Providence, R.I., 1977; MR **57** #9839 (pp. 27–28)].

As in the case of Thompson's factorization, the group Σ_4 violates the hypothesis and the conclusion of the author's theorem. Moreover, it can be shown that for $p = 2$, conditions (A) and (B) are equivalent to the conditions (I) H is Σ_4-free (i.e., Σ_4 is not involved in H) and (II) S is a Sylow 2-subgroup of H and $C_H(O_2(H)) \leq O_2(H)$. Thus, Σ_4 is the "root" of all counterexamples.

The main part of the paper proves the conclusion under the assumptions (I), (II), and an additional condition: (III) Every non-abelian simple section of H

is isomorphic to $\mathrm{Sz}(2^m)$ or $\mathrm{PSL}_2(3^m)$ for some odd m. A similar approach was followed in the reviewer's triple factorization paper. As in that paper [op. cit., p. 29], one easy consequence of the main part and of a theorem of D. Goldschmidt (about groups containing a strongly closed abelian 2-subgroup) is the classification of all Σ_4-free non-abelian simple groups. (However, this is not mentioned by the author.) This, in turn, shows that the only Σ_3-free non-abelian groups are the groups in (III). Thus, since $\mathrm{SL}(2,2) \cong \Sigma_3$, condition (III) follows from (B) and is finally seen to be superfluous. A further consequence is that the Suzuki groups are the only non-abelian simple groups of order not divisible by 3.

The proof is largely inspired by the "amalgam method", although it does not explicitly use it. For example, to define $W(S)$, the author embeds a certain family of groups H as above for which $J(S) \lhd H$ into one huge group G_1 (their amalgamated product over S), which may be infinite. He then defines $\Omega_1(Z(S))$ to be the group of all $x \in Z(S)$ for which $x^2 = 1$ (as usual) and $W(S)$ to be the group generated by all conjugates of $\Omega_1(Z(S))$ in G_1. Then $W(S)$ is a characteristic subgroup of S contained in $Z(J(S))$, and he must prove that $W(S) \lhd H$ for every group H satisfying (I), (II), and (III), regardless of whether $J(S) \lhd H$.

The main preliminary results about a group H as above are given in Section 2. By exploiting (III) and properties of $\mathrm{Sz}(2^m)$ and $\mathrm{PSL}_2(3^m)$, the author finds and analyzes a useful subgroup L of H such that $L/O_2(L)$ is dihedral or a Suzuki group. Most of Section 2 is devoted to studying the action of H or of L on a GF(2)-module V. An essential tool in earlier work was that $|V/C_V(A)| > |A/C_A(V)|$ if $O_2(H/C_H(V)) = 1$, A is a subgroup of S, and $A/C_A(V)$ is a non-trivial elementary abelian 2-group. But this did not cover some important situations in which $O_2(H/C_H(V)) > 1$. Here, the author extends this inequality to these situations by using an earlier result, (2.3), and a remarkable new result, (2.9).

The proof proper appears in the last section (Section 3), and can only be described as a tour de force. The author assumes the existence of a group H in which $W(S)$ is not normal. By using his information about the amalgam construction, the subgroup L, and GF(2)-modules to their fullest extent, he eventually obtains some delicate inequalities that lead to a contradiction. Somewhat surprisingly, the case in which H is solvable is much more difficult than the non-solvable case.

{Reviewer's remarks: The author neglected to give a choice of $J(S)$. One may take Thompson's definition above or the subgroup generated by the elementary abelian subgroups of maximal rank in S. The author has a correction for page 378: Replace "≤ 2, and again 2.6" by "≤ 4 and $|B/(B \cap O_2(U))| = |B/(B \cap X_1)| = 2$, and again 2.6 applied to L and X/X_1".} *George Glauberman* (Chicago, IL)

98j:20022 20D30 06B15 20E22

Baddeley, Robert (Leicester); **Lucchini, Andrea** (Brescia)

On representing finite lattices as intervals in subgroup lattices of finite groups. (English summary)

J. Algebra **196** (1997), *no.* 1, 1–100.

This paper is a major contribution to the most intriguing problem about subgroup lattices of finite groups: What are the possible interval sublattices in the subgroup lattices of finite groups? (An interval sublattice $[G/T]$ is the lattice formed by all intermediate subgroups K, $G \ge K \ge T$, where G is a group and T is a fixed subgroup of G.) Baddeley and Lucchini reduce the problem to a collection of questions about finite simple groups. Although these questions are quite hard, eventually they might be resolved using the classification of finite simple groups.

The problem originates from universal algebra. In 1963 G. Grätzer and E. T. Schmidt [Acta Sci. Math. (Szeged) **24** (1963), 34–59; MR **27** #1391] gave an abstract characterization of congruence lattices of algebras, namely these are the so-called algebraic lattices, i.e., complete lattices in which every element is a join of compact elements. Since every finite lattice trivially satisfies this condition, every finite lattice can be represented as a congruence lattice of a suitable algebraic structure. However, the construction of Grätzer and Schmidt (as well as the more recent ones by Pudlák and by Tůma) is an infinite procedure, yielding an infinite algebra even if the lattice to be represented is finite. Hence there naturally arose the problem whether every finite lattice can occur as the congruence lattice of a finite algebra.

P. Pudlák and the reviewer [Algebra Universalis **11** (1980), no. 1, 22–27; MR 82g:08003] proved that this question about general algebras is in fact equivalent to a problem about finite groups: Does every finite lattice occur as an interval in the subgroup lattice of a finite group? Later results showed that this is not just a formal equivalence, but that indeed the crux of the problem is group-theoretic. With an ingenious construction J. Tůma [J. Algebra **125** (1989), no. 2, 367–399; MR 91e:20020] proved that every algebraic lattice (so every finite lattice, in particular) can be represented as an interval in the subgroup lattice of an infinite group. Thus the difficulty is caused by requiring the finiteness of the group.

Most of the attention has been focused on the very simple lattices $\mathcal{M}_n$ consisting of a largest, a smallest and n pairwise incomparable elements: not only because this is a more easily accessible case, but also because it can be the first step towards understanding the local structure of the subgroup lattice of a finite group. For $n =$ 1, 2, $q+1$ (where q is a prime power) it is easy to find examples of $\mathcal{M}_n$ as intervals in subgroup lattices. However, in the subgroup lattices of finite solvable groups no other $\mathcal{M}_n$'s can occur. So the finite nonabelian simple groups inevitably play a major role in this problem. After the first interesting examples had been found by W. Feit [Algebra Universalis **17** (1983), no. 2, 220–221; MR 85d:20022], Lucchini [J. Algebra **164** (1994), no. 1, 85–90; MR 95c:06016] discovered new series of examples, namely for $n = q+2$ and for $n = (q^t+1)/(q+1)+1$ (where t is an odd prime).

So at the present time the only values of n for which an occurrence of $\mathcal{M}_n$ as an interval in the subgroup lattice of a finite group has been found are $n = 1$, 2, $q+1$, $q+2$, $(q^t+1)/(q+1)+1$, the smallest undecided cases being $n = 16$, 23, 35. The authors are cautious, but their ultimate goal seems to be to show that the present list is complete.

Now let G be a finite group and T a subgroup of G such that $[G/T] \cong \mathcal{M}_n$. Assume that n is not on the above list of known representations of $\mathcal{M}_n$'s, n is large enough ($n > 50$), and G has minimal order among all groups for which $\mathcal{M}_n$ is an interval in the subgroup lattice of G. By a result of P. Köhler, G has a unique minimal normal subgroup M, and M is nonabelian. Hence M is the direct product of isomorphic simple groups; let F denote one of these factors. The case $M \cap T \neq 1$ has been dealt with in an earlier paper of Lucchini, where it was shown that under the said assumptions M itself is a nonabelian simple group, so G is an almost simple group. Hence in this case the problem is reduced to a question about almost simple groups, where an expected classification of maximal subgroups would lead to the solution.

The present paper of Baddeley and Lucchini deals with the case $M \cap T = 1$. Now either $MT \neq G$ or $MT = G$. In the first case M is also complemented, i.e., there exists a subgroup H with $M \cap H = 1$, $MH = G$ and $H > T$. Then G has the structure of a twisted wreath product of F and H. The main thread in the paper is the investigation of covering relations in subgroup lattices of twisted wreath products. In the present case it leads to the following description of the structural ingredients of G: F is a nonabelian simple group, H is an almost simple group, T is a core-free maximal subgroup of H, $Q = \mathbf{N}_T(F)$ is a core-free subgroup of H satisfying $H = QT$, the homomorphism $\varphi\colon Q \to \operatorname{Aut} F$ satisfies $\varphi(Q \cap T) \geq \operatorname{Inn} F$ and its restriction to $Q \cap T$ has no extension to any subgroup of T properly containing $Q \cap T$. Furthermore, $n-1$ is the number of those homomorphisms $\varphi'\colon Q \to \operatorname{Aut} F$ for which both $\varphi'|_{Q\cap T} = \varphi|_{Q\cap T}$ and $\widetilde{\varphi}' = \widetilde{\varphi}$ hold, where $\widetilde{\varphi}$ denotes the composition of φ with the natural homomorphism onto $\operatorname{Out} F = \operatorname{Aut} F/\operatorname{Inn} F$. However, no such example is known with $n \geq 3$.

The case $MT = G$ leads to even more complex technical conditions that cannot be reproduced in full detail here. This case is further divided into two subcases according to whether $\mathbf{N}_T(F)$ induces all inner automorphisms of F or not. It turns out that T always has a unique minimal normal subgroup N, which is a direct product of isomorphic copies of a nonabelian simple group E. The crucial observation is that F is isomorphic to a section of E. The bulk of the paper is then concerned with obtaining as far as possible a characterization of the pair (G, T) in terms of E, F, the representation of F as a section of E, and other pieces of information as appropriate. Overall, the paper is highly technical, but a handy "results diagram" and a transparent cross-referencing system significantly help the reader.

The authors finish the paper by formulating two general problems on which the successful resolution of the remaining questions depends: (i) Describe the maximal nonabelian simple sections of the nonabelian simple groups. (ii) Describe all pairs (F, L) where F is a nonabelian simple group and L is a group of automorphisms of F such that there is exactly one proper nontrivial L-invariant subgroup of F.

P. P. Pálfy (Budapest)

99a:20038 20F32 20F05
Ol′shanskiĭ, A. Yu. (Moscow)
On the distortion of subgroups of finitely presented groups. (Russian. Russian summary)
Mat. Sb. **188** (1997), *no.* 11, 51–98; *translation in Sb. Math.* **188** (1997), *no.* 11, 1617–1664.

In this deep and important paper the author obtains a characterization of the possible metric distortion of finitely presented subgroups in finitely presented groups, answering the question raised by M. L. Gromov [in *Geometric group theory, Vol. 2 (Sussex, 1991)*, 1–295, Cambridge Univ. Press, Cambridge, 1993; MR 95m:20041]. Informally speaking, subgroup distortion measures how much the intrinsic word metric of a finitely generated subgroup differs from the restriction of the word metric from the ambient (finitely generated) group to this subgroup.

The main result of the paper under review is the following. Theorem 2: Let G be a group (not even necessarily finitely generated). Let $f: G \to \mathbf{N}$ be a computable function satisfying the following restrictions: $f(g) = f(g^{-1})$ for every $g \in G$; $f(g) = 0$ iff $g = 1$; $f(gh) \leq f(g) + f(h)$ for every $g, h \in G$; there exists $a > 0$ such that $\#\{g \in G \mid f(g) \leq r\} \leq a^r$ for any $r \in \mathbf{N}$ (computable here means that there is an algorithm which for every word in the generators of G computes the value of f on the element of g corresponding to this word).

Then there exist an embedding of G into a finitely presentable group Q and a finite generating set of Q inducing the word length l_Q on Q, such that the function f is equivalent (in some natural sense) to $l_Q|_G$. This implies that any "reasonable" function can occur as the distortion function of an embedding of G in a finitely presented group.

In particular, there exist embeddings of the infinite cyclic group $\langle t \rangle$ into finitely presented groups such that the length of t^n, measured in the ambient group, grows as n^θ, where $\theta \geq 1$ is any "computable" real number, such as $\sqrt{2}$ or π. One of the limitations of Theorem 2, however, is that the existence of a computable function on G implies (and for finitely generated G is equivalent to the fact) that G has solvable word problem. Thus Theorem 2 does not apply to all finitely presented groups.

The proof of this remarkable result consists of two major steps. In the first step, carried out by the author in a previous paper ["Distortion functions for subgroups", in *Proc. Conf. on Geometric Group Theory (Canberra, 1996)*, de Gruyter, to appear], it is proved that Theorem 2 holds if one requires Q to be just finitely generated, rather than finitely presentable, and drops the assumption that f be computable. In the present article the author shows that if f is computable then the group Q does turn out to be recursively presentable.

The main part of the article consists in proving a geometric version of the Higman embedding theorem. Together with step one, described above, this implies Theorem 2.

Theorem 3: Let R be a finitely generated recursively presentable group with word length l_R on R corresponding to some finite generating set of R. Then there exists a finitely presentable group Q, containing R as a subgroup, and a finite generating set of Q inducing the word length l_Q on Q such that $l_R(r) = l_Q(r)$ for every $r \in R$.

Thus Theorem 3 says that a finitely generated recursively presentable group can be isometrically embedded into a finitely presented group. This result is certainly remarkable in its own right. The proof of Theorem 3 is very complicated. One of the main ingredients is to find the right modification of the presentation for Q from S. Aanderaa's proof of the Higman embedding theorem [in *Word problems: decision problems and the Burnside problem in group theory (Conf. on Decision Problems in Group Theory, Univ. California, Irvine, Calif., 1969; dedicated to Hanna Neumann)*, 1–18, Studies in Logic and the Foundations of Math., 71, North-Holland, Amsterdam, 1973; MR **54** #412].

In the present paper (as well as in Aanderaa's proof) the Turing machine associated with a recursive presentation of R is rather directly encoded in a certain finite group presentation for Q. The presentation turns out to have many nice "aspherical properties", similar to those of HNN-extensions. This allows the author to conduct a very careful analysis of van Kampen diagrams for Q and to conclude finally that R isometrically embeds in Q. The particular methods used in the present paper are reminiscent of the geometric machinery developed by the author earlier [see *Geometry of defining relations in groups*, English translation, Kluwer Acad. Publ., Dordrecht, 1991; MR 93g:20071; *The geometry of defining relations in groups* (Russian), "Nauka", Moscow, 1989; MR 91i:20035] as well as of the methods used by J. C. Birget, E. Rips and M. V. Sapir in their recent preprint ["Dehn functions of groups", Preprint, 1997; per bibl.] where groups with exotic isoperimetric functions are constructed. Since the present paper appeared, its technique has been used in a joint paper of Ol′shanskiĭ, Birget, Rips and Sapir, in which the Higman embedding theorem was generalized in yet another direction. Namely, they proved that for a group G the word problem is solvable by a non-deterministic Turing machine in polynomial time if and only if the group G is embeddable in a finitely presentable group with a polynomial isoperimetric function.

Also, in the reviewer's opinion, there is another direction of investigation, associated with Theorem 2, which is of considerable interest. It would be very interesting to understand what kind of distortion can occur for word hyperbolic (or just finitely presented) subgroups of word hyperbolic groups. It is well known that a finitely generated subgroup of a word hyperbolic group is either linearly distorted (the generic case) or has at least exponential distortion. There are many examples of exactly exponential distortion for word hyperbolic subgroups of word hyperbolic groups. Also, M. Mitra pointed out how to construct examples of hyperbolic subgroups of hyperbolic groups having iterated exponential distortion and distortion of the type $f(n) = \exp(\exp(\cdots \exp(n))\cdots)$ (n compositions). However, to the best of the reviewer's knowledge, there are no other examples available at this time.

Ilya Kapovich (Piscataway, NJ)

98i:20039 20F36 20J05 57M07

Bestvina, Mladen (Salt Lake City, UT); **Brady, Noel** (Salt Lake City, UT)

Morse theory and finiteness properties of groups. (English summary)

Invent. Math. **129** (1997), *no.* 3, 445–470.

In this paper, Bestvina and Brady present two important results and introduce the use of Morse-theoretic techniques into the study of infinite groups. First, this paper resolves a 20-year-old question in the study of infinite groups. There has been an ongoing refrain in articles and books on finiteness properties of infinite groups which goes something like: "Every finitely presented group is FP_2, but it is not known if the converse holds." Bestvina and Brady establish that the converse is false. Second, they introduce Morse-theoretic techniques into the study of groups acting on non-positively curved spaces, techniques which have already been applied by other authors to other situations. Finally, in their concluding section, they establish that either the Eilenberg-Ganea conjecture or the Whitehead conjecture (or possibly both) is false. We will briefly discuss each of these contributions in turn.

A group G is of type $\mathcal{F}_n$ if it admits a $K(G,1)$ complex $\mathcal{X}$ with finite n-skeleton. This definition was introduced by C. T. C. Wall [Ann. of Math. (2) **81** (1965), 56–69; MR **30** #1515], and it directly extends the two most common finiteness properties of groups: A group is $\mathcal{F}_1$ if and only if it is finitely generated, and it is $\mathcal{F}_2$ if and only if it is finitely presented. J. Stallings ($n = 3$) and R. Bieri ($n \geq 4$) introduced the following family of examples: Let $G_n = \underbrace{F_2 \times \cdots \times F_2}_{n \text{ copies}}$ be the direct product of n free groups, each of rank 2. Then the kernel of the map taking each generator to $1 \in$

$\mathbf{Z}$ is $\mathscr{F}_{n-1}$ but not $\mathscr{F}_n$. (Stallings' original example is in [Amer. J. Math. **85** (1963), 541–543; MR **28** #2139] and Bieri's work is in [J. Pure Appl. Algebra **7** (1976), no. 1, 35–51; MR **52** #10904].)

While the $\mathscr{F}_n$ property is a nice generalization of finite generation and finite presentability, and it has concrete geometric meaning, it is not very tractable in terms of using algebraic machinery. This is very similar to the fact that homology groups of topological spaces are easier to work with than homotopy groups. Bieri introduced a different set of finiteness conditions, the FP_n properties, which are considerably more amenable to algebraic techniques: A group G is FP_n if and only if $\mathbf{Z}$ has a projective resolution as a trivial $\mathbf{Z}G$-module that is finite through dimension n [R. Bieri, *Homological dimension of discrete groups*, Mathematics Department, Queen Mary College, London, 1976; MR **57** #6224]. If $\mathscr{X}$ is a $K(G,1)$ complex, then the action of G on the universal cover $\widetilde{\mathscr{X}}^{(n)}$ induces such a resolution, hence the property $\mathscr{F}_n$ implies the FP_n property. It is relatively simple to prove that $\mathrm{FP}_1 \Rightarrow \mathscr{F}_1$, and a lot of interesting work has appeared in the intervening years, trying to prove or disprove the equivalence of these finiteness properties for $n \geq 2$.

Bestvina and Brady begin by clarifying the connection between the $\mathscr{F}_n$ and the FP_n properties. If G admits a cocompact action on an n-acyclic space $\mathscr{X}$, meaning that $\widetilde{H}_i(\mathscr{X}) = 0$ for $0 \leq i \leq n$, then the action of G still induces a resolution of $\mathbf{Z}$ as a $\mathbf{Z}G$-module. So even if $\mathscr{X}$ is not n-connected, G is FP_n. Bestvina and Brady say that G is FH_n if G acts freely and cocompactly on an n-acyclic space. It is immediate that $\mathscr{F}_n \Rightarrow \mathrm{FH}_n \Rightarrow \mathrm{FP}_n$, and Bestvina and Brady establish that $\mathscr{F}_n \not\Leftarrow \mathrm{FH}_n$ for $n \geq 2$. As the authors indicate, this leaves open the intriguing question: Are the FP_n and FH_n properties the same?

Bestvina and Brady's study concentrates on the topology of right-angled Artin groups, or what used to be called "graph groups". Let L be a finite flag complex, that is, L is a simplicial complex where every complete graph contained in $L^{(1)}$ is the 1-skeleton of a simplex in L. The barycentric subdivision of any simplicial complex is a flag complex, so any topology that can be presented in the simplicial setting also occurs in the flag complex setting. The right-angled Artin group associated to L has a presentation where the generators correspond to the vertices of L, and the defining relations state that two generators commute when their affiliated vertices are joined by an edge. Denote this group by G_L. If L is a simplex, then G_L is free abelian, while if L is a null graph, G_L is a finitely generated free group. More interestingly, direct products of finitely generated free groups are right-angled Artin groups.

Main theorem: Let L be a finite flag complex, let G_L be the associated right-angled Artin group, and let H_L be the kernel of the map sending every generator of G_L to $1 \in \mathbf{Z}$. Then (1) H_L is FP_n if and only if L is n-acyclic. (2) H_L is finitely presented if and only if L is simply connected.

It follows that if L is not simply connected, but has trivial reduced homology in dimensions 1 and 2, then H_L will be FP_2 but it is not finitely presented. Further, by carefully choosing the complex L one can arrange to have the kernel be of type FP (meaning that $\mathbf{Z}$ has a finite projective resolution as a trivial $\mathbf{Z}G$-module) but not be finitely presentable.

The techniques used in proving this theorem are as intriguing as the result. Bestvina and Brady introduce Morse-theoretic techniques into the study of CAT(0) spaces. These techniques have already been used in several other situations to establish important results: Brady ["Branched coverings of cubical complexes and subgroups of hyperbolic groups", Proc. London Math. Soc., to appear] has used them to establish the existence of finitely presented subgroups of hyperbolic groups which are not themselves hyperbolic. Bestvina and M. E. Feighn ["The topology at infinity of $\mathrm{Out}(F_n)$", Preprint; per revr.] have used similar techniques to establish connectivity at infinity properties of $\mathrm{Aut}(F_n)$ and $\mathrm{Out}(F_n)$ which imply that these groups are duality groups.

Every right-angled Artin group G_L has a canonical non-positively curved space $\widetilde{K}_L$ on which it acts; the space is constructed by identifying tori of appropriate dimensions in a manner prescribed by the complex L. For each simplex $\sigma \subset L$, let T_σ be a $(|\sigma|+1)$-torus. The space $\widetilde{K}_L$ is then the disjoint union of the T_σ's, modulo the identifications $T_\sigma \cap T_\tau = T_{\sigma\cap\tau}$, with the convention that $T_\varnothing$ is the basepoint.

Since $H_L < G_L$, H_L also acts on $\widetilde{K}_L$. The trick is to understand the topology of a level set inside of $\widetilde{K}_L$ on which the kernel H_L acts with finite quotient. In general terms, this is essentially the same strategy that is employed by Bieri, Neumann, Strebel and Renz in their theory of Σ-invariants for infinite groups [see R. Bieri, W. D. Neumann and R. Strebel, Invent. Math. **90** (1987), no. 3, 451–477; MR 89b:20108; R. Bieri and B. Renz, Comment. Math. Helv. **63** (1988), no. 3, 464–497; MR 90a:20106]. In fact, Bestvina and Brady's results have been extended to give a complete description of the Σ-invariants for right-angled Artin groups. It follows from this work that it is possible to determine the finiteness properties of any subgroup above the commutator subgroup of a right-angled Artin group. This was first done by the reviewer, H. Meinert and L. VanWyk [Comment. Math. Helv. **73** (1998), no. 1, 22–44; MR 99f:57002], and an independent proof, based more closely on Bestvina and Brady's work, has been presented by K.-U. Bux and C. Gonzales ["The Bestvina-Brady construction revisited—geometric computation of Σ-invariants for right angled Artin groups", J. London Math. Soc., to appear].

The key technical result of this paper is that inside the level set of $\widetilde{K}_L$ associated with the kernel H_L one can find arbitrarily large copies of the defining flag complex L. A careful argument then shows that the homology and the homotopy type of this level set are determined by the homology and the homotopy type of L.

Finally, the article concludes with a very interesting discussion of the example when L is a flag triangulation of the spine of the Poincaré homology sphere. In this case, either the kernel H_L is a counterexample to the Eilenberg-Ganea conjecture, or there must be a counterexample to the Whitehead conjecture. The argument is roughly as follows: The cohomological dimension of H_L is two. S. Eilenberg and T. Ganea [Ann. of Math. (2) **65** (1957), 517–518; MR **19**, 52d] proved that for $n \geq 3$ the cohomological and the geometric dimension of a group are the same; the Eilenberg-Ganea conjecture claims that this also holds in dimension two. If one assumes that the Eilenberg-Ganea conjecture is true, then the kernel H_L does have geometric dimension two, and there is some contractible 2-complex Y on which the kernel H_L acts freely and with finite quotient. This space is then quasi-isometric to the level set on which H_L acts inside of $\widetilde{K}_L$, which implies that Y contains connected subcomplexes which are not aspherical. This contradicts the Whitehead conjecture. In a remark at the end of the paper the authors indicate their own conjecture, which would imply that H_L is a counterexample to the Eilenberg-Ganea conjecture.

John Meier (Easton, PA)

97j:20037 20F50 20F05 20F06

Lysënok, I. G. (Moscow)

Infinite Burnside groups of even period. (Russian. Russian summary)

Izv. Ross. Akad. Nauk Ser. Mat. **60** (1996), *no.* 3, 3–224; *translation in Izv. Math.* **60** (1996), *no.* 3, 453–654.

In 1902 W. Burnside posed a question on the finiteness of a group that has finitely many generators and satisfies the identity $X^n = 1$. This is the well-known Burnside problem on periodic groups. Each m-generated group satisfying $X^n = 1$ is a quotient of the group $B(m,n) = F_m/F_m^n$, where F_m denotes a free group of rank m and F_m^n is the subgroup in F_m generated by all nth powers of its elements. Thus the equivalent question is whether $B(m,n)$ is finite. The results due to Burnside, Sanov and M. Hall show that $B(m,n)$ is finite for the cases $n = 2,3,4,6$. In 1968 P. S. Novikov and S. I. Adyan published the proof that $B(m,n)$ $(m > 1)$ is infinite provided $n \geq 4381$ is an odd integer. In 1975 Adyan improved the estimate down to odd $n \geq 665$. To investigate the structure of $B(m,n)$, Novikov and Adyan gave a construction of this group in terms of generators and defining relations. The set of defining relations (each of them has the form $A^n = 1$) is subdivided into subsets enumerated by a natural parameter α, which is called the rank. To describe the group $B(m,n)$, it suffices to study the groups $B_\alpha(m,n)$ that are given by all defining relations of rank $\leq \alpha$. The proof involves many basic concepts and consists of hundreds of lemmas proved by simultaneous induction on α. In 1982 A. Yu. Ol′shanskiĭ gave a significantly shorter proof of the Novikov-Adyan theorem for the case of odd $n > 10^{10}$ using the geometric approach of van Kampen-Lyndon diagrams. The case of even exponents remained open. There was a natural suggestion that the structure of $B(m,n)$ for even n should be more complicated than for the odd case. The results that $B(m,n)$ is infinite for sufficiently large exponents were independently announced by S. V. Ivanov [Bull. Amer. Math. Soc. (N.S.) **27** (1992), no. 2, 257–260; MR 93a:20061] and the author [Uspekhi Mat. Nauk **47** (1992), no. 2(284), 201–202; MR 93g:20077]. The full version of Ivanov's result [Internat. J. Algebra Comput. **4** (1994), no. 1-2, ii+308 pp.; MR 95h:20051] includes a description and several properties of $B(m,n)$ for any $n \geq 2^{48}$ divisible by 2^9, in particular, a description of finite subgroups in $B(m,n)$. A consequence is infiniteness of $B(m,n)$ for $m > 1$ and any $n \geq 2^{48}$. The proof is based on the geometric method of graded diagrams and uses the approach of the above-mentioned paper by Ol′shanskiĭ. Let us formulate the main results obtained by the author of the paper under review.

Theorem 1: For any $m \geq 2$ and $n \geq 8000$ the group $B(m,n)$ is infinite. Theorem 2: Let $n = 16k \geq 8000$ and $m \geq 2$. (i) Any non-empty word X that is reduced in a free group and equals 1 in $B(m,n)$ contains a non-empty subword of the form $A^{n/2-1240}$. (ii) The group $B(m,n)$ has exponential growth. (iii) The group $B(m,n)$ cannot be presented by a finite number of defining relations. (iv) Any finite subgroup in $B(m,n)$ is embeddable into a finite direct product $D_n \times D_{\overline{n}} \times \cdots \times D_{\overline{n}}$, where D_r is the dihedral group of order $2r$, and $\overline{n}$ is the maximal power of 2 that divides n.

The description of $B(m,n)$ for $n \geq 8000$, n divisible by 16, given in the paper also allows one to solve the word problem for this case (recall that to solve the word problem in a group means to give an algorithm that decides whether a word is equal to 1 in the group). The author mentions that if n is even but not divisible by 16 then the structure of $B(m,n)$ becomes more complicated. Although we know that $B(m,n)$ is infinite for sufficiently large n, we do not have a proof that the word problem in $B(m,n)$ is decidable for all $n \gg 1$. Let us characterize briefly the contents of the paper. The author's basic tool is a modified version of the Novikov-Adyan method. Some geometric constructions and ideas of Ol′shanskiĭ are also used in the proof. To give a presentation of $B(m,n)$, the author constructs a sequence of groups $B_\alpha(m,n)$ by induction on the rank α. §3 and §4 of the paper contain some basic information on Cayley graphs of groups and van Kampen-Lyndon diagrams. The main definitions of basic concepts and the construction of $B_\alpha(m,n)$ are given in §5. From §6 to §16, all statements are proved by simultaneous induction on α. §6 contains a series of lemmas on periodic paths on the Cayley graph of $B_\alpha(m,n)$. These statements play the same role as the so-called periodization in the Novikov-Adyan method. One can treat these facts as a generalization of well-known properties of periodic words. In §7 some important finite subgroups of $B_\alpha(m,n)$ are introduced. §8 consists of essentially technical statements on almost periodic paths on Cayley graphs. §9 and §10 deal with a special class of van Kampen-Lyndon diagrams. The estimates obtained are used in §11 to analyze relations in $B_\alpha(m,n)$ in terms of diagrams. One of the main results of §12 is the statement that any element of $B_\alpha(m,n)$ can be presented by a word reduced in rank α, and the analogue of it for cyclic words. §13 deals with conjugacy and normal forms of elements of finite order in $B_\alpha(m,n)$. The concept of a complete finite subgroup in $B_\alpha(m,n)$ is introduced in §14. It is shown there that any finite subgroup in $B_\alpha(m,n)$ can be embedded into a complete subgroup. §15 contains the description of finite subgroups in $B_\alpha(m,n)$. The goal of §16 is to prove the following important statement: If G is a finite subgroup in $B_\alpha(m,n)$ and $x,w \in B_\alpha(m,n)$ are such that $x^{-i}wx^i \in G$ for $i=0,1,\cdots,5$, then $x^{-i}wx^i \in G$ for any i. Let us mention that the divisibility of n by 16 is used only in the proof of the following lemma: If $G \subseteq H$ are finite subgroups in $B_\alpha(m,n)$ and $|H:G| \le 2$, then $x^4 \in N(G)$ for any $x \in N(H)$ (here N denotes the normalizer of a subgroup in $B_\alpha(m,n)$). Note that §16 is one of the most important parts of the paper. Finally, §17 contains proofs of Theorem 1 and Theorem 2.

The paper has a self-contained proof of a very important group-theoretic result. It introduces many significant ideas that can be used for constructing new, interesting groups. *Victor S. Guba* (Vologda)

99m:20088 20F55 20C05 20F36

Broué, Michel (Paris); **Malle, Gunter** (Heidelberg); **Rouquier, Raphaël** (Paris)

Complex reflection groups, braid groups, Hecke algebras. (English summary)

J. Reine Angew. Math. **500** (1998), 127–190.

The theory of finite real linear groups generated by reflections is very beautiful and very well known. Here, a reflection is a linear nonidentity transformation r of a vector space V fixing a hyperplane pointwise and stabilizing a 1-dimensional complement. The fixed hyperplane is called the reflection hyperplane and nonzero vectors in the 1-dimensional complement are called its roots.

Suppose that W is a such a group, acting on the real vector space V. Then there is a Euclidean inner product preserved by W such that V decomposes into mutually orthogonal irreducible components. Moreover, the nontrivial components can be chosen so as to be generated by roots; a direct product decomposition of W into reflection subgroups (i.e., subgroups generated by reflections) results. Suppose (zooming in to a single component) that W acts irreducibly on V. The inner product can be used to provide a presentation of W as follows: take $v \in V$ to be a regular vector (i.e., not contained in any reflection hyperplane of [a reflection in] W), and consider the reflection hyperplanes "closest" (in angle, defined by the inner product) to v. There are exactly $\dim(V)$ such hyperplanes; denote by R the set of corresponding reflections. Then W is generated by R. For $r, s \in R$, the angle between the corresponding reflection hyperplanes, or, which amounts to the same, the angle between their roots, determines the order $m_{r,s}$ of rs; now $\{(rs)^{m_{r,s}} \mid r, s \in R\}$, where $m_{r,r} = 1$, is a set of defining relations for a presentation of W on the generator set R. Coxeter has classified the real reflection groups by determining which matrices $(m_{r,s})_{r,s\in R}$ may occur. The data is usually given in terms of a labeled graph, known as the Coxeter diagram.

For $r = s$ the relations express that real reflections have order 2. Those for $r \neq s$ can be written

$$(1)\qquad \underbrace{rsr\cdots}_{m_{r,s}\text{ factors}} = \underbrace{srs\cdots}_{m_{r,s}\text{ factors}} .$$

Now consider the connected topological space M obtained by first complexifying V and next removing the complexified reflection hyperplanes of W. Its fundamental group P is called the pure braid group related to W, whereas the fundamental group of the quotient space M/W is called the braid group B of W. By Steinberg's theorem, which asserts that any vector of V fixed by a nontrivial element of W belongs to a reflection hyperplane of W, the subgroup P is normal in B with quotient W. A presentation for B can be given on generators $\widehat{r}$ (mapping onto r) by taking as relations the hatted versions of (1). In case W is the symmetric group on $n+1$ letters, in its irreducible reflection representation on n-dimensional space, the braid group is the original braid group as defined by E. Artin [Ann. of Math. (2) **48** (1947), 101–126; MR **8**, 367a].

There is another, "bigger" object of interest related to W, namely its Hecke algebra. This associative algebra over the ring $S = \mathbf{Z}[\{u_{i,0}, u_{i,1}, u_{i,0}^{-1}, u_{i,1}^{-1}\}_i]$, where i indexes the conjugacy classes of reflections in W, is obtained from the above presentation of W by taking the quotient of the free algebra on the generators $\overline{r}$ (corresponding to $r \in R$) by the ideal generated by the barred versions of LHS-RHS of (1) and (2) $(\overline{r} - u_{i,0})(\overline{r} - u_{i,1})$. This algebra is a deformation of the group algebra $\mathbf{Z}[W]$ in that the latter occurs as the specialization $u_{i,j} = \det(r_i)^j$, where r_i is a reflection of the conjugacy class indexed by i.

Besides the morphism $B \to W$, a morphism $B \to (\mathbf{C}[W])^*$ (the group of invertible elements of the group algebra of W) can be defined by means of monodromy. By deformation, this monodromy morphism can be viewed as a specialization and a restriction to B of a morphism $T\colon A[B] \to A[W]$ of (group) algebras, where A is the ring of holomorphic functions in the variables $\{t_{i,0}, t_{i,1}\}_i$. Now T factors through the Hecke algebra $A \otimes_S H$.

If W is a Weyl group, it is involved in Lie groups, algebraic groups and groups of Lie type and the Hecke algebra is important for many aspects of their representation theory. Nevertheless, the constructions and results mentioned are also valid for the other finite real reflection groups.

How much of the above holds if we consider finite complex reflection groups (i.e., linear groups on a complex vector space V generated by reflections—defined as above) instead of the real ones? Ever since the classification of these groups in 1954

[G. C. Shephard and J. A. Todd, Canadian J. Math. **6** (1954), 274–304; MR **15**, 600b], there have been arguments in favor of and against the point of view that the finite complex reflection groups would behave just as nicely and would be just (or almost) as important. To begin, thanks to work of Chevalley, the reflection groups (real and complex) are characterized by the fact that the algebras of W-invariant polynomials on V are themselves polynomial algebras. Clearly an argument in favor. But the fact that certain complex reflection groups (even amongst those all of whose reflections have order 2) cannot be generated by $\dim(V)$ reflections is an argument against.

Lehrer and Springer showed that if one takes an arbitrary element in W, its centralizer induces a reflection group on each eigenspace. Even if W is taken to be real, the induced reflection group may be a complex one; many, but not all, irreducible reflection groups can be found in this way from the real ones. To alternate this nice intricate liaison between real and complex groups with a negative aspect (at least up till now) we mention that no analog of the natural Coxeter presentation of the real groups has been found. A natural procedure, starting from a regular vector v, using a Hermitian inner product, and taking all reflections with hyperplanes nearest to v is not known (to me). It is clear though that more complicated relations, involving at least three reflections at a time, are needed.

The present paper is an attempt to show that there is an analog to most of the above observations, even though it may be difficult to find. The authors, Broué, Malle and Rouquier, take the viewpoint that a natural presentation is one where the factoring of the monodromy morphism through the Hecke algebra comes about "naturally". Observe that the braid group and the pure braid group can again be defined in terms of topology (Steinberg's theorem is also valid for complex reflection groups). The resulting morphism now maps B onto the opposite group W^{op} of W. In deviation from the above description of the real case (depending on a presentation of W), the three authors define the Hecke algebra H for complex W as a quotient algebra of the group algebra of B by relations generalizing (2) to products of as many linear factors as the order of the corresponding reflections. They find (this is their main theorem) that the factorization of the monodromy morphism (defined as above, but now from B to $(\mathbf{C}[W^{\mathrm{op}}])^*$) through $A \otimes_S H$ exists.

But then, to push the analogy with the real case, explicit generators and relations for the braid group and the Hecke algebra associated with those of W are called for. The authors succeed in finding such a presentation for W with interpretations for B and H for many finite irreducible complex reflection groups W, but not for all. The proofs consist of mainly case-by-case analysis.

Besides this impressive collection of facts, which seems to indicate that an overall theory leading to general proofs (and statements of results!) may someday become available, several other properties of (some of) the groups involved, generalizing known results for the real case, are proved (e.g., a description of the center of B).

Arjeh M. Cohen (Eindhoven)

97d:20056 20G15 20G40 22E40
Serre, Jean-Pierre (Paris)
Exemples de plongements des groupes $\mathrm{PSL}_2(\mathbf{F}_p)$ dans des groupes de Lie simples.
[Examples of embeddings of the groups $\mathrm{PSL}_2(\mathbf{F}_p)$ into simple Lie groups]
Invent. Math. **124** (1996), *no.* 1-3, 525–562.

Generalizing a conjecture of Kostant's, the author proves the following results for p a prime number and G a simple linear algebraic group over an algebraically closed field k with Coxeter number h: (i) Suppose $p = h + 1$; if $\operatorname{char} k \neq 2$ or $G \not\cong \mathrm{PGL}_2$, then $\mathrm{PGL}_2(\mathbf{F}_p)$ embeds in $G(k)$; (ii) if $p = 2h + 1$, then $\mathrm{PSL}_2(\mathbf{F}_p)$ embeds in $G(k)$.

In particular, for $G(k) = \mathrm{PGL}_2(\mathbf{C})$ (with Coxeter number 2) one finds the well-known octahedral and icosahedral subgroups $\mathrm{PGL}_2(\mathbf{F}_3)$ and $\mathrm{PSL}_2(\mathbf{F}_5)$. Result (ii) has been established previously, albeit with the help of computers for some of the exceptional Lie types. Also, part of (i) was known, but (for instance) not for G of type E_7.

The (unified) proof uses Testerman's results on the existence of special classes of subgroups of $G(\mathbf{F}_p)$ isomorphic to $\mathrm{PSL}_2(\mathbf{F}_p)$. A self-contained description of these subgroups, containing many principal elements, is given. Next a suitable ring R of integers in a local field with residue field $\mathbf{F}_p$ is taken and the morphism $G(R) \to G(\mathbf{F}_p)$ is considered. Its kernel is a profinite p-group. The crux of the proof is the fact that the inverse image of the special subgroup isomorphic to $\mathrm{PSL}_2(\mathbf{F}_p)$ is a split extension of the subgroup $\mathrm{PSL}_2(\mathbf{F}_p)$ with a normal p-subgroup (part of the above kernel). In other words, the special subgroup of $G(\mathbf{F}_p)$ "lifts" to an isomorphic subgroup in $G(R)$. In case (i) this is established by showing that $H^*(\mathrm{PGL}_2(\mathbf{F}_p), L) = 0$, where L is the restriction to $\mathrm{PGL}_2(\mathbf{F}_p)$ of the adjoint $G(\overline{\mathbf{F}_p})$-module. In case (ii), $H^2(\mathrm{PSL}_2(\mathbf{F}_p), L)$ has dimension 1 and so a more intricate study of the action of subgroup elements is carried out to show that the standard Borel subgroup of $\mathrm{PSL}_2(\mathbf{F}_p)$ lifts. Since this Borel subgroup contains a Sylow p-subgroup of $\mathrm{PSL}_2(\mathbf{F}_p)$, it follows that the whole group lifts. Subsequently, from the existence of a subgroup of $G(R)$, passing to local fields K, the result is deduced for arbitrary algebraically closed fields k. A step of more general interest is here the existence, for any local field K and finite subgroup L of $G(K)$, of a finite totally ramified extension K' of K such that a $G(K')$-conjugate of L embeds in a hyperspecial maximal compact subgroup of $G(K')$ (so that L maps onto a subgroup of $G(F)$, where F is the residue field of K' and K).

Many details regarding the subgroups are discussed. For instance, in case (ii) with k of characteristic 0, the minimal field K for which there exists a form G_K of G such that $\mathrm{PSL}_2(\mathbf{F}_p)$ embeds in $G_K(K)$ is shown to contain $\mathbf{Q}(\sqrt{p^*})$, where $p^* = \pm p$ and $p^* \equiv 1 \pmod 4$; the question is raised whether K coincides with $\mathbf{Q}(\sqrt{p^*})$.

The only embeddings of $L(2,q)$ in Lie groups that are conceivable but have not yet been established in the literature are $L(2,29)$ for type E_7 and $L(2,32)$, $L(2,41)$, and $L(2,49)$ for type E_8, but there are two preprints by Griess and Ryba proving existence for all four cases by computer. *Arjeh M. Cohen* (Eindhoven)

99g:20085 20G20 20C15 20E32 20G05

Griess, Robert L., Jr. (Ann Arbor, MI); **Ryba, A. J. E.** (Milwaukee, WI)

Finite simple groups which projectively embed in an exceptional Lie group are classified! (English summary)

Bull. Amer. Math. Soc. (N.S.) **36** (1999), *no.* 1, 75–93.

This paper gives an overview of some work which began in the 80's and has recently been completed by the authors. This is the determination of which simple or quasisimple groups are subgroups of the exceptional Lie groups over the complex numbers. These are the complex Lie groups of type G_2, F_4, E_6, E_7, E_8. The question for the classical Lie groups, i.e. the general linear, orthogonal, and symplectic groups, can be determined by ordinary character theory and the Frobenius-Schur indicator applied to the defining vector space. This leaves the exceptional groups to be investigated separately.

The classification is for simple groups or quasisimple groups. A group is quasisimple if it is its own commutator and modulo the center is simple. Such a group is called a covering group of the simple group. For example, the group of 2×2 unimodular matrices contains a group of order 120. This group modulo its center, $\pm I$, is the alternating group on 5 letters. The alternating group itself is not a subgroup. These quasisimple groups are interesting subgroups of Lie groups, and often appear unrelated to other natural subgroups of the Lie group. Because of the classification of the simple groups, it is possible to try to determine which groups occur.

The paper is divided into several parts which give an introduction to the topic, a history of the developments, and brief sketches of the methods. There is an excellent bibliography and there are sections discussing other subgroup embeddings. The techniques all use one or more of the natural representations of the groups. In particular, the adjoint representation on the Lie algebras plays a pivotal role, as well as other smaller-dimensional representations. The adjoint representations have degrees $14, 52, 78, 133, 248$ for G_2, F_4, E_6, E_7, E_8, respectively. There are additional representations of degrees $7, 26, 27, 56$ for G_2, F_4, E_6, E_7, respectively. The traces are known for elements of small order acting on these spaces. To check if a particular group might possibly be a subgroup a search is made for a sum of characters for the purported group for which the sum of the characters has the correct value on these elements of small order. At this stage it is not always possible to tell that the character is correct on all classes, but only on certain ones. A theorem of Jordan says that for any given integer, say 248, there are only a finite number of quasisimple groups which have a nontrivial representation of that degree or smaller. The classification of finite simple groups allows the determination of all these groups. Many groups occur in a closed subgroup of positive dimension. These are known from the classical theory or induction. There are very restrictive character considerations for the remaining ones, called Lie primitive. It is then necessary to examine each remaining group to try to decide if it is a subgroup. There are many techniques for doing this. More about this later.

The most important section, perhaps, is Section 3, which gives a table of the simple groups for which they or their covering groups are subgroups of an exceptional Lie group. The table and much of the work in this field make use of the inclusions

$$G_2 < F_4 < 3E_6 < 2E_7 < E_8.$$

Here the 3 in front of E_6 and the 2 in front of E_7 mean there are centers of orders $3,2$, respectively, and the quotient is the group described. These groups are their own commutators. Each row of the table is indexed by a quasisimple group which occurs in E_8. Elsewhere in the row, the smallest of the above groups in which it first appears is indicated. There is much more information, including the order of the center and references to the constructions in each case. There are instances when a given group occurs with a completely different action on the modules. These are not given here but some information is given.

The work naturally started with G_2 in the early 80's [A. M. Cohen and D. B. Wales, Comm. Algebra **11** (1983), no. 4, 441–459; MR 85b:20010] using the 7-dimensional representation, which was small enough that all its subgroups were known. It continued with the groups F_4 and E_6 [A. M. Cohen and D. B. Wales, Israel J. Math. **82** (1993), no. 1-3, 45–86; MR 94h:20020; Proc. London Math. Soc. (3) **74** (1997), no. 1, 105–150; MR 97k:20078]. The work for E_7 and E_8 was started by A. M. Cohen and the first author [in *The Arcata Conference on Representations of Finite Groups (Arcata, Calif., 1986)*, 367–405, Proc. Sympos. Pure Math., 47, Part 2, Amer. Math. Soc., Providence, RI, 1987; MR 90a:20089], and finished by the authors, with constructions by them, A. M. Cohen, Bert Lisser, Peter Kleidman, and J.-P. Serre. The final group to be constructed was the smallest Suzuki group, Sz(8), by the authors. They have constructed several of the groups in E_7 and E_8, including the projective special linear group, $\mathrm{PSL}_2(31)$, the group of 2×2 unimodular matrices over Z_{31} modulo the center. Here Z_{31} is the integers mod 31. The constructions of these two groups are described briefly. These constructions, indeed, many of the constructions, use computers. Up to algebraic equivalence there are unique ways to write the character in the 248-dimensional adjoint module. This means there is a way to write down matrices that give the group needed. In each case the work was done over a finite field whose characteristic is relatively prime to the order of the group. For $\mathrm{PSL}_2(31)$ it is Z_{2791} and for Sz(8), Z_{29}. The existence of a subgroup in $E_8(p)$ for these primes p implies the existence over the complex numbers. There are two different methods illustrated by these cases. In the first case the matrices for a subgroup are determined in E_8 and then another element found also in E_8 which together with the subgroup generates the desired group. In the other, the matrices for the whole group are given and then a Lie algebra structure on the 248-dimensional space is found preserved by the matrices. Computers are used for both of these. In these cases and very often in these hard embeddings the computer work gives uniqueness, or perhaps a small number of possibilities.

Many constructions for $\mathrm{PSL}_2(p)$, p a prime, were done by Serre [Invent. Math. **124** (1996), no. 1-3, 525–562; MR 97d:20056, see p. B104 in this volume]. In these cases he lifted a subgroup isomorphic to $\mathrm{PSL}_2(p)$ by ingenious means from the corresponding Lie group over Z_p. This gave the first construction of some of these for E_7 and E_8 and constructions for many of the earlier groups which had been done by computer. In these cases, however, the uniqueness is not determined, as it often is by the computer construction.

This is a nicely written paper with many gems hidden throughout. To the extent possible it is written in an elementary way, with notation and references carefully given. Even the character table of the symmetric group on 4 letters is given, with illustrations of some of the arguments in this case. There are sketches of much more than can be included in this review about the methods and techniques used, including a discussion of the Kostant conjecture. The overview obtained by reading the paper gives a very good introduction to these developments. *D. Wales* (Pasadena, CA)

97f:20059 20G40 03C40 03C45 03C60 20A15
Hrushovski, E. (Jerusalem); **Pillay, A.** (Notre Dame, IN)
Definable subgroups of algebraic groups over finite fields.
J. Reine Angew. Math. **462** (1995), 69–91.

Theorem: Let p be a prime number and G a subgroup of $\mathrm{GL}_n(\mathbf{F}_p)$ generated by elements of order p. Then $G = \langle x_1\rangle \cdots \langle x_k\rangle$ for suitable elements $x_1, \cdots, x_k$ of G of order p, where the natural number k depends only on n and not on p. This attractive result is obtained here by a model-theoretic argument. It is related to Nori's Theorem B [M. V. Nori, Invent. Math. **88** (1987), no. 2, 257–275; MR 88d:20068], which is also derived here. In addition the authors give elegant model-theoretic proofs for the results of C. R. Matthews, L. N. Vaserstein and B. Weisfeiler [Proc. London Math. Soc. (3) **48** (1984), no. 3, 514–532; MR 85d:20040] and Nori [op. cit.] on the reduction modulo large p of certain finitely generated subgroups of $\mathrm{GL}_n(\mathbf{Q})$. The paper contains other nice results, but this review will focus on the proof of the theorem stated above, to illustrate the novel methods used.

First some background on "definable groups" and "pseudo-finite fields", notions which play a key role in the paper.

Definable groups: In the last decade groups finally began to play a pervasive role in model theory and its applications. The model-theoretic notion of stability gave rise to a theory of stable groups [see B. Poizat, *Groupes stables*, Poizat, Lyon, 1987; MR 89b:03056]. Algebraic groups and also differentially algebraic groups are stable, and some of their characteristic properties are consequences of that fact alone.

Many structures in mathematics—the real and p-adic fields, for example—are not stable, but are nevertheless model-theoretically "nice" in the sense that one can determine more or less explicitly the category of definable sets and maps associated to the structure, and introduce a well-behaved dimension function on the class of definable sets. (Example: The definable sets associated to the real field $\mathbf{R}$ are the semialgebraic subsets of $\mathbf{R}^n$ for $n \in \mathbf{N}$.) In that case the definable groups (the group objects in the category of definable sets and maps) are especially rewarding objects of study, and behave much like stable groups. "Definability" has the virtue of flexibility and ease in handling compared to more delicate geometric notions, while often being of comparable scope. Moreover, various model-theoretic generalizations by Hrushovski of Weil's theorem on recovering algebraic groups from generic data provide powerful tools in constructing definable groups. Several recent applications of model theory, including the one under discussion, depend in fact on rather basic results on definable groups in certain "nice" mathematical structures.

Pseudo-finite fields: Another part of the background to the paper under review concerns pseudo-finite fields: (infinite) fields that satisfy all elementary statements that are true in all finite fields. J. Ax [Ann. of Math. (2) **88** (1968), 239–271; MR **37** #5187] showed that the pseudo-finite fields are exactly the perfect fields F that have exactly one extension of each positive degree n in their algebraic closure $\widetilde{F}$, and such that each absolutely irreducible affine curve defined over F has an F-rational point. Pseudo-finite fields have certain advantages over finite fields, especially because of the last property, which is called PAC (= pseudo-algebraically closed). At the same time, results on pseudo-finite fields often imply asymptotic properties of finite fields, a sort of Lefschetz principle that is put to good use in this article. The pseudo-finite fields are unstable but "nice" in the (admittedly vague) sense above, by taking the dimension of a definable set $S \subseteq F^n$ to be the algebraic-geometric dimension of its Zariski-closure in $\widetilde{F}^n$ [see Z. M. Chatzidakis, L. P. D. van den Dries and A. J. Macintyre, J. Reine Angew. Math. **427** (1992), 107–135; MR 94c:03049].

Let F be a pseudo-finite field. Early on in the paper the authors use simple dimension arguments to establish a result of which the following is a special case.

Proposition: Let $(G_x)_{x\in X}$ be a family of connected algebraic subgroups of $\mathrm{GL}_n(\widetilde{F})$ defined over F. Then the subgroup of $\mathrm{GL}_n(F)$ generated by $\bigcup_x G_x(F)$ is of the form $G_{x_1}(F)\cdots G_{x_k}(F)$ for certain $x_1,\cdots,x_k \in X$.

(It might be mentioned here that each algebraic group G defined over F gives rise to the definable group $G(F)$.) The theorem stated at the beginning of this review is an easy consequence of the proposition. To see this, let $\mathrm{char}(F)=p>n$ or $\mathrm{char}(F)=0$. An element $x\in \mathrm{GL}_n(F)$ is said to be unipotent if $x=1_n+y$ with $y^n=0$. Each unipotent $x=1_n+y$ gives rise to a one-parameter subgroup of $\mathrm{GL}_n(\widetilde{F})$:

$$t\mapsto x^t=\sum_{i=0}^{n-1}\binom{t}{i}y^i\colon \widetilde{F}\to \mathrm{GL}_n(\widetilde{F}) \text{ (binomial expansion)},$$

whose image is the connected algebraic subgroup $x^{\widetilde{F}}$ of $\mathrm{GL}_n(\widetilde{F})$, which is defined over F and satisfies $x^{\widetilde{F}}(F)=x^F$. Now apply the proposition to the case where X is any set of unipotent elements of $\mathrm{GL}_n(F)$, with $G_x=x^{\widetilde{F}}$. Because of the uniformity in the definitions of the groups $G_x(F)$, a routine model-theoretic compactness argument shows that in this case the number k in the proposition can be taken to depend only on n and to be independent of F. The theorem now follows from the observation that if $\mathrm{char}(F)=p>n$, then x is unipotent if and only if $x^p=1_n$.
Lou van den Dries (Urbana, IL)

97e:20106a 20P05 20E32 20F05
Liebeck, Martin W. (London); **Shalev, Aner** (Jerusalem)
Classical groups, probabilistic methods, and the (2, 3)-generation problem. (English summary)
Ann. of Math. (2) **144** (1996), *no.* 1, 77–125.

97e:20106b 20P05 20D06 20F05
Liebeck, Martin W. (London); **Shalev, Aner** (Jerusalem)
Simple groups, probabilistic methods, and a conjecture of Kantor and Lubotzky. (English summary)
J. Algebra **184** (1996), *no.* 1, 31–57.

These two papers by Liebeck and Shalev constitute a big step forward in probabilistic group theory, and demonstrate the power of probabilistic methods in the solution of long-standing non-probabilistic problems.

The foundations of the subject lie in a series of papers by P. Erdős and P. Turán [see, e.g., Z. Wahrsch. Verw. Gebiete **4** (1965), 175–186 (1965); MR **32** #2465] where they investigated random permutations, and created a statistical theory for the symmetric group S_n. J. D. Dixon [Math. Z. **110** (1969), 199–205; MR **40** #4985] used this theory to settle a 19th-century conjecture of Netto, proving that two randomly chosen elements of the alternating group A_n generate A_n with probability $\to 1$ as $n \to \infty$. He then conjectured that a similar result holds for arbitrary finite simple groups. To be precise, let $P(G)$ denote the probability that two randomly chosen elements of a group G generate G; Dixon conjectured that, if G is a finite simple group, then $P(G) \to 1$ as $|G| \to \infty$.

In 1990 W. M. Kantor and the reviewer proved Dixon's conjecture for classical (and small-rank exceptional) groups [Geom. Dedicata **36** (1990), no. 1, 67–87; MR 91j:20041]; and in 1995 Liebeck and Shalev completed the proof of Dixon's conjecture by dealing with the remaining exceptional groups of Lie type [Geom. Dedicata **56** (1995), no. 1, 103–113; MR 96h:20116]. While the arguments of Dixon are elementary, the proofs of Kantor, Lubotzky, Liebeck and Shalev rely heavily on the classification of finite simple groups, and the information on their maximal subgroups. The relevance of maximal subgroups to this subject can be seen from the following simple observation (made by Kantor and the reviewer, and used in suitable variations in all subsequent papers in the area): suppose the elements x, y chosen at random from a group G do not generate G; then they both lie in some maximal subgroup M of G, and given M this holds with probability $|G{:}M|^{-2}$; therefore $1 - P(G) \leq \sum_M |G{:}M|^{-2}$, where M ranges over all maximal subgroups of G. This observation led to questions concerning enumeration of maximal subgroups and the estimation of their indices. These questions were solved using results by M. Aschbacher [Invent. Math. **76** (1984), no. 3, 469–514; MR 86a:20054] and others on the subgroup structure of groups of Lie type. In some cases, special tricks had to be employed in order to deal with "unknown" maximal subgroups.

The above description summarizes the situation before the appearance of the papers under review. These papers deal with more delicate problems, such as random generation by elements of given order. The main motivation behind the study of such questions is the long-standing problem of determining the finite simple images of the classical modular group $\mathrm{PSL}_2(\mathbf{Z})$. A group G is an image of the modular group if and only if it is $(2,3)$-generated, namely, can be generated by elements x, y satisfying $x^2 = y^3 = 1$. Partial results on the $(2,3)$-generation problem were obtained throughout the century, starting with Miller (1901) and Barahana (1928). More recent results have been obtained by Tamburini, Wilson, Di Martino, Vavilov, Malle and others, who showed (using explicit constructions) that many groups of Lie type are $(2,3)$-generated. However, the $(2,3)$-generation problem remained open for various infinite families of classical (and exceptional) groups, including some projective special linear groups.

Instead of explicitly constructing generators of order 2 and 3, as done by other authors, Liebeck and Shalev ask the following question: suppose we choose at random an element x of order 2 and an element y of order 3 from the simple group G; what is the probability $P_{2,3}(G)$ that x, y generate G? The main result of the first paper under review shows that, for G classical, $P_{2,3}(G) \to 1$ unless $G = \mathrm{PSp}_4(p^k)$ with $p \geq 5$, in which case $P_{2,3}(G) \to 1/2$, or $G = \mathrm{PSp}_4(p^k)$ with $p = 2, 3$, in which case $P_{2,3}(G) = 0$. Consequently, the authors deduce that except for $\mathrm{PSp}_4(2^k)$, $\mathrm{PSp}_4(3^k)$, and finitely many other groups, all finite simple classical groups are images of $\mathrm{PSL}_2(\mathbf{Z})$. This is a very powerful result, and the discovery that $\mathrm{PSp}_4(2^k)$ and $\mathrm{PSp}_4(3^k)$ are not images of $\mathrm{PSL}_2(\mathbf{Z})$, being unexpected, is also quite remarkable.

The proof starts with bounding $1 - P_{2,3}(G)$ from above by a certain expression involving the maximal subgroups of G. But this expression now does not depend on the indices of the maximal subgroups, but rather on the number of elements of order 2 and 3 in them. By counting elements of orders 2 and 3 in G and in its maximal subgroups (a rather painstaking task), and assuming $G \neq \mathrm{PSp}_4(q)$, the authors derive another upper bound on $1 - P_{2,3}(G)$, which has the form $\sum_M |G{:}M|^{-s}$ for some s slightly bigger than 1. Then they show that, for every fixed $s > 1$, the expression $\sum_M |G{:}M|^{-s}$ tends to 0 as $|G| \to \infty$, and thus they deduce that $P_{2,3}(G) \to 1$.

Another problem solved in the papers of Liebeck and Shalev was raised by Kantor and the reviewer. In our above-mentioned 1990 paper we conjectured that a randomly chosen involution and a randomly chosen additional element of a finite simple group G generate G with probablity $\to 1$ as $|G| \to \infty$. The techniques developed in the first paper under review enable Liebeck and Shalev to prove this conjecture for classical (and alternating) groups. However, the case of exceptional groups of Lie type required different methods. This case is solved in the second paper under review, where the proof of the conjecture is completed. It is interesting that $(2,3)$-generation of classical groups plays a key role in proving the Kantor-Lubotzky conjecture for exceptional groups. This demonstrates the beautiful interplay between probabilistic and deterministic results in the papers under review, which is really two-sided: on the one hand probabilistic methods are used to prove deterministic results, such as $(2,3)$-generation; on the other hand, the $(2,3)$-generation results, when applied to the maximal subgroups M of an exceptional group G, are used to prove that G is randomly generated by an involution and another element. Another important ingredient in that proof is a recent (yet unpublished) work by Liebeck and Seitz on the subgroup structure of exceptional groups. The second paper under review also contains remarkably precise estimates for the generation probability $P(G)$ discussed in the first part of this review; extending an estimate of Babai concerning $P(A_n)$, the authors show that $P(G)$ is almost determined by the minimal index of a proper subgroup of G.

Some of the results just described have interesting combinatorial applications [see M. W. Liebeck and A. Shalev, J. Combin. Theory Ser. A **75** (1996), no. 2, 341–352; MR 98b:20005]. They may also find applications in computational group theory, since probabilistic results are instrumental in many group-theoretic algorithms.

Let me conclude this review with a historical comment. The theory of random graphs was created in order to prove various existence theorems in graph theory which could not be proved by constructive methods. The papers under review bring this idea to group theory, and will hopefully have a similar impact.

Alexander Lubotzky (Jerusalem)

C. REVIEWS IN ANALYSIS

97g:26016 26D10 52A40
Burchard, Almut (Princeton, NJ)
Cases of equality in the Riesz rearrangement inequality.
Ann. of Math. (2) **143** (1996), *no.* 3, 499–527.

The Riesz rearrangement inequality of the title is the assertion that for nonnegative measurable functions f, g, h in $\mathbf{R}^n$, we have

$$\int_{\mathbf{R}^n \times \mathbf{R}^n} f(y)g(x-y)h(x)dx\,dy \leq \int_{\mathbf{R}^n \times \mathbf{R}^n} f^*(y)g^*(x-y)h^*(x)dx\,dy,$$

where * denotes symmetric decreasing rearrangement in $\mathbf{R}^n$. For $n = 1$ the inequality is due to F. Riesz (1930), and for general n to Sobolev (1938).

The paper under review is devoted to the identification of the triples f, g, h for which equality holds. The author gives a complete description when f, g, h are characteristic functions of sets A, B, and C of finite positive measure: If the volume radius of one of the sets, say C, is $\geq$ the sum of the volume radii of A and B, then equality holds iff $A + B \subset C$, modulo sets of measure zero. But if the volume radius of each of the sets is less than the sum of the volume radii of the other two, then A, B, and C must be suitable similitudes of a single ellipsoid. The former case follows fairly easily from known results related to the Brunn-Minkowski inequality, but proof of the latter case requires a good deal of inventiveness with symmetrization techniques and some new hands-on criteria for deciding when certain types of sets are ellipsoids.

The author deduces, via “layer-cake” representations [see, e.g., E. H. Lieb and M. Loss, *Analysis*, Amer. Math. Soc., Providence, RI, 1997; MR 98b:00004] and her characterization in the characteristic function case, that for general functions f, g, h, if at least two of the symmetric decreasing rearrangements are strictly decreasing, then, except in trivial cases, equality holds in Riesz's inequality if and only if there are a measure-preserving linear transformation T and constants a, b, c with $c = a + b$, such that $f = f^* \circ (T + a)$, $g = g^* \circ (T + b)$, and $h = h^* \circ (T + c)$. As a corollary, the author is able to identify the optimizers in a weak-type Young inequality.

In the reviewer's opinion, this paper constitutes a very significant contribution to the theory of symmetrization. Some related uniqueness problems are still open. For example, the author notes that her proof does not identify optimizers in the generalization of Riesz's inequality by H. J. Brascamp, Lieb and J. M. Luttinger [J.

Functional Analysis **17** (1974), 227–237; MR **49** #10835]:

$$\int_{\mathbf{R}^{nk}} \prod_{i=1}^{m} f_i\Big(\sum_{j=1}^{k} a_{ij}x_j\Big)\, dx_1 \cdots dx_k \le \int_{\mathbf{R}^{nk}} \prod_{i=1}^{m} f_i^{*}\Big(\sum_{j=1}^{k} a_{ij}x_j\Big)\, dx_1 \cdots dx_k.$$

Albert Baernstein, II (St. Louis, MO)

99i:28012 28A75 30C65 46E35 49Q15
David, Guy (Orsay); **Semmes, Stephen** (Houston, TX)
Quasiminimal surfaces of codimension 1 and John domains. (English summary)
Pacific J. Math. **183** (1998), *no.* 2, 213–277.

This paper deals with the following general question: how does the topology of a set restrict its measure and geometry? Here we consider closed subsets of Euclidean space only. Questions like this are of course old, arising for example in the theory of minimal surfaces, but they have recently gained new prominence in the hands of the two authors. A concrete problem of this type was posed by Semmes in his 1994 ICM address [in *Proceedings of the International Congress of Mathematicians, Vol. 1, 2 (Zürich, 1994)*, 875–885, Birkhäuser, Basel, 1995; MR 97j:28010]: If K is a compact set in $\mathbf{R}^n$, separating two concentric spheres of radii 1 and 2, and having Hausdorff $(n-1)$-measure bounded by a constant C, is it then true that K has a substantial piece of a Lipschitz graph? More precisely, is it true that there is a constant $M = M(n, C) > 0$ and a Lipschitz graph Γ such that the Hausdorff $(n-1)$-measure of $K \cap \Gamma$ is at least M? The point is that the assumptions on K only involve one global condition on topology, and one on measure, as opposed to the more standard local and scale-invariant hypotheses. The paper under review answers this question affirmatively. There is indeed a more substantial answer, valid in a larger context of quasiminimizing sets, which will be discussed below. As stated above, Semmes's problem was independently solved by P. W. Jones, N. H. Katz and A. M. Vargas [Rev. Mat. Iberoamericana **13** (1997), no. 1, 189–210; MR 98j:49057].

The paper contains several interesting results, but one can highlight three issues as follows. First, sets that are quasiminimizers for a kind of obstacle problem (presenting the topological nontriviality) are shown to possess several nice geometric and measure-theoretic properties; second, some of the above properties are shown to imply, outside a variational context, further restrictions on the geometry, namely John-type conditions; third, it is shown how the particular problem mentioned at the beginning of this review fits in this framework. Now let us proceed to examine these three steps in more detail.

Fix two concentric cubes in $\mathbf{R}^n$, $n \ge 2$, and consider measurable sets V that contain the smaller cube, are contained in the larger, and whose characteristic function 1_V is a function of bounded variation, or a BV-function. Thus, V is a set of finite perimeter in the language of geometric measure theory. The authors call such a set a quasiminimizer if, roughly, the BV-norm of its characteristic function is smaller than that of any other similar set, up to a fixed error. (As is carefully explained in the paper, a corresponding quasiminimizing requirement in terms of the Hausdorff $(n-1)$-measure of the (Euclidean) boundary ∂V, in place of the BV-norm of 1_V, would lead to similar results.) The authors show that each quasiminimizing set has an open representative W such that the measure $\mu = |\nabla 1_W|$, which is supported on

∂W, satisfies $\mu(B_r) \approx r^{n-1}$ on boundary balls of radii $0 < r < \operatorname{diam} W$. It follows that a similar statement about the local mass of ∂W is true when μ is replaced by the Hausdorff $(n-1)$-measure. Thus, using the standard terminology of the genre, ∂W is an Ahlfors regular set of dimension $n-1$ for each quasiminimizer W. Moreover, ∂W satisfies a type of porosity condition, called Condition B by the authors, meaning that each ball centered at ∂W, with radius no more than $\operatorname{diam} W$, contains balls of comparable size in two different components of $\mathbf{R}^n \setminus \partial W$. Finally, it is established that quasiminimizing domains are domains of isoperimetry, which means that subdomains satisfy a relative version of the classical isoperimetric inequality. It follows in particular that $\mathbf{R}^n \setminus W$ has precisely two complementary components. The verification of these facts is a beautiful play with definitions, Sobolev-type inequalities, and direct geometric arguments.

The second step is to verify that a domain of isoperimetry in $\mathbf{R}^n$ is a John domain if its boundary satisfies Condition B and is Ahlfors regular of dimension $n-1$. B. Bojarski [in *Complex analysis, Joensuu 1987*, 52–68, Lecture Notes in Math., 1351, Springer, Berlin, 1988; MR 90b:46068] had earlier established that John domains are domains of isoperimetry. The authors display examples showing that a full converse to Bojarski's theorem cannot be true. Recall that a John domain is a domain with the property that each pair of points in it can be joined by a curve that escapes from the boundary at a fixed linear rate in comparison to the length travelled (from either point). It is curious that this innocent looking definition, given by F. John [Comm. Pure Appl. Math. **14** (1961), 391–413; MR **25** #1672] in connection with the rigidity of nearly isometric mappings, keeps showing up in ever new contexts in geometric analysis.

At this point it follows that a quasiminimizing domain has Ahlfors regular boundary of dimension $n-1$, and that there are precisely two complementary components of its boundary which are both John domains. The authors next show that this result is sharp in the sense that every such domain (subject to similar topological nontriviality conditions) arises as a quasiminimizer. This is achieved by showing that the domain is actually a minimizer for a weighted problem, where one puts weight 1 on the boundary and a large constant weight $A \gg 1$ on the two domains. A similar weighted problem is considered when the authors return to the original query mentioned at the beginning of this review: one penalizes sets that miss a lot of the separating compact set K, and the minimizer to this weighted problem is a quasiminimizer to the corresponding unweighted problem. Thus, the above results apply. Finally, to obtain a Lipschitz graph Γ that contains a substantial piece of K, the authors invoke some of their earlier results about uniformly rectifiable sets [*Analysis of and on uniformly rectifiable sets,* Amer. Math. Soc., Providence, RI, 1993; MR 94i:28003]. Condition B and the Ahlfors regularity of the boundary are central conditions here. Besides this point, the authors' use of their weak geometric lemma, which also occurs in the proof of the John property as explained in the previous paragraph, the paper is essentially self-contained. It even includes a nice discussion of functions of bounded variation and minimization problems, which goes beyond the particular needs of the paper.

The authors have a real gift for making hard things look easy. There is rarely anything technically difficult going on locally, arguments are discussed from several angles, down to the last detail, and prerequisites are kept at the minimum. This paper is a pleasure to read. Read it! *Juha Heinonen* (Ann Arbor, MI)

97f:28040 28D05 43A25

Akcoglu, Mustafa (Toronto, ON); **Bellow, Alexandra** (Evanston, IL);
Jones, Roger L. (Chicago, IL); **Losert, Viktor** (Vienna);
Reinhold-Larsson, Karin [Reinhold, Karin] (Albany, NY); **Wierdl, Máté** (Evanston, IL)

The strong sweeping out property for lacunary sequences, Riemann sums, convolution powers, and related matters. (English summary)

Ergodic Theory Dynam. Systems **16** (1996), *no.* 2, 207–253.

This article gives us a complete description of the best of the worst behavior in a number of different contexts. When almost everywhere convergence fails, one can ask for a clearer understanding of how badly it fails. The best way it can fail, in that it generally implies all other possible fashions of failure, is to be strongly sweeping out. Specifically, suppose that (T_n) is a sequence of positive linear operators acting on $L_1(X)$ of a probability space $(X, \mathcal{B}, m)$. Suppose that $T_n 1 = 1$ for all $n \geq 1$. Bellow was the first to define strong sweeping out for this sequence to be the condition that, for all $\varepsilon > 0$, there exists a measurable set E such that $m(E) < \varepsilon$, and $\limsup_{n\to\infty} T_n 1_E = 1$ a.e. $[m]$. It follows (by a Baire category argument or a direct construction along such a line of reasoning) that many measurable sets E (including ones of arbitrarily small measure) have the property that $\limsup_{n\to\infty} T_n 1_E = 1$ a.e. $[m]$ and at the same time $\liminf_{n\to\infty} T_n 1_E = 0$ a.e. $[m]$. This property is then another way to describe strong sweeping out. This clearly shows that $(T_n f)$ fails to converge a.e. for some function $f \in L_1(X)$, specifically even for a characteristic function.

The original motivation for the study in this paper was to prove that strong sweeping out occurs for averages along lacunary powers in the ergodic theorem. That is, fix a lacunary sequence (m_k) and some invertible ergodic transformation τ of $(X, \mathcal{B}, m)$. Let $T_n f(x) = n^{-1} \sum_{k=1}^n f(\tau^{m_k} x)$ for a function $f \in L_1(X)$. It was shown by Bellow that such averages fail to converge a.e., giving a specific instance of a universally bad sequence as first discovered by U. Krengel. Later, the reviewer showed that this sequence of averages is bad for characteristic functions too. However, strong sweeping out for a general lacunary sequence of powers was not known to hold; this article proves that this is the case. Moreover, the authors show this by looking at an interesting combinatorial property of lacunary sequences originally studied by A. Pollington in relation to the density of sequences of the form $(m_k \alpha)$. Actually, the method used in this paper shows that one cannot avoid the strong sweeping out behavior by replacing (T_n) by any nontrivial averaging method along the lacunary sequence (m_k).

With this very successful investigation, the authors of this article looked further for proving strong sweeping out for other operator sequences (T_n) arising in ergodic theory and real analysis. One example from ergodic theory that is shown to have the strong sweeping out property is where (T_n) is given by the convolution powers of a probability measure whose Fourier transform has appropriate properties. This includes any probability measure μ whose Fourier transform $\widehat{\mu}(\gamma)$ approaches 1 tangentially as γ approaches 1; such behavior occurs for instance when the measure has finite second moment and mean not zero. Another example from real analysis that is shown to have the strong sweeping out property is the sequence of Riemann sums taken as a sequence of operators. That is, consider $T_n f(x) = n^{-1} \sum_{k=1}^n f(x + k/n)$, for a periodic Lebesgue integrable function f defined on $\mathbf{R}$. Then this sequence was shown by W. Rudin to fail to converge a.e. even for characteristic functions, but the authors of this article show that indeed the sequence of Riemann sums as operators is strongly sweeping out.

This article has many very carefully crafted theorems along the lines of proving the best of the worst. It is a very readable paper, well worth study by a wide audience. Few articles in ergodic theory in recent years have been so thorough and so definitive in what is proven in the article. Shall we rate articles like restaurants, since they are often places to have an intellectual feast? If so, this article deserves a 6-star rating. *Joseph Max Rosenblatt* (Urbana, IL)

98m:28034 28D05 60J15
den Hollander, Frank (Nijmegen); **Steif, Jeffrey E.** (Göteborg)
Mixing properties of the generalized T, T^{-1}-process. (English summary)
J. Anal. Math. **72** (1997), 165–202.

The setup in this paper is as follows. S_n is a bi-infinite random walk in $\mathbf{Z}^d$ with i.i.d. stepsizes X_i, $i \in \mathbf{Z}$. In addition, $\{C_z,\ z \in \mathbf{Z}^d\}$ is an i.i.d. colouring of $\mathbf{Z}^d$; the C_z's take values $+1$ and -1 with probability $1/2$ each. The walk and the colouring are independent. The process of interest is $Z_i = (X_i, C_{S_i})$, that is, we observe both the current step of the random walk and the colour of the newly arrived position. S. A. Kalikow [Ann. of Math. (2) **115** (1982), no. 2, 393–409; MR 85j:28019] studied this process in $d = 1$ with simple random walk. He showed that the process is not Bernoulli in this case. Since it is well known that $\{Z_i\}$ is K, Kalikow's result gives a "natural" example of a process which is K but not Bernoulli. The paper under review deals with mixing properties of the $\{Z_i\}$ process for general random walks. Since these mixing properties are intimately related to intersection properties of random walks, much of the paper is devoted to properties of random walks.

The main results of the paper are summarised in (i)–(iii) below: (i) The process $\{Z_i\}$ is weak Bernoulli (WB) if and only if the future and the past of the random walk have only finitely many intersections a.s. (ii) If S_n is transient, then $\{Z_i\}$ is very weak Bernoulli (VWB). (iii) If (1) S_n is recurrent, (2) the step-sizes X_i have a finite δ-moment for some positive δ and (3) a weak technical condition ($(*)$, say) is satisfied, then $\{Z_i\}$ is not VWB. A few comments are appropriate. It turns out that for the $\{Z_i\}$ process, VWB is equivalent to Bernoulli. The technical condition $(*)$ mentioned in (iii) is probably not a condition at all: the authors conjecture that every random walk S_n satisfies it. The condition in (i) above can be related to the Green's function G. It is shown in the paper that if $\sum_{x \in \mathbf{Z}^d} G(x)G(-x) < \infty$, then the condition holds. (The Green's function condition is not necessary though, as the authors show with a counterexample.) It is then shown that this Green's function condition actually holds for an arbitrary random walk in $d \geq 5$, but fails in $d \leq 4$ when the random walk has zero mean and finite variance. A very neat generalisation of Kalikow's result can now be formulated. Consider a simple random walk in $\mathbf{Z}^d$. When $d = 2$, we have from (iii) (plus the fact that the simple random walk satisfies condition $(*)$) that $\{Z_i\}$ is not Bernoulli. When $d = 3$ or 4, (ii) tells us that $\{Z_i\}$ is Bernoulli. Since the Green's function condition fails here, $\{Z_i\}$ is not WB. Finally, from (i) together with the fact that the Green's function condition holds, we see that now $\{Z_i\}$ is WB when $d \geq 5$.

A few words about the proofs. The proof of (i) proceeds via the coupling characterisation of WB. (ii) is a consequence of the fact that a transient random walk has so-called zero density intersection, a fact that is easy to prove. Most of the work in the paper under review is in the proof of (iii). This is of course no surprise since this is the point where Kalikow's result is extended and people who are familiar with Kalikow's proof will appreciate its ingenuity. In some sense, the authors more or less follow the ideas of Kalikow. However, the fact that the step-sizes are unbounded results in great technical complications.

Apart from the results mentioned above, the authors go into some depth to investigate the relevant conditions on the random walk. We already mentioned the relation between the Green's function and the condition in (i). They also show that condition $(*)$ holds when the random walk is in a certain domain of attraction of a stable law. As such, the paper is as much about random walks as it is about the process $\{Z_i\}$. The classification of the different mixing properties is surprisingly complete. Probabilists will learn a lot about ergodic-theoretic notions by studying this paper. It is a very fine piece of work. *Ronald Meester* (Utrecht)

99c:28055 28D05
Rudolph, Daniel J. (College Park, MD)
Fully generic sequences and a multiple-term return-times theorem. (English summary)
Invent. Math. **131** (1998), *no.* 1, 199–228.

Let $(X,\mathcal{F},\mu,T)$ be a measure-preserving system on the probability measure space $(X,\mathcal{F},\mu)$. A bounded sequence (a_n) is said to be a good universal weight for the pointwise ergodic theorem if the averages

$$\text{(i)} \qquad M_{(a_n)}(g)(y)=\frac{1}{N}\sum_{n=1}^{N}a_n g(S^n y)$$

converge a.e. ν for all measure-preserving systems $(Y,\mathcal{G},\nu,S)$ and all functions $g\in L^1(\nu)$. In an unpublished preprint J. Bourgain announced the solution to a quite long-standing problem [cf. A. Bellow and V. Losert, Trans. Amer. Math. Soc. **288** (1985), no. 1, 307–345; MR 86c:28035]; he stated that given $f\in L^\infty$ the sequence $a_n=(f(T^n x))$ is μ-a.e. a good universal weight. Shortly after, Bourgain published a shorter proof jointly with H. Furstenberg, Y. Katznelson and D. S. Ornstein in an appendix to a paper of his [Inst. Hautes Études Sci. Publ. Math. No. 69 (1989), 5–45; MR 90k:28030]. Already at this stage Rudolph [Ergodic Theory Dynam. Systems **14** (1994), no. 1, 197–203; MR 95e:28015] rewrote their proof in terms of joinings. (The reader may also consult [K. Collins, "Bourgain's return times theorem", Master's project, Univ. North Carolina, Chapel Hill, NC, 1995; per revr.] for more details on both proofs.) In both published proofs the main path was to decompose a bounded function f into the sum of two functions, one being its projection onto the Kronecker factor $\mathcal{K}$, then deal with both functions separately. Furthermore, one could characterize the orthocomplement of the Kronecker factor in the following way: for μ-a.e. x the averages $M_{(f(T^n x))}(g)(y)$ converge a.e. ν to 0 for all dynamical systems $(Y,\mathcal{G},\nu,S)$ and all functions g in $L^2(\nu)$ if and only if the function f belongs to the orthocomplement of $\mathcal{K}$.

Quite impressed by this return-time phenomenon where a single sampling is "good" for an uncountable number of functions and dynamical systems, in 1991 the reviewer asked the following questions: (1) Can this return time result be extended to finitely many products of functions and dynamical systems? In other words, can we have a.e. convergence in the same universal sense for averages of the form

$$\text{(ii)} \qquad \frac{1}{N}\sum_{n=1}^{N}\prod_{i=1}^{H} f_i(T_i^n x_i)?$$

(universal sense meaning that the sets of full measure X_{f_i} associated with each bounded function f_i depend only on those j with $1 \leq j < i$). (2) Can such a result, if true, help characterize higher-order algebras than the Kronecker factor or their complement? (3) What are the exact spaces of functions for which such a result could be true? Could it be true if all functions f_i are in L^1?

A promising step towards an answer to (1) and (2) was reached with the Wiener-Wintner generalization of Bourgain's return-time theorem obtained by E. Lesigne, Rudolph and the reviewer [Israel J. Math. **92** (1995), no. 1-3, 375–395; MR 97e:28011]. The algebra characterized was exactly the one created by J.-P. Conze and Lesigne [Bull. Soc. Math. France **112** (1984), no. 2, 143–175; MR 86i:28019] while proving the convergence in L^1 norm of the averages

$$\text{(iii)} \qquad \frac{1}{N}\sum_{n=1}^{N} f_1 \circ T^n f_2 \circ T^{2n} f_3 \circ T^{3n}.$$

Thus a deep connection was established between multiple recurrence problems, introduced by H. Furstenberg, and return-time phenomena. This connection is important as it is not known whether the averages

$$\text{(iv)} \qquad \frac{1}{N}\sum_{n=1}^{N}\prod_{i=1}^{4} f_i \circ T^{in}$$

converge in L^1 norm for bounded functions.

The convergence of the averages in (ii) was obtained for all positive integer H and L^1 i.i.d. random variables by the reviewer [Duke Math. J. **88** (1997), no. 2, 217–246; MR 98g:60059]. It was the first multiple-term return-time theorem obtained for all positive integer H. It answered partially questions (1) and (3).

The paper under review answers the first question in the affirmative. In a remarkable way the author proves his result without having to create higher-order algebras. Several elegant key concepts are introduced in the paper: pointwise genericity, fully generic sequences, thick Følner sequences, to name a few. The most important concept introduced is in our view the notion of fully generic sequences.

Briefly, the path followed by the author is the following. First he defines what it means for an infinite sequence of values m from a compact metric space to be fully generic for some shift-invariant measure on M^N. Then he shows that, for any shift-invariant measure μ, μ-a.e. sequence is fully generic. Theorem 1, the goal of the paper and the key to the induction steps on H, proves that if m is fully generic, M_1 is some other metric space and ν is some shift invariant measure on M_1^N, then for ν-a.e. sequence m_1 the sequence of pairs (m, m_1) is again fully generic. This being done, it "only" remains to show that fully generic sequences are universally good. This involves to some extent joinings (or couplings).

In spite of the technical definitions and arguments the paper is well written. For clarity the author treats first the case where all the transformations T_i are weakly mixing (the case considered independently by the reviewer). In this case the notion of full genericity coincides with the classical notion of genericity introduced by H. Furstenberg some years ago.

One example is given at the end of the paper to show that there is a necessary modification of the sets X_{f_i} when one moves from the case $H = 2$ to the general case. The generalization obtained by the author cannot be reached by a trivial induction argument in the case $H = 2$. This is another indication of the depth of the author's main result. Some natural questions are raised at the end of the paper. They deal with the refinement of the notion of full genericity to k-fold genericity.

This paper is a very fine piece of work not only because of its results but also because of the techniques introduced. It provides important criteria for good randomly generated universal weights and opens the door to many multiple-term results. It also gives a tool to study the higher-order algebras that the author skillfully avoids in this paper. *Idris Assani* (Chapel Hill, NC)

97c:28041 28D15 11K55 22D40 28D20
Schmidt, Klaus [Schmidt, Klaus[1]] (Vienna)
★**Dynamical systems of algebraic origin. (English summary)**
Progress in Mathematics, 128.
Birkhäuser Verlag, Basel, 1995. xviii+310 pp. $89.00. ISBN 3-7643-5174-8

This award-winning monograph is about the dynamics and ergodic theory of actions (particularly $\mathbf{Z}^d$ actions) by group automorphisms on a compact metric group X. It is a marvelous display of the richness and depth of the interplay between dynamical systems on one hand and commutative algebra and number theory on the other. This study was initiated by an example due to F. Ledrappier which illustrates a phenomenon of breakdown of higher order mixing for $\mathbf{Z}^2$ actions.

The author systematically develops the underlying theory of actions by countable groups via group automorphisms of a compact group X. The foundations are laid down in the first chapter by developing basic results related to ergodicity and mixing. Three basic finiteness conditions (expansiveness, conjugacy to Lie subshifts and the descending chain condition (d.c.c.)) are introduced and their relationship with one another is examined. These conditions play a role throughout the text. The second chapter introduces the fundamental way in which the dynamics of the $\mathbf{Z}^d$ action by group automorphisms and commutative algebra interact. Every $\mathbf{Z}^d$ action α_d, $d \in \mathbf{Z}^d$, where $\alpha_d \in \operatorname{Aut}(X)$, determines a module $\mathscr{M}$ over the ring $\mathscr{R}_d = \mathbf{Z}[u_1^{\pm 1}, \cdots, u_d^{\pm 1}]$ of Laurent polynomials in (commuting) variables $u_1, \cdots, u_d$ with coefficients in $\mathbf{Z}$, and conversely. The module $\mathscr{M}$ is just the dual group $\widehat{X}$ (written additively) with the module structure given by $(f, a) = \sum_{n \in \mathbf{Z}^d} C_f(n) \widehat{\alpha}_n(a)$, where f is a finite sum $f = \sum C_f(n) u^n \in \mathscr{R}_d$, $C_f(n) \in \mathbf{Z}$, $a \in \widehat{X}$ and $\widehat{\alpha}_n \in \operatorname{Aut} \widehat{X}$ is the induced automorphism on $\widehat{X}$. Then throughout the book a "dictionary" is built that translates the dynamical properties (e.g., ergodicity, mixing, higher-order mixing, positive entropy, intrinsic ergodicity, etc.) of the action into algebraic properties of the module $\mathscr{M}$. These algebraic properties are expressed in terms of algebraic conditions on the prime ideals associated with $\mathscr{M}$.

Chapter III develops a structure theory of group automorphisms which in particular yields entropy as a complete invariant for topological conjugacy of expansive and ergodic group automorphisms of compact, zero-dimensional groups. The fourth chapter deals with periodic points for the $\mathbf{Z}^d$ actions by group automorphisms. It is shown that the periodic points are dense provided that the action satisfies the d.c.c. condition and the connected component of the identity in X is abelian.

The next three chapters are about entropy. A higher-dimensional analogue of the "entropy addition formula" is proved. Then the author starts developing tools for computing the entropy of the automorphism corresponding to the quotient module $\mathcal{M} = \mathcal{R}_d/(f)$ where $f \in \mathcal{R}_d$. This is done by introducing the notion of the "Mahler measure" of a Laurent polynomial $f \in \mathcal{R}_d$. Then pieces are put together to develop an expression for the entropy of any $\mathbf{Z}^d$ action by automorphisms on a compact abelian group.

Next, the question of characterization of positive (and complete positive) entropy is addressed. This involves higher-dimensional generalization of a theorem of Kronecker that characterizes Laurent polynomials with Mahler measure 1 in terms of being able to write such a polynomial as a product of cyclotomic polynomials. For certain $\mathbf{Z}^d$ actions, this also leads to some explicit entropy formulae in terms of the Dirichlet L-function and Mahler measure of certain polynomials. These are some of the new and amazing connections between dynamics and number theory. A dynamical interpretation of Lehmer's conjecture concerning polynomials with minimal Mahler measure is also discussed. For finite entropy, ergodic $\mathbf{Z}^d$ actions by automorphisms on compact abelian groups, complete positive entropy is characterized in terms of uniqueness of measure with maximal entropy.

The relation of entropy to growth rates of certain quantities related to periodic points is discussed. In general, an ergodic $\mathbf{Z}^d$ action by automorphisms on a compact abelian group can have uncountably many periodic points with a given period and hence a meaningful growth rate is obtained in terms of the number of connected components in the set of periodic points with a given period. The distribution of periodic points is also studied. For expansive actions, it is shown that all the limits of the distributions of periodic points are measures of maximal entropy.

The relation between entropy and dimension is examined in the class of zero-entropy $\mathbf{Z}^d$ actions by group automorphisms. Specific examples of zero entropy $\mathbf{Z}^2$-shift actions on closed, shift-invariant subgroups of $(\mathbf{Z}/p\mathbf{Z})^{\mathbf{Z}^2}$ are studied (where p is a rational prime). This study leads to certain measurable conjugacy invariants for more general $\mathbf{Z}^d$ actions and these invariants are related to relative entropy and certain residual sigma-algebras.

Chapter VIII deals with the breakdown of higher order mixing—a phenomenon discovered by Ledrappier. This phenomenon is again governed by the prime ideals associated with the $\mathcal{R}_d$ module $\mathcal{M}$. Another highlight of this work is the connection between mixing properties and solutions to certain arithmetical equations studied by Mahler, Masser, van der Poorten and Schlickewei. Using this connection the author proves that mixing $\mathbf{Z}^d$ actions by automorphisms on a compact, connected abelian group X is mixing of all orders. If X is not connected then the action can have non-mixing sets, and these are governed by the factorization of the prime ideals associated with $\mathcal{M}$ in certain rings of Laurent polynomials in fractional powers of the variables $u_1, \cdots, u_d$. This study leads to further measurable conjugacy invariants which distinguish between $\mathbf{Z}^d$ actions that cannot be distinguished by the earlier invariants.

The final chapter discusses various rigidity results for zero entropy $\mathbf{Z}^d$ actions by group automorphisms. Invariant measures for almost minimal, mixing actions are analyzed. It is shown that if such an action on a compact, abelian, zero-dimensional group admits a "sufficiently mixing" invariant probability measure then that measure is the normalized Haar measure. Next, the question of cohomological rigidity for $\mathbf{Z}^d$ actions (by group automorphisms) in the class of Hölder continuous, real-valued cocycles is investigated. When $d=1$, for Anosov actions the classical result of Livshits completely describes the obstructions to being a coboundary. Recently, A. Katok and R. Spatzier (and subsequently many others) have discovered a phenomenon which basically says that for $d>1$ with "sufficient partial hyperbolicity" in the dynamics, things are even better, i.e., Hölder cohomology with coefficients in $\mathbf{R}$ is trivial. The author proves this type of result for expansive and mixing $\mathbf{Z}^d$ actions by automorphisms of a compact abelian group. The final section deals with another manifestation of rigidity—the isomorphism rigidity. This says that every measurable self-conjugacy of the action (modulo null sets) is a continuous group automorphism. This type of result is established for a specific $\mathbf{Z}^2$ action.

This book is a very valuable additon to the literature on dynamical systems and ergodic theory. Further development of this theory and in particular several questions posed in this book will be a significant source of interesting and challenging problems in the years to come. *Mahesh Nerurkar* (Camden, NJ)

98b:28027 28D15

den Hollander, Frank (Nijmegen); **Steif, Jeffrey E.** (Göteborg)

On K-automorphisms, Bernoulli shifts and Markov random fields. (English summary)

Ergodic Theory Dynam. Systems **17** (1997), *no.* 2, 405–415.

Let $d>0$ be an integer. In what follows, a "dynamical system" will be a measure-preserving $\mathbf{Z}^d$-action on a finite standard measure space. If F is a finite set and if $X_F := F^{\mathbf{Z}^d}$ is given some finite $\mathbf{Z}^d$-invariant measure, then the resulting dynamical system will be called a "stationary random field" or an "SRF". A Markov random field (MRF) is a kind of SRF and we refer the reader to the paper for the definition of this and other terms.

The three main theorems of this paper are as follows: (1) If an MRF is very weak Bernoulli (VWB), then it is Følner independent (FI). (2) If an MRF is K, then it has a trivial full tail (TFT). (3) If an SRF is FI, then it has a TFT.

In what follows, we will say "is TFT" rather than "has a TFT", though the former is, strictly speaking, not quite grammatical. With regrets, we will also need to use these abbreviations quite frequently.

The technical definitions of the various terms are either given in the paper or referenced and probably would not enlighten anyone who does not already know, in principle, what they are. Instead, I will try to describe the context.

First, the collection of MRFs is a well-studied and natural class of SRFs. Another extremely important class of dynamical systems is the class called "Bernoulli" systems. It should be pointed out that Bernoulli systems are ubiquitous and are, at the same time, completely classified, so it is natural that much work should go into figuring out exactly which systems are Bernoulli and which are not.

Second, VWB and K are adjectives that modify dynamical systems, while FI and TFT are adjectives that modify SRFs. Each of these terms describes some type of measure-theoretic "chaos".

The misleading terminology notwithstanding, it is known, for any dynamical system, that VWB is equivalent to Bernoulli. For $d=1$, the statement that Bernoulli implies VWB is due to D. Ornstein and B. Weiss, while the converse is due to Ornstein alone. The result for $d > 1$ is proved by generalizing the ideas in the proofs in the $d = 1$ case.

Because VWB is equivalent to Bernoulli, much work goes into verifying the VWB condition for many dynamical systems. It is also known, for all SRFs, that FI implies VWB, but that the converse is not true. The FI condition is somewhat simpler to study, so, in trying to verify VWB for a general SRF, one might first try to verify FI. If that turns out to be impossible, then one faces the somewhat more complicated VWB definition. The content of (1), then, is that, for MRFs, there is no need to look at the VWB condition; the FI condition will do.

In fact, one might hope, for MRFs, that all these adjectives are equivalent: FI, VWB, Bernoulli, K, and TFT. If this were true, then, to prove that a given MRF is Bernoulli, one would have many different things to try. From this point of view, (1) and (2) can be seen as partial results. By (1), we know that FI and VWB are equivalent. By classical results, VWB is equivalent to Bernoulli and Bernoulli implies K. By (2), K implies TFT. It is not hard to show, for any SRF, that TFT implies K (see the well-known Theorem 1.14 on p. 409 of the paper under review).

So the only missing implication in the aforementioned "hope" is that, for MRFs, K implies Bernoulli. However, in a note added in proof, the authors mention that C. Hoffman has constructed a counterexample to this. Nevertheless, it is nice that the various implications between these terms are now completely established.

Note that the paper collects together enough results to show, for MRFs, that FI implies TFT. In this context, then (3) becomes interesting, since it says, for all SRFs, and not just MRFs, that FI implies TFT.

So, for SRFs, we can say that FI implies TFT and that TFT implies K. We also know that FI implies VWB and that VWB implies K. The converses to these four statements are all false. As pointed out in the paper (p. 409, l.+18 to l.+19), we also know that VWB does not imply TFT and that TFT does not imply VWB. Finally, as mentioned above, VWB is equivalent to Bernoulli. So, using (3), all possible implications have been established also for SRFs. *Scot Adams* (Minneapolis, MN)

97e:31003 31B15 35Q60 49J10 52A20 78A30
Jerison, David (Cambridge, MA)
A Minkowski problem for electrostatic capacity. (English summary)
Acta Math. **176** (1996), *no.* 1, 1–47.

In this article, the author introduces and solves, in complete generality, a Minkowski problem for electrostatic capacity. The classical Minkowski problem consists in finding a convex polyhedron from data consisting of normals to their faces and their surface areas. The corresponding problem for convex bodies with smooth boundaries is to find the convex body given the Gauss curvature of its boundary, as a function of the unit normal. More specifically, let Ω be a bounded, convex, open subset of $\mathbf{R}^N$, and let $n = N - 1$, $g: \partial\Omega \to S^n$ be the Gauss map, i.e., the mapping from $x \in \partial\Omega$ to the outer unit normal at x (g is defined almost everywhere with respect to surface measure $d\sigma$ on $\partial\Omega$). Consider the measure $d\mu = g^*(d\sigma)$ defined on the unit sphere by $\mu(E) = \int_{g^{-1}(E)} d\sigma$ for any Borel set $E \subset S^n$. When Ω is a polyhedron, the measure is a sum of point masses at the normals to each of the faces, and the coefficient at a normal is the surface area of that face. The classical Minkowski problem consists in describing which measures arise in this way. The result of a long line of investigations is that, if $N \geq 2$ and μ is a positive measure on S^n, there exists a bounded, convex domain Ω such that $g^*(d\sigma) = d\mu$ if and only if μ satisfies $\int_{S^n} \xi_k d\mu(\xi) = 0$, $k = 1, \cdots, N$, and μ is not supported on any equator. Ω is unique up to translation. In addition, if k is a non-negative integer, $0 < \alpha < 1$, and $d\mu = (1/K)d\xi$ for some strictly positive function $K \in C^{k,\alpha}(S^n)$, then Ω is $C^{k+2,\alpha}$. The case of polyhedra was solved by Minkowski, and existence in general was proved by Aleksandrov [see T. Bonnesen and W. Fenchel, *Theory of convex bodies*, Translated from the German and edited by L. Boron, C. Christenson and B. Smith, BCS Assoc., Moscow, ID, 1987; MR 88j:52001]. In the smooth case, the function K is the Gauss curvature of $\partial\Omega$, which explains the second phrasing of the Minkowski problem. C^∞ regularity was proved by Pogorelov, Nirenberg, and Cheng and Yau [see S. Y. Cheng and S. T. Yau, Comm. Pure Appl. Math. **29** (1976), no. 5, 495–516; MR **54** #11247]. The exact gain of two derivatives, and the treatment of small values of k, are due to L. A. Caffarelli [see Ann. of Math. (2) **130** (1989), no. 1, 189–213; MR 90i:35046; Ann. of Math. (2) **131** (1990), no. 1, 129–134; MR 91f:35058; Ann. of Math. (2) **131** (1990), no. 1, 135–150; MR 91f:35059; Comm. Pure Appl. Math. **44** (1991), no. 8-9, 965–969; MR 92h:35088].

The point of departure of the present article is the following: for $N \geq 3$, let Ω' be the complement of the closure of Ω, and let U be the equilibrium potential of Ω, i.e., the continuous function satisfying $\Delta U = 0$ in Ω, $U \equiv 1$ on $\partial\Omega$ and such that U tends to 0 at ∞. It is not difficult to see that $U(x) = \gamma a_N |x|^{2-N} + O(|x|^{1-N})$ as $x \to \infty$, where $a_N = 1/((N-2)\mathrm{vol}(S^n))$, and $\gamma > 0$. The constant γ is known as the electrostatic capacity of Ω, $\mathrm{cap}\,\Omega$. A formula, due to H. Poincaré [*Figures d'équilibre d'une masse fluide*, Naud, Paris, 1902; JFM 34.0757.05], states that $\mathrm{cap}\,\Omega = (N-2)^{-1} \int_{\partial\Omega} u|\nabla U|^2 d\sigma$, where u is the Minkowski support function of Ω, i.e., the function $u(\xi) = x \cdot \xi$, where ξ is the unit normal to a support plane P of Ω, pointing away from Ω, and $x \in P \cap \partial\Omega$ (a support plane P is a hyperplane tangent to $\partial\Omega$). Jerison points out the resemblance between this formula and the formula for the volume of Ω, $\mathrm{vol}\,\Omega = N^{-1} \int_{\partial\Omega} u d\sigma$, and also a resemblance between their first variations. He is thus led to considering the analogue of the Minkowski problem, with the density $|\nabla U|^2 d\sigma$ in place of $d\sigma$. He considers the measure $g^*(|\nabla U|^2 d\sigma)$ on S^n, and proves that, for any $N \geq 4$, if μ is a positive measure on S^n, there exists a bounded convex body such that $g^*(|\nabla U|^2 d\sigma) = d\mu$ if and only if μ satisfies $\int_{S^n} \xi_k d\mu(\xi) = 0$, $k = 1, \cdots, N$, and μ is not supported on any equator. Then the body Ω is unique up to translation. In addition, if k is a non-negative integer and $0 < \alpha < 1$, $d\mu = R d\xi$ for some strictly positive density $R \in C^{k,\alpha}(S^n)$, then Ω is $C^{k+2,\alpha}$. (If $N = 3$, the problem is also dilation invariant, and the answer is slightly different, and also described in full in the paper.) As in the classical Minkowski problem, the proof consists of three parts: existence, uniqueness and regularity. However, the details of the proof of each part are substantially different from those in the classical case.

In the proof of existence, an important role is played by a Brunn-Minkowski inequality for capacity, due to C. Borell [Math. Ann. **263** (1983), no. 2, 179–184; MR 84e:31005]. This, coupled with the formula for the first variation of capacity, due to Hadamard [see P. R. Garabedian and M. Schiffer, Proc. Amer. Math. Soc. **5** (1954), 206–211; MR **16**, 36c], yields an "isoperimetric inequality for capacity", which is a key ingredient in the a priori estimates needed for existence. The validity of Hadamard's formula for general convex bodies (instead of just smooth ones) and the continuity of the first variation of capacity, in the Minkowski metric, are established by Jerison. This requires a delicate argument, which relies on estimates for harmonic functions on Lipschitz domains, due to B. E. J. Dahlberg [Arch. Rational Mech. Anal. **65** (1977), no. 3, 275–288; MR **57** #6470], and on C^1 domains, due to Jerison and Kenig [Trans. Amer. Math. Soc. **273** (1982), no. 2, 781–794; MR 83k:31004]. In addition, explicit calculations for ellipsoids and semi-infinite strips are used, in conjunction with the "isoperimetric inequality" mentioned above and Dahlberg's "comparison principle", to obtain a priori estimates on eccentricity. Existence follows, in the case of polyhedra, by considering a suitable variational problem, and, in the general case, by weak continuity, using the a priori bounds, and the continuity result mentioned above. In analogy with the classical case, uniqueness follows by identifying the case of equality in Borell's Brunn-Minkowski inequality for capacity. In the present case, this was done in a joint paper of Caffarelli, Jerison and E. H. Lieb [Adv. Math. **117** (1996), no. 2, 193–207]. Finally, the regularity part of the proof, as in the classical case, depends on the ideas of Caffarelli for regularity of solutions to Monge-Ampère equations. The new twist here is that the "doubling estimates on slices" for $|\nabla U|^2$, needed for applying Caffarelli's results, may fail. (The connection between doubling and Caffarelli's results was first discovered by the author in his related work on bounded domains [Invent. Math. **105** (1991), no. 2, 375–400; MR 92k:31003].) Slightly weaker versions of these estimates do hold, however, and are proved by Jerison by an ingenious argument. He then shows that this weaker version of doubling suffices to apply Caffarelli's method for regularity. This is extremely fortunate, because no estimates weaker than the ones obtained would have sufficed, as Jerison shows by appropriate examples.

This paper is an important contribution to potential theory and the regularity theory of fully nonlinear equations. It formulates and completely solves an interesting potential-theoretic problem for convex bodies, amalgamating a collection of difficult and important results in quite different areas of analysis, and showing how they can be used together. Along the way, fundamental new estimates for harmonic measure on convex domains are also obtained.

The paper is dedicated to the author's father, Meyer Jerison, whose untimely passing we continue to mourn. *Carlos E. Kenig* (Chicago, IL)

99a:32018a 32G07 32F20 32F40 35P15
Epstein, Charles L. (Philadelphia, PA)
A relative index on the space of embeddable CR-structures. I. (English summary)
Ann. of Math. (2) **147** (1998), *no.* 1, 1–59.

99a:32018b 32G07 32F20 32F40 35P15
Epstein, Charles L. (Philadelphia, PA)
A relative index on the space of embeddable CR-structures. II. (English summary)
Ann. of Math. (2) **147** (1998), *no.* 1, 61–91.

Let $(M, \bar{\partial}_b)$ denote a compact strongly pseudoconvex three-dimensional CR manifold. It is well known that the question of embeddability of such structures is far more subtle in three dimensions than in higher dimensions, and that the set of embeddable structures is a small subset of the space of all structures. In this fundamental paper, Epstein studies the structure of the space of embeddable structures, defines a relative index on this space, and defines operators which "track" the space of CR functions from one structure to another. These operators are also used to relate jumps in the relative index to the presence of small eigenvalues for the associated $\Box_b$ operator. In Part II, he collects analytic results which are required for the proof of the theorems in Part I, including the regularity properties of the operators which he uses in Part I and their smooth dependence on the underlying structure.

The heuristics of the approach can be summarized as follows. Let $\bar\partial_b^0$ denote a reference CR structure which is embeddable. If $\bar\partial_b^\varphi$ is a small perturbation of $\bar\partial_b^0$, then we should be able to "track" the CR functions from one structure to the next. If only a finite-dimensional subspace of CR functions disappears, then the new structure should also be embeddable; on the other hand, this should also be a necessary condition, since regularity properties for the $\Box_b$ operator suggest that there can be at most a finite number of "small" nonzero eigenvalues for $\Box_b$ for embeddable structures. This "tracking" of the CR functions can be done using the Szegő projector ${}^0\colon \ker\bar\partial_b^\varphi \to \ker\bar\partial_b^0$, even though the definition of ^{0} involves making choices. These heuristics are made precise in Theorem A in Part I of this paper: The deformed structure $\bar\partial_b^\varphi$ is embeddable if and only if the operator ${}^0\colon \ker\bar\partial_b^\varphi \to \ker\bar\partial_b^0$ is Fredholm. The proof of this theorem relies upon the functional analytic criteria for embeddability due to J. J. Kohn [Duke Math. J. **53** (1986), no. 2, 525–545; MR 87m:32041] and L. Boutet de Monvel [in *Séminaire Goulaouic-Lions-Schwartz 1974–1975; Équations aux derivées partielles linéaires et non linéaires*, Exp. 9, 14 pp., Centre Math., École Polytech., Paris, 1975; MR **53** #13645]: a compact strongly pseudoconvex three-dimensional CR manifold is embeddable if and only if the range of $\bar\partial_b$ is closed.

The proof of Theorem A involves having a precise understanding of the Szegő projector. However, the Szegő projector is a very delicate operator with which to work, since it is definitely not Fredholm on L^2. To circumvent this difficulty, Epstein defines the auxiliary operator $B := I + @\bar\partial_b^*\varphi\partial_b$, an operator that arises naturally if one tries to solve the $\bar\partial$ equations for the deformed structure. Working formally, it is easy to see that if $h \in \ker\bar\partial_b^0$ and $u \in (\ker\bar\partial_b^0)^\perp$, then $\bar\partial_b^\varphi(u+h) = 0$ implies that $B(u+h) = h$. More generally, $B = @\bar\partial_b^*(\bar\partial_b^\varphi) + {}^0$, so $B|_{\ker\bar\partial_b^\varphi} = {}^0$. It follows that Theorem A can be rephrased as: A structure $\bar\partial_b^\varphi$ is embeddable if and only if $B^\varphi\colon \ker\bar\partial_b^\varphi \to \ker\bar\partial_b^0$ is Fredholm.

The operator B is much easier to analyse than the Szegő projector since for a small deformation $\bar\partial_b^\varphi$ of $\bar\partial_b^0$, B^φ is a small perturbation of the identity; in particular, the operator B is close to being invertible as an operator on L^2. In Part II of this paper, Epstein makes this precise by constructing a parametrix for B^φ for arbitrary deformations $\bar\partial_b^\varphi$ of the reference structure $\bar\partial_b^0$; from this construction follows Corollary 3.1: The operator $B\colon L^2 \to L^2$ is a Fredholm operator of index zero. This fact is particularly important for two reasons: first, the statement does not involve restricting the operator to a subspace of L^2, and second, it is a statement for the full deformation space of the reference structure which does not depend upon the embeddability of the deformed structure. Using this fundamental result, Epstein is able to complete the proof of Theorem A.

In light of Theorem A, one can define the relative index between two embeddable CR structures by $\operatorname{Ind}(\bar\partial_b^0, \bar\partial_b^\varphi) = \operatorname{ind}({}^0\colon \ker\bar\partial_b^\varphi \to \ker\bar\partial_b^0)$; this definition can be extended to arbitrary CR structures by defining $\operatorname{Ind}(\bar\partial_b^0, \bar\partial_b^\varphi) = -\infty$ when $\bar\partial_b^\varphi$ is not embeddable. (Here, the reference structure $\bar\partial_b^0$ is assumed to be embeddable.) Theorem B states that the relative index is well defined; that is, it is independent of the volume form used to define the Szegő projection as well as being invariant under the action of the contact diffeomorphism group. To establish the latter half of this result, Epstein first shows that the relative index satisfies a cocycle formula; that is, if $\bar\partial_b^0, \bar\partial_b^\varphi, \bar\partial_b^\psi$ are embeddable CR structures, and $\bar\partial_b^\psi$ is sufficiently close to $\bar\partial_b^\varphi$, then $\operatorname{Ind}(\bar\partial_b^0, \bar\partial_b^\varphi) = \operatorname{Ind}(\bar\partial_b^0, \bar\partial_b^\psi) + \operatorname{Ind}(\bar\partial_b^\psi, \bar\partial_b^\varphi)$. Indeed, this formula provides a crucial tool to pass from local statements about the deformation space to global statements.

The remainder of Part I is devoted to giving a more precise description of the space of CR structures, and to giving some applications of these results. Let φ be a deformation such that B^φ is invertible. Define the operator $A := \overline{\quad}^{\,0}\varphi\partial_b B^{-1}\quad{}^{0}$, where $\overline{\quad}^{\,0}$ denotes the orthogonal projection onto $\ker(\bar\partial_b^0)^*$. This operator has the property that $A = \bar\partial_b^\varphi B^{-1}\quad{}^{0}$. Thus, if B^φ restricts to an invertible transformation $B^\varphi\colon \ker\bar\partial_b^\varphi \to \ker\bar\partial_b^0$ between the spaces of CR functions, then A is identically zero. More generally, Epstein shows that there is a neighbourhood of $\bar\partial_b^0$ such that $\operatorname{Rank} A = -\operatorname{Ind}(\bar\partial_b^0, \bar\partial_b^\varphi)$. Thus, the operator A "tracks" the CR functions which "disappear" under small deformations, and the rank of A is precisely the dimension of this space. Theorem D makes this description of disappearing CR functions more precise as follows: There is a neighbourhood of $\bar\partial_b^0$ such that if $\bar\partial_b^\varphi$ is in this neighbourhood and $\operatorname{Rank} A^\varphi = k$, then there exist precisely k "small" eigenvalues for the perturbed $\Box_b^\varphi$ operator.

The relative index defines a stratification of the space of CR structures, where the strata are defined by $\mathcal{G}_n = \{\varphi\colon \|\varphi\|_{L^\infty} < 1,\ \operatorname{Ind}(\bar\partial_b^0, \bar\partial_b^\varphi) \geq -n\}$. Theorem E states that the strata are closed.

Epstein provides several applications of his results. First, if two CR structures are embeddable in a common Stein space X, and their images are smoothly isotopic as strongly pseudoconvex hypersurfaces within the ambient space X, then the relative index is zero. This makes precise the intuitive idea that two isotopic hypersurfaces should have the same "number" of CR functions. Next, combining this with a result of Eliashberg on the filling of embeddable CR structures on the three-sphere (private communication), he establishes the following stability result: Let $(S^3, \bar\partial_b^\varphi)$ be an embeddable CR structure on S^3; then $\operatorname{Ind}(\bar\partial_b^0, \bar\partial_b^\varphi) = 0$. Finally, he provides a class of examples where the index takes on arbitrarily large negative values.

In Part II of this paper, Epstein establishes the very precise analytic results which are necessary to prove the results in Part I. Using the Heisenberg calculus as developed by R. W. Beals and P. C. Greiner [*Calculus on Heisenberg manifolds*, Ann. of Math. Stud., 119, Princeton Univ. Press, Princeton, NJ, 1988; MR 89m:35223] and M. E. Taylor [Mem. Amer. Math. Soc. **52** (1984), no. 313, iv+182 pp.; MR 86f:58156; *Noncommutative harmonic analysis*, Amer. Math. Soc., Providence, RI, 1986; MR 88a:22021], Epstein follows the procedure outlined by Beals and Greiner to construct an approximate partial inverse and Szegő projector for $\bar\partial_b^\varphi$, and shows that these operators depend smoothly upon the CR structure. Of particular importance is that this analysis holds for the entire deformation space, including the non-embeddable structures, while exact operators with estimates exist only in case the structure is embeddable. He next shows that the operator B is invertible up to a smoothing operator, and he constructs a parametrix for B; again, the parametrix depends smoothly upon the CR structure. These precise analytic results set the groundwork necessary to effectively analyse the $\overline{\partial}_b$ operator for a general compact strongly pseudoconvex CR manifold. *John Bland* (Toronto, ON)

99m:32028 32H25 14J29 32H30
McQuillan, Michael (Oxford)
Diophantine approximations and foliations. (English summary)
Inst. Hautes Études Sci. Publ. Math. No. 87 (1998), 121–174.

In 1925, R. Nevanlinna proved his "Second Main Theorem" for value distribution of holomorphic maps from $\mathbf{C}$ to a smooth complex projective curve (i.e., complete Riemann surface). Except for refinements in the error term, his inequality is best possible. Attempts to extend this result to higher dimensions, however, have not been nearly as successful. It has been conjectured (but not universally accepted) that the Second Main Theorem should still hold, with the same formal inequality, for holomorphic curves $f:\mathbf{C}\to X$ with Zariski-dense image, where X is a smooth projective variety and the divisor in the inequality is a normal crossings divisor. A special case of this conjecture is the Green-Griffiths conjecture, which asserts that if X is a smooth complex projective variety of general type of (complex) dimension 2, then any holomorphic curve $f:\mathbf{C}\to X$ cannot have Zariski-dense image [M. Green and P. Griffiths, in *The Chern Symposium 1979 (Proc. Internat. Sympos., Berkeley, Calif., 1979)*, 41–74, Springer, New York, 1980; MR 82h:32026].

If the Green-Griffiths conjecture is true, then the closure of the image of any holomorphic curve $f:\mathbf{C}\to X$ must be either a single point or a curve of geometric genus 0 or 1. Thus, the Green-Griffiths conjecture is related to the conjecture in algebraic geometry that a smooth projective surface of general type over a field of characteristic zero has only finitely many curves of geometric genus ≤ 1. This question is still open, but F. Bogomolov proved that it holds for all surfaces of general type with $c_1^2>c_2$ [M. Deschamps, in *Séminaire Bourbaki, 30e année (1977/78)*, Exp. 519, pp. 233–247, Lecture Notes in Math., 710, Springer, Berlin, 1979; MR 82e:14045].

The present paper extends Bogomolov's work to the case of holomorphic curves, thus proving the Green-Griffiths conjecture for surfaces of general type with $c_1^2>c_2$. Moreover, the paper goes further by proving a Nevanlinna-like inequality. Specifically, let X be a smooth projective surface of general type with $c_1^2>c_2$, let H be an ample divisor on X, and let Z be the union of the (finitely many) curves on X of geometric genus 0 or 1. Let $p\colon Y\to\mathbf{C}$ be a finite ramified cover, and let $f\colon Y\to X$ be a holomorphic map whose image is not contained in Z. Then there exist an étale map $h:\mathbf{C}\to\mathbf{C}$ and a constant α, depending on f, p, and h, such that $T_{\mathcal{K}_X,g}(r)\le\alpha d_g(r)+o(T_{H,g}(r))$ for all real $r>0$ outside a set of finite Lebesgue measure. Here $Y'=Y\times_h\mathbf{C}$, $g\colon Y'\to X$ is the lifting of f, $\mathcal{K}_X$ is the canonical sheaf on X, and $d_g(r)$ is the difference $N_{\mathrm{Ram},p}(r)-N_{\mathrm{Ram},f}(r)$ of the counting functions for the ramification of p and f.

Briefly, Bogomolov's proof proceeds as follows. Let $f\colon Y\to X$ be the normalization of a curve on X of genus 0 or 1. This lifts to a map $f'\colon Y\to\mathbf{P}(\Omega_X^1)$. (Here, as in the paper itself, we use Grothendieck's conventions on $\mathbf{P}(\cdot)$ of a vector sheaf.) Bogomolov then constructed a divisor D on $\mathbf{P}(\Omega_X^1)$ such that if the image of f' is not contained in D, then f cannot exist, by a simple degree-counting argument. If, on the other hand, the image of f' is contained in D, then f is a leaf of a foliation, and he showed that there can be only finitely many such curves. This can be done either by applying a general result of Seidenberg, or more directly via Hodge theory.

The present paper follows the above outline of Bogomolov's proof. Let now $f\colon Y \to X$ be a holomorphic map as discussed previously. If its lifting $f'\colon Y \to \mathbf{P}(\Omega^1_X)$ is not contained in the above divisor D, then the paper proves a "tautological inequality" (Theorem A) to replace the degree-counting argument in Bogomolov's proof. Let $\mathcal{O}(1)$ denote the tautological line sheaf on $\mathbf{P}(\Omega^1_X)$; then (with other notations as above) Theorem A says that $T_{\mathcal{O}(1),f'}(r) \le d_f(r) + O(\log r + \log T_{H,f}(r))$ for all $r > 0$ outside a set of finite Lebesgue measure. The proof of this theorem involves an (unstated) intermediate result related to the classical lemma on the logarithmic derivative. Theorem A follows from this result by an argument using Fulton's deformation to the normal cone.

The other possibility is that the image of f' is contained in D, so that f is a leaf of the associated foliation. In that case the paper builds on Seidenberg's argument to show that f must satisfy the desired inequality if its image is Zariski-dense. In Bogomolov's case, one uses Seidenberg to find a blowing-up of X such that all curves of genus 0 or 1 in the foliation are disjoint; then the argument follows easily from finite-dimensionality of the Néron-Severi group. In the context of holomorphic curves, however, these algebraic facts cannot be used directly; hence the author relies instead on a detailed analysis of the singularities of the foliation, combined with a refinement of his Theorem A.

Having read this far, the reader may be wondering what the words "Diophantine approximation" are doing in the title. It has been observed by C. F. Osgood, the reviewer, and others that Nevanlinna theory bears a close (but formal) resemblance to Diophantine approximations. In previous papers, the author has used this observation to obtain proofs in Nevanlinna theory using constructions of the auxiliary line sheaf in Thue's method (a prevalent method in Diophantine approximation). He uses this method (which he calls "dynamic Diophantine approximation") in this paper as well, in his proof of Theorem A.

This paper is noteworthy not only because it makes progress on the Green-Griffiths conjecture, but also because it gives perhaps the first example of how to take a purely algebraic proof (Bogomolov's theorem in this case) and carry it over to the context of holomorphic curves. Central to this is the paper's Theorem A, about which the author is justifiably proud in noting that "to our knowledge, Theorem A implies all '[Second] Main Theorem type' results...." Theorem A has one minor shortcoming, however: it should have been stated for liftings to $\mathbf{P}(\Omega^1_X(\log D))$ for a normal crossings divisor D on X (which would have necessitated an additional truncated counting function term $N^{(1)}_f(D,r)$ on the right-hand side of the inequality). This flaw is very minor, though, and does not detract from the fact that the paper breaks a lot of new ground. *Paul Vojta* (Berkeley, CA)

99i:32035 32L10 32G13 32J18
Siu, Yum-Tong (Cambridge, MA)
Invariance of plurigenera.
Invent. Math. **134** (1998), *no.* 3, 661–673.

Starting with the works of Euler and Abel on integrals on algebraic curves it has become clear that all the analytic information about a compact Riemann surface is contained in global holomorphic 1-forms. These are objects that locally can be written as $f(z)dz$, where z is a local coordinate and f is a holomorphic function. Riemann established that the dimension of the vector space of global holomorphic 1-forms is the same as the topological genus of the underlying surface. This is one of the earliest results establishing the topological nature of certain analytically defined numbers.

In the past 100 years much effort was directed towards finding higher-dimensional versions of the result of Riemann. One direction is to replace the dimension of the vector space of global holomorphic 1-forms with the Euler characteristic of the line bundle of holomorphic n-forms where n is the dimension. The Hirzebruch-Riemann-Roch theorem expresses this number in terms of the Chern classes of the complex manifold. The situation becomes murkier if we would like to study the vector space of global holomorphic n-forms. Hodge theory shows that this is naturally a subspace of one of the topological cohomology groups, but it is still unknown how to convert this information into an explicit formula. The theory does, however, imply the following weaker statement: If X_t is a family of smooth, complex, projective varieties depending continuously on a parameter t then the dimension of the vector space of global holomorphic n-forms on X_t is a locally constant function of t.

It has also been realized that in higher dimensions this single number does not carry enough information about a variety. Instead, one should look at global sections of tensor powers of the line bundle of n-forms. This line bundle is frequently denoted by $K_X^{\otimes m}$ and its local sections can be written as

$$f(z_1,\cdots,z_n)(dz_1\wedge\cdots\wedge dz_n)^{\otimes m}.$$

The dimension of the vector space of global sections of $K_X^{\otimes m}$ is called the mth plurigenus of X and it is denoted by $P_m(X)$. A very natural question is: are the numbers $P_m(X)$ topological in nature? Very little is known about this problem. The following weaker conjecture received much attention because it is especially important in moduli problems: Let X_t be a family of smooth, complex, projective varieties depending continuously on a parameter t. Are the $P_m(X_t)$ locally constant functions of t?

The traditional approach is to convert $P_m(X_t)$ into an Euler characteristic and then use the Hirzebruch-Riemann-Roch theorem. Unfortunately, the higher cohomology groups of $K_X^{\otimes m}$ are usually nonzero and they vary wildly within the birational equivalence class of X. This, however, led to the first approaches to the conjecture. In many cases one can find a suitable birational model X^* such that $P_m(X) = P_m(X^*) = \chi(K_{X^*}^{\otimes m})$. If such a model can be found for every X_t in a family in a continuous manner then we arrive at a solution of the conjecture. This approach has been successfully carried out for surfaces [S. Iitaka, J. Math. Soc. Japan **22** (1970), 247–261; MR **41** #6252] and for threefolds [J. Kollár and S. Mori, J. Amer. Math. Soc. **5** (1992), no. 3, 533–703; MR 93i:14015]. Some cases were settled in all dimensions by M. N. Levine [Invent. Math. **74** (1983), no. 2, 293–303; MR 85d:32054].

The paper under review approaches the question differently. Instead of changing the variety X, we would like to change the line bundle $K_X^{\otimes m}$ in such a way that we do not change the space of global sections but we do eliminate higher cohomology groups. (In general this is only possible if $K_X^{\otimes m}$ is replaced by a sheaf which is not locally free.) The key step is to introduce a metric on the line bundle $K_X^{\otimes m}$ which blows up along a subvariety in a carefully controlled manner. These ideas were introduced in algebraic geometry by Iitaka and his school [cf. Y. Kawamata, K. Matsuda and K. Matsuki, in *Algebraic geometry, Sendai, 1985*, 283–360, Adv. Stud. Pure Math., 10, North-Holland, Amsterdam, 1987; MR 89e:14015] and in complex manifold theory by J.-P. Demailly [in *Complex algebraic varieties (Bayreuth, 1990)*, 87–104, Lecture Notes in Math., 1507, Springer, Berlin, 1992; MR 93g:32044].

In the paper under review the author proves the deformation invariance of plurigenera for varieties where sections of $K_X^{\otimes m}$ separate points over a dense open set. These are usually called varieties of general type. For such varieties there is a natural choice of the singular metric. Let g_i be a basis of the global sections of $K_X^{\otimes m}$; then we can declare that $(\sum |g_i|^2)^{1/2m}$ is a section of K_X which has norm 1 everywhere. We get a singular metric which depends on the choice of the basis and m. The proof now depends on two key observations. First, all global sections of $K_X^{\otimes m}$ are L^2 in these metrics and this remains so after small perturbations of the metric. In particular, the dependence on m is not a serious problem. Second, the extension theorems of T. Ohsawa and K. Takegoshi [Math. Z. **195** (1987), no. 2, 197–204; MR 88g:32029] and L. Manivel [Math. Z. **212** (1993), no. 1, 107–122; MR 94e:32050] can be globalized to extend L^2 sections from X_0 to nearby X_t with a suitable perturbation of the metric. The perturbations required for the second part are just small enough that they do not matter if all choices are made carefully.

Besides solving a long-standing problem, the author's method is applicable to several other problems concerning deformations of varieties. Two such results are in papers by Kawamata [J. Amer. Math. Soc. **12** (1999), no. 1, 85–92; MR 99g:14003] and N. Nakayama ["Invariance of the plurigenera of algebraic varieties", Preprint, Res. Inst. Math. Sci., Kyoto Univ., Kyoto, 1998; per revr.].

János Kollár (Salt Lake City, UT)

98i:32051 32L30 14J99 32J15
Brunella, Marco (Dijon)
Feuilletages holomorphes sur les surfaces complexes compactes. (English and French summaries) [Holomorphic foliations on compact complex surfaces]
Ann. Sci. École Norm. Sup. (4) **30** (1997), *no.* 5, 569–594.

In this remarkable, carefully written and deep work, the classification of (non-singular) holomorphic foliations on compact complex surfaces is achieved. The first two sections of the paper recall and exploit the basic and essential material needed, especially P. Baum and R. Bott's formulae in the two-dimensional case [see J. Differential Geometry **7** (1972), 279–342; MR **51** #14092], as well as behaviour under blowing-up. For instance, a nice and simple application of the contents of Section 1 is the following: Suppose X is a compact complex surface and $\mathcal{F}$ and $\mathcal{G}$ are non-singular holomorphic foliations on X which are transverse. Let $S \subset X$ be a curve which is neither $\mathcal{F}$-invariant nor $\mathcal{G}$-invariant. It follows that $S \cdot S \geq \chi(S)$ and hence, that X does not contain holomorphic spheres with negative self-intersection; in particular, X is minimal.

The remainder of the work can be divided into two parts, one dealing with surfaces of non-general type, the other with surfaces of general type.

In the first part, Sections 3, 4 and 5, foliations on surfaces of non-general type are studied. The Enriques-Kodaira classification is used throughout [see W. P. Barth, C. A. M. Peters and A. J. H. M. Van de Ven, *Compact complex surfaces*, Springer,

Berlin, 1984; MR 86c:32026]. General arguments are developed in Sections 3 and 4, especially exploiting the existence of fibrations, over "most" algebraic surfaces X, with generic fibre a rational or an elliptic curve. These fibrations are cleverly used as "reference fibrations", with which a foliation $\mathcal{F}$ in X is compared, that is, either $\mathcal{F}$ coincides with the fibration, is transverse to it, or else is a "turbulent" foliation. The concept of turbulent foliation is due to Reeb and was generalized, in the complex context, by É. Ghys [Ann. Fac. Sci. Toulouse Math. (6) **5** (1996), no. 3, 493–519; MR 98d:32037]. Inspired by the definition of Ghys, the author defines a turbulent foliation on a compact complex surface X as a foliation $\mathcal{F}$ with the following property: there is a regular elliptic fibration of X such that: (i) a finite number of fibres are $\mathcal{F}$-invariant and (ii) all other fibres are transverse to $\mathcal{F}$. The results in the case of algebraic surfaces can be summarized as: If X is a compact complex algebraic surface and $\mathcal{F}$ is a non-singular holomorphic foliation on X, then we have one of the following (non-exclusive) possibilities: (1) $\mathcal{F}$ is a fibration; (2) $\mathcal{F}$ is transverse to a fibration; therefore, it is a suspension of an automorphism group of an algebraic curve; (3) $\mathcal{F}$ is a linear foliation on a complex torus; (4) $\mathcal{F}$ is a transversely hyperbolic foliation with dense leaves whose universal cover is a fibration over the disc, with the disc as typical fibre; (5) $\mathcal{F}$ is a turbulent foliation over an elliptic surface.

Section 5 of the paper displays the structure of foliations over surfaces of non-general type, and the results are: Let $\mathcal{F}$ be a non-singular holomorphic foliation on a compact complex surface X of Kodaira dimension < 2. Then $\mathcal{F}$ is one of the following objects (observe that such surfaces satisfy $c_1^2(X) = 2c_2(X)$): (1) an elliptic or rational fibration; (2) a foliation transverse to an elliptic or rational fibration; (3) a linear foliation on a complex torus; (4) certain (precisely described) foliations on Hopf and Inoue surfaces; (5) a turbulent foliation.

In the second part, Sections 6 and 7, the case of foliations on surfaces of general type is treated by considering dynamical properties of the foliation; essentially, a transverse invariant metric is constructed, and the arguments here are quite involved and make use of a result of J.-P. Demailly [in *Complex algebraic varieties (Bayreuth, 1990)*, 87–104, Lecture Notes in Math., 1507, Springer, Berlin, 1992; MR 93g:32044]. The main result of Sections 6 and 7 is: If $\mathcal{F}$ is a non-singular foliation on a surface of general type, then $\mathcal{F}$ is either a fibration or a transversely hyperbolic foliation with dense leaves whose universal cover is a fibration over the disc, with the disc as typical fibre.

Let us point out some amazing consequences of the results above: (i) The only non-minimal surface which admits a non-singular foliation is $\mathbf{P}^2_{\mathbf{C}}$ blown up at a point, and this foliation is the blow-up of the radial foliation centered at this point. (ii) If a surface X admits a regular foliation, then its signature is non-negative, that is, $\frac{1}{3}(c_1^2(X) - 2c_2(X)) \geq 0$. In particular, the only complete intersection which admits a non-singular foliation is $\overline{\mathbf{C}} \times \overline{\mathbf{C}}$.

In conclusion, Brunella has synthesized and clearly presented a key and very important result. This is a very substantial contribution to the area.

M. G. Soares (Belo Horizonte)

97h:33034 33E05 11F27 33C05 33D10

Berndt, Bruce C. (Urbana, IL); **Bhargava, S. [Bhargava, Srinivasamurthy]** (Mysore); **Garvan, Frank G.** (Gainesville, FL)

Ramanujan's theories of elliptic functions to alternative bases.

Trans. Amer. Math. Soc. **347** (1995), *no.* 11, 4163–4244.

In his celebrated 1914 paper on modular equations and approximations to π, Ramanujan offers several fantastic series representations for $1/\pi$ which he claims are derived from "corresponding theories" for theta functions in which the classical base q is replaced by one of three other bases. The most famous example of these series representations is

$$\frac{1}{\pi}=\frac{2\sqrt{2}}{9801}\sum_{n=0}^{\infty}\frac{(1103+26390n)(4n)!}{396^{4n}(n!)^4},$$

each term of which provides roughly eight additional digits of precision.

The formulas for $1/\pi$ were only recently proved by J. M. and P. B. Borwein in 1987 [*Pi and the AGM*, Wiley, New York, 1987; MR 89a:11134], and inspired infinite classes of new results in the same vein. For example, in 1988 D. V. and G. V. Chudnovsky [in *Ramanujan revisited (Urbana-Champaign, Ill., 1987)*, 375–472, Academic Press, Boston, MA, 1988; MR 89f:11099] discovered

$$\frac{1}{\pi}=12\sum_{n=0}^{\infty}(-1)^n\frac{(13591409+545140134n)(6n)!}{(640320^3)^{n+1/2}(n!)^3(3n)!},$$

which they used to compute π to over two billion digits. The latter formula is based on the hypergeometric function ${}_2F_1(\frac{1}{6},\frac{5}{6};1;x)$ and Klein's absolute invariant J, while the classical theory concerns ${}_2F_1(\frac{1}{2},\frac{1}{2};1;x)$ and the modular function λ.

Aside from series representations for $1/\pi$, Ramanujan's notebooks contain approximately fifty additional results without proofs in the so-called "alternative-base" theories. The paper under review develops these theories, proving Ramanujan's claims and establishing several further results as well.

The most striking of the three alternative theories, that of signature 3, arises from the hypergeometric function ${}_2F_1(\frac{1}{3},\frac{2}{3};1;x)$, in which Ramanujan's beautiful cubic transformation formula

$$(1)\qquad {}_2F_1\left(\tfrac{1}{3},\tfrac{2}{3};1;1-\left(\tfrac{1-x}{1+2x}\right)^3\right)=(1+2x)\,{}_2F_1\left(\tfrac{1}{3},\tfrac{2}{3};1;x^3\right)$$

plays a pivotal role. Identifiable functions with cubic functional equations are rare flowers. The other two theories are more easily extracted from the classical theory, that is, the theory of signature 2, and so are of less interest.

In the classical theory of elliptic functions, the processes of duplication and dimidiation, which rest upon Landen's transformation, are very useful in obtaining formulas from previously derived formulas when q is replaced by q^2 or $\sqrt{q}$, respectively. They lead naturally to quadratically convergent iterations for elliptic functions and directly for π. Correspondingly, Ramanujan's cubic transformation (1) can be employed to devise the new processes of triplication and trimidiation. These processes were exploited by the Borweins in their cubically convergent algorithm for π. Although E. Goursat provided several cubic transformations for hypergeometric functions in his classical memoir [Ann. Sci. École Norm. Sup. (2) **10** (1881), suppl., 3–142; JFM 13.0267.01], (1) apparently cannot be deduced from Goursat's results. In fact, (1) remained unproved until 1991, when the Borweins deduced it in connection with their cubic analogue of the arithmetic-geometric mean [J. M. Borwein and P. B. Borwein, Trans. Amer. Math. Soc. **323** (1991), no. 2, 691–701; MR 91e:33012].

The paper under review develops transformation formulas (e.g. the so-called transformation principle) for relating hypergeometric functions of various signatures, thus enabling one to transform formulas in one theory to those of another. These transformations are used to establish Ramanujan's modular equations in the alternative theories and, as a general principle, allow the authors to employ formulas in the classical theory to prove corresponding theorems in the alternative theories. There is an enormous amount of hard, subtle, sophisticated analysis involved, and it is a tribute to the authors' exposition that the paper is accessible and highly readable. The reviewers would like to encourage those interested in Ramanujan's work to consult Volumes I–IV of *Ramanujan's notebooks* [*Part I*, Springer, 1985; MR 86c:01062; *Part II*, 1989; MR 90b:01039; *Part III*, 1991; MR 92j:01069; *Part IV*, 1994; MR 95e:11028; *Part V*, 1998; MR 99f:11024], for which Bruce C. Berndt was awarded the 1996 Leroy P. Steele Prize of the AMS for Exposition. The citation [Notices Amer. Math. Soc. **43** (1996), no. 11, 1340–1347; MR 97k:01066 (p. 1340)] states: "To Bruce C. Berndt for the four volumes *Ramanujan's notebooks, Parts I, II, III* and *IV*, Springer, 1985, 1989, 1991, and 1994. In recognition of Berndt's heroic and extraordinary achievement in exposing to the general mathematical researcher a trove of results that were utterly inaccessible before, the AMS decided this year, exceptionally, to broaden the standard interpretation of 'exposition'. In an impressive scholarly accomplishment spread out over 20 years, Berndt has provided a readable and complete account of the Notebooks, making them accessible to other mathematicians. Ramanujan's enigmatic, unproven formulas are now readily available, together with context and explication, often after the most intense and clever research efforts on Berndt's part."

To conclude we offer a quote of G. H. Hardy concerning Ramanujan's work on elliptic and modular functions: "It is here that both the profundity and the limitations of Ramanujan's knowledge stand out most sharply...." In just that light, the present authors have made remarkably complete the fragmentary insights left to us by Ramanujan; fragments which many speculated would never be truly limned.

J. Borwein and David Bradley (Burnaby, BC and Orono, ME)

97k:41019 41A15 05B35 15A04

de Boor, Carl (Madison, WI); **Ron, Amos** (Madison, WI); **Shen, Zuowei** (Madison, WI)

On ascertaining inductively the dimension of the joint kernel of certain commuting linear operators. (English summary)

Adv. in Appl. Math. **17** (1996), *no.* 3, 209–250.

The paper is concerned with the dimension of the joint kernel of certain commuting linear operators. Let $\mathbf{X}$ be an index set and associate each $x \in \mathbf{X}$ with a linear map l_x on a fixed linear space S. The maps $\{l_x\}_{x\in\mathbf{X}}$ are assumed to commute with one another: $l_x l_y = l_y l_x$ for all $x, y \in \mathbf{X}$. Thus, for each subset A of $\mathbf{X}$, the product $l_A := \prod_{x\in A} l_x$ is well-defined. Let $\mathscr{B}$ be a collection of subsets of $\mathbf{X}$ having the same cardinality s. The elements of $\mathscr{B}$ are called bases. For $y \in \mathbf{X}$, $\mathscr{B}_{\setminus y}$ denotes the collection $\{B \in \mathscr{B}\colon\ y \notin B\}$. A subset C of $\mathbf{X}$ is said to be independent if C is a subset of a basis. Once $\mathscr{B}$ is selected, the joint kernel $K = K(\mathscr{B})$ is defined as $K(\mathscr{B}) := \bigcap_{A\in\mathbf{A}(\mathscr{B})} \ker(l_A)$, where $\mathbf{A} = \mathbf{A}(\mathscr{B})$ is the collection of all subsets of $\mathbf{X}$ which meet every $B \in \mathscr{B}$. The goal of this paper is to establish the dimension formula (1)

$\dim K(\mathscr{B}) = \sum_{B\in\mathscr{B}} \dim K(\{B\})$ under appropriate conditions on $\mathscr{B}$ and the linear maps $\{l_x\}_{x\in\mathbf{X}}$.

If $\mathscr{B}$ is matroidal, i.e., $\mathscr{B}$ is the collection of bases for a matroid defined on a subset of $\mathbf{X}$, then it was proved by W. Dahmen and C. A. Micchelli [Adv. Math. **76** (1989), no. 1, 33–93; MR 90m:41020] that (2) $\dim K(\mathscr{B}) \le \sum_{B\in\mathscr{B}} \dim K(\{B\})$. A stronger result is established in this paper. Let $\mathscr{E} = \mathscr{E}(\mathscr{B})$ be the collection of all those independent subsets C of $\mathbf{X}$ which either are in $\mathscr{B}$, or else there is some $b \in \mathbf{X}\setminus C$ which satisfies the following two conditions: (i) $C \cup b \in \mathscr{E}$; (ii) if $\mathscr{B}_{\setminus b} \ne \varnothing$, then $C \in \mathscr{E}(\mathscr{B}_{\setminus b})$. The definition of $\mathscr{E}$ seems rather complicated, but the recursion required in it does terminate after finitely many steps. If $\varnothing \in \mathscr{E}(\mathscr{B})$, then $\mathscr{B}$ is said to satisfy the $\mathscr{E}$-condition. It is shown in this paper that the inequality in (2) holds true provided $\mathscr{B}$ satisfies the $\mathscr{E}$-condition, which is weaker than the condition that $\mathscr{B}$ is matroidal.

The dimension formula in (1) was obtained by the reviewer, S. D. Riemenschneider and Shen [Amer. J. Math. **114** (1992), no. 1, 157–184; MR 93b:47002] under the conditions that $\mathscr{B}$ is matroidal and all the linear maps $\{l_x\}$ lie in a semigroup G of linear operators with s-additivity. The s-additivity is connected to the solvability of systems of the form (C,φ): $l_c? = \varphi_c$, $c \in C$, with $C \subseteq \mathbf{X}$ and φ a map into S and defined (at least) on C. The system (C,φ) is said to be (i) special if $\varphi_c \in K(\mathscr{B}_{\setminus c})$ for all $c \in C$, (ii) compatible if $l_c\varphi_b = l_b\varphi_c$ for all $c, b \in C$, and (iii) basic if $C \in \mathscr{B}$. The following characterization for s-additivity is established in this paper: G is s-additive if and only if, for every matroidal $\mathscr{B}$ and every $l\colon \mathbf{X} \to G$ with $\dim K < \infty$, every special compatible basic system is solvable. The main result of this paper states that if $\mathscr{B}$ satisfies the $\mathscr{E}$-condition, and if the linear maps $\{l_x\}$ satisfy the solvability condition that any special compatible basic system is solvable, then the dimension formula in (1) holds true.

The paper is an important contribution to approximation theory and related areas. It is amazing that the authors use elementary methods to obtain such deep results. The paper should be of interest to both analysts and algebraists.

The flavor of this paper is algebraic and combinatorial. In the same spirit, Dahmen, A. Dress and Micchelli [Adv. in Appl. Math. **17** (1996), no. 3, 251–307; MR 97m:41008, see p. C27 in this volume] have derived similar results by using homological algebra. In the same issue [Adv. in Appl. Math. **17** (1996), no. 3, 308–336; MR 98h:13027] the reviewer obtained related results on the codimension of a multivariate polynomial ideal by using the perturbation technique from analysis. Among many applications of research in this direction the reviewer mentions his solution [Linear Algebra Appl. **250** (1997), 69–103; MR 97k:41024] of the long-standing conjecture of R. P. Stanley [Duke Math. J. **43** (1976), no. 3, 511–531; MR **56** #2865] on symmetric magic squares. *Rong Qing Jia* (Edmonton, AB)

97m:41008 41A15 05B35 13D03 41A63
Dahmen, Wolfgang (Bielefeld); **Dress, Andreas** (Bielefeld);
Micchelli, Charles A. (Bielefeld)
On multivariate splines, matroids, and the Ext-functor. (English summary)
Adv. in Appl. Math. **17** (1996), *no.* 3, 251–307.

In this article the authors study the problem of determining the dimension of the joint kernel of commuting linear operators. This fascinating problem originated in spline theory, more specifically in the area of box splines, and has intimate connections with elements of algebra and combinatorics (e.g., matroids and polynomial ideals). An overview of the problem, its origin and its brief history occupies the first part of this review. The actual contribution of this particular paper is analysed in the second part of the review.

A box spline is a compactly supported smooth piecewise-polynomial function in $d \geq 1$ variables. In approximation theory, the space S spanned by the integer translates of the box spline is considered as a potential source of approximants. Approximation theory basics tell us that the subspace $\mathcal{D}$ of all entire functions in S contains critical information concerning approximation properties of S. I recommend the monograph by C. de Boor, K. H. Höllig and S. D. Riemenschneider [*Box splines*, Springer, New York, 1993; MR 94k:65004], for further reading in this direction.

Each box spline M_Ξ is defined by a multiset Ξ of directions, i.e., a spanning multiset of nonzero vectors in $\mathbf{Z}^d$. De Boor and Höllig established [J. Analyse Math. **42** (1982/83), 99–115; MR 86d:41008] that the space $\mathcal{D} := \mathcal{D}(\Xi)$ is the joint kernel of the differential operators (1) $p_Y := \prod_{\xi \in Y} D_\xi$, where Y varies over all subsets of Ξ that satisfy $Y \cap B \neq \varnothing$, for all $B \in \mathbf{B}(\Xi)$, with D_ξ the directional derivative in the ξ-direction, and with $\mathbf{B}(\Xi)$ the collection of all bases (for $\mathbf{R}^d$) made of the directions in Ξ. Dahmen and Micchelli [Studia Math. **82** (1985), no. 3, 243–263; MR 87k:41008] proved that the dimension of $\mathcal{D}(\Xi)$ equals the cardinality of $\mathbf{B}(\Xi)$. They established this celebrated result with the aid of two separate arguments, a complicated one for the upper bound inequality ($\dim \mathcal{D}(\Xi) \leq \#\mathbf{B}(\Xi)$), and a more complicated one for the complementary lower bound inequality. Subsequently, they introduced a discrete version of $\mathcal{D}(\Xi)$ (where the directional derivative is replaced by a directional difference), and proved that the dimension of the discrete analogue of $\mathcal{D}(\Xi)$ is $\sum_{B \in \mathbf{B}(\Xi)} |\det B|$. (This expression is also the volume of $\operatorname{supp} M_\Xi$, and is connected to the problem of tiling zonotopes by parallelepipeds, but that is a different story.)

The problem was revived with the introduction of exponential box splines [A. Ron, Constr. Approx. **4** (1988), no. 4, 357–378; MR 89j:41017]. In the exponential case, each of the directional derivatives in (1) is replaced by a first-order differential operator $D_\xi - c_\xi$, $c_\xi \in \mathbf{C}$. The problem of finding $\dim \mathcal{D}(\Xi)$ for this more general case was attacked independently by Ben-Artzi and Ron [Trans. Amer. Math. Soc. **309** (1988), no. 2, 683–710; MR 89m:41008] and by Dahmen and Micchelli [Adv. Math. **76** (1989), no. 1, 33–93; MR 90m:41020]; both papers showed that the dimension is independent of the values of constants (c_ξ), hence remains $\#\mathbf{B}(\Xi)$. However, those two papers approached the problem from completely different angles. The first focused primarily on the lower bound inequality. It was observed that this estimate is quite trivial for a generic selection of the constants $(c_\xi)_\xi$ (which is the case when $\mathcal{D}(\Xi)$ is spanned by pure exponentials), and this was used to prove the lower bound in general. This approach was refined later by de Boor and Ron [J. Math. Anal. Appl. **158** (1991), no. 1, 168–193; MR 93a:41014], who used lower-order perturbations of the generators of polynomial ideals as a painless approach to establishing the lower bounds. They proved that if $\mathbf{B}(\Xi)$ is replaced in (1) by any subset $\mathbf{B}_1$ of it, the corresponding joint kernel $\mathcal{D}_{\mathbf{B}_1}(\Xi)$ still satisfies $\dim \mathcal{D}_{\mathbf{B}_1}(\Xi) \geq \#\mathbf{B}_1(\Xi)$. This result was extended later by Jia to differential operators of higher degree, but not to any other type of operators.

In contrast, Dahmen and Micchelli [op. cit.; MR 90m:41020] attacked the problem via the upper bound inequality. They first observed that the equality $\dim \mathscr{D}(\Xi) = \#\mathbf{B}(\Xi)$ can be rewritten as (2) $\dim \mathscr{D}(\Xi) = \sum_{B\in\mathbf{B}(\Xi)} \dim \mathscr{D}(B)$ (since $\mathscr{D}(B)$ can be easily shown to contain constants only), and they further observed that the dimension formula for the discrete case can also be rewritten in a similar manner. With these observations, they generalized the problem to the following set-up: Let Ξ be a matroid, $\mathbf{B}(\Xi)$ its set of bases. Let $(D_\xi)_\xi$ be a set of endomorphisms on some linear space , which pairwise commute. Defining $\mathscr{D}(\Xi)$ as in (1), prove the dimension formula (2). Using a clever linear map on $\mathscr{D}(\Xi)$, which subsequently was dubbed the "DM map", Dahmen and Micchelli showed that the upper bound estimate can be reduced to analogous bounds on certain submatroids, and eventually to rudimentary linear algebra, which requires no information on the nature of $(D_\xi)_\xi$. Once the upper bound is in place, the complementary lower bound is determined by the ability to find solutions for certain linear systems which involve those D_ξ operators. This approach, which was subsequently studied also by Jia, Riemenschneider, Z. Shen, de Boor, Ron and others, consists thus of two basic ingredients: (i) find structural conditions on the subset $\mathbf{B}(\Xi)$ that allow the deduction of the upper bound, regardless of the nature of the operators $(D_\xi)_\xi$, and (ii) find, in addition, the smallest set of "solvability conditions" that the operators (D_ξ) should satisfy to obtain equality in (2) (and not only an inequality $\leq$).

The present authors recast this problem in algebraic terms and then use homological algebra tools for its solution. The basic approach is as follows: Let $(D_\xi)_{\xi\in\Xi}$ be pairwise commuting operators on a linear space . With R the polynomial ring in the variables Ξ, the operators $(D_Y)_Y$ (cf. (1)) can be viewed as pure monomials, which generate an ideal I in R. This leads to an exact sequence of R-modules, $0 \to I \to R \to R/I \to 0$. The two underlying observations are then that the dimension sought for is $\dim \operatorname{Hom}_R(R/I, \quad)$ and that the required "solvability conditions" are equivalent to $\operatorname{Ext}^1_R(R/I, \quad) = 0$. While this does not simplify the problem, it allows someone with a good grasp of homological algebra to examine new approaches for this problem. Overall, there are important new observations and results in this paper, but I was unable to determine whether the algebraic tools were essential or merely convenient for these achievements. Also, the authors do not connect their results to concrete unsolved problems in approximation theory, and it takes a great deal of effort to judge the relevance of their various results by converting them back into approximation theory language.

The article is divided into two different, almost disjoint, studies: Sections 4–6 on the one hand, and Sections 7–8 on the other. Section 4 treats the simple case of "pure exponentials", when the block spaces $(\mathscr{D}(B))_{B\in\mathbf{B}(\Xi)}$ form a direct sum. Section 6 highlights the case of matroid selection rules (for the basis collection). Section 5, while making in its results assumptions that do not hold in most of the applications, is intriguing in its generalization of the sought-for equality: it attempts to establish an equality of the form $\dim \mathscr{D}(\Xi) = \sum_{B\in\mathbf{B}(\Xi)} m_B \dim \mathscr{D}(B)$, where m_B is a multiplicity of B in $\mathbf{B}(\Xi)$ (and which is determined by the structure of $\mathbf{B}(\Xi)$). This generalization is important, since there are recent examples of dimension problems in box spline theory where $\dim \mathscr{D}(\Xi)$ exceeds $\sum_{B\in\mathbf{B}(\Xi)} \dim \mathscr{D}(B)$. In Sections 7 and 8, a structural assumption on $\mathbf{B}(\Xi)$ ($\mathbf{B}$-purity), which implies the upper bound inequality, is made, and conditions that ensure the matching lower bound are sought. The results there, among other things, capture and extend the s-additivity notion of Shen [Proc. Amer. Math. Soc. **112** (1991), no. 2, 381–390; MR 91j:35069].

I could not locate in the article results where a lower bound is obtained without a matching upper bound, and thus was unable to confirm the authors' contention that they extend and cover all previous results from the literature. At the same time, the paper breaks new ground and changes permanently the way we view the dimension problem. *Amos Ron* (Madison, WI)

97b:42022 42B08 42B05

Bourgain, Jean (Princeton, NJ)

Spherical summation and uniqueness of multiple trigonometric series.

Internat. Math. Res. Notices **1996,** *no.* 3, 93–107.

The author proves the following theorem: Let $\sum_{\xi\in\mathbf{Z}^d} a_\xi e^{i\langle x,\xi\rangle}$ be a trigonometric series which converges spherically to zero at every point x of the d-torus $\mathbf{T}^d$, i.e., $\lim_{R\to\infty}\sum_{|\xi|<R} a_\xi e^{i\langle x,\xi\rangle}=0$ for all $x\in\mathbf{T}^d$. Then $a_\xi=0$ for all ξ.

This masterful paper solves what had been the most important open problem in this area of harmonic analysis. The special case when $d=1$ was done in 1870 by Georg Cantor, who made good use of an idea of Riemann to study the original series by determining properties of the associated Riemann function, $a_0x^2/2-\sum_{\xi\neq 0}(a_\xi/\xi^2)e^{i\langle x,\xi\rangle}$. V. L. Shapiro [Ann. of Math. (2) **66** (1957), 467–480; MR **19**, 854d] had solved the $d=2$ case, subject to a side condition that was later shown by R. Cooke to follow from the original hypothesis [Proc. Amer. Math. Soc. **30** (1971), 547–550; MR **43** #7847]. After B. Connes [C. R. Acad. Sci. Paris Sér. A-B **283** (1976), no. 4, Aii, A159–A161; MR **54** #10975] extended Cooke's work to higher dimensions, a higher-dimensional uniqueness result became possible, but for many years it seemed too difficult to show that the Riemann function was continuous. This is exactly what Bourgain does. His methodology is a beautiful blend of classical harmonic analysis and Brownian motion theory.

To prove the theorem, first note that the hypothesis implies that the sequence $\sum_{|\xi|=R} a_\xi e^{i\langle x,\xi\rangle}$ tends to 0 as R tends to ∞ for all $x\in\mathbf{T}^d$, from whence the aforementioned theorem of Connes yields $\lim_{R\to\infty}\sum_{|\xi|=R}|a_\xi|^2=0$. This is enough to show that the series $a_0|x|^2/(2d)-\sum_{\xi\neq 0}(a_\xi/|\xi|^2)e^{i\langle x,\xi\rangle}$ defining the Riemann function converges to, and is also the Fourier series of, an everywhere finite $L^2(\mathbf{T}^d)$ function F. The calculations $\Delta|x|^2=2d$ and $\Delta e^{i\langle x,\xi\rangle}=-|\xi|^2e^{i\langle x,\xi\rangle}$ suggest that we ought to have $\Delta F=0$ and consequently F harmonic. Indeed, a summation by parts shows that the generalized Laplacian of F, defined to be $\lim_{r\to 0} c(\mathrm{av}(F)-F)/r^2$ for an appropriate constant c, where $\mathrm{av}(F)$ is the average of F over the solid ball of radius r, is everywhere 0. Temporarily assume that F is continuous. Continuous functions with generalized Laplacian 0 are easily seen to be harmonic, hence C^∞, so that the coefficients must decay to 0 rapidly. Given any η in $\mathbf{Z}^d$, multiply the original series by $e^{-i\langle x,\eta\rangle}$, integrate in x over $\mathbf{T}^d$, and interchange integral and sum (justified by the rapid decay) to get that $a_\eta=0$, which completes the proof. Thus the essence of the problem lies in showing F to be continuous.

Here, in most heuristic terms, is the logical flow of Bourgain's argument. Suppose that F is not continuous and let Z be the nonempty set of discontinuities of F. A Baire category argument on Z produces a point $p\in Z$ and a solid ball B about

p such that if $Z \cap B$ is "thin" (is of measure 0 with respect to a certain harmonic measure), then Bourgain shows that F is harmonic and hence continuous on B, contrary to the definition of Z, while if $Z \cap B$ is "thick", then Bourgain shows that F must be continuous at p, contrary to $p \in Z$. Bourgain achieves the first contradiction using a balayage argument that seems to use the axiom of choice in an unavoidable way. His other contradiction is reached by hard analysis, harmonic measure, and capacity theory. The power and originality used here prompted Victor Shapiro to say to me that he had casually mentioned this 40-year-old problem to Jean Bourgain, but if he had not it might very well have gone unsolved for another century.

Here is one typographical correction: The function F_R appearing on the right hand side of equation (1.14) should be replaced by F_R^*, the strong maximal function of F_R.

Spherical summation is probably the most important way of summing a multiple trigonometric series. Four other common ways are iterated summation, unrestricted rectangular summation, restricted rectangular summation, and square summation. Correspondingly, there are four statements analogous to Bourgain's uniqueness theorem. If the word "spherically" is replaced by either "iteratedly" or "unrestricted rectangularly", the corresponding statements are indeed theorems, while if it is replaced by "restricted rectangularly" or "squarely", open questions result, even in two dimensions. For an expansion of the somewhat tersely written proof of Bourgain, as well as a summary of what is known when spherical summation is replaced by other modes of summation, see [J. M. Ash and G. Wang, in *Harmonic analysis and nonlinear differential equations (Riverside, CA, 1995)*, 35–71, Contemp. Math., 208, Amer. Math. Soc., Providence, RI, 1997; MR 99f:42019].

J. Marshall Ash (Chicago, IL)

99f:42026 42B10 42B25

Tao, Terence (Los Angeles, CA); **Vargas, Ana** (Madrid); **Vega, Luis** (Bilbao)

A bilinear approach to the restriction and Kakeya conjectures. (English summary)

J. Amer. Math. Soc. **11** (1998), *no.* 4, 967–1000.

In this paper, bilinear versions of two of the most classical problems of modern harmonic analysis are considered, and results about them are then applied to the usual versions. These two problems are the restriction problem for the Fourier transform and the Kakeya maximal problem. The (dual) form of the restriction problem asks for which $1 \le p, q \le \infty$ one has the estimate

$$\|\widehat{g\,d\sigma}\|_{L^q(\mathbf{R}^n)} \le C\|g\|_{L^p(d\sigma)} \tag{1}$$

(say for all test functions g), where $d\sigma$ denotes surface measure on, for example, the unit sphere or the base of a paraboloid P^{n-1} in $\mathbf{R}^n$. The Kakeya maximal problem asks for the precise $L^p \to L^q$ mapping properties (in terms of the parameter δ) of the mapping

$$f \mapsto \sup_{x \in R} \frac{1}{|R|} \int_R |f|, \tag{2}$$

where the sup is taken over (say) the family of all $\delta \times \delta \times \cdots \times \delta \times 1$ rectangles in $\mathbf{R}^n$.

A great deal about these problems is known, especially in low dimensions, as is the fact that the two problems are inextricably linked and also provide a quantitative framework for estimating lower bounds on dimensions of Kakeya sets (that is, sets containing unit line segments in each direction) in $\mathbf{R}^n$.

The bilinear restriction problem is the problem of determining those p_1, p_2 and q such that

$$\|\widehat{gd\sigma}\widehat{hd\sigma}\|_{L^q(\mathbf{R}^n)} \leq C\|g\|_{L^{p_1}(d\sigma)}\|h\|_{L^{p_1}(d\sigma)} \tag{3}$$

for all g and h supported on 1-separated 1-cells on P^{n-1}, and the ideas connected with this problem have a long (if rather unexplicit) history going back to works of Sjölin, Beals and the present reviewer in the early 1980s.

The first result presented by the authors is that problems (1) and (3) are essentially equivalent in the range where it is conjectured that (1) holds. This is essentially an advanced form of a scaling argument. On the other hand, the range of indices for which one might expect (3) to be valid is larger than what simple considerations of (1) might lead one to imagine. Indeed, the next result characterises the p's for which (3) holds for $q=2$ and $p_1=p_2=p$ as being $[4n/(3n-2),\infty]$; this is based upon "a routine modification" of a previous argument (in the case $n=3$) of Moyua, Vargas and Vega. Examples giving necessary conditions on $p_1=p_2=p$ and q such that (3) holds are also given. An immediate consequence of the aforementioned results is the first sharp (dual) restriction theorem on the line $1/q=(n-1)/(n+1)p'$ when $n=3$ which goes beyond the Stein-Tomas point $q=4$, $p=2$. A further improvement on this occurs later in the paper.

With a view to this further improvement, the authors next turn to formulating and proving bilinear variants of the dual of the Kakeya problem (2) with a suitable separation hypothesis on the supports of the pair of functions. Once again they prove an equivalence between the usual and the bilinear forms of the problem in the natural ranges of indices. They then prove a bilinear estimate which in the case $n=3$ improves upon Wolff's Kakeya maximal theorem—but only on the scale of bilinear, not linear estimates.

In the last part of the paper the authors obtain their sharpest results for the restriction problem by modifying the "Bourgain machine" to make it suitable for the bilinear estimates considered earlier. This is by far the most technical part of the paper, but the final result is that when $n=3$, (1) holds for $p>\frac{170}{77}$ and $q>\frac{34}{9}$ while on the line $1/q=(n-1)/(n+1)p'$ $(n=3)$ (1) holds for $q>4-\frac{5}{27}$. The authors also show that the so-called Bochner-Riesz conjecture holds when $n=3$ and $p>\frac{34}{9}$ (≥ 2).

This paper represents an exciting new development in the study of the restriction phenomenon. *Anthony Carbery* (Edinburgh)

99d:42026 42B20 45P05 49L20

Nazarov, F. L. (East Lansing, MI); **Treĭl′, S. R.** (East Lansing, MI)

The hunt for a Bellman function: applications to estimates for singular integral operators and to other classical problems of harmonic analysis. (Russian. Russian summary)

Algebra i Analiz **8** (1996), *no.* 5, 32–162; *translation in St. Petersburg Math. J.* **8** (1997), *no.* 5, 721–824.

The work reviewed here was announced in a 6-page note [C. R. Acad. Sci. Paris Sér. I Math. **323** (1996), no. 7, 717–722; MR 99j:42010], where an elementary proof of the Hunt-Muckenhoupt-Wheeden theorem was given. The theorem states that the Hilbert transform is bounded in the weighted space $L^2(w)$ if and only if the weight w satisfies the (A_2) condition: $\sup\langle w\rangle_I\langle 1/w\rangle_I<\infty$ (the supremum taken over all intervals I, brackets denoting the mean over an interval). The authors admitted that the proof looked like a clever trick but claimed that there are deep ideas behind it. This 131-page hunt for a Bellman function explains these ideas.

The main theorem of the paper is a generalization of the above-mentioned theorem to the case of weighted L^p spaces of vector-valued functions. Specifically, let E be a finite-dimensional Hilbert space and let the weight W be a function from $\mathbf{R}$ into the cone of positive definite selfadjoint operators on E. Then $L^p(W)$, $1<p<\infty$, consists of the functions $\varphi\colon\mathbf{R}\mapsto E$ for which the norm $\|\varphi\|_{L^p(W)}=(\int_{\mathbf{R}}\|W(t)^{1/p}\varphi(t)\|_E^p dt)^{1/p}$ is finite. The Hilbert transform, defined first for smooth compactly supported E-valued functions on $\mathbf{R}$ by $H\varphi(t)=\lim_{\varepsilon\to 0+}\int_{|s-t|<\varepsilon}(\varphi(s)/(s-t))ds$ and extended to a unitary operator on $L^2(\mathrm{Id})$, is said to operate on $L^p(W)$, $1<p<\infty$, if $\|H\varphi\|_{L^p(W)}\le C\|\varphi\|_{L^p(W)}$ for every $\varphi\in L^p(W)\cap L^2(\mathrm{Id})$, where the constant C does not depend on φ. First a necessary condition for the boundedness of the Hilbert transform on $L^p(W)$ in terms of W is obtained by a rather standard procedure using harmonic extension to the upper half-plane, and to no surprise it is an expected generalization of Muckenhoupt's (A_p) condition; it is called the $(\mathscr{A}_{p,q})$ condition. The sufficiency is proved in two stages. In the first it is shown that the Haar system is an unconditional basis for $L^p(W)$ if W satisfies the $(\mathscr{A}_{p,q})$ condition, and the space of coefficients is identified. The matrix of the Hilbert transform (actually of any classical Calderón-Zygmund operator) with respect to the Haar basis is then shown to give a bounded operator on the space of coefficients.

This is the main result but the main theme of this work is the method by which it is obtained. The method is applicable to a wide range of problems and has the following steps. The first step is to identify the so-called Bellman function (the name comes from the theory of optimal control) associated with the problem at hand. Then, by the method of wishful thinking, one finds the properties which the Bellman function must have in order for the problem to have the expected solution. As a rule one of the properties is a convexity-type condition. Usually one does not have an explicit expression for the Bellman function so one cannot check that it has the desired properties. An exception is the case of unweighted scalar L^p spaces, for which D. L. Burkholder computed the sharp unconditionality constant of the Haar basis in a first occurrence of this method [see Ann. Probab. **12** (1984), no. 3, 647–702; MR 86b:60080]. Luckily one does not really have to know the Bellman function explicitly. To solve the problem it may be enough to construct any function with the wanted properties, a function of Bellmanian type, in the authors' terminology.

The authors admit that building functions of Bellmanian type is a kind of trade and go to great lengths in explaining the secrets of the trade. They first illustrate the method by proving Carleson's imbedding theorem and the maximal function inequality (in both cases the dyadic version), and go on to prove the unconditionality of the Haar basis gradually: first for the scalar weighted $L^2(w)$, next for $L^2(W)$, then scalar L^p without a weight, and finally $L^p(W)$. Then they show that the space of coefficients of expansions of $L^p(W)$ functions with respect to the Haar basis is a certain Triebel-Lizorkin space, independent of W, and finish off with proving boundedness of matrices of Calderón-Zygmund operators on these spaces.

The paper is self-contained; it even contains a chapter on the theory of bases in Banach spaces. It is very pleasant to read because it radiates excitement and enthusiasm for a new approach which may very well become one of the more powerful methods in classical harmonic analysis. *Aleš Založnik* (Ljubljana)

99i:42018 42B20 30E20
David, Guy (Orsay)
Unrectifiable 1-sets have vanishing analytic capacity. (English and French summaries)
Rev. Mat. Iberoamericana **14** (1998), *no.* 2, 369–479.

In this remarkable paper the author completes the proof of a 32-year-old conjecture of Vitushkin. The conjecture states that, among sets of finite one-dimensional Hausdorff measure (i.e. length), those that are removable for bounded analytic functions are precisely those projecting in sets of zero length in almost all directions (otherwise called Besicovitch irregular). A compact set is said to be removable for bounded analytic functions whenever it cannot be the set of singularities of a bounded analytic function (an elementary example is provided by a point which is removable according to Riemann). Thus the result describes in metric terms an analytic property of sets of finite length which is extremely difficult to handle in the most general case.

It turns out that the eventual proof of Vitushkin's conjecture is the result of contributions of several people at different stages. We wish to describe them briefly. First of all let us point out that the critical dimension for Painlevé's problem (that is, the problem of describing removable sets) is one. Indeed, Painlevé proved more than a century ago that zero length implies removability. That removability is not described by length was discovered by A. G. Vitushkin [Dokl. Akad. Nauk SSSR **127** (1959), 246–249; MR **22** #9607], who constructed a removable set with positive length. This set is Besicovitch irregular and thus Vitushkin's conjecture becomes natural.

Besicovitch's theory of sets of positive finite length shows that Vitushkin's conjecture can be reformulated by saying that a set of finite length is removable if and only if it intersects each rectifiable curve in a set of zero length. In 1977 A.-P. Calderón [Proc. Nat. Acad. Sci. U.S.A. **74** (1977), no. 4, 1324–1327; MR **57** #6445] proved that the Cauchy integral operator is bounded on L^2 with respect to the arc length measure on a Lipschitz graph with a small Lipschitz constant. On the other hand, work of Garabedian in the 50's established the equivalence between removability for H^∞ functions and removability for functions in the Hardy space H^2. Combining Calderón's theorem with Garabedian's results one verifies readily the Denjoy conjecture, namely, that sets of positive length on any rectifiable curve are non-removable. Therefore one direction of Vitushkin's conjecture is proved.

Proving the other direction took 20 more years. M. Christ [Colloq. Math. **60/61** (1990), no. 2, 601–628; MR 92k:42020] found a way to modify a function with nonremovable singularities on our given set to construct a non-trivial piece of the set where the Cauchy integral is bounded on L^2. This required proving a local $T(b)$-theorem and making an assumption of homogeneity on the length measure of the set to meet the natural (at that time) conditions for Calderón-Zygmund theory.

Combining deep work of P. W. Jones [Invent. Math. **102** (1990), no. 1, 1–15; MR 91i:26016] on the traveling salesman problem, and of David and S. Semmes [Astérisque No. 193 (1991), 152 pp.; MR 92j:42016] on uniform rectifiability, P. Mattila, M. S. Mel′nikov and the reviewer were able to show in 1996 [Ann. of Math. (2) **144** (1996), no. 1, 127–136; MR 97k:31004] that the subset provided by Christ's work is indeed contained in a countable union of rectifiable curves. The argument used a new technical device called Menger curvature, which was not powerful enough to allow us to dispense with the classical homogeneity assumption of Calderón-Zygmund theory. This is what the author does in the paper under consideration. Dwelling on previous work by J. C. Leger on one hand and by David and Mattila on the other, the author, surmounting enormous difficulties, is able to prove a $T(b)$-theorem without assuming doubling conditions on the underlying measure. This completes the last missing step in the proof of Vitushkin's conjecture.

The interested reader is invited to consult the survey paper by David [Publ. Mat. **43** (1999), no. 1, 3–25], in which the background of the conjecture is presented and the main steps of its proof are described with a broad audience in mind.

Joan Verdera (Bellaterra (Barcelona))

97b:46016 46Bxx

Johnson, W. B. (College Station, TX); **Lindenstrauss, J.** (Jerusalem);
Schechtman, G. (Rehovot)

Banach spaces determined by their uniform structures. (English summary)

Geom. Funct. Anal. **6** (1996), *no.* 3, 430–470.

In this important article, the authors provide a deep analysis of the following general question: Let X and Y be two Banach spaces which are uniformly homeomorphic. Are they linearly isomorphic? This turns out to be a rather delicate question, which remains unsolved even for some classical spaces such as c_0 or l_1. This work contains, however, positive or negative results on several classes of Banach spaces and introduces powerful new techniques, thus opening the way to further research in this direction.

The first of these techniques is a principle attributed by the authors to the late E. Gorelik [Israel J. Math. **87** (1994), no. 1-3, 1–8; MR 95f:46028], to whom the article is dedicated. The Gorelik principle, which relies on a nice application of Brouwer's fixed-point theorem, can be stated as follows: A uniform homeomorphism between Banach spaces cannot take a large ball of a finite-codimensional subspace into a small neighborhood of a subspace of infinite codimension. This principle is used, together with linear results, to show (Corollary 1.7) that if a space X has an unconditional basis which is "not too close to l_2" and if Y is uniformly homeomorphic to X, then no linear quotient of Y is isomorphic to l_2. An important consequence of this result is (Theorem 2.1) that if p is finite and strictly greater than 1, then any space which is uniformly homeomorphic to l_p is actually isomorphic to l_p. In other words, the reflexive l_p spaces are determined by their uniform structures. Section 2 also contains sufficient conditions on a Banach space which ensure that it is uniformly homeomorphic to at most two isomorphically distinct Banach spaces. The proof of this statement (Proposition 2.9) uses two basic tools from this area: ultrapowers and differentiation of Lipschitz maps. Section 3 deals with c_0. It is shown there that if a complemented subspace of a $C(K)$ space is uniformly homeomorphic to c_0, then it is isomorphic to c_0. It is an important open problem to know whether c_0 is in fact determined by its uniform structure. This result suffices, however, to clear up the situation for $C(K)$ spaces, even in the nonseparable case: this is explained in Section 6.

Section 4 deals with uniform homeomorphisms between unit balls, which turn out to be considerably easier to establish than those between the whole spaces: for instance, a result which goes back to Mazur asserts that the unit balls of two arbitrary separable l_p spaces are uniformly homeomorphic. Generalized Mazur maps were recently provided by E. Odell and T. Schlumprecht [Acta Math. **173** (1994), no. 2, 259–281; MR 96a:46031]. These maps are used in Sections 4 and 5 to provide pairs of Banach spaces which are uniformly homeomorphic. A striking consequence of the results from Sections 2, 4 and 5 is the existence of a Banach space which is uniformly homeomorphic to exactly two nonisomorphic Banach spaces. Actually, the result is shown not only for two but for any power of two (Theorem 5.8). The examples happen to be variants of a Tsirelson space.

Section 7 investigates how well one can identify a Banach space by looking at its finite-dimensional subspaces. Kwapień's theorem implies that this is indeed so for Hilbert space, and it is natural to conjecture that l_2 is somehow the only space that can be determined by its finite-dimensional subspaces. However, the authors show the existence of non-Hilbertian spaces which are determined by their pavings, where a paving is defined to be an increasing sequence of finite-dimensional subspaces whose union is dense in the space (Corollary 7.7). Such spaces nevertheless have to be very close to Hilbert space (Proposition 7.4). Section 8, finally, contains a list of open problems. It is not known, for instance, whether all classical function spaces are determined by their uniform structure, nor whether two separable Banach spaces which are Lipschitz homeomorphic to each other are in fact isomorphic.

This work clearly shows the depth and complexity now reached by the theory called geometry of Banach spaces. However, it shows equally well that simple questions on simple spaces are still unsolved. It therefore provides the reader with fascinating challenges. *Gilles Godefroy* (Paris)

99g:46010 46B15 46B07 46B20 46B28
Terenzi, Paolo (Milan)
A positive answer to the basis problem. (English summary)
Israel J. Math. **104** (1998), 51–124.

One of the fundamental problems in Banach space theory was the question about the existence of a Schauder basis in every separable Banach space. This problem was solved in the negative by P. Enflo [Acta Math. **130** (1973), 309–317; MR **53** #6288], who gave an example of a separable Banach space which does not even possess the approximation property. Nevertheless, modified versions of this question still persist, namely what kind of "basislike" systems one can find in an arbitrarily given separable Banach space X.

For example, there always exists a uniformly minimal M-basis, i.e. a sequence of elements $x_n \in X$ together with $x_n^* \in X^*$ such that $x_n^*(x_m) = 1$ when $n = m$, $x_n^*(x_m) = 0$ for $n \neq m$, $\overline{\operatorname{span}}\{x_n\}_{n=1}^{\infty} = X$, $x = 0$ whenever $x_n^*(x) = 0$ for all n, and $\sup \|x_n\| \cdot \|x_n^*\| < \infty$. This is due to R. I. Ovsepian and A. Pełczyński [Studia Math. **54** (1975), no. 2, 149–159; MR **52** #14942]. Here, every $x \in X$ is uniquely represented by $\sum_{n=1}^{\infty} x_n^*(x)x_n$; however, this series need not converge in norm. In a previous publication [Studia Math. **111** (1994), no. 3, 207–222; MR 95i:46013] the author of the current paper even showed that every X has a uniformly minimal strong M-basis, i.e. where in addition $x \in \overline{\operatorname{span}}\{x_n^*(x)x_n\}_{n=1}^{\infty}$ for all $x \in X$.

If one wants to ensure norm convergence of the "basis expansion" of every $x \in X$ one has to allow blockings and permutations: A uniformly minimal M-basis $\{x_n\}_{n=1}^{\infty}$ is called a uniformly minimal basis with quasi-fixed brackets and permutations if there exists a fixed increasing sequence of indices $q(m)$ such that, for each $x \in X$, there is a sequence of indices $q'(m)$ with $q(m) < q'(m) \leq q(m+1)$, and a permutation π of $\mathbf{N}$ such that $\{\pi(n)\}_{n=q(m)+1}^{q(m+1)}$ is a permutation of $\{n\}_{n=q(m)+1}^{q(m+1)}$, and we have

$$x = \sum_{m=1}^{\infty} \sum_{n=q'(m)+1}^{q'(m+1)} x_{\pi(n)}^* x_{\pi(n)}$$

with norm convergence. The nicest special case of such a system is the case where $\pi = \mathrm{id}$ and $q'(m) = q(m)$ for all m. Then $\{x_n\}_{n=1}^{\infty}$ is called a uniformly minimal basis with fixed brackets. We obtain that X has a uniformly minimal basis with fixed brackets if and only if X is the direct sum of a sequence of finite-dimensional subspaces. Hence Enflo's counterexample shows that there are Banach spaces even failing to have a uniformly minimal basis with fixed brackets.

Now in the current paper the author shows that if we allow $q'(m)$ to depend on $x \in X$ (within the bounds of $q(m)$ and $q(m+1)$) and if we allow permutations then such a basis always exists.

He proves that any subspace Y of a separable Banach space X has a uniformly minimal basis with quasi-fixed brackets and permutations which can be extended to a uniformly minimal basis with quasi-fixed brackets and permutations of X. In particular, every X has such a basis.

In a complement to the current paper [Israel J. Math. **108** (1998), 345–348; MR 99m:46025] the author observes that, by repeating essential portions of his proof, it is even possible to show that every separable Banach space X has a uniformly minimal basis with fixed brackets and quasi-fixed permutations, i.e. here $q(m)$ and $q'(m)$ are prefixed with $q(m) < q'(m) \leq q(m+1)$ such that, for every $x \in X$, one has $x = \sum_{m=1}^{\infty} \sum_{n=q(m)+1}^{q(m+1)} x_{\pi(n)}^* x_{\pi(n)}$, and only the permutations $\{\pi(n)\}_{n=q'(m)+1}^{q'(m+1)}$ of $\{n\}_{n=q'(m)+1}^{q'(m+1)}$ depend on x.

The idea of the proof is to find well-positioned finite-dimensional subspaces in X which have "nice" M-bases and then to start constructing a common M-basis from them. If X has type >1 then it contains uniformly complemented copies of l_2^n. Otherwise l_1 is finitely represented in X. In the latter case the author shows that X is "finitely transformable in c_0", i.e. for every finite-dimensional subspace Y of c_0

and every $\varepsilon > 0$ there is a subspace X_0 of X such that X/X_0 is $(1+\varepsilon)$-isomorphic to Y. Using this fact one can find enough suitable M-bases for finite-dimensional subspaces of X even in this case.

Thus, uniformly minimal bases with fixed brackets and quasi-fixed permutations are the sequences closest to Schauder bases which so far have been proven to exist in every separable Banach space. Moreover, the paper gives an important contribution to the solution of the following variants of the problem about finding basislike structures which are still open: Does every separable Banach space have a uniformly minimal basis with quasi-fixed brackets (where $\pi = \mathrm{id}$)? The author points out that, using his ideas of the proofs in the current paper and from previous papers, this question can be reduced to Banach spaces with finite cotype.

Another related problem is the existence of a uniformly minimal basis $\{x_n\}_{n=1}^{\infty}$ with individual permutations, i.e. where, for every $x \in X$, there is a permutation π of $\mathbf{N}$ with $x = \sum_{n=1}^{\infty} x^*_{\pi(n)}(x)x_{\pi(n)}$. The result of the current paper includes a weak positive answer to this: For a uniformly minimal basis with quasi-fixed brackets and permutations one always has $x^*(x) = \sum_{n=1}^{\infty} x^*_{\pi(n)}(x)x^*(x_{\pi(n)})$ for each $x^* \in X^*$ where π depends on x and x^*. *Wolfgang Lusky* (Paderborn)

99h:46009 46B20 46B28 46E30 46M05

Alspach, Dale E. (Stillwater, OK)

Tensor products and independent sums of $\mathcal{L}_p$-spaces, $1 < p < \infty$. (English summary)

Mem. Amer. Math. Soc. **138** (1999), *no.* 660, viii+77 pp.

Let $1 < p < \infty$, $p \neq 2$. An $\mathcal{L}_p$-space is a Banach space isomorphic to some complemented subspace of a function space $L_p(\mu)$, yet not isomorphic to a Hilbert space. (This is actually not the original definition due to Lindenstrauss and Pełczyński, but an equivalent reformulation.) The problem of studying the different types of separable $\mathcal{L}_p$-spaces has been the subject of many papers, and this memoir makes a significant contribution to those questions.

When the theory of $\mathcal{L}_p$-spaces was developed, up to isomorphism only four infinite-dimensional separable $\mathcal{L}_p$-spaces were known: $L_p[0,1]$, l_p, $l_p \oplus l_2$, and $l_p(l_2)$. Then, in a seminal work from 1970, H. P. Rosenthal added some more spaces to this list, notably a space X_p of $\{0,1,-1\}$-valued random variables on $[0,1]$; here $p > 2$. The question of whether there exist, in fact, infinitely many isomorphically distinct $\mathcal{L}_p$-spaces was settled by G. Schechtman, who considered tensor products $X_p^n = X_p \otimes \cdots \otimes X_p$ in which the norm on the tensor product is the one canonically inherited from $L_p([0,1]^n)$. He showed that these spaces are pairwise nonisomorphic $\mathcal{L}_p$-spaces [Israel J. Math. **22** (1975), no. 2, 138–147; MR **52** #11545]. Another approach was launched by J. Bourgain, Rosenthal and Schechtman [Ann. of Math. (2) **114** (1981), no. 2, 193–228; MR 83j:46031], who exhibited even uncountably many different $\mathcal{L}_p$-spaces. They defined an $\mathcal{L}_p$-space R_p^α for each ordinal $\alpha < \omega_1$ inductively; the inductive step involves embedding a sequence of subspaces of $L_p[0,1]$ onto independent coordinates in $L_p([0,1]^{\mathbf{N}})$. They went on to show that uncountably many of the spaces R_p^α are pairwise nonisomorphic although it remained open precisely which ones are isomorphic and which ones are not.

In the present memoir the relationship between these two methods of constructing infinitely many $\mathcal{L}_p$-spaces is investigated, and it is shown that they are essentially incompatible. More precisely, the author proves that $X_p \otimes X_p$ is not isomorphic to a complemented subspace of any R_p^α. A great deal of technical work is required to reach this point; for instance, a method for dealing with gliding-hump arguments in spaces with bases which have many subsequences spanning an isomorph of the space is developed. Another major result of this paper is the complete isomorphic classification of the spaces R_p^α: namely, R_p^α is isomorphic to R_p^β, $\beta > \alpha$, if and only if $\beta = \alpha + n$ for some $n \in \mathbf{N}$. *Dirk Werner* (Berlin)

98a:46034 46E20 30D50 30H05 47A15 47B38
Aleman, A. [Aleman, Alexandru] (Hagen);
Richter, S. [Richter, Stefan] (Knoxville, TN);
Sundberg, C. [Sundberg, Carl] (Knoxville, TN)
Beurling's theorem for the Bergman space.
Acta Math. **177** (1996), *no.* 2, 275–310.

A celebrated theorem in operator theory is A. Beurling's description of the invariant subspaces in H^2 in terms of inner functions [Acta Math. **81** (1949), 239–255; MR **10**, 381e]. To do the same thing for the Bergman space L_a^2 has been deemed virtually impossible by many analysts, in view of the fact that the lattice of invariant subspaces is so large, and that the invariant subspaces may have weird properties as viewed from the H^2 perspective. The size of the lattice can be appreciated from the known fact that essentially every operator on separable Hilbert space can be realized as the compression of the Bergman shift on $M \ominus N$, where M and N are invariant subspaces, $N \subset M$.

But a Beurling-type theorem is precisely what the present paper delivers. Given an invariant subspace M in L_a^2, consider the subspace $M \ominus TM$, where T stands for multiplication by z. This makes sense because TM is a closed subspace of M. In Beurling's H^2 case, $M \ominus TM$ is one-dimensional and spanned by an inner function. In the L_a^2 setting, the dimension of $M \ominus TM$ may be arbitrarily large, even infinite. However, with the correct analogous definition of inner functions in L_a^2, all vectors of unit norm in $M \ominus TM$ are L_a^2-inner. Following Halmos, the subspace $M \ominus TM$ is called the wandering subspace of M.

Given an invariant subspace, a natural question is: which collections of elements generate it? In particular, one can ask for the least number of elements in a set of generators. It is known that the dimension of the wandering subspace represents a lower bound for the least number of generators. In the paper, it is shown that any orthonormal basis in the wandering subspace (which then consists of L_a^2-inner functions) generates M as an invariant subspace. This settles the issue of the minimal number of generators.

Let P be the orthogonal projection $M \to M \ominus TM$, and let $L\colon M \to M$ be the operator such that TL is the orthogonal projection $M \to TM$. Then, for $f \in M$, $f = Pf + TLf$. If we do the same for $Lf \in M$, we get $Lf = PLf + TL^2 f$ and, inserting it into the original relation for f, we get $f = Pf + TPLf + T^2L^2f$. As we go on repeating this process, we get $f = Pf + TPLf + T^2PL^2f + \cdots + T^{n-1}PL^{n-1}f + T^nL^nf$. The point with this decomposition is that, apart from the last term, each term is of the form T to some power times an element of $M \ominus TM$, so that $Pf + TPLf + T^2PL^2f + \cdots + T^{n-1}PL^{n-1}f$ is in $[M \ominus TM]$, the invariant subspace generated by $M \ominus TM$. If the operators T^nL^n happened to be uniformly bounded, as they are in the case of H^2, T^nL^nf would tend to 0 in the weak topology, and f

would be in the weak closure of $[M \ominus TM]$, which by standard functional analysis coincides with $[M \ominus TM]$. However, for the Bergman space, it seems unlikely that the T^nL^n are uniformly bounded for all possible invariant subspaces M, although no immediate counterexample comes to mind.

For this reason, the authors try Abel summation instead, and consider for $0 < s < 1$ the operators $R_s\colon M \to [M \ominus TM]$ given by $R_s = P + sTPL + s^2T^2PL^2 + s^3T^3PL^3 + \cdots$; it is easy to show that the series converges in norm. To prove that $M = [M \ominus TM]$, it suffices to check that (a) $\|R_s\| \le C$ for some constant independent of s, and (b) $R_sf \to f$ in the topology of uniform convergence on compacts in the unit disk. This is so because $R_sf \in [M \ominus TM]$ then converges weakly to $f \in M$ as $s \to 1$. Part (b) is easy; the trick is to obtain (a). The authors show at an early stage that for $\lambda \in \mathbf{D}$, $M = (M \ominus TM) + (T - \lambda)M$, and that the sum is direct (in the Banach space sense). They write $Q_\lambda f$ for the skewed projection operator $M \to M \ominus TM$ associated with this decomposition. The operator Q_λ has a convergent Taylor series expansion $Q_\lambda = P + \lambda PL + \lambda^2PL^2 + \lambda^3PL^3 + \cdots$, which very much resembles the expression for R_s. In fact, one sees that $R_sf(w) = Q_{sw}f(w)$. Now one observes that $Q_\lambda f(\lambda) = f(\lambda)$, because the element of $(T - \lambda)M$ which one has to add to $Q_\lambda f$ to obtain f vanishes at the point λ. It follows that $R_sf(w) \to f(w)$ normally in $\mathbf{D}$ as $s \to 1$.

We come to the hard part: obtaining the uniform boundedness of the operators R_s. What is needed is a concrete representation formula for the norm in M which makes it possible to compare the norms of f and R_sf. Here, the factorization theory due to the reviewer [J. Reine Angew. Math. **422** (1991), 45–68; MR 93c:30053], and subsequent improvements by Duren-Khavinson-Shapiro-Sundberg, come to assistance. Let φ be an L^2_a-inner function, and let $f \in [\varphi]$, the invariant subspace generated by φ. The norm of f can then be expressed in terms of the quotient f/φ. That formula suggests the norm identity (1) $\|f\|^2 = \pi^{-1}\int_{\mathbf{D}} \|Q_wf\|^2 dA(w) + \pi^{-2}\int_{\mathbf{D}\times\mathbf{D}} \Gamma(z,w)\Delta_z\Delta_w|Q_wf(z)|^2\,dA(z)\,dA(w)$, which turns out to be valid for general invariant subspaces M, where $\Gamma(z,w)$ is the biharmonic Green function for the unit disk (suitably normalized), which is known to be positive. One first checks that the formula (1) holds for all f in $[M \ominus TM]$. Second, a related integral formula for the norm in M, in terms of integrals along concentric circles, is established for all $f \in M$. Then an intricate argument, involving Green's formula and rather subtle analysis of signs of functions, shows that for $f \in M$, we have at least a $\ge$ inequality in (1). To get the identity (1) for general $f \in M$, it is necessary to show first that $M = [M \ominus TM]$.

Formula (1), which we know to hold for $f \in [M \ominus TM]$ and with $\ge$ for general $f \in M$, applies to $R_sf \in [M \ominus TM]$ with equality, and if we notice that $Q_wR_sf(z) = Q_{sw}f(z)$, we get (2) $\|R_sf\|^2 = \pi^{-1}\int_{\mathbf{D}} \|Q_{sw}f\|^2 dA(w) + \pi^{-2}\int_{\mathbf{D}\times\mathbf{D}} \Gamma(z,w)\Delta_z\Delta_w|Q_{sw}f(z)|^2\,dA(z)\,dA(w)$. An almost radial monotonicity property of $\Gamma(z,w)$ in the w variable then shows that the last term on the right-hand side of (2) has a lim sup as $s \to 1$ bounded by twice the value one gets when s is set equal 1. Moreover, the first term on the right-hand side of (2) has a lim sup as $s \to 1$ which equals what one gets when $s = 1$ is plugged in. Thus, $\limsup_s \|R_sf\| \le 2\|R_1f\| \le 2\|f\|$. It follows that $R_sf \to f$ weakly, which completes the proof. In the paper, it is even shown that $R_sf \to f$ in norm as $s \to 1$. Incidentally, the proof also answers affirmatively one of the conjectures raised by the reviewer concerning polynomial approximation in certain weighted Bergman spaces [in *Linear and com-*

plex analysis. Problem book 3. Part II, 114, Lecture Notes in Math., 1574, Springer, Berlin, 1994].

The above-mentioned theorem is new also in the case when the wandering subspace $M \ominus TM$ is one-dimensional.

The theorem seems to represent a breakthrough in our understanding of the invariant subspaces of the Bergman space. What is desirable and remains to be developed is a better understanding of the wandering subspaces.

Håkan Hedenmalm (Lund)

97j:46033 46E35 42B20 53C23 57N15

Semmes, S. (Houston, TX)

Finding curves on general spaces through quantitative topology, with applications to Sobolev and Poincaré inequalities. (English summary)

Selecta Math. (N.S.) **2** *(1996), no.* 2, 155–295.

It is a little appreciated feature of Euclidean n-space, $n \geq 2$, that every pair of points a, b in it can be joined not only by a straight line segment $[a, b]$—the shortest curve between the points—but by a large family of curves with length comparable to the distance between the points. One such family is obtained when we draw two line segments symmetrically from the points a and b to an $(n-1)$-disk of radius $|a-b|$ that meets the segment $[a, b]$ perpendicularly at its midpoint. From the existence of this curve family one deduces the important Sobolev and Poincaré inequalities that are valid in $\mathbf{R}^n$. This deduction is an abstract procedure that only uses some basic harmonic analysis, and in which the Euclidean structure no longer plays a role.

In this very interesting paper, the author addresses the issue of finding plentiful curve families joining pairs of points in a metric space. This study provides important new examples of metric spaces that support Sobolev-Poincaré inequalities similar to those in $\mathbf{R}^n$. These spaces need not be smooth in any classical sense of the word, yet for many analytic problems they are indistinguishable from Euclidean space.

It is worthwhile to describe more precisely what is meant by plentiful curve families in a metric space. Let (X, d, μ) be a (complete, separable) metric measure space; for simplicity, assume here that X has Hausdorff dimension $n > 0$ and that μ is the Hausdorff n-measure $\mathscr{H}_n$, finite on bounded sets. A family Γ of curves in X is said to be ν-controlled, where ν is a Borel measure on X, if there is a probability measure $d\gamma$ on Γ such that ν dominates the measure $A \mapsto \int_\Gamma \operatorname{length}(\gamma \cap A)\, d\gamma$, for $A \subset X$ a Borel set. Ideally, Γ should be a family of curves joining the given points a and b in X and ν should be absolutely continuous with respect to the background measure $\mu = \mathscr{H}_n$ with a nice density. If Γ is the curve family in $\mathbf{R}^n$ as described in the first paragraph, then Γ is ν-controlled with $d\nu \approx (|a-x|^{(1-n)} + |b-x|^{(1-n)})d\mathscr{H}_n$; the probability measure on Γ is an appropriately normalized angular measure. The main result of the paper gives conditions on X that guarantee the existence of a ν-controlled family Γ of curves from a point a to a point b with all curves contained in a ball of radius comparable to $d(a, b)$ and with $d\nu \approx (d(a, x)^{(1-n)} + d(b, x)^{(1-n)})d\mathscr{H}_n$.

One set of sufficient conditions given in the paper is as follows: X is an orientable topological n-manifold whose local covering and contractibility properties are uniformly controlled. I forgo the precise definitions of the last two conditions here. The first is satisfied if the Hausdorff n-measure in X is nicely distributed, say it obeys the Ahlfors regularity condition: $\mathscr{H}_n(B_R) \approx R^n$ for all balls B_R of radius $R < \operatorname{diam} X$.

The second is satisfied if each ball can be contracted to a point inside a concentric ball with radius enlarged by a fixed factor. These conditions are already quite general (yet more general ones are given in the paper), leading to significant applications. However, the author is careful in pointing out that his methods are quite flexible and that there is room for weakening the assumptions while retaining the same conclusions. It would be a nice task to find such weaker assumptions in an elegant, concise formulation.

There would have been a good alternative for the (purely qualitative) assumption that X be a manifold, for this property is never really used in the paper. In all the arguments, one only needs that X has the local (co-)homological properties of an n-manifold, that is, X is "algebraically" a manifold of correct dimension (cf. Remark A.35 in the paper). This distinction is important both for conceptual reasons and for applications. Many spaces that naturally arise in mathematics, e.g. in the classical decomposition space theory or in the theory of transformation groups, cannot be expected to be manifolds although they often are so-called cohomology manifolds. These examples are relevant in this analytical context as well. For example, the author [Rev. Mat. Iberoamericana **12** (1996), no. 1, 187–275; MR 97e:57025] showed recently that many well-known spaces of the classical decomposition space theory can be realized geometrically so as to satisfy the hypotheses required by this paper.

The curve family Γ joining the fixed points a and b is constructed from the fibers of an appropriate Lipschitz map F from the product space $X \times (0, d(a,b))$ to the n-sphere. The map $F_t = F(\cdot, t)$ should be constant outside a ball $B(a, Cd(a,b))$ for t near 0, and outside a ball $B(b, Cd(a,b))$ for t near $d(a,b)$; moreover, none of the maps F_t should be homotopic to a constant. The required Lipschitz bound is: $\operatorname{Lip} F|_{X\times(\varepsilon,d(a,b)-\varepsilon)} \leq C\varepsilon^{-1}$. The fibers of F are projected to X, and under the Lipschitz conditions, the co-area formula essentially guarantees that a desired curve family Γ can be found inside the image of the projection.

The main task of the paper is to construct the Lipschitz map described above. This is a beautiful, albeit lengthy, story, and the reader is rewarded by numerous interesting details and techniques, many important in their own right and certain to be useful in other contexts. The three assumptions on X mentioned above are used roughly speaking as follows. The controlled covering function allows one to reduce the problem to the case where X lies in some large-dimensional Euclidean space, or to assure that X is well approximated by polyhedra in the spirit of the classical dimension theory; an embedding theorem of P. Assouad [Bull. Soc. Math. France **111** (1983), no. 4, 429–448; MR 86f:54050] is put to use here in an interesting manner. After this, the controlled contractibility is used in a familiar way together with the polyhedral structure of the surrounding or approximating rectilinear space, for maps from polyhedra into (locally) uniformly contractible spaces easily allow for quantitative extensions. The local homological conditions are used to secure a degree theory which then can be used to assure that the obtained maps are homotopically nontrivial.

Arguments of this type were used earlier by G. David and the author [in *Analysis and partial differential equations*, 101–111, Dekker, New York, 1990; MR 91c:30037] in the special case of a quasisymmetrically deformed $\mathbf{R}^n$. In a companion paper the author [Publ. Mat. **40** (1996), no. 2, 411–430; MR 97j:46034] discusses anew this important special case, where a more straightforward approach is possible.

The analytic inequalities that are derived from these considerations have to be reinterpreted, for there is no underlying smooth structure on X. One can define a generalized gradient of a continuous function u in a metric space to be a nonnegative Borel function g such that $|u(x)-u(y)| \le \int_\gamma g$ for each pair of points x and y in X and for each rectifiable curve γ joining these points. Various Poincaré-type inequalities can be formulated in terms of generalized gradients, and their validity is implied by the existence of a thick family of curves between pairs of points as discussed above. The procedure is familiar from $\mathbf{R}^n$: first one averages the inequality $|u(x)-u(y)| \le \int_\gamma g$ over all curves in the family, and then over the points x and y. (I am brushing aside here some interesting technical points related to weak- versus strong-type estimates for function/gradient pairs; cf. Appendix B in the paper under review.)

Of late, several authors have isolated the validity of a Poincaré inequality in a metric space as an important sufficient condition which leads to nontrivial analytic statements without a priori smoothness. The assumptions usually are accompanied by certain mass bounds or minor topological stipulations. For example, Poincaré inequalities without smooth structure play an important role in the theory of quasiconformal mappings, in potential theory on manifolds, and in the study of limit spaces of Riemannian manifolds with curvature bounds. The paper by Semmes is necessary reading for anyone interested in this type of geometric analysis. The reader should not fear the daunting length of the paper, much of which is caused by extremely careful exposition. *Juha Heinonen* (Ann Arbor, MI)

98d:46034 46E35 28A80

Hajłasz, Piotr (Warsaw); **Martio, Olli** (Helsinki)

Traces of Sobolev functions on fractal type sets and characterization of extension domains. (English summary)

J. Funct. Anal. **143** (1997), *no.* 1, 221–246.

The authors investigate trace theory and extension domains for Sobolev spaces in the light of the following result obtained recently by Hajłasz [Potential Anal. **5** (1996), no. 4, 403–415; MR 97f:46050]: If $\Omega \subset \mathbf{R}^n$ is a bounded domain with a Lipschitz boundary and $1 < p \le \infty$ then $u \in W^{1,p}(\Omega)$ if and only if there exists $g \in L^p(\Omega)$, $g \ge 0$, such that

$$|u(x)-u(y)| \le |x-y|(g(x)+g(y))$$

for almost all $x, y \in \Omega$. Because of this characterization, it is possible to extend the notion of a Sobolev space on $\mathbf{R}^n$ to an arbitrary metric space with a measure μ. Thus, if (X,d,μ) is such a metric space with $\mu(X) < \infty$ and $\operatorname{diam} X < \infty$, for $1 < p \le \infty$ the Sobolev space $W^{1,p}(X,d,\mu)$ is defined as the space of all functions $u \in L^p(X,\mu)$ such that $|u(x)-u(y)| \le d(x,y)(g(x)+g(y))$ μ-a.e. for some function $g \in L^p(X,\mu)$. The function g can be regarded as the generalized gradient of u. In the case when Ω is a Lipschitz domain, with $X = \Omega$ and $\mu =$ Lebesgue measure, we have by the above result that $W^{1,p}(X,d,\mu) = W^{1,p}(\Omega)$. However, for an arbitrary Ω we have only the inclusion $W^{1,p}(X,d,\mu) \subset W^{1,p}(\Omega)$. There are some domains Ω for which strict inclusion occurs.

The authors show that if Ω is a bounded Lipschitz domain, then

$$W^{1-1/p,p}(\partial\Omega) \subset W^{1,p}(\partial\Omega, |\cdot|^{1-1/p}, H^{n-1}) \subset W^{1-1/(p-\varepsilon),p-\varepsilon}(\partial\Omega),$$

where H^{n-1} denotes the $(n-1)$-dimensional Hausdorff measure. Along with classical results, this leads to the bounded trace (i.e. restriction) operator

$$T\colon W^{1,p}(\mathbf{R}^n) \to W^{1,p}(\partial\Omega, |\cdot|^{1-1/p}, H^{n-1}).$$

In general, the authors obtain the following trace result. If $0<\lambda<1$, $1\le(n-d)\lambda<p\le n/\lambda$ and μ is a Borel measure supported on a compact set $K\subset\mathbf{R}^n$ such that $\mu(B(x,r))\le Cr^d$ for all $x\in\mathbf{R}^n$ and all $r>0$, then there is a bounded trace operator $T\colon W^{1,p}(\mathbf{R}^n)\to W^{1,dp/(n-\lambda p)}(K,|\cdot|^{1-\lambda},\mu)$, where $|\cdot|^{1-\lambda}(x,y)=|x-y|^{1-\lambda}$. While this theorem yields a very general trace result, the authors point out that when applied to the situation of a bounded Lipschitz domain, it is slightly weaker than the classical result. Indeed, in this case one is led to the trace operator

$$T\colon W^{1,p}(\mathbf{R}^n) \to W^{1,p}_{\mathrm{w}}(\partial\Omega, |\cdot|^{1-1/p}, H^{n-1}),$$

where $W^{1,p}_{\mathrm{w}}$ is defined as above with $L^p(\partial\Omega, H^{n-1})$ replaced by weak-$L^p(\partial\Omega, H^{n-1})$.

The authors also obtain several results on extension operators, the following serving as an important representative. A domain Ω is said to satisfy the $A(c)$-condition, $0<c<1$, if for each $x\in\partial\Omega$ and every $0<r<\operatorname{diam}\Omega$ there exists $y\in\Omega$ such that $B(y,cr)\subset\Omega\cap B(x,r)$. With this definition in place, it is shown that when $1<p\le\infty$ and when a bounded domain $\Omega\subset\mathbf{R}^n$ has the $A(c)$ property, then $u\in W^{1,p}(\Omega)$ is a trace of a $W^{1,p}(\mathbf{R}^n)$ function if and only if $u\in W^{1,p}(\Omega,|\cdot|,H^n)$. The results in this paper concerning extension domains give a different proof of the Jones extension theorem when $p\ne 1$ [P. W. Jones, Acta Math. **147** (1981), no. 1-2, 71–88; MR 83i:30014]. *W. P. Ziemer* (Bloomington, IN)

97h:46078 46H30 46H15 46J05 47A53 47B35 47G10
Böttcher, A. [Böttcher, Albrecht] (Chemnitz); **Gohberg, I.** (Tel Aviv); **Karlovich, Yu.** (Chemnitz); **Krupnik, N.** (Ramat Gan); **Roch, S.** (Leipzig); **Silbermann, B.** (Chemnitz); **Spitkovsky, I.** (Williamsburg, VA)
Banach algebras generated by N idempotents and applications. (English summary)
Singular integral operators and related topics (*Tel Aviv*, 1995), 19–54, *Oper. Theory Adv. Appl.*, 90, *Birkhäuser, Basel*, 1996.

Recall that the effective known symbol calculi for the Fredholm property for algebras of operators, such as the algebras of singular integral operators, of Wiener-Hopf operators, of Toeplitz operators, etc., emerged essentially from local techniques such as the Allan-Douglas local principle, which reads as follows. Let $\mathcal{A}$ be a Banach algebra with identity e, and let $\mathcal{C}$ be a subalgebra of the center of $\mathcal{A}$ which contains e. For each maximal ideal x of the Banach algebra $\mathcal{C}$, let I_x denote the smallest closed two-sided ideal of $\mathcal{A}$ which contains x. Then an element a of $\mathcal{A}$ is invertible if and only if the cosets $a+I_x$ are invertible in the quotient algebra $\mathcal{A}/I_x$ for all maximal ideals of $\mathcal{C}$.

To examine the Fredholm property of operators, one applies the local principle to the quotient algebra modulo the ideal of compact operators. In some cases Banach algebras generated by two or more idempotents may be used instead of local algebras $\mathcal{A}/I_x$; hence it is convenient when these algebras have a symbol.

A Banach algebra $\mathcal{B}$ is said to possess a matrix symbol of order n if there is a family $(f_t)_{t\in T}$ of continuous algebra homomorphisms f_t from $\mathcal{B}$ into the algebra $\mathbf{C}^{l(t)\times l(t)}$, with $l(t)\le n$, such that an element $x\in\mathcal{B}$ is invertible in $\mathcal{B}$ if and only if the matrices $f_t(x)$ are invertible for all $t \in T$. Algebras generated by two idempotents have a rather simple structure: they always have a matrix symbol of order 2 [S. Roch and B. Silbermann, Integral Equations Operator Theory **11** (1988), no. 3, 385–419; MR 89d:47116]. On the other hand, if the number of generating idempotents is greater than two, the algebra may have no symbol. The existence of a matrix symbol for such algebras is a consequence of certain additional hypotheses.

The main result of the article under review consists in the construction of a matrix symbol of order $2N$ for the Banach algebras generated by idempotents $p_1, p_2, \cdots, p_{2N}, P$ such that $p_1 + p_2 + \cdots + p_{2N} = e$, $p_i p_j = \delta_{ij} p_i$, $P(p_{2i-1} + p_{2i})P = (p_{2i-1} + p_{2i})P$, $Q(p_{2i} + p_{2i+1})Q = (p_{2i} + p_{2i+1})Q$, where $Q = e - P$. These algebras naturally arise as local when one investigates the essential spectrum of the Cauchy singular integral operator S_Γ,

$$(S_\Gamma f)(z) := \lim_{\varepsilon \to 0} (\pi i)^{-1} \int_{\Gamma \setminus \Gamma(z,\varepsilon)} f(\zeta)(\zeta - z)^{-1} d\zeta$$

($z \in \Gamma$), which is well defined and bounded in $L^p(\Gamma)$ $(1 < p < \infty)$ if and only if the curve Γ is a Carleson curve, i.e. $\sup_{z \in \Gamma} \sup_{\varepsilon > 0} |\Gamma(z,\varepsilon)|/\varepsilon < \infty$, where $|\Gamma(z,\varepsilon)|$ denotes the measure of $\Gamma(z,\varepsilon) := \{\zeta \in \Gamma\colon\ |\zeta - z| < \varepsilon\}$ [G. R. David, Ann. Sci. École Norm. Sup. (4) **17** (1984), no. 1, 157–189; MR 85k:42026]. Investigating the properties of S_Γ, the authors further suppose that for every point z of the curve Γ there exists $\varepsilon > 0$ such that the connected component of $\Gamma(z,\varepsilon)$ containing z is a substar of some Carleson flower. The number N of petals in the flower determines the number of idempotents $p_1, p_2, \cdots, p_{2N}, P$ in the local algebra, with respect to the point $z \in \Gamma$. The operator S_Γ is realized in this algebra as an abstract analogue of a singular integral operator.

The results obtained by the authors are an important contribution both to the theory of Banach algebras generated by a finite number of idempotents and to Fredholm theory, since they facilitate a wider application of the local principle.

{For the entire collection see MR 97d:47002.} *A. G. Myasnikov* (Moscow)

99k:46094a 46Lxx 11B85 22D25 39B12 42C15 47D25
Bratteli, Ola; Jorgensen, Palle E. T.
Iterated function systems and permutation representations of the Cuntz algebra. (English summary)
Mem. Amer. Math. Soc. **139** (1999), *no.* 663, x+89 pp.

99k:46094b 46Lxx 22D25 39B12 42C15 46N40 47D25
Bratteli, Ola (Oslo); **Jorgensen, Palle E. T.** (Iowa City, IA)
Isometries, shifts, Cuntz algebras and multiresolution wavelet analysis of scale N. (English summary)
Integral Equations Operator Theory **28** (1997), *no.* 4, 382–443.

The main theme of these two articles is the study of some representations of the Cuntz algebra $\mathcal{O}_N$, coming from suitable dynamical systems in the first one and from wavelets in the second one. These studies constitute a generalization and an improvement of work by the authors and G. L. Price [in *Quantization, nonlinear partial differential equations, and operator algebra (Cambridge, MA, 1994)*, 93–138, Proc. Sympos. Pure Math., 59, Amer. Math. Soc., Providence, RI, 1996; MR 97h:46107] and Jorgensen and S. Pedersen [Constr. Approx. **12** (1996), no. 1, 1–30; MR 97c:46091], among others.

Before discussing the content of the articles, let us recall that $\mathscr{O}_N$ is the C^*-algebra generated by N isometries $s_0, s_1, \cdots, s_{N-1}$ satisfying (1) $s_i^* s_j = \delta_{ij}$ and (2) $\sum_{i=0}^{N-1} s_i s_i^* = 1$. It is a simple C^*-algebra, and every system of operators $\{S_0, S_1, \cdots, S_{N-1}\}$ on a Hilbert space $\mathscr{H}$ satisfying relations (1) and (2) determines a representation of $\mathscr{O}_N$.

The study of the representations of $\mathscr{O}_N$ on $\mathscr{H}$ is not only interesting in itself, but also provides endomorphisms of $B(\mathscr{H})$: If $S_0, \cdots, S_{N-1} \in B(\mathscr{H})$ satisfy (1) and (2), then the map $\alpha: A \mapsto \sum_i S_i A S_i^*$ is an endomorphism of $B(\mathscr{H})$, and, conversely, every endomorphism is of this form. Moreover, such an endomorphism is a shift in the sense of Powers if and only if the corresponding representation is irreducible when restricted to the subalgebra $\mathrm{UHF}_N = \{a \in \mathscr{O}_N;\ \gamma_z(a) = a \text{ for all } z \in \mathbf{T}\}$, where γ_z is the automorphism of $\mathscr{O}_N$ defined by $\gamma_z(s_i) = z s_i$.

All representations dealt with are particular cases of the following scheme: $\mathscr{H} = L^2(\Omega, \mu)$, where Ω is a measure space and μ is a probability measure on Ω. Assume that there are N maps $\sigma_i: \Omega \to \Omega$ with the property that (3) $\mu(\sigma_i(\Omega) \cap \sigma_j(\Omega)) = 0$ for $i \neq j$, (4) $\mu(\sigma_i(\Omega)) = 1/N$, so that $\{\sigma_0(\Omega), \cdots, \sigma_{N-1}(\Omega)\}$ is a partition of Ω up to measure zero. Assume furthermore that (5) $\mu(\sigma_i(Y)) = 1/N$ for every measurable subset Y of Ω. Then the σ_i's are injections up to measure zero, and hence it is possible to define an N-to-1 map $\sigma: \Omega \to \Omega$ (well defined up to measure zero) by $\sigma \circ \sigma_i = \sigma_i$ for $i \in \mathbf{Z}_N = \{0, \cdots, N-1\}$. Finally, the announced representations $s_i \mapsto S_i$ of $\mathscr{O}_N$ on $L^2(\Omega, \mu)$ are defined by using N measurable functions $m_0, \cdots, m_{N-1}: \Omega \to \Omega$ with the property that the $N \times N$ matrix (6) $N^{-1/2}(m_i(\sigma_j(x)))_{0 \leq i,j \leq N-1}$ is unitary for almost all $x \in \Omega$. Then, setting (7) $(S_i \xi)(x) = m_i(x) \xi(\sigma(x))$, one gets a representation π of $\mathscr{O}_N$.

For instance, take $\Omega = \mathbf{T}$ with its normalized Haar measure, and set $\sigma_k(e^{2\pi i \theta}) = \exp(2\pi i(\theta + k)/N)$, so that $\sigma(z) = z^N$. Moreover, choose integers $r_0, \cdots, r_{N-1}$ that are pairwise incongruent mod N and define (8) $(S_k \xi)(z) = z^{r_k} \xi(z^N)$, for $\xi \in L^2(\mathbf{T})$. With respect to the natural basis of $L^2(\mathbf{T})$, the corresponding representation of $\mathscr{O}_N$ is permutative in the following general sense: There exists an orthonormal basis $(e_n)_{n \in \mathbf{N}}$ of $\mathscr{H}$ such that (9) $S_k e_n \in \{e_m;\ m \in \mathbf{N}\}$.

The first article under review is mainly devoted to the study of general permutative representations of $\mathscr{O}_N$; it contains the construction of a universal permutative (nonseparable) representation, a detailed analysis of the case $N = 2$ based on arithmetic and combinatorial properties of $\mathbf{Z}$ and other classes of representations associated to pairs $(\mathbf{N}, D)$ where $\mathbf{N}$ is a suitable integer $(\nu \times \nu)$-matrix and where $D \subset \mathbf{Z}^\nu$ plays the role of the r_k's in the above; the associated representation acts on $L^2(\mathbf{T}^\nu)$ and is defined as in (8). Their study requires a self-similar compact subset $\Omega \in \mathbf{R}^\nu$, and several examples are treated.

We now review the second article. The authors start by studying isometries on $L^2(\mathbf{T})$ of the form (10) $(S_m \xi)(z) = m(z) \xi(z^N)$. (Such an operator is an isometry if and only if $N^{-1} \sum_{w: w^N = z} |m(w)|^2 = 1$.) Using the so-called Wold decomposition (into a unitary part and a "shift" part), they prove that the unitary part of S_m is one- or zero-dimensional. Moreover, it is one-dimensional if and only if $|m(z)| = 1$ a.e. and there exist a measurable $\xi: \mathbf{T} \to \mathbf{T}$ and $\lambda \in \mathbf{T}$ such that $m(z)\xi(z^N) = \lambda \xi(z)$ a.e. Furthermore, they are able to characterize representations of $\mathscr{O}_N$ generated by isometries S_{m_i} as in (10), where $m_i = \sqrt{N} \chi_{A_i} u$ with suitable measurable $A_0, \cdots, A_{N-1} \subset$

$\mathbf{T}$ and $u:\mathbf{T}\to\mathbf{T}$: these are representations π^u on $L^2(\mathbf{T})$ for which the elements of $\pi^u(\mathscr{D}_N)''$ are multiplication operators by functions in $L^\infty(\mathbf{T})$ (recall that $\mathscr{D}_N$ is the closed linear span of $\{s_I s_I^*;\ I$ a multi-index set$\}$). Moreover, for specific A_i's, they classify these representations: π^u is equivalent to $\pi^{u'}$ if and only if there exists a measurable function $\Delta:\mathbf{T}\to\mathbf{T}$ such that (11) $\Delta(z)u(z)=u'(z)\Delta(z^N)$ a.e.

Finally, a connection is made between the above representations and wavelets; for instance, it is proved that if S_{m_0} (as in (10)) is a shift in the sense that $\bigcap_{n\geq 1} S_{m_0}^n L^2(\mathbf{T})=0$, then S_{m_0} is a compression of the scaling operator $(U_N\xi)(x)=N^{-1/2}\xi(x/N)$ which acts on $L^2(\mathbf{R})$ and which appears in wavelet analysis of scale N. Unfortunately, it is impossible to give more details on that interesting construction here because in order to do so we would have to rewrite some parts of the article.

Paul Jolissaint (Neuchâtel)

98c:46121 46L05 15A57 47A58 47B15 47C15

Lin, Huaxin (Eugene, OR)

Almost commuting selfadjoint matrices and applications. (English summary)

Operator algebras and their applications (*Waterloo, ON*, 1994/1995), 193–233, *Fields Inst. Commun.*, 13, *Amer. Math. Soc.*, *Providence*, *RI*, 1997.

In this paper Lin solves the problem that an approximate normal matrix is close to a normal matrix in the affirmative. This problem has been open for almost 30 years and has been studied extensively from an operator theory point of view as well as in the context of operator algebras. More precisely, Lin shows the following. Theorem: For any $\varepsilon>0$ there is a $\delta>0$ such that, for any matrix A in the algebra $M_n(\mathbf{C})$ of $n\times n$ complex matrices with $\|AA^*-A^*A\|<\delta$, there is a normal matrix N with $\|N-A\|<\varepsilon$. There is an equivalent similar statement about almost commuting Hermitian matrices: For any two selfadjoint matrices $H,S\in M_n(\mathbf{C})$ with $\|H\|,\|S\|\leq 1$ and $\|HS-SH\|<\delta$ there exist commuting selfadjoint matrices H_1, S_1 with $\|H-H_1\|+\|S-S_1\|<\varepsilon$.

The importance of this result lies in the fact that it holds uniformly in n. Thus it is in fact a statement about operators on infinite-dimensional spaces. There is a simple example in the form of a weighted shift with weights $\min(k/n,1)$, n fixed, which shows that the result does not hold in $B(l^2)$. This suggests the study of this problem in the framework of the C^*-algebra $\mathscr{A}$ that is the weak or c_0-sum of matrix algebras $M_{k(n)}$, $\mathscr{A}=\sum'\oplus M_{k(n)}$. A bounded sequence of nearly normal matrices $A_n\in M_{k(n)}$ with $\|A_nA_n^*-A_n^*A_n\|\to 0$ defines an element in the multiplier algebra $M(\mathscr{A})$ whose canonical image $\pi((A_n))$ in the corona algebra $M(\mathscr{A})/\mathscr{A}$ is normal. Thus this problem is equivalent to the statement that any extension $0\to\mathscr{A}\to\mathscr{E}\to C(X)\to 0$ is splitting. In this paper Lin proves that any normal element $\pi(\mathscr{A})$ in the corona algebra $M(\mathscr{A})/\mathscr{A}$ can be approximated by a normal element with finite spectrum.

This is rather technical and draws heavily on a wide arsenal of C^*-algebra methods; in particular techniques which have been developed for C^*-algebras of real rank zero. Meanwhile Lin's proof has been greatly simplified and extended by P. Friis and M. Rørdam [J. Reine Angew. Math. **479** (1996), 121–131; MR 97i:46097].

Lin's theorem has a number of important consequences. One of these is that any normal element in an AF algebra can be approximated by a normal element with finite spectrum. It also implies I. D. Berg's conjecture which is essentially the above result on a separable Hilbert space with the additional requirement that $A - \lambda$ be of index 0 for all λ [Canad. J. Math. **30** (1978), no. 6, 1152–1165; MR 81a:47029]. If, additionally, $AA^* - A^*A$ is compact, one can even find a normal operator N with $\|A - N\| < \varepsilon$ and $N - A$ compact.

Both viewpoints have been unified in the recent paper by Lin [J. Funct. Anal. **150** (1997), no. 1, 65–74; MR 98g:46084].

{For the entire collection see MR 97f:46002.} *Horst Behncke* (Osnabrück)

99c:46066 46L05 20C99 22D20 46J10 47D25

Pisier, G. (College Station, TX)

The similarity degree of an operator algebra. (English summary)

Algebra i Analiz **10** (1998), *no.* 1, 132–186; *reprinted in St. Petersburg Math. J.* **10** (1999), *no.* 1, 103–146.

This important paper develops several deep themes about the similarity of representations of operator algebras on Hilbert spaces related to Pisier's recent solution of the Halmos problem on polynomially bounded operators (in the negative) and to his paper on polynomially bounded operators. The main idea introduced in this paper is a "similarity setting". Some of the applications of this are mentioned in this review; however, it is essential to read the paper to appreciate its full scope.

In 1950 J. Dixmier [Acta Sci. Math. Szeged **12** (1950), Leopoldo Fejer et Frederico Riesz LXX annos natis dedicatus, Pars A, 213–227; MR **12**, 267a] and M. M. Day [Trans. Amer. Math. Soc. **69** (1950), 276–291; MR **13**, 357e] proved the following result concerning continuous (nonunitary) representations of an amenable group G; B. Szőkefalvi-Nagy [Acta Univ. Szeged. Sect. Sci. Math. **11** (1947), 152–157; MR **9**, 191b] had dealt with the case $G = \mathbf{Z}$ though in different notation. Let G be an amenable locally compact group and let π be a continuous uniformly bounded representation from G into the group of continuous invertible operators on H, with $\|\pi\| = \sup\{\|\pi(g)\|\colon g \in G\}$ the uniform bound on π. Then there is an invertible operator S on H, the "similarity", such that $\theta(g) = S^{-1}\pi(g)S$ $(g \in G)$ is a unitary representation of G on H and $\|S^{-1}\| \cdot \|S\| \le \|\pi\|^2$. Dixmier asked the question whether amenable groups are the only ones in which each representation is similar to a unitary representation. This question is still open; however, Pisier (Theorem 0.1) proves the following strong converse to the full Day-Dixmier theorem for discrete groups G. If there are positive constants K and $\alpha < 3$ such that for each uniformly bounded group representation π of G on H there is an invertible operator S with $\theta(g) = S^{-1}\pi(g)S$ $(g \in G)$ a unitary representation of G and $\|S^{-1}\| \cdot \|S\| \le K\|\pi\|^{\alpha}$, then G is amenable ($K = 1$, $\alpha = 2$ is Day-Dixmier).

"Converses" to the Day-Dixmier theorem for various specific groups were studied by L. Ehrenpreis and F. I. Mautner, R. A. Kunze and E. M. Stein, M. Cowling and others (see Cowling [in *Harmonic analysis and group representations*, 49–128, Liguori, Naples, 1982; MR 86h:22012] for a discussion of these results).

To handle representations of discrete groups and operator algebras with a single set of technical theorems, Pisier introduces a "similarity setting". This is a triple (i, E, A) where E is an operator space, A is an operator algebra in $B(H)$ and i is an injective linear map from E into A with $\|i\|_{\mathrm{cb}} \le 1$ such that A is generated by $i(E)$ as a Banach algebra. The similarity degree d is the greatest lower bound of all numbers $\alpha \ge 0$ satisfying the following condition: there is a constant K such that for each unital homomorphism $\pi: A \to B(H)$ with πi completely bounded from E to $B(H)$ there is an invertible linear operator $S: H \to H$ for which $\theta(e) = S^{-1}\pi i(e)S$ $(e \in E)$ is completely contractive on E and $\|S^{-1}\| \cdot \|S\| \le K\|\pi i\|_{cb}^{\alpha}$. Pisier proves that d is an integer (or infinity), using analyticity and operator-space and -algebra techniques. This leads to the converse of Day-Dixmier mentioned above, to a characterisation of proper uniform algebras in terms of the similarity degree (Theorem 5.1) and to the following result concerning C^*-algebras (Theorem 6.1). If A is a C^*-algebra of similarity degree $d \le 2$, and if π is a representation of A such that $\pi(A)''$ (the von Neumann algebra generated by $\pi(A)$) is semifinite, then $\pi(A)''$ is injective.

This paper is essential reading for those who wish to understand operator space techniques and their links with other parts of analysis.

Allan M. Sinclair (Edinburgh)

99f:46096 46L55 46L40

Katayama, Yoshikazu (Ikeda); **Sutherland, Colin E.** (Sydney); **Takesaki, Masamichi** (Los Angeles, CA)

The characteristic square of a factor and the cocycle conjugacy of discrete group actions on factors.

Invent. Math. **132** (1998), *no.* 2, 331–380.

This very interesting paper is the last step in providing a complete classification of actions of any countable amenable group on any injective separable factor, up to cocycle conjugacy. This subject has a long history, beginning with the pioneering work of A. Connes [Ann. Sci. École Norm. Sup. (4) **8** (1975), no. 3, 383–419; MR **52** #15031; Acta Sci. Math. (Szeged) **39** (1977), no. 1-2, 39–66; MR **56** #6411] on the outer classification of single automorphisms of the type-II_1 injective factor R and the type-II_∞ injective factor R_∞. Later, important progress was achieved by V. F. R. Jones, who found the complete classification of actions of finite groups on R, by introducing a very elaborate cohomological invariant [Mem. Amer. Math. Soc. **28** (1980), no. 237, v+70 pp.; MR 81m:46094]. A further fundamental step is due to A. Ocneanu [*Actions of discrete amenable groups on von Neumann algebras*, Lecture Notes in Math., 1138, Springer, Berlin, 1985; MR 87e:46091]. He succeeded in classifying actions of countable amenable groups on R by a clever use of paving techniques. Afterwards, using groupoid and duality techniques, Sutherland and Takesaki [Pacific J. Math. **137** (1989), no. 2, 405–444; MR 90k:46142] extended this classification result to actions of countable amenable groups on all type-III injective factors except the type-III_1 case, and partial results were obtained by these authors together with Y. Kawahigashi [Acta Math. **169** (1992), no. 1-2, 105–130; MR 93j:46068] in the type-III_1 case. The paper under review now gives the complete solution for the injective type-III_1 case.

Let M be a factor. The authors begin by constructing a universal invariant associated to M, called the characteristic square of M. It is an $\mathrm{Aut}\, M \times \mathbf{R}$ commutative diagram involving the cohomology of the (smooth) flow of weights $(C(M), \theta, \mathbf{R})$ and the subgroups in $\mathrm{Aut}\, M$ of inner automorphisms and centrally trivial ones. An equivalent version, called the intrinsic invariant, is also formulated. It is a cohomology class which originates in spirit from the thesis of Jones.

This universal object is used to construct functorially the invariants associated to a given action of a countable group G on M. A major feature of the cocycle conjugacy classification problem is that it is not smooth. More precisely, the quotient of the Polish group $\mathrm{Act}(G, M)$ of all actions of G on M by the equivalence relation $\mathscr{R}$ defined by cocycle conjugacy is not a standard Borel space. In this paper, the authors adopt a categorical point of view to overcome this difficulty. Instead of working at the level of $\mathrm{Act}(G, M)/\mathscr{R}$, they construct a groupoid desingularisation of this space and a groupoid desingularisation of the space of invariants, and they carefully define the (lifted) invariant associated to a given action, in a Borel and functorial way. In the same spirit, when G is amenable, they construct model actions on the injective type-III_1 factor realising particular given invariants. Then they use Brauer group techniques to compare a given action with a model action.

Finally, combining the type-III_1 classification result with the results mentioned at the beginning of this review (after an appropriate restatement), it is now possible to give the following complete and satisfactory statement: Let G be a countable amenable group and M a separable injective factor; denote by $\mathrm{Aut}_{\mathbf{R}}(C(M))$ the group of automorphisms of $C(M)$ commuting with the flow of weights. Then each action α of G on M determines a normal subgroup $N(\alpha)$ of G, a homomorphism $\operatorname{mod}\alpha$ from G into $\mathrm{Aut}_{\mathbf{R}}(C(M))$ and an element $(\chi(\alpha), \nu(\alpha))$ obtained from the intrinsic invariant of M by pull-back. The element $(N(\alpha), \operatorname{mod}\alpha, (\chi(\alpha), \nu(\alpha)))$ (up to the natural action of $\mathrm{Aut}_{\mathbf{R}}(C(M))$) is a complete invariant for cocycle conjugacy. Moreover, the invariants that occur are identified. Here $N(\alpha)$ is the subgroup of elements $g \in G$ such that α_g is centrally trivial, and $\operatorname{mod}\alpha$ is the Connes-Takesaki module. *Claire Anantharaman-Delaroche* (Orléans)

97a:47006 47-02 34L05 47B25 47E05 81Q10
Simon, Barry (Pasadena, CA)
Operators with singular continuous spectrum. VII. Examples with borderline time decay. (English summary)
Comm. Math. Phys. **176** (1996), *no.* 3, 713–722.

This is the seventh, but certainly not the last, article of Simon on the role of the singular continuous (s.c.) spectrum in quantum mechanical problems. These papers deal essentially with one-dimensional systems, though the main result of the present paper shows that a naive extension of the Simon-Spencer result of high barriers [B. Simon and T. C. Spencer, Comm. Math. Phys. **125** (1989), no. 1, 113–125; MR 91g:81018] is not possible. In all of these papers the following questions on the s.c. spectrum are central: (i) How and when does it arise? (ii) How typical is it? (iii) How does it behave under rank-one or boundary perturbations? (iv) What is the spatial or temporal behaviour of states and how are they localized?

Even though the inverse spectral theorem of I. M. Gel′fand and B. M. Levitan [Amer. Math. Soc. Transl. (2) **1** (1955), 253–304; MR **17**, 489c] allows almost any type of measure as a spectral measure of a Sturm-Liouville problem, s.c. measures have largely been ignored by mathematical physicists until recently. Indeed, at first glance the case against s.c. measures seems rather convincing. The corresponding states have no physical interpretation [M. Reed and B. Simon, *Methods of modern mathematical physics. I. Functional analysis,* Academic Press, New York, 1972; MR **58** #12429a]. They defy scattering theory, because even a tiny rank-one perturbation will move it to an orthogonal position or turn it partially into a point spectrum. Just this for a science where stability properties occupy an eminent position and which has turned perturbation theory into a high art. Their applicability to disordered media or quasicrystals seems rather artificial. Finally, their indirect definition by properties which they do not have does not endear them to the analyst. In this respect they share the fate of s.c. measures of probability theory, which occupy a vast but mostly uncharted terrain.

Thus mathematical physicists have avoided them for a long time and tried rather hard to extend the class of "nice" potentials increasingly further instead of taking all these odds as a challenge, as is done in these papers. A notable exception and forerunner of these articles was the paper of D. B. Pearson [Comm. Math. Phys. **60** (1978), no. 1, 13–36; MR **58** #4076], who gave a rather intuitive way of constructing s.c. measures with sparse bump type potentials. Two typical quantum effects are considered responsible for such strange effects. Besides tunnelling it is essentially the partial reflection of a wave, which even works for wave-particles travelling energetically well above such bumps. A careful arrangement of such bumps can therefore be responsible for ultimate complete reflection—bound states—dispersion and partial reflection, i.e. a plethora of spectral types and behaviour. S.c. spectra can thus be expected essentially from bumpy type potentials, i.e. for almost periodic potentials, random potentials, high barrier and wide barrier potentials. All these classes may lead to s.c. spectra. These properties and considerations apply likewise to the discrete version of the Schrödinger equation, the Jacobi matrices, and there is a well-established correspondence of going from one class to the other.

The most important paper in this series is certainly the first [Part I, Ann. of Math. (2) **141** (1995), no. 1, 131–145; MR 96a:47038], because it introduces the methods of the Baire category theorem to this field and establishes the necessary techniques for handling such problems. In a single sweep Simon shows that an s.c. spectrum arises typically for Jacobi matrices and Schrödinger operators. These Baire category results have been extended to other settings, requiring different approaches and proofs. In the second paper of this series [Part II, Comm. Math. Phys. **165** (1994), no. 1, 59–67; MR 97a:47002] R. Del Rio, N. Makarov and Simon show, for example, that typically rank-one perturbations lead to operators without eigenvalues. Thus if A has no absolutely continuous (a.c.) spectrum, the spectrum of $A_\lambda = A + \lambda P$ will typically be s.c. For quasiperiodic potentials V_ω from the hull Ω of an even quasiperiodic function V the corresponding Hamiltonian will likewise typically have no eigenvalues. Thus again a purely s.c. spectrum is typical, once one member of this class does not have an a.c. spectrum [Part III, S. Jitomirskaya and B. Simon, Comm. Math. Phys. **165** (1994), no. 1, 201–205; MR 97a:47003].

These Baire category results are nonconstructive, however. Thus there is still need to consider specific examples, in particular if one wants to establish additional properties. Simon and G. Stolz [Part V, Proc. Amer. Math. Soc. **124** (1996), no. 7, 2073–2080; MR 97a:47004] studied sparse bump type potentials where the s.c. spectrum is stable with respect to compactly supported perturbations or changes of the boundary condition.

Since s.c. measures occupy this vast no-man's land between discrete and a.c. measures, one can expect examples exhibiting almost either of these types. In the present paper Simon constructs even sparse potentials V with purely s.c. spectral measures ν, which have Hausdorff dimension 1 and are almost a.c. This is done by carefully adjusting the spacing between the bumps. In this way he obtains Hamiltonians H such that $|\langle\varphi, e^{itH}\varphi\rangle| \leq Ct^{-1/2}\ln|t|$ holds for a dense set of states

φ in $\mathscr{L}^2(\mathbf{R})$. With such a potential it is then natural to look at $H=-\Delta+V_1(x_1)+\cdots+V_n(x_n)$ on $\mathscr{L}(\mathbf{R}^n)$. The corresponding spectral measure for simple tensor states $\varphi(x)=\varphi_1(x_1)\cdots\varphi_n(x_n)$ is then the nth convolution power of ν. For $n\geq 3$ one has therefore high barrier Hamiltonians with a purely transient spectrum.

Since rank-one perturbations can destroy s.c. spectra, such spectra can also be created in this way. Thus the analysis of rank-one perturbations has always been of central importance. One reason for this is certainly also that it allows the use of standard mathematical machinery for analyzing measures like Borel transforms or Herglotz functions. While such problems crop up in most of these papers, maybe even only for changes of the boundary condition, they are studied in detail in the second [Part II, op. cit.] and the fourth [Part IV, R. Del Rio et al., Preprint; per bibl.] paper of this series.

{For Part VI see MR 97a:47005.} *Horst Behncke* (Osnabrück)

98d:47002 47-02 43A65 46L05 47Axx 47D25
Pisier, Gilles (College Station, TX)
★Similarity problems and completely bounded maps.
Lecture Notes in Mathematics, 1618.
Springer-Verlag, Berlin, 1996. viii+156 pp. \$38.95. ISBN 3-540-60322-0

This book is devoted to three similarity problems, which we explain below, concerning unitary group representations, contractions on Hilbert space, and $*$-representations on C^*-algebras. For each of them, the solutions that were known (at the time of the writing of the book) are given with complete proofs and the necessary background. Furthermore, many important related questions are discussed throughout the book. As the title suggests, the author emphasizes the link between these problems and the theory of completely bounded maps on operator algebras (or operator spaces). The book is truly remarkable for at least three reasons. First, it gathers in one volume almost all the important results on similarity problems and is therefore a valuable tool for anybody working on similarity problems. Second, the book contains several new or simplified proofs which shed some new light on several aspects of the problems. Third, the reader can find here several new results and complements not available elsewhere. It will undoubtedly become a reference book on the subject.

The first problem concerns group representations. Given a Hilbert space H, we denote by $B(H)$ the C^*-algebra of all bounded linear operators on H. Let G be a locally compact group. Consider a representation $\pi\colon G\to B(H)$ and assume that π is continuous (with respect to the strong operator topology) and uniformly bounded (i.e. $\sup_{g\in G}\|\pi(g)\|<\infty$). Then the problem is the following: does there exist an isomorphism $S\in B(H)$ such that the representation $g\mapsto S^{-1}\pi(g)S$ is unitary? In this case we say that π is unitarizable. The two main results are the following. First, any uniformly bounded continuous representation on G is unitarizable if G is amenable (Dixmier). Second, there exist counterexamples on some nonamenable groups (Ehrenpreis-Mautner). However, the following problem remains open: is G amenable if any uniformly bounded continuous representation on G is unitarizable? After a short proof of Dixmier's theorem, the reader in Chapter 2 finds a construction of non-unitarizable representations on the free group with at least two generators, as well as classical criteria of amenability related to this first similarity problem.

The second problem concerns operator theory and requires more background. This includes completely bounded maps and Wittstock's factorization theorem, dealt with in Chapter 3, as well as classical dilation theory, starting with von Neumann's inequality. This famous inequality is discussed at length in Chapter 1, where three different proofs and related results are provided. Let $T: H \to H$ be a bounded operator on Hilbert space. We say that T is similar to a contraction if there exists an isomorphism $S \in B(H)$ such that $\|S^{-1}TS\| \leq 1$, and the general problem here is to find simple conditions for T to have this property. By von Neumann's inequality, similarity to contraction implies that T is polynomially bounded, i.e. there exists a constant $C > 0$ such that $\|P(T)\| \leq C\|P\|$ $(= C \sup\{|P(z)|\colon\ |z| \leq 1\})$ for any complex polynomial P. In particular, if T is similar to a contraction, then T is power bounded, i.e. $\sup_{n\geq 1} \|T^n\| < \infty$. A longstanding open problem known as the Halmos problem has been: is T similar to a contraction whenever it is polynomially bounded? This question was still open when the book under review was completed but it was recently solved by the author [see J. Amer. Math. Soc. **10** (1997), no. 2, 351–369; MR 97f:47002, see p. C62 in this volume]. In the book, the reader finds examples of power bounded operators not similar to a contraction (the first example is due to Foguel) and proofs of old classical results (Rota, Sz.-Nagy). The main positive result, due to Paulsen, definitely shows the connection between similarity problems and completely bounded maps. Let $A(D)$ be the disc algebra. Assume that T is a polynomially bounded operator and let $u_T: A(D) \to B(H)$ be the (unique) bounded unital homomorphism mapping z to T. Paulsen's theorem asserts that T is similar to a contraction if and only if u_T is completely bounded. This result (and some generalizations) is treated in Chapter 4. Finite-dimensional estimates (Bourgain's theorem) are also considered without proofs.

The third problem concerns operator algebras. Let A be a unital C^*-algebra and let $u: A \to B(H)$ be a bounded unital homomorphism. The problem is the following: is u similar to a $*$-representation, i.e. does there exist an isomorphism $S \in B(H)$ such that the mapping $a \mapsto S^{-1}u(a)S$ is a $*$-representation from A into $B(H)$? This problem, known as the Kadison problem, is still open in general. As above, it is strongly connected to complete boundedness. Namely, u is similar to a $*$-representation if and only if it is completely bounded. The main cases where Kadison's problem has a positive solution are discussed in Chapter 7. It is shown that a bounded unital homomorphism $u: A \to B(H)$ is completely bounded (and hence similar to a $*$-representation) if A is nuclear (Bunce, Christensen) or if u is cyclic (Haagerup). The latter is an outstanding result involving the noncommutative Grothendieck inequality, which is included in the book, and shows that Kadison's problem has a positive solution if $A = B(\mathscr{H})$ for some Hilbert space $\mathscr{H}$.

As we mentioned above, the book also contains results on connected problems. For instance, (bounded) Schur multipliers are studied and characterized in connection with completely bounded maps and Grothendieck's inequality. A whole chapter is also devoted to a generalization of completely bounded maps to a Banach space setting. Analogues of the theorems of Wittstock and Paulsen, due to Pisier himself, are proved. *Christian Le Merdy* (Besançon)

99f:47001 47-02 42A82 46E22 47A45 47A57 47B25 47E05 47F05
Gorbachuk, M. L. (Kiev); **Gorbachuk, V. I.** (Kiev)
★**M. G. Krein's lectures on entire operators. (English summary)**
Operator Theory: Advances and Applications, 97.
Birkhäuser Verlag, Basel, 1997. x+220 pp. \$98.50. ISBN 3-7643-5704-5

In this book the authors present a detailed exposition of M. G. Kreĭn's remarkable theory of entire operators, including applications to a number of classical problems in analysis as well as to the spectral theory of ordinary and partial differential equations. In particular, the Hamburger moment problem, and the Kreĭn problem of extending two important classes of continuous functions from a finite interval to the full line, are discussed in detail.

The book consists of three chapters and three appendices. The three chapters are based primarily on notes of Kreĭn's lectures on the subject which he delivered at the Odessa Pedagogical Institute in 1961. The notes were written by V. M. Adamyan, A. A. Nudel′man and D. Z. Arov, who recalls that about twenty lectures, two hours each, were given to an audience of approximately ten collaborators and graduate students. Kreĭn was invited to give this series of lectures by M. S. Brodskiĭ, who was the chairman of the Mathematics Department of the Odessa Pedagogical Institute at that time.

The authors also draw on other sources, including a number of published papers by Kreĭn himself, two notes by Kreĭn and Saakyan, as well as related papers by M. S. Livšic, H. Langer, Yu. L. Shmul′yan and the authors themselves. In particular, the third appendix is a translation of handwritten notes taken from the Kreĭn archives that are based on his lecture at the Jubilee Session of the Moscow Mathematical Society in 1964.

Chapter 1 is devoted to preliminary material on symmetric operators and the theory of functions. Chapter 2 is devoted to entire operators with deficiency index $(1,1)$. A number of applications are presented in Chapter 3. Appendix 1 treats entire operators with arbitrary deficiency indices (m,m), $1 \le m \le \infty$. Appendix 2 presents applications of the theory of entire operators to the spectral theory of ordinary and partial differential equations.

1. Entire operators: Let A be a closed symmetric operator with dense domain $\mathcal{D}(A)$ in a separable complex Hilbert space $\mathfrak{H}$. Let $\mathfrak{M}_z = (A - zI)\mathcal{D}(A)$ and $\mathfrak{N}_{\bar z} = \mathfrak{H} \ominus \mathfrak{M}_z$. It is well known that the dimension m_+ [resp. m_-] of the so-called defect spaces $\mathfrak{N}_z$ is constant for z in the open upper half-plane $\mathbf{C}_+$ [resp. the open lower half-plane $\mathbf{C}_-$]. The ordered pair (m_+, m_-) is called the deficiency index of A.

A point $z \in \mathbf{C}$ is said to be a point of regular type of an arbitrary operator A with dense domain $\mathcal{D}(A)$ if there exists a constant $k_z > 0$ such that $\|(A - zI)f\| \ge k_z \|f\|$ for every $f \in \mathcal{D}(A)$. It is well known that if A is a closed symmetric operator with dense domain, then every point $z \notin \mathbf{R}$ is a point of regular type of A and hence $\mathfrak{M}_z$ is closed in $\mathfrak{H}$ (i.e., $\overline{\mathfrak{M}_z} = \mathfrak{M}_z$) for every $z \notin \mathbf{R}$. If $a \in \mathbf{R}$ is a point of regular type of A, then every point in the full interval $(a - k_a, a + k_a)$ is a point of regular type of A and $m_+ = m_-$, i.e., A has deficiency index (m,m). Some useful facts on the spectrum of selfadjoint extensions of a symmetric operator A with finite deficiency index (m,m) in such a spectral lacuna $(a - k_a, a + k_a)$ are presented in Section 4 of Chapter 1.

A symmetric operator A is said to be regular if every point $z \in \mathbf{C}$ is a point of regular type of A. Kreĭn introduced and investigated the following very nice and important subclass of symmetric regular operators, which he termed entire operators: A is called an entire operator if A is a closed symmetric operator with dense domain $\mathscr{D}(A)$ for which there exists a fixed subspace M of $\mathfrak{H}$ such that

$$\overline{\mathfrak{M}_z} = \mathfrak{M}_z \quad \text{and} \quad \mathfrak{H} = \mathfrak{M}_z \dotplus M \tag{1}$$

for every point $z \in \mathbf{C}$. Here M is a fixed subspace of $\mathfrak{H}$ which is called a gauge. Clearly, every entire operator is regular and has equal deficiency indices (m,m), and $\dim M = m > 0$ (i.e., A is not selfadjoint).

If $\widetilde{A}$ is a selfadjoint extension of A onto the Hilbert space $\widetilde{\mathfrak{H}} \supset \mathfrak{H}$ and if $\widetilde{E}_\lambda$ is the spectral function of $\widetilde{A}$, then $E_\lambda = P_{\mathfrak{H}} \widetilde{E}_\lambda | \mathfrak{H}$ and $R_z = P_{\mathfrak{H}} (\widetilde{A} - zI)^{-1} | \mathfrak{H}$ are called the generalized spectral function and the generalized resolvent of the symmetric operator A. If the defect numbers m_+ and m_- of A are both positive, then there exist many nonunitarily equivalent selfadjoint extensions $\widetilde{A}$ of A, and a basic problem in spectral theory is to describe the set $\{E_\lambda\}$ or, equivalently, the set $\{R_z\}$.

2. Deficiency index $(1,1)$: If $\dim M = 1$, then one can choose a so-called gauge vector $u \in M$, $u \neq 0$. Then for each $f \in \mathfrak{H}$, there exists a unique scalar function $f_u(z)$ such that

$$f - f_u(z)u \in \mathfrak{M}_z \quad \text{and} \quad f_M(z) = f_u(z)u \in M, \tag{2}$$

i.e., $f_M(z)$ is the projection of f onto M parallel to $\mathfrak{M}_z$. If A is an entire operator, then the function $f_u(z)$ is entire with bounded characteristic in both $\mathbf{C}_+$ and $\mathbf{C}_-$. A number of Kreĭn's results on this class of entire functions are presented in Section 4 of Chapter 1. On the basis of these results, Kreĭn obtained the estimate

$$\int_{-\infty}^{\infty} \int_{-\infty}^{\infty} \frac{\ln^+ \|f_M(z)\|}{(1+|z|)^4} dx\, dy \le \Gamma < \infty$$

for every $f \in \mathfrak{H}$ with $\|f\| = 1$ (see Lemma 2.1 in Chapter 2). He then showed that

$$\nabla_M(z) = \sup \left\{ \frac{\|f_M(z)\|}{\|f\|} \colon f \in \mathfrak{H} \right\}$$

is a log-subharmonic function with growth of at most exponential type. Moreover, its indicator diagram is the smallest interval on the imaginary axis $i\mathbf{R}$ which contains the indicator diagrams of all the functions $f_u(z)$ (see Theorem 5.1 in Chapter 2). The numerical value of the function $\nabla_M(z)^{-1}$ has the following geometric meaning:

$$\nabla_M(z)^{-1} = \min\{\|g-h\| \colon g \in M,\ \|g\| = 1 \text{ and } h \in \mathfrak{M}_z\}.$$

For the considered entire operator A with generalized spectral function $\{E_\lambda\}$,

$$E_\Delta f = \int_a^b f_u(\lambda) dE_\lambda \tag{3}$$

for each $f \in \mathfrak{H}$ and each finite interval $\Delta = [a,b]$ and hence, for every $f, g \in \mathfrak{H}$,

$$(f,g) = \int_{-\infty}^{\infty} f_u(\lambda) \overline{g_u(\lambda)}\, d\sigma(\lambda), \quad \text{where } \sigma(\lambda) = (E_\lambda u, u). \tag{4}$$

Thus the formula $\Phi f = f_u(z)$ serves to define an isometric operator $\Phi \colon \mathfrak{H} \to L_2(\mathbf{R}, d\sigma)$ such that $(\Phi A f)(\lambda) = \lambda (\Phi f)(\lambda)$ for every $f \in \mathscr{D}(A)$; Φ maps $\mathfrak{H}$ onto $L_2(\mathbf{R}, d\sigma)$ if and only if E_λ is an orthogonal spectral function of A, i.e., if and only if the corresponding $\widetilde{A}$ is a selfadjoint extension of A within the Hilbert space $\mathfrak{H}$ ($\widetilde{\mathfrak{H}} =$

$\mathfrak{H}$). Formulas (3) and (4) were obtained by Livšic under the weaker assumption that the function $f_u(x)$ is continuous on $\mathbf{R}$ [C. R. (Doklady) Acad. Sci. URSS (N.S.) **44** (1944), 3–7; MR **6**, 131b].

One of the main problems in the spectral theory of an entire operator A with gauge vector u is to describe the set V_u of all nondecreasing bounded functions $\sigma(\lambda)$ on $\mathbf{R}$ for which formula (4) is valid. For each such σ, the function

$$\omega(z)=\int_{-\infty}^{\infty}\frac{d\sigma(\lambda)}{\lambda-z}=(R_z u,u) \tag{5}$$

belongs to the (Nevanlinna) class R of functions which are holomorphic in $\mathbf{C}_+$ with nonnegative imaginary part in $\mathbf{C}_+$. But more is true: ω belongs to $R_0=\{h\in R\colon \sup_{y>0} y|h(iy)|<\infty\}$.

Kreĭn's formula for the generalized resolvents R_z of a symmetric operator A with deficiency $(1,1)$ is established in Section 6 of Chapter 2. This formula is then used in Section 7 to obtain a description of the set of $\omega(z)=(R_z u,u)$ in terms of a linear fractional transformation

$$T_W[\tau]=(w_{11}\tau+w_{12})(w_{21}\tau+w_{22})^{-1}$$

which maps $\{\tau\in R\cup\{\infty\}\}$ onto $\{\omega(z)=(R_z u,u)\colon R_z$ is a generalized resolvent of $A\}$. This description is valid for symmetric operators which satisfy (1) in some region of the complex plane as well as for entire operators. The entries w_{ij} in the matrix-valued function

$$W(z)=\begin{bmatrix} w_{11}(z) & w_{12}(z)\\ w_{21}(z) & w_{22}(z)\end{bmatrix}$$

are holomorphic in the set of points for which the condition (1) holds. Thus, if A is an entire operator, then $W(z)$ is entire. Moreover, it is $\mathcal{J}_2$-inner with respect to the signature matrix

$$\mathcal{J}_2=\begin{bmatrix} 0 & i\\ -i & 0\end{bmatrix},$$

i.e., $W(z)^*\mathcal{J}_2W(z)\le\mathcal{J}_2$ for $z\in\mathbf{C}_+$ and $W(z)^*\mathcal{J}_2W(z)=\mathcal{J}_2$ for $z\in\mathbf{R}$. Concrete formulas for the elements of $W(z)$ in terms of the functions $\mathscr{D}_k(z)=(d_k)_u(z)$ and

$$\mathscr{E}_k(z)=\int_{-\infty}^{\infty}\frac{\mathscr{D}_k(\lambda)-\mathscr{D}_k(z)}{\lambda-z}d\sigma(\lambda)$$

are given, where $\{d_k\}_{k=1}^{\infty}$ is an arbitrary orthonormal basis for $\mathfrak{H}$. Proofs of these facts and many other nice results are presented in Chapter 2. In particular, this chapter includes an exposition of Kreĭn's famous method of directing functionals. Recall that a function $\Phi(f,\lambda)$ which is linear with respect to f in a linear space $\mathscr{L}\supset\mathscr{D}(A)$ $(\mathscr{L}\ne\mathscr{D}(A))$ is called a directing functional for A if: (1) $\Phi(f,\lambda)$ is a holomorphic function on $\mathbf{R}$ for each fixed $f\in\mathscr{L}$, (2) $\Phi(\cdot,\lambda)\not\equiv 0$ for some $\lambda\in\mathbf{R}$, and (3) the equation $(A-\lambda_0 I)x=f_0$ is solvable for $f_0\in\mathscr{L}$ and $\lambda_0\in\mathbf{R}$ if and only if $\Phi(f_0,\lambda_0)=0$. If Φ is a directing functional for A that is defined on $\mathscr{L}=\mathfrak{H}$, then from (3) it is easy to infer that

$$f_u(\lambda)=\Phi(f,\lambda)/\Phi(u,\lambda),$$

$$(f,g)=\int_{-\infty}^{\infty}\Phi(f,\lambda)\overline{\Phi(g,\lambda)}\frac{d\sigma(\lambda)}{|\Phi(u,\lambda)|^2}.$$

Section 9 of Chapter 2 considers generalized gauge vectors $\widetilde{u}$ which belong to the space $\mathfrak{H}_-$ of continuous linear functionals on the Hilbert space $\mathfrak{H}_+$ of vectors in $\mathscr{D}(A^*)$ with the graph metric scalar product in $\mathscr{D}(A^*)$: $(f,g)_+ = (A^*f, A^*g) + (f,g)$ for $f,g \in \mathscr{D}(A^*)$. Then, the natural extensions $\widehat{R}_z$ of R_z are defined by the relation $(R_z f, g) = (f, \widehat{R}_{\bar{z}} g)$ for $f \in \mathfrak{H}$ and $g \in \mathfrak{H}_-$ and the functions

$$\omega_{\widetilde{u}}(z) = \int_{-\infty}^{\infty} \frac{1+\lambda z}{\lambda - z} \frac{d\sigma(\lambda)}{1+\lambda^2} = (\widehat{R}_z \widetilde{u}, \widetilde{u}) \in R \tag{6}$$

come into play in place of the functions $\omega_u(z) = (R_z u, u) \in R_0$. The nondecreasing function $\sigma(\lambda)$ considered in (6) is now subject to the constraint $\int_{-\infty}^{\infty}(1+\lambda^2)^{-1} d\sigma(\lambda) < \infty$; it may be unbounded.

A parametrization formula in terms of a linear fractional transformation T_W which maps $R \cup \{\infty\}$ onto $\{\omega_{\widetilde{u}}(z) = (\widehat{R}_z \widetilde{u}, \widetilde{u})\} \subset R$ is given for this setting also. Here too, the 2×2 matrix function $W(z)$ will be entire and $\mathscr{J}_2$-inner whenever the set $M \subset \mathfrak{H}_-$ is such that the operator A is entire, in the appropriate sense for this setting.

3. Deficiency index (m,m), $1 \leq m \leq \infty$: The theory of entire operators for the general case when $m = \dim M \leq \infty$ and M is either an ordinary gauge (i.e., $M \subset \mathfrak{H}$) or a generalized gauge (i.e., $M \subset \mathfrak{H}_-$) is presented in Appendix 1. The analysis for $M \subset \mathfrak{H}$ is based on a fundamental 1949 paper by Kreĭn [Ukrain. Mat. Žurnal **1** (1949), no. 2, 3–66; MR **14**, 56d; translated in *American Mathematical Society Translations. Series 2. Vol. 97*, Amer. Math. Soc., Providence, R.I., 1971; MR **42** #5740] which develops the theory for $m < \infty$, and on two short notes by Kreĭn and his student Sh. N. Saakyan which treat the case $m = \infty$. The discussion also exploits the investigations of Langer and Shmul′yan on the integration of operator functions with respect to an operator measure. Finally, the analysis of generalized gauges $M \subset \mathfrak{H}_-$ rests on the work of Shmul′yan on the theory of representations of symmetric operators A with arbitrary deficiency indices (m,m), $1 \leq m \leq \infty$, with respect to a generalized gauge $M \subset \mathfrak{H}_-$.

Let A be an entire operator with gauge $M \subset \mathfrak{H}$ and let $\mathscr{P}(z)$ be the projection operator from $\mathfrak{H}$ onto M that is parallel to $\mathfrak{M}_z$. Then it is shown that

$$E_\Delta f = \int_a^b dE_s \mathscr{P}(s) f, \quad f \in \mathfrak{H} \text{ and } \Delta = [a,b], \quad \text{and}$$

$$(f,g) = \int_{-\infty}^{\infty} (d\sigma(\lambda) f_M(\lambda), g_M(\lambda)),$$

where $f_M(z) = \mathscr{P}(z) f$ and

$$(\sigma(\Delta) f, g) = (E_\Delta f, g) \quad \text{for } f,g \in M. \tag{7}$$

Let $C(M)$ denote the class of holomorphic operator-valued functions $\Omega(z)$ in $\mathbf{C}_+$ whose values are contractive linear operators acting in the Hilbert space M and let $S(A,M)$ denote the class of bounded nondecreasing functions $\sigma(\lambda)$ whose values are bounded linear operators acting in M and which relate to the generalized spectral functions E_λ of the given symmetric operator A by formula (7). Then the set $S(A,M)$ is given by the parametrization formula

$$\int_{-\infty}^{\infty} \frac{d\sigma(\lambda)}{\lambda - z} = [A(z)\Omega(z) + B(z)][C(z)\Omega(z) + D(z)]^{-1}, \quad \Omega \in C(M), \tag{8}$$

where $$\begin{bmatrix} A(z) & B(z) \\ C(z) & D(z) \end{bmatrix} = \begin{bmatrix} w_{11}(z) & w_{12}(z) \\ w_{21}(z) & w_{22}(z) \end{bmatrix} \begin{bmatrix} I_M & I_M \\ iI_M & -iI_M \end{bmatrix}.$$

In Kreĭn's investigations, the block operator function $W(z) = [w_{ik}(z)]$ is called the resolvent matrix of the entire operator under consideration. Formula (8) is obtained for symmetric operators which satisfy property (1) in some region of the complex plane as well as for entire operators. In this more general setting, formula (8) holds for all points z in this region, as does the following important identity:

(9) $$J_M - W(z)J_M W(\zeta)^* = -i(z-\overline{\zeta})G(z)G(\zeta)^*|_M,$$

where $$J_M = \begin{bmatrix} 0 & iI_M \\ -iI_M & 0 \end{bmatrix}, \quad G(z) = \begin{bmatrix} \mathscr{P}(z) \\ -\mathscr{Q}(z) \end{bmatrix}$$

and $\mathscr{Q}(z) = P_M(A-zI)^{-1}(I-\mathscr{P}(z))$. If a point $a \in \mathbf{R}$ belongs to the domain under consideration, then $W(z)$ can be normalized by the condition $W(a) = I_{M\oplus M}$. Then, by formula (9),

$$W(z) = I_{M\oplus M} + i(z-a)G(z)G(a)^*J_M.$$

In particular, this formula holds for every point $a \in \mathbf{R}$ when A is an entire operator, and then choosing $a = 0$, we see that $W(z)$ is an entire J_M-inner operator function which is normalized by the condition $W(0) = I_{M\oplus M}$.

We remark that the cited papers of Kreĭn and Saakyan also consider the connection between the resolvent matrix and the characteristic function of the nonselfadjoint operator which is obtained from the symmetric operator when the point $a \in \mathbf{R}$ is a point of regular type for A. If $a = 0$ is such a point, then $W(1/z)$ is the characteristic function of such an operator. When A is an entire operator and $W(z)$ is normalized by the condition $W(0) = I_{M\oplus M}$, then $W(1/z)$ is the characteristic function of a simple, selfadjoint bounded operator B for which $(I-\lambda B)^{-1}$ is an entire function, i.e., B has only one point of spectrum, at $\lambda = 0$, and this point is not an eigenvalue.

4. Applications: A number of applications of the theory of entire operators with $\dim M = 1$ (both for $M \subset \mathfrak{H}$ and $M \subset \mathfrak{H}_-$) to problems in analysis and prediction theory for (weakly) stationary stochastic processes and processes with (weakly) stationary increments are presented in Chapter 3. Additional applications to the spectral theory of ordinary and partial differential equations, which serve to illustrate the usefulness of the theory in the case that $\dim M = \infty$ and/or $M \subset \mathfrak{H}_-$, are presented in Appendix 2.

The first application in Chapter 3 is to the classical Hamburger moment problem, in which a sequence of numbers $\{s_k\}_{k=0}^{\infty}$ is given and the objective is to find a nondecreasing scalar function $\sigma(\lambda)$ on $\mathbf{R}$ such that

$$s_k = \int_{-\infty}^{\infty} \lambda^k d\sigma(\lambda), \quad k \geq 0.$$

For this problem, a Hilbert space $\mathfrak{H}$ is obtained from the closure of the linear space $\mathscr{L}$ of polynomials $f(\lambda) = \sum_{k=0}^{n} \xi_k \lambda^k$ with respect to the metric based on the (possibly degenerate) inner product

$$(f,g) = \sum_{j=0}^{n} \sum_{k=0}^{n} s_{j+k}\xi_k\overline{\eta}_j \quad \left(g = \sum_{j=0}^{n} \eta_j \lambda^j\right).$$

The operator A of multiplication by λ in $\mathscr{L}$ is symmetric. If the moment problem has more than one solution, then A is an entire operator with $\dim M = 1$. Moreover, $\nabla_M(z)$ is of zero exponential type. Therefore, the resolvent matrix $W(z)$ of the problem is an entire $\mathscr{J}_2$-inner matrix function of minimal exponential type. Thus the general theory of entire operators includes the Hamburger moment problem as a special case and yields the basic results which are known for this problem, including information on the associated orthogonal polynomials.

The second section of Chapter 3 considers the classical Kreĭn problem of extending a continuous function $f(t)$ that is given on the interval $[-a,a]$ with positive-definite kernel $F(t,s) = f(t-s)$ on $[0,a]\times[0,a]$ from $[-a,a]$ to a continuous function $\widetilde{f}(t)$ on the full line with positive-definite kernel $\widetilde{F}(t,s) = \widetilde{f}(t-s)$ on $[0,\infty)\times[0,\infty)$. The section begins by establishing the Kreĭn integral representation formula

$$f(t) = \int_{-\infty}^{\infty} e^{i\lambda t}\,d\sigma(\lambda), \quad |t| \le a, \tag{10}$$

for functions in the given class, where $\sigma(\lambda)$ is a bounded nondecreasing function on $\mathbf{R}$.

A Hilbert space $\mathfrak{H}$ is obtained by completing the linear space $\mathscr{L}$ of functions $\varphi(t)$ which are continuous on $[0,a]$ and vanish on a left neighborhood of a with respect to the metric based on the (possibly degenerate) inner product

$$(f,g) = \int_0^a \int_0^a F(t-s)f(s)ds\overline{g}(t)dt.$$

The operator A_0 that is defined on the set $\mathscr{D}(A_0)$ of all continuously differentiable functions $\varphi \in \mathscr{L}$ with $\varphi(0) = 0$ by the formula $(A_0 f)(t) = if'(t)$ is symmetric. In the case when the Kreĭn problem has more than one solution, the operator A (which is built from A_0 in a standard way) is entire with $\dim M = 1$ and $f_M(z) = \int_0^a e^{izt} f(t)dt$ for $f \in \mathscr{L}$. Moreover, in this instance, the resolvent matrix $W(z)$ for this problem has exponential type a. In particular it is shown that the continuation problem has a nonunique solution if there exists an integral representation (10) for $f(t)$ based on a nondecreasing function $\sigma(\lambda)$ such that the derivative $\sigma'(\lambda)$ of the absolutely continuous component of $\sigma(\lambda)$ meets the condition

$$\int_{-\infty}^{\infty} \frac{\log \sigma'(\lambda)}{1+\lambda^2} d\lambda > -\infty.$$

A number of concrete examples are then considered and another set of conditions (that are both necessary and sufficient) for the existence of a nonunique solution of this extension problem are also presented in this section.

The third section of Chapter 3 presents Kreĭn's results on the extension of unitary and spiral curves in Hilbert space. In Kreĭn's terminology, a continuous vector function x_t, $t \in [a,b]$, in a Hilbert space is called a spiral arc if $B(s,t) = (x_{r+s} - x_r, x_{r+t} - x_r)$ is independent of r for every choice of r, s and t such that $r+s$, r and $r+t$ all belong to $[a,b]$; the vector function x_t is called a unitary arc if the scalar product $F(t,s) = (x_t, x_s)$ depends only on the difference $t-s$ for every choice of s and t in $[a,b]$. If the stated conditions for x_t are met when $[a,b]$ is replaced by $(-\infty,\infty)$, then x_t is referred to as a spiral line in the first case and as a unitary line in the second. Clearly, a stochastic process x_t with weakly stationary increments and correlation function $B(t,s)$ (of the increments) generates a spiral arc or line, whereas a weakly stationary stochastic process with correlation function $f(t) = F(t,0)$ generates a unitary arc or line. The problem of extending the correlation function $f(t)$ of a weakly stationary process from a finite interval to the full line was considered in the preceding section. This section focuses on the problem of extending continuous functions $g(t) = g(-t)$ that are given on the interval $[-a,a]$ with positive definite kernel

$$B(t,s) = g(t-s) - g(t) - g(-s) + g(0)$$

on $[0,a] \times [0,a]$ to the full line, in such a way that the corresponding kernel $B(t,s)$ is positive-definite on $[0,\infty) \times [0,\infty)$. It is shown that every such g admits an integral representation of the form

$$g(t) = \alpha + i\beta t + \int_{-\infty}^{\infty} \left(e^{i\lambda t} - 1 - \frac{i\lambda t}{1+\lambda^2} \right) \frac{d\sigma(\lambda)}{\lambda^2},$$

where $\sigma(\lambda)$ is a nondecreasing function on $\mathbf{R}$ such that $\int_{-\infty}^{\infty} (1+\lambda^2)^{-1} d\sigma(\lambda) < \infty$. Then

$$B(t,s) = \int_{-\infty}^{\infty} \left(\frac{e^{i\lambda t} - 1}{\lambda} \right) \left(\frac{e^{-i\lambda s} - 1}{\lambda} \right) d\sigma(\lambda).$$

In Appendix 2, the authors present their own results on the spectral theory for minimal symmetric operators generated by the second order differential expression

$$(L[u])(x) = u''(x) + (\mathscr{A}u)(x) - q(x)u(x), \quad x \in [0,b], \tag{11}$$

where for each fixed x, $u(x)$ belongs to a separable Hilbert space $\mathscr{U}$, $\mathscr{A}$ is a selfadjoint operator on $\mathscr{U}$ that is bounded from below and $q(x) = q(x)^*$ is a strongly continuous function on $[0,b]$ whose values are bounded linear operators on $\mathscr{U}$. In this application $m = \dim M$ may be finite or infinite and the gauge M can be either ordinary (i.e., $M \subset \mathfrak{H}$) or generalized (i.e., $M \subset \mathfrak{H}_-$). The formulas for the (operator-valued) blocks of the resolvent matrix $W(z)$ are obtained in terms of two special solutions of the equation $L[y] = \lambda y$. When $m = 1$ and $\mathscr{A} = 0$, then the differential expression (11) reduces to the ordinary Sturm-Liouville operator. Under appropriate assumptions on q, the partial differential expression of hyperbolic type,

$$(L[u])(x,y) = \frac{\partial^2 u}{\partial x^2} - \frac{\partial^2 u}{\partial y^2} - q(x,y)u,$$

for $(x,y) \in [0,b] \times \mathbf{R}$, can also be fit into this framework, with $m = \infty$ and $\mathscr{U} = L_2(\mathbf{R})$.

Kreĭn's theory of entire operators is a special branch of his general theory of representations for symmetric operators with deficiency index (m,m). It is intimately connected with the theory of the reproducing kernel Hilbert spaces of entire functions that was developed independently by de Branges in the late fifties and early sixties [see, e.g., L. de Branges, *Hilbert spaces of entire functions*, Prentice Hall, Englewood Cliffs, N.J., 1968; MR **37** #4590], somewhat after Kreĭn's main activities in this direction had stopped. Kreĭn discussed some connections between these two theories in his talks at the Jubilee Session of the Moscow Mathematical Society in 1964 and at the International Congress in Moscow in 1966. The first of these talks is reproduced in Appendix 3. It is an English translation of handwritten notes (by an unknown member of the audience) that were taken from the Kreĭn archives. This appendix gives an excellent picture of the development of the theory of entire operators and the place of this theory in general operator theory. Connections with a number of different problems of mathematical analysis and its applications are discussed, including spectral theory for the string equation and canonical differential systems, prediction theory for weakly stationary stochastic processes and processes with weakly stationary increments, and relations to de Branges' theory of Hilbert spaces of entire functions.

Kreĭn's theory begins with a given symmetric operator A and a fixed gauge subspace M. If A is an entire operator, then the space $\{f_M(z)\colon f \in \mathfrak{H}\}$ is a Hilbert space of entire functions and the resolvent matrix $W(z)$ describes the selfadjoint extensions of the given operator A.

In de Branges' work the central objects of study are the reproducing kernel Hilbert spaces of entire functions $\mathscr{H}(E)$ and $\mathscr{H}(W)$, among others, not symmetric operators. However, it turns out that if P_M denotes the orthogonal projection from $\mathfrak{H}$ onto M and if $\mathscr{P}(z)f = f_M(z)$ and $\mathscr{Q}(z)f = P_M(A-zI)^{-1}(f - f_M(z))$, then, with appropriate choice of inner product (and basis in M), the space $\{\mathscr{P}(z)f\colon f \in \mathfrak{H}\}$ is a de Branges space $\mathscr{H}(E)$ and the space of pairs $\left\{\begin{bmatrix} \mathscr{P}(z)f \\ -\mathscr{Q}(z)f \end{bmatrix} : f\ in \mathscr{H}\right\}$ is a de Branges space $\mathscr{H}(W)$ based on a matrix-valued function $W(z)$ which turns out to coincide with Kreĭn's resolvent matrix. This was shown using methods based on the theory of de Branges spaces in a paper of D. Alpay and H. Dym [Integral Equations Operator Theory **8** (1985), no. 2, 145–180; MR 87h:47022b].

One of the basic problems in de Branges' investigations is the connection between nested sets of spaces $\mathscr{H}(W_t)$ and the theory of canonical differential systems of the form

$$\frac{dW_t(\lambda)}{dt} = i\lambda W_t(\lambda)H(t)\mathscr{J}_2 \quad (H(t) \geq 0)$$

on the finite interval $[0, a]$ and the infinite intervals $[0, \infty)$ or $(-\infty, \infty)$. Kreĭn characterized de Branges' result on the uniqueness of the real normalized solution $H(t)$ (i.e. $\overline{H(t)} = H(t)$ and trace $H(t) = 1$ a.e.) of the inverse monodromy problem for a given real entire $\mathscr{J}_2$-inner monodromy matrix $W(\lambda) = W_a(\lambda)$ as "brilliant".

5. Conclusions: In our opinion, the authors have done an excellent job in presenting a coherent and comprehensive introduction to Kreĭn's fundamental and striking results in the theory of selfadjoint extensions of symmetric operators and their applications. The book is highly recommended to everyone who has an interest in operator theory and classical analysis. It is an ideal source book for a "Topics" course or a student seminar. The prerequisites are modest: some familiarity with Hilbert spaces (at the level of the first few chapters of the well-known book by Akhiezer and Glazman) and complex function theory, and the rewards are manifold. There are a number of misprints, mostly minor, but fortunately they do not detract from the value of this important book. Most of this material was either not available or not easily accessible before. The authors deserve our thanks for bringing it to a wide audience and sharing with us one more facet of the remarkable legacy of M. G. Kreĭn. *Damir Z. Arov and Harry Dym* (Odessa and Rehovot)

97f:47002 47Axx 42B30 47Bxx 47C15
Pisier, Gilles (College Station, TX)
A polynomially bounded operator on Hilbert space which is not similar to a contraction. (English summary)
J. Amer. Math. Soc. **10** (1997), *no.* 2, 351–369.

This paper solves a long-standing problem in operator theory that resolves an investigation begun over fifty years ago. This problem, known as the polynomially bounded operator problem, was the most famous open problem in operator theory except for the invariant subspace problem.

An operator A acting on a Hilbert space $\mathcal{H}$ is called a contraction provided that $\|Ax\| \le \|x\|$ for every vector $x \in \mathcal{H}$; i.e. $\|A\| := \sup_{\|x\|\le 1} \|Ax\| \le 1$. An operator T is similar to a contraction if there exists an invertible operator S on $\mathcal{H}$ such that $\|STS^{-1}\| \le 1$. Contractive linear maps play a special role in operator theory, and thus it is natural to ask for simple analytic criteria for an operator which are equivalent to being similar to a contraction.

In 1947, B. Sz.-Nagy [Acta Univ. Szeged. Sect. Sci. Math. **11** (1947), 152–157; MR **9**, 191b] showed that an operator is similar to a unitary operator if and only if $\sup_{n\in\mathbf{Z}} \|T^n\|$ is finite. If $\|STS^{-1}\| \le 1$, it is evident that $\sup_{n\ge 0} \|T^n\| \le \|S\|\,\|S^{-1}\|$. Operators with this property are said to be power bounded. Subsequently, Sz.-Nagy [Magyar Tud. Akad. Mat. Kutató Int. Közl. **4** (1959), 89–93; MR **21** #7436] showed that every power bounded compact operator is similar to a contraction. This led him to conjecture that every power bounded operator is similar to a contraction.

It turned out that this conjecture is incorrect. S. R. Foguel [Proc. Amer. Math. Soc. **15** (1964), 788–790; MR **29** #2646] soon found a counterexample. Let S be the unilateral shift, that is, the isometry defined on an orthonormal basis e_k, $k \ge 0$, by $Se_k = e_{k+1}$ for all $k \ge 0$. Foguel's operator has the form $R(Q) := \begin{bmatrix} S^* & Q \\ 0 & S \end{bmatrix}$ acting on $\mathcal{H} \oplus \mathcal{H}$, where Q is the projection onto the span of the lacunary sequence $\{e_{3^k}\}$. Operators of this form play an important role in the rest of the story.

To obtain a better conjecture, it is necessary to delve more deeply into the structure of a contraction. The first important step in this direction is a basic inequality due to J. von Neumann [Math. Nachr. **4** (1951), 258–281; MR **13**, 254a]. If T is a Hilbert space operator such that $\|T\| \le 1$, then $\|p(T)\| \le \|p\|_\infty := \sup_{|z|\le 1} |p(z)|$ for every polynomial p. It follows easily that if $\|STS^{-1}\| \le 1$, then $\|p(T)\| \le \|S\|\,\|S^{-1}\|\,\|p\|_\infty$ for every polynomial p. Consequently, any operator similar to a contraction is polynomially bounded, meaning that there is a constant c such that $\|p(T)\| \le c\|p\|_\infty$ for every polynomial p.

A. Lebow [Michigan Math. J. **15** (1968), 397–399; MR **38** #5047] showed that Foguel's operator, which is power bounded but not similar to a contraction is, in fact, not even polynomially bounded. Then P. R. Halmos, in his famous "Ten problems in Hilbert space" paper [Bull. Amer. Math. Soc. **76** (1970), 887–933; MR **42** #5066], reformulated Sz.-Nagy's question as: Is every polynomially bounded operator similar to a contraction? The problem survived in this form for twenty-five years.

Returning to the structure of contractions themselves, we go back to Sz.-Nagy's seminal work on unitary dilations [Acta Sci. Math. Szeged **15** (1953), 87–92; MR **15**, 326d]. He showed that if T is a contraction on a Hilbert space $\mathcal{H}$, then there is a (unique minimal) unitary operator U acting on a Hilbert space $\mathcal{K}$ containing $\mathcal{H}$ such that $\mathscr{K}$ decomposes as a direct sum $\mathscr{K} = \mathscr{H}_+ \oplus \mathscr{H} \oplus \mathscr{H}_-$ so that the corresponding matrix for U has the form $\begin{bmatrix} * & * & * \\ 0 & T & * \\ 0 & 0 & * \end{bmatrix}$.

Clearly, any polynomial $p(U)$ of U has $p(T)$ as its 2, 2 entry. Since von Neumann's inequality follows easily from the spectral theorem for unitary operators, this yields a simple proof for arbitrary contractions.

Von Neumann's inequality and the unitary dilation theorem have important algebraic reformulations. The closure of the polynomials in the sup-norm over the unit disk is the disk algebra $A(\mathbf{D})$, consisting of continuous functions on the closed disk which are analytic on the interior. Now von Neumann's inequality shows that every contraction T yields a contractive representation of $A(\mathbf{D})$ by $\rho_T(h) = h(T)$. Thus the contractive representations of $A(\mathbf{D})$ are in a one-to-one correspondence with the set of contractions by identifying a representation ρ with the operator $\rho(z)$.

A unitary operator determines a C^*-representation of $C(\mathbf{T})$, the space of continuous functions on the unit circle by $\sigma_U(f) = f(U)$ using the continuous functional calculus for normal operators. Again we obtain a one-to-one correspondence between the $*$-representations of $C(\mathbf{T})$ and the set of unitary operators. Thus a unitary dilation U of a contraction T yields a $*$-representation σ_U of $C(\mathbf{T})$ which dilates ρ_T in the sense that ρ_T is recovered by restricting the domain to $A(\mathbf{D})$ and compressing down to the Hilbert space $\mathscr{H}$.

This viewpoint was initiated by W. B. Arveson [Acta Math. **123** (1969), 141–224; MR **40** #6274]. He observed that every representation ρ of an operator algebra $\mathscr{A}$ determines a representation $\rho^{(n)}$ of $\mathscr{M}_n(\mathscr{A})$ into $\mathscr{M}(\mathscr{B}(\mathscr{H})) \simeq \mathscr{B}(\mathscr{H}^{(n)})$ by the natural formula $\rho^{(n)}([A_{ij}]) = [\rho(A_{ij})]$. It is evident that if ρ has a $*$-dilation σ, then $\rho^{(n)}$ has a $*$-dilation $\sigma^{(n)}$. But every $*$-representation is automatically contractive—so it would follow that $\|\rho^{(n)}\| \leq 1$. Define $\|\rho\|_{\rm cb} := \sup_{n\geq 1} \|\rho^{(n)}\|$. Arveson's fundamental dilation theorem states that a representation ρ of a unital subalgebra of a C^*-algebra has a $*$-dilation if and only if $\|\rho\|_{\rm cb} \leq 1$.

Arveson noted that any operator T which is similar to a contraction satisfies $\|\rho_T\|_{\rm cb} < \infty$. For if $\|STS^{-1}\| \leq 1$, then the contractive representation of $\rho_{STS^{-1}}$ is completely contractive. Therefore, if $[p_{ij}]$ is any matrix of polynomials, then

$$\|[p_{ij}(T)]\| = \|S^{-1(n)}[p_{ij}(STS^{-1})]S^{(n)}\| \leq \|S\|\,\|S^{-1}\|\,\|[p_{ij}]\|_\infty,$$

where $S^{(n)}$ denotes an $n \times n$ diagonal matrix with all diagonal coefficients equal to S, and the polynomial matrix is endowed with the sup-norm over the unit circle: $\|[p_{ij}]\|_\infty = \sup_{|z|=1} \|[p_{ij}(z)]\|$. Consequently, we say that an operator T is completely polynomially bounded if there is a constant c such that $\|[p_{ij}(T)]\| \leq c\|[p_{ij}]\|_\infty$ for every matrix of polynomials. In other words, the representation ρ_T of the disk algebra is completely bounded.

V. I. Paulsen [J. Funct. Anal. **55** (1984), no. 1, 1–17; MR 86c:47021] established the first (and now we believe the best) quantitative criterion for an operator T to be similar to a contraction: An operator T is similar to a contraction if and only if it is completely polynomially bounded. Moreover, the similarity S may be chosen so that $\|S\|\,\|S^{-1}\| = \|\rho_T\|_{\rm cb}$.

This result establishes an excellent criterion for similarity that is consistent with a variety of other results in operator theory and operator algebras which have found completely bounded maps to be a fundamental tool. However, it was left open whether such a strong condition needs to be verified. The Halmos question remained open for more than a dozen more years. The remaining story focuses on the search for counterexamples.

An important class of operators related to Foguel's example was introduced by C. Foiaş and J. P. Williams ["On a class of polynomially bounded operators", Preprint; per bibl.] and V. V. Peller [J. Operator Theory **7** (1982), no. 2, 341–372; MR 83i:47019]. These are the operators of the form $R(\Gamma_f)$, where Γ_f is the Hankel operator $[a_{i-j}]$, where $f(z)=\sum_{k\geq 0}a_k z^k$ is an analytic function on the disk. Such operators satisfy the relation $S^*\Gamma_f=\Gamma_f S=\Gamma_{S^*f}$. Using this identity, it is routine to verify that the functional calculus for these operators is given by $p(R(\Gamma_f))=\begin{bmatrix} p(S^*) & \Gamma_{p'(S^*)f} \\ 0 & p(S)\end{bmatrix}$. The diagonal entries are contractive maps. So norm estimates may be confined to the $1,2$ entry. Norms of Hankel operators may be computed using function-theoretic tools since $\|\Gamma_f\|$ equals the distance of f to the coanalytic functions in the L^∞ norm on the circle (this is known as the BMOA norm).

Peller showed that a sufficient condition for $R(\Gamma_f)$ to be polynomially bounded is that $\Gamma_{f'}$ be bounded (f' is in BMOA). J. Bourgain [Israel J. Math. **54** (1986), no. 2, 227–241; MR 88h:47024] showed that this condition implied complete polynomial boundedness, and thus similarity to a contraction. Very recently, A. B. Aleksandrov and Peller [Internat. Math. Res. Notices **1996**, no. 6, 263–275; MR 97c:47006b, see p. C72 in this volume] have completed this picture by showing that this condition is also necessary. This means that there are no counterexamples of this form.

Pisier's crucial insight is that operator-valued functions could yield counterexamples. He considers operators of the form $R(\Gamma_F)=\begin{bmatrix} S^{*(\infty)} & \Gamma_F \\ 0 & S^{(\infty)}\end{bmatrix}$, where $F(z)=\sum_{k\geq 0}A_k z^k$ is an analytic function on the disk with operator-valued Fourier coefficients A_k and $A^{(\infty)}$ denotes the direct sum of countably many copies of A. The Hankel property yields the relation $S^{*(\infty)}\Gamma_F=\Gamma_F S^{(\infty)}$.

He makes use of important properties of the generators of the CAR C^*-algebra. These are operators C_i for $i\geq 1$ satisfying (i) $C_iC_j+C_jC_i=0$ and (ii) $C_iC_j^*+C_j^*C_i=\delta_{ij}I$ for $i,j\geq 1$. It is well known that these equations have a unique solution up to a $*$-isomorphism between the C^*-algebras generated by the solutions. In particular, the norms of matrices with coefficients in the C_i's are independent of this choice. There is a map Λ from l^2 into $\mathscr{B}(\mathscr{H})$ given by $\Lambda(a)=\sum_{i\geq 1}a_iC_i$. This map is isometric but is not completely bounded. This will be exploited in the construction of a counterexample.

Pisier considers the function $F(z)=\sum_{n\geq 1}2^{-n}C_nz^{2^n}$ and other similar functions with lacunary behaviour. By considering the $0,0$ entry of the $1,2$ matrix coefficient of $p(R(\Gamma_F))$, one fairly easily can show that this map is not completely bounded. Thus $R(\Gamma_F)$ is not similar to a contraction. The difficult part is establishing that it is polynomially bounded. An important simplification is to consider the vector-valued function $G(z)=\sum_{n\geq 1}2^{-n}e_nz^{2^n}$, where e_n is an orthonormal basis for l^2 considered as operators from the scalars into Hilbert space. It suffices to show that $R(\Gamma_G)$ is polynomially bounded because of the identity $\Lambda(G)=F$. The reason is that norm estimates involving these functions are related because Λ is bounded, but similarity to a contraction may differ because Λ is not completely bounded.

Pisier's methods are probabilistic and make use of the martingale version of BMO functions. The result is based on inequalities due to Doob, Burkholder, Garsia and others and the Fefferman and Stein theorem on the duality between H^1 and BMO. Recent preprints by J. E. McCarthy ["On Pisier's construction", Preprint; per revr.] and S. V. Kislyakov [Zap. Nauchn. Sem. S.-Petersburg. Otdel. Mat. Inst. Steklov. (POMI) **247** (1997), Issled. po Linein. Oper. i Teor. Funkts. 25, 79–95, 300] provide classical function-theoretic proofs of the same key estimate.

The reviewer and Paulsen [J. Reine Angew. Math. **487** (1997), 153–170; MR 98d:47003, see p. C65 in this volume] have a different approach that establishes the result by explicitly computing a similarity that conjugates $R(\Gamma_G)$ to a contraction. Moreover, the Aleksandrov-Peller results are generalized and simplified to give a complete answer to when $R(\Gamma_G)$ is similar to a contraction if G takes values in $\mathcal{B}(\mathcal{H},\mathcal{K})$ provided that one of $\mathcal{H}$ or $\mathcal{K}$ is finite-dimensional. It turns out that for these examples, polynomial boundedness does imply similarity to a contraction.

Because a finite number of CAR operators can be represented in a matrix algebra, Pisier is also able to obtain quantitative lower bounds for norms of polynomials acting on matrices. Bourgain obtained upper bounds for the complete bound norm of a polynomially bounded matrix in terms of the polynomial bound norm and the dimension of the form $C\|\rho_T\|\log(n+1)$. Pisier's examples provide lower bounds on the order of $c(\|\rho_T\|-1)(\log(n+1))^{1/2}$.

So a famous conjecture in operator theory has finally been laid to rest. These results have further implications, especially related to the structure of operator spaces and to the factorization of operator-valued analytic functions. The interested reader is referred to the papers by Pisier, Aleksandrov and Peller, and the reviewer and Paulsen for further detail. Two excellent references for background on this problem are the monographs by Paulsen [*Completely bounded maps and dilations*, Longman Sci. Tech., Harlow, 1986; MR 88h:46111] and Pisier [*Similarity problems and completely bounded maps*, Lecture Notes in Math., 1618, Springer, New York, 1995; MR 98d:47002, see p. C52 in this volume].

{The result reviewed here was announced in a brief note [C. R. Acad. Sci. Paris Sér. I Math. **322** (1996), no. 6, 547–550; MR 97c:47006a, see p. C72 in this volume].}

Kenneth R. Davidson (Waterloo, ON)

98d:47003 47Axx 47B35 47B37 47Bxx 47C15

Davidson, Kenneth R. (Waterloo, ON); **Paulsen, Vern I.** (Houston, TX)

Polynomially bounded operators. (English summary)

J. Reine Angew. Math. **487** (1997), 153–170.

The paper deals with the similarity problem for bounded operators on Hilbert space and, more particularly, with the celebrated Halmos problem recently solved by Pisier. Let H be a Hilbert space and let $T\in B(H)$ be a bounded operator on H. One says that T is similar to a contraction provided that there exists an invertible operator $S\in B(H)$ such that $\|S^{-1}TS\|\leq 1$. The problem of classifying such operators came to light in a paper from 1955 where Sz.-Nagy showed that if T is power bounded (i.e. $\sup_{n\geq 0}\|T^n\|<\infty$) and compact, then it is similar to a contraction. A few years later, answering a question of Sz.-Nagy, Foguel devised an operator T which is power bounded although not similar to a contraction. Then Sz.-Nagy's question was revised in 1970 by Halmos as follows. First note that if $T\in B(H)$ is similar to a contraction, then there exists a constant $K\geq 1$ such that for any complex polynomial P, (1) $\|P(T)\|\leq K\|P\|$ (with $\|P\|=\sup\{|P(z)|\colon |z|\leq 1\}$). Indeed, this follows from von Neumann's inequality, which asserts that (1) holds with $K=1$ when T is a contraction. An operator T satisfying (1) is said to be polynomially bounded, and Halmos asked whether every polynomially bounded operator is similar to a contraction (see the well-known paper by P. R. Halmos [Bull. Amer. Math. Soc. **76** (1970), 887–933; MR **42** #5066]). This question, known as the Halmos problem, was solved very recently in the negative by G. Pisier, who produced counterexamples [see J. Amer. Math. Soc. **10** (1997), no. 2, 351–369; MR 97f:47002, see p. C62 in this volume].

Let $\mathscr{H}_1, \mathscr{H}_2$ be Hilbert spaces and let S_1, S_2 be the shift operators on the Hardy spaces $H^2(\mathscr{H}_1)$ and $H^2(\mathscr{H}_2)$, respectively. The class of operators on $H = H^2(\mathscr{H}_2) \oplus_2 H^2(\mathscr{H}_1)$ of the form (2) $T = \begin{pmatrix} S_2^* & \Gamma \\ 0 & S_1 \end{pmatrix}$ has played a central role in the study of similarity to a contraction. Indeed, both Foguel's counterexample to Sz.-Nagy's problem and Pisier's counterexample to Halmos's problem are of this form and, in the meantime, many important contributions to the similarity problem were given for operators of that form by Foias and Williams, Peller, Bourgain, and Aleksandrov and Peller. The study of (2) is especially attractive when $\Gamma\colon H^2(\mathscr{H}_1) \to H^2(\mathscr{H}_2)$ is a Hankel operator (i.e. $\Gamma S_1 = S_2^*\Gamma$) since in that case one simply has $P(T) = \begin{pmatrix} P(S_2^*) & \Gamma P'(S_1) \\ 0 & P(S_1) \end{pmatrix}$ for any polynomial P. From now on, T will be an operator of the form (2), with the assumption that Γ is a Hankel operator.

Basically, the paper under review is devoted to such operators in the case when either $\mathscr{H}_1$ or $\mathscr{H}_2$ is finite-dimensional. The great interest of the paper lies in the tractable nature of the examples that are considered and in the remarkable simplicity of the proofs. That simplicity comes partly from the following key idea, explained in the paper. Let D be the (unbounded) derivation operator on $H^2(\mathscr{H}_1)$. Then if ΓD is bounded, the corresponding operator T is similar to a contraction. The two main points of the paper are a new proof of Pisier's counterexample to the Halmos problem which is simpler than the original one, and a generalization of some recent work of Aleksandrov and Peller.

In 1995 A. B. Aleksandrov and V. V. Peller [Internat. Math. Res. Notices **1996**, no. 6, 263–275; MR 97c:47006b, see p. C72 in this volume] showed that in the scalar case (i.e. when $\mathscr{H}_1 = \mathscr{H}_2 = \mathbf{C}$), T is similar to a contraction if and only if it is polynomially bounded. Moreover, if F is a symbol of Γ, this holds if and only if the Hankel operator with symbol F' is bounded (the "if" part is due to Bourgain). In the paper under review, that result is fully extended to the case when $\mathscr{H}_1$ and $\mathscr{H}_2$ are finite-dimensional. Moreover, it is proved that if only one of the two spaces $\mathscr{H}_1$ or $\mathscr{H}_2$ is finite-dimensional, it remains true that T is similar to a contraction if and only if it is polynomially bounded. However, examples (given in the paper) show that in that case, T can be similar to a contraction although the Hankel operator with symbol F' is unbounded. In some sense, these results are optimal since Pisier's counterexample is an operator of the form considered here, but with $\mathscr{H}_1$ and $\mathscr{H}_2$ infinite-dimensional. The following more specific class of operators is considered in the paper. Let $\mathscr{H}_1 = \mathbf{C}$ and let $\mathscr{H}_2 = l^2$, with basis $(e_n)_{n\geq 0}$. Let $(\alpha_n)_{n\geq 0}$ be a sequence in l_2 and let Γ be the Hankel operator with matrix $[\alpha_{i+j}e_{i+j}]_{i,j\geq 0}$. Then it is shown that the corresponding operator T is similar to a contraction (if and only if it is polynomially bounded) if and only if $\sup_{k\geq 0}(k+1)(\sum_{i\geq k}|\alpha_i|^2)^{1/2} < \infty$. Counterexamples to the Halmos problem are then derived from that result by means of intricate manipulations. *Christian Le Merdy* (Besançon)

98b:47011a 47A13 11K70 20C35 28A80 42A20 42C10 46N20 47F05
Pedersen, Steen (Dayton, OH)
Spectral sets whose spectrum is a lattice with a base. (English summary)
J. Funct. Anal. **141** (1996), *no.* 2, 496–509.

98b:47011b 47A13 11K70 20C35 28A80 42A20 42C10 46N20 47F05
Lagarias, Jeffrey C. (Murray Hill, NJ); **Wang, Yang [Wang, Yang[3]]** (Atlanta, GA)
Spectral sets and factorizations of finite abelian groups. (English summary)
J. Funct. Anal. **145** (1997), *no.* 1, 73–98.

Both of these papers deal with a special case of a long-standing conjecture of B. Fuglede concerning tilings of $\mathbf{R}^n$ and basis properties of $L^2(\Omega)$, where Ω is some subset in $\mathbf{R}^n$ of finite positive Lebesgue measure. The papers also both acknowledge suggestions from the reviewer. Fuglede conjectured [J. Functional Analysis **16** (1974), 101–121; MR **57** #10500] that a subset $\Omega\subset\mathbf{R}^n$, of finite positive Lebesgue measure, tiles $\mathbf{R}^n$ by translations by some subset $S\subset\mathbf{R}^n$ if and only if $L^2(\Omega)$ has an orthogonal basis of the form $\{e^{i\lambda\cdot x}\colon \lambda\in\Lambda\}$ for some subset $\Lambda\subset\mathbf{R}^n$. (The functions $e_\lambda(x)=e^{i\lambda\cdot x}$ are restricted to Ω and then Λ is said to be a spectrum.) The tiling is up to measure zero, i.e., it is assumed that the distinct translates $s+\Omega$ and $s'+\Omega$ for $s,s'\in S$ overlap on a set of at most Lebesgue measure zero.

To give the basis for the conjecture, we provide some history: Fuglede's paper resulted from a question raised much earlier (1957) by I. E. Segal, who in turn (later) suggested to the reviewer that the origin is from the desire to understand the candidates for spectral transforms of the partial derivatives $\partial/\partial x_j$, $j=1,\cdots,n$, on bounded open sets $\Omega\subset\mathbf{R}^n$, going back to von Neumann. Segal asked which Ω admit commuting selfadjoint extension operators H_j of the partial derivatives $(1/\sqrt{-1})\partial/\partial x_j$ on the dense subspace $C_c^\infty(\Omega)$ in $L^2(\Omega)$. If commuting extensions exist, we say that Ω has the extension property. In that case, we have the identity $\sqrt{-1}H_jf=(\partial/\partial x_j)f$, $\forall f\in C_c^\infty(\Omega)$, $\forall j=1,\cdots,n$. But Fuglede also showed that this property is very restrictive: the triangle and the disk in $\mathbf{R}^2$ do not have the extension property. Fuglede further showed that the extension property is closely connected to the spectral property, i.e., the presence of a subset $\Lambda\subset\mathbf{R}^n$ such that the exponentials $\{e^{i\lambda\cdot x}\colon \lambda\in\Lambda\}$ form an orthogonal basis for $L^2(\Omega)$. If that is the case, then commuting selfadjoint extensions H_j may be defined by $H_je_{(\lambda_1,\cdots,\lambda_n)}=\lambda_je_{(\lambda_1,\cdots,\lambda_n)}$, $j=1,\cdots,n$, where $e_\lambda:=e^{i\lambda\cdot x}$, $x\in\Omega$. If Ω is further assumed to be connected, the converse also holds. (This involved earlier work of Fuglede, Pedersen, and the reviewer (1982).) The original motivation and justification for the conjecture came from the von Neumann-Stone spectral theorem. To see this, note that if Ω has the extension property then we may define a unitary representation U of the group $\mathbf{R}^n$, acting on the Hilbert space $L^2(\Omega)$, via $U(t_1,\cdots,t_n)=\exp(i\sum_{j=1}^n t_jH_j)$, $t\in\mathbf{R}^n$; and there is a projection-valued measure E on $\mathbf{R}^n$ with values in projections on $L^2(\Omega)$ such that $U(t)=\int_{\mathbf{R}^n}e^{i\lambda\cdot t}E(d\lambda)$. The support of E is the spectrum Λ of U, and the above-mentioned result states that this spectrum Λ must be discrete, the spectral measure E must be atomic, and $E(\{\lambda\})$ must be the one-dimensional projection onto $\mathbf{C}e_\lambda$ if $\lambda\in\Lambda$, and zero if not.

While the spectral theorem had been used this way earlier for one dimension, there was, prior to Fuglede's paper, very little, or nothing, for higher dimensions. In making the connection between (i) the extension property, (ii) the spectral property, and (iii) the tile property for some subset $\Omega\subset\mathbf{R}^n$ (as specified), Fuglede also noticed that (ii)–(iii) make sense for other groups, and are both interesting and nontrivial for finite abelian groups, and even that the problem for the finite cyclic groups $\mathbf{Z}_k = \mathbf{Z}/k\mathbf{Z}$ is closely connected to the problem for $\mathbf{R}^n$; and he stated two results (without proofs) to that effect. The proofs have now been given by Lagarias and Wang in the second paper under review.

It is not immediately clear to the uninitiated that there is any even intuitive connection between properties (ii) and (iii) for general sets $\Omega \subset \mathbf{R}^n$. To make this connection (the conjecture is that they are the same!), note that, in one dimension, the selfadjoint extensions H of $(1/i)\frac{d}{dx}$ on $C_c^\infty(0,1)$ in $L^2(0,1)$ are parameterized by $\mathbf{T}$; specifically, any extension H must have spectrum of the form $a+\mathbf{Z}$ for some $a \in [0,1\rangle$; in fact, a given $H = H_a$ is specified by its domain $\mathscr{D}(H_a) = \{f \in C^1(0,1)\colon f(0) = e^{i2\pi a} f(1)\}$, and moreover, the unitary one-parameter group $U_a(t) = e^{itH_a}$ is induced in the sense of Mackey by the one-dimensional representation $n \mapsto e^{i2\pi a\cdot n}$ of $\mathbf{Z}$. The reviewer [Adv. in Math. **44** (1982), no. 2, 105–120; MR 84k:47024] showed in one dimension that the unitary representation U of $\mathbf{R}$ on $L^2(\Omega)$ arising from a general domain $\Omega \subset \mathbf{R}$ with the extension property must be induced (in Mackey's sense) not just from a character of some lattice Γ in $\mathbf{R}$, but from a representation of Γ which is possibly of higher (finite) dimension. He had an analogous result for $\mathbf{R}^n$, when $n > 1$, but then the rank of Γ could be smaller than n. However, geometric conditions were given, in the 1982 paper, under which $\operatorname{rank}(\Gamma) = n$. But those were restricting assumptions, and the general case for $n > 1$ is to this day not well understood. If $\operatorname{rank}(\Gamma) = n$, then the spectrum of the corresponding commuting tuple $(H_1, \cdots, H_n)$ of selfadjoint operators must have the form $A + \Gamma^\circ = \{a + \xi\colon a \in A,\ \xi \in \Gamma^\circ\}$, where Γ° is the dual lattice and A is a finite subset of $\mathbf{R}^n$ such that distinct points in A correspond to different points in the torus $\mathbf{R}^n/\Gamma^\circ \simeq (\Gamma)\hat{}$ (the compact dual group), under the quotient morphism $\mathbf{R}^n \to \mathbf{R}^n/\Gamma^\circ$. After changing basis in $\mathbf{R}^n$, we note that sets in $\mathbf{R}^n$ of the form $A + \mathbf{Z}^n$ are natural candidates for Λ (spectrum) when Ω has the spectral property with spectrum Λ, and when Ω is a tile with translation set S, i.e., the candidates for translation vectors would have the same form.

While the nature of more general examples of (ii)–(iii) is not known, this setting of quasiperiodicity is reasonable, and it is adopted in the two papers under review. The above-mentioned 1982 paper by the reviewer also showed that, if some open $\Omega \subset \mathbf{R}^n$ has the extension property, then the corresponding unitary representation $U(t)$ of $\mathbf{R}^n$ acts locally on $L^2(\Omega)$ by translation, i.e., $x \mapsto x + t$, when both x and $x + t$ are in Ω. When the point $x + t$ exits from Ω, it must return to Ω through a boundary operator. But the commutativity of the extension operators H_j suggests that the $\mathbf{R}^n$-translations will "wrap around" Ω in a way that makes Ω a tile for some set S of translations, and that S should be related to the spectrum Λ through the boundary operators. Unfortunately the latter are poorly understood, and the two papers under review are based instead on methods from harmonic analysis and finite abelian group combinatorics.

Both papers have a result to the effect that finite sets $A \subset \mathbf{R}^n$, which embed in the torus $\mathbf{R}^n/\mathbf{Z}^n$, will automatically have $\{e_\lambda\colon \lambda \in A + \mathbf{Z}^n\}$ as an orthogonal basis for $L^2(\Omega)$ if they are only assumed orthogonal, and it is further given that the cardinality of A, $\#A$, satisfies $\#A \geq \lambda_n(\Omega)$ (> 0), where $\lambda_n(\Omega)$ is the Lebesgue measure of Ω. The viewpoints and proofs in the two papers are very different, and they are both valuable and independent contributions.

Both papers provide a classification of the subsets $\Omega \subset \mathbf{R}^n$ of positive Lebesgue measure with spectrum Λ of the form $A + \mathbf{Z}^n$. The Pedersen classification is in terms of certain bundles over the torus $\mathbf{R}^n/\mathbf{Z}^n$, whereas the Lagarias-Wang classification is given in terms of an interesting new notion of universal spectrum. Their starting point is sets $\Omega \subset \mathbf{R}^n$ which tile with translation vectors $B + \mathbf{Z}^n$, where B is a subset of $(1/k_1)\mathbf{Z} \times (1/k_2)\mathbf{Z} \times \cdots \times (1/k_n)\mathbf{Z}$, i.e., $B \subset \mathbf{R}^n$ is rational with respect to the unit lattice $\mathbf{Z}^n$. But they also give a necessary and sufficient condition for such a rational periodic set to be a universal spectrum, and it is expressed in terms of set-factorizations $X \oplus Y = G$, where $G = \mathbf{Z}_{k_1} \times \mathbf{Z}_{k_2} \times \cdots \times \mathbf{Z}_{k_n}$ and $X = B \pmod{\mathbf{Z}^n}$.

Palle E. T. Jorgensen (Iowa City, IA)

99f:47008 47A13 46L99 47B20 58B25

Martin, Mircea; Salinas, Norberto (Lawrence, KS)

Flag manifolds and the Cowen-Douglas theory. (English summary)

J. Operator Theory **38** (1997), *no.* 2, 329–365.

One can generalize the classical Grassmann and flag manifolds of subspaces of $\mathbf{C}^n$ to the infinite-dimensional setting by considering closed subspaces of Hilbert spaces, thus obtaining important examples of Banach manifolds. More generally, any C^*-algebra A contains the set $E(A) = \{e \in A\colon\ e^2 = e\}$ of idempotents, and the set $P(A) = \{p \in E(A)\colon\ p = p^*\}$ of projections, as submanifolds under the norm topology. For $A = \mathscr{L}(H)$ (all bounded operators on a Hilbert space H) one recovers the Grassmann manifold over H. The concept of flag can also be adapted to the C^*-algebra setting by considering increasing n-tuples $E_n(A) = \{(e_1, \cdots, e_n) \in E(A)^n\colon\ e_1 \le \cdots \le e_n\}$ and $P_n(A) := E_n(A) \cap P(A)^n$. Here $e_1 \le e_2$ means $e_1 e_2 = e_1 = e_2 e_1$. A crucial fact is that the real Banach manifolds $P(A)$ and $P_n(A)$ possess a natural complex structure, essentially given as a quotient manifold under the "Gram-Schmidt orthonormalization" map $\pi\colon E_n(A) \to P_n(A)$, which becomes a holomorphic submersion. Thus one may consider holomorphic $P_n(A)$-valued maps on a domain $\Omega \subset \mathbf{C}^m$, and one of the goals of the paper is the classification of such maps up to an equivalence relation called congruence. The main motivation is the well-known Cowen-Douglas theory (for $A = \mathscr{L}(H)$), which studies m-tuples of Hilbert space operators in terms of their kernel and range projections depending holomorphically on spectral parameters [M. J. Cowen and R. G. Douglas, Acta Math. **141** (1978), no. 3-4, 187–261; MR 80f:47012].

The present authors consider elements with closed range in a C^*-algebra A, and study the associated kernel and range projections. Given holomorphic families of such elements, the associated projection-valued maps are characterized in terms of a nonlinear first-order differential equation related to the underlying complex structure ($\overline{\partial}$-operator). Next, the Cowen-Douglas class is generalized to commuting m-tuples of elements in A, denoted by $\mathscr{A}_k(\Omega, A)$ ($k \ge 1$ an integer), giving rise to holomorphic $P(A)$-valued maps. It is shown that equivalence in $\mathscr{A}_k(\Omega, A)$ can be reduced to congruence $P \sim Q$ for the corresponding projection-valued maps, i.e., the existence of a unitary $u \in A$ such that $uP(\omega)u^* = Q(\omega)$ for all $\omega \in \Omega$. As for $A = \mathscr{L}(H)$, congruence is closely related to (arbitrarily high) order of contact at each $\omega \in \Omega$, and the main point is to prove that order of contact $\le k$ implies congruence and thus unitary equivalence. This is shown here for "inner" C^*-algebras (including $A = \mathscr{L}(H)$).

In the second part of the paper, these concepts and results are generalized to $P_n(A)$-valued maps with $n \geq 1$. This leads to multi-order classes $\mathscr{A}_{\langle k \rangle}(\Omega, A)$, $\langle k \rangle = (k_1, \cdots, k_n)$, and the flag-type congruence theorem states that, again for "inner" C^*-algebras, two holomorphic families in $P_n(A)$ are (componentwise) congruent if and only if they have multi-order of contact $\leq \langle k \rangle$ on Ω. The natural question of whether such flag-valued maps yield (Cowen-Douglas type) invariants for operator m-tuples is answered in the last section, where it is shown that commuting m-tuples having the holomorphic spanning property with respect to domains $\Omega', \Omega'' \subset \mathbf{C}^m$ [M. J. Cowen, Indiana Univ. Math. J. **35** (1986), no. 4, 855–895; MR 88d:47037] lead to holomorphic 2-flag-valued maps (on $\Omega' \times \Omega''$), and that unitary equivalence between two such m-tuples can be treated by the geometric methods developed here.

Harald Upmeier (Marburg)

98m:47004 47A15
Read, C. J. (Cambridge)
Quasinilpotent operators and the invariant subspace problem. (English summary)
J. London Math. Soc. (2) **56** (1997), *no.* 3, 595–606.

There are a number of theorems and open problems concerning invariant subspaces of bounded linear operators. Studies of the invariant subspaces of certain particular operators (such as Beurling's treatment of the unilateral shift) have interesting connections to classical analysis. The behavior of some operators with respect to some of their invariant subspaces is the essence of structure theorems such as the Jordan canonical form theorem in finite dimensions and the spectral theorem for normal operators on Hilbert space. Von Neumann's double commutant theorem can be viewed as a statement about invariant subspaces of self-adjoint algebras of operators; the study of reflexive algebras provides analogues in the setting of non-self-adjoint algebras.

More specifically, questions concerning the existence of invariant subspaces for particular classes of operators have produced a wealth of interesting theorems and examples. It has now been a quarter of a century since Victor Lomonosov proved his remarkable theorem that an operator has a nontrivial invariant subspace if it commutes with a non-scalar operator which commutes with a nonzero compact operator. This is the main affirmative result for operators on general Banach spaces; on Hilbert space there is also Scott Brown's beautiful theorem establishing the existence of nontrivial invariant subspaces for subnormal operators, which has been much generalized. There are many other existence theorems that have been proven using a variety of techniques.

It is still not known whether there is a bounded linear operator on a Hilbert space which has only the trivial invariant subspaces: this is the famous invariant subspace problem. On Banach spaces, however, there have been several extraordinary constructions of counterexamples. An operator is said to be transitive if it has only the trivial invariant subspaces. The first example was found by P. Enflo [Acta Math. **158** (1987), no. 3-4, 213–313; MR 88j:47006], who established the existence of a transitive operator on a Banach space he constructed as he constructed the operator. Enflo's construction was ingenious and very difficult; its publication was delayed for a number of years as Enflo refined his presentation and referees struggled to check it.

The author of the present paper, C. J. Read, produced a transitive operator on the well-known sequence space l^1 [Bull. London Math. Soc. **17** (1985), no. 4, 305–317; MR 87e:47013]. His construction was also ingenious. It was simpler than Enflo's, and was on the familiar space l^1. The author subsequently made an even more remarkable construction, of an operator that is transitive in a stronger sense than that defined above: he produced a bounded linear operator on l^1 which has no closed invariant sets (much less subspaces) except the trivial two [Israel J. Math. **63** (1988), no. 1, 1–40; MR 90b:47013]. This means that the operator has the property that the orbit of every vector other than 0 under the action of the operator is dense in l^1.

In the present paper the author considers the problem of whether every quasinilpotent operator on a Banach space has a nontrivial invariant subspace. By Gel′fand's spectral radius formula, a quasinilpotent operator can be defined either as one having only 0 in its spectrum or as an operator such that the sequence of nth roots of norms of its nth powers converges to 0. The proof of existence of invariant subspaces for compact operators immediately reduces to the quasinilpotent case (non-quasinilpotent compact operators have eigenspaces). Many people tried to extend M. Hilden's simplified proof of (a slightly weaker form of) Lomonosov's theorem so it would apply to all quasinilpotent operators; Hilden's proof seemed to suggest that such an extension might be possible. On the other hand, most other known techniques for producing invariant subspaces require the operator to have more than one point in its spectrum (and satisfy other conditions, such as growth restrictions on the resolvent). Thus the question of the existence of a transitive quasinilpotent operator on any Banach space has been of interest for quite a while.

This question is answered in the paper under review. Using techniques he introduced in a previous paper [J. London Math. Soc. (2) **34** (1986), no. 2, 335–348; MR 87m:47020], the author constructs a transitive quasinilpotent operator on l^1. As in the previous work of Enflo and of the author, this is a very delicate construction of a sequence of vectors $\{e_i\}$ such that the operator T defined by $Te_i = e_{i+l}$ is the example. Though the construction is not at all easy to verify in detail, the exposition contains a heuristic guide to the proof which is very informative. The proof is nicely divided into a series of lemmas whose statements provide a clear outline of the approach.

Several existence problems remain in addition to the invariant subspace problem on Hilbert space. As the author reminds us in his introduction, it is not known whether there is a transitive operator on a reflexive Banach space. More generally, Lomonosov has pointed out that the examples of Enflo and Read are not adjoints of any operators, so it is therefore conceivable that every operator which is an adjoint has a nontrivial invariant subspace. Also, there are still no counterexamples to other stronger assertions than existence of invariant subspaces for all operators on Hilbert space, such as the hyperinvariant subspace, transitive algebra and reductive algebra problems.

Moreover, it would be very interesting to find a transitive operator which can be more easily described on a familiar space of sequences or functions. The classical Weierstrass theorem on uniform approximation of continuous functions by polynomials is the assertion that the constant function 1 is not in any proper invariant subspace of the operator consisting of multiplication by the independent variable on the space of continuous functions on $[0, 1]$. If T is a transitive operator, then, whenever f is a nonzero vector, for each vector g and $\varepsilon > 0$ there is a polynomial in T

whose value at f is within ε of g. If such a T has any computable form this would yield very interesting approximation theorems.

At the end of the paper the author discusses the relationship between his construction and the open problem of whether there is a commutative Banach algebra which has no nontrivial closed ideals. He states that it can be shown that every ideal other than $\{0\}$ in the uniformly closed algebra of operators generated by the quasinilpotent operator he constructed has strong closure which includes the entire algebra. This operator algebra would have been a counterexample to the Banach algebra question if this result had been obtained for the uniform (rather than strong) closure of every ideal other than $\{0\}$. The author states that this is not the case for his present construction, but that he thinks that some variant should have the stronger property. This seems a very plausible conjecture. *P. Rosenthal* (Toronto, ON)

97c:47006a 47A20 47B35
Pisier, Gilles (College Station, TX)
Un opérateur polynomialement borné sur un Hilbert qui n'est pas semblable à une contraction. (English and French summaries) [A polynomially bounded operator on Hilbert space which is not similar to a contraction]
C. R. Acad. Sci. Paris Sér. I Math. **322** (1996), *no.* 6, 547–550.

97c:47006b 47A20 47B35
Aleksandrov, A. B. (St. Petersburg); **Peller, V. V.** (St. Petersburg)
Hankel operators and similarity to a contraction.
Internat. Math. Res. Notices **1996,** *no.* 6, 263–275.

Pisier has obtained a deep counterexample to the longstanding problem of whether a polynomially bounded operator is similar to a contraction. This resolution is striking both for its contacts with abstract operator space theory (matricial norms and all that) and for its use of hard analysis in the setting of H^1–BMOA duality and martingale Hardy spaces.

It was P. R. Halmos who popularised the problem in 1970 as one of the Ten Problems in Hilbert Space [Bull. Amer. Math. Soc. **76** (1970), 887–933; MR **42** #5066], but the germ of the issue goes back to 1947 at least, when Sz.-Nagy showed that the uniform boundedness of the powers of an invertible operator R and its inverse guarantees similarity to a contraction: $\|S^{-1}RS\| \leq 1$ for some S. Although the condition of one-sided boundedness ($\|R^n\| \leq C$, for $n \geq 0$) gives the same conclusion for compact operators, S. Foguel showed, with a 2×2 upper triangular operator matrix example, that the condition is inadequate in general. However, by virtue of von Neumann's inequality for contractions, operators that are similar to contractions are also polynomially bounded, that is, $\|p(R)\| \leq C\|p\|_\infty$ with $\|p\|_\infty$ the disc algebra norm. Also, A. Lebow has shown that Foguel's example is not polynomially bounded and so Halmos' problem was the natural one.

The polynomial boundedness of an operator R means precisely that it determines a bounded representation ρ of the disc algebra A as an operator algebra. This viewpoint is useful since if R is similar to a contraction then the representation ρ has a stronger property, namely it is similar to a contractive representation. By elementary dilation theory, contractive representations are completely contractive (induced maps on matrix algebras over A are contractive) and so ρ is completely bounded. Drawing on W. B. Arveson's important work on complete boundedness,

V. I. Paulsen [J. Funct. Anal. **55** (1984), no. 1, 1–17; MR 86c:47021] showed that this latter condition—R being "completely polynomially bounded"—is in fact equivalent to similarity to a contraction. This was the first significant involvement of the theory of matricial norms in the problem.

The papers under review focus on the so-called Foguel-Hankel operators (the terminology is due to K. R. Davidson and Paulsen), a class of operators R for which the other side of Halmos' problem, polynomial boundedness (PB), simplifies. These operators are upper triangular 2×2 operators with a Hankel operator Γ in the $1,2$ position and it is simply the boundedness of ΓD, with D differentiation, that is equivalent to PB. The boundedness properties of a (classical) Hankel operator Γ are intimately related to the function theory of its defining function φ, and Peller showed in 1982 that if the derivative φ' lies in BMOA then R is PB. Subsequently J. Bourgain seemed to reduce hope of finding a counterexample to the Halmos question in this direction by showing that this derivative condition implies Paulsen's complete boundedness condition. (Bourgain gave the inequality $\|\Gamma_\varphi D\| \leq C\|\Gamma_{\varphi'}\|$.) And hope seemed extinguished when Aleksandrov and Peller obtained, in the second paper under review, new weak factorisations for analytic functions which lead quickly to the equivalence of φ' in BMOA with the PB condition.

Pisier's fundamental idea was that a counterexample should exist amongst infinite-dimensional variants of the Foguel-Hankel operators, these being analogues which act on operator-valued Hardy spaces. This might not seem particularly hopeful at first sight inasmuch as the matricial and operator-valued theory of Hankel operators echoes the scalar case. However, in the theory of operator spaces it is known that (even Hilbert) spaces can be isometrically isomorphic without being completely isometrically isomorphic, and Pisier was able to embed this discrepancy into the Foguel-Hankel operators to show that there is indeed a gap between polynomial boundedness and complete polynomial boundedness. The particular operator-valued defining function $\varphi(z)$ for Γ has the lacunary form $\sum_{k\geq 0} 2^{-k} z^{2^k} C_k$, where $C_1, C_2, \cdots$ is a familiar sequence of operators satisfying the canonical anticommutation relations (the CAR relations). It is through the use of H^1–BMOA duality (in martingale versions of H^p) and considerable hard analysis that the polynomial boundedness condition is established for the associated operator R. (The details are announced in the first article under review.) The failure of complete polynomial boundedness turns on the fact that whilst the operators $\{C_k\}$ span a Hilbert space ($\|\sum \alpha_k C_k\| = (\sum |\alpha_k|^2)^{1/2}$), the matricial norm structure of this space differs from that of the usual row (or column) Hilbert space. This bad matricial norm structure is reflected when matrix-valued polynomials (whose Fourier coefficients involve matrix realisations of CAR matrices!) are applied to Pisier's Foguel-Hankel operator.

Subsequent to Pisier's breakthrough, Davidson and Paulsen [J. Reine Angew. Math. **487** (1997), 153–170; MR 98d:47003, see p. C65 in this volume] have now given a beautifully clear operator-theoretic analysis of higher-dimensional Foguel-Hankel operators. Moreover, they have quantified and widened the gap (a little) and have obtained further explicit counterexamples with φ non-lacunary. Curiously, however, they show that the operator version of Bourgain's inequality is false.

Returning to the scalar case: the condition of polynomial boundedness for the operator R, which is equivalent to the boundedness of $\Gamma_\varphi D$, reformulates as an inner product condition: for analytic polynomials p, q the quantity $|\langle p'q, \varphi\rangle|$ is bounded by $C\|p\|_\infty \|q\|_1$ for some constant C. It is a familiar fact (Nehari's theorem) that the boundedness of a Hankel operator Γ_ψ is equal to the norm of its defining function ψ as an element of the dual of the Hardy space H^1 (this dual is BMOA). This comes about from the Riesz factorisation $f = gh$ of H^1 functions by H^2 functions. Moreover, even the property of weak factorisation ($f = \sum_k g_k h_k$ with control of norms) leads to the equivalence of Hankel-operator norms with dual space norms, and this idea has been exploited in higher-dimensional situations. Aleksandrov and Peller obtain a remarkable family of seemingly classical weak factorisations for spaces of derivatives of H^p functions, and it is the case $p = 1$ that unlocks the equivalence of the inner product condition with the condition $\varphi' \in$ BMOA. They show that if f has a derivative in H^1 then, with the usual control of norms, $f = g_1'h_1 + \cdots + g_4'h_4$, with $g_k \in H^\infty$, $h_k \in H^1$. Through such weak factorisations Aleksandrov and Peller obtain symbol function criteria for the boundedness of Hankel operators between various Banach spaces of analytic functions. *Stephen Power* (Lancaster)

98j:47017 47A45 46E20 46H25 46M20 47B35
Ferguson, Sarah H. (West Lafayette, IN)
Backward shift invariant operator ranges. (English summary)
J. Funct. Anal. **150** (1997), *no.* 2, 526–543.

The author obtains significant generalizations of results of J. F. Carlson and the reviewer [J. Funct. Anal. **128** (1995), no. 2, 278–306; MR 96a:46129] on the groups $\mathrm{Ext}(H, K)$, where H and K are Hilbert modules over the disk algebra $A(\mathbf{D})$. She proves that if H is an isometric $A(\mathbf{D})$-module and $H_1 = H/H_0$, then H_1 is similar to an isometric module if and only if $\mathrm{Ext}(H_1, H_0) = \{0\}$.

If $\Theta(z)\colon \mathscr{E} \to \mathscr{D}$ is the characteristic operator function of a completely nonunitary contraction operator and if $S(\Theta)$, acting on $K(\Theta)$, is the corresponding Sz.-Nagy-Foias model, this implies that $\mathrm{Ext}(K(\Theta), H^2_{\mathscr{E}}) = \{0\}$ if and only if $S(\Theta)$ is similar to an isometry, and it generalizes a result of Carlson and the reviewer [op. cit.] for the case $\dim \mathscr{E} < \infty$. By a theorem of B. Sz.-Nagy and C. Foias [Acta Sci. Math. (Szeged) **35** (1973), 225–254; MR **53** #3737], the vanishing of $\mathrm{Ext}(K(\Theta), H^2_{\mathscr{E}})$ is thus equivalent to left invertibility of Θ. (The author also asserts that this is equivalent to similarity of a different operator $S(\Theta)$ to an isometry in Corollary 1.5, but its proof is not clear to the reviewer.) The results on the Sz.-Nagy-Foias model show that a Hilbert module K is projective in the category of contractive Hilbert modules over $A(\mathbf{D})$ if and only if K is similar to an isometric module. The "if" portion was proved by Carlson et al. [New York J. Math. **1** (1994/95), 26–38, electronic; MR 95g:46136].

The "invariant operator ranges" in the title, denoted $\mathscr{H}(W; B)$, refer to the ranges of operators $(T_W^{1/2}, \Gamma_B)\colon H^2_{\mathscr{D}} \oplus H^2_{\mathscr{E}} \to H^2_{\mathscr{D}}$, where T_W is a selfadjoint (operator-valued) Toeplitz operator and Γ_B is a bounded Hankel operator. The space $\mathscr{H}(W; B)$ is endowed with the norm $\|(T_W^{1/2} f, \Gamma_B g)\| = \|P(f, g)\|$, where P is the orthogonal projection onto the orthogonal complement of the kernel of $T_W^{1/2} \oplus \Gamma_B$. The author proves that the vanishing of $\mathrm{Ext}(\overline{H^2_{\mathscr{E}}}, \mathscr{H}(W; B))$ is equivalent to extension of the Hankel operator Γ_B as a module map from $L^2_{\mathscr{E}}$ to $\mathscr{H}(W; B)$.

In case $W = I - BB^*$, where B is a contractive, analytic operator function, $\mathscr{H}(W; B)$ is a de Branges-Rovnyak space. In this case, the extension property of Γ_B is equivalent to the existence of analytic operator functions C and D such that $CA + DB = I$, where A is an operator-valued, outer function associated with $I - B^*B$. *Douglas N. Clark* (Athens, GA)

99g:47018 47A45 30E05 46B15 47A15 47A57

Treil, S. (East Lansing, MI)

Unconditional bases of invariant subspaces of a contraction with finite defects. (English summary)

Indiana Univ. Math. J. **46** (1997), *no.* 4, 1021–1054.

The problem investigated in the present paper originates in a celebrated result of L. Carleson about free interpolation in H^∞, which concerns the determination of the sequences $(\lambda_n)_{n\in\mathbf{N}}$ for which the restriction operator $f\mapsto(f(\lambda_n))_n$ maps H^∞ onto the sequence space l^∞. Carleson [Ann. of Math. (2) **76** (1962), 547–559; MR **25** #5186] proved that they are characterized by a simple analytic condition, namely

$$\inf_{n\in\mathbf{N}}\prod_{m\neq n}\frac{\lambda_m-\lambda_n}{1-\bar\lambda_m\lambda_n}>0.$$

This fact can be rephrased in the language of the geometry of Banach spaces. Here there are several notions that generalize linear independence to the infinite-dimensional case. Suppose (x_n) is a sequence of vectors in a Banach space X; it is called an unconditional basis if any vector $x\in X$ can be uniquely written as an unconditionally convergent series $x=\sum_n a_nx_n$. In a Hilbert space (which is the situation we will be interested in) this notion is equivalent to that of Riesz basis, which is the transform of an orthogonal basis by an invertible operator. A weaker condition that can be imposed on the linearly independent sequence (x_n) is the requirement that the projections P_k, defined on the linear span of (x_n), and which map $\sum_n a_nx_n$ to a_kx_k, are uniformly bounded. Such a sequence is said to be uniformly minimal, and it is easy to show that, although any sequence that is a Riesz basis in its linear span is uniformly minimal, the converse is not necessarily true.

Now, Carleson's result can be rephrased by stating that the last difference does not exist in the case of a special type of sequence, formed by reproducing kernels in the Hilbert space H^2. More precisely, defining

$$k_\lambda(z)=\frac{(1-|\lambda|^2)^{1/2}}{1-\bar\lambda z},$$

the interpolation theorem stated above is equivalent to the fact that, if $(\lambda_n)_n$ is a sequence of complex numbers in the open unit disc, then the sequence (k_{λ_n}) of elements of H^2 is uniformly minimal if and only if it is a Riesz basis in its span.

The functions $k_\lambda(z)$ happen to be also eigenvectors of the adjoint S^* of the operator S of multiplication by the variable z. This opens a path for the generalization of Carleson's theorem. One can consider, instead of one-dimensional invariant spaces, general invariant subspaces; there exist immediate extensions of the notions of Riesz basis and of uniform minimality to sequences of subspaces. Also, since the condition actually is to be verified on the span of the functions, not on the whole space, we may replace H^2 with a subspace that is invariant under the adjoint of multiplication.

It is at this point that the function context may be forgotten, and all results can be stated using only the language of contractions on Hilbert spaces. The restriction of S^* to an invariant subspace is a rather general type of completely nonunitary contraction with defect indices $(1,1)$. V. I. Vasjunin [Trudy Mat. Inst. Steklov. **130** (1978), 5–49, 223; MR 80b:47041] proved that, for an operator with defect indices $(1,1)$, a complete system of invariant subspaces is a Riesz basis in its span if and only if it is uniformly minimal. The extension to larger defect spaces has been investigated by the author in a series of papers [Dokl. Akad. Nauk SSSR **288** (1986), no. 2, 308–312; MR 87g:47023; in *Toeplitz operators and spectral function theory*, 209–280, Birkhäuser, Basel, 1989; MR 91j:47036; Algebra i Analiz **1** (1989), no. 6, 200–234; MR 91f:47014]. The present one solves the problem completely, by proving the following theorem: Suppose T is a completely nonunitary Hilbert space contraction with finite defect indices. Then a system of T-invariant subspaces is a Riesz basis in its linear span if and only if it is uniformly minimal.

This result seems to be definitive, since the absence of the unitary part, and the finite dimensionality of the defect spaces, are necessary conditions. Although function theory has somehow disappeared in the statement, it is strongly used in the proof. A number of highly nontrivial analytic facts appear as ingredients; they are related to the fine structure of the space of bounded analytic functions on the unit disc and its vector-valued generalizations. One may expect that some of the techniques developed by the author might find further applications. As it stands, the paper is a masterly piece of work; the mixture of operator theory and function theory gives it a distinctive flavour. *Dan Timotin* (Atlanta, GA)

99i:47023 47A55 47B25 47N50 81Q10 81Q15
Agmon, Shmuel (Jerusalem)
A perturbation theory of resonances.
Comm. Pure Appl. Math. **51** (1998), *no.* 11-12, 1255–1309.

Quantum resonances describe states with certain rest energies and rates of decay, consequently providing a more realistic model than bound states, which have no decay. They arise in many settings, from the dynamical study of chemical reactions to the theory of automorphic forms. Although the study of resonances has a very long tradition, many basic problems remain open (see for instance the reviewer's elementary survey [Notices Amer. Math. Soc. **46** (1999), no. 3, 319–328] and references cited there).

The definition of resonances usually proceeds via some form of analytic Fredholm theory, and the ideal solution is an "effective meromorphic continuation" where, due to some deformation or change of Hilbert space, the original operator becomes Fredholm in some range of complex energies z, that is, $P-z$ has finite-dimensional kernels and cokernels. In other words, $(P-z)^{-1}$ is meromorphic with poles of finite rank. Perturbation of resonances can then be studied in the framework of perturbation theory of non-self-adjoint operators. In many situations such "effective continuation" is not available even though we can continue the resolvent meromorphically as an operator $R(\lambda)=(P-\lambda)^{-1}\colon B_0\to B_1$, where B_0 is some smaller and B_1, some larger space; for instance, for perturbations on $\mathbf{R}^n$, we can often take $B_0 = e^{-|x|^2}L^2(\mathbf{R}^n)$, and $B_1 = e^{|x|^2}L^2(\mathbf{R}^n)$.

Professor Agmon introduces a brilliantly simple new method for turning P into an operator with discrete spectrum, under the assumption of the meromorphy of $R(\lambda)$ alone. Roughly speaking, this is done as follows. Let Γ be the boundary of Ω and let

$$B_0 \subset B_\Gamma \stackrel{\text{def}}{=} \left\{g + \int_\Gamma R(\zeta)\Phi(\zeta)d\zeta\colon\ g\in B_0,\ \Phi\in C(\Gamma;B_0)\right\} \subset B_1,$$

which, equipped with a natural norm, is a Banach space. Then $P: B_\Gamma \to B_\Gamma$ is an unbounded operator with discrete spectrum in Ω coinciding, with multiplicities, with the poles of $R(\lambda)$. This is so because the continued resolvent, $R(\lambda)$, can be applied to the elements of B_Γ through a formal use of the resolvent identity $(R(\zeta)R(\lambda) = (R(\lambda) - R(\zeta))/(\lambda - \zeta)$, which is, however, illegal as $R(\lambda): B_0 \to B_1$ and $R(\zeta)$ is not defined on B_1):

$$R(\lambda)\left(g + \int_\Gamma R(\zeta)\Phi(\zeta)d\zeta\right) = R(\lambda)g + \int_\Gamma \frac{R(\lambda) - R(\zeta)}{\lambda - \zeta}\Phi(\zeta)d\zeta,$$

where the right-hand side is well defined. One can then show that $R(\lambda): B_\Gamma \to B_\Gamma$ and consequently that it is the resolvent of $P: B_\Gamma \to B_\Gamma$ with an appropriately chosen domain. Since we assumed the meromorphy of $R(\lambda)$ with poles of finite rank, we obtain the statement about the spectrum.

If the operator P is changed by a "small" operator which takes values in B_0, then the same space B_Γ can be used for the perturbed operator (the definition of B_Γ depends on P), which then has discrete spectrum and a meromorphic resolvent. Hence the perturbation theory of resonances is reduced to the perturbation theory of eigenvalues of non-self-adjoint operators. This allows an immediate application of the results already available in the classical monograph of T. Kato [*Perturbation theory for linear operators*, Reprint of the 1980 edition, Springer, Berlin, 1995; MR 96a:47025]. It also allows a new, abstract approach to the celebrated "Fermi golden rule" which describes the (second) variation of an embedded eigenvalue dissolving into a resonance.

An example of a new application of Agmon's approach is given by considering generic simplicity of resonances. When "effective meromorphic continuation" is achieved by complex scaling (for the most up-to-date version see J. Sjöstrand [in *Microlocal analysis and spectral theory (Lucca, 1996)*, 377–437, Kluwer Acad. Publ., Dordrecht, 1997; MR 99e:47064, see p. C103 in this volume] and references cited there), a generic compactly supported perturbation makes the resonances simple, as shown by F. Klopp and the reviewer [Helv. Phys. Acta **68** (1995), no. 6, 531–538; MR 97j:81067]. This however does not apply to the case of "conformally compact manifolds", which are a generalization of infinite-volume hyperbolic quotients, despite the meromorphy of the resolvent (established in greatest generality by R. R. Mazzeo and R. B. Melrose [J. Funct. Anal. **75** (1987), no. 2, 260–310; MR 89c:58133]). The methods of the paper under review allow a resolution of this problem in this and many other situations: a generic compactly supported metric or potential perturbation makes all the resonances simple.

In the case of asymptotically hyperbolic manifolds this was in fact achieved in a recent preprint of D. Borthwick and P. Perry ["Scattering poles for asymptotically hyperbolic manifolds", http://xxx.lanl.gov/abs/math.SP/9903118].

{Editorial note: A brief erratum concerning the proof of Theorem 6.2 has appeared [S. Agmon, Comm. Pure Appl. Math. **52** (1999), no. 12, 1617–1618].}

Maciej Zworski (Berkeley, CA)

97c:47018 47A57 44A60 47B20
Putinar, Mihai (Riverside, CA)
Extremal solutions of the two-dimensional L-problem of moments. (English summary)
J. Funct. Anal. **136** (1996), *no.* 2, 331–364.

Let f be a measurable function on $\mathbf{R}$ satisfying $0 \le f \le L$ a.e. with respect to Lebesgue measure and such that its (power) moments

$$a_n := \int_{\mathbf{R}} x^n f(x) dx \tag{1}$$

exist for every $n \ge 0$. Consider the formal power series identity $\exp(L^{-1}(a_0 X + a_1 X^2 + \cdots)) = 1 + b_0 X + b_1 X^2 + \cdots$. A straightforward calculation shows that the sequence $\{b_n\}_{n=0}^{\infty}$ is positive semi-definite on $\mathbf{R}$; conversely, the positive semi-definiteness of $\{b_n\}_{n=0}^{\infty}$ is sufficient to guarantee the existence of a measurable function f with $0 \le f \le L$ and whose moments are given by the a_n's.

When the sequence $\{b_n\}$ is degenerate, i.e., $\det((b_{i+j})_{0 \le i,j \le N}) = 0$ for some $N \ge 1$, the associated solution of the L-moment problem (1) is the characteristic function of a union of at most $(N-1)$ disjoint, bounded intervals. These degenerate solutions are the only functions on $\mathbf{R}$ which are characterized by finitely many moments.

The analogous L-moment problem in two dimensions requires consideration of monomials in z and $\bar{z}$, that is,

$$a_{mn} := \int_{\mathbf{R}} z^m \bar{z}^n f(z, \bar{z}) dA(z) \quad (m, n \ge 0), \tag{2}$$

where dA denotes planar Lebesgue area measure. Degeneracy here is detected by $\det((b_{mn})_{m,n=0}^{N}) = 0$, where the b_{mn}'s are related to the a_{mn}'s by the formal exponential transform $\exp(-(\pi L)^{-1} \sum_{m,n=0}^{\infty} a_{mn} X^{m+1} Y^{n+1}) = 1 - \sum_{m,n=0}^{\infty} b_{mn} X^{m+1} Y^{n+1}$. Previously [J. Funct. Anal. **94** (1990), no. 2, 288–307; MR 91m:47018], the author showed that any degenerate solution of (2) is of the form $f = L\chi_D$, where χ_D is the characteristic function of a bounded set of the complex plane with real analytic boundary. In the paper under review, the following refinement is obtained.

Theorem: A solution f of (2) is degenerate if and only if $f = \chi_\Omega$, where Ω is a bounded open set with real-analytic boundary and such that there exists a meromorphic function m with poles in Ω and continuous up to the boundary $\partial\Omega$, with $\bar{z} = m(z)$ for every $z \in \partial\Omega$.

The proof of this theorem is based on a careful analysis of the kernel

$$E_L(z, w) := \exp\left(-\frac{1}{\pi L} \int_{\mathbf{C}} \frac{f(\varsigma) dA(\varsigma)}{(\varsigma - w)(\bar{\varsigma} - \bar{z})} \right)$$

($|z|, |w|$ large), which has previously been considered by several authors in connection with the theory of hyponormal operators with rank-one self-commutator. It is precisely the interplay between such an operator T, its characteristic kernel E and the associated principal function f/L that establishes the theorem.

As a direct consequence, the author obtains as a corollary that the following conditions on Ω are equivalent: The closure of Ω is the support of a degenerate solution of (2); the kernel $E_L(z, w)$ is rational with denominator of the form $P(w)\overline{P(z)}$, P a polynomial, for large values of $|z|$ and $|w|$; there exist a pure hyponormal operator T with rank-one self-commutator and a polynomial P, such that $[T^*, P(T)^* T] = 0$ and $\sigma(T) = \overline{\Omega}$.

In addition, the author sheds light on a beautiful connection between the domains described in the above theorem and its corollary and the theory of Schwarz functions for a domain. Namely, if $\Omega = \operatorname{int}(\overline{\Omega})$ fulfills the conditions in the theorem, then Ω is simply connected if and only if there is a rational conformal map from the unit disk onto Ω; and Ω is one-connected if and only if it is the image of a conformal map of an annulus $\{z\colon r < |z| < 1\}$ by the transformation $f(\log z)$, where f is an elliptic function with periods $2\pi i$, $2\log r$.

The author has been successful at combining tools and techniques from function theory, perturbation theory for pairs of selfadjoint operators, and the classical theory of moments, to describe all degenerate solutions of the L-problem of moments with compact support in $\mathbf{R}^2$. An important application is also made to the so-called truncated moment problem, $a_{mn} = \int_{\overline{\mathbf{D}}} z^m \overline{z}^n d\mu(z)$ $(0 \le m, n \le N)$, where μ is a positive measure supported in the closed unit disk $\overline{\mathbf{D}}$ in the complex plane. By considering the associated generating function $G(z,w) := \sum_{m,n=0}^{N} a_{mn} w^{-m-1} \overline{z}^{-n-1}$, the author is able to show that if the truncated moment problem has a solution, then there are finitely many points λ_j in $\overline{\mathbf{D}}$, $1 \le j \le M$, and a positive definite matrix $(c_{ij})_{i,j=1}^{M}$ such that

$$G(z,w) = \sum_{i,j=1}^{M} c_{ij} (\lambda_j - \overline{z})^{-1} (\lambda_i - w)^{-1} \mod(\overline{z}^{-N-2}, w^{-N-2}).$$

In particular, if $M \le N$, then the solution to the truncated moment problem is unique; this is also the case when $\det((a_{mn})) = 0$.

It is worth noting that more moments than the number of nodes are needed to determine uniqueness, a special feature of multivariable moment problems, and one that poses serious difficulties in terms of constructing one-step matrix extensions [cf. R. E. Curto and L. A. Fialkow, Mem. Amer. Math. Soc. **119** (1996), no. 568, x+52 pp.; MR 96g:47009]. The matrix (a_{mn}) considered by the author corresponds to the analytic part of the matrix $M(N)$ in the Fialkow-Curto paper cited above; in particular, the above-mentioned condition $\det((a_{mn})) = 0$ means that the analytic columns of $M(N)$ are linearly dependent, so Corollary 5.15 of the Fialkow-Curto paper cited above gives a concrete description of the unique measure which solves the truncated moment problem. For more on truncated moment problems, the reader is referred to recent work by L. A. Fialkow and the reviewer [Mem. Amer. Math. Soc. **136** (1998), no. 648, x+56 pp.; MR 99d:47015, see p. C80 in this volume].

The article concludes with a section consisting of specific examples of degenerate moment sequences and their exponential transforms. The domains in question are "quadrature domains", those that in addition to the properties described in the author's theorem are compact semi-algebraic. For instance, the disk of center a and radius r is determined by the moments $a_{00} = \pi r^2$, $a_{01} = \overline{a_{10}} = \pi a r^2$, and $a_{11} = \pi(r^2/2 + |a|^2 r^2)$. Domains that are not finitely determined by their moments include those that do not have real algebraic boundary; contain a straight segment in their boundaries; are multiply connected and allow $\overline{z}$ to be extended from the exterior boundary to the whole polynomial hull; have a connected component of the boundary containing two circular arcs, of different radii, or intersecting at an angle different from π; allow for meromorphic extensions of $\overline{z}$ from different connected components of the boundary, but which take different values inside the domain, e.g., annuli.

Raúl E. Curto (Iowa City, IA)

99d:47015 47A57 15A57 30E05 44A60

Curto, Raúl E. (Iowa City, IA); **Fialkow, Lawrence A.** (New Paltz, NY)

Flat extensions of positive moment matrices: recursively generated relations. (English summary)

Mem. Amer. Math. Soc. **136** (1998), *no.* 648, x+56 pp.

One of the oldest and most thoroughly studied inverse problems in mathematics is the moment problem, with all its wide ramifications. To fix ideas, we recall the refined dictionary between four categories of objects arising in the study of the moments $(s_k)_{k=0}^{\infty}$ of a positive, rapidly decreasing measure $d\nu$ on the real line. They are: the sequence $(P_k)_{k=0}^{\infty}$ of orthogonal polynomials, the Jacobi matrix representing in the basis (P_k) the multiplication M_x by the real variable x, the symmetric operator M_x regarded on a domain of the Lebesgue space $L^2(d\nu)$ and the Hankel matrix $H=(s_{k+l})_{k,l=0}^{\infty}$. The basic relations between these objects have shaped functional analysis and probability theory for more than a century. It is important to remark that most of these relations are constructive, and therefore applicable via algorithms. In particular, one of these relations lies at the root of the spectral theorem for self-adjoint operators [N. I. Akhiezer, *The classical moment problem and some related questions in analysis*, Translated by N. Kemmer, Hafner Publishing Co., New York, 1965; MR **32** #1518].

The truncated moment problem, asking roughly, for a fixed integer N, that one describe the multivalued correspondence $(s_k)_{k=0}^{N} \mapsto d\nu$, is at the origins of modern convexity theory and has close connections with classical interpolation problems for spaces of analytic functions. In the light of modern operator theory, the truncated moment problem is reduced to the analysis of all selfadjoint extensions/dilations of the $(N+1)\times(N+1)$-upper left corner J_N of the Jacobi matrix J, or equivalently to the analysis of all nonnegative Hankel extensions of the corresponding truncated Hankel matrix H_N. This latter point of view is amply commented upon in the fundamental work of N. I. Akheizer and M. Kreĭn [*Some questions in the theory of moments*, translated by W. Fleming and D. Prill, Translations of Mathematical Monographs, Amer. Math. Soc., Providence, R.I., 1962; MR **29** #5073]. An inspired and comprehensive survey of the present status of moment problems on the line and their applications has been given by B. Simon [Adv. Math. **137** (1998), no. 1, 82–203].

In the multivariable case ($d\nu$ a measure on $\mathbf{R}^n$) there are still many gaps in understanding the parallel relations between moments, orthogonal polynomials, Hankel matrices and systems of commuting Jacobi matrices. The authors of the present Memoir began half a decade ago a unique and sustained program of elucidating the truncated moment problem in two variables from the perspective of the dilation theory of finite-dimensional Hankel matrices. They have already published a series of interesting articles and a preceding Memoir [Mem. Amer. Math. Soc. **119** (1996), no. 568, x+52 pp.; MR 96g:47009].

The present paper is a direct continuation of the previous works. It has two main themes: a careful parallel between the one-variable case and the two-dimensional truncated moment problem (written in a new language introduced by the authors) and an analysis of the one-step extension of Hankel matrices with additional structure and some rank constraints. The first theme incites the reader to consider the numerous conjectures raised by Curto and Fialkow; the second subject is the motivation of many numerical experiments and examples which justify the correctness of this approach to moment problems. It is worth remarking that the authors focus on extensions of moment sequences of even degree to the next even degree, contrary to the customary odd-degree extension analysis familiar to experts in cubature formulas [cf. I. P. Mysovskikh, *Interpolational cubature formulas* (Russian), "Nauka", Moscow, 1981; MR 83i:65025].

In vague terms, the main results of this Memoir revolve around the following idea: if the truncated moments reflect some algebraic conditions implied by the existence of a representing measure with support on a proper real algebraic subset in the plane, then the one-step extension process can be fully understood, and such a representing measure (even atomic) can be recursively constructed. The reader will find the precise statements in the ample, ten-page, introduction.

Needless to say, the paper under review will be of interest and have an impact in operator theory, probability theory and numerical analysis. Although the final touch to this theory is still far away, the results of Curto and Fialkow are profound and original. Beginners can approach this well-written volume directly, without consulting all the preceding works. *Mihai Putinar* (Santa Barbara, CA)

99i:47027 47A57 30E05 47A20 47N70 93B36

Foias, C. (Bloomington, IN); **Frazho, A. E.** (West Lafayette, IN); **Gohberg, I.** (Tel Aviv); **Kaashoek, M. A.** (Amsterdam)

★Metric constrained interpolation, commutant lifting and systems. (English summary)

Operator Theory: Advances and Applications, 100.

Birkhäuser Verlag, Basel, 1998. xii+587 pp. $152.00. ISBN 3-7643-5889-0

The simplest Nevanlinna-Pick interpolation problem is to find a function analytic and bounded by 1 on the unit disk which takes prescribed values at finitely many prescribed points in the disk ($f(z_j)=w_j$ for $j=1,\cdots,n$ where $z_1,\cdots,z_n$ are prescribed points in the unit disk and $w_1,\cdots,w_n$ are prescribed complex values). The solution criterion is simply the positivity of the associated Pick matrix $\Lambda=[(1-w_iw_j)/(1-z_iz_j)]_{i,j=1,\cdots,n}$ built directly out of the interpolation data. Recently a number of authors have studied a "time-varying" (also called nonstationary) generalization of this problem; roughly, the unit disk $\mathbf{D}$ is replaced by the set of diagonal matrices Z (thought of as operators on $l^2(\mathbf{Z})$, the l^2-space over the integers $\mathbf{Z}$) such that the weighted shift operator U^*Z (where U is the usual bilateral shift operator on $l^2(\mathbf{Z})$) has spectral radius strictly less than 1. The function f analytic and bounded by 1 on the disk is replaced by a bi-infinite lower triangular matrix giving rise to an operator L on $l^2(\mathbf{Z})$ required to have operator norm at most 1. The value w for an analytic function f at a point z in the unit disk is replaced by a diagonal matrix W (thought of as an operator on $l^2(\mathbf{Z})$). The interpolation conditions assume the form $L^\wedge(Z_j)=W_j$ for $j=1,\cdots,n$ for given "time-varing points" Z_j and "time-varying" values W_j, where $Z\to L^\wedge(Z)$ is the time-varying version of "point-evaluation" introduced by D. Alpay, P. Dewilde and H. Dym. Various teams of researchers have now developed a whole theory of nonstationary Nevanlinna-Pick interpolation, including canonical generalizations of the entire conventional Nevanlinna-Pick theory. A sample point from the conventional theory covered by the more general theory is the simple Pick matrix criterion for the existence of solutions. Motivation for conventional Nevanlinna-Pick interpolation theory (especially its generalizations to matrix- and operator-valued functions and the the quest for more practical solution algorithms) has become particularly intense in the last 20 years or so, due to its penetrating connections with problems in time-invariant system theory, particularly, in prediction theory and H^∞-H^2-control theory. It is therefore particularly satisfying for those working with the nonstationary generalization of Nevanlinna-Pick interpolation that exactly the same connections occur, but with time-varying systems in place of time-invariant systems.

The present book provides a grand, unified commutant lifting formalism for understanding the conventional time-invariant interpolation theory and the more recent nonstationary version of the theory, along with how the two versions are connected with each other. Part A of the book focuses on the time-invariant case only, but in a very general setting; this part can be viewed as an update of two recent books, namely, *The commutant lifting approach to interpolation problems* by Foias and Frazho [Birkhäuser, Basel, 1990; MR 92k:47033] and *Interpolation of rational matrix functions* by the reviewer, Gohberg and L. Rodman [Birkhauser, Basel, 1990; MR 92m:47027]. The presentation is well written with the constraint in mind that the present book should be self-contained. The aggregation of tangential Hermite-Fejér interpolation conditions (both one-sided and two-sided versions) into three matrix residue interpolation conditions (or a single matrix residue interpolation condition in the case of one-sided Hermite-Fejér interpolation conditions), introduced in the above-cited book of the reviewer, Gohberg and Rodman, is now viewed as a set of interpolation conditions of functions of more traditional type on a function of a complex variable but now with the interpolation conditions involving a matrix-valued argument. The authors are then able to extend these ideas with ease to the infinite-dimensional setting where one has possibly an operator-valued argument. As was done by the reviewer, Gohberg and Rodman [op. cit] for the matrix-valued case, it is shown how all the classical problems and their analogues for operator-valued functions (tangential Nevanlinna-Pick interpolation, tangential Hermite-Fejér interpolation, and Sarason interpolation) can all be viewed as particular cases of a standard left Nevanlinna-Pick interpolation problem (involving just one interpolation condition but with operator-valued argument). It is also detailed explicitly how the two-sided Sarason interpolation problem is equivalent to a standard two-sided Nevanlinna-Pick interpolation problem (left and right standard Nevanlinna-Pick interpolation conditions combined with a coupled left-right interpolation condition)—here called the Nudelman interpolation problem. In addition, the Nehari extension theorem is discussed in detail in this framework. A separate chapter presents the system-theoretic background, including a system-theoretic interpretation of point evaluation (with operator argument) of an analytic function. Unlike the approach of the reviewer, Gohberg and Rodman [op. cit.], which focused on reduction to homogeneous interpolation conditions and construction of matrix functions having given directional zero and pole structure, the present monograph reduces the proofs of all interpolation theorems to an application of the Sz.-Nagy-Foias commutant lifting theorem, the approach to interpolation problems originating with D. Sarason in 1967. The proof of the commutant lifting theorem itself is organized as an application of the Parrott lemma characterizing contractive completions of 2×2 block matrices with one unspecified matrix extry. The proof works with the Sz.-Nagy-Schäffer choice of coordinates for the minimal isometric dilation of a contraction operator. In this way the authors are able to arrive at an explicit formula for the so-called "central" lifting, and an explicit parametrization (in terms of a free "Schur contraction" parameter) for the set of all contractive solutions of the commutant (or more properly, intertwining) lifting problem. The authors are able to then convert these operator-theoretic formulas to state-space coordinates roughly by simply working with Taylor coefficients. The central lift, in the abstract and in the context of the many, varied applications, has the interpretation as the unique contractive solution of the lifting problem which maximizes a quantity connected with entropy; the exposition here provides a good, general setting for understanding in a unified way the many recent papers on various particular instances of this theme (i.e., the "maximum entropy" or "central" solution of a control or interpolation problem). Using the central lift, the authors also give quantitative estimates on the trade-off between 2-norm and infinity-norm performance for lifts (or interpolants). Applications to prediction theory and mixed H^2/H^∞ control problems (for discrete-time systems only, with the associated function theory on the unit disk rather than on a half plane) are discussed in some detail, with inclusion of graphical plots to illustrate 2-norm–infinity-norm performance tradeoffs for a particular example.

Part B of the book is the time-varying parallel of Part A. The nonstationary versions of all the interpolation problems and the Nehari extension problem, of all the solution criteria and of all the system-theoretic background and applications, are formulated. Two methods of proof of the nonstationary interpolation theorems are provided, one via reduction to a standard, stationary interpolation problem but for an operator-valued function, and a second via an application of a nonstationary version of the commutant lifting theorem itself. The "sparse embedding" operator and the "diagonal compression" operator give the precise correspondence between a nonstationary interpolation problem and an associated stationary interpolation problem (but on a much bigger space). The nonstationary commutant lifting theorem used in the second method of proof is due to the authors, although a closely related theorem, formulated in terms of representations of a nest algebra, is due earlier to V. Paulsen and S. Power. Three proofs are provided for the nonstationary commutant lifting theorem: (1) a proof via reduction to the classical (stationary) commutant lifting theorem (a more abstract version of the first method of proof for nonstationary interpolation problems), (2) a direct proof via the Parrott lemma, a nonstationary analogue of the proof given in Part A for the standard commutant lifting theorem, and (3) a proof by way of an explicit formula for the central intertwining lifting. Included also are parametrization formulas for the set of all solutions in terms of an appropriate nonstationary Schur contraction free parameter.

In conclusion this book provides an excellent exposition of a rich, multi-layered theory. The layers proceed from the most concrete to the most abstract, from the most specific to the most general. Parallel but seemingly different layers are seen to be equivalent by simple transformation of the data sets. More shocking (at least at first) is that higher (more general) layers (namely, nonstationary interpolation and nonstationary commutant lifting) can be seen as special cases of lower (more specific) layers (namely, stationary interpolation and stationary commutant lifting). Of course this is possible only with a cost—the stationary problem associated with a given nonstationary problem has data arranged very inefficiently with lots of sparsity. Nevertheless, this idea serves to unify conceptually all of interpolation and commutant lifting. The book certainly provides a definitive, state-of-the-art look at interpolation and commutant lifting, along with its connections with and applications to system and control theory. *Joseph A. Ball* (Blacksburg, VA)

97d:47023 47A60 42B15 47B44 47D03
Cowling, Michael (Sydney); **Doust, Ian** (Sydney); **McIntosh, Alan** (North Ryde); **Yagi, Atsushi** (Himeji)
Banach space operators with a bounded H^∞ functional calculus. (English summary)
J. Austral. Math. Soc. Ser. A **60** (1996), *no.* 1, 51–89.
Unbounded operators whose spectrum is contained in a sector of the complex plane, and whose resolvents satisfy suitable growth conditions, have been intensively studied in operator theory. Examples are m-accretive and m-sectorial operators treated in detail in the classical monograph of T. Kato [*Perturbation theory for linear operators*, Springer-Verlag New York, Inc., New York, 1966; MR **34** #3324]. Natural and important classes of differential operators fall into this category. If T is an m-sectorial operator on a Hilbert space with vertex at 0, then $-T$ is the generator of a holomorphic semigroup. It is well known that fractional powers can be defined in a natural way for accretive operators [K. Yosida, *Functional analysis*, Academic Press, New York, 1965; MR **31** #5054; B. Sz.-Nagy and C. Foiaş, *Harmonic analysis of operators on Hilbert space*, English translation, revised, North-Holland, Amsterdam, 1970; MR **43** #947]. A list of examples and newer applications can be found for instance in an article of J. Prüss and H. Sohr [Math. Z. **203** (1990), no. 3, 429–452; MR 91b:47030].

Let T be a closed operator on a Banach space X, and let $0 \le \omega < \pi$ be a real number. Then T is called an operator of type ω if the spectrum of T is contained in the closed sector S_ω consisting of all complex numbers z with $|\arg z| \le \omega$ and if, for each $\omega < \mu < \pi$, the growth of the resolvent of T outside the larger sector S_μ is restricted by $\|R(z,T)\| \le C|z|^{-1}$ with a constant C depending on μ. The Riesz-Dunford functional calculus allows one to define a bounded operator $f(T)$ for each function f analytic in a neighbourhood of $\sigma(T)$ including the point ∞ when T is unbounded. Denote by $S = S_\mu^0$ $(0 \le \omega < \mu < \pi)$ the interior of the sector S_μ. There are natural subalgebras $\Psi(S)$ and $\mathcal{F}(S)$ of the algebra $\mathcal{O}(S)$ of all analytic functions on S (defined by suitable growth conditions at 0 and at ∞) such that $\Psi(S) \subset H^\infty(S) \subset \mathcal{F}(S)$ and such that, for each one-to-one operator T of type ω with dense domain and dense range, the Riesz-Dunford functional calculus of T extends to an algebra homomorphism from $\Psi(S)$ to $L(X)$, and to a map which associates with each function f in $\mathcal{F}(S)$ a closed densely defined operator $f(T)$ on X. In this way, questions about powers or holomorphic semigroups associated with T become questions concerning the boundedness properties of a suitable functional calculus of T.

For the Hilbert-space case, this observation is contained in an earlier paper of McIntosh [in *Miniconference on operator theory and partial differential equations (North Ryde, 1986)*, 210–231, Austral. Nat. Univ., Canberra, 1986; MR 88k:47019]. The central problem studied in the paper under review is to find necessary and sufficient conditions on T which guarantee that $f(T)$ is a bounded operator on X whenever $f \in H^\infty(S)$ and that the resulting algebra homomorphism $H^\infty(S) \to L(X)$ is continuous. In the above-cited paper McIntosh proved that a one-to-one operator of type ω on a Hilbert space has a bounded H^∞-functional calculus on S_μ^0 for every $\mu > \omega$, or equivalently for some $\mu > \omega$, if and only if the imaginary powers T^{is} $(s \in \mathbf{R})$ of T form a (C_0)-group which for each $\omega < \mu < \pi$ satisfies an estimate of the form $\|T^{is}\| \le C\exp(\mu|s|)$, where the constant C depends on μ. In the same paper he proved that these conditions are equivalent to the fact that both T and its adjoint T^* satisfy suitable square function estimates.

It is the aim of the present paper to clarify to which extent these Hilbert-space results remain true in a Banach-space setting, and to show that more classical functional calculi such as the L^p-multiplier theory on $\mathbf{R}^n$ fit into this more general framework. For instance, a result of Stechkin on L^p-multipliers is used to show that in the Banach-space case the boundedness of the imaginary powers is not enough to guarantee the existence of a bounded H^∞-functional calculus. In the case of L^p-spaces it is proved that suitable square-function estimates for T and T' are equivalent to the existence of a bounded H^∞-functional calculus. The paper gives a well-written introduction to the general area which contains a survey on some of the Hilbert-space results obtained so far, and which gives new insight into the behaviour of Banach-space operators with spectrum contained in a sector. *Jörg Eschmeier* (Saarbrücken)

99g:47034 47A68 39A10 42A75 45E10 47B35
Bastos, M. A. (Lisbon); **Karlovich, Yu. I.** (Chemnitz);
Spitkovsky, I. M. (Williamsburg, VA); **Tishin, P. M.** (Odessa)
On a new algorithm for almost periodic factorization. (English summary)
Recent progress in operator theory (*Regensburg*, 1995), 53–74, *Oper. Theory Adv. Appl.*, 103, *Birkhäuser, Basel*, 1998.

Probably the most fruitful idea in recent work on almost periodic factorization is the so-called Portuguese transformation. The reviewer has had several opportunities to write about the fascinating results of this transformation and about how this transformation has been cultivated in the meantime. Now, in contrast to the chronology, the reviewer has the pleasure of reviewing just the paper the story started with.

Let AP be the algebra of all continuous almost periodic functions on $\mathbf{R}$. Convolution equations over the interval $(0,\lambda)$ lead to the consideration of matrix functions of the form

$$G_f^{(\lambda)}(x)=\begin{pmatrix} e^{i\lambda x} & 0 \\ f(x) & e^{-i\lambda x}\end{pmatrix}, \quad x\in\mathbf{R},$$

and a central problem consists in finding an AP-factorization of $G_f^{(\lambda)}(x)$, that is, a representation

$$G_f^{(\lambda)}(x)=G_+(x)\mathrm{diag}(e^{i\mu x},e^{-i\mu x})G_-(x)$$

where $G_+^{\pm 1}$ and $G_-^{\pm 1}$ are matrix functions whose entries belong to AP and have their Bohr-Fourier spectrum in $\mathbf{R}_+=[0,\infty)$ and $\mathbf{R}_-=(-\infty,0]$, respectively.

The notion of AP-factorization was introduced in 1985 by Karlovich and Spitkovsky, who also constructed AP-factorizations of $G_f^{(\lambda)}$ in the cases where f is an almost periodic monomial or binomial. Moreover, they discovered almost periodic trinomials f for which $G_f^{(\lambda)}$ does not have an AP-factorization. In a series of subsequent papers, the class of almost periodic polynomials f giving rise to AP-factorable matrix functions $G_f^{(\lambda)}$ has been understood better and better, but up to the present a complete description of this class is unknown (even for trinomials).

The idea of the paper under review consists in multiplying $G_f^{(\lambda)}$ from the left and the right by appropriate matrix functions so that the resulting matrix function is of the form $G_{f_1}^{(\lambda_1)}$ with $\lambda_1<\lambda$ and with an f_1 whose structure is comparable with the structure of f and so that $G_f^{(\lambda)}$ is AP-factorable if and only if $G_{f_1}^{(\lambda_1)}$ is also. If necessary, one can repeat this procedure to produce a sequence of matrix functions $G_{f_1}^{(\lambda_1)}, G_{f_2}^{(\lambda_2)}, \cdots$. The authors show that in this way one can reduce the problem of constructing an AP-factorization for several classes of almost periodic polynomials f to cases which were already studied in earlier papers.

This paper is heavily based on joint work of its authors in Lisbon, by virtue of which the passage from $G_f^{(\lambda)}$ to $G_{f_1}^{(\lambda_1)}$ is now called the Portuguese transformation. The number $-\lambda_1$ is the left-most point of the intersection of the Bohr-Fourier spectrum of f and the interval $(-\lambda, \lambda)$. In order to carry out the Portuguese transformation, one has to find almost periodic polynomials M_1^+, M_2^+ with nonnegative Bohr-Fourier spectrum such that

$$M_1^+(x)f(x)e^{i\lambda_1 x} + M_2^+(x)e^{i(\lambda_1+\lambda)x} = 1, \quad x \in \mathbf{R}. \tag{1}$$

Once M_1^+, M_2^+ are available, one can construct the desired f_1.

The authors are able to solve (1) in various interesting cases, and these cases constitute the contents of the paper. Only once this paper was written was it realized that (1) is actually a corona problem whose data $f(x)e^{i\lambda_1 x}$ and $e^{i(\lambda_1+\lambda)x}$ are analytic almost periodic functions and whose solution $M_1^+(x), M_2^+(x)$ is sought in the set of analytic almost periodic functions. This insight gave birth to a new development in the factorization business. Bastos, Karlovich, A. F. dos Santos and Tishin [J. Math. Anal. Appl. **223** (1998), no. 2, 494–522; MR 99f:47022a; J. Math. Anal. Appl. **223** (1998), no. 2, 523–550; MR 99f:47022b] further enlarged the range of applicability of the Portuguese transformation, while D. Quint, L. Rodman and Spitkovsky [Michigan Math. J. **45** (1998), no. 1, 73–102; MR 99c:47020] essentially cultivated the approach and showed that many (or even virtually all) known results on AP-factorization can be obtained by clever and possibly repeated application of the Portuguese transformation. Finally, the Portuguese transformation has resulted in a revival of the interest in corona theorems: see, for example, J. A. Ball, Karlovich, Rodman and Spitkovsky [Integral Equations Operator Theory **32** (1998), no. 3, 243–281; MR 99k:47033], Rodman and Spitkovsky [Indiana Univ. Math. J. **47** (1998), no. 4, 1243–1256], and the reviewer [Integral Equations Operator Theory **33** (1999), no. 3, 253–272].

In summary, the ball is rolling. Although we now can do some things better than in the paper under review, we should not forget that this paper showed us how to do things at all and that this paper alone started the ball rolling.

{For the entire collection see MR 99d:47002.} *A. Böttcher* (Chemnitz)

98i:47019 47B20 30C40 30C85 30D30 47A60 47B47

McCarthy, John E. (St. Louis, MO); **Yang, Liming [Yang, Li Ming]** (St. Louis, MO)

Subnormal operators and quadrature domains.

Adv. Math. **127** (1997), *no.* 1, 52–72.

An operator S on a Hilbert space $\mathscr{H}$ is called subnormal if S is the restriction to $\mathscr{H}$ of a normal operator N acting on a Hilbert space $\mathscr{K} \supseteq \mathscr{H}$. The self-commutator of S is $[S^*, S] := S^*S - SS^*$, which is always a positive operator when S is subnormal. S is said to be cyclic [resp. rationally cyclic] if there exists a vector $\xi \in \mathscr{H}$ such that every other vector in $\mathscr{H}$ can be approximated in the Hilbert space norm by vectors of the form $r(S)\xi$, where r is a polynomial [resp. a rational function with poles off $\sigma(S)$, the spectrum of S]. R. Olin, J. Thomson and T. Trent ["Subnormal operators with finite-rank self-commutators", Trans. Amer. Math. Soc., to appear] have obtained a complete classification of cyclic subnormal operators with finite-rank self-commutators, by first classifying the irreducible ones, and then showing when a finite direct sum of such operators is again cyclic. (Independently of this work, an attempt was made by D. Xia in 1987 to characterize all pure subnormal operators with finite-rank self-commutators using a suitable function model, and concrete results were obtained for the polynomially cyclic case [D. Xia, Integral Equations Operator Theory **10** (1987), no. 2, 258–289; MR 89e:47034; Integral Equations Operator Theory **10** (1987), no. 6, 880–903; MR 89e:47035; errata; MR 90i:47025; Integral Equations Operator Theory **24** (1996), no. 1, 106–125; MR 96k:47043].)

First, we need some notation. The open unit disk will be denoted by $\mathbf{D}$ and its boundary, the unit circle, by $\mathbf{T}$. For a compact subset $K \subseteq \mathbf{C}$ and a finite positive Borel measure μ supported on K we let $S_{K,\mu}$ denote the operator of multiplication by the coordinate function on the space $R^2(K,\mu)$, the closure in $L^2(\mu)$ of the space of rational functions with poles off K. As usual, we assume that $\sigma(S_{K,\mu}) = K$.

Theorem [R. Olin, J. Thomson and T. Trent, op. cit.]: Let S be an irreducible cyclic subnormal operator. Then S has finite-rank self-commutator if and only if there exist a rational function r and a measure ν such that: (i) r is bounded in $\mathbf{D}$; (ii) r is a weak* generator of $H^\infty(\mathbf{D})$; (iii) ν is the sum of Haar measure on $\mathbf{T}$ and a finite number of point masses in $\mathbf{D}$; and (iv) S is unitarily equivalent to $r(S_{\mathbf{D},\nu})$.

To obtain the analogue of this theorem for rationally cyclic operators, one first needs to discuss quadrature domains in $\mathbf{C}$. A subset $\Omega \subseteq \mathbf{C}$ is called a quadrature domain if there exist points $z_1, \cdots, z_N \in \mathbf{C}$ and complex numbers $a_{m,n}$ such that every analytic function $f \in L^1(\Omega)$ satisfies the identity $\int_\Omega f(z)\,dA = \sum_{j=1}^N \sum_{i=0}^{N_j} a_{i,j} f^{(i)}(z_j)$; the order of Ω is $n := \sum_{j=1}^N (N_j+1)$. Quadrature domains are naturally associated with a function R_Ω, the so-called Schwarz function, which is meromorphic in Ω, continuous up to $\partial\Omega$, and such that $R_\Omega(z) = \bar{z}$ on $\partial\Omega$. Inside Ω, the equation $R_\Omega(z) = \bar{z}$ admits finitely many solutions, which we collect in a set $E(\Omega)$. Bounded simply connected domains are quadrature domains precisely when they are conformal images of $\mathbf{D}$ under a rational function. The theorem of Olin, Thomson and Trent says that $S_{K,\mu}$ is irreducible, cyclic, and with finite-rank self-commutator if and only if K is polynomially convex, $\Omega := \operatorname{int} K$ is a quadrature domain, and μ is mutually absolutely continuous with respect to the sum of harmonic measure for Ω plus a finitely atomic measure supported inside Ω.

In the article under review, the authors establish that (a) every quadrature domain gives rise to a rationally cyclic subnormal operator with finite-rank self-commutator; (b) every rationally cyclic subnormal operator with finite-rank self-commutator "comes from" a quadrature domain; and (c) the set of unitary equivalence classes of irreducible rationally cyclic subnormal operators with spectrum K and self-commutator of rank $s \geq n$ is parametrized by $[0,1)^t \times \{\Omega \times (0,\infty)\}^{s-n} \times \mathbf{R}_+^{E(\Omega)}$, where $t+1$ is the connectivity of Ω.

The precise phrasing for (b) above is as follows. Theorem: Let $S_{K,\mu}$ be a rationally cyclic subnormal operator with spectrum K, and let $\Omega := \operatorname{int} K$. Then $S_{K,\mu}$ is irreducible and has finite-rank self-commutator if and only if: (i) $K = \overline{\Omega}$; (ii) Ω is a quadrature domain; (iii) $\mu|_{\partial\Omega}$ is absolutely continuous with respect to harmonic measure ω for Ω, and $\int_{\partial\Omega} \log(d\mu/d\omega)d\omega > -\infty$; and (iv) $\mu|_\Omega$ is finitely atomic.

In the special case when Ω is simply connected, that is, $\Omega = r(\mathbf{D})$, where r is conformal and rational, the Schwarz function is given by $R_\Omega(z) = \overline{r}(1/\overline{\varphi}(z))$, where $\varphi := r^{-1}$; this observation leads to a new proof of Theorem 2 in Olin, Thomson and Trent's paper [op. cit.].

Along the way, the authors obtain a number of significant auxiliary results. We mention here the following. Lemma 1.6: Let G be the set of analytic bounded point evaluations for $R^2(K,\mu)$, let k_λ denote the kernel function for $R^2(K,\mu)$, let $\{\lambda_j\}$ be a sequence in G that converges to some point $\lambda_0 \in \partial G$, and let Δ be a disk centered at λ_0. Then $\lim_{j\to\infty}\int_\Delta(|k_{\lambda_j}|^2/\|k_{\lambda_j}\|^2)\,d\mu = 1$. Theorem 2.2 (ii): All irreducible rationally cyclic subnormal operators with spectrum K and finite-rank self-commutator are similar, so the spectrum is a complete similarity and quasi-similarity invariant for pure rationally cyclic subnormal operators with finite-rank self-commutator.

The article concludes with a section on the geometric connections between subnormal operators and quadrature domains. As is well known, two pure subnormal operators S_1 and S_2 are unitarily equivalent if and only if there is a unitary operator from $\mathscr{L}_1 := \mathrm{Ran}[S_1^*, S_1]$ to $\mathscr{L}_2 := \mathrm{Ran}[S_2^*, S_2]$ that simultaneously intertwines both $S_1^*|_{\mathscr{L}_1}$ with $S_2^*|_{\mathscr{L}_2}$ and $[S_1^*, S_1]|_{\mathscr{L}_1}$ with $[S_2^*, S_2]|_{\mathscr{L}_2}$. It follows that each quadrature domain of order n gives rise to two $n\times n$ matrices; the specific collection of $n\times n$ matrices which can arise in this way are then described in terms of extendibility properties to bounded rationally cyclic subnormal operators with non-atomic spectral measure, thus providing, in an abstract sense, a parametrization of bounded quadrature domains.

Related to the article under review is recent work of M. Putinar, and of B. Gustafsson and Putinar, on a different approach to quadrature domains, which exploits the moment-theoretic properties of subnormal operators. In this context, quadrature domains arise as degenerate solutions of the so-called L-problem of moments, and are semi-algebraic sets uniquely determined by finitely many of their (area) moments. (For details on this new viewpoint, the reader is referred to the articles by Gustafsson and Putinar ["Linear analysis of quadrature domains. II", Preprint, 1997; per revr.] and Putinar [J. Funct. Anal. **136** (1996), no. 2, 331–364; MR 97c:47018, see p. C78 in this volume; J. Approx. Theory **92** (1998), no. 1, 38–58; MR 99d:47017; Ark. Mat. **33** (1995), no. 2, 357–376; MR 96m:47025], and to the featured review of Putinar's 1996 paper [op. cit.].) *Raúl E. Curto* (Iowa City, IA)

97b:47024 47B35 46M20 47A45

Yakubovich, D. V. (St. Petersburg)

Dual piecewise analytic bundle shift models of linear operators. (English summary)

J. Funct. Anal. **136** (1996), *no.* 2, 294–330.

The structure of linear transformations of a (topological) vector space has attracted and intrigued mathematicians, with waves of fashion or resentment, for almost a century. Even the classification of linear transformations of a finite-dimensional space, up to similarity or unitary equivalence, is rather complex. (Don't we feel this daily when teaching linear algebra?) By now, a large amount of information and techniques has accumulated; accordingly, we can be optimistic and regard classes of linear operators as accessible (e.g. selfadjoint, normal, scalar, dissipative, decomposable operators) or we resign when we are faced with intractable classification and structure problems as in the case of quasi-nilpotent operators or pairs of noncommuting operators. On the other hand, we have well understood the power of studying algebras of operators rather than single operators.

In the history of this subject the functional models occupy a distinguished place. To be more specific, whenever a class of operators was represented in explicit form as transformations of a function space, more light was shed on the situation thanks to an import of methods from contingent fields (typically function theory of a real or a complex variable). Classical examples are: the operators which admit spectral measures [see N. Dunford and J. T. Schwartz, *Linear operators. Part III*, Interscience Publishers [John Wiley & Sons, Inc.], New York, 1971; MR **54** #1009], then the Sz. Nagy and Foiaş functional model for contractive Hilbert space operators [see B. Sz.-Nagy and C. Foiaş, *Harmonic analysis of operators on Hilbert space*, English translation, revised, North-Holland, Amsterdam, 1970; MR **43** #947] or de Branges-Rovnyak-type models [see N. K. Nikol′skiĭ and V. I. Vasyunin, in *The Gohberg anniversary collection, Vol. II (Calgary, AB, 1988)*, 405–434, Birkhäuser, Basel, 1989; MR 91c:47017]. We need not mention the wide-spread impact of these functional models on quantum mechanics, statistics or the control theory of linear systems.

One step further, to give a more precise example: the study of Toeplitz operators alone has led to several remarkable geometric models. (A Toeplitz operator is represented as an infinite matrix with equal entries on the directions parallel to the main diagonal.) Two notable contributions which bring together Toeplitz operators and function theory on complex manifolds are the paper of D. N. Clark [J. Reine Angew. Math. **320** (1980), 6–31; MR 82h:47025] and, respectively, the paper of M. J. Cowen and R. G. Douglas [Acta Math. **141** (1978), no. 3-4, 187–261; MR 80f:47012]. In these and many other similar works it was understood that it is very natural to represent Toeplitz operators as multiplication operators on the section space of an appropriate analytic vector bundle. Then invariants of the operator (index, spectrum, etc.) have direct geometric meaning, and this dictionary turns out to be beneficial in both directions. For other illustrations of the same principle see the recent monograph by J. Eschmeier and M. Putinar [*Spectral decompositions and analytic sheaves*, Oxford Univ. Press, New York, 1996; MR 98h:47002].

The present paper by Dmitry Yakubovich is an inspired attempt at unifying, in a simple scheme, several vector bundle models for Toeplitz-like operators. The starting point in this axiomatic treatment is a bounded linear operator T acting on a reflexive Banach space X and satisfying two basic conditions: the point spectrum of T is empty and the eigenvalues of T^* span the space X^*. The standard example (as in most other similar models in operator theory) is the unilateral shift on $l^2(\mathbf{N})$. Under these and several additional natural assumptions a finite-rank (or in general "small") operator $J\colon X^* \to \mathbf{R}$ produces the analytic vector bundle $J\operatorname{Ker}(T^* - \lambda)$, $\lambda \in \sigma(T) \setminus \sigma_{\rm ess}(T^*)$. With the aid of this object the author diagonalizes simultaneously T and T^* on two analytic vector bundles in duality. Several simple formulae, reminiscent of Grothendieck-Köthe duality, relate the original operators to their functional models. As a validation of the whole construction, the refined model of D. N. Clark for rational Toeplitz operators cited above as well as certain subtle computations due to the former Leningrad school of functional analysis (to which the author belonged) are incorporated and simplified in the new framework.

The paper represents a lucid symbiosis between classical and modern chapters of operator theory. Besides its clear value as a new tool of constructing functional models for general classes of operators, it raises several challenging open questions for function theorists. *Mihai Putinar* (Santa Barbara, CA)

98m:47031 47B35 45E10 47A50
Böttcher, Albrecht (Chemnitz); **Karlovich, Yuri I. [Karlovich, Yu. I.]** (Odessa)
★**Carleson curves, Muckenhoupt weights, and Toeplitz operators. (English summary)**
Progress in Mathematics, 154.
Birkhäuser Verlag, Basel, 1997. xvi+397 pp. \$75.00. ISBN 3-7643-5796-7

The theory of Toeplitz operators has always been a richly productive meeting ground of operator theory and function theory. Deep ideas and results from function theory and harmonic analysis provide powerful tools for studying both individual operators and the algebras they generate. The operator theory leads to deep and natural questions in function theory. This book in that tradition is a presentation of the work of the authors and others in recent years on Toeplitz operators with piecewise continuous symbols acting on weighted Lebesgue spaces on Carleson curves.

Suppose Γ is a simple closed curve in the complex plane and w is a strictly positive weight function on Γ. If Γ is smooth we can consider the Cauchy projection, the bounded linear operator S on $L^2(\Gamma, w(z)|dz|) = L^2(\Gamma, w)$ which takes f to the boundary values on Γ of the Cauchy integral of f along Γ. Thus $Sf(z) = \lim_{w\to z}(\pi i)^{-1}\int_\Gamma f(\zeta)(\zeta - z)^{-1}d\zeta$. Let $H^2 = H^2(\Gamma, w)$ be the range of the Cauchy projection $P := (I+S)/2$. Given a bounded function a on Γ define T, the Toeplitz operator with symbol a, by $Tf = P(af)$. The theory of such operators when Γ, w, and a are sufficiently smooth is classical. For instance, if Γ and w are C^2 and a is continuous then T_a is Fredholm if and only if a never vanishes. In that case the Fredholm index of the operator is the negative of the winding number of the curve $w(\Gamma)$ [I. C. Gohberg, Uspehi Matem. Nauk (N.S.) **7** (1952), no. 2(48), 149–156; MR **14**, 54a]. However, for rough Γ, unbounded w, or discontinuous a, fundamental and difficult questions arise at once; for instance, is S bounded? However, for a very natural class of curves, weights, and symbol functions the situation is now well understood on both L^2 and L^p, $1 < p < \infty$. This book is a self-contained presentation of those results.

For $t \in \Gamma$, $\varepsilon > 0$ set $\Gamma(t,\varepsilon) = \Gamma \cap \{|z-t| < \varepsilon\}$ and denote its arc-length by $|\Gamma(t,\varepsilon)|$. A curve is said to be a Carleson curve (or Ahlfors-David regular) if $\sup_{t\in\Gamma}\sup_{\varepsilon>0}\varepsilon^{-1}|\Gamma(t,\varepsilon)| < \infty$. If $1 < p, q < \infty$, $1/p + 1/q = 1$ then a weight w defined on Γ is said to satisfy the Muckenhoupt condition $A_p(\Gamma)$ if

$$\sup_{t\in\Gamma}\sup_{\varepsilon>0}\left(\varepsilon^{-1}\int_{\Gamma(t,\varepsilon)} w(s)^p|ds|\right)^{1/p}\left(\varepsilon^{-1}\int_{\Gamma(t,\varepsilon)} w(s)^{-q}|ds|\right)^{1/q} < \infty.$$

The first three chapters of the book develop the basics about such curves and weights. The next two chapters present a detailed proof that S is bounded on $L^p(\Gamma, w)$ if and only if Γ is a Carleson curve and w satisfies $A_p(\Gamma)$. This result is the culmination of the deep work of many people in the 70's and 80's. The presentation here of a self-contained proof for the non-expert is a welcome addition. (The analogous question when Γ is a compact set rather than a curve is still being actively studied [see, e.g., P. Mattila, M. S. Mel′nikov and J. Verdera, Ann. of Math. (2) **144** (1996), no. 1, 127–136;

MR 97k:31004; F. Nazarov, S. Treil and A. L. Vol′berg, Internat. Math. Res. Notices **1997**, no. 15, 703–726; MR 99e:42028].)

In the smooth category the essential spectrum of a Toeplitz operator with continuous symbol is the range of the symbol. It was known in the 60's how to find the essential spectrum of T_a if a is continuous except for a single jump discontinuity. Chapter 6 contains that and other Toeplitz algebra basics, including a localization principle which reduces many questions about the spectrum of a Toeplitz operator to the study of local behavior of the symbol functions in a neighborhood of each point. Of the next chapter the authors say, "This chapter is the heart of the book.... We know from the preceding chapter that the essential spectrum of a classical Toeplitz operator is the union of the essential range of the symbol and line segments joining the endpoints of each jump. We will show that in the general case [of piecewise continuous symbol] these line segments metamorphose into circular arcs, logarithmic double spirals, horns, spiralic horns, and eventually into what we call leaves." The beginning of the story is old, the basics having been worked out in the 60's. However, much of this chapter is work of the 90's and much is work of the authors. The detailed description of the essential spectrum comes down to describing the "leaves". The shape of the leaf at, say, 0 is determined by knowing the set $\{\gamma\colon\ |z^\gamma|w$ satisfies $A_p(\Gamma)\}$. The details of the analysis of this set and the passage from this set to the leaf are technical and complicated but they are satisfyingly concrete and geometric. The general theory is augmented with detailed examples and more than two dozen figures of various geometric objects that arise. In addition to their mathematical interest the figures are visually enchanting.

Chapters 8 and 9 extend the results to algebras of operators generated by P and multiplication by piecewise continuous functions and also from simple curves to more complicated, but still locally finite, geometries. For a Jordan curve the analysis begins by using the localization principle. Spectral analysis in the "local algebra" rests on the fact that the algebra is generated by its identity and two other idempotents. Roughly, one idempotent distinguishes the two different one-sided limits and the other distinguishes locally between holomorphy and antiholomorphy. This approach goes back to [R. G. Douglas, Proc. London Math. Soc. (3) **36** (1978), no. 2, 243–272; MR **58** #2421]; the more general version here is due to a number of people working in the 90's. If N arcs cross at a point then the attempt to localize leads to technical questions about the possibility of extending Carleson curves and Muckenhoupt weights. After those are resolved using rather delicate arguments from geometric function theory, the local algebra is seen to be generated by the identity and $2N$ idempotents which satisfy additional relations. The spectral analysis is then completed using Banach algebraic techniques developed recently for working with these more complicated algebras [A. Böttcher et al., in *Singular integral operators and related topics (Tel Aviv, 1995)*, 19–54, Birkhäuser, Basel, 1996; MR 97h:46078, see p. C44 in this volume].

The final chapter has a brief look at very recent work on related topics, including Toeplitz operators acting on spaces of vector-valued functions, the analysis of Wiener-Hopf operators, and the use of Mellin analytical techniques.

{See also the following review [MR 98m:47032].}

Richard Rochberg (St. Louis, MO)

97h:47023 47B38 30D05 30D55 47A15
Bourdon, Paul S. (Lexington, VA); **Shapiro, Joel H.** (East Lansing, MI)
Cyclic phenomena for composition operators. (English summary)
Mem. Amer. Math. Soc. **125** (1997), *no.* 596, x+105 pp.

Let **D** be the unit disc in the complex plane **C** and X a linear topological space of analytic functions on **D**. For each analytic $\varphi\colon \mathbf{D}\to\mathbf{D}$ there is an induced composition operator $C_\varphi(f)=f\circ\varphi$ on X. These operators have been studied on classical Banach spaces of analytic functions, almost continuously since the 1960s, focussing on questions like boundedness, compactness, spectra, semigroups, and subnormality.

In the present monograph the authors undertake a systematic study of yet another direction: that of cyclicity of composition operators, on the Hardy space H^2 as a model space. A linear operator T is said to be cyclic if there is a vector x (called a cyclic vector) such that the orbit $O(T,x)=\{T^n(x)\colon n=0,1,2,\cdots\}$ has dense linear span. If the orbit itself is dense then T is said to be hypercyclic. Because the closed linear span of $O(T,x)$ is the smallest closed T-invariant subspace containing x, the concept of cyclicity is closely connected with the study of invariant subspaces. The study of cyclicity is especially natural for composition operators because the nth power of C_φ is the composition operator induced by the nth iterate of φ. This suggests that cyclic properties of C_φ should depend heavily on the dynamical behavior of the inducing map.

The first part of the monograph is devoted to the study of cyclicity when the inducing map φ is linear fractional. For such symbols the cyclic behavior of C_φ is completely determined and is found to depend on the location of the fixed points of φ relative to the interior and the boundary of the disc. All possible cases occur: C_φ can be hypercyclic, cyclic but not hypercyclic, or not at all cyclic.

For the general case the authors use a convenient classification of arbitrary self-maps of **D** based on the concept of Denjoy-Wolff point and suggested by an analogous classification of linear fractional maps. Maps φ with a fixed point in **D** are of dilation type. Maps with no fixed point in **D** and derivative <1 at the Denjoy-Wolff point are of hyperbolic type. The remaining case is when there is no fixed point in **D** and the derivative at the Denjoy-Wolff point is $=1$. These are maps of parabolic type, and they fall into two subclasses: those of automorphic type for which the orbits $\{\varphi_n(z)\}$ $(z\in\mathbf{D})$ are separated, i.e. the hyperbolic distance of successive points in $\{\varphi_n(z)\}$ stays bounded away from zero, and those of nonautomorphic type for which no orbit is separated.

Only univalent φ need be considered for cyclicity because nonunivalent symbols induce noncyclic composition operators. A major objective of the entire study is to determine the extent to which cyclic behavior of operators with linear fractional symbol in each of the classes above is transferred to operators induced by arbitrary φ in the same class. This connection is possible by the linear fractional model theorem. In the univalent case the theorem states that there is a univalent map σ on **D** and a linear fractional self-map ψ of **D** such that $\sigma\circ\varphi=\psi\circ\sigma$, and φ and ψ are of the same classification type. The functional equation then relates orbits of C_φ and C_ψ (with the function σ creeping in) and, from this, transference of cyclicity is possible.

For cyclicity of C_φ it is necessary that polynomials in φ be dense in H^2, but the transference technique requires that polynomials in σ be dense in H^2. This polynomial approximation property, however, does not pass down from φ to σ in the linear fractional model theorem. Much of the work of the authors is devoted to showing that for sufficiently regular maps φ the corresponding function σ has the desired approximation property. As a result, for such φ the operator C_φ inherits in full the cyclic behavior of the linear fractional composition operator C_ψ.

The regularity imposed on φ is rather strong. It requires it to be continuous on the closure of $\mathbf{D}$ with $\varphi(\mathbf{D})$ making contact with $\partial\mathbf{D}$ in at most one point, plus some smoothness conditions on the Denjoy-Wolff point. Under such conditions the authors expand on the linear fractional model theorem, formulated in the simpler setting of the right half-plane, to derive delicate asymptotic representations of the intertwining map σ in both the hyperbolic and the parabolic case (Section 4). This permits them to obtain the desired polynomial approximation property and to derive criteria for distinguishing maps φ of parabolic automorphic type from those of parabolic nonautomorphic type. The cyclicity of operators with parabolic nonautomorphic symbol is treated in Section 5. The last section contains open problems related to cyclicity.

This work continues and expands the study of cyclicity initiated by Shapiro [*Composition operators and classical function theory*, Springer, New York, 1993; MR 94k:47049]. The results obtained show that composition operators exhibit diverse types of cyclic behavior. In addition these operators possess their cyclic or noncyclic properties in an extreme form. For example, the authors observe that a cyclic linear fractional composition operator has a dense set of cyclic vectors, while if the operator is not hypercyclic then only constant functions can be limit points of orbits. The work uncovers new links between classical function theory and operator theory, by connecting the cyclicity problem with polynomial approximation and analytic functional equations. The work is very readable. It is accessible to anyone with familiarity with elementary complex analysis and elementary operator theory.

Aristomenis Siskakis (Thessaloniki)

98a:47035 47B38 30D05 30D55 47A10

Poggi-Corradini, Pietro (Charlottesville, VA)

The Hardy class of geometric models and the essential spectral radius of composition operators. (English summary)

J. Funct. Anal. **143** (1997), *no.* 1, 129–156.

Let Ω be a simply connected region in the complex plane and let ψ be a Riemann mapping of the open unit disk U onto Ω. The Hardy number $h(\Omega)$ of Ω is defined by $h(\Omega) = \sup\{p\colon \psi \in H^p\}$, where H^p denotes the usual Hardy space of U. Observe that the definition of $h(\Omega)$ is independent of the particular Riemann mapping chosen; also observe that via Littlewood's subordination principle, every holomorphic mapping on U with range contained in Ω will belong to H^p for $p < h(\Omega)$.

In the paper under review, Poggi-Corradini obtains upper and lower bounds for $h(\Omega)$ that involve the hyperbolic metric on Ω, ρ_Ω. Specifically, he shows that for any $z \in \Omega$, $h(\Omega)$ lies between the lim sup and lim inf of the quotient $\rho_\Omega(z, \alpha\partial U)/\log(\alpha)$ as $\alpha \to \infty$. In addition, he shows that if Ω is a part of a "geometric model" (Ω, λ) for a univalently induced composition operator on the Hardy spaces, then the lim sup and lim inf coincide, yielding $h(\Omega) = \lim_{\alpha\to\infty} \rho_\Omega(z, \alpha\partial U)/\log(\alpha)$, the preceding limit being termed the hyperbolic stricture of Ω.

The pair (Ω, λ) is a geometric model for a composition operator provided that $0 \in \Omega$, that $0 < |\lambda| < 1$, and that $\lambda\Omega \subset \Omega$. When (Ω, λ) is a model, Poggi-Corradini also relates hyperbolic stricture to the Hardy numbers for the connected components of the invariant subset $\operatorname{int}(\bigcap_{n\geq 0} \lambda^n \Omega)$ of Ω, establishing interesting ties between his work and complex dynamics (these connections are explored further in his paper [Proc. Amer. Math. Soc. **126** (1998), no. 6, 1697–1708; MR 98g:30049]).

Poggi-Corradini's work has important applications to the theory of composition operators on Hardy spaces; in fact, these applications motivated his work. Every holomorphic φ mapping U into itself induces a bounded composition operator C_φ on each of the Hardy spaces H^p, $0 < p \leq \infty$, where, of course, $C_\varphi f = f \circ \varphi$. If φ fixes the origin and $\varphi'(0) \neq 0$, then classical work of Koenig shows that there is a holomorphic function σ on U such that $C_\varphi \sigma = \varphi'(0)\sigma$; moreover, σ is univalent whenever φ is univalent, and σ is uniquely determined up to a constant multiple as long as φ is not an automorphism of U. When σ is univalent (and φ is not an automorphism), setting $\Omega = \sigma(U)$ and $\lambda = \varphi'(0)$ yields a geometric model (Ω, λ) for C_φ: this terminology is used because the action of φ on U is conjugate to the action of multiplication by λ on Ω (since $\varphi = \sigma^{-1} \circ (\lambda\sigma)$). Poggi-Corradini's formula for $h(\Omega)$ in terms of hyperbolic stricture thus provides information about the Hardy classes to which the eigenfunction σ belongs.

The reviewer and Joel H. Shapiro [J. Amer. Math. Soc. **10** (1997), no. 2, 299–325; MR 97h:30040] obtained information about the Hardy classes to which the eigenfunctions σ belong, proving that $|\varphi'(0)|^{p/2} > r_{\rm e}(C_\varphi: H^2 \to H^2)$ is sufficient to ensure that $\sigma \in H^p$, where $r_{\rm e}$ denotes the essential spectral radius. They obtained necessity under the assumption that φ be analytic on the closed disk. Poggi-Corradini's work with hyperbolic stricture yields necessity under the hypothesis that φ be univalent. He has recently established necessity with no additional hypotheses on φ.

One final remark: when (Ω, λ) is a geometric model Poggi-Corradini's formula $h(\Omega) = \lim_{\alpha\to\infty} \rho_\Omega(z, \alpha\partial U)/\log(\alpha)$ may be viewed as a generalization of a formula for the Hardy number of a star-like region obtained by L. J. Hansen [Michigan Math. J. **17** (1970), 235–248; MR **41** #7118]. *Paul Bourdon* (Lexington, VA)

98f:47043 47B40 47A11 47B38

Albrecht, Ernst (Saarbrücken); **Eschmeier, Jörg** (Saarbrücken)

Analytic functional models and local spectral theory.

Proc. London Math. Soc. (3) **75** (1997), *no.* 2, 323–348.

A bounded linear operator $T \in L(X)$ on a complex Banach space X is said to have Bishop's property (β) if, for every open subset U of $\mathbf{C}$ and every sequence of analytic functions $f_n: U \to X$ for which $(T - \lambda) f_n(\lambda) \to 0$ as $n \to \infty$ locally uniformly on U, it follows that $f_n(\lambda) \to 0$, again locally uniformly on U. In an equivalent version, this condition was introduced by E. Bishop [Pacific J. Math. **9** (1959), 379–397; MR **22** #8339] in an attempt to develop a general duality theory for operators on Banach spaces. The paper under review completes the aspirations of the original Bishop inquiry, and provides a remarkably strong connection with the theory of decomposable operators in the sense of C. Foias.

An operator $T \in L(X)$ is said to be decomposable if, for every open cover $\{U, V\}$ of $\mathbf{C}$, there exist T-invariant closed linear subspaces Y and Z of X such that $Y + Z = X$, $\sigma(T|Y) \subseteq U$, and $\sigma(T|Z) \subseteq V$. The class of decomposable operators contains the spectral operators and, in particular, the normal operators. Moreover, all operators with a non-analytic functional calculus are decomposable; see F.-H. Vasilescu [*Analytic functional calculus and spectral decompositions*, Translated from the Romanian, Reidel, Dordrecht, 1982; MR 85b:47016].

For an open set U of $\mathbf{C}$, let $H(U,X)$ denote the Fréchet space of all X-valued analytic functions on U, and, for arbitrary $T \in L(X)$, let T_U denote the operator on $H(U,X)$ given by $(T_U f)(\lambda) := (T-\lambda)f(\lambda)$ for all $\lambda \in U$. It is easily seen that T has property (β) precisely when, for each open set $U \subseteq \mathbf{C}$, the operator T_U is injective and has closed range. The paper under review provides an important new characterization which lies much deeper: an operator has property (β) if and only if it is similar to the restriction of a decomposable operator to one of its closed invariant subspaces.

To introduce the dual notion of property (β), let us consider the local spectral subspaces of an operator $T \in L(X)$ given by $X_T(F) := X \cap \operatorname{ran} T_{\mathbf{C}\setminus F}$ for all closed sets $F \subseteq \mathbf{C}$. The operator T is said to have the decomposition property (δ) if $X = X_T(\overline{U}) + X_T(\overline{V})$ for every open cover $\{U,V\}$ of $\mathbf{C}$. It is well known that T is decomposable if and only if T has both properties (β) and (δ). In the paper under review, it is shown that property (δ) characterizes the quotients of decomposable operators with respect to closed invariant subspaces, and that (β) and (δ) are completely dual to each other, in the sense that an operator T has one of these properties exactly when its adjoint T^* on the dual space X^* has the other. In particular, it follows that T is decomposable if and only if both T and T^* have property (β).

These results make essential use of the theory of vector-valued analytic functions and topological tensor products. The decomposable extension of an operator with property (β) follows from a powerful functional model in the spirit of M. Putinar [J. Operator Theory **12** (1984), no. 2, 385–395; MR 85h:47027] which may be sketched as follows.

For an arbitrary operator $T \in L(X)$ and a bounded open neighborhood Ω of $\sigma(T)$, let $\Phi\colon H(\Omega,X) \to X$ denote the local analytic functional calculus given by $\Phi(f) := (2\pi i)^{-1}\int_\Gamma (T-\lambda)^{-1} f(\lambda)\,d\lambda$ for all $f \in H(\Omega,X)$, where Γ is a contour which surrounds $\sigma(T)$ in Ω. Let $A(\Omega,X)$ and $W(\Omega,X)$ denote the Banach spaces of all $f \in L^2(\Omega,X)$ for which the distributional derivative $\overline{\partial} f$ satisfies $\overline{\partial} f = 0$ and $\overline{\partial} f \in L^2(\Omega,X)$, respectively, and let M denote the operator of multiplication by the independent variable on these spaces. Then M is decomposable on $W(\Omega,X)$ and on $L^2(\Omega,X)$ since there is an obvious functional calculus on $C^\infty(\mathbf{C})$ for M. Moreover, because Φ maps $A(\Omega,X)$ onto X, it follows that T is similar to the operator induced by M on the quotient space $A(\Omega,X)/\ker(\Phi|A(\Omega,X))$. Hence, if $S \in L(Y)$ denotes the operator induced by M on $Y := W(\Omega,X)/(\ker\Phi|A(\Omega,X))$, then X may be regarded as a subspace of Y, and T may be identified with the restriction of S to X. This shows that every operator is similar to the restriction of a quotient of a decomposable operator. On the other hand, it follows from the Cauchy-Pompeiu formula that the $\overline{\partial}$-operator maps $W(\Omega,X)$ onto $L^2(\Omega,X)$, which leads to an isomorphism from the space $Z := W(\Omega,X)/A(\Omega,X)$ onto $L^2(\Omega,X)$. Let $R \in L(Z)$ denote the operator induced by M on $L^2(\Omega,X)$. Then R is decomposable, and there is an exact sequence $0 \to X \xrightarrow{i} Y \xrightarrow{q} Z \to 0$ for which $iT = Si$ and $qS = Rq$.

This analytic functional model yields, for every open set U in $\mathbf{C}$, an exact sequence

$$0 \to H(U,X) \xrightarrow{I\otimes i} H(U,Y) \xrightarrow{I\otimes q} H(U,Z) \to 0$$

for which $(I\otimes i)T_U = S_U(I\otimes i)$ and $(I\otimes q)S_U = R_U(I\otimes q)$, where I is the identity operator on $H(U)$. A simple diagram chase now shows that property (β) carries over from T to S. Hence it follows that S is a decomposable extension of T provided that T has property (β).

The case of decomposable liftings requires a certain reformulation of property (δ). For arbitrary $T\in L(X)$ and an open subset U of the Riemann sphere $\widehat{\mathbf{C}} := \mathbf{C}\cup\{\infty\}$ with $\infty\in U$, let $P(U,X)$ denote the Fréchet space of all X-valued analytic functions on U which vanish at ∞, and let $(T^U f)(\lambda) := (T-\lambda)f(\lambda) + \lim_{\mu\to\infty}\mu f(\mu)$ for all $f\in P(U,X)$ and $\lambda\in U\setminus\{\infty\}$. For a closed set $F\subseteq\widehat{\mathbf{C}}$ with $\infty\in F$, let $P(F,X)$ denote the inductive limit of the spaces $P(U,F)$, as U ranges over all open neighborhoods of F, and let T^F denote the operator on $P(F,X)$ which is canonically induced by the operators T^U. The authors establish that T has property (δ) precisely when T^F is surjective for every closed set $F\subseteq\widehat{\mathbf{C}}$ that contains ∞. This characterization is one of the tools in the construction of a decomposable lifting for an operator with property (δ). Another ingredient is a functional model which represents an arbitrary operator as the quotient of a restriction of a decomposable operator.

The heart of the complete duality between (β) and (δ) is the classical Grothendieck-Köthe duality principle, which provides an isomorphism from $P(F)$ onto the strong dual of $H(\widehat{\mathbf{C}}\setminus F)$ through the formula $\langle g,f\rangle = (2\pi i)^{-1}\int_\Gamma \widetilde{g}(\lambda)f(\lambda)d\lambda$ for all $f\in H(\widehat{\mathbf{C}}\setminus F)$ and $g\in P(F)$, where $\widetilde{g}$ is a representative of the equivalence class g and Γ is a contour surrounding the complement of the domain of $\widetilde{g}$ in $\widehat{\mathbf{C}}\setminus F$. Since an operator between two Fréchet spaces is injective and has closed range precisely when its adjoint is surjective, it follows from the Grothendieck-Köthe duality principle and the preceding characterizations of (β) and (δ) that an operator T has property (β) if and only if T^* has property (δ). For operators on reflexive Banach spaces, it is immediate that here the roles of (β) and (δ) may be interchanged. The authors establish the latter characterization also in the much more involved non-reflexive case, and provide a number of further interesting results.

The paper shows that restrictions and quotients of decomposable operators provide a natural and elegant framework in axiomatic spectral theory. It seems remarkable that, as we know now, Bishop dealt with these classes of operators long before decomposable operators were introduced by Foias in 1963.

Since a preprint of the paper under review has been around for quite some time, its important impact on further developments in local spectral theory has already become visible. For the close connections to sheaf theory, we refer to Eschmeier and Putinar [*Spectral decompositions and analytic sheaves*, Oxford Univ. Press, New York, 1996; MR 98h:47002]. Also, Eschmeier and B. Prunaru [J. Funct. Anal. **94** (1990), no. 1, 196–222; MR 91m:47008] have employed the theory of this paper to extend the results on the existence of invariant subspaces in the spirit of S. Brown from the Hilbert to the Banach space setting. This confirms the intuition that restrictions of decomposable operators are the proper generalizations of subnormal and hyponormal operators in the general context of Banach spaces, but the details are far from obvious.

In a different direction, it was shown by Putinar [Integral Equations Operator Theory **15** (1992), no. 6, 1047–1052; MR 94a:47036] and, independently, by L. M. Yang [J. Funct. Anal. **112** (1993), no. 1, 204–217; MR 94c:47033] that quasisimilar operators with property (β) have the same essential spectrum, thus providing a general positive answer to a question which had been open for hyponormal operators for a long time. The corresponding result for property (δ) was recently established by T. L. Miller and V. G. Miller [Glasgow Math. J. **38** (1996), no. 1, 21–28; MR 96m:47009]. For further applications to the theory of spectral inclusions, we refer to K. B. Laursen and the reviewer [Glasgow Math. J. **36** (1994), no. 3, 331–343; MR 95k:47002].

In connection with concrete classes of operators, we finally point out that properties (β) and (δ) play a natural role in the spectral theory of convolution operators on group and measure algebras; see Eschmeier, Laursen and the reviewer [J. Funct. Anal. **138** (1996), no. 2, 273–294; MR 97g:46066].

Michael M. Neumann (Mississippi State, MS)

98h:47048 47B40 47A60 47C05
Gillespie, T. A. (Edinburgh)
Boundedness criteria for Boolean algebras of projections. (English summary)
J. Funct. Anal. **148** (1997), *no.* 1, 70–85.

Ever since N. Dunford's original conception of the notion of a scalar-type spectral operator in a Banach space was introduced (in the late 1940s and early 1950s), one of the fundamental questions has been to determine criteria which ensure that the sum and product of two such commuting operators are again scalar-type spectral operators. An example of C. A. McCarthy [Pacific J. Math. **11** (1961), 295–307; MR **23** #A2750] shows, even in a separable reflexive Banach space, that this is not always the case. The above question is intimately related to the problem of determining criteria which guarantee that the Boolean algebra generated by two commuting, bounded Boolean algebras of projections (say, $\mathcal{A}$ and $\mathcal{B}$) is again bounded. An account of what was known up to 1970 in regard to such questions can be found in the book by N. Dunford and J. T. Schwartz [*Linear operators. Part III*, Interscience Publishers [John Wiley & Sons, Inc.], New York-London-Sydney, 1971; MR **54** #1009 (pp. 2098–2101)]. Since then, not too many general results have appeared in this direction, since it is difficult to identify geometric conditions on a Banach space which ensure that sums and products of commuting scalar-type spectral operators are again scalar-type spectral operators. Two positive results in this regard show that the class of Grothendieck spaces with the Dunford-Pettis property and the class of hereditarily indecomposable Banach spaces (recently identified by W. T. Gowers and B. Maurey) have this property [see W. J. Ricker, Glasgow Math. J. **28** (1986), no. 2, 215–222; MR 87i:47047; Acta Sci. Math. (Szeged) **59** (1994), no. 3-4, 475–488; MR 96j:47027].

The paper under review is the first significant paper (since the 1970s) which systematically identifies large (and new) classes of Banach spaces X with the property that the Boolean algebra generated by $\mathcal{A}$ and $\mathcal{B}$ is again bounded. It is shown that this is always the case for arbitrary (complex) Banach lattices and for closed subspaces of p-concave Banach lattices for some $1 \le p < \infty$ (actually, it is also shown that a quantitative refinement of this property characterizes p-concave Banach lattices within the class of all Banach lattices). These results are used to give similar boundedness results for other related classes of Banach spaces, such as subspaces of an $\mathcal{L}_p$ space, $p \in [1,\infty)$, complemented subspaces of $\mathcal{L}_\infty$ and Banach spaces admitting local unconditional structure. In the final section, two applications of the boundedness property are presented. One of these involves multiplicative representations of $C(K)$ spaces (with K a compact Hausdorff space), whilst the other is concerned with the original motivating question about sums and products of commuting scalar-type spectral operators. It should be noted that in most cases the author is also able to give an explicit bound on the Boolean algebra generated by $\mathcal{A}$ and $\mathcal{B}$ in terms of the individual bounds for $\mathcal{A}$ and $\mathcal{B}$. For details we refer the reader to this very interesting paper. *W. Ricker* (Sydney)

97a:47055 47D06 34C35 34G10 47B38 58F17
Chicone, C. (Columbia, MO); **Latushkin, Yu.** (Columbia, MO)
Quadratic Lyapunov functions for linear skew-product flows and weighted composition operators. (English summary)
Differential Integral Equations **8** (1995), *no.* 2, 289–307.

In this nice article the authors extend the method of Yu. L. Daletskiĭ and M. G. Kreĭn [*Stability of solutions of differential equations in Banach space*, English translation, Amer. Math. Soc., Providence, RI, 1974; MR **50** #5126] to the case of unbounded operators D that generate a C_0-group $(e^{tD})_{t\in\mathbf{R}}$. This method gives a quadratic Lyapunov function for the equation $u'(t)=Du(t)$, $t\in\mathbf{R}$, and permits one to characterize the hyperbolicity of the solution $u(t)=e^{tD}u(0)$. This was already observed by A. M. Lyapunov (1892), who characterized the stability of (e^{tD}) (this means $\lim_{t\to\infty}\|e^{tD}\|=0$) in terms of the eigenvalues of D in the finite-dimensional case.

To obtain the main result of this paper the authors observe that the Daletskiĭ-Kreĭn technique remains valid if the operator $(i\xi+d)(i\xi-D)^{-1}$ is replaced by e^D. This extension is applied to obtain a quadratic Lyapunov function for the case of a group of weighted composition operators generated by a flow on a compact metric space and this characterizes the hyperbolicity of linear skew-product flows. In the last part of the paper the authors illustrate the interrelations of the abstract results with Schrödinger equations, Riccati equations, Hamiltonian equations, and Anosov geodesic flows on two-dimensional Riemannian manifolds.

Let us briefly give some of the results of the article. Let D be a generator of a C_0-group $(e^{tD})_{t\in\mathbf{R}}$ on a Hilbert space H with scalar product $(\cdot,\cdot)$. The operator e^{tD}, $t\neq 0$, is said to be hyperbolic if its spectrum $\sigma(e^{tD})$ does not intersect the unit circle. It is proven that e^{tD}, $t\neq 0$, is hyperbolic if and only if there exists an invertible Hermitian operator W such that D is uniformly W-dissipative, i.e., there is some $\delta>0$ such that $((WD+D^*W)f,f)\le -\delta\|f\|^2$ for each $f\in H$. Moreover, W has the same properties as in the bounded case [see op. cit.]. This result is applied to characterize the hyperbolicity of a linear skew-product flow $\widehat{\varphi}:=\widehat{\varphi}^t$, where, for simplicity, $\widehat{\varphi}^t(x,v):=(\varphi^t x,\Phi(x,t)v)$ for $v\in H$ and x in a compact metric space X. Here $(\varphi^t)_{t\in\mathbf{R}}$ is a continuous flow on X and $\Phi: X\times\mathbf{R}\to \mathrm{GL}(H)$ is a continuous cocycle over $(\varphi^t)_{t\in\mathbf{R}}$. The linear skew-product flows are one of the major objects in studying the asymptotics of the variational differential equation $(*)$ $u'=A(\varphi^t x)u$, $x\in X$, where $A: X\to\mathcal{L}(H)$ (see the work of J. K. Hale [*Asymptotic behavior of* *dissipative systems*, Amer. Math. Soc., Providence, RI, 1988; MR 89g:58059], R. A. Johnson, K. J. Palmer and G. R. Sell [SIAM J. Math. Anal. **18** (1987), no. 1, 1–33; MR 88a:58112] and R. Sacker and Sell [J. Differential Equations **113** (1994), no. 1, 17–67; MR 96k:34136; J. Differential Equations **27** (1978), no. 3, 320–358; MR **58** #18604] and the references therein).

The main idea to characterize the hyperbolicity of $\widehat{\varphi}$ is to consider the group of weighted composition operators $(T_\Phi(t))_{t\in\mathbf{R}}$ on $L^2(X,\mu,H)$ defined by

$$(T_\Phi(t)f)(x)=\left(\frac{d(\mu\circ\varphi^{-t})}{d\mu}\right)^{1/2}\Phi(\varphi^{-t}x,t)f(\varphi^{-t}x),\quad t\in\mathbf{R},$$

where μ is a Borel φ^t-invariant measure on X with $\operatorname{supp}\mu=X$.

It is well known that the hyperbolicity of $(T_\Phi(t))_{t\in\mathbf{R}}$ is equivalent to the hyperbolicity of the linear skew-product flow $\widehat{\varphi}$ [see Yu. D. Latushkin and A. M. Stěpin, Uspekhi Mat. Nauk **46** (1991), no. 2(278), 85–143, 240; MR 92k:47062; R. Raŭ, "Hyperbolic evolution semigroups", Dissertation, Univ. Tübingen, Tübingen, 1992; per bibl.]. Therefore it follows from the first result that $\widehat{\varphi}$ is hyperbolic if and only if there exists an Hermitian invertible operator-valued function $W(\cdot)$ defined on X and continuously differentiable along the trajectories of the flow such that $\dot{W}(x):=(d/dt)W(\varphi^t x)|_{t=0}$ satisfies

$$((\dot{W}(x)+A^*(x)W(x)+W(x)A(x))v,v)\le-\delta\|v\|^2$$

for $v\in H$, $x\in X$ and some constant $\delta>0$. This is a generalization of a classical result of A. M. Samoĭlenko [*Elements of the mathematical theory of multifrequency oscillations. Invariant tori* (Russian), "Nauka", Moscow, 1987; MR 89d:34071] on the exponential dichotomy for the trivial torus for equation $(*)$ in case X is an m-dimensional torus and $\dim H<\infty$.

The techniques used in this paper are very elegant and the article is written in a very clear style. *Abdelaziz Rhandi* (Marrakech)

98m:47067 47D25 42C15 46B28 47A99

Dai, Xingde (Charlotte, NC); **Larson, David R.** (College Station, TX)

Wandering vectors for unitary systems and orthogonal wavelets. (English summary)

Mem. Amer. Math. Soc. **134** (1998), *no.* 640, viii+68 pp.

This memoir connects the areas of operator theory and orthonormal wavelets. Let $\mathscr{U}$ be a set of unitary operators acting on a Hilbert space $\mathscr{H}$ such that $\mathscr{U}$ contains the identity operator I of $\mathscr{B}(\mathscr{H})$. An element x of $\mathscr{H}$ is said to be a complete wandering vector for $\mathscr{U}$ if the set $\mathscr{U}x\equiv\{Ux\colon U\in\mathscr{U}\}$ is an orthonormal basis for $\mathscr{H}$. The set of all complete wandering vectors for $\mathscr{U}$ is denoted by $\mathscr{W}(\mathscr{U})$. An orthonormal wavelet is a function ψ in $L^2(\mathbf{R})$ such that the set $\{2^{j/2}\psi(2^jt-k)\colon j,k\in\mathbf{Z}\}$ is an orthonormal basis of $L^2(\mathbf{R})$. Taking the unitary operators on $\mathscr{H}=L^2(\mathbf{R})$ defined by $(Tf)(t)=f(t-1)$, $(Df)(t)=\sqrt{2}f(2t)$, $f\in L^2(\mathbf{R})$, $t\in\mathbf{R}$, it is easy to see that the set of orthonormal wavelets is the set of complete wandering vectors for the set of unitaries $\mathscr{U}_{D,T}\equiv\{D^jT^k\colon j,k\in\mathbf{Z}\}$. This provides the connection mentioned at the beginning of this review.

Chapter 1 presents the notion of local commutant $C_x(\mathscr{U})$ at $x\in\mathscr{H}$ for a set of operators $\mathscr{U}\subset\mathscr{B}(\mathscr{H})$:

$$C_x(\mathscr{U})\equiv\{A\in\mathscr{B}(\mathscr{H})\colon (AU-UA)x=0\text{ for all }U\in\mathscr{U}\}.$$

The main result of this chapter (Proposition 1.3) provides a parametrization for the set $\mathscr{W}(\mathscr{U})$ in terms of the unitary operators belonging to the local commutant at an element $\psi \in \mathscr{W}(\mathscr{U})$.

The case $\mathscr{U} = \mathscr{U}_1\mathscr{U}_0$, where $\mathscr{U}_0$ and $\mathscr{U}_1$ are groups of unitaries, is studied in Chapter 2. (Notice that this is the situation for orthonormal wavelets.) In this case the authors present some structural theorems for the local commutant. In particular, they give a description of the set $C_\psi^P\mathscr{U} \equiv C_\psi(\mathscr{U}) \cap \{P_\psi\}'$ (Lemma 2.15), where P_ψ is the projection operator from $\mathscr{H}$ into the linear span of $\mathscr{U}_0\psi$ and S' denotes the commutator of a set of operators S, and show that this set is a von Neumann algebra (Theorem 2.9).

The remaining chapters treat the case of the wavelet system $\mathscr{U}_{D,T}$. The results of the previous chapters are applied in Chapter 3, in which the most striking result is that the set $\mathscr{W}(\mathscr{U}_{D,T})$ is far from empty. More precisely, the linear span of the set $\mathscr{W}(\mathscr{U}_{D,T})$ is dense in $L^2(\mathbf{R})$. To prove this result the authors use the fact that if $\psi \in \mathscr{W}(\mathscr{U}_{D,T})$, the function $\eta = \mathscr{F}^{-1}(hg\widehat{\psi})$ is also an element of $\mathscr{W}(\mathscr{U}_{D,T})$, where h and g are unimodular, h is 2π-periodic and g is 2-dilation-periodic. (Here $\widehat{}$ denotes the Fourier transformation and $\mathscr{F}^{-1}$ its inverse.)

Chapter 4 deals with the case of orthonormal wavelets ψ_E whose Fourier transforms satisfy $\widehat{\psi_E} = \frac{1}{\sqrt{2\pi}}\chi_E$, where E is a measurable set contained in $\mathbf{R}$. In this case the wavelet ψ_E is called an s-elementary wavelet and the set E is said to be a wavelet set. (Such wavelets are called minimally supported frequency (MSF) wavelets in [E. Hernández, X. Wang and G. Weiss, J. Fourier Anal. Appl. **2** (1996), no. 4, 329–340; MR 97h:42015; J. Fourier Anal. Appl. **3** (1997), no. 1, 23–41; MR 98b:42049].) The authors show, with a different proof, the already known result that E must be both translation congruent to $[0, 2\pi)$ modulo 2π and dilation congruent to $[-2\pi, -\pi) \cup [\pi, 2\pi)$ modulo 2 [see Y.-H. Ha et al., Michigan Math. J. **41** (1994), no. 2, 345–361; MR 95g:42050; X. Fang and X. Wang, J. Fourier Anal. Appl. **2** (1996), no. 4, 315–327; MR 97d:42030]. Using this result many examples of s-elementary wavelets are given.

The notion of operator interpolation of wavelets is developed in Chapter 5 in the abstract setting and applied mainly when the wavelets are s-elementary. Suppose that E and F are two wavelet sets and let $\sigma\colon E \to F$ be the one-to-one and onto map implementing the 2π-translation congruence. Since E and F generate a partition of $\mathbf{R} \setminus \{0\}$ under dilation by powers of 2, σ can be extended to a one-to-one map of $\mathbf{R}$ onto $\mathbf{R}$ which is 2-homogeneous. When the map σ is involutive, that is, $\sigma^2 =$ identity, the wavelets ψ_E and ψ_F admit operator interpolation and it can be proved (Proposition 5.4) that

$$\widehat{\psi}(s) = h_1(s)\widehat{\psi_E} + h_2(s)\widehat{\psi_F}$$

is the Fourier transform of an orthonormal wavelet whenever h_1 and h_2 are 2-dilation-periodic functions on $\mathbf{R}$ such that

$$\begin{pmatrix} h_1(s) & h_2(s) \\ h_2 \circ \sigma(s) & h_1 \circ \sigma(s) \end{pmatrix}$$

is unitary for almost every $s \in \mathbf{R}$. With this criterion the authors are able to obtain in a very easy way all the Meyer wavelets starting from two very simple s-elementary wavelets.

Although the case of wavelets in $\mathbf{R}^n$ is not specifically treated, it is remarked that many of the results proved apply in this multidimensional situation, and examples are given of single wavelets on $\mathbf{R}^n$.

This is a carefully written memoir. The authors should be congratulated for shedding some needed light on the mathematical theory of wavelets using notions and results from the abstract area of operator theory. *Eugenio Hernández* (Madrid)

97a:47068 47D30 47B38

de Pagter, B. (Delft); **Ricker, W. J.** (Sydney)

Bicommutants of algebras of multiplication operators.

Proc. London Math. Soc. (3) **72** (1996), *no.* 2, 458–480.

The introduction of this paper describes its content so clearly and adequately that we cannot avoid the temptation of using parts of it for our report. Moreover, in writing this report we were in touch with the authors, and we use this opportunity to thank them for answering some of our questions and for sending us a preprint of a sequel to the paper under review. This sequel is entitled "Algebras of multiplication operators in Banach function spaces", and we will briefly mention its content later on.

We begin by reminding the reader of the classical bicommutant theorem due to von Neumann, which is a starting point of this work. According to this theorem, for each unital selfadjoint algebra $\mathscr{B}$ of operators on a Hilbert space, the bicommutant $\mathscr{B}^{cc}$ of $\mathscr{B}$ coincides with the closure of $\mathscr{B}$ in the weak operator topology τ_w. This result was an important part of von Neumann's program of the axiomatization of quantum physics. This and the intrinsic beauty of von Neumann's theorem have made the questions associated with this theorem an object of intense study in "pure" functional analysis. The general question of describing conditions on an operator algebra under which the bicommutant coincides with the closure of the algebra in the weak operator topology is far from being answered. It is known [D. Sarason, Pacific J. Math. **17** (1966), 511–517; MR **33** #590] that the hypothesis of the selfadjointness of $\mathscr{B}$ is essential for the validity of von Neumann's theorem (even if the algebra $\mathscr{B}$ is commutative). This fact explains one of the serious problems arising when one wants to generalize this theorem to algebras of operators on Banach spaces: it is not clear which algebraic properties should replace selfadjointness.

On the other hand, the paper begins with some encouraging examples that suggest the possibility of extending the von Neumann theorem. We need some notation to describe these examples. The symbols $C[0,1]$ and $L^p[0,1]$, where $1 \le p \le \infty$, have their usual meaning, and we consider $C[0,1]$ as a subalgebra of $L^\infty[0,1]$. Each function $\varphi \in L^\infty$ generates a natural multiplication operator $M_\varphi \colon f \mapsto \varphi f$ which is bounded on each L^p. Following the authors we denote by $\mathfrak{M}_p$ the unital algebra $\{M_\varphi \colon \varphi \in C[0,1]\}$ of all such multiplication operators on L^p generated by the multipliers from $C[0,1]$. It is possible to verify that for each finite p the bicommutant $\mathfrak{M}_p^{cc}$ coincides with the algebra $\{M_\varphi \colon \varphi \in L^\infty[0,1]\}$ and that $\mathfrak{M}_\infty^{cc} = \{M_\varphi \colon \varphi \text{ is Riemann integrable}\}$.

These results (of which the latter is much more surprising) can certainly be considered as the desired extensions of von Neumann's theorem, and they lead the authors to the next more general question for the following algebras of multiplication operators. Replace $C[0,1]$ by an arbitrary closed subalgebra $\mathfrak{A}$ of $L^\infty(\mu)$, and consider in the space $\mathscr{L}(L^p)$ of all continuous operators on $L^p(\mu)$ the algebra $\mathfrak{M}_p(\mathfrak{A}) = \{M_\varphi\colon \varphi \in \mathfrak{A}\}$. The measure μ here can be any non-pathological measure; to be exact, the authors deal with the class of Maharam measures. What is the bicommutant of this algebra? The purpose of this work is to give a complete answer to this question. At this point it is important to mention that this answer depends heavily on whether p is finite or not, and also on whether one deals with real or complex algebras. The first distinction is more important, as the solution for infinite p was hard to guess, while the finite p case has its "antecedents", at least for the algebra $L^\infty(\mu)$ over a σ-finite measure (for $p = 2$ this is Theorem IX.6.6 in J. B. Conway's book, *A course in functional analysis* [Springer, New York, 1985; MR 86h:46001], and for other finite p the same proof can easily be adapted). We will confine our attention mostly to spaces over the reals, mentioning the complex case only briefly. Theorem 1.1 handles the case when p is finite and states that $\mathfrak{M}_p(\mathfrak{A})^{cc} = \tau_w - \mathrm{cl}(\mathfrak{M}_p(\mathfrak{A}_1))$, where $\mathfrak{A}_1$ denotes the smallest unital subalgebra of $L^\infty(\mu)$ containing $\mathfrak{A}$.

To formulate the answer for $p = \infty$ we need an additional concept of upper and lower elements. Namely, if $\mathfrak{U}$ is a unital subalgebra of $L^\infty(\mu)$ and $f \in L^\infty(\mu)$ then consider $f^\uparrow = \sup\{u \in \mathfrak{U}\colon\ u \leq f\}$ and $f^\downarrow = \inf\{u \in \mathfrak{U}\colon\ f \leq u\}$, where the sup and inf above are taken in the Dedekind complete vector lattice $L^\infty(\mu)$ and should not be confused with the pointwise sup and inf. The collection $\mathfrak{R}(\mathfrak{U}) = \{f \in L^\infty\colon\ f^\uparrow = f^\downarrow\}$ is called the Dedekind closure and was introduced by S. Kaplan. The Dedekind closure is a closed unital subalgebra of $L^\infty(\mu)$ and this object is crucial for formulating the answer for $p = \infty$ in Theorem 1.2, which states that $\mathfrak{M}_\infty(\mathfrak{A})^{cc} = \mathfrak{M}_\infty(\mathfrak{R}(\mathfrak{A}_1))$. In the case of the complex field, the algebra $\mathfrak{A}_1$ in the statement of Theorem 1.1 should be replaced by a "selfadjoint" hull $\mathfrak{A}^\# = \{\sum f_j \bar{g}_j\colon\ f_j, g_j \in \mathfrak{A}\}$ and in the statement of Theorem 1.2 by a naturally defined Dedekind closure of $\mathfrak{A}^\#$ in $L^\infty_{\mathbb{C}}(\mu)$.

Section 2 contains many well-chosen examples illustrating the main results, the proofs of which are given in Sections 3 and 4. The concluding section, Section 5, has further examples and relevant remarks. A few words about the proofs. They are rather involved but neatly presented, and show that the problem at hand has a much more intimate relationship with the theory of vector and Banach lattices than one might suspect. The main ingredients of the proofs are the operator of the conditional expectation and a detailed analysis of the structure of the Dedekind closure.

The moment one starts reading through the proofs the question arises as to what extent the concrete geometry of L^p spaces is essential for the results obtained in the paper. The authors themselves answer this question in the above-mentioned preprint, in which—among many other things—they extend the results of the paper under review to a much more general class of Banach spaces, in particular to (L^1, L^∞) interpolation (fully symmetric) Banach function spaces with order continuous norm. Notice that from an example by J. Dieudonné [Portugal. Math. **14** (1955), 35–38; MR **17**, 1228b] it follows that the interpolation condition above is essential. For the details, more connections and, of course, the proofs, we refer the reader to these interesting papers.

Yu. A. Abramovich and A. Kitover (Indianapolis, IN and Philadelphia, PA)

99e:47064 47F05 35A27 35P25 47A40 47A55 81Q10
Sjöstrand, J. (Palaiseau)
A trace formula and review of some estimates for resonances. (English summary)
Microlocal analysis and spectral theory (*Lucca*, 1996), 377–437, *NATO Adv. Sci. Inst. Ser. C Math. Phys. Sci.*, 490, *Kluwer Acad. Publ.*, *Dordrecht*, 1997.

Formulas relating the scattering resonances and the trace of some functions of the perturbed and unperturbed operators play an important role in scattering theory and in the analysis of the distribution of resonances. For compact perturbations L of the Laplacian $-\Delta$ in $\mathbf{R}^n$, n odd, such a formula was established in increasing order of generality by Lax and Phillips, by Bardos, Guillot and Ralston, by Melrose, and by Sjöstrand and Zworski. In this case the scattering operator $S(\lambda)$ admits a meromorphic continuation to the complex plane $\mathbf{C}$ with scattering resonances λ_j and we have (1) $u(t)=$

$$2\mathrm{tr}(\cos(t\sqrt{L})-\mathbf{1}\restriction_{\mathbf{R}^n\setminus B(0,R_0)}\cos(t\sqrt{-\Delta})\mathbf{1}\restriction_{\mathbf{R}^n\setminus B(0,R_0)})=\sum_j e^{i\lambda_j|t|},$$

$t\neq 0$, where the summation is over all resonances λ_j counted with their multiplicities and the ball $B(0,R_0)$ contains the perturbation.

For noncompact perturbations the resonances can only be defined in some neighbourhood of the real axis. The situation becomes more complicated when we study long-range perturbations for which a scattering theory is not always available. In the paper under review the author proves a trace formula covering a very large class of perturbations including even-dimensional cases and long-range perturbations for Schrödinger operators. Moreover, the main result holds for a general semiclassical framework concerning two unbounded selfadjoint operators $P_i(h)$, $i=0,1$, in a Hilbert space $\mathscr{H}_i=\mathscr{H}_{i,R_0}\oplus L^2(\mathbf{R}^n\setminus B(0,R_0))$, depending on $h\in(0,h_0]$, $h_0>0$. The operators $P_i(h)$ satisfy some general assumptions formulated in the framework of complex scaling developed by Sjöstrand and M. Zworski [J. Amer. Math. Soc. **4** (1991), no. 4, 729–769; MR 92g:35166]. The resonances λ_j of $P_i(h)$ are defined in a sector

$$S_{\theta_0}=\{z\in\mathbf{C}\setminus\{0\}\colon\ 0\leq-\arg z<2\theta_0\}$$

as the eigenvalues of $P_i(h)$ defined on a suitable contour in $\mathbf{C}^n$. To state the main result of the paper let us consider a domain Ω independent of h, and let $W\subset\subset\Omega$ be an open relatively compact subset of $e^{i(-2\theta_0,\varepsilon_0]}[0,+\infty)$, where ε_0 can be taken arbitrarily small. Let $I=\mathbf{R}^+\cap W$, $J=\mathbf{R}^+\cap\Omega$ be intervals and let $\chi\in C_0^\infty(J)$ be h-independent and such that $\chi=1$ near $\overline{I}$. Finally, let Ω_- [resp., W_-] be the intersections of Ω [resp., W] with $e^{i(-2\theta_0,0]}(0,+\infty)$ and let $f(z;h)$ be holomorphic in z on Ω and satisfy $|f(z;h)|\leq 1$ for $z\in\Omega\setminus W$. The author proves the following formula:

$$(2)\quad \mathrm{tr}[(\chi f)(P_1(h))-(\chi f)(P_0(h))]=$$
$$\Big[\sum_{\lambda_{\bullet,j}\in W_-}f(\lambda_{\bullet,j};h)-\sum_{\mu\in\sigma(P_\bullet)\cap(-\infty,0)\cap W_-}f(\mu;h)\Big]_0^1+O(h^{-n_{\max}}),$$

where $n_{\max}$ is an integer, $\lambda_{\bullet,j}$ denotes the resonances of $P_\bullet(h)$, the dot stands for $i=0,1$ and $[a_\bullet]_0^1$ is equal to a_1-a_0. The proof of (2) is quite ingenious; it is based on reduction to a suitable Grushin problem for the scaled operator as well as on trace class estimates for the inverse of the Grushin problem. The approach of the author does not use scattering theory and this is the main difference with the previous proofs of (1). The integer $n_{\max}$ is connected with the assumptions for $P_i(h)$ but in many cases we have $n_{\max} = n$. It is important to note that the trace formula (2) is local; this is very useful for the analysis of the distribution of the resonances in some bounded domain.

The author presents several applications based on his trace formula. First he compares the trace formulas (1) and (2) in the case when P_i are independent on h. More precisely, he obtains an asymptotic version of (1) saying that for every $\chi \in C_0^\infty(\mathbf{R}^+)$ and $\lambda \in \mathbf{R}$ we have $\widehat{\chi u}(\lambda) = \sum_j \hat{\chi}(\lambda - \lambda_j) + O(|\lambda|^{-\infty})$. Next, lower bounds of resonances near the real axis are discussed and some previous results are extended to the even-dimensional case. The author reviews some upper bounds on the resonances near the real axis, treating operators depending on h. Recently, he obtained an impressive result for semiclassical Schrödinger operators based on the formula (2) [see in *Séminaire sur les Équations aux Dérivées Partielles, 1996–1997*, Exp. No. II, 17 pp., École Polytech., Palaiseau, 1997; MR 98k:35143]. Under general assumptions he proved that in a complex neighbourhood W of a curve $\gamma\colon [0,1] \to \mathbf{C}$, related to the analytic singularities on the negative axis, we have

$$[\#(\mathrm{Res}(P.(h)) \cap W)]_0^1 \geq \frac{1}{C(W)} h^{-n}, \quad 0 < h \leq h_0(W).$$

The paper is a very important contribution to the theory of resonances.

{For the entire collection see MR 97k:35002.} *Vesselin M. Petkov* (Talence)

97a:47077a 47G10 42A50 45P05 47A99 47B38
Rochberg, Richard (St. Louis, MO)
Size estimates for eigenvalues of singular integral operators and Schrödinger operators and for derivatives of quasiconformal mappings. (English summary)
Amer. J. Math. **117** (1995), *no.* 3, 711–771.

97a:47077b 47G10 42A50 45P05 47A99 47B38
Rochberg, Richard (St. Louis, MO)
A singular integral operator with a large eigenvector.
Function spaces (*Edwardsville, IL,* 1994), 347–351, *Lecture Notes in Pure and Appl. Math.,* 172, *Dekker, New York,* 1995.

97a:47077c 47G10 42A50 45P05 47A99 47B38
Rochberg, Richard (St. Louis, MO)
Size estimates for eigenvectors of singular integral operators with $T1 = T^*1 = 0$. (English summary)
Interaction between functional analysis, harmonic analysis, and probability (*Columbia, MO,* 1994), 409–420, *Lecture Notes in Pure and Appl. Math.,* 175, *Dekker, New York,* 1996.

The word "eigenvalues" in the title of the first paper ([a]) is an annoying misprint: the paper is devoted to estimates of eigenfunctions rather than those of eigenvalues. The latter topic is comparatively well studied; see e.g. the survey paper of M. Birman and the reviewer [Uspekhi Mat. Nauk **32** (1977), no. 1(193), 17–84, 271; MR **55** #11104], and the books of H. H. König [*Eigenvalue distribution of compact operators*, Birkhäuser, Basel, 1986; MR 88j:47021] and A. Pietsch [*Eigenvalues and* *s-numbers*, Cambridge Univ. Press, Cambridge, 1987; MR 88j:47022b]. The problem for the eigenfunctions is much harder. The main reason for this is that, unlike eigenvalues, the eigenfunctions are not unitarily invariant objects, and therefore the powerful Hilbert space technique does not apply.

The paper [a] is basic in the cycle under review; the short notes [b] and [c] complement its contents. The author starts with the operators

$$T_\Lambda f = \sum_Q \lambda_Q \Big(|Q|^{-1} \int_Q f(t)\,dt \Big) \chi_Q(x). \tag{1}$$

In (1) Q runs over the set of all dyadic cubes in $\mathbf{R}^d$, $|Q|$ and χ_Q denote the volume and the characteristic function of Q, and $\Lambda = \{\lambda_Q\}$ is the family of nonnegative scalars, which specifies the operator. It was proved by the author and S. Semmes [J. Funct. Anal. **86** (1989), no. 2, 237–306; MR 90k:47047] that, for any $0 < p < \infty$, $\Lambda \in l_p$ implies $T_\Lambda \in S_p$. The following estimate for the normalized eigenfunctions of T_Λ, $T_\Lambda f = \lambda f$, is established in [a], Theorem A: There exist positive constants c_1, c_2 and c_3 such that for any dyadic cube Q and $\lambda |Q|^{1/2} \mu > c_1$ one has (2)

$$|Q|^{-1} \mathrm{meas}(Q \cup \{x \colon |f(x)| > \mu\}) \le \exp(-c_2 \log(c_3 \lambda |Q|^{1/2} \mu)^{\frac{p}{p-1}}).$$

In [b] it is shown that the estimate is sharp, within the values of c_j.

The author mentions in [a] that (2) actually remains valid under the milder assumption $\Lambda \in l_{p,\,\mathrm{weak}}$. This is a useful improvement; in my opinion, it would be worthwhile to state the result for Λ in this class at once: the preference of l_p to $l_{p,\,\mathrm{weak}}$ is nothing but customary practice.

The result for singular integral operators is much harder to understand and explain because many technical parameters, introduced in the course of the proof, appear in the final formulation of Theorem C (p. 747). Basically, it says that any normalized eigenfunction f, $Tf = \lambda f$, corresponding to $\lambda \ne 0$, on any dyadic cube W is majorized by $C(W) \exp(R_W(1))$, where $C(W)$ is a constant and $R_W(1)$ is a function whose size is under control. To establish this estimate, the author splits T into the sum $T = L + R_1 + R_2$, where R_1 and R_2^* are paraproducts (not dyadic ones), whose construction makes use of the functions $u := T(1)$ and $v := T^*(1)$. The assumptions on T are stated in terms of u, v, and the integral kernel of L. If $u = v = 0$, the result of Theorem C can be considerably improved; this is done in [c].

Eigenvalues of the Schrödinger equation on $\mathbf{R}^3$, $-\Delta f + vf = \lambda f$, are classical objects, and a lot is known about their behaviour [see, e.g., H. L. Cycon et al., *Schrödinger operators with application to quantum mechanics and global geometry*, Springer, Berlin, 1987; MR 88g:35003]. However, usually some specific requirements on v are imposed, like its membership in the Kato class. In (1) the only general assumption is $v \in A_\infty$. Theorem D (p. 755) basically states that under one more assumption of a somewhat more technical nature (depending on the value of λ), the eigenfunction on any dyadic cube W is majorized by $C(W) \exp((\Delta - \lambda)^{-1}(v\chi_W))$.

A result of similar nature (Theorem E) is obtained for derivatives of quasiconformal mapping. Here we are dealing with the equation $f_{\bar z} = \mu f_z$, where μ is a function rather than a constant. Nevertheless, there is some resemblance to an eigenvalue problem, and the same approach applies.

The proofs cannot be called easy, though each step is rather elementary. Numerous informal explanations by the author simplify the comprehensive reading (even in spite of a good bunch of misprints). The main idea is a suitable representation of an operator as a sum of pieces, which are well localized or at least "well concentrated". For the operators (1) the wanted decomposition is given directly by the definition. This case is the simplest one and serves the author as a model. In other cases the author uses wavelets and their technical generalization, called "molecules". The heart of the proof is a careful combinatorial analysis of a sum each of whose terms involves a product of these pieces. This sum appears when the resolvent of an operator is expanded into the Neumann series.

The paper [a] is interesting as the first attempt to develop a unified approach to the study of eigenfunction behaviour for a wide class of operators on $\mathbf{R}^d$.

{For the entire collection containing the second paper see MR 96c:46001; for the entire collection containing the third paper see MR 96f:00024.}

Michael Z. Solomyak (Rehovot)

98k:47097 47G10 15A52 30E25 34A55 34E20 41A60 82B23 82B44

Deift, Percy A. (New York, NY); **Its, Alexander R. [It·s, A. R.]** (Indianapolis, IN); **Zhou, Xin** (Durham, NC)

A Riemann-Hilbert approach to asymptotic problems arising in the theory of random matrix models, and also in the theory of integrable statistical mechanics.

Ann. of Math. (2) **146** (1997), *no.* 1, 149–235.

The central problem considered in this paper is the $x \to +\infty$ asymptotic analysis of a Fredholm determinant $P_x = \det(1 - K_x)$, where K_x is the integral operator with kernel $\sin x(z - z')/\pi(z - z')$ acting on the disjoint union of intervals $J = \bigcup_{j=1}^n (a_j, b_j)$. In random matrix theory P_x arises as the bulk scaling limit of the probability that no eigenvalues lie in the interval collection $(x/\pi)J$ for the Gaussian unitary ensemble (GUE). The principal result in the paper is a theorem that gives the asymptotics of $(d/dx)\log P_x$ as $x \to \infty$: Theorem 1.31. Let J be the union of disjoint intervals given above. Then for any $j = 1, 2, \cdots$ one has the $x \to \infty$ asymptotic development

$$\frac{d}{dx}\log P_x = -2\alpha x + \frac{d}{dx}\log\theta(xV) + \frac{G_1(x)}{x} + \cdots + \frac{G_j(x)}{x^j} + O\left(\frac{1}{x^{j+1}}\right),$$

where $\alpha > 0$ is a positive constant, θ is a theta function, V is a vector in $\mathbf{R}^n$, and the functions $G_j(x)$ are bounded quasiperiodic functions of x.

The quantities α, θ, V, and G_j are explicitly described in the paper, and this theorem is a refinement of a result originally proved by H. Widom [Comm. Math. Phys. **171** (1995), no. 1, 159–180; MR 96i:47050; J. Approx. Theory **77** (1994), no. 1, 51–64; MR 95f:42041; Advances in Math. **3** (1969), 127–232; MR **39** #418]. Widom's results are, in turn, a mathematically rigorous generalization of asymptotics obtained by J. des Cloizeaux and M. L. Mehta [J. Mathematical Phys. **14** (1973), 1648–1650; MR **48** #6500] and F. J. Dyson [Comm. Math. Phys. **47** (1976), no. 2, 171–183; MR **53** #9993] for the single interval case, $n = 1$. In this case the theta function, θ, degenerates and one has $(d/dx)\log P_x = -x - 1/4x + O(1/x^2)$, as a special case of Theorem 1.31. Widom's analysis, which is a tour de force of classical analysis and approximation theory, makes it clear that the rigorous investigation of such asymptotics is a difficult and subtle problem. The main point of the paper under review is that the general collection of techniques which one might collectively refer to as "Riemann-Hilbert steepest descent methods" have an illuminating application to the study of such asymptotics. Riemann-Hilbert problems involve the characterization of a (matrix-valued) analytic function defined on the complement of some collection of contours in the complex projective plane in terms of (matrix-valued) multipliers which relate the boundary values of the analytic function on the two sides of the contours. That there is a Riemann-Hilbert characterization of the logarithmic derivative of $\det(1-K_x)$ goes back to M. Jimbo et al. [Phys. D **1** (1980), no. 1, 80–158; MR 84k:82037], where it appears as an early instance of what later became a general theory of τ-functions. A simplified version of this characterization is the starting point in the Deift-It·s-Zhou analysis, which then proceeds by introducing a cleverly chosen multiplier which at least formally reduces the Riemann-Hilbert problem to an asymptotically simpler and "explicitly solvable" form. Both the choice of the multiplier and the explicit solution of the asymptotic Riemann-Hilbert problem involve detailed function theory on the hyperelliptic curve associated with the equation $y^2=\prod_{j=1}^n(z-a_j)(z-b_j)$ (the critical function appearing in the multiplier is also a central figure in Widom's analysis). These developments formally produce the asymptotics in Theorem 1.31 but some non-uniformity in the convergence stops matters short of a proof.

When a system of contours has finite length and the multipliers are uniformly close to the identity it is a relatively elementary fact that one can solve an integral equation to find an analytic function that solves the Riemann-Hilbert problem which has this system of multipliers. It is satisfying that this simple result is the principal tool used to justify the approximations that lead to the asymptotics in Theorem 1.31. However, it is by no means a trivial matter to see how to find the approximations to which this result applies. Deift, It·s, and Zhou employ ideas from steepest descent methods, monodromy preserving deformation theory, WKB asymptotics, and turning point analysis just to arrive at an appropriate approximation. All the while the focus of the method is on producing an approximation characterized by Riemann-Hilbert data—for example, in the course of the analysis, a linear differential equation is first approximated, then asymptotically solved near a turning point and finally discarded when only a certain aspect of the local behavior of the solution is required for the Riemann-Hilbert data.

We will conclude this review with a few words about the scope of Riemann-Hilbert methods in the analysis of nonlinear "exactly solvable" problems. The success of these methods includes a growing number of problems such as: the analysis of connection problems for Painlevé transcendents [A. R. It·s, A. S. Fokas and A. A. Kapaev, Nonlinearity **7** (1994), no. 5, 1291–1325; MR 96c:34011], an analysis of the asymptotics of solutions to equations like the mKdV equation [P. A. Deift and X. Zhou, Ann. of Math. (2) **137** (1993), no. 2, 295–368; MR 94d:35143], examples of universality in bulk scaling for non-Gaussian matrix models [A. R. It·s and P. Bleher, "Semiclassical asymptotics of orthogonal polynomials, Riemann-Hilbert problem, and universality in the matrix model", Preprint 97-2, Dept. Math. Sci., Indiana Univ.–Purdue Univ. Indianapolis (IUPUI), Indianapolis, IN; per revr.], and the analysis of the asymptotics of correlation functions in some solvable models in statistical mechanics [A. R. It·s et al., Internat. J. Modern Phys. B **4** (1990), no. 5, 1003–1037; MR 91k:82009]. These examples and more are surveyed in the introduction to the paper under review and in the extensive references there. The method itself is evolving. In this paper the "method of steepest descent" differs from the earlier Deift-Zhou triangular factorization-contour shifting scheme and the role of WKB methods is also changing. Considerable art is required to apply the method to problems as subtle as the one considered here but in the hands of skilled practitioners the technique is opening a mathematical window on a wide range of difficult asymptotic problems with importance in both mathematics and physics. *John N. Palmer* (Tucson, AZ)

99d:47074 47H30 30C20
Lanza de Cristoforis, Massimo (Padua)
A functional decomposition theorem for the conformal representation.
J. Math. Soc. Japan **49** (1997), *no.* 4, 759–780.
This paper deals with the nonlinear operator which maps an injective function φ defined on the unit disk in $\mathbf{R}^2$ with values in $\mathbf{R}^2$, of class $C^{m,\alpha}$, and with nonvanishing Jacobian, into the unique holomorphic, one-to-one map $g[\varphi]$ of the unit disk $\mathscr{D}$ onto the Jordan domain $\varphi(\mathscr{D})$, normalized by $g[\varphi](0)=\varphi(0)$, $g'[\varphi](0)\in(0,+\infty)$. There are some papers which have investigated the dependence of the conformal representation upon the domain [see, e.g., T. Radó, Acta Sci. Math. (Szeged) **1** (1923), 180–186; JFM 49.0247.03; A. Rosenblatt, C. R. Acad. Sci. Paris **202** (1936), 1398–1400; JFM 62.0384.02; C. R. Acad. Sci. Paris **202** (1936), 1832–1834; JFM 62.0384.03; D. Zeitlin, "Behavior of conformal maps under analytic deformation of the domain", Thesis, Univ. Minnesota, 1957; per bibl.; H. Yoshikawa, J. Math. Soc. Japan **12** (1960), 174–186; MR **23** #A1786]. The methods of the above-mentioned papers are different from those in the work under review.

A new crucial idea of the paper is that the nonlinear operator $\varphi\mapsto g[\varphi]$ can be decomposed as $\varphi\mapsto\varphi\circ S[\varphi]^{(-1)}$, where $S[\varphi]^{(-1)}$ denotes the inverse function of $S[\varphi]$, and the operator $\varphi\mapsto S[\varphi]$ is analytic from a set of "admissible" φ's in $C^{m,\alpha}(\mathrm{cl}\,\mathscr{D},\mathbf{R}^2)$ to $C^{m,\alpha}(\mathrm{cl}\,\mathscr{D},\mathbf{R}^2)$. This result allows one to give precise information on the regularity of $\varphi\mapsto g[\varphi]$.

The analyticity statement for $S[\cdot]$ is completely new and is in contrast with the nondifferentiability of the map $\varphi\mapsto g[\varphi]$ from the set of "admissible" φ's in $C^{m,\alpha}(\mathrm{cl}\,\mathscr{D},\mathbf{R}^2)$ to $C^{m,\alpha}(\mathrm{cl}\,\mathscr{D},\mathbf{R}^2)$ (see a related topic in L. V. Kantorovich and V. I. Krylov [*Approximate methods of higher analysis* (Russian), Fifth corrected edition, Gosudarstv. Izdat. Fiz.-Mat. Lit., Moscow-Leningrad, 1962; MR **27** #4338]). The author shows, in particular, that a stronger expansion theorem for S (not for g) holds when no analyticity assumption on the dependence of φ on the variable in $\partial\mathscr{D}$ is made, thus confirming the remarkable difference between $\varphi\mapsto g[\varphi]$ and $\varphi\mapsto g[\varphi]^{(-1)}\circ\varphi$.

The proof of the main result is based on the following key idea: having observed that $g[\varphi]$ is the unique solution to a nonlinear elliptic system for each φ, the author studies this system locally by means of an implicit function theorem. By using the regularity theorem for $\varphi\mapsto S[\varphi]$, together with two "higher-order" differentiability results for the superposition and for the inversion operator, he proves that $\varphi\mapsto g[\varphi]$ is of class C^k, $k\in\mathbf{N}$, if the derivative loss between the domain and the range of g is k, and if the domain of $g[\cdot]$ is restricted to the closure of the smooth functions in the norm of the domain. This statement is shown to be sharp in the case $k=1$.

The author explains briefly how such differentiability statements can be applied to study the dependence of the spectrum of the Laplacian upon perturbations of the domain. Besides this, he notes that "functional regularity" theorems for $g[\cdot]$ and for $S[\cdot]$ find applications in fluid-solid interaction problems where the conformal representation can be used to handle domain-dependent exterior boundary-value problems.

Theorem 3.10 of the paper covers a recent result of S. J. Wu [Comm. Pure Appl. Math. **46** (1993), no. 10, 1303–1326; MR 96e:30020].

S. K. Vodop′yanov (Novosibirsk)

D. REVIEWS IN GEOMETRY AND TOPOLOGY

98f:22018 22E40 11F41 22E46

Farb, Benson (Chicago, IL); **Schwartz, Richard [Schwartz, Richard Evan]** (Chicago, IL)

The large-scale geometry of Hilbert modular groups. (English summary)

J. Differential Geom. **44** (1996), *no.* 3, 435–478.

The word metric on a finitely generated group supplies the latter with the structure of a path metric space. A quasi-isometry between two metric spaces is a map that distorts distances by a bounded factor, above a given scale. M. Gromov's program of studying infinite discrete groups includes, in particular, classifying the groups up to quasi-isometry [cf. in *Geometric group theory, Vol. 2 (Sussex, 1991)*, 1–295, Cambridge Univ. Press, Cambridge, 1993; MR 95m:20041]. A uniform lattice in a Lie group is quasi-isometric to the ambient group, so the question reduces to studying quasi-isometries of Lie groups. The first breakthrough in classifying non-uniform lattices in semisimple Lie groups was made by Schwartz [Inst. Hautes Études Sci. Publ. Math. No. 82 (1995), 133–168 (1996); MR 97c:22014]. He proved that two non-uniform irreducible lattices in a semisimple group G of $\mathbf{R}$-rank 1, where G is not isomorphic locally to $\mathrm{PSL}(2,\mathbf{R})$, are quasi-isometric iff they are commensurable (up to conjugation in G). Also he conjectured that the same holds for higher-rank groups (that is, Mostow rigidity for lattices should hold if instead of isomorphisms between lattices one considers quasi-isometries).

The paper under review is the first attempt to understand quasi-isometries of lattices in higher-rank groups. Here we have $G=\mathrm{PSL}(2,\mathbf{R})\times\mathrm{PSL}(2,\mathbf{R})$ and Γ is a non-uniform irreducible lattice in G. An example of such a lattice comes from the ring of integers O_d in the real quadratic field $\mathbf{Q}(\sqrt{d})$, $d>0$ being a square-free integer. If $M\in\mathrm{PSL}(2,O_d)$ is a matrix and $\overline{M}$ stands for the matrix obtained by replacing each entry of M by its Galois conjugate, then the map $M\to(M,\overline{M})$ sends $\mathrm{PSL}(2,O_d)$ to an irreducible lattice in G, called a Hilbert modular group. It is known that any irreducible lattice in G is commensurable to one of the Hilbert modular groups.

The idea behind the proof of the conjecture is to replace Γ with its geometric model Ω, called a neutered space. Namely, to get Ω one takes the product $X=H^2\times H^2$ of two hyperbolic planes and deletes a Γ-equivariant disjoint union of horoballs in X. Then Γ acts properly and cocompactly on Ω and hence is quasi-isometric to Ω. In the course of their study of quasi-isometries between two neutered spaces, the authors develop coarse metrical versions of theorems from algebraic topology, such as Alexander duality and the Jordan separation theorem.

The main result is that any quasi-isometry between two nonuniform irreducible lattices $\Gamma_1,\Gamma_2\subset G$ is given by an element $g\in G$ which "commensurates" Γ_1 and Γ_2, i.e. $\Gamma_2\cap g\Gamma_1 g^{-1}$ is of finite index in Γ_2. In particular, two distinct Hilbert modular groups cannot be quasi-isometric. As a consequence, one proves that a finitely generated group quasi-isometric to Γ is a finite extension of a lattice in G commensurable with Γ. Another application is a complete quasi-isometry classification of lattices in G. Namely, there are three quasi-isometry classes of lattices in G. The first consists of all uniform lattices in G. The second consists of two subclasses: products of two nonuniform lattices in $\mathrm{PSL}(2,\mathbf{R})$ and products of nonuniform and uniform lattices therein. The last one consists of infinitely many subclasses given by Hilbert modular groups.

The results of the paper have since been generalized by Schwartz to non-uniform irreducible lattices in a product of rank-1 semisimple groups [Acta Math. **177** (1996), no. 1, 75–112; MR 97m:53093]. The same question for irreducible lattices in products of higher-rank semisimple groups was recently solved by A. Eskin [J. Amer. Math. Soc. **11** (1998), no. 2, 321–361; MR 98g:22005, see p. D6 in this volume]. This completed the quasi-isometry classification for lattices in semisimple Lie groups.

Alexander Starkov (Moscow)

98g:22005 22E40 20F32 53C35

Eskin, Alex (Chicago, IL)

Quasi-isometric rigidity of nonuniform lattices in higher rank symmetric spaces.

J. Amer. Math. Soc. **11** (1998), *no.* 2, 321–361.

This paper represents the culmination of Gromov's program for classifying finitely generated groups by their geometry, for the class of irreducible lattices acting on symmetric spaces. The program has progressed through different classes of symmetric spaces, with a growing body of tools and techniques (see the survey in [B. Farb, Math. Res. Lett. **4** (1997), no. 5, 705–717; MR 98k:22044]). The current paper handles nonuniform lattices acting on symmetric spaces of rank ≥ 2, with many beautiful new ideas and several clever new applications of well-known tools.

Two metric spaces X, Y are said to be quasi-isometric if there exists a map $\varphi\colon X \to Y$, called a quasi-isometry between X and Y, such that, for some constants $K \geq 1$, $C \geq 0$, the inequality $K^{-1}d(x,y) - C \leq d(f(x), f(y)) \leq Kd(x,y) + C$ holds for all $x, y \in X$, and moreover each point of Y is within distance C of some point of $f(X)$. Any finitely generated group Γ has a well-defined notion of geometry, namely the quasi-isometric geometry of the word metric on Γ. Gromov proposed the problem of classifying all finitely generated groups up to quasi-isometry. This problem is motivated by a principle discovered by J. Milnor, A. S. Schwarz and others—any geodesic metric space X on which Γ acts properly discontinuously and cocompactly is quasi-isometric to the group Γ endowed with its word metric.

Quasi-isometries play an important role in the famous rigidity theorem of G. D. Mostow, extended by G. Prasad and G. A. Margulis: if G is a semisimple Lie group with finite center and associated symmetric space X, if Γ_1, Γ_2 are lattices in G, i.e. discrete subgroups with cofinite volume, and if Γ_1 is isomorphic to Γ_2, then Γ_1, Γ_2 are conjugate by an element of G. When the lattice is uniform, i.e. cocompact, the proof was given by Mostow; for nonuniform lattices, the case when G has rank 1 was done by Prasad and the higher-rank case by Margulis. The key idea in these proofs is to use the isomorphism between Γ_1 and Γ_2 to induce a quasi-isometry on X, and show that this induces a map on the Furstenberg boundary of X.

In studying quasi-isometric rigidity, the main problem is: given a finitely generated group Λ, describe all finitely generated groups that are quasi-isometric to Λ. The quasi-isometric rigidity theorem for uniform lattices says that if G, X are as above, and if Λ is a finitely generated group quasi-isometric to (a uniform lattice acting on) X, then there is a short exact sequence $1 \to F \to \Lambda \to \Gamma \to 1$, where F is finite and Γ is a uniform lattice acting on X. For hyperbolic 3-space the proof was done by Gromov and D. Sullivan; higher-dimensional real hyperbolic spaces by P. Tukia; complex hyperbolic spaces by R. Chow; quaternionic hyperbolic spaces and the Cayley hyperbolic plane by P. Pansu; and higher-rank symmetric spaces by B. Kleiner and B. Leeb. The latter case was re-proved by Eskin and Farb using different methods which are relevant to the current paper. (For the hyperbolic plane the result has its own long history, culminating independently in work of D. Gabai and joint work of A. Casson and D. Jungreis.)

R. Schwartz conjectured a surprisingly strong quasi-isometric rigidity theorem for Γ an irreducible, nonuniform lattice in a semisimple Lie group G with finite center. Schwartz' conjecture says that if Λ is a finitely generated group quasi-isometric to Γ, then there is a short exact sequence $1 \to N \to \Lambda \to \Gamma' \to 1$, where N is finite and Γ' is a lattice in G commensurable to Γ. The conjecture also says that two nonuniform irreducible lattices in G are quasi-isometric if and only if they are commensurable in G, up to conjugation. This conjecture was verified by Schwartz when G has rank 1, by Farb and Schwartz for $G = \mathrm{SL}(2, \mathbf{R}) \times \mathrm{SL}(2, \mathbf{R})$, and by Schwartz for products of $\mathrm{SL}(2, \mathbf{R})$ and $\mathrm{SL}(2, \mathbf{C})$.

The current paper completes the picture by verifying Schwartz' conjecture for any G of rank ≥ 2. To accomplish this, the author verifies another conjecture of Schwartz: if Γ is an irreducible, nonuniform lattice in G, and if φ is any quasi-isometry of Γ to itself, then there exists an element $g \in G$ such that $g\Gamma \cap \Gamma$ has finite index in Γ—that is, g is a commensurator of Γ—and such that the actions of φ and of g on Γ are within a bounded distance of each other.

Here is an overview of the proof. The strategy is to show that the quasi-isometry φ induces a bi-Hölder map on the Furstenberg boundary $\widehat{X}$, which in turn induces an order preserving bijection of the Tits building of X. By the Tits theorem, φ is within a bounded distance of an isometry of X. A final argument shows that this isometry is the action of a commensurator of Γ.

Obtaining the action of φ on $\widehat{X}$ is an extremely delicate matter, requiring technical finesse and ingenious application of a wide range of powerful tools. It is worthwhile to first review the problem for the case of uniform lattices. If φ is a quasi-isometry of hyperbolic space, say, and if γ is a geodesic, then $\varphi(\gamma)$ is a quasigeodesic, and Morse's lemma says that $\varphi(\gamma)$ is contained in a bounded neighborhood of a geodesic. By considering the ends of this geodesic one obtains the induced action of φ on the boundary of hyperbolic space. For uniform lattices in a higher-rank symmetric space X, geodesics are replaced by flats in X—isometrically embedded copies of $\mathbf{R}^n$ where n is the rank. Ends of geodesics are replaced by Weyl chambers of flats, to obtain the Furstenberg boundary $\widehat{X}$. Unfortunately, as discovered by Kleiner and Leeb, a quasiflat in X is not contained in a bounded neighborhood of a flat; however, it is contained in a bounded neighborhood of a finite union of flats, as discovered by Eskin and Farb.

Consider now the case at hand, where φ is a quasi-isometry of an irreducible, nonuniform lattice Γ in G with rank ≥ 2. What needs to be shown is that φ acts on the set of flats in X. Unfortunately, φ does not even act naturally on X, only on a Γ-invariant cocompact subset of X called the neutered space, which we will denote N. In the cases handled by Schwartz and by Farb-Schwartz, a reasonably natural extension of φ to all of X was constructed, but those methods do not seem suitable when the $\mathbf{Q}$-rank of the lattice is ≥ 1.

To overcome this obstacle, the author abandons the idea of extending φ to X, and instead he applies an array of tools to obtain an action of φ on almost all of $\widehat{X}$. First, the Lubotzky-Mozes-Raghunathan theorem says that the intrinsic, geodesic metric on N is quasi-isometric to the metric induced from X. Given a flat F in X, the intersection $F \cap N$ is a subset of F called a "flat with holes". Next, the proof combines the Moore ergodicity theorem with results of Eskin and Farb on flats with holes to prove that for almost all flats F, the set $\varphi(F \cap N)$ is contained in a bounded neighborhood of a finite union of flats.

To complete the proof there is still much hard work to be done. The author proves that for almost all flats F, the set $\varphi(F \cap N)$ is actually contained in a bounded neighborhood of a single flat, and one obtains an action of φ on almost all of $\widehat{X}$. Then he shows that this partially defined map is bi-Hölder, and so extends to a bi-Hölder map defined on all of $\widehat{X}$. Finally, as indicated above, he shows that this induces an order preserving bijection of the Tits building, and the associated isometry is a commensurator of Γ.

The paper is well written and merits close study, for anyone interested in the rich and rapidly evolving field of quasi-isometric rigidity. *Lee Mosher* (Newark, NJ)

98b:22034 22E50 20G25 22E35
Lusztig, George (Cambridge, MA)
Classification of unipotent representations of simple p-adic groups.
Internat. Math. Res. Notices **1995,** *no.* 11, 517–589.

Let G be the group of rational points of a connected reductive group defined over a p-adic field F (not assumed to be of characteristic 0). If P is a parahoric subgroup of G with pro-unipotent radical U the quotient M is the group of rational points of a reductive group defined over the residue field $\mathbf{F}_q$ of F. If $(\pi, \mathscr{V})$ is an irreducible smooth representation of G we say that it is of level 0 if, when restricted to some P as above, the finite-dimensional subspace $\mathscr{V}^U$ of U-fixed vectors is nonzero and contains a cuspidal representation of M. We remark that $\mathscr{V}^U$ is typically zero; if it is not, then, after changing P if necessary, $\mathscr{V}^U$ must contain a cuspidal representation. We say that $(\pi, \mathscr{V})$ is unipotent if it is of level 0 and if, in addition, it contains a unipotent cuspidal representation of M. The best known and most common example of a unipotent representation is the case when $(\pi, \mathscr{V})$ contains a nontrivial Iwahori fixed vector: in other words, the parahoric in question is an Iwahori subgroup I so that M is a torus and the unipotent cuspidal representation is the trivial character.

Here are two facts concerning level-0 representations: (i) If σ is an irreducible cuspidal representation of M as above one can consider the intertwining algebra $\mathscr{H}(\sigma)$ of the smooth representation c-$\mathrm{Ind}_P^G(\check{\sigma})$, where "$\vee$" denotes the contragredient. The structure of this algebra was investigated by the reviewer [Invent. Math. **114** (1993), no. 1, 1–54; MR 94g:22035]; roughly speaking, it is the tensor product of a twisted group algebra by an affine Iwahori-Hecke algebra with not necessarily equal parameters; this is quite analogous to the situation for parabolic induction in groups over finite fields, and indeed the proof uses the theory of R. B. Howlett and G. I. Lehrer [Invent. Math. **58** (1980), no. 1, 37–64; MR 81j:20017] in a crucial way. The group algebra may exhaust the intertwining algebra; however, if σ is unipotent the classification of such representations implies that the group algebra piece is very small and comes from automorphisms of P, and the 2-cocycle is trivial. (ii) The category of smooth representations of G which are generated by their σ-isotypic vectors is equivalent to the category of nondegenerate $\mathscr{H}(\sigma)$-modules. This was proved by the reviewer [Compositio Math. **118** (1999), no. 2, 135–157], and by A. Moy and G. Prasad [Comment. Math. Helv. **71** (1996), no. 1, 98–121; MR 97c:22021]; although these proofs are different, each relies to some extent on the theory of types due to C. J. Bushnell and P. Kutzko [Proc. London Math. Soc. (3) **77** (1998), no. 3, 582–634].

This generalises a well-known result of A. Borel [Invent. Math. **35** (1976), 233–259; MR **56** #3196] for the Iwahori case. It is worth pointing out, however, that under quite general hypotheses one can show that there are bijections between the isomorphism classes of irreducible objects on each side here. This does not depend on knowing a categorical equivalence, and proofs can be found in the paper by the reviewer cited above, as well as in Bushnell and Kutzko [*The admissible dual of* $\mathrm{GL}(N)$ *via compact open subgroups*, Ann. of Math. Stud., 129, Princeton Univ. Press, Princeton, NJ, 1993; MR 94h:22007].

Now assume G is split simple adjoint, and write ${}^LG$ for the connected component of a dual group for G—we are deliberately vague here—so that ${}^LG$ is simply connected. In the present paper the author establishes a bijection between the set of isomorphism classes of irreducible unipotent representations of G and the set of triples (s, N, ρ), where $s \in {}^LG$ is semisimple, $N \in \mathrm{Lie}\,{}^LG$ is nilpotent, with $\mathrm{Ad}(s)N = qN$, where q is the order of the residue field of F, and ρ is an irreducible representation of the component group $Z_{{}^LG}(s, N)/Z^0_{{}^LG}(s, N)$ which is trivial on the image of the centre $Z_{{}^LG}$. He obtains variants for the inner forms of G, by placing no restriction on ρ. This completely proves an earlier conjecture of the author's [Trans. Amer. Math. Soc. **277** (1983), no. 2, 623–653; MR 84j:22023]. The Iwahori part of the conjecture was established earlier by D. Kazhdan and the author [Invent. Math. **87** (1987), no. 1, 153–215; MR 88d:11121] using equivariant K-homology: they showed that the equivariant K-homology tensored by $\mathbf{C}$ of the variety of triples $\{(u, B, B')|\ B, B' \text{ Borel groups in } {}^LG,\ u \text{ unipotent} \in B \cap B'\}$ provides a model for the regular representation of the Iwahori-Hecke algebra $\mathscr{H}$. They were then able to construct a "standard" $\mathscr{H}$-module for each triple (s, N, ρ); this module has a unique irreducible quotient, and this provides the desired correspondence, by Borel's version of (ii) above. Unfortunately, the equivariant K-homology method runs into serious difficulties if one tries to extend it to the general conjecture. In the paper under review, following a strategy outlined in his address at the 1990 International Mathematics Congress in Kyoto [in *Proceedings of the International Congress of Mathematicians, Vol. I, II (Kyoto, 1990)*, 155–174, Math. Soc. Japan, Tokyo, 1991; MR 93e:20059], the author proves his conjecture essentially by replacing affine Hecke algebras by graded versions and using his parametrisation of the irreducible modules of these latter algebras. Indeed, in a series of previous papers he has used (ordinary) equivariant homology and perverse sheaves in the equivariant derived category to construct and parametrise such modules. In particular this method provides a new proof of the Iwahori case.

We now try to provide a rough, and sometimes heuristic, sketch of some of the steps in the bijection itself, beginning with the geometric (complex) side.

(a) First, given a nilpotent orbit $\mathscr{C}$ of some Levi component L of a connected reductive complex algebraic group ${}^LG$ and an irreducible equivariant cuspidal local system $\mathscr{F}$ of this orbit the author has constructed a graded Hecke algebra $\overline{H}$ over $\mathbf{C}[r]$ (r an indeterminate) with typically unequal parameters [see Inst. Hautes Études Sci. Publ. Math. No. 67 (1988), 145–202; MR 90e:22029; in *Representations of groups (Banff, AB, 1994)*, 217–275, Amer. Math. Soc., Providence, RI, 1995; MR 96m:22038 (with errata for Part I)]. For example, if the Levi component is a maximal torus with the trivial nilpotent orbit this algebra is a graded algebra associated to a filtration of the standard affine Hecke algebra over $\mathbf{C}(v, v^{-1})$. In any case the graded Hecke algebra can be usefully identified with the full endomorphism ring of a certain equivariant perverse sheaf associated to $\mathcal{F}$ in the equivariant derived category $\mathcal{D}_{^LG\times\mathbf{C}^*}\mathrm{Lie}^LG$, or again, as a graded vector space, with the full equivariant homology of a generalised variety of triples with values in an associated local system.

(b) Now fix ^LG and consider the equivalence relation on the triples $(L,\mathcal{C},\mathcal{F})$ in (a) obtained via the action of ^LG. This provides a finite collection of graded Hecke algebras $\overline{H}_1,\cdots,\overline{H}_k$. Let $\mathrm{Irr}_{r_0}\overline{H}_i$ denote the set of isomorphism classes of irreducible modules of $\overline{H}_i$ on which the indeterminate r acts via the real number r_0. The author has shown that there is a bijection from $\coprod_i \mathrm{Irr}_{r_0}\overline{H}_i$ to the set whose elements consist of triples (s,N,ρ) as above except that s is now in Lie^LG and q is replaced by $2r_0$. We refer the reader to the author's two-part paper [op. cit.; MR 90e:22029; op. cit.; MR 96m:22038], especially Part II, Section 8.

(c) In particular, if ^LG is simply connected and almost simple with maximal torus $^LT={}^LT_c\times{}^LT_h$ decomposed into its compact and hyperbolic parts, and $t'\in{}^LT_c$, we may apply (a) and (b) above to the connected reductive group $Z_{^LG}(t')$.

(d) Fix $(^LG,{}^LT)$ as in (c) and let I be an index set in one-to-one correspondence with a set of simple affine roots (or the extended Dynkin diagram) for this pair. To $J\subsetneq I$ one can associate a connected reductive group $^LG_J\subsetneq{}^LG$. Fix a nilpotent orbit $\mathcal{C}$ on Lie^LG_J and an irreducible ^LG_J-equivariant cuspidal local system $\mathcal{F}$ on $\mathcal{C}$. In Section 5 of the paper under review the author associates an affine Hecke algebra $H(^LG,{}^LG_J,\mathcal{C},\mathcal{F})$ with unequal parameters starting from a certain root datum $(\mathcal{R},\mathcal{L},\mathcal{R}',\mathcal{L}')$ arising from the triple $(^LG_J,\mathcal{C},\mathcal{F})$. Let $\mathcal{W}$ denote the Weyl group associated to this root datum; let $\mathcal{T}$ denote the complex algebraic torus whose cocharacter group is $\mathcal{L}'$. The centre $\mathcal{Z}$ of $H(^LG,{}^LG_J,\mathcal{C},\mathcal{F})$ can be identified with the algebra of regular functions of the quotient $\mathcal{W}\backslash\mathcal{T}\times\mathbf{C}^*$; thus a character of $\mathcal{Z}$ can be identified with a pair $(\mathcal{W}t,v_0)$, $t\in\mathcal{T}$. Let $\mathrm{Irr}_{\mathcal{W}t,v_0}H(^LG,{}^LG_J,\mathcal{C},\mathcal{F})$ denote the set of isomorphism classes of irreducible $H(^LG,{}^LG_J,\mathcal{C},\mathcal{F})$ modules which transform on $\mathcal{Z}$ by the character $(\mathcal{W}t,v_0)$.

(e) Take a pair $(\mathcal{W}t,v_0)$ as in (d) above and decompose $t=t_ct_h$ into its compact and hyperbolic parts. We assume v_0 is a positive real number. To t_c the author associates an element $t_c^\psi\in T_c$, and ^LG_J is a Levi component of $Z_{^LG}(t_c^\psi)$. One then has the graded Hecke algebra $\overline{H}(Z_{^LG}(t_c^\psi),{}^LG_J,\mathcal{C},\mathcal{F})$ as in (a). The element $\log t_h$ lies in $\mathrm{Lie}\,\mathcal{T}$ and this can be identified with $\mathrm{Lie}\,Z_{^LG_J}\subset\mathrm{Lie}^LT$. Just as in (d) one has the set $\mathrm{Irr}_{\mathcal{W}_{t_c}\log t_h,r_0}\overline{H}(Z_{^LG}(t_c^\psi),{}^LG_J,\mathcal{C},\mathcal{F})$, where $r_0=\log v_0$ and the Weyl group $\mathcal{W}_{t_c}$ may be identified with $N_{Z_{^LG}(t_c^\psi)}{}^L(G_J)/{}^LG_J$. The author then applies the results of his paper [J. Amer. Math. Soc. **2** (1989), no. 3, 599–635; MR 90e:16049] to establish a bijection $\mathrm{Irr}_{\mathcal{W}t,v_0}H(^LG,{}^LG_J,\mathcal{C},\mathcal{F})\to\mathrm{Irr}_{\mathcal{W}_{t_c}\log t_h,r_0}\overline{H}(Z_{^LG}(t_c^\psi),{}^LG_J,\mathcal{C},\mathcal{F})$.

(f) Fix a positive number v_0, and let $\overline{\mathfrak{G}}$ denote the collection of triples $(J,\mathcal{C},\mathcal{F})$. By varying the orbits of $\mathcal{T}$ under $\mathcal{W}$, and combining the results in (c)–(e), the author obtains a bijection of $\coprod_{(J,\mathcal{C},\mathcal{F})\in\overline{\mathfrak{G}}}\mathrm{Irr}\,H_{v_0}(^LG,{}^LG_J,\mathcal{C},\mathcal{F})$ with the set $\mathfrak{G}(v_0)$ of triples in the introduction, except that q is replaced by v_0^2. (See Theorem 5.21 of the paper.) Here $H_{v_0}(\cdots)$ denotes the algebra $H(\cdots)$ specialised at v_0. Note that if one takes $v_0=1$, the left side of the bijection becomes the collection of irreducible representations of a certain finite set of generalised affine Weyl groups, while the right side becomes the collection of pairs (g,ρ): $g\in{}^LG$, ρ an irreducible representation of the component group of $Z_{^LG}(g)$.

(g) Let $S'(I)$ denote the (finite) group of characters of $Z_{^LG}$. If $\chi\in S'(I)$ then in (f) above we can consider the sets $\overline{\mathfrak{G}}_\chi,\mathfrak{G}_\chi(v_0)$: the first set denotes triples $(J,\mathcal{C},\mathcal{F})$ where $Z_{^LG}$ acts on each stalk of $\mathcal{F}$ by χ; the second denotes triples (s,N,ρ) where

$\rho|$image $Z_{^LG}$ acts by χ. The bijection in (f) respects this decomposition. The elements $\mathfrak{G}_\chi$ are called by the author geometric diagrams of type χ.

We remark that much of §§ 2–5 of the paper is concerned with carefully establishing the bijections in (d)–(f) above. We now pass to the p-adic (arithmetic) side.

(h) Let $\widetilde{F}$ denote the unramified closure of the field F. We take a split adjoint simple F-group $\mathbf{G}_0$ and fix an inner form $\mathbf{G}$ of $\mathbf{G}_0$; the Bruhat-Tits classification implies that $\mathbf{G}$ splits over $\widetilde{F}$ and we identify $\mathbf{G}$ with its $\widetilde{F}$-points. Take the complex group ^LG in (a)–(g) above to be the connected component of the dual group of $\mathbf{G}$. (We are being careless here.) The Galois action on $\mathbf{G}$ is determined by a Frobenius element Φ which induces a permutation u on the vertices $\mathbf{I}$ of the ($\widetilde{F}$-) local Dynkin diagram of $\mathbf{G}$. This permutation can be identified with an element χ of the group $S'(I)$ in (g) by a result of Kneser-Kottwitz.

(i) Let P be a parahoric subgroup of $G=\mathbf{G}^\Phi$; we write $\widehat{P}$ for the full centraliser of the facet in the affine enlarged building of G fixed by P. Up to G-conjugacy P corresponds to a u-stable subset $\mathbf{J}$ of $\mathbf{I}$, and we suppose that its reductive quotient M admits a unipotent cuspidal representation $\mathbf{E}$, with contragredient $\check{\mathbf{E}}$. This forces strong restrictions on both M and P. Let ψ be a character of the "component" group $\widehat{P}/P$. The triple $(\mathbf{J},\mathbf{E},\psi)$ is called by Lusztig an arithmetic diagram of type u; let $\overline{\mathfrak{A}}_u$ denote the collection of such diagrams. Write $\mathscr{H}(\mathbf{I},\mathbf{J},u,\mathbf{E})$ for the intertwining algebra End_G (c-$\mathrm{Ind}_P^G(\check{\mathbf{E}})$) (compact induction); the irreducible unital modules of this algebra are in one-to-one correspondence with the irreducible admissible representations of G which contain $\mathbf{E}$ by (ii) above. Now, this algebra can be decomposed into a direct sum of two-sided ideals $\mathscr{H}^\psi$ corresponding to central idempotents parametrised by the ψ above, and $\mathscr{H}^\psi$ is isomorphic with the algebra End_G (c-$\mathrm{Ind}_{\widehat{P}}^G(\check{\mathbf{E}}_\psi)$), where $\check{\mathbf{E}}_\psi$ is an extension of $\check{\mathbf{E}}$ to $\widehat{P}$, parametrised by ψ. This last assertion depends on the fact that $\mathbf{E}$ is unipotent. Each of the algebras $\mathscr{H}^\psi$ is noncanonically isomorphic to a fixed algebra $\mathscr{H}'(\mathbf{I},\mathbf{J},u,\mathbf{E})$ which has the same presentation as End_G (c-$\mathrm{Ind}_P^G(\check{\mathbf{E}})$) except that the group algebra part can be identified with the group algebra of $N_G(P)/\widehat{P}$. Let $\mathscr{U}(G)$ denote the set of isomorphism classes of irreducible unipotent representations of G, and let $\mathrm{Irr}\,\mathscr{H}'(\mathbf{I},\mathbf{J},u,\mathbf{E})$ denote isomorphism classes of irreducible unital modules of the algebra $\mathscr{H}'(\mathbf{I},\mathbf{J},u,\mathbf{E})$. Combining these facts one obtains a bijection $\mathscr{U}(G)\to\coprod_{(\mathbf{J},\mathbf{E},\psi)\in\overline{\mathfrak{A}}_u}\mathrm{Irr}\,\mathscr{H}'(\mathbf{I},\mathbf{J},u,\mathbf{E})$.

(j) Fix $\chi\in S'(I)$ corresponding to u. The desired bijection

$$\coprod_{(\mathbf{J},\mathbf{E},\psi)\in\overline{\mathfrak{A}}_u}\mathrm{Irr}\,\mathscr{H}'(\mathbf{I},\mathbf{J},u,\mathbf{E})\to\coprod_{(J,\mathscr{C},\mathscr{F})\in\overline{\mathfrak{G}}_\chi}\mathrm{Irr}\,H_{v_0}(^LG,{}^LG_J,\mathscr{C},\mathscr{F})$$

is realised in §7 of the paper under review by matching families of arithmetic diagrams with families of geometric diagrams case by case, and specialising the bijection in (g) to $v_0=\sqrt{q}$, where q is the order of the residue field of F. For this the author is guided by earlier bijections he has established in the classification of character sheaves. (The interested reader may wish to look at the reviewer's paper [Ann. Sci. École Norm. Sup. (4) **29** (1996), no. 5, 639–667; MR 97g:22013] for a special case.)

This is an important paper which will repay close reading.

Lawrence Morris (Worcester, MA)

98i:22026 22E55 20G35

Prasad, Gopal (Ann Arbor, MI); **Rapinchuk, Andrei S.** (Charlottesville, VA)

Computation of the metaplectic kernel.

Inst. Hautes Études Sci. Publ. Math. No. 84 (1996), 91–187 (1997).

Let F denote a global field with ring of adèles $\mathbf{A}$. If S is a finite, possibly empty, set of places of F we write $\mathbf{A}_S$ for the ring of S-adèles: it is the restricted direct product of those F_v for $v \notin S$. Now let G be an absolutely simple simply connected algebraic group defined over F, and write $I = \mathbf{R}/\mathbf{Z}$ for the circle group. The paper under review is concerned with the description of topological central extensions of the type $1 \to I \to E \to G(\mathbf{A}_S) \to 1$ which split over $G(F)$. Thus E is locally compact second countable, I embeds as a closed central subgroup of E, the natural map $E/I \to G(\mathbf{A}_S)$ is a homeomorphism, and there is a continuous section over $G(F)$; such extensions are classified by the S-metaplectic kernel $M(S,G) = \ker(H^2(G(\mathbf{A}_S), I) \to H^2(G(F), I))$ (abstract cochains on the right, measurable cochains on the left). What is new and difficult here is that the authors systematically allow G to be F-anisotropic. But before discussing the nuts and bolts of the paper, we shall briefly describe the framework and some origins of the problem.

To begin, in the theory of automorphic forms there are the well-known two-fold metaplectic covers of $\mathrm{Sp}_{2n}(\mathbf{A})$; these were introduced by Weil to explain theta functions in a group-theoretic context with the goal of recovering an adelic group-theoretic explanation of the Siegel formula. In another direction, T. Kubota introduced generalised theta functions which are automorphic under more general metaplectic covers of $\mathrm{SL}_2(\mathbf{A})$, defined by the nth power norm residue symbol of F, when F contains the nth roots of unity. The functions in this case are proportional to the residues of certain Eisenstein series. The simplest example is $F = \mathbf{Q}(\omega)$, where $\omega^2 + \omega + 1 = 0$. Here S. J. Patterson studied the associated Eisenstein series and showed that their Fourier coefficients are cubic Gauss sums; from this via a "converse theorem" he deduced that the Fourier coefficients of the associated theta functions, called cubic theta functions, were also cubic Gauss sums. Starting from this he and Heath-Brown were able to determine the distribution of cubic Gauss sums at prime arguments, thus answering a celebrated question of Kummer.

The relation between these metaplectic covers and the group $M(S,G)$ can be explained as follows. First, if H, A are second countable locally compact groups and A is abelian with a continuous H action, C. C. Moore [Trans. Amer. Math. Soc. **113** (1964), 40-63; ibid. **113** (1964), 64–86; MR **30** #2106; Trans. Amer. Math. Soc. **221** (1976), no. 1, 1–33; MR **54** #2867; Trans. Amer. Math. Soc. **221** (1976), no. 1, 35–58; MR **54** #2868] showed how to define cohomology groups $H^*(H,A)$ with good properties, based on Borel measurable cochains; these groups can themselves be topologised (in several ways). In particular, $H^2(H,A)$ classifies topological extensions of H by A and if the H action on A is trivial then such extensions are of course central. Second, Moore's work implies that if $H^2(G(\mathbf{A}_S), I)$ is locally compact and Hausdorff and if $[G(\mathbf{A}_S), G(\mathbf{A}_S)]$ is dense in $G(\mathbf{A}_S)$ then $G(\mathbf{A}_S)$ admits a universal central extension $1 \to \pi_1(G(\mathbf{A}_S)) \to \widetilde{G}_S \to G(\mathbf{A}_S) \to 1$: thus, given any other central extension $1 \to C' \to \widehat{G}' \to G(\mathbf{A}_S) \to 1$ there is a unique continuous homomorphism $f\colon \widetilde{G}_S \to \widehat{G}'$ which makes the diagram obtained from the two exact sequences commutative. Moreover, $\pi_1(G(\mathbf{A}_S))$ is isomorphic to the Pontryagin dual of the topological group $H^2(G(\mathbf{A}_S), I)$. Third, it is elementary that if $G(F)$ is perfect, then since it is discrete, it has a universal cover as abstract group with central kernel $\pi_1(G(F))$, and the map $i: G(F) \to G(\mathbf{A}_S)$ provides a map $i_*: \pi_1(G(F)) \to \pi_1(G(\mathbf{A}_S))$. Finally, if we let D be the closure of $i_*(\pi_1(G(F)))$ in $\pi_1(G(\mathbf{A}_S))$ then the central extension $1 \to \pi_1(G(\mathbf{A}_S))/D \to \widetilde{G}_S/D \to G(\mathbf{A}_S) \to 1$ splits over $G(F)$ and is universal with respect to this property. The group $\pi_1(G(\mathbf{A}_S))/D$ is called by Moore the relative fundamental group $\pi_1(G(\mathbf{A}_S), G(F))$; thus under these assumptions $M(S,G)$ is just the Pontryagin dual of this group and measures the number of finite central extensions of $G(\mathbf{A}_S)$ splitting over $G(F)$. A central extension of $G(\mathbf{A})$ by a finite group which splits over $G(F)$ (case $S = \varnothing$) is traditionally called a metaplectic cover of $G(\mathbf{A})$. We remark that as a consequence of the local results to be discussed below the group $H^2(G(\mathbf{A}_S), I)$ is indeed locally compact and Hausdorff, as it is a direct product of finite groups; on the other hand it is known that $G(F)$ is perfect (which implies that $G(\mathbf{A})$ is perfect as well) except possibly for some F-rank-one forms of E_6, and some anisotropic forms of A_n, D_4, E_6.

One may view the S-metaplectic kernel as a cohomological formulation of the S-congruence kernel for appropriate S. To describe this it is convenient to write V for the set of places of F, and V_∞ for the set of Archimedean places of F. If $v \in V$ we write F_v for the associated local field, and $\mathfrak{o}_v$ for the local ring of integers if v is finite. If $S \subset V$ is finite containing V_∞ we write $\mathfrak{o}_S$ for the ring of S-integers: $\mathfrak{o}_S = \{x \in F \mid x \in \mathfrak{o}_v,\ v \notin S\}$. We also write SL_n for the special linear group, viewed as an algebraic group defined over F.

Now with G as above, let $\rho: G \to \mathrm{SL}_n$ be a faithful representation; the problems we shall consider will not depend on the choice of ρ. If $S \supset V_\infty$ we set $\Gamma = \rho(G(F)) \cap \mathrm{SL}_n(\mathfrak{o}_S)$; a subgroup of $G(F)$ that is commensurable with Γ is called an S-arithmetic subgroup. One can construct a plentiful supply of such subgroups as follows. Let $\mathfrak{a}$ be any nonzero ideal of $\mathfrak{o}_S$; there is then a reduction map $\pi_\mathfrak{a}: \mathrm{SL}_n(\mathfrak{o}_S) \to \mathrm{SL}_n(\mathfrak{o}_S/\mathfrak{a})$ and we set $\Gamma_\mathfrak{a} = (\ker)\pi_\mathfrak{a} \cap \Gamma$, which we call the principal S-congruence subgroup of level $\mathfrak{a}$. An S-arithmetic subgroup that contains a principal congruence subgroup of some level is called an S-congruence subgroup. The congruence subgroup problem asks whether every S-arithmetic subgroup is an S-congruence subgroup. For $G = \mathrm{SL}_n$, $F = \mathbf{Q}$, $S = V_\infty$ the answer is negative if $n = 2$ and this has been known since the late 19th century; the response is positive if $n > 2$; this was proved by J. L. Mennicke [Ann. of Math. (2) **81** (1965), 31–37; MR **30** #2083] and H. Bass, M. Lazard and J.-P. Serre [Bull. Amer. Math. Soc. **70** (1964), 385–392; MR **28** #5117].

Serre reformulated this problem in the following suggestive fashion. First, we can put a topology on $G(F)$ by taking as a base of neighbourhoods of 1 the S-congruence subgroups; the strong approximation theorem (Kneser, Platonov, Prasad) implies that the completion of $G(F)$ with respect to this topology is just the group $G(\mathbf{A}_S)$. Second, we can put a topology on $G(F)$ by taking as a base of neighbourhoods of 1 the S-arithmetic subgroups; we denote the completion of $G(F)$ with respect to this topology by $\widehat{G}_S$. Since every S-congruence subgroup is an S-arithmetic subgroup the universal property of the completion map $G(F) \to \widehat{G}_S$ implies there is a unique continuous homomorphism $\widehat{G}_S \to G(\mathbf{A}_S)$ which splits over $G(F)$. By examining the closure of Γ_S in the completed groups it follows that this homomorphism is onto, and that the kernel $C(S,G)$ is a profinite group, which we call the S-congruence kernel. The S-congruence subgroup problem is equivalent to determining when this kernel is trivial. For $G = \mathrm{SL}_n$, computing it is closely related to computing certain algebraic K_2-groups, and to the computation of Whitehead torsion in topology.

We can now describe the advertised relation between the groups $C(S,G)$ and $M(S,G)$ when $S \supset V_\infty$. First, the S-congruence kernel $C(S,G)$ is part of a topological extension $1 \to C(S,G) \to \widehat{G}_S \to G(\mathbf{A}_S) \to 1$ splitting over $G(F)$, but it may not be central. If it is, and if $G(F)$ is perfect, then one can show this extension is universal. From this one deduces as above that $M(S,G)$ is the Pontryagin dual of $C(S,G)$. H. Matsumoto [Ann. Sci. École Norm. Sup. (4) **2** (1969), 1–62; MR **39** #1566] proved that $C(S,G)$ is central in $\widehat{G}_S$ for all (simply connected) Chevalley groups of F-rank ≥ 2, and it has been conjectured that this is so whenever $\sum_{v\in S} F_v$-rank $(G) \geq 2$ and G is F-isotropic for all finite $v \in S$. For the considerable progress on this conjecture we refer the reader to the survey by Prasad [in *Proceedings of the International Congress of Mathematicians, Vol. I, II (Kyoto, 1990)*, 821–832, Math. Soc. Japan, Tokyo, 1991; MR 93c:22019] or the article by Rapinchuk [Astérisque No. 209 (1992), 11, 73–84; MR 94b:20047].

The first results of a general nature on the computation of the S-metaplectic kernel were due to Moore [Inst. Hautes Études Sci. Publ. Math. No. **35** (1968), 157–222; MR **39** #5575]. Starting from the results of Steinberg on the generators and relations for the abstract universal central extension of a simply connected split group over any field F, he found a relation between the norm power residue symbols of local class field theory and the topological central extensions of $\mathrm{SL}_2(F)$ when F is a local non-Archimedean field. He then proved that $\pi_1(G(F))$ is a quotient of $\pi_1(\mathrm{SL}_2(F))$ via the image of the SL_2 coming from any long root in G when F is local non-Archimedean. (Matsumoto [op. cit.] later proved that this map is an isomorphism.) These results imply that $\pi_1(G(\mathbf{A}_S), G(F))$ is a quotient of $\pi_1(\mathrm{SL}_2(\mathbf{A}(S)), \mathrm{SL}_2(F))$ when F is global. In the same paper he proved the "uniqueness of reciprocity" of global class field theory; he used this to compute $M(S,G)$ (actually the relative fundamental group) for SL_2, and by the same reduction trick as above showed that for any simply connected split group and any finite $S \subset V$, $M(S,G)$ is trivial if S contains a noncomplex place and in any case it is isomorphic to a subgroup of the dual of the group of roots of unity of F. Shortly afterwards, Matsumoto [op. cit.] explicitly constructed certain topological central extensions of $G(\mathbf{A}_S)$ when G is a Chevalley group and from this was able to prove that in fact $M(S,G)$ is isomorphic to $\widehat{\mu}(F)$ if S does not contain any noncomplex place. In the same spirit, V. V. Deodhar [Amer. J. Math. **100** (1978), no. 2, 303–386; MR 80c:20058] extended these results to quasisplit groups. These methods are "local to global" starting from generators and relations.

For non-(quasi)-split G the generator and relation approach is not feasible. Prasad and M. S. Raghunathan [Ann. of Math. (2) **119** (1984), no. 1, 143–201; MR 86e:20051a; Ann. of Math. (2) **119** (1984), no. 2, 203–268; MR 86e:20051b] replace it by Bruhat-Tits theory to compute $H^2(G(F), I)$ when F is non-Archimedean (in which case one replaces measurable cochains by continuous cochains) when G is F-isotropic. These ideas deserve a bit more explanation as they are quite intricate, and different from the techniques above. For the time being F will denote a local non-Archimedean field, with ring of integers $\mathfrak{o}$ and residue field $\mathbf{F}_q$ of characteristic p.

(a) Let X be any locally compact abelian group viewed as a trivial $G(F)$-module; we can then consider the group $C^i(X)$ of X-valued oriented simplicial i-cochains of the Bruhat-Tits building of $G(F)$. There is a spectral sequence abutting to $H^*(G(F), X)$ (continuous cochains) whose $E_1^{i,j}$-term is $H^j(G(F), C^i(X))$, and since

X is a trivial $G(F)$-module one quickly finds that $E_2^{i,0}=0$, $i>0$, and $H^i(G(F),X)=0$, $i\geq 1$, when $X=\mathbf{R}$ or $\mathbf{Q}$. From the exact sequences $0\to\mathbf{Z}\to\mathbf{R}\to\mathbf{R}/\mathbf{Z}\to 0$, $0\to\mathbf{Z}\to\mathbf{Q}\to\mathbf{Q}/\mathbf{Z}\to 0$ one then deduces that $H^i(G(F),\mathbf{Q}/\mathbf{Z})\cong H^i(G(F),\mathbf{R}/\mathbf{Z})$ for any $i\geq 1$. Applying this to $i=2$ one then breaks up $H^2(G(F),\mathbf{Q}/\mathbf{Z})$ as a direct sum $H^2(G(F),\mathbf{Q}/\mathbf{Z})=H^2(G(F),J)\oplus H^2(G(F),J')$ where J [resp. J'] is the p-primary part [resp. sum of the prime-to-p-primary components]. Prasad and Raghunathan then compute each of these pieces separately.

(b) For the tame part $H^2(G(F),J')$ one carefully analyses certain exact sequences arising from the spectral sequence; by making use of known results on Schur multipliers of finite groups of Lie type (which appear here as reductive quotients of parahoric subgroups) one can show that $H^2(G(F),J')$ has the same order as the tame part of $\widehat{\mu}(F)$.

(c) For the wild part one again employs the spectral sequence, as well as the previously known results of Moore and Deodhar, and some case-by-case arguments. The most difficult part of the paper is to show that $H^2(G(F),J)$ is a cyclic group of order at most that of the wild part of $\widehat{\mu}(F)$. We refer the reader to the introduction of Prasad and Raghunathan's work for a lucid description of the arguments.

(d) Finally, using (b) and (c) the authors find $H\subset G$ with $H\cong R_{L/F}(\mathrm{SL}_2)$ over F, where L is a finite cyclotomic extension of F, such that the restriction $H^2(G(F),I)\to H^2(H(F),I)\cong\widehat{\mu}(F)$ is an isomorphism onto a subgroup of index at most 2. We remark that if G is anisotropic, the authors obtain information on $H^2(G(F),I)$ as well.

Building upon the above results, Prasad and Raghunathan [Invent. Math. **71** (1983), no. 1, 21–42; MR 84f:20051] proved the following: Let F be global, $S\subset V$ finite, and G F-isotropic. Then $M(S,G)$ is trivial if S contains a non-Archimedean place or if it contains a real place v such that the group $G(F_v)$ is not topologically simply connected; in general it is a subgroup of $\widehat{\mu}(F)$. For this they used the local injectivity results above to reduce the problem to F-rank-one groups and then to restriction of scalars of SL_2, at which point one can apply the results of Moore. We note that A. Bak and U. Rehmann [J. Algebra **78** (1982), no. 2, 475–547; MR 85j:11041] used K-theory techniques to prove these results for $\mathrm{SL}_{n,D}$, where $n\geq 2$ and D is a division algebra.

When G is F-anisotropic, several problems present themselves. One can still use the local results above to describe the topological central extensions of $G(\mathbf{A}_S)$ by I, but it seems to be difficult to determine when such an extension splits over $G(F)$, and it is not obvious how one might reduce the problem to a smaller-rank group. Thus new ideas are necessary, and this brings us to the paper under review.

One is supposing that G is absolutely simple, simply connected but not necessarily F-isotropic; moreover, if G is F-isotropic (in particular, split or quasisplit), one can ignore it if it is convenient to do so in view of known results. The main theorem essentially says that modulo conjecture (U) (page 111) for certain forms of type A_n (see below), $M(S,G)$ is isomorphic to a subgroup of $\widehat{\mu}(F)$. The authors first give a (new) proof that $M(S,G)$ is always finite (Theorem 2.7). A key role is then played by an elementary three-step reduction process:

(e) If $M(S\cup\{v\},G)$ is trivial for almost all v, then $M(S,G)$ is isomorphic to a subgroup of $\widehat{\mu}(F)$. This essentially follows by choosing a v such that G_v is F_v-isotropic (so that $H^2(G(F_v),I)$ is isomorphic to a subgroup of $\widehat{\mu}(F_v)$) and $M(S\cup$

$\{v\}, G)$ is trivial, showing that $M(S,G)$ embeds into $H^2(G(F_v),I)$ and applying Chebotarev's density theorem. See 1.13 of the paper under review.

(f) As a consequence of (e), it is enough to suppose that S contains a place v which is non-Archimedean such that G is F_v-isotropic, or contains a real place such that $G(F_v)$ is not topologically simply connected, and prove that in these cases $M(S,G)$ is trivial.

(g) Suppose that for any finite $U \subset V$ the group $M_U(G) = \ker(H^2(G(U),I) \to H^2(G(F),I))$ is trivial; here $G(U) = \prod_{v \in U} G(F_v)$. Suppose that for all but finitely many $v \notin S$ there is an F-subgroup $H \subset G$ such that (i) $M(S,H)$ is trivial, and (ii) the restriction map $H^2(G(F_v),I) \to H^2(H(F_v),I)$ is injective. Then $M(S,G)$ is trivial. This is Lemma 6.1 in the paper under review.

The toughest part of the paper is devoted to proving (f) for groups of type A_n; in fact the authors are unable to prove what they need for those groups of type A which require a division algebra whose centre is a quadratic extension of F: for these groups (called "special" in the paper under review) they assume a conjecture to make a uniform statement. For the other classical groups one uses their geometric realisations to find a subgroup H of type A_n satisfying the conditions in (g); this is relatively easy. Fortunately one can avoid special groups in the reduction process. For exceptional groups one again employs the same strategy; this time it is necessary to study the Galois cohomology of these groups. For groups of type G_2, F_4, E_7, E_8 the authors show that such forms split over a suitable quadratic extension of F; work of Weisfeiler then enables them to construct an anisotropic subgroup H which is either an inner form of A_1 or a (non-special!) special unitary group in 3 variables. Forms of type E_6 and D_4 are technically more troublesome but the reduction strategy is the same.

We briefly sketch some of the ideas for inner forms of type A_n, for which partial results when $n = 1$ were obtained by J. Klose [Math. Z. **193** (1986), no. 4, 625–649; MR 88c:11068] and Rapinchuk [Izv. Akad. Nauk SSSR Ser. Mat. **51** (1987), no. 5, 1033–1064, 1118; MR 89h:11074]. For the group G of elements of norm 1 in a quaternion central F-algebra D the authors show that it is enough to prove the following (Theorem 3.4): If S contains a noncomplex place v_0 such that D_{v_0} splits, and q is a prime ($= 2$, if v_0 is real), then $M(S,G)$ contains no element of order q. (Note that at this point one knows that $M(S,G)$ is finite.) To prove this they take such an element x which corresponds to a central extension of $G(\mathbf{A}_S)$ by I splitting over $G(F)$. Now, there is a finite set of places S' containing S, all Archimedean places and all places v where G_v is F_v-anisotropic such that if $v \notin S'$ then for any maximal compact subgroup K of $G(F_v)$, $H^1(K,I)$ is trivial, and the map $H^2(G(F_v),I) \to H^2(K,I)$ is trivial. This implies that it is enough to show that $x_v = 0$ (this has a meaning) for all $v \notin S'$; for this the authors use the local description: it is enough to show that the corresponding character $\chi_v = 0$ (in $\widehat{\mu}(F_v)$). They proceed by choosing a separable quadratic extension L/F which is a subfield of D with the property that the degree $[L_{v_i} : F_{v_i}] = 1$ for $i = 0, 1$; this corresponds to a maximal anisotropic mod centre torus T in $\mathrm{GL}_{1,D}$. By means of T and explicit local commutator calculations in $\mathrm{GL}_{1,D}$ the authors show the desired result. The dévissage for inner forms of type A_n, $n > 1$, is similar in spirit once the authors have made a reduction to the case where the division algebra has prime degree.

In Section 8 of the paper the authors use the results they have obtained (with the same caveat for special groups) to show that $M(\varnothing, G) \cong \widehat{\mu}(F)$. They get this by showing that the central extension of $G(\mathbf{A})$ by $\widehat{\mu}(F)$ splitting over $G(F)$, constructed by P. Deligne [Inst. Hautes Études Sci. Publ. Math. No. 84 (1996), 35–89 (1997); MR 98d:14056], corresponds to an element of order $|\widehat{\mu}(F)|$ in $M(\varnothing, G)$. The claim then follows since they have already established that $M(\varnothing, G)$ is isomorphic to a subgroup of $\widehat{\mu}(F)$. They then use this result to complete the earlier local calculations of Prasad and Raghunathan: they show that if G is absolutely simple, simply connected and F-isotropic for F local non-Archimedean then $H^2(G(F), I) \cong \widehat{\mu}(F)$.

This review has only sketched some of the principal contributions to this area. For surveys on this and related topics, and further references, the interested reader can consult the previously cited surveys by Prasad and Rapinchuk as well as the article by Raghunathan [in *Proceedings of the Hyderabad Conference on Algebraic Groups (Hyderabad, 1989)*, 465–494, Manoj Prakashan, Madras, 1991; MR 92i:20048].

Lawrence Morris (Worcester, MA)

97b:51009 51H10 51-02 51H25

Salzmann, Helmut [Salzmann, Helmut R.] (Tübingen); **Betten, Dieter** (Kiel); **Grundhöfer, Theo** (Würzburg); **Hähl, Hermann** (Kiel); **Löwen, Rainer** (Braunschweig); **Stroppel, Markus** (Darmstadt)

★**Compact projective planes.**

With an introduction to octonion geometry.

de Gruyter Expositions in Mathematics, 21.

Walter de Gruyter & Co., Berlin, 1995. xiv+688 pp. DM 258.00.

ISBN 3-11-011480-1

This is the book that everybody interested in topological (incidence) geometry has been waiting for. It is the most comprehensive account of all important results relating to compact projective planes and for years to come it will be the standard reference and a must read for everybody who wants to learn about or work in topological geometry. The material in the book originated from a multitude of sources scattered in the literature and has been completely reworked, streamlined and complemented to produce a self-contained whole.

Although the book is of a specialist nature, the authors have made every effort to make it accessible and of interest to as broad a section of the mathematical community as possible. Mathematicians interested in a modern, highly readable and self-contained introduction to the classical and general affine and projective planes and their collineation groups should read the first two chapters of the book. Mathematicians interested in topology and Lie group theory will be interested to read about the interplay of these two disciplines with incidence geometry, which is one of the central themes of the book.

Topological geometry is the study of incidence geometries whose object sets are endowed with topologies such that the natural geometric operations associated with the geometries are continuous maps with respect to these topologies. A topological projective plane, for example, is a projective plane such that the operations of joining two points and intersecting two lines are continuous with respect to the topologies on the point and line sets. Topological projective planes play a central role in topological geometry since many other important topological geometries, such as topological projective spaces, topological circle planes, topological generalized polygons and buildings admitting topological projective planes as substructures, are generalizations of these geometries or are connected with them in some other way. As a consequence, results about topological projective planes often play an important part in proving results about other kinds of topological geometries.

Not much can be said about topological geometries in general. Still, quite weak and natural assumptions about the topologies on the object sets of topological geometries often force the geometries to be very close to some classical examples. In the case of topological projective planes, requiring the point and line sets to be locally compact, connected and of finite dimensions yields four natural classes of topological projective planes, in the following referred to as compact projective planes, that are closely related to the four classical examples of such planes, that is, the projective planes over the reals, the complex numbers, the quaternions and the octonions. In particular, the two object sets of a compact projective plane turn out to be compact topological spaces of dimensions 2, 4, 8 or 16 and these dimensions are the same for both its point and its line set. Depending on the dimension of its object sets a compact projective plane is called 2-, 4-, 8- or 16-dimensional, respectively. Although not quite accurate, it helps to think of a compact projective plane as a deformed classical example, that is, as a geometry that shares the point set with one of the classical examples and whose line set is a distorted version of the classical line system. In fact, many of the known non-classical compact projective planes have been constructed by distorting one of the classical line systems. Famous examples are the first example of a non-Desarguesian plane constructed by D. Hilbert and the Moulton planes. Despite this closeness in the topological sense, two planes in one of the four classes can be very different in the purely incidence-geometric sense.

Almost forty years ago Salzmann set out to study and classify the compact projective planes. One of the most fundamental results in the study of compact projective planes is the fact that, under reasonable conditions, the group of continuous collineations of a compact projective plane is a Lie group of finite dimension. This makes it possible to study and classify these planes using a highly appealing and effective mix of tools, techniques and results borrowed from topology, the theory of Lie groups and incidence geometry. One of the highlights of the theory of compact projective planes is the classification in the spirit of F. Klein's Erlangen program of the most homogeneous compact projective planes. Here the dimension of the automorphism group turned out to be an appropriate measure for the degree of homogeneity of such planes. In this sense the classical planes are the most homogeneous planes and can be characterized by having automorphism groups of maximal dimensions. It took more than 30 years of concerted effort by many outstanding mathematicians, first and foremost Salzmann himself, to complete this truly remarkable classification. One of the main objectives of the book is to give a complete account of this classification.

Although there are still many tantalizing open problems connected with compact projective planes, in view of the classification mentioned above, it is justified to say that today we understand these geometries very well. The book under review can serve as a proof of this fact.

Chapter 1 is devoted to a detailed study of the classical compact projective planes. This includes a description of the full automorphism groups of these geometries together with some of their important subgroups, like motion groups with respect to polarities. This chapter comprises the first 130 pages and is a book in itself. The material of this chapter is classical and well known but is presented in a manner that sets it apart from other texts dedicated to similar subjects. It can serve as a highly accessible introduction to affine and projective planes and their collineation groups in general and some important topological and Lie group-theoretical aspects of these classical objects in particular. A special emphasis, which is also reflected in the subtitle of the book, is placed on a careful elementary study of the projective plane over the octonions, its group of collineations, which is the real exceptional simple Lie group $E_6(-26)$, and of certain subgroups of this collineation group such as the elliptic motion group, which is an exceptional simple Lie group of type F_4. The exposition owes much to some papers by J. Tits on exceptional Lie groups. It appears to be the first time that an approach influenced by those papers has been elaborated in a book.

Chapter 2 is a brief summary of notions and results concerning general affine and projective planes, coordinatizations and collineations of such planes.

Chapter 3 is another good starting point for a leisurely browse through the book. It starts with a detailed study of geometries whose point set is the plane $\mathbf{R}^2$ and whose lines are curves homeomorphic to $\mathbf{R}$ such that two points are contained in exactly one of the lines. If such a geometry satisfies the parallel axiom, that is, if it is an affine plane, then its projective closure can be made into a flat projective plane, that is, a compact projective plane of dimension 2. Every flat projective plane arises in this manner. Several characterizations of the real projective plane among the flat projective planes are given. After showing that the automorphism group of a 2-dimensional projective plane is a Lie group of finite dimension, the classification of all flat projective planes whose collineation group has dimension ≥ 3 is given.

Flat projective planes are a showcase for what to expect in arbitrary compact projective planes, but, of course, not everything that works for these special compact projective planes works for compact projective planes in general. Chapter 4 introduces general topological and compact projective planes and a transfer of results from the flat to the general compact case is begun. In particular, it is shown that the classical planes are precisely the compact connected Moufang planes and that the group of all continuous collineations of a compact projective plane is always a locally compact transformation group.

In Chapter 5 topological spaces associated with compact projective planes are investigated from the point of view of algebraic topology. For compact projective planes of finite dimensions, many important results are deduced in this chapter. For example, given such a plane, its point space has topological dimension 2, 4, 8 or 16, and a line is homotopy-equivalent to a sphere and has topological dimension 1, 2, 4 or 8, respectively. Other interesting facts that can be found in this chapter are about closed, connected proper subplanes and topological ovals in compact projective planes of finite dimensions. The latter results are especially important for the theories of finite-dimensional circle planes and compact generalized quadrangles.

Chapter 6 deals with various notions of homogeneity. It is shown that compact projective planes with point-transitive collineation groups are classical. Furthermore, groups of axial collineations and transitivity conditions for these groups are studied and characterizations of the classical planes are given with respect to these groups. This chapter closes with a short summary of what is known about groups of projectivities of compact projective planes.

In Chapters 7 and 8 a detailed account of the celebrated classification of the compact projective planes mentioned above is given. The classification of the 2-

dimensional projective planes has already been given in Chapter 3. Chapter 7 deals with the 4-dimensional case and Chapter 8 with the 8- and 16-dimensional cases. We only mention that special classes of compact projective planes, like translation planes, shift planes and Hughes planes, which provide the examples for the most homogeneous non-classical planes, are dealt with in great detail, thereby providing the reader with a solid basis for developing a good feeling for compact projective planes. Many formerly unpublished results about shift planes and Hughes planes are contained in Sections 74 and 86, respectively.

Chapter 9 is an appendix. In this chapter many results from topology and the theory of Lie groups are gathered, especially the topological characterization of Lie groups (Hilbert's fifth problem), the structure and the classification of (simple) Lie groups and real linear representations of almost simple Lie groups. In particular, all irreducible representations of almost simple Lie groups on real vector spaces of dimension at most 16 are listed.

The book closes with an extensive bibliography and an index.

Burkard Polster (Adelaide)

98g:52005 52A10 52A37

Rychlik, Marek R. (Tucson, AZ)

A complete solution to the equichordal point problem of Fujiwara, Blaschke, Rothe and Weitzenböck. (English summary)

Invent. Math. **129** (1997), *no.* 1, 141–212.

The well-known equichordal problem from geometry asks whether there exists an equichordal curve in the plane. The problem was posed by Fujiwara in 1916 and, independently, by Blaschke, Rothe and Weitzenböck in 1917. Here are all the definitions needed to understand the problem: (1) An equichordal point P of a Jordan curve (a simply closed curve) is a point inside the curve such that all chords (line segments) connecting two points of the curve and passing through P have the same length. (2) An equichordal curve is a Jordan curve that has two distinct equichordal points. The definition of an equichordal curve originally required it to be convex but this property was dropped later.

Since the equichordal problem is so easy to formulate and so accessible even to non-mathematicians, a large number of problem solvers tried their luck with it. But most of them quickly found out that the problem cannot be solved by just drawing a picture of an "equichordal curve" and arriving at a contradiction by means of elementary geometry. Some "solutions" were obtained in this way, though. The problem can even claim to be of practical importance because the reviewer knows an engineer who asked for a formula of an equichordal curve in order to develop a better Wankel engine. It was quite clear for some time that equichordal curves do not exist, but a proof was missing. At first sight it may appear strange that many people spent so much effort and time on showing that something does not exist. But on the other hand, work on challenging problems has often led to exciting new developments in mathematics.

In spite of the fact that the problem appeared in book collections of open problems and in unsolved problem columns of journals, the number of published papers on the subject remained remarkably small. An important contribution was made by E. Wirsing [Arch. Math. **9** (1958), 300–307; MR **21** #2205], who was the first to use analytical methods to attack the problem. His work was continued by R. Schäfke and the reviewer [J. Reine Angew. Math. **425** (1992), 9–60; MR 93d:52003]. Using methods from asymptotic analysis, they showed that equichordal curves of small excentricities do not exist. The excentricity of an equichordal curve is the ratio of the distance between its equichordal points and the common length of the chords through them. Equichordal curves of larger excentricities could then be excluded by improving results of G. Michelacci and A. Volčič [Arch. Math. (Basel) **55** (1990), no. 6, 599–609; MR 91m:52002] with the assistance of a computer.

The paper under review proves the nonexistence of equichordal curves by applying methods from the theory of dynamical systems and Riemann surfaces. It is shown that an equichordal curve, if it existed, would be a heteroclinic connection of a map T naturally associated with the equichordal problem. A heteroclinic connection of T is an invariant curve connecting two fixed points of T. The map is then complexified. The main idea of the proof is to consider a Riemann surface associated with the heteroclinic connection. This surface would have to be compact, which would force an equichordal curve to be algebraic. It is then quite straightforward to show that T does not admit an invariant algebraic curve. As one would expect, the paper is long (72 pages) and involved. The reading is easy at the beginning but becomes more difficult towards the end of the paper.

The paper represents an extraordinary effort of the author that, without doubt, will find high appreciation by everyone in the mathematical world.

Hans W. Volkmer (Milwaukee, WI)

98h:53045 53C15 53C40 53C65 57R15 58F05
Donaldson, S. K. (Oxford)
Symplectic submanifolds and almost-complex geometry.
J. Differential Geom. **44** (1996), *no.* 4, 666–705.

This important paper develops a general procedure for producing symplectic submanifolds of any even codimension within a given compact symplectic manifold (V,ω) of dimension $2n$, $n \geq 2$. The main construction of the paper is the following deep existence theorem for codimension 2 symplectic submanifolds. Assume that the class $[\omega/2\pi] \in H^2(V;\mathbf{R})$ is "rational", i.e. there is a lift of $[\omega/2\pi]$ to an integral class $h \in H^2(V;\mathbf{Z})$. Then for sufficiently large integers k, the Poincaré dual of kh in $H_{2n-2}(V;\mathbf{Z})$ may be represented by a symplectic submanifold $W \subset V$.

In order to obtain the symplectic submanifold W the author first endows V with a compatible almost complex structure and considers a complex line bundle $L \to V$ with first Chern class $c_1(L) = [\omega/2\pi]$. In the integrable case, when V is a complex manifold, since L is a positive line bundle one may apply the Kodaira embedding theorem and conclude that for k sufficiently large, $L^{\otimes k}$ gives rise to a projective embedding of V in $\mathbf{CP}^{\mathrm{N}}$. If $s_k\colon V \to L^{\otimes k}$ is a holomorphic section which is transverse to the zero section then $W_k = \{x \in V\colon\ s_k(x) = 0\}$ is a complex submanifold and hence a symplectic submanifold (equivalently, W is given by a hyperplane section of $V \subset \mathbf{CP}^{\mathrm{N}}$). In the nonintegrable case this result is significantly more difficult. In particular, holomorphic sections will not exist in general. One must instead try to find sections, transverse to the zero section, which are as near to holomorphic as possible. The author shows that there is a transverse section s of $L^{\otimes k}$ which satisfies $|\bar\partial s| < (C/\sqrt{k})|\partial s|$ on the zero set of s, where ∂s and $\bar\partial s$

are the complex linear and antilinear parts of the derivative ∇s and C is a positive constant. It is easy to see that this inequality ensures that the submanifold $W = \{x \in V\colon\ s(x) = 0\}$ is symplectic. Proving the existence of a transverse section which satisfies this inequality is a delicate analytic task which makes up the core of the paper. Roughly speaking, the proof consists of two parts. The first is an analytic construction of "approximately holomorphic" sections s with $\bar\partial s$ small. The second, more difficult, part is to show that among these sections one can find ones where ∂s is not small on the intersection of s with the zero section. The local part of the proof of this "quantitative transversality" result is accomplished by adapting a technique from real algebraic geometry developed by Y. Yomdin [Math. Ann. **264** (1983), no. 4, 495–515; MR 86f:58017].

For large values of k these symplectic submanifolds are "approximately pseudo-holomorphic". This is the key point that allows the author to remove the condition that the class $[\omega/2\pi]$ be rational. If one begins with an arbitrary compact symplectic manifold (V, ω) then one can find an arbitrarily close symplectic structure ω' on V which is rational. If J' is an almost complex structure which is compatible with ω' then the approximately J'-pseudo-holomorphic submanifolds will also be symplectic with respect to ω. By replacing V with W and iterating the construction one produces submanifolds of arbitrary even codimension.

In the final section the author establishes a number of related geometric results. He proves an asymptotic result which shows that the submanifolds W are quite complicated, essentially filling out all of V as $k \to \infty$. In particular, when viewed as currents, the sequence $k^{-1} W_k$ converges to a multiple of the symplectic form ω.

By adapting the main construction to the integrable case, the author is able to obtain bounds on the first fundamental form and curvature of complex hypersurfaces of high degree (representing the Poincaré dual of $k[\omega]$ for large k) in a Kähler manifold. These estimates are optimal in terms of their dependence on k. The author raises the interesting problem of trying to find the optimal constant when $V = \mathbf{CP}^2$; in other words one would like to understand which curves of high degree are the "smoothest" (or rather "flattest") among all curves representing the same homology class.

In addition the author proves an analogue of the Lefschetz hyperplane theorem which states that for k sufficiently large the inclusion of W_k in V induces an isomorphism on the homotopy groups π_p for $p \le n-2$ and a surjection on π_{n-2}.

Prior to the theory developed here, general methods for producing symplectic submanifolds were available only in either codimension $d \ge 4$ or dimension 4 (i.e. $n = 2$). In the former case, under certain natural topological assumptions, M. Gromov's h-principle may be used to produce symplectic submanifolds of codimension $d \ge 4$ [see *Partial differential relations*, Springer, Berlin, 1986; MR 90a:58201]. In the latter case one has the theory of pseudo-holomorphic curves, also introduced by Gromov. This has been an extremely important tool in the study of symplectic four-manifolds and has recently seen remarkable applications, primarily through the work of C. H. Taubes [see, e.g., J. Differential Geom. **44** (1996), no. 4, 818–893; MR 97k:58029, see p. D73 in this volume]. The paper under review presents the only general existence result known for codimension 2 symplectic submanifolds in dimensions greater than 4. It is clearly the beginning of a very rich theory. *Daniel Pollack* (Seattle, WA)

98d:53052 53C21 53C20 57R20
Gromov, M. [Gromov, Mikhael L.] (Bures-sur-Yvette)
Positive curvature, macroscopic dimension, spectral gaps and higher signatures. (English summary)
Functional analysis on the eve of the 21*st century, Vol. II* (*New Brunswick, NJ*, 1993), 1–213, *Progr. Math.*, 132, *Birkhäuser Boston, Boston, MA*, 1996.

The author wrote in an earlier paper that "the curvature tensor of a Riemannian manifold is a little monster of (multi)linear algebra whose full geometric meaning remains obscure" [Rend. Sem. Mat. Fis. Milano **61** (1991), 9–123 (1994); MR 95j:53055]. Invariant theory and our limited grasp on the full curvature tensor lead to the study of certain components of curvature; the paper under review emphasizes scalar curvature, probably the most elusive of the basic notions of curvature.

Many of the pieces covered by featured reviews in Mathematical Reviews are suitable for a wide audience. This article will be consulted by a large number of geometers, topologists, and operator algebraists, but most readers outside these specialties would do better to begin with the recent survey literature since this paper reports on several of the author's long-standing interests and research programs and assumes a moderate level of familiarity with the problems and methods discussed. (Two other readers whose work is in those specialties were consulted in the preparation of this review; their views are reported in separate paragraphs below.) The author's Italian survey paper cited above is an admirable overflight of the terrain of modern geometry for a nonexpert audience. Collections from the Fields Institute [G. Besson et al., *Riemannian geometry (Waterloo, ON, 1993)*, Amer. Math. Soc., Providence, RI, 1996; MR 96i:53001] and MSRI [*Comparison geometry (Berkeley, CA, 1993–94)*, Cambridge Univ. Press, Cambridge, 1997; MR 97m:53001] offer convenient access to other views of that rich landscape. Marcel Berger's historical brief ["Riemannian geometry during the second half of the twentieth century", Preprint, Inst. Hautes Études Sci., Bures-sur-Yvette, 1997; per revr.] is recommended for long-range perspective and a massive bibliography, while one of A. L. Besse's books [*Einstein manifolds*, Springer, Berlin, 1987; MR 88f:53087] is an excellent reference for tools, terminology, and Einstein metrics (those whose Ricci curvature is a scalar multiple of the metric tensor). Finally, J. Rosenberg and S. Stolz are expected to follow their survey [in *Algebraic topology and its applications*, 241–267, Springer, New York, 1994; see MR 95b:55001] with a new one for the DMV seminar series.

Sectional curvature is based on the Gaussian curvature of a surface and is therefore well adapted to comparison arguments, especially with two-dimensional model spaces such as triangles. This is a strong invariant, at least by one of the basic geometric tests: if all the sectional curvatures of a Riemannian manifold vanish (so the manifold is "Riemannian flat") then this space is everywhere locally isometric to Euclidean space.

The Ricci curvature tensor is a contraction of the full curvature tensor which has close ties to the analysis of the Laplace-Beltrami operator. While comparison is a well-established technique for sectional curvature, comparison arguments for Ricci curvature have seen considerable recent progress (see the MSRI volume cited above) and Ricci-flat manifolds are richer and more numerous than Riemannian-flat manifolds. Of particular interest since the rise of string field theory are the Calabi-Yau manifolds, a large family of compact complex manifolds with Ricci-flat Kähler metrics.

As the author describes it on p. 3 of the paper under review, the scalar curvature of a Riemannian metric g is a real-valued function Sc_g which is obtained by applying to g the (unique up to scale) second order differential operator from metrics to functions which is (i) invariant under the natural action of diffeomorphisms on metrics and functions, and (ii) linear in the second derivatives of g. Scalar-flat metrics exist on many manifolds which are far from admitting a Riemannian-flat metric [see A. L. Besse, op. cit. (Chapter 4D)]; while scalar curvature is simple enough to generate many interesting problems in geometry and differential equations, its relationship to the underlying manifold remains mysterious. Sections 1 and 3 of this paper describe the scalar curvature and some of its properties with a richness which even experts may find provocative.

Sign constraints on Ricci and scalar curvatures lead to strangely asymmetric results. J. Lohkamp has shown that in most cases, Riemannian metrics of negative scalar or even negative Ricci curvature form a dense subset of the space of all metrics. (See his paper in the Fields Institute volume [in *Riemannian geometry (Waterloo, ON, 1993)*, 23–51, Amer. Math. Soc., Providence, RI, 1996; MR 97b:53037] for the topologies and arguments involved in this claim. Some of his original papers are [Ann. of Math. (2) **142** (1995), no. 3, 457–498; MR 96m:53040; Ann. of Math. (2) **140** (1994), no. 3, 655–683; MR 95i:53042; Invent. Math. **110** (1992), no. 2, 403–407; MR 93h:58025].) In sharp contrast, many manifolds do not admit any Riemannian metrics of positive scalar curvature. The analysis of this situation is not complete and our current understanding rests on a variety of supports: index arguments applied to spinor bundles by Lichnerowicz and Hitchin, minimal surface arguments of Schoen and Yau, cobordism and index arguments due to Gromov and Lawson, and, most recently, Seiberg-Witten invariants.

The present reviewer's assessment of this paper after a few weeks spent with it was that the main idea offered is a notion of K-area, initially presented as a sort of stable operator norm of curvatures for unitary bundles with nontrivial Chern numbers. (The K-area invariant is defined more than once and occupies much of Sections 4–6. Redefinition is a feature of the author's style which readers new to his work should watch for.) Bounds on K-area lead to new proofs of some of the Gromov-Lawson results which have inspired Rosenberg, Stolz, and others. Features of K-area on symplectic manifolds are emphasized here and have also been studied by L. Polterovich [Geom. Funct. Anal. **6** (1998), no. 4, 726–739; MR 97e:58031].

An expert on the Novikov conjecture and Novikov's theorem on invariance of rational Pontryagin classes reported a quite different weighting of the elements of this paper. After the K-area is introduced and applied to the positive scalar curvature problem, those subjects and their relationship to the signature of a manifold are reviewed. The second reader's attention was seized by a sketched proof of Novikov's theorem which is remarkably efficient (that reader suggests that it would be a good argument to present in a graduate class) and by remarks on the Novikov conjecture (these elements appear in Sections 7 and 8). Here as in some of his other work the author is heavily influenced by G. Lusztig's contribution to the Novikov conjecture and index theory [J. Differential Geometry **7** (1972), 229–256; MR **48** #1250].

Another reader was consulted who works with the geometry of foliations and believes that the last section (Section 9) of this paper could be its most important contribution. Here the author puts "macroscopic" or "asymptotic" notions considered in Section 2 and partially familiar from his work on geometric group theory to work on the Novikov conjecture, index theory for foliated manifolds, and other issues. The philosophy at work here is that well-tuned targets for indicial invariants might not see the familiar aspects of an open manifold's topology and geometry, but could depend only on the topology and geometry of its ends.

The paper concludes with a brief section on "perspectives, problems, omissions". Almost every reader is more likely to find an excess of riches than omissions here, and the interviews reported above suggest that mathematicians of varied interests and backgrounds will find something useful.

{For the entire collection see MR 96k:00022.}

Christopher W. Stark (Gainesville, FL)

98k:53044 53C21 53C20

Cheeger, Jeff (New York, NY); **Colding, Tobias H.** (New York, NY)

On the structure of spaces with Ricci curvature bounded below. I.

J. Differential Geom. **46** (1997), *no.* 3, 406–480.

The paper under review is the first in a series of papers by Cheeger and Colding on the geometry and topology of spaces with a lower Ricci curvature bound. Typically, by a space with a lower Ricci curvature bound they mean the Gromov-Hausdorff limit of a sequence of manifolds with the same lower Ricci curvature bound. Several of the tools were developed by Cheeger and Colding in their earlier paper [Ann. of Math. (2) **144** (1996), no. 1, 189–237; MR 97h:53038]. In addition to applying to singular limits, the results of these papers also have important applications to manifolds with lower Ricci curvature bounds. These papers, and many of these results, were announced in another paper by the authors [C. R. Acad. Sci. Paris Sér. I Math. **320** (1995), no. 3, 353–357; MR 96f:53047].

Manifolds with lower Ricci curvature bounds have been extensively studied, and it would be impossible to do justice to all of the work in this area. We will, however, attempt to briefly survey some of the main tools for studying manifolds M^n with $\mathrm{Ric}_M \geq 0$. The first is the Bochner formula: if u is a harmonic function, i.e. $\Delta u = 0$, then

$$\Delta|\nabla u|^2 \geq 2|\nabla^2 u|^2 + 2|\nabla u|^2 \inf_{x\in M} \mathrm{Ric}_x \geq 0. \tag{1}$$

Second, the Laplacian comparison theorem states that $\Delta r \leq (n-1)/r$, where r is the distance function. This can be used to derive the relative volume comparison theorem: Given $x \in M^n$ with $\mathrm{Ric}_M \geq 0$, then

$$\frac{d}{ds}\left(\frac{\mathrm{Vol}\,(B_s(x)\subset M)}{\mathrm{Vol}\,(B_s\subset \mathbf{R}^n)}\right) \leq 0, \tag{2}$$

and, in particular, $\mathrm{Vol}\,(B_s(x)\subset M) \leq \mathrm{Vol}\,(B_s \subset \mathbf{R}^n)$ (Bishop's volume comparison theorem). If M admits a line then, by the splitting theorem of Cheeger and D. Gromoll [J. Differential Geometry **6** (1971/72), 119–128; MR **46** #2597], the associated Busemann function is affine (and hence M splits isometrically). More recently, U. Abresch and Gromoll [J. Amer. Math. Soc. **3** (1990), no. 2, 355–374; MR 91a:53071] used a similar analysis to study thin triangles and the excess. Another important classical result is the gradient estimate of S. Y. Cheng and S. T. Yau [Comm. Pure Appl. Math. **28** (1975), no. 3, 333–354; MR **52** #6608] which gives an a priori estimate for the gradient of a harmonic function.

M. Gromov's compactness theorem gives a useful framework for studying these manifolds [*Structures métriques pour les variétés riemanniennes,* Edited by J. Lafontaine and P. Pansu, CEDIC, Paris, 1981; MR 85e:53051]. The Riemannian distance function dist allows us to view Riemannian manifolds as metric spaces (in fact, length spaces). Recall that if (X,d) is a metric space then the Hausdorff distance d_H between $A,B\subset X$ is less than ε if

$$A\subset\{x\in X|\ d(x,B)<\varepsilon\}\text{ and }B\subset\{x\in X|\ d(x,A)<\varepsilon\}.$$

The Gromov-Hausdorff distance d_{GH} between metric spaces X and Y generalizes this notion by defining $d_{\mathrm{GH}}(X,Y)=\inf_{(f,g,Z)} d_H(f(X),g(Y))$ where $f\colon X\to Z$ and $g\colon Y\to Z$ are isometric. Gromov's compactness theorem is the assertion that the set of metric spaces

$$\{(M^n,\mathrm{dist})|\ \mathrm{Ric}_M\geq -1,\ \mathrm{diam}_M\leq D\}$$

is precompact in the Gromov-Hausdorff topology. Similarly, if $p_j\in M_j$ is a sequence of manifolds with $\mathrm{Ric}_{M_j}\geq 0$ a subsequence, $B_1(p_{j'})\subset M_{j'}\to B_1(p)\subset X$ in the pointed Gromov-Hausdorff topology where X is a length space. In the important special case when M_j is the rescaling $(M,r_j^{-2}g)$ of a fixed manifold (M,g) with $\mathrm{Ric}_M\geq 0$, such a limit X is called a tangent cone. If $r_j\to 0$ we get the tangent cone at a point, and when $r_j\to\infty$ we get the tangent cone at infinity. Since this metric space X is generally not a Riemannian manifold, this raises the question of what, if any, additional structure it has.

When the manifolds M_j have sectional curvature bounded from below then the triangle comparison theorems can be used to show that the limit X has a great deal of structure [Yu. D. Burago, M. L. Gromov and G. Ya. Perel′man, Uspekhi Mat. Nauk **47** (1992), no. 2(284), 3–51, 222; MR 93m:53035]. In the case of Ricci curvature, these pointwise estimates fail and much less was known. The first *integral* estimates on distances and angles were developed by Colding [Invent. Math. **124** (1996), no. 1-3, 175–191; MR 96k:53067; Invent. Math. **124** (1996), no. 1-3, 193–214; MR 96k:53068; Ann. of Math. (2) **145** (1997), no. 3, 477–501; MR 98d:53050]. This led to the solutions of well-known conjectures of Anderson-Cheeger, Gromov, and Fukaya-Yamaguchi.

The following example illustrates the various viewpoints. Let M^n be a closed manifold with $\mathrm{Ric}_M\geq 0$. By integrating (1), Bochner showed that $b_1(M^n)\leq n$ with equality for a flat torus. Applying (2) to the universal cover of M, J. Milnor showed that $\pi_1(M)$ has polynomial growth [J. Differential Geometry **2** (1968), 1–7; MR **38** #636]. The splitting theorem of Cheeger and Gromoll implies that $\pi_1(M)$ is almost crystallographic. Suppose now that $\mathrm{diam}_M=1$ and $\mathrm{Ric}_M\geq-\varepsilon$. Gromov showed that there is some $\varepsilon_1>0$ so that $\varepsilon<\varepsilon_1$ implies that $b_1(M^n)\leq n$. Settling a conjecture of Gromov, Colding showed that there is some $\varepsilon_2>0$ such that $\varepsilon<\varepsilon_2$ and $b_1(M^n)=n$ implies that M is a torus.

In [op. cit., 1996], Cheeger and Colding proved quantitative generalizations of several rigidity theorems for Riemannian manifolds of positive (or nonnegative) Ricci curvature. Included in this were the volume cone implies metric cone theorem, the maximal diameter theorem, and the splitting theorem. It is perhaps useful to recall the first and third above.

From the volume comparison theorem, if $\mathrm{Ric}_M\geq 0$ and the ratio in (2) is constant for $s<r<t$ then $B_t\setminus B_r\subset M$ is a metric cone. That is, there is a closed manifold $(N^{n-1},d\omega^2)$ such that $B_t\setminus B_r$ is isometric to $(N\times[s,t],dr^2+r^2\,d\omega^2)$. Cheeger and

Colding proved that if the ratio is almost constant then $B_t \setminus B_r \subset M$ is Gromov-Hausdorff close to a metric cone. As a consequence, they showed that if M^n has $\mathrm{Ric}_M \geq 0$ and Euclidean volume growth (namely, $\mathrm{Vol}(B_r) \geq cr^n$ for all $r > 0$ and some $c > 0$) then every tangent cone at infinity M_∞ is a metric cone of Hausdorff dimension n.

It follows from their almost splitting theorem that if $M_i^n \to Y$ in the pointed Gromov-Hausdorff distance where $\liminf \mathrm{Ric}_{M_i} \geq 0$ and Y contains a line then Y splits isometrically as $\mathbf{R} \times X$ for some length space X of Hausdorff dimension at most $n-1$. This is important for the structure theory of singular limits, but it also has implications for classical geometry. For instance, Cheeger and Colding used this to show that there exists some $\delta > 0$ such that if $\mathrm{diam}(M) \leq 1$ and $\mathrm{Ric}_M \geq -(n-1)\delta$ then $\pi_1(M^n)$ contains a nilpotent subgroup of finite index (answering a conjecture of Gromov).

In the present paper, the authors combine the tools developed in their first paper with techniques similar to those used in geometric measure theory to analyze the structure of spaces with Ricci curvature bounded from below. Suppose now that a length space (Y, d) is the pointed Gromov-Hausdorff limit $(M_i^n, p_i) \to (Y, p)$ where $\mathrm{Ric}_{M_i} \geq (n-1)\Lambda$ for all i. We recall next some of their definitions (which differ from those of the announcement). Define the weakly k-Euclidean set by $y \in \mathscr{W}_k$ if *some* tangent cone at y splits off a factor $\mathbf{R}^k$. If moreover *every* tangent cone at y is isometric to $\mathbf{R}^k$ then y is in the k-regular set $\mathscr{R}_k$. Let $\mathscr{R} = \bigcup_k \mathscr{R}_k$ denote the set of regular points. Define the k-degenerate set $\mathscr{D}_k$ by $\mathscr{D}_k = Y \setminus \mathscr{R}_{k+1}$, $\ _k$ to be $\mathscr{D}_k \setminus \mathscr{R}$, and the singular set by $= \bigcup \ _k = Y \setminus \mathscr{R}$. The authors show first that the set of regular points $\mathscr{R} \subset Y$ is dense. Similar definitions are very important in regularity theory for harmonic maps and minimal surfaces (see, for instance, H. Federer [*Geometric measure theory*, Springer-Verlag New York Inc., New York, 1969; MR **41** #1976], R. Schoen and K. Uhlenbeck [J. Differential Geom. **17** (1982), no. 2, 307–335; MR 84b:58037a; correction MR 84b:58037b], or the more recent survey of L. M. Simon [in *Nonlinear partial differential equations in differential geometry (Park City, UT, 1992)*, 185–223, Amer. Math. Soc., Providence, RI, 1996; MR 96k:58060]).

In the first four sections, the authors study the general case (which includes the possibility that $\lim \mathrm{Vol}(B_1(p_i)) = 0$, i.e., the sequence is collapsing). They construct renormalized limit measures ν on Y which they show satisfy the appropriate volume comparison and $\nu(\quad) = 0$.

If in addition $\mathrm{Vol}(B_1(p_i)) \geq v > 0$ for all i (the sequence is noncollapsing) then they show that Y has Hausdorff dimension n and the renormalized limit measure is a multiple of n-dimensional Hausdorff measure ("volume convergence"). This implies that the relative volume comparison theorem holds for Y. Furthermore, they show that every tangent cone is a metric cone (note that they give examples showing that the tangent cones need not be unique). They also show that $\dim \mathscr{D}_k \leq k$.

In the noncollapsing case, they define the ε-regular set by $y \in \mathscr{R}_\varepsilon$ if the Gromov-Hausdorff distance between $B_1(0) \subset \mathbf{R}^n$ and $B_1(y_\infty) \subset Y_y$ is less than ε for every tangent cone Y_y at y. In other words, $y \in \mathscr{R}_\varepsilon$ if all sufficiently small balls at y are (d_{GH}) close to balls in $\mathbf{R}^n$, and hence $\mathscr{R}_\varepsilon$ contains an open neighborhood of the regular set $\mathscr{R}$. They show that if $\varepsilon < \varepsilon(n)$ is sufficiently small then the interior of $\mathscr{R}_\varepsilon$ is homeomorphic to a smooth manifold. In fact, the metric is bi-Hölder equivalent to a smooth metric. The idea is that if $y \in \mathscr{R}_\varepsilon$ then since $B_1(y_\infty)$ is close to $B_1(0) \subset \mathbf{R}^n$ it must have almost maximal volume. The relative volume comparison then implies that all sufficiently small subballs have almost maximal volume. It then follows (see the earlier discussion) that these subballs are also Gromov-Hausdorff close (on their own scales) to balls in $\mathbf{R}^n$. In other words, we have a "Reifenberg" condition [cf. E. R. Reifenberg, Acta Math. **104** (1960), 1–92; MR **22** #4972], which allows the authors to construct the bi-Hölder homeomorphism.

In this noncollapsing case, they also show that $\ \subset\ {}_{n-2}$ and hence the singular set has codimension at least two (which is optimal). In the collapsed case there are examples with codimension-one singular sets. The authors also obtain various stronger results under various additional assumptions on the sequence (two-sided Ricci curvature bounds, Einstein manifolds); this uses a result of M. T. Anderson [Invent. Math. **102** (1990), no. 2, 429–445; MR 92c:53024]. A future paper of the authors with G. Tian will contain further results in this direction (including sharper results for Kähler manifolds); see their announcement [C. R. Acad. Sci. Paris Sér. I Math. **324** (1997), no. 6, 645–649; MR 98g:53078].

William P. Minicozzi, II (Baltimore, MD)

99e:53049 53C21 53C20 57R99 58G11
Hamilton, Richard S. (La Jolla, CA)
Four-manifolds with positive isotropic curvature.
Comm. Anal. Geom. **5** (1997), *no.* 1, 1–92.

This remarkable paper uses the flow of metrics by their Ricci curvature to prove the following result: Suppose M is a compact four-dimensional manifold carrying a Riemannian metric with non-negative curvature on isotropic 2-planes (that is, for any orthonormal 4-frame, $R_{1313}+R_{1414}+R_{2323}+R_{2424}\geq 2R_{1234}$), and M does not contain any essential incompressible space-form (that is, there is no submanifold N^3 diffeomorphic to S^3/Γ in M with fundamental group Γ injecting into $\pi_1(M)$ and not equal to $\{1\}$ or Z_2). Then M is diffeomorphic to one of S^4, the real projective space $\mathbf{RP}^4$, $S^3\times S^1$, the non-oriented twisted product $S^3\tilde{\times}S^1$, or a connected sum of these. In particular, if M is simply connected then M is diffeomorphic to S^4; if $\pi_1(M)=Z_2$ then M is diffeomorphic to $\mathbf{RP}^4$, and if $\pi_1(M)=Z$ then M is diffeomorphic to $S^3\times S^1$ if M is oriented, or to $S^3\tilde{\times}S^1$ if not. The result is related to earlier work by M. J. Micallef and J. D. Moore [Ann. of Math. (2) **127** (1988), no. 1, 199–227; MR 89e:53088] which showed (also in higher dimensions) that a compact simply connected manifold with positive isotropic curvature is homeomorphic to the sphere. The techniques used here are quite different, and quite new: As in previous work by the author using the flow by Ricci curvature [J. Differential Geom. **17** (1982), no. 2, 255–306; MR 84a:53050; J. Differential Geom. **24** (1986), no. 2, 153–179; MR 87m:53055], a key step is to establish curvature inequalities which are preserved or improve under the flow. In previous work, the inequalities were sufficient to rule out the formation of any singularity in the flow, except that of the entire manifold contracting to zero size in finite time, approaching constant curvature in the process. In contrast, in the present paper the inequalities allow the formation of "neck-like" singularities which look like $S^3\times\mathbf{R}$ (or a quotient of this) in the limit of rescaling. The author introduces a new ingredient by showing that the flow can be "stopped" before such a singularity forms, with sufficient control to allow surgery to be performed to avoid the singularity. Successive such surgeries gradually decompose the manifold into a connected sum of simple pieces. This requires a detailed understanding of the geometry of the manifold in a neck-like region produced by an approaching singularity, and of the changes in curvature produced during the surgery process. *Ben Andrews* (Canberra)

97d:53052 53C25 53C20
Joyce, D. D. (Oxford)
Compact 8-manifolds with holonomy Spin(7).
Invent. Math. **123** (1996), *no.* 3, 507–552.

Half a century ago, S. Bochner [Bull. Amer. Math. Soc. **52** (1946), 776–797; MR **8**, 230a] proved that any Ricci-flat Riemannian metric on an n-dimensional torus must be flat, and this prompted him to ask whether every compact Ricci-flat manifold is flat. It was not until the late 1970s that S. T. Yau [Comm. Pure Appl. Math. **31** (1978), no. 3, 339–411; MR 81d:53045], building upon the work of Bochner's student Calabi, established that the answer is emphatically *no*. Calabi had observed that the problem simplifies dramatically if one chooses to consider only Kähler metrics, and Yau combined these geometric insights with new analytic ideas to prove the existence of Ricci-flat Kähler metrics on all compact Kähler-type complex manifolds with trivial canonical class.

Joyce's recent work represents the most spectacular advance in the theory of Ricci-flat manifolds since the work of Yau. In the paper under review, Joyce constructs the first examples of non-Kähler Ricci-flat metrics on simply connected even-dimensional compact manifolds. In two closely related previous papers [J. Differential Geom. **43** (1996), no. 2, 291–328; J. Differential Geom. **43** (1996), no. 2, 329–375] the author constructed analogous odd-dimensional examples.

Kähler geometry deals with those Riemannian manifolds which admit a parallel nondegenerate 2-form. A nice generalization, essentially due to M. Berger [Bull. Soc. Math. France **83** (1955), 279–330; MR **18**, 149a], is to consider manifolds which admit a nontrivial parallel differential form. The classification is simplest and clearest if one sticks to metrics which are neither locally symmetric nor locally split as Riemannian products. Then in dimension 8, for example, there are exactly 5 kinds of metrics with parallel 4-forms, corresponding to the holonomy groups U(4), SU(4), Sp(4), Sp(4)Sp(1), and Spin(7). Among such metrics, the Ricci-flat ones are exactly those with holonomy SU(4), Sp(4), or Spin(7). And of these, the only ones which are not Kähler are those with holonomy Spin(7).

Note that while Spin(7) is just the double cover of SO(7), it acts on $\mathbf{R}^8$ in a slightly non-obvious way—namely, by the spin representation. As a subgroup of SO(8), it can also be understood as the subgroup which fixes a generic right-handed spinor. Thus a Riemannian 8-manifold with holonomy Spin(7) can also be characterized by the fact that it is a spin manifold with a parallel non-null spinor field, so that in the compact case the $\widehat{A}$ genus must be 1. By contrast, holonomy SU(4) and Sp(2) would respectively force the $\widehat{A}$ genus to be 2 or 3, since these geometries are characterized by the existence of 2 or 3 linearly independent parallel (null) spinors.

It was R. L. Bryant [Ann. of Math. (2) **126** (1987), no. 3, 525–576; MR 89b:53084; cf. R. L. Bryant and S. M. Salamon, Duke Math. J. **58** (1989), no. 3, 829–850; MR 90i:53055] who first constructed metrics of holonomy Spin(7) on noncompact manifolds, and Bryant's viewpoint is exploited systematically throughout Joyce's proof. Bryant's trick is to concentrate on the parallel 4-form rather than the metric. The

4-form associated with any Spin(7) metric can be equated with a certain standard 4-form on $\mathbf{R}^8$ by choosing a suitable basis for the tangent space. A 4-form with this non-generic property is called an admissible 4-form. Any admissible 4-form determines a reduction of the structure group of the tangent bundle to Spin(7), and so in particular a metric. The metric in question then has holonomy Spin(7) iff the admissible 4-form is closed. The ultimate source of this trick can be traced to the pioneering work of R. Harvey and H. B. Lawson, Jr. [Acta Math. **148** (1982), 47–157; MR 85i:53058].

Joyce's approach to constructing compact 8-manifolds with holonomy Spin(7) consists of a geometric part and an analytic part. The geometric part constructs compact 8-manifolds with admissible 4-forms for which the exterior derivative is small. The analytic part then deforms these admissible forms to obtain closed admissible 4-forms—and thus metrics with holonomy Spin(7).

The geometric construction is modeled on the Kummer construction of $K3$ surfaces. One considers the orbifold T^8/Γ, where $\Gamma \cong (Z_2)^{\oplus 4}$ is generated by reflections through suitable 4-tori $T^4 \subset T^8$. Provided these 4-tori are chosen correctly, the standard constant-coefficient admissible 4-form on T^8 can be pulled back to a desingularization of the orbifold and then perturbed so as to yield a family of admissible 4-forms in a family of cohomology classes. This depends in part on the observation that the Riemannian product $TS^2 \times T^4$ of the Eguchi-Hanson gravitational instanton and a flat torus has holonomy $SU(2) \subset \text{Spin}(7)$, and rescaling the Eguchi-Hanson factor leads to a degeneration of this model to the flat orbifold $(\mathbf{R}^4/Z_2) \times T^4$. However, this only tells one what to do at the codimension-4 singularities of T^8/Γ. There are also codimension-8 singularities because the 4-tori in question are not all disjoint, and one needs additional prescriptions for resolving these. The same strategy also works for certain other finite groups of automorphisms of T^8.

The analytic part of the paper deals with the problem of perturbing an admissible 4-form with small exterior derivative so as to obtain a closed admissible 4-form. With suitable gauge fixing, this can be made into a nonlinear elliptic PDE and solved in a Sobolev space by a Picard iteration scheme provided that certain estimates hold. In practice, these are provided by the fact that the examples degenerate to flat orbifolds; even though the injectivity radius goes to zero and the curvature blows up in the process, they do so at estimable rates that allow one to uniformly control the operator norms of the appropriate differential operators in a uniform manner.

For Riemannian geometry, this paper and its companions must surely represent one of the major achievements of the decade. A new and unimagined landscape has been opened up for a new generation of explorers, and in the coming years we can look forward to learning much more about the geography of these Einstein manifolds and the field theories that inhabit them. *Claude LeBrun* (Stony Brook, NY)

97m:53084 53C25
Joyce, Dominic D. (Oxford)
Compact Riemannian 7-manifolds with holonomy G_2. I, II.
J. Differential Geom. **43** (1996), *no.* 2, 291–328, 329–375.

The Cheeger-Gromoll splitting theorem [J. Cheeger and D. Gromoll, J. Differential Geometry **6** (1971/72), 119–128; MR **46** #2597] implies that any compact Ricci-flat manifold is finitely covered by the product of a flat torus and a compact, simply connected Ricci-flat manifold. Examples of the latter, however, are rather difficult to produce. Indeed, until very recently, the only examples available to us were those given by S.-T. Yau's solution [Comm. Pure Appl. Math. **31** (1978), no. 3, 339–411; MR 81d:53045] of the Calabi conjecture. These Calabi-Yau manifolds, however, are Kähler by construction, and in particular are all even-dimensional. The papers under review remedy this by constructing odd-dimensional simply connected compact Ricci-flat manifolds. This constitutes the most important advance in the subject in nearly two decades.

Although the Ricci-flat manifolds of Joyce are superficially quite different from those of Calabi-Yau, there is a deep analogy between them. Kähler manifolds are exactly the Riemannian manifolds which admit nondegenerate parallel 2-forms. Which other manifolds admit nontrivial parallel differential forms? The rough answer to this question was given by M. Berger [Bull. Soc. Math. France **83** (1955), 279–330; MR **18**, 149a] in the language of holonomy groups—i.e. groups of linear transformations arising from Riemannian parallel transport around loops. The only "nontrivial" odd-dimensional holonomy geometry lives in dimension 7, and corresponds to the holonomy group G_2. Riemannian 7-manifolds of this type are characterized by the existence of a parallel 3-form [cf. R. Harvey and H. B. Lawson, Jr., Acta Math. **148** (1982), 47–157; MR 85i:53058]. Like Calabi-Yau manifolds, they are also characterized by the fact that they carry parallel spinor fields. In particular, a metric of holonomy G_2 is automatically Ricci-flat.

While Berger's classification allowed for the possible existence of metrics of holonomy G_2, nearly three decades elapsed until R. L. Bryant settled the matter by constructing some examples [Ann. of Math. (2) **126** (1987), no. 3, 525–576; MR 89b:53084]. Beautiful and explicit complete examples were later constructed by Bryant and S. M. Salamon [Duke Math. J. **58** (1989), no. 3, 829–850; MR 90i:53055].

Joyce's assault on the compact case systematically exploits the point of view developed by Bryant: forget the metric, and focus on the 3-form. A generic 3-form on an oriented 7-manifold determines a (pseudo-Riemannian) metric by requiring that the metric times its volume form is an irreducible component of the tensor-cube of the form. Taking the Hodge star of the 3-form with respect to this metric defines a nonlinear map from 3-forms to 4-forms. A closed 3-form is then parallel with respect to the corresponding metric iff its dual 4-form is also closed.

Joyce begins by constructing compact 7-manifolds equipped with closed 3-forms whose dual 4-forms have small exterior derivative. This is done by smoothing the singularities of certain flat orbifolds whose singular sets are flat tori of codimension 4. It is then shown that these 3-forms can be perturbed into 3-forms which are both closed and co-closed; this is set up as an elliptic PDE, and solved by Picard iteration in a suitable Sobolev space. Joyce later used a similar approach to construct compact 8-manifolds with holonomy Spin(7) [Invent. Math. **123** (1996), no. 3, 507–552; MR 97d:53052, see p. D29 in this volume].

As one more indication of the importance of Joyce's results, recall that no compact manifold previously known to admit a Ricci-flat metric could also admit a metric of positive scalar curvature. Any scalar-flat metric on such a manifold is thus Ricci-flat and has special holonomy [cf. A. Futaki, Invent. Math. **112** (1993), no. 1, 23–29; MR 94f:53072]. By contrast, a result of M. Gromov and Lawson [Ann. of Math. (2) **111** (1980), no. 3, 423–434; MR 81h:53036] tells us that every simply connected compact 7-manifold admits metrics of positive scalar curvature. In particular, Joyce's manifolds carry many scalar-flat metrics which are *not* Ricci-flat. Now just this behavior is also expected for Ricci-flat manifolds of $\mathrm{SO}(n)$-holonomy. Thus, Joyce's examples would seem to be, in an important sense, much more "typical" than their Calabi-Yau cousins. *Claude LeBrun* (Stony Brook, NY)

98m:53068 53C35 20E42 20F32 53C23 57M07

Kleiner, Bruce (Salt Lake City, UT); **Leeb, Bernhard** (Mainz)

Rigidity of quasi-isometries for symmetric spaces and Euclidean buildings.

Inst. Hautes Études Sci. Publ. Math. No. 86 (1997), 115–197 (1998).

One of the key tools used in studying the asymptotic geometry of finitely generated groups and the spaces on which they act is the notion of a quasi-isometry between metric spaces, a map $\Phi: X \to X'$ such that for some $L, C \geq 0$ we have $L^{-1}d(x_1, x_2) - C \leq d(\Phi(x_1), \Phi(x_2)) \leq Ld(x_1, x_2) + C$ for all $x_1, x_2 \in X$, and for all $x' \in X'$ we have $d(x', \mathrm{Im}(\Phi)) \leq C$. For example, if X, X' are two length spaces, meaning that distances are realized by lengths of shortest paths, and if some group G acts cocompactly and properly discontinuously on X and on X', then there exists a quasi-isometry $\Phi: X \to X'$; this quasi-isometry is moreover G-equivariant.

An important conjecture of Margulis from the 1970s says that if X, X' are irreducible symmetric spaces of noncompact type, each non-flat and of rank at least 2, and if $\Phi: X \to X'$ is a quasi-isometry, then the metric on X may be rescaled, multiplying by a constant, so that Φ is a bounded distance from some isometry. One of the key steps in the Mostow rigidity theorem in rank ≥ 2 is a proof of Margulis' conjecture under the stronger hypothesis that Φ is an equivariant quasi-isometry. Margulis' conjecture drops the equivariance hypothesis, producing a pure statement about the asymptotic geometry of symmetric spaces. Such statements have strong repercussions in geometric group theory, via Gromov's program of quasi-isometric classification of finitely generated groups equipped with the word metric.

In this paper the authors prove Margulis' conjecture. The paper also gives a complete classification of symmetric spaces of noncompact type up to quasi-isometry, without assuming irreducibility, but instead assuming that the de Rham factors all have rank ≥ 2: If X, X' are two such spaces, and if they are quasi-isometric, then X, X' become isometric after the metrics on their de Rham factors are suitably rescaled. The latter result is obtained by combining Margulis' conjecture, earlier results of G. Mostow and of M. Anderson and V. Schroeder, and another new theorem of the paper under review: a quasi-isometric rigidity theorem for de Rham splittings. Applications of these theorems to geometric group theory are given in a paper by Kleiner and Leeb ["Groups quasi-isometric to symmetric spaces", Preprint, 1996; per bibl.].

All the main theorems of the paper are stated and proved in a more general setting in which the class of symmetric spaces of noncompact type is enlarged to include certain Euclidean buildings, common examples of which occur under the guise of "p-adic symmetric spaces".

The proofs involve elegant applications of the asymptotic cone methods of L. van den Dries and A. J. Wilkie [J. Algebra **89** (1984), no. 2, 349–374; MR 85k:20101] and of M. Gromov [in *Geometric group theory, Vol. 2 (Sussex, 1991)*, 1–295, Cambridge Univ. Press, Cambridge, 1993; MR 95m:20041]. The asymptotic cone of a metric space X captures the intuitive notion of looking at X from infinitely far away.

Starting with a sequence of real numbers $\lambda_i \to 0$, one chooses a nonprincipal ultrafilter ω and applies the ultralimit construction to the sequence of rescaled metric spaces $\lambda_i X$, thereby producing a metric space X_ω called the asymptotic cone of X. A similar ultralimit construction applied to rescalings of a quasi-isometry $\Phi\colon X \to X'$ yields a bi-Lipschitz homeomorphism of asymptotic cones $\Phi_\omega\colon X_\omega \to X'_\omega$.

In order to understand the asymptotic cone of a symmetric space of noncompact type and rank ≥ 2, the authors develop a beautiful new theory of indiscrete buildings, which are related to discrete Tits buildings as **R**-trees are related to simplicial trees. They prove that the asymptotic cone of a symmetric space of noncompact type and rank $r \geq 2$ is a Euclidean building of rank r satisfying a strong indiscreteness property (for example, the isometry group of this building acts transitively) and a branching property (the building is "thick"), and if the symmetric space is irreducible then so is the building. Next they study the topology of Euclidean buildings of rank ≥ 2, proving for example that a homeomorphism between two of them becomes an isometry after the metric is scaled by an appropriate factor, assuming irreducibility and the indiscreteness and branching properties. In the reducible case they prove that a building homeomorphism preserves a natural "de Rham splitting". The remaining work is to translate results about homeomorphisms between asymptotic cones back to results about quasi-isometries between symmetric spaces.

Another proof of Margulis' conjecture was given later by A. Eskin and B. Farb [J. Amer. Math. Soc. **10** (1997), no. 3, 653–692; MR 98e:22007], who proved a theorem analyzing quasi-isometric embeddings of Euclidean spaces into symmetric spaces, and used this theorem to derive Margulis' conjecture; another proof of this theorem of Eskin and Farb appears in the paper under review.

The paper is carefully and clearly written, with a self-contained exposition of building theory, including spherical buildings as well as the new theory of indiscrete Euclidean buildings. This theory is presented in a geometric manner that contrasts with more standard algebraic presentations, and should be useful to geometers. The authors' use of asymptotic cone methods should be studied carefully by anyone interested in large-scale geometry and quasi-isometric rigidity.

Lee Mosher (Newark, NJ)

99j:53083 53C42 32G10 32J18 53C25 58D27

McLean, Robert C.

Deformations of calibrated submanifolds. (English summary)

Comm. Anal. Geom. **6** (1998), *no.* 4, 705–747.

Complex submanifolds of Kähler manifolds are possibly the most readily available source of minimal submanifolds. The reason for their minimality is a piece of linear algebra—the Wirtinger inequality—which says that complex subspaces are those for which the induced volume form is a minimum. Many years ago R. Harvey and H. B. Lawson, Jr. [Acta Math. **148** (1982), 47–157; MR 85i:53058] generalized the Wirtinger inequality to encompass not only Kähler manifolds but so-called calibrated manifolds, differential geometric structures given by a Riemannian metric and a covariant constant differential form. At the time the only available stock of examples was through Yau's proof of the Calabi conjecture, which showed that any Kähler manifold with trivial canonical bundle had a metric with a covariant constant holomorphic form in the top degree. Perhaps for this reason, Harvey and Lawson concentrated on using their methods to produce minimal varieties in Euclidean space.

More recently, however, the work of D. D. Joyce [J. Differential Geom. **43** (1996), no. 2, 291–328, 329–375; MR 97m:53084, see p. D30 in this volume; Invent. Math. **123** (1996), no. 3, 507–552; MR 97d:53052, see p. D29 in this volume] has shown that there are compact examples of the other classes of geometrical structures—those with holonomy G_2 and Spin(7). Not only that, but Joyce's construction comes together with some distinguished submanifolds. When we add to this the interest of physicists in D-branes and calibrations, it becomes clear that there needs to be more investigation of calibrated submanifolds, and McLean's paper provides just such a start.

If we think of calibrated submanifolds as a generalization of complex submanifolds then we might want to study the space of all such submanifolds in a particular class, the analogue of the basic local question of whether, given one compact complex submanifold, we can find its nearby deformations as a complex submanifold. This is a well-developed subject. One hopes that an infinitesimal deformation will correspond to a holomorphic section of the normal bundle, and in good cases this is true, but in general such deformations may be obstructed. McLean tackles the parallel question for deformations of calibrated submanifolds. These are of different types. In a Calabi-Yau manifold of complex dimension n there are of course the complex submanifolds, but also the special Lagrangian submanifolds which are characterized by the vanishing of both the Kähler form and the imaginary part of the holomorphic n-form. In a G_2 manifold, which is of dimension 7, there are two types of calibrated submanifolds—the four-dimensional coassociative ones and the three-dimensional associative submanifolds. In an 8-manifold with a Spin(7) structure there are the four-dimensional Cayley submanifolds.

What McLean does is first to identify the space of (potentially) infinitesimal deformations—what for a complex submanifold would be the holomorphic sections of the normal bundle. Here for the special Lagrangian case it turns out to be the space of harmonic 1-forms, and for the coassociative submanifolds, the space of self-dual harmonic 2-forms. In the other two cases it is the space of solutions of a twisted Dirac operator. The key point here is that the space has a topological interpretation in the first two cases, and this enables the author to prove that the deformation theory there is unobstructed. The proof is relatively simple: using the exponential map to identify a neighbourhood of the zero section with a tubular neighbourhood of the submanifold, the equations become nonlinear equations for a section of the normal bundle, and Banach space implicit function theorems give the result that each harmonic form of the appropriate type is actually tangential to an actual deformation. At the same time this result produces a manifold structure on the moduli space of deformations, and also the $\mathscr{L}^2$ norm gives a natural metric on it.

This is a very interesting extension of deformation theory for complex submanifolds and it is one which has already been put to use in the approach to mirror symmetry by A. Strominger, S.-T. Yau and E. Zaslow [Nuclear Phys. B **479** (1996), no. 1-2, 243–259; MR 97j:32022; see also N. J. Hitchin, Ann. Scuola Norm. Sup. Pisa Cl. Sci. (4) **25** (1997), no. 3-4, 503–515 (1998)]. In this situation one hypothesizes that a Calabi-Yau manifold of dimension n should be fibred generically by special Lagrangian n-tori (forgetting the metric, this is just a completely integrable Hamiltonian system). Since in this case the first Betti number of the special Lagrangian torus is n, by McLean's result the base space of the fibration can be identified with the moduli space of deformations. The picture is somewhat like an elliptic fibration of a $K3$ surface, and McLean mentions this idea in the paper, written before these ideas came to prominence in physics, but there is now an expanding literature in the area.

In fact there are a number of interesting problems which the author raises at the end of his paper, on some of which light can now be shed. In the special Lagrangian case, whatever the dimension of the moduli space, there is always a natural n-form, since each tangent vector is represented by a harmonic 1-form θ, and integrating the exterior product $\theta_1 \wedge \cdots \wedge \theta_n$ over the submanifold defines an n-form on the moduli space. McLean asks if it is closed (it is) and what its relevance is (not yet known). He also asks about the metric structure on the moduli space (constant curvature, Einstein?). In this case the guiding principle for our current knowledge is the hyper-Kähler case where the metric is a special Kähler metric (in the terminology of Strominger [Comm. Math. Phys. **133** (1990), no. 1, 163–180; MR 92a:53068] and D. S. Freed [Comm. Math. Phys. **203** (1999), no. 1, 31–52]), and some properties of this are known by the work of Z. Lu [Math. Ann. **313** (1999), no. 4, 711–713]. In particular the natural metric has nonnegative Ricci curvature and is flat if the scalar curvature ρ vanishes. It follows from the inequality $\Delta\rho \geq 3\rho^2/n$ in Lu's paper that the metric is not in general Einstein since then ρ would be constant, hence zero by the inequality and hence the metric would be flat.

McLean's paper has been highly influential already. Perhaps the only disappointment is that we have no good means to construct such calibrated submanifolds except in very special circumstances, but one hopes that things will improve in the near future. *N. J. Hitchin* (Oxford)

98f:54002 54-02 18B30 54Bxx 54Cxx 54Exx

Lowen, R. [Lowen, Robert] (Wilrijk)

★**Approach spaces.**

The missing link in the topology-uniformity-metric triad.

Oxford Mathematical Monographs.

Oxford Science Publications.

The Clarendon Press, Oxford University Press, New York, 1997. x+253 pp. $110.00.

ISBN 0-19-850030-0

This book is a landmark in the history of general topology. The author developed approach spaces in a series of papers from 1987 onwards and the book contains the results from those papers, presented in a systematic and definitive manner. Approach spaces form a supercategory of the categories of topological spaces and metric spaces, and the main purpose for their introduction was to fill some gaps concerning categorical aspects of metrizable spaces (see below for details). They not only solved these fundamental problems but also turned out to be interesting objects of study for their own sake. Moreover, they are useful in many areas such as probability theory, functional analysis, function spaces, hyperspaces, and probabilistic metric spaces. They also provide a proper setting for a unified treatment of certain topological and metric properties in the theory of approximation.

Let us discuss a motivation. Suppose (X, d) is a metric space and $\delta_d(x, A)$ is the distance between $x \in X$ and $A \subset X$ defined by $\inf\{d(x,a)\colon\ a \in A\}$. We say that x is near A if $\delta_d(x, A) = 0$. About ninety years back, F. Riesz showed that the concept of a continuous function can be defined by using this nearness relation between points and subsets. A function f from one metric space X to another metric space Y is continuous at $c \in X$ if whenever c is near $A \subset X$ in X, $f(c)$ is near $f(A)$ in Y. Most teachers and texts of calculus motivate continuity by an intuitive statement such as "when x is near c, $f(x)$ is near $f(c)$". This intuitive statement, which is covariant, is not precise. To make it precise with the use of metrics, one arrives at the usual contravariant ε-δ definition. But this contravariant definition is neither easy to teach nor easy for a student to understand. This concern has been debated in recent issues of the Notices of the AMS. When we understand that continuity is a topological concept and not a metric one, then the situation becomes simple. Riesz's definition is unique in mathematics; it is both intuitive and rigorous at the same time. Of course, the metric definition is useful and students need to understand it [see P. Cameron, J. G. Hocking and S. A. Naimpally, Amer. Math. Monthly **81** (1974), 739–745; MR **50** #2403]. From the above discussion, it is clear that the distance δ_d is more valuable than the metric d. The same conclusion is reached by Lowen from the categorical point of view.

The category **TOP**, of topological spaces and continuous maps, is well behaved with respect to subspaces, products, quotients, coproducts, etc. On the other hand, the category **MET**, of metric spaces and nonexpansive maps, is stable only under the formation of subspaces and finite products. Even when metric spaces are used in analysis (e.g. Banach spaces), topology provides the framework for most of the basic concepts such as continuity, convergence, compactness, etc. This is partially remedied by the use of the extended pseudometrics, i.e. the category $p\mathbf{MET}^\infty$. The new category still does not behave well with respect to initial structures. Weil introduced the concept of a uniform space, which makes it possible to extend some metric concepts such as total boundedness and completeness. But we lose numerical information that was available earlier. Consider the following diagram:

$$\begin{array}{ccccc}
p\mathbf{MET}^\infty & \to & ? & \to & ? \\
\downarrow & & \downarrow & & \downarrow \\
pm\mathbf{UNIF} & \to & \mathbf{UNIF} & \to & q\mathbf{UNIF} \\
\downarrow & & \downarrow & & \downarrow \\
pm\mathbf{TOP} & \to & \mathbf{CREG} & \to & \mathbf{TOP}
\end{array}$$

In the second row, we have q**UNIF**, the category of quasi-uniform spaces and uniformly continuous maps, **UNIF**, of uniform spaces, and pm**UNIF**, of pseudometrizable uniformities. In the third row, we have **TOP**, and its subcategories **CREG**, of completely regular spaces, and pm**TOP**, of pseudometrizable topological spaces. The first row gives numerical information, the second uniform information (entourages) and the third local information (neighborhoods). The horizontal arrows stand for embedding functors and the vertical arrows for the forgetful functors. The two question marks show that the diagram is incomplete; do there exist structures which fill the gaps such that the two new vertical arrows represent forgetful functors and the two new horizontal arrows represent the embedding functors? Lowen shows that approach spaces fill the gaps in an admirable manner. In addition to the above motivation, Lowen points out several other considerations which call for a new structure: (1) products of metric spaces, (2) analysis versus numerical analysis and approximation theory, (3) topological properties versus metric properties, (4) quantification of topological properties such as the Kuratowski or Hausdorff measure of non-compactness.

The central idea in *Approach spaces* is that of a distance δ, which is a function on $X \times 2^X$ to $[0,\infty]$. And the interesting fact is that a distance can be defined not only in a metric space, as we saw above, but also in a topological space, a uniform space, etc. Of course, the abstract distance to be defined should satisfy some very simple general conditions: (D1) For each x in X, $\delta(x,\{x\})=0$. (D2) For each x in X, $\delta(x,\varnothing)=\infty$. (D3) For each x in X and for all A, B in 2^X, $\delta(x, A\cup B)=\min\{\delta(x,A),\delta(x,B)\}$. (D4) For each x in X, for each A in 2^X and each ε in $[0,\infty]$, $\delta(x,A)\le\delta(x,A^{(\varepsilon)})+\varepsilon$, where $A^{(\varepsilon)}=\{x\in X\colon\ \delta(x,A)\le\varepsilon\}$. We have already seen how a distance arises from a metric. Lowen shows how we can define a distance in a topological space or a uniform space. For example, in a topological space, we define $\delta(x,A)=0$ if $x\in \mathrm{cl}(A)$, $\delta(x,A)=\infty$ otherwise. The distance δ so defined satisfies (D1)–(D4) and is compatible with the topology. If δ is a distance on X, then (X,δ) is called an approach space. If (X,δ) and (X,δ') are approach spaces, a function $f\colon X\to Y$ is called a contraction if for each x in X and for each $A\subset X$, $\delta(f(x),f(A))\le\delta(x,A)$. The approach spaces form the objects and contractions form the morphisms of a construct denoted by **AP**. A topological space can be described in several equivalent ways: open sets, closed sets, Kuratowski closure operator, etc. Similarly, approach spaces can be defined not only in terms of a distance but also in terms of limit operators, approach systems, gauges, towers, hull operators, regular function frames, etc. Lowen describes these and shows their equivalence. A clear diagram is provided showing various connections. The first chapter ends by showing that **AP** is a topological construct.

The second chapter deals with topological approach spaces. In these spaces, distances arise from topologies as described above. The functor **TOP** $\to$ **AP** is a full concrete embedding. Moreover, **TOP** is simultaneously bireflectively and bicoreflectively embedded in **AP**. In the third chapter, we find metric approach spaces. Here the distances arise from extended pseudo-quasimetrics. The category of extended pseudo-quasimetric spaces as objects and nonexpansive maps as morphisms *pq*$\mathbf{MET}^\infty$ is also nicely embedded in **AP**. Uniform approach spaces are dealt with in the fourth chapter. These are subspaces of products of ∞p-metric approach spaces in **AP**. They form the epireflective hull, denoted by **UAP**, of $p\mathbf{MET}^\infty$ in **AP**. Their convergence in **UAP** and their relationship with **UNIF** are discussed. The two gaps (question marks) in the diagram given above are now filled. The new first row looks like

$$p\mathbf{MET}^\infty \to \mathbf{UAP} \to \mathbf{AP}$$

and the diagram is complete!

Chapter five gives canonical examples: spaces of measures, function spaces, hyperspaces, probabilistic metric spaces and spaces of random variables. Subsequent chapters give approach properties (compactness, connectedness, completeness), completion (construction, and an example of a function space) and compactification (construction, βX).

For purposes of reference, Appendix A provides information on basic categorical concepts, topological constructs, reflective and coreflective subcategories. Appendix B contains (probabilistic) metric spaces, convergence spaces, quasi-uniform and uniform spaces and proximity spaces. There is a bibliography of 117 items and an index.

The book is very carefully written, and, with hindsight, the author has presented the material in a simpler manner than that used in his original papers. Although the subject matter goes deep into abstract territory, several examples are provided which make the book easy to read. The book deserves a place on the shelves of mathematicians, especially topologists, who are interested in fundamentals.

S. A. Naimpally (Nepean, ON)

99c:55009 55N91 14F25 14F32 16E99 18G10 55N33

Goresky, Mark (Princeton, NJ); **Kottwitz, Robert** (Chicago, IL); **MacPherson, Robert** (Princeton, NJ)

Equivariant cohomology, Koszul duality, and the localization theorem.

Invent. Math. **131** (1998), *no.* 1, 25–83.

The idea that in order to study cohomological properties of group actions on spaces, it is best to consider a variant of the usual cohomology that incorporates the group action into its very definition has been around for nearly half a century. The Borel style equivariant cohomology of a space X with respect to the action of a group G is defined as the non-equivariant cohomology of $EG \times_G X$. Here $EG \to BG$ is a principal G-fibration, EG is acyclic, has a free action by G and BG is the classifying space of the group G. In a book by A. Borel [*Seminar on transformation groups*, Princeton Univ. Press, Princeton, N.J., 1960; MR **22** #7129] many remarkable properties of this equivariant cohomology theory were established, typical of which are localization theorems for actions of tori and a spectral sequence relating the group cohomology of G, the (non-equivariant) cohomology of X and the equivariant cohomology of X with respect to G. That this theory could be used in computing the cohomology of geometric quotients in algebraic geometry became clear from the work of M. F. Atiyah and R. H. Bott [Topology **23** (1984), no. 1, 1–28; MR 85e:58041] in the early 1980s, which was systematically explored by F. C. Kirwan [*Cohomology of quotients in symplectic and algebraic geometry*, Princeton Univ. Press, Princeton, NJ, 1984; MR 86i:58050]. That a variant of this theory could be defined for studying group actions on singular spaces was observed independently in the middle 1980s by J.-L. Brylinski and the present reviewer [J.-L. Brylinski, in *Kazhdan-Lusztig theory and related topics (Chicago, IL, 1989)*, 5–32, Amer. Math. Soc., Providence, RI, 1992; MR 94c:55010; R. Joshua, Math. Z. **195** (1987), no. 2, 239–253; MR 88g:14048], who then went on to define equivariant intersection cohomology. The equivariant intersection cohomology complexes form the simple objects in the heart of the equivariant derived category, which is the natural generalization of the usual derived category to study group actions.

In the paper under review, the authors obtain several interesting results in this area. Several results are established on what are called equivariantly formal spaces. Real coefficients are used throughout the paper when defining cohomology and all complexes of sheaves are complexes of sheaves of real vector spaces. According to the authors' definition, a space X with the action by a compact connected group G is equivariantly formal if the spectral sequence in equivariant cohomology degenerates; as examples of equivariantly formal spaces, one has symplectic manifolds with Hamiltonian G-action, any space with a G-invariant CW-decomposition, or any G-space whose ordinary cohomology vanishes in odd degrees. One of the main results in this paper concerns the case in which an algebraic torus acts on a complex projective algebraic variety with finitely many fixed points and finitely many 1-dimensional orbits. Assuming in addition that the equivariant cohomology of X is free over the cohomology ring of G, the authors show that the restriction map from the G-equivariant cohomology of the space X to the G-equivariant cohomology of the set of fixed points is an injection, and they relate the image of this map with the finitely many 1-dimensional G-orbits.

The next main topic discussed in the paper is one that involves Koszul duality [J. N. Bernstein, I. M. Gel′fand and S. I. Gelfand, Funktsional. Anal. i Prilozhen. **12** (1978), no. 3, 66–67; MR 80c:14010a; A. A. Beilinson, V. A. Ginzburg and W. Soergel, J. Amer. Math. Soc. **9** (1996), no. 2, 473–527; MR 96k:17010] in the following context. If G denotes a connected compact group, the Koszul duality that is discussed here concerns the duality between modules over the cohomology ring $H^*(BG)$ (identified with the symmetric algebra of polynomial functions on the Lie algebra of G) and modules over $H^*(G)$ (identified with the exterior algebra of the Lie algebra of G). Observe that if X is a G-space, the equivariant cohomology $H^*_G(X)$ is a module over the former while the ordinary cohomology of X is a module over the latter. While these two are not Koszul dual to each other, it is shown that one may lift the action of $H^*(BG)$ [resp. $H^*(G)$] to a co-chain complex representing $H^*_G(X)$ [resp. $H^*(X)$]. Now it is shown that the Koszul duality functor sends the former complex to the latter and that it defines an equivalence of the derived category of bounded-below complexes on $H^*(BG)$ with the derived category of bounded-below complexes on $H^*(G)$. It is also shown that a G-space X is equivariantly formal if and only if the chain complex representing the ordinary cohomology of X is quasi-isomorphic to its cohomology in the derived category over $H^*(G)$ and the action of $H^*(G)$ on $H^*(X)$ is trivial.

The authors also obtain extensions of the above results when the equivariant cohomology is replaced by equivariant hypercohomology with respect to any complex in the equivariant derived category of bounded-below complexes on a G-space X. If G is a connected compact group acting on a reasonable space X, a complex K in the G-equivariant derived category of X is called equivariantly formal if the corresponding spectral sequence for G-equivariant hypercohomology of X degenerates. The authors also establish the existence of certain cohomology operations on $H^*(X; K)$ whose vanishing is shown to be equivalent to K being equivariantly formal. In addition, various other results that are of secondary importance are also established; for example, a cycle-theoretic definition of equivariant intersection cohomology paralleling the original definition of intersection cohomology is given. While they do not discuss other applications here, it is to be expected that the authors have several applications to problems in representation theory in mind.

The reviewer has not stated all the results contained in this very fine piece of work. The reader is encouraged to look at the paper. *Roy Joshua* (Columbus, OH)

97d:57004 57M25
Bar-Natan, Dror (Cambridge, MA)
On the Vassiliev knot invariants.
Topology **34** (1995), *no.* 2, 423–472.
This paper is one of the standard introductions to the subject of Vassiliev invariants. It synthesises the axiomatic approach to Vassiliev's work à la Birman and Lin with the theory of quantum knot invariants of Reshetikhin and Turaev.

J. S. Birman and X. S. Lin [Invent. Math. **111** (1993), no. 2, 225–270; MR 94d:57010] characterized Vassiliev's invariants as finite-type invariants. A knot invariant which takes values in an abelian group can be extended to singular knots—that is, immersions of the circle in 3-space which have a finite number of transversal self-intersections. This extension is done by considering a singular knot as an alternating sum of proper knots obtained by resolving each of the double points in the two possible directions. A knot invariant is said to be of finite type or of type n if, for some $n \in \mathbf{N}$, it vanishes on knots with more than n double points.

The Reshetikhin-Turaev invariant τ_V [see, e.g., Comm. Math. Phys. **127** (1990), no. 1, 1–26; MR 91c:57016] is a (framed) knot invariant which depends on a semi-simple Lie algebra $\mathfrak{g}$ and a representation V, and it takes values in the complex power series in a formal parameter h. For example, the Jones polynomial with the substitution $q = e^h$ gives a power series in h which is actually the invariant coming from the defining representation of SU(2). The author was working on perturbative Chern-Simons theory and noticed connections with diagrams in the Vassiliev theory. This led to the realization that the coefficient of h^n in τ_V is a type-n invariant. This added impetus to both approaches.

The Vassiliev theory is underpinned by certain combinatorial objects. To a knot with n double points can be associated a chord diagram—that is, an oriented circle with chords marked on it—each chord corresponding to a double point, the chord ends signifying the points on the circle which meet at the double point. It is not difficult to show that the value that a type-n Vassiliev invariant takes on a singular knot with n double points depends only on the underlying chord diagram of the singular knot. Thus a type-n invariant naturally determines a function on the set of chord diagrams with n chords. This function can be thought of as the leading term or "symbol" of the invariant. From topological considerations each of these functions can be shown to satisfy two combinatorially defined relations—known as the four-term (4T) and one-term (1T) relations (the latter also referred to as "isolated chord" and as "framing independence"). Functions of this form satisfying 4T and 1T are called weight systems.

The invariants in this paper take values in a field $\mathbf{F} = \mathbf{Q}$, $\mathbf{R}$ or $\mathbf{C}$. Let $\mathscr{A}^r$ be the graded vector space over $\mathbf{F}$ which has as the homogeneous components of degree n linear combinations of chord diagrams with n chords, modulo the ideal generated by the 1T and 4T relations. Note that an element of $\mathscr{A}^r$ is not necessarily a finite linear combination of chord diagrams. Although this is a simple space to define, the 1T and 4T relations make it difficult to figure out even the dimensions of the homogeneous parts; up to now the dimensions are known only up to degree 9.

M. Kontsevich's theorem (also known as the fundamental theorem) [in *I. M. Gel′fand Seminar*, 137–150, Amer. Math. Soc., Providence, RI, 1993; MR 94k:57014] says that over $\mathbf{F}$, the structure of the Vassiliev invariants is the same as the structure of $\mathscr{A}^r$; the first rigorous proof (due to Kontsevich) is presented in the paper under review. The theorem consists of the transcendental construction of a function $\widetilde{Z}$, the Kontsevich integral, from the set of knots to $\mathscr{A}^r$, with the property that a knot with n double points is mapped to its underlying chord diagram plus higher order terms (i.e. diagrams with more than n chords). $\widetilde{Z}$ is universal in the sense that all Vassiliev invariants factor through it. Dualizing this means that, modulo invariants of lower order, a Vassiliev invariant is determined by the weight system it defines.

Actually, the construction of $\widetilde{Z}$ given in the paper under review, although the first rigorous one, is not very satisfactory as it is not easy to do any calculations with. Other constructions have since appeared but nothing entirely satisfactory—see a paper by the author and A. Stoimenow [in *Geometry and physics (Aarhus, 1995)*, 101–134, Dekker, New York, 1997] for an overview of some of these.

The set of knots admits the operation of connected sum, which makes it into a commutative monoid, and the set of **F**-valued knot invariants has a commutative product coming from the multiplication in **F**; similarly $\mathscr{A}^r$ can be given a natural Hopf algebra structure with the product and coproduct having simple combinatorial definitions, so that $\widetilde{Z}$ respects these structures. There is a different normalization of $\widetilde{Z}$ which is relevant for the Reshetikhin-Turaev invariants mentioned below.

The structure theory of Hopf algebras implies that $\mathscr{A}^r$ is actually a polynomial algebra generated by its so-called primitive elements. Again the structure of these primitive elements is not properly understood, but $\mathscr{A}^r$ has an alter ego in which they are easier to spot: it is sometimes more convenient to consider the space of Chinese character diagrams which have extra trivalent vertices and a relation STU replacing 4T. The author proves that this space is isomorphic to $\mathscr{A}^r$.

The quantum invariants are really naturally defined for framed knots, and one can repeat the above theory for framed knots. One obtains a Hopf algebra $\mathscr{A}$ which is chord diagrams modulo just the 4T relation (or Chinese character diagrams modulo just STU); and as an algebra it is generated by one more primitive element than $\mathscr{A}^r$—the diagram with one chord. There is also a natural Hopf algebra map $\mathscr{A}^r \to \mathscr{A}$ which corresponds to assigning the zero framing to a knot.

Recall that for V a representation of the semi-simple Lie algebra $\mathfrak{g}$, the coefficient, τ_V^n, of h^n in τ_V is a type-n invariant; so the τ_V^n define weak weight systems (i.e. they satisfy 4T but not 1T). These weak weight systems have a simple description: a chord diagram (or Chinese character diagram) gives a recipe for "glueing together" various tensors associated to $\mathfrak{g}$ to obtain an element of the center of the universal enveloping algebra of $\mathfrak{g}$; taking the trace of this in the representation V gives a number—this is the value of the weight system on the diagram.

For the classical Lie algebras with certain representations, the author gives computationally easier ways of evaluating these weight systems. This is possible because of the behaviour of these weight systems with respect to some operations in the representation ring of a fixed $\mathfrak{g}$. On the level of (framed) knots one can define various cabling operations. One is the ith disconnected cabling, which results in an i-component link consisting of i parallel (along the framing) copies of the original knot. Evaluating $\tau_{V\otimes W}$ on a knot gives the same result as evaluating τ on the second disconnected cabling with (in the language of quantum invariants) one component coloured by V and one by W—this generalizes in the obvious way to higher-order cabling. The disconnected cabling operation descends to a map on the chord diagram level which can be utilised to evaluate $\tau_{V\otimes W}$.

Also very interesting is the ith connected cabling operation on knots which results in a knot which wraps i times around the old knot. This operation is known to be related to the Adams operation in the representation ring of $\mathfrak{g}$. On the level of chord diagrams one obtains a family of combinatorially defined maps $\psi^i : \mathcal{A} \to \mathcal{A}$ satisfying $\psi^i \psi^j = \psi^{ij}$, and indeed these are adjoint to the Adams operations, $\widehat{\psi}^i$, in the representation ring of $\mathfrak{g}$, i.e. for a diagram D, $\tau^n_{\widehat{\psi}^i V}(D) = \tau^n_V(\psi^i D)$. Further, each homogeneous part of $\mathcal{A}$ splits naturally into a direct sum of simultaneous eigenspaces for these operations, and spanning sets for these eigenspaces are given by the Chinese characters which are introduced.

A question raised in this paper is, do all weight systems come from semisimple Lie algebras? This has been answered in the negative by P. Vogel ["Algebraic structures on modules of diagrams", Preprint, Inst. Math. Jussieu, Paris; per revr.].

The author also presents another method for constructing weight systems which is at least as strong as the semisimple Lie algebra method. It involves the use of marked surfaces but why marked surfaces should lead to knot invariants is not clear. This construction is conjectured to give all weight systems.

Many of the subsidiary results in the paper are left as exercises for the reader, and several outstanding problems are listed. All in all, this paper is a good introduction to a subject in the intersection of many areas of current mathematical interest.

Simon Willerton (Strasbourg)

97h:57028 57M50 57N10

Gabai, David (Pasadena, CA)

On the geometric and topological rigidity of hyperbolic 3-manifolds.

J. Amer. Math. Soc. **10** (1997), *no.* 1, 37–74.

In this important and impressive paper, the author considers rigidity problems for 3-manifolds. He considers a closed hyperbolic 3-manifold N which satisfies a certain condition which he calls the insulator condition. The precise statement of this condition is technical and will be discussed later, but very many hyperbolic 3-manifolds satisfy this condition. In particular, if the shortest closed geodesic in N is very short, or is very long, then N satisfies the insulator condition. Precise bounds for these lengths are given in the paper. The main theorem of this paper is as follows: Theorem: Let N be a closed hyperbolic 3-manifold satisfying the insulator condition. (1) If M is a closed irreducible 3-manifold and $f : M \to N$ is a homotopy equivalence, then f is homotopic to a homeomorphism. (2) If f and g are homotopic homeomorphisms of N, then they are isotopic. (3) The space of hyperbolic metrics on N is path connected. The author conjectures that all closed hyperbolic 3-manifolds satisfy the insulator condition. In March 1996, Gabai, Meyerhoff and N. Thurston announced that they had proved this conjecture.

Before the review itself, here is a brief discussion of how these results relate to previous results. For brevity, I will restrict attention to the situation where M and N are both closed, orientable and irreducible with infinite fundamental group. In [Ann. of Math. (2) **87** (1968), 56–88; MR **36** #7146], F. Waldhausen showed that Parts (1) and (2) hold under the additional assumption that N is Haken. (Actually he used the term "sufficiently large" but this term has been replaced by "Haken" these days.) In [Ann. of Math. (2) **117** (1983), no. 1, 35–70; MR 84c:57008; Topology **24** (1985), no. 3, 341–351; MR 87c:57012], the reviewer showed that (1) and (2) hold under the additional assumption that N is a Seifert fibre space. In [Topology **31** (1992), no. 3, 493–517; MR 94g:57021; Comment. Math. Helv. **68** (1993), no. 3, 341–364; MR 94f:57016], the reviewer and J. Hass showed that (1) and (2) hold if N admits a π_1-injective map f of a closed orientable surface, not S^2, which satisfies some technical conditions which mean that f is not very singular. Also Mostow's rigidity theorem implies that (1) holds if both M and N are assumed to be hyperbolic. Finally, the author of the paper under review showed in a recent paper [J. Amer. Math. Soc. **7** (1994), no. 1, 193–198; MR 94b:57016] that under the hypotheses of Part (1) of the above theorem, and without any insulator condition, there is a common finite cover of M and N. Both Waldhausen's result and this result are used in the proof of the paper under review. It should be pointed out that Part (3) does not follow from Mostow's rigidity theorem, as that result shows only that given two metrics on a closed hyperbolic 3-manifold N, there is an isometry between them which is homotopic to the identity, whereas the result of Part (3) asserts that the isometry can be chosen to be isotopic to the identity.

Now we turn to the author's proof of his main theorem. This is a major piece of work. I will attempt to describe the ideas in simple terms, but several parts of the argument are very difficult to carry out. The initial idea for the proof is that one should choose a simple closed curve γ in M in such a way that f sends γ to a simple closed curve δ in N by a homeomorphism, sends a regular neighbourhood $N(\gamma)$ of γ to a regular neighbourhood $N(\delta)$ of δ by a homeomorphism, and in addition f maps $M - N(\gamma)$ into $N - N(\delta)$. If one can do this, then Waldhausen's theorem on Haken manifolds implies that the restricted map $f\colon M - N(\gamma) \to N - N(\delta)$ is homotopic to a homeomorphism by a homotopy fixing the boundary of $N(\gamma)$, and hence that f itself is homotopic to a homeomorphism. This is not a new idea, but the problem was that there seemed no way of choosing γ and δ appropriately. In [op. cit.], the author showed that these choices can be made if one is prepared to replace M and N by appropriate finite covers. In this paper, he shows that these choices can be made if one has the insulator condition.

As there is no natural way to choose any particular simple closed curve in M, one starts by considering the choice of δ in N. Here one has the slight advantage that there is a canonical choice for δ within any homotopy class. Namely one can choose δ to be the closed geodesic in that class. It is possible that this geodesic might be singular, but this can be avoided by choosing δ to be the shortest closed geodesic in N, as this geodesic is automatically simple. For the moment, let δ denote any simple closed geodesic in N and consider the problem of choosing γ correctly. The choice of δ determines the homotopy class of the curve γ in M, but there are infinitely many isotopy classes to choose from. If M and N are indeed homeomorphic there is exactly one "correct" isotopy class of γ, and we can think of all the other isotopy classes of homotopic curves as being knotted in some sense. Thus the problem can be thought of as the problem of choosing the unknotted isotopy class for γ. But as we do not yet know that M and N are homeomorphic, it is not at all clear that there is any isotopy class of γ which will have the required property.

Given a simple closed geodesic δ in the hyperbolic 3-manifold N, the author describes a canonical solid torus neighbourhood W of δ, which he calls the Dirichlet neighbourhood of δ, as its definition is analogous to that of the Dirichlet fundamental region of a discrete group acting on hyperbolic space. Consider the picture in the universal cover $\widetilde{N}$ of N. There will be an infinite family Δ of disjoint geodesics forming the pre-image of δ. Fix one of these geodesics δ_0. Given any other geodesic δ_i in this family, there is a shortest line segment d_i between δ_0 and δ_i, and d_i is itself a geodesic segment and is perpendicular to both δ_0 and δ_i. Let D_i denote the hyperbolic plane which meets d_i orthogonally at its midpoint. Thus D_i is the plane halfway between δ_0 and δ_i. This gives rise to an infinite family D_0 of planes in $\widetilde{N}$ which will be invariant under the stabiliser of the geodesic δ_0 and will cut $\widetilde{N}$ into regions. Let W_0 denote the region which contains δ_0. This region has closure homeomorphic to $\mathbf{R} \times D^2$ and its image in N has closure W which is a solid torus neighbourhood of the closed geodesic δ. Thus the canonical family D_0 of planes in $\widetilde{N}$ defines a canonical solid torus neighbourhood of δ in N. The key idea of the whole paper is that one should construct a family of planes with similar properties in the universal cover of M, but without having first chosen γ. If this can really be done, then there is a natural choice of γ, namely the core of the solid torus obtained in M. The construction of a suitable family of planes is described in the next few paragraphs.

At this point one uses the earlier result of the author which states that M and N must have a common finite cover. This makes it easy to identify the universal covers of M and N with each other and with hyperbolic 3-space H^3 in such a way that there is a lift of $f: M \to N$ to a map $\widetilde{f}: H^3 \to H^3$, which extends to the identity map on the sphere at infinity. The fundamental groups of M and N both act on H^3 as covering groups, and both actions extend to the same action on the sphere at infinity. (The actions are the same when one identifies the two groups via the isomorphism given by the map f.) There is no natural metric on $\widetilde{M}$, but if one chooses a metric on M, this will yield a metric on $\widetilde{M}$, and $\pi_1(M)$ will act by isometries. Of course, this metric will almost certainly not be hyperbolic.

Now consider the family Δ of geodesics in $\widetilde{N}$ and the family D_0 of planes midway between δ_0 and its translates. One wants to construct a similar picture in $\widetilde{M}$. As these two universal covers have already been identified, it is tempting to simply consider the images of these geodesics and planes under this identification. But this family need not be invariant under the action of $\pi_1(M)$. However, if we consider only the sphere at infinity of H^3, the endpoints of the geodesics forming Δ will be invariant under the action of $\pi_1(M)$, as the actions of $\pi_1(M)$ and $\pi_1(N)$ are the same. Similarly, the circles at infinity of the planes forming D_0 are also invariant under the action of $\pi_1(M)$. The idea now is to span these circles by planes invariant under the action of $\pi_1(M)$ which are area minimising using the metric in the universal cover of M. The first difficulty here is that it is not clear that there exists an area-minimising plane spanning each circle. The author needs to work very hard to get round this problem and shows instead that there is an area-minimising lamination spanning each circle. In what follows, I will ignore this problem and continue to refer to planes for simplicity of exposition.

Suppose now that we can span one of these circles by an area-minimising plane. By taking translates of this plane, we will obtain a family E_0 of planes invariant under the action of $\pi_1(M)$ which are area minimising using the metric in the universal cover $\widetilde{M}$ of M. Standard cut-and-paste arguments show that two of these planes are disjoint if they have disjoint boundary circles. The converse is trivial because any two of the circles being considered in the sphere at infinity must be disjoint or cross transversely. As two of the hyperbolic planes in D_0 are also disjoint if and only if their boundary circles are disjoint, it follows that two of the area-minimising planes in E_0 are disjoint if and only if the corresponding planes of the family D_0 are disjoint. Thus the way in which the new planes intersect each other is identical to that of the original hyperbolic planes if we only consider pairs of planes. However, there is no obvious information about how triples of planes should meet. The family of planes E_0 might divide $\widetilde{M}$ into regions in quite a different way from the way in which the hyperbolic planes in D_0 divided $\widetilde{N}$. We are interested in the component of the complement of the planes of E_0 which contains the two points at infinity a and b which form the endpoints of the geodesic δ_0. We would like this region to be homeomorphic to $\mathbf{R} \times D^2$ and to cover a solid torus in M, as for the family of hyperbolic planes D_0. The main problem is that there may be no such component. In other words, the two endpoints a and b of δ_0 might lie in different components of the complement of the family of planes E_0. This is where the insulator condition is used.

Now I will discuss what the insulator condition is about. First we consider a condition which the author calls "tri-linking". Recall that we are interested in a pair of points $\{a, b\}$ in the sphere, namely the endpoints of δ_0, and a family of circles in the sphere with the property that each circle separates the pair $\{a, b\}$ from some translate of $\{a, b\}$. Thus each individual circle does not separate a from b. This family of circles is said to exhibit tri-linking if there are three of them whose union separates a from b. (It is clearly impossible for the union of two of them to do this.) If tri-linking occurs, then it is possible to isotop the three planes so as to separate a from b in the 3-ball. Thus there is no reason why this should not occur within the family of planes E_0. However, the author proves that if no tri-linking occurs, then a and b must lie in one component V_0 of the complement of the planes in E_0, and that V_0 covers a solid torus V in M. A special case of the insulator condition occurs when there is some simple closed geodesic δ in N such that if one proceeds as above to choose a geodesic δ_0 and the family of hyperbolic planes D_0, then no tri-linking occurs. The most general form of the insulator condition asserts the existence of a family of circles on the sphere at infinity which again satisfy the no tri-linking condition but need not be the boundaries of hyperbolic planes, but I will say no more about that.

The conclusion so far is that if no tri-linking occurs in the family of planes D_0 in $\widetilde{N}$, then the corresponding family of planes E_0 in $\widetilde{M}$ yields a solid torus embedded in M, so that we can define γ to be the core of this solid torus. However, this is not the end of the argument by any means. First note that there is a worry here, as we know that there is only one correct isotopy class for γ, but the above construction seems as if it might yield different isotopy classes as one alters the choices of the planes E_0. And we still need to prove that f can be homotoped to send a regular neighbourhood $N(\gamma)$ of γ to a regular neighbourhood $N(\delta)$ of δ in N by a homeomorphism and in addition to send $M - N(\gamma)$ into $N - N(\delta)$. It is trivial to arrange that f is a homeomorphism between regular neighbourhoods of γ and δ. The difficulty is to arrange that f maps $M - N(\gamma)$ into $N - N(\delta)$. The author deals with these problems using some more impressive arguments. First he shows that the isotopy class of γ is independent of the choice of planes forming E_0 and is also independent of the choice of metric on M. Next he applies this result with the manifold M replaced by the common finite cover M_1 of M and N. This implies that the lifts of γ and δ to M_1 must be isotopic. It follows that the link Γ in H^3 formed by the pre-image of γ is equivalent to that formed by the pre-image

Δ of δ, meaning that there is a homeomorphism of H^3 which sends one link to the other and is the identity on the sphere at infinity. In the last part of the argument, the author shows that this implies that f can be homotoped in the required way. Note that there are many homeomorphisms from $N(\gamma)$ to $N(\delta)$, but probably only one which extends to a homeomorphism from M to N. Thus there is a problem of choosing the correct homeomorphism from $N(\gamma)$ to $N(\delta)$. The details of this part of the argument are interesting but technical, and I will say nothing more about them. *G. Peter Scott* (Ann Arbor, MI)

97b:57017 57P99
Bryant, J. [Bryant, John Logan] (Tallahassee, FL); **Ferry, S.** (Binghamton, NY); **Mio, W.** (Tallahassee, FL); **Weinberger, S.** (Chicago, IL)
Topology of homology manifolds.
Ann. of Math. (2) **143** (1996), *no.* 3, 435–467.
Homology manifolds are topological spaces with the local homology properties of manifolds. More formally, an n-dimensional homology manifold is a topological space X such that at each point $x \in X$,

$$H_r(X, X \setminus \{x\}) = \begin{cases} \mathbf{Z} & \text{if } r = n, \\ 0 & \text{if } r \neq n. \end{cases}$$

Homology manifolds are also known as generalized manifolds.

Topological manifolds are themselves examples of homology manifolds. A resolution (M, f) of a space X is defined to be an n-dimensional topological manifold M with a surjection $f: M \to X$ such that each inverse image $f^{-1}(x)$ $(x \in X)$ is contractible in each of its neighbourhoods in M, in which case X is an n-dimensional homology manifold. Does every homology manifold admit a resolution? The paper under review constructs compact ANR homology manifolds in dimensions ≥ 5 which do not admit a resolution. Nonresolvable homology manifolds are called exotic. The exotic homology manifolds are important new objects in high-dimensional manifold theory. Their construction may be compared to the observational discovery of a new planet whose existence was predicted by calculations. In order to properly appreciate the significance of the exotic homology manifolds, it is best to trace the history of homology manifolds.

Homology manifolds were introduced in 1933 independently by Čech and Lefschetz. In the first instance, homology manifolds were artefacts of algebraic topology, serving to clarify the homological properties of manifolds. Much of the algebraic topology of manifolds works just as well for homology manifolds. In particular, the cohomology and homology of a compact oriented n-dimensional ANR homology manifold X are related by Poincaré duality isomorphisms $H^{n-*}(X) \cong H_*(X)$. Homology manifolds also became artefacts of geometric topology. The decomposition theory of manifolds pioneered in the 1920's and 1930's by R. L. Moore and G. T. Whyburn was refashioned in terms of homology manifolds by R. Wilder in the 1930's and 1940's, and carried forward by R. H. Bing in the 1950's and 1960's. All this led to the use of homology manifolds in the geometric study of the connectivity and separation properties of manifolds.

Homology manifolds play an important role in the recognition problem for topological manifolds: formulate a general position property on a homology manifold X

which is necessary and sufficient for X to be a topological manifold. The beautiful paper by J. W. Cannon "The recognition problem: what is a topological manifold?" [Bull. Amer. Math. Soc. **84** (1978), no. 5, 832–866; MR **58** #13043] was written at a previous turning point in the history of the problem, shortly after the work of Cannon and R. D. Edwards on the double suspension of homology spheres. The paper included the formulation of the "disjoint disk property", and the statement of a theorem of Edwards that for $n \geq 5$ a resolvable n-dimensional ANR homology manifold with this property is an n-dimensional topological manifold. A proof of the theorem may be found in the book of R. J. Daverman [*Decompositions of manifolds*, Academic Press, Orlando, FL, 1986; MR 88a:57001]. The theorem reduces the recognition problem for topological manifolds to the problem of characterizing the ANR homology manifolds which admit resolutions. With hindsight, it is clear that this was the best result geometric topology could achieve without algebraic surgery theory.

The Browder-Novikov-Sullivan-Wall surgery theory was developed in the 1960's, solving the homotopy-theoretic version of the recognition problem: when is a space with n-dimensional Poincaré duality homotopy equivalent to an n-dimensional manifold? For $n \geq 5$ an n-dimensional Poincaré duality space X is homotopy equivalent to an n-dimensional topological manifold if and only if the Spivak normal fibration of X admits a topological reduction with 0 surgery obstruction. Since homology manifolds are Poincaré duality spaces it is natural to ask if they are homotopy equivalent to topological manifolds. D. E. Galewski and R. J. Stern [Invent. Math. **39** (1977), no. 3, 277–292; MR **56** #3847] showed that the Spivak normal fibration ν_X of a polyhedral homology manifold X admits a canonical topological reduction with 0 surgery obstruction, so that if $\dim(X) \geq 5$ then X is homotopy equivalent to a topological manifold. The triangulability of X was an essential ingredient of the method. The results of Cannon and Edwards allowed Galewski and Stern [Ann. of Math. (2) **111** (1980), no. 1, 1–34; MR 81f:57012] to prove that in fact every polyhedral homology manifold of dimension ≥ 5 admits a resolution by a (triangulated) topological manifold. With hindsight, it is clear that this was the best result algebraic topology could achieve without geometric surgery theory.

Further progress depended on the development of the controlled topology pioneered in the 1970's by T. A. Chapman, Ferry, and F. Quinn. Controlled topology grew out of the use of noncompact manifolds in Novikov's proof of the topological invariance of rational Pontryagin classes and in the Kirby-Siebenmann structure theory of compact topological manifolds. The methods of combinatorial topology can be extended to non-triangulable compact spaces by the systematic use of metric conditions to limit the size of allowed operations. From this point of view a resolution is a sufficiently controlled homotopy equivalence.

Controlled topology was first applied to the resolution problem for homology manifolds by Quinn [Invent. Math. **72** (1983), no. 2, 267–284; MR 85b:57023]. The paper had the following programme for proving that every ANR homology manifold X can be resolved: (1) view X as an X-controlled Poincaré duality space, meaning that the Poincaré duality chain equivalence $[X] \cap - : C(X)^{n-*} \to C(X)$ has "small point inverses" when measured in X, and (2) extend in that case Quinn's methods [Ann. of Math. (2) **110** (1979), no. 2, 275–331; MR 82k:57009] to develop just enough controlled surgery theory to prove that the controlled Poincaré space in (1) is controlled homotopy equivalent to a topological manifold, and hence resolvable. In

Quinn's 1983 paper [op. cit.; MR 85b:57023] it was claimed that the programme works, and that every ANR homology manifold X of dimension ≥ 5 admits a resolution. Unfortunately, a potential obstruction in (2) had been overlooked. The result was withdrawn, and replaced by Quinn's 1987 study [Michigan Math. J. **34** (1987), no. 2, 285–291; MR 88j:57016]. That paper introduced an invariant $I(X) \in \mathbf{Z}$ (for connected X) such that $I(X) = 0$ if and only if X admits a resolution. Roughly speaking, $8I(X)+1$ is the number of "points" in the generic "point inverse" of $[X] \cap -: C(X)^{n-*} \to C(X)$, with $I(X) = 0$ for resolvable X by topological transversality. Either every X was resolvable or else the invariant was realized by an exotic X. The hunt was on.

The main new results in the paper under review are: (i) for every compact simply-connected n-dimensional topological manifold M with $n \geq 6$ and every integer $k \in \mathbf{Z}$ there exists a compact n-dimensional ANR homology manifold X homotopy equivalent to M with $I(X) = k$, and (ii) a non-simply-connected surgery classification theory for compact ANR homology manifolds of dimension ≥ 6, including analogues of the Sullivan-Wall surgery exact sequence and the reviewer's total surgery obstruction.

The simply-connected examples in (i) are obtained by an ingenious infinite process which constructs a sequence of ever more controlled Poincaré duality spaces, converging to a compact ANR homology manifold X in the homotopy type of M realizing the prescribed resolution obstruction $I(X) = k$. Unfortunately, Quinn did not work out all the controlled surgery theory necessary for the classification of homology manifolds. The surgery classification in (ii) uses the bounded surgery theory of Ferry and E. K. Pedersen [in *Novikov conjectures, index theorems and rigidity, Vol. 2 (Oberwolfach, 1993)*, 167–226, Cambridge Univ. Press, Cambridge, 1995; MR 97g:57044] instead. A controlled surgery problem determines a bounded surgery problem over the open cone. A solution of the bounded surgery problem determines a solution of the original controlled surgery problem by codimension-1 splitting, but it is not clear if the solution is sufficiently controlled for the application to (ii). The proof of such a result is to appear in a paper by Ferry and Pedersen ["Squeezing structures", to appear]. Until this paper or some appropriate substitute becomes available the surgery classification in (ii) must be regarded as somewhat provisional—although there is little doubt among the experts that it is correct.

Even though the exotic homology manifolds have now been hunted down, much remains to be done, apart from completing the surgery classification programme.

An n-dimensional topological manifold is homogeneous, since every point has a neighbourhood homeomorphic to the Euclidean space $\mathbf{R}^n$. Is there a corresponding homogeneity for exotic homology manifolds with the disjoint disk property? The paper formulates the intriguing conjecture that there exist spaces $\mathbf{R}^4_k$ ($k \in \mathbf{Z}$) such that every connected ANR homology manifold X_n, $n \geq 5$, with the disjoint disk property and $I(X) = k$ is locally homeomorphic to $\mathbf{R}^4_k \times \mathbf{R}^{n-4}$. The hunt is now on for $\mathbf{R}^4_k$!

Homology manifolds may well feature in work on the Borel conjecture for Poincaré duality groups, and the closely related Novikov conjecture on the homotopy invariance of the higher signatures. The classifying space of an n-dimensional Poincaré duality group π is an aspherical n-dimensional Poincaré duality space $B\pi$. The Borel conjecture for π is that $B\pi$ is homotopy equivalent to a compact n-dimensional topological manifold. The conjecture has been verified for a large class of Poincaré duality groups. The hunt is also on for aspherical nonresolvable homology manifolds, since their fundamental groups would be counterexamples to the Borel conjecture.

The report of Weinberger [in *Proceedings of the International Congress of Mathematicians, Vol. 1, 2 (Zürich, 1994)*, 637–647, Birkhäuser, Basel, 1995; MR 97g:57028] explores the analogy between homology manifolds and orbifolds, pointing towards new applications of homology manifolds in the study of group actions.

In conclusion, it should be noted that the discovery of the exotic homology manifolds provides a remarkable example of collaboration between algebraic and geometric topologists. The reviewer hopes that collaborations of this type will lead to further progress into this fascinating realm of topology.

A. A. Ranicki (Edinburgh)

97a:57033 57R57 53C15 58D10 58D27 58G30
Taubes, Clifford H. (Cambridge, MA)
SW ⇒ Gr: **from the Seiberg-Witten equations to pseudo-holomorphic curves.**
J. Amer. Math. Soc. **9** (1996), *no.* 3, 845–918.

Background. In the fall of 1994 N. Seiberg and E. Witten introduced their monopole equations in the study of smooth 4-manifolds [cf. Nuclear Phys. B **426** (1994), no. 1, 19–52; MR 95m:81202a; erratum; MR 95m:81202b; Nuclear Phys. B **431** (1994), no. 3, 484–550; MR 95m:81203]. This discovery led to significant simplifications of many earlier results by Donaldson and others, which were based on Donaldson's polynomial invariants and used anti-self-dual Yang-Mills equations with nonabelian structure groups SU(2) and SO(3). The new Seiberg-Witten equations reduced the structure group to U(1) and it was soon realized by Kronheimer and Mrowka that the notorious compactness problem all but disappeared. As a result they were able to prove the Thom conjecture about the genus minimizing property of embedded complex curves in $\mathbf{CP}^2$ [cf. P. B. Kronheimer and T. S. Mrowka, Math. Res. Lett. **1** (1994), no. 6, 797–808; MR 96a:57073]. Taubes [Math. Res. Lett. **1** (1994), no. 6, 809–822; MR 95j:57039; Math. Res. Lett. **2** (1995), no. 1, 9–13; MR 96a:57075] established that symplectic 4-manifolds have nontrivial Seiberg-Witten invariants and, in particular, do not admit metrics of positive scalar curvature unless $b^+ = 1$. He thus obtained large classes of smooth 4-manifolds which do not admit symplectic structures although there are no obvious obstructions (the simplest example is a connected sum of three copies of $\mathbf{CP}^2$).

From an entirely different direction Gromov had in the mid 1980s introduced pseudo-holomorphic curves in symplectic geometry [M. L. Gromov, Invent. Math. **82** (1985), no. 2, 307–347; MR 87j:53053] and showed how they can be used to prove a number of surprising and remarkable new theorems, including his celebrated nonsqueezing theorem and a uniqueness result for symplectic structures on $\mathbf{CP}^2$ under the assumption of the existence of pseudo-holomorphic spheres. This line of research was taken up by McDuff, who showed, for example, that every symplectic 4-manifold which is minimal (no symplectically embedded sphere with self-intersection -1) and contains an embedded pseudo-holomorphic sphere with nonnegative self-intersection must be rational or ruled [cf. D. McDuff, J. Amer. Math. Soc. **3** (1990), no. 3, 679–712; MR 91k:58042; erratum; MR 93k:58098]. Y. Ruan [in *Geometry and analysis on complex manifolds*, 171–197, World Sci. Publishing, River Edge, NJ, 1994; MR 98h:32050] extended Gromov's construction by defining a version of the Gromov invariant which counts pseudo-holomorphic curves in a given homology class. One of the serious obstacles for further developments along these lines was the problem of establishing the existence of pseudo-holomorphic curves. An important step in this direction was the work of S. K. Donaldson ["Symplectic submanifolds and almost complex geometry", Preprint, Univ. Oxford, Oxford, 1995; per revr.], where he proved the existence of very ample divisors (codimension-2 symplectic submanifolds) representing the class $k[\omega]$ for large k provided that the cohomology class of the symplectic form ω is rational.

Taubes [Math. Res. Lett. **2** (1995), no. 2, 221–238; MR 96a:57076] brought these developments together with the announcement of his striking new theorem about the equivalence of the Seiberg-Witten and the Gromov invariants for symplectic 4-manifolds. The story has three parts. The paper under review proves the existence of pseudo-holomorphic curves from solutions of the Seiberg-Witten equations. In a subsequent paper ["Counting pseudo-holomorphic submanifolds in dimension 4", Preprint, Harvard Univ., Cambridge, MA, 1995; per revr.] Taubes extends the construction of Ruan to include the counting of disconnected pseudo-holomorphic curves in his definition of the Gromov invariant [see also D. McDuff, in *Gauge theory and symplectic geometry (Montreal, PQ, 1995)*, 175–210, Kluwer Acad. Publ., Dordrecht, 1997; MR 98j:58018]. The third part is the forthcoming paper by Taubes ["Gr $\Rightarrow$ SW: from pseudoholomorphic curves to the Seiberg-Witten invariants", Preprint], which constructs solutions of the Seiberg-Witten equations from pseudo-holomorphic curves.

The main theorem. The Seiberg-Witten invariants of a compact oriented smooth 4-manifold take the form of an integer-valued function SW: ${}^c(X) \to \mathbf{Z}$ defined on the set ${}^c(X)$ of isomorphism classes of spin^c structures on TX. A spin^c structure can be thought of as a rank-4 Hermitian vector bundle $W = W^+ \oplus W^-$ over X which carries an action $\Gamma\colon TX \to \mathrm{End}(W)$ of the tangent bundle, called Clifford multiplication, which interchanges W^+ and W^- and satisfies $\Gamma(v)^*\Gamma(v) = |v|^2 1$. A spin^c connection on W is a Hermitian connection which is compatible, via Clifford multiplication, with the Levi-Civita connection on X. Any such connection preserves the splitting and is uniquely determined by the induced connection A on the "square root" of the line bundle $L = L_\Gamma = \det(W^+)$. In turn the spin^c connection determines a Dirac operator $D_A\colon C^\infty(X, W^+) \to C^\infty(X, W^-)$. The Seiberg-Witten equations have the form

$$D_A\Phi = 0, \quad F_A^+ + \eta = \sigma^+((\Phi\Phi^*)_0) \tag{1}$$

for a connection $A \in \mathscr{A}(L^{1/2})$ and a section $\Phi \in C^\infty(X, W^+)$. Here $\eta \in \Omega^{2,+}(X, i\mathbf{R})$ is a perturbation, F_A^+ is the self-dual part of the curvature of the connection on $L^{1/2}$, and $\sigma^+\colon \mathrm{End}_0(W^+) \to \Lambda^+T^*X \otimes i\mathbf{R}$ is the natural isomorphism, induced by Γ, between traceless Hermitian endomorphisms of W^+ and imaginary-valued self-dual 2-forms on X. The term $(\Phi\Phi^*)_0 \in \mathrm{End}_0(W^+)$ denotes the traceless part of the Hermitian endomorphism $\Phi\Phi^*$ of W^+. The Seiberg-Witten invariant $\mathrm{SW}(X, \Gamma)$ is defined, roughly speaking, as the algebraic number of U(1)-gauge equivalence classes of solutions of (1) for a regular perturbation η. In the case $b^+ \geq 2$ this number is independent of the choice of the perturbation and the metric. (See a book by the reviewer [*Spin geometry and Seiberg-Witten invariants*, Birkhäuser, to appear] for more details.)

If (X, ω) is a symplectic 4-manifold then there exists an almost complex structure J on TX which is compatible with ω, meaning that the formula $\langle v, w\rangle = \omega(v, Jw)$ defines a Riemannian metric. The set of such almost complex structures is contractible and hence the corresponding canonical spin^c structures $\Gamma_J\colon TX \to \mathrm{End}(W_J)$ with $W_J = \Lambda_J^{0,*}T^*X$ are all isomorphic. In Taubes' first two papers cited above, he proved that the Seiberg-Witten invariant of Γ_J is $\mathrm{SW}(X, \Gamma_J) = \pm 1$. Taubes also proved that if $E \to X$ is a Hermitian line bundle such that the twisted spin^c structure $\Gamma_E\colon TX \to W_E$ with $W_E = W_J \otimes E$ has nonzero Seiberg-Witten invariants then

$$0 \le c_1(E)\cdot[\omega] \le c_1(K)\cdot[\omega],$$

where $K = K_J = \Lambda_J^{2,0}T^*X$. Equality holds if and only if $E = \mathbf{C}$ or $E = K$. In the paper under review Taubes proves the following theorem.

Theorem 1. Let (X, ω) be a compact symplectic 4-manifold with $b^+ \ge 2$. Suppose that $E \to X$ is a complex line bundle such that $\mathrm{SW}(X, \Gamma_E) \ne 0$. Then there exists an embedded symplectic submanifold $C \subset X$ which represents the class $[C] = \mathrm{PD}(c_1(E))$. Moreover, every component C_i of C satisfies

$$c_1(K)\cdot C_i \le g(C_i) - 1 \le C_i \cdot C_i.$$

If $E = \mathbf{C}$ is the trivial bundle then C is the empty curve.

Taubes actually proves the existence of a J-holomorphic curve representing the Poincaré dual of $c_1(E)$. But to obtain embedded curves it is necessary to choose a generic (possibly nonintegrable) almost complex structure and, even in the generic case, it may be necessary to allow for higher multiplicity in the case of tori with zero self-intersection.

Although the result is formulated for the case $b^+ \ge 2$ the proof obviously extends to the case $b^+ = 1$. In that case there are two Seiberg-Witten invariants $\mathrm{SW}^{\pm}(X, \Gamma_E)$ for every spin^c structure and, with a suitable choice of sign, Taubes' theorem continues to hold under the assumption $\mathrm{SW}^+(X, \Gamma_E) \ne 0$.

Corollaries. Theorem 1 has many important consequences. For example, the Poincaré dual of the canonical class $c_1(K) \in H^2(X, \mathbf{Z})$ in a symplectic 4-manifold (X, ω) with $b^+ \ge 2$ can be represented by a symplectic submanifold. Another consequence is that if a homology class $e \in H_2(X, \mathbf{Z})$ with self-intersection $e \cdot e = -1$ can be represented by a smoothly embedded sphere then either e or $-e$ can be represented by a symplectically embedded sphere. Thus smooth blowup is equivalent to symplectic blowup. Taubes also shows that minimal symplectic 4-manifolds with $b^+ \ge 2$ satisfy $c_1(K)\cdot c_1(K) \ge 0$. In the "elliptic case" $c_1(K)\cdot c_1(K) = 0$ he shows that if $\mathrm{SW}(X, \Gamma_E) \ne 0$ then the Poincaré dual of $c_1(E)$ can be represented by a disjoint union of symplectically embedded tori with self-intersection zero. Another consequence is that symplectic 4-manifolds with $b^+ \ge 2$ have simple type, meaning that $\mathrm{SW}(X, \Gamma_E) = 0$ unless the Seiberg-Witten moduli space is zero-dimensional, i.e. $c_1(E)^2 = c_1(E)\cdot c_1(K)$. Finally, Taubes settles a long-standing open question by proving the uniqueness of symplectic structures on $\mathbf{CP}^2$ (up to scaling and symplectomorphism). All these results are proved in the paper under review.

Taubes' existence theorem for J-holomorphic curves was quickly taken up by other researchers. For example A. Liu [Math. Res. Lett. **3** (1996), no. 5, 569–585; MR 97k:57038] disposes of a conjecture by Gompf which asserts that every minimal symplectic 4-manifold with $c_1(K)\cdot c_1(K) < 0$ is a ruled surface. It was shown by T. J. Li and Liu [Math. Res. Lett. **2** (1995), no. 4, 453–471; MR 96m:57052] and H.

Ohta and K. Ono [in *Moduli of vector bundles (Sanda, 1994; Kyoto, 1994)*, 209–212, Dekker, New York, 1996; MR 97d:57043; Internat. J. Math. **7** (1996), no. 6, 755–770; MR 97k:57042] that minimal symplectic 4-manifolds which either satisfy $c_1(K)\cdot[\omega]<0$ or admit a metric of positive scalar curvature are rational or ruled. Li-Liu and Lalonde-McDuff proved a uniqueness theorem for symplectic structures on ruled surfaces and Kotschick observed that simply connected symplectic 4-manifolds are irreducible. These results constitute first steps towards a classification of symplectic 4-manifolds in analogy to algebraic surfaces. The crucial geometric ingredient in all this is Taubes' existence theorem for pseudo-holomorphic curves.

About the proof. In the symplectic case with spin^c structure Γ_E, the perturbed Seiberg-Witten equations, after suitable rescaling, take the form

$$\begin{gathered}\bar\partial_B\psi_0+\bar\partial_B^*\psi_2=0,\quad \frac{1}{\lambda}2F_B^{0,2}=\overline{\psi}_0\psi_2,\\ \frac{1}{\lambda}4i(F_B)_\omega=4\pi+|\psi_2|^2-|\psi_0|^2.\end{gathered}\tag{2}$$

Here B is a connection on E, ψ_0 is a section of E, and ψ_2 is a $(0,2)$-form on X with values in E. In a first step Taubes shows that the solutions of (2) satisfy the L^2-estimates $\|\psi_2\|^2+\lambda\|\bar\partial_B\psi_0\|^2\le c/\lambda^2$ and

$$\|\partial_B\psi_0\|^2+\frac{\lambda}{2}\int_X\left(4\pi-|\psi_0|^2\right)^2\le 8\pi^2[\omega]\cdot c_1(E)+\frac{c}{\lambda}.$$

This shows that $|\psi_0|^2$ converges to 4π almost everywhere as $\lambda\to\infty$. This convergence cannot be uniform unless the bundle $E\to X$ is trivial, and the goal of the proof is in fact to show that the zero-sets $C_\lambda={\psi_0}^{-1}(0)$ will converge to a J-holomorphic curve. The argument can be viewed as an almost complex version of the correspondence between holomorphic sections and divisors in Kähler surfaces.

To make this precise Taubes begins, in Section 2, by establishing the pointwise estimates for the solutions of (2) such as

$$\begin{gathered}|\psi_0|^2\le 4\pi+c/\lambda^2,\quad \lambda|\psi_2|^2\le c(4\pi+c/\lambda^2-|\psi_0|^2),\\ |F_B^+|\le\frac{\lambda}{\sqrt8}(4\pi-|\psi_0|^2)+c,\ |F_B^-|\le\frac{\lambda}{\sqrt8}\left(1+\frac{c}{\sqrt\lambda}\right)(4\pi-|\psi_0|^2)+c.\end{gathered}$$

The proof of these inequalities is based on the maximum principle and the operator $d^*d+\lambda|\psi_0|^2$ plays a fundamental role.

The heart of the proof of Theorem 1 is the crucial monotonicity estimate for the solutions of (2): Taubes shows, in Section 3, that there are constants $c>0$, $\lambda_0>0$, and $\delta>0$ such that if $\lambda\ge\lambda_0$, $1/4\lambda\le r^2\le\delta$, and $|\psi_0(x)|\le\pi$ then

$$\frac{r^2}{c}\le\frac{\lambda}{2}\int_{B_r(x)}|4\pi-|\psi_0|^2|\le cr^2.$$

The proof makes use of the precise factor $\lambda/\sqrt8$ in the curvature estimate. A consequence of this monotonicity is that the set $|\psi_0|^2\le\pi$ can be covered by approximately $N_\rho\le c/\rho^2$ balls of radius ρ whenever $1/\sqrt\lambda\le\rho\le\rho_0$ (for suitable constants c and $\rho_0>0$). This is used subsequently to prove that the zero sets of ψ_0 converge to a set of Hausdorff dimension 2.

The next step (in Section 4) is a kind of localization result which asserts that the important quantities are concentrated near the zero set of ψ_0 for large λ.

More precisely, the function $u = |F_B|/\lambda + |4\pi - |\psi_0|^2| + |\nabla_B\psi_0|/\lambda + \lambda|\psi_2|^2 + |\nabla_B\psi_2|^2$ satisfies the exponential decay estimate $u(x) \le ce^{-\sqrt{\lambda}d(x,C_\lambda)/c}$, $C_\lambda = \psi_0{}^{-1}(0)$. The proof is based on comparison with model solutions on $\mathbf{C}^2$.

In Section 5 Taubes establishes the convergence of the curvature forms as currents. For a sequence $\lambda_n \to \infty$ and a corresponding sequence $B_n, \psi_{n0}, \psi_{n2}$ of solutions of (2) with $\lambda = \lambda_n$, consider the currents $\mathcal{F}_n\colon \Omega^2(X) \to \mathbf{R}$ defined by

$$\mathcal{F}_n(\tau) = \frac{i}{2\pi}\int_X F_{B_n} \wedge \tau.$$

It follows from the pointwise estimates that there exists a subsequence which converges in the weak*-topology to a current $\mathcal{F}\colon \Omega^2(X) \to \mathbf{R}$ which is continuous with respect to the L^∞-norm. Taubes uses the monotonicity and localization estimates to show that the limit current $\mathcal{F}$ has type $(1,1)$ and is supported in a set C of finite 2-dimensional Hausdorff measure. The crucial point is that C has a well-defined intersection number with every local J-holomorphic disc whose boundary is disjoint from C and that this intersection number is positive whenever the intersection is nonempty. The final step of the proof (in Section 6) is a general result which asserts that under these conditions (finite 2-dimensional Hausdorff measure and positive intersection with local J-holomorphic discs) the set C must be a J-holomorphic curve. *Dietmar A. Salamon* (Coventry)

98h:57058 57R57 53C15 57R15 58D27 58G10

Kronheimer, P. B. (Cambridge, MA); **Mrowka, T. S.** (Cambridge, MA)

Monopoles and contact structures.

Invent. Math. **130** (1997), *no.* 2, 209–255.

The monopole, or Seiberg-Witten, invariants of a smooth, closed oriented 4-manifold were introduced by E. Witten [Math. Res. Lett. **1** (1994), no. 6, 769–796; MR 96d:57035], and proved immediately to be very powerful (see S. K. Donaldson's survey paper [Bull. Amer. Math. Soc. (N.S.) **33** (1996), no. 1, 45–70; MR 96k:57033]). In the reviewer's opinion this paper represents a major contribution towards the understanding of the relationship between the Seiberg-Witten equations and various geometric structures on 3- and 4-dimensional manifolds. Some of the most far-reaching applications of the Seiberg-Witten invariants are due to Cliff Taubes, and were obtained by considering the Seiberg-Witten equations on closed symplectic 4-manifolds. This paper can be viewed as a natural generalization of Taubes' work [Math. Res. Lett. **1** (1994), no. 6, 809–822; MR 95j:57039; Math. Res. Lett. **2** (1995), no. 1, 9–13; MR 96a:57075] to the case of symplectic 4-manifolds with boundary. More precisely, the paper shows that an appropriate category in this respect is the category of 4-manifolds with contact boundary carrying a compatible symplectic form (see below for the definition).

Before we can describe the content of the paper, we need to recall some of the basic ideas and problems concerning contact structures. Given an oriented 3-manifold Y, a field of 2-planes $\xi \subset TY$ is a contact structure if it is the kernel of a smooth 1-form θ on Y such that $\theta \wedge d\theta \neq 0$ at every point of Y. For an introduction to contact structures and a guide to the existing literature the reader is referred to [D. Bennequin, in *Third Schnepfenried geometry conference, Vol. 1 (Schnepfenried, 1982)*, 87–161, Astérisque, 107-108, Soc. Math. France, Paris, 1983; MR 86e:58070; Y. M.

Eliashberg, Ann. Inst. Fourier (Grenoble) **42** (1992), no. 1-2, 165–192; MR 93k:57029; E. Giroux, Astérisque No. 216 (1993), Exp. No. 760, 3, 7–33; MR 94k:57040]. When the form $\theta \wedge d\theta$ is positive [resp. negative] the contact structure ξ is called positive [resp. negative]. Eliashberg defined a special class of contact structures, which he called overtwisted, and proved that in any homotopy class of oriented 2-plane fields on an oriented 3-manifold there exists a unique positive overtwisted contact structure up to isotopy. The non-overtwisted contact structures are called tight, and are quite mysterious objects at present. For example, the questions of existence and uniqueness in a given homotopy class are known to have a negative answer, in general, and a complete classification is known for just a few 3-manifolds. Eliashberg proved that the Euler classes of tight contact structures (as oriented 2-plane bundles) satisfy the same inequalities as the Euler classes of taut foliations, which are highly reminiscent of the "adjunction" inequalities satisfied by the Seiberg-Witten basic classes of a closed 4-manifold. Such inequalities force the Euler classes of the tight contact structures on a given 3-manifold to belong to a finite set. The only tight contact structures known at present are fillable in one sense or another, i.e., roughly speaking, they constitute a 3-dimensional phenomenon induced by a 4-dimensional one. A 4-manifold with contact boundary is a pair (X,ξ), where X is a connected, oriented smooth 4-manifold with boundary and ξ is a contact structure on ∂X (positive with respect to the boundary orientation). A compatible symplectic form on (X,ξ) is a symplectic form ω on X such that $\omega|_\xi > 0$ at every point of ∂X. A contact 3-manifold (Y,ζ) is called symplectically fillable if ζ is positive and there exists a 4-manifold with contact boundary (X,ξ) carrying a compatible symplectic form ω and an orientation-preserving diffeomorphism φ from Y to ∂X such that $\varphi_*(\zeta) = \xi$. The triple (X,ξ,ω) is called a symplectic filling of Y. More generally, (Y,ζ) is called symplectically semi-fillable if the diffeomorphism φ sends Y onto a connected component of ∂X. In this case the triple (X,ξ,ω) is called a symplectic semi-filling of Y. If (Y,ζ) is symplectically semi-fillable, then ζ is tight by a theorem of Eliashberg and Gromov.

We shall now outline the results of the paper. Given a 4-manifold with contact boundary (X,ξ), the contact structure ξ determines a Spin^c structure (see below for the definition) $\mathbf{s}_\xi$ on a neighborhood of the boundary. Let Spin$^c(X,\xi)$ be the set of isomorphism classes of extensions of $\mathbf{s}_\xi$ to all of X. The authors define a monopole invariant of the pair (X,ξ) as a map $\mathrm{SW}_{(X,\xi)}\colon \mathrm{Spin}^c(X,\xi) \to \mathbf{Z}$, sending a Spin^c structure to the "number of points" of a certain moduli space, whose orientation determines the sign of the map (see below for details). This is the first main result of the paper. The set Spin$^c(X,\xi)$ is a principal homogeneous space for the relative cohomology $H^2(X,\partial X;\mathbf{Z})$, so that if $\mathbf{s}_1,\mathbf{s}_2 \in \mathrm{Spin}^c(X,\xi)$, there is a difference element $\mathbf{s}_1 - \mathbf{s}_2 \in H^2(X,\partial X;\mathbf{Z})$. We shall now state the other two main results of the paper. Suppose that (X,ξ) is equipped with a compatible symplectic form ω. The first result (Theorem 1.1) says that there are natural choices of an element $\mathbf{s}_\omega \in \mathrm{Spin}^c(X,\xi)$ and of an orientation of the corresponding moduli space so that $\mathrm{SW}_{(X,\xi)}(\mathbf{s}_\omega) = 1$. The second result (Theorem 1.2) states that if $\mathbf{s} \in \mathrm{Spin}^c(X,\xi)$ and $\mathrm{SW}(\mathbf{s}) \neq 0$, then $[\omega] \cup (\mathbf{s} - \mathbf{s}_\omega) \geq 0$ in $H^4(X,\partial X;\mathbf{R})$, with equality only if $\mathbf{s} = \mathbf{s}_\omega$. In particular, if ω is exact then $\mathbf{s}_\omega$ is the only element of Spin$^c(X,\xi)$ for which the monopole invariant is nonzero.

The paper contains two applications of these results. The first one (Theorem 1.3) is that for any closed 3-manifold, there are only finitely many homotopy classes of 2-plane fields which are realized as semi-fillable contact structures. Together with a theorem of Eliashberg and W. P. Thurston [*Confoliations*, Amer. Math. Soc., Providence, RI, 1998; MR 98m:53042], this implies that on any closed 3-manifold the number of homotopy classes of 2-plane fields which are realized as the tangent distributions to smooth, taut foliations is finite. The second application (Corollary 1.6) was proved earlier by Lisca and G. Matić [Invent. Math. **129** (1997), no. 3, 509–525; MR 98f:57055]. It states that if two Stein structures with boundary on a 4-manifold X induce isotopic contact structures on the boundary, then the two Spin^c structures induced in the interior are isomorphic. As shown in the above-mentioned paper, this result allows the construction of interesting examples of contact structures which are homotopic as oriented 2-plane fields but not homotopic through contact structures.

Now we come to the proofs of the main results. A Spin^c structure on a smooth 4-manifold X is a triple (W^+,W^-,ρ), where W^+ and W^- are Hermitian rank-2 bundles over X, called respectively the positive and negative spinor bundle, and $\rho\colon T^*X\to \mathrm{Hom}(W^+,W^-)$ is a linear map satisfying the Clifford relation: $\rho(\theta)^*\rho(\theta)=|\theta|^2\mathrm{Id}_{W^+}$ for every $\theta\in T^*X$. The map ρ extends to a linear embedding $\rho\colon \Lambda^*T^*X\to \mathrm{End}(W^+\oplus W^-)$. A Spin connection A is a unitary connection on $W=W^+\oplus W^-$ such that the induced connection on $\mathrm{End}(W)$ agrees with the Levi-Civita connection on the image of ρ. To any Spin connection A is associated, via ρ, a twisted Dirac operator $D_A^+\colon \Gamma(W^+)\to\Gamma(W^-)$. Given a 4-manifold with contact boundary (X,ξ), let X^+ be the smooth manifold obtained from X by attaching the open cylinder $[1,+\infty)\times\partial X$ along ∂X. Up to some choices which can be shown to be irrelevant, the contact structure ξ determines on $[1,+\infty)\times\partial X$ a metric g_0 and a self-dual 2-form ω_0 of constant length $\sqrt{2}$. ω_0 determines on $[1,+\infty)\times\partial X$ a Spin^c structure $\mathbf{s}_0=(W^+,W^-,\rho)$ and a unit section Φ_0 of W^+. Moreover, there is a unique Spin connection A_0 such that $D_{A_0}^+(\Phi_0)=0$. Choose an arbitrary extension of g_0 to all of X^+. The triple (X^+,ω_0,g_0) is an example of an AFAK (asymptotically flat almost Kähler) manifold. Although it is important that the fundamental properties of the moduli spaces hold for a general AFAK 4-manifold, in order to limit the amount of technical discussions involved, we shall be vague about this. Define $\mathrm{Spin}^c(X,\xi)$ to be the set of isomorphism classes of Spin^c structures on X^+ whose restriction to $[1,+\infty)\times\partial X$ is isomorphic to $\mathbf{s}_0$. Given $\mathbf{s}=(W^+,W^-,\rho)\in\mathrm{Spin}^c(X,\xi)$, extend Φ_0 and A_0 arbitrarily to all of X^+. Let L_l^2 and L_{l,A_0}^2, $l\geq 4$, be, respectively, the standard Sobolev spaces of Spin connections and sections of W^+, and let $\mathscr{C}$ be the space of pairs (A,Φ) such that $A-A_0\in L_l^2$ and $\Phi-\Phi_0\in L_{l,A_0}^2$. Then, $\mathscr{G}=\{u\colon X^+\to\mathbf{C}\colon |u|=1,\ 1-u\in L_{l+1}^2\}$ is a Hilbert Lie group acting freely on $\mathscr{C}$. Let $\eta\in L_{l-1}^2(i\mathfrak{su}(W^+))$. Given a Spin connection A, let $\widehat{A}$ be the induced U(1) connection on $\det(W^+)$. Let $M_\eta(\mathbf{s})$ be the quotient, under the action of $\mathscr{G}$, of the set of pairs $(A,\Phi)\in\mathscr{C}$ which satisfy the η-perturbed monopole equations

$$\begin{cases}\rho(F_{\widehat{A}}^+)-\{\Phi\otimes\Phi^*\}=\rho(F_{\widehat{A}_0}^+)-\{\Phi_0\otimes\Phi_0^*\}+\eta\\ D_A^+(\Phi)=0,\end{cases}$$

where $\{\Phi\otimes\Phi^*\}$ denotes the traceless part of the endomorphism $\Phi\otimes\Phi^*$. The result which constitutes the core of the paper is the proof that, for η in a Baire set of perturbing terms exponentially decaying along the end, $M_\eta(\mathbf{s})$ is a smooth, compact orientable manifold having (if non-empty) dimension $d(\mathbf{s})=\langle e(W^+,\Phi_0),[X,\partial X]\rangle$, the obstruction to extending Φ_0 to all of X as a non-vanishing section of W^+. Assuming that an orientation for $M_\eta(\mathbf{s})$ has been chosen, the authors define $\mathrm{SW}_{(X,\xi)}(\mathbf{s})$ to be the number of points of $M_\eta(\mathbf{s})$ counted with signs when $d(\mathbf{s}) = 0$, and zero when $d(\mathbf{s}) \neq 0$.

The proof that, for a generic exponentially decaying η, the moduli space $M_\eta(\mathbf{s})$ is a smooth orientable manifold of expected dimension $d(\mathbf{s})$ may be considered more standard compared with the proof of its compactness. In fact, once the natural map between appropriate Banach manifolds has been identified, the first part amounts to establishing that the map is Fredholm and then proving the dimension formula. In the present situation the problem is subtler than usual, because the underlying 4-manifold is non-compact. The authors show that the operator $\mathscr{D}$ obtained by linearizing the equations and its adjoint $\mathscr{D}^*$ are "admissible". This is a technical property which is sufficient to establish the required Fredholmness via an argument essentially due to M. Gromov and H. B. Lawson, Jr. (who used such an argument to prove the Fredholmness of operators "positive at infinity" [Inst. Hautes Études Sci. Publ. Math. No. 58 (1983), 83–196 (1984); MR 85g:58082]). The dimension formula is proved by first showing that the expected dimension $d(\mathbf{s})$ of the moduli space is zero for a Spin^c structure $\mathbf{s}$ arising from an almost complex structure, and then using an excision argument. To see that $d(\mathbf{s})$ is zero when $\mathbf{s}$ comes from an almost complex structure one uses the following facts: (i) $d(\mathbf{s})$ can be interpreted as the index of a Fredholm operator $D(A,\Phi)$, which is invariant under deformation and is therefore independent of the metric and the configuration (A,Φ); (ii) since $\mathbf{s}$ is induced by an almost complex structure, there is a configuration (A_0,Φ_0) with Φ_0 never vanishing; (iii) the operators $D(A_0,\Phi_0)$ and its adjoint $D(A_0,\Phi_0)^*$ have no kernel for a suitably rescaled metric, because of inequalities coming from the Weitzenböck formulae and the fact that Φ_0 is never vanishing. Let us turn to the proof of the compactness of $M_\eta(\mathbf{s})$. If the underlying manifold were compact, as in the case of the original Seiberg-Witten invariants, this would be an immediate consequence of the Weitzenböck formula. Because of non-compactness, the authors need to prove that the solutions of the monopole equations satisfy a uniform exponential decay estimate, which is not at all obvious. In order to do this, they adapt and use a series of differential inequalities for solutions of the equations from Taubes [J. Amer. Math. Soc. **9** (1996), no. 3, 845–918; MR 97a:57033, see p. D49 in this volume].

Once the compactness of the moduli space has been established, the verification that the map SW is a well-defined invariant of the pair (X,ξ) is not difficult. The proofs of the other two main results described above go along the lines of the analogous ones established by Taubes for a closed symplectic 4-manifold. There is only one last step before one can apply Taubes' arguments, a technical lemma saying that if (X,ξ) carries a compatible symplectic form ω, then the restriction of ω to the complement of a neighborhood of the boundary of X extends to X^+ and defines there, together with a compatible metric, an AFAK structure. The above properties of the moduli space still hold when X^+ is given this AFAK structure, and the existence of the symplectic form on all of X^+ allows one to apply Taubes' arguments.

We finish with a sketch of the proof of the first application (the proof of the second application is essentially the same as the earlier proof mentioned before). Suppose that a 4-manifold with contact boundary (X,ξ) carries a compatible symplectic form ω. Then $\mathrm{SW}(\mathbf{s}_\omega) = 1$, which implies that the moduli space $M_\eta(\mathbf{s})$ is non-empty for any metric on X^+. By considering metrics which are more and more stretched in the direction normal to the boundary of X one sees that certain L^2 moduli spaces on 4-manifolds with cylindrical ends are non-empty. Using results from J. W. Morgan, Mrowka and D. Ruberman [*The L^2-moduli space and a vanishing theorem for Donaldson polynomial invariants*, Internat. Press, Cambridge, MA, 1994; MR 95h:57039] and K. A. Frøyshov [Math. Res. Lett. **3** (1996), no. 3, 373–390; MR 97i:57037], such L^2 moduli spaces can be described to an extent which is sufficient for the application. In particular, their dimensions can be expressed in terms of the limit values of their elements along the cylindrical ends, and the homotopy class of ξ. The results of Frøyshov imply that for suitable perturbations of the equations there are only a finite number of possible limit values, and the non-negativity of the dimensions constrains the homotopy class of ξ within a finite set.

Paolo Lisca (Pisa)

98m:57041 57R57 53C25 57N10 58D30 58G18 81Q30 81T18
Rozansky, L. [Rozansky, Lev] (Chicago, IL); **Witten, E.** (Princeton, NJ)
Hyper-Kähler geometry and invariants of three-manifolds. (English summary)
Selecta Math. (N.S.) **3** (1997), *no.* 3, 401–458.

This paper utilises the geometry of hyper-Kähler manifolds to generate a new family of $(2+1)$-dimensional quantum field theories, indexed by such manifolds, whose partition function gives a finite-type invariant of 3-manifolds. Although the actual invariants of 3-manifolds obtained are not new, the paper does provide a new construction of weight systems, from compact hyper-Kähler manifolds rather than Lie algebras, and also enables the authors to make a very natural proposal for the $\mathrm{SU}(n)$ generalisation of the Casson-Walker invariant of rational homology 3-spheres.

A $(d+1)$-dimensional topological quantum field theory [M. F. Atiyah, Inst. Hautes Études Sci. Publ. Math. No. 68 (1988), 175–186 (1989); MR 90e:57059] associates, to any closed d-dimensional manifold, a vector space, and, to any $(d+1)$-dimensional manifold with boundary, a vector in the vector space of the boundary, this assignment being done in such a way that the whole collection of assignments behaves naturally under gluing operations on the manifolds, that is, it defines a functor from a suitable category of manifolds to the category of vector spaces. If $S(A)$ is the action functional associated with the theory, then the partition function associated to a closed $(d+1)$-dimensional manifold gives a topological invariant $Z(M)=\int_{\mathscr{A}}\exp(-S(A)/\hbar)\mathscr{D}A$ as a Feynman integral over some space $\mathscr{A}$ of fields, where $\hbar$ is considered as a perturbative parameter.

The classical example is that of Chern-Simons theory [E. Witten, Comm. Math. Phys. **121** (1989), no. 3, 351–399; MR 90h:57009], in which a $(2+1)$-dimensional field theory is constructed with the Chern-Simons action $S(A)=\int_M \operatorname{tr}(A\wedge dA+\frac{2}{3}A\wedge A\wedge A)$, where $A\in\mathscr{A}$ is a G-connection on a chosen 3-manifold M, and G is a fixed Lie group. The partition function of the theory on a closed 3-manifold then gives a 3-manifold invariant, dependent on the parameter $\hbar=4\pi/K$, where K is the level of the theory (assumed to be an integer), and the choice of Lie group G. This may also be extended to produce an invariant of the pair of a link L contained in a 3-manifold M, by computing the expectation value $Z(M,L)=\int_{\mathscr{A}} W_L(A)\exp(-S(A)/\hbar)\mathscr{D}A$ of a certain observable known as the Wilson loop W_L, which is the product over components of the link of the trace of the holonomy of the connection A around that component. These invariants are the so-called quantum invariants of knots and 3-manifolds, and are believed to be identical with those constructed (rigorously) from quantum groups by N. Yu. Reshetikhin and V. G. Turaev [Invent. Math. **103** (1991), no. 3, 547–597; MR 92b:57024], while containing in the case of $G = \mathrm{SU}(2)$ the Jones polynomial [V. F. R. Jones, Bull. Amer. Math. Soc. (N.S.) **12** (1985), no. 1, 103–111; MR 86e:57006] invariant of links.

When the parameter $\hbar$ is small, it is possible to perform a stationary phase expansion, to obtain an asymptotic expansion decomposition of the partition function $Z(M) = \sum_c Z^c(M)$, whose terms are of the form $Z^c(M) = \exp(-S_c/\hbar)\sum_{n=0}^{\infty} W_n(c)\hbar^{n+d}$, indexed by equivalence classes c of elements of $\mathscr{A}$ at which the action $S(A)$ is stationary. While the whole Feynman integral $Z(M)$ does not yet have a rigorously-defined meaning for the general case, owing to the non-existence of a measure $\mathscr{D}A$ on the space $\mathscr{A}$, the coefficients $W_n(c)$ are well defined, and expressible as a combinatorial sum of a finite number of Feynman diagrams, each described by a finite-dimensional integral, indexed by graphs with "size" n, and of a type determined by the nature of the action S.

In the case of Chern-Simons theory, the contribution from the trivial connection, $Z^0(M)$, provides a formal power series invariant of M, and the coefficient of $\hbar^n$ is an invariant of finite type [T. Ohtsuki, J. Knot Theory Ramifications **5** (1996), no. 1, 101–115; MR 97i:57019] of order n (or $2n$ or $3n$, depending on the definition of "finite type" employed). Similarly, the associated link invariants may be expanded in powers of $\hbar$ and the coefficients in such an expansion are Vassiliev link invariants.

An invariant $I(M)$, of rational homology 3-spheres M, is said to be of finite type n if for any link L_0 and sublink L_1 of more than n components, the alternating sum of $(-1)^{|L|}I(S^3_L)$ over all links $L_1 \subseteq L \subseteq L_0$ vanishes. It is a notion closely related to Vassiliev invariants, e.g. any invariant $I(M)$ of manifolds of finite type gives rise to a Vassiliev invariant of links, $V(L) = I(S^3_L)$, namely the evaluation of I on the manifold obtained from S^3 by surgery along L. The Casson invariant is the simplest nontrivial finite-type invariant.

The condition for a finite-type invariant is similar to that of the vanishing of a multi-dimensional derivative, so that such an invariant is to be compared with a multi-variable polynomial. For a given degree, it is thus determined, up to the addition of lower order terms, by the coefficients of the top degree monomials. Such data in the context of finite-type invariants is known as a weight system; it assigns a scalar (the weight) to any trivalent graph, in such a way that the IHX relation, a certain condition amongst weights on diagrams which differ only locally in a prescribed manner [D. Bar-Natan, Topology **34** (1995), no. 2, 423–472; MR 97d:57004, see p. D40 in this volume], is satisfied. Weight systems for Vassiliev invariants are similar, the difference being that a weight is assigned for any trivalent graph with additional external legs.

It is known that from any Lie algebra, a weight system can be generated, using contractions on the structure constants of the Lie algebra, the Jacobi identity guaranteeing the IHX relation. Indeed these weight systems are precisely those which correspond to the manifold invariants produced by Chern-Simons theory. However, it has been of interest to attempt to construct weight systems which do not come from Lie algebras.

The current paper investigates a family of topological quantum field theories, indexed by a hyper-Kähler manifold X (dimension $4n$), rather than by a Lie algebra, obtained by twisting a reduction of a supersymmetric sigma-model naturally definable in six dimensions, to three. An element of the space $\mathcal{A}$ of fields now consists of a map $M \to X$ (or $4n$ scalar-valued fields φ on M) along with a scalar η and a one-form χ on M, while the action of the theory is given by $S = \int_M (L_1 + L_2)\sqrt{h}d^3x$, where L_1 and L_2 are two contributions to the Lagrangian: $L_1 = (d\varphi, d\varphi) + \chi \cdot D\eta$ and $L_2 = \chi D\chi + \Omega\chi\chi\chi\eta$, where we have omitted all indices and scaling, so as to only give the form of the terms. Here Ω is essentially the Riemann curvature tensor, and plays a rôle analogous to that of the Lie algebra structure constants in Chern-Simons theory. Although the theory is defined with the use of a choice of metric, the metric-dependence is eliminated by restricting consideration to a specific class of observables, the expectation values of which are therefore topological invariants. The fact that Ω is symmetric, while Lie algebra structure constants are anti-symmetric, is reflected in a reversal of statistics for the gauge field. This change of statistics, as the authors explain, results in there being only a finite number of nontrivial Feynman diagrams in this theory, η playing the rôle of a coupling parameter for χ but "fermionic-valued". In particular, for a hyper-Kähler manifold X of dimension $4n$, and a 3-manifold M which is a rational homology sphere, the only nontrivial Feynman diagram evaluations come from trivalent graphs with $2n$ vertices. The effect of a larger Betti number is a simplification in diagrams which contribute: $b_1 = 1$ gives only collections of circles with vertex insertions, $b_1 = 2$ gives collections of intervals, while from $b_1 = 3$ the invariant reduces just to information about the cubic intersection form on $H^1(M, \mathbf{Z})$, and for higher b_1 the partition function of the theory vanishes entirely. Hence the partition function of the theory is a finite-type invariant of 3-manifolds.

Although the way in which the Feynman diagrams are computed relies on well-worn principles in quantum field theory, it is possible, once the weight system is derived, to perfectly rigorously define a manifold invariant directly from it, without further reference back to the Feynman integral, once the IHX relation is checked. In this theory, it is shown that the IHX relation follows from the Bianchi identity for the Riemann curvature tensor. An analogue of Wilson lines can also be defined here and the authors state that they expect the corresponding link invariants to be Vassiliev invariants.

The invariant associated with the Atiyah-Hitchin manifold, the reduced moduli space of a pair of BPS monopoles on $\mathbf{R}^3$ [M. F. Atiyah and N. J. Hitchin, *The geometry and dynamics of magnetic monopoles*, Princeton Univ. Press, Princeton, NJ, 1988; MR 89k:53067], is shown to be the Casson-Walker invariant of rational homology 3-spheres. It is proposed that a reasonable candidate for the SU(n)-version of the Casson invariant would be the invariant associated in this theory to the hyper-Kähler manifold described by the configuration space of n magnetic monopoles.

For the case when X is four-dimensional, the authors conclude by advancing the construction of the topological field theory further, to obtain a complete description, including the Hilbert spaces associated with a sphere and a torus, along with an explicit action of the mapping class group of the torus.

Ruth Lawrence (Ann Arbor, MI)

99j:57033 57R57 57M25 57N13 57R55
Fintushel, Ronald (East Lansing, MI);
Stern, Ronald J. [Stern, Ronald John] (Irvine, CA)
Knots, links, and 4-manifolds.
Invent. Math. **134** (1998), *no.* 2, 363–400.

The history of the invariants of smooth four-dimensional manifolds constructed via gauge theory goes back to the seminal work of Simon Donaldson at the beginning of the 1980s. Since then, our knowledge about smooth 4-manifolds has gradually increased, with a leap in the mid 1990s when Seiberg and Witten introduced their celebrated equations, easier to handle than the anti-self-duality equations. The paper under review extracts, in a way which is clever, intriguing and elegant at the same time, a large amount of topological information from the latest gluing theorems for Seiberg-Witten invariants. Using such information the authors are able to uncover and display quite spectacularly the amazing (and, to such an extent, previously unknown) complexity of the set of smooth structures on a four-dimensional manifold. One of the most esthetically appealing features of the paper is the authors' ability to recast the gluing properties of the Seiberg-Witten invariants using the formalism of skein relations, thus bringing invariants of links in S^3 into the picture, perhaps almost suggesting the possibility of an axiomatic approach to gauge theory. The authors acknowledge that it was G. Meng and C. H. Taubes [Math. Res. Lett. **3** (1996), no. 5, 661–674; MR 98j:57049] who first discovered the relationship between Seiberg-Witten type invariants and the Alexander polynomial, but in the reviewer's opinion this does not detract from the beauty of the present paper.

The Seiberg-Witten invariant of a smooth closed oriented 4-manifold X with $b_2^+(X)>1$ and no 2-torsion in $H_1(X;\mathbf{Z})$ (the case considered in the paper) can be viewed as an integer-valued map SW_X defined on the set of characteristic 2-dimensional cohomology classes of X. A class β such that $\mathrm{SW}_X(\beta)\neq 0$ is called a basic class. The set of non-zero basic classes of X is of the form $\{\pm\beta_1,\cdots,\pm\beta_n\}$, and the information contained in the map SW_X can be collected in a Laurent polynomial

$$\mathscr{W}_X=b_0+\sum_{j=1}^n b_j(t_j+(-1)^{(1/4)(\chi(X)+\sigma(X))}t_j^{-1}),$$

where $t_j=\exp(\beta_j)$, $b_0=\mathrm{SW}_X(0)$, $b_j=\mathrm{SW}_X(\beta_j)$ and $\chi(X)$ and $\sigma(X)$ are respectively the Euler characteristic and the signature of X. A cusp neighborhood is the 4-manifold obtained by performing 0-framed surgery on a right-handed trefoil knot on the boundary of the 4-ball. Suppose that X is a smooth 4-manifold containing a smoothly embedded 2-torus of self-intersection zero inside a cusp neighborhood (a cusp-embedded torus, for short), with $\pi_1(X)=\pi_1(X\setminus T)=\{1\}$. Then, given a knot K in S^3, it is possible to construct a smooth 4-manifold X_K homotopy equivalent to X by removing a neighborhood of the torus T inside X and replacing it with the complement of the knot K in S^3 crossed with S^1. Denote by $[T]$ the Poincaré dual of the homology class of T inside X. The first main result of the paper says that

$$\mathscr{W}_{X_K}=\mathscr{W}_X\cdot\Delta_K(t),\tag{1}$$

where $t=\exp(2[T])$, and $\Delta_K(t)$ is the symmetrized Alexander polynomial of the knot K. This result implies, for instance, that there are at least as many smoothly distinct 4-manifolds homeomorphic to the $K3$ surface ($\mathscr{W}_{K3}=1$) as there are Alexander polynomials of knots! In fact, the authors conjecture that if K and K' are two knots (or, more generally, n-component links—see below for the definition of X_K in this case) then $K3_K$ and $K3_{K'}$ are diffeomorphic if and only if K and K' are isotopic. If $\Delta_K(t)$ is not monic it follows from (1) and the work of Taubes [Math. Res. Lett. **1** (1994), no. 6, 809–822; MR 95j:57039] that X_K does not admit a symplectic structure. Under some additional hypotheses—satisfied e.g. by the homotopy $K3$ surfaces mentioned above—if $\Delta_K(t)$ is not monic, X_K does not admit a symplectic structure with either orientation. On the other hand, when the knot K is fibered ($\Delta_K(t)$ is known to be monic in this case), X is symplectic and the torus T is symplectically embedded, the construction can be made in the symplectic category, and due to further work of Taubes [J. Differential Geom. **44** (1996), no. 4, 818–893; MR 97k:58029, see p. D73 in this volume] formula (1) gives the Gromov invariants of X_K.

The idea to prove (1) is to use some recent gluing theorems for Seiberg-Witten invariants due to J. Morgan, T. S. Mrowka, Z. Szabó and Taubes together with work of J. Hoste on sewn-up link exteriors in order to express the invariants of X_K in terms of those of X by performing a series of topological log-transforms on null-homologous tori in X_K which reduce it to X. This turns out to be the same algorithm used to compute the Alexander polynomial $\Delta_K(t)$. There are two gluing theorems needed to prove this. The first one, due to the above-mentioned authors, computes the change in the Seiberg-Witten invariants of a 4-manifold X containing two cusp-embedded tori under the operation of "internal fiber sum" on X, which involves removing a neighborhood of the tori and gluing together the resulting two boundary components. The second one is due to Morgan, Mrowka and Szabó [Math. Res. Lett. **4** (1997), no. 6, 915–929; MR 99f:57039] and computes the change in $\mathcal{W}_X$ when a neighborhood of a null-homologous torus $T \subset X$ of self-intersection zero is cut out and reglued via a certain diffeomorphism of the boundary—the resulting manifold $X(p/q)$ is called a generalized (p/q)-log transform of X along T.

In order to sketch the line of the proof of (1) we need to give a more general definition of X_K. Let K be either an oriented knot or an oriented 2-component link in S^3. When K is a link there is a canonical way to build a three-manifold M_K by removing a neighborhood of K and gluing together the two boundary components of the resulting 3-manifold via a diffeomorphism determined by the linking number of the two components of K. When K is a knot, let M_K be the result of performing 0-surgery on K. Then, if m is a meridian of any component of K viewed as a loop inside M_K, $T_m = m \times S^1 \subseteq M_K \times S^1$ is a 2-torus of self-intersection zero. If X is a simply connected 4-manifold containing a cusp-embedded torus T with $\pi_1(X \setminus T) = \{1\}$, one can define X_K as the fiber sum $X \#_{T=T_m} (M_K \times S^1)$ (there is an ambiguity in the choice of the gluing diffeomorphism, but it does not affect the resulting invariants). Fix a projection of K to the plane, and let $K_+ = K$. Let K_- be the result of changing a single oriented positive crossing in K_+ to a negative crossing, and let K_0 be the result of resolving the crossing. At the crossing of K_- that is in question, there is an unknotted circle U linking K_- algebraically 0 times, so that the result of $+1$ surgery on U turns K_- into K_+. In terms of X_K this translates into the fact that X_{K_+} is a $(1/1)$-log transform of X_{K_-} along $U \times S^1$, and the second gluing theorem mentioned before says that $\mathcal{W}_{X_{K_-}(1/1)} = \mathcal{W}_{X_{K_-}} + \mathcal{W}_{X_{K_-}(0/1)}$.

On the other hand, the work of Hoste [Pacific J. Math. **112** (1984), no. 2, 347–382; MR 86f:57004] taken together with the first gluing theorem mentioned above implies that $\mathcal{W}_{X_{K_-}(0/1)} = (t^{1/2} - t^{-1/2})^{2(c-1)} \mathcal{W}_{X_{K_0}}$, where c is the number of components of K. Putting these two relations together yields

$$\mathcal{W}_{X_{K_+}} = \mathcal{W}_{X_{K_-}} + (t^{1/2} - t^{-1/2})^{2(c-1)} \mathcal{W}_{X_{K_0}}. \tag{2}$$

Now, given an oriented knot or 2-component link K, it is easily checked that if one defines $\Theta_K = (t^{1/2} - t^{-1/2})^{1-c} \mathcal{W}_{X_K} / \mathcal{W}_X$, it follows from (2) that Θ satisfies the skein relation

$$\Theta_{K_+} = \Theta_{K_-} + (t^{1/2} - t^{-1/2}) \cdot \Theta_{K_0}. \tag{3}$$

It is a well-known fact that, starting with a knot K, one can perform skein moves so as to build a resolution tree starting from K and at each stage adding the bifurcation determined by K_+, K_- and K_0, where each of them is either a knot or a 2-component link, so that at the bottom of the tree one obtains only unknots and split links. Moreover, for the unknot U the manifold X_U is just $X\#_T(S^2 \times T^2) = X$, while for a split link S, since X_S contains an essential 2-sphere of self-intersection zero, we have $\mathcal{W}_{X_S} = 0$. Hence $\Theta_U = 1$, while $\Theta_S = 0$. Since these initial values, together with the resolution tree and the analogue of the skein relation (3), uniquely determine the symmetrized Alexander polynomial $\Delta_K(t)$ for any knot K, it follows that Θ_K is a Laurent polynomial in a single variable t, and $\Theta_K(t) = \Delta_K(t)$, so we have (1).

The second main result of the paper is a consequence of the fact that the techniques used to prove (1) can be generalized to the case of multi-component links in S^3. Let $L = \{K_1, \cdots, K_n\}$ be an $(n \geq 2)$-component oriented ordered link in S^3 and suppose that X_j, $j = 1, \cdots, n$, are simply connected smooth 4-manifolds with $b_2^+ \geq 1$ and each containing a smoothly embedded torus T_j of self-intersection 0 with $\pi_1(X_j \setminus T_j) = \{1\}$ and representing a nontrivial homology class. Then one can construct a smooth 4-manifold $X(X_1, \cdots, X_n; L)$ from the disjoint union of the X_j by removing a neighborhood of the tori T_j and replacing it with the complement of the link L in S^3 crossed with S^1. In the special case when all the X_j are the same manifold X, denote the resulting construction by X_L. The second main result of the paper states that if $E(1)$ is the rational elliptic surface and $\Delta_L(t_1, \cdots, t_n)$ denotes the symmetric multivariable Alexander polynomial of the $n \geq 2$ component link L, then

$$\mathcal{W}_{E(1)_L} = \Delta_L(t_1, \cdots, t_n), \tag{4}$$

where, if $[T_j]$ denotes the Poincaré dual of the homology class of T_j, $t_j = \exp(2[T_j])$. The authors observe that when L is a certain k-twisted Whitehead link $W(k)$ and $k \neq \pm 1$, the manifolds $E_{W(k)}$ are diffeomorphic to the first examples of nonsymplectic simply connected irreducible smooth 4-manifolds constructed by Szabó [Invent. Math. **132** (1998), no. 3, 457–466; MR 99f:57033]. Indeed, by looking at the Alexander polynomial of $W(k)$ it follows as before from (4) and the work of Taubes that $E_{W(k)}$ does not admit a symplectic structure with either orientation. Formula (4) is proved by showing that a suitable element of $\mathbf{Q}(t_1, \cdots, t_n)$ depending on K, and defined in a way similar to Θ_K, satisfies the axioms which according to V. G. Turaev [Uspekhi Mat. Nauk **41** (1986), no. 1(247), 97–147, 240; MR 87i:57009] characterize the Alexander polynomial of a link. Apart from a few technical difficulties which are quite easily dealt with, the key ingredients of the proof of (4) are the same as before, i.e. gluing theorems for Seiberg-Witten invariants and 3-dimensional results on sewn-up link exteriors.

In the last section of the paper the authors discuss examples with $b_2^+ = 1$ given by the same construction. An analogue of formula (1) allows them to state conditions under which a 4-manifold of the form X_K with $b_2^+ = 1$ does not admit a symplectic structure. This implies, for example, that for any knot K in S^3 whose Alexander polynomial $\Delta_K(t)$ is not monic, the manifolds $E(1)_K$ admit no symplectic structure with either orientation, and this remains true even after an arbitrary number of blowups. *Paolo Lisca* (Pisa)

99k:57070 57R57 11F11 57N13 58D29 81T60
Moore, Gregory [Moore, Gregory W.] (New Haven, CT);
Witten, Edward (Princeton, NJ)
Integration over the u-plane in Donaldson theory. (English summary)
Adv. Theor. Math. Phys. **1** (1997), *no.* 2, 298–387.

In 1988 Witten [Comm. Math. Phys. **117** (1988), no. 3, 353–386; MR 89m:57037] rewrote the Donaldson invariants of smooth 4-manifolds as correlation functions in a certain supersymmetric four-dimensional gauge theory. This work introduced a new technique—topological twisting—which allows one to formulate certain supersymmetric theories on curved manifolds. A mathematical interpretation of this work was quickly worked out by M. F. Atiyah and L. C. Jeffrey [J. Geom. Phys. **7** (1990), no. 1, 119–136; MR 92f:58025], who explained that the path integral in this problem is a formal expression for the Euler class of the instanton moduli space. This view of Donaldson invariants as an infinite-dimensional Euler class was well known by geometers, and it was very unclear what, if anything, the rewriting as a formal integral over an infinite-dimensional space added to our understanding. Furthermore, there was no indication that the quantum field theory expression was sensitive to subtle points in the theory, such as the difference between manifolds with $b^+ > 1$ and manifolds with $b^+ = 1$. (The second Betti number of a compact oriented 4-manifold decomposes as a sum $b^+ + b^-$ via the nondegenerate intersection form.) For manifolds with $b^+ > 1$ Donaldson invariants are independent of the metric, whereas if $b^+ = 1$ there is a mild—that is, finite-dimensional—metric dependence which can be tracked by the position of the line of self-dual forms in the real second cohomology.

The situation changed radically in the mid 1990s due to new developments in quantum field theory. One can broadly say that up to that point the applications of quantum field theory to mathematics involved mostly formal aspects of field theory: the equivalence of Lagrangian and Hamiltonian, the method of steepest descent and perturbation theory, etc. In fact, dynamical or nonperturbative aspects of quantum field theory were only understood physically in special cases—mostly in dimensions less than four—and did not find mathematical application. The work of N. Seiberg on four-dimensional supersymmetric gauge theories, and in particular the work of Seiberg and Witten [Nuclear Phys. B **426** (1994), no. 1, 19–52; MR 95m:81202a; erratum; MR 95m:81202b] on the model relevant to Donaldson invariants, gave for the first time a detailed description of the dynamics. (See lectures of Seiberg and Witten in [*Quantum fields and strings: a course for mathematicians*, edited by P. Deligne et al., Amer. Math. Soc., Providence, RI, 1999] for an account of these advances in quantum field theory written for mathematicians.) Witten [Math. Res. Lett. **1** (1994), no. 6, 769–796; MR 96d:57035] quickly applied this work to the topologically twisted version of the theory, and so expressed Donaldson invariants for manifolds with $b^+ >$

1 in terms of simpler "monopole" equations, now called Seiberg-Witten equations. This revolutionized the mathematical study of these invariants [see S. K. Donaldson, Bull. Amer. Math. Soc. (N.S.) **33** (1996), no. 1, 45–70; MR 96k:57033]. The paper under review treats the more subtle case $b^+ = 1$ and also fills in some details about the $b^+ > 1$ case.

The Seiberg-Witten "solution" to the pure $N = 2$ supersymmetric Yang-Mills theory with gauge group SU(2), the theory relevant to Donaldson invariants, is described in terms of the modular curve for $\Gamma_0(4)$, the so-called u-plane. Here "solution" means "description of the low-energy behavior". A fundamental idea is that since the Donaldson invariants are topological invariants, they can be computed at any energy scale. The low-energy limit is achieved by scaling the metric of a curved Riemannian 4-manifold to be large. In that limit the path integral over the infinite-dimensional space of fields reduces to an integral over the u-plane. Now the modular curve in question has three cusps—one at infinity and two in the finite u-plane—and the measure against which one integrates has point masses at the two finite cusps. The contribution of these points to Donaldson invariants was computed by Witten [op. cit.; MR 96d:57035] in terms of the Seiberg-Witten equations. For $b^+ > 1$ this is the only contribution, an assertion simply stated in Witten's paper [op. cit.; MR 96d:57035] and demonstrated in detail here. The main point of this paper is the analysis of the continuous part of the u-plane integral for $b^+ = 1$, both general properties and detailed formulas. A schematic summary of the structure is $(*)$ $I_{\rm D} = I_{\rm SW} + I_u$. That is, the Donaldson invariant is a sum of contributions from the Seiberg-Witten equations (the two finite cusp points) and the u-plane integral. As stated above, $I_u = 0$ for manifolds with $b^+ > 1$. From now on we discuss manifolds with $b^+ = 1$.

The first goal of the paper is to find the correct integrand for I_u. That is, the infinite-dimensional path integral must be rewritten as an integral over the u-plane plus the contributions at the finite cusps. The derivation involves some detailed points of quantum field theory which are carefully explained in the paper. Then the authors turn to a verification of the general properties: homotopy invariance, wall crossing formulas, vanishing in certain chambers, and behavior under blow-up. Some of these are quite evident from the explicit integrand, while others require more detailed arguments. Note that "wall crossing formulas" refers to the change in I_u under a change of metric which switches chambers. (Recall that $I_{\rm D}$ is not metric-independent, but has a mild metric dependence. So does $I_{\rm SW}$, and one must verify that the metric dependence of all three quantities in $(*)$ is compatible.) These general properties in principle determine I_u in all cases, but in any case it is much more useful to have explicit expressions in special cases, which the authors derive. These include $\mathbf{P}^1 \times \mathbf{P}^1$ and $\mathbf{P}^2$. The results obtained from quantum field theory agree with those obtained previously by Göttsche and collaborators [L. Göttsche, J. Amer. Math. Soc. **9** (1996), no. 3, 827–843; MR 96k:57029; G. Ellingsrud and L. Göttsche, Quart. J. Math. Oxford Ser. (2) **49** (1998), no. 195, 307–329; MR 99g:57034; L. Göttsche and D. B. Zagier, Selecta Math. (N.S.) **4** (1998), no. 1, 69–115; MR 99d:57022]. (In fact, the comparison has some interesting number-theoretic consequences.)

There is much more in this paper. One significant physical prediction is the assertion that all 4-manifolds with $b^1 = 0$ and $b^+ > 1$ have "generalized simple type". The authors also include results about correlation functions on manifolds with $b^1 > 0$. Finally, there are extensions to u-plane integrals in related theories.

An expository version of the results of this paper appears in [M. Mariño and G. W. Moore, Nuclear Phys. B Proc. Suppl. **68** (1998), 336–347]. A continuation of this line of investigation for higher-rank gauge groups was carried out by Mariño and Moore [Comm. Math. Phys. **199** (1998), no. 1, 25–69].

Besides the intrinsic interest of the results obtained here, the paper under review is noteworthy as an example of quantum field theory techniques applied to a mathematics problem. Very detailed properties of the particular field theory at hand impressively match intricate formulas in geometry. *Daniel S. Freed* (Austin, TX)

99m:57029 57R57 58D10 58D29
Taubes, Clifford Henry (Cambridge, MA)
The structure of pseudo-holomorphic subvarieties for a degenerate almost complex structure and symplectic form on $S^1 \times B^3$. (English summary)
Geom. Topol. **2** (1998), 221–332 (electronic).

This paper constitutes another step in Taubes' program that aims to provide a precise relation between Seiberg-Witten invariants and the Gromov invariants. The SW-invariants provide, for any smooth 4-manifold with non-vanishing β_2^+, a way to associate an integer to each Spin^c class (which is the same as a class in $H_2(M, \mathbf{Z})$, at least up to some identification). That integer counts the number of solutions to the gauge-theoretical Seiberg-Witten equations. On the other hand, given a symplectic manifold (M, ω), the Gromov invariant counts the number of pseudo-holomorphic curves in a given homology class in $H_2(M, \mathbf{Z})$, that is to say, real closed surfaces whose tangent spaces are everywhere J-invariant and which realise that homology class (when the space of such curves has dimension larger than 0, one actually counts the number of such J-curves that meet some generic set of points in M). Here J is any almost complex structure compatible with ω. It turns out that that number is independent of the choice of the ω-compatible structure J. Taubes has shown in previous papers [see, e.g., Math. Res. Lett. **2** (1995), no. 2, 221–238; MR 96a:57076; J. Amer. Math. Soc. **9** (1996), no. 3, 845–918; MR 97a:57033, see p. D49 in this volume] that both invariants agree on symplectic manifolds. The problem is then the following: since the SW-invariants are well-defined for all smooth 4-manifolds with $\beta_2^+ \neq 0$, is there some way to extend the definition of the Gromov invariants in such a manner that the equality continues to hold between them?

In the paper under review, Taubes recalls that any closed smooth 4-manifold, which is not homeomorphic to a multiple connected sum of S^4 with some number of $\mathbf{CP}^2$'s having the anti-complex orientation, can be endowed with a closed 2-form ω which is nondegenerate everywhere except on a finite set S of circles. Thus ω is symplectic everywhere except on each circle $Z \in S$ (where it actually vanishes). He then claims that, up to some change of coordinates (which is unfortunately not of C^∞-class—the paper does not specify which class), one can take as a model of (ω, J) in some deleted neighbourhood $U = S^1 \times (B^3 - \{0\})$ of the circle $Z = S^1 \times \{0\}$ the following:

$$\omega = dt \wedge (x\,dx + y\,dy - 2z\,dz) + x\,dy \wedge dz - y\,dx \wedge dz - 2z\,dx \wedge dy,$$

and some J-structure that blows up near Z and which is such that $\sqrt{2}\omega(\cdot, J(\cdot))/|\omega|$ gives back the usual flat metric on $S^1 \times (B^3 - \{0\})$. Therefore, in order to extend Gromov's invariants to these manifolds, one must first study the way that J-holomorphic curves approach the singular circle Z in the "standard" model. This is the goal of the paper, which forms a nice piece of work.

The main theorem of the paper states that if C is a finite energy, pseudo-holomorphic subvariety of $S^1 \times (B^3 - \{0\})$, then each point in its closure in $S^1 \times B^3$ has a well-defined tangent cone. Furthermore, there is a finite set $Z_s \subset Z$ of singular points and an integer N such that any sufficiently small neighbourhood of a point $z \in Z - Z_s$ intersects C in a disjoint union of N embedded components. The closure of each such component is the image of a standard half-disc by a real-analytic embedding that sends the straight edge to Z and the interior to the complement of Z. Moreover, they can be explicitly described as perturbations of simple "ribbons" that rotate round the Z-axis.

The entire paper is devoted to a proof of that theorem.

At the singular points, that is to say, those in Z_s, two types of events can occur: either there is a cone with the top point lying on Z, or there is a piece of a ribbon avoiding Z which slowly approaches Z in a way described in the paper as Example 1.6. In any case, the precise behaviour of finite energy pseudo-holomorphic subvarieties near singular points is postponed to another paper.

The paper is written in a clear and straightforward style. There are a few misprints, and some statements are false as stated, like Proposition 3.3 (but, given the context, one easily sees how to correct them). *François Lalonde* (Montreal, QC)

99b:58003 58-02 28A78 54F45 54H20 58F11 58F15 58F99
Pesin, Yakov B. (University Park, PA)
★**Dimension theory in dynamical systems. (English summary)**
Contemporary views and applications.
Chicago Lectures in Mathematics.
University of Chicago Press, Chicago, IL, 1997. xii+304 pp. \$56.00; \$19.95 paperbound. ISBN 0-226-66221-7; 0-226-66222-5

This book is the first one to offer a unified approach to many different notions of dimension, as well as important methods and results previously scattered throughout the literature, at the interface between dimension theory and the theory of dynamical systems—an exciting area that has rapidly developed over the last decade. In fact, it is an excellent and highly enjoyable book, self-contained, and clearly written from the first to the last page. It puts many ideas and concepts in a higher and more general framework without introducing cumbersome technicalities, and still touches on the pathologies arising in connection with the dimensions and the tantalizing questions in the field. As a result, there is a smoother ride to the core of the subject than experienced by many of us when the area was in its early stages.

In a nutshell, some main themes of the book are the following. At the heart of the theory is the C-structure of the underlying space, and the elaboration of the Carathéodory construction. Via an analogue of the free energy in statistical physics, the Carathéodory dimension and the lower and upper Carathéodory capacities of sets and measures are defined, which can be specialized to the Hausdorff dimension and the lower and upper box (or Minkowski) dimensions, respectively. A natural question concerns the existence of sufficient conditions that guarantee equalities between these quantities. In fact, for measures, certain conditions are given, whereas for a set, the phenomenon of equality is declared (by the author) to be rare.

Increasingly interesting seems the chapter about the "dimensional" view of the thermodynamical formalism (including the classical one of dynamical systems due to Bowen, Ruelle, Sinaĭ and Walters), topological pressure and topological and measure-theoretic entropies, the extension to non-compact or non-invariant sets, along with the variational principle for topological pressure and the so-called Bowen equation.

The second part of the book is devoted to applications to dynamical systems. A prominent role is given to the Moran geometric constructions relying on the property of self-similarity, which is crucial to the fundamental relations between dimension, Lyapunov exponents, and metric entropy. Another example concerns the regular geometric constructions. The discussion continues on to some examples revealing the pathological properties of the pointwise dimension, and to examples that have strictly different Hausdorff dimension and box dimension. Further objects of study are repellers for smooth expanding maps (including hyperbolic Julia sets, repellers for one-dimensional Markov maps, and limit sets for Schottky groups), basic sets of Axiom A diffeomorphisms, and Smale-Williams solenoids. A highlight of the book is the multifractal analysis of Gibbs measures for smooth conformal dynamical systems, and the multifractal study of conformal repellers and conformal hyperbolic sets associated with the pointwise dimension, local entropy, and Lyapunov exponents. The culmination of this dimension story is the presentation of a pearl to a still young field, the recently achieved affirmative solution of the Eckmann-Ruelle conjecture.

For dozens of years to come this book will be a standard reference and rich source for graduate students, researchers and anyone interested in dimension theory of dynamical systems. *Irene Hueter* (Gainesville, FL)

99b:58023 58B30 81T30

Connes, Alain (Bures-sur-Yvette); **Douglas, Michael R.** (Piscataway, NJ); **Schwarz, Albert** (Davis, CA)

Noncommutative geometry and matrix theory: compactification on tori. (English summary)

J. High Energy Phys. **1998**, *no.* 2, Paper 3, 35 pp. (electronic).

The goal of this very promising paper is to formulate the Ishibashi-Kawai-Kitazawa-Tsuchiya (IKKT) and Banks-Fischler-Shenker-Susskind (BFSS) matrix models so as to make more precise the relation between them and to describe, thanks to methods of noncommutative geometry, toroidal compactifications of them. Roughly speaking, the Euclidean version of the IKKT model consists of the data of a complex supermanifold (the configuration space) and an action functional S considered as a holomorphic function. S is invariant with respect to the action of the complex orthogonal group $SO(10, \mathbf{C})$ and some infinitesimal gauge transformations; moreover S is connected with the action functional of the Green-Schwarz superstring.

The compactification of the IKKT model on a circle arises in the following way: one wants to restrict S to a subspace consisting of points that remain in the same gauge class after a shift by a real number $2\pi R_0$ in the direction X_0 where $X_0, X_1, \cdots$ are elements of some complex Lie algebra. So for these points there should exist an invertible matrix U such that $X_0 + 2\pi R_0 = UX_0U^{-1}$, $X_1 = UX_1U^{-1}, \cdots$. It is much easier to find such operators by considering the Hilbert space $L^2(S^1)$. By means of Wick rotation, one can obtain corresponding results in the BFSS matrix model.

When the authors want to construct such a compactification for a noncommutative torus, they obtain precisely a projective module over it and a connection. In this context, the dimension of a vector bundle is a real number (think of the trace II_1 of a projection). The authors then use tools of noncommutative geometry to study the moduli space for the noncommutative torus; it is endowed with an action of $\mathrm{SL}_2(\mathbf{Z})$. The moduli space of constant curvature connections associated with odd cohomology plays the role of space-time.

Then the authors discuss gauge theory on the noncommutative torus and, moreover, propose an interpretation for the BFSS model compactified on T^θ.

Schwarz [Nuclear Phys. B **534** (1998), no. 3, 720–738] has subsequently shown that compactifications on Morita equivalent tori are (in some sense) physically equivalent.

{See also the following review [MR 99b:58024].} *Éric Leichtnam* (Paris)

99k:58015a 58B30 53C55 58G30 81Q60 83C99
Fröhlich, Jürg (Zürich); **Grandjean, Olivier** (Zürich); **Recknagel, Andreas** (Zürich)
Supersymmetric quantum theory, non-commutative geometry, and gravitation.
Symétries quantiques (*Les Houches*, 1995), 221–385, *North-Holland, Amsterdam*, 1998.

99k:58015b 58B30 53C55 58G30 81Q60 83C99
Fröhlich, J. [Fröhlich, Jürg] (Zürich); **Grandjean, O.** (Zürich); **Recknagel, A.** (Zürich)
Supersymmetric quantum theory and differential geometry. (English summary)
Comm. Math. Phys. **193** (1998), *no.* 3, 527–594.

Non-commutative geometry is a far-reaching generalisation of classical differential geometry developed by A. Connes [*Noncommutative geometry*, Academic Press, San Diego, CA, 1994; MR 95j:46063] which permits mathematical analysis of objects beyond the reach of classical methods. Quantum theory provides much of the motivation and many of the examples and tools of noncommutative geometry. On the other hand, E. Witten's supersymmetric approach to the Morse inequalities [J. Differential Geom. **17** (1982), no. 4, 661–692 (1983); MR 84b:58111] initiated extensive use of supersymmetry in geometry, with the de Rham complex forming the canonical example. The papers under review aim at a synthesis of these ideas, with supersymmetric quantum mechanics providing the algebraic framework for both classical real and complex differential geometry and also their noncommutative analogues. The motivation of the authors is quantum gravity, that is, a theory combining general relativity and quantum theory. Underlying their approach is the belief that space-time is noncommutative in some sense not yet understood (although concrete suggestions have been formulated [S. Doplicher, K. Fredenhagen and J. E. Roberts, Comm. Math. Phys. **172** (1995), no. 1, 187–220; MR 96f:81068]).

The basic concept in both papers under review is that of a spectral triple. This consists of a *-algebra $\mathscr{A}$, a Hilbert space $\mathscr{H}$ which forms a module over this algebra and an unbounded self-adjoint operator D on $\mathscr{H}$ such that the commutators $[D,a]$ are bounded for all elements a of $\mathscr{A}$. This basic structure is formulated in terms of supersymmetric quantum mechanics in these papers. Roughly speaking, supersymmetric quantum mechanics requires additional operators Q, Q^* with $D = Q+Q^*$, $Q^2=(Q^*)^2=0$ and with $H=D^2=QQ^*+Q^*Q$ being proportional to the Hamiltonian. In general one also imposes a Z_2 grading with a grading operator γ which anticommutes with D and commutes with $\mathscr{A}$. When the algebra $\mathscr{A}$ is commutative, for example $C^\infty(M)$, where M is a compact spin^c manifold, then this supersymmetric data can be realised classically with D being the Dirac operator.

To characterise noncommutative geometries amongst all spectral triples one imposes additional axioms on the spectral triple. These are discussed in [A. Connes, Comm. Math. Phys. **182** (1996), no. 1, 155–176; MR 98f:58024]. These additional axioms enable one to describe the metric (or Riemannian) aspects of noncommutative geometry and to characterise spin geometry. On restriction to an abelian algebra $\mathscr{A}$ it is claimed by Connes [op. cit.; MR 98f:58024] that one can show from the axioms that the spectrum of $\mathscr{A}$ is a smooth manifold. In the Comm. Math. Phys. paper under review $\mathscr{A}$ is always assumed to be $C^\infty(M)$ for some compact manifold M.

The Les Houches paper takes a more general view analogous to that of Connes [op. cit.; MR 98f:58024]. An important difference is in the notion of integration, where the Dixmier trace is replaced by a "KMS-style" construction of a trace. The paper starts with general relativity and the difficulty of reconciling it with quantum theory, progresses through supersymmetric quantum mechanics based around Pauli's theory of the electron (and the electron/positron pair) and then begins a discussion of noncommutative geometry. The approach diverges from that of Connes [op. cit.; MR 98f:58024] in Section 5.2 with the introduction of the notion of supersymmetric spectral data as an approach to noncommutative geometry. The supersymmetric algebras which characterise spin and Riemannian geometry form the most extensively developed examples in this exposition. These algebraic structures do not distinguish between classical and fully noncommutative geometry (the distinction occurs at the level of the algebra $\mathscr{A}$, which is $C^\infty(M)$ in the former case and a noncommutative *-algebra of bounded operators in the latter). Examples of the fully noncommutative theory are given by the noncommutative torus. Further details on supersymmetric noncommutative spectral data and their application in quantum field theory are deferred to a third, as yet unpublished, manuscript. The Les Houches paper concludes with a discussion of string theory, matrix models and conformal field theory and their relationship to noncommutative geometry.

The Comm. Math. Phys. paper extends the material in the Les Houches review by developing classical complex differential geometry as the theory of commutative supersymmetric spectral triples possessing additional algebraic structure. This additional structure takes the form of further operators on the Hilbert space of the triple and their algebraic relations. What is the point of this reformulation? This supersymmetric algebraic approach is not sensitive to the commutativity of $\mathscr{A}$. Thus the noncommutative analogues of complex geometries are also defined in this approach. Just as many highly singular (and classically intractable) objects such as the dual of a discrete group, Penrose tilings or quantum groups are encoded by spectral triples and may be analysed by applying cyclic cohomology, K-theory and other tools of noncommutative geometry, so too we anticipate successful analysis of noncommutative complex geometries.

The purpose of the Comm. Math. Phys. paper is to give a comprehensive exposition of the supersymmetric algebraic formulation of Riemannian, spin, symplectic, complex, Hermitian, Kähler and hyper-Kähler geometry. To give the flavour of the results let us focus on the spectral data associated to a complex manifold. We assume first that we have the commutative $N = (1,1)$ structure given, that is, a spin manifold M, the spinor bundle S over M, and the *-algebra $\mathscr{A}$ as the smooth functions $C^\infty(M)$ acting on the Hilbert space $L^2(\quad)$, which consists of the square integrable sections of the bundle $\quad = S \otimes \overline{S}$ (n even) or $\quad = S \otimes \mathbf{C}^2 \otimes \overline{S}$ (n odd), where $\overline{S}$ is the "conjugate spinor bundle" constructed using the complex conjugates of the transition functions of S. Acting on smooth sections of we have both the Dirac operator D and its "conjugate" $\overline{D}$. These data, together with the grading, are referred to as $N = (1,1)$ supersymmetry. From this supersymmetry algebra one may recover the graded algebra of differential forms on M and in particular the exterior differential d. The additional information needed to specify complex spectral data is an operator T such that $G := [T,d]$ satisfies $G^2 = 0$, $[T,G] = G$, $[T,f] = 0$ for all $f \in C^\infty(M)$ and $[T,\omega]$ is a smooth one-form on M (acting on $L^2(\quad)$) for all smooth one-forms ω. The final element of structure is the Dolbeault complex, and for this we set $\overline{G} := d - G$ and define

$$\Omega^{(0,1)}(M) = \Big\{ \sum_{i=1}^{N} f_0^i [G, f_1^i] \,|\, f_j^i \in C^\infty(M),\ N \in \mathbf{Z} \Big\},$$

$$\Omega^{(1,0)}(M) = \Big\{ \sum_{i=1}^{N} f_0^i [\overline{G}, f_1^i] \,|\, f_j^i \in C^\infty(M),\ N \in \mathbf{Z} \Big\}$$

and the anti-linear map

$$\sharp : \Omega^{(1,0)}(M) \to \Omega^{(0,1)}(M)$$

defined by

$$(f_0[G, f_1])^\sharp = \overline{f}_0[\overline{G}, \overline{f}_1].$$

One should recognise T as the operator which counts the holomorphic degree p of a (p,q)-form in the Dolbeault complex. The proof that this algebraic data does indeed specify a complex manifold in the usual sense uses the algebraic data to show that the complex structure J which is multiplication by i on $\Omega^{(1,0)}(M)$ and $-i$ on $\Omega^{(0,1)}(M)$ is integrable.

It is not essential to have all of the algebraic structure of $N = (1,1)$ supersymmetry to characterise Spin^c manifolds; indeed the spectral triple $(C^\infty(M), L^2(S), D)$ suffices to recover spin^c geometry. The complex geometries (Hermitian, Kähler and hyper-Kähler) however do require higher supersymmetry in which one introduces additional "chiral supercharges", so-called $N = (n,n)$ supersymmetry with $n = 1, 2$ and 4. This raises the obvious questions: what are the relationships between one supersymmetry algebra and another, and how do we pass between them? A discussion of these issues forms the third section of the Comm. Math. Phys. paper and, utilising the notion of broken supersymmetry, part of Section 5 of the Les Houches paper.

Readers interested in the fully noncommutative theory can find some discussion in the Les Houches paper (but with a more extensive treatment deferred to a subsequent, unpublished manuscript). The papers are very accessible and clearly written and will be a valuable source for both newcomers and experienced researchers.

{For the entire collection containing the Les Houches paper see [MR 99k:81008].}

Alan L. Carey (Adelaide)

97k:58024 58C27 32S15 57R45

du Plessis, Andrew (Aarhus); **Wall, Terry [Wall, Charles T. C.]** (Liverpool)

★**The geometry of topological stability.**

London Mathematical Society Monographs. New Series, 9.

Oxford Science Publications.

The Clarendon Press, Oxford University Press, New York, 1995. viii+572 pp.

ISBN 0-19-853588-0

Whitney can be said to have founded the field of singularity theory with his work in the 1940s and 1950s on approximating differentiable maps by maps having especially simple properties: a C^∞-map $f\colon N\to P$ between smooth manifolds of dimensions n and p can be approximated by embeddings if $n<2p$, by immersions with normal crossings if $n=2p$, by immersions with normal crossings and isolated "Whitney umbrella" singularities if $n=2p-1$, and by maps with fold and cusp singularities if $n=p=2$. He formulated the crucial notion of stability, and hypothesized that stable maps can be characterized by their transversality to some finite collection of submanifolds of a jet space; this would allow one to use the Thom transversality theorem to prove that one can approximate smooth maps by stable maps, and perhaps allow one to classify the singularities of stable maps. Thom, however, showed by example that one can't always approximate by stable maps, and conjectured that one could approximate by C^0-stable maps. A map f is stable [resp. C^0-stable] if it has a neighborhood W in the Whitney topology such that, for any $g\in W$, there exist C^∞-diffeomorphisms [resp. homeomorphisms] r and l such that $g=l\circ f\circ r$. In the 1960s Mather, in a remarkable series of papers, characterized stability in terms of transversality, calculated the "nice range of dimensions" for which stable maps are dense, and made considerable progress on the classification of germs of stable maps. Also he defined the so-called Thom-Mather canonical stratification of $J^k(n,p)-W^k(n,p)$ ($W^k(n,p)$ is a certain $\mathscr{K}$-invariant closed algebraic subset whose codimension goes to ∞ with k) such that multitransversality to this stratification while avoiding W^k implies C^0-stability, proving Thom's conjecture.

An important notion in both theories is that of contact (or $\mathscr{K}$-) equivalence: germs f and g are $\mathscr{K}$-equivalent if there is a diffeomorphism of the ambient space of the graphs sending the graph of f to that of g, leaving the domain invariant, and commuting with projection to the domain; in particular, the sets $f^{-1}(0)$ and $g^{-1}(0)$ are diffeomorphic. This defines equivalence classes called contact classes in jet space. For proper maps, stability is equivalent to multitransversality to contact classes; outside the nice dimensions, there are continuously varying families of contact classes forming sets (called "moduli") of codimension less than or equal to n; hence one can't apply Thom's transversality theorem to prove density (and density fails in fact). A key property of the Thom-Mather stratification is that it has only finitely many strata, which are $\mathscr{K}$-invariant, and Whitney's conditions (a) and (b) hold for it.

The Thom-Mather theory of C^0-stability has two principal defects. The first is that it is very difficult to explicitly calculate this stratification. The second is that Mather's sufficient condition is not a necessary condition for C^0-stability.

The goal of the authors is to correct this situation by finding an explicit $\mathscr{K}$-invariant stratification $\mathscr{B}^k$ of $J^k(n,p)-W^k(n,p)$ such that multitransversality to it is equivalent to C^0-stability. The authors have not yet reached this goal, which is extremely difficult because of the inherent difficulties in studying C^0-equivalence of smooth maps. Nevertheless, their partial successes presented in this book are beautiful achievements, using techniques which should command wide interest.

For simplicity, suppose N is compact. The book's main result is that there are a $\mathcal{K}$-invariant subset $Y^k(n,p)$ of $J^k(n,p)$ with $W^k(n,p) \subset Y^k(n,p)$ and a $\mathcal{K}$-invariant stratification $\mathcal{B}^k$ of $J^k(n,p) - Y^k(n,p)$ such that: (a) if f is C^0-stable, then f is multitransverse to $\mathcal{B}^k$, and misses Y^k if $n < \operatorname{codim} Y^k(n,p)$; (b) if f is multitransverse to $\mathcal{B}^k$ and misses Y^k, then f is C^0-stable. Thus, in the range of dimensions for which $n < \operatorname{codim} Y^k(n,p)$, this gives a necessary and sufficient characterization of C^0-stability. This range of dimensions is approximately the same as those for which $\mathcal{K}$-orbits have at most one-dimensional moduli; in particular it is considerably larger than Mather's nice range of dimensions. Furthermore, the strata are explicitly defined, so that one knows a great deal about the C^0-stable maps in these dimensions. Two important consequences of this result are: in the dimensions for which this classification works, C^0-stability is shown to be a local condition; C^0-stability is equivalent to stability in the nice range of dimensions.

There is much in this book which should appeal to a larger audience than those interested in the central problem we have described. The study of diversity, quasi-diversity and tame retractions should appeal to topologists. The algorithm for computing discriminant matrices and instability ideals is a beautiful application of commutative algebra and computer algebra. There is much of interest to students of singularities of varieties, in particular: the $\mathcal{K}$-classification of finite map-germs (i.e. of "fat points") and the study and calculation of real multiplicity and of mod 2 Milnor numbers.

Chapters 1–3 present a very readable introduction to the problem and a review of background material. Chapter 4 discusses the possible definitions of C^0-stability of maps and of map-germs, and the interrelationships between these. In the non-compact case this is quite complicated; this is illustrated by May's example of the C^0-stable function from **R** to **R** whose germ at 0 has an x^3 singularity, which in no case can be considered a C^0-stable germ. The authors fix upon an appropriate notion of locally C^0-stable; still it has the unfortunate property that it is not an open condition; near stable germs are non-stable ones. In Chapter 5 a stronger condition, disruptiveness (and the closely related quasi-disruptiveness), is studied; this is open, and if f is quasi-disruptive it is not a C^0-stable germ and cannot be a germ of a C^0-stable map.

$\mathcal{K}$-sufficient k-jets are ST-equivalent (ST stands for "stably topologically") if representatives of them have C^0-equivalent C^∞-stable unfoldings. The ST-equivalence classes partition $J^k(n,p) - W^k(n,p)$; the authors conjecture that they will serve as the sought-for $\mathcal{B}^k$ in arbitrary dimensions. This cannot now be proven. However, several procedures are developed for ST-distinguishing two given strata, and these are used inductively to define and calculate a stratification into "fine LST-invariants". This is done in Chapter 6, where it is also proven that non-transversality to these fine LST-invariants (plus the "immersion condition", which is very weak) implies quasi-disruptiveness. Hence transversality to them is a necessary condition for C^0-stability.

The next two chapters are concerned with explicit calculations. Chapter 7 deals with the case $n > p$ (complete intersection singularities). Considerable classification up to $\mathcal{K}$-type has been done. The Milnor number is not a C^0-invariant over the reals, but is when reduced mod 2. While this sounds like a weak invariant, the pattern of mod 2 Milnor numbers appearing in adjacent strata is a very important invariant, and is used to ST-distinguish collections of $\mathscr{K}$-classes, and hence to organize them into fine LST-invariants. Chapter 8 considers the case $n \leq p$ (fat points). Here the real multiplicity replaces the mod 2 Milnor numbers. Also a substantial $\mathscr{K}$-classification is carried out.

Chapter 9 begins the study of sufficient conditions. Suppose a map-germ f has a stable unfolding F. If F is a C^0-trivial unfolding of f, then there are C^0-retractions R and S such that $SF = fR$. Then S is a V-tame retraction if, for every embedding-germ j from the target of f to that of F which is transverse to the parameter space, Sj is a homeomorphism. This implies that f is a C^0-stable germ. Also one can piece together such germs to get maps which are globally C^0-stable; the key step in the globalization is an application of the Kirby-Edwards theorem on spaces of embeddings. Using the notion of V-tameness, civilized submanifolds of jet space are defined with the property that multitransversality to them implies C^0-stability. Finally, weighted homogeneity allows the construction of V-tame retractions of stable unfoldings when appropriate multitransversality conditions are met.

Chapter 10 is mostly commutative algebra. A module is constructed which allows one to determine the instability locus of an unfolding. This module allows one to check the multitransversality condition just referred to. To carry out these calculations, efficient algorithms are developed, which have been implemented on a computer. Also, many examples are presented for which the results can be determined without a computer. The results of the computer calculations are described in Chapter 11. This provides extensive lists of civilized submanifolds.

The main result is proved in Chapter 12. Fine LST-invariants which are also civilized are the $\mathscr{B}^k$-strata of the theorem. The book closes with a long list of open problems. *Leslie Charles Wilson* (Honolulu, HI)

97k:58029 58D10 57N13 57R15 57R57 58D27 58D29

Taubes, Clifford Henry (Cambridge, MA)

Counting pseudo-holomorphic submanifolds in dimension 4.

J. Differential Geom. **44** (1996), *no.* 4, 818–893.

This is the second in a series of papers in which Taubes proves the equivalence of the Seiberg-Witten and Gromov invariants for compact symplectic 4-manifolds (X, ω) [see also Math. Res. Lett. **1** (1994), no. 6, 809–822; MR 95j:57039]. This paper is devoted to the definition of the relevant Gromov invariants. It is self-contained and independent of SW theory.

Every (X, ω) admits a tamed almost complex structure J, unique up to homotopy. Fix a class in $H^2(X; \mathbf{Z})$, or equivalently its Poincaré dual homology class A. A J-holomorphic A-curve is an embedded (not necessarily connected) submanifold C representing A whose tangent space is J-invariant. The set of such curves forms a moduli space $\mathscr{M}_A$ with even formal dimension $d_A = A^2 - K \cdot A$, where K is the canonical class. Now choose $\frac{1}{2}d_A$ points in X and consider the "cutdown moduli space" $\mathscr{M}'$ of curves in $\mathscr{M}_A$ that pass through those points. The formal tangent space to $\mathscr{M}'$ at a curve C is the kernel of a real linear deformation operator D_C; this is a compact perturbation of the $\overline{\partial}$ operator with values in the normal bundle, and depends on the 1-jet of J along C.

Preparatory results prove that for generic $(J, \{\text{pts}\})$, $\mathcal{M}'$ is a finite set of curves at which D_C is invertible. Each $C \in \mathcal{M}'$ then has an associated sign ± 1 according to the parity of the spectral flow from D_C to any $\overline{\partial}$ operator which commutes with J. Counting the points in $\mathcal{M}'$ with sign (incorporating the subtlety described below) gives an invariant Gr: $H^2(X;\mathbf{Z}) \to \mathbf{Z}$. The main result of this paper is that Gr is independent of $(J, \{\text{pts}\})$ and is invariant under isotopies of the symplectic form.

Isotopy invariance is not a priori clear. The space of operators D_C is divided into chambers by codimension-one walls consisting of those operators with one-dimensional kernel. As $(\omega, J, \{\text{pts}\})$ varies the operators D_C can cross walls, changing the invariant. This turns out to be a problem only when counting embedded tori with $A^2 = K \cdot A = 0$. Here a surprising subtlety arises: to get an invariant one must also count the multiple covers of these "toric" curves.

There are eight possible types of toric curves, determined by the sign of D_C on C and its double covers. For each type, there is a generating function f_i that describes the total contribution of the covers of the curve to the invariant. At a wall, curves of different types can coalesce and bifurcate, changing sign and type. By studying several such wall crossings and demanding that the invariant be unchanged, the author shows that the f_i are determined by f_1, which he fixes to agree with the conventions of Seiberg-Witten theory.

The final section works out a specific example (the product of a torus with a Riemann surface), showing how to perturb J and identify the resulting toric curves.

Thomas H. Parker (East Lansing, MI)

98m:58017 58D27 53C07 57R57 58E15 58G05

Mrowka, Tomasz (Cambridge, MA); **Ozsváth, Peter**; **Yu, Baozhen** (Pasadena, CA)

Seiberg-Witten monopoles on Seifert fibered spaces.

Comm. Anal. Geom. **5** (1997), *no.* 4, 685–791.

Since their appearance on the mathematical scene, the Seiberg-Witten equations have had a dramatic impact in low-dimensional topology. Besides producing simpler arguments for earlier results obtained via Donaldson's theory of instantons, they have uncovered previously unsuspected relationships between the topology and the differential geometry of low-dimensional manifolds.

There are two reasons these equations look more attractive than the Yang-Mills equations. Firstly, the gauge group of this new theory is abelian, and secondly, they possess remarkable positivity features which, among many other things, guarantee the compactness of the moduli spaces of solutions.

However, one has to pay a price for these simplifications, and perhaps the best illustration of the new difficulties involved is what occurs when one tries to produce gluing formulae for the new invariants.

Suppose M is a closed, smooth, oriented 4-manifold decomposed into two manifolds with boundary M_i, $i = 1, 2$, by an oriented hypersurface $N \subset M$. As in Donaldson theory, an important amount of information is contained in "tunnelings", i.e. finite energy solutions of the Seiberg-Witten equations on the infinite cylinder $\mathbf{R} \times N$. If one regards the longitudinal coordinate as a temporal variable t, one can formally regard these equations as describing the gradient flow of a certain energy functional $\mathcal{E}$ on N. The techniques of Morgan-Mrowka-Ruberman in the instanton case extend with no conceptual changes to the Seiberg-Witten case and one can prove that, much as in a finite-dimensional situation, the above tunnelings have asymptotic limits as t goes to $\pm\infty$ which are critical points of the energy functional $\mathscr{E}$. In the instanton case these critical points can be described as flat SU(2)-connections, which are objects quite well understood topologically. Unfortunately, in the Seiberg-Witten case, there is no hope for a simple description, mainly because these critical points are extremely sensitive to the Riemannian metric on the 3-manifold N. It is reasonable to try first to look at special examples.

A large and important class of examples is supplied by the Seifert fibrations. These can be viewed as circle bundles over 2-orbifolds. They are very well understood topologically and, moreover, the work of Thurston shows that they possess certain (almost) canonical locally homogeneous metrics. This paper is devoted precisely to the study of tunnelings on $\mathbf{R}\times N$ where N is a Seifert manifold equipped with the Thurston geometry.

The first part of the paper is devoted to the study of the nature of the critical points of $\mathscr{E}$. These can be alternatively described as solutions of the three-dimensional Seiberg-Witten equations. As is known, these involve Dirac equations which traditionally are defined using connections induced on spinor bundles by the Levi-Civita connection of the metric. The first important point of this paper is that the Levi-Civita connection may not be the most revealing choice. Instead, the authors employ as start-up connection a metric connection which has torsion but is compatible with the fibration structure. These perturbations of the Seiberg-Witten equations can be solved quite explicitly following an approach very similar to the one employed for solving the Seiberg-Witten equations on Kähler manifolds. The solutions can be described as vortices on the base of the Seifert fibrations, which is an orbifold. Moreover, the part of the critical set consisting of irreducible solutions is Morse-Bott nondegenerate, while the degeneracies of the reducible part are quite well understood. These results were also independently obtained by the present reviewer using adiabatic techniques [see L. I. Nicolaescu, Comm. Anal. Geom. **6** (1998), no. 2, 331–392].

The next part in this paper is heavily analytical and is devoted to a detailed study of tunnelings. The infinite cylinder $\mathbf{R}\times N$ can be compactified to a ruled complex orbifold X. Using techniques pioneered by G. Y. Guo [Comm. Math. Phys. **179** (1996), no. 3, 737–775, 777–788; MR 98a:32040] the authors identify these tunnelings with certain divisors (vortices) on X with prescribed behavior near the zero and infinity sections of X. They go a step further than Guo. They show that the deformation complex associated to a tunneling is isomorphic to the deformation complex associated to the corresponding vortex on X.

This result opens the possibility of computing the virtual dimensions of the spaces of tunnelings by computing the index of the deformation complex for vortices on X. However, the computation of this index is by no means a trivial job. One source of difficulties is the presence of singularities on X, which is only an orbifold. The third part of the paper is devoted precisely to this problem. The problem is reduced to a computation on a smooth manifold using the Hirzebruch-Jung desingularization procedure, and the end result is an explicit description of the virtual dimensions of the spaces of tunnelings in elementary, albeit quite elaborate, terms. An immediate consequence of these computations is an explicit description of the (irreducible) Seiberg-Witten-Floer homology of the Brieskorn homology spheres with at most four singular fibers, equipped with the Thurston metric.

The computation of these virtual dimensions is not the only consequence of this work. The identification between the above two deformation complexes plays an important part in the recent proof of the Thom conjecture by Ozsváth and Z. Szabó ["The symplectic Thom conjecture", Preprint]. The key point in their proof is a certain gluing formula which requires an explicit knowledge of certain obstruction bundles. This is where the above identification between the two deformation complexes enters crucially.

It is very difficult to capture in this review the remarkable complexity of the problem solved in this paper. Its complete resolution required a large amount of nontrivial input from differential geometry, complex algebraic geometry and partial differential equations. The results and techniques in this paper have already had a substantial impact in the development of Seiberg-Witten theory, and I am convinced this work has many other potential applications that are yet to be discovered.

Liviu I. Nicolaescu (Notre Dame, IN)

97d:58038 58D29 14N10 20F55 32G34 58F07 81T40

Dubrovin, Boris (Trieste)

Geometry of 2D topological field theories.

Integrable systems and quantum groups (*Montecatini Terme*, 1993), 120–348, *Lecture Notes in Math.*, 1620, *Springer*, *Berlin*, 1996.

In assembling the material in these lecture notes, the author brings to a wider public a series of ideas in which he has been fundamentally involved over a period of years. The subject is a crossroads of many disciplines—it is one of those pieces of mathematics (like Dynkin diagrams) which have a long history in one form or another, and which one encounters in many walks of life.

In the current incarnation of the subject, the focal point is the WDVV (Witten-Dijkgraaf-Verlinde-Verlinde) equation. This is a simply stated problem: we need to find a function $F(t_1,\cdots,t_n)$ such that the third derivative $c_{ijk}=\partial^3F/\partial t_i\partial t_j\partial t_k$ defines the structure constants of a commutative, associative algebra (when we raise one index with the quadratic form c_{1jk}). The function F is also required to be quasi-homogeneous. We obtain in this way an n-dimensional space of algebras, and the modern interest in the subject derives from the fact that conformal field theories come in families with moduli spaces, and in particular 2-dimensional topological field theories. The axiomatic approach to these as in [M. F. Atiyah, *The geometry and physics of knots*, Cambridge Univ. Press, Cambridge, 1990; MR 92b:57008] tells us to associate a vector space to a circle and various operations to cobordisms. In particular, a pair of pants describes an algebra structure—that of a commutative Frobenius algebra. The general structure of field theories tells us that the vector space itself parametrizes deformations of this structure, and that is how the WDVV equations arise. In Dubrovin's geometric version of this theory this is known as a Frobenius manifold.

Questions on such topological field theories abound. For example, there are many enumerative problems in algebraic geometry which are answerable by using generating functions for Gromov-Witten invariants. These functions satisfy the WDVV equations. Ask how many rational curves of degree d in the plane pass through $3d-1$ generic points, and you are asking for a solution of a nonlinear differential equation. It is important, then, to approach these equations from the right point of view, or, even better, several points of view.

From Dubrovin's vantage point, we are looking at an n-dimensional manifold (usually complex) which has the structure of a Frobenius algebra on each tangent space. This is a commutative algebra with a linear form θ such that $\theta(xy)$ is a nondegenerate inner product. The manifold thus has a metric $g(x,y)=\theta(xy)$, but $\theta(xyz)$ also defines a rank-3 tensor $c(x,y,z)$, which is required to satisfy an integrability condition. With this structure, one finds in the generic semisimple case that the common eigenspaces of all the tangent vectors give distinguished directions which can be integrated to an orthogonal coordinate system. The algebraic structure on the manifold is then little more than such a coordinate system. A Frobenius manifold is, however, defined to be flat and this provides the nonlinear differential equation. Put another way, we have two sets of coordinates on $\mathbf{C}^n$: flat ones $t_1,\cdots,t_n$, in which the associativity of the algebra is described by a nonlinear equation for a function $F(t)$, and orthogonal ones $x_1,\cdots,x_n$, in which the algebra structure is trivial, but flatness gives a nonlinear equation for the coefficients g_{ii} of the metric. It is in this context that the equations, and some of the solutions, derive from the work of Darboux and Egorov at the turn of the century.

In fact, this respectable pedigree goes further. As Dubrovin shows, the problem can be reformulated in the language of isomonodromic deformations, another 19th century preoccupation. These are the equations whose solution requires one to pass from the monodromy of a first order system of ODEs with regular singular points to its coefficients. For Frobenius manifolds, the monodromy group is generated by reflections. It is here that one first begins to appreciate the seriousness of the equations, for in the lowest-dimensional interesting case—three dimensions—these are Painlevé equations, for which we need in general Painlevé transcendants to find the solutions.

In specific cases, the geometry saves the day, however, and a whole section of the lectures is devoted to a beautiful set of concrete examples of Frobenius structures, defined on the orbit spaces of Coxeter groups. These examples grew out of some novel observations of K. Saito in 1979 on deformations of singularities.

Yet another aspect of these fascinating objects is their relationship to nonlinear partial differential equations of "hydrodynamic type"—in fact, the one-dimensional isentropic flow of an ideal gas is the general 2-dimensional Frobenius manifold in this language. The lectures deal extensively with these ideas and the derivation of the corresponding structures as limits of integrable hierarchies, where one ignores higher nonlinearities.

Dubrovin's lecture notes provide a comprehensive treatment of a fascinating and important subject. The organization of the material is as nonlinear as the subject itself, but the text makes for challenging and surprising reading and will repay any attempt to probe its depths.

{For the entire collection see MR 97b:58001.} *N. J. Hitchin* (Oxford)

97h:58032 58D29 55N91 58D27 81T40 81T70
Cordes, Stefan (New Haven, CT);
Moore, Gregory [Moore, Gregory W.] (New Haven, CT);
Ramgoolam, Sanjaye (New Haven, CT)
Lectures on 2D Yang-Mills theory, equivariant cohomology and topological field theories.
Géométries fluctuantes en mécanique statistique et en théorie des champs (*Les Houches*, 1994), 505–682, *North-Holland, Amsterdam*, 1996.
The very genesis of string theory is in attempts to formulate a theory of the strong interactions. In this sense, the idea that Yang-Mills theory is a string theory is over 20 years old. The idea is attractive but has proven evasive. Recently, D. J. Gross [Nuclear Phys. B **400** (1993), no. 1-3, 161–180; MR 94h:81139] initiated a program to find a string-theory interpretation of 2d YM. One of the first goals of Gross' program was to find a string action of YM_2. The significance of the results is that they provide a bridge between YM_2 and cohomological field theory. The two themes just mentioned, 2d Yang-Mills as a string theory and cohomological QFT as a method of deriving an action for it, are the focus of this set of lectures, delivered by Moore at the Les Houches session LXII in 1994. Accordingly, the lecture notes are split into two parts, dealing with the respective theories.

Part I, published in the proceedings of the 1994 Trieste Spring School on superstrings [Nuclear Phys. B Proc. Suppl. **41** (1995), 184–244; MR 96m:81174], deals with Yang-Mills as a string theory, in particular in the case $d=2$. After discussing some general pros and cons of formulating YM as a string theory, the authors move to more detailed topics. There is an exact solution of YM_2, for which (a) the Hilbert space is the space of class functions on the gauge group G, and (b) the amplitudes are known exactly. For example, the authors derive the famous result which states that if a closed oriented space-time Σ_T has area a, then the partition function of Yang-Mills theory for gauge group G is

$$Z=\sum_R (\dim R)^{\chi(\Sigma_T)} e^{-e^2 a C_2(R)},$$

where the sum is over the unitary irreducible representations of G, $\chi(\Sigma_T)$ is the Euler character of Σ_T, and $C_2(R)$ is the quadratic Casimir in the representation R. Similar results hold for all amplitudes. Furthermore, the beautiful calculations of Gross and W. Taylor, IV of the $1/N$ expansions of amplitudes in YM_2 are described. The outcome is that the $1/N$ expansion may be identified with a sum over branched covers or simple generalizations of these. Moreover, the branched covers are weighted, and it is shown that these weights are Euler characteristics of the moduli space of holomorphic maps from Riemann surfaces to other Riemann surfaces.

Part II, which is reviewed here, develops in great detail the so-called cohomological field theories, a special case of the more general notion of topological field theories. The latter may be grouped into two classes: "Schwarz type" and "cohomological type". The former have Lagrangians which are metric independent and hence—formally—the quantum theory is expected to be topological. Examples of such theories include the $e^2=0$ YM_2 theory with Lagrangian $I=\int \mathrm{Tr}(\varphi F)$ or 3d Chern-Simons theory $I=\int \mathrm{Tr}(AdA+\frac{2}{3}A^3)$. Mathematically such theories are related to flat connections, knot invariants, Knizhnik-Zamolodchikov equations, etc. Physically they are related to rational conformal field theories, the fractional quantum Hall effect, anyons, etc. In the lectures under review here, the authors focus on the latter type of TFT's, those of cohomological type. These have a very different flavor. They are not manifestly metric independent, but have a "BRST operator", that is, an odd nilpotent operator Q. Physical observables are metric independent because of the decoupling of BRST trivial degrees of freedom. In fact, cohomological field theory (CoFT) is of very broad significance—each of the examples of topological field theories has important applications to both mathematics and physics. It is enough to mention Donaldson theory and its recent successor, Seiberg-Witten theory.

The topics involved in this presentation include equivariant cohomology and its relation to universal bundles, as well as to Lie algebra cohomology. Here it is described that "equivariant cohomology of a Lie algebra $\mathfrak{g}$ is the same as a 'supersymmetrized' Lie algebra cohomology of a corresponding graded Lie algebra $\mathfrak{g}_{\text{super}}$". The amplitudes of a TFT are defined as intersection numbers in certain moduli spaces. Since these are often infinite-dimensional, some work is required in defining them. The approach applied by the authors is based on the following paradigm of describing a cohomological FT by three basic data: symmetries, fields, and equations. The symmetries will generally be characterized by an infinite-dimensional Lie group, for example a group of gauge transformations, or a diffeomorphism group. Having chosen the symmetry group G one studies the G-equivariant cohomology. To study equivariant cohomology one must choose a model for the cohomology. A choice of differential complex amounts to a choice of fields. Mathematically, the ultimate objects of study are intersection numbers. The equations are used to isolate interesting subspaces of the space of fields. One writes Poincaré duals to these subspaces using Mathai-Quillen representatives of Thom forms. The equations are viewed as the vanishing of a section of a vector bundle.

In the cases treated in these lectures, the equations define a finite-dimensional subspace M of the relevant moduli space, which is a priori infinite-dimensional. A typical example (and chronologically one of the first) is the moduli space of self-dual SU(2)-connections on a 4-manifold (Donaldson theory). The authors then apply a kind of Lefschetz fixed point argument to give a formula for intersection numbers on the finite-dimensional space M. It is based on the construction of the Thom class of a vector bundle (whose direct image on the base gives the Euler class of that bundle). For this the Mathai-Quillen formalism is applied [see V. Mathai and D. G. Quillen, Topology **25** (1986), no. 1, 85–110; MR 87k:58006]. This formalism is described in detail, giving a systematic description from a unified point of view for all the different CoFT's. As a first example, supersymmetric quantum mechanics is discussed in detail, following the original presentation [E. Witten, J. Differential Geom. **17** (1982), no. 4, 661–692 (1983); MR 84b:58111]; here the mathematical theory is de Rham cohomology as well as Morse theory.

Next the authors discuss topological sigma models, specifying the necessary data to get a CoFT. Also, the relation to the physical sigma model is discussed. In the following section the more difficult case is considered, where the topological theory has local symmetries. The essential principle motivating this is that of space-time locality. Examples of this are given by topological YM, 2d topological gravity and, finally, topological strings. Each of these are considered in detail, and it is here that the string action for YM_2 is derived. In a last section the possibility of extending these results to $d=4$ is discussed, and an appendix contains some generalities of differential geometry.

Although by now no longer timely, these nicely written lecture notes give a comprehensive introduction to these theories, suitable for both a mathematical and a physical audience.

{For the entire collection see MR 97f:81002.} *Bruce Hunt* (Offenbach)

99e:58038 58E05 58D29 58F09

Fukaya, Kenji (Kyoto); **Oh, Yong-Geun** (Madison, WI)

Zero-loop open strings in the cotangent bundle and Morse homotopy.

Asian J. Math. **1** (1997), *no.* 1, 96–180.

Given a closed smooth manifold M with a metric g as an auxiliary structure, there are two—a priori different—methods to investigate the topology of M, namely Morse theory on M and a variant of Lagrangian intersection Floer homology on T^*M. The main result of the paper is that they are in fact equivalent. Let us first describe both theories in more detail and then give a summary of contents.

Morse theory: Recall the construction of the Morse complex of M as (re)discovered by E. Witten in 1982 [J. Differential Geom. **17** (1982), no. 4, 661–692 (1983); MR 84b:58111]. Let f be a smooth Morse function on M such that its negative gradient flow is Morse-Smale (i.e. stable and unstable manifolds intersect transversely). The Morse chain groups are generated freely (over the integers or Z_2 if M is not orientable) by $\mathrm{Crit}\, f$, the set of critical points of f, and graded by the Morse index $\mathrm{Ind}_f(\cdot)$. The matrix elements of the boundary operator are defined by counting (with appropriate signs or mod 2 for Z_2 coefficients) connecting trajectories between critical points of index difference 1. The homology of this complex computes the singular homology of M.

Now generalize this construction by using several—say k—Morse functions, and study their gradient flows along the edges of some tree T. This way one obtains the cup product and higher (co)homology operations on Morse homology [M. Betz and R. L. Cohen, Turkish J. Math. **18** (1994), no. 1, 23–41; MR 95i:58037]. The present approach is slightly different and was outlined by one of the authors in an earlier paper [K. Fukaya, in *Geometric topology (Athens, GA, 1993)*, 409–440, Amer. Math. Soc., Providence, RI, 1997; MR 98i:57061]. Pick k distinct points $\overrightarrow{p}=(p_1,\cdots,p_k)$ in M and k smooth functions $\overrightarrow{f}=(f_1,\cdots,f_k)$ on M such that the difference between any two of them is a Morse function and the Morse-Smale property holds. A ribbon tree $t=(T,\iota,v_1)$ is a tree T together with an embedding $\iota\colon T\to D^2$ such that no vertex of T has two edges and precisely the exterior vertices of T are mapped to ∂D^2. v_1 is a distinguished exterior vertex of T. We consider two ribbon trees to be equivalent if their trees are isomorphic and the embeddings are isotopic, transforming the distinguished points into one another. Let l be a function associating to each interior edge of T a positive number (its length); then (t,l) is called a metric ribbon tree. The result of Stasheff that the set Gr_k of metric ribbon trees is homeomorphic to $\mathbf{R}^{k-3}$ [J. D. Stasheff, Trans. Amer. Math. Soc. **108** (1963), 275-292; ibid. **108** (1963), 293–312; MR **28** #1623] is recovered here by proving the existence of a diffeomorphism between both spaces. Now an element of the moduli space $\mathcal{M}_g(M;\overrightarrow{f},\overrightarrow{p})$ of graph flows is a pair $((t,l),I)\in \mathrm{Gr}_k\times C^0(T,M)$ such that I restricted to edges of T is a trajectory of the negative gradient flow of the difference of two of the functions, e.g. on the ith exterior edge (which is identified with $\mathbf{R}_0^-$) a trajectory of $-\nabla(f_{i+1}-f_i)$ which emanates from the zero p_i. It is shown that for generic $\overrightarrow{f}$ this moduli space of graph flows is a smooth manifold of dimension

$$\sum_{i=1}^{k} \mathrm{Ind}_{f_{i+1}-f_i}(p_i)-(k-1)n+(k-3).$$

The last term reflects the dimensionality of Gr_k and the other terms may be heuristically obtained by summing up the codimensions of the unstable manifolds of the p_i to get the codimension of their intersection, which can be identified with $\mathcal{M}_g(M;\overrightarrow{f},\overrightarrow{p})$.

Lagrangian intersection Floer theory: Any function f_i gives rise to a Lagrangian submanifold $\Lambda_i=\mathrm{Graph}(df_i)$ of T^*M and any (nondegenerate) critical point p_i of $f_{i+1}-f_i$ corresponds one-to-one to a point $x_i=df_i(p_i)$ where Λ_{i+1} and Λ_i intersect (transversely).

In the case $k=2$ A. Floer constructed an algebraic chain complex [J. Differential Geom. **28** (1988), no. 3, 513–547; MR 90f:58058] whose chain groups are generated freely by the (finite) set $\Lambda_1\cap\Lambda_2$ and whose boundary operator is given by counting J-holomorphic disks in T^*M with finite energy and Lagrangian boundary conditions (i.e. $u\colon D^2\setminus\{\pm1\}\mapsto T^*M$ is required to satisfy $\partial D^2\to\Lambda_1\cup\Lambda_2$ and $\pm1\mapsto\Lambda_1\cap\Lambda_2$). Its homology, called Lagrangian Floer homology, computes the singular homology of Λ_i, $i=1,2$, and hence that of M. Note that any two disks with two marked points on the boundary (each) are equivalent (via a biholomorphic automorphism φ of D^2 respecting marked points) and so we can choose the points ±1 above without loss of generality.

For $k=3$ this still holds true (there is a unique φ), but for $k>3$ there is a $(k-3)$-dimensional moduli space $\mathcal{T}_{0,k}$ of equivalence classes of disks with k marked (distinct) points on the boundary (respecting the cyclic order of ∂D^2). Multiplication by $\varepsilon>0$ yields Lagrangians $\Lambda_i^\varepsilon=\mathrm{Graph}(\varepsilon df_i)$ and intersection points $x_i^\varepsilon=\varepsilon df_i(p_i)$, $p_i\in\mathrm{Crit}(f_{i+1}-f_i)$. Let $J=J_g$ denote the almost complex structure on T^*M induced by the metric g on M and define the moduli space $\mathcal{M}_J(T^*M;\overrightarrow{\Lambda}^\varepsilon,\overrightarrow{x}^\varepsilon)$ to be the set of pairs $([z_1,\cdots,z_n],w)\in\mathcal{T}_{0,k}\times\mathrm{Map}(D^2,T^*M)$ such that w is J-holomorphic, $w(z_i)=x_i^\varepsilon$ and $w(\partial D^2)\subset\Lambda_1^\varepsilon\cup\cdots\cup\Lambda_k^\varepsilon$.

Main theorem: For generic $\overrightarrow{f}$ and $\varepsilon>0$ sufficiently small, there exists an orientation preserving diffeomorphism

$$\Phi^\varepsilon\colon\mathcal{M}_g(M;\overrightarrow{f},\overrightarrow{p})\to\mathcal{M}_J(T^*M;\overrightarrow{\Lambda}^\varepsilon,\overrightarrow{x}^\varepsilon).$$

An immediate consequence is that $\mathcal{M}_J(T^*M;\overrightarrow{\Lambda}^\varepsilon,\overrightarrow{x}^\varepsilon)$ is a smooth manifold for generic $\overrightarrow{f}$ and its dimension is given by the same formula stated above. Moreover, for any Riemannian manifold (M,g) there is a topological A^∞-category $\mathcal{MS}(M)$ whose objects are the set of smooth functions on M. To any (generic) choice of two of them one associates the Morse-Witten complex of their difference, and these morphisms are related by a certain composition map [cf. K. Fukaya, op. cit.]. The main theorem then establishes the existence of an A^∞-category in the realm of Lagrangian intersection theory; its objects are the graphs of exact 1-forms on M and to any two of them (generic choice) one associates their Floer complex. The authors remark that these A^∞-structures determine the rational homotopy type of M, but regrettably do not shed any further light on this statement.

The paper is devoted to the proof of the main theorem and consists of two parts. As $\mathcal{T}_{0,3}$ and Gr_3 consist of a single point each, the structure of the J-holomorphic disk and graph flow moduli spaces simplifies essentially for $k=3$ and the main theorem is proved in this case in great analytical detail in Part I. The method of proof is a version of Floer's glueing construction extended to the study of singular perturbation problems in symplectic geometry by S. Dostoglou and D. A. Salamon in their proof of the Atiyah-Floer conjecture [Ann. of Math. (2) **139** (1994), no. 3, 581–640; MR 95g:58050]. The idea is to construct an approximate zero $\widetilde{w}^\varepsilon$ of the Cauchy-Riemann operator (in appropriately ε-weighted Banach space norms), given an element of $\mathcal{M}_g(M;\overrightarrow{f},\overrightarrow{p})$, and then to find nearby a true zero $w^\varepsilon\in\mathcal{M}_J(T^*M;\overrightarrow{\Lambda}^\varepsilon,\overrightarrow{x}^\varepsilon)$ using Newton's iteration method. Whereas usually a candidate for $\widetilde{w}^\varepsilon$ is pretty much dictated by the geometric situation at hand, this is not at all the case here, due to the degeneration of a domain of dimension 2 into one of dimension 1. The authors overcome this problem by an ingenious rescaling construction. Part II deals with the general case $k\geq 3$. The main difficulty appearing here is that the moduli spaces are now stratified, because Gr_k is stratified by cells consisting of all metric ribbon trees of the same tree type—so what varies within one cell are the lengths of interior edges. Instead of constructing the required diffeomorphism on each stratum separately and then showing that their images fit together appropriately, approximate solutions are defined on each stratum, then glued together, and finally J-holomorphic disks are found using Newton's method. The crucial point to make this construction work is the existence of natural cell decompositions of the compactifications of Gr_k and $\mathcal{T}_{0,k}$, which are dual to each other. To compactify Gr_k, use the natural $\mathbf{R}^+$-action on (the lengths of) interior edges. Compactification of $\mathcal{T}_{0,k}$ is defined as taking the closure of $\mathcal{T}_{0,k}$ considered as a subset of the space of stable curves of genus zero with k marked points. In that way collections of disks intersecting at isolated boundary points arise—similarly to the familiar bubble trees. *Joa Weber* (Stony Brook, NY)

99m:58047 58E05 55N35 57R15 57R70 58F05
Liu, Gang (Cambridge, MA); **Tian, Gang** (Cambridge, MA)
Floer homology and Arnold conjecture.
J. Differential Geom. **49** (1998), *no.* 1, 1–74.
The conjecture mentioned in the title has had a major influence in the creation of symplectic topology. The prehistory of the subject begins in 1912 with the statement by H. Poincaré of a theorem on fixed points of area-preserving maps of the annulus, which was proved in 1913 by G. D. Birkhoff. In the 1960s and 1970s, V. I. Arnol′d conjectured symplectic generalizations of this result. In particular [cf. F. E. Browder, in *Mathematical developments arising from Hilbert problems (De Kalb, IL, 1974)*, 35–79, Amer. Math. Soc., Providence, RI, 1976; see MR **54** #7158 (p. 66)]: Conjecture A. Let (V,ω) be a closed symplectic manifold, and let φ be a Hamiltonian diffeomorphism of V. Then the number of fixed points of φ is bounded from below by the number of critical points of a function, "both algebraically and geometrically".

Here a Hamiltonian diffeomorphism is a map of the form $\varphi=\varphi_{t_0}$ where (φ_t) is the solution of a non-autonomous Hamilton equation $d\varphi_t/dt=X_{H_t}(\varphi_t)$. The algebraic case is the most interesting: it means that if the fixed points are nondegenerate, then $\#\mathrm{Fix}(\varphi)$ should be bounded below by the minimal number of critical points of a Morse function on V. Using Morse theory, this conjecture has usually been interpreted as follows: Conjecture B. If the fixed points of φ are nondegenerate, then $\#\mathrm{Fix}(\varphi) \geq \mathrm{SB}(V)$, where $\mathrm{SB}(V)$ is the sum of Betti numbers over $\mathbf{Z}$ or a field.

The conjecture is valid for $\mathscr{C}^1$-small transformations since then the fixed points are in one-to-one correspondence with the critical points of a suitable generating function. Actually this argument applies to intersections of Lagrangian submanifolds under Hamiltonian deformations, and leads to a generalization of the conjecture to that case, but that would take us too far.

Let us give some history about this conjecture. In 1978, a still unpublished topological proof of Y. Eliashberg for the case of surfaces (other than S^2, for which it follows from Hopf) attracted some attention, but the first important result was the theorem of C. C. Conley and E. J. Zehnder [Invent. Math. **73** (1983), no. 1, 33–49; MR 85e:58044], who proved Conjecture B (and Conjecture A) for any standard symplectic torus $V = T^{2n}$. The proof involved the study of the action functional

$$\mathscr{A}_H(x) = \int_x p \cdot dq - H_t dt$$

defined on the space $\mathscr{L}(V)$ of contractible loops (one can assume that (H_t) is defined on $V \times S^1$ and is smooth). By Hamilton's principle the critical points of $\mathscr{A}_H$ are in bijection with the fixed points of φ such that the loop $(\varphi_t(x))$ is contractible. Critical points of a_H were found using a Galerkin-type method to reduce the problem to a finite-dimensional one. Some time afterwards, this method was extended independently by A. Floer and the reviewer to cover the case of some symplectic manifolds of nonpositive curvature, including surfaces.

The most important breakthrough is due to Floer. After having studied the general case of Lagrangian intersections, he adapted his method to the case of fixed points of symplectomorphisms [Comm. Math. Phys. **120** (1989), no. 4, 575–611; MR 90e:58047], defining what became known as Floer (co-)homology of symplectic manifolds. His construction was based on the application to $\mathscr{A}_H$ of the description of Morse theory given by E. Witten to define a complex based on $\mathrm{Fix}(\varphi)$, whose differentials are obtained by counting gradient lines joining two fixed points. A crucial point is that these gradient lines are maps $f\colon \mathbf{R} \times S^1 \to V$ which are solutions of a nonhomogeneous Cauchy-Riemann equation $\bar{\partial}_{J,H} f = 0$, where J is an almost complex structure compatible with ω. This enabled him to use the newly created theory of pseudoholomorphic curves of M. L. Gromov [Invent. Math. **82** (1985), no. 2, 307–347; MR 87j:53053].

In the simplest case when $[\omega]$ vanishes on $\pi_2(V)$, Floer cohomology is defined as follows. For $x, y \in \mathscr{L}(V)$, denote by $\widetilde{\mathscr{M}}(J, H, x, y)$ the space of gradient lines connecting x and y. Then one can define a relative Morse (or Conley-Zehnder) index $\mu(x)$ (i.e. $\mu(y) - \mu(x)$ is well-defined), and for a generic choice of (H, J) one has the following properties: (o) $\widetilde{\mathscr{M}}(J, H, x, y)$ is a smooth oriented manifold of dimension $\mu(y) - \mu(x)$, with a free action of $\mathbf{R}$ of quotient $\mathscr{M}(J, H, x, y)$. Thus it is empty if $\mu(y) - \mu(x) \leq 0$. (i) $\mathscr{M}(J, H, x, y)$ is compact if $\mu(y) - \mu(x) = 1$. (ii) $\mathscr{M}(J, H, x, y)$ can be compactified by adding the oriented boundary

$$\bigcup_{\mu(z) = \mu(y) + 1} \mathscr{M}(J, H, x, z) \times \mathscr{M}(J, H, z, y)$$

if $\mu(y) - \mu(x) = 2$.

To prove this Floer used Gromov's compactness theorem for pseudoholomorphic curves and his newly created technique of gluing. The complex giving Floer cohomology $FH^*(V,\omega;J,H)$ is free over $\mathrm{Fix}(\varphi)$, graded by μ, and $\partial x = \sum_{\mu(y)=\mu(x)+1} n(x,y)y$, where $n(x,y)$ is the algebraic number of points of $\mathscr{M}(J,H,x,y)$. Then Floer proved that $FH^*(V,\omega;J,H)$ is independent of the generic choice of (J,H), and that for H small enough it can be identified with the standard cohomology $H^*(V,\mathbf{Z})$ given by the Morse complex of H. This implies Conjecture B for (V,ω). Actually, Floer proved that all this holds in the "positive" (or "monotone") case when $[\omega]_{|\pi_2(V)} = \lambda c_{1|\pi_2(V)}$ with $\lambda \geq 0$.

Floer's result was afterwards generalized by H. H. W. Hofer and D. A. Salamon [in *The Floer memorial volume*, 483–524, Progr. Math., 133, Birkhäuser, Basel, 1995; MR 97f:57032] and by K. Ono [Invent. Math. **119** (1995), no. 3, 519–537; MR 96a:58046] to the "semi-positive" (or "weakly monotone") case when $c_1 \cdot A + 3 - n \geq 0$ for every $A \in \mathrm{Im}(\pi_2(V) \to H_2(V))$ such that $\omega \cdot A > 0$ $(2n = \dim(V))$. Here one needs to work on a suitable covering $\Gamma \to \widetilde{\mathscr{L}}(V) \to \mathscr{L}(V)$ of the space of contractible loops so that the action functional is well defined. The complex giving Floer cohomology is then free over some Novikov ring Λ_ω (a suitable completion of $\mathbf{Z}[\Gamma]$), and one has $FH^*(V,\omega,J;H) \approx H^*(V,\omega) \otimes \Lambda_\omega$. This again implies Conjecture B for the semipositive case.

To understand the significance of semipositivity, note that $c_1 \cdot A + 3 - n$ is the formal dimension (over $\mathbf{C}$) of the space of unparametrized J-holomorphic spheres in the class A. If J is generic, this is always ≥ 0 for all *simple* such spheres, and the semipositivity implies that it will also be true for multiply covered spheres. This is the key to obtain properties (i) and (ii), and more generally to prove that the Gromov-Floer compactification of every space $\mathscr{M}(J,H,x,y)$ contains only strata of lower codimension.

If (V,ω) is not semipositive, the Gromov-Floer compactification may involve multiply covered spheres with negative dimension, which will give strata having higher dimension than the moduli space itself. In particular one cannot guarantee that properties (i) and (ii) hold, thus the general definition of Floer homology is not clear. This was the situation in 1993, and a similar difficulty appeared in the definition of quantum cohomology and Gromov-Witten invariants.

A solution to this problem has been announced by several authors in the context of Gromov-Witten invariants [J. Li and G. Tian, in *Topics in symplectic 4-manifolds (Irvine, CA, 1996)*, 47–83, Internat. Press, Cambridge, MA, 1998; K. Fukaya and K. Ono, Topology **38** (1999), no. 5, 933–1048; B. Siebert, "Gromov-Witten invariants of general symplectic manifolds", Preprint, http://xxx.lanl.gov/abs/dg-ga/9608005] as well as in that of Floer homology (see the present paper, and Fukaya-Ono [op. cit.]). In all these works one constructs virtual moduli cycles, using ideas inspired by excess intersection theory in algebraic geometry.

The construction of such a virtual moduli cycle (or rather "chain" here in general!) is the main result of the paper under review. This is highly technical, but written very clearly. Roughly, the construction proceeds as follows.

First one defines the space $\mathscr{B}(\widetilde{x},\widetilde{y})$ of *stable* maps connecting $\widetilde{x}$ to $\widetilde{y}$, where $\widetilde{x}$ and $\widetilde{y}$ are two critical points of a_H on $\mathscr{L}(V)$. This is the idea of Kontsevich to reformulate Gromov's compactness theorem, and involves maps defined on various domains. This space is equipped with an infinite-dimensional "bundle" $\mathscr{L}$ with fiber $\mathscr{L}_f = \Omega^{0,1}(f^*TV)$ and a section $\overline{\partial}_{J,H}\colon \mathscr{B}(\widetilde{x},\widetilde{y}) \to \mathscr{L}$, so that the Gromov-

Floer compactification $\overline{\mathscr{M}}(J,H,\widetilde{x},\widetilde{y})$ is $\overline{\partial}_{J,H}^{-1}(0)$. Since it is compact and $\mathscr{B}(\widetilde{x},\widetilde{y})$ is a Banach orbifold, it is covered by finitely many W_i with isotropy groups contained in the finite groups Γ_i (the isotropy groups arise because of multiply covered maps).

Then one modifies locally the equation $\overline{\partial}_{J,H}f=0$ by adding generic perturbations v_i which are not Γ_i-invariant to obtain transversality. One thus obtains local moduli spaces $\widetilde{\mathscr{M}}_i^{v_i}=(\overline{\partial}_{J,H}+v_i)^{-1}(0)$ which are smooth of the expected dimension. This involves a suitable generalization of the gluing construction.

Finally, one globalizes this construction by "averaging". One finally obtains $C(\mathscr{M}^v(\widetilde{x},\widetilde{y}))$, which is a singular chain with rational coefficients (because of the averaging) in $\mathscr{B}(\widetilde{x},\widetilde{y})$, of dimension $\mu(\widetilde{x})-\mu(\widetilde{y})-1$. Moreover, one has

$$\partial C(\mathscr{M}^v(\widetilde{x},\widetilde{y}))=\bigcup_{\widetilde{z}} C(\mathscr{M}^v(\widetilde{x},\widetilde{z}))\times C(\mathscr{M}^v(\widetilde{z},\widetilde{y})).$$

It follows that if $\mu(x)-\mu(\widetilde{y})=1$ one can define $n(\widetilde{x},\widetilde{y})\in\mathbf{Q}$ as before, and thus define the Floer (co-)homology $FH^*(V,\omega;J,H,v)$. A parametrized version of all this proves the independence with respect to (J,H,v). Finally, one proves that it is isomorphic to $H^*(V,\mathbf{Q})$ for H small.

As a corollary, the authors obtain the proof of Conjecture B with rational coefficients for every closed symplectic manifold.

{Reviewer's remarks: One may wonder whether it is really necessary to go to rational coefficients. Also, in Arnol′d's original statement, the lower bound for the nondegenerate case would seem to be the minimal number of critical points of a Morse function on V. This might be much bigger than $\mathrm{SB}(V;\mathbf{Z})$ if $\pi_1(V)$ is complicated. For instance, one should have the lower bound $\min(p+q)$ over all presentations of $\pi_1 V$ with p generators and q relations. It seems unclear how to obtain such a result with Floer homology.} *Jean-Claude Sikorav* (Toulouse)

98g:58030a 58E12 35J60 53A10

Lancaster, Kirk E. (Wichita, KS); **Siegel, David** (Waterloo, ON)

Existence and behavior of the radial limits of a bounded capillary surface at a corner. (English summary)

Pacific J. Math. **176** (1996), *no.* 1, 165–194.

98g:58030b 58E12 35J60 53A10

Lancaster, Kirk E.; Siegel, David

Correction to: "Existence and behavior of the radial limits of a bounded capillary surface at a corner".

Pacific J. Math. **179** (1997), *no.* 2, 397–402.

In this remarkable and in some ways definitive paper, the authors characterize the behavior of radial limits $Rf(\theta)=\lim_{r\to 0^+}f(r\cos\theta,r\sin\theta)$, $-\alpha<\theta<\alpha$, for solutions of the capillary problem (1) $\nabla\cdot Tf=\kappa f+\lambda$ in Ω, $\nu\cdot Tf=\cos\gamma$ on $\partial\Omega$, $Tf=\nabla f/\sqrt{1+|\nabla f|^2}$, in a two-dimensional domain Ω with a corner point $-\alpha<\theta<\alpha$. Here κ and λ are constants, ν is an exterior unit normal, and γ is the "contact angle" on $\partial\Omega$, prescribed as a function of distance s from the vertex O. It can happen that no such radial limit exists for any θ in that range, as the authors demonstrate. However, they give natural conditions ensuring the existence and also the continuity of $Rf(\theta)$, and providing a priori information on possible behavior.

In a case of principal interest, for which there exist $\lim_{s\to 0+}\gamma^{\pm}(s)=\gamma_0^{\pm}\neq 0,\pi$, it is shown that there are fans of constant radial limits adjacent to each corner side, and explicit estimates are obtained for the size of the fans (in some cases, a "central fan" covering a half space and separated from the boundary fans by fans of monotone change can also exist). These estimates are later used in some striking ways; notably, conditions are given under which the fans overlap, thus ensuring continuity at O. These conditions are identical to the ones obtained by another procedure in a joint work of J. T. Chen, E. Miersemann and the reviewer ["Capillary surfaces in wedge domains", in preparation]. It is not yet known whether the conditions are sharp.

The demonstrations for the theorems are based in part on estimates for conformal parametrizations and in part on blow-up procedures. A rather lengthy technical effort is needed to complete the discussion; however, the material is broken down into individual steps that are very carefully presented, with scrupulous attention to detail.

The paper is perhaps as important for questions to which it calls attention as for those it answers. Notably, no general conditions are presented under which a central fan will appear. The paper does contain an example in which such behavior occurs; it is Example 2 on p. 184, for which the existence of the central fan follows as in the proof of Corollary 2 just above it.

Additionally, the necessary conditions for continuity of f at the vertex, and the possible behaviors when $\gamma_0^{\pm}=0$ or π, remain open questions, as does the question of finding general asymptotic expansions at the vertex points. This last question was settled by Miersemann [Pacific J. Math. **157** (1993), no. 1, 95–107; MR 93m:35039] for "subcritical" opening angles (the singular case, not addressed in the present paper) in "positive" gravity fields. To the reviewer's knowledge, all other cases remain open.

The sufficient conditions obtained for continuity of f require $\alpha<\pi/2$. If $\alpha>\pi/2$, general sufficiency conditions are yet to be established. In the example mentioned above, $\alpha>\pi/2$ and f is discontinuous at the vertex, although the prescribed data are constant.

Unfortunately, due to a printer's error, the figures appear without captions and labels; this error is rectified in the correction (cited in the heading of this review).

The results of the paper relate closely to and complement independent work by P. Concus and the reviewer [SIAM J. Math. Anal. **27** (1996), no. 1, 56–69; MR 96m:76006] and by the reviewer [Calc. Var. Partial Differential Equations **4** (1996), no. 4, 305–322; MR 97f:53005].

In a later work [Z. Anal. Anwendungen **15** (1996), no. 4, 819–850; MR 97m:53011], the authors carry the investigation further, with estimates on the rate of approach to the limit values. *R. Finn* (Stanford, CA)

97i:58048 58F05 70F25 70H33 70H35 70Q05 93B29 93C85

Bloch, Anthony M. (Ann Arbor, MI); **Krishnaprasad, P. S.** (College Park, MD); **Marsden, Jerrold E.** (Pasadena, CA); **Murray, Richard M.** (Pasadena, CA)

Nonholonomic mechanical systems with symmetry. (English summary)

Arch. Rational Mech. Anal. **136** (1996), *no.* 1, 21–99.

Reducing ODEs or PDEs which are invariant under the action of a Lie group is at the same time a very old and a very recent problem. Euler, Lagrange, Poisson, Jacobi, Kovalevskaya, and especially E. Noether, fully understood the interplay between symmetry and integrals of motion. Recently, the combination of abstract nonsense and computer algebra has allowed a systematic and algorithmic approach—doing tedious calculations by hand is nowadays a matter of personality deviation. Such powerful tools would certainly be of much delight to these men and women of mathematics, but more importantly, are useful now for applied scientists and engineers. For ODEs, the Hamiltonian case is of special significance, as each one-parameter subgroup of symmetry, roughly speaking, does it twice: it allows reducing the dimension by two. In the PDE context, symmetry and reduction are associated to "conservation laws" and PDEs of evolution often have a Hamiltonian structure.

On the other hand, until a few years ago non-holonomic systems were essentially left behind in the chest of lost items. With some indulgence, they were regarded as curious if not bizarre objects. Certain stones, known to the Celts and supposedly possessing magic properties, prefer to rotate in one direction; rolling balls on vertically rotating cylinders defy gravity; billiard professionals perform amazing tricks. Notwithstanding this somewhat dubious reputation, non-holonomic systems had attracted the interest of important scientists. Hertz, while struggling against a terminal disease, devoted himself to the foundations of mechanics: he advocated replacing forces by geometric constraints. É. Cartan made an interesting address to the 1928 International Congress of Mathematicians, a work that certainly should be useful in the present new developments.

For the mechanical systems encountered in engineering, one should take d'Alembert's principle as the leitmotiv—an idea cherished by Sommerfeld. Instead, if one uses the rules of the calculus of variations, a completely wrong set of equations obtains, as Hertz already realized. That non-integrable constraints produce systems which are not described by the calculus of variations is a good reason for the old prejudice mainstream physicists have against them. At any rate, the "wrong" set of equations, using the non-holonomic variational calculus (called by Arnol′d and Kozlov "vakonomic"), also has a very beautiful structure, being related to sub-Riemannian geometry (and, when quantized, to hypoelliptic operators). Recently it was realized that the same mathematical framework governs the motion of a classical particle in a Yang-Mills field, and the motion of a microorganism at low Reynolds number. The proper terminology for distinguishing the two different structures may be lacking. Tentatively, we call the former "non-vakonomic systems".

The paper under review concerns the non-vakonomic, non-holonomic mechanics, but with a novel twist: the systems depend on parameters ("shape variables") that can be used as control variables. This new feature paves the way to future developments and potentially many important practical applications in robotics, especially in biomechanics. At a glance, the equations of motion have the form

$$g^{-1}\dot{g} = -A(r)\dot{r} + B(r)p, \tag{1}$$

$$\dot{p} = \dot{r}^{\mathsf{T}}(\alpha(r)\dot{r} + \beta(r)p) + p^{\mathsf{T}}\gamma(r)p, \tag{2}$$

$$M(r)\ddot{r} = -C(r,\dot{r}) + N(r,\dot{r},p) + \tau, \tag{3}$$

where g denotes the symmetry group variables, p the momentum (not conserved in the general case), and r the reduced variables which describe the shape of the system.

Sometimes one has full control of the shape r, so (3) can be abandoned and one simply imposes a prescribed $r(t)$ on (1)–(2), or, equivalently, one assumes $\ddot{r} = u$.

In intrinsic terms, the authors use Ehresmann connections to model the constraints, and show how the curvature appears as a forcing into Lagrange equations. The momentum equation, which plays an important role in control applications, involves parallel transport operators. The cases where the momentum is conserved vs. the cases where it is not can be distinguished by geometric considerations. The momentum equation decouples from the group variables. The latter allow full reconstruction of the overall notion.

A delightful example is given, the snakeboard, which is a souped-up version of a skate. The front and back pairs of wheels are independently actuated by the rider—but he or she does not need to put a foot on the ground! Forward motion is generated by the interplay among these extra degrees of freedom, the non-holonomic constraints, and the shape variations determined by the twists of the rider. It is worth mentioning some of the findings. Simulations and experiments have detected several "gaits" involving periodic inputs at integrally related frequencies. For instance, 1 : 1 yields forward motion, 1 : 2 rotation about a fixed point and 3 : 2 sideways motion. These features could be explained theoretically.

In summary, the long-term goal of the whole project is to develop a control theory for Lagrangian mechanical systems using modern geometric tools. One hopes that before too long most control engineers will feel comfortable enough to use this language quite easily in their control recipes. *Jair Koiller* (Petrópolis)

98k:58095 58F05 17B99 46L89 58F06

Weinstein, Alan (Berkeley, CA)

The modular automorphism group of a Poisson manifold. (English summary)

J. Geom. Phys. **23** (1997), *no.* 3-4, 379–394.

Pursuing the philosophy that Poisson manifolds appear as semiclassical limits of operator algebras, the author shows that Poisson algebras, like von Neumann algebras, have modular automorphism groups, and geometric consequences of this observation as well as the links between Poisson geometry and operator algebras are studied. This philosophy was in fact stimulated by work of A. Lichnerowicz, to whom the article is dedicated (as well as by that of his collaborators).

Thus, let P be a Poisson manifold and, for a function f on P, write H_f for its Hamiltonian vector field. Given a density μ, consider the operator φ_μ, which, on a function f on P, is defined by the identity $(\varphi_\mu f)\mu = \mathscr{L}_{H_f}\mu$, where $\mathscr{L}_\xi$ is the Lie derivative of a vector field ξ. The antisymmetry of the Poisson tensor entails that φ_μ is a derivation and hence a vector field on P, called the modular vector field of P with respect to the density μ; this vector field represents a cohomology class (which does not depend on the choice of density) in Poisson cohomology, called the modular class. Further, the modular vector field φ_μ is precisely zero when μ is an invariant density for the flows of all Hamiltonian vector fields; a Poisson manifold with such an invariant density is called unimodular, and the modular class of such a Poisson manifold is plainly zero. The modular vector field was introduced without a name by J.-L. Koszul [Astérisque **1985**, Numéro Hors Série, 257–271; MR 88m:17013] and studied for the dual of a Lie algebra with its Lie-Poisson structure. For a general Poisson manifold, integration of a (particular complete) modular vector field yields a 1-parameter group of Poisson diffeomorphisms. When the density is changed, the generator of the group changes by a Hamiltonian vector field, so a 1-parameter group of "outer automorphisms" is intrinsically attached to the Poisson manifold, this group being trivial in the unimodular case. The resulting flow is called the modular flow. For example, for the structure $\{x,y\}=x^2+y^2$ in the plane, the modular flow is shown to consist of rotations around the origin; for a 3-manifold of the type $\mathrm{PSL}(2,\mathbf{R})/\Gamma$, where Γ is a surface group, for a suitable Poisson structure (defined in terms of the standard basis of left invariant vector fields on $\mathrm{PSL}(2,\mathbf{R})$), the modular flow is shown to be the horocycle flow. Many more examples are given in the paper—for instance the Reeb foliation on the 3-sphere has nonzero modular class while its Godbillon-Vey class is zero—and much new insight into Poisson geometry is offered by means of the modular flow, e.g. how global conditions can cause the modular class of a regular Poisson manifold to be nonzero. Further, the corresponding theory for regular Poisson manifolds is shown to admit generalizations to foliations, even to general Lie algebroids. For example, the Bott connection for the conormal bundle of a foliation is shown to yield precisely that structure which generalizes to foliations the modular vector field of a Poisson manifold. It is moreover indicated what the theory signifies for Lie groupoids: For example, given a density on a Poisson manifold admitting a symplectic groupoid, its modular flow lifts in a natural way on the symplectic groupoid by symplectic groupoid automorphisms. This lifted flow is Hamiltonian and is generated by the logarithm of what is called the modular function.

The notion of modular flow in Poisson geometry is also shown to be a classical limit of that of the modular automorphism group for von Neumann algebras. The modular theory of von Neumann algebras has its origin in the KMS theory in quantum mechanics [cf., e.g., A. Connes, *Noncommutative geometry*, Academic Press, San Diego, CA, 1994; MR 95j:46063], where a 1-parameter group of automorphisms is related to a certain type of positive linear functional, called a weight. In this theory, the modular automorphism group measures the extent to which the weight fails to be a trace. In the paper, weights on the Poisson algebra of a Poisson manifold (or what corresponds to them) are positive Borel measures, and the defining properties of a modular vector field are shown to entail that, likewise, such a vector field measures the extent to which integration with respect to a positive density fails to be a Poisson trace (i.e. a trace that vanishes on functions of the kind $\{f,h\}$ where $\{\cdot,\cdot\}$ denotes the Poisson bracket). Indeed, the similarities between KMS theory and symplectic geometry appearing throughout J. N. Renault's book [*A groupoid approach to C^*-algebras*, Lecture Notes in Math., 793, Springer, Berlin, 1980; MR 82h:46075] prompted the author to develop the theory of symplectic groupoids, [cf., e.g., A. D. Weinstein and P. Xu, J. Reine Angew. Math. **417** (1991), 159–189; MR 92k:58094].

For an orientable Poisson manifold P, a somewhat more abstract definition of the modular vector field is this: From the Poisson structure, the cotangent bundle of P inherits a Lie algebroid structure, and the assignment to an n-form μ of the Lie derivatives $\mathscr{L}_{H_f}\mu$, where f runs through the functions on P, turns the space of n-forms $\Omega^n P$ into a module for this Lie algebroid; a modular vector field then completely describes this module structure. In the non-orientable case, the structure just defined is still available but yields a different notion of modular structure. These observations have certain formal consequences, of cohomological and other nature.

Some of them are mentioned in the paper, and a hint is given that a suitable Poincaré duality theory should in particular explain the obvious pairing between the zeroth Poisson homology and the invariant densities—the Poisson traces cited above.

{Reviewer's remarks: Other formal consequences of the last-mentioned observations have been worked out by S. Evens, J.-H. Lu and the author ["Transverse measures, the modular class, and a cohomology pairing for Lie algebroids", Preprint, http://xxx.lanl.gov/abs/dg-ga/9610008, Quart J. Math., to appear], by Xu ["Gerstenhaber algebras and BV-algebras in Poisson geometry", Preprint, http://xxx.lanl.gov/abs/dg-ga/9703001], and by the reviewer [Ann. Inst. Fourier (Grenoble) **48** (1998), no. 2, 425–440; MR 99b:17021; J. Reine Angew. Math. **510** (1999), 103–159]; in the last reference, a precise notion of Poincaré duality for Lie algebroids, or more generally Lie-Rinehart algebras, is given. For the special case of a Poisson manifold, the appropriate notion of trace in this theory is substantially different from that of Poisson trace, though.} *Johannes Huebschmann* (Villeneuve d'Ascq)

99m:58089 58F05 58D10 58F22

Hofer, H. [Hofer, Helmut H. W.] (New York, NY); **Wysocki, K.** (Parkville); **Zehnder, E.** (Zürich)

The dynamics on three-dimensional strictly convex energy surfaces. (English summary)

Ann. of Math. (2) **148** (1998), *no.* 1, 197–289.

Periodic solutions have always been lying at the core of Hamiltonian dynamical systems. Throughout the last decade the existence problem for solutions of period 1 known as the Arnol′d conjecture [V. Arnol′d, C. R. Acad. Sci. Paris **261** (1965), 3719–3722; MR **33** #1861] has led to very fruitful advances in symplectic topology, mainly based on the theory known as Floer homology. It becomes an even more intricate question if one considers an autonomous, i.e. time-independent, Hamiltonian and asks for periodic solutions of any period on a prescribed level set, called an energy surface. Without further geometric assumptions on the energy surface there need not be any periodic solutions at all. A major breakthrough by P. H. Rabinowitz [J. Differential Equations **33** (1979), no. 3, 336–352; MR 81a:58030a] established an existence result on star-shaped regular energy surfaces. A related result by A. Weinstein under the assumption of convexity for the regular energy surface led to the following seminal conjecture [A. Weinstein, J. Differential Equations **33** (1979), no. 3, 353–358; MR 81a:58030b]: Let M^{2n-1} be a $(2n-1)$-dimensional closed manifold with a 1-form λ and a non-singular smooth vector field X such that $\lambda \wedge d\lambda^{n-1}$ is a volume form, $d\lambda(X,\cdot)=0$ and $\lambda(X)>0$. Then X has a periodic orbit. λ is called a contact form on the manifold M and the hyperplane distribution given by the kernel of λ is called its contact structure. If the regular energy surface $S=H^{-1}(1)$ in $\mathbf{R}^{2n}$ is strictly convex and encloses the origin, it carries such a contact structure, where the Hamiltonian vector field gives X, which after the normalization $\lambda(X)=1$ is known as the Reeb vector field. Note that the geodesic flow for a Riemannian manifold N is such a Hamiltonian dynamical system, where the unit (co-)tangent bundle S^*N is the energy surface of contact type and its periodic solutions correspond exactly to the nontrivial closed geodesics on N. The paper under review considerably extends the understanding of the dynamics on a 3-dimensional strictly convex energy surface beyond a mere existence result for any periodic solution.

An important tool in the study of periodic solutions is the Poincaré section map on a surface of section. The main result of the current paper is the existence of a global surface of section Σ of disk type in a 3-dimensional strictly convex energy surface S. A global surface of section is a compact embedded surface bounded by a closed orbit of X such that every other flow line of X intersects Σ in forward and backward time. Thus, a groundbreaking result by J. M. Franks [New York J. Math. **2** (1996), 1–19, electronic; MR 97c:58123] can be applied to the area-preserving Poincaré map on the disk Σ, which for obvious reasons has at least one fixed point. The consequence is the alternative that the prescribed energy surface S carries either 2 or infinitely many periodic solutions. The core of this result is the existence of an open book decomposition of the given energy surface which, due to the assumption of strict convexity, has to be a 3-sphere with its unique so-called tight contact structure. An open book decomposition is a fibration $M \setminus P_0 \to S^1$ where the binding is given by a particular periodic solution P_0 of X and the pages by the fibres, in this case of disk type $\mathbf{C}$. Every page of this open book decomposition provides a global surface of section.

The central theorem of the paper proves the existence of an open book decomposition of the spherical contact threefold $S = (M, \lambda)$ under the assumption of what the authors define as dynamical convexity. This is a condition in addition to the existence of a contact structure which is expressed in terms of Maslov-type indices associated to periodic orbits of the Reeb vector field. The index of a periodic solution is essentially defined by the winding number of the linearized flow along the orbit. A 3-dimensional contact manifold (M, λ) is dynamically convex if all indices of periodic solutions are at least 3. The authors show that strict convexity of the energy surface S enclosing the origin implies dynamical convexity. The method for obtaining the open book decomposition is based on a concept of moduli spaces of punctured finite energy surfaces in the so-called symplectization $M \times \mathbf{R}$. This is a 4-dimensional manifold inheriting a symplectic structure from the contact form λ. Choosing a suitable compatible almost complex structure the authors employ a powerful theory for holomorphic punctured surfaces in $M \times \mathbf{R}$ which are asymptotic to cylinders over periodic orbits in M near the punctures. Such holomorphic surfaces of genus 0 with exactly one puncture asymptotic to a periodic solution are called finite energy planes. This concept stands in close analogy to M. Gromov's seminal theory of pseudoholomorphic curves [Invent. Math. **82** (1985), no. 2, 307–347; MR 87j:53053], used with great success for 4-dimensional symplectic manifolds by D. McDuff [J. Amer. Math. Soc. **3** (1990), no. 3, 679–712; MR 91k:58042; erratum; MR 93k:58098]. The main step at the heart of the current paper is the work carried out by the same three authors in [Duke Math. J. **81** (1995), no. 1, 159–226 (1996); MR 97a:53043; correction; MR 98j:53037]. A periodic solution P of index 3 in a dynamically convex contact-threefold provides an embedded spanning disk which is transversal to the flow and stems from the projection of a finite energy plane in $M \times \mathbf{R}$ asymptotic to the periodic orbit P. As a consequence the moduli space of all such finite energy planes asymptotic to P provides an open book decomposition of M. The major work of the paper under review is first to obtain the existence of such an index minimizing orbit together with the property of dynamic convexity from the assumption of a strictly convex energy surface enclosing the origin, which implies that it has to be the 3-sphere with its unique tight contact structure. Behind this matter lies the fact that suitable combinatorics for the complex of periodic solutions involving the indices form an invariant of the contact structure which is known in the present case of the 3-sphere and invariant under continuous deformations of λ. Analogously to the proof of the Arnol′d conjecture via Floer homology this proves a posteriori the existence of the required periodic solution of index 3. Moreover, since the underlying work in their earlier paper [op. cit.; MR 97a:53043] regards the generic situation of nondegenerate periodic orbits, the authors have to show that the general degenerate case can be suitably approximated by nondegenerate Reeb flows. Much of the analysis for the moduli spaces of finite energy surfaces can be carried out in arbitrary dimensions. It is, however, in particular the positivity of intersections of holomorphic curves in dimension 4 (e.g. studied by McDuff [J. Differential Geom. **34** (1991), no. 1, 143–164; MR 93e:53050]) which provides the crucial embeddedness of the spanning disk and which ties this result to the 3-dimensional framework.

The paper presents a profound result about Hamiltonian dynamics based on an extremely powerful theory in contact topology which was developed by Hofer and his coauthors in a series of papers. This theory of holomorphic curves in symplectizations is largely inspired by Gromov's work on pseudoholomorphic curves and Floer's homology theory in symplectic topology. It requires an extensive analysis of moduli spaces related to Cauchy-Riemann type Fredholm operators and so-called bubbling-off analysis providing essential (non-)compactness results. Although the paper under review relies on results obtained in previous papers (in particular [op. cit.; MR 97a:53043]) it nevertheless offers also the non-expert a very intelligible self-contained account of the employed theory of finite-energy surfaces in symplectizations. For an additional excellent expository article on this subject of Hamiltonian dynamics and holomorphic curves, the reader may be referred to Hofer's plenary lecture at the ICM 1998 [Doc. Math. **1998**, Extra Vol. I, 255–280 (electronic); MR 99k:58028]. Moreover, holomorphic curves in symplectizations also find seminal application in terms of invariants for contact structures. This is now known as contact homology, partly similar to Floer homology for symplectic manifolds. For a superb overview the reviewer recommends Y. Eliashberg's invited lecture at the ICM 1998 [Doc. Math. **1998**, Extra Vol. II, 327–338 (electronic)]. *Matthias Schwarz* (Leipzig)

98e:58087 58F06 55N91 57R20 57S25 58G05 58G10
Vergne, Michele (Paris)
Multiplicities formula for geometric quantization. I, II.
Duke Math. J. **82** (1996), *no.* 1, 143–179, 181–194.

Let M be a compact quantizable symplectic manifold (or more generally, orbifold) acted upon in a Hamiltonian fashion by a compact connected Lie group G preserving the structure and suppose that 0 is a regular value of the moment map so that the reduced space M_{red} acquires the structure of a symplectic orbifold. Let $\mathscr{L}$ be a G-equivariant prequantum bundle for M, the G-action being compatible with the moment map, and denote by $\mathscr{L}_{\text{red}}$ the corresponding orbifold line bundle on the reduced space. Pick an almost complex structure on M which is compatible with the other data; it induces a compatible almost complex structure on M_{red}. In these two papers, the quantization $Q(M,\mathscr{L})$ of these data is defined as the direct image of $[\mathscr{L}]\in K_G(M)$ in the complex K_G-theory of a point, that is, as a virtual complex representation of G or as an element of the complex representation ring $R(G)$ of

G. This notion of direct image is a special case of a more general one due to M. F. Atiyah and F. Hirzebruch [in *Proc. Sympos. Pure Math., Vol. III*, 7–38, Amer. Math. Soc., Providence, R.I., 1961; MR **25** #2617], who used it to prove the differentiable Riemann-Roch theorem; it involves a Dirac operator with reference to the almost complex structure (or rather the corresponding Riemannian structure). The resulting virtual representation does not depend on the choice of almost complex structure. This hinges on the fact that the group coming into play is a torus; there are in fact counterexamples to the quantization conjecture when the group is nonabelian and the almost complex structure is non-positive. Likewise, the quantization $Q(M_{\text{red}}, \mathscr{L}_{\text{red}})$ of the reduced data is defined as an element of $K(\text{point})$ which is just a copy of the integers. In the two papers, two arguments are given establishing equality of the two numbers $Q(M, \mathscr{L})^G$ and $Q(M_{\text{red}}, \mathscr{L}_{\text{red}})$ under suitable circumstances, $Q(M, \mathscr{L})^G$ meaning the multiplicity of the trivial representation in $Q(M, \mathscr{L})$. In the first part, I, this is done for the case where G is a torus by means of a method which is based on localization in equivariant cohomology [cf. L. C. Jeffrey and F. C. Kirwan, Topology **34** (1995), no. 2, 291–327; MR 97g:58057; M. Vergne, Topology **35** (1996), no. 1, 243–266; MR 97g:58058]. In the second part, II, a proof of the equality of the two numbers is given for the case where M is an orbifold arising from an action of a torus on a manifold and where G is a torus, by means of the author's extension to almost complex manifolds of the index theorem for transversely elliptic operators [Duke Math. J. **82** (1996), no. 3, 637–652; MR 97j:58150]. Under the circumstances outlined, this establishes what is referred to as the statement of the quantization conjecture, raised [Invent. Math. **67** (1982), no. 3, 515–538; MR 83m:58040] by V. Guillemin and S. Sternberg (but neither phrased nor named in this way) and verified for general compact G but under the additional stronger conditions that M be a positive Kähler manifold and that the line bundle $\mathscr{L}$ be holomorphic and sufficiently positive.

Several other papers have recently appeared or will appear which extend the Guillemin-Sternberg result, and in which the main tool is localization; more information may be found in the reviews of the two papers of Jeffrey and Kirwan [Quart. J. Math. Oxford Ser. (2) **47** (1996), no. 186, 165–185; MR 97i:58064; Topology **36** (1997), no. 3, 647–693; MR 98e:58088, see p. D93 in this volume; see the following review]. *Johannes Huebschmann* (Villeneuve d'Ascq)

98e:58088 58F06 55N91 57R20 57S25 58G05 58G10

Jeffrey, Lisa C. (Princeton, NJ); **Kirwan, Frances C.** (Oxford)

Localization and the quantization conjecture. (English summary)

Topology **36** (1997), *no.* 3, 647–693.

Let M be a compact quantizable symplectic manifold acted on in a Hamiltonian fashion by a compact connected Lie group K preserving the structure and suppose that 0 is a regular value of the moment map so that the reduced space M_{red} acquires the structure of a symplectic orbifold. Pick a compatible almost complex structure on M—this is always possible; it descends to a compatible almost complex structure on M_{red}. Let $\mathscr{L}$ be a K-equivariant prequantum bundle for M, the K-action being compatible with the moment map, and denote by $\mathscr{L}_{\text{red}}$ the corresponding orbifold line bundle on the reduced space. When the almost complex structure is a genuine complex structure and $\mathscr{L}$ a holomorphic line bundle, geometric quantization endows the cohomology groups $H^j(M,\mathscr{L})$ with structures of representation spaces for K, and the Riemann-Roch numbers $\mathrm{RR}^K(\mathscr{L})$ and $\mathrm{RR}(\mathscr{L}_{\mathrm{red}})$ are defined as usual by $\mathrm{RR}^K(\mathscr{L})=\sum(-1)^j \dim H^j(M,\mathscr{L})^K$ and $\mathrm{RR}(\mathscr{L}_{\mathrm{red}})=\sum(-1)^j \dim H^j(M_{\mathrm{red}},\mathscr{L}_{\mathrm{red}})$. When the structure is only almost complex, the Riemann-Roch numbers $\mathrm{RR}^K(\mathscr{L})$ and $\mathrm{RR}(\mathscr{L}_{\mathrm{red}})$ can still be defined by means of a direct image construction in K-theory involving a Spin^c Dirac operator [cf. J. J. Duistermaat, *The heat kernel Lefschetz fixed point formula for the spin-c Dirac operator*, Birkhäuser Boston, Boston, MA, 1996; MR 97d:58181; V. W. Guillemin, in *Lie theory and geometry*, 305–334, Progr. Math., 123, Birkhäuser Boston, Boston, MA, 1994; MR 96m:58095; M. Vergne, Duke Math. J. **82** (1996), no. 1, 143–179, 181–194; MR 98e:58087, see p. D92 in this volume; see the preceding review]. This direct image construction has already been used by M. F. Atiyah and F. Hirzebruch in the proof of the differentiable Riemann-Roch theorem [Bull. Amer. Math. Soc. **65** (1959), 276–281; MR **22** #989]; it involves, in the smooth context, a suitable Dirac operator as a replacement for the algebro-geometric Dolbeault operator; cf., e.g., D.18 in the book by H. B. Lawson, Jr. and M.-L. Michelsohn [*Spin geometry*, Princeton Univ. Press, Princeton, NJ, 1989; MR 91g:53001]. What is referred to as the quantization conjecture in the title is the statement that, under appropriate circumstances, the two Riemann-Roch numbers coincide. V. Guillemin and S. Sternberg [Invent. Math. **67** (1982), no. 3, 515–538; MR 83m:58040] verified equality of the two numbers under the additional condition that M be a positive Kähler manifold and that the line bundle $\mathscr{L}$ be holomorphic and sufficiently positive.

There are counterexamples showing that the quantization conjecture cannot hold in full generality. In the paper under review, which extends earlier results of the authors for the special case where K has rank one [Quart. J. Math. Oxford Ser. (2) **47** (1996), no. 186, 165–185; MR 97i:58064], the quantization conjecture is shown to hold in many circumstances, including more or less those of the cited result of Guillemin and Sternberg. (More precisely, one of the sufficient hypotheses in the paper says that the power of the requisite line bundle coming into play should be sufficiently high, but it is not clear whether the given bound coincides with the bound required in the Guillemin-Sternberg result.) Among the tools is the authors' residue formula [Topology **34** (1995), no. 2, 291–327; MR 97g:58057], applied to the Todd class of M_{red}, the Todd class being given by Kawasaki's Riemann-Roch formula for orbifolds when K does not act freely on the zero locus. This suffices when M is actually a complex manifold; in the more general case where M carries only an almost complex structure, Vergne's extension [Duke Math. J. **82** (1996), no. 3, 637–652; MR 97j:58150] of Kawasaki's Riemann-Roch theorem is used, with the appropriate K-theory interpretation of the Riemann-Roch numbers. A formula for the K-invariant Riemann-Roch number is obtained by means of the holomorphic Lefschetz formula, which is a special case of the Atiyah-Segal-Singer equivariant index formula. The resulting formulas for the two Riemann-Roch numbers are then shown to coincide in many circumstances.

Several other papers have recently appeared or will appear which extend the Guillemin-Sternberg result, and in which the main tool is localization. See the review of the cited paper of Jeffrey and Kirwan [op. cit.; MR 97i:58064]. By means of a different method based on E. Lerman's technique of symplectic cutting [Math. Res. Lett. **2** (1995), no. 3, 247–258; MR 96f:58062], Duistermaat et al. proved a version of the quantization conjecture for circle actions [Math. Res. Lett. **2** (1995), no. 3,

259–266; MR 96g:58178]. The same technique enabled E. Meinrenken [Adv. Math. **134** (1998), no. 2, 240–277; MR 99h:58179] to establish the quantization conjecture for arbitrary nonabelian groups provided the almost complex structure is positive with respect to the symplectic structure. In the paper under review, positivity is not assumed and the quantization conjecture is shown to hold in some situations where the almost complex structure is not necessarily positive, for a torus and for certain nonabelian groups. Essentially the same result for the case of a torus action may be found in the already cited papers of Vergne [op. cit.; MR 98e:58087, see p. D92 in this volume]. The known counterexamples to the quantization conjecture are in fact for the case of an action of a nonabelian group and non-positive almost complex structure. In a somewhat different direction, the Guillemin-Sternberg result has been extended by R. Sjamaar [Ann. of Math. (2) **141** (1995), no. 1, 87–129; MR 96a:58098] for the case where 0 is not a regular value, under certain additional hypotheses which ensure that the reduced space acquires the structure of a Kähler quotient. A generalization thereof, based on the technique of symplectic cutting, has recently been obtained by Meinrenken and Sjamaar [Topology **38** (1999), no. 4, 699–762] for genuinely singular quotients which are not necessarily Kähler.

Johannes Huebschmann (Villeneuve d'Ascq)

98b:58096 58F10

Hayashi, Shuhei (Shinjuku)

Connecting invariant manifolds and the solution of the C^1 stability and Ω-stability conjectures for flows.

Ann. of Math. (2) **145** (1997), *no.* 1, 81–137.

This paper contains both a very powerful new perturbation result (the connecting lemma) and an application to prove the long-outstanding conjecture on the converse to structural stability for flows. These results are in the nature of determining which properties of dynamical systems persist under perturbation. This subject has a long history going back to the work of H. Poincaré and A. A. Andronov. In the 1960's this perspective on dynamical systems was resurrected with the work of M. Peixoto about the structural stability of certain flows on compact two-manifolds. Peixoto interested S. Smale in the problem and an outburst of results followed. By the early 1970's, the sufficient conditions for structural stability were established for both diffeomorphisms and flows with the results of Smale, J. Palis and Smale, D. V. Anosov, J. Moser, J. Robbin, and Robinson. This led to the obvious question of whether these same conditions (of hyperbolicity and transversality) were also necessary for structural stability. This question became know as the stability conjecture.

After some early work by V. A. Pliss and J. Franks, R. Mañé and S. T. Liao led the way with the best results. By 1988, Mañé had proved that these conditions were necessary for the structural stability of diffeomorphisms [Inst. Hautes Études Sci. Publ. Math. No. 66 (1988), 161–210; MR 89e:58090]. Along the way he had developed several perturbation results, including the ergodic closing lemma and a technique to create new homoclinic connections. These technical aspects of the proofs were very complicated although the general outline was clear. Finally Palis proved that any Ω-stable diffeomorphism has a hyperbolic chain recurrent set [Inst. Hautes Études Sci. Publ. Math. No. 66 (1988), 211–215; MR 89e:58091].

It was felt that the general approach for diffeomorphisms should apply to flows (differential equations) but there were several new aspects which had to be overcome. S. Hu [Trans. Amer. Math. Soc. **342** (1994), no. 2, 753–772; MR 94f:58075] was able to prove the conditions were necessary for flows in three dimensions. Again, to prove the conjecture in higher dimensions, Hayashi developed a new perturbation result, the connecting lemma. This theorem not only allows him to prove the stability conjecture for flows in all dimensions, but also simplifies the technical details of the proof for diffeomorphisms. Finally, it has been proved that for both diffeomorphisms and flows, necessary and sufficient conditions for structural stability are the hyperbolicity of the nonwandering set and transversality of stable and unstable manifolds.

One indication of the importance of this result of Hayashi is that there has already been another proof by L. Wen which gives a new complete presentation of the proof using the connecting lemma [J. Differential Equations **129** (1996), no. 2, 334–357; MR 97j:58082].

Hayashi's contribution is not only the final important step in finishing the proof of the stability conjecture, but also the development of the new perturbation technique, the connecting lemma. This method should prove very important in determining more precisely what dynamics is robust under perturbations. The connecting lemma is in the spirit of the closing lemma by C. Pugh [C. C. Pugh and C. Robinson, Ergodic Theory Dynam. Systems **3** (1983), no. 2, 261–313; MR 85m:58106]. In fact, the connecting lemma is a major extension of the closing lemma and uses the closing lemma in its proof.

The simplest form of the connecting lemma to understand relates to closing stable and unstable manifolds. Suppose a diffeomorphism or flow f has two periodic points p and q such that the unstable manifold of the orbit of p accumulates on the stable manifold of the orbit of q, $\mathrm{closure}(W^u(\mathcal{O}(p),f))\cap W^s(\mathcal{O}(q),f)\neq\varnothing$. The connecting lemma says that there is a C^1 approximation of f by g such that there is an actual homoclinic intersection of the manifolds, $W^u(\mathcal{O}(p),g)\cap W^s(\mathcal{O}(q),g)\neq\varnothing$.

The connecting lemma has already been crucial in recent unpublished work by C. A. Morales, M. J. Pacífico, and E. Pujals for studying robust invariant sets for flows in three dimensions. These robust systems are shown to have a partially hyperbolic structure and some examples resemble the Lorenz attractor. Z. Xia has also adapted the connecting lemma to symplectic diffeomorphisms and Hamiltonian flows to obtain new generic properties for these systems [Comm. Math. Phys. **177** (1996), no. 2, 435–449; MR 97d:58154]. In the future, other such applications of the connecting lemma should be abundant. *Clark Robinson* (Evanston, IL)

99c:58122 58F15 58F03

Nowicki, Tomasz (Warsaw); **Sands, Duncan** (Stony Brook, NY)

Non-uniform hyperbolicity and universal bounds for S-unimodal maps. (English summary)

Invent. Math. **132** (1998), *no.* 3, 633–680.

In dynamical systems, stochastic behavior is usually associated with positive Lyapunov exponents. For unimodal maps, it is enough to assume that the lower Lyapunov exponent at the critical value is positive (the Collet-Eckmann condition) in order to obtain stochastic properties for the map, such as the existence of an absolutely continuous invariant probability measure with exponential decay of correlations. The Collet-Eckmann condition is abundant in families of one-dimensional maps (see the reference list in the paper along with [M. Tsujii, Invent. Math. **111** (1993), no. 1, 113–137; MR 93j:58081]). The authors prove a very important result about the Collet-Eckmann condition for unimodal maps with negative Schwarzian derivative. Namely, the condition is equivalent to a positive lower bound on Lyapunov exponents for all periodic orbits, the so-called uniform hyperbolic structure. Over the course of the decade of research, a number of properties of S-unimodal maps were shown to follow from the Collet-Eckmann condition. Many of them, in turn, were shown to imply the uniform hyperbolic structure. The result of the paper implies that all these conditions are in fact equivalent.

Shortly after the results of this paper became known, it was observed by Nowicki and F. Przytycki [Fund. Math. **155** (1998), no. 1, 33–43; MR 99a:58058] that this family of conditions includes a purely topological one. Hence, the topological invariance of the Collet-Eckmann condition has followed for unimodal maps.

From the technical standpoint, the paper is based on a refined technique. A certain sequence of iterations, that of so-called cutting and co-cutting times, is considered in the vicinity of the critical value of the map. Using the cross-ratio technique, precise geometric estimates are derived, often within a factor of 2 or so from observed ratios. This outstanding technique has been used by Sands to obtain an estimate of the so-called "a priori bounds", crucial in applications of complex analytic-methods (see a derived argument in [J. Graczyk and G. Świątek, *The real Fatou conjecture*, Ann. of Math. Stud., 144, Princeton Univ. Press, Princeton, NJ, 1998]).

In addition to the weight of the result and its technical significance, the quality of the paper is further enhanced by the careful way in which it is written, with all details included or clearly quoted from other sources. Hence any patient reader, not necessarily a specialist, should be able to read it. The paper is among the most important and best written in one-dimensional dynamics in the present decade.

Grzegorz Świątek (University Park, PA)

99h:58140 58F15 58F11
Young, Lai-Sang (Los Angeles, CA)
Statistical properties of dynamical systems with some hyperbolicity.
Ann. of Math. (2) **147** (1998), *no.* 3, 585–650.

This paper contains a breakthrough result in the study of the decay of correlations for hyperbolic systems with singularities. Hyperbolic systems (characterized by the absence of zero Lyapunov exponents) include many classical and modern models of statistical physics: Lorentz gases, hard ball systems, certain chaotic attractors. While smooth uniformly hyperbolic systems are very well understood, this is not the case for many physical models, which are not uniformly hyperbolic and contain singularities in the phase space. One of the most important issues here is the behavior of the time correlation function, which is directly related to the basic laws of statistical mechanics.

It was long conjectured that singularities in hyperbolic systems, e.g., in Lorentz gases and other Sinaĭ billiards, might slow down the decay of correlations from exponential to subexponential. That would imply that statistical properties of these models are weak and physical laws could be hard to prove or even observe experimentally. Many researchers have computed the correlation function numerically, but the data obtained were compatible with both exponential and subexponential laws, and the controversy persisted for more than a decade. This paper proves that correlations actually decay exponentially for hyperbolic maps with singularities under quite general assumptions. The main result of this paper covers certain Sinaĭ billiards, including planar periodic Lorentz gases with finite horizon, and this is also shown in the paper.

The content of the main theorem is the following, technical details being left out. Let $f: M \to M$ be a nonuniformly hyperbolic map, possibly with singularities. Let $\Gamma \subset M$ be a "hyperbolic product structure", i.e. the intersection between a family of unstable manifolds and a family of stable manifolds. The set Γ is iterated under f^n, $n \geq 1$, until its images properly cross Γ itself (in the sense that the intersection stretches completely across Γ in the unstable direction). Every time a proper intersection occurs, the points of $f^n\Gamma$ coming back to Γ "make a return". The points that make a return are taken out of circulation, and the rest of $f^n\Gamma$ is iterated further, until proper intersections with Γ occur again, etc. The main assumption of the theorem is that the fraction of Γ that does not make a return until time n is exponentially small in n (with respect to the Lebesgue measure on unstable manifolds). Under this assumption, the author proves that an f-invariant Sinaĭ-Ruelle-Bowen measure on M exists, enjoys exponential decay of correlations and satisfies a central limit theorem.

The author applies the main theorem to dispersing Sinaĭ billiards, Axiom A attractors and logistic maps of the interval. The theorem has potentially wide applications to other chaotic deterministic systems of physical origin.

Nikolai Chernov (Birmingham, AL)

98d:58141 58F17 53C22

Mañé, Ricardo

On the topological entropy of geodesic flows.

J. Differential Geom. **45** (1997), *no.* 1, 74–93.

Ricardo Mañé was born in Montevideo on January 14, 1948 and did most of his mathematical work at IMPA in Rio de Janeiro. Among his outstanding contributions to dynamical systems were his solution of the stability conjecture and the book *Ergodic theory and differentiable dynamics* [English translation, Springer, Berlin, 1987; MR 88c:58040]. He died in Montevideo on March 9, 1995 (while the present paper was being refereed).

When Mañé was a student at the University of Uruguay, Jorge Lewowicz got him interested in mathematics, particularly dynamical systems, and also put him in contact with Jacob Palis at IMPA in Rio, whose student Mañé then became. He was to stay at IMPA for his entire career. His 1973 thesis was a precursor to his main achievement; it showed that normal hyperbolicity, which was known to be sufficient for persistence of invariant manifolds, is also necessary [Bull. Amer. Math. Soc. **80** (1974), 90–91; MR **49** #4043; Trans. Amer. Math. Soc. **246** (1978), 261–283; MR 80c:58019]. That year C. Robinson [J. Differential Equations **22** (1976), no. 1, 28–73; MR **57** #14051] sharpened the hypotheses of a result of J. W. Robbin [Ann. of Math. (2) **94** (1971), 447–493; MR **44** #4783], showing that hyperbolicity of recurrent orbits combined with transversality conditions on transient orbits is sufficient for structural stability of a dynamical system (i.e., the rigidity of the topological type of the orbit structure under C^1 perturbation). This settled half of the 1968 Palis-Smale stability conjecture that these conditions are necessary and sufficient for structural stability. Until 1986 only various partial results towards necessity were known. Then Mañé proved the stability conjecture in a masterpiece [Topology **17** (1978), no. 4, 383–396; MR 84b:58061; Inst. Hautes Études Sci. Publ. Math. No. 66 (1988), 161–210; MR 89e:58090] which certainly qualifies as one of the crowning achievements of the theory of smooth dynamics and which he described as the focal point of his contributions to dynamical systems. These range widely with over 40 papers and a major book [*Introduction to ergodic theory* (Portuguese), Inst. Mat. Pura Apl., Rio de Janeiro, 1983; MR 87d:58085; op. cit.; MR 88c:58040] in twenty years.

Further highlights of his work are "An ergodic closing lemma" [Ann. of Math. (2) **116** (1982), no. 3, 503–540; MR 84f:58070], finite-dimensionality of invariant sets of maps of Banach spaces whose noncontracting part is compact [in *Dynamical systems and turbulence, Warwick 1980 (Coventry, 1979/1980)*, 230–242, Lecture Notes in Math., 898, Springer, Berlin, 1981; MR 84k:58119], elements of Pesin theory in Banach spaces [in *Geometric dynamics (Rio de Janeiro, 1981)*, 522–577, Lecture Notes in Math., 1007, Springer, Berlin, 1983; MR 85j:58126], and prevalence of zero Lyapunov exponents outside of partially hyperbolic systems, a work which was never finished (it was announced in his 1983 ICM address [in *Proceedings of the International Congress of Mathematicians, Vol. 1, 2 (Warsaw, 1983)*, 1269–1276, PWN, Warsaw, 1984; MR 87e:58127 (Theorem C and its corollary)] and he wrote an outline for a seminar in Paris [in *International Conference on Dynamical Systems (Montevideo, 1995)*, 110–119, Longman, Harlow, 1996; MR 98d:58105]). Characteristically, his only foray into one-dimensional dynamics produced a famous result, remarkable for its generality; it says that a compact set forward invariant under a C^2 interval map and not containing critical points, stable or indifferent periodic orbits, is hyperbolic, i.e., expanding [Comm. Math. Phys. **100** (1985), no. 4, 495–524; MR 87f:58131; erratum; MR 88i:58139]. In complex dynamics he distinguished himself by joint work with P. Sad and D. P. Sullivan [Ann. Sci. École Norm. Sup. (4) **16** (1983), no. 2, 193–217; MR 85j:58089] which essentially set the stage for using complex dynamics to settle the density of Axiom A for quadratic maps. A joint result with A. Freire [Invent. Math. **69** (1982), no. 3, 375–392; MR 84d:58063] which extended A. Manning's result [Ann. of Math. (2) **110** (1979), no. 3, 567–573; MR 81e:58044] about equality of topological entropy and volume growth entropy of geodesic flows on nonpositively curved manifolds to the situation with no conjugate points, is interesting in many regards, for example with respect to G. Besson, G. Courtois and S. Gallot [Geom. Funct. Anal. **5** (1995), no. 5, 731–799; MR 96i:58136]. The present paper re-proves it as a corollary of its main results. Finally, his 1983 book [op. cit.; MR 87d:58085] was the first introduction to smooth ergodic theory and remains important today.

At IMPA Mañé directed eleven doctoral theses in various areas of dynamics and from there he visited several of the leading research instutions of North America and Europe, such as the University of North Carolina, Northwestern University, IAS Princeton, MSRI Berkeley, IHES Bures-sur-Yvette. He gave invited addresses at the International Congresses in Warsaw and Zürich.

In this paper, Mañé examines the number $n_T(x,y)$ of geodesic segments of length at most T with given endpoints on a compact C^∞ Riemannian manifold M and shows that the average N_T over $M \times M$ has exponential growth rate equal to the topological entropy h of the geodesic flow. In the absence of conjugate points there is no need to take the average and one furthermore gets coincidence with the exponential rate of the volume growth of balls on the universal cover. Even in the presence of conjugate points, $\limsup_T T^{-1} \log n_T(x,y) \le h$ for every $x \in M$ and a.e. $y \in M$ due to an inequality of G. P. Paternain [Ergodic Theory Dynam. Systems **12** (1992), no. 1, 109–121; MR 93g:58117]. This leads Mañé to ask whether $\lim_T T^{-1} \log n_T(x,y) = h$ for a.e. $(x,y) \in M \times M$, at least for a generic metric on a surface M. (That "a.e." is necessary is due to the possibility of conjugate points and even away from these the growth can be very bad [K. Burns and G. P. Paternain, in *International Conference on Dynamical Systems (Montevideo, 1995)*, 7–20, Longman, Harlow, 1996; MR 98g:58137].) This question has been answered in the negative by Burns and Paternain [Ergodic Theory Dynam. Systems **17** (1997), no. 5, 1043–1059], who further develop the subject in a forthcoming paper ["Geodesic arcs and hyperbolic measures", in preparation].

To obtain the main result, comparisons with $\mathrm{Ex}(g^T) := \int_{SM} \mathrm{ex}(d_\theta g^T)\, d\theta$, where $\mathrm{ex}(L) := \max_S |\det(L \restriction S)|$ (the maximum taken over subspaces), are made. The proof shows $\liminf_T T^{-1} \log \mathrm{Ex}(g^T) \le \liminf_T T^{-1} \log N_T$ and $\limsup_T T^{-1} \log \mathrm{Ex}(g^T) \le \limsup_T T^{-1} \log N_T \le h$. This implies the result by F. Przytycki's inequality $h \le \liminf_T T^{-1} \log \mathrm{Ex}(g^T)$ [Invent. Math. **59** (1980), no. 3, 205–213; MR 82a:58032].

The third of these inequalities is proved using a method of Paternain [op. cit.; MR 93g:58117] which invokes Y. Yomdin's solution [Israel J. Math. **57** (1987), no. 3, 285–300; MR 90g:58008] of the Shub entropy conjecture [M. Shub, Bull. Amer. Math. Soc. **80** (1974), 27–41; MR **48** #12603]; this is where the C^∞ assumption is, as yet, indispensable. The preceding two inequalities are the main novelty of this paper. To obtain them, a preliminary step is (roughly) to replace N_T by a number $A(T)$ appearing in work of M. Berger and R. Bott [Topology **1** (1962), 301–311; MR **26** #4296], then to identify a Lagrangian subspace R where the max defining ex is attained, and finally to show that when $g^T R$ is sufficiently transverse to the vertical subspace then A majorizes ex. The core idea is that the required transversality can be achieved by easily controlled small changes in time because the set of times of nontransversality is discrete. *Boris Hasselblatt* (Medford, MA)

98a:58128 58F18 32L30 34A20 34C05

Berthier, M. (Rennes); **Cerveau, D.** (Rennes); **Lins Neto, A.** (Rio de Janeiro)

Sur les feuilletages analytiques réels et le problème du centre. [Real-analytic foliations and the problem of the center]

J. Differential Equations **131** (1996), *no.* 2, 244–266.

This remarkable paper concerns a subject that is traditionally important in the local analytic theory of planar vector fields. There is a general rule that algebraic properties of the projective (also known as hidden or vanishing) holonomy group of a singular point of an analytic vector field reflect analytic properties of that field, and the same applies to singular analytic foliations of the complex plane. The authors have found yet another spectacular manifestation of this principle; in particular, they show how integrability in a special class of functions and reversibility can affect the projective holonomy.

Formally the paper is devoted to the single goal of constructing an example of a center (a real-analytic foliation in a neighborhood of an isolated singular point with identity return map) that admits neither a Liouville first integral nor a nontrivial symmetry. This example is important since in all known cases when a singular point of a planar vector field is a center, it either possesses a Liouville-type first integral, or is reversible and hence possesses an axis of symmetry. (Some time ago H. Żołądek conjectured that this is always the case, at least for polynomial vector fields, but later he retracted his claim.)

To construct the required example, the authors first show that any group generated by an even number of germs of holomorphisms that come in symmetric pairs can be realized as a projective holonomy group of a real monodromic foliation with an isolated singularity. Unfortunately, this theorem is wrong in such generality (more about the gap below). Still, a weaker fact is true: if in addition at least one germ among the generators of the group is analytically linearizable, then the group can be realized as a projective holonomy. With this modification, the further construction goes through with no changes and the other results of the paper are not affected by this error. I am grateful to Yu. Il′yashenko, who pointed me to his recent paper [Proc. Steklov Inst. Math. **213** (1996), 6–29] revealing the error.

The two claims that follow the realizability theorem are very natural. It is proved that a vector field that admits a Liouville function as a first integral must have a solvable holonomy group. Next, the authors show that multipliers of generators of the holonomy group of a reversible (or, more precisely, foldable) field must exhibit multiplicative integer dependencies. Using these necessary conditions, the authors construct an example of a group that cannot be the projective holonomy of a Liouville integrable or reversible vector field. The field itself is then constructed using (the corrected form of) the realization theorem.

The paper is so elegant that its basic notions and ideas can be explained in the limited space allowed for this review. Below follows a brief description of various stages of the construction.

The main objects of study are a real-analytic 1-form $\omega = f(x,y)dx + g(x,y)dy$ having an isolated singularity at the origin on the plane $\mathbf{R}^2$, and a foliation $\mathcal{F}_\omega$ whose leaves are tangent to the distribution $\{\omega = 0\}$. This form can be complexified. Consider the blow-up π of the origin on the complex plane $\mathbf{C}^2$: if the form is not diacritical, then the exceptional divisor (the complex projective line $\pi^{-1}(0)$) becomes a leaf $\mathcal{L}$ of the pullback form $\pi^*\omega$ after a finite number of points (the singularities of the blown-up form) are deleted. Once an arbitrary transversal T_a to $\mathcal{L}$ at some point $a \in \mathcal{L}$ is chosen, the procedure of analytic continuation of leaves of the foliation $\mathcal{F}_{\pi^*\omega}$ over different loops in $\mathcal{L}$ maps the fundamental group of the leaf $\mathcal{L}$ into a group of holomorphic automorphisms of (T_a, a): the holonomy transformation corresponding to a loop γ is a holomorphic germ $h_\gamma\colon (T,a) \to (T,a)$. This group, defined up to a holomorphic conjugacy, is the projective holonomy of the foliation $\mathcal{F}_\omega$ (or the form ω itself). Without loss of generality one can always assume that the projective holonomy group is generated by the holonomy transformations of small loops around all deleted points (the only relation between them being that the composition of all germs is the identity).

The first problem addressed by the authors is the realization problem in the real category: when is a group generated by finitely many holomorphic germs $h_k\colon (\mathbf{C},0) \to$

$(\mathbf{C},0)$, $k=1,\cdots,n$, satisfying the identity $h_1\circ\cdots\circ h_n=\mathrm{id}$ the projective holonomy group of an appropriate real form ω?

A prototype for this problem is the question of the existence of a (nondegenerate) singular point with specified monodromy (the holonomy of the analytic invariant curve tangent to an eigenvector). For example, any linear map $z\mapsto vz$, $v=2\pi i\alpha$, for any choice of $\alpha\in\mathbf{C}$ is the monodromy map of the axis $\{w=0\}$ for the foliation $\{\alpha w\,dz-z\,dw=0\}$ (note that α can be replaced by $\alpha+k$ for any $k\in\mathbf{Z}$). As a corollary, one can easily construct an analytic singular foliation having an invariant analytic curve with any specified linearizable monodromy.

Essentially, nonlinear maps tangent to the linear map $z\mapsto vz$ with $v^n=1$ for some n (the resonant case) can be realized as monodromies by the now-classical result of J. Martinet and J.-P. Ramis [Ann. Sci. École Norm. Sup. (4) **16** (1983), no. 4, 571–621 (1984); MR 86k:34034]; see also the paper by P. M. Elizarov and Il′yashenko [Mat. Sb. (N.S.) **121(163)** (1983), no. 1, 111–126; MR 84i:58084].

In the remaining case of nonresonant maps that are not analytically linearizable ("wild"), the realization theorem was obtained by R. Pérez Marco and J.-C. Yoccoz [Astérisque No. 222 (1994), 7, 345–371; MR 96b:58090]. Another approach to this problem was used by Elizarov, Il′yashenko, A. A. Shcherbakov and S. M. Voronin [in *Nonlinear Stokes phenomena*, 57–105, Amer. Math. Soc., Providence, RI, 1993; MR 94e:32055]. Notice that in this case the vector field tangent to the foliation must be of saddle type with eigenvalues $(1,\alpha)$, with α real negative (otherwise the field is analytically linearizable and hence its monodromy cannot be wild by the Poincaré linearization theorem).

One can obtain the global realization theorem by gluing the singular foliations for each generator of the group, thus obtaining an abstract line bundle over $\mathbf{CP}^1$ and a singular foliation on it with the specified holonomy group (the construction actually goes back to G. D. Birkhoff [*Collected mathematical papers (in three volumes). Vol. I*, Dover, New York, 1968; MR **38** #4271a; *Vol. II*; MR **38** #4271b; *Vol. III*; MR **38** #4271c]). To make sure that this bundle is isomorphic to the canonical line bundle of the exceptional divisor, one has to verify that the Chern class of this constructed bundle is indeed -1, as it should be for the canonical bundle. This is wrong in general, as noticed by Il′yashenko (see below). However, if at least one of the generators is linearizable, one can use the freedom of choice of the eigenvalue α to satisfy this topological condition. This construction was implemented in detail by Lins Neto [J. Differential Geom. **26** (1987), no. 1, 1–31; MR 88f:32047]. The result is the theorem claiming that any group generated by germs $h_1,\cdots,h_k$ satisfying the condition $h_1\circ\cdots\circ h_k=\mathrm{id}$ can be realized as the holonomy of an analytic singular foliation $\{\omega=0\}$, provided that at least one germ h_j is analytically linearizable. An example showing that a group generated by three wild germs cannot be realized as a projective holonomy was given by Il′yashenko [op. cit., 1996]. Alternatively, one can add to the list of generators the identical monodromy germ realized by the nontrivial foliation $\{nw\,dz-z\,dw=0\}$, $n\in\mathbf{Z}$, thus forcing the Chern class of the constructed abstract bundle to become -1, by a proper choice of n.

As soon as the topological obstruction is removed, one can blow down the resulting singular foliation and construct the form ω that would realize the specified monodromy. However, without additional considerations the form ω may be non-real and hence will not define any (singular) foliation on the real plane $\mathbf{R}^2$. The first basic result of the authors is a (necessary and sufficient) condition for realization of a group of germs as the projective holonomy of a real form.

If the original form ω is real, then the leaf $\mathscr{L}$ should be symmetric with respect to the complex conjugacy $z \mapsto \overline{z}$ (z being the chart on the exceptional divisor), which means that all deleted points come in pairs. Moreover, provided that the transversal T was chosen real, the holonomy maps should also come in pairs: if γ is a small loop around a deleted point $p \in \mathbf{CP}^1$ and $\overline{\gamma}$ a symmetric loop around $\overline{p}$, then $h_\gamma(z) = \overline{h_{\overline{\gamma}}(\overline{z})}$. Finally, for a monodromic real singularity (centro-focal configuration) the real line $\mathbf{RP}^1 \subset \mathbf{CP}^1$ on the exceptional divisor should be free of singular points after the blow-up, and one can easily see that the composition of all holonomy operators corresponding to singular points in each hemisphere should be a germ h that is real and such that $h'(0) < 0$ (the real monodromy of the singular point is its square $h \circ h$). These necessary conditions are also sufficient for realizability in the real-analytic context, as the following result shows. Theorem. Any group generated by $2p$ analytic germs $\{h_1(z), \cdots, h_p(z), \overline{h_1(\overline{z})}, \cdots, \overline{h_p(\overline{z})}\}$, such that the composition $h = h_1 \circ \cdots \circ h_p$ is real and $h'(0) < 0$, can be realized as the projective holonomy group of a real-analytic form ω with an isolated monodromic singularity, provided that at least one of the germs h_k is analytically linearizable.

To prove this theorem, the authors first construct a non-real foliation with the specified holonomy, as described above. Then they construct an antiholomorphic involution of a neighborhood of the exceptional divisor (first outside the singular points, and then showing that the result can be extended onto these points). The claim that this involution is analytically conjugate to the standard one, $z \mapsto \overline{z}$, is very simple.

The second part of the paper describes the projective holonomy of a form that admits a Liouville first integral. Recall that a Liouville function is a function that can be obtained from meromorphic function(s) by a finite number of algebraic operations, integration and taking exponentials. The following is a result in the spirit of differential Galois theory.

Theorem. If a form ω possesses a Liouville first integral, then the projective holonomy group of this form is solvable.

This theorem is proved by reasoning with formal power series. More precisely, one can prove that if the form ω admits a Liouville first integral, then after blow-up and division by an appropriate product of the form $\exp h \times \prod_{i=1}^{p} f_i^{\lambda_i}$, $\lambda_i \in \mathbf{C}$, f_i, h holomorphic, and restriction on the transversal T equipped with the chart z, this form is formally conjugate to the exact form dG, where $G(z) = z^\gamma \exp(V(z)/z^p)$ and V is a formal series in z. One can show that all holonomy operators must preserve G up to a multiplicative constant. The corollary is that any element from the holonomy group that is tangent to identity (e.g., any commutator) must preserve the form $z^{-(p+1)}(\gamma z^p + 1)\, dz$. It is known then that such groups are solvable [D. Cerveau and R. Moussu, Bull. Soc. Math. France **116** (1988), no. 4, 459–488 (1989); MR 90m:58192]. It is interesting to compare this result with the old (1971) result by A. Khovanskiĭ, recently published in detail [J. Dynam. Control Systems **1** (1995), no. 1, 91–123; MR 96c:34012] and which asserts that the monodromy group of a Liouville function in one variable is necessarily solvable.

The next step is to establish the necessary condition for reversibility of a vector field tangent to a foliation. The usual definition is that the foliation is symmetric with respect to some line or, more generally, analytically equivalent to a symmetric one. The authors extend the class of non-symmetric foliations by introducing non-foldable foliations (we prefer this to the authors' term "irreducible foliations" since the latter is burdened by irrelevant associations). By definition, this means that if $\mathcal{F}_\omega$ is a pullback of some other foliation $\mathcal{F}'$ by a holomorphic map $H: (\mathbf{C}^2, 0) \to (\mathbf{C}^2, 0)$, then H must necessarily be locally analytically invertible, unless it squeezes down entire curves to points (a foliation symmetric by the y-axis in $(\mathbf{C}^2, 0)$ can be obtained as a pullback of another foliation by the non-invertible fold $(x, y) \mapsto (x^2, y)$).

It is rather clear that if a foliation with isolated singularities is foldable and has analytic separatrices through the singularity, then these separatrices when folded must be taken into separatrices of the folded foliation and conversely, the preimages of separatrices also must be separatrices. Moreover, it is obvious that if some separatrix is folded into another separatrix, then its monodromy (not to be confused with the generators of the projective holonomy: this is not a "hidden" but rather a "visible" map) should be a superpositional root of the monodromy of the corresponding image. Thus, if two different separatrices are folded into the same curve, then their multipliers λ, μ should be multiplicatively dependent: $\lambda^n = \mu^m$ for some natural n, m. As a conclusion, one has the following condition: if a foliation possesses at least three separatrices with different limit tangencies and multiplicatively independent linearizations of the corresponding monodromies, then the foliation is not foldable, and hence cannot have a symmetry axis.

It finally remains to produce the counterexample. The authors use a very difficult algebraic theorem by S. D. Cohen [Quart. J. Math. Oxford Ser. (2) **46** (1995), no. 181, 21–93; MR 96e:20033] claiming that the shift $z \mapsto z+1$ and the power $z \mapsto z^p$, $1 \neq p \in \mathbf{N}$, generate a free group. From this theorem it follows that the two analytic germs $h_1: z \mapsto z/(z+1)$ and $h_2: z \mapsto z/(1+z^p)^{1/p}$ generate a free group and hence any group containing these two germs cannot be solvable. This implies that the germs $\lambda_1 h_1, \lambda_2 h_2$ also generate the free group of rank 2 for almost all $\lambda_1, \lambda_2 \in \mathbf{C}$. We add the third germ $h_3 = -h_2^{-1} h_1^{-1}$ and then add three more germs symmetric to the previous three. Then this group generated by three pairs of germs can be realized by a real-analytic foliation, since for a generic choice of λ_1 the first germ will be linearizable. By construction, the projective holonomy is non-solvable and hence there cannot be any Liouville first integral. The (Poincaré) monodromy of this foliation is the square of $z \mapsto -z$, i.e. the configuration is central. Finally, one can easily verify that the separatrices corresponding to the singularities on the exceptional divisor have independent multipliers for generic values of λ_i. Hence the foliation cannot be foldable, and the construction is finished. *Sergeĭ Yu. Yakovenko* (Rehovot)

99a:58130 58F20 30D05 58F03 58F23
Pérez-Marco, Ricardo (Orsay)
Fixed points and circle maps.
Acta Math. **179** (1997), *no.* 2, 243–294.

The problem of the center, that is, the so-called "Zentrumsproblem", deals with holomorphic dynamics near an indifferent fixed point. The first contributions were made by H. Cremer in 1938 and, later, by C. L. Siegel in 1942. As V. I. Arnol′d pointed out in 1961, there is a heuristic relation between this problem and the dynamics of analytic circle homeomorphisms: "The problem of the center is a singular case of the problem of a map of the circumference into itself whose radius, in the singular case, is equal to zero...." The main purpose of the present article is to make rigorous the heuristic relation which has long been known and was recently illustrated by the work of Fields medallist J.-C. Yoccoz. Actually, the author presents a dictionary of the connections between the two problems.

Contributions to both areas are typically long and technical, and deal with difficulties associated with "small divisors". In order to avoid repetition of proofs, the author has developed a geometric construction which turns out to be a powerful tool.

Theorem. Let $f(z)=\exp(2\pi i\alpha)\cdot z+O(z^2)$, where $\alpha\in\mathbf{R}$, be a local holomorphic diffeomorphism. Let U be a Jordan neighborhood of the indifferent fixed point 0. Assume that f and f^{-1} are defined and univalent on a neighborhood of the closure of U. Then there exists a set K such that (i) K is compact, connected and full, (ii) $0\in K\subset\overline{U}$, (iii) $K\cap\partial U\neq\varnothing$, and (iv) K is totally invariant with respect to f. Moreover, if f is not of finite order, f is linearizable at 0 if and only if $0\in\operatorname{int}(K)$.

The set K is called a "Siegel compactum" and should be viewed as a degenerate Siegel disc, just as Aubry-Mather sets are related to KAM invariant curves for a twist map of the annulus. If the fixed point is irrationally indifferent and K is *not* contained in the closure of a linearization domain, then K is called a "hedgehog".

The existence of forward (or backward) invariant sets was established by Birkhoff in 1922. The advantage of the above theorem is that it allows one to obtain a totally invariant compactum. Using this compactum, Carathéodory's extension theorem, and Schwarz' reflection principle, the author associates an analytic circle diffeomorphism g to an indifferent fixed point. Finally he proves the crucial fact that the rotation number of g equals α.

Included in this article are further substantial contributions, e.g. a proof of Dulac-Moussu's conjecture, as well as a number of new and more natural proofs of known results, e.g. Naishul's theorem and the snail lemma.

The reviewer has not stated all the results contained in this fine piece of work. Instead the reader is strongly encouraged to look at the paper.

Hartje Kriete (Göttingen)

98e:58145 58F23 30C10 30D05

Lyubich, Mikhail (Stony Brook, NY)

Dynamics of quadratic polynomials. I, II.

Acta Math. **178** (1997), *no.* 2, 185–247, 247–297.

A famous Polish mathematician and one-time editor of Fundamenta Mathematicae is sometimes quoted as saying that he had never accepted an article more than eight pages in length, because a longer proof could not possibly be correct. The reviewer has never written an article which was less than nine pages, so needless to say, he doesn't agree. But as many a provocative overstatement, this one has some truth to it, particularly if eight is replaced with eighty. From writing a few papers longer even than that, I recall the certain numbness of mind which might set in on page seventy-seven, when I no longer could remember notations or definitions from the first ten pages, and I didn't care enough to look them up. If one believes in miracles, such a paper may still be right. But will it ever find a devoted enough reader to be *read*, not just skimmed or quoted?

The paper by Lyubich has 113 pages and while lengthy by any standard it is hardly an exception in the field of conformal dynamics. In fact, some recent works in this area have never been published as articles, but appeared straightaway as books. Great complexity seems to be the staple of works in one-dimensional dynamics. Some papers from back in the 1980s, to mention just one [M. V. Jakobson, Comm. Math. Phys. **81** (1981), no. 1, 39–88; MR 83j:58070], were long and complicated when written, have not been much simplified since, and yet have been read and have given inspiration for further research. One could speculate about reasons for this state of affairs. First of all, proofs in dynamical systems tend to be complicated. They depend on the "picture" of the dynamics which exists in the author's mind and is typically not easy to explain to the reader. When such an explanation is undertaken it often leads to lengthy combinatorial, topological or geometric constructions. Somewhat paradoxically, the technical side of one-dimensional proofs also tends to be very complicated. There are a number of "principles" that have underlain the development of the 1990s, and in particular the present paper. Examples of these are "treat one-dimensional real and complex systems jointly, looking to explore the real and complex facets of the problem at hand", "use quasiconformal deformations and pullback", "try inducing or renormalization". Without getting into the exact meaning of these principles, they are not theorems or even well-defined techniques, but merely indications of directions in which to search. For reasons not completely understood, they work very well on concrete problems, but require work in order to be made into proofs. In addition to these general factors, the paper is also made lengthy by its ambitious scope. The four theorems stated in the introduction together with Corollary 1.1 cover a large part of the development of the field in the 1990s.

Among these results, the density theorem is probably the most famous. It states that in the logistic family $x \to ax(1-x)$ an attracting periodic orbit exists for values of a in a dense subset of $(0,4)$. The complex bounds theorem is a long-sought-for and powerful technical result. The statements of Theorem III and the rigidity theorem cannot be understood without going more deeply into the paper. Roughly, Theorem III means that high iterates of the quadratic polynomial become increasingly expanding (an example of the "inducing principle" first used by Jakobson [op. cit.]), and the rigidity theorem is an attempt at a generalization of the density theorem to a class of complex polynomials. Such a generalization is not likely to be relevant in the study of complex quadratic polynomials. The approach used in the paper requires complex bounds. These are true for real polynomials, but are generally not satisfied for infinitely renormalizable complex quadratic polynomials. At the top of page 188, the author observes that the Hausdorff dimension of the complex parameters to which his rigidity theorem applies is at least 1, the Hausdorff dimension of the boundary of the Mandelbrot set is 2, and 1 is one-half of 2. This is not convincing. Hausdorff dimension 1 is attained on the real line and being a half of the full dimension is hardly the sign of the set being robust, let alone typical.

The density theorem has another proof by J. Graczyk and the reviewer [see Ann. of Math. (2) **146** (1997), no. 1, 1–52; MR 99b:58079]. This surely affects the way I look at the present paper, thinking first of all similarities with and differences from my own work. The approaches of both proofs are very similar. Our work also uses a version of Theorem III (Theorem C in [J. Graczyk and G. Świątek, Ann. Sci. École Norm. Sup. (4) **29** (1996), no. 4, 399–482; MR 98d:58152]) and includes a proof of the complex bounds theorem, together with quasiconformal pullback techniques.

There are two main differences. The first concerns the technique of proof of the growth of moduli (Theorem III in the present paper). Both approaches use an invariant based on "separating annuli" but these invariants are quite different. The second difference occurs in the way partial conjugacies obtained on subsequent levels of renormalization for an infinitely renormalizable polynomial are pasted together. Our technique uses the fact that polynomials are real. Since the author's method covers some complex polynomials, he has to use a more subtle complex construction.

To read the paper critically will not be an easy task. The size is one obvious problem. Sometimes, a very lengthy paper may not be so hard to go through if its structure is modular, more like a sequence of related papers which can be checked independently. Such is not the case with the paper at hand. The proof depends on a number of constructions, techniques and combinatorial schemes which are the basis of a fairly sophisticated language developed by the author. A simple attempt of going through the statements of the paper named "theorems", and presumably representing the main steps of the proof, reveals that most of them cannot be understood by an expert without extensive reading through earlier parts of the paper. For example, we find Theorem 8.1 on page 241, which is attributed to Martens. The problem is, Martens never states any such result. He does not talk of "tails of long central cascades" or intervals I_m. As far as I understand the author's intention and Martens' paper, I believe that a proof of the statement needed for the present paper can be recovered from Martens' work. However, a critical reader has no choice but to go through Martens' work in some detail. Besides, the statement of Theorem 8.1 uses colloquial language and is not clear at all (are the bounds referred to uniform in some sense?). The impossibility of skimming through the paper without reading the whole might be particularly annoying for an expert, and imprecise statements will bother a devoted non-expert reader.

Nevertheless, it may be rewarding to go through the paper. The density theorem alone is worth 113 pages. Other results included in the paper (complex bounds and local connectivity) make the work more interesting still. After all, the density problem can be traced over seventy years back to Fatou, is actually relevant even to non-mathematicians, and represents a very rare circumstance in which we can make a statement about what is typical and what is not in a *whole class* of non-hyperbolic dynamical systems. It is true that most of the results of this paper can also be found in works by other authors [J. Graczyk and J. Świątek, op. cit., 1996; op. cit., 1997; G. Levin and S. J. van Strien, "Local connectivity of Julia sets of real polynomials", Preprint No. 5, Inst. Math. Sci., SUNY, Stony Brook, NY, 1995; per bibl.], but those are technically involved as well. The present paper represents a great deal of work and some elements of its approach are different and original.

The progress in conformal dynamics of the 1990s has been impressive if measured by problems solved. The price of these results has been the incredible complexity of proofs. For this reason, the achievements are less than completely satisfying. Hopefully, the works of the 1990s will help others to find better proofs as well as contribute to nonlinear dynamics in general.

Grzegorz Świątek (University Park, PA)

99e:58143 58F23 30D05 58F03

Levin, Genadi (Jerusalem); **van Strien, Sebastian** (Coventry)

Local connectivity of the Julia set of real polynomials. (English summary)

Ann. of Math. (2) **147** (1998), *no.* 3, 471–541.

If z^l+c is a unimodal polynomial with locally connected Julia set J then the dynamics on J has a particularly simple representation: it is semi-conjugate to the multiplication by l modulo 1 on $[0,1]$. The quest for local connectivity of Julia sets dates back to the work of A. Douady and J. H. Hubbard [*Étude dynamique des polynômes complexes. Partie I*, Univ. Paris XI, Orsay, 1984; MR 87f:58072a; *Étude dynamique des polynômes complexes. Partie II*, Univ. Paris XI, Orsay, 1985; MR 87f:58072b]. It was shown there that the Julia sets of complex unimodal polynomials with a parabolic or attracting cycle are locally connected. The breakthrough contribution came from J.-C. Yoccoz, who proved local connectivity of the Julia sets for all complex finitely renormalizable quadratic polynomials without neutral cycles; see the exposition by Hubbard [in *Topological methods in modern mathematics (Stony Brook, NY, 1991)*, 467–511, Publish or Perish, Houston, TX, 1993; MR 94c:58172]. In his proof, Yoccoz uses dynamically generated partitions (known as Yoccoz partitions) which nest to points and have connected intersections with the Julia set. This gives a base of locally connected topology for the corresponding Julia sets. However, a part of Yoccoz's reasoning, based on calculation of certain dynamically defined annuli, breaks down when applied to infinitely renormalizable polynomials or non-renormalizable polynomials of degree higher than 2. J. Hu and J. Jiang made progress in the specific case of real infinitely renormalizable quadratic polynomials of bounded combinatorial type ["The Julia set of the Feigenbaum quadratic polynomial is locally connected", Preprint, 1993; per bibl.]. Eventually, in the presence of additional symmetry (real unimodal polynomials), Levin and van Strien proved that Yoccoz partitions nest to points for all nontrivial combinatorics. This leads immediately to the first main result of the paper under review (Main Theorem): the Julia sets of real unimodal polynomials z^l+c are either locally connected or totally disconnected. The main part of the proof is carried out on the real line. By producing strong numerical estimates on the geometry of some dynamically defined points on the real line (real bounds), the authors obtain various types of so-called complex bounds. This and a careful analysis of Yoccoz partitions, specifically their real traces, conclude the proof. It should be mentioned that there are examples of complex quadratic polynomials with connected but not locally connected Julia sets due to Douady and D. Sullivan; see the work of D. E. K. Sørensen [Ergodic Theory Dynam. Systems **16** (1996), no. 3, 555–590; MR 97h:58135].

The property of complex bounds was introduced by Sullivan [in *American Mathematical Society centennial publications, Vol. II (Providence, RI, 1988)*, 417–466, Amer. Math. Soc., Providence, RI, 1992; MR 93k:58194] as a key point in renormalization theory. In general, the role of the property of complex bounds is twofold. Firstly, it allows one to derive properties of the holomorphic extension of the real system, as the local connectivity for Julia sets of real unimodal polynomials, and secondly, the complex extension can then be used to provide more information about the original real system, as happens in the theory of renormalization.

The second main result (Theorem A) of Levin and van Strien's paper states that every real infinitely renormalizable unimodal polynomial has complex bounds. It shows also how to find explicit real bounds. In the real unimodal case renormalization captures the following property of the dynamics: there is an interval (a so-called restrictive interval) which is mapped into itself by an iterate and this transformation is unimodal. If this process can be repeated infinitely many times then a map is infinitely renormalizable. In this case, f has complex bounds if every renormalization can be defined as a polynomial-like map in the sense of Douady and Hubbard [Ann. Sci. École Norm. Sup. (4) **18** (1985), no. 2, 287–343; MR 87f:58083], that is, there exists a planar domain U with the range $f(U) \supset U$ so that $f: U \mapsto f(U)$ is a proper holomorphic map, and moreover, $f(U) \setminus \overline{U}$ is an annulus with modulus bounded from below by a positive constant independent from the choice of renormalization.

In Sullivan's work, complex bounds were proved for the class of real infinitely renormalizable quadratic polynomials of bounded combinatorial type. This proof was based on a subtle estimate known as the "sector lemma". The result implies both the compactness of the renormalization operator as well as a quasi-conformal equivalence, through the work of Douady and Hubbard [op. cit., 1985], between the renormalizations of combinatorically bounded type quadratic polynomials and prototype mappings $z^2 + c$.

In Levin and van Strien's paper, the bounds depend only on the degree of the polynomial. This gives a significant extension of Sullivan's result. Moreover, Theorem A assures that one can choose the domain U of the polynomial-like renormalization so that $f(U)$ has a universal shape, up to rescaling, with the diameter comparable to the corresponding restrictive interval. In fact, in many cases, $f(U)$ is a Euclidean disc. Also, Theorem 11.1 states that Theorem A can be extended to the class of real-analytic maps which are unimodal and infinitely renormalizable. The proof of the complex bounds uses the technique of the "shortest interval" and cross-ratio estimates. The explicit real bounds are of crucial importance. Domains in the complex plane are constructed directly and the reasoning does not require the sector lemma.

The property of complex bounds for real unimodal polynomials with quadratic singularity has independent proof by G. Świątek and the reviewer [Topology Proc. **21** (1996), 33–112; MR 99k:58156] and constitutes an important technical part in the proof of the dense hyperbolicity conjecture for real quadratic polynomials [J. Graczyk and G. Świątek, Ann. of Math. (2) **146** (1997), no. 1, 1–52; MR 99b:58079]. Later, another proof of Theorem A for polynomials with quadratic singularity was provided by M. Lyubich and M. Yampolsky [Ann. Inst. Fourier (Grenoble) **47** (1997), no. 4, 1219–1255; MR 98m:58113] with a generalization to an uncountable class of complex quadratic polynomials by Lyubich [Acta Math. **178** (1997), no. 2, 185–247, 247–297; MR 98e:58145, see p. D105 in this volume]. Finally, a short proof was presented by D. Sands ["Complex bounds", manuscript, 1996; per revr.]; for the exposition see also Graczyk and Świątek [*The real Fatou conjecture*, Ann. of Math. Stud., 144, Princeton Univ. Press, Princeton, NJ, 1998]. The technique is purely real and is based on cutting time formalism. Explicit real bounds which are similar to those used by Levin and van Strien are obtained and give complex bounds immediately. It should be emphasized that the property of complex bounds is generally not satisfied by complex quadratic polynomials. The examples are based on constructions by Douady [see D. E. K. Sørensen, op. cit.].

When a unimodal map is not renormalizable then the notion of complex bounds has to do with the behavior of holomorphic extensions of induced real maps. Consider a real-analytic unimodal map f with the critical point at 0 and a sequence of the first return maps to regularly returning intervals $I_n \ni 0$, for all $k > 0$, $f^k(\partial I_n) \cap I_n = \varnothing$ (Martens' property), which form a base of neighborhoods for 0. By Martens' property, the first return maps are proper. We look for extension domains Ω_n of I_n, $\partial\Omega_n \supset \partial I_n$, which satisfy the following conditions: Ω_n are Jordan discs and the first return maps F_n to Ω_n are proper holomorphic maps. The complex bounds for F_n mean that every connected component of the domain of F_n is separated from the boundary of Ω_n by an annulus with the conformal modulus greater than a positive number $m(n)$. Consequently, we say that f satisfies complex bounds if we can show sequences I_n and Ω_n as described above so that the first return maps F_n satisfy the bound with a constant m which is independent of n. As stated, the process of proving bounds involves two stages: guessing how to find the right Ω_n and then actually proving the bound.

Theorems B and C of the paper under review deal with the non-renormalizable case. Theorem B emphasizes the relative simplicity of constructing holomorphic extensions of the first return maps in some special cases. Theorem C states that every real non-renormalizable and unimodal polynomial $f = z^l + c$, $l \geq 4$, $\omega(c)$ minimal, has complex bounds which depend only on l. Moreover, I_n are chosen as the intersections of the real line with Yoccoz pieces.

The minimality of the ω-limit set $\omega(c)$ implies that the first return maps F_n have finitely many branches. In this case, we say that F_n are generalized polynomial-like maps. If we allow the number of the branches of F_n to be infinite then we obtain the more general concept of a box mapping. A closely related result to Theorem C was obtained by Świątek and the reviewer [in *Dynamical systems and applications*, 315–328, World Sci. Publishing, River Edge, NJ, 1995; MR 97b:58121]. The assumption about the minimality of $\omega(c)$ was dropped and the notion of a polynomial-like map was replaced by a box mapping.

As listed above, the property of complex bounds is used frequently in the real systems. Levin and van Strien provided another example of its usefulness in ["Julia sets of real unimodal polynomials carry no invariant line-fields", Preprint, Univ. Amsterdam, Amsterdam, 1995; per bibl.]. The property of complex bounds also played an important role in O. Kozlovski's Ph.D. thesis ["Structural stability in one-dimensional dynamics", Univ. Amsterdam, Amsterdam, 1998; per revr.], where the dense hyperbolicity conjecture for smooth unimodal maps was reported. Other important applications are likely to come.

In the 1990s significant progress was achieved in the understanding of one-dimensional dynamical systems. The award of two Fields medals in the field is a token of the attention it has received in the last decade. The paper under review is an important contribution to the subject. The rapid progress in the field has brought with it a great technical complexity of proofs. The paper of Levin and van Strien is not an exception to this trend. *Jacek Graczyk* (Stockholm)

97e:58202 58F27 34D20 58F05 58F10 58F30
Delshams, Amadeu (Barcelona); **Gutiérrez, Pere** (Barcelona)
Effective stability and KAM theory. (English summary)
J. Differential Equations **128** (1996), *no.* 2, 415–490.

This long paper is devoted to the classical problem of stability of the action variables I for nearly-integrable Hamiltonian systems governed by Hamilton functions of the form $H(\varphi, I) = h(I) + f(\varphi, I)$. Here, φ varying in $\mathbf{T}^n = (\mathbf{R}/2\pi\mathbf{Z})^n$ and I ranging in a domain $\mathcal{G} \subset \mathbf{R}^n$ are, respectively, the angles and actions, while f is a small perturbation (of size ε) of the integrable Hamilton function h. The main stability results for such systems are provided by the celebrated KAM (Kolmogorov-Arnol′d-Moser) and Nekhoroshev theorems. According to the KAM theorem, most of the trajectories of a perturbed system lie on invariant n-tori which are close to the unperturbed invariant n-tori $\{I = \text{const}\}$ and carry quasi-periodic motions (provided that h meets suitable nondegeneracy conditions). The Nekhoroshev theorem states that all the solutions $(\varphi(t), I(t))$ of a perturbed system satisfy the estimate $|I(t) - I(0)| < c_1\varepsilon^b$ for $|t| \le T(\varepsilon)$ with $T(\varepsilon) = c_2 \exp(c_3\varepsilon^{-a})$ exponentially large in $1/\varepsilon$ (provided that H is analytic and h meets some nondegeneracy conditions known as steepness). This property is called effective stability of I, the positive constants a and b being called the stability exponents. Thus, the KAM theory deals with eternal stability of the action variables, but on some subset of the phase space only, whereas the Nekhoroshev theory concerns stability of I in the whole phase space, but over finite time intervals only. Lately, a close relationship between these two different types of stability has begun to be revealed. Important steps in this direction were made by A. D. Perry and S. Wiggins [Phys. D **71** (1994), no. 1-2, 102–121; MR 95a:58104] and A. Morbidelli and A. Giorgilli [Phys. D **86** (1995), no. 3, 514–516; MR 96f:58133; J. Statist. Phys. **78** (1995), no. 5-6, 1607–1617; MR 96h:58154], who studied, in particular, the so-called stickiness of invariant tori. One more step is taken in the present paper. The authors give detailed and complete proofs for both theorems outlined above (assuming H to be analytic). Their proofs are by no means new; moreover, they are quite standard. What is new in the paper is that the authors prove the two theorems in parallel, and a large common part of the two proofs is emphasized explicitly. This part consists in bringing the system to a normal form via successive canonical transformations constructed by the Lie series method.

More precisely, given a module $\mathcal{M} \subset \mathbf{Z}^n$, an integer $K > 0$, and a number $\alpha > 0$, consider the subdomain $G = G_{\mathcal{M}} \subset \mathcal{G}$ consisting of all the points I such that $|k \cdot \omega(I)| \ge \alpha$ for all $k \in \mathbf{Z}^n \setminus \mathcal{M}$, $|k| = \sum_{j=1}^n |k_j| \le K$ (here $\omega = \partial h/\partial I$ is the unperturbed frequency map). For any analytic Hamilton function $h(I) + Z(\varphi, I) + R(\varphi, I)$ defined in the set $\mathbf{T}^n \times G$, the authors construct an analytic canonical transformation reducing this function to the form $h(I) + Z^{\text{new}}(\varphi, I) + R^{\text{new}}(\varphi, I)$ with R^{new} much smaller than R. Here the Fourier expansions of Z and Z^{new} involve only resonant harmonics $\exp(ik \cdot \varphi)$ with $k \in \mathcal{M}$, $|k| \le K$, whereas R and R^{new} are remainders. This procedure is common to both the results. The rest of the proofs proceeds as follows.

For the Nekhoroshev theorem, the whole domain $\mathcal{G} \subset \mathbf{R}^n$ is covered by a family of sets $G_{\mathcal{M}}$ associated to different modules $\mathcal{M} \subset \mathbf{Z}^n$, with suitably chosen parameters K and α. On each set $\mathbf{T}^n \times G_{\mathcal{M}}$, the perturbed Hamilton function $h(I) + f(\varphi, I)$ is brought, by a finite series of canonical transformations, to a normal form with a remainder exponentially small in K. Then an estimate on the deviations $|I(t) - I(0)|$ is obtained for this normal form under the assumption that h is quasiconvex, i.e., there exists $m > 0$ such that $|v \cdot (\partial\omega/\partial I)v| \ge m|v|^2$ for each vector $v \in \mathbf{R}^n$ orthogonal to $\omega = \partial h/\partial I$ and each point $I \in \mathcal{G}$. In fact, quasiconvexity is the simplest and most important particular case of steepness.

For the KAM theorem, the authors always take $\mathcal{M} = \{0\}$ and assume that h is isoenergetically nondegenerate, i.e., $(\partial\omega/\partial I)v + \lambda\omega \ne 0$ everywhere in $\mathcal{G}$ for any number $\lambda \in \mathbf{R}$ and any vector $v \in \mathbf{R}^n \setminus \{0\}$ orthogonal to ω. An infinite series of normalizing canonical transformations is constructed, with orders $K_1, K_2, \cdots$ increasing to infinity. The definition domains of these transformations shrink down to a Cantorian set which is the union of the desired perturbed invariant tori, the final Hamilton function becoming integrable on this set. The frequency vectors ω^* of the tori obtained satisfy the Diophantine condition: there exist $\tau, \gamma > 0$ such that $|k \cdot \omega^*| \geq \gamma |k|^{-\tau}$ for each $k \in \mathbf{Z}^n \setminus \{0\}$.

Another new ingredient of the paper is the concept of a nearly-invariant torus, which constitutes a further bridge between the KAM and Nekhoroshev theorems. Namely, under the same hypotheses as for the KAM theorem and by a slight modification of the proof, the authors arrive at the following nice result: The unperturbed invariant tori whose frequency vectors ω^* meet the Diophantine condition approximately (up to a given precision ρ which means that the products $k \cdot \omega^*$ with $|k| \leq c_4 \rho^{-c_5}$ only are taken into account) survive a perturbation in the form of nearly-invariant tori. The trajectories starting on such a torus remain in its close vicinity (of a width exponentially small in $1/\rho$) up to a time exponentially long in $1/\rho$. The union of all the nearly-invariant tori contains inner points, in contrast to that of the genuinely invariant tori.

The authors also use the normalizing procedure with $\mathscr{M} = \{0\}$ to obtain effective stability in a perturbation of a system of n harmonic oscillators with a Diophantine frequency vector.

The paper is very well written, involved calculations alternating with informal discussions. The coefficients in all the estimates and bounds are explicitly expressed in terms of the parameters characterizing the system at hand. Many of the estimates proven are optimal. In particular, the values $a = b = 1/2n$ are obtained for the stability exponents in the Nekhoroshev theorem. Some indications of further or related results are presented. To summarize, the paper is a substantial contribution to the theory of Hamiltonian dynamical systems. *Mikhail B. Sevryuk* (Moscow)

99b:58210 58F30 34G20 35B40 47H20 58F15 58F39

Bates, Peter W. (Provo, UT); **Lu, Kening** (Provo, UT); **Zeng, Chongchun** (Provo, UT)

Existence and persistence of invariant manifolds for semiflows in Banach space. (English summary)

Mem. Amer. Math. Soc. **135** (1998), *no.* 645, viii+129 pp.

Two of the fundamental problems in dynamical systems are the persistence of invariant manifolds under perturbations and the qualitative properties of the behaviour near invariant manifolds. For finite-dimensional systems, normal hyperbolicity of invariant manifolds is exactly the right concept for dealing with the question of persistence. Roughly speaking, this requires the tangent bundle of the phase space restricted to the manifold to split into three subbundles one of which is the tangent bundle of the manifold, and the linearized flow either expands or contracts the other bundles at rates greater than any expansion or contraction in the tangent bundle of the manifold. Fenichel, and Hirsch, Pugh and Shub proved that normal hyperbolicity is a sufficient condition for persistence in the finite-dimensional case, while Mañé proved that this condition is also necessary.

In the present paper, the sufficiency result is extended to the case of C^1 semiflows in general Banach spaces. For example, this extension applies to infinite-dimensional systems generated by certain parabolic partial differential equations and other infinite-dimensional evolution equations. Also, the case of discrete systems generated by iterating a map in a Banach space is covered. Besides proving persistence of normally hyperbolic, C^2 compact invariant manifolds under small perturbations of the semiflow with respect to the C^1 norm, the authors establish in that context existence results for local center-unstable and center-stable manifolds in a neighborhood of the unperturbed invariant manifold. The latter provides information about the qualitative behaviour of the perturbed semiflow near the persistent invariant manifold. In fact, the local center-stable manifold consists of points that approach the invariant manifold under the forward semiflow, while the center-unstable manifold consists of points for which backwards orbits which approach the invariant manifold exist. Hence, these manifolds can be regarded as the stable and unstable manifolds of the invariant manifold, respectively. Invariant foliations of a whole neighborhood of the invariant manifold are not constructed in this paper, but by the theory presented the stage is set for doing that in a forthcoming paper. In any case, a theory as general as that presented here was missing so far, and only results for more specific classes of infinite-dimensional systems and invariant manifolds such as equilibria or periodic orbits have been established before.

As the basic technical tool, Hadamard's graph transform is used, which is the standard method for constructing invariant manifolds besides the Lyapunov-Perron method. In order to construct such a graph transform for flows, usually the reversed flow is used. One of the main difficulties in the situation studied in the present paper is the fact that for general semiflows in Banach spaces, the associated time-t maps are not invertible. As a consequence, a new technique had to be developed in order to establish the existence of a center-stable manifold. Another significant difficulty which had to be overcome is that in general, a trivialization of the normal bundle of an invariant manifold in an infinite-dimensional phase space is not available.

The paper is an important contribution to the theory of infinite-dimensional dynamical systems, and it will have numerous applications to a wide range of problems from various areas of science and engineering. *Jürgen Scheurle* (Munich)

99d:58148 58F32 34K15

Walther, Hans-Otto (Giessen); **Yebdri, Mustapha** (Giessen)

Smoothness of the attractor of almost all solutions of a delay differential equation. (English summary)

Dissertationes Math. (*Rozprawy Mat.*) **368** (1997), 72 pp.

In recent years there has been a great deal of research on questions related to the qualitative behaviour of solutions of delay differential equations. Such equations arise in the modeling of diverse physical, biological and ecological phenomena. These models often give rise to equations which are simple to state but nevertheless exhibit complicated behaviour, and results of a nonlinear and global nature are difficult to obtain.

The equation $\dot{x}(t) = -\mu x(t) + f(x(t-1))$ is the simplest differential equation for a state variable under the combined influence of friction, i.e., $\mu \geq 0$, and delayed negative feedback ($\xi f(\xi) < 0$ for $\xi \neq 0$). This equation generates a semiflow F in the phase space $C = C([-1,0], \mathbf{R})$, given by $F(t, \varphi) = x_t$ for all $t \geq 0$. Here $x_t(\theta) =$

$x(t+\theta)$, $-1 \leq \theta \leq 0$, and $x: [-1, \infty) \to \mathbf{R}$ is the solution of the equation with $x_0 = \varphi$.

A solution x is called slowly oscillating if there exists a $t \in \mathbf{R}$ such that $|z - z'| > 1$ for any pair of zeros $z \neq z'$ of x in $[t, \infty)$. J. L. Kaplan and J. A. Yorke [J. Differential Equations **23** (1977), no. 2, 293–314; MR **55** #807] conjectured that the initial data for slowly oscillating solutions form an open and dense set in C, provided that the nonlinearity f satisfies the stronger condition that f is C^1-smooth, $f(0) = 0$ and $f'(\xi) < 0$ for all $\xi \neq 0$ and that f is bounded from above or bounded from below. Under these additional conditions on the nonlinearity f, it was recently shown by J. Mallet-Paret and Walther ["Rapid oscillations are rare in scalar systems governed by monotone negative feedback with a time tag", Preprint, Univ. Giessen, Giessen, 1994; per bibl.] that all other trajectories constitute a closed graph of codimension 2 in C. In particular, it follows that the conjecture by Kaplan and Yorke is correct. Furthermore, the phase curves x_t of slowly oscillating solutions x enter the set $S \subset C$ of initial data $\varphi \neq 0$ with at most one change.

Thus, under the semiflow F, an open and dense set of initial data is absorbed into the set S. It turns out that the general position of the set S in the phase space C can be described in terms of spectral properties of the linearization of F at 0. The infinitesimal generator of this linearization has a leading pair of eigenvalues. The associated generalized eigenspace $L \subset C$ is two-dimensional and a subset of $\overline{S}$. The complementary subspace Q, $C = L \oplus Q$, has the property that $S \cap Q = \varnothing$. The set S is positively invariant under the semiflow F and the restricted semiflow to the closure of S, $\overline{S} = S \cup \{0\}$, has a global attractor $A \subset \overline{S}$. Every $\varphi \in A \setminus \{0\}$ uniquely determines a solution $x = x^{\varphi}$ which is defined on $\mathbf{R}$, satisfies $x_0 = \varphi$, is bounded, and is slowly oscillating.

In earlier work by Walther [Mem. Amer. Math. Soc. **113** (1995), no. 544, vi+76 pp.; MR 95f:58070], it was shown that the attractor A can be written as a Lipschitz map a from a subset $L_a \subset L$ into Q, i.e., $A = \{\chi + a(\chi) |\ \chi \in L_a\}$. The aim of the present work is to further elaborate smoothness properties of the map a. The authors prove that in case $A \neq \{0\}$, the map $a: L_a \to Q$ is not only Lipschitz but actually a continuously differentiable map. This proof of the smoothness of the attractor A is long and complicated. The main ingredients are Poincaré-Bendixson arguments and a detailed analysis of Poincaré maps.

This beautiful result completes a long journey with important intermediate steps by Walther [J. Reine Angew. Math. **414** (1991), 67–112; MR 91k:34112; in *Oscillation and dynamics in delay equations (San Francisco, CA, 1991)*, 177–239, Amer. Math. Soc., Providence, RI, 1992; MR 93i:34137]. *Sjoerd M. Verduyn Lunel* (Amsterdam)

97g:58162 58G10 57R57 58G25

Cappell, Sylvain E. (New York, NY); **Lee, Ronnie** (New Haven, CT);
Miller, Edward Y. (Brooklyn, NY)

Self-adjoint elliptic operators and manifold decompositions. I. Low eigenmodes and stretching. (English summary)

Comm. Pure Appl. Math. **49** (1996), *no.* 8, 825–866.

This is the first in a series of three articles. It contains a number of results concerning the effect of cut-and-paste and stretching constructions on the eigenvalues of a first order self-adjoint elliptic operator on a closed manifold.

To set notation, let M be a closed smooth manifold, $\Sigma \subset M$ a separating codimension-1 submanifold decomposing M as $M = M_1 \cup_\Sigma M_2$. Let D be some first order elliptic self-adjoint operator on M of "Atiyah-Patodi-Singer type" on a neighborhood $[-1,1]\times\Sigma$ of Σ; thus (1) $D = \sigma(\partial/\partial t + \widehat{D})$ in this neighborhood. Let $M(r)$ denote the manifold obtained by replacing the neighborhood of Σ by $[-r,r]\times\Sigma$. The operator D has a natural extension to $M(r)$ using (1) to define D on the neighborhood of Σ. Define $M_1(r) = M_1 \cup ([0,r]\times\Sigma)$, $M_1(\infty) = M_1 \cup ([0,\infty)\times\Sigma)$; note that D determines an operator on $M_1(r)$ and $M_1(\infty)$ using (1) again. Similarly define $M_2(r)$ and $M_2(\infty)$.

With this notation in place, the results of this paper concern constructing eigenvectors with low eigenvalues of D on $M(r)$ in terms of corresponding eigenvectors of D on $M_i(r)$ and L^2 eigenvectors of D on $M_i(\infty)$.

The main theorem of this article roughly states that for r large enough, the eigenvectors of D on $M(r)$ whose eigenvalues lie in $(-r^{3/2+\varepsilon}, r^{3/2+\varepsilon})$ can be approximated by splicing together L^2 eigenvectors on $M_1(\infty)$ and $M_2(\infty)$. (If the kernel of $\widehat{D}$ is non-empty, one needs also to "splice in" certain extended L^2 solutions to $Du=0$ on $M_i(\infty)$.) Moreover, any eigenvalue in the range $(-r^{3/2+\varepsilon}, r^{3/2+\varepsilon})$ actually lies in the range $(-\exp(cr), \exp(cr))$ for some $c>0$; i.e. "small eigenvalues are exponentially small in r". This exponential control over the eigenvalues is used in the second article [S. E. Cappell, R. Lee and E. Y. Miller, Comm. Pure Appl. Math. **49** (1996), no. 9, 869–909; see the following review; MR 97g:58163, see p. D115 in this volume] in the situation when $D(u)$ is a 1-parameter family of such operators to prove a Mayer-Vietoris theorem for the spectral flow of the family $D(u)$ in terms of the decomposition $M = M_1 \cup M_2$. *Paul A. Kirk* (Bloomington, IN)

97g:58163 58G10 57R57 58G25

Cappell, Sylvain E. (New York, NY); **Lee, Ronnie** (New Haven, CT); **Miller, Edward Y.** (Brooklyn, NY)

Self-adjoint elliptic operators and manifold decompositions. II. Spectral flow and Maslov index. (English summary)

Comm. Pure Appl. Math. **49** (1996), *no.* 9, 869–909.

This is the second in a series of three articles. It contains the proofs of several formulas expressing the spectral flow of a 1-parameter family of self-adjoint elliptic differential operators on a closed manifold as a sum of terms computed from a decomposition of the manifold into two submanifolds.

Let the closed manifold M be decomposed along a separating codimension-1 submanifold Σ as $M = M_1 \cup_\Sigma M_2$. Let $D(u)$, $u \in [0,1]$, be a continuous family of first-order, self-adjoint elliptic operators on M. The spectral flow of the family $D(u)$ is the integer defined to be the net number of eigenvalues which cross from negative to positive along the path (a slight modification of this definition is needed if $D(0)$ or $D(1)$ has a nontrivial kernel).

The operator $D(u)$ is assumed to be of "Atiyah-Patodi-Singer type" near Σ, so that in a collar $[-1,1]\times\Sigma$, $D(u)$ essentially takes the form $D(u) = \partial/\partial t + \widehat{D}(u)$ with $\widehat{D}(u)$ elliptic, self-adjoint on Σ.

The restrictions of $D(u)$ to M_1 and M_2 can be made self-adjoint in several ways, depending on the properties of the tangential operator $\widehat{D}(u)$. In the general case one must make a choice of Lagrangian subspace $L_i(u)$ of $H(u)$, where $H(u)$ is a finite-dimensional symplectic subspace of $L^2(\Sigma)$ spanned by those eigenvectors of $\widehat{D}(u)$ whose eigenvalues are small. The choice is used to provide $D(u)_{|M_i}$ with self-adjoint boundary conditions.

With this setup in place, the main result of this paper states that, after replacing the collar neighborhood of Σ by a larger collar $[-r,r]\times\Sigma$ for r large enough, the spectral flow of $D(u)$ on M (properly defined) is equal to the sum of three terms: (1) the spectral flow of $D(u)|_{|M_1}$ and $D(u)_{|M_2}$ with respect to the boundary conditions induced by the $L_i(u)$, (2) the Maslov index of the pair of Lagrangians $(L_1(u), L_2(u))$, and

$$(3) \qquad \tfrac{1}{2}(\dim\ker\widehat{D}(1)-\dim\ker\widehat{D}(0)).$$

(The precise formula is slightly more complicated if the spaces $H(u)$ jump in dimension.)

The arguments are based on the results of the first article [S. E. Cappell, R. Lee and E. Y. Miller, Comm. Pure Appl. Math. **49** (1996), no. 8, 825–866; see the preceding review; MR 97g:58162, see p. D114 in this volume], which establishes exponential control of the small eigenvalues of the operators $D(u)$ and $D(u)_{|M_i}$ and defines splicing constructions to create eigenvectors on M from eigenvectors on the M_i.

Paul A. Kirk (Bloomington, IN)

99a:58144 58G10

Melrose, Richard B. (Cambridge, MA); **Piazza, Paolo** (Rome)

Families of Dirac operators, boundaries and the b-calculus. (English summary)

J. Differential Geom. **46** (1997), *no.* 1, 99–180.

The authors solve the index problem for families of Dirac operators on compact, even-dimensional manifolds with boundary. This problem had been treated previously by J.-M. Bismut and J. Cheeger [J. Funct. Anal. **89** (1990), no. 2, 313–363; MR 91e:58180; J. Funct. Anal. **90** (1990), no. 2, 306–354; MR 91e:58181], who derived an index formula under the assumption that all boundary Dirac operators of the family are invertible, or at least of constant rank. This assumption is too restrictive for some potential applications. The main achievement of Melrose and Piazza is to remove this restriction. The central problem that arises is, then, that the boundary condition traditionally used (the Atiyah-Patodi-Singer (APS) boundary condition) may vary discontinuously among the family, leaving the index bundle ill-defined. Melrose and Piazza deal with this problem by introducing a class of boundary conditions, containing the APS boundary condition, which is general enough to overcome this obstacle and specific enough so that established methods can still be used, with only small modifications. Another aspect in which this article generalizes previous work is the consideration of generalized Dirac operators, so the existence of a spin structure is not needed.

Before stating the main ideas and results, we will give a short overview of the historical context in which this article is placed. The classical Atiyah-Singer (AS) index theorem solves in great generality the problem of calculating the index $\operatorname{ind} P = \dim\ker P - \dim\operatorname{coker} P$ of an elliptic partial differential operator P on a compact manifold in terms of the coefficients of P, i.e. the symbol of P. The case where P is a Dirac operator on a Riemannian manifold has special significance since (a) it is well-motivated from geometric and topological contexts (signature, Gauss-Bonnet,

Dolbeault operators), (b) it allows, by its additional structure, a richer theory, and (c) it represents, up to stable homotopy, the general case.

From now on we will restrict our attention to Dirac operators. Various generalizations of the setup above were also treated by Atiyah and Singer, and later by many other authors. Those of interest here are the "family" case and the case of a compact manifold with boundary. In the family problem one is given a smooth fibration $\varphi\colon M \to B$ with compact fibers $M_z = \varphi^{-1}(z)$ (to be considered as members of a family of manifolds), $z \in B$, and a smoothly varying family $\{D_z\}_{z\in B}$ of Dirac operators on the fibers; then the formal difference of vector spaces $\ker D - \operatorname{coker} D$ can be made sense of as a virtual vector bundle, again called $\operatorname{ind} D$, over the base B. The family index theorem calculates the Chern class $\operatorname{ch}(\operatorname{ind} D) \in H^*(B)$. The definitive form of this theorem, which even calculates the Chern character of $\operatorname{ind} D$ on the differential form level, for a suitable connection on this bundle (at least when $\dim \ker D_z$ is constant in z), was given by Bismut [Invent. Math. **83** (1985), no. 1, 91–151; MR 87g:58117], who found the "correct" way to use D. Quillen's idea [Topology **24** (1985), no. 1, 89–95; MR 86m:58010] of a superconnection. A detailed exposition, with simplifications of some proofs, was written by N. Berline, E. Getzler and M. Vergne [*Heat kernels and Dirac operators,* Springer, Berlin, 1992; MR 94e:58130]; Melrose and Piazza rely heavily on this work.

An index theorem for Dirac operators on manifolds with boundary was first proved by M. F. Atiyah, V. K. Patodi and I. M. Singer [Math. Proc. Cambridge Philos. Soc. **77** (1975), 43–69; MR **53** #1655a], who introduced their now famous APS boundary condition; this condition defines the domain of the Dirac operator to consist of those sections whose restriction to the boundary lies in the negative spectral subspace of the boundary Dirac operator. Some elliptic boundary condition is necessary to make the operator Fredholm (so the index problem makes sense), and this choice was motivated by the case of the signature operator, where the resulting index is the topological signature. The remarkable new feature that occurs in the APS index formula is the appearance of a non-local, spectral-theoretic term, the so-called η-invariant of the boundary operator, in addition to the "local" term that appears already in the boundaryless case (i.e., an integral over the manifold, of an expression computed locally from the symbol of D); there are no local boundary terms, like the geodesic curvature in the Gauss-Bonnet theorem, since the metric is assumed to have product form near the boundary. More recent treatments include Cheeger's [J. Differential Geom. **18** (1983), no. 4, 575–657 (1984); MR 85d:58083] (see also A. W. Chou [Trans. Amer. Math. Soc. **289** (1985), no. 1, 1–40; MR 86i:58124]) and Melrose's [*The Atiyah-Patodi-Singer index theorem*, A K Peters, Wellesley, MA, 1993; MR 96g:58180]. Cheeger "closes-off" the manifold by attaching a cone to the boundary and then uses his cone calculus. Melrose proves the APS theorem in the context of his b-calculus, carrying through systematically the philosophy that the APS theorem is "just" the classical AS theorem in the b-category (instead of the category of closed manifolds).

Since this philosophy is also central to this article, we will explain it a little more. In the b-category, instead of closed manifolds and the pseudodifferential calculus on them one considers compact manifolds with boundary and the b-pseudodifferential calculus, which is tailored to deal with operators arising from Riemannian metrics (on the interior) having complete cylindrical ends (i.e. essentially of the form $x^{-2}dx^2 + g_N$, where $M = [0,1)_x \times N$ near the boundary $\partial M = \{0\} \times N$, and g_N is a Riemannian metric on N). There is a simple correspondence between compact manifolds with boundary, $\widetilde{M}$, with incomplete metrics that have product form $d\widetilde{x}^2 + g_N$ near the boundary, and b-manifolds M with b-metrics: attach an infinite cylinder $[-\infty, 0] \times N$ to $\widetilde{M}$, and change variables $x = e^{\widetilde{x}}$. Then, sections on $\widetilde{M}$ satisfying the APS condition correspond to square-integrable sections on M, and invertibility of the boundary Dirac operator $\widetilde{M}$ is equivalent to Fredholmness of the Dirac operator on M. Although this causes the slight difficulty that one has to deal with non-Fredholm operators occasionally, this approach has the great virtue that the APS theorem can be proved in a way largely parallel to (one version of) the proof of the classical AS theorem, if only one replaces "usual" differential geometric notions by the corresponding b-notions.

The present article combines the two settings described above: families and boundaries. However, when considering families of manifolds with boundary one encounters new difficulties which are not present in either problem separately, both of a conceptual and of a technical nature. The main conceptual problem is the possible discontinuity of the APS boundary projection operator as a function of the base point; Melrose and Piazza solve this elegantly by introducing the notion of a spectral section; see below. As to the technical problems, the b-perspective turns out to be useful also for suggesting natural solutions to them, for example when constructing perturbations of the operators to make the family Fredholm and the index bundle well-defined. An outline of the version of the b-calculus needed here is given in the appendix of the paper.

We will now state the main results precisely. First, we quote the definition of a spectral section: Given a smooth fibration $M' \to B$ with closed Riemannian manifolds as fibers and a smooth family $\{D'_z\}_{z\in B}$ of selfadjoint differential operators on the fibers, a spectral section for D' is a smooth family $\{P_z\}_{z\in B}$ of pseudodifferential projection operators on $L^2(M'_z)$ which act as the identity on all eigenspaces of D'_z with sufficiently large positive eigenvalue, and annihilate all eigenspaces with sufficiently large negative eigenvalue.

Now assume that the following data are given: (a) A smooth fibration $\varphi\colon M \to B$, with compact, even-dimensional fibers $M_z = \varphi^{-1}(z)$, $z \in B$, diffeomorphic to a fixed manifold with boundary. (b) For each $z \in B$, the following data, varying smoothly with z: (b1) a Riemannian metric on M_z, of product type near the boundary, and (b2) a Hermitian Clifford module E_z, with unitary Clifford connection.

These data determine a family of selfadjoint generalized Dirac operators $\{D_z\}_{z\in B}$ acting on sections of E_z. Given a spectral section P for the boundary fibration $\partial M \to B$ and the family of boundary Dirac operators, one obtains a smooth family of Fredholm operators D_z^P by using the boundary condition $P_z(u_{|\partial M_z}) = 0$ for the operator D_z. The Z_2-graded kernels of D_z^P define a virtual vector bundle $\operatorname{ind} D^P$ over B.

Main theorem: In the cohomology of the base, one has

$$\operatorname{ch}(\operatorname{ind} D^P) = (2\pi i)^{-n/2} \int_{M/B} \widehat{A}(M/B)\operatorname{Ch}'(E) - \tfrac{1}{2}\widehat{\eta}_P.$$

Here, $\widehat{A}(M/B)$ is the $\widehat{A}$-genus of the vertical tangent bundle, and $\operatorname{Ch}'(E)$ is the twisted Chern character of $E \to M$. The form $\widehat{\eta}_P$ on B is determined from the boundary operators and the spectral section P; it corresponds to the η-invariant in the APS theorem. In fact, Melrose and Piazza also state a slightly more general theorem, in the language of b-geometry, from which this is an immediate consequence.

To complete this result, one needs to know that a spectral section exists. Melrose and Piazza prove this by first showing that, for a general family $\{D'_z\}$ on $M' \to B$, the existence of a spectral section is equivalent to the vanishing of the index of this family in the K^1-theory of the base, as defined by Atiyah and Singer; then they use a "classical" version of the index theorem to show that this K^1-index vanishes if D' arises as boundary of a family of Dirac operators. To round things off, Melrose and Piazza also analyze the change of $\operatorname{ind} D^P$ under a change of P.

Daniel Grieser (Berlin)

97a:58179 58G12 58F06
Fedosov, Boris (Dolgoprudnyĭ)
★Deformation quantization and index theory. (English summary)
Mathematical Topics, 9.
Akademie Verlag, Berlin, 1996. 325 pp. DM 130.00. ISBN 3-05-501716-1

One of the main results in the mathematics of quantization is the theorem that every symplectic manifold M has a formal deformation quantization. More precisely, this means that there exists a (noncommutative) product $*$ on the space $\mathscr{C}^\infty(M)[\![\hbar]\!]$ of formal power series with coefficients in $\mathscr{C}^\infty(M)$ such that for $f,g\in\mathscr{C}^\infty(M)$ the following conditions hold: (1) The product $f*g$ has an expansion of the form

$$f*g=\sum_{n=0}^{\infty} c_n(f,g)\hbar^n,$$

where the c_n are bidifferential operators and $c_0(f,g)=fg$. (2) The Heisenberg quantization condition

$$[f,g]=f*g-g*f=i\hbar\{f,g\}+o(\hbar^2)$$

is satisfied, where $\{\ ,\ \}$ is the Poisson bracket of smooth functions on M.

A first proof of this fact was given by M. De Wilde and P. B. A. Lecomte [Lett. Math. Phys. **7** (1983), no. 6, 487–496; MR 85j:17021]. It is based on extensive cohomological considerations. Fedosov gave an independent and more geometric proof [J. Differential Geom. **40** (1994), no. 2, 213–238; MR 95h:58062]. The book under review now provides a detailed account of Fedosov's constructions for deformation quantization. Preparatory material such as symplectic geometry and Weyl quantization on Euclidean space is provided.

The main idea of Fedosov's proof consists in first quantizing the bundle WM of Weyl algebras over M by applying the Moyal product. Then, after the choice of a symplectic connection on M, a unique flat connection on $WM\otimes\Lambda M$ is inductively constructed. It has the property that the space of global parallel sections of $WM\otimes\Lambda M$ is canonically isomorphic to the space $\mathscr{C}^\infty(M)[\![\hbar]\!]$ and that it is closed under the Moyal product on the Weyl bundle. Thus one derives the desired deformed product on $\mathscr{C}^\infty(M)[\![\hbar]\!]$ by applying the canonically constructed isomorphism.

Another major result in analysis is the Atiyah-Singer index theorem, which asserts that the analytic and the topological index of an elliptic operator coincide. The book under review aims at interpreting the Atiyah-Singer index theorem within the theory of deformation quantization. Again the author gives an introduction to all the needed preparatory material such as characteristic classes, K-theory with compact support, pseudodifferential operators and index theory.

Unfortunately the book contains several misprints. Nevertheless it is very recommendable for anybody looking for a concise and (almost) self-contained introduction to deformation quantization and its application to index theory. Moreover, for experts in the field it provides stimulating ideas for further developments, especially in the last three chapters about asymptotic operator expansion, symplectic reduction and a trace formula for the Schrödinger operator. *Markus J. Pflaum* (Berlin)

99d:58168 58G12 19K56 46L87 58G15

Leichtnam, Eric (Paris); **Piazza, Paolo** (Rome)

The b-pseudodifferential calculus on Galois coverings and a higher Atiyah-Patodi-Singer index theorem. (English and French summaries)

Mém. Soc. Math. Fr. (N.S.) No. 68 (1997), iv+121 pp.

One of the basic goals of index theory is to compute the index of an elliptic differential (or pseudodifferential) operator D on a compact manifold M, i.e. the dimension of the nullspace of D minus the dimension of its cokernel. It is more realistic to try to compute this quantity rather than, for example, the dimension of the nullspace alone (which might well be what one is really interested in) because the index is stable under perturbations. The famous theorem of Atiyah and Singer equates this index with an explicit characteristic number on M, $\int_M$ AS. The integrand AS is an explicit differential form associated to D.

To prove this theorem for general differential (or pseudodifferential) operators it suffices to prove it only for Dirac-type operators. The most efficient means to do this uses the heat equation [cf. N. Berline, E. Getzler and M. Vergne, *Heat kernels and Dirac operators*, Springer, Berlin, 1992; MR 94e:58130]. Briefly, if the Dirac-type operator D is Z_2-graded, $D = D^+ \oplus D^-$, one may then identify its index with $\dim\ker D^-D^+ - \dim\ker D^+D^-$, and then with the limit as $t \to \infty$ of the so-called supertrace of the heat kernel, $\mathrm{Str}(\exp -tD^2)$. On the other hand, $\int_M$ AS is the limit as $t \to 0$ of this supertrace. In fact, this works even on the local level: the limit as $t \to 0$ of the pointwise supertrace equals AS exactly; this is known as the local index formula, and is the most analytically difficult of the steps. The theorem follows because one computes that the t derivative of this supertrace is the supertrace of a (super)commutator, which therefore vanishes.

I have sketched this proof in detail because all the generalizations discussed here follow the same general outline. One important generalization is the Atiyah-Patodi-Singer index theorem. Here M is a compact manifold with boundary, with all structures, such as the metric, spin structure, etc., of product type near the boundary. One introduces non-local boundary conditions involving a spectral decomposition of the induced Dirac-type operator D_0 on ∂M. The final formula equates the index of D with these boundary conditions with the difference of two terms. The first is $\int_M$ AS as before, but note that this is no longer a characteristic number because the manifold is not closed. The second is the eta invariant, a global spectral invariant of D_0 which is a regularized measure of its spectral asymmetry. Melrose gave a reformulation of the proof of this theorem using the b-calculus of pseudodifferential operators. This is a class of operators adapted to the geometry of manifolds with boundary, or more correctly with infinite asymptotically cylindrical ends, which provides a convenient and functorial way to construct parametrices for D or D^2. The attractive feature of this reformulation is that the argument proceeds precisely as in the boundaryless case. One difficulty is that the heat kernels are no longer trace class, and for this Melrose introduced the b-trace, an extension of the trace functional, and so all terms are at least defined. However, the t-derivative of the b-supertrace does not vanish because the b-trace does not vanish on commutators; one readily computes, however, that the integral from 0 to ∞ of this derivative is precisely the eta invariant, and so this represents the difference of the supertrace at $t = 0$ and at $t = \infty$.

The next step in this story is the index theorem of Connes and Moscovici. This is a "noncommutative" index theorem in the following sense. Starting with a compact spin manifold and Dirac operator D as before, let $\widetilde{M} \to M$ be any Galois covering with group Γ. If $\nu: M \to B\Gamma$ is the associated classifying map and β any element of $H^*(B\Gamma)$, then we consider the so-called "higher" characteristic numbers

$$\int_M \mathrm{AS} \wedge \nu^* \beta.$$

For example, when D is the signature operator, these are the higher signatures appearing in the famous conjecture of Novikov, which has been the focus of much research. The theorem equates these with a more complicated sort of analytic index constructed as follows. The lift $\widetilde{D}$ of D to $\widetilde{M}$ is a Γ-invariant operator, and in particular its nullspace and cokernel are modules over the group algebra $\mathbf{C}\Gamma$. If $C_r^*(\Gamma)$ denotes the reduced C^*-algebra of Γ (which is closely related to a Banach completion of $\mathbf{C}\Gamma$), then $\widetilde{D}$ is Fredholm in the sense that its kernel and cokernel are finitely generated projective $C_r^*(\Gamma)$ modules. The formal difference of these is now an element of the K-group $K_0(C_r^*(\Gamma))$. To obtain a number from this, one uses β to produce an element τ_β of cyclic cohomology, which can then be paired with the element of $K_0(C_r^*(\Gamma))$. The Connes-Moscovici theorem equates these two numbers. The original proof of this result is of a different nature than we have been describing, but Lott found a heat equation proof using Bismut's theory of superconnections.

In a later paper Lott used this same formalism to define a higher eta invariant $\widetilde{\eta}$, which is a noncommutative topological de Rham class over $\mathscr{B}^\infty$, a "smoothing" of $C_r^*(\Gamma)$. He showed that the integral defining $\widetilde{\eta}$ converges, assuming that Γ is virtually nilpotent and that $\widetilde{D}$ has a bounded inverse. It is natural to speculate, as he did, that this form appears in a higher Atiyah-Patodi-Singer theorem. This is exactly the accomplishment of the paper under review. The main result of Leichtnam and Piazza is that if M is a compact manifold with boundary, and if $\widetilde{M} \to M$ is a Galois covering, then one may equate the Chern character of the index class, which is a noncommutative topological de Rham class over $\mathscr{B}^\infty$, with the difference of two terms. The first is $\int_M \mathrm{AS} \wedge \omega$, where ω is a biform valued in this de Rham theory, while the second is the higher eta invariant $\widetilde{\eta}$. For this one needs to assume that Γ is virtually nilpotent and that $\widetilde{D}$ has a gap in its spectrum around zero. Using the notion of spectral sections introduced by Melrose and Piazza, and extended to the noncommutative setting by Wu, the authors intend to prove the theorem in generality in a later paper, without making assumptions on the spectrum of $\widetilde{D}$.

Although the ramifications of this rather complicated result are not fully transparent, this theorem once again unifies several areas of research and is a milestone in bringing together a number of diverse methods in its proof. Indeed, it is a testament to the development of these various methods that the same general scheme of proof may be used here, so that at a first glance the proof appears to be not too different from that of Melrose, for example. The technical details, needless to say, require substantial diligence. One of the main new ingredients required in the proof is an integration of the Mishchenko-Fomenko theory of pseudodifferential operators on covering manifolds with the calculus of b-pseudodifferential operators. The authors develop all these tools and ideas both carefully and succinctly, starting from the boundaryless case. The discussion of the Mishchenko-Fomenko calculus is particularly good. Altogether, this paper provides a very thorough and readable treatment of this area of active research, and the techniques here will without doubt prove important in future developments. *Rafe Mazzeo* (Stanford, CA)

99m:58186 58G12 16W30 19D55 58B30

Connes, A. (Bures-sur-Yvette); **Moscovici, H.** (Columbus, OH)

Hopf algebras, cyclic cohomology and the transverse index theorem. (English summary)

Comm. Math. Phys. **198** (1998), *no.* 1, 199–246.

In this paper the authors solve a very important (internal) problem of noncommutative geometry. The authors introduce a new Hopf algebra H_n, and define and compute its cyclic cohomology. These tools and results allow them to compute the local index formula for tranversally hypoelliptic operators on foliations.

Description of the problem: The spaces of leaves of foliations are basic examples of noncommutative spaces. For such a foliation V one has the assembly map μ from a certain geometric group to the K-theory of the leaf space $K(V/F)$. The analogue for foliations of property T (in the group case) gives an obstruction to tentative proofs of the surjectivity of μ in general. One way to test the K-group $K(V/F)$ is to use its natural pairing with the K-homology groups of the C^*-algebra of the foliation. So one has to construct cycles (i.e. elements of the K-homology groups). This is a quite elaborate problem that the authors settled [Geom. Funct. Anal. **5** (1995), no. 2, 174–243; MR 96e:58149] by constructing (motivated by the work of Hilsum-Skandalis) a spectral triple (A, H, D), where D is a suitable differential hypoelliptic operator. So for any element E of $K(V/F)$ one gets an index

$$\langle \mathrm{ch}_*(D), \mathrm{ch}^*(E)\rangle = \mathrm{Index}\, D_E$$

where $\mathrm{ch}_*(D)$ is a cyclic cocycle that needs to be computed. The explicit computation is a formidable task.

The fundamental discovery of the authors is that there exists a Hopf algebra H_n (n is the codimension of the foliation) which acts on the C^*-algebra of the transverse frame bundle of the foliation and that the index computation takes place within the cyclic cohomology of H_n and thus is doable.

This Hopf algebra H_n is essentially the dual of the enveloping algebra of the Lie algebra of vector fields:

$$f(x)\frac{\partial}{\partial x}, \quad f(0) = 0 = f'(0) \text{ modulo } x^{n+2}\frac{\partial}{\partial x}.$$

In this paper, for any Hopf algebra H (such as H_n) satisfying a suitable unimodularity condition the authors develop a cyclic cohomology theory. They use the natural cosimplicial space, but the point is that it has a hidden cyclic structure determined by τ_n acting on $H^{\otimes n}$ as

$$\tau_n(h^1 \otimes \cdots \otimes h^n) = (\Delta^{n-1}\widetilde{S}(h_1))h^2 \otimes \cdots \otimes h^n \otimes 1$$

where $\widetilde{S}$ is the twisted antipode. See page 219. For instance, if H is the enveloping algebra of a Lie algebra $\mathscr{G}$ then the periodic cyclic cohomology of H is isomorphic to the Lie algebra homology $H_*(\mathscr{G}, C)$.

Moreover, the first important fact is that there exists an isomorphism θ between the Lie algebra cohomology of formal vector fields and the periodic cyclic cohomology of the corresponding Hopf algebra H.

Now the Hopf algebra H acts on the algebra A of the above spectral triple (defining the transverse fundamental class). The authors construct a characteristic map from $HC^*(H)$ to $HC^*(A)$. By composing the map θ above with this latter map, one gets a map, denoted by $\widetilde{\theta}$, from $H^*(W\mathrm{SO}_n)$ to the cyclic homology of A such that for each n there exists a universal polynomial $L_n \in H^*(W\mathrm{SO}_n)$ such that $\mathrm{Ch}_*(Q) = \widetilde{\theta}(L_n)$.

The results of this paper should have important developments and motivate interesting research. *Éric Leichtnam* (Paris)

97f:58131 58G25 58F17 58G15 58G18
Guillemin, Victor (Cambridge, MA)
Wave-trace invariants.
Duke Math. J. **83** (1996), *no.* 2, 287–352.

This major paper contains the proof of a remarkable inverse result in spectral geometry which was inspired in part by a deformation result of Y. Colin de Verdière [in *South Rhone seminar on geometry, III (Lyon, 1983)*, 122–139, Hermann, Paris, 1984; MR 86a:58078], and by a conjecture of A. Weinstein. The wave group $e^{it\sqrt{\Delta}}$ is the solution operator for the wave equation associated to the Laplacian on a (compact) Riemannian manifold. (Any positive first-order elliptic pseudodifferential operator A could replace $\sqrt{\Delta}$.) Its distribution trace $\mathrm{Tr}(e^{it\sqrt{\Delta}})$ is, perhaps unexpectedly, smooth except for a discrete set of values $\{\pm L_\gamma\} \cup \{0\}$, where L_γ varies over the lengths of all closed geodesics (or closed orbits of the Hamiltonian flow corresponding to the principal symbol of A in the more general setting). The singularity at zero was analyzed by L. Hörmander in his famous paper [Acta Math. **121** (1968), 193–218; MR **58** #29418] introducing the local theory of Fourier integral operators, while the other singularities, and in particular the relationship between the spectrum of the Laplacian and the lengths of closed geodesics, was discovered by Colin de Verdière, Chazarain, and by Guillemin and Duistermaat. Each of these singularities is classical, i.e. represented by an asymptotic expansion in powers of $(t - L_\gamma)$, and the wave-trace invariants are the coefficients of the singular terms in these asymptotic expansions. The information given by the coefficients for the singularity at zero is local and equivalent to the information given by the better-known heat invariants. However, the coefficients at the other singularities contain some sort of infinitesimal information about the closed geodesic of that length. The theorem is that the wave invariants at γ determine the complete Birkhoff normal form of the geodesic flow at γ. In fact, they determine the microlocal analogue of the Birkhoff normal form. Unfortunately, because of their nonlocal nature, these wave invariants are very difficult to compute. One of the big achievements in this paper is a method to filter out such meaningful information from the messy higher coefficients in these expansions, at least at nondegenerate elliptic closed geodesics. Here ellipticity is the condition that the eigenvalues of the Poincaré return map around the closed geodesic are all of unit modulus, and the nondegeneracy condition is that the exponents of these eigenvalues be rationally independent. Subsequently, an independent proof, as well as details also for the general nondegenerate hyperbolic, loxodromic and mixed cases, was given by S. Zelditch [Geom. Funct. Anal. **7** (1997), no. 1, 145–213; MR 98f:58191; Geom. Funct. Anal. **8** (1998), no. 1, 179–217; MR 98m:58136].

Let us be more precise about the definition of the classical Birkhoff normal form. Associated to the operator A is its symbol, a real-valued function on $T^*X \setminus \{0\}$. The principal symbol is homogeneous of degree one, and the closed orbits for its bicharacteristic (Hamiltonian) flow correspond to the closed geodesics in the main example. The classical Birkhoff canonical form theorem states that this principal symbol may be carried by a canonical transformation (i.e. a homogeneous symplectomorphism) to a simpler function (up to an error which vanishes to infinite order at the closed orbit γ), the quadratic part of which incorporates the symbol of a quantum harmonic oscillator acting in the normal bundle of γ. The microlocal version of this theorem, one of the main results of this paper, is a quantized form of this statement. More precisely, the canonical transformation corresponds to a Fourier integral operator, and the result is that the operator may be conjugated by this FIO to a rather simple sort of operator built out of these transverse quantum harmonic oscillators, plus an operator with symbol vanishing to infinite order at γ. The classical normal form is the principal symbol of the quantum normal form.

The use of microlocal normal forms is an important and old technique in linear PDE, having been used in significant and essential ways by Boutet de Monvel, Grushin, Sjöstrand, Melrose, and others, starting from the early 1970s. The microlocal Birkhoff normal form here is also related to the by-now classical constructions of quasi-modes, particularly the work of Babich and Lazutkin. The use in such a substantial way of a normal form in spectral geometry is quite original, though.

The main ideas and steps in this proof, then, may be summarized as follows. First the operator is conjugated to its microlocal Birkhoff normal form around a closed bicharacteristic γ. The wave invariants at γ are then expressed as noncommutative residues of the wave group generated by the quantum normal form. As Guillemin acknowledges, this relationship was discovered by Zelditch, and is one of the main ingredients which makes the computations tractable. Finally, the inverse problem is solved, which recovers the normal form coefficients from the wave invariants. Because the microlocal Birkhoff normal form is obtained in a sense directly from the classical Birkhoff normal form, and thus the classical Birkhoff normal form invariants appear directly in the microlocal normal operator, this gives the conjectured relationship.

There are, of course, many technicalities along the way, and many important parts of the proof which I am omitting here. But, as usual with Guillemin's work, the writing is lucid, and the paper is a pleasure to read. *Rafe Mazzeo* (Stanford, CA)

98g:58181 58G25 11M36 35P20 35P25
Guillopé, Laurent (Nantes); **Zworski, Maciej** (Toronto, ON)
Scattering asymptotics for Riemann surfaces.
Ann. of Math. (2) **145** (1997), *no.* 3, 597–660.
This paper is a very important study of scattering asymptotics for a class of non-compact Riemannian manifolds X that includes quotients of real hyperbolic two-dimensional space by a finitely generated discrete group of hyperbolic isometries. The authors prove two main results: (i) sharp lower bounds on the counting function for scattering resonances, and (ii) a Weyl law for the scattering phase for a relative scattering operator. The authors build on their own previous studies of scattering theory and scattering resonances for hyperbolic manifolds [J. Funct. Anal. **129** (1995), no. 2, 364–389; MR 96b:58116], R. R. Mazzeo and R. B. Melrose's resolvent parametrix construction [J. Funct. Anal. **75** (1987), no. 2, 260–310; MR 89c:58133] and the second author's joint work with J. Sjöstrand [J. Funct. Anal. **123** (1994), no. 2, 336–367; MR 95g:35149] on lower bounds for the counting function of scattering poles for compactly supported perturbations of the Laplacian in $\mathbf{R}^n$. As the authors remark, their techniques and results depend heavily on an exact (metric) description of the geometry at infinity: their Riemannian manifolds have finitely many ends isometric to hyperbolic cusps or cylinders, so that model resolvents and scattering operators for the ends can be computed explicitly.

Scattering resonances are poles of the meromorphically continued resolvent operator for the Laplacian on X; for an excellent review of the resonance-counting question in both classical and geometric scattering theory, see the second author's survey [in *Spectral and scattering theory (Sanda, 1992)*, 301–331, Dekker, New York, 1994; MR 95i:35210]. They are discrete data associated to the continuous spectrum of the Laplacian which appear to play a role similar to eigenvalues of the Laplacian on a compact manifold; resonances are much harder to analyze in practice than L^2 eigenvalues since they arise as eigenvalues of a non-selfadjoint problem. In the case under study, the resolvent $(\Delta_X - s(1-s))^{-1}$ is defined as an operator from $L^2(X)$ to itself in the half-plane $\operatorname{Re}(s) > \frac{1}{2}$ (possibly with finitely many poles due to L^2-eigenvalues) and admits a meromorphic continuation to the complex plane as an operator from $C_0^\infty(X)$ to $C^\infty(X)$. The counting function for scattering poles, $N(r)$, counts the number of scattering poles with multiplicity in the region $\{s \in \mathbf{C}\colon |s| < r\}$. The authors show that $N(r)$ is bounded below by constants times r^2 (their previous paper [op. cit.] shows an upper bound of the same form); this result is remarkable for being the first instance of sharp upper and lower bounds for a large class of elliptic operators in the absence of symmetries.

Weyl laws for the scattering phase date back to A. Selberg's 1956 analysis of spectral theory for finite-volume hyperbolic surfaces [J. Indian Math. Soc. (N.S.) **20** (1956), 47–87; MR **19**, 531g]; for such surfaces, Selberg showed that the sum of the counting function for eigenvalues of the Laplacian and the "winding number of the scattering operator" obeys an asymptotic law analogous to Weyl's law for the eigenvalues of the Laplacian on a compact surface. For the manifolds X considered here, the scattering operator $S_X(s)$ is not a matrix as in the Selberg case but a pseudodifferential operator on the geometric boundary of X. The authors define a renormalized scattering operator using a trick that appears, implicitly at least, in S. J. Patterson's study of the Laplacian on infinite area surfaces [Compositio Math. **31** (1975), no. 1, 83–107; MR **52** #5575; Compositio Math. **32** (1976), no. 1, 71–112;

MR **54** #7385; Compositio Math. **33** (1976), no. 3, 227–259; MR **58** #10750]: they consider a comparison scattering operator which uses scattering on hyperbolic half-cylinders as a "model" and thus obtain a relative scattering operator $\quad_X(s)$ which is the identity plus a trace-class perturbation. Thus one can define a scattering phase $\sigma_X(s)=(i/2\pi)\log\det \quad_X(s)$. One obtains

$$\sigma_X(s)=0\text{-vol}(X)|s|^2-(M/\pi)|s|\log|s|+O(|s|)$$

for a surface with M cuspidal ends; the 0-volume is a natural renormalized volume which in the case of hyperbolic quotients is exactly the volume of the Fenchel-Nielsen region.

Both results depend on a Poisson formula which relates the renormalized trace of the wave group to the scattering resonances ρ:

$$0\text{-tr}\left(\cos\left(\sqrt{\Delta_X-\frac{1}{4}}\right)\right)=\frac{1}{2}\sum_\rho e^{(\rho-1/2)|t|}m_\rho,$$

where m_ρ is the multiplicity of the resonance, and 0-tr denotes a regularized trace over the infinite volume. It is significant that, in contrast to previous studies, the Poisson formula is derived without the use of the Lax-Phillips scattering theory. The authors compute the small-time behavior of the wave trace using results of V. Ya. Ivriĭ [Funktsional. Anal. i Prilozhen. **14** (1980), no. 2, 25–34; MR 82m:58057] and explicit calculations for the model spaces. They estimate the counting function for resonances from below by using the small-time behavior of the wave trace, the Poisson formula, and techniques previously used by the second author and Sjöstrand [op. cit.]. Weyl's law for the scattering phase is obtained from the short-time singularities of the wave trace, a Birman-Kreĭn type trace formula relating the wave trace to the derivative of the scattering phase, and a modified Tauberian argument used by Melrose [Comm. Partial Differential Equations **13** (1988), no. 11, 1431–1439; MR 90a:35183] to study Weyl asymptotics for the scattering phase in exterior obstacle scattering in Euclidean space. *Peter A. Perry* (Lexington, KY)

97i:58177 58G26 46L05 57Q10 58G12

Burghelea, D. (Columbus, OH); **Friedlander, L.** (Tucson, AZ);
Kappeler, T. (Columbus, OH); **McDonald, P. [McDonald, Patrick]** (Columbus, OH)

Analytic and Reidemeister torsion for representations in finite type Hilbert modules. (English summary)

Geom. Funct. Anal. **6** (1996), *no.* 5, 751–859.

The classical history of torsion invariants is covered in J. Milnor's article [Bull. Amer. Math. Soc. **72** (1966), 358–426; MR **33** #4922]. This theory is purely combinatorial, associating to the complex given by a triangulation or cell decomposition of a Riemannian manifold an invariant which is defined using determinants of the boundary operator of the complex. A new twist was added by D. B. Ray and I. M. Singer [Advances in Math. **7** (1971), 145–210; MR **45** #4447], who introduced an analytic version of torsion using the zeta function definition of the determinant of the Laplacian on differential forms. They conjectured equality of the analytic and combinatorial torsions and this was eventually proved, independently, by J. Cheeger [Ann. of Math. (2) **109** (1979), no. 2, 259–322; MR 80j:58065a] and W. Müller [Adv. in Math. **28** (1978), no. 3, 233–305; MR 80j:58065b]. The paper under review is concerned with the von Neumann or L^2 version of these invariants. The matter really starts with the L^2 index theorem of Atiyah and Singer (see, for example, a paper by M. F. Atiyah [in *Colloque "Analyse et Topologie" en l'Honneur de Henri Cartan (Orsay, 1974)*, 43–72, Astérisque, 32-33, Soc. Math. France, Paris, 1976; MR **54** #8741] in which the L^2 Betti numbers were introduced). They were investigating a generalization of the Atiyah-Singer index theorem to (non-compact) covering spaces of compact Riemannian manifolds. The idea was to use von Neumann's notion of "continuous" dimension to replace the ordinary dimension. J. Dodziuk [Topology **16** (1977), no. 2, 157–165; MR **56** #3898] proved the Hodge-de Rham isomorphism for the L^2 cohomology, thus establishing the homotopy invariance of the L^2 Betti numbers. The L^2 cohomology was exploited in several places, for example in a paper by Atiyah and W. Schmid [Invent. Math. **42** (1977), 1–62; MR **57** #3310; erratum; MR 81d:22015], and also in the proof of the Hopf conjecture [M. L. Gromov, J. Differential Geom. **33** (1991), no. 1, 263–292; MR 92a:58133] and in the vanishing of the Euler characteristic of classifying spaces of discrete groups which contain an infinite amenable normal subgroup [J. Cheeger and M. L. Gromov, Topology **25** (1986), no. 2, 189–215; MR 87i:58161]. In an earlier paper and in its predecessors [J. Cheeger and M. L. Gromov, in *Differential geometry and complex analysis*, 115–154, Springer, Berlin, 1985; MR 86h:58131; J. Differential Geom. **21** (1985), no. 1, 1–34; MR 87d:58136] the Atiyah-Singer theory was extended to covering spaces of finite-volume manifolds with bounded geometry. Results on L^2 Betti numbers on residually finite covering spaces were proved by W. Lück [Geom. Funct. Anal. **4** (1994), no. 4, 455–481; MR 95g:58234] and, on amenable covering spaces, by Dodziuk and V. Mathai [in *Lipa's legacy (New York, 1995)*, 151–167, Contemp. Math., 211, Amer. Math. Soc., Providence, RI, 1997; MR 98j:58106; J. Funct. Anal. **154** (1998), no. 2, 359–378; MR 99e:58201].

In the late 1980s, Novikov and Shubin [D. V. Efremov and M. A. Shubin, in *Séminaire sur les Équations aux Dérivées Partielles, 1988–1989*, Exp. No. VIII, 19 pp., École Polytech., Palaiseau, 1989; MR 91b:58258] defined some new numerical invariants using the spectral density function of the Laplacian, which measures the von Neumann multiplicity. The motivation for defining these invariants was as follows. By Hodge theory, we know that the multiplicity of the zero eigenvalue of the Laplacian on a compact manifold is a homotopy invariant. What Dodziuk proved was that the von Neumann multiplicity of the zero eigenvalue of the Laplacian on a covering space is again a homotopy invariant. Novikov and Shubin reasoned that the germ of the spectral density function near zero should also be a homotopy invariant and this was established by Gromov and Shubin [Geom. Funct. Anal. **1** (1991), no. 4, 375–404; MR 92i:58184; erratum; MR 96f:58161] and J. Lott [J. Differential Geom. **35** (1992), no. 2, 471–510; MR 93b:58140]; see also a paper by J. Block, Mathai and S. Weinberger [Proc. Amer. Math. Soc. **125** (1997), no. 12, 3757–3762; MR 98d:58175]. There is an outstanding conjecture that the Novikov-Shubin invariants are always positive.

The notion of L^2 analytic torsion for L^2 acyclic covering spaces with positive Novikov-Shubin invariants was first studied by Mathai [J. Funct. Anal. **107** (1992), no. 2, 369–386; MR 93g:58156] and Lott [op. cit.], and L^2 Reidemeister-Franz torsion was first studied by Carey and Mathai [J. Funct. Anal. **110** (1992), no. 2, 377–409; MR 94a:58211]. These L^2 torsion invariants were subsequently generalised by Lück and M. G. Rothenberg [*K*-Theory **5** (1991), no. 3, 213–264; MR 93g:57025]. Further generalisation occurs in the paper under review and more recently using the theory of determinant lines of finitely generated Hilbertian modules over finite von Neumann algebras initiated by Carey, M. Farber and Mathai [J. Reine Angew. Math. **484** (1997), 153–181; MR 98c:58175]. These L^2 torsion invariants specialise to the Ray-Singer-Quillen torsion and the classical Reidemeister-Franz torsion respectively, in the case of finite-dimensional representations of the fundamental group.

The paper under review considers unitary representations of the fundamental group on a Hilbert $\mathcal{A}$ module, where $\mathcal{A}$ is a finite von Neumann algebra, and proves in that setting an L^2 version of the Cheeger-Müller theorem (including their result in the case $\mathcal{A} = \mathbf{C}$). The proof of this equality of the combinatorial and analytic L^2 torsions is obtained whenever the covering space is of so-called "determinant class". This latter condition is a restriction which arises because zero is, in general, in the spectrum of the Laplacian and hence the L^2 zeta function is not a priori well defined. The determinant class condition is satisfied for example whenever the Novikov-Shubin invariants are positive.

By this result and calculations done independently by Mathai [op. cit.] and Lott [op. cit.], one obtains an alternate simplicial description of hyperbolic volume for a closed 3-dimensional hyperbolic manifold in terms of the L^2 Reidemeister-Franz torsion. It is also known that residually finite covering spaces are of determinant class, a result going back to Lück [op. cit.]. A complete proof of this fact is given in a paper by Burghelea, Friedlander and Kappeler ["Torsions for manifolds with boundary and glueing formulas", Preprint, http://xxx.lanl.gov/abs/dg-ga/9510010] (and extended to amenable covering spaces by Dodziuk and Mathai [op. cit.]).

The main tool in the proof of equality of the two L^2 torsions is an L^2 version of the Witten deformation [E. Witten, J. Differential Geom. **17** (1982), no. 4, 661–692 (1983); MR 84b:58111] following ideas of B. Helffer and J. Sjöstrand [Comm. Partial Differential Equations **10** (1985), no. 3, 245–340; MR 87i:35162] and Shubin [Geom. Funct. Anal. **6** (1996), no. 2, 370–409; MR 97i:58171].

For the strategy of the proof one needs to refer to earlier work of Burghelea, Friedlander and Kappeler [J. Funct. Anal. **137** (1996), no. 2, 320–363; MR 97b:58147; J. Funct. Anal. **107** (1992), no. 1, 34–65; MR 93f:58242; op. cit., Preprint] and to papers by J.-M. Bismut and W. P. Zhang [Astérisque No. 205 (1992), 235 pp.; MR 93j:58138; Geom. Funct. Anal. **4** (1994), no. 2, 136–212; MR 96f:58179]. These papers present alternative new proofs of the Cheeger-Müller theorem. It was Bismut and Zhang who first exploited the Witten deformation in the context of the Cheeger-Müller result, with the Burghelea-Friedlander-Kappeler papers contributing a different approach using a Mayer-Vietoris glueing argument for torsion. It is this second proof which is generalised to the L^2 setting in the paper under review. This necessitates a somewhat lengthy excursion in which the Mayer-Vietoris technique is extended to the von Neumann module situation. This is accomplished in Section 3 and depends on von Neumann module versions of some asymptotics à la Seeley [R. T. Seeley, in *Singular Integrals (Proc. Sympos. Pure Math., Chicago, Ill., 1966)*, 288–307, Amer. Math. Soc., Providence, R.I., 1967; MR **38** #6220; Amer. J. Math. **91** (1969), 963–983; MR **42** #877] which are discussed in Section 2 although full details are not provided. This early part of the paper describes previous contributions to the theory of L^2 torsion from the viewpoint of the authors and introduces a novel way of handling operators of determinant class. The Witten deformation in the L^2 setting occupies Section 4. The main proofs appear in Section 5 (On the L^2 Witten deformation of the de Rham complex) and Section 6 (On the asymptotic analysis of the deformation of the torsion).

The idea of the proof (if not the details) can be briefly explained. One applies the work of Witten [op. cit.] and Helffer and Sjöstrand [op. cit.] extended to the L^2 Witten deformation of the de Rham complex. This deformed complex is split into a direct sum of "small" and "large" subcomplexes with the small complex being of finite type (in the sense of being a finite von Neumann module). The small complex is the key one and is asymptotically isometric to the combinatorial complex arising from a triangulation of the covering space. The Witten deformation localises the torsion of the small complex to a neighbourhood of the critical points but, in the approach adopted by the authors, one needs to glue via the Mayer-Vietoris argument onto the complex on the complement of this neighbourhood. This is where the detailed analysis (of Sections 2 and 3) of L^2 torsion on manifolds with boundary is exploited. Once one handles, by this method, the error in the torsion introduced by using only the small complex, the equality of the two torsions may be established. *Alan L. Carey* (Adelaide)

E. REVIEWS IN DIFFERENTIAL EQUATIONS AND APPLIED MATHEMATICS

99h:34003 34A20 34E05

Costin, Ovidiu (Chicago, IL)

On Borel summation and Stokes phenomena for rank-1 nonlinear systems of ordinary differential equations.

Duke Math. J. **93** (1998), *no.* 2, 289–344.

The author studies analytic nonlinear (systems of) ODE's at an irregular singularity of rank 1 under nonresonance conditions. He starts with a prepared form for such equations:

$$y' = f_0(x) - \Lambda y - x^{-1}By + g(x,y),$$

where f_0 and g are analytic at $x=\infty$, $y=0\in\mathbf{C}^n$, $\Lambda=\operatorname{diag}(\lambda_j)$ with the λ_j linearly independent over $\mathbf{Z}$, $B=\operatorname{diag}(\beta_j)$, $f_0(x)=O(x^{-M-1})$, $g(x,y)=O(y^2,x^{-M-1}y)$, $x\to\infty$, $y\to 0$. He assumes that $\arg\lambda_h\neq\arg\lambda_j$ if $h\neq j$, and $\operatorname{Re}\beta_h<0$. The condition on the β_h can always be obtained by a suitable substitution.

There exists an n-parameter family of transseries solutions (formal exponential series)

$$\widehat{y} = \widehat{y}_0 + \sum_{k\in\mathbf{N}^n} C^k e^{-(k,\lambda)x} x^{(k,m)} \widehat{y}_k$$

where $\widehat{y}_k = x^{-(k,\beta+m)}\sum_{l=0}^{\infty} a_{k,l}x^{-l}$, $m_j=-1-[\beta_j]$, $j=1,\cdots,n$, and the parameter $C\in\mathbf{C}^n$ [cf. M. Iwano, Ann. Mat. Pura Appl. (4) **44** (1957), 261–292; MR **20** #3320; J. Écalle, *Les fonctions résurgentes. Tome I*, Univ. Paris XI, Orsay, 1981; MR 84h:30077a; *Tome II*, 1981; MR 84h:30077b; *Tome III*, 1985; MR 87k:32009; W. Balser, Asymptot. Anal. **15** (1997), no. 3-4, 261–282; MR 98g:34015b, see p. E8 in this volume]. Here $\widehat{y}_0$ is a formal power series solution of the given ODE and $\widehat{y}_k$, $k\neq 0$, is a formal series solution of an associated linear ODE. Generically all these formal series are divergent.

The author considers the subset of transseries which are asymptotic expansions in a direction d, so $C_j \neq 0$ implies $|\arg \lambda_j + d| < \pi/2$. In general these expansions are not classical since the terms in the transseries are beyond all orders of each other. He shows that there is a one-to-one correspondence between these transseries and actual solutions $y \sim \widehat{y}_0$ on d. This correspondence is obtained by means of a generalization of the Borel transform.

It is known that $\widehat{y}_k$ is Borel-summable in all directions except the so-called Stokes directions. These Stokes directions contain singularities of the usual Borel transform Y_k of $\widehat{y}_k$. The Y_k satisfy a convolution equation (E_k) and they have infinitely many singularities (in general branchpoints) due to the nonlinearity of the ODE. The singularities become less regular with respect to integrability the further away they are from the origin. Therefore the inverse of the Borel transform—the Laplace transform—of Y_k over a Stokes line does not exist in the usual sense. So the main problem is to sum the transseries in a Stokes direction.

The author considers without loss of generality the case in which 1 is the first singularity of Y_0 on $\mathbf{R}^+$ and considers the Stokes line $\mathbf{R}^+$. Then there are also singularities at the points $h \in \mathbf{N}$. To overcome the integrability problems at these singularities he introduces so-called staircase distributions on $\mathbf{R}^+$ which are of order mh on $(0, h)$ for all $h \in \mathbf{N}$ and some $m \in \mathbf{N}$, and a suitable convolution subalgebra of these which forms a Banach algebra $\mathcal{D}'_m$. He shows that Y_0 can be extended to a one-parameter family of solutions in $\mathcal{D}'_m$ of (E_0) which are obtained from the analytic continuations of Y_0 in the upper and in the lower half-plane. If $k \neq 0$ then Y_k satisfies a linear convolution equation (E_k) depending on the $Y_{k'}$ with $k' \prec k$ and it has singularities at all positive integers. It appears that generically Y_k is uniquely determined by Y_0. This is an example of the general phenomenon of resurgence.

There exists a one-parameter family of weighted averages of the analytic continuations of Y_k which are utilized to obtain maps of the formal solution $\widehat{y}_k$ to solutions in $\mathcal{D}'_m$ of (E_k). These maps commute with complex conjugation and with differentiation and they are the required generalized Borel transforms. Application of these transforms to the terms $\widehat{y}_k$ in the transseries followed by the Laplace transform maps this transseries into an actual solution y of the given ODE depending on the parameter $C \in \mathbf{C}^n$ and with $y \sim \widehat{y}_0$ on $\mathbf{R}^+$. In this way all actual solutions with this asymptotic behavior are obtained. The summation algorithm uses the formal solutions alone and not the differential equations they solve. The sums thus obtained can be written as Laplace integrals over special paths avoiding the singularities.

Finally the author considers local Stokes transitions: he calculates the change in the coefficient $C \in \mathbf{C}^n$ in the transseries at the Stokes directions. This is different from the classical Stokes phenomenon where one considers these changes at the anti-Stokes directions.

The paper elucidates the general theory of Écalle of analyzable functions, resurgence and medianization [cf. J. Écalle, *Introduction aux fonctions analysables et preuve constructive de la conjecture de Dulac*, Hermann, Paris, 1992; MR 97f:58104] for the special class of nonlinear ODE of rank 1 under consideration. Due to this specialization the author only needs a subset of Écalle's median paths in his summation procedure. The resurgence relations which he obtains correspond to Écalle's general bridge equations. In a previous paper [Internat. Math. Res. Notices **1995**, no. 8, 377–417; MR 96m:34103] the author studied the same ODE but with further restrictions on the transseries and with $\beta_j \in [0, 1)$. Compared with the previous paper the use of the convolution algebra of staircase distributions $\mathscr{D}'_m$ is an important new tool: it remedies the nonintegrability properties of the Borel transforms. Resurgence properties of the type studied here occur also in the work of M. V. Berry on hyperasymptotics [Proc. Roy. Soc. London Ser. A **422** (1989), no. 1862, 7–21; MR 90h:34084], where the first asymptotic series reappears in various shapes in the process of computing higher terms in the expansions. *B. L. J. Braaksma* (Groningen)

98g:34015a 34A25 34A20 34A34 34C20 34E05
Balser, Werner (Ulm)
Formal solutions of non-linear systems of ordinary differential equations. (English summary)
The Stokes phenomenon and Hilbert's 16*th problem* (*Groningen*, 1995), 25–49, *World Sci. Publishing, River Edge, NJ,* 1996.

98g:34015b 34A25 34A20 34A34 34C20 34E05
Balser, Werner (Ulm)
Existence and structure of complete formal solutions of non-linear meromorphic systems of ordinary differential equations. (English summary)
Asymptot. Anal. **15** (1997), *no.* 3-4, 261–282.

For systems of linear, homogeneous, meromorphic differential equations, there is a well-established theory which constructs formal fundamental solutions in the vicinity of an irregular singularity and then gives their asymptotic interpretation. For analogous nonlinear systems, a "complete" and general theory has been slow to develop and either has involved various kinds of restrictive assumptions on the equations or else has only been concerned with certain types (e.g., formal power series) of solutions. The purpose of this paper is to lay the groundwork for a complete and general theory of formal solutions for systems of nonlinear (meromorphic) differential equations of the form $z^{r+1}dx/dz = g(z,x)$, where r is a non-negative integer and g is a formal power series in the $n+1$ variables $(z, x_1, x_2, \cdots, x_n)$. By this, the author means constructing an expression of the form $\hat{x}(z) = \sum_{|p|\geq 1} \hat{x}_p(z)c^p$ which formally satisfies the differential equation, where $c = (c_1, c_2, \cdots, c_n)$ contains the n free constants one would expect in a "complete" solution. Here, p is a multi-index, and the $\hat{x}_p(z)$ are vectors of formal log-exponential sums of the form $\hat{f}(z)z^{\lambda'} \log^k z \cdot \exp\{p(z)\}$, where $p(z)$ (not to be confused with the multi-index p) is a polynomial in some root of $1/z$, k is a positive integer, λ' is a complex constant, and $\hat{f}(z)$ ("hat" always means formal) is a formal power series in a root of z. The coefficients can be formally, recursively calculated by formally solving systems of nonhomogeneous linear differential equations, obtained by substituting the above series into the differential equation and equating like powers of the multi-index p. While in special cases such formal solutions have been constructed, it is rather surprising that such expressions suffice, in general.

In order to prove such a far-reaching result, the author first investigates the structure of formal analytic and meromorphic transformations of the nonlinear system. These are natural analogues of ones used in the linear case and while others have used similar transformations, here the treatment is especially elegant, making very effective and efficient use of both notation and terminology. (A good knowledge of the linear theory is especially useful here, say either from W. R. Wasow [*Asymptotic expansions for ordinary differential equations*, Reprint of the 1976 edition, Dover, New York, 1987; MR 88i:34003] or from W. B. Jurkat [*Meromorphe Differentialgleichungen*, Lecture Notes in Math., 637, Springer, Berlin, 1978; MR 82a:34004].)

The transformations can be used to simplify the differential equation and its solutions in much the same way as block-diagonalization according to distinct eigenvalues is used in the linear case. For this purpose, a very useful new concept introduced here is that of the system being "reduced of type $Q(z)$", where Q is a specified diagonal matrix polynomial. This generalizes the block diagonal structure and allows certain kinds of exponential transformations to be made that are essential for the reduction process.

A very important consequence of the reduction procedure is that the reduced equations have solutions which are explicit, elementary functions. This was not the case in earlier treatments of nonlinear systems, expecially in the so-called "regular singular" cases where $r = 0$. Here is where the formal log-exponential transformations are especially useful because they are capable of removing many terms from the nonlinear part which previous treatments could not handle.

Finally, the "detailed" structure of the complete formal solutions is discussed, analogously to the linear theory (e.g., as given in Jurkat [op. cit.]). First a "monodromy function" is introduced, which plays the role for nonlinear systems of the monodromy factor in the linear case. Then, in preparation for an eventual application to an asymptotic theory for the formal solutions, a factorization (composition) theorem shows that the formal transformations can be split according to levels induced by the exponential polynomials present in the formal solution.

{Reviewer's remark: I believe that this seminal work lays an excellent foundation for a new and potentially very useful theory for nonlinear meromorphic systems that will bring the understanding of their solutions to a new level.}

{For the entire collection containing the first paper (which is a preliminary version of the second paper) see MR 97k:34002.} *Donald A. Lutz* (San Diego, CA)

98h:34009a 34A25 12H20 14H20 14P05 30B10 30C55 34A20 34C05
Roytwarf, N. (Rehovot); **Yomdin, Y.** (Rehovot)
Bernstein classes. (English and French summaries)
Ann. Inst. Fourier (Grenoble) **47** (1997), *no.* 3, 825–858.

98h:34009b 34A25 12H20 14P05 32B05 34A20 34C05
Briskin, M.; Yomdin, Y. (Rehovot)
Algebraic families of analytic functions. I.
J. Differential Equations **136** (1997), *no.* 2, 248–267.

98h:34009c 34A25 12H20 14P05 32B05 34A20 34C05
Francoise, J.-P. (Paris); **Yomdin, Y.** (Rehovot)
Bernstein inequalities and applications to analytic geometry and differential equations. (English summary)
J. Funct. Anal. **146** (1997), *no.* 1, 185–205.

General background: These three papers are devoted to different aspects of the same problem: establishing global finiteness properties for some natural families of analytic functions. The following families serve as main examples: algebraic functions of a given degree, solutions of algebraic differential equations of a given degree, and

Poincaré "first return mappings". By "finiteness" the authors mean a variety of topological, metric, algebraic and analytic properties illustrated for polynomials by two basic ones: Bézout's theorem, which bounds the number of zeroes of a polynomial by its degree, and Bernstein's inequality [S. Bernstein, Proc. Kharkov Math. Soc. (2) **14** (1913), 1–6; per bibl.], which bounds the ratio of the maxima of a polynomial on two cofocused ellipses (in terms of its degree and the ratio of the semiaxes of the ellipses). Traditionally these two properties are investigated along two quite separate directions.

In the realm of algebraic geometry, starting with polynomials (i.e. with a finite-dimensional effectively given "input") one can obtain, at least in principle, a finite-dimensional effectively given answer to any algebraic or topological question concerning the set of common zeroes of these polynomials.

In turn, Bernstein's inequality is basic for approximation theory (and, more generally, "constructive function theory"), where the interrelations among analytic, approximative and metric properties of functions are investigated.

Extension of finiteness properties of polynomials to wider classes of functions is in no way straightforward. Integration of algebraic functions still can be treated to some extent within the algebraic geometry framework (although some basic finiteness questions, like bounding the number of zeroes of Abelian integrals, become highly nontrivial). However, solving algebraic differential equations seems to destroy finiteness properties completely. Virtually any global phenomenon which can be encountered for general analytic functions can be reproduced also for solutions of algebraic differential equations.

Besides the classical examples of differential equations generating special functions, only very few situations are known to which finiteness results can be extended. Most of them have been found within the framework of transcendental number theory [see C. L. Siegel, *Transcendental numbers*, Ann. of Math. Stud., 16, Princeton Univ. Press, Princeton, N. J., 1949; MR **11**, 330c; A. J. van der Poorten, Enseignement Math. (2) **23** (1977), no. 1-2, 19–38; MR **57** #16221]. Another important class is formed by Pfaffian functions [A. G. Khovanskiĭ, *Fewnomials*, Translated from the Russian by Smilka Zdravkovska, Amer. Math. Soc., Providence, RI, 1991; MR 92h:14039].

As far as Bernstein's inequality is concerned, the problem of its generalization even to algebraic functions remained open for a long time. Recently a generalization has been obtained by C. L. Fefferman and R. Narasimhan [Ann. Inst. Fourier (Grenoble) **44** (1994), no. 4, 1091–1179; MR 95k:32011] and (for a closely related Markov inequality) by L. Bos et al. [Indiana Univ. Math. J. **44** (1995), no. 1, 115–138; MR 96i:41009], and improved by A. Y. Brudnyi [Indiana Univ. Math. J. **46** (1997), no. 1, 93–116; MR 98k:32008; Ann of Math. (2) **149** (1999), no. 2, 511–533].

Hilbert's problem: Probably the most prominent finiteness problem, related to a fairly general class of algebraic differential equations, is the second part of Hilbert's sixteenth problem, which asks whether the number of limit cycles of a polynomial vector field is always bounded in terms of its degree. This is one of the few Hilbert problems which remain unsolved.

The following two discoveries were probably the most important in the long line of research related to Hilbert's sixteenth problem: 1. The first is based on the results of classical investigations of H. Dulac [J. École Polytech. (2) **9** (1904), 5–125; JFM 35.0331.02], A. Kapteyn, and N. Bautin [C. R. (Doklady) Acad. Sci. URSS (N. S.) **24** (1939), 669–672; MR **2**, 49a; translated in American Math. Soc. Translation **1954**

(1954), no. 100, 19 pp.; MR **15**, 527h] and allows one to bound (in terms of the degree only) the number of limit cycles which can appear locally in small perturbations of certain degenerate situations.

2. The second is based on the much more recent results of Yu. S. Il′yashenko [Uspekhi Mat. Nauk **45** (1990), no. 2(272), 143–200, 240; MR 92a:58110] on one hand, and R. Bamón and J. P. Écalle et al. [C. R. Acad. Sci. Paris Sér. I Math. **304** (1987), no. 13, 375–377; MR 89i:58121a] on the other, and gives a global finiteness of the number of limit cycles for each individual vector field (though it leaves open the question of the existence of a bound, depending on degrees only, even for the vector fields of degree two).

The main tool in these investigations was the study of the first return (or Poincaré) mapping associated with a closed trajectory (or a "polycycle" of the vector field).

In the second case, which presents the main difficulties in establishing global finiteness, the Poincaré mapping usually has a very complicated singularity at the origin, and highly sophisticated asymptotic expansions have been used to prove that its fixed points cannot accumulate at the origin. The main open problem in this direction is a bound on the number of the fixed points only in terms of the degree.

In contrast with the latter, in the classical Bautin case, the Poincaré mapping is regular at the origin, and its Taylor expansion can be used. Bautin discovered that the finite dimensionality of the "input", seemingly lost in the trajectories of the vector field, returns as the algebra of these Taylor coefficients (which turn out to be polynomials in the parameters of the field). The main open problem here is how to globalize the local bounds. This problem naturally splits into two parts: 1. Globalization to the disk of convergence of the Taylor series. 2. Globalization beyond this disk (and, in particular, beyond singularities of the Poincaré mapping).

While the first part can, one hopes, still be treated by a careful investigation of the algebra of the Taylor coefficients, the second one requires analytic continuation of the Poincaré mapping, and probably encounters at least some of the difficulties which appear in the study of polycycles.

Papers under review: In the framework given above, the papers under review can be considered as a continuation of Bautin's line of research, which relates the finiteness properties of functions with the algebraic properties of their Taylor coefficients.

The main problem treated is that of a globalization of the results produced by Bautin's approach. In the case of solutions of algebraic ODE's and of their Poincaré mappings, the authors obtain a partial answer: globalization to a disk of a size explicitly given in terms of the parameters of the problem. In the case of algebraic functions the authors obtain a globalization to the disk of convergence.

Two important new tools are introduced. On the algebraic side the authors give an analogue of the convergent Hironaka division theorem, which turns out to be directly related to a globalization of Bautin's results, and the recurrence relations that preserve ideals, which allow one to compute Bautin ideals for solutions of algebraic ODE's. On the analytic side the authors show the role of Bernstein's inequality as an adequate analytic tool. In particular, it brings together the two lines of investigation mentioned above.

Finally, the authors state and discuss various aspects of the problem of globalization to the disk of convergence, i.e. that of the existence for a given family of the structural Bernstein inequality (as it is called in the papers under review).

Let us now describe in greater detail the content of each of the papers. The main purpose of the first paper is to relate properties of the Taylor coefficients of a function, which satisfies the Bernstein-type inequalities, with the number of its zeroes, i.e., the number of points at which f vanishes.

By definition, an analytic function f on a disk D_{2R} of radius $2R$ in $\mathbf{C}$ is said to belong to the Bernstein class B_K^1 if

$$(\max |f| \text{ on } D_{2R})/(\max |f| \text{ on } D_R) < K.$$

On the other hand f is said to belong to the Bernstein class $B_{N,C}^2$ on D_R if the Taylor coefficients a_j of f satisfy $|a_j|R^j < C\max\{|a_i|R^i;\ i = 0, 1, \cdots, N,$ for any $j > N\}$. Polynomials of degree d belong to $B_{d,0}^2$, and to B_K^1, with $K = 2^d$. The function $\exp \lambda x$ belongs to B_K^1, with $K = \lambda R$, and to $B_{1,C}^2$, with C roughly $\exp \lambda R$, or, alternatively, to $B_{d,1}^2$, with d roughly λR.

The first main result of the first paper is that the classes B^1 and B^2 essentially coincide: if f belongs to B_K^1, it also belongs to $B_{N,C}^2$, with N roughly $\log K$ and C explicitly given in terms of K, and if f belongs to $B_{N,C}^2$, it also belongs to B_K^1, with K explicitly given in terms of N and C.

The relation between the number of zeroes and the behavior of Taylor coefficients is known classically. Call f as above p-valent if it attains any value at most p times in D_R. The classical result is that if f is p-valent, its Taylor coefficients satisfy an inequality $|a_j|R^j < (A/p)^{2p} j^{2p} \max\{a_i|R^i;\ i = 0, 1, \cdots, N\}$, for any $j > p$ [M. Biernacki, C. R. Acad. Sci. Paris **203** (1936), 449–451; Zbl 014.31904]. In particular, f belongs to B_p^2 on any smaller disk. (For $p = 1$, as we know now by the de Branges theorem, the correct answer is $|a_j|R^j < j|a_1|R$.)

On the other hand, another well-known result [A. J. van der Poorten, op. cit.] is that for f in B_K^1 the number of zeroes of f in D_R does not exceed $\log K$.

Combining the results above, the authors show that the properties of f to be p-valent and of f' to belong to B_p^2 (and of f and f' to belong to a corresponding B_K^1 class) are essentially equivalent.

Next the authors use the above approach to prove Bernstein-type inequalities for algebraic functions. Recently such inequalities found applications in the study of pseudodifferential operators [C. Fefferman and R. Narasimhan, op. cit.], in potential theory [L. Bos et al., op. cit.; A. Brudnyi, op. cit., 1997; op. cit., 1999] and in dynamical systems [Y. Yomdin, Israel J. Math. **57** (1987), no. 3, 285–300; MR 90g:58008; addendum; MR 90g:58009; Ergodic Theory Dynam. Systems **11** (1991), no. 3, 583–602; MR 92m:58076]. Since the valency of an algebraic function is bounded in terms of its degree by the Bézout theorem, the authors obtain an explicit (and essentially sharp) Bernstein inequality for algebraic functions of any number of variables. In particular, this improves the results of Fefferman and Narasimhan [op. cit.] (and of Bos et al. [op. cit.]) and answers one of the questions posed in Fefferman and Narashimhan's paper.

In one variable the result is that if a certain branch $y(x)$ of an algebraic function, satisfying $P(x, y) = 0$, with P of degree d in x, is regular in D_R, then for any $R_1 < R$, $y(x)$ belongs to $B_{d,C}^2$ with $C = A(3R/R_1)^{3d}$, where A is an absolute constant.

Let us stress that the inequality obtained is valid in the maximal disk of regularity of the algebraic function considered, and the bounds depend only on the degree of the function and on the relative position of the disks inside the maximal disk of regularity. The authors call the Bernstein-type inequality satisfying these conditions a structural Bernstein inequality, and discuss the problem of for which families of analytic functions the structural Bernstein inequality is valid (e.g. it is not valid for the family $\exp \lambda x$).

As is explained above these results are directed toward extending the finiteness properties of polynomials to wider classes of functions. The first paper also briefly presents the main results of the second and third papers and thus can be viewed as a survey of some of these methods and results.

In the second paper solutions of algebraic differential equations, and, in particular, algebraic functions, are studied by means of a detailed analysis of the recurrence relations for their Taylor coefficients. By means of certain variants of the Cauchy majorization technique the authors prove that starting with some index, each Taylor coefficient is bounded in terms of the initial ones. Consequently, the functions considered belong to B^2 classes, and the results of the first paper imply both the Bernstein-type inequalities and the bounds on the number of zeroes (and, in particular, a generalization of the Fefferman-Narasimhan inequality to solutions of algebraic differential equations). The remainder of the second paper is devoted to the study of the ideals generated by the Taylor coefficients in the ring of polynomials in the parameters of the problem (cf. the discussion below).

In the third paper the authors consider a family of first return (or Poincaré) functions for a polynomial vector field of a given degree on a plane. Zeroes (or, more accurately, fixed points) of these functions correspond to closed trajectories of the vector field (limit cycles, if they are isolated), and the possibility of a bound on their number in terms of the degree only is (roughly) the content of the second part of Hilbert's sixteenth problem. Not much is known about this. One of the very important classical results is a theorem of Bautin [op. cit.; MR **2**, 49a; op. cit.; MR **15**, 527h]. It bounds by 3 the maximal number of limit cycles that may appear in a small perturbation of a center for a quadratic vector field.

One of the main difficulties is that the recurrence relation for the Taylor coefficients of the Poincaré function involves integration in time. As a result, it does not produce pointwise inequalities between the Taylor coefficients, and the direct approach of the second paper does not work. It was Bautin's discovery that another important property of these Taylor coefficients can be used: they turn out to be polynomials in the parameters of the initial equation. Denote by I the ideal generated by all Taylor coefficients. By Hilbert's finiteness theorem, I is generated by a finite number d of these coefficients. Using Rolle-lemma-type arguments, Bautin has shown that d bounds the number of zeroes of the Poincaré function that may appear locally. He also succeeded in computing the number d for vector fields of degree 2. This remains the only case where this number is known.

In the third paper the authors combine Bautin's approach with that of the convergent Hironaka division theorem to provide explicit (although only in principle) bounds for the size of admissible perturbation in Bautin's theorem. An important observation here is that the structure of the ideal I is directly related to the property of the corresponding function to belong to the class B^2. Indeed, assume that $a_0, \cdots, a_d$ generate the ideal I. Then for any a_j with $j > d$, $a_j = \sum_{i=0}^d p_i^j a_i$. Assuming that we can estimate the coefficients p_i^j in this expression, it will provide (for any value of the parameters) the estimate for a_j, which is necessary to show that the Poincaré function belongs to B_d^2. But a calculation similar to the one in the convergent Hironaka division theorem allows one to estimate p_i^j assuming that the degrees and the norms of a_j are known.

In the second part of the second paper the authors introduce a class of analytic families, which they call an A_0-series. The approach of the third paper can be naturally extended to these families. An A_0-series consists of a power series with polynomial coefficients parametrized by a finite-dimensional parameter λ, such that the degrees of these polynomials grow at most linearly, and their norms at most exponentially with their index. The Bautin ideal I is generated by all coefficients, and the Bautin index d is the minimal number of coefficients that generate I. The authors of the second paper prove (generalizing the result of the third paper) that for any value of the parameter λ, the number of zeroes of the A_0-series in the disk of radius $r(\lambda)$ is at most d, where $r(\lambda)$ is an explicitly given function (tending to zero as λ tends to infinity). As an example, the authors compute the Bautin ideals and the Bautin index for the A_0-series given by algebraic functions and, also, by solutions of algebraic differential equations. (For the equation $y' = P(x, y)$, d is the degree in x of $P(x, 0)$ and the generators of I are written explicitly in terms of the coefficients of P.) The computation utilizes recurrence relations, which, in contrast to the case of the Poincaré mapping, preserve the ideals. *Pierre D. Milman* (Toronto, ON)

98e:34029 34Bxx 34E10

Kozlov, Vladimir [Kozlov, Vladimir Arkad′evich] (Linköping); **Maz′ya, Vladimir** (Linköping)

★**Theory of a higher-order Sturm-Liouville equation.**

Lecture Notes in Mathematics, 1659.

Springer-Verlag, Berlin, 1997. xii+140 pp. $33.00. ISBN 3-540-63065-1

The ordinary differential equation under consideration is

$$\{(\partial + k_+)^m(-\partial - k_-)^n - q(t)\}w(t) = f(t),$$

mainly on $\mathbf{R}$ but also on $\mathbf{R}_+$ or a bounded interval. Here $\partial = d/dt$, k_+ and k_- are real constants with $k_+ > k_-$, m and n are positive integers, and q is measurable and non-negative. The special case where $k_+ = -k_- = k$ and $m = n = 1$ is the Sturm-Liouville equation $-w''(t) + \{k^2 - q(t)\}w(t) = f(t)$. More detailed conditions imposed on q are that either

$$\sup q(t) < m^m n^n \{(k_+ - k_-)/(m+n)\}^{m+n} \quad \text{or}$$

$$\int_{\mathbf{R}} q(t)dt < \{(m-1)!(n-1)!/(m+n-2)!\}(k_+ - k_-)^{m+n-1},$$

which, in the Sturm-Liouville case, become $\sup q(t) < k^2$ or $\int_{\mathbf{R}} q(t)dt < 2k$. Thus the book covers basically an extension of the hyperbolic situation for the second-order equation.

The central theme of the book is the Green's function defined initially for $q(t) = 0$ and $f(t) = \delta(t)$ and then for general q by the Neumann series. Later the definition in terms of the adjoint equation is recovered. Detailed estimates on the size of the Green's function as $t \to \pm\infty$ are obtained, as well as positivity and monotonicity properties. These results lead to existence and uniqueness theorems for solutions $w(t)$ subject to growth conditions at $\pm\infty$. As may be expected, the exponentials

$\exp(\pm k_{\pm}t)$ feature in these results, and both necessary and sufficient conditions are given.

Detailed asymptotic formulae with error term are also obtained for the homogeneous equation with $f=0$, and one example is as follows. Theorem. Let σ and κ be integers such that $0\le\kappa\le\sigma\le m-1$. Then the condition $\int^{\infty}t^{m+\kappa-1}q(t)dt<\infty$ is necessary and sufficient for the existence of a solution $w(t)$ such that

$$w(t)=t^{\sigma}\exp(-k_{+}t)\{1+o(t^{-\kappa})\}\quad(t\to\infty).$$

There is a similar result for k_{-} and n in place of k_{+} and m, and these results reduce to known ones for the Sturm-Liouville equation when $m=n=1$, as shown in the book of P. Hartman [*Ordinary differential equations*, Reprint of the second edition, Birkhäuser, Boston, Mass., 1982; MR 83e:34002 (p. 381)], although this standard reference is not cited.

The authors moot the possibility of generalizing their differential equation so that the differential operator is $M(\partial)=\sum a_s\partial^s$, a finite sum with constant coefficients for which the roots of the characteristic equation $M(z)=0$ are real. However, this matter is not pursued. In an appendix, the authors give applications of their theory to ordinary differential equations with operator coefficients, together with the promise of more to follow in another forthcoming book.

In the book under review, the authors have identified a special type of higher-order analogue of the hyperbolic Sturm-Liouville equation and they have developed a coherent theory based on the Green's function. The account is concise but full and written with clarity. *M. S. P. Eastham* (Cardiff)

99f:34029 34B15 34A26 47H15 58C27

Malta, Iaci (Rio de Janeiro); **Saldanha, Nicolau C.** (Rio de Janeiro); **Tomei, Carlos** (Rio de Janeiro)

Morin singularities and global geometry in a class of ordinary differential operators.

Topol. Methods Nonlinear Anal. **10** (1997), *no.* 1, 137–169.

The authors consider the nonlinear ordinary differential equation $(*)$ $u'(t)+f(t,u(t))=g(t)$, where t is in the circle S^1 (i.e. t is periodic), u is real-valued, and f maps $S^1\times\mathbf{R}$ to $\mathbf{R}$. This differential equation defines an operator $F\colon B^1\to B^0$, with $F(u)[t]=u'(t)+f(t,u(t))$, where the domain B^1 is either $C^1(S^1)$, the Banach space of periodic functions with continuous derivatives, or $H^1(S^1)$, the Hilbert space of periodic functions with derivatives in $L^2(S^1)$. They seek (topological or smooth) global coordinate changes in both the domain B^1 and range B^0 so that F has some simple standard form Θ. In this case we say that F is (topologically or smoothly) globally equivalent to Θ. For E any Banach space, the fold map $\Phi_1\colon\mathbf{R}\times E\to\mathbf{R}\times E$ is defined by $\Phi_1(t,v)=(t^2,v)$, and the cusp map $\Phi_2\colon\mathbf{R}^2\times E\to\mathbf{R}^2\times E$ is defined by $\Phi_2(s,t,v)=(s^3+ts,t,v)$. If an operator is globally equivalent to Φ_1 [resp., Φ_2], we say that it is a global fold [resp., cusp].

In an influential paper A. Ambrosetti and G. Prodi [Ann. Mat. Pura Appl. (4) **93** (1972), 231–246; MR **47** #9377] proved that a certain nonlinear Dirichlet problem led to an operator on Hölder spaces with properties of Φ_1, and M. S. Berger and the reviewer [Indiana Univ. Math. J. **28** (1979), no. 6, 935–952; MR 80k:58097] proved that this operator on Sobolev spaces was a global fold. This seems to be the first proof, for a differential operator that is neither linear nor bijective, of global equivalence to a standard form. H. P. McKean, Jr. and J. C. Scovel [Ann. Scuola Norm. Sup. Pisa Cl. Sci. (4) **13** (1986), no. 2, 299–346; MR 88e:58103] then showed that, for convex nonlinearities f like $f(t,x)=x^2$, the operator F resulting from equation $(*)$ is a global fold. The reviewer, E. N. Dancer and J. G. Timourian [Trans. Amer. Math. Soc. **338** (1993), no. 1, 1–42; MR 93j:35076] proved that for a certain nonlinear Dirichlet problem, the operator was a topological global cusp, and B. Ruf [Ann. Scuola Norm. Sup. Pisa Cl. Sci. (4) **17** (1990), no. 1, 1–33; MR 92d:35107] proved that for the corresponding Neumann problem the operator had many of the properties of the cusp map Φ_2. I. B. Mandhyan [Nonlinear Anal. **13** (1989), no. 9, 1057–1066; MR 90i:58011] showed that a certain integral operator was a topological global cusp. Two extensive surveys of differential equations leading to folds, cusps and even other Morin singularities (see below) are [in *Topological nonlinear analysis, II (Frascati, 1995)*, 109–160, Birkhäuser Boston, Boston, MA, 1997; MR 98e:58039; ibid., 161–245; MR 98e:58040] by the reviewer and Timourian, and [in *Topological nonlinear analysis, II (Frascati, 1995)*, 315–395, Birkhäuser Boston, Boston, MA, 1997; MR 98a:58023] by Ruf.

In Section 1 of the present paper the authors use a global Lyapunov-Schmidt decomposition of F for $(*)$, writing each $u\in B^i$ $(i=0,1)$ as $\widetilde{u}+\overline{u}\in\widetilde{B}^i+\overline{B}^i$, where $\widetilde{u}$ has average 0 and $\overline{u}$ is constant, and $F(u)=v$. They prove (Theorem 1.2) that for each $\overline{u}$, the map $\widetilde{u}\mapsto\widetilde{v}$ is a global diffeomorphism, thereby giving new coordinates in the domain and writing F in these adapted coordinates as $(\widetilde{v},\overline{u})\mapsto(\widetilde{v},\overline{v})$. They then prove that, for example for $f(t,x)=x^2$, the map F of $(*)$ has a fold on each fiber and thus is itself a global fold. In general, properness of f implies the properness of F (Proposition 1.4). These adapted coordinates are an important tool in the sequel. Furthermore, if 0 is a regular value of Σ defined by $\Sigma(u)=\int D_2f(t,u(t))\,dt$ (i.e. if f is generic), then there is a global homeomorphism Ξ of B^1 taking the critical set S_1F to a closed linear subspace of B^1 of codimension 1; if $B^1=H^1$, then Ξ can be chosen to be a diffeomorphism.

The Morin singularity of order k $(k=1,2,\cdots)$ is given by $\Phi_k\colon\mathbf{R}^k\times E\to\mathbf{R}^k\times E$, $(x_1,x_2,\cdots,x_k,v)\mapsto(x_1^{k+1}+x_2x_1^{k-1}+\cdots+x_kx_1,x_2,\cdots,x_k,v)$, for any Banach space E, so that cases $k=1$ and 2 are the fold and cusp, respectively. In Section 2 the authors prove that at generic critical points of F, the operator F is (locally) a Morin singularity, and they give considerable information, especially (Section 3) in the autonomous case $f(x)$ (independent of t). In Section 4 the adapted coordinates and fibration of Theorem 1.2 are used to study several classes of smooth functions $f\colon S^1\times\mathbf{R}\to\mathbf{R}$. For example, suppose that f is proper and either (a) $D_2^2f(t,x)>0$ or (b) 0 is a regular value of Σ and all singularities of F are (local) smooth folds; then F is a smooth global fold.

In their paper cited above McKean and Scovel asked about the global nature of the operator F for $(*)$ with $f(t,x)=x^3-x$. V. V. Cafagna and F. Donati [C. R. Acad. Sci. Paris Sér. I Math. **300** (1985), no. 15, 523–526; MR 86g:58114] and the reviewer and Timourian [Nonlinear Anal. **20** (1993), no. 11, 1319–1343; MR 94h:58043] considered a generalized form of this question with $f(t,x)=ax+bx^2+cx^{2k+1}$, where $k=1,2,\cdots$, and $a\geq 0$, $a^2+b^2>0$ and $c>0$, and they proved that F has many properties of the global cusp Φ_2. (This is a natural approach since a map satisfying certain listed properties is a topological global cusp [P. T. Church and J. G. Timourian, op. cit.; MR 94h:58043 (p. 1323)].) In Section 5 the present authors prove that F is a global cusp (Corollary 5.5) as a consequence of Theorem 5.1: Let

$f:\mathbf{R}\to\mathbf{R}$ be a proper function such that (a) $f'''(x)\geq 0$ [resp., $f'''(x)\leq 0$] for all real x, (b) $f'''(x)$ has isolated roots, and (c) $f'(x)$ assumes both signs. Then F is a topological global cusp, and in the case $H^1(S^1)$ it is smooth. The argument in part (Lemma 5.4) uses a sketched global parametrized version of Whitney's proof of the standard coordinates for a cusp in dimension 2 [H. Whitney, Ann. of Math. (2) **62** (1955), 374–410; MR **17**, 518d].

If $f(t,x)$ is a polynomial in x of degree $d\leq 3$ with coefficients in t and nonzero coefficient of highest degree, then from the results of Sections 4 and 5 the operator F associated to $(*)$ is a diffeomorphism, a global fold or a global cusp. In particular, equation $(*)$ has at most d periodic solutions. In contrast, A. Lins Neto [Invent. Math. **59** (1980), no. 1, 67–76; MR 81i:34009] had proved that for such a polynomial f with degree $d=4$ the number of solutions can be arbitrarily large. In Section 6 the authors obtain such a polynomial $f(x)$ of degree 4 for which the operator F given by $(*)$ at a point u_b has a Morin singularity of order $k=4$ (a butterfly). It follows that there are smooth periodic functions v near $F(u_b)$ with 6 solutions.

This paper makes a substantial contribution to a difficult and interesting area.

P. T. Church (Syracuse, NY)

98i:34051 34C05 58F21
Żołądek, Henryk (Warsaw)
Algebraic invariant curves for the Liénard equation. (English summary)
Trans. Amer. Math. Soc. **350** (1998), *no.* 4, 1681–1701.

The paper contains a thorough study of algebraic invariant curves for the polynomial Liénard equation $\dot{x}=y$, $\dot{y}=-f_m(x)y-g_n(x)$ for all degrees (m,n) of f_m and g_n respectively, except $(m,n)=(1,3)$. The Liénard equation is important because it is equivalent to the second order differential equation.

It is now well known that invariant algebraic curves are fundamental in the study of polynomial systems. G. Darboux [Bull. Sci. Math. Astronom. (2) **2** (1878), 60–96, 123–124, 151–200; JFM 10.0214.01] was the first to remark that if a system has sufficiently many invariant algebraic curves, then this allows one to find an integrating factor or a first integral. The work of Darboux was forgotten for many years until D. I. Schlomiuk [Trans. Amer. Math. Soc. **338** (1993), no. 2, 799–841; MR 93j:34037] noticed that algebraic invariant curves control the bifurcation diagram of quadratic systems with centers. Also, in the presence of algebraic invariant curves, the number of limit cycles of a polynomial vector field drops significantly.

There exists an explicit algorithm to find the invariant algebraic curves of a given degree. However, looking for an invariant curve $F(x,y)=\sum_{0\leq i+j\leq n}a_{ij}x^iy^j=0$ leads to solving a system of polynomial equations whose variables are the coefficients a_{ij} of F. Although the tool of Gröbner bases exists for such a problem, in practice the complexity of the problem grows so rapidly with the degree of F that the method is inapplicable without new theoretical ingredients.

K. Odani started the study of algebraic invariant curves of the Liénard equation and in his paper [Differential Equations Dynam. Systems **5** (1997), no. 3-4, 347–354; MR 99j:34039; see also J. Differential Equations **115** (1995), no. 1, 146–152; MR 95i:34051] analyzed the case $n\leq m$, in which case he concluded that if both polynomials f_m and g_n do not vanish and their quotient is not a constant, then the system has no invariant algebraic curve. The results of Żołądek apply to the remaining case $n > m$, where he shows that for any given (m, n) there always exist systems with an invariant algebraic curve. He also shows that generic systems different from the linear one have no invariant algebraic curve. Moreover he shows that for $n > m + 1 > 2$ there exist systems with an algebraic limit cycle.

The ingredients of the proof of Żołądek are the following. He considers three kinds of invariant curves, which are the only ones which can occur for a Liénard equation: rational curves $y = R(x)$, hyperelliptic curves $(y + P(x))^2 = Q(x)$ (P, Q rational functions) and d-fold curves ($d \geq 3$). The last ones are obviously the difficult ones to investigate but fortunately their occurrence is very limited. To prove the existence of systems with invariant algebraic curves, Żołądek starts by deriving the general form of a system with a given invariant algebraic curve and he constructs systems with rational invariant curves or hyperelliptic invariant curves or limit cycles. In the converse part of the theorem he needs to discuss the absence of invariant algebraic curves other than the ones already found. For that purpose he needs to introduce two other ingredients. The first one is the study of singular points at infinity (in the compactification of the system to the projective space), since algebraic invariant curves pass through these points. This yields a severe restriction to the kind of d-fold curves that can occur. The second ingredient is the asymptotic expansion of the characteristic curves (just a few of them can be branches of algebraic curves) of the singular points at infinity and the introduction of local first integrals around these points. In particular, in the case of the existence of a d-fold invariant algebraic curve two of the singular points at infinity are formally linearizable and he derives the existence of a formal local Darboux first integral. The absence of such a local first integral is hence an obstruction to the existence of such a curve.

The paper is self-sufficient, in the sense that it presents a new proof of Odani's result as well as a simple presentation of A. I. Yablonskiĭ's example of a quadratic system with a quartic limit cycle.

{See also M. Hayashi [Funkcial. Ekvac. **39** (1996), no. 3, 403–408; MR 97m:34056].} *Christiane Rousseau* (Montreal, QC)

98m:34053 34C05 34D30 58F21

Artés, Joan C.; Kooij, Robert E.; Llibre, Jaume

Structurally stable quadratic vector fields. (English summary)

Mem. Amer. Math. Soc. **134** (1998), *no.* 639, viii+108 pp.

The description of the topology of the phase portraits of polynomial vector fields in the plane is a classical problem of the theory of differential equations. It was studied by Poincaré when he created the geometric theory of differential equations. The problem of the number and location of limit cycles for these fields was included by Hilbert in his famous list, as the second part of the sixteenth problem. At the present time it is known that the number of limit cycles of any planar polynomial vector field is finite. Each of the two known independent proofs of this theorem occupies a whole book [Yu. S. Il′yashenko, *Finiteness theorems for limit cycles,* Translated from the Russian by H. H. McFaden, Amer. Math. Soc., Providence, RI, 1991; MR 92k:58221; J. P. Écalle, *Introduction aux fonctions analysables et preuve constructive de la conjecture de Dulac,* Hermann, Paris, 1992; MR 97f:58104].

The existence of a uniform estimate of the number of limit cycles for a planar polynomial vector field depending on the degree of polynomials only is strongly suggested but not yet proved even in the quadratic case (i.e., for polynomials of degree two).

The problem discussed in the present paper is in a sense converse to the previous one. The authors study quadratic vector fields which have no limit cycles at all.

Let $\mathbf{RP}^2$ be a projective closure of $\mathbf{R}^2$, S be the line at infinity, and $\mathrm{Hom}(\mathbf{RP}^2, S)$ be the space of all homeomorphisms of $\mathbf{RP}^2$ to itself which preserve S. Definition: Let $\mathscr{P}$ be the space of all planar quadratic vector fields. A vector field X on $\mathbf{R}^2$ is said to be structurally stable in $\mathscr{P}$ if there exist a neighborhood $\mathscr{N}$ of X in $\mathscr{P}$ and a continuous map $h: \mathscr{N} \to \mathrm{Hom}(\mathbf{RP}^2, S)$ such that $h(X) = \mathrm{id}$, and $h(Y)$ transforms the phase portrait of X into that of Y. Theorem 1: Quadratic vector fields without limit cycles which are structurally stable in the family $\mathscr{P}$ are topologically equivalent to one of 44 phase portraits drawn in the paper.

The problem solved by this theorem has a long history. At the beginning of the work on the paper under review, a list of 59 phase portraits was known, obtained through the efforts of many authors. It was established that the phase portrait of any quadratic vector field without limit cycles is topologically equivalent to one from the list. It was further established that 43 phase portraits from the list may be realized by a quadratic vector field. The contribution to the proof of Theorem 1 made in the article under review is to show that from the remaining 16 phase portraits only one may be realized by a quadratic vector field.

Another important statement is that if we shrink all the limit cycles of the quadratic vector field together with their intrinsic domains in $\mathbf{R}^2$ to points, then we will get a portrait from the same list. The precise statement is the following. Theorem 2: If a quadratic vector field with limit cycles is structurally stable in $\mathscr{P}$ and we identify the region bounded by the outmost limit cycle surrounding an antisaddle point with a single point, then we obtain the phase portrait of some quadratic vector field without limit cycles.

Two other theorems give the list of the phase portraits under a milder equivalence relation: topological equivalence in $\mathbf{R}^2$ that may not in general be extended to the projective plane. *Yu. S. Il′yashenko* (Moscow)

97a:34121 34C35 34C11 34D08 47N20

Chow, Shui-Nee (Atlanta, GA); **Leiva, Hugo** (Atlanta, GA)

Existence and roughness of the exponential dichotomy for skew-product semiflow in Banach spaces. (English summary)

J. Differential Equations **120** (1995), *no.* 2, 429–477.

Exponential dichotomies provide a means for formulating and investigating the counterpart of hyperbolic behavior, in particular robustness to small perturbations, in nonautonomous systems. The evolution of such a system on its state space X, $x \to \Phi(\theta, t)x$, obeys a cocycle (rather than a group or semigroup) property: $\Phi(\theta, t+s) = \Phi(\theta \cdot t, s)\Phi(\theta, t)$ with respect to a flow σ on an associated parameter space Θ where $\sigma(\theta, t)$ is written $\theta \cdot t$. For example, in a periodic system with period 1, $\Theta \simeq S^1$ and $\theta \cdot t = \theta + t \bmod 1$. Obviously, $\Phi(\theta, 0)x \equiv x$ for all θ, x is required here. The pair $\pi = (\Phi, \sigma)$ then defines a flow or semiflow on $X \times \Theta$ by $\pi(x, \theta, t) = (\Phi(\theta, t)x, \theta \cdot t)$, which is called a skew-product flow or semiflow. In this paper the authors consider linear skew-product semiflows on a Banach space X with Θ a compact Hausdorff space. Their motivation comes from parabolic partial differential equations and functional-differential equations, examples of which are considered at the end of the paper. Their aim is to establish necessary and sufficient conditions for the existence of an exponential dichotomy, as defined by them, for a linear skew-product semiflow and then to prove that such an exponential dichotomy is not destroyed by a small perturbation of the skew-product semiflow. They assume that the $\Phi(\theta, t)$, $t \geq 0$, are bounded linear operators, and continuous in t and in $(\theta, x) \rightarrow \Phi(\theta, t)x$. The terminology of Banach bundles is used throughout, with the trivial bundle $\mathscr{E} = X \times \theta$ being considered in particular.

The concept of an exponential dichotomy introduced in the paper is an extension of that used by D. Henry [*Geometric theory of semilinear parabolic equations*, Lecture Notes in Math., 840, Springer, Berlin, 1981; MR 83j:35084] for evolution operators. Call a mapping $\mathbf{P}: \mathscr{E} \rightarrow \mathscr{E}$ a projector if it is continuous and has the form $\mathbf{P}(x, \theta) = (P(\theta)x, \theta)$, where $P(\theta)$ is a bounded linear projection on the fibre $\mathscr{E}(\theta)$, and $\mathscr{N}(P(\theta)) = \{x \in \mathscr{E}(\theta)\colon\ P(\theta)x = 0\}$. Such a projector $\mathbf{P}$ is said to be invariant with respect to a linear skew-product semiflow $\pi = (\theta, \sigma)$ if $P(\theta \cdot \iota)\Phi(\theta, t) = \Phi(\theta, t)P(\theta)$, for all $t \geq 0$, $\theta \in \Theta$. The authors say that a linear skew-product semiflow $\pi = (\Phi, \sigma)$ on $\mathscr{E}$ has an exponential dichotomy over an invariant set $\widehat{\Theta}$, where $\widehat{\Theta} \subset \Theta$, if there exist constants $k \geq 1$, $\beta > 0$ and an invariant projector $\mathbf{P}$ such that for all $\theta \in \widehat{\Theta}$ the following properties hold: (1) $\Phi(\theta, t)\colon \mathscr{N}(P(\theta)) \rightarrow \mathscr{N}(P(\theta \cdot t))$, $t \geq 0$, is an isomorphism with inverse $\Phi(\theta \cdot t, -t)\colon \mathscr{N}(P(\theta \cdot t)) \rightarrow \mathscr{N}(P(\theta))$, $t \geq 0$; (2) $\|\Phi(\theta, t)P(\theta)\| \leq ke^{-\beta t}$, $t \geq 0$; (3) $\|\Phi(\theta, t)(I - P(\theta))\| \leq ke^{\beta t}$, $t \leq 0$. Note that the $\Phi(\theta, t)$ in (1) are well defined for $t \leq 0$ on the appropriately restricted subspaces by backwards continuation.

As an intermediate tool, the authors also introduce the concept of a discrete skew-product flow $\widehat{\pi}$ on $\mathscr{E}$ by

$$\widehat{\pi}(x, \theta, n) = (\Phi_n(\theta)x, \theta \cdot n), \quad \text{for all } (x, \theta) \in \mathscr{E}, n \in \mathbf{Z},$$

where the $\Phi_n(\theta) \in L(X)$ are uniformly bounded and $(x, \theta) \rightarrow \Phi_n(\theta)x$ are continuous. The $\Phi_n(\theta)$ are not necessarily related or invertible on the whole space X. Pointwise and uniform (with respect to $\theta \in \Theta$) discrete dichotomy is shown to be equivalent to the existence of a unique bounded solution $\{x_n\}$ of $x_{n+1} = \Phi_n(\theta) + f_n$ for each $\{f_n\} \in l_\infty(\mathbf{Z}, X)$; additional conditions on the Φ_n for a pointwise dichotomy to be uniform are considered; and the roughness of such discrete dichotomies is investigated. These results are then transformed to the original linear skew-product semiflow $\pi = (\Phi, \sigma)$ through the associated discrete skew-product flow $\widehat{\pi}$ with $\Phi_n(\theta) = \Phi(\theta \cdot n, 1)$.

The paper is a substantial conceptual and technical contribution with potential for far-reaching applications. It differs from a related paper of R. J. Sacker and G. R. Sell [J. Differential Equations **113** (1994), no. 1, 17–67; MR 96k:34136] in that the unstable spaces need not be finite-dimensional.

Peter E. Kloeden (Frankfurt am Main)

98m:34095 34C35 34A26 34C23 34C37 58F12 58F13
Gedeon, Tomáš (Bozeman, MT)
Cyclic feedback systems.
Mem. Amer. Math. Soc. **134** (1998), *no.* 637, viii+73 pp.
Context. Solutions of differential equations can exhibit simple and ordered behavior, but also very complicated structures. The classical Poincaré-Bendixson theorem shows that two-dimensional autonomous vector fields do not admit chaotic dynamics; the main reason for this fact is that a solution curve locally separates $\mathbf{R}^2$ into two components and thus, since different solutions cannot cross, restricts the possibilities for other solutions. A higher-dimensional system is said to have cyclic feedback (CF) if it has the form $\dot{x}_i = f_i(x_i, x_{i-1})$, $i = 1, \cdots, n$ (where i is taken mod n), and if there exist $\delta_i \in \{-1, 1\}$ such that $\delta_i f_i(0, y)y > 0$ for all $y \in \mathbf{R}$, $y \neq 0$, and $\delta_i \partial_2 f_i(x, 0) > 0$ for all $x \in \mathbf{R}$, $i = 1, \cdots, n$. The system is said to have monotone cyclic feedback (MCF) if even $\delta_i \partial_2 f_i(x, y) > 0$ for all $x, y \in \mathbf{R}$. Systems with CF or MCF appear, e.g., as models for neural networks. A main tool for the analysis of such systems is an integer-valued Lyapunov function N, which for $x \in \mathbf{R}^n$ with $x_i \neq 0$ for all $i \in \{1, \cdots, n\}$ is defined by $N(x) = \#\{i \mid \delta_i x_i x_{i-1} < 0\}$. (It is an analogue of the zero number known from scalar parabolic equations.) For MCF systems, arguments in the spirit of the Poincaré-Bendixson theorem can be applied to certain two-dimensional projections of solutions, which do not intersect for such systems [see J. Mallet-Paret and H. L. Smith, J. Dynam. Differential Equations **2** (1990), no. 4, 367–421; MR 91k:58098]. For CF equations, it is still possible to prove results which express simplicity of the dynamics: There exists a global attractor that has a Morse decomposition, i.e., a decomposition into finitely many invariant sets with a certain order, and connecting orbits from "higher" to "lower" sets in this hierarchy. If one neglects the dynamics within these Morse sets, the flow on the attractor looks like a gradient flow, consisting only of equilibria and connecting solutions. Results of Gedeon and Mischaikow (Theorems 1.3, 1.5 and 1.7 in the work under review) describe the dynamics of CF systems by a semiconjugacy to a model flow on a disk in $\mathbf{R}^k$. This semiconjugacy maps Morse sets of the original flow onto stationary points or periodic orbits of the model flow. It is possible, in principle, that complicated dynamics take place within some Morse set, and this is the point where the new results in Gedeon's work start.

Generally, differential equations in dimension three or greater can generate chaotic motions. The Bernoulli shift $\sigma: \{0, 1\}^{\mathbf{Z}} \to \{0, 1\}^{\mathbf{Z}}$ is a standard model for complicated dynamics, with countably many periodic points which are dense, and with dense orbits. The following result, in weaker form due to Smale and Birkhoff, is meanwhile classical [see, e.g., K. J. Palmer, in *Dynamics reported, Vol. 1*, 265–306, Wiley, Chichester, 1988; MR 89j:58060]: Let the vector field $\dot{x} = f(x)$ in $\mathbf{R}^n$ $(n \geq 3)$ have an unstable periodic orbit γ, with global stable and unstable manifolds W^{s}, W^{u} which intersect transversely at some point (not on γ). Then there exist a Poincaré map π associated with γ and an extension Π of π which has an invariant set S such that the restriction of Π to S is conjugate to the shift σ by a homeomorphism. (Trajectories $(x_n)_{n \in \mathbf{Z}}$ of Π correspond to solutions $\xi: \mathbf{R} \to \mathbf{R}^n$ of $\dot{x} = f(x)$.) In this sense, the differential equation contains the Bernoulli shift as a subsystem.

The notion of "topological crossing", due to K. Burns and H. N. Weiss [Comm. Math. Phys. **172** (1995), no. 1, 95–118; MR 96e:58120], is a weaker property than transverse intersection. If W^{s} and W^{u} cross topologically, theorems of Burns and

Weiss and Gedeon, Mischaikow and McCord (Theorem 2.8 in the work under review) show that one can still construct a semiconjugacy between the restriction of some iterate Π^d to an invariant set S and the shift σ. The semi-conjugating map $\rho: S \to \{0,1\}^{\mathbf{Z}}$ is surjective, but not necessarily injective. Hence the result, which does not require transversality, gives a "lower bound" on the complexity of the dynamics of Π, and thereby on the dynamics of the differential equation. This theorem is basic for Gedeon's work.

The main result (Theorem 1.11), which for simplicity is described here only for the case $n=3$, is the following: One starts with an MCF system in $\mathbf{R}^3$ of the form $\dot{x}_1 = -a_1x_1 - b_1 f(x_3)$, $\dot{x}_i = -a_i x_i + b_i x_{i-1}$ $(i=2,3)$, where only the first equation is nonlinear. Assume that the system has an unstable periodic orbit γ with one simple, unstable multiplier (and therefore a two-dimensional unstable manifold). Then one can deform the monotone function f outside the range of γ_3 (the third component of γ) to a nonmonotone function g with only one extremum, such that the modified system contains chaotic dynamics in the sense of Theorem 2.8 (see above). The modified equation still is a CF system, and the complicated dynamics take place in the "lowest", most attractive Morse set. (A related result for scalar delay equations with negative feedback and one extremum of the nonlinearity was obtained by the reviewer.)

Main ideas. Let γ be the unstable periodic orbit, and let $p \in \gamma$ be the point with maximal value M_3 of the third component. For $\delta \to 0$, one considers families of functions $g_\delta: \mathbf{R} \to \mathbf{R}$ which coincide with f on $(-\infty, M_3+\delta)$ and which decrease to values less than $f(M_3+\delta)$ on an interval $(M_3+\delta, M_3+\delta+\eta(\delta))$, where $\eta(\delta) = O(\delta^q)$ for some $q>2$. It is essential that when f is replaced by g_δ, the flow Φ changes to Φ_δ only above the plane $J_\delta := \{x \mid x_3 = M_3+\delta\}$; in particular, the periodic orbit and local unstable and stable manifolds are preserved. Using information about the values of the Lyapunov functional N on the tangent spaces $T_pW^{\rm s}$, $T_pW^{\rm u}$, Gedeon shows that the global invariant manifolds $W^{\rm u}$ and $W^{\rm s}$ intersect the region above J_δ, where the flows Φ and Φ_δ differ.

Solutions in $W^{\rm u}$ spend (within one turn along γ) only a short time interval above J_δ, and during this time they hit the hyperplane $H := \{x \mid -a_3x_3 + b_2x_2 = 0\}$, where the third component has a maximum. After hitting H, points move towards J_δ again (the third coordinate decreases) and, after a short time, hit J_δ. The motion from J_δ to H is described by a map $\widehat{\varphi}_\delta$, and the second part (from H to a different portion of J_δ) by a map $\widehat{\psi}_\delta$. These maps are carefully studied, using linearization along γ and Taylor expansion of solutions with respect to time, and measuring the distance of points x in H from the tangent space $T_pW^{\rm s}$ by the scalar product $\langle x-p, \alpha\rangle$ with a normal vector α. It is also important that solutions in $W^{\rm u}$ spiral away from γ at increasing distances after each turn. With these techniques, the author proves topological intersection of the global invariant manifolds for sufficiently small δ. We pick out a central part of the proof: Using the information on N, it can be shown that the first component of α satisfies $\alpha_1 \neq 0$. The fact that the nonlinearity appears in the first equation of the system, combined with the cyclic structure, implies that the difference between Φ and Φ_δ is larger in the first component than in the other two. Together with the property $\alpha_1 \neq 0$, one obtains a lower estimate on the size of the change perpendicular to $W^{\rm s}$ of trajectories starting in $W^{\rm u} \cap J_\delta$, when f is changed to g_δ. It is not a priori clear that this change goes in the right direction, i.e., that $W^{\rm u}$ moves towards $W^{\rm s}$ under this change, so that finally an intersection takes place. Here is the key argument: One could also deform f to $2f - g_\delta$ on the interval $[M_3 + \delta, \infty)$, and then obtain a change of trajectories of the same size, but in the opposite direction. Now, contrary to g_δ, the function $2f - g_\delta$ is monotone, and an intersection of W^u and W^s is impossible for this nonlinearity: For an MCF system, the values of N on the stable and unstable eigenspaces and the fact that N is nonincreasing along solutions exclude homoclinic solutions. Thus the deformation of f to g_δ must automatically yield the "right" motion of W^u (towards W^s).

With the topological intersection proved, the result on erratic dynamics follows from the already mentioned Theorem 2.8. The whole construction can be carried out with g_δ arbitrarily C^0-close to f, if one allows that g also has a minimum (at a value larger than the maximum). The last section shows that the starting point, a hyperbolic unstable periodic orbit with one unstable multiplier, can actually be obtained with an MCF system via Hopf bifurcation. This is done with methods from the paper by S.-N. Chow and Mallet-Paret [J. Differential Equations **26** (1977), no. 1, 112–159; MR **58** #7718] in combination with some skillfully performed explicit calculations.

Comment. Gedeon's work is an important result that complements the Poincaré-Bendixson-type theorems on simple dynamics for monotone cyclic feedback systems. It is clear from the construction and from the Hopf bifurcation result of the last section that the examples given here are not exotic, but in some sense typical.

The author gives an introduction to the framework of related publications, and a good exposition of the new construction. Although proofs are not always fully formalized, the nice geometric ideas become very clear. The interplay between geometric objects in state space, the cyclic feedback structure, the integer-valued Lyapunov function and analytical estimates on local solution behavior is mathematically very attractive—one hopes that this work will find many readers.

Bernhard Lani-Wayda (Giessen)

99b:34082 34C37 35Q51 58F39 76B15

Lombardi, Eric (Valbonne)

Orbits homoclinic to exponentially small periodic orbits for a class of reversible systems. Application to water waves. (English summary)

Arch. Rational Mech. Anal. **137** (1997), *no.* 3, 227–304.

This paper is about the existence of homoclinic orbits associated to periodic orbits in 1-parameter families of reversible evolution equations, close to a fixed point where the linearized operator has a pair of simple purely imaginary eigenvalues and a double, nonsemisimple zero eigenvalue for some critical value of the parameter. A strong motivation for this work has been the water wave problem, where one considers the irrotational flow of an inviscid fluid layer under the influence of gravity as well as surface tension. Under the principle of spatial dynamics for this problem, a homoclinic orbit connecting the rest state to itself corresponds to a solitary wave, and a homoclinic orbit associated to a small amplitude periodic perturbation of the rest state corresponds to a so-called generalized solitary or quasi-solitary wave, i.e. a solitary wave with ripples at infinity. For Froude number F equal to 1 and small surface tension (Bond number less than $1/3$), a singularity of the described type occurs in this problem, and the existence of a true solitary wave close to this singularity is still a major open problem [see G. Iooss and K. Kirchgässner, Proc.

Roy. Soc. Edinburgh Sect. A **122** (1992), no. 3-4, 267–299; MR 94b:76012]. Hence, it makes sense to ask the question whether there exist generalized solitary waves whose ripples at infinity are as small as possible. Generalized solitary wave solutions were established for the first time by J. K. Hunter and the reviewer in [Phys. D **32** (1988), no. 2, 253–268; MR 89j:76031], where a fifth-order perturbed Korteweg-de Vries equation was considered as an asymptotic model for the full water wave problem. In the present paper, as an application of the general theory developed, the author establishes the existence of generalized solitary wave solutions in which the size of the ripples at infinity tends to zero exponentially fast as the parameter $\mu = F^{-2} - 1$ tends to zero from below. Before, such a result had been suggested only by means of formal asymptotics in [J. T. Beale, Comm. Pure Appl. Math. **44** (1991), no. 2, 211–257; MR 92d:76011a] and [S. M. Sun and M. C. Shen, J. Math. Anal. Appl. **172** (1993), no. 2, 533–566; MR 94c:76011]. Thus, the issue of asymptotics beyond all orders or rather of exponentially small asymptotics, which has become a major research topic in recent years, is involved here [*Asymptotics beyond all orders (La Jolla, CA, 1991)*, edited by H. Segur, S. Tanveer and H. Levin, Plenum, New York, 1991; MR 93i:00016]. Regarding this issue, a fundamental contribution of the present paper is to provide rigorous exponentially small upper bounds for oscillatory integrals related to solutions of analytic evolution equations in a quite general, infinite-dimensional framework. These estimates are based on the analytic continuation of real solutions to the complex time plane [cf. P. J. Holmes, J. E. Marsden and J. Scheurle, in *Hamiltonian dynamical systems (Boulder, CO, 1987)*, 213–244, Contemp. Math., 81, Amer. Math. Soc., Providence, RI, 1988; MR 90b:58217]. The author proves the existence, near a singularity of the type mentioned above, of reversible homoclinic orbits associated to periodic solutions with amplitudes that are exponentially small in terms of the distance of the parameter from its critical value. In particular, this result applies to the water wave problem to yield the above-mentioned generalized solitary wave solutions.

The main result, concerning the infinite-dimensional case, is derived in Section 4. As a first step, the 4-dimensional case is considered in Section 3. One should note, though, that it is not possible to use center manifold theory to reduce the infinite-dimensional case to the 4-dimensional one, since the technique used to obtain exponentially small upper bounds heavily relies on the analyticity of the system. In fact, in the nonanalytic case that is discussed in Section 2, only homoclinic orbits corresponding to periodic solutions with amplitudes of algebraically small order are found.

The ideas developed in this paper might also be useful in other situations involving exponentially small asymptotics related to solutions of infinite-dimensional evolution equations. As a matter of fact, in the meanwhile, the author of the present paper himself has succeeded in obtaining sharp, exponentially small upper bounds for certain oscillatory integrals by a refined version of his method. This has enabled him to prove generic nonexistence results for true solitary waves in the present context. *Jürgen Scheurle* (Munich)

98k:34084 34D05 58F10
Cima, Anna (Barcelona); **van den Essen, Arno** (Nijmegen);
Gasull, Armengol (Bellaterra (Barcelona)); **Hubbers, Engelbert** (Nijmegen);
Mañosas, Francesc (Bellaterra (Barcelona))
A polynomial counterexample to the Markus-Yamabe conjecture. (English summary)
Adv. Math. **131** (1997), *no.* 2, 453–457.

In this paper the authors give explicit and simple 3-dimensional polynomial counterexamples to both the continuous and the discrete Markus-Yamabe conjectures. The continuous Markus-Yamabe conjecture was already proved to be true in two dimensions: first for polynomial 2D-vector fields in 1987 by the reviewer and C. Olech [in *Analyse mathématique et applications*, 373–381, Gauthier-Villars, Montrouge, 1988; MR 90b:58135]; and then for all $\mathcal{C}^1$ 2D-vector fields, independently by Fessler and Gutiérrez, at Marco Sabatini's 1993 Trento conference [R. Fessler, "A solution of the global asymptotic stability Jacobian conjecture and a generalization", in *Recent results on the global asymptotic stability Jacobian conjecture (Trento, 1993)*, Univ. Trento, Trento; per revr.; C. Gutiérrez, "A solution to the bidimensional global asymptotic stability conjecture", ibid.; per bibl.]. Revised versions of these appeared as [R. Fessler, Ann. Polon. Math. **62** (1995), no. 1, 45–74; MR 96k:34099a; C. Gutiérrez, Ann. Inst. H. Poincaré Anal. Non Linéaire **12** (1995), no. 6, 627–671; MR 96k:34099b]. A. A. Glutsyuk [Funktsional. Anal. i Prilozhen. **29** (1995), no. 4, 17–30, 95; MR 96k:34099c] then gave a 2D proof shorter than those of Fessler and Gutiérrez, but the injectivity results in the latter two are more general. Already in 1988 ideas for a 4-dimensional $\mathcal{C}^1$ counterexample were sketched, but with faulty and obscure arguments, in a paper by N. E. Barabanov [Sibirsk. Mat. Zh. **29** (1988), no. 3, 3–11, 219; MR 89g:93077]. Nevertheless, that paper inspired J. Bernat and J. Llibre to carry out a careful and rigorous construction of analytic (but not polynomial) counterexamples to the Kalman and Markus-Yamabe conjectures in dimensions $\geq$ 4 which exhibit nontrivial periodic solutions [Dynam. Contin. Discrete Impuls. Systems **2** (1996), no. 3, 337–379; MR 98c:34079, see p. E29 in this volume]. Thus the beautiful 3-dimensional polynomial counterexamples given in the paper under review end a 35-year quest in which the reviewer was personally involved. Among those present when the original version of this conjecture was presented in the fall of 1960 in a lecture given by Lawrence Markus to the differential equations group at RIAS (Research Institute for Advanced Study) in Baltimore, Maryland were Philip Hartman, Czesław Olech, and the reviewer. The mathematics section of RIAS was led by Solomon Lefschetz [1884–1972], Joe LaSalle [1916–1983], and Jack Hale. In that fall lecture of 1960 Larry Markus presented the results of his freshly written joint paper with the Japanese-American mathematician Hidehiko Yamabe [Osaka Math. J. **12** (1960), 305–317; MR **23** #A3316; see also M. Goto, Osaka Math. J. **13** (1961), i–ii; MR **23** #A3658; H. Yamabe, *Collected works of Hidehiko Yamabe*, Gordon and Breach, New York, 1967; MR **36** #6255]. But behind the paper under review is *𝒜 𝒯ale of 𝒯wo 𝒞onjectures & erendipity*. The fortunate discovery of these beautiful polynomial counterexamples was made accidentally by the authors while they were in pursuit of a counterexample to a different but intimately related conjecture—namely, the Jacobian conjecture (JC) of Ott-Heinrich Keller: For polynomial maps $f:\mathbf{C}^n \to \mathbf{C}^n$, is the nonvanishing of the Jacobian determinant $\det f'(x)$ on $\mathbf{C}^n$ sufficient for the (global) invertibility of the mapping f? The final sequence of events was roughly as follows: Gaetano Zampieri was visiting Bo Deng and the reviewer at the University of Nebraska during the academic year 1993–94. At lunch one day, early in March 1994, Deng and Zampieri suggested to me that perhaps the Jacobian conjecture could be proved by means of a group conjugation. Let $f\colon \mathbf{C}^n \to \mathbf{C}^n$ be a nonlinear polynomial mapping of the form $f(x) = x - g(x)$, $g(0) = 0$, and suppose with Keller that $\det f'(x) = 1$ for all x in $\mathbf{C}^n$. They asked whether it might be possible, by introducing a parameter s, to find, for some appropriate values of s, some s-family of analytic (if not actually polynomial) automorphisms $x \to h(s,x)$, so that the mappings $x \to s\,f(x)$ would satisfy the conjugation equation $h(s, s\,f(x)) = s\,h(s,x)$ for all x in $\mathbf{C}^n$. Because of the Poincaré-Siegel theory we knew that such analytic conjugations must exist at least locally. However, at that time they had computed no examples, and so I was skeptical; but I spent the next few days working out several examples to see if global conjugations might be feasible; and indeed it worked in all the cases I checked! Since I had been invited to give an hour talk at the upcoming Curaçao Conference on Polynomial Automorphisms, I developed as many examples of this phenomenon as time permitted and presented them at van den Essen's Curaçao Conference on July 4, 1994 [in *Automorphisms of affine spaces (Curaçao, 1994)*, 67–87, Kluwer Acad. Publ., Dordrecht, 1995; MR 97m:14020]. This got van den Essen and his colleagues busy examining these and many other examples of polynomial and analytic conjugations; but we were unable to prove that it would work in general [B. Deng, G. H. Meisters and G. Zampieri, Z. Angew. Math. Phys. **46** (1995), no. 6, 872–882; MR 96m:32022]. What was needed was a criterion for such conjugations to work. Then Deng discovered just such a criterion which actually characterized linearizable dilations of analytic maps ["Analytic conjugation, global attractor, and the Jacobian conjecture", Preprint, Univ. Nebraska, Lincoln, NE, 1995; per revr.]. Deng proved that the dilation λf of an analytic map f of $\mathbf{C}^n$ into itself with $f(0) = 0$, $f'(0) = I$, and $|\lambda| > 1$, has an analytic conjugation to its linear part λx if and only if f is a holomorphism (i.e., a holomorphic automorphism) of $\mathbf{C}^n$ and $x = 0$ is a global attractor for the inverse of λf. This is a corollary of the more general theorem: Suppose $F\colon \mathbf{C}^n \to \mathbf{C}^n$ is an analytic map, $F(0) = 0$, $\det F'(0) \neq 0$, and all eigenvalues of $F'(0)$ are in the open unit disk and not resonant; then F has an analytic conjugation to its linear part $F'(0)$ iff F is an analytic automorphism of $\mathbf{C}^n$ and the fixed point $x = 0$ is a global attractor, i.e., for all $x \in \mathbf{C}^n$, $F^k(x) := F \circ F^{k-1}(x) \to 0$ as $k \to \infty$. This was just the kind of criterion that van den Essen and his colleagues needed. Soon they had counterexamples to global conjugation [A. R. P. van den Essen and E. Hubbers, Adv. in Appl. Math. **18** (1997), no. 3, 382–388; MR 98b:14011] and they finally noticed that one of the polynomial maps they had generated was a counterexample, not to Keller's JC, but to the Markus-Yamabe conjecture, with which by this time they had also become familiar. Later it was kindly pointed out to us (by Franc Forstnerič, January 10, 1996) that Deng's discovery was, in fact, a rediscovery! Deng's criterion is essentially the same as a theorem which occurs in the appendix of the 1988 paper by J.-P. Rosay and W. Rudin [Trans. Amer. Math. Soc. **310** (1988), no. 1, 47–86; MR 89d:32058]. A more detailed exposition of this twisted *Tale of Two Conjectures* was given as an hour talk by the reviewer [in *Proceedings of the Conference on Aspects of Mathematics—Algebra, Geometry, and Several Complex Variables (Hong Kong, 1996)*; per revr.]. But the JC of Keller remains open; in fact, there is a restricted form of the Markus-Yamabe conjecture (rmyc) equivalent to it: JC is false iff $\neg$ (rmyc): there exist n and an $n \times n$ complex matrix A such that $[\mathrm{diag}(Ax)]^2 A$ is nilpotent for all $x \in \mathbf{C}^n$ and the system

$\dot{x} = G_A(x) := c - x + [\mathrm{diag}(Ax)]^2 Ax$, for some constant vector c, has two distinct rest points x_1, x_2. Note that $G'_A(x)$ has all eigenvalues -1. The 3D polynomial counterexample to the Markus-Yamabe conjecture given in the paper under review is not of this form! So rmyc $\equiv$ JC is still open! *Gary H. Meisters* (Lincoln, NE)

99d:34088 34D05 34C27 34C35 35K55 47H20 54H20 58F11

Shen, Wenxian (Auburn, AL); **Yi, Yingfei** (Atlanta, GA)

Almost automorphic and almost periodic dynamics in skew-product semiflows. (English summary)

Mem. Amer. Math. Soc. **136** (1998), *no.* 647, x+93 pp.

The monograph under review contains a study of almost automorphic and in particular almost periodic limit solutions of broad and important classes of differential equations with Bohr-almost periodic coefficients. These include cooperative systems of ordinary and functional-differential equations, and nonlinear parabolic partial differential equations. An almost periodic solution of an almost periodic differential equation is analogous to a fixed point of a differential equation with constant coefficients, or a T-periodic solution of a differential equation with T-periodic driving terms. Almost automorphic, non-almost periodic solutions have no true analogues in the constant coefficient and periodic cases, and their appearance in almost periodic systems is a remarkable phenomenon.

The study of the limit, or asymptotic, behavior of almost periodic differential equations was initiated by H. Bohr himself and by Favard in the 1920s. Among the themes which attracted the attention of researchers before (and after) 1975 was that of the fundamental relation between the uniform asymptotic stability of a solution and the asymptotic almost periodicity of that solution. Among many noteworthy contributions, we mention only [R. K. Miller, J. Differential Equations **1** (1965), 337–345; MR **32** #2690] and [G. R. Sell, *Topological dynamics and ordinary differential equations,* Van Nostrand Reinhold, London, 1971; MR **56** #1283]. For an elegant summary of results up to 1976 concerning uniformly stable and uniformly asymptotically stable solutions of almost periodic differential equations, see the monograph by R. J. Sacker and Sell [Mem. Amer. Math. Soc. **11** (1977), no. 190, iv+67 pp.; MR **56** #6632]. Particularly noteworthy in this paper is the systematic use of methods of topological dynamics. In fact, an explicit connection is made between the uniform stability of a solution and the "distallity" of its ω-limit set. Distallity is a basic concept in topological dynamics: it has been studied by R. Ellis [*Lectures on topological dynamics,* W. A. Benjamin, Inc., New York, 1969; MR **42** #2463], by H. Furstenberg [Amer. J. Math. **85** (1963), 477–515; MR **28** #602] and in numerous other notable papers.

One can approach the Shen-Yi monograph by asking for properties of the ω-limit set of a solution to an almost periodic differential equation when the solution is not uniformly stable. One of Shen and Yi's main results can be paraphrased as follows. Suppose that the solutions of a given almost periodic differential equation determine a strongly monotone (skew-product) semiflow. Suppose further that the solution operator is compactifying for large times. Finally, replace the uniform stability condition by the much weaker hypothesis that the ω-limit set of a given solution is linearly stable. Then (1) all minimal subsets of that ω-limit set are almost automorphic; (2) almost automorphic but non-almost periodic minimal sets frequently exist. Thus almost automorphy plays a striking role in characterizing the limit behavior of solutions of strongly monotone almost periodic systems. Just as did Sacker and Sell (and other authors since 1976), Shen and Yi avail themselves of the machinery of topological dynamics. In particular, they use facts about almost automorphic functions and almost automorphic minimal sets due to S. Bochner [Proc. Nat. Acad. Sci. U.S.A. **48** (1962), 2039–2043; MR **26** #2816], W. A. Veech [Bull. Amer. Math. Soc. **83** (1977), no. 5, 775–830; MR **57** #7558] and others.

We give an overview of the contents of the monograph. It contains three parts. The first recounts basic facts from topological dynamics about minimal flows, distal flows, almost automorphic flows, almost periodic flows, and related matters. In Part Two, differential equations with almost periodic or, more generally, Birkhoff recurrent coefficients are considered. General results are proved under the assumption that the solutions of the differential equation define a skew-product semi-flow which preserves a partial order. The result alluded to in the previous paragraph is precisely formulated and proved.

In Part Three, these results are applied to cooperative systems of ordinary or functional-differential equations and to parabolic PDEs. For brevity, we describe only the Shen-Yi theory of the following parabolic PDE in one space dimension: (1) $\partial u/\partial t = \partial^2 u/\partial x^2 + f(t,x,u,u_x)$, $0<x<1$, $u(0,x)=u_0(x)$, with Dirichlet or Neumann boundary conditions at $x=0$ and $x=1$. The function f is almost periodic in t and is C^2-smooth. One introduces the hull Y of f and the corresponding almost periodic minimal flow on Y. The solutions of (1) induce a skew-product semiflow on $Y \times X$, where X is an appropriate fractional power space.

A bounded solution u of (1) has a compact ω-limit set $\Omega = \Omega(u)$, and Ω admits a two-sided flow extension. If f is periodic in t, then results of P. Brunovský, P. Poláčik and B. Sandstede [Nonlinear Anal. **18** (1992), no. 3, 209–215; MR 93b:35064] imply that Ω reduces to a periodic orbit. If f is almost periodic but not periodic in t, then Shen and Yi prove that Ω contains at most two minimal subsets, and these are almost automorphic extensions of Y. Almost automorphic, non-almost periodic minimal sets do occur, and examples are given. No linear stability hypothesis on Ω is required to prove these results. If a minimal subset of Ω happens to be uniformly stable or hyperbolic, then it is actually almost periodic. The zero number of H. Matano [J. Fac. Sci. Univ. Tokyo Sect. IA Math. **29** (1982), no. 2, 401–441; MR 84m:35060] is used in the proofs of these results, as is the theory of Part Two.

The authors pose several intriguing problems; here is one. It is known that an almost automorphic minimal set can have positive topological entropy [N. G. Markley and M. E. Paul, Israel J. Math. **34** (1979), no. 3, 259–272 (1980); MR 81f:28024]. The problem is to determine a driving function f for which such a minimal set exists in the corresponding skew-product semiflow. This would surely be a remarkable "route to chaos" in equation (1).

Part Three also has many significant results about almost automorphy in cooperative systems and in parabolic equations of space dimension greater than one. For reasons of space, we forego the pleasure of discussing them here. We conclude by stating our belief that Shen and Yi have fully justified their assertion: "almost automorphy ... is essential and fundamental in the qualitative study of almost periodic differential equations." *Russell A. Johnson* (Florence)

98c:34079 34D20 34C25 34H05 58F10 93D20

Bernat, Josep (Bellaterra (Barcelona)); **Llibre, Jaume** (Bellaterra (Barcelona))

Counterexample to Kalman and Markus-Yamabe conjectures in dimension larger than 3. (English summary)

Dynam. Contin. Discrete Impuls. Systems **2** (1996), *no.* 3, 337–379.

A matrix is called Hurwitz if all its eigenvalues have negative real parts. An n-dimensional $\mathscr{C}^1$-system $\dot{x} = F(x)$ of ordinary differential equations is called Hurwitz if its Jacobian matrix $F'(x)$ is Hurwitz at each $x \in \mathbf{R}^n$. This paper was inspired by, and benefited from, a paper of N. E. Barabanov [Sibirsk. Mat. Zh. **29** (1988), no. 3, 3–11, 219; MR 89g:93077]. The paper under review accomplishes, by different methods and different arguments, what Barabanov had tried but, unfortunately, had failed to accomplish. Namely, Bernat and Llibre carefully construct a 4-dimensional, Hurwitz analytic system $\dot{x} = F(x)$, with $F(0) = 0$, but having a non-constant periodic solution—thus providing the first proved counterexample to the Markus-Yamabe conjecture [L. Markus and H. Yamabe, Osaka Math. J. **12** (1960), 305–317; MR **23** #A3316]. One aspect of great merit of the paper under review is that, although the construction is extremely difficult, Bernat and Llibre provide all the necessary details in a clear and orderly manner, written out in full for anyone to check; even though there are complicated formulas involving many specific numbers, all the numbers involved are given in exact form (not mysterious long decimals)—so the calculations can be easily checked. Not very long after this important and amazing paper was accepted for publication, a 3-dimensional polynomial counterexample to the Markus-Yamabe conjecture was ferreted out (in a surprising and serendipitous manner) by A. Cima, A. R. P. van den Essen, A. Gasull, E. Hubber and F. Mañosas [Adv. Math. **131** (1997), no. 2, 453–457; MR 98k:34084, see p. E25 in this volume]. Namely, $(\dot{x}, \dot{y}, \dot{z}) = (-x + z(x+yz)^2, -y - (x+yz)^2, -z)$ has the unbounded orbit $(x(t), y(t), z(t)) = (18e^t, -12e^{2t}, e^{-t})$, even though $F'(x, y, z)$ has -1 for all its eigenvalues. Nevertheless, the Bernat-Llibre system $\dot{\mathbf{x}} = (x_2, -x_4, x_1 - 2x_4 - k_1\psi(x_4), x_1 + x_3 - x_4 - k_2\psi(x_4))$, where $k_1 = 9131/900$, $k_2 = 1837/180$, and $\psi\colon \mathbf{R} \to \mathbf{R}$ can be chosen analytic, given in the paper under review, remains important because of its periodic orbit; such a non-constant periodic orbit is surely impossible for polynomial Hurwitz systems. (This remains to be proved.) An important precursor which led to the accidental discovery of the 3-dimensional polynomial counterexample is a paper by van den Essen and Hubbers [Adv. in Appl. Math. **18** (1997), no. 3, 382–388; MR 98b:14011]. In a 1997 paper [Publ. Mat. **41** (1997), no. 1, 85–100; MR 98h:34088], Cima, Gasull and Mañosas studied the dynamics of the polynomial example of Cima et al.'s Advances in Mathematics paper [op. cit.] (see also the above-cited paper by Cima, Gasull and Mañosas). In the 2-dimensional case, the Markus-Yamabe conjecture is true. It was proved for 2-dimensional polynomial systems by the reviewer and C. Olech in their joint paper [in *Analyse mathématique et applications*, 373–381, Gauthier-Villars, Montrouge, 1988; MR 90b:58135]. Interesting variations and generalizations were given by Nguyễn Văn Châu [Ann. Polon. Math. **62** (1995), no. 2, 143–154; MR 97a:34140]. However, Nguyễn Văn Châu's results did not contain the earlier results of Olech, and Meisters and Olech, because both his assumptions and his conclusions are different. Nguyễn Văn Châu's paper included two new injectivity results, which are now of even more interest since S. Pinchuk has given a counterexample to the strong real Jacobian conjecture [Math. Z. **217** (1994), no. 1, 1–4; MR 95g:14018]. In 1993 the Markus-Yamabe conjecture was proved for all class $\mathscr{C}^1$ vector fields on $\mathbf{R}^2$, simultaneously but independently, by R. E. Fessler and C. Gutiérrez, both of whom presented their proofs at M. Sabatini's conference [R. Fessler, "A solution to the global asymptotic stability conjecture and a generalization", in *Recent results on the global asymptotic Jacobian conjecture (Trento, 1993)*, Publication UTM 429, Dept. Mat., Univ. Trento, Trento, 1994; per revr.; C. Gutiérrez, "A solution to the bidimensional global asymptotic stability conjecture", ibid.; per revr.]. Their proofs, both of which are long and difficult, as well as a third proof by A. A. Glutsyuk, shorter and somewhat easier to read, appeared in 1995 [R. E. Fessler, Ann. Polon. Math. **62** (1995), no. 1, 45–74; MR 96k:34099a; C. Gutiérrez, Ann. Inst. H. Poincaré Anal. Non Linéaire **12** (1995), no. 6, 627–671; MR 96k:34099b; A. A. Glutsyuk, Funktsional. Anal. i Prilozhen. **29** (1995), no. 4, 17–30, 95; MR 96k:34099c; see also R. E. Fessler, in *Automorphisms of affine spaces (Curaçao, 1994)*, 127–135, Kluwer Acad. Publ., Dordrecht, 1995; MR 96i:26013]. In a related paper [C. Gutiérrez and M. A. Teixeira, Bol. Soc. Brasil. Mat. (N.S.) **26** (1995), no. 1, 57–66; MR 96c:58099] it is shown that if a planar vector field F has at least one rest point, $\det F'(x) > 0$ on $\mathbf{R}^2$, and $\operatorname{trace} F'(x) < 0$ in a neighborhood of ∞, then ∞ is a repellor [resp. attractor] provided that $\int_{\mathbf{R}^2} \operatorname{trace} F'(x)dx < 0$ [resp. > 0]. There are still more proofs appearing. But neither the Bernat-Llibre 4-dimensional analytic counterexample presented in the paper under review, nor the 3-dimensional polynomial counterexample of Cima et al., ends the story of the Markus-Yamabe conjecture. For there still remains the weak Markus-Yamabe conjecture, which is equivalent to the still open Jacobian conjecture (JC) [O.-H. Keller, Monatsh. Math. Phys. **47** (1939), 299–306; Zbl 021.15303 (p. 301, table, items 6 and 7); H. Bass, E. H. Connell and D. Wright, Bull. Amer. Math. Soc. (N.S.) **7** (1982), no. 2, 287–330; MR 83k:14028]. The weak Markus-Yamabe conjecture: Every system of the special form $\dot{x} = G(x) := c - x + H(x)$, with $H(tx) = t^3H(x)$ and $H'(x)^n \equiv 0$, has at most one rest point x_0. Note that all eigenvalues of $G'(x)$ are -1. By the reduction-of-degree theorem of A. V. Yagzhev, and Bass-Connell-Wright [op. cit.], JC is false if and only if there is a polynomial map $F(x) = x - H(x)$, with $H(tx) = t^3H(x)$ and $H'(x)$ nilpotent for all x, and yet the system $\dot{x} = G(x) = F(x_1) - F(x) = c - x + H(x)$ has two distinct rest points x_1, x_2 (that is, $F(x)$ is not injective). The 3-dimensional polynomial example found by Cima et al. is not of this form.

{See also the following review [MR 98c:34080, see p. E30 in this volume].}

Gary H. Meisters (Lincoln, NE)

98c:34080 34D20 26B10 58F10

Parthasarathy, Thiruvenkatachari (New Delhi); **Sabatini, Marco** (Trento)

Some new results on the global asymptotic stability Jacobian conjecture. (English summary)

Bull. Polish Acad. Sci. Math. **41** (1993), *no.* 3, 221–228 (1994).

The global asymptotic stability Jacobian conjecture is concerned with the problem of determining whether the critical point $0 = F(0)$ of a system of differential equations (S) $\dot{x} = F(x)$, with $x \in \mathbf{R}^n$, $F \in \mathscr{C}^1(\mathbf{R}^n, \mathbf{R}^n)$, is globally asymptotically stable, provided at every point x of $\mathbf{R}^n$ every eigenvalue of the Jacobian matrix $F'(x)$ has negative real part. This question was brought into prominence in the United States by the 1960 paper of L. Markus and H. Yamabe [Osaka Math. J. **12** (1960), 305–317; MR **23** #A3316]. At that time the most important result on this question had been obtained by C. Olech [Contributions to Differential Equations **1** (1963), 389–400; MR **26** #5248], who showed that, in the two-dimensional case, such systems are globally asymptotically stable if and only if the mapping $F:\mathbf{R}^2 \to \mathbf{R}^2$ is injective (univalent, one-to-one). The paper under review is concerned only with the two-dimensional case and establishes two partial results. The authors' first result, which extends a result of Olech [op. cit.], is that a sufficiently smooth F satisfying the Markus-Yamabe hypotheses will be injective if at least one of the diagonal entries of $F'(x)$ never changes sign; and this is an application of a perturbation technique (interesting in itself) which was used by S. Rădulescu and M. Rădulescu [Nonlinear Anal. **13** (1989), no. 5, 539–553; MR 90d:58017]. It consists in deducing injectivity of F from that of each term in a sequence of maps F_k converging uniformly to F on compact sets. The second result of the authors is that F is injective (and the origin is globally asymptotically stable) if, in addition to the other hypotheses, at least one of the components of F is a rational function. At the time the paper under review was published, the Markus-Yamabe conjecture had been proved only for two-dimensional polynomial systems by the reviewer and Olech [in *Analyse mathématique et applications,* 373–381, Gauthier-Villars, Montrouge, 1988; MR 90b:58135], and had been asserted to be false for certain class-$\mathscr{C}^1$ systems in dimension four by N. E. Barabanov [Sibirsk. Mat. Zh. **29** (1988), no. 3, 3–11, 219; MR 89g:93077]. But in the time since the paper under review was published, the Markus-Yamabe conjecture has been settled in all dimensions as follows: (1) in 1993 it was proved true, for all class-$\mathscr{C}^1$ two-dimensional vector fields $F(x)$, simultaneously and independently by R. E. Fessler and C. Gutiérrez, at a conference organized by Sabatini and held in Trento, Italy, September 14–17, 1993 [R. E. Fessler, "A solution to the global asymptotic stability conjecture and a generalization", in *Recent results on the global asymptotic stability Jacobian conjecture (Trento, 1993),* Publication UTM 429, Dept. Mat., Univ. Trento, Trento, 1994; per revr.; C. Gutiérrez, "A solution to the bidimensional global asymptotic stability conjecture", ibid; per revr.], and a bit later also proved by A. A. Glutsyuk, and subsequently published in three separate papers [R. E. Fessler, Ann. Polon. Math. **62** (1995), no. 1, 45–74; MR 96k:34099a; C. Gutiérrez, Ann. Inst. H. Poincaré Anal. Non Linéaire **12** (1995), no. 6, 627–671; MR 96k:34099b; A. A. Glutsyuk, Funktsional. Anal. i Prilozhen. **29** (1995), no. 4, 17–30, 95; MR 96k:34099c]; see the featured review of these articles by Olech; (2) the Barabanov claim of a four-dimensional $\mathscr{C}^1$-counterexample with a periodic-orbit was rigorously proved in clear detail [J. Bernat and J. Llibre, Dynam. Contin. Discrete Impuls. Systems **2** (1996), no. 3, 337–379; MR 98c:34079, see p. E29 in this volume; see the preceding review]; (3) finally, an idea of Gaetano Zampieri for attacking the related Jacobian conjecture led to a series of papers which culminated in the discovery of the three-dimensional polynomial vector field $\dot{x}_1 = -x_1 + x_3(x_1 + x_2x_3)^2$, $\dot{x}_2 = -x_2 - (x_1 + x_2x_3)^2$, $\dot{x}_3 = -x_3$, which satisfies the hypotheses of the Markus-Yamabe conjecture, but nevertheless possesses unbounded orbits which can be explicitly written [see G. H. Meisters, in *Automorphisms of affine spaces (Curaçao, 1994),* 67–87, Kluwer Acad. Publ., Dordrecht, 1995; MR 97m:14020; B. Deng, G. H. Meisters and G. Zampieri, Z. Angew. Math. Phys. **46** (1995), no. 6, 872–882; MR 96m:32022; B. Deng, "Analytic conjugation, global attractor, and the Jacobian conjecture", Preprint, 1995; per revr.; A. R. P. van den Essen and E. Hubbers, Adv. in Appl. Math. **18** (1997), no. 3, 382–388; MR 98b:14011; A. Cima, A. R. P. van den Essen, A. Gasull, E. Hubbers and F. Mañosas, Adv. Math. **131** (1997), no. 2, 453–457; MR 98k:34084, see p. E25 in this volume]. For additional references and an exposition of these recent developments, see the reviewer's paper ["A biography of the Markus-Yamabe conjecture", in *Proceedings of the Conference on Aspects of Mathematics—Algebra, Geometry and Several Complex Variables (Hong Kong, 1996)*; per revr.]. See also the recent paper by van den Essen and E. Hubbers [J. Algebra **187** (1997), no. 1, 214–226; MR 97m:14019]. *Gary H. Meisters* (Lincoln, NE)

97a:34186a 34K15 34C20 58F36
Faria, Teresa (Lisbon); **Magalhães, Luis T.** (Lisbon)
Normal forms for retarded functional-differential equations and applications to Bogdanov-Takens singularity. (English summary)
J. Differential Equations **122** (1995), *no.* 2, 201–224.

97a:34186b 34K15 34C20 58F36
Faria, Teresa (Lisbon); **Magalhães, Luis T.** (Lisbon)
Normal forms for retarded functional-differential equations with parameters and applications to Hopf bifurcation. (English summary)
J. Differential Equations **122** (1995), *no.* 2, 181–200.

Almost twenty years have now passed since the first rigorous proofs of the existence of Hopf bifurcations in functional-differential equations (FDEs). Since then, many authors have contributed to the computational issue of calculating the direction of bifurcation/stability of such bifurcations. Normal forms have played a fundamental role in this problem, dating from the work of S. N. Chow and J. Mallet-Paret [J. Differential Equations **26** (1977), no. 1, 112–159; MR **58** #7718].

A variety of alternate approaches have appeared in the literature, often requiring the approximation of the center manifold before evaluating the normal form of the ordinary differential equation on the manifold. (See, for example, the book by B. D. Hassard, N. D. Kazarinoff and Y. H. Wan [*Theory and applications of Hopf bifurcation*, Cambridge Univ. Press, Cambridge, 1981; MR 82j:58089], where the Poincaré normal form is used.) Frequently, the relationship between the coefficients in the normal form and the coefficients in the original FDE is not easily determined, obscuring the effects of model parameters on system behavior.

In the papers under review, the authors consider the computation of normal forms for ordinary differential equations describing the flow of FDEs in finite-dimensional invariant spaces associated with finitely many characteristic values for the linearized equation. The method, which does not involve center manifold approximation, is applied to the classic Hopf bifurcation problem, providing a useful comparison to other known methods. New ground is broken in its application to the Bogdanov-Takens singularity of delay-difference equations, including a detailed analysis of the scalar equation $\dot{x}(t) = f(x(t), x(t-1))$.

The phase space chosen in this investigation is that of the Chow and Mallet-Paret paper, cited above. However, the methods developed here require a more thorough examination of the direct sum decomposition of the phase space. This is obtained by extension of the classical bilinear form for retarded FDEs and the use of an associated adjoint theory. These results are of independent interest, and perhaps as important as the normal form calculations to which they are applied. Although not discussed here, the algorithms of these papers raise the prospect that symbolic manipulation software can be applied to general FDE normal form calculations, as has been previously demonstrated in the case of Hopf bifurcations [J. M. Franke and H. W. Stech, in *Delay differential equations and dynamical systems (Claremont, CA, 1990)*, 161–175, Lecture Notes in Math., 1475, Springer, Berlin, 1991; MR 92k:34092].

Harlan W. Stech (Duluth, MN)

97a:34193a 34K15 34C35 58F32

Mallet-Paret, John (Providence, RI); **Sell, George R.** (Minneapolis, MN)

Systems of differential delay equations: Floquet multipliers and discrete Lyapunov functions. (English summary)

J. Differential Equations **125** (1996), *no.* 2, 385–440.

97a:34193b 34K15 34C35 58F32

Mallet-Paret, John (Providence, RI); **Sell, George R.** (Minneapolis, MN)

The Poincaré-Bendixson theorem for monotone cyclic feedback systems with delay. (English summary)

J. Differential Equations **125** (1996), *no.* 2, 441–489.

This remarkable pair of companion papers has been anticipated for more than a decade and comes as a surprise to virtually no one working in the area of dynamical systems who has listened to either author speak at conferences over this period of time. The long delay for this work to appear in print is more than offset by the carefully crafted presentation of deep results.

The central idea of this work, the notion that certain systems of delay differential equations possess an integer-valued Lyapunov function, that is, a function that measures roughly the number of sign changes in the state of the system, and which, as time progresses, cannot increase, has a substantial history in dynamical systems theory. K. Nickel [J. Reine Angew. Math. **211** (1962), 78–94 (1 insert); MR **26** #4056] discovered the so-called crossing number or lap number for scalar reaction diffusion equations, which was later used by H. Matano [J. Fac. Sci. Univ. Tokyo Sect. IA Math. **29** (1982), no. 2, 401–441; MR 84m:35060] to show that solutions of such systems converge to an equilibrium, and by D. B. Henry [J. Differential Equations **59** (1985), no. 2, 165–205; MR 86m:58080] to show the transversality of stable and unstable manifolds of hyperbolic equilibria. The first author and B. Fiedler [Arch. Rational Mech. Anal. **107** (1989), no. 4, 325–345; MR 90j:35116] used a similar function to prove the Poincaré-Bendixson theorem for scalar reaction-diffusion equations on the circle with periodic boundary conditions. J. Smillie [SIAM J. Math. Anal. **15** (1984), no. 3, 530–534; MR 85g:58054] appears to be the first to have used such a Lyapunov function for tridiagonal competitive and cooperative systems of ordinary differential equations to show that every bounded semi-trajectory converges to equilibrium. Later, G. Fusco and W. M. Oliva [J. Dynam. Differential Equations **2** (1990), no. 1, 1–17; MR 91d:58116] used a modified version of this Lyapunov function to establish transversality of stable and unstable manifolds for hyperbolic critical elements of autonomous tridiagonal systems of ordinary differential equations, and the first author and the reviewer [J. Dynam. Differential Equations **2** (1990), no. 4, 367–421; MR 91k:58098] used a related Lyapunov function to establish the Poincaré-Bendixson theorem for monotone cyclic feedback systems of ordinary differential equations. A. D. Myschkis [*Lineare Differentialgleichungen mit nacheilendem Argument*, Deutscher Verlag Wissensch., Berlin, 1955; MR **17**, 497b] first discovered an integer-valued Lyapunov function for scalar delay differential equations, which was later used by the first author [J. Differential Equations **72** (1988), no. 2, 270–315; MR 89m:58182] to obtain a Morse decomposition of the attractor which reflects the different rates of oscillation of the globally bounded solutions of the scalar delay equation (3) below. The discrete Lyapunov function exploited in this pair of papers extends and unifies analogous Lyapunov functions used for monotone cyclic feedback systems of ordinary differential equations and for scalar delay equations.

In this review, to keep the notation to a minimum, only the special class of "unidirectional systems" considered by the authors will be featured. More general systems allowing two-sided feedback are treated in their work. The papers focus on the asymptotic behavior of solutions of a monotone cyclic feedback system with delays somewhat more general than the following:

$$\dot{x}^i(t) = f^i(x^i(t), x^{i+1}(t-\beta)), \tag{1}$$

where $0 \le i \le N$, $N \ge 0$, $\beta \ge 0$, indices are taken modulo $N+1$, and $f^i \in C^1$ enjoys the monotonicity property:

$$\frac{\partial f^i}{\partial x^{i+1}} > 0, \quad i \ne N, \quad \text{and} \quad \delta^* \frac{\partial f^N}{\partial x^0} > 0, \tag{2}$$

where the constant $\delta^* \in \{-1, 1\}$ determines whether the system has positive ($\delta^* = +1$) or negative ($\delta^* = -1$) feedback. When $\beta = 0$, the results of these papers contain those of the first author and this reviewer mentioned above. When $N = 0$ and $\beta = 1$, the results apply to the well-studied scalar delay equation

$$\dot{x}(t) = f(x(t), x(t-1)). \tag{3}$$

The $(N+1)$st order scalar delay equation of the form $P(D)x(t) = f(x(t-1))$, $D = d/dt$, where p is an $(N+1)$st degree polynomial all of whose roots are real, and where $f'(x) \ne 0$ for all x, can readily be reduced to (1) above. Systems of the type (1) appear in many mathematical models, especially those motivated by the classical Goodwin model of cellular control of enzyme synthesis and in neural networks. Equation (3) is the most well-studied delay differential equation, containing the famous Wright equation and the Mackey-Glass equation as special cases.

The first paper is devoted to the construction of the integer-valued Lyapunov function for a more general system of delay differential equations than (1) which satisfy Carathéodory conditions and feedback inequalities, rather than the monotonicity condition (2). These more general systems arise as the linearization of (1) along a solution or as the equation satisfied by the difference between two different solutions of (1). The Lyapunov function is then used to study the Floquet theory of periodic linear systems of delay differential equations of the form that would result from linearizing (1) about a periodic or equilibrium solution. It is shown that the sum of the real parts of the Floquet eigenspaces corresponding to all Floquet multipliers of a given modulus is at most two-dimensional and, when zero is removed, belongs to a level set of the Lyapunov function. Roughly, solutions which decay to zero at the same exponential rate have the same number of sign changes and it is shown that faster rates of decay correspond to a greater number of sign changes. The sharpest results are obtained for the autonomous linear system. Issues dealing with so-called superexponential solutions are put off to a forthcoming paper.

The second paper, although its main results are strongly based on the ideas in the first, is presented in a self-contained form so that the reader who is mainly interested in the Poincaré-Bendixson theorem can begin with it. The main result of the second paper is contained in its title: the omega limit set of a bounded forward orbit of (1) is either (a) a nonconstant periodic orbit, or (b) a compact invariant set with the property that the omega and alpha limit sets of any of its points are contained in the set of equilibria. This result follows from the fact that the omega limit set of a bounded forward orbit projects homeomorphically onto a subset of the (x^i, x^{i+1})-plane in a natural way. This, in turn, makes possible the same arguments used in the proof of the classical Poincaré-Bendixson theorem.

In addition to the results announced in these works, many conjectures are made and many directions for future work are outlined. The authors view their work as "opening the door to a general inquiry into the structure of the attractor of the system (1)". Obtaining a nontrivial Morse decomposition of the attractor compatible with the Lyapunov function, describing connecting orbits between Morse sets, and transversality of stable and unstable manifolds of hyperbolic critical elements are suggested as future topics of research. Clearly, we may expect further progress on this interesting class of systems in the future. *Hal Leslie Smith* (Tempe, AZ)

97f:34067 34L05 47H12 58F07 58F19
Tkachenko, V. [Tkachenko, Vadim Aleksandrovich] (Be'ér Sheva')
Spectra of non-selfadjoint Hill's operators and a class of Riemann surfaces.
Ann. of Math. (2) **143** (1996), *no.* 2, 181–231.

Hill's operator $L = -d^2/dx^2 + q(x)$ with real potential of period π and of class $L^2[0, \pi]$ has been the object of intensive study since the early 1970's owing to its connection with the periodic KdV equation. As an operator in $L^2(\mathbf{R})$, it is self-adjoint with band spectrum σ marked off by its periodic/anti-periodic eigenvalues. After a (simple) periodic ground state λ_0, these come in separated pairs $\lambda_n^- \leq \lambda_n^+$, of alternating anti-periodic/periodic character, located at n^2 or nearly so for $n \uparrow \infty$. Normally, every "gap" $[\lambda_n^-, \lambda_n^+]$ is open. The intervening intervals are the "bands" cited above.

In more detail, let Δ be Hill's discriminant formed as in $\Delta(\lambda) = (1/2)[y_1(1, \lambda) + y_2'(1, \lambda)]$ from the solutions of $Ly = \lambda y$ with $y_1(0) = 1$, $y_1'(0) = 0$, $y_2(0) = 0$, $y_2'(0) = 1$. $Lf = \lambda f$ has a solution with $f(x + \pi) = mf(x)$ precisely when $m^2 - 2\Delta m + 1 = 0$, i.e., for $m = \Delta \pm \sqrt{\Delta^2 - 1}$; in particular, the periodic/anti-periodic eigenvalues are the roots of $\Delta(\lambda) = +1/-1$. The formula for the "multiplier" m prompts the introduction of the Riemann surface M of the irrationality $\sqrt{\Delta^2 - 1}$, of which more below [see, e.g., S. P. Novikov, Funkcional. Anal. i Priložen. **8** (1974), no. 3, 54–66; MR **52** #3760; H. P. McKean and P. van Moerbeke, Invent. Math. **30** (1975), no. 3, 217–274; MR **53** #936; H. P. McKean and E. Trubowitz, Comm. Pure Appl. Math. **29** (1976), no. 2, 143–226; MR **55** #761].

The periodicity of q requires that the $2\infty + 1$ points of the periodic/anti-periodic spectrum satisfy ∞ conditions, expressible by means of Abelian integrals on M. The matter is most neatly described by the recipe of V. A. Marchenko and I. V. Ostrovskiĭ [Mat. Sb. (N.S.) **97(139)** (1975), no. 4(8), 540–606, 633–634; MR **53** #13717]. The map $\lambda \to -(1/\pi)\int_{\lambda_0}^{\lambda} \Delta^\bullet(1 - \Delta^2)^{-1/2} d\lambda$ sends the cut plane $\mathbf{C} - [\lambda_0, \infty)$ to the upper half-plane with an infinite number of vertical "spikes" $(n + ih = 0 \leq h \leq h_n)$ removed, one to each $n \neq 0$, these being the double images of the gaps $[\lambda_n^-, \lambda_n^+]$. The spike heights h_n: $n \geq 1$ serve as free moduli, subject only to $\sum_1^\infty n^2 h_n^2 < \infty$, the periodicity being built into the evaluation

$$(1/\pi)\int_{\lambda_0}^{\lambda_n^-} \Delta^{\bullet}(1-\Delta^2)^{-1/2}dx = \pm n.$$

Naturally, this picture breaks down for complex potentials. The form of the spectrum is much more various and the periodic/anti-periodic spectrum cannot determine it fully anymore. Moreover, the Riemann surface of $\sqrt{\Delta^2-1}$ is not the right object of study, and the whole question seemed intractable until Tkachenko had the excellent idea to look not at $\sqrt{\Delta^2-1}$ but at $\cos^{-1}\Delta$ instead. The present paper explains the beautiful description of complex spectra that he has obtained by this means.

The spectrum σ is the inverse image in $\mathbf{C}$ of $-1 \le \Delta \le +1$ [see V. A. Tkachenko, Dokl. Akad. Nauk SSSR **155** (1964), 289–291; MR **29** #468; and references cited therein]. Next, Δ is a (complex) Hill discriminant if and only if it is an even integral function of $k=\sqrt{\lambda}$, of exponential type π, capable of the development $\Delta(\lambda) = \cos(\pi k) + ck^{-1}\sin(\pi k) - (c^2/2)k^{-2}\cos(\pi k) + k^{-2}r(k)$, in which c is a (complex) constant and r is an integral function, of type $\le \pi$, with $\int_0^\infty |r|^2 dk < \infty$ and $\sum_1^\infty |r(n)| < \infty$ [see V. A. Tkachenko, in *Spectral operator theory and related topics*, 41–71, Amer. Math. Soc., Providence, RI, 1994; MR 95i:34157]. It follows that σ is comprised of a countable family of analytic arcs; these may intersect in a tame manner.

Now come the Riemann surfaces. The "sheet" S_n $(n \ne 0)$ is a copy of the $\sqrt{\lambda}$ plane provided with a system of (mostly two but sometimes more) disjoint cuts in the form of mild deformations of rays, each bank of every cut being paired to a bank of an identical cut on another sheet and pasted to form a Riemann surface S, with projection $S \to \mathbf{C}$ via the natural identifications $S_n = \mathbf{C}$. The following conditions are imposed on S: (1) simple connection; (2) all ramifications of index $< \infty$; (3) the natural identification $S_n \to \mathbf{C} \to S_{-n}$ is a holomorphic involution; (4) for all but a finite number of $n \ge 1$, S_n has just two cuts starting at ramifications of S, one ramified point of index 1 in common with S_{n-1}, one such with S_{n+1}, and no other ramifications at all; (5) the projections γ_{n-1} and γ_n of these ramifications are alternately close to ± 1 in the sense that $\sum n^2|(-1)^n - \gamma_n|^2 < \infty$. This type of surface is automatically conformally equivalent to the plane. If the uniformizing map is $\Phi\colon S \to \mathbf{C}$, then $\Delta = \text{projection} \circ \Phi^{-1}$ is an integral function. Φ is unique if you require that the sheet label n of $\Phi^{-1}(r)$ increases to $+\infty$ as r traverses $[0,\infty)$.

The main result may now be stated: $\Delta = \text{projection} \circ \Phi^{-1}$ is a Hill discriminant, and every discriminant arises in this way; in particular, the general complex Hill spectrum σ is obtained by application of Φ^2 to the points of S which project to $[-1,+1]$. It is a source of extra satisfaction that all this is conformally invariant: two conformal surfaces produce the same spectrum.

A few examples will clarify what is happening. Think, first, of the Riemann surface of $(1/\pi)\cos^{-1}$ covering the plane $\mathbf{C} = \mathbf{C}_-$. It is, itself, a plane $\mathbf{C}_+$ divided into vertical strips of unit width, indexed by $n \ne 0$ in the obvious way. What Tkachenko does is simple but decisive: it is to place, between $\mathbf{C}_+$ and $\mathbf{C}_-$, an intermediate surface S of which the nth sheet is the image under the map $k \to \cos \pi k$ of the nth strip of $\mathbf{C}_+$, with the label n still attached. These sheets are stacked up one over the other, cut along $(-\infty,-1]$ and $[1,\infty)$, and then pasted bank to bank in conformity with the covering $\mathbf{C}_+$ so that $\mathbf{C}_+$ and S are conformally the same. The uniformizer is

$\Phi = (1/\pi)\cos^{-1}$, with branch determined by the sheet label, and $\Delta =$ projection $\circ$ $\Phi^{-1} = \cos \pi k$ so that $\Delta^{-1}[-1, +1] = \mathbf{R}$ and $\sigma = [0, \infty)$. It is Tkachenko's insight that small modifications of the intermediate surface serve to describe the general complex Hill spectrum. The full picture begins to emerge with the help of three simple modifications of the surface just described. (1) Take two paired cuts $[1, \infty)$, say, at level n, together with their companions at level $-n$, and move the point 1 a little to the right. This produces a "gap" in the spectrum $[0, \infty)$. (2) Moving 1 a little to the left produces instead a "spine", this being a small segment cutting $(0, \infty)$ at 90°. (3) Now take the sheets as for $\cos^{-1}$ with four exceptions: for $n = \pm 2$, cut along $(-\infty, 0]$ and $[1, \infty)$; for $n = \pm 1$, cut along $(-\infty, 0]$ only. Then paste everything in the old way except for the four left-hand cuts $(-\infty, 0]$, which are pasted in a helical manner, as for the Riemann surface of the fourth root. This produces a mild deformation of $\sigma = [0, \infty)$ with a new structure attached, viz., a "barb" comprised of six segments meeting the deformed (open) ray at a single point, the eight subtended angles being 45° each. Clearly, variants abound: the rays $(-\infty, -1]$ and $[1, \infty)$ can be (mildly) deformed; the exceptional rays $(-\infty, 0]$ can also be varied both in number and in their mode of pasting. In this way, various numbers of gaps, spines, and/or barbs may be introduced, spines can sprout further spines, etc. That is the story Tkachenko tells in this distinctive paper.

There remain the (to my knowledge) open question of what are the correct auxiliary spectral data to specify the Hill operator fully, and the subsequent question of describing the class of all complex potentials with fixed discriminant. In the real case, the auxiliary data take the form of a divisor on the Riemann surface M of $\sqrt{\Delta^2 - 1}$ encoding the roots of $y_2(1, \lambda) = 0$ and the signature of $\sqrt{\Delta^2 - 1} = (1/2)[y_2'(1, \lambda) - y_1(1, \lambda)]$ at each of these. The divisor has just one point on each of the real ovals of M, permitting the identification of the spectral class with (the real part of) the associated Jacobi variety; in particular, the spectral class is a (real) ∞-dimensional torus as mentioned before [cf. H. P. McKean and E. Trubowitz, op. cit.]. B. Birnir [Comm. Pure Appl. Math. **39** (1986), no. 1, 1–49; MR 87f:58061; Comm. Pure Appl. Math. **39** (1986), no. 3, 283–305; MR 87k:58109] showed that the same description applies to complex potentials having real Hill-type spectrum with a finite number of open gaps: The spectral class imitates the full complex Jacobi variety of the desingularized curve M provided one admits potentials with singularities of the type x^{-2}. Presumably, the full Jacobi variety is what is needed here as well.

Addendum: M. Schmidt has brought to the reviewer's attention two papers, closely allied to Tkachenko's, on the complex periodic nonlinear Schrödinger equation and the allied Dirac-type spectral problem. P. G. Grinevich and M. U. Schmidt [Phys. D **87** (1995), no. 1-4, 73–98; MR 96k:58113] have described the corresponding class of compact (finite genus) Riemann surfaces. A partial discussion of the general (infinite genus) case for smooth potentials is given in [M. U. Schmidt, Mem. Amer. Math. Soc. **122** (1996), no. 581, viii+111 pp.; MR 96k:58123], in which it is proved that the usual map, from potentials with fixed discriminant to divisors on the Riemann surface, is injective and that the image of such divisors in the associated Jacobi variety is open and dense. These results may be adapted to KdV and the allied Hill-type spectral problem studied by Tkachenko. *H. P. McKean, Jr.* (New York, NY)

98a:35008 35B25 35Q30 76D05 76D10 76D30
Temam, Roger (Bloomington, IN);
Wang, Xiaoming [Wang, Xiao Ming[1]] (New York, NY)
Asymptotic analysis of Oseen type equations in a channel at small viscosity. (English summary)
Indiana Univ. Math. J. **45** (1996), *no.* 3, 863–916.

The main object of research in this paper is a system of Navier-Stokes equations linearized around a fixed-velocity flow. The authors study the behavior of its solutions as the viscosity parameter tends to zero. This article extends previous results of the authors [Differential Integral Equations **8** (1995), no. 7, 1591–1618; MR 96j:35194] concerning the Stokes equations, i.e. the Navier-Stokes equations linearized around the state of rest. In these two works the authors develop their own promising method to investigate the problem of vanishing viscosity, based, firstly, on specially constructed boundary layer functions which they call correctors, and secondly, on original a priori estimates.

Properly speaking, these two points, namely, consideration of types of boundary layer functions and a way to estimate the difference between explicit and asymptotic solutions, are crucial in any boundary layer function method. They are characterized by the fact that a solution of a singularly perturbed equation is typically sought as a sum of a solution of the corresponding degenerate equation and a function that vanishes far from the boundary. This approach seems common and natural now but it has been achieved through the efforts of many scientists. In addition to those cited in the paper under review, the names of A. B. Vasil′eva, V. F. Butuzov and M. I. Imanaliev are appropriate to recall here. However, the name of such methods comes from continuum mechanics, where the theory of boundary layers is rapidly developing.

But although the majority of works in this sphere are based on the Navier-Stokes equations at small viscosity, as a rule the functions used there for the description of boundary layers are not boundary layer functions in the above-mentioned mathematical sense for the Navier-Stokes equations.

A number of scientists working on the problem of small (or vanishing) viscosity have obtained many results for different formulations. In particular, the weak convergence of solutions for the linearized three-dimensional Navier-Stokes equations with convective terms was established by the reviewer [S. N. Alekseenko, in *Studies in integro-differential equations, No. 16*, 243–257, "Ilim", Frunze, 1983; MR 85k:35190; in *Studies in integro-differential equations, No. 19 (Russian)*, 288–304, 320, "Ilim", Frunze, 1986; MR 89c:76045]. Interesting results for the two-dimensional case were obtained by G. V. Alekseev [Dinamika Splošn. Sredy Vyp. 10 (1972), 5–25, 246; MR **56** #17400]. There are also other results on convergence of some asymptotic expansions arising in this problem.

But an approach giving a common solution of singularly perturbed boundary value problems has not yet been found. Indeed, this is not a single problem, rather there are many problems differing from one another. And while some of these problems have been solved successfully still others are far from being solved.

In the article under review the authors study the asymptotic behavior of periodic solutions in a two-dimensional channel. The boundary layer phenomena occur near the side walls.

The main problem is formulated as follows. Let $\Omega_\infty = \mathbf{R}^1 \times (0,1) \equiv \{(x,y)\colon x \in \mathbf{R}^1,\ y \in (0,1)\}$, $\Omega = (0,2\pi) \times (0,1) \equiv \{(x,y)\colon x \in (0,2\pi),\ y \in (0,1)\}$; $H^s(\Omega) = \{v \in H^s_{\rm loc}(\Omega_\infty),\ v$ is periodic in x with period $2\pi\}$, $s \geq 0$, $H^1_0(\Omega) = \{v \in H^1(\Omega),\ v|_{y=0,1} = 0\}$, $H = \{v\colon v \in (L^2(\Omega))^2,\ {\rm div}\, v = 0,\ v_2|_{y=0,1} = 0\}$, $V = \{v\colon v \in (H^1_0(\Omega))^2,\ {\rm div}\, v = 0\}$. One needs to find an unknown vector function $u^\varepsilon \equiv u(t,x,y,\varepsilon) = (u_1^\varepsilon, u_2^\varepsilon)$ satisfying the evolutionary Oseen vector equation with initial-boundary conditions, namely (1) $\partial_t u^\varepsilon - \varepsilon\Delta u^\varepsilon + U_\infty D_1 u^\varepsilon + \nabla p^\varepsilon = f$, $u^\varepsilon|_{t=0} = u_0$, $u^\varepsilon \in V$, where D_1 denotes the partial derivative with respect to x, and U_∞ is constant.

The corresponding problem for the inviscid equation has the form (2) $\partial_t u^0 + U_\infty D_1 u^0 + \nabla p^0 = f$, $u^0|_{t=0} = u_0$, $u^0 \in H$. In addition, let θ^ε be a solution of the problem (3) $\partial_t \theta^\varepsilon - \varepsilon\Delta\theta^\varepsilon + U_\infty D_1 \theta^\varepsilon = 0$, $\theta^\varepsilon|_{t=0} = 0$, $\theta^\varepsilon|_{y=0,1} = -u^0|_{y=0,1}$. Since $u_2^0|_{y=0,1} = 0$, and thus $\theta_2^\varepsilon = 0$, then (3) is in fact an equation for θ_1^ε only.

The main result of this article is the following. Theorem 3.1: Suppose $u_0 \in V$, u_0 and f are sufficiently smooth; then the function θ^ε defined by equation (3) is a zeroth-order corrector for problems (1) and (2) in the sense that there exists a generic constant k, depending on the data but independent of ε, such that

$$\|u^\varepsilon - (u^0 + \theta^\varepsilon)\|_{L^\infty(0,T;(L^2(\Omega))^2)} \leq k\varepsilon^{1/2},$$

$$\|u^\varepsilon - (u^0 + \theta^\varepsilon)\|_{L^\infty(0,T;(H^s(\Omega))^2)} \leq k\varepsilon^{(1-s)/2} \text{ for } s \in [0,1],$$

$$\|p^\varepsilon - p^0\|_{L^2(0,T;(H^1(\Omega)))} \leq k\varepsilon^{1/2}.$$

On the whole, the corrector θ^ε is of parabolic type and depends on $y/\sqrt{\varepsilon t}$ or $(1-y)/\sqrt{\varepsilon t}$, although the presence of the convective term brings about a certain intricacy.

Besides this main result, the authors establish the following result on the rate of convergence of u^ε in $L^2(\Omega)$. Proposition 3.1: Suppose u_0 and f are sufficiently smooth (but do not necessarily vanish on the boundary); then $\|u^\varepsilon - u^0\|_{L^\infty(0,T;H)} \leq k\varepsilon^{1/4}$, where k is a generic constant independent of ε.

The larger part of the article is devoted to determining the asymptotic expansion of the corrector. To this end the authors begin with the investigation of the vector equations $-\varepsilon\Delta v^\varepsilon + U_\infty D_1 u^\varepsilon + \nabla p^\varepsilon = f$, for which they establish asymptotic expansions and the form of the correctors. Here they use a decisive decomposition of the space H, namely $H = K \oplus K^\perp$, where $K = \{v \in H\colon v$ is independent of $x\}$ and $K^\perp$ is the orthogonal complement of K in H. The authors show that if $f \in K$ then $v^\varepsilon = \varepsilon^{-1}A^{-1}f$ and no boundary layer phenomena exist, but when $f \in K^\perp$ then $|v^\varepsilon| = (1/|U_\infty|)|f|$ and $\|v^\varepsilon - (v^0 - \theta^\varepsilon)\|_{(L^2(\Omega))^2} \leq K\varepsilon^{1/2}$, where v^0 is a solution of the system $U_\infty D_1 v^0 = f$, ${\rm div}\, v^0 = 0$, v^0 is periodic in x with period 2π, and θ^ε is a solution of $-\varepsilon\Delta\theta^\varepsilon + U_\infty D_1\theta^\varepsilon = 0$, $\theta^\varepsilon|_{y=0,1} = -v^0|_{y=0,1}$.

Further the authors study the asymptotic behavior of the problem (4) $\partial_t v^\varepsilon - \varepsilon\Delta v^\varepsilon + U_\infty D_1 v^\varepsilon = f$, $v^\varepsilon|_{t=0} = v_0$, $v^\varepsilon|_{y=0,1} = 0$, where v is periodic in x with period 2π. In the proof the authors derive the form and asymptotic behavior of the corrector that they use later to solve the main problem (1). Although the behavior of solutions to equations of type (4) for $\varepsilon \to 0$ has been studied by other specialists too, the authors follow their own original method of solving the problem, adapted to their stated aims. They give an explicit approximate formula for the description of the corrector, using the function ${\rm erf}(y) = (1/\sqrt{2\pi})\int_0^y \exp(-z^2/2)dz$, among other tools. The investigation of equation (4) is done as a straight continuation of the authors' previous article mentioned above. Moreover, the authors improve some previous results here, even in the case of the heat equation or the Stokes equations.

The proofs of all results are based on sharp and efficient a priori estimates, using a functional-analytic global treatment of the pressure term. These a priori estimates are of independent interest as ones giving tools for explorations of similar tasks.

Besides the already reviewed results the authors obtain asymptotic expansions for the Navier-Stokes equations linearized around $(U(y),0)$, namely $\partial_t u^\varepsilon - \varepsilon\Delta u^\varepsilon + U'(y)u_2^\varepsilon e_x + U(y)D_1 u^\varepsilon + \nabla p^\varepsilon = f$, where e_x, e_y is the canonical basis of $\mathbf{R}^2$. Also, they study the large-time behavior of the solutions to the evolutionary Oseen equation. It turns out to be very interesting because if $f \in K$ then the solution blows up as $\varepsilon \to 0$ and $t \geq 1/\varepsilon$, and conversely if $f \in K^\perp$ then the solution exhibits typical boundary layer phenomena.

The paper is written carefully so as to make it easy for the reader to follow and is distinguished by an exhaustive analysis of the problems considered. The exposition of the material is truly absorbing and expressive, as far as this is possible for a mathematical paper. The article could serve as a model of how scientific works should be written. *Sergey Nikolaevich Alekseenko* (Bishkek)

98j:35026 35C20 35B25 35L60

Grenier, Emmanuel (Paris); **Guès, Olivier** (Nice)

Boundary layers for viscous perturbations of noncharacteristic quasilinear hyperbolic problems. (English summary)

J. Differential Equations **143** (1998), *no.* 1, 110–146.

Singular perturbation problems in differential equations arise when the perturbed equation ($\varepsilon \neq 0$) contains derivatives of order higher than those in the unperturbed equation ($\varepsilon = 0$). As a result, the perturbed equation allows more boundary or initial conditions to be imposed, and consequently, solutions of the perturbed equation cannot be uniformly close to those of the unperturbed equation, unless the boundary conditions are chosen in a special way.

In many cases, this problem can be resolved by introducing boundary layers. The solution of the perturbed problem is approximated by a superposition of two terms: one is the solution of the reduced equation for $\varepsilon = 0$, while the other is a contribution which is confined to a small neighborhood of the boundaries where the additional boundary conditions are required. Typically, the construction of this boundary layer solution involves rescaling one of the independent variables by ε or a power of ε.

While the formal construction of solutions involving boundary layers is well known and widely applied, the rigorous justification of such asymptotic solutions is not easy, especially in the context of partial differential equations.

Here the authors solve this problem for a broad and important class of equations. They consider a symmetrizable hyperbolic system with a small viscosity of order ε: $\partial u/\partial t + \sum_{j=1}^n A_j(t,x,u)\partial u/\partial x_j + B(t,x,u)u - \varepsilon\Delta u = f(t,x)$. The problem is posed for $t > 0$ and $x_n > 0$, and for simplicity of exposition the initial and boundary conditions are chosen to be $u = 0$. It is assumed that the boundary is noncharacteristic. The authors characterize the boundary conditions which survive in the limit $\varepsilon \to 0$. They then show how a formal approximate solution can be constructed using matched asymptotics; this can be carried out up to any order in ε. Let u_n be the formal asymptotic solution, including terms up to order ε^n. It is not difficult to see that

u_n will satisfy the equation up to a residual term R_n which is of order ε^n. The heart of the proof, and the main difficulty, is in deriving energy estimates which yield a bound of the error of the solution $u - u_n$ in terms of the residual R_n.

Contrary to what one might naively expect, the existence of a formal boundary layer profile combined with an existence result for the full partial differential equations is not enough to yield the result that the formal solution approximates the true solution. In addition, the authors require a smallness condition on the boundary layer profile. The fact that such a smallness condition is necessary is illustrated by the counterexample $v_t + v_x - v w_x - \varepsilon v_{xx} = f$, $w_t - w_x - \varepsilon w_{xx} = 0$ $(t > 0,\ x > 0)$, with the boundary condition $v = w = 0$ and initial conditions $v = 0$, $w = \lambda(1 - \exp(-x/\varepsilon))$. One can solve successively for w and then for v, and it is clear that smooth solutions exist for every $t > 0$. It is also easy to construct a formal approximate solution for $\varepsilon \to 0$; indeed, it is given by $w = \lambda(1 - \exp(-x/\varepsilon))$ and v satisfies $v_t + v_x = f$, $v(0,t) = v(x,0) = 0$. If λ is sufficiently large, however, the difference between this approximation and the true solution is not of order ε; indeed, due to an instability in the boundary layer, the difference becomes infinite as $\varepsilon \to 0$ if f is suitably chosen.

The paper generalizes earlier work on semilinear equations [O. Guès, Ann. Inst. Fourier (Grenoble) **45** (1995), no. 4, 973–1006; MR 97i:35103] and one-dimensional problems [M. Gisclon, J. Math. Pures Appl. (9) **75** (1996), no. 5, 485–508; MR 97f:35129; M. Gisclon and D. Serre, C. R. Acad. Sci. Paris Sér. I Math. **319** (1994), no. 4, 377–382; MR 95e:35119]; the authors also cite several recent works addressing the important case where the boundary is characteristic for the hyperbolic system.

Michael Renardy (Blacksburg, VA)

98f:35038 35J40 35A20 35S15 46N20 47F05

Kozlov, V. A. [Kozlov, Vladimir Arkad′evich]; Maz′ya, V. G.; Rossmann, J.

★Elliptic boundary value problems in domains with point singularities. (English summary)

Mathematical Surveys and Monographs, 52.

American Mathematical Society, Providence, RI, 1997. x+414 pp. $99.00.

ISBN 0-8218-0754-4

The goal of this book is a systematic and self-contained exposition of the theory of elliptic boundary value problems in smooth domains and domains with isolated singularities on the boundary. The book consists of three parts.

In Part I (Chapters 1–4) the authors begin with consideration of ordinary differential equations on the semi-axis; then they move on to boundary problems in a half-space (periodic in the tangential variables) and in bounded smooth domains.

The important difference from other books devoted to elliptic partial differential equations is the form of the boundary conditions: throughout the book they are written as $Bu + C\underline{u} = g$, where $\underline{u}$ is an additional unknown vector-valued function on the boundary and C is a matrix of differential operators along the boundary. Thus the theory developed in this part contains the classical theory of elliptic differential boundary value problems (without C) and is contained in the Boutet de Monvel theory, where B and C can be pseudodifferential and the elliptic equation can also be pseudodifferential and can contain additional (Green and coboundary) terms.

The advantage in comparison with the classical theory consists in the fact that a modification of the classical Green formula is always valid and the formally adjoint problem exists and belongs to the same class of boundary value problems even if the normality condition is violated. Using the formally adjoint problem, the authors are able to consider their problems in a complete scale of Sobolev L^2 spaces of any integer order, to obtain the Fredholm property by proving a priori estimates and regularity theorems without constructing the parametrix and to describe the cokernel of the given boundary problem as the kernel of the formally adjoint boundary problem. (To some extent a similar point of view is present in the books of J.-L. Lions and E. Magenes [*Problèmes aux limites non homogènes et applications. Vol. 1*, Dunod, Paris, 1968; MR **40** #512] and Y. A. Roĭtberg [*Elliptic boundary value problems in the spaces of distributions*, Translated from the Russian by Peter Malyshev and Dmitry Malyshev, Kluwer Acad. Publ., Dordrecht, 1996; MR 97m:35055], devoted to classical differential elliptic boundary problems.)

The theory is constructed carefully and without too great generality. The Green function is constructed, the integral representation of solutions is obtained, and elliptic problems with parameter are discussed in Chapter 3; however, more special questions, such as boundary conditions with terms of higher order, boundary value problems for elliptic systems, boundary problems in the variational form, the use of Sobolev spaces of real order, L^p and Hölder spaces, etc., are left for Chapter 4. The whole book has such a structure, and this, along with numerous examples, makes it convenient for beginners.

Part II (Chapters 5–8) is devoted to elliptic boundary value problems in cylinders, cones and bounded domains with conical points (angular points if $n = 2$). At first the authors consider problems in an infinite cylinder $\{(x, t) \in \Omega \times \mathbf{R}\}$, where Ω is a smooth bounded n-dimensional domain, with coefficients not depending on t, in Sobolev spaces of positive and negative integer order with an exponential weight $e^{\beta t}$. The weight is chosen so that the corresponding operator pencil (obtained by the Laplace transform $v(\lambda) = \int_{-\infty}^{\infty} e^{\lambda t} u(t) dt$) has no eigenvalues on the corresponding line parallel to the imaginary axis. The authors obtain conditions necessary and sufficient for unique solvability and determine the asymptotics of the solutions at infinity. For boundary value problems with t-dependent coefficients, under a stabilization condition at infinity, the authors obtain the Fredholm property and generalize the asymptotic results.

Then the authors move on to model boundary value problems in an infinite cone, find conditions for unique solvability of the problem in Sobolev spaces with a power weight and determine the asymptotics of the solution near a conical point. Again, Sobolev spaces of positive and negative order are used. After this the authors consider elliptic boundary value problems in bounded domains with conical singularities, prove the Fredholm property, investigate the dependence of the index on the choice of spaces and construct the asymptotics of the solutions. They also investigate elliptic problems with parameter in domains with conical points; such problems arise when one considers elliptic problems in domains with higher-order singularities.

The weights originally considered do not contain the case of the absence of a weight. In Chapter 7 the authors consider boundary value problems in weighted Sobolev spaces with nonhomogeneous norms containing the usual Sobolev spaces. Again, more special questions are concentrated in the last chapter of this part.

Part III (Chapters 9 and 10) is devoted to elliptic boundary value problems in quasicylindrical domains and domains with cuspidal points. The cusps may be exterior or interior. The results include the solvability (in special weighted Sobolev spaces depending on the geometry of the domain near the singular points of the boundary) and assertions on the regularity and on the structure of the solutions.

Each part is supplemented by references to the bibliography, which includes 256 items.

It is widely known that the authors have obtained many deep results for elliptic boundary value problems in domains with singularities. Here their experience permits them to present the results in a very complete and accessible form. Without doubt, the book will be interesting for many mathematicians working with elliptic boundary problems in smooth and nonsmooth domains, and it would be frequently used in any mathematical library. *M. S. Agranovich* (Moscow)

98e:35091 35K57 47H15 47N20 58F39

Brunovský, P. (Bratislava)**; Poláčik, P.** (Bratislava)

The Morse-Smale structure of a generic reaction-diffusion equation in higher space dimension.

J. Differential Equations **135** (1997), *no.* 1, 129–181.

The authors consider reaction-diffusion equations of the form $u_t = \Delta u + f(x,u)$, $t > 0$, $x \in \Omega$, with Dirichlet boundary conditions on $\partial\Omega$, where Ω is a bounded domain in $\mathbf{R}^n$ with smooth boundary and f is a sufficiently regular function on $\overline{\Omega} \times \mathbf{R}$. This problem defines a semiflow on an appropriate Banach space and is gradient-like there. Brunovský and S.-N. Chow [J. Differential Equations **53** (1984), no. 1, 1–23; MR 86a:35072] have shown that generically with respect to f in the C^r-topology all equilibria are hyperbolic. In the one-dimensional case D. B. Henry [J. Differential Equations **59** (1985), no. 2, 165–205; MR 86m:58080] and S. B. Angenent [J. Differential Equations **62** (1986), no. 3, 427–442; MR 87e:58115] have shown that stable and unstable manifolds of any two hyperbolic equilibria always intersect transversally. This is proved via nodal properties of the solutions and the Jordan curve theorem. In higher dimensions Poláčik [Differential Integral Equations **7** (1994), no. 5-6, 1527–1545; MR 94m:35161] has given examples with nontransversal intersections.

The authors prove in the present article that there is a residual set in C^k with the C^r-Whitney topology such that for any f in this set all equilibria are hyperbolic and the stable and unstable manifolds intersect transversally. This is the Morse-Smale property.

For the proofs they consider the operator $u \mapsto u_t - \Delta u + f(x,u)$ and formulate the transversality of stable and unstable manifolds in terms of regular values of this operator. The generic hyperbolicity of the equilibria was first proved via an abstract transversality theorem by F. Quinn [in *Global Analysis (Proc. Sympos. Pure Math., Vol. XV, Berkeley, Calif., 1968)*, 213–222, Amer. Math. Soc., Providence, R.I., 1970; MR **41** #9304].

Then they apply the abstract transversality theorem (Theorem 4.c.1) for equations of the form $(*)$ $u_t + Au = f(u,\lambda)$, where A is a sectorial operator on a real Banach space X with compact resolvent, $f\colon G \times Y \to X$ is of class C^1, where G is an open set in X^α, the domain of the fractional power A^α, $0 \le \alpha < 1$, Y is open in a separable Banach space Λ and if u, λ is a solution of $(*)$ on an interval (t_1, t_2) then the evolution operator $T(t,s)$ of the variational equation $v_t + Av - f_u(u(t),\lambda)v = 0$ is injective with dense range in X. Under these assumptions and some further technical hypotheses the theorem states that there is a residual subset $\Gamma \subset Y$ such that for any $\lambda \in \Gamma$ any heteroclinic solution of $(*)$ is transverse. The main theorem is proved in Section 5 by verifying the hypotheses of Theorem 4.c.1.

Since all the results on the Morse-Smale property for parabolic equations are somehow related to the result of Angenent and Henry, this is really a new result. It must be compared to the Kupka-Smale theorem for finite-dimensional dynamical systems. In my opinion the methods of this paper will certainly be applied to other infinite-dimensional dynamical systems; in particular, Theorem 4.c.1 is of independent interest.

In conclusion, I think the paper by Brunovský and Poláčik gives an important result with interesting methods, which will be fruitful in other problems.

Gunter Bengel (Münster)

99m:35157 35L67 35B35 35L65 47D06 47N20

Zumbrun, Kevin (Bloomington, IN); **Howard, Peter** (Bloomington, IN)

Pointwise semigroup methods and stability of viscous shock waves. (English summary)

Indiana Univ. Math. J. **47** (1998), *no.* 3, 741–871.

The stability of shock profile solutions for systems of viscous conservation laws of the form $u_t + f(u)_x = (B(u)u_x)_x$ is a fundamental issue in the study of nonlinear waves and in particular for a better understanding of some crucial topics such as convergence of numerical schemes or convergence in the inviscid limit. However, present-day theory of stability is essentially weakly nonlinear, since it requires small-perturbation assumptions about the approximate decoupling of the linearized equations [see A. Szepessy and Z. P. Xin, Arch. Rational Mech. Anal. **122** (1993), no. 1, 53–103; MR 93m:35125; T.-P. Liu, Comm. Pure Appl. Math. **50** (1997), no. 11, 1113–1182; MR 98j:35121; and references therein].

In this paper the authors present a new approach to shock wave stability, partially inspired by the functional-analytic methods used by D. H. Sattinger in the scalar case [Advances in Math. **22** (1976), no. 3, 312–355; MR **55** #8561], which makes no special assumption on the shock strength, but puts more essential conditions on the point spectrum of the linearized operator around the wave.

The first difficulty for this program is that there is no separation between the top eigenvalue $\lambda = 0$ of the linearized operator about the wave, given by $L(u)v = -(f'(u)v)_x + (B(u)v_x)_x$, and its essential spectrum, and therefore it is impossible to use the standard semigroup theory to yield stability. In the scalar and in the totally compressive case this problem was solved by the use of weighted norms to shift the essential spectrum to the left and obtain a spectral gap, but similar techniques do not work for more general situations.

In this paper the authors use analytic extension into the essential spectrum of L of the resolvent kernel, formally defined by $G_\lambda(\cdot,y) := (L-\lambda I)^{-1}\delta_y$. This extension leads to the notion of "effective" point spectrum, which obeys a modified version of the Fredholm theory.

Applying these ideas to the problem of shock stability gives a simple condition for linear stability in terms of the zeroes of the Evans function $D_L(\lambda)$. This is an analytic function that plays a role for unbounded operators analogous to that played by $\det(L-\lambda I)$ for finite-dimensional operators. This function was already carefully considered in a recent paper by R. A. Gardner and Zumbrun [Comm. Pure Appl. Math. **51** (1998), no. 7, 797–855; MR 99c:35152]. Actually it turns out that the zeroes of the Evans function correspond, with their algebraic multiplicity, to the effective eigenvalues of L inside some sector of the essential spectrum. Using some quite detailed, and truly nontrivial, estimates on the linearized operator about the wave, it is possible to establish one of the main results of the paper, namely the Evans function criterion: The L^p linear (orbital) stability of a (stationary) wave u with respect to L^1-perturbations, $p>1$, is equivalent to the fact that the Evans function $D_L(\lambda)$ has precisely k zeroes in $\{\operatorname{Re}\lambda\geq 0\}$, where k is the dimension of the stationary manifold.

Using some very precise bounds on the Green function of the linearized operator, it is possible to establish an analogous result for nonlinear stability using in an essential way the ideas of Liu [op. cit.] and in particular for Lax and overcompressive shocks with perturbations with a suitable decay at infinity.

This theory is now in a rapid evolution and many problems could be handled by the basic theory of the present work: non-strictly hyperbolic systems, sonic shocks, dispersive undercompressive shocks, multidimensional planar shocks.

With the exception of the results about nonlinear stability, this valuable paper is completely self-contained and provides at the same time a fundamental step forward in the theory of nonlinear waves and an excellent introduction to the subject; it is possible that these ideas could be of interest even to researchers working outside of the area of conservation laws. *Roberto Natalini* (Rome)

99a:35196 35Qxx 35B10 35B35

Gardner, Robert A. (Amherst, MA)

Spectral analysis of long wavelength periodic waves and applications. (English summary)

J. Reine Angew. Math. **491** (1997), 149–181.

The stability of periodic travelling wave solutions of nonlinear PDE is a classical problem of fundamental importance. The L^∞-spectrum of a periodic wave consists of bands or loops, rather than isolated eigenvalues. This complication can be avoided by restricting the analysis to a class of perturbations with the same periodicity as that of the underlying wave.

A natural question then is whether the restricted problem correctly predicts the stability of the wave under all bounded perturbations. This issue has been addressed extensively in several celebrated yet problem-specific applications, including Benjamin-Feir instability of periodic deep water waves [T. B. Benjamin and J. E. Feir, J. Fluid Mech. **27** (1967), 417–430; Zbl 144.47101; T. B. Benjamin, Proc. Royal Soc. London Ser. A **299** (1967), no. 1456, 59–75] and the closely related Eckhaus instability for periodic solutions of the Ginzburg-Landau equation [W. Eckhaus, *Studies in non-linear stability theory*, Springer-Verlag New York, New York, Inc., 1965; MR **33** #4490]. In both cases, a small-amplitude periodic wave of frequency ω which is stable under perturbations of frequency ω can sometimes be destabilized by perturbations of frequency $\omega\pm\varepsilon$ for sufficiently small ε. These examples illustrate the importance of considering the stability of a periodic wave relative to all bounded perturbations.

The spectral analysis of large-amplitude wave solutions of general systems in the absence of additional structure such as a variational principle is a difficult problem. In the case of large-amplitude pulse or front solutions, the Evans function $D(\lambda)$ has proven to be an effective tool in stability calculations [J. W. Evans, Indiana Univ. Math. J. **22** (1972/73), 577–593; MR **52** #14697; errata; MR **53** #8686; C. K. R. T. Jones, Trans. Amer. Math. Soc. **286** (1984), no. 2, 431–469; MR 86b:35011; J. C. Alexander, R. A. Gardner and C. K. R. T. Jones, J. Reine Angew. Math. **410** (1990), 167–212; MR 92d:58028; R. L. Pego and M. I. Weinstein, Philos. Trans. Roy. Soc. London Ser. A **340** (1992), no. 1656, 47–94; MR 93g:35115]. The roots of the analytic function $D(\lambda)$ coincide with the spectrum of the associated eigenvalue problem in a region Ω of the spectral plane outside the essential spectrum. Such investigations usually begin with a crude estimate to determine the approximate location of eigenvalues. A convenient approach in this case is the use of an associated topological invariant called the stability index, which was introduced by Alexander, Gardner and Jones [op. cit.]. In situations of strong instability, crude estimates suffice; however, in other situations the wave may have spectrum close to the imaginary axis, and in such cases a finer analysis of the Evans function is usually required in order to determine the stability of the wave.

In an earlier paper [J. Math. Pures Appl. (9) **72** (1993), no. 5, 415–439; MR 94j:35072], Gardner constructed both the Evans function and the stability index for periodic travelling waves. In this setting, the Evans function $D(\lambda, \gamma)$ depends on a second free parameter $\gamma = e^{i\theta}$ along the unit circle, which arises as the Floquet multiplier of the monodromy matrix of the linearized equations about the wave. For fixed γ the roots of $D(\lambda, \gamma)$ are called the γ-eigenvalues of the wave. A crucial result proved by Gardner [op. cit.] is that all eigenvalues of a periodic wave have finite L^∞-multiplicity even though they occur in 1-parameter loops in the spectral plane and are therefore never isolated. More precisely, all (generalized) L^∞-eigenfunctions can be expressed as a finite superposition of (generalized) γ-eigenfunctions to yield Jordan chains of finite ascent. The (discrete) γ-spectrum of the wave therefore determines the entire L^∞ spectrum as γ varies over S^1. The γ-eigenvalue index counts γ-eigenvalues inside a fixed contour K in the spectral plane that is disjoint from the spectrum of the wave.

In the paper under review, Gardner implements his earlier machinery in a novel approach to the stability analysis of periodic travelling waves subject to another frequently encountered limiting regime: that in which a 1-parameter family of periodic waves $u_\alpha(\xi)$ tends to a homoclinic (solitary) wave $u_0(\xi)$ as $\alpha \to 0$ in the state space of the travelling wave ODE. The only other structural assumptions are that u_0 decays exponentially in the far field, and that the location and multiplicities of the isolated eigenvalues λ_0 of the solitary wave u_0 in the domain Ω are known.

Under the above assumptions, the main result of this paper is that the multiplicity of γ-eigenvalues of the periodic wave u_α in a small but fixed neighborhood of λ_0 is equal to the multiplicity of λ_0 as an eigenvalue of u_0, for all sufficiently small $\alpha > 0$ and for all $\gamma \in S^1$. If λ_0 is a simple eigenvalue of u_0, the periodic wave u_α then has a continuous loop $\lambda(\gamma)$ of simple γ-eigenvalues, parametrized by $\gamma \in S^1$ in a small neighborhood of λ_0. While the result appears straightforward on heuristic grounds, the limit in question is highly singular since the topology of the spectrum changes for $\alpha > 0$. This result is a deep and original contribution to the stability analysis of periodic travelling waves.

The proof uses the γ-eigenvalue index to count γ-eigenvalues of u_α interior to a small contour K enclosing λ_0 in its interior. A preliminary step consists of a delicate analysis of certain Poincaré maps defined for associated systems of ODE induced by the original linearized system on spaces of projectivized differential forms. The analysis reveals that the Poincaré maps have hyperbolic fixed points which determine stable and unstable subspaces of the associated monodromy matrix for the problem. This information is then used to analyze both the growth and the decay properties of solutions of the original linear equations, and to show that the Evans function $D_\alpha(\lambda, \gamma)$ of the periodic wave essentially decouples into a product consisting of the Evans function $D_0(\lambda)$ of the homoclinic limit with a second, trivial factor. The proof is completed with a topological continuation argument.

The last section of the paper presents various applications of the main result to systems arising in several physical models, including periodic travelling profile solutions of nonclassical systems of viscous conservation laws, various dispersive equations such as the generalized KdV (gKdV) equation, and reaction-diffusion systems such as the FitzHugh-Nagumo and the Gray-Scott models. In certain parameter regimes the homoclinic wave is itself unstable, and the instability of large-wavelength periodic waves follows from the main theorem. It is interesting to note that for the gKdV equation with nonlinear flux $f(u) = u^{p+1}$, the solitary wave is unstable for $p > 4$ [J. L. Bona, P. E. Souganidis and W. A. Strauss, Proc. Roy. Soc. London Ser. A **411** (1987), no. 1841, 395–412; MR 88m:35128]. Hence, by Gardner's result large-amplitude periodic wave solutions of the gKdV equation are also unstable for $p > 4$.

Gardner also notes that solitary wave solutions $u_0(\xi)$ of the gKdV can sometimes be unstable for certain $p < 4$ when $u_0(\pm\infty) = b$ with $b < 0$. This is interesting since it follows from a theorem of J.-C. Saut [J. Math. Pures Appl. (9) **58** (1979), no. 1, 21–61; MR 82m:35133] that global solutions of the gKdV equation exist whenever $p < 4$. Hence there is a regime in which finite-time blowup does not occur, yet periodic travelling waves of the gKdV equation are unstable. Since the equation determines an infinite-dimensional Hamiltonian system, the global attractor in such situations is likely to be a rather complicated object.

In situations in which the homoclinic wave is stable, such as the FitzHugh-Nagumo system and, occasionally, the Gray-Scott pulse solutions, the main theorem demonstrates that the long-wavelength periodic waves will necessarily have a loop of simple γ-eigenvalues passing through the origin. A finer analysis of the location of the marginally stable eigenvalues is required in order to resolve fully the periodic wave's stability. While this issue is not directly tackled in the present paper, the innovative methodology developed in the proof of the main theorem presents a suggestive first step in this direction.

This paper is a deep and highly original contribution to the stability problem for travelling waves of nonlinear PDE. *Joel Smoller* (Ann Arbor, MI)

98m:35167 35Q30 35Q53 76B15
Wu, Sijue (Evanston, IL)
Well-posedness in Sobolev spaces of the full water wave problem in 2-D. (English summary)
Invent. Math. **130** (1997), *no.* 1, 39–72.

Everyone is familiar with the motion of water waves in everyday experience, and there has been an extremely rich variety of phenomena observed in the motion of such waves. However, the full equations governing the motion of the waves are notoriously difficult to work with because of the free boundary and the inherent nonlinearity, which are nonstandard and nonlocal. Although many approximate treatments, such as linear theory and shallow-water theory as well as numerical computations, have been used to explain many important phenomena, it is certainly of importance to study the solutions of the equations which include the effects neglected by approximate models. The well-posedness of the fully nonlinear problem is one of the main mathematical problems in fluid dynamics. Here, the motion of two-dimensional irrotational, incompressible, inviscid water waves under the influence of gravity is considered. There is a free surface separating the air and water, and the surface tension is zero. The existence of progressing periodic or solitary waves of small and large amplitude has been obtained in many works using different methods. For the unsteady water-wave problem, T. Kano and T. Nishida [J. Math. Kyoto Univ. **19** (1979), no. 2, 335–370; MR 82d:76012] showed the local existence in time for the general initial value problem with analytic data. For initial data in Sobolev spaces, V. I. Nalimov [Dinamika Splošn. Sredy Vyp. 18 Dinamika Židkost. so Svobod. Granicami (1974), 104–210, 254; MR **58** #29458], and subsequently H. Yosihara [Publ. Res. Inst. Math. Sci. **18** (1982), no. 1, 49–96; MR 83k:76017], gave the well-posedness locally in time for the full equations of water waves of infinite and finite depth, respectively, if the initial free surface is a localized perturbation of the static state, while W. Craig [Comm. Partial Differential Equations **10** (1985), no. 8, 787–1003; MR 87f:35210] used a similar approach to obtain the well-posedness for the equations in Boussinesq scaling and K-dV scaling. In the paper under review the author proves the local existence in time for the full equations with initial data in some Sobolev spaces, if the initial free surface is non-self-intersecting.

It is well known from linear theory that weakly hyperbolic equations behave badly under perturbation and in general the mapping from the initial data to the solution at a later time loses derivatives. In a nonlinear problem, this loss can have quite serious implications for an existence theory. However, in the case of water waves, it was first observed by Nalimov that in Lagrangian variables a cancellation for second-time derivations of the solution in an equation occurs, which leads to control of the loss of derivatives and reduces the system to a quasi-linear system. The existence of solutions for the initial value problem is obtained by solving the quasi-linear system using an iteration method. Based on this special structure of the equations, the local well-posedness of the problem in time was obtained for various physical problems if the initial data are near the static state. Moreover, for a linearized problem about an arbitrary solution of the full equations, it was observed by J. T. Beale, T. Y. Hou and J. S. Lowengrub [Comm. Pure Appl. Math. **46** (1993), no. 9, 1269–1301; MR 95c:76016] that the linearized problem is well posed if a certain pointwise condition on the free surface holds. The condition is essentially that the surface is not accelerating downward, normal to itself, as rapidly as the normal acceleration of gravity.

In the paper under review the author studies the fully nonlinear problem with arbitrary initial data. It is shown that the condition mentioned above always holds as long as the free surface is non-self-intersecting. Then, the existence and uniqueness of solutions in certain Sobolev spaces for the full water-wave problem, local in time, are proved if the initial free surface is non-self-intersecting. The basic approach in solving the problem is similar to that of Nalimov and many others. However, the author applies Riemann mapping to express an important operator on the free surface and uses the reparametrization by the Riemann mapping to derive a new and neat equivalent quasi-linear system which holds for every non-self-intersecting free surface of water waves. The paper is certainly of interest to researchers who study mathematical problems in water waves. *Shu Ming Sun* (Blacksburg, VA)

98f:35121 35Q35 35A20 35B99 65M99 76C05 76V05
Caflisch, Russel E. (Los Angeles, CA); **Ercolani, Nicholas** (Tucson, AZ); **Steele, Gregory**
Geometry of singularities for the steady Boussinesq equations. (English summary)
Selecta Math. (*N.S.*) **2** (1996), *no.* 3, 369–414.

The Euler equations describe the motion of an inviscid, incompressible fluid. One of the outstanding open questions of mathematical fluid dynamics is whether a singularity can develop in solutions of the three-dimensional Euler equations starting from smooth initial data, in the case when no boundaries are present (i.e., for periodic solutions or solutions decaying at infinity). The answer to the corresponding question for two-dimensional flow is clear-cut: no singularities will develop if the initial data are smooth.

The difference between two and three dimensions is best understood by writing the Euler equations in terms of the vorticity $\boldsymbol{\omega} = \nabla \times \mathbf{u}$, where $\mathbf{u}$ is the velocity field. The equations then take the form $\partial\boldsymbol{\omega}/\partial t + (\mathbf{u}\cdot\nabla)\boldsymbol{\omega} = (\boldsymbol{\omega}\cdot\nabla)\mathbf{u}$, $\nabla\cdot\mathbf{u} = 0$. The velocity field is reconstructed from the vorticity field using the Biot-Savart law. In two dimensions, $(\boldsymbol{\omega}\cdot\nabla)\mathbf{u} = 0$ and the first equation above immediately implies that the vorticity field satisfies $\boldsymbol{\omega}_t(\Phi_t(x)) = \boldsymbol{\omega}_0(x)$, where $\Phi_t(x)$ denotes the fluid particle trajectory starting from x at $t = 0$, and $\boldsymbol{\omega}_t(x)$ is the vorticity at position x and time t. In simple terms, the vorticity field is convected along particle path lines, and singularities cannot develop from smooth initial data. On the other hand, for three-dimensional flow the term $(\boldsymbol{\omega}\cdot\nabla)\mathbf{u}$ describes vortex stretching. Heuristic arguments suggest that the presence of this term could lead to infinite vorticity in finite time. However, despite much attention, neither a global existence theorem nor an example which shows the development of a singularity in finite time is known.

The simplest form of the Euler equations that capture nontrivial vortex stretching occurs for axisymmetric flow with swirl. Here, the governing equations involve only two independent spatial variables, radius r and axial position z. A. Pumir and E. D. Siggia [Phys. Fluids A **4** (1992), no. 7, 1472–1491; MR 93c:76014] and S. Childress et al. [J. Fluid Mech. **203** (1989), 1–22; MR 90e:76054] have pointed out that the Boussinesq equations describing stratified, incompressible flow are very similar to the Euler equations for axisymmetric flow with swirl. Thus, an understanding of singularity formation in the 2-D Boussinesq equations can lead to advances in these other problems. The paper under review presents the first rigorous analysis (supported by numerical computation) of generic singularity types for the steady Boussinesq equations. This is an important development in a widely followed area.

The phrase "singularity type" refers to the equivalence class of singularities that are preserved under analytic change of variables. "Generic" means that a singularity is stable with respect to perturbations, and that a solution with a singularity type that is not stable may be perturbed into a solution having only stable singularity types. In the paper under review, complex-valued solutions to the steady Boussinesq equations are considered, and the singularities are allowed to have complex spatial positions. That is, if x is a suitably chosen spatial coordinate, then the singularity may occur at complex values of x and for complex values of the velocity $\mathbf{u} = (u, v)$, density ρ, and vorticity ω.

There are several reasons for studying singularities of this type. For one, the authors believe that the form of steady complex singularities may be indicative of the form of real singularities in the unsteady, real-valued Boussinesq equations. This belief is supported by the fact that special "traveling wave" solutions can be obtained from the steady solutions by a simple Galilean transformation. (These traveling wave solutions have complex wave speed and actually correspond to solutions that are growing in amplitude.) In the special solutions the complex singularities move in the complex space plane $(x, y) \in \mathbf{C}^2$ (where x and y are appropriate spatial variables) toward the real plane. The situation is analogous to the behavior of complex singularities in the inviscid Burgers equation, $w_t + ww_x = 0$. Here a complex conjugate pair of singularities forms off the real x-axis and propagates unchanged toward the real line. The singularities change type upon colliding on the real axis, but the kinds of singularities that form after the collision are influenced by possible types of complex singularities. It is reasonable to assume that such a scenario occurs in the time-dependent Euler and Boussinesq equations. Consequently, the singularities for the steady and traveling wave solutions analyzed in the present paper are perhaps relevant in the general time-dependent problem.

The main feature of the analysis in the paper under review is a geometric approach to the study of differential equations. In this approach the graph of a solution is regarded as a fundamental object. Here the graph is situated in the space with coordinates $(x, y, u, v, \omega, \rho)$, where $\mathbf{x} = (x, y)$ are independent variables and the other variables are dependent. The PDE's provide constraints which enable one to reduce the space to (x, y, u, v, ψ), where ψ is the stream function. The solution surface is then the lifting of the graph of ψ onto this space. This surface can be interpreted as a multi-valued classical (i.e. smooth) solution—the Legendrian singularities are the branch curves of the solution surfaces under projection to $\mathbf{x}$. An outgrowth of the present application is a novel extension of Legendrian singularity theory to problems in which the dependent variables have infinite values at the singularities.

The singularity types uncovered by the analytical theory are supported by the results of accurate numerical computations. The computations are performed for the exact Euler equations with (complex) initial data that are upper analytic in the axial variable z. This means that the Fourier expansion of the data contains only nonnegative wavenumbers in z. As a result, the coefficient of the mode with wavenumber k depends only on those modes with wavenumber less than k. The advantage of this structure is that no truncation error is introduced by the restriction to a finite set of wavenumbers. This facilitates computations of sufficiently high accuracy to be able to deduce singularity properties from the decay of the Fourier coefficients. Two of the three analytically predicted singularity types were found in the computations. Thus, the computational results demonstrate the genericity of the singularity types established analytically.

In summary, this paper is pioneering in its use of geometric methods to analyze singularities of partial differential equations arising in problems from fluid dynamics. This includes the steady 2-D Boussinesq equations for stratified flow, and the Euler equations for axisymmetric flow with swirl, to which extension of the Boussinesq results is straightforward. The paper utilizes an impressive combination of analytical and numerical techniques. The results may serve as a guide in the analysis of time-dependent singularities in these and related problems. *Michael Siegel* (Newark, NJ)

99k:35166 35Q55

Kenig, Carlos E. (Chicago, IL); **Ponce, Gustavo** (Santa Barbara, CA); **Vega, Luis** (Bilbao)

Smoothing effects and local existence theory for the generalized nonlinear Schrödinger equations.

Invent. Math. **134** (1998), *no.* 3, 489–545.

Nonlinear Schrödinger equations of the form

$$\partial_t u = i\mathscr{L}u + P(u, \nabla_x u, \overline{u}, \nabla_x \overline{u}), \quad u(x,0) = u_0, \tag{1}$$

have been studied [T. Kato, in *Spectral theory and differential equations (Proc. Sympos., Dundee, 1974; dedicated to Konrad Jörgens)*, 25–70, Lecture Notes in Math., 448, Springer, Berlin, 1975; MR **53** #11252; S. Klainerman, Arch. Rational Mech. Anal. **78** (1982), no. 1, 73–98; MR 84b:35015; S. Klainerman and G. Ponce, Comm. Pure Appl. Math. **36** (1983), no. 1, 133–141; MR 84a:35173; J. Shatah, J. Differential Equations **46** (1982), no. 3, 409–425; MR 84g:35036] when $\mathscr{L} = \Delta$ and the nonlinear term P has a special form, which allows one to use the standard energy estimates. Another approach used to treat the general nonlinear terms of derivative type is restricted to suitable analytic function spaces [cf. N. Hayashi, Duke Math. J. **62** (1991), no. 3, 575–591; MR 92f:35127]. Recently, there has been much interest in the study of nonlinear Schrödinger equations with general derivative nonlinear terms in the usual order Sobolev space $H^s = \{f \in L^2;\ \|(1-\Delta)^{s/2} f\|_{L^2} < \infty\}$. In [C. E. Kenig, G. Ponce and L. Vega, Ann. Inst. H. Poincaré Anal. Non Linéaire **10** (1993), no. 3, 255–288; MR 94h:35238] local well-posedness for "small data" in H^s for s large enough and $P(z) = P(z_1, \cdots, z_{2n+2}) = \sum_{l_0 \le |\alpha| \le d} a_\alpha z^\alpha$ for $l_0 \ge 3$ was shown by using the so-called local smoothing effect of Kato type [T. Kato, in *Studies in applied mathematics*, 93–128, Academic Press, New York, 1983; MR 86f:35160]. In the one-dimensional case, the smallness assumption on the size of the data was removed in the paper by Hayashi and T. Ozawa [Differential Integral Equations **7** (1994), no. 2, 453–461; MR 94j:35165] by the gauge transformation technique. For the elliptic case $\mathscr{L} = \Delta$, H. Chihara [Math. Japon. **42** (1995), no. 1, 35–51; MR 96d:35128] removed the size restriction on the data in any dimension by making use of a modified version of the classical pseudo-differential operator of order 0 which was discovered by S. Doi [J. Math. Kyoto Univ. **34** (1994), no. 2, 319–328; MR 95g:35190], and the diagonalization method which depends on the ellipticity of $\mathscr{L}$.

In this article the authors show the local well-posedness of problem (1) in the nonelliptic case $\mathscr{L}=\sum_{j\le k}\partial_{x_j}^2-\sum_{j>k}\partial_{x_j}^2$, where $k\in\{1,\cdots,n\}$, which has several physical applications, for example, in water wave problems [see, e.g., A. Davey and K. Stewartson, Proc. Roy. Soc. London Ser. A **338** (1974), 101–110; MR **50** #1620; V. E. Zakharov and E. I. Shul′man, Phys. D **1** (1980), no. 2, 192–202; MR 81j:35077]. The main contribution of the paper is to remove the size restriction on the data in any dimension for general $\mathscr{L}$ when the data are sufficiently regular. The authors reduce the problem to that of L^2 well-posedness of the linear Schrödinger equation of the form

$$\partial_t v = i\mathscr{L}v + b_1(x)\cdot\nabla_x v + b_2(x)\cdot\nabla_x\overline{v} + F(x,t), \tag{2}$$

which is considered as a linearization of problem (1), where the coefficients $b_j(x)$ depend on the data of problem (1).

The L^2 well-posedness of (2) for $\mathscr{L}=\Delta$ and $b_2(x)=0$ has been obtained in [S. Doi, op. cit.] by making use of a 0th-order classical pseudo-differential operator which yields smoothing effects of solutions to (2). However, this operator does not work for (2) for general $\mathscr{L}$ and $b_2(x)$. The interesting idea of this article is to introduce a different type of operator C such that C is invertible, the difference between the commutator $i[\mathscr{L},C]$ and $Cb_1(x)\cdot\nabla_x$ essentially vanishes, and the complex conjugate of Cv is equal to $C\overline{v}$, which allows one to use the energy method in the remaining term $b_2(x)\cdot\nabla_x\overline{v}$. The operator C is not a classical pseudo-differential operator of order 0 but in the class of 0th-order operators studied by W. Craig, T. Kappeler and W. A. Strauss [Comm. Pure Appl. Math. **48** (1995), no. 8, 769–860; MR 96m:35057], who showed smoothing effects of solutions to linear Schrödinger equations with $b_2(x)=0$ and $\mathscr{L}=\mathscr{L}(x)$ elliptic with variable coefficients. The authors show the existence of the operator C with the properties mentioned above by the modified calculus for the pseudo-differential operators of Craig, Kappeler and Strauss.

Nakao Hayashi (Shinjuku)

97b:35174 35Q99 35B25 35C20 35K55 76D10 80A22

Alikakos, Nicholas D. (Knoxville, TN); **Bates, Peter W.** (Provo, UT); **Chen, Xinfu** (Pittsburgh, PA)

Convergence of the Cahn-Hilliard equation to the Hele-Shaw model. (English summary)

Arch. Rational Mech. Anal. **128** (1994), *no.* 2, 165–205.

In this paper, the authors carry out a program of rigorous justification of the connection between the Cahn-Hilliard equation and the Hele-Shaw problem in an appropriate asymptotic limit which corresponds to a sharpening of the transition layer between phases. The Cahn-Hilliard equation, a fourth-order nonlinear partial differential equation, serves as a well-regarded model for the process of phase separation and coarsening in a melted alloy. It has been the subject of much work over the past ten to fifteen years. In particular, formal asymptotic analysis [R. L. Pego, Proc. Roy. Soc. London Ser. A **422** (1989), no. 1863, 261–278; MR 90j:80010] established the connection between the level sets of the solution to the Cahn-Hilliard equation and interfacial motion in the Hele-Shaw problem. In the latter problem, the normal velocity of a propagating interface depends on the jump across the interface in the normal derivative of a function which is harmonic on either side and which equals the local mean curvature on the propagating interface. For the time interval in which the Hele-Shaw problem possesses a classical solution, the present paper now makes this formal connection rigorous.

To set this result in context, one can compare it to the large body of work devoted to an asymptotic analysis of the (second-order) Allen-Cahn equation—which tends to dissipate the same energy as the Cahn-Hilliard equation but which, unlike the Cahn-Hilliard equation, does not respect conservation of mass. The connection between the level sets of solutions to the Allen-Cahn equation and the problem of motion by mean curvature was made rigorous by L. C. Evans, H. M. Soner and P. E. Souganidis [Comm. Pure Appl. Math. **45** (1992), no. 9, 1097–1123; MR 93g:35064], using a weak formulation of the curvature flow via viscosity theory. At roughly the same time, P. de Mottoni and M. Schatzman ["Geometrical evolution of developed interfaces", unpublished] used a detailed spectral analysis to justify the formal asymptotics for the Allen-Cahn equation for as long as the mean curvature flow retained a classical solution. The paper under review is in spirit the Cahn-Hilliard cousin of this last reference. (We should note that more recently Chen [J. Differential Geom. **44** (1996), no. 2, 262–311; MR 98a:35059] developed a weak formulation of the Hele-Shaw problem using varifolds and carried out the asymptotic connection in this setting as well.) The procedure combines a delicate and nonstandard application of the method of matched asympotic expansions with detailed spectral analysis for the Cahn-Hilliard equation, previously worked out by N. D. Alikakos and G. Fusco [Indiana Univ. Math. J. **42** (1993), no. 2, 637–674; MR 95b:35089] in two dimensions and by Chen [Comm. Partial Differential Equations **19** (1994), no. 7-8, 1371–1395; MR 95f:35023] in higher dimensions. With regard to the matched asymptotics, one complicating feature is the necessity for a nontrivial boundary layer in addition to the interior layer marking the separation of phases.

This work is a true tour-de-force and successfully closes a chapter in the development of Cahn-Hilliard theory. At the same time, it serves as an impressive example of how careful spectral analysis can be combined with the tool of matched asymptotic expansions to yield rigorous results in challenging areas of nonlinear differential equations. *Peter J. Sternberg* (Bloomington, IN)

97g:35169 35Q99 35M20 35Q35 76H05 76X05 82D99
Gamba, Irene M. [Martínez-Gamba, Irene] (New York, NY); **Morawetz, Cathleen S.** (New York, NY)
A viscous approximation for a 2-D steady semiconductor or transonic gas dynamic flow: existence theorem for potential flow. (English summary)
Comm. Pure Appl. Math. **49** (1996), *no.* 10, 999–1049.

This paper proves a beautiful (even if not the ultimate) existence theorem for a difficult class of equations. Two important applications are named in the title and I believe the method introduced here will apply even more widely.

In the last decade, modeling of semiconductors has advanced from what are called drift-diffusion models to "hydrodynamic" models, which look like the equations of compressible fluid flow, coupled to electric field equations. The former models appear to be satisfactory for large devices and low electric fields, but fail to capture the behavior of submicron systems. (The reader is referred to the monograph of J. W. Jerome [*Analysis of charge transport*, Springer, Berlin, 1996; MR 98b:35206] for more detail.) Continuum models of the sort analyzed in this paper take as variables the density of a charge carrier (free electrons, for example) and its velocity, along with the electric field created by and influencing the motion of the charged fluid. A simplification in the present work is that the flow field is irrotational and hence the velocity is the gradient of a potential.

The mathematics is difficult for several linked reasons. First, the partial differential equation is of mixed type. If it were written as a system of first-order equations and linearized, there would be five equations and hence at least one real characteristic field. The authors introduce an "artificial" viscosity and show that then the system can be cast in the form of a pair of second-order elliptic equations for the electric potential and the velocity potential, coupled to a transport equation for the density.

If it were not for the viscous regularization, the system would exhibit a second difficulty: the velocity potential equation would actually exhibit change of type, in the manner of steady transonic flow, which is a special case of the model considered here. A theory for quasilinear equations which change type in a region which is not known a priori remains elusive; this question motivated early work of K. O. Friedrichs [Comm. Pure Appl. Math. **11** (1958), 333–418; MR **20** #7147] and Morawetz [Comm. Pure Appl. Math. **11** (1958), 315–331; MR **20** #3375] and also R. J. DiPerna [Trans. Amer. Math. Soc. **292** (1985), no. 2, 383–420; MR 87g:35148], and more recent work of Morawetz [Comm. Pure Appl. Math. **38** (1985), no. 6, 797–817; MR 87c:76087]. Questions which are still open include correct specification of boundary conditions for a well-posed problem, and even, to some extent, the definition of a well-posed problem in this case. It is a belief among practitioners in the field that the ultimate criterion for well-posedness is evolution of the time-dependent problem to a steady state. However, there is as yet no satisfactory theory for the time-dependent problem, even for the case of ideal gas flow, in more than a single space dimension. The paper at hand does not discuss time-dependent problems. In fact, there may be little that current knowledge of time-dependent problems can contribute to the theory of steady flows, though the authors were influenced by an important paper of B. Zhang [Comm. Math. Phys. **157** (1993), no. 1, 1–22; MR 95e:82080] on a one-dimensional unsteady version of the hydrodynamic model. The unsteady system does not show transonic change of type, but is still of mixed type since it couples the hyperbolic hydrodynamic equations to a potential equation for the electromagnetic field. (A fully time-dependent problem would replace electrostatics by electrodynamics, coupling Maxwell's equations to the hydrodynamic equations, as in the magnetohydrodynamic model for a plasma. This refinement does not appear to be necessary—nor would the analysis be easy—for devices on the scale of transistors.) Zhang's approach to treating the mixed problem presaged the method adopted in the present paper. Zhang himself built on a steady one-dimensional problem solved by Gamba in a paper [Comm. Partial Differential Equations **17** (1992), no. 3-4, 553–577; MR 93f:35186] which introduced the artificial viscosity approach to this class of problems.

The authors use a variant of the Schauder fixed point theorem, the Leray-Schauder fixed point theorem, an elegant technique which was developed to prove existence of solutions to elliptic boundary value problems. Since the problem at hand is not elliptic, it is impressive that the estimates needed to show compactness of the mapping still hold. The technical problem necessitating use of the more elaborate fixed point theorem is that the velocity and density must both be bounded away from zero for the estimates to hold, and hence the domain of the problem is not convex. In this connection it should be noted that while positive density is natural in transonic flow, the velocity is zero at stagnation points, which occur in some geometries. Thus, the approach of this paper would have to be modified in order to include some interesting problems in transonic aerodynamics. The specific boundary value problem considered here is an interior domain with corners. In classical elliptic theory, a nonsmooth domain complicates the analysis. However, for this problem of decidedly mixed type it is not at all clear what boundary conditions will lead to a well-posed problem. Since the velocity potential gives rise to streamlines, it seems natural to choose the sort of domain used in this paper: one with side boundaries consisting of streamlines, and "upstream" and "downstream" ends. Bounds on the density and velocity are obtained by controlling the inflow conditions. A difficulty which can be anticipated—that streamlines may cross the downstream boundary more than once—is handled by an artificial extension of the boundary.

The fixed-point theorem is applied to a nonlinear mapping T of the pair (ρ,φ), the density and the velocity potential. These satisfy the equations $\nabla\cdot(\rho\nabla\varphi)=0$ and $\nu\Delta\varphi=f(\nabla\varphi,\rho,\Phi)$; the electrostatic potential is governed by an equation of the form $\Delta\Phi=g(\rho)$, where f and g are given functions and ν is the viscosity coefficient. The authors ingeniously cast the first two equations as a coupling of a Poisson equation for φ and a transport equation for ρ and from this are able to pull out an extra derivative and obtain compactness of T. (Credit is assigned to unpublished work of R. V. Kohn and Morawetz for this clever idea.) The estimates, though in large part standard, are still quite lengthy. (To simplify this summary, I have not described several steps—solving the electrostatic equation, modifying the friction term outside the range of interest, and conformally mapping the domain—which influence the bounds.)

The main theorem proves existence of a classical solution to the boundary value problem for (ρ,φ,Φ) for a fixed, positive ν. The data are not small, and definitely include the transonic regime. In spite of achieving uniform bounds for ρ, Φ and $\nabla\varphi$ for small values of the artificial viscosity, this paper does not allow one to take the limit $\nu\to 0$, because the estimates are not good enough to show that the convergence is strong. The one-dimensional problem solved by Gamba [op. cit.] did include this limit, but the authors state that a "major gap" still exists in the two-dimensional problem. The source of the difficulty is not really the dimension of the physical space. In fact, for a scalar equation in any dimension, vanishing viscosity limits enjoy strong convergence properties as a consequence of the maximum principle for scalar parabolic or elliptic equations. For some systems of conservation laws, the celebrated technique of compensated compactness uses the nonlinear structure of the flux functions to achieve the same effect. It is not clear how big a step is needed to extend this technique to the current set of problems, or even whether it can be done at all.

Despite this limitation, the authors have introduced an important method by showing how a fixed-point procedure can be applied to mixed-type equations. Obtaining uniform small-viscosity bounds for a transonic equation is another major contribution. The model in this paper interweaves fluid dynamic and electrostatic fields; in this it is similar to other complex fluid dynamic models, such as viscoelastic, multiphase, and porous medium flows, which are at the heart of much modern technology. Computational interest in problems of this sort has established a need to understand

and solve systems of partial differential equations of mixed and changing type, both unsteady and steady. This paper provides a valuable tool.

Barbara Lee Keyfitz (Houston, TX)

98k:49002 49-02 06A15 46A40 46N10 47N10 49N15 52A01 90C26

Singer, Ivan (Bucharest)

★**Abstract convex analysis.**

With a foreword by A. M. Rubinov.

Canadian Mathematical Society Series of Monographs and Advanced Texts.

A Wiley-Interscience Publication.

John Wiley & Sons, Inc., New York, 1997. xxii+491 pp. $79.95. ISBN 0-471-16015-6

Since the appearance of the classical book by R. T. Rockafellar [*Convex analysis*, Princeton Univ. Press, Princeton, N.J., 1970; MR **43** #445], convex analysis has been a well-established theory. Its applications cover a broad range of mathematical subjects, optimization theory being the one in which it plays the most important role. The power of convex analytic methods in the study of convex optimization problems has for a long time suggested extending the notion of convexity in order to get more general concepts applicable to the study of nonconvex optimization problems. As a result, the research field of so-called generalized convexity grew rapidly starting in the seventies, and it continues to develop today. However, in spite of the considerable number of existing papers devoted to the subject, before the appearance of the book under review only a very small number of monographs on abstract or generalized convexity were published, that by M. Avriel et al. [*Generalized concavity*, Plenum, New York, 1988; MR 89e:90155] being the closest to the applications to optimization (and to mathematical economics as well). In particular, the study of duality theory has not been covered by any of the previously published books.

Duality theory is a very important tool in the analysis of optimization problems. By associating to a given problem another one, its dual, in a meaningful way, one gains insight into the structure of the original problem, from which one is often able to derive optimality conditions, and to devise numerical methods for solving it. In real world (e.g. economic) applications, the dual problem usually is of interest in its own right, and admits its own (economic) interpretation. In classical convex optimization, duality is based on the theory of conjugate functions. It is therefore a natural approach to base duality theories for specific families of nonconvex optimization problems on suitable generalizations of the notion of conjugation. One of the main contributors to the development of this line of research in the last two decades has been the author, who provides us with a deep study of the topic in this monograph.

The book is written in a very systematic and careful way. A long introductory chapter describes the fundamental concepts of abstract convex analysis and some of their main applications to optimization theory. The next three chapters deal with generalizations of the notion of convexity. Chapter 1 presents the general framework of complete lattices, within which abstract convexity is studied in this monograph. The basic definition says that an element in a complete lattice is convex with respect to a subset if it is the supremum of its minorants in the subset. This general notion is specialized in Chapters 2 and 3 to the cases when the complete lattice is the power set of a given set X, ordered by containment, or the set of all extended real-valued functions on X, with the pointwise ordering, respectively. This approach permits one to study abstract convexity for subsets of an arbitrary set X (without assuming any of the usual algebraic or topological structures) and functions on an arbitrary set X, generalizing the classical representation of closed convex sets [resp., lower semicontinuous convex functions] as intersections of closed half-spaces [resp., suprema of continuous affine functions], and to apply generalizations of separation theorems. Abstract quasi-convex functions (i.e., functions with abstract convex level sets) are discussed in Chapter 4. Chapters 5–9 are devoted to duality theory. The general theory for arbitrary complete lattices is presented in Chapter 5, whereas the more specific theories dealing with the particular complete lattices considered in Chapters 2 and 3 are developed in Chapters 6 and 7, respectively. The starting point in Chapter 5 is the definition of a duality (also called "polarity" by some authors), a mapping from a complete lattice into another, such that the image of the infimum of any set is the supremum of the images of its elements. In particular, given two sets, dualities relate the subsets of [resp., the functions on] the first set to the subsets of [resp., the functions on] the second set. A duality between two complete lattices of extended real-valued functions is called a conjugation when the image of the sum of a function with a constant is the image of the function minus the constant. It turns out that such an operator is none other than the Fenchel-Moreau conjugation associated to a coupling function. Among dualities, conjugations are those appearing the most frequently in the definition of dual optimization problems; their study is the subject of Chapter 8. Chapter 9 deals with $\vee$-dualities and $\perp$-dualities in the sense of [J.-E. Martínez-Legaz and I. Singer, Optimization **22** (1991), no. 4, 483–511; MR 92e:06017], two special classes of dualities arising in some applications. The last chapter is devoted to abstract subdifferentials: the definition of subdifferential with respect to a duality between two complete lattices of extended real-valued functions, as introduced in [J.-E. Martínez-Legaz and I. Singer, ZOR—Math. Methods Oper. Res. **42** (1995), no. 1, 109–125; MR 96d:49024], is first presented and then specialized to several particular cases. In particular, the subdifferential with respect to the Fenchel conjugation is just the usual subdifferential. The book ends with a long (73 pages) notes and remarks section that contains interesting historical details and other complements to the theory developed in the text. The list of references includes 317 items. There are also detailed notation, author and subject indexes.

As one can infer from the above description of the contents, the emphasis of the book is on the duality features of convexity. Actually, a significant portion of the material is based on the author's research papers on the subject.

The book is highly recommended to anybody wishing to get a deep knowledge of the abstract foundations of optimization duality theory. It clearly shows the essence of duality, and provides a unified presentation of virtually all the generalized conjugation concepts that have been proposed in the nonconvex context. The book is suitable also for students: the abstract concepts and results are exemplified by numerous particular cases, and the results are given with complete proofs.

Generalized convexity is not aimed at creating generalizations of the notion of convexity just for the sake of generalization. Convexity can be regarded as a sort of bridge between the ("easy") linear and the (difficult) nonlinear worlds, and the aim of generalized convexity is to find hidden convexity features in nonconvex problems, in order to better analyze them. So, generalized convexity is far from being an exhausted field of research. This book provides us with solid ground on which future research on this topic can be based. *Juan-Enrique Martínez-Legaz* (Bellaterra)

98m:49001 49-02 46N10 47N10 49J52 49K40 90C30

Rockafellar, R. Tyrrell (Seattle, WA); **Wets, Roger J.-B.** (Davis, CA)

★**Variational analysis.**

Grundlehren der Mathematischen Wissenschaften [Fundamental Principles of Mathematical Sciences], 317.

Springer-Verlag, Berlin, 1998. xiv+733 pp. $129.00. ISBN 3-540-62772-3

Nonsmooth and set-valued analysis, for which the authors propose the all-encompassing term variational analysis, constitute the subject matter of this book. Nonsmooth analysis is that body of theory that is concerned with the calculus of functions and sets that do not admit linear approximations, in the sense of the customary derivative or the usual tangent space. This area of research has been a very active one in its own right in the last quarter-century, and its applications to other fields have greatly multiplied as well. This fact reflects the growing appreciation of the essential need to grapple with nonsmoothness (together with absence of convexity) in such topics as optimization, control, nonlinear partial differential equations, optimal design, elasticity, variational methods, and, in fact, in nonlinear analysis generally. In nonsmooth analysis there intervene certain set-valued functions which play the role of (for example) generalized derivatives or tangents. The study of multifunctions such as these and others, together with related and more general issues of convergence (e.g., graph or gamma-convergence), has given rise to that subject known as set-valued analysis.

Certain characteristics have now become the norm in all types of nonsmooth analysis: the presence of a calculus of subgradients and normal vectors, in tandem with constructs dual to these (directional derivatives, tangents), and especially the placement on an equal footing of sets and of functions, which lends to the subject a distinctly geometric flavor. As pointed out by the authors, this pattern was established in F. H. Clarke's work in the early 1970s, which featured the calculus of proximal normals and generalized gradients. The principal applications of the theory were initially in optimization and control; the state of the subject in the early 1980s is summarized in the monograph by Clarke [*Optimization and nonsmooth analysis*, Wiley, New York, 1983; MR 85m:49002; second edition, SIAM, Philadelphia, PA, 1990; MR 91e:49001]. Another reference of note is the monograph by J.-P. Aubin and H. Frankowska [*Set-valued analysis*, Birkhäuser Boston, Boston, MA, 1990; MR 91d:49001].

In recent years, the subject has known a period of intense abstract development which has led to a rather bewildering array of competing and unclearly related theories. It is the authors' contention that it is possible to adduce a relatively small number of basic and interrelated concepts that will serve as the foundation of a complete and satisfying theory. In this they are certainly correct, and they present a detailed and convincing account of such an approach, one to which they have themselves contributed. As regards nonsmooth analysis, it is a good sign for the subject that the choice of basic constructs is much the same as in another recent and related monograph [F. H. Clarke et al., *Nonsmooth analysis and control theory*, Springer, New York, 1998; MR 99a:49001, see p. E59 in this volume]. The chapter headings of the book under review give an idea of its coverage: 1. Max and Min, 2. Convexity, 3. Cones and cosmic closure, 4. Set convergence, 5. Set-valued mappings, 6. Variational geometry, 7. Epigraphical limits, 8. Subderivatives and subgradients, 9.

Lipschitzian properties, 10. Subdifferential calculus, 11. Dualization, 12. Monotone mappings, 13. Second-order theory, 14. Measurability.

In preparing this systematic exposition of the basics, the authors have fashioned a great deal of new nomenclature and notation. Neologisms abound, and despite the authors' detailed arguments in the many pages of "Commentary", the choices do not always seem to be the happiest. The attributions, comments and historical remarks in the Commentaries are occasionally puzzling or misleading as well, tending as they do to examine certain trees in minute detail to the detriment of the forest. More serious shortcomings of the book, however, are its limitation to finite dimensions and its lack of motivating applications. It is true, as the authors say, that the line must be drawn somewhere. But in tracing it where they have, they have placed out of reach many interesting connections with functional analysis, together with most of the currently active research directions. Despite the book's considerable length, it does not offer a complete treatment of even the finite-dimensional theory; important topics such as fuzzy proximal calculus, multidirectional mean-value theorems and viscosity subdifferentials are not discussed. And the nonexpert will have difficulty grasping what all this theory is meant to be good for.

Despite these shortcomings, this volume is undoubtedly an important contribution to the subject. The authors have provided a very thorough, closely-reasoned synthesis of many of the basics of set-valued analysis. Great pains have been taken with the book: the writing and organization are very clear and remarkably error-free, and excellent figures accompany the text. The vision of the subject put forward here will influence it for many years to come. *Francis H. Clarke* (Villeurbanne)

99a:49001 49-02 49J52 58C20 93-02

Clarke, F. H. (Villeurbanne); **Ledyaev, Yu. S.** (Moscow);
Stern, R. J. [Stern, Ronald Jay] (Montreal, QC); **Wolenski, P. R.** (Baton Rouge, LA)

★Nonsmooth analysis and control theory. (English summary)

Graduate Texts in Mathematics, 178.

Springer-Verlag, New York, 1998. xiv+276 pp. $49.95. ISBN 0-387-98336-8

Nonsmooth analysis concerns the local description of nondifferentiable functions and sets lacking smooth boundaries, in terms of generalizations of classical concepts of derivatives, normals and tangents. Initial impetus for nonsmooth analysis came from optimization: how can we usefully adapt classical optimality conditions involving gradients of Hamiltonian functions or solutions to Hamilton-Jacobi equations to cover optimization problems of modern advanced engineering design, where these functions are no longer differentiable? It has been highly successful in addressing this specific question. But, as is now clear, it has broader implications as a general methodology. Just as in proving trigonometric identities over the real field we might benefit from an excursion into the complex field, there can be advantages to studying functions (even smooth ones) via the local properties of related, nonsmooth functions; nonsmooth analysis alerts us to this possibility.

F. H. Clarke's book [*Optimization and nonsmooth analysis,* Wiley, New York, 1983; MR 85m:49002] had a major role in winning an audience for nonsmooth analysis, techniques from which are now used, regarded as "part of the furniture" even, by practitioners in optimal control, mathematical programming, games theory and other fields. The period since its publication has been one of technical progress, fresh applications and changing perspectives—advances in the field which until now have been largely inaccessible, one suspects, to many who could profit from them, for lack of a good expository monograph.

Nonsmooth analysis and control theory has been written to fill this gap. It is a largely self-contained text in which the authors, taking as their starting point merely some basic functional analysis, build from scratch a theory of nonsmooth analysis, centered on properties of proximal normals, and follow up with numerous applications.

Care is taken to motivate the analysis and to convey the essential ideas. Often, proof of a general theorem is preceded by the simple proof of an important special case (as with the density theorem for proximal normals) which clarifies essential ideas. Anyone already familiar with the subject matter of this book will be surprised that it covers so much ground in its 276 pages. The authors have achieved this partly by finding shortcuts to some complicated proofs in the literature (their short, transparent proof of the Borwein-Preiss variational principle is particularly welcome) and partly by summarizing many routine steps as exercises, with accompanying hints. In some books, "exercises" are used to camouflage gaps in an argument, but not here: they serve genuinely both to clarify and to educate. The aim is to help readers understand techniques of nonsmooth analysis, rather than provide an inventory of sharpest results. "Give a man a fish and you feed him for a day; teach him how to fish and feed him for a lifetime."

And now to subject matter. This divides into basic theory and applications.

In the key first chapter, a theory of nonsmooth analysis is constructed around proximal subgradients and normals in Hilbert space. Why this particular framework? Because the authors wish to combine the generality of an infinite-dimensional theory with the simplicity of an approach based on a few intuitive geometric ideas, and because proximal normals feature so prominently in later applications. Most applications of nonsmooth analysis hinge on the properties of "subgradients" to certain "derived" functions (functions obtained by composing other "basic" functions); these are analysed with the help of a "nonsmooth calculus" which relates the subgradients of the derived and basic functions. In contrast to the "exact" calculus that can be constructed around some alternative nonsmooth constructs, the calculus of proximal subgradients is "fuzzy", i.e. it provides only approximate relationships, but ones whose precision is controlled by a parameter which can be taken arbitrarily small. This might seem like a disadvantage. But the strength of the fuzzy rules is that, in infinite dimensions, they often are valid under less stringent hypotheses than their exact counterparts. (The extra hypotheses of "exact" calculus are usually introduced to compensate for the lack of strong compactness of closed, bounded sets in infinite-dimensional spaces.) In infinite-dimensional applications, it is often simpler to apply the fuzzy calculus of proximal subgradients and to use the special nature of the application at hand to justify subsequent limit taking, than to try to match up the hypotheses of an exact calculus rule to the application.

Chapter 2, also on the basics of nonsmooth analysis, focuses on the generalized gradient of a locally Lipschitz function. This set-valued derivative, introduced by Clarke in 1975, inaugurated nonsmooth analysis by providing the first convincing answer to the question: how should one define a "generalized" gradient for a large class of functions, which reduces to the classical gradient for smooth functions and the subdifferential (in the sense of convex analysis) for convex functions and is accompanied by a useful calculus? Generalized gradients are not the only choice and alternative constructs are potentially the source of more information in certain applications. (For example, the variously named "limiting subdifferential" or "codifferential" has advantages in finite-dimensional optimization, as stressed by R. T. Rockafellar and B. Mordukhovich.) Yet the theory of the original generalized gradients remains the best-known and most widely applied area of nonsmooth analysis. One explanation is that, unlike some of its "rivals" (such as proximal analysis) which have a more restrictive setting, the theory of generalized gradients is a Banach space theory; this provides greater flexibility in certain applications. Another is that (in finite-dimensional spaces) the generalized gradients are often easy to calculate, thanks to an important representation theorem which tells us that the generalized gradient of a locally Lipschitz function at a point coincides with the convex hull of all limits of gradients of the function at neighbouring points, outside an arbitrary nullset. This is the rationale behind having a chapter on generalized gradients. The material here refines that in *Optimization and nonsmooth analysis* and augments it with more detailed information about the interrelation of generalized gradients and other tools of nonsmooth analysis.

The remaining two chapters introduce some new constructs, but are largely devoted to applications of the preceding theory. In Chapter 3, the emphasis is on refinements of classical theorems in nonlinear analysis, as well as new types of results, obtainable with the help of nonsmooth analysis. Specific topics covered include solvability of equations, sensitivity and the mean value theorem. Proximal subgradients are estimated in cases of special interest (proximal subgradients of integral functionals, for example). Where there is overlap with material in the earlier *Optimization and nonsmooth analysis*, the use of proximal gradients typically provides simplified proofs, under reduced hypotheses. The extension of the classical mean value theorem proved here, a "multidirectional", nonsmooth mean value inequality, to be precise, is something of a showpiece. The well-known classical theorem concerns the relationship between the values of a Gateaux differentiable function at the endpoints of a given line segment in its domain and the gradient at some point along the line segment. It is to be expected that this, like so many classical theorems involving derivatives, has a nonsmooth analogue. But what is surprising is that nonsmooth analysis can be used to prove a generalization for general lower semicontinuous functions in Hilbert spaces which, when specialized to the smooth case, is in some respects a significant improvement on the classical theorem—if equality in the classical relationship is replaced by inequality, then inequality holds *uniformly* with respect to the location of one of the endpoints of the line segment involved. Introducing uniformity or multidirectionality into the mean value theorem has a number of interesting consequences, a sample of which are pursued in Chapter 3. The significance of results like these is that they give evidence that nonsmooth analysis can be used to do more than solve problems of its own creation.

The final chapter is on control applications. The area of control traditionally associated with nonsmooth analysis is necessary conditions for local optimality. Few developments reflect the expanding range of applications so strikingly as recent applications of nonsmooth analysis to control theory and, in particular, to studying the qualitative properties of nonlinear control systems. These take up more than half of the chapter. It begins with the study of invariance properties of differential inclusions. When does there exist a solution arc which satisfies a given pathwise state constraint? This territory is already well explored. Aubin's viability theory provides a necessary and sufficient condition in terms of the "contingent cone" to the state constraint set, namely the requirement that at every boundary point there exists a velocity pointing into the contingent cone. The analysis then establishes equivalences of the contingent cone condition with other, seemingly less stringent, criteria for "viability". This book uses alternative methods for characterizing invariance, centered on conditions involving the proximal normal. The merits of the proximal normal approach are: first, existence of invariant trajectories is established by a simple constructive procedure called "proximal aiming" (traditional proofs, based on an abstract compactness argument, are not constructive); second, the proximal normal characterization is the "weakest", so equivalences with other criteria (including the contingent cone criterion) follow more or less directly from the definitions. However, one of the most striking benefits of studying invariance and proximal normals is its implications for control system design. Techniques involving "proximal aiming" have recently been used to resolve a long-standing open question regarding the existence of stabilizing feedback controllers for open-loop stabilizable plants. Proximal normals so perfectly match this application area that familiarity with proximal normals would seem indispensable for anyone seeking to understand it. While material in this book does not take the reader all the way to characterizing systems stabilizable by state feedback (the details are rather complicated), it is very good preparation for reading the research literature on this topic. In the concluding sections of the book, the authors return to familiar territory for applications of nonsmooth analysis: optimal control. Dynamic programming is briefly touched on (for free right endpoint problems value functions are characterized as unique "proximal" solutions to the Hamilton-Jacobi equation). However, the main focus of attention is necessary conditions of optimality. The many technical advances that have been made in this area since publication of *Optimization and nonsmooth analysis* have distracted attention from the simple proofs which are now available of some basic results in this area. Consistent with their expository objectives, the authors restrict attention to necessary conditions in the form of a Hamiltonian inclusion (a generalization of Hamilton's system of equations from the calculus of variations). A short, accessible proof of this condition is given, based on estimates of proximal subgradients of integral functionals and a sensitivity analysis of the value function associated with the endpoint constraints.

Writing this book is an impressive achievement. Besides flair, it has a streamlined feel to it, the result of multiple redrafting and extensive test-runs on graduate students. Newcomers to the field can, with the help of this book, acquire a perspective on proximal normals and other important tools of modern-day nonsmooth analysis, and the skills to use them. Specialists will find here a convenient reference and lucid proofs of many important results. It is an invaluable addition to the literature on nonsmooth analysis and its applications, and will be widely read.

Richard B. Vinter (London)

99g:49002 49-02 49K15 49L05 93-02

Milyutin, A. A.; Osmolovskii, N. P.

★Calculus of variations and optimal control. (English summary)

Translated from the Russian manuscript by Dimitrii Chibisov.

Translations of Mathematical Monographs, 180.

American Mathematical Society, Providence, RI, 1998. xii+372 pp. $129.00.

ISBN 0-8218-0753-6

The first author is one of the patriarchs of the theory of optimal control who was most instrumental in introducing functional-theoretic methods into the theory at an early stage of its development. The book under review summarizes the results of some rather recent studies which are much less known in the West. This is a highly original treatise full of new results and ideas which should be recommended to all interested in the theory of necessary and sufficient conditions in the calculus of variations and optimal control: strongly to specialists and users, with some reservation to students, firstly because the presentation is far from elementary (sometimes even superfluously complicated) and, secondly, because the organization of the book is not perfect, which adds to the difficulties of reading the text. On the other hand, the collection of examples and applications in the book seems to be unrivaled in the literature, as regards both the number and the thoroughness of the analyses.

The monograph is devoted to a systematic study of a class of problems that can be equally interpreted and treated as problems of the calculus of variations and/or optimal control. The class is fairly general although it does not embrace the entire complexity of either of these fields. Specifically, the general statement of the problem is the following: (P) Minimize $J_0(x(t_0),x(t_1),t_0,t_1)$ subject to $\dot{x}=f(t,x,u)$, $g(t,x,u)=0$, $(t,x,u)\in\Omega\subset\mathbf{R}^{1+m+n}$, $t_0\le t\le t_1$; $J_i(x(t_0),x(t_1),t_0,t_1)\le 0$, $i=1,\cdots,k$; $K_j(x(t_0),x(t_1),t_0,t_1)=0$, $j=1,\cdots,r$ (with Ω being an open set). Typical for the calculus of variations in this formulation are (a) the absence of unilateral and non-functional constraints on dynamic variables and (b) functional requirements on the data, which are usually assumed as smooth as necessary; typical for optimal control is the appearance of a "control system" with two types of variables, state variable x and control variable u, and functional requirements for the admissible controls which are allowed to be bounded measurable functions of time.

The circle of questions discussed in the monograph is rather traditional for books on the calculus of variations: extremals, Hamiltonian systems, Hamilton-Jacobi theory, field theory, invariance of extremals, quadratic problems and second-order conditions, conjugate points. Most of these questions have not been discussed earlier (even in the specialized literature) for variational problems of such generality. But the main novelty of the authors' approach is the extensive use of methodology and techniques of optimal control theory. The most visible manifestation of this approach is the systematic use of a concept of minimality which is intermediate between the natural extensions to (P) of the standard concepts of weak and strong minima. This concept, called the Pontryagin minimum by the authors, was introduced more than 30 years ago by Milyutin and Egorov but still is not sufficiently known and appreciated, even by specialists. Meanwhile, it is very natural for optimal control. Its main distinction (from the concepts of weak and strong minimum) is that it is not associated with any topology: e.g. for the problem with fixed end-points, $(\overline{x}(\cdot),\overline{u}(\cdot))$ is a Pontryagin minimum in (P) if for any $K>0$ it is a local minimum, with respect to the product of the uniform topology (for the x-component) and the L^1-topology (for the u-component) on the set of admissible $(x(\cdot), u(\cdot))$ satisfying the additional constraint $\|u(\cdot)\|_{L^\infty} \leq K$.

It is with respect to the Pontryagin minimum that the maximum principle is the main first order necessary condition. In the first part of the book the authors introduce and study a concept of an extremal associated with the maximum principle and develop for it the entire theory involving extensions of Hamiltonian equations, Weierstrass field theory, etc. They go even further and present in the course of the book a fairly complete set of first and second order necessary and sufficient conditions for the Pontryagin minimum. Among them are conditions for which no analogues have been known even for the simplest problem of the calculus of variations, e.g. the set of second order conditions for piecewise smooth extremals in which the connection between necessary and sufficient conditions is very similar to what we have in the classical case, when one type of inequalities is replaced by another (say $\geq$ by $>$ for quadratic functionals or $<$ by $\leq$ for the location of conjugate points). The proof of these conditions, by the way, occupies an entire chapter almost 100 pages long. The proof may be considered as evidence that the result is highly nontrivial, but can hardly be accepted as the "right" way to obtain the result. A search for an alternative proof is the most exposed but not the only open problem suggested by the book. Other types of minima (weak, strong and some variants of the latter) are also considered in detail, and differences between conditions of various types are carefully analyzed. Here too, an attentive reader will find many new observations and results practically in every chapter of the book.

As was mentioned at the beginning, the book is not easy to read, and not only because of the complexity of the problem and the sophistication of the techniques. Such things as the absence of an index, and not always the best choice of notation, which is sometimes unusual and awkward, should be mentioned among the contributing factors. The main shortcoming, however, is the references: one can find very few in the text, and the list of references is extremely poor: the latest references to anything not written by the authors are from 1980 as regards publications in Russian, and from 1975 otherwise. The absence of any attempt to make a comparison or even any mention of ongoing research on the maximum principle, Hamilton-Jacobi theory, and second order conditions is quite embarrassing. In spite of all that, the book is certainly at the very forefront of the developments and may, in many respects, determine the state-of-the-art and directions for future studies in the areas to which it relates. *Alexander D. Ioffe* (Haifa)

97k:49032 49K15 49J52 90C25

Rockafellar, R. Tyrrell (Seattle, WA)

Equivalent subgradient versions of Hamiltonian and Euler-Lagrange equations in variational analysis. (English summary)

SIAM J. Control Optim. **34** (1996), *no.* 4, 1300–1314.

This paper proves a formula in finite-dimensional nonsmooth calculus of great importance for the study of optimality conditions in nonsmooth problems of the calculus of variations and optimal control.

Given are an open set Ω in $\mathbf{R}^m$ and a lower semicontinuous function ("the Lagrangian") $L = L(x, v)$ defined on $\Omega \times \mathbf{R}^n$ with values in $\mathbf{R} \cup \{+\infty\}$. For each fixed x in Ω, the mapping $v \mapsto L(x,v)$ is convex. Associated with L is the function $H(x,p) = \sup\{\langle p,v\rangle - L(x,v)\colon\ v \in \mathbf{R}^n\}$ ("the Hamiltonian"): note that H is convex in p because L is convex in v. Under the hypotheses to be described below, both functions L and H turn out to be lower semicontinuous, and hence to lie in the domain of the operators ∂, which produces the set of "subgradients in the general sense" (also known as the set of limiting proximal subgradients), and ∂^∞, which produces the set of "subgradients in the horizon sense" (also known as the set of singular limiting proximal subgradients). (These sets will be familiar to everyone interested in nonsmooth optimization; neither must be convex in general.) The central issue is the interpretation and validity of the following identity involving fixed elements v and p of $\mathbf{R}^n$:

$$(*) \qquad \operatorname{co}\{w\colon\ (w,p) \in \partial L(x,v)\} = \operatorname{co}\{w\colon\ (-w,v) \in \partial H(x,p)\}.$$

Rockafellar's main result says that under mild hypotheses, one has that (i) for any point (x,v) in $\Omega \times \mathbf{R}^n$ where $L(x,v) < \infty$ and $\partial^\infty L(x,v) \cap [\mathbf{R}^n \times \{0\}] = \{(0,0)\}$, identity $(*)$ holds for all p in $\mathbf{R}^n$, and, similarly, (ii) for any point (x,p) in $\Omega \times \mathbf{R}^n$ where $H(x,p) < \infty$ and $\partial^\infty H(x,p) \cap [\mathbf{R}^n \times \{0\}] = \{(0,0)\}$, identity $(*)$ holds for all v in $\mathbf{R}^n$.

He shows that the left-hand side of $(*)$ can be nonempty only if $p \in \partial_v L(x,v)$, while the right-hand side can be nonempty only if $v \in \partial_p H(x,p)$: thus $(*)$ implies both of these (equivalent) partial subgradient conditions.

The hypotheses under which these results are obtained do not go far beyond the partial convexity of L and H mentioned above. Rockafellar requires only joint lower semicontinuity of the functions L and H on the set $\Omega \times \mathbf{R}^n$, and the continuity of one of the two set-valued mappings $x \mapsto \operatorname{epi} L(x,\cdot)$ or $x \mapsto \operatorname{epi} H(x,\cdot)$. (As the paper's Proposition 2.1 shows, the continuity of either implies that of the other.)

The proof is a tour de force of convex analysis, using Fenchel duality, Moreau regularization, Wijsman's epi-continuity thoerem, and Attouch's theorem on convergence of subgradients.

The importance of this result comes from its application to the theory of necessary conditions for optimality in nonsmooth problems of the calculus of variations, where it can be used to unify and extend the Euler-Lagrange and Hamiltonian approaches to writing the necessary conditions for optimality. In a problem of minimizing an extended-valued functional $\int_0^1 L(x(t),\dot{x}(t))\,dt$ over curves $x\colon [0,1] \to \mathbf{R}^n$ with $x(0) = 0 = x(1)$, the sharpened Euler-Lagrange inclusion involving a costate arc p, namely,

$$\dot{p}(t) \in \operatorname{co}\{w\colon\ (w,x(t)) \in \partial L(x(t),\dot{x}(t))\} \text{ a.e. } t \in [0,1],$$

turns out to be equivalent, under very modest qualification conditions, to a new form of the Hamiltonian inclusion involving the same p, namely,

$$\dot{p}(t) \in \operatorname{co}\{w\colon\ (-w,\dot{x}(t)) \in \partial H(x(t),p(t))\} \text{ a.e. } t \in [0,1].$$

The paper contains a thorough introduction to this motivating problem, but the full implications of the identity $(*)$ for the variational setting are elaborated elsewhere [see P. D. Loewen and R. T. Rockafellar, SIAM J. Control Optim. **34** (1996), no. 5, 1496–1511; MR 97d:49021]. *Philip D. Loewen* (Vancouver, BC)

99b:49038 49Q05 49Q20 53C42 58E20
White, Brian [White, Brian Cabell] (Stanford, CA)
Stratification of minimal surfaces, mean curvature flows, and harmonic maps.
J. Reine Angew. Math. **488** (1997), 1–35.

Much of the most interesting work in geometric analysis involves using local information to produce global conclusions. Often new results of this type require great technical effort and the resulting complication may well obscure the underlying insights. Apropos of this is F. J. Almgren, Jr.'s 1700-page preprint ["Q-valued functions minimizing Dirichlet's integral and the regularity of area-minimizing rectifiable currents up to codimension two", Preprint; per bibl.], in which the stratification theorem is proved for mass-minimizing integral currents. In the paper under review, the author unifies the treatment of Almgren's stratification theorem and a number of other stratification results in geometric analysis by proving some abstract theorems for closed sets and for upper semicontinuous functions. This is work of fundamental importance and should help in clarifying the direction of future efforts. Some special cases of the results of this paper are the following: (1) For a stationary or bounded mean curvature integral varifold in $\mathbf{R}^N$, the set of points at which no tangent cone can be rotated to have the form $\mathbf{R}^{k+1} \times C$ has Hausdorff dimension at most k. (2) Let U be a domain in $\mathbf{R}^N$ and let M be a Riemannian manifold. For a map $u\colon U \to M$ that minimizes $\int |Du|^p$ (for some $p \in (1,\infty)$), the set of points at which no tangent map has more than k independent directions of translational invariance has Hausdorff dimension at most k. (3) For a mean curvature flow in the sense of Brakke of an m-dimensional integral varifold in $\mathbf{R}^N$, the set of points in spacetime at which no tangent flow is a static plane (with constant multiplicity for all time, past and future) has parabolic Hausdorff dimension at most m.

The abstract theorems are set in a finite-dimensional real vector space with a translation invariant metric d that induces the usual topology. It is not required that the function $\|x\| := d(x,0)$ satisfy $\|rx\| = |r|\,\|x\|$, where r is a scalar and x is a vector. Similarly, the notion of dilation is allowed to be more general than the standard scalar multiplication, and a type of generalized cone is defined in terms of this generalized dilation. The author defines an important quantity associated with a generalized cone C called the "building dimension". The building dimension of C is the supremum of the Hausdorff dimension of subsets S of the vector space for which $\{x-y\colon\ x,y \in S\} \subset C$. One of the preliminary theorems, of interest in its own right, states that if $\mathscr{C}$ is a compact family of closed generalized cones, each of building dimension less than or equal to k, and if S is a subset of the vector space such that any limit of blow-ups of S at any $x \in S$ is a subset of some generalized cone in $\mathscr{C}$, then S has Hausdorff dimension less than or equal to k. An analogous theorem applies to upper semicontinuous functions.

The actual stratification theorem (Theorem 3.2) applies to an upper semicontinuous function and its tangent cone functions, and states that if Σ_k is the set of points at which no tangent cone function has more than k independent directions of translation invariance, then $\dim(\Sigma_k) \le k$ and Σ_0 is countable. The remainder of the paper is then devoted to showing how the stratification results are applied to various problems in geometric analysis. *Harold Parks* (Corvallis, OR)

98e:49102 49Q20 28A99 35Q99 60B99 65K10
Gangbo, Wilfrid (Atlanta, GA); **McCann, Robert J.** (Providence, RI)
The geometry of optimal transportation.
Acta Math. **177** (1996), *no.* 2, 113–161.

The Monge-Kantorovich mass transfer problem concerns the question of finding an optimal transportation between two mass distributions P_1 and P_2. In its original formulation due to Monge in 1782, the cost of transportation of unit mass between two domains in Euclidean space is given by the Euclidean distance. P_1, P_2 are the uniform masses on two subsets U_1, U_2 and admissible plans are those mappings between U_1 and U_2 which preserve uniform masses. It was conjectured by Monge that an optimal transport $S(x)$ satisfies the condition that $(S(x)-x)/|S(x)-x|$ is given by the gradient Du of a scalar potential. Appel (1928) gave a formal argument based on the corresponding Euler-Lagrange equation.

Real progress on the transportation problem is due to Kantorovich's 1940 relaxation of the transportation problem by allowing the class of all doubly stochastic measures as admissible transportation plans. This relaxation introduces convexity of the domain and linearity of the transportation functional. The basic result then is a duality theorem. A solution of the dual problem can be interpreted as Monge's transport potential and also as the Lagrange multiplier of the original mass transfer problem. Versions of the Kantorovich duality theorem for general cost functions $c(x,y)$ on abstract measure spaces and general probability distributions have been established by Kellerer (1984) and recently by Ramachandran and Rüschendorf (1994). As a consequence, characterizations of optimal transport measures in terms of subgradients of convex functions (equivalently, in terms of cyclically monotone support) in the case of Euclidean squared distance were given by Knott and Smith (1984), Rüschendorf and Rachev (1990) and Brenier (1991). This characterization of optimal solutions also implies existence of a Monge solution if one of the probability measures is Lebesgue-continuous. An extension of these results to general cost functions was given by Rüschendorf (1991, 1995). The characterization of optimal plans is obtained in terms of generalized c-subgradients of c-convex functions.

Gangbo and McCann work out a different approach to existence results as well as uniqueness results for optimal Monge functions. Their approach is no longer based on duality theory but uses directly the c-cyclical monotonicity condition of the support. Optimal maps are characterized by corresponding c-convex potentials. This leads to existence results without any moment conditions as in the duality approach. A related existence result was also obtained independently by Caffarelli (1996). To establish uniqueness, Aleksandrov's proof for surfaces of prescribed integral curvature is generalized to the transportation problem. Along the way the authors establish several important properties of c-concave functions, like the local Lipschitz property, and investigate the related Legendre transforms. The case of costs given by convex functions of the difference and by concave functions of the absolute difference is discussed in detail. In the strictly convex case the optimal transfer plan is uniquely determined by the potential; for radial costs of the form $h(|x-y|)$ the potential specifies the direction of transport but not the distance. Evans and Gangbo (1996) established in the case of the Euclidean cost $|x-y|$ an ordinary differential equation for the missing piece of information. The paper of Gangbo and McCann gives not only a new and sound basis for a general solution of the Monge-Kantorovich problem and finds a general solution in two important cases, but also contains some methodological insights and developments on the underlying class of c-concave functions and generalized Legendre transforms.

It should be mentioned that the problem treated in this paper is related to several important mathematical problems, and the methods developed in this paper are of interest for those problems, too. Brenier has discussed in several papers relations of the Monge-Kantorovich problem to rearrangement theorems for vector fields (polar decomposition) as well as to generalized solutions of a least action problem describing the motion of an ideal incompressible fluid. The Euler-Lagrange equation of this variational problem is identical to the Euler equations of incompressible flows in Lagrangian form. A discrete version of the least action principle leads to a Monge-Kantorovich-type problem with a fixed finite number of marginals.

It is also known that the solution of the transportation problem for the quadratic cost $c(x,y)=\|x-y\|^2$ is equivalent to generalized solutions of the Monge-Ampère PDE; for a general cost function c one obtains a correspondence with c-concave solutions of a generalized Monge-Ampère equation. Thus, numerical results and regularity theory can be transferred from one problem to the other. There are only partial results for these problems up to now. *Ludger Rüschendorf* (Freiburg)

98i:70004 70F07 57R99 58F05 70F15 70G35

McCord, Christopher K. (Cincinnati, OH); **Meyer, Kenneth R.** (Cincinnati, OH); **Wang, Quidong [Wang, Qiu Dong]** (Cincinnati, OH)

The integral manifolds of the three body problem. (English summary)

Mem. Amer. Math. Soc. **132** (1998), *no.* 628, viii+90 pp.

This is a paper of historical value and fundamental importance. It finally elucidates a question that has its roots in the qualitative methods promoted by Henri Poincaré toward the end of the 19th century. In his attempts to understand the three-body problem of celestial mechanics, Poincaré encountered insurmountable difficulties. The mathematical techniques he needed had not yet been developed, so the French mathematician came up with a different approach. The quantitative methods had proved limited and shortsighted. Progress in the three-body problem had been minimal. Poincaré's fundamental idea was to tackle problems qualitatively. His main goal was to understand the global geometry of the phase space for a given system of ordinary differential equations.

But before explaining the qualitative behavior of orbits in phase space, one has to understand its topology. Poincaré's prize paper of 1889, as well as his three-volume *Les méthodes nouvelles de la méchanique céleste* (1892–1899), emphasized the importance of this issue and opened new directions of mathematical thinking, among them being the well-established contemporary field of algebraic topology, which includes homotopy, homology, and cohomology theory. A full cohomological description of the phase space counts today as a pertinent understanding of the phase-space topology.

More than one century after Poincaré asked the question of the qualitative behavior of the three-body problem, the authors have finally understood the cohomology of the phase space associated to this problem. This is remarkable not only for the problem itself, but for the whole history of mathematics. When unable to solve a problem, mathematicians open new directions and create new fields, which they later develop, often forgetting the initial question. We now face an example in which the development of a new field has helped answer the initial question more than a century later. A rare event, indeed!

The paper under review describes the cohomology of the integral manifolds that foliate the phase space of the three-body problem—the integral manifold of the energy and angular momentum, as well as that of the Hill regions, in their full and reduced form (the reduction being done via rotations). The authors are the ones to reach the top of a pyramid that has been conquered through the work of several generations of mathematicians. Direct contributions in this sense were previously made by Albouy, Birkhoff, Cabral, Chen, Easton, Saari, Simó, Smale, and others. The techniques used in the present article are those of cohomology theory and the qualitative theory of dynamical systems. The paper itself is a mathematical tour de force that has been possible thanks to a close collaboration between three distinguished mathematicians: a topologist with a keen interest in celestial mechanics, McCord, and two top specialists in celestial mechanics and dynamical systems, Meyer and Wang.

Florin N. Diacu (Victoria, BC)

97h:70014 70F10 58F27 70F15 70H05
Moeckel, Richard (Minneapolis, MN)
Transition tori in the five-body problem. (English summary)
J. Differential Equations **129** (1996), *no.* 2, 290–314.

This is an exceptional paper in the direction of the theory of dynamical systems, in general, and of celestial mechanics, in particular, in which progress has been very slow and very difficult to make. As a response to a question posed by Kolmogorov in connection with what he called the theory of invariant tori (today—KAM theory, after Kolmogorov, Arnol′d, and Moser), V. I. Arnol′d showed in 1963 that there exist stable integrable Hamiltonian systems which, if perturbed properly, give rise to a special kind of chaos, a phenomenon that was subsequently called Arnol′d diffusion. This occurs when, under perturbation, resonant tori develop stable and unstable manifolds that intersect transversely connecting the tori in a chaotic chain that allows certain orbits to travel from torus to torus in an unpredictable manner. However, Arnol′d's example was somewhat artificial, in the sense that his system of differential equations has no (up to now) known physical meaning. In the same paper Arnol′d conjectured that this phenomenon takes place in a natural system: the n-body problem. A proof of this fact would have been of particular importance for astronomy, since the existence of the Kirkwood gaps in the belt of asteroids, which had been mathematically tackled as a resonance phenomenon, might have found a complete explanation.

The only important progress on the Arnol′d diffusion since Arnol′d's original work was made by Holmes and Marsden in 1982, but not in connection with the n-body problem. The first to prove the existence of Arnol′d's diffusion in the planar three-body problem was Xia, in 1994, who extended an earlier construction he had made in the restricted three-body problem. Moeckel's paper comes as the next step towards the understanding of the transition-tori phenomenon in the n-body problem.

The author considers a planar five-body problem with three equal masses, called primaries, and two bodies of small mass (which depends on the cube of a small parameter ε) that comprise the binary. The primaries move on nearly elliptic orbits in a nearly equilateral configuration (in the unperturbed case, i.e. for $\varepsilon = 0$, the primaries describe the well-known Lagrangian relative-equilibrium orbit), while in the center the two small masses move on nearly circular orbits (which are circular in the unperturbed case). Such a solution can be described by two angles (one for the primaries, by fixing the equilateral triangle, and the other for the binary), so it lives on a torus. By varying data, a family of tori is generated. Perturbing the system, a laborious analysis puts into evidence a chaotic Arnol′d-diffusion-type behavior. However, the transition-tori phenomenon obtained by Moeckel here is different from the one of Arnol′d. In Arnol′d's example the tori have one-half the dimension of the phase space (the maximum possible), whereas in this case the tori are low-dimensional and normally hyperbolic before perturbation.

To obtain his main result, the author uses modern tools in dynamical systems: KAM theory, the Poincaré-Mel′nikov method, the theory of normally hyperbolic invariant manifolds, etc. *Florin N. Diacu* (Victoria, BC)

98m:70017 70F10 70F15
Albouy, Alain (Paris); **Chenciner, Alain** (Paris)
Le problème des n corps et les distances mutuelles. [The n-body problem and mutual distances]
Invent. Math. **131** (1998), *no.* 1, 151–184.

A fundamental question in celestial mechanics is how best to take account of the symmetry. The Newtonian potential function of the n-body problem depends only on the mutual distances between the masses and is therefore invariant under translations and rotations in $\mathbf{R}^3$. One approach is to introduce the mutual distances themselves as coordinates. In the present paper, Albouy and Chenciner carry out this program in an elegant way for general potentials and for any space dimension.

The configuration of n bodies in $\mathbf{R}^3$ can be represented by a $3\times n$ matrix, X, whose jth column is the position of the jth mass. Forming the $n\times n$ Gram matrix $G=X^t\cdot X$ gives a rotation invariant representation. The restriction of the quadratic form determined by G to the plane $\xi_1+\cdots+\xi_n=0$ in $\mathbf{R}^n$ is also translation invariant. This restricted quadratic form, which can be expressed entirely in terms of the mutual distances, represents the relative configuration of the bodies. After introducing this idea (in coordinate-free style) the authors proceed to write Newton's laws of motion, obtaining equations generalizing those found by Lagrange for $n=3$.

Writing the equations in this form is particularly useful for studying homographic motions—solutions for which the configuration changes only by rotation and scaling. Special cases include pure scaling (leading to central configurations) and pure rotation. The paper contains a short proof of the fact that configurations admitting homographic motions which are not pure rotations are central configurations. For pure rotations in dimension at least 4 a wider class of configurations, called "configurations équilibrées", occurs. For example, the only noncollinear central configuration of $n=3$ masses is the equilateral triangle, but if the masses are equal, every isosceles triangle is a "configuration équilibrée" (they admit rotational solutions in $\mathbf{R}^4$).

The concept of "configuration équilibrée" has turned out to be useful even for studying the classical problem of central configurations in $\mathbf{R}^3$. Every central configuration is also a "configuration équilibrée", and the equations characterizing such configurations provide relations which must be satisfied by the mutual distances of the central configuration. These relations are the key to Albouy's proof that every central configuration of four equal masses has a symmetry [C. R. Acad. Sci. Paris Sér. I Math. **320** (1995), no. 2, 217–220; MR 95k:70023]. It is likely that other applications of the ideas and equations in this paper can be found, particularly in problems with few bodies. *Richard Moeckel* (Minneapolis, MN)

97e:70025 70H35 34C23 34D05 58F05 70K20

Bolotin, Sergey [Bolotin, Sergei V.] (Moscow); **Negrini, Piero** (Rome)

Asymptotic solutions of Lagrangian systems with gyroscopic forces. (English summary)

NoDEA Nonlinear Differential Equations Appl. **2** (1995), *no.* 4, 417–444.

The subject of this paper is the exploration of the stability problem of equilibrium solutions of Lagrangian systems in the presence of gyroscopic forces.

Let M be the n-dimensional configuration manifold of a system described in local coordinates $q \in \mathbf{R}^n$ by the Lagrange equations

$$\frac{d}{dt}\frac{\partial T}{\partial \dot{q}} - \frac{\partial T}{\partial q} = \frac{\partial U}{\partial q} + F. \tag{1}$$

Here, T is the kinetic energy, which defines a Riemannian metric on M and is given in local coordinates by $T = \frac{1}{2}\langle B(q)\dot{q}, \dot{q}\rangle$, where $B(q)$ is a symmetric positive definite matrix, U is the potential field and F is a gyroscopic force field. This means that $F = A(q)\dot{q}$, where the matrix $A(q)$ is antisymmetric and the differential 2-form $\Gamma = \frac{1}{2}A(q)dq \wedge dq$ is closed: $d\Gamma = 0$.

The system with gyroscopic forces (M, T, U, Γ) is invariantly defined by T, U, Γ as a second-order differential equation on the tangent bundle TM, independently of the local coordinates.

Let $O = \{q = 0\} \in M$ be an equilibrium position of (1): $\nabla U(0) = 0$. If O is a point of strict relative maximum for U, the Lagrange-Dirichlet theorem guarantees that the corresponding equilibrium solution is stable (according to Lyapunov) independently of the presence of the gyroscopic forces. As is well known, the Lagrange-Dirichlet theorem is not invertible in general; i.e., the strict maximum property is a sufficient but not necessary condition for the stability of O. For example, this is shown by the simple example with $n = 1$, $T(q, \dot{q}) = |\dot{q}|^2/2$ and $U(q) = \cos(q^{-1})\exp(-q^{-2})$ if $q \neq 0$, $U(0) = 0$ [P. Painlevé, C. R. Acad. Sci. Paris **138** (1904), 1555–1557; JFM 35.0704.02].

The search for additional conditions which ensure the instability of O when U does not have a strict relative maximum at O is a challenging problem, whose formulation goes back to Lyapunov (1892). Known in the literature as the inversion problem of the Lagrange-Dirichlet theorem, it has been widely investigated, particularly in the case of reversible systems (namely, when $A(q) = 0$). For such a case, the instability has been established in several works, for several classes of potentials. For a survey of results in this direction we refer to a paper by L. Salvadori [Atti Accad. Sci. Torino Cl. Sci. Fis. Mat. Natur. **126** (1992), suppl. 2, 151–168; MR 94m:70047; see also N. Rouche, P. Habets and M. Laloy, *Stability theory by Liapunov's direct method*, Springer, New York, 1977; MR **56** #9008]. Here, we think it worthwhile mentioning that, in spite of the efforts of several mathematicians throughout the century, only quite recently was a complete answer given to the problem in the reversible case. Indeed, V. P. Palamodov [in *Dynamical systems in classical mechanics,*

5–20, Amer. Math. Soc. Transl. Ser. 2, 168, Amer. Math. Soc., Providence, RI, 1995; MR 97b:34055] proved, for the case of natural mechanical systems, the instability for analytic U such that O is not a maximum point.

In the paper under review, the inversion problem is studied for nonreversible systems, describing mechanical systems on which gyroscopic forces are also acting. The presence of the gyroscopic forces further contributes to the difficulty and subtleness of the question at hand. Recall in this connection Kelvin's theorem, according to which for nondegenerate critical points of the potential U, nondegenerate gyroscopic forces may have a stabilizing effect.

The authors consider systems for which the functions A, B, U are of class $C^N(M)$ with N suitably large. Specifically, they study equilibria which are degenerate critical points of U, i.e. such that the Taylor expansion of the function U near $q = 0$ takes the form $U(q) = U_k(q) + o(|q|^k)$ with $k \geq 3$, where $U_k(q)$ is the first nonzero homogeneous form. And they require the gyroscopic forces to be nondegenerate, namely such that $\det A(0) \neq 0$. So, the dimension n must be even: $n = 2m$.

For such systems, the authors establish an algebraic criterion for the instability. More precisely, they prove the existence of a motion which asymptotically tends in the past [respectively, in the future] to the equilibrium whenever the "eigenvalue problem"

$$\nabla U_k(a) = \lambda A(0)a, \quad a \in \mathbf{R}^n \setminus \{0\}, \tag{2}$$

has a solution with $\lambda < 0$ [respectively, $\lambda > 0$]. Together with the conservation of the Liouville measure in the phase space this obviously implies the instability.

Several corollaries follow, which provide concrete conditions for the existence of asymptotic solutions. In particular, it is shown that, if $n = 2$ and O is an isolated critical point of U_k, the stability of the equilibrium holds if and only if O is a strict maximum or minimum for U_k, while asymptotic orbits exist if and only if the critical point is of saddle type. Related results for systems with two degrees of freedom are contained in a paper by S. D. Furta [Vestnik Moskov. Univ. Ser. I Mat. Mekh. **1987**, no. 5, 100–101; MR 89f:70031].

While carrying out the proof, the authors also establish the following result: In the presence of nondegenerate gyroscopic forces, the formal stability of the degenerate equilibrium O holds whenever O is a point of strict local minimum for U which is $(2k-3)$-decidable. (We recall that the formal stability of the equilibrium O amounts to the existence of a formal integral $F = F_2 + \cdots$ such that for N large enough the polynomial $F = F_2 + \cdots + F_N$ is positive definite.)

In contrast with this situation, when O is a point of strict local minimum for U and the gyroscopic forces are absent, for any point q_0 in a neighborhood of $O \in M$ an asymptotic solution $q(t)$ is known to exist such that $q(0) = q_0$ and $q(t) \to O$ as $t \to \infty$. So, one could wonder what happens to the asymptotic solutions existing in the reversible case, when a small nondegenerate gyroscopic force $F = \varepsilon A(q)\dot{q}$ is switched on. The authors investigate this question and show that a bifurcation phenomenon takes place: for small positive ε there exists a compact invariant set Σ (subset of the tangent bundle), to which the instability properties of O are transferred, which collapses at O (the equilibrium solution $q \equiv 0$, $\dot{q} \equiv 0$) when the gyroscopic forces are switched off.

The proof of this bifurcation result relies on the use of global variational methods. In fact, the set Σ is defined as the union of supports of action-minimizing invariant measures in the sense of J. N. Mather. Then, the problem is tackled by means of variational techniques developed in papers by Mather [Math. Z. **207** (1991), no. 2, 169–207; MR 92m:58048] and Bolotin [Vestnik Moskov. Univ. Ser. I Mat. Mekh. **1992**, no. 6, 34–41, 96; MR 94c:58064; in *Dynamical systems in classical mechanics*, 21–90, Amer. Math. Soc. Transl. Ser. 2, 168, Amer. Math. Soc., Providence, RI, 1995; MR 97b:58124]. In contrast, the proof of the existence of asymptotic solutions mainly develops according to the pattern of V. V. Kozlov and Palamodov [Dokl. Akad. Nauk SSSR **263** (1982), no. 2, 285–289; MR 83h:58056] and Kozlov [J. Appl. Math. Mech. **50** (1986), no. 6, 719–725 (1987); MR 89c:70046] and exploits some modifications suggested by Negrini [J. Differential Equations **115** (1995), no. 2, 350–367; MR 95m:34085] for the case of nondegenerate gyroscopic forces. Specifically, this proof involves the construction of a suitable normal form and fixed point arguments: the equations (1) are written in Hamiltonian form, by using a twisted symplectic structure in the phase space; then, a formal symplectic transformation is shown to exist, which allows one to construct a formal invariant manifold. As for the asymptotic solutions, first, the proof is carried out for the case in which $A, B, U \in C^\infty(M)$. For the restriction of the equations of motion to the formal invariant manifold, a formal asymptotic solution is found by induction. At this step, the application of a theorem by A. N. Kuznetsov [Funkcional. Anal. i Priložen. **6** (1972), no. 2, 41–51; MR **47** #3738] provides the existence of an asymptotic solution of the system (1). Second, the case in which $A, B, U \in C^N(M)$ with N suitably large is also considered. Specifically, N is required to satisfy the inequality $N > (k-2)\Lambda + k - 1$, where $\Lambda = \max \operatorname{Re} \operatorname{Spec}(S)$, $S = -A^{-1}(0)\nabla^2 U_k(a)$ and (a, λ) is a solution of (2). In this case, some suitable coordinate transformations are performed and the original problem is reformulated as a fixed point problem for a map on a Banach space Z. Under the postulated assumptions, this map is seen to be a contraction of the unit ball in Z. This gives the solution. For the applications concerning the case $n = 2$, the KAM theory also comes into play.

Very briefly, the article is an excellent contribution in the field of qualitative theory of the equations of classical mechanics. It addresses long-standing and hard questions, employing a variety of techniques from advanced variational theories, classical perturbation theory, symplectic geometry and local analysis.

Maria Letizia Bertotti (Trento)

97m:73001 73-02 49Q20 58E50 73B99 73C50 73V25 76A15 82D30

Virga, Epifanio G. (Pisa)

★**Variational theories for liquid crystals. (English summary)**

Applied Mathematics and Mathematical Computation, 8.

Chapman & Hall, London, 1994. xiv+375 pp. ISBN 0-412-39880-X

The discovery of liquid crystals at the end of the nineteenth century initiated an era of scientific research devoted to understanding the fascinating properties of the new phases of matter intermediate between liquid and solid. The discovery is often associated with a letter that, in March of 1888, the botanist Friedrich Reinitzer wrote to Otto Lehmann, Professor of Physics at Aachen. He described observations that esters of cholesterol appeared to have two melting points between which the liquid showed striking colors. The collaboration that followed between the two scientists set important foundations for the science of liquid crystals.

After 1888, the research efforts very quickly extended beyond Germany, especially to France and Holland. It was towards the end of the century that chemists such as Vorlander in Halle uncovered the basic molecular structure responsible for liquid crystalline behavior. At the same time, Georges Friedel in Strasbourg identified new types of liquid crystal phases, and cleared up the terminology of the subject.

From a different point of view, it was after Lehmann's death in 1922 that Staudinger established the notion of long-chain molecules. Vorlander conceived the notion of liquid crystal polymer through the systematic synthesis and study of increasingly long molecules.

Important theoretical advances took place in the 1920's with the formulation of the continuum model of the static theory by Oseen, Zocher, and later by Frank. The first molecular field theory for nematics was proposed by Born (1916) with a permanent dipole model. Statistical theories based on rigid-rod macromolecular models began with the work by Onsager in 1949. Such theories contributed to the understanding of how liquid crystals interact with electric fields. The electro-optical properties of liquid crystals have been studied very actively since the beginning of the century, and many important phenomena such as the Freedericksz transition, field-induced domain patterns and spatially periodic structures have been discovered. The wealth of experimental and theoretical knowledge resulted in important technological advances such as the manufacturing in 1968 of the first liquid crystal device capable of switching from cloudy to clear by application of an electric field.

Since the manufacture of the first display device, numerous researchers emerging from diverse scientific disciplines have collaborated in experimental and theoretical endeavors to elucidate different aspects of the liquid crystal phenomenology. A result of such work is manifested in the important technological discoveries that have already taken deep roots in everyday life.

The formulation of general conservation laws and constitutive equations for nematic flow by Ericksen and Leslie marked the entrance of the mathematical community into the field of liquid crystal research. Indeed, their pioneering work has made deep inroads into the fields of applied mathematics and analysis of the last two decades. However, outstanding textbooks on liquid crystal science such as those by P.-G. de Gennes and J. Prost [*The physics of liquid crystals*, Second edition, Oxford Univ. Press, Oxford, 1993] and by S. Chandrasekhar [*Liquid crystals*, Second edition, Cambridge Univ. Press, Cambridge, 1992] have remained largely inaccessible to mathematicians.

With *Variational theories for liquid crystals*, Epifanio Virga reaches out to the mathematical community, from the point of view of liquid crystal research as well as education. It is perhaps the first book capable of making inroads into senior undergraduate and graduate education in applied mathematics. Researchers coming into the field from different disciplinary directions will also draw substantial support from this book.

The first chapter of the book presents the physical background of liquid crystal phases, and defines mathematical tools to measure properties related to molecular orientation. It also presents a survey of fundamental concepts in optics necessary to interpret and understand the liquid crystal phenomenology. This is one of the strongest points of the book, since, very often, applied mathematicians encounter a conceptual gap between the analysis of the equations and the ensuing physical properties. The first chapter of the book does an outstanding job of breaking the cross-disciplinary ice, while paving the path of the mathematician who wishes to analyze models of liquid crystals.

The second chapter offers very clearly formulated and rigorous background material on continuum mechanics. It casts the liquid crystal theory within the framework of rational mechanics. That chapter itself contains basic and independent materials for a course in continuum mechanics.

Chapter 3 presents a survey of the static theory by Frank, Oseen and Zocher, casting it within the framework of the calculus of variations. It is well known that one of the main difficulties in the modeling of complex physical systems is the problem of finding a suitable mathematical interpretation of the boundary effects, namely, the formulation of boundary conditions. The author has skillfully dealt with such issues by giving an excellent mathematical interpretation of strong anchoring.

The description of elastic deformations of liquid crystals together with the careful presentation of tensor analysis of Chapter 2 suggests that the book could be used in a vector calculus classroom for the third-semester sequence. In fact, students could be introduced to the geometric aspects of differential operators such as divergence and curl through the liquid crystal plates of Lavrentovich and Luskin as presented in the book.

The study of the Freedericksz transition in terms of the nonlinear stability analysis of solutions of ordinary differential equations is elegantly carried out in Chapter 4. The author succeeds in developing a valuable theme for an advanced course in ordinary differential equations and bifurcation theory. Again, we are presented with a topic where students can visualize solutions of problems through experimental pictures, and also generate them while playing with their own watch or calculator.

Liquid crystals are very sensitive to their boundary conditions due to the flexible orientational ordering of their constituent molecules. The efficient manufacturing of high-quality liquid crystal displays requires an elaborate control of boundary alignment and the development of appropriate alignment techniques. These have prompted a lot of research activity on liquid crystal boundary effects such as epitaxy on crystal surfaces, prevalence of perpendicular alignment to a treated glass plate, and influence of unidirectional rubbing.

The book presents an outstanding mathematical treatment of boundary effects. In fact, Chapter 5 is entirely devoted to the study of free boundary problems such as droplet configurations. The author now discusses weak anchoring boundary conditions, and their interpretation in terms of interfacial energy. At this point, the book seems to make a departure from the general background approach of the first four chapters, and deals with more specialized research topics. It offers an outstanding presentation of the physical challenges in the study of free surfaces separating material phases with different degrees of order. At the same time, the author proceeds with the mathematical description of surface energy functionals appropriate for modeling the nematic-isotropic phase contact. The material presented in Chapter 5 may also prove to be very relevant in mathematical studies of liquid crystal encapsulated systems. The study of such systems belongs to the mainstream of research into the future generation of display devices.

The last chapter of the book is devoted to the study of defects that may arise in minimizer configurations of the free energy. First, the author addresses the classical theory of Oseen, Frank and Zocher, presenting a thorough survey of the theory of regularity of the minimizers and the dimension of the singular sets. The study of the harmonic mappings that result as a special case of the former model occupies a distinguished place in the survey.

The goal of modeling line and plane defects motivated Ericksen to propose a theory for uniaxially nematic liquid crystals that takes into account an order parameter in the model. The author devotes the second part of Chapter 5 to a presentation of Ericksen's new theory for liquid crystals with variable degree of orientation. The book concludes with an expository account of the regularity and the properties of the singular sets within the framework of the new theory. The author has mastered the unusual skill of focusing on the fundamental ideas and theorems as well as their physical interpretation. Although the presentation does not include proofs of theorems, he successfully brings out the main ingredients and underlines the mathematical difficulties.

Although the book does not deal with flow phenomena, it nevertheless puts forward many of the elements that a researcher may need for setting up and analyzing dynamical and flow problems of nematic liquid crystals. The following historical observation may support such a point of view. The first successful commercialization of liquid crystal polymers was achieved by DuPont in the late 1960's with the fiber marketed under the name of Kevlar. It has strength comparable to steel but much lower density. Since then many different polymeric liquid crystal compounds have been synthesized and studied. One of the common features of such polymeric materials is that their processing flows present many more, and more persistent, defects than the small-molecule liquid crystal counterparts. The new theory of Ericksen that Epifanio Virga describes may indeed prove valuable in modeling such complex flow behavior.

Variational theories for liquid crystals is a very clear and well-written book, with a very rich bibliography. It deals with historical issues with extreme care. In addition to being a very valuable resource for applied mathematicians and researchers in the field, it is a lot of fun to read. *Maria Carme Calderer* (University Park, PA)

98j:73001 73-02 73A05 73B30 73Bxx 76A02 80-02

Šilhavý, Miroslav (Prague)

★**The mechanics and thermodynamics of continuous media. (English summary)**

Texts and Monographs in Physics.

Springer-Verlag, Berlin, 1997. pp. xiv+504 pp. $89.50. ISBN 3-540-58378-5

Both in depth and in its encyclopedic coverage, this treatise by Šilhavý evokes memory of the justly famous *Handbuch* articles of C. A. Truesdell, III and R. Toupin [in *Handbuch der Physik, Bd. III/1,* 226–793; appendix, pp. 794–858, Springer, Berlin, 1960; MR **22** #8778] and Truesdell and W. Noll [*The non-linear field theories of mechanics,* Springer, Berlin, 1965; MR **33** #2030]. Indeed it may be regarded as an ambitious attempt on the part of the author to write a sequel to the *Handbuch* articles, covering a large portion of the major developments in continuum thermostatics and thermomechanics in the three decades since Truesdell and Noll's account of their revival in the 1959–65 papers of Bernard D. Coleman and his coworkers.

Commenting on the axioms of continuum physics, Truesdell and Noll, in their preface to *The non-linear field theories* [op. cit.], spoke about "a principle of irreversibility", the "true form" of which "is not yet known". Search for the true form of this principle occupied the center stage in thermomechanics for the first half of the three decades since then. Miroslav Šilhavý made his first impact as a researcher in thermomechanics by creating, concurrently with and independently of James B. Serrin, what is now sometimes called the Serrin-Šilhavý approach to thermodynamics.

In the work of Coleman and his coworkers in the sixties, the entropy and the absolute temperature were taken as given a priori, and the Clausius-Duhem inequality (revived by Truesdell and Toupin [op. cit.]) was postulated as the principle of irreversibility or the Second Law in continuum thermomechanics. Perhaps arising partly as attempts to allay criticisms from proponents of other approaches, there soon followed a stream of foundational studies (the efforts of W. A. Day and of Coleman and Owen being good representatives) examining the general validity of the Clausius-Duhem inequality and investigating the existence and uniqueness of entropy, or the lack thereof, for various material bodies.

These efforts set the stage for the work of Serrin and of Šilhavý in the late seventies and early eighties.

Following the pioneers of classical thermodynamics, Serrin and Šilhavý take the concepts of hotness, heat and work as primitives. As was intuitively clear to Carnot and was explicitly pointed out by Gibbs, "in thermodynamic problems, heat received at one temperature is by no means the equivalent of the same amount of heat received at another temperature.... This is a result of the general law, that heat can only pass from a hotter to a colder body...." In other words, we should consider in thermodynamics not only the quantity but also the quality of heat as characterized by the hotness at which the heat is received or given out. If a body receives a quantity of heat over a range of temperatures, just knowing the total amount received will not suffice for us to carry out a full thermodynamic analysis; in addition, the quality of the heat received must be specified by a detailed breakdown of the total according to the temperatures of receipt.

It will be a problem for future historians of thermodynamics to explain why a suitable mathematical expression to capture both the quantitative and qualitative aspects of heat was not created until Serrin and Šilhavý, around 1978, independently proposed their "accumulation function" and "heat distribution measure", respectively, so that once and for all classical thermodynamics could be set on a firm mathematical foundation. A general theory of thermodynamics was subsequently developed. I would call this theory neoclassical thermodynamics, for it is best described as the classical thermodynamics of Clausius and Kelvin put in the most general setting (plus a careful delineation of the required axioms and logical arguments). In neoclassical thermodynamics, the existence of the mechanical equivalent of heat and that of an absolute temperature scale follow as general theorems. Furthermore, whenever the existence of a local entropy function can be demonstrated for a classical continuous body, the Clausius-Duhem inequality will indeed be the "true form" of the principle of irreversibility. For simple cases such as the thermoelastic solid and the heat-conducting Navier-Stokes fluid, it has been explicitly shown that the entropy does exist as a local state function.

The theory of neoclassical thermodynamics is presented in Part II of the present treatise. It is a masterful account and is more comprehensive than either the outline by Serrin [in *New perspectives in thermodynamics*, 3–32, Springer, Berlin, 1986; see MR 87h:80002] or the elementary exposition by D. R. Owen [*A first course in the mathematical foundations of thermodynamics*, Springer, New York, 1984; MR

85m:80001], although these three presentations each carry distinguishing personal touches of their authors and none of them can be regarded as complete. In particular, discussion of applications in Šilhavý's book is mainly restricted to systems with perfect accessibility.

Occupying less than one-tenth of the page space notwithstanding, this part puts an unmistakable imprint on the rest of the book. It affects the overall organization and the selection of topics. First and foremost, the Clausius-Duhem inequality is taken throughout as the principle of irreversibility. Secondly, the constitutive equations treated in this book are restricted mainly to elastic materials with heat conduction and viscosity. Restricting attention to this special class of materials will soften any objection to taking the Clausius-Duhem inequality as the Second Law, for it is only the particular form of this law for the specific bodies in question that is at issue. For elastic materials with heat conduction and viscosity, most thermomechanical theories will have their principle of irreversibility equivalent to the Clausius-Duhem inequality. On the other hand, even within the Serrin-Šilhavý approach itself, I am not aware of a published proof that such bodies in the most general case do obey the accessibility axiom, although hardly anyone would doubt the validity of this assertion for them.

The present treatise is certainly not meant to be a survey of thermomechanical theories. As for other approaches in thermomechanics, apart from a brief mention of Müller's entropy inequality (p. 165), three short sections (§§12.2–12.4) in small print are devoted to linear irreversible thermodynamics, the dissipation potential, and relaxation models (extended linear irreversible thermodynamics), respectively, which are all "completely compatible with the Clausius-Duhem inequality".

Šilhavý's treatise is divided into five parts. Like the article of Truesdell and Noll, it begins with a chapter on tensor algebra and analysis. This chapter together with Chapter 8 on isotropic functions in Part III gives a valuable compilation of formulae on tensor functions, many of which were discovered after the publication of Truesdell and Noll's *Handbuch* article. In Part I (entitled "Balance equations"), besides the standard fare, there are well-written sections on Bravais lattices (§1.5), compatibility of deformations at the interface (§2.3), rank-1 connections (§2.4) and twins (§2.5), which will help prepare the reader for later studies of phase transitions in crystals. There are also appendices on piecewise smooth objects (§2.6) and on the Gauss-Green theorem (§3.9); a brief review of sets of finite perimeter and functions of bounded variation is given in the latter.

In Part II, "Foundations", there is also a chapter in which Šilhavý gives an exposition of his own work (1989) to derive Cauchy's equations of motion from the balance of energy and the principle of material frame-indifference. In his approach mass is a derived concept, and the classical splitting of the total energy into the kinetic energy plus the objective internal energy is a consequence, not a presumption. Interestingly enough, it was also Serrin who, basing his ideas partly on those of Šilhavý, came up with a related approach to these concepts at about the same time.

Part III is devoted to the constitutive equations of elastic materials with heat conduction and viscosity. Special cases include Navier-Stokes-Fourier fluids, Kelvin-Voigt solids, thermoelastic materials, and ideal dissipationless materials. The restrictions placed on the response functions by frame-indifference, symmetry, and the Clausius-Duhem inequality are derived.

Part IV, which runs to about two hundred pages long, treats the theory of thermodynamic equilibrium. It is a celebration of the point of view championed by J. L. Ericksen since the late sixties, namely that thermodynamics as taught by Gibbs is not only the theory of heat, but also a theory of equilibrium and stability, with the main tool the extremum principles and the calculus of variations. This part begins with chapters describing different types of environments, the equilibrium states of a body in a given environment, and the extremum principles. Then various notions of convexity (including quasiconvexity, rank-1 convexity, and polyconvexity) are presented in two chapters entitled "Convexity" and "Constitutive inequalities", respectively. There is a section (§17.4) on Maxwell's relation, in which the continuity of the normal component of the Eshelby energy-momentum tensor across the static phase interface is established as the generalization of the equality of chemical potentials, which has been known since the days of Gibbs as a condition of equilibrium for the special case of fluids. After "Constitutive inequalities" follow a chapter on the thermostatics of fluids, and one on the linearized approach to the equilibrium of solids, which includes classical linear elasticity and the linearized elasticity of stressed bodies. The final chapter of this part is devoted to direct methods in equilibrium theory. It begins with a section which describes the main mathematical ingredients, namely weak convergence, Young measures, and the lower semicontinuity of integral functionals. The analysis of Ball (1977) and some further developments on solutions to extremum problems for rubber-like bodies, which have polyconvex stored energy functions, are presented. The final section gives an exposition of the work of Chipot and Kinderlehrer (1988) on Young measure minimizers and the equilibrium configurations of crystals.

Part V, entitled "Dynamics", has its emphasis placed on moving singular surfaces, which include propagating phase boundaries and shock waves. Here the Clausius-Duhem inequality gives us the entropy admissibility criterion for jumps. This criterion, however, is not strong enough to secure uniqueness of solutions for the initial-value problem. In this part, extra conditions for evolving phase boundaries and various admissibility criteria for shocks are reviewed. Besides shock waves, various types of elastic waves (surface waves, acceleration waves, etc.) are also discussed, and there is a chapter devoted to adiabatic fluid dynamics (shock waves, shock layers). The book ends with a chapter in which the properties of a linearized system of equations for a viscous solid with heat conduction are examined.

The treatment given to the topics covered is encyclopedic. It includes accounts of much that is of secondary or even tertiary importance to the main theme of the book. Those accounts are often given in small print but are supported by a full bibliography.

While this treatise is impressive in its scope of coverage, it does have major omissions. Particularly notable among these are the gradient theory of phase transformations and M. E. Gurtin's theory of configurational forces and phase interfaces with structure (i.e., phase boundaries that are stressed and carry energy and entropy). Both of these topics are compatible with the general theme of this book; in fact, they are mentioned briefly in small print, and the main references are listed in the bibliography. On the other hand, exclusion of these topics from detailed discussion is also understandable. All the topics treated in the present book fit together nicely within a relatively simple framework: the balance equations and the form of the Clausius-Duhem inequality are classical; the existence of internal energy and entropy as local state functions is not in doubt for the special classes of materials in question. In this sense the aforementioned omitted topics are outcasts; besides foundational issues on the existence and uniqueness of internal energy and entropy as local state functions, which remain to be clarified as the author should reexamine these issues from the standpoint of the Serrin-Šilhavý approach, these theories will also require significant modification of the structure of the balance equations.

Most chapters of this treatise carry a bibliographical note, in which the author comments on the literature and provides information on the history of and other approaches to the subject. I find these bibliographical notes informative and helpful.

This book has an excellent bibliography, comprising twenty-two pages. The subject index, however, can be improved substantially. Since the whole treatise is roughly the same size as Truesdell and Noll's book [op. cit.], which has a subject index about three times as long, this shortcoming sometimes renders it difficult to locate quickly items of secondary or tertiary importance.

This treatise is clearly not directed to beginners but to scholars, specialists, and researchers who are already active in continuum thermomechanics, and to advanced students who desire to begin research on the subject. Its encyclopedic coverage, exhaustive bibliography, and helpful bibliographical notes make it a valuable book of reference. Supplemented by the original papers, this book can also serve as the basic reference or roadmap for several topic courses or seminars for advanced graduate students. I used it profitably in a seminar in which students learnt about various notions of convexity, Young measures, and the equilibrium theory of crystals and rubber-like bodies.

I strongly recommend this book to research libraries and to all practitioners of continuum thermomechanics. *Chi-Sing Man* (Lexington, KY)

98c:73007 73B27 73K20 73V25 78A99

Lipton, Robert (Worcester, MA)

Variational methods, bounds, and size effects for composites with highly conducting interface. (English summary)

J. Mech. Phys. Solids **45** (1997), *no.* 3, 361–384.

Bounds on effective properties of composites are important in many fields of applied and industrial mathematics. One of the first major results is that of Z. Hashin and S. Shtrikman [J. Appl. Phys. **33** (1962), 3125–3131; Zbl 111.41401; J. Mech. Phys. Solids **11** (1963), 127–140; MR **28** #2676]. They introduced new variational principles and found optimal bounds on the effective conductivity and elastic moduli of two-phase isotropic composites. Since that time, several powerful methods (such as the Hashin-Shtrikman method, the translation method, the Beran method, and others) were developed to attack the problem, and a number of questions were answered. Most of the works were devoted to mixtures with perfect bonding between the phases.

However, in many cases of practical interest, the interface conditions are far from being perfect: interfacial conductivity or resistance, cracks, and other interfacial defects may play a crucial role in defining effective properties of composites. Hashin [J. Mech. Phys. Solids **40** (1992), no. 4, 767–781; MR 93b:73006], Lipton [SIAM J. Appl. Math. **57** (1997), no. 2, 347–363; MR 98c:73008], Lipton and B. Vernescu [Math. Models Methods Appl. Sci. **5** (1995), no. 8, 1139–1173; MR 96m:73005; Proc. Roy. Soc. London Ser. A **452** (1996), no. 1945, 329–358; MR 97a:73008], and S.

Torquato and M. D. Rintoul [Phys. Rev. Lett. **75** (1995), no. 22, 4067–4070] used variational definitions of effective conductivity and effective elastic moduli to study composites with imperfect interface. The paper under review develops a new, rigorous variational method to study such composites.

New variational principles are developed for the effective conductivity tensor of anisotropic conductors. These variational principles are in fact a nontrivial generalization of the Hashin-Shtrikman variational principles to describe composites with highly conducting interface. Extra geometric interface information is encoded into principles through the solution operator of a simpler transport problem. These operators can be expressed as gradients of simple layer potentials with densities supported on the phase interface or in terms of simple Dirichlet problems inside each phase region.

By using these new variational principles, new bounds on the effective conductivity are found. The upper bounds depend upon phase volume fractions, a surface energy tensor and a scale-free matrix of parameters equal to the effective conductivity of the composite with the same microgeometry but with superconducting inclusions. New lower bounds are given in terms of phase volume fractions, interfacial geometric parameters, and two-point correlation functions. The bounds are found to be optimal for certain choices of interfacial parameters. They are equivalent to the Hashin-Shtrikman bounds in case of perfect interface.

The new bounds can be used to estimate the composite properties in terms of the size distribution of the particles. As an example, for polydisperse suspensions of spheres it is shown that the effective conductivity is greater than that of the matrix when the harmonic mean of the sphere radii is greater than the ratio of interfacial conductance divided by the contrast between matrix and particle conductivities.

The paper is an important contribution to the theory of composites. The approach developed in the paper will have a wide range of applications, in particular, to elastic composites with imperfect interface. The paper is self-contained, well written, and easy to follow. *Leonid V. Gibiansky* (Kensington, MD)

97c:73018 73C50 35Q72 73S10 73V25
Dolzmann, Georg (Freiburg); **Müller, Stefan** (Freiburg)
Microstructures with finite surface energy: the two-well problem. (English summary)
Arch. Rational Mech. Anal. **132** (1995), *no.* 2, 101–141.

What is microstructure and why should we be concerned with it? Microstructure or mesoscale structure typically occurs over micron or submicron scales in solids, but can concern us at finer or coarser scales as well. Microstructural characteristics influence the function of the material as a whole and, correspondingly, understanding restrictions on microstructure imposed by the macroscopic environment offers the possibility of improving and optimizing behavior. "Microstructure is at the heart of materials science" [W. W. Mullins, "Remarks on the evolution of materials science", Materials Res. Soc. Bull. **21** (1996), no. 7, 20]. Materials with mobile defect structures and whose function is dictated by diffusionless phase transformations are of special interest. Such "active materials" include shape memory and magnetostrictive materials and are an area of intense research and application [e.g., *Materials for smart systems (Boston, MA, 1994)*, Materials Res. Soc., Pittsburgh, PA, 1995]. Many of the mathematical questions and some of the interesting issues arising from the analysis of microstructure are embodied in this paper. As part of a broader perspective, methods of nonlinear analysis have been developed in the past ten years to describe the interplay of crystallography and energy and their effects across disparate length scales. This has come about of necessity. Crystallography implies that energy densities of ordered solids are invariant under discrete groups, which, in turn, means that they define functionals in the calculus of variations which lack lower semicontinuity. These densities attain their minima on a discrete set of copies of $SO(n)$, often referred to as potential wells. Minimizing sequences generally do not converge strongly, turning the applecart on the traditional pastime of analysis of proving that weak things are really strong ones. In the present enterprise, the oscillatory behavior of minimizing sequences can be exploited to give deep characterizations of limit configurations. As such a sequence typically involves successively higher frequencies, it mimics successively distant microscopic views of a structure, and so accomplishes a scale change from mesoscopic to macroscopic. The technical device for this is the Young measure of L. C. Young. In the multidimensional setting, this often assures that what are classical kinematical coherence conditions at the macroscopic level remain properties of the deformation gradients in the support of the Young measure at the microscopic level. Profound results concerning the characterization of microstructure then follow, but many difficult questions arise as well.

The coarse graining accomplished by passing to a weak limit within the Young measure framework necessarily forgets and neglects some energetic features of the system. As a consequence, the possible microstructures which result from theory are far too vast in number to represent a physically reasonable characterization. Among the neglected effects that come to mind, the first is the interfacial energy between different potential wells. A cubic centimeter of some shape memory alloys may contain a football field of surface, and yet the bulk may be easily deformed by hand, so this energy must be extremely small. Nonetheless, we may entertain interfacial energy as a possible sensible selection criterion and incorporate it in the system. This is the question treated by Dolzmann and Müller in the paper under review. They consider, in addition, a related situation which arises in the study of magnetostrictive materials.

The situation encountered most frequently is that in which deformation gradients in equilibrium populate two coherent energy wells. The simplest equilibrium configuration is then a simple laminate, or variation thereof, where the lamella, finite or infinite in number, have a common normal direction. Thus the interfaces between them form a collection of parallel plane segments. The result of Dolzmann and Müller is, in essence, that this is the only situation in which surface energy proportional to surface area remains finite and the system is at minimum energy. Thus many common microstructures, including those encountered in magnetic materials, do not meet their criterion, and the question of selection criteria or the question of equilibrium remains open.

The proof is a tour de force of methods related to the calculus of variations and geometric measure theory. The authors first show an interesting Liouville type of result valid for deformations whose gradient lies in $SO(n)$ on a set of finite perimeter. These deformations must be affine, which generalizes a well-known result of Reshetnyak. Unlike the more usual problems of nonlinear elasticity theory, working in the reference configuration alone is not always convenient; inverses must be used, so invertibility of functions in Sobolev spaces and degree theory are invoked (cf., e.g., the book by I. Fonseca and W. Gangbo [*Degree theory in analysis and applications*, Oxford Univ. Press, New York, 1995; MR 96k:47100] for the most up-to-date treatment). Both the approach of the authors and their results will be important reference points for the future development of the field.

{References include [J. M. Ball and R. D. James, Arch. Rational Mech. Anal. **100** (1987), no. 1, 13–52; MR 89c:80005; Philos. Trans. Roy. Soc. London Ser. A **338** (1992), no. 1650, 389–450; A. Cellina, Nonlinear Anal. **20** (1993), no. 4, 337–341; MR 94b:49036a; M. Chipot and D. Kinderlehrer, Arch. Rational Mech. Anal. **103** (1988), no. 3, 237–277; MR 90a:73037; I. Fonseca, Arch. Rational Mech. Anal. **107** (1989), no. 3, 195–223; MR 90k:73023; R. D. James and D. Kinderlehrer, Philos. Mag. B **68** (1993), no. 2, 237–274; A. G. Khachaturyan, *Theory of structural transformations in solids*, Wiley, New York, 1983; R. V. Kohn and S. Müller, Philos. Mag. A **66** (1992), no. 5, 697–715].} *David Kinderlehrer* (Pittsburgh, PA)

98f:73025 73K10 35C20 35Q72 47N20 73C99 73K12 73K20

Campbell, A. [Campbell, Alain] (Caen);
Nazarov, S. [Nazarov, Sergei A.] (St. Petersburg)

Une justification de la méthode de raccordement des développements asymptotiques appliquée à un problème de plaque en flexion. Estimation de la matrice d'impédance. (English and French summaries) [A justification of the method of matching asymptotic expansions applied to a bending plate problem. Estimation of the impedance matrix]

J. Math. Pures Appl. (9) **76** (1997), *no.* 1, 15–54.

A very important application of the singular perturbation theory to continuum mechanics is the asymptotic behaviour of solutions for domains with small inclusions or defects. The basic tool for such investigations is the method of matched asymptotic expansions dating back to the early work of M. I. Vishik and L. A. Lyusternik [Russian Math. Surveys **15** (1960), no. 3, 1–73; MR **23** #A1920] and M. Van Dyke [*Perturbation methods in fluid mechanics*, Academic Press, New York, 1964; MR **31** #974]. This method is based on two asymptotic expansions and the corresponding limit problems, an outer problem which gives asymptotically precise solutions farther away from the inclusion, and an inner problem which approximates the solution in the vicinity of the inclusion. Both solutions are matched so that they coincide in some intermediate zone. In this article, the authors tackle the problem of a vibrating Kirchhoff plate with a small clamped region of diameter ε which is located at a fixed place, distant from the outer lateral boundary. For $\varepsilon \to 0$, this is a singular perturbation problem since the degenerate outer problem becomes a so-called Sobolev problem, that is, the problem of a bending plate with fixed displacements at the origin—but no restrictions on the derivatives. The inner problem, valid in the vicinity of the inclusion and characterized by the stretched variable $\xi = \varepsilon^{-1}x$, becomes an exterior boundary value problem in $\mathbf{R}^2 \setminus \omega$, where $\omega_\varepsilon = \varepsilon\omega$ is the original, small inclusion. Exterior problems of this type in classical acoustics have already been analyzed by R. C. MacCamy [Quart. Appl. Math. **23** (1965), 247–255; MR **32** #6823].

Also, the two-dimensional Stokes expansion of exterior Navier-Stokes flows leads to singular perturbation problems of similar character [G. C. Hsiao and R. C. MacCamy, in *Theory and applications of singular perturbations (Oberwolfach, 1981)*, 229–244, Lecture Notes in Math., 942, Springer, Berlin, 1982; MR 84b:76028]. Campbell and Nazarov use for the outer as well as for the inner problem formulations as operator equations on weighted Sobolev spaces, the so-called Kondrat′ev spaces [V. A. Kondrat′ev, Trudy Moskov. Mat. Obšč. **16** (1967), 209–292; MR **37** #1777].

These operators are Fredholm if the frequency parameter does not coincide with an eigenvalue of the Sobolev problem. The Fredholm indices depend on the weight indices, i.e. the strength of the weight of the corresponding weighted Sobolev spaces. This dependence and precise relations between the kernels and cokernels have been found in the pioneering work of V. G. Maz′ya, Nazarov and B. A. Plamenevskiĭ [in *Partial differential equations. Spectral theory (Russian)*, 72–153, 222–223, Leningrad. Univ., Leningrad, 1981; MR 84c:35037] and Maz′ya and Plamenevskiĭ [Math. Nachr. **76** (1977), 29–60; MR **58** #29176]. Using these results, the general solutions and the solvability conditions for the inner and outer problems are derived. Appropriate matching of the outer and inner solutions provides the construction of the uniquely determined asymptotic expansion of the plate vibrations.

This representation of the solution defines an approximate solution operator whose norm turns out to be of the order $\varepsilon^{-\kappa}$ with $\kappa > 0$. The results are applied to the calculation of an asymptotic formula for the impedance matrix which describes the reaction of the plate to a rigid motion of the clamped small zone.

The paper is a beautiful application of the method of matched asymptotic expansions to plate theory. The use of the general Fredholm theory in the weighted Sobolev spaces gives a very elegant method of derivation and also a rigorous justification of the asymptotic expansions.

The reviewer acknowledges the substantial help provided by M. Bach and C. Eck for this review. *Wolfgang L. Wendland* (Stuttgart)

97i:73071a 73K15

Ciarlet, Philippe G. (Paris); **Lods, Véronique** (Paris)

Asymptotic analysis of linearly elastic shells. I. Justification of membrane shell equations. (English summary)

Arch. Rational Mech. Anal. **136** (1996), *no.* 2, 119–161.

97i:73071b 73K15

Ciarlet, Philippe G. (Paris); **Lods, Véronique** (Paris);
Miara, Bernadette (Noisy-le-Grand)

Asymptotic analysis of linearly elastic shells. II. Justification of flexural shell equations. (English summary)

Arch. Rational Mech. Anal. **136** (1996), *no.* 2, 163–190.

97i:73071c 73K15

Ciarlet, Philippe G. (Paris); **Lods, Véronique** (Paris)

Asymptotic analysis of linearly elastic shells. III. Justification of Koiter's shell equations. (English summary)

Arch. Rational Mech. Anal. **136** (1996), *no.* 2, 191–200.

97i:73071d 73K15

Ciarlet, Philippe G. (Paris); **Lods, Véronique** (Paris)

Asymptotic analysis of linearly elastic shells: "generalized membrane shells". (English summary)

J. Elasticity **43** (1996), *no.* 2, 147–188.

The asymptotic justification when the shell thickness goes to zero of Koiter's two-dimensional shell model in three different cases is presented in these four papers. Rigorous and detailed proofs are given. A survey of recent investigations concerning the low-dimensional elastic bodies covers nearly 150 items.

A family of linearly elastic isotropic shells of thickness 2ε, all having the same middle surface $S=\varphi(\overline{\omega})\subset\mathbf{R}^3$, where $\omega\subset\mathbf{R}^2$ is a bounded and connected open set with a Lipschitz-continuous boundary $\gamma=\partial\omega$, is considered. The shells are clamped along a portion of their lateral face, whose middle line is $\varphi(\gamma_0)$, where γ_0 is a portion of $\partial\omega$ with length $\gamma_0>0$. The authors consider the three-dimensional variational problem

$$(1)\qquad \int_{\Omega^\varepsilon} A^{ijkl}e_{kl}(\mathbf{u}^\varepsilon)e_{ij}(\mathbf{v}^\varepsilon)\sqrt{g}\,dx=\int_{\Omega^\varepsilon} f^i v_i^\varepsilon\sqrt{g}\,dx \quad \text{for all } \mathbf{v}^\varepsilon\in\mathbf{V}_{3d}(\Omega^\varepsilon),$$

where $\Omega^\varepsilon=\omega\times[-\varepsilon,\varepsilon]\subset\mathbf{R}^3$, x_α $(\alpha=1,2)$ are the curvilinear coordinates in ω and x_3 is the transverse variable, $\mathbf{u}^\varepsilon(x_j)=(u_i^\varepsilon(x_j))$ is the unknown vector field of body displacements, e_{ij} and A^{ijkl} are the components of the strain tensor and of the three-dimensional elasticity tensor, respectively. They also consider the corresponding two-dimensional variational problem

$$(2)\quad \varepsilon\int_\omega a^{\alpha\beta\sigma\tau}\gamma_{\sigma\tau}(\boldsymbol{\zeta}^\varepsilon)\gamma_{\alpha\beta}(\boldsymbol{\eta})\sqrt{a}\,dy + \frac{\varepsilon^2}{3}\int_\omega a^{\alpha\beta\sigma\tau}\rho_{\sigma\tau}(\boldsymbol{\zeta}^\varepsilon)\rho_{\alpha\beta}(\boldsymbol{\eta})\sqrt{a}\,dy=\int_\omega p^{i,\varepsilon}\eta_i\sqrt{a}\,dy$$
$$\text{for all } \boldsymbol{\eta}=(\eta_i)\in\mathbf{V}_{2d}(\omega),\ p^{i,\varepsilon}=\frac{1}{2\varepsilon}\int_{-\varepsilon}^{\varepsilon} f^i\,dx_3,$$

where $\boldsymbol{\zeta}^\varepsilon(x_\alpha)=(u_i^\varepsilon(x_\alpha))$ is the unknown vector field of the middle surface displacements, $\gamma_{\alpha\beta}$, $\rho_{\alpha\beta}$, and $a^{\alpha\beta\sigma\tau}$ are the components of the linearized change of metric and change of curvature tensors and of the two-dimensional elasticity tensor, respectively. The functional spaces $\mathbf{V}_{3d}(\Omega^\varepsilon)$ and $\mathbf{V}_{2d}(\omega^\varepsilon)$ in (1) and (2) depend on the particular case under consideration. The average $\overline{\mathbf{u}}^\varepsilon=(2\varepsilon)^{-1}\int_{-\varepsilon}^{\varepsilon}\mathbf{u}^\varepsilon\,dx_3$ is compared as $\varepsilon\to 0$ with the solution $\boldsymbol{\zeta}$ of (2). The semi-norm $|\cdot|_\omega^M$: $\boldsymbol{\eta}\to|\boldsymbol{\eta}|_\omega^M=\{\sum_{\alpha,\beta}\|\gamma_{\alpha,\beta}\|_{0,\omega}^2\}^{1/2}$ over the space $\mathbf{V}(\omega)=\{\boldsymbol{\eta}\in\mathbf{H}^1(\omega);\ \boldsymbol{\eta}=\mathbf{0}$ on $\gamma_0\}$ is introduced.

If the shell edge is clamped ($\gamma_0=\partial\omega$) and the Gaussian curvature of the middle surface S is positive then as $\varepsilon\to 0$ problem (2) is reduced to the membrane problem (the ε^2 term in the left-hand side of equation (2) is neglected). In this case $\mathbf{V}_{3d}=\mathbf{V}(\Omega^\varepsilon):=\{\mathbf{v}^\varepsilon\in\mathbf{H}^1(\Omega^\varepsilon);\ \mathbf{v}^\varepsilon=\mathbf{0}$ on $\Gamma^\varepsilon=\gamma\times[-\varepsilon,\varepsilon]\}$ and $\mathbf{V}_{2d}=\mathbf{V}_M(\omega):=H^1(\omega)\times H^1(\omega)\times L^2(\omega)$. The semi-norm $|\cdot|_\omega^M$ is a norm on $\mathbf{V}_M(\omega)$. By using the generalized Korn inequality it is proved that $u_\alpha^\varepsilon\to\zeta_\alpha$ in $H^1(\Omega^\varepsilon)$, $u_3^\varepsilon\to\zeta_3$ in $L^2(\Omega^\varepsilon)$ and $\overline{u}_\alpha^\varepsilon\to\zeta_\alpha$ in $H^1(\omega^\varepsilon)$, $\overline{u}_3^\varepsilon\to\zeta_3$ in $L^2(\omega^\varepsilon)$ as $\varepsilon\to 0$.

If the space of inextensional displacements, $\mathbf{V}_F(\omega)=\{\boldsymbol{\eta}=(\eta_i)\in H^1(\omega)\times H^1(\omega)\times H^2(\omega);\ \eta_i=\partial_\nu\eta_3=0$ on γ_0, $\gamma_{\alpha\beta}(\boldsymbol{\eta})=0$ in $\omega\}$, is not empty, then the order of the flexural displacements is ε^{-2} times larger than for the membrane shell equation. It is proved that $\varepsilon^{-2}\mathbf{u}^\varepsilon \to \widehat{\zeta}$ in $\mathbf{H}^1(\Omega^\varepsilon)$ as $\varepsilon \to 0$. Here $\widehat{\zeta}$ is the solution of (2) with $\varepsilon = 1$ and $\mathbf{V}_{2d} = \mathbf{V}_F(\omega)$ (the first term of the left side of (2) vanishes).

The remaining cases (the so-called generalized membrane shells) are intermediate between the membrane and the flexural shells and the results are more complex. At first the body forces are not arbitrary but admissible: there exist functions $\varphi^{\alpha\beta}(\varepsilon) \in L^2(\omega)$ and $F_i(\varepsilon) \in L^2(\Omega^\varepsilon)$ such that the right-hand side of (2) is of the form $\int_\omega \varphi^{\alpha\beta}\gamma_{\alpha\beta}(\mathbf{v})dy + \int_{\Omega^\varepsilon} F^i \partial_3 v_i \, dx$ for all $\mathbf{v} \in \mathbf{V}(\Omega^\varepsilon)$. It is proved that if the body forces are admissible then the convergence $\mathbf{u}^\varepsilon \to \zeta$ as $\varepsilon \to 0$ occurs in the abstract Hilbert space $\dot{\mathbf{V}}_M^\#(\omega)$, the completion of $\mathbf{V}(\omega)/\mathbf{V}_0(\omega)$ with respect to the norm $|\cdot|_\omega^M$, where $\mathbf{V}(\omega) = \{\boldsymbol{\eta} \in \mathbf{H}^1(\omega);\ \boldsymbol{\eta} = \mathbf{0} \text{ on } \gamma_0\}$, $\mathbf{V}_0(\omega) := \{\boldsymbol{\eta} \in \mathbf{V}(\omega),\ \gamma_{\alpha\beta}(\boldsymbol{\eta}) = 0 \text{ in } \omega\}$.

The authors state that these results "thus constitute an asymptotic analysis of linearly elastic shells in all possible cases". It seems hard to fully agree with this statement if we think about applications. These results are obtained under some restrictions which may not be fulfilled. (i) Only the results for two extreme cases (membrane and flexural shells) are carried to completion. For the generalized membrane shell it is not clear what occurs when the external forces are not admissible. (ii) Only the clamped and the free parts of the edges are examined. With respect to equation (2) a lot of various intermediate boundary conditions are used. With respect to equation (1) it is interesting to examine the case when a shell is attached to an elastic body. (iii) Under the assumption that the given functions are sufficiently smooth, by using formal asymptotic expansions of three-dimensional equations one can obtain the orders of errors of two-dimensional models [A. L. Gol′denveĭzer, Mech. Solids **31** (1996), no. 4, 124]. There are sufficient distinctions between the expansions for inner points and near the edge. It is interesting to derive rigorous estimations for errors of two-dimensional models (the statement that the error goes to zero with ε is deficient for applications). (iv) A large variety of formal asymptotic expansions of solutions of two-dimensional equations, for statics [A. L. Gol′denveĭzer, *Theory of elastic thin shells* (Russian), Second edition, Izdat. "Nauka", Moscow, 1976; MR **56** #1867], dynamics [A. L. Gol′denveĭzer, V. B. Lidskiĭ and P. E. Tovstik, *Free oscillations of thin elastic shells* (Russian), "Nauka", Moscow, 1979; MR 80i:73051] and buckling [P. E. Tovstik, *Stability of thin shells: asymptotic methods* (Russian), Fizmatlit "Nauka", Moscow, 1995; MR 96j:73001], have been constructed. The question is how to estimate rigorously the errors for them for both three-dimensional equations and two-dimensional models.

To conclude, these papers by Ciarlet, Lods, and Miara contain a promising method of investigation and very important results in two-dimensional mathematical theory of thin elastic shells. But there are in this field a lot of mathematical problems which are imposed by applications and which are to be solved by similar (or by other) methods. Some of these problems were indicated above.

Petr E. Tovstik (St. Petersburg)

98a:73046 73K40 49Q15 65K10 73B27 73V05 73V25
Allaire, Grégoire (Gif-sur-Yvette); **Bonnetier, Eric** (Palaiseau);
Francfort, Gilles (Villetaneuse); **Jouve, François** (Palaiseau)
Shape optimization by the homogenization method. (English summary)
Numer. Math. **76** (1997), *no.* 1, 27–68.

This paper addresses both theoretical and algorithmic aspects of the "homogenization method" for shape optimization. The viewpoint is similar to that of Allaire and R. V. Kohn [European J. Mech. A Solids **12** (1993), no. 6, 839–878; MR 97a:73062]. The present paper extends that earlier work to 3D and provides a much-improved numerical algorithm. Perhaps the most important contribution, however, is its systematic examination of "penalization schemes" for obtaining classical rather than relaxed designs.

The context is shape optimization for minimal compliance. Thus the goal is to choose the shape of a body so as to optimize its rigidity under specified mechanical loads. An optimal shape may not exist, because an optimizing sequence can develop microstructure. The homogenization method "relaxes" the problem by permitting (physically realizable) composites as structural components. This has both theoretical and practical advantages. There is a satisfying mathematical theory, based on ideas from homogenization (optimal bounds on the effective moduli of composites) and the calculus of variations (quasiconvexification). Moreover, the method is more global than those based on free boundaries and front-tracking.

The paper starts with a detailed introduction to the homogenization method for structural optimization. The viewpoint is not new, but the exposition is very clear and fully informed by recent developments. An oversight: it would have been natural to cite the work of L. Gibiansky and A. Cherkaev on optimal bounds for 3D composites with applications to structural optimization [in *Topics in the mathematical modelling of composite materials*, 273–317, Birkhäuser Boston, Boston, MA, 1997; per revr.].

The paper's main contributions are as follows: (1) Justification of the shape-optimization limit. In considering shape optimization one is effectively considering mixtures of a given elastic material with void. Homogenization cannot be applied directly, because it requires a hypothesis of nondegeneracy. A natural idea for resolving this difficulty is to replace the void with a nearly-degenerate elastic material, take the relaxation, then pass to the limit. Allaire and Kohn proved the correctness of this procedure in a special case, using explicit knowledge of the relaxed problem. The present work proves its correctness in great generality (for compliance optimization).

(2) Comparison of the 2D and 3D problems with one applied load. There are some surprising differences. For example, in 2D the relaxation is achieved by a rank-2 laminate, but the associated Hooke's law is degenerate; in 3D one must use rank-3 laminates, and the associated Hooke's law is nondegenerate.

(3) An efficient solution algorithm. The relaxed design problem can be formulated as a variational problem involving (statically admissible) stresses and (sequentially laminated) microstructures. This paper uses an alternating direction method, taking full advantage of that structure. Each iteration must solve an elasticity problem then a microstructure optimization problem. The former is done using a standard, displacement-based finite element scheme; the latter is done analytically (for the case of one load). This approach is similar in concept to the "optimality criteria" method widely used in the structural optimization community; however, it has the advantage

of being monotone. Its efficiency is demonstrated by an impressive example of 3D shape optimization.

(4) Penalization schemes for obtaining classical designs. The homogenization method has one overwhelming disadvantage: it usually delivers a "relaxed" design rather than a "classical" one. Several investigators have explored methods for obtaining nearly-optimal classical designs automatically, by penalizing the presence of composites. A few have advocated simplifying the relaxation procedure by introducing convenient "fictitious composites", arguing that the composites will be eliminated by penalization so their realizability is not important. The present work provides an extremely clear summary of this discussion. It shows that the alternating direction method is especially well-suited to the introduction of penalization. And it gives evidence that use of fictitious composites + penalization leads to worse results than realizable composites + penalization. The authors correctly note that little is known with rigor about these penalization schemes and what they produce. (There is an interesting analogy with image segmentation. The idea of penalization has much in common with the work of A. Blake and A. Zisserman [*Visual reconstruction*, MIT Press, Cambridge, MA, 1987; MR 89k:92086]; the present paper's combination of alternating-direction and penalization resembles the work of Charbonnier et al. [IEEE Trans. Image Process. **6** (1997), no. 2, 298–311]; see also A. Chambolle ["Image segmentation: variants of the graduated nonconvexity algorithm", Preprint; per revr.].) *Robert V. Kohn* (New York, NY)

98i:73049 73M25 35J99 35Q72 47F05 73C02 73C50
Nazarov, S. A.; Polyakova, O. R.
Rupture criteria, asymptotic conditions at crack tips, and selfadjoint extensions of the Lamé operator. (Russian)
Tr. Mosk. Mat. Obs. **57** (1996), 16–74; *translation in Trans. Moscow Math. Soc.* **1996,** 13–66 (1997).

There exist different approaches to obtain a condition under which a crack in a brittle elastic 2-d body starts to propagate. Such conditions are called rupture criteria and they are based on various physical considerations, e.g., the energy criterion of A. A. Griffith, the force criterion of G. Irwin, the deformation criterion of D. S. Dugdale, M. Ya. Leonov, V. V. Panasyuk, and others. Those criteria are usually connected with certain assumptions on the mechanical structure of the crack mouth (a small zone around the crack tip).

In the paper under review mathematical tools are developed to model cracks with structured mouths and, in particular, to describe nonlinear effects at crack tips. The boundary value problem of the elasticity theory is considered in a weighted space with detached asymptotics which consists of vector functions with a prescribed asymptotic form, including usual square-root singularity terms as well as terms with higher-order singularities. The formulation of the elasticity problem in these spaces involves additionally a system of algebraic equations for the coefficients of the singularity terms mentioned above. Solvability and uniqueness results for a problem with such asymptotic conditions at the tips are obtained with the help of a generalized Green's formula which, in addition, contains a quadratic form in the coefficients of the singularity terms. If the asymptotic conditions are homogeneous and the corresponding Green's formula is symmetric, one can interpret the operator of the problem as a selfadjoint extension of the Lamé operator in a proper Hilbert space that gives rise to the energy functional and then reformulate the problem as a minimization one (note that the common elastic energy functional becomes infinite due to the higher-order singularity terms). In the latter framework, nonlinear asymptotic conditions are posed at the crack tips and the problems obtained are investigated.

Based on the method of matched asymptotic expansions, a rigorous asymptotic analysis of the elasticity problem with small singular perturbations of the crack near its tips is performed. It is shown that the "far-field" asymptotic decomposition can be interpreted as a solution to the problem with suitable asymptotic conditions at the tips. For a long list of mechanical models of crack mouths, explicit forms of asymptotic conditions are calculated together with the corresponding energy functionals, fragments of which depend on so-called stress intensity factors and are related to the energy stored by the crack mouths. The most interesting and unexpected physical inference is the asymptotic equivalence of the following couple of rupture criteria: a natural mechanical criterion for the assumed structure of the crack mouth and the Griffith criterion with an energy functional which has been derived by asymptotic treatment of the crack mouth and formulation of asymptotic conditions at the tips.

The paper is an important contribution to the theory of boundary value problems on singular domains. *Vladimir A. Kondratiev* (Moscow)

99m:76001 76-02 35-02 35Q30 76N10
Lions, Pierre-Louis (Paris)
★**Mathematical topics in fluid mechanics. Vol. 2. (English summary)**
Compressible models.
Oxford Lecture Series in Mathematics and its Applications, 10.
Oxford Science Publications.
The Clarendon Press, Oxford University Press, New York, 1998. xiv+348 pp.
ISBN 0-19-851488-3

This book is the second of two volumes that P.-L. Lions dedicated to fluid mechanics. This one emphasizes the compressible side, with various contexts: viscous isentropic flows, viscous non-isentropic flows and inviscid flows. The author focuses on the large time, large data Cauchy problem. This a priori choice may be a matter of discussion. It is justified by the fact that large-time existence for small smooth data is useless for applications, since realistic data are large. Similarly, small-time existence for smooth data does not apply to the range of times which are considered in practical applications.

The main part of the book concerns the isentropic viscous case, the so-called "compressible Navier-Stokes equation". For a long time, the theory of the Cauchy problem with large data has been done only in one space dimension, using the order structure of the real line and some special transport properties along the particle path [see A. V. Kazhikhov and V. V. Shelukhin, J. Appl. Math. Mech. **41** (1977), no. 2, 273–282; translated from Prikl. Mat. Meh. **41** (1977) , no. 2, 282–291; MR **57** #8425; A. V. Kazhikhov, Comm. Math. Phys. **82** (1981/82), no. 2, 37–62, 175; MR 83e:76055; D. Hoff, Proc. Roy. Soc. Edinburgh Sect. A **103** (1986), no. 3-4, 301–315; MR 88b:35156; Arch. Rational Mech. Anal. **114** (1991), no. 1, 15–46; MR 92e:76052; D. Serre, C. R.

Acad. Sci. Paris Sér. I Math. **303** (1986), no. 13, 639–642; MR 88a:35193; C. R. Acad. Sci. Paris Sér. I Math. **303** (1986), no. 14, 703–706; MR 88a:35194]. From those old times (say the eighties), it was known by a dozen people that the trace of the stress tensor plays a special rôle. In particular, it remains smoother than each of its two components, the pressure and the strain. It also plays a prominent rôle in the study of the propagation of oscillations [D. Serre, Phys. D **48** (1991), no. 1, 113–128; MR 92b:76069; A. A. Amosov and A. A. Zlotnik, Dokl. Akad. Nauk **342** (1995), no. 3, 295–299; MR 96e:76101]. Therefore it was clear that a higher-dimensional proof would involve a regularity result about this quantity. In spite of this knowledge, specialists were unable to step beyond the 1-D case.

The reason for this failure is that two other important tricks were needed, which eventually were found by Lions. One of them uses the notion of renormalisation already introduced by R. J. DiPerna and Lions [Invent. Math. **98** (1989), no. 3, 511–547; MR 90j:34004] in the study of the Boltzmann equation. However, contrary to the Boltzmann context, the renormalisation is not used as a new definition of (very weak) solutions; actually the solutions constructed in this book are genuine solutions of the mass and momentum conservation laws. Renormalisation is merely a tool which helps one to extract some compactness property from the equations, or from approximate equations. Actually, this argument was devised independently by Hoff [Arch. Rational Mech. Anal. **132** (1995), no. 1, 1–14; MR 96j:76127; see also V. A. Vaĭgant and A. V. Kazhikhov, Sibirsk. Mat. Zh. **36** (1995), no. 6, 1283–1316, ii; MR 97b:35142; V. A. Vaĭgant, Siberian Adv. Math. **6** (1996), no. 2, 103–150; MR 97f:35169; V. A. Vaĭgant and A. V. Kazhikov, Dokl. Akad. Nauk **357** (1997), no. 4, 445–448; MR 99c:35189]. The second trick is a rather delicate estimate of commutators involving Riesz transforms and variable coefficient operators.

The presentation of this memoir pays attention to pedagogy, in the sense that stability of weak solutions is studied first, although the target of establishing existence requires the stability of approximate solutions, which is a more technical task. Similarly, stationary problems precede time-dependent ones, since one strategy consists in semi-discretizing the system in time, so that steady problems become intermediate steps. Actually, these take up the main part (110 pages) of the book, while the final proof of existence of time-dependent solutions is rather short.

In the last part, the author discusses several complementary aspects of the general problem, in particular the heat-conducting compressible viscous fluid. For this last model, the results are by no means complete, but some compactness properties may be given. Other interesting models are: an elliptic-hyperbolic problem, where the acceleration of the fluid is neglected, a Stokes-like model, where only the convective term of the acceleration is neglected, a shallow water model, a low Mach number model, and inviscid barotropic fluids. The latter case behaves very much differently from the general one, since it corresponds to a hyperbolic system, without dissipative term. Therefore almost no smoothing effect is expected. However, interesting regularization properties were derived by the author and his colleagues B. Perthame, T. Souganidis and E. Tadmor, using their kinetic formulation [see P.-L. Lions, B. Perthame and E. Tadmor, Comm. Math. Phys. **163** (1994), no. 2, 415–431; MR 95c:76093; P.-L. Lions, B. Perthame and P. E. Souganidis, Comm. Pure Appl. Math. **49** (1996), no. 6, 599–638; MR 97e:35107]. A new estimate of $\rho^{1+\gamma}$ (γ being the adiabatic constant) is obtained through the very elegant Theorem 8.13. The existence result now allows all values of γ in $(1,\infty)$, thus completing the seminal work by DiPerna [Comm. Math. Phys. **91** (1983), no. 1, 1–30; MR 85i:35118]. Let us note however that global existence for the Cauchy problem for the full gas dynamics remains open. The method employed here (compensated compactness) does not extend to this case, due to the lack of conserved quantities, the so-called entropies. We should also point out that the result obtained in the barotropic case is rather weak, in the sense that there is a huge gap between the known regularity of the solution and what would ensure its uniqueness.

In conclusion, one can say that this book is a milestone in the mathematics of fluid dynamics. Perhaps no other before it offered so many novelties in one handful. The advances it makes in the field are such that specialists will need a decade to exploit them in their full range. It opens so many questions that the field will be active for a while. Certainly, every researcher on compressible fluids will wish to add this text to his private bookshelf.

{Vol. 1 has been reviewed [MR 98b:76001].} *Denis Serre* (Lyon)

97k:76018 76B15 35B32 35Q35 47H15 47N20 58C27

Okamoto, Hisashi (Kyoto)

Interfacial progressive water waves—a singularity-theoretic view. (English summary)

Tohoku Math. J. (2) **49** (1997), *no.* 1, 33–57.

The paper presents rigorous results, using functional analysis and equivariant singularity theory, on an interesting open question in the theory of water waves. The water-wave problem studied involves waves at the surface of an inviscid, irrotational, constant-density fluid with gravitational and surface tension forces present. Previous numerical studies of the problem, for the case of an infinite-depth fluid, had found complicated bifurcation diagrams which implied that there might be a singularity in parameter space. Analysis of the finite-depth problem introduced a new parameter, but this parameter introduced no new singularities and so the question remained unsolved. In this paper, the author is able to explain the anomalous bifurcation diagrams as follows.

The idea of the author is to embed the problem in a larger one: in this case, the problem of interfacial progressive waves. In the interfacial wave problem, the interface lies between two fluids of differing density. In the limit as the density of the upper fluid goes to zero, the classic water-wave problem is recovered. A parameter, denoted b, is introduced which is a measure of the distance from the classic water-wave problem; in the limit $b \to 0$, the water-wave problem is recovered. The author then shows that there is a critical value of b at which indeed a degeneracy does occur, and a rigorous analysis, using equivariant singularity theory, of the unfolding of this degeneracy produces bifurcation diagrams that qualitatively explain the observed bifurcations of previous work on the classic water-wave problem.

An overview of the analysis is as follows. The interfacial-wave problem is formulated as an operator equation using a modification of the set of equations for interfacial waves due to N. E. Kochin [Math. Ann. **98** (1928), no. 3-4, 582–615]. It is proved that the resulting operator equation is well-defined as a mapping on a Hilbert space, that the operator equation is O(2)-equivariant and that the limit $b \to 0$ recovers the water-wave problem. The operator equation is then analyzed using the equivariant Lyapunov-Schmidt method. It is proved that at each point of a three-dimensional manifold, parametrized by b and two other parameters, the linearization of the operator equation about the trivial state has a four-dimensional kernel. The equivariant Lyapunov-Schmidt method then reduces the analysis to a parameter-dependent O(2)-equivariant mapping from $\mathbf{R}^4$ to $\mathbf{R}^4$. The main results of the paper are then deduced by applying O(2)-equivariant singularity theory to this reduced mapping.

The discovery of the singularity, combined with an embedding of the bifurcation equation in the singularity theory framework of M. Golubitsky, I. N. Stewart and D. G. Schaeffer [*Singularities and groups in bifurcation theory. Vol. II*, Springer, New York, 1988; MR 89m:58038], then completes the proof, and resolves the open question. There is some discussion of the implications for water waves and also the possibility of further singularities of interest if b is allowed to take negative values. *Thomas J. Bridges* (Guildford)

97d:76007 76C05 35-02 35Q35 76-02

Chemin, Jean-Yves

Fluides parfaits incompressibles. (French summary) [Incompressible perfect fluids]

Astérisque No. 230 (1995), 177 pp.

The theory of the Euler equation for an incompressible inviscid fluid has been revived in the early nineties, mainly by researchers from the École Polytechnique. After a 28-year-long wake, two main open questions were solved in 1991: the persistence of smooth vortex patches [J.-Y. Chemin, in *Séminaire sur les Équations aux Dérivées Partielles, 1990–1991*, Exp. No. XIII, 11 pp., École Polytech., Palaiseau, 1991; see MR 92f:35007; Ann. Sci. École Norm. Sup. (4) **26** (1993), no. 4, 517–542; MR 94j:35141] and the existence of solutions for vortex sheet initial data [J.-M. Delort, C. R. Acad. Sci. Paris Sér. I Math. **312** (1991), no. 1, 85–88; MR 91i:35149; J. Amer. Math. Soc. **4** (1991), no. 3, 553–586; MR 92f:76019]. Since these results deal with planar motions, the last big problem to elucidate is whether smooth solutions of 3D flows survive, or blow up in finite time.

The book under review has two main goals, namely to present the classical theory (Wolbiner, Yudovich and others) on the one hand, and the new theory on the other hand. The true quality of this text is that it unifies the past and the present by an intensive use of Sobolev and Hölder estimates of the paraproduct given by the Littlewood-Paley decomposition (see Chapters 2 and 3). In these, the space C^1_* plays the role of a Hölder space instead of C^1. Although the former is larger than C^1, it still fits well enough with the theory of ODEs so that a flow may be defined for such velocity fields. Thus Lagrangian properties may be addressed.

The Euler equation is introduced in Chapter 1 as a consequence of the least action principle. This is a true mathematical definition, which is not fully convincing from a mechanical point of view for several reasons. First, it cannot be generalized to real (that is, viscous) fluids. Secondly, although it may be adapted to the compressible case, the variational principle does not distinguish a time arrow, so that it cannot select a physically admissible solution from among others. One might argue that incompressible inviscid flows do not need an "entropy condition", since they admit unique global smooth solutions, but this is not entirely true: the 3D flows might not be smooth beyond some finite time; even in 2D, vortex sheet initial data would yield very weak solutions for which we are not able to prove uniqueness.

Nevertheless, regarding the mathematical scope of this book, this definition of a fluid is certainly acceptable. It is followed in Chapter 2 by a very neat presentation of the Littlewood-Paley theory. This part could be of great help, even for graduate students. It ends with a study, via paradifferential calculus, of the quadratic operator $v = \text{velocity} \overset{\Pi}{\mapsto} \text{pressure} = p := -\Delta^{-1}\partial_i\partial_j(v_i v_j)$.

Chapter 3 is devoted to estimates of the velocity in terms of estimates of the vorticity $\Omega(v)$. A special emphasis is given to the notion of tangential regularity with respect to a geometrical structure, anticipating the vortex patch problem.

Chapter 4 presents a new approach to Wolibner's classical solutions (1933). The Euler equation is seen as an abstract evolution equation with a quadratic operator Π satisfying tame estimates. In dimension two, one also assumes the transport equation for the vorticity, but in all cases, one does not make use of energy conservation.

Chapter 5 considers first Yudovich's theory [V. I. Yudovich, Ž. Vyčisl. Mat. i Mat. Fiz. **3** (1963), 1032–1066; MR **28** #1415] for initial data with bounded vorticity in 2D. The solution is still unique and its structure is preserved. However, the Hölder regularity of the flow may decay as time increases and an explicit example shows (Section 5.3) that it does decay in general. The especially important Section 5.5 proves the persistence of 2D vortex patches. This result, due to the author (see above) and independently to P. Serfati [C. R. Acad. Sci. Paris Sér. I Math. **318** (1994), no. 6, 515–518; MR 95g:35160], seemed unbelievable when announced in 1991. Indeed, numerical experiments of N. J. Zabusky [J. Comput. Phys. **30** (1979), no. 1, 96–106; MR 80g:76016] suggested that singularities of the boundary of the patches would develop, presumably in finite time. A. J. Majda [Comm. Pure Appl. Math. **39** (1986), no. S, suppl., S187–S220; MR 87j:76041] even conjectured, concerning the boundary of piecewise constant vortex patches, that there are smooth initial curves such that the curve becomes nonrectifiable in finite time.

In the same (wrong) direction, S. Alinhac [J. Funct. Anal. **98** (1991), no. 2, 361–379; MR 92g:35172] studied a quadratic approximation of the equation governing the evolution of the boundary and found some evidence of finite time breakdown. Thus it is fair to say that the author's result is one of the biggest achievements of the last decade in the mathematical theory of fluid flows.

The next chapter is devoted to the problem of vortex sheets in two space dimensions. Intermediate results were obtained by R. Di Perna and Majda [Comm. Pure Appl. Math. **40** (1987), no. 3, 301–345; MR 88e:35149]. It is the weakest of the 2D theories, in the sense that it provides only the existence of a weak solution and says nothing about regularity and uniqueness. Since the Birkhoff-Rott equation, which formally governs the evolution of Lipschitz vortex sheets, is linearly ill-posed, we do not expect the geometric structure of the solution to persist for positive time in general. In particular, the flow should not be well-defined as a measure-preserving one-to-one mapping.

The key point in Delort's analysis is the study of a singular integral. It assumes that the singular part of the initial vorticity is nonnegative (or nonpositive). This severe restriction seems to be justified by numerical experiments of R. Krasny, but who knows?

Chapters 7 (wave front, Gevrey regularity) and 9 (vortex patches with nonsmooth boundaries) are less fundamental. However, Chapter 8 presents recent results about the time regularity of the flow which are far from obvious. In short, the flow is analytic with respect to time in the classical theory (velocity in C^r, $r > 1$). It is still

Gevrey-3 in Yudovich's theory ($\Omega(v)$ bounded). Nevertheless, the space regularity is much weaker, since the velocity field is not smoothed out. The results are due to the author [J. Math. Pures Appl. (9) **71** (1992), no. 5, 407–417; MR 94b:35223].

Overall, this book is a wonderful piece of mathematics and offers an almost complete overview on the field of the mathematical theory of incompressible inviscid fluids (although one might wish to learn about initial-boundary value problems). It clearly shows how big the difference is between the 2D and the 3D cases (the reader should be aware of the fact that a few theorems and sections make sense only in 2D, although the author does not say so). This book fills a wide gap. As far as the reviewer knows, there has not existed a detailed overview on this topic. Such a new book in a classical field is a rare event and most of the researchers in theoretical fluid mechanics will want to have it in their private libraries. *Denis Serre* (Lyon)

98m:76030 76C05 35Q35 49J40 76M30

Brenier, Yann (Paris)

A homogenized model for vortex sheets.

Arch. Rational Mech. Anal. **138** (1997), *no.* 4, 319–353.

It is well known that the Euler equations of incompressible flows can be derived by minimizing an action functional. So, in geometrical terms (following Arnol′d), the problem arises of finding a minimal geodesic on the group of all volume-preserving diffeomorphisms of the domain D occupied by the flow. In 1970, D. G. Ebin and J. Marsden [Ann. of Math. (2) **92** (1970), 102–163; MR **42** #6865] solved this problem locally when the fluid motion connects two states which are close in a Sobolev norm. In 1985, A. I. Shnirel′man [Mat. Sb. (N.S.) **128(170)** (1985), no. 1, 82–109, 144; MR 87d:58034] constructed an example showing that two arbitrarily chosen fluid states cannot be connected by a smooth flow. This negative result stimulated a search for a more general framework. It was found by Shnirel′man and by Brenier independently [J. Amer. Math. Soc. **2** (1989), no. 2, 225–255; MR 90g:58012]. The idea (following L. C. Young) was to treat each admissible fluid flow as a generalized flow, a probability measure on the set of particle trajectories.

In the current paper the author makes a number of substantial contributions to this area. The new mathematical model which he presents generalizes the Euler equations for the case of a finite number of fluid phases. Each phase is equipped with a volume concentration field $\rho_\alpha > 0$, called density ($\sum_\alpha \rho_\alpha = 1$), and a velocity field v_α. All the phases have the same pressure field p and they obey the equations $\partial_t \rho_\alpha + \nabla\cdot(\rho_\alpha v_\alpha) = 0$ and $\partial_t v_\alpha + (v_\alpha\cdot\nabla)v_\alpha + \nabla p = 0$. Each velocity field is considered to be potential, $v_\alpha = \nabla\Phi_\alpha$. The Euler equations are formally included if each ρ_α takes values in $\{0,1\}$. First, the author proves that the new model can be obtained from a variational principle, with $\frac{1}{2}\sum\int_0^T\int_D \rho_\alpha v_\alpha^2$ taken as an action. Then considerable work goes into showing that the generalized flow (or the measure-valued solution) which minimizes the action satisfies the equations in the distributional sense. The main results are obtained under the assumption that $\inf_{\alpha,t,x}\rho_\alpha(t,x) \ge \delta > 0$. (Such solutions are called strongly homogenized.) In this case, the regularity properties are improved, namely, the solution is shown to be weak. Moreover, uniqueness and stability theorems are proved.

To solve the least action problem, the author formulates it in a dual form and uses his concept of generalized flow. In this way the existence of a minimizer is ensured by Rockafellar's theorem. To improve the regularity of the minimizer, its Lagrangian deformations are studied. In particular, a refined version of Moser's lemma due to B. Dacorogna and J. K. Moser [Ann. Inst. H. Poincaré Anal. Non Linéaire **7** (1990), no. 1, 1–26; MR 91i:58148] is used to prove weak compactness of the minimizing sequence and the existence of the pressure field. The Lagrangian deformations are also used to obtain a priori estimates of the first derivatives of velocity and of the pressure gradient. As for the uniqueness question, it is reduced to that of the uniqueness of weak solutions of the linear transport equation. The latter is studied with the help of the notion of renormalized solution developed by R. J. DiPerna and P.-L. Lions.

There is one more indication of the importance of Brenier's results. He proves that the initial boundary-value problem has a local analytic strongly homogenized solution. On the other hand, it follows from his estimates that any strongly homogenized solution of the initial boundary-value problem cannot be global in time. This phenomenon is indirectly revealed through the analysis of the time boundary-value problem, when a final time $T>0$ is prescribed as well as the values of ρ_α at $t=0$ and $t=T$. Thus, we see that it is not the Cauchy problem but the time boundary-value problem that has the properties of global existence, uniqueness, and stability. *Vladimir V. Shelukhin* (Rio de Janeiro)

98e:76024 76C99 35Q35 58F40

Shnirelman, A. (Tel Aviv)

Evolution of singularities, generalized Liapunov function and generalized integral for an ideal incompressible fluid. (English summary)

Amer. J. Math. **119** (1997), *no.* 3, 579–608.

The understanding of fluid flow remains one of the great challenges in the theory of PDE. In this paper, the author starts from the classical concept of generalized Lyapunov function (GLF) for a dynamical system, and extends it to functions with values in some Fréchet space with a partial order given by a positive cone. He then constructs such a GLF for the 2D incompressible fluid flow, and outlines some of the numerous consequences of this construction: a precise version of propagation of singularities of the flow is obtained (Theorem 1.5); each point of a dense open set of states has a wandering neighbourhood, that is, the state is never seen again in this neighbourhood for large times; there exist generalized integrals of the motion, etc.

The approach here is by writing the Euler equations on the two-dimensional torus as a dynamical system on (g,ω), respectively the diffeomorphism telling where the fluid particles are, and the vorticity of the fluid in Lagrangian coordinates. The main feature of Shnirel′man's approach is the use of microlocal analysis. Of course, microlocal analysis had been used before to investigate the singularities of the flow [see, e.g., J.-Y. Chemin, Astérisque No. 230 (1995), 177 pp.; MR 97d:76007, see p. E92 in this volume], which was actually the original purpose of the author. But here, going beyond this so to speak "standard use", and completing the existent tools by new ones, the author displays monotonicity properties of the fluid flow that allow him to construct the GLF. The idea is as follows: using the machinery of paraproduct (introduced by Bony) and paracomposition (introduced by the reviewer), the author displays some monotonicity up to some additional terms of better regularity; without these terms, the Lyapunov function would be constructed

(Lemma 6.1 and the formula immediately above). He then introduces the definition of microlocal scalar product: this product associates to two functions (u, v) a (class of) measure(s) on the cotangent bundle (in a quotient space). For $u = v$ in a certain Besov space B_s, the quadratic form corresponding to this product measures, in the various directions of the cotangent bundle, the amount of norm of u at the exact level of regularity s. The effect of this microlocal product is that it will cancel the contribution of the additional terms of better regularity, making the monotonicity property exact for the GLF with values in the appropriate quotient space of measures (the positive cone being the positive measures). In fact, the analysis of the author leads to an even more precise result (Theorem 1.4), which yields generalized integrals (GI), also with values in a (quotient) space of measures. From these GLF and GI, one can of course obtain numerous scalar LF and integrals.

This may all seem very complicated, but it is not, the technical "complications" arising naturally in the course of a basically simple argument. Moreover, the new tool (the microlocal scalar product) introduced by the author is certainly likely to have many other applications, just as similar tools (microlocal defect measures, Young measures, etc.) already have in the theory of weak solutions of nonlinear PDE, in homogenization, in control theory, etc. Finally, we would like to emphasize what we believe can be learned from the approach of the author: considering such an old problem as the fluid flow, it is likely that not many new mathematical results are going to be obtained by nineteenth-century PDE methods. Even researchers oriented towards applications will have to incorporate Shnirel′man's results into their research, just because these results, far from being some (irrelevant) refinement of basically well understood things, are the first rigorous ones in the subject.

Serge Alinhac (Orsay)

98b:76024a 76D05 35Q30
Frehse, Jens (Bonn); **Růžička, Michael** (Bonn)
Existence of regular solutions to the stationary Navier-Stokes equations.
Math. Ann. **302** (1995), *no.* 4, 699–717.

98b:76024b 76D05 35Q30
Struwe, Michael (Zürich)
Regular solutions of the stationary Navier-Stokes equations on $\mathbf{R}^5$.
Math. Ann. **302** (1995), *no.* 4, 719–741.

One of the most challenging problems in applied mathematical analysis is the problem of regularity of the solutions to the three-dimensional Navier-Stokes equations. Since Leray's fundamental paper of 1934 many mathematicians have tried to prove that not only for small data but also for arbitrary initial values there exists a unique regular solution for all times. So far all attempts to establish a complete answer (without a hidden smallness or symmetry assumption) have failed.

One idea towards a better understanding is to study the stationary Navier-Stokes equations $-\Delta u + (u \cdot \nabla)u + \nabla p = f$ and $\operatorname{div} u = 0$ in Ω, either with $u = 0$ on $\partial\Omega$ (Dirichlet problem) or with periodic boundary conditions and $\int u dx = 0$, if Ω is a cube. By dimensional analysis the three-dimensional nonstationary problem is related to the five-dimensional stationary problem; hence one is led to assume that $\Omega \subset \mathbf{R}^5$. On the other hand one has to keep in mind that a positive answer in the stationary case is only a point in favour for believers in the regularity of nonstationary problems—it does not lead to a proof and has no direct physical significance. Nevertheless there was (and is) considerable interest in understanding the stationary Navier-Stokes equations in higher dimensions. The term "higher dimensions" in this context means that $\Omega \subset \mathbf{R}^N$, $N \geq 5$—the case $N=3$ is standard, and for $N=4$ see, for example, a paper by C. Gerhardt [Math. Z. **165** (1979), no. 2, 193–197; MR 81h:35039].

Now in May 1993, Frehse and Růžička presented a preprint (which later appeared [J. Frehse and M. Růžička, Ann. Scuola Norm. Sup. Pisa Cl. Sci. (4) **21** (1994), no. 1, 63–95; MR 95e:35158]), where they proved the existence of certain weak solutions, fulfilling additional (weighted) estimates and integrability conditions, respectively. The background was the experience of the first author with regularity questions for elliptic systems ("hole-filling technique"). The crucial result was the existence of a weak solution with a maximum estimate (M) $\frac{1}{2}|u|^2 + p \leq c(\Omega')$ on $\Omega' \subsetneq \Omega$ for $N=5$, which reflects the fact that the above quantity is formally a subsolution of a certain elliptic equation. (By the way, this fact is also an important ingredient in the exclusion of certain self-similar blow-up solutions proposed by Leray [see J. Nečas, M. Růžička and V. Šverák, Acta Math. **176** (1996), no. 2, 283–294; MR 97f:35165]. Furthermore the "nearly optimal" estimate

$$(*) \qquad \sup_{x_0 \in \Omega} \int_\Omega \frac{(u\cdot(x-x_0))^2}{|x-x_0|^N} dx \leq c(f)$$

holds.

At this point, Struwe (in the second paper under review) succeeded in improving $(*)$ to a uniform local smallness condition: $(**)$ For $\varepsilon > 0$ there is $R > 0$ with

$$\sup_{x_0} \int_{B_R(x_0)} \frac{(u\cdot(x-x_0))^2}{|x-x_0|^N} dx \leq \varepsilon$$

(here with $N=5$). This "little (but decisive!) step" in turn was enough for him to continue and to establish the smoothness of this solution, where he restricted himself to the Cauchy and the periodic problem. Thus the first general result for regular existence in five dimensions was proved.

After this Frehse and Růžička (in the first paper under review) continued to higher dimensions. They showed $(**)$ by starting with $(***)$ $\int_\Omega (\frac{1}{2}|u|^2 + p)_+^q dx \leq c$ for all $q < \infty$ if $N=6$, and for $q = 2N/(N-6)$ if $6 < N < 10$, and established Hölder estimates, from which again the existence of a smooth solution follows.

In the meantime, Frehse and Růžička have improved their results in several respects. First they proved the existence of a smooth periodic solution in a cube, if $N \leq 15$. Next, in the Dirichlet case, they proved the existence of a regular solution for $N=5$ [Arch. Rational Mech. Anal. **128** (1994), no. 4, 361–380; MR 95j:35170], and recently they improved $(***)$ to (M) in the case $N=6$ [Ann. Scuola Norm. Sup. Pisa Cl. Sci. (4) **23** (1996), no. 4, 701–719 (1997); MR 98i:35142], thus solving the Dirichlet problem smoothly in six dimensions.

The series of results mentioned above and the techniques used have added considerably to our understanding of the Navier-Stokes equations and represent an important development in mathematical analysis. *Michael Wiegner* (Aachen)

98c:76023 76D05 35Q35 76E15 76E99

Kagei, Yoshiyuki (Fukuoka); **von Wahl, Wolf** (Bayreuth)

The Eckhaus criterion for convection roll solutions of the Oberbeck-Boussinesq equations. (English summary)

Internat. J. Non-Linear Mech. **32** (1997), *no.* 3, 563–620.

The theory of sideband instability was developed by physicists to study the stability of a spatially periodic steady state of a partial differential equation on an unbounded domain with respect to general (spatially non-periodic) perturbations. The term sideband instability relates to the fact that often the most dangerous perturbations are those which are linear combinations of periodic functions with slightly perturbed periods. This effect was first explained in 1965 in W. Eckhaus' pioneering book [*Studies in non-linear stability theory*, Springer, New York, 1965; MR **33** #4490] using physically motivated multiple scaling arguments. In problems $\partial_t u = A_R u + N(u)$ with one unbounded spatial direction and with reflection symmetry, generically a family of periodic states bifurcate from the homogeneous solution if the bifurcation parameter R (e.g. the Rayleigh number) is moved through a critical value R_0. Small nontrivial periodic solutions $u_{R,k}$ with period $2\pi/k$ exist whenever $R > R_0 + c_0(k-k_0)^2 + \mathcal{O}(|k-k_0|^3)$, where k_0 is the critical wave length and $c_0 > 0$. The Eckhaus criterion states that $u_{R,k}$ is unstable if $R < R_0 + 3c_0(k-k_0)^2 + \mathcal{O}(|k-k_0|^3)$ and is stable otherwise.

The first mathematical proofs of the Eckhaus instability were given for the model problem of the Swift-Hohenberg equation in the following works [P. Collet and J.-P. Eckmann, *Instabilities and fronts in extended systems*, Princeton Univ. Press, Princeton, NJ, 1990; MR 92e:58063; A. Mielke, in *Nonlinear dynamics and pattern formation in the natural environment (Noordwijkerhout, 1994)*, 206–222, Longman, Harlow, 1995; MR 97e:76039; M. Kuwamura, SIAM J. Math. Anal. **27** (1996), no. 5, 1311–1335; MR 97h:76044]. The latter two works also include the zigzag instability, which appears in systems with two unbounded directions. The linearized stability problem is equivalent to the spectral theory for the associated elliptic operator $L_{R,k}\colon v \mapsto A_R v + DN(u_{R,k}(x))v$ with periodic coefficients on an unbounded domain $x \in \mathbf{R}^d$. For Schrödinger operators with a periodic potential there exists a well-known theory using the Bloch decomposition: Instead of $L_{R,k}$ one considers the Bloch operators

$$B_{R,k}(\sigma)\colon w \mapsto e^{-i\sigma\cdot x} L_{R,k}[e^{i\sigma\cdot x} w]$$

acting on functions w having the same period as $u_{R,k}$. The major result of the Bloch theory states that

$$\operatorname{spec}_{\mathcal{L}^2}(L_{R,k}) = \operatorname{cl}\Big(\bigcup_{\sigma\in\mathbf{R}^d} \operatorname{spec}(B_{R,k}(\sigma))\Big). \tag{1}$$

Here the spectrum of each $B_{R,k}(\sigma)$ is discrete while the $\mathcal{L}^2$-spectrum of $L_{R,k}$ is purely continuous. This theory was generalized to non-selfadjoint elliptic operators by B. Scarpellini [NoDEA Nonlinear Differential Equations Appl. **1** (1994), no. 3, 281–311; MR 95g:35095] and Mielke [Comm. Math. Phys. **189** (1997), no. 3, 829–853; MR 98i:76037].

The work under review, as well as those of Scarpellini [Analysis **15** (1995), no. 4, 359–391; MR 97i:35141] and Mielke [J. Nonlinear Sci. **7** (1997), no. 1, 57–99; MR 99a:76063], considers the associated spectral problem for spatially periodic steady states of the Navier-Stokes equation or the Oberbeck-Boussinesq equation (OBE) which includes the coupling to the temperature field, often also called the Rayleigh-Bénard problem. In contrast to the case of selfadjoint or proper elliptic operators, formula (1) does not hold in general but needs a modification at $\sigma = 0$ [see B. Scarpellini, op. cit., 1995]. This is due to the Stokes projector, which in Fourier coordinates has a singularity at $\sigma = 0$. Most of these complications disappear for the two-dimensional OBE, since only one unbounded direction is present, that is, $\sigma \in \mathbf{R}^1$.

The general idea is to control the dangerous part of the spectrum of the Bloch operators $B_{R,k}(\sigma)$ by using the fact that $B_{R,k}(\sigma) - B_{R_0,k_0}(\sigma)$ is small for $(R,k) \approx (R_0,k_0)$ and that $B_{R_0,k_0}(\sigma)$ can be calculated explicitly. The latter fact depends crucially on the choice of the boundary conditions, namely the stress-free ones. The analysis shows that dangerous eigenvalues occur only for $\sigma \approx 0 \bmod k$ and that there are exactly two of them. The Taylor expansion of these two eigenvalues as functions of σ, R and k can be calculated up to any desired order using a perturbation argument in the flavor of the Lyapunov-Schmidt reduction.

The proof of instability is much easier than the proof of stability since it is sufficient to find one σ such that $B_{R,k}(\sigma)$ has an eigenvalue with positive real part. In this sense it was possible in [A. Mielke, op. cit., MR 99a:76063] to treat the three-dimensional OBE and find additional instabilities. For the stability proof for $u_{R,k}$ one has to show that all the eigenvalues of all $B_{R,k}(\sigma)$ have a stable spectrum. This important point is settled in the present paper for the two-dimensional case, but the three-dimensional case remains open.

The authors employ the decomposition of solenoidal vector fields into the poloidal and the toroidal vector field and the mean flow [see B. J. Schmitt and W. von Wahl, in *The Navier-Stokes equations II—theory and numerical methods (Oberwolfach, 1991)*, 291–305, Lecture Notes in Math., 1530, Springer, Berlin, 1992; MR 94j:35150].

Alexander Mielke (Hannover)

98g:76020 76D05 35K55 35Q30 35S50

Danchin, Raphaël (Palaiseau)

Poches de tourbillon visqueuses. (English and French summaries) [Viscous vortex patches]

J. Math. Pures Appl. (9) **76** (1997), *no.* 7, 609–647.

This paper is devoted to the study of viscous approximation of vortex patches in the plane. Let us first recall some basic facts about the Euler equation. Denote by $v = (v_1, v_2)$ a vector field on $\mathbf{R}^2$, depending on time $t \in \mathbf{R}$. It has been known since the work of W. Wolibner [Math. Z. **37** (1933), 698–726; Zbl 008.06901] that the Euler equations for v: $\partial v/\partial t + (v \cdot \nabla)v = -\nabla p$, $\operatorname{div} v = \partial v_1/\partial x_1 + \partial v_2/\partial x_2 \equiv 0$, have a smooth global solution if $v|_{t=0}$ is smooth and divergence free. This result is proven using in an essential way the fact that the vorticity of v, $\omega(t,x) = (\partial v_1/\partial x_2 - \partial v_2/\partial x_1)(t,x)$, satisfies the transport equation $(\partial\omega/\partial t + v \cdot \nabla)\omega = 0$. In 1963, V. I. Yudovich [Ž. Vyčisl. Mat. i Mat. Fiz. **3** (1963), 1032–1066; MR **28** #1415] studied global existence for a class of nonsmooth Cauchy data, namely those divergence-free vector fields $v_0(x)$ whose vorticity ω_0 is a compactly supported L^∞ function. He was able to get global existence and uniqueness of the solution $v(t,x)$, in spite of the fact that $x \mapsto v(t,x)$ is not Lipschitz continuous for such data. His result implies that if ω_0 is the characteristic function of a bounded domain Ω_0, the vorticity at time t is the characteristic function of a bounded domain $\Omega(t)$. In 1986, A. J. Majda [Comm. Pure Appl. Math. **39** (1986), no. S, suppl., S187–S220; MR 87j:76041] conjectured that if $\partial\Omega_0$ is smooth, $\partial\Omega(t)$ is also smooth for t small enough, but develops singularities in finite time. J.-Y. Chemin proved that actually $\partial\Omega(t)$ stays smooth for any $t \in \mathbf{R}$ [Ann. Sci. École Norm. Sup. (4) **26** (1993), no. 4, 517–542; MR 94j:35141] (see also the proof of A. L. Bertozzi and P. Constantin [Comm. Math. Phys. **152** (1993), no. 1, 19–28; MR 94b:35221]). In fact, the result of Chemin applies to more general situations, where one assumes that ω_0 is L^∞, and that for a convenient vector field X_0, $X_0 \cdot \nabla\omega_0$ is in the negative Hölder space C^{s-1} ($s \in (0,1)$). One then proves that the vorticity at any time has the same type of regularity.

In this paper, the author studies Navier-Stokes equations $\partial v_\nu/\partial t + v_\nu \cdot \nabla v_\nu = -\nabla p_\nu + \nu\Delta v_\nu$, $\operatorname{div} v_\nu = 0$, in two space dimensions, for Cauchy data v_0 whose vorticity ω_0 is the characteristic function of a smooth bounded domain Ω_0. Of course, due to the diffusion term $\nu\Delta v_\nu$ of the equation, the solution is smooth at positive times, and the aim of the paper is to study the limit v of v_ν when $\nu \to 0+$. The main theorem asserts that if $\Omega_\nu(t)$ [resp. $\Omega(t)$] is the image of Ω_0 under the flow of the vector field v_ν [resp. v], and if $\partial\Omega_0$ is of class $C^{1+\varepsilon}$ ($\varepsilon \in (0,1)$), then $\partial\Omega_\nu(t)$ is of class $C^{1+\varepsilon'}$ for any positive t ($\varepsilon' < \varepsilon$) and converges to $\partial\Omega(t)$ in $C^{1+\varepsilon'}$ when $\nu \to 0$. The author gives also a more general statement, where the assumption $\omega_0 = \mathbf{1}_{\Omega_0}$, with $\partial\Omega_0$ smooth, is replaced by the conditions $\omega_0 \in L^\infty$, $(X_0 \cdot \nabla)\omega_0 \in B_a^{s-1}$, for any X_0 describing a convenient finite family of vector fields (B_a^{s-1} being a Besov space, $s \in (2/a, 1)$, $a \in (2, +\infty)$).

Most of the interest of the paper under review comes from the proofs of the results, which rely on a new, remarkable estimate for diffusion equations. To give the flavor of the arguments, let us first recall the case $\nu = 0$: one is given an initial vorticity $\omega_0 \in L^\infty_{\text{comp}}$ such that $(X_0 \cdot \nabla)\omega_0 \in C^{s-1}$ ($s \in (0,1)$) for a given vector field X_0 with vanishing (or at least smooth) divergence. The solution v of the Euler equation with Cauchy data corresponding to ω_0 has a vorticity $\omega(t,\cdot)$ satisfying $\partial_t\omega + v \cdot \nabla\omega = 0$. If $X(t)$ is the vector field transported from X_0 by the flow of v at time t, $[\partial_t + v(t,\cdot) \cdot \nabla, X(t)] = 0$, whence one has a transport equation $\partial_t(X(t) \cdot \nabla\omega) + v \cdot \nabla(X(t) \cdot \nabla\omega) = 0$. One deduces from that an inequality for $\|X(t) \cdot \nabla\omega(t,\cdot)\|_{C^{s-1}}$ in terms of a time integral of the exponential of $\|\nabla v(t,\cdot)\|_{L^\infty}$. Moreover, the main inequality of the work of Chemin cited above gives a control of $\|\nabla v(t,\cdot)\|_{L^\infty}$ in terms of the logarithm of $\|X(t) \cdot \nabla\omega(t,\cdot)\|_{C^{s-1}}$ (and of other well-estimated quantities). Both inequalities together give a control of $\|X(t) \cdot \nabla\omega(t,\cdot)\|_{C^{s-1}}$ by the Cauchy data. This implies that the type of smoothness given by the condition $X_0 \cdot \nabla\omega_0 \in C^{s-1}$ is propagated for any positive time.

When one considers a solution to the Navier-Stokes equation, the vorticity $\omega_\nu(t,x)$ of the solution satisfies $\partial_t\omega_\nu + v_\nu \cdot \nabla\omega_\nu = \nu\Delta\omega_\nu$. If the vector field $X_\nu(t)$ is obtained from X_0 through transport by the flow of v_ν, $X_\nu \cdot \nabla\omega_\nu(t,\cdot)$ satisfies a diffusion equation $\partial_t(X_\nu(t) \cdot \nabla\omega_\nu) + v_\nu \cdot \nabla(X_\nu(t) \cdot \nabla\omega_\nu) - \nu\Delta(X_\nu(t) \cdot \nabla\omega_\nu) = \nu[\Delta, X_\nu \cdot \nabla]\omega_\nu$. One would like to estimate this quantity as in the case $\nu = 0$. The trouble comes from the right-hand side of the preceding equation, which will be, in any case, two orders less regular than ω_ν. The idea of the author is to use the parabolic nature of the equation to compensate for this loss of smoothness, uniformly in ν. As a model problem, remark that if u satisfies $\partial_t u - \nu\Delta u = \nu g$, $u|_{t=0} = 0$ and $\operatorname{Supp}\widehat{u}(t,\xi) \subset \{\lambda \le |\xi| \le 2\lambda\}$ one has $\|u(t,\cdot)\|_{L^2} \le C\lambda^{-2}\sup_s \|g(s,\cdot)\|_{L^2}$, i.e. one has a gain of two derivatives uniformly in ν. The main part of the proof in this paper is to get a similar estimate for $u = X_\nu(t) \cdot \nabla\omega_\nu$ when $g = [\Delta, X_\nu(t) \cdot \nabla]\omega_\nu$, not only for L^2-norms but also for L^a ones ($a > 2$). Two difficulties occur: the first one is that a quantity like $[\Delta, X(t) \cdot \nabla]\omega$ has no meaning at all, because of the low smoothness of X and ω. This is easily resolved through the use of Bony's paradifferential calculus. A more serious problem is the following: to estimate in L^{2p} the solution of $\partial_t v - \nu\Delta v = \nu g$, the author writes the L^2 inequality satisfied by v^p considered as a solution to an equation with left-hand side $\partial_t v^p - \nu\Delta v^p$. The difficulty is that for $p > 1$, if $\widehat{v}$ is supported in an annulus $\lambda \leq |\xi| \leq 2\lambda$, $\widehat{v^p}$ does not vanish close to 0, which prevents one from getting any gain on the regularity as above. The very clever idea of the author is to use a refined microlocal dyadic decomposition of v, designed in such a way that pth powers of each of its terms have Fourier transforms supported far away from 0. This method allows one to control Besov norms of $X_\nu(t) \cdot \nabla\omega_\nu(t, \cdot)$, which is the key point in the conclusion of the proof. *Jean-Marc Delort* (Villetaneuse)

99j:76033 76D10 35Q35

Олейник, О. А. [Oleĭnik, O. A.]; Самохин, В. Н. [Samokhin, V. N.]

★Математические методы в теории пограничного слоя. (Russian. Russian summary) [Mathematical methods in boundary-layer theory]

Fizmatlit "Nauka", Moscow, 1997. 512 pp. ISBN 5-02-015202-1

This book is a vast treasury of rigorous mathematical results about the Prandtl systems of PDE's in fluid dynamics. Oleĭnik and her many collaborators are responsible for developing most of this theory over the last 4 decades.

The Prandtl equations were devised early in this century as a simpler replacement for the Navier-Stokes equations to describe viscous laminar fluid flows near boundaries to which the fluid adheres. These replacement equations have a quite different mathematical character than the Navier-Stokes equations, and as one can easily see from this book, the theory goes along very different lines.

The Prandtl equations come in several versions, and the following are considered here: stationary and unstationary problems, 2D flows and flows around axisymmetric bodies. In addition, versions of these equations exist and are treated here for non-Newtonian fluids and for magnetohydrodynamic problems.

The two most commonly occurring mathematical (and applied) problems are: (1) The problem of the continuation downstream of a given velocity profile for the stationary equations. The velocity "outside" the boundary layer is assumed given, as a function of position on the surface of the body. One cannot expect the continuation to be always possible arbitrarily far downstream, because of the well-known boundary layer separation effect. That fact is reflected in the mathematical theory given here. (2) The initial-value problem for the nonstationary equations, with data given at time 0 and also for all time at a given point on the body over which the fluid flows.

These basic problems and many variations of them as well are given rigorous theories. An example of a variation is the problem in magnetohydrodynamics when the magnetic field oscillates "rapidly" (though the oscillation is in space). Then one is able (as is proved) to approximate the solution by the solution of a suitable averaged equation.

Other "spin-off" problems include (1) the initial development of boundary layers in cases when the fluid, initially at rest, begins moving, either gradually or impulsively;

(2) solutions periodic in time (when the given data are); (3) two fluids, each satisfying the stationary Prandtl equations, separated by an unknown surface; (4) the analog of the Prandtl system for non-Newtonian fluids, the viscosity depending on velocity gradients.

Generally, more than one method is presented for the solution of these problems. For example, existence theorems are presented using the method of lines, finite difference methods, and successive approximations. Moreover, the problem is generally reduced by means of the nonlinear transformations of von Mises or Crocco to a simpler form. Both options are considered in this book.

The book is exceedingly detailed, since full proofs are generally given.

There are also rigorous results, presented in much less detail in this book, relating to the connection between solutions of the Prandtl system and those of the Navier-Stokes equations in situations when the former are thought to approximate the latter. These results are stated as theorems in the introductory chapter. The same is true of the asymptotic approximation of solutions by self-similar solutions.

The book concludes with a list of 15 suggested future research directions (some rather generally defined) and unsolved problems. *P. C. Fife* (Salt Lake City, UT)

97c:76028 76E30 35B35 35Q35 58F39
Bridges, Thomas J. (Guildford); **Mielke, Alexander** (Hannover)
A proof of the Benjamin-Feir instability. (English summary)
Arch. Rational Mech. Anal. **133** (1995), *no.* 2, 145–198.

This paper is an important contribution to the study of nonlinear waves, treating a basic instability mechanism that comes up in fluid mechanics (surface waves, Rayleigh-Bénard convection, Wilton ripples, etc.), nonlinear optics (resonant wave mixing) and many other areas. The paper is specifically concerned with a fundamental mechanism for instability that goes by the names of the Benjamin-Feir instability, the modulational instability or sideband instabilities.

The discovery of this type of instability goes back to the work of T. B. Benjamin [Proc. Roy. Soc. London Ser. A **299** (1967), 59–75; per bibl.] and Benjamin and J. E. Feir [J. Fluid Mech. **27** (1967), 417–430; Zbl 144.47101], G. B. Whitham [J. Fluid Mech. **27** (1967), 399–412; MR **34** #8711] and V. E. Zakharov [J. Appl. Mech. Tech. Phys. **9** (1968), no. 2, 190–194]. The original context of Benjamin and Feir concerned the stability of the gravity wavetrains that were found by Stokes in 1849. Since this original work, many others have contributed significantly to the problem, such as Saffman, Yuen, etc., and through this work, deep connections with integrable Hamiltonian systems and solitons were found.

A. C. Newell, on page 45 of his book [*Solitons in mathematics and physics*, SIAM, Philadelphia, PA, 1985; MR 87h:35314], explained what the instability mechanism is quite succinctly: "The Benjamin-Feir (or, as it is sometimes called, the modulational) instability is widespread and plays an important role in various nonlinear wave phenomena. Simply put, if dispersion and nonlinearity act against each other, monochromatic wave trains do not wish to remain monochromatic. The sidebands of the carrier wave can draw on its energy via a resonance mechanism, with the result that the envelope becomes modulated. In one space dimension, this envelope modulation continues to grow until the soliton shape is reached. At this point, nonlinearity and dispersion are in exact balance and no further distortion occurs."

This reference as well as the book of Whitham [*Linear and nonlinear waves*, Wiley-Intersci., New York, 1974; MR **58** #3905] contains many more references, insights, and interesting comments on the literature. In particular, the basic mechanism that starts the instability is a linear instability and, as the above quotation suggests, there is much interest in what happens after the instability sets in, due to nonlinear effects.

The development of the instability involves very important resonant wave phenomena and it is one of many reasons why the three- and four-wave resonant interaction models, common in fluid mechanics, optics and plasma physics, etc., are interesting. These resonant wave interactions come up in some amazing and unexpected ways, including the macroscopic quasicrystal patterns seen in the forced surface waves studied by Gollub and his collaborators [see J. P. Gollub, Proc. Nat. Acad. Sci. U.S.A. **92** (1995), no. 15, 6705–6711; and references therein]. The resonant wave equations themselves are particularly fascinating completely integrable finite-dimensional Hamiltonian systems. (Their canonical and Lie-Poisson structure relative to $\mathrm{su}(n)^*$ is studied in a preprint of the reviewer, Alber and Luther ["On the Hamiltonian structure of the 3-wave interaction", Preprint; per revr.].)

The paper under review studies the basic linear instability in a rigorous fashion. Many of the original arguments, while perfectly correct, can be hard to justify in a proper nonlinear PDE–functional analysis setting. One of the several difficulties, but a crucial one, that must be overcome is that the Stokes wavetrain is spatially periodic and yet the instability is to perturbations that could be quasiperiodic; thus, for example, Banach spaces of periodic functions on the line would not capture the instability, and because of this and the spatial noncompactness, setting up a proper dynamical function space is not so straightforward.

This paper provides a setting in which these problems can be posed and the desired results proved. This is done in a way that maintains the computational efficacy of the original methods. A basic mechanism is to reformulate the problem using a Hamiltonian formulation of both the spatial and the temporal evolution structure of the waves. The key technical tool used is a Hamiltonian center manifold theory developed by Mielke [*Hamiltonian and Lagrangian flows on center manifolds*, Lecture Notes in Math., 1489, Springer, Berlin, 1991; MR 93f:58080] combined with Hamiltonian formulations of the water wave problem developed by Bridges [see, e.g., J. Nonlinear Sci. **4** (1994), no. 3, 221–251; MR 95g:58077]. These techniques allow one to reduce the problem to a finite-dimensional one where the final calculations are carried out. The techniques do leave some important issues open, such as the complete analysis of the well-posedness of the initial value problem and the full structure of the spectrum of the linearized problem. This is presumably much more difficult and will involve small divisor problems. However, linear instability is one thing that the authors can rigorously deduce.

In a related preprint [J. Nonlinear Sci. **7** (1997), no. 1, 57–99; MR 99a:76063], Mielke uses another method for rigorously detecting these instabilities based on the direct use of the Lyapunov-Schmidt procedure and applies it to a number of additional fluid instability problems, including the instabilities analyzed by Busse and Bernoff in Rayleigh-Bénard convection. One should note that these results provide a rigorous and computationally efficient scheme for the linear instability problem. As was indicated, the classical literature contains many other interesting problems too that could also benefit from an airtight mathematical framework. One such problem would be to obtain comparative estimates between the unstable solution and a solution of a corresponding resonant three- or four-wave equation.

J. E. Marsden (Pasadena, CA)

98g:76037a 76E99 58F40 76M99
Benjamin, T. Brooke; Bridges, Thomas J. (Guildford)
Reappraisal of the Kelvin-Helmholtz problem. I. Hamiltonian structure. (English summary)
J. Fluid Mech. **333** (1997), 301–325.

98g:76037b 76E99 58F40 76M99
Benjamin, T. Brooke; Bridges, Thomas J. (Guildford)
Reappraisal of the Kelvin-Helmholtz problem. II. Interaction of the Kelvin-Helmholtz, superharmonic and Benjamin-Feir instabilities. (English summary)
J. Fluid Mech. **333** (1997), 327–373.

The intrinsic appeal of fluid dynamics to generations of applied mathematicians and physicists is arguably due to the fact that while most fundamental hydrodynamic phenomena can be observed by the naked eye both in nature and in experiment, their adequate modeling requires the application of the most sophisticated and advanced mathematical techniques.

The instability of a plane vortex sheet, also known as the "wind over water" instability, is undeniably such a phenomenon. It is named after Kelvin and Helmholtz (KH), who developed the linearized theory describing the early evolution of the sheet. Accounts of their theory are included in most texts on hydrodynamics; of these, H. Lamb [*Hydrodynamics*, sixth edition, Cambridge Univ. Press, Cambridge, 1932; reprint, 1993; MR 96f:76001] and P. G. Drazin and W. H. Reid [*Hydrodynamic stability*, Cambridge Univ. Press, Cambridge, 1981; MR 82h:76021] have the most lucid expositions.

In the KH model the vorticity is localized at the initially horizontal interface between two ideal incompressible irrotational fluids with different densities ρ and ρ' and different horizontal velocities U and U'. The density in the lower (unprimed) layer is higher than the density of the upper (primed) layer: $\rho > \rho'$. Thus gravity (which is directed downward) has a stabilizing impact. In order to suppress spurious short-wavelength instabilities a nonzero surface tension on the interface between two fluids is introduced. The disturbed interface is characterized by the function $\eta(x,y,t)$ while the velocities below and above the interface are characterized by velocity potentials $\varphi(x,y,z,t)$ for $z \le \eta(x,y,z)$, and $\varphi'(x,y,z,t)$ for $z \ge \eta(x,y,z)$, respectively. The objective is to describe the dynamics of these quantities.

Although the original linearized theory gives an adequate description of the early motions of the fluid interface arising as a result of the development of the KH instability, it cannot be used in order to study many complicated features which the motion exhibits in later stages of evolution, when the vortex sheet rolls up into billows. An adequate analysis of the fully developed KH instability can only be achieved in the framework of the nonlinear theory. Various techniques needed for such an analysis were proposed in the past, notably in a seminal paper by P. G. Drazin [J. Fluid Mech. **42** (1970), 321–335; Zbl 213.54403]. In the present series of papers Benjamin (who died before its publication) and Bridges make several very important contributions to the nonlinear theory of the KH instability. Their approach is based on the novel Hamiltonian formulation of the KH problem.

In the first paper, the authors show how the pair of spatial velocity potentials φ, φ' can be replaced by a single surface function $\zeta(x, y, t)$ and express the perturbed total energy E in terms of η and ζ. They show that E is the Hamiltonian governing the dynamics of η and ζ. The authors also extend their analysis to the case when the vortex sheet becomes folded, so that it can no longer be described by a single-valued function η, and propose a non-canonical Hamiltonian formulation of the problem. Finally, they revisit the linearized problem from the Hamiltonian viewpoint. The formalism developed in the first paper can be considered as a natural generalization of the Hamiltonian formalism developed in the works of V. E. Zakharov in the late 1970s and others for the water wave problem, which is a degenerate case of the KH problem.

In the second and more technically demanding paper the authors apply their theory to the study of nonlinear periodic traveling waves, and present several new results on their instabilities and bifurcations. In particular, they find a pervasive superharmonic instability when the density ratio $r = (\rho - \rho')/(\rho + \rho')$ exceeds a critical value r_c and the velocity difference $u = U - U'$ approaches its KH threshold u_c. The authors also provide an interesting discussion of the interaction among the superharmonic, KH and Benjamin-Feir instabilities.

The papers are written in a clear and inviting style and repay the careful reader with a lot of unexpected information on the nonlinear KH problem. They are a fitting monument to the giant of the British school of fluid dynamics.

Alexander Lifschitz (New York, NY)

98d:76079 76F20 58F40 76D05 76M25

Holmes, Philip (Princeton, NJ); **Lumley, John L.** (Ithaca, NY);
Berkooz, Gal (Ithaca, NY)

★**Turbulence, coherent structures, dynamical systems and symmetry. (English summary)**

Cambridge Monographs on Mechanics.
Cambridge University Press, Cambridge, 1996. xviii+420 pp. \$69.95.
ISBN 0-521-55142-0

Turbulence in fluids is one of the great challenges to physics, mathematics, and modern engineering science. The revolution in classical mechanics during the twentieth century—most notably the last generation's developments in dynamical systems—has generated new hope that the complex motions characterizing turbulence may be understood rationally and productively within a consistent mathematical framework. Soon after the groundbreaking computational, analytical and experimental developments of the 1960s through the 1980s, attention focused on the most fundamental question: Is turbulence chaos? This book goes a long way toward answering that question by providing a thorough presentation of an effort to develop a dynamical systems approach to modeling turbulence based on empirically determined "coherent structures".

In principle, the dynamics of turbulent fluid motions is contained in the Navier-Stokes equations, the nonlinear partial differential equations constituting Newton's laws of motion for incompressible continuum systems with a linear relation between the local stress and rate-of-strain tensors. The universality of this mathematical description for a tremendous range of physical systems makes these equations of paramount importance in the applied and fundamental sciences. The phenomenon of turbulence, however, simultaneously involves the full complications of both the nonlinearity and the continuum nature of the equations: the many degrees of freedom inherent in the continuum interact strongly and effectively inseparably via the equations' nonlinearities. Hence direct numerical simulation of intense turbulence necessarily requires prohibitively large computational efforts in terms of memory size and raw computing speed, and exact analytical results are essentially nonexistent. In order to reduce the mathematical complexity of the most fundamental description, traditional turbulence modeling efforts have been directed toward statistical models and so-called closure approximations to produce more tractable lower equations to solve for average quantities.

This book is devoted to an alternative approach to turbulence modeling, that of developing (relatively) low-dimensional ordinary differential equation models for the evolution of the amplitudes of coherent structures embedded in the turbulence. Many turbulent flows display identifiable and recurrent structures in the flow field, and the first step is to objectively identify and rationally order a basis for the system's phase space composed of these structures. This is accomplished here via the "proper orthogonal decomposition", also widely referred to as the Karhunen-Loève method, of extracting an orthogonal set of basis functions from an ensemble of functions complemented with an appropriate probability distribution. In practice one extracts the eigenfunctions of a time-averaged correlation function measured in an experiment or via numerical simulation, with the key observation for the problem at hand being that they are ordered in the spectrum according to the average kinetic energy in the modes. Then the projection of the Navier-Stokes equations onto these basis functions, followed by a truncation of the resulting infinite set of coupled ordinary differential equations, produces a reduced model of the dynamics of the dominant energy-containing modes. As fully explained in the book, complete implementation of this procedure also involves many more modeling decisions and techniques, but in the end it has proven successful in helping to identify key dynamical structures and mechanisms in a number of turbulent flows. Symmetry is a central tool used to identify robust features of the models, and the attention paid to symmetries inherited by the truncated model is particularly productive in this respect: symmetry arguments are essential to establishing the convincing connection between turbulence and chaotic dynamics in the reduced models.

This book is an up-to-date research monograph combined with a large amount of background material in turbulence theory and phenomenology, Karhunen-Loève methods, and both deterministic and stochastic dynamical systems. It will be accessible and of interest to a wide audience, ranging from engineering and applied mathematics graduate students to active researchers in nearly all of the applied sciences impacted by turbulence in fluids (e.g., astrophysics, geophysics, atomospheric and ocean sciences, the many engineering sciences). The authors make an impressive effort to introduce nearly all the necessary background material, but this approach draws on so many areas that it will almost certainly be necessary for dedicated readers to consult at least some of the references on the particular subfields. The authors of this book have played key roles in the development of these methods during the last decade—indeed, they have been at the forefront of the effort—and the first two authors previously wrote definitive texts on, respectively, dynamical systems and turbulence theory. Hence this book benefits not only from the authors' authoritative insights and intuition regarding the various physical theories and mathematical techniques employed in this project, but also from the high level of clarity of their presentation. *Charles R. Doering* (Ann Arbor, MI)

99k:76082 76F99 35Q99 60K40 76M35 82C99
Turkington, Bruce (Amherst, MA)
Statistical equilibrium measures and coherent states in two-dimensional turbulence. (English summary)
Comm. Pure Appl. Math. **52** (1999), *no.* 7, 781–809.

The statistical theory of two-dimensional incompressible turbulence (2DT) has attracted a lot of attention in recent years for several reasons: two-dimensional turbulence is of interest for its own sake because of possible applications to geophysical flows; its theory may be of use in setting up a theory for (the very different) three-dimensional turbulence, and it is also an interesting paradigm of statistical theories for general partial differential equations. L. Onsager [Nuovo Cimento (9) **6** (1949), Supplemento, no. 2 (Convegno Internazionale di Meccanica Statistica), 279–287; MR **12**, 60f] observed that the statistical theory of 2DT is dominated by coherent vortices and gave a qualitative explanation. The effort to provide a quantitative theory started with the work of D. Montgomery and G. Joyce [Phys. Fluids **17** (1974), 1139–1145; MR **52** #4912] in the 70's and has produced a large and interesting literature.

A salient property of the equations of motion of fluid mechanics in two dimensions in the inviscid limit is the existence of an infinite sequence of vorticity invariants. A key problem in the theory of 2DT is to find maximum entropy states compatible with the equations of motion and thus with the invariants; these states are the coherent structures of 2DT. Because of mixing, not every invariant of the equations of motion becomes an effective constraint on the maximum entropy state. Furthermore, to be compatible with turbulent mixing, the resulting theory should be free of ultraviolet cut-offs. One then has to wonder which invariants should be taken into account, and how to make sure that there is no effective cut-off. The earliest theories generally used two invariants and did have an effective cut-off; there exist a number of outstanding papers with rigorous results in various limiting situations; a beautiful construction due to R. Robert and J. Sommeria [J. Fluid Mech. **229** (1991), 291–310; MR 93d:76034] and J. Miller [Phys. Rev. Lett. **65** (1990), no. 17, 2137–2140; MR 92a:82126] even takes *all* invariants into account but still, as it turns out, contains an effective cut-off. The present remarkable paper elegantly cuts the Gordian knot: It weakens all the constraints on the maximum entropy states produced by the invariants into inequality constraints, and then imposes as effective constraints only those inequalities which reach their maxima. The resulting theory has no cut-off and is dynamically self-consistent, at least to the extent that can be checked at present (issues of ergodicity remain largely open); it is derived by taking a limit of a sequence of lattice models defined by Gibbs measures. In addition, the author suggests, on the basis of earlier work, a priori ways of finding vorticity invariants that are likely to be effective.

In the present paper the theory is outlined and applied to some commonly studied problems, in particular the patch problem (in which the microscopic vorticity can take only one of two values). The results make a strong case for the new theory.

A mathematically more complete presentation will be found in a later paper. The practical uses of the theory are unclear—simpler theories give satisfactory results in many applications—but the present paper provides an intellectually satisfying resolution of an old conundrum, and may lead to more intelligent thinking about the mysteries of turbulence in other contexts as well.

Alexandre J. Chorin (Berkeley, CA)

97b:76084 76M10 35L65 65M15 65M60 76N10

Johnson, Claes (Göteborg); **Szepessy, Anders** (Stockholm)

Adaptive finite element methods for conservation laws based on a posteriori error estimates. (English summary)

Comm. Pure Appl. Math. **48** (1995), *no.* 3, 199–234.

One of the fundamental open problems of computational fluid dynamics concerns the construction of an algorithm for the numerical solution of the Navier-Stokes equations such that the error between the exact and the computed solution, measured in a given norm, is below a prescribed tolerance. Solutions to the Navier-Stokes equations exhibit a rich collection of small-scale features and localised structures, including boundary layers and shock waves, whose accurate resolution necessitates the use of adaptively refined computational meshes.

This article is concerned with the derivation of a posteriori error bounds for Galerkin finite element approximations of scalar nonlinear conservation laws, and systems of conservation laws in one space dimension, with the aim of constructing reliable and efficient adaptive finite element algorithms for the numerical solution of these equations, and of providing an appropriate framework for the development of a similar analysis for the compressible Navier-Stokes equations in several space dimensions. The term "a posteriori" signifies the fact that, upon performing a computation, the computed solution U is used to quantify the size of the error between itself and the unknown analytical solution u. This is achieved by inserting U into the partial differential equation and monitoring the size of the resulting residual $R(U)$, which measures the extent to which U fails to satisfy the differential equation; an a posteriori bound provides an estimate of the error $u-U$ in terms of this (computable) residual. While for finite element approximations to partial differential equations of elliptic type there has been a considerable amount of research on a posteriori error estimation, relatively little is known about the validity of analogous results for hyperbolic equations and parabolic problems with dominant hyperbolic behaviour [B. J. Lucier, Math. Comp. **46** (1986), no. 173, 59–69; MR 87m:65141; H. Nessyahu and T. Tassa, SIAM J. Numer. Anal. **31** (1994), no. 3, 628–654; MR 95g:65125]. The work under review represents a significant attempt to bridge this gap.

The first part of the paper is concerned with the derivation of an a posteriori error bound for scalar viscous conservation laws of the form $u_t + f(u)_x - \varepsilon u_{xx} = 0$, for $x \in \mathbf{R}$ and $t > 0$, subject to the initial condition $u(x,0) = u_0(x)$ for $x \in \mathbf{R}$ and the decay condition $u(x,t) \to 0$ as $|x| \to 0$, for $t > 0$; more specifically, the analysis applies to the Burgers equation corresponding to $f(u) = u^2/2$. The authors replace the conservation law by an analogous problem, with viscosity coefficient $\widehat{\varepsilon}$ instead of ε, and denote its solution by $\widehat{u}$; in practice $\widehat{\varepsilon} = \max(\varepsilon, O(h^\alpha))$, where h is the mesh size in the discretisation and α is a suitably chosen positive constant depending on the approximation properties of the method (with piecewise linear basis functions, typically, $\alpha = 1$ or $\alpha = \frac{3}{2}$). It is this modified partial differential equation that is approximated by the finite element method, with U signifying the corresponding numerical solution. The total error $u - U$ is split up as $(u - \widehat{u}) + (\widehat{u} - U)$, where $u - \widehat{u}$ can be thought of as a perturbation error caused by changing the viscosity coefficient in the differential equation from ε to $\widehat{\varepsilon}$. With $\widehat{e} = \widehat{u} - U$, the main result in this part of the paper is the a posteriori error bound on $\widehat{e}$ stated in Theorem 3.6, which is of the following generic form:

$$|||\widehat{u} - U|||_Q \leq C|||h^2 R(U)|||_Q;$$

here $|||\cdot|||_Q$ is a weighted space-time L_2 norm over $(-\infty, \infty) \times (0, T)$, with T the final time. The selection of the weight function ψ involved in the definition of the norm is driven by the admissibility criterion formulated in Lemma 3.3 and is related to the strong stability of the linearised dual problem; it is shown that suitable choices of the weight function can be made according to the local features that are present in the analytical solution (e.g., $\psi \equiv 1$ in the case of a shock wave, $\psi = (t/T)^{1+\gamma}$ with $0 \leq \gamma \leq 1$ in the case of a rarefaction wave, and $\psi = \exp(3\beta(t - T))$, with β denoting the supremum of $|\frac{1}{2}(\widehat{u} + U)_x|$, in the case of a smooth solution).

The second part of the paper deals with systems of conservation laws in one space dimension; it is assumed that the reduced problem (corresponding to $\varepsilon = 0$) is strictly hyperbolic and genuinely nonlinear. Supposing that the computational domain $(-\infty, \infty) \times (0, T)$ can be split into a shock region, a simple wave region and a regular region, and that the shocks and simple waves are weak, the authors study the stability of the linearised dual problem in a weighted L_2 norm. Again, the weight depends crucially on the structure of the solution, but this time it is a matrix-valued function. Once the strong stability of the linearised dual problem has been established, in Theorem 4.4, the a posteriori error bound is arrived at as in the scalar case.

This article presents an important contribution to the a posteriori error analysis of finite element approximations of nonlinear conservation laws; it is written in a self-contained and clear style which allows the reader to follow the exposition with ease, and will, undoubtedly, generate further research into this exciting and significant subject area. *Endre E. Süli* (Oxford)

98d:76140 76N10 35L65 35Q35
Sideris, Thomas C. (Santa Barbara, CA)
Delayed singularity formation in 2D compressible flow. (English summary)
Amer. J. Math. **119** (1997), *no.* 2, 371–422.

The compressible Euler equations in two space dimensions (for isentropic gas) form a 3×3 strictly hyperbolic symmetric system of conservation laws in the unknown $\rho > 0$ (the density) and u (the speed of the fluid) (see A. J. Majda [*Compressible fluid flow and systems of conservation laws in several space variables*, Springer, New York, 1984; MR 85e:35077] for an introduction to the multidimensional aspects of conservation laws and historical references). The biggest and the smallest eigenvalues are genuinely nonlinear, while the middle one is linearly degenerate. If the initial data are irrotational (that is, the vorticity of the initial speed field is zero), the system reduces to a scalar second order quasilinear wave equation, which is also a (second order) conservation law. On the other hand, for constant initial density and divergence-free initial velocity fields, the system reduces to the *in*compressible Euler equation. Hence the full system can be viewed as a coupling between these two special forms. In this paper, T. Sideris considers smooth Cauchy data of small size ε, and discusses the life span of the corresponding smooth solutions, extending previous work by the reviewer [Invent. Math. **111** (1993), no. 3, 627–670; MR 94a:35089]. For the incompressible Euler equation in 2D, existence of global classical solutions is known; however, only a poor control of the derivatives of these solutions for large time is available (for instance, the spatial gradient of the velocity may grow as a double exponential in time). For a rotationally symmetric initial velocity, the solution does not depend on time, in which case the above-mentioned difficulty disappears. For the quasilinear wave equations, the method of Klainerman of combining energy inequalities and commutations of Lorentz vector fields [see, e.g., L. Hörmander, *Lectures on nonlinear hyperbolic differential equations*, Springer, Berlin, 1997; MR 98e:35103] yields good lower bounds for the life span, which turns out to have the magnitude of C/ε^2.

The author proves first the lower bound C/ε of the life span (for *any* C) under very general conditions, a result which can be rescaled to a statement about the "incompressible limit" problem [see also A. J. Majda, op. cit.]. He then verifies the magnitude C/ε^2 in the irrotational case, thus preparing for the principal result of the paper, which concerns rotationally symmetric initial data, small in an appropriate weighted Sobolev space; the author obtains the life span $(\ln(1+\varepsilon^{-2}))^{-1}C/\varepsilon^2$ (C/ε^2 is the correct one for compactly supported initial data). It is hard to guess whether C/ε^2 is also the correct life span for general data such as those considered by Sideris.

The method of proof uses two main ingredients: (i) a rough approximate solution constructed by just superposition of a (global) incompressible flow (solution of the incompressible Euler equations for the divergence-free part of the data) and a gradient flow (solution of a quasilinear wave equation for the irrotational part of the data), (ii) new decay estimates in the spirit of S. Klainerman [Comm. Pure Appl. Math. **38** (1985), no. 3, 321–332; MR 86i:35091].

Better approximations than the one used here are available in the reviewer's paper [op. cit.], but they require much more work; it is not clear, beside the interest that these approximations may have as an explicit description of the solution, whether they are needed to obtain the correct life span. The remarkable fact is that the rough approximation (i) is sufficient to obtain essentially the good life span: this shows that the middle mode is better decoupled from the two other modes than it seems at first sight. The crucial point in obtaining decay estimates for the solutions of compressible Euler equations is that one cannot use the full set of the fields $\partial_j, x_i\partial_j - x_j\partial_i, x\nabla_x + t\partial_t, t\partial_j + x_j\partial_t$ as in the case of wave equations; the last ones (the "hyperbolic rotations") are forbidden. The whole point is to adapt the usual Klainerman inequalities (using the full set of fields) to the case where some fields are missing (the situation is analogous to the situation of nonlinear elasticity, with the difference that the part of the solution governed by the incompressible Euler equations has no decay at all). Such inequalities are obtained by the techniques developed in Klainerman and Sideris [Comm. Pure Appl. Math. **49** (1996), no. 3, 307–321; MR 96m:35231]. This is the heart of the proof, which yields a control of the correction between the solution and the approximation by using an induction on time. Without any symmetry assumption on the data, the problem of obtaining relevant lower bounds and upper bounds of the life span would certainly require a deeper understanding of the coupling between the middle mode of the system and the others, and is completely open. *Serge Alinhac* (Orsay)

98j:76127 76P05 35Q35 45K05 82C40
Liu, Jian-Guo (Philadelphia, PA); **Xin, Zhouping** (New York, NY)
Boundary-layer behavior in the fluid-dynamic limit for a nonlinear model Boltzmann equation. (English summary)
Arch. Rational Mech. Anal. **135** (1996), *no.* 1, 61–105.

The Boltzmann equation for the one-particle distribution function $f(\mathbf{x},\xi,t)$ of a gas of molecules is given by

$$\frac{\partial f^\varepsilon}{\partial t}+\xi\cdot\nabla f^\varepsilon=\frac{Q(f^\varepsilon,f^\varepsilon)}{\varepsilon} \tag{1}$$

where Q is the collision operator, ξ is the particle velocity, $\mathbf{x}\in\mathbf{R}^3$, $t>0$, and $\varepsilon>0$ is the Knudsen number which captures the space-time scaling. For $\varepsilon\sim O(1)$ we are in the Boltzmann regime where the mesoscopic Boltzmann description is correct; if $0<\varepsilon\ll 1$, we are in the macroscopic fluid regime where continuum theory should be appropriate.

Continuum theory is derivable from the Boltzmann equation by a straightforward though formal procedure. Simply multiply the Boltzmann equation by 1, ξ, $|\xi|^2$ and integrate over $\xi\in\mathbf{R}^3$ to obtain the balance laws of mass, momentum, and energy:

$$\frac{\partial f}{\partial t}+\frac{\partial}{\partial x_i}(\rho u_i)=0, \tag{2}$$

$$\frac{\partial}{\partial t}(\rho u_j)+\frac{\partial}{\partial x_i}(\rho u_i u_j+T_{ij})=0, \tag{3}$$

$$\frac{\partial}{\partial t}\left[\rho\left(\frac{1}{2}|\mathbf{u}|^2+e\right)+\frac{\partial}{\partial x_i}\left[\rho u_i\left(\frac{1}{2}|\mathbf{u}|^2+e\right)+T_{ij}u_j+q_i\right]=0.\right. \tag{4}$$

$$\rho=\int f\,d\xi,\quad \rho\mathbf{u}=\int \xi f\,d\xi,\quad \rho e=\frac{1}{2}\int|\xi-\mathbf{u}|^2 f\,d\xi, \tag{5}$$

$$\mathbf{T}=\int(\xi_i-u_i)(\xi_j-u_j)f\,d\xi,\quad \mathbf{q}=\frac{1}{2}\int(\xi_i-u_i)|\xi-\mathbf{u}|^2 f\,d\xi.$$

Equations (2), (3), (4) describe the evolution of the fluid variables ρ (density), $\mathbf{u}$ (velocity), e (internal energy). System (2),(3),(4) is not closed: expressions for the Cauchy stress $\mathbf{T}$ and heat flux $\mathbf{q}$ depend on the unknown mesoscopic dependent variable f^ε. However, if one formally multiplies (1) by ε and lets $\varepsilon\to 0+$ it follows that f^0 is a Maxwellian distribution dependent only on the macroscopic variables ρ, $\mathbf{u}$, e and hence $\mathbf{T}$ and $\mathbf{q}$ depend only on the macroscopic variables in the $\varepsilon\to 0+$ limit. In fact the expressions for $\mathbf{T}$, $\mathbf{q}$ in this limit reduce equations (2)–(4) to the classical inviscid Euler equations for an ideal gas.

Unfortunately the rigorous validity of this limiting process has never been proven. In fact, should one be able to prove the rigorous validity of the process one will have proven global existence of the equations of gas dynamics in three space dimensions.

Since this is as yet an open problem even in one space dimension, the difficulty of the problem is apparent.

One approach to this limit problem, which is really one of compactness of solutions $\{f^\varepsilon\}$ to the Boltzmann equation, is to alter the point of view. Instead of compactness one may pose the question of approximation: Can the solutions $(\rho, \mathbf{u}, e)$ of the limit equation be approximated by solutions of the Boltzmann equation via the identification (5)? Notice that compactness implies approximation but not vice versa, so the approximation question should hopefully be more accessible to analysis. In fact this is indeed the case. The main idea is to construct via matched asymptotic expansion a solution f^ε to the Boltzmann equation in terms of powers of ε so that its associated fluid variables (via (5)) ρ^ε, $\mathbf{u}^\varepsilon$, e^ε approximate the given solution to the inviscid Euler equation $(\mathbf{u}^0, \rho^0, e^0)$. This approach was pioneered by R. E. Caflisch [Comm. Pure Appl. Math. **32** (1979), no. 4, 521–554; MR 80g:76040; Comm. Pure Appl. Math. **33** (1980), no. 5, 651–666; MR 81j:76072] and more recently applied by Xin [Comm. Pure Appl. Math. **44** (1991), no. 6, 679–713; MR 92f:76078].

In the paper under review Liu and Xin have returned to the approximation issue but now in the presence of a boundary. The authors consider a simple model problem: the Broadwell system, which is a discrete (three) velocity model for the kinetic theory of gases. (This is the model considered by Caflisch in his original 1979 paper cited above.) What makes the Liu-Xin paper interesting is the necessity of dealing with diffusive and reflective boundary conditions. Now there are two approximations that must be made to the smooth limit flow (a smooth solution exists for the limit fluid like equation for some finite time t_0). One approximation like Caflisch's is constructed away from the boundary but now an additional approximation to the Broadwell system must be made in a neighborhood of the boundary. To obtain a uniform approximation to the smooth solution a matching procedure is employed. The procedure given here is employed only for a smooth solution of the limit equation but one may surmise that, based on Xin's success [op. cit.] in constructing solutions of the Broadwell equation which approximate a solution of the limit system possessing a shock, the shock case can be handled as well. *Marshall Slemrod* (Madison, WI)

99a:76143 76S05 35B27 35K55 35Q35 73B27 76T05

Bourgeat, Alain (Saint-Étienne); **Luckhaus, Stephan** (Bonn); **Mikelić, Andro** (Villeurbanne)

Convergence of the homogenization process for a double-porosity model of immiscible two-phase flow. (English summary)

SIAM J. Math. Anal. **27** (1996), *no.* 6, 1520–1543.

This paper deals with the homogenization problem for the system of equations governing two-phase flows in a porous medium. This system is a main tool in mathematical research on oil recovery problems and, therefore, its properties are studied intensively. However, the results pertaining to the existence of its solutions were obtained only quite recently because the system of equations in question does not fit in the ordinary framework. It relates to the class of parabolic equations supplemented by nonlinearity and degeneration. The homogenization of ordinary, i.e. linear and nondegenerate, parabolic operators with spatially oscillating coefficients was investigated by many authors [see, e.g., V. V. Zhikov, S. M. Kozlov and O. A. Oleĭnik, Trudy Moskov. Mat. Obshch. **45** (1982), 182–236; MR 85k:35024].

Usually the homogenized operator has the same form as the original one, and it becomes more simple because its coefficients do not oscillate. A result of this kind had been established by Bourgeat, Kozlov and Mikelić [Calc. Var. Partial Differential Equations **3** (1995), no. 3, 385–406; MR 97e:76089] for the system governing two-phase flows in porous media.

In the paper under review a special situation is considered. Here the homogenized operator happens to be non-local with respect to the time variable. That non-locality describes memory effects. The appearance of memory as a result of the homogenization procedure is the central point of the paper. A similar result for the ordinary heat-conduction equation was obtained by E. Khruslov ["Averaged model of non-stationary diffusion in a fractured-cellular medium", Preprint No. 50, Kharkov Inst. Low Temp., Kharkov, 1988; per revr.]. Time non-locality arises due to the specific choice of porous microstructure. Namely, the authors consider a reservoir with a periodic set of isolated "bubbles". The values of the operator's coefficients inside the bubbles differ substantially from the external ones. The orders of the coefficients are correlated with the spatial scale of the microstructure in such a way that the relaxation time period for a single periodicity cell is equal to the typical time of the process in the whole domain. To this end the permeability coefficient inside the bubbles has to be proportional to ε^2, where ε is the size of the periodicity cell.

To prove the central result the authors have successfully applied the method of two-scale convergence. This was suggested by G. Nguetseng [SIAM J. Math. Anal. **20** (1989), no. 3, 608–623; MR 90j:35030] and G. Allaire [SIAM J. Math. Anal. **23** (1992), no. 6, 1482–1518; MR 93k:35022] in connection with the homogenization of periodic structures. In this asymptotic problem one deals with expansions of the solution into a sum of terms like $u(x,y)$ where $y=x/\varepsilon$ is the so-called "fast" variable on the periodicity torus. Roughly, the two-scale convergence of a sequence $u_\varepsilon(x)$ to a function $u(x,y)$ means that $u_\varepsilon(x)$ is close to $u(x,x/\varepsilon)$ for small values of ε. With the concept of two-scale convergence it is possible to avoid complicated computations pertaining to asymptotic expansions, and all limit properties of the solution could be obtained in the most straightforward way.

Thus, the paper under review develops the method of two-scale convergence with respect to the most complicated homogenization problem it was applied to before. Any analysis of the system for two-phase porous flows is a subject of great interest both in the theory and in applications, and this contribution seems to be an important part of that research. *Vasily V. Zhikov* (Vladimir)

99c:90002 90-02 49J24 90A16 93A30
Aubin, Jean-Pierre (Paris)
★**Dynamic economic theory.**
A viability approach.
Studies in Economic Theory, 5.
Springer-Verlag, Berlin, 1997. xxviii+510 pp. $96.00. ISBN 3-540-62687-5

A long-standing concern of mathematical economics centers around the familiar Adam Smith "invisible hand" story whereby market pressures and best response actions by agents drive prices to a Walrasian equilibrium. While fixed-point theorems establish the existence of such equilibria, no realistic formulation of the price dynamics can ensure convergence of prices to one of the equilibria. Instead, examples can be created which generate any desired form of chaotic price dynamics [see D. G. Saari, Notices Amer. Math. Soc. **42** (1995), no. 2, 222–230; MR 95m:90029].

Attempts to correct this failure of price theory and related economic concerns include imposing market restrictions and concepts of bounded rationality. Aubin explores a fundamentally different, novel approach which leans heavily on the realism that most of us make our economic decisions only when we must. Contrast Aubin's approach with basic assumptions of typical economic models which require everyone to respond instantly, like brokers on the stock market, to even the slightest of changes in news and events.

Technically, Aubin's analysis of his theme borrows heavily from the "viability theory" he and his coauthors developed over the last two decades. While related to control theory, viability theory differs in that instead of optimizing a performance function, controls are sought to minimize effort. Conflict occurs when the dynamics collides with the boundary of the constraint set (e.g., a scarcity of certain commodities, etc.), finally forcing a decision to be made, a decision determined by the theory.

Using his characteristically entertaining and intuitive writing style, full of historical asides, Aubin develops his theme while describing (sometimes critically) traditional approaches. In the process he generates several interesting mathematical results that should be of value even to readers not interested in economics.

The first couple of chapters provide a sufficiently complete introduction of viability theory and appropriate economic concepts so that the book is accessible to readers unfamiliar with either topic. The second part of the book shows how to use the main viability theorems to understand the allocation of scarce resources, planning mechanisms, and the design of dynamical economies which support "viability". He then describes how to obtain adjustment procedures from a regulation mechanism; this leads, in a natural manner, to questions of dynamical games and inequalities.

The third part of the book is of independent interest to mathematicians. It covers a surprisingly wide variety of topics including pseudo-inverses of linear operators, set-valued functions and differential inclusions, convex optimization, Wiener processes, the Ky Fan inequality, stochastic epiderivatives, and on and on.

I expect this book to generate controversy, thought, and insight.

Donald G. Saari (Evanston, IL)

98e:90168 90C30 14Q15 52A41 90C60

Porkolab, Lorant (New Brunswick, NJ); **Khachiyan, Leonid** (New Brunswick, NJ)

On the complexity of semidefinite programs. (English summary)

J. Global Optim. **10** (1997), *no.* 4, 351–365.

Let $\mathbf{R}^d$, $d = n(n+1)/2$, be the space of all symmetric $n \times n$ matrices with the scalar product $\langle A, B\rangle = \operatorname{tr}(AB)$. We write $A \succeq 0$ provided A is a positive semidefinite matrix. There is a theory of linear inequalities and linear programming, called semidefinite programming, in the space of symmetric matrices with the order $\succeq$, which is somewhat parallel to the theory of "standard" linear inequalities and "standard" linear programming in $\mathbf{R}^n$. At the same time, semidefinite programming can be considered as a particular case of linear programming in a topological vector space with a fixed cone of non-negative vectors, a point of view which goes back to L. V. Kantorovich, a founder of linear programming. Semidefinite programming, long before it acquired its modern name, attracted the attention of many researchers. For example, the geometry of the cone of positive semidefinite matrices was studied by G. F. Voronoï in connection with problems of the geometry of numbers. C. Davis related geometry of the order $\succeq$ and quadratic convexity, notably the Toeplitz-Hausdorff convexity theorem [see Canad. Math. Bull. **14** (1971), 245–246; MR **47** #850]. L. Lovász used what is now called semidefinite programming in coding theory [see IEEE Trans. Inform. Theory **25** (1979), no. 1, 1–7; MR 81g:05095]. Applications of semidefinite programming to optimization as well as the history of the subject are surveyed in [M. X. Goemans, Math. Programming **79** (1997), no. 1-3, Ser. B, 143–161; MR 98g:90028].

Semidefinite programming is different from the standard linear programming (linear programming of the non-negative orthant) in some essential aspects. First, the cone of positive semidefinite matrices is not polyhedral (its extreme rays are spanned by positive semidefinite matrices of rank 1). Therefore, the image of the cone under a linear transformation is not always closed. Therefore, the duality theory is not as straightforward and needs to be modified. Second, there are essential numerical problems. Suppose we want to find a positive semidefinite solution to a system of linear equations in the space of symmetric matrices. It may happen that the complexity (bit size) of such a solution is inevitably exponentially large in the bit size of the input. For example, we can encode a "pathological" system of quadratic inequalities $x_0 \geq 2$, $x_1 \geq x_0^2, \cdots, x_{k+1} \geq x_k^2, \cdots, x_n \geq x_{n-1}^2$ by a system of linear inequalities with respect to the order $\succeq$: to say that $x_{k+1} \geq x_k^2$ is to say that the 2×2 symmetric matrix having 1 and x_k in the first row and x_k and x_{k+1} in the second row is positive semidefinite. For any solution of such a system, we will have $x_n \geq 2^{2^n}$.

In the paper under review, an important result is achieved. The authors consider testing the feasibility of a system of linear inequalities $\langle A_i, X\rangle \leq b_i$, $i = 1, \cdots, m$, $X \succeq 0$, where A_i are given symmetric $n \times n$ integral matrices and b_i are given integers. They present an algorithm for testing the consistency of the system, whose complexity is $mn^{O(\min\{m,n^2\})}$ (in terms of the number of arithmetic operations used) and in which the bit size of all the numbers used is $ln^{O(\min\{m,n^2\})}$, where l is the maximum binary size of the input coefficients. Also, the authors show that any such feasible system has a solution whose bit size is at most $ln^{O(\min\{m,n^2\})}$. In particular, if the size n of the matrices or the number m of the inequalities is fixed, we get a polynomial time algorithm. If n is fixed, the existence of a polynomial time algorithm follows from general complexity results of the first-order theory of the reals (for a survey see, for example, J. Renegar [J. Symbolic Comput. **13** (1992), no. 3, 255–299; MR 93h:03011a]). That is, if the number of variables is fixed, the validity of any statement about real numbers which uses arithmetic operations, comparisons and quantifiers can be checked in polynomial time. If the number m of inequalities is fixed, the idea is to use convex duality: in the dual problem, m becomes the number of variables (Lagrange multipliers). The dual problem can be understood as a formula of the first-order theory of the reals for the Lagrange multipliers. Since the number of variables is fixed, we get a polynomial time algorithm. We also note that once a polynomial bound is proved for the bit size of a possible solution, the feasibility of the system can be tested in polynomial time by the ellipsoid method. The authors use a subtle analysis to get their complexity bound, which indeed looks optimal.

The existence of a polynomial time algorithm when the number of inequalities is fixed may be related to the reviewer's result on polynomial time solvability of systems of real quadratic equations when the number of equations is fixed although the number of variables is allowed to grow [see Discrete Comput. Geom. **10** (1993), no. 1, 1–13; MR 94f:14051]. Indeed, the condition $X \succeq 0$ can be replaced by writing X as the sum $X = \sum_{k=1}^n x_k x_k^t$, where $x_1, \cdots, x_n \in \mathbf{R}^n$ are $n \times 1$ columns of new real variables. Then we get a system of m real quadratic inequalities, which, of course, can be transformed into a system of m quadratic equations by adding slack variables. Next, the standard trick of "homogenization" reduces the problem to finding a common nontrivial zero of $m+1$ real quadratic forms.

Alexander I. Barvinok (Ann Arbor, MI)

97f:90132 90D20 90D06

Simon, R. S. (Bielefeld); **Spież, S.** (Warsaw); **Toruńczyk, H.** (Warsaw)

The existence of equilibria in certain games, separation for families of convex functions and a theorem of Borsuk-Ulam type. (English summary)

Israel J. Math. **92** (1995), *no.* 1-3, 1–21.

Repeated games model ongoing relationships. Their equilibria reflect phenomena such as altruism, revenge, and the gradual building of trust; in the incomplete information case, also signalling and the strategic revelation and concealment of information. In a 1968 technical report that was widely circulated—but not formally published until recently—the reviewer and M. B. Maschler (with the collaboration of R. E. Stearns) [*Repeated games with incomplete information*, MIT Press, Cambridge, MA, 1995; MR 96k:90001] conjectured that any two-person infinitely repeated game with complete information for one player (but not necessarily the other) possesses an equilibrium in the sense of J. Nash [Ann. of Math. (2) **54** (1951), 286–295; MR **13**, 261g]. In spite of concerted attempts, no headway was made on this conjecture until fifteen years later, when S. Sorin [Internat. J. Game Theory **12** (1983), no. 4, 193–205; MR 85c:90104] proved it for the case in which the informed player has just two "types"; i.e., when the uncertainty of the uninformed player is concentrated on just two possibilities. The current paper proves the general conjecture, for arbitrarily many types.

Formally, let $G^1, \cdots, G^k$ be k bimatrix games, all with the same number of rows and columns. Chance chooses G^i with probability p^i; the row player, but not the column player, is informed of the choice, and the game is then played infinitely often. After each play of the chosen game, both players are informed of the "moves"—i.e., the pure strategies picked—by both in that play, but not of the payoff; information about the payoff might tip off the uninformed player about which game was chosen. The whole scenario, including the numerical values of the probabilities p^i, is assumed commonly known between the players.

An equilibrium is a pair σ of mixed strategies in the repeated game such that, roughly speaking, neither side can improve its long-term average expected payoff by changing its strategy, when the other side does not change; in other words, that for some sequence $(\varepsilon_1, \varepsilon_2, \cdots)$ tending to 0, the n-stage truncation of σ is an ε_n-equilibrium in the n-stage truncation of the repeated game, for each n (an ε-equilibrium is a pair of strategies such that neither side can gain more than ε by changing when the other does not change).

The current paper shows that with these definitions, an equilibrium always exists. In fact, an equilibrium of a particularly simple kind—called a joint plan—exists. One can think of a joint plan as follows: The players enter into a "contract", in which the informed player undertakes to give a one-time signal to the uninformed player, which may be correlated with his information; each player then plays an "agreed" sequence of moves, which may depend on the signal. If either player i deviates from the "agreed" move sequence, the other "punishes" him by switching to a sequence of moves that brings i's payoff sufficiently far down so as to render the deviation unprofitable. The plan is set up so that the informed player has no incentive to give a signal that differs from the one that the plan prescribes.

Joint plans were first defined by the reviewer and Maschler [op. cit.], who also showed that many equilibria are *not* of this form. When $k = 2$, the equilibria of Sorin [op. cit.] are joint plans of a particular sort, called independent and 2-safe; this also holds for the equilibria in the current paper. Here "independent" means that for each signal of the row player, the agreed sequence of move-pairs has a time distribution that is a product of two distributions (one for each of the players); "2-safe" means that the column player's strategy in the repeated game guarantees him his maxmin.

The main mathematical tool in the proof is the following theorem from algebraic topology, established in the paper: Let F be a convex-valued upper semicontinuous correspondence from a compact set C in $\mathbf{R}^{n+1}$ to a polyhedron built up of simplexes of dimension $\leq n$, and let $x \in C$. Then the boundary of C has a subset D whose convex hull contains x and such that $\bigcap_{y \in D} F(y)$ is non-empty. In particular, when F is a function (i.e., point-valued), then all points of D are mapped into the same point. This is related to a theorem of K. Borsuk [Fund. Math. **20** (1933), 177–190; Zbl 006.42403] which says that if f is a continuous function from the n-sphere S^n into $\mathbf{R}^n$, then there are antipodal points y, z with $f(y) = f(z)$; in that case $D = \{y, z\}$.

This beautiful paper, which solves a long-standing conjecture of central importance, is a major contribution to mathematical game theory, one of the two or three most significant in the last decade. *Robert J. Aumann* (Jerusalem)

98g:93003 93B05 93D15

Clarke, Francis H. (Montreal, QC); **Ledyaev, Yuri S. [Ledyaev, Yu. S.]** (Moscow); **Sontag, Eduardo D.** (New Brunswick, NJ); **Subbotin, Andrei I.** (Ekaterinburg)

Asymptotic controllability implies feedback stabilization. (English summary)

IEEE Trans. Automat. Control **42** (1997), *no.* 10, 1394–1407.

A long-standing, central question in nonlinear control has been: "Does asymptotic controllability imply feedback stabilizability?" This article combines sophisticated tools from positional differential games and nonsmooth analysis with the evolving theory of control Lyapunov functions to authoritatively and affirmatively settle this question.

A major contribution of this paper concerns the precise meaning of "solution" for discontinuous controllers. Various notions of solution for discontinuous dynamical systems have been proposed in the existing literature, yet only this article brings together deep insights from the respective areas to develop a coherent larger framework. The notion of solution given here is an accurate model of the process used in computer control ("sample and hold"). It gives rise to a very natural definition of s-stabilizing feedback (with "s" for sampling), characterized by uniform attractiveness, overshoot boundedness, and Lyapunov stability. This article complements these natural concepts of solution and s-stability with the appropriate notion of derivative of the control Lyapunov functions: the proximal subgradient.

The theory developed here exploits a deep and intrinsic connection between invariance properties (of the control Lyapunov function) with respect to trajectories, and the characterization of these properties in terms of proximal subgradients. This connection is reflected by the agreement of the local viscosity supersolution and the upper minimax solution of the corresponding Hamilton-Jacobi equation.

The extremely well written introduction of this outstanding article lays out, in very accessible terms, the technical difficulties that for so long have stood in the way of an affirmative answer to the central problem. This is complemented with a self-contained exposition of major tools from nonsmooth analysis utilized in the construction. The stabilization strategy is based on a pointwise optimization of a smoothed version of a control Lyapunov function. Intermediate stages that achieve semiglobal practical stabilization are combined iteratively, and, with appropriate smoothing, eventually yield the desired global result.

In general, one cannot expect that discontinuous controllers exhibit the robustness properties against perturbations that make feedback control so attractive in the first place (as opposed to open-loop control). However, as the authors prove in the last few sections, the s-stabilizing feedback control constructed here is indeed insensitive to small perturbations in the system's dynamics. This highly nontrivial theorem reaffirms that these notions of solution and s-stability are indeed the right ones for control. The general robustness result is formulated in terms of differential inclusions, which makes it suitable for applications dealing with various kinds of disturbances. (It allows for actuator errors—but no claims are made about measurement errors. The latter can be dealt with nicely using robust *dynamic* feedback controllers.) A corollary is that asymptotic controllability is structurally stable with respect to a natural class of perturbations. *Matthias Kawski* (Tempe, AZ)

98f:93003 93B18 93B29 93B52 93C10

Pomet, Jean-Baptiste (Nantes)

On dynamic feedback linearization of four-dimensional affine control systems with two inputs. (English summary)

ESAIM Control Optim. Calc. Var. **2** (1997), 151–230 (electronic).

A useful tool in the analysis of a nonlinear control system of the form $\dot{x} = f(x,u)$ (where the state x is in $\mathbf{R}^n$, the control u is in $\mathbf{R}^m$, and the mapping $f\colon \mathbf{R}^n \times \mathbf{R}^m \to \mathbf{R}^n$ is smooth or C^∞) is the (feedback) coordinate transformation $z = \varphi(x)$, $v = \psi(x,u)$ to obtain a transformed system $\dot{z} = \widetilde{f}(z,v)$ which (one hopes) has a simpler form than the original system. Of particular interest is the case where the transformed system can be made linear, i.e., of the form $\dot{z} = Az + Bv$, where A and B are constant matrices of the appropriate dimensions, for then one can apply the formidable body of knowledge concerning linear control systems to the original nonlinear system. Necessary and sufficient conditions for so-called static feedback linearization have been known for some time [see L. R. Hunt, R. J. Su and G. Meyer, in *Differential geometric control theory (Houghton, Mich., 1982)*, 268–298, Progr. Math., 27,

Birkhäuser, Boston, Mass., 1983; MR 84k:93010; and B. Jakubczyk and W. Respondek, Bull. Acad. Polon. Sci. Sér. Sci. Math. **28** (1980), no. 9-10, 517–522 (1981); MR 82k:93020]. As would be expected, these conditions are rather restrictive and one is led naturally to examine other types of coordinate transformations which might produce linear systems from nonlinear ones. A useful generalization of static feedback transformations is the so-called dynamic feedback transformations, which have the general form $z=\varphi(x,\xi)$, $\dot{\xi}=\gamma(x,\xi,u)$, $v=\psi(x,\xi,u)$. In this case the transformed system $\dot{z}=\widetilde{f}(z,v)$ is defined on a state space of dimension strictly larger than that of the original system.

In contrast with the static problem, relatively little is known about necessary and/or sufficient conditions for dynamic feedback linearization. The author makes a notable contribution to the understanding of the dynamic feedback linearization problem by providing necessary and sufficient conditions for dynamic feedback linearization in the simplest nontrivial cases, namely fully nonlinear systems in dimension three and affine two-input systems in dimension four (the situation for planar systems and single-input systems is well understood). The author's results are fairly general in the sense that the only restrictive assumption requires that the "linearizing outputs" be functions of x and u only. At the same time his results are constructive and the requisite computations can be implemented via a symbolic manipulation software package. *Kevin A. Grasse* (Norman, OK)

F. REVIEWS IN PROBABILITY AND MATHEMATICAL PHYSICS

97h:60016 60E15 60D05

Talagrand, Michel (Paris)

Concentration of measure and isoperimetric inequalities in product spaces. (English summary)

Inst. Hautes Études Sci. Publ. Math. No. 81 (1995), 73–205.

Given a metric space (Ω, d) and a probability measure μ on its Borel σ-field, one can pose the following two problems. (i) The isoperimetric problem: Given $\varepsilon > 0$, among all (Borel) subsets of Ω of measure $1/2$, find the sets A for which the measure of the set $A_\varepsilon = \{\omega \in \Omega;\ d(\omega, A) \leq \varepsilon\}$ is minimal. (ii) Concentration: Find a good lower bound on $\inf\{\mu(A_\varepsilon);\ \mu(A) = 1/2\}$. Of course, a solution to (i) also solves (ii) (provided the minimal set is relatively simple). However, in many specific situations one can reach a satisfactory lower bound to (ii) not through a solution to (i) and a lot of times in situations in which the answer to (i) is not known.

The importance of the concentration inequality stems from the fact, first noticed by P. Lévy, that, if the lower bound in (ii) is $1 - \delta$, then every real-valued Lipschitz function, f, on (Ω, d) is "well concentrated" near its median: The measure of the set of points in which f deviates from its median by more than ε/L is smaller than or equal to 2δ, where L denotes the Lipschitz constant of f. Lévy solved (i) for the sphere in n-dimensional Euclidean space and used it in the way described above. It turns out that in this case δ is practically $e^{-\varepsilon^2 n}$, which leads to the surprising fact that a Lipschitz function on the sphere is practically a constant on all but an exponentially small portion of the sphere. This fact and similar ones in other natural metric probability spaces are extremely useful. (See V. D. Milman and G. Schechtman [*Asymptotic theory of finite-dimensional normed spaces*, Lecture Notes in Math., 1200, Springer, Berlin, 1986; MR 87m:46038] for applications in the direction of the local theory of Banach spaces.)

Proving concentration inequalities in interesting situations and applying them to problems in various areas has become quite an active area of research in the last few years. The proofs are geometrical, combinatorial or probabilistic. Elsewhere [Proc. Amer. Math. Soc. **104** (1988), no. 3, 905–909; MR 90h:60016] Talagrand introduced a relatively simple but quite powerful new method which works in many situations in which the probability space is a product space with many components. The proofs, as naive as they may look, are by induction on the number of components. This is not always as easy as it sounds, mostly because the right induction hypothesis is sometimes hard to guess.

The present paper can be viewed as a comprehensive account of this very successful method for proving concentration inequalities in product spaces. A portion of the paper is devoted to previous results of Talagrand (usually with new proofs or presentations) but most of it is new. About 40% of the paper is devoted to applications. An attempt was made to get the best results the method can yield in each instance. Another new feature is that sometimes Talagrand uses various non-standard notions of "closeness", not always defined by a metric. Some of the applications show that these notions are really necessary. Let us review in more detail some of the results presented here. Section 2 deals with the basic situation: The probability space is a product space (Ω^n, μ^n) and the "distance" between a point $x = (\omega_1, \cdots, \omega_n) \in \Omega^n$ and a set $A \subseteq \Omega^n$ is $f_h(A, x) = \inf\{\sum h(\omega_i, \omega_i');\ (\omega_1', \cdots, \omega_n') \in A\}$, where h is some non-negative function on $\Omega \times \Omega$ with $h(x, x) = 0$. The section starts with the simple case of the Hamming metric (where $h(\omega, \omega') = \delta_{\omega, \omega'}$) and ends with a result of the author [Geom. Funct. Anal. **1** (1991), no. 2, 211–223; MR 92j:60004] (see also B. Maurey [Geom. Funct. Anal. **1** (1991), no. 2, 188–197; MR 92g:60024] for a much simpler proof) which generalizes the Gaussian concentration inequality. Section 3 returns to another beautiful idea of the author: For $q \geq 2$ and $x = (\omega_1, \cdots, \omega_n)$, define $f(A, x) = \inf\{\#\{i \leq n;\ \omega_i \notin \{\omega_i^1, \cdots, \omega_i^q\}\};\ (\omega_1^j, \cdots, \omega_n^j) \in A,\ 1 \leq j \leq q\}$. This (as well as a generalization of it) serves now as the distance between a point and a set. It turns out that on one hand one gets much better concentration here than in the case $q = 1$ (which reduces to the Hamming metric) and on the other hand this is the right definition for some applications. The idea of considering this measure of "closeness" was first introduced in an earlier paper by the author [Ann. Probab. **17** (1989), no. 4, 1546–1570; MR 91e:60054]. The proof here is much simpler than the original one.

If Ω is a subset of $\mathbf{R}$, say, then one can consider as a distance between a point $x \in \Omega^n$ and a set $A \subseteq \Omega^n$ the Euclidean distance between x and the convex hull of A. This distance (in the special case of the Hamming cube) and the relevant concentration inequality appeared in another paper by the author [Proc. Amer. Math. Soc. **104** (1988), no. 3, 905–909; MR 90h:60016], where the method of proof presented throughout the present paper was first used. The point here is that for convex Lipschitz functions one gets much better concentration than for general Lipschitz functions. In Section 4 of the paper under review a far-reaching generalization of this distance is considered for general product spaces. It turns out that this generalization has a lot of applications as well as at least one analogous result, which is presented in the next section. This generalization was motivated in part by a slight generalization of the original result of the author [W. B. Johnson and G. Schechtman, in *Functional analysis (Austin, TX, 1987/1989)*, 72–77, Lecture Notes in Math., 1470, Springer, Berlin, 1991; MR 92m:60017]. As Talagrand puts it: "The author's desire to have the last word prompted the discovery of the abstract setting of Theorem 4.1.1."

Section 5 deals with the symmetric group S_n equipped with its natural probability measure, μ. Although this is not a product space one can deal with it in a similar manner. The main result in this section is an improvement on a result of Maurey [C. R. Acad. Sci. Paris Sér. A-B **288** (1979), no. 14, A679–A681; MR 80c:46020] obtained by considering the Euclidean distance from a point to a convex hull of an appropriate set, in the spirit of the result of the previous section. Let us be more specific here: For $\sigma \in S_n$ and $A \subseteq S_n$ let

$$f(A, \sigma) = \inf\Big\{\sum_{i=1}^{n} s_i^2;\ (s_1, \cdots, s_n) \in V_A(\sigma)\Big\},$$

where $V_A(\sigma)$ is the convex hull of the set $\{(s_1, \cdots, s_n) \in \{0,1\}^n;\ \exists \tau \in A$ such that $\forall i \leq n,\ s_i = 0 \Rightarrow \tau(i) = \sigma(i)\}$. The concentration inequality proved here is: $\mu(A_t) \geq 1 - \mu(A)^{-1} e^{-t/16}$ for every $A \subset S_n$, $t > 0$, where $A_t = \{\sigma;\ f(A, \sigma) \leq t\}$.

The rest of the paper is devoted to applications of the concentration inequalities discussed in the first five sections. There are applications to computer science and combinatorial problems: statistical properties of bin packing (where the sizes of the items to be packed are random), of the longest increasing sequence of a random sequence, and of the chromatic number of a random graph, along with problems related to the traveling salesman problem, minimal spanning tree and the assignment problem. There are applications related to problems in mathematical physics: concentration of the first-passage time in percolation on graphs and concentration of the energy in the Sherrington-Kirkpatrick spin glass model. Not surprisingly, there are also applications to probability theory, particularly to deviation inequalities for norms of sums of Banach-space-valued independent random variables.

To a large extent the paper is self-contained. Also, the individual sections are, to a large extent, logically independent. This paper is a cornerstone in this important and rapidly developing area of probability and its applications. It should be read by anyone interested in this area. *G. Schechtman* (Rehovot)

98i:60021 60F05 58F15 60J05 60J99

Kifer, Yuri (Jerusalem)

Limit theorems for random transformations and processes in random environments. (English summary)

Trans. Amer. Math. Soc. **350** (1998), *no.* 4, 1481–1518.

This basic paper is within the following framework: Let θ be a measure-preserving invertible transformation of a probability space $(\Omega, \mathcal{G}, P)$. Further, let $\{f_\omega: \Xi^\omega \to \Xi^{\theta\omega}\}_{\omega\in\Omega}$ be a family of measurable maps between the ω-sections of a measurable set $\Xi \subset \mathbf{\Xi} \times \Omega$, where $(\mathbf{\Xi}, \mathcal{F})$ is another measurable space. Finally, let $\{\mu^\omega \in \mathcal{P}(\Xi^\omega)\}_{\omega\in\Omega}$ be a family of probability measures on the ω-sections which satisfies $f_\omega\mu^\omega = \mu^{\theta\omega}$. This setup is sometimes called a bundle random dynamical system (RDS) with fibers Ξ^ω and fiber maps f_ω.

If φ is a measurable function on Ξ then for each fixed $\omega \in \Omega$ the sequence $\{\varphi(f_{\theta^{n-1}\omega} \circ \cdots \circ f_\omega\xi, \theta^n\omega)\}_{n\in\mathbf{N}}$ can be considered as a stochastic process on (Ξ^ω, μ^ω). The author provides conditions under which for P-almost all ω this process satisfies a central limit theorem (CLT) and a law of the iterated logarithm (LIL). His results are deep generalizations of parts of the corresponding theory concerned with deterministic dynamical systems [cf. M. Denker, in *Dynamical systems and ergodic theory (Warsaw, 1986)*, 33–62, PWN, Warsaw, 1989; MR 92d:28007]—the fundamental difference being that the process under consideration is no longer stationary. The basic approach is to modify the existing probabilistic tools which are used in the deterministic theory (uniform mixing condition and representation by martingale differences), and which are formulated for stationary processes, in such a way that they become applicable in the present random context.

Let us sketch a sample result: Consider the RDS which is obtained by inducing θ on an arbitrary measurable set $Q \subset \Omega$ with $P(Q) > 0$. One can also define some uniform mixing coefficient φ_j which takes into account only the state of mixing of the RDS at hitting times, i.e., when $\theta^k\omega \in Q$. Under some summability conditions on φ_j and the second moments of the induced process one obtains that for P-almost all $\omega \in \Omega$ the limit

$$\sigma^2 := \lim_{n\to\infty} \frac{1}{n} \mathbf{E}_{\mu^\omega} \left(\sum_{i=0}^{n-1} \psi_{\theta^i\omega} \circ f_{\theta^{i-1}\omega} \circ \cdots \circ f_\omega \right)^2$$

exists and is independent of ω. Here $\psi_\omega(\xi) := \varphi(\xi, \omega) - \mathbf{E}_{\mu^\omega}(\varphi(\cdot, \omega))$. The CLT statement then is that the process

$$\left(\frac{1}{\sqrt{n}} \left(\sum_{i=0}^{n-1} \psi_{\theta^i\omega}(f_{\theta^{i-1}\omega} \circ \cdots \circ f_\omega\xi) \right) \right)_{n\in\mathbf{N}}$$

on (Ξ^ω, μ^ω) converges in law to $\mathcal{N}(0, \sigma^2)$.

To apply such a result to some more concrete situation one has to find a good set Q such that one has control over the φ_j's. This is done for Markov chains in random environments, random subshifts of finite type, and random expanding in average transformations. At the end, the method of random suspension is introduced to produce continuous time RDSs. It is shown that the general discrete-time result can also be applied here to obtain continuous-time versions of the CLT and the LIL. *Thomas Bogenschütz* (Essen)

98i:60060 60H15 35K99 35R60 58G11 58G32

Sowers, Richard B. (Urbana, IL)

Short-time geometry of random heat kernels. (English summary)

Mem. Amer. Math. Soc. **132** (1998), *no.* 629, viii+130 pp.

Sowers' memoir studies fundamental solutions of the stochastic partial differential equation (SPDE)

$$(*) \qquad dp = \frac{1}{2}\Delta p\, dt + (\sigma, \nabla p) \circ dW(t).$$

Here, W is a standard Brownian motion, $\circ\, dW$ represents the Stratonovich differential, and $p(t,x)$ is a function on $(0,\infty) \times M$, where M is a Riemannian manifold; Δ and ∇ are respectively the Laplace-Beltrami and gradient operators associated to the Riemannian structure on M, and σ is a vector field on M. The equation for p is a generalization of the deterministic heat equation $\partial u/\partial t = \frac{1}{2}\Delta u + (\overline{\sigma}, \nabla u)$ in which $\overline{\sigma}$ is replaced by a random field generated by a white noise in the temporal variable. It appears in several applications, and it is, in some sense, the simplest SPDE in which the stochastic differential is essential. Of course, a simpler equation is obtained if the term $(\sigma, \nabla p) \circ dW(t)$ is replaced by $hp \circ dW$, where h is a function; but in this case, the stochastic differential can be factored out by writing $p = e^{h(x)W(t)} q$ and deriving a partial differential equation for q with random coefficients but no stochastic differentials.

The utility of studying the usual, deterministic heat equation in the setting of Riemannian manifolds is a formula expressing the dominant short-time asymptotics of fundamental solutions in terms of geometric quantities. The goal of Sowers' work is to derive the analogous formula for the stochastic case. Let p_y denote the fundamental solution of the stochastic equation tending to a delta function at y as $t \downarrow 0$. Sowers' main result states that, when M is compact, $p_y(t,x) = P_y(t,x)(1 + \varepsilon(t,x))$, where $\varepsilon(t,x)$ is of the order of some positive power of t as $t \downarrow 0$ and

$$P_y(t,x) = (2\pi t)^{-d/2}\Theta_y(x) \exp\left\{-\frac{d^2(x,y)}{2t} - R(t,x) - S(t,x)\right\}.$$

In this formula, $\Theta_y(x)$ is a deterministic normalization constant, and $d(x,y)$ is the Riemannian distance between x and y. To define the term R, introduce the geodesic curve γ which connects y to x in time t. Then

$$R(t,x) = \int_0^t (\gamma'(s), \sigma(\gamma(s))) \circ dW(s).$$

These first two terms in the exponent replicate, with a stochastic twist, the short-time asymptotics

$$\ln u_y(t,x) = -\frac{d^2(x,y)}{2t} - \int_0^t (\gamma'(s), \sigma(\gamma(s)))\, ds + O(t), \quad t \downarrow 0,$$

of fundamental solutions to the deterministic equation $\partial u/\partial t = \frac{1}{2}\Delta u + (\overline{\sigma}, \nabla u)$.

The third term S contains a more subtle contribution of randomness. It can be expressed as a twice-iterated stochastic integral of Jacobi vector fields, which are vector fields along geodesics expressing the first variation of a geodesic η by a flow along geodesics transverse to η. Thus S expresses a correction term revealing the contribution of the random input to variation about the geodesic. The Jacobi vector fields in S are defined along the geodesic γ connecting y to x. S is of order $O(\ln\ln t^{-1})$, and hence truly differs from the $O(t)$ correction in the deterministic case.

The second main result of the memoir further refines the formula $p_y(t,x) = P_y(t,x)(1+\varepsilon(t,x))$ by providing an expansion of $\varepsilon(t,x)$ as an infinite series of multiple Stratonovich integrals with deterministic kernels. The author calls this a Minakshisundaram-Pleijel expansion. Error bounds incurred by truncating the series at a finite number of terms are provided as part of the analysis. Extensions of the results, for example, to noncompact M, are also stated.

All the main results are motivated and summarized in the first chapter of the monograph and illustrated with simple examples, before any technical details are introduced. The rest of the exposition builds the machinery for the proofs. The analysis required to prove the main results is extensive, deep, and involves considerable calculation. Despite this, the author does an excellent job of guiding the reader through the calculations and supplying intuition. The second chapter presents an interesting heuristic derivation of the leading order term $P_y(t,x)$ via a rescaling argument and an application of large deviation theory. The properties of the formally derived P_y are then discussed rigorously in Chapter 3; in particular, it is shown that P_y satisfies a SPDE, which differs from that for p_y only by the appearance of an additional and explicitly defined potential term. The fourth chapter then develops an equation for the ratio of p_y to P_y and performs a multiple Wiener expansion of this ratio, the Minakshisundaram-Pleijel expansion. Finally, the fifth chapter establishes error estimates on the previously developed formulae in order to establish the short-time asymptotic equivalence of p_y and P_y.

This memoir is a mathematically rich and difficult achievement in SPDE theory and contains many technical innovations. The results are clearly presented at the outset for the reader wishing to understand only the main theorems, while the structure of the proofs is well charted for those wishing to delve deeper into the technical aspects. *Daniel Ocone* (New Brunswick, NJ)

99c:60168a 60J60 60J45 60J55 60J80
Dynkin, E. B. (Ithaca, NY); **Kuznetsov, S. E.** (Moscow)
Natural linear additive functionals of superprocesses. (English summary)
Ann. Probab. **25** (1997), *no.* 2, 640–661.

99c:60168b 60J60 35K55 60J55 60J80
Dynkin, E. B. (Ithaca, NY); **Kuznetsov, S. E.** (Moscow)
Nonlinear parabolic P.D.E. and additive functionals of superdiffusions. (English summary)
Ann. Probab. **25** (1997), *no.* 2, 662–701.

99c:60168c 60J60 35J60 58G32 60J80
Dynkin, E. B. (Ithaca, NY); **Kuznetsov, S. E.** (Moscow)
Solutions of nonlinear differential equations on a Riemannian manifold and their trace on the Martin boundary. (English summary)
Trans. Amer. Math. Soc. **350** (1998), *no.* 11, 4521–4552.

These three papers are part of a long effort by the authors and others to extend some of the most fruitful results linking Brownian motion to solutions of Laplace's or Poisson's equation to non-linear analogs that arise in the study of the family of superprocesses. These three papers extend many previously observed relationships to far more general settings. In summary, the authors identify conditions that are necessary for solutions to exist, and conditions that are sufficient. The key tool for the three papers is a new construction of additive functionals of superprocesses. These additive functionals yield probabilistic representations of solutions. These papers come amidst a large stream of related works primarily but not solely by the authors. Although the papers are not the final word on the subject, they are of great interest in their own right.

The story for the linear equations is familiar. Let L be a second-order elliptic operator. If E is a bounded smooth domain, f a continuous function in E and φ a bounded continuous function on ∂E, then the solution of the boundary value problem $Lh=-f$ in E, $h=\varphi$ on ∂E can be expressed classically in terms of the Green's function and Poisson kernel by $h(x)=\int_E g(x,y)f(y)+\int_{\partial E} k(x,y)\varphi(y)\sigma(dy)$. A generalization to the boundary value problem with measures is the equation $Lh=-\eta$ in E and $h=\nu$ on ∂E. A mild solution is then given by the formula $h(x)=\int_E g(x,y)\eta(dy)+\int_{\partial E} k(x,y)\nu(dy)=G\eta+K\nu$.

Pioneering work on the connections between these representations and probabilistic representations was done by Kakutani, Kac, Doob and Hunt, starting some 50 years ago. Probabilistically, if ξ is a diffusion with generator L and τ is its exit time from a regular domain E, then a solution to the first equation is also available through the following formula involving the expectation of the process started from x: $h(x)=E_x(\int_0^\tau f(\xi_s)+\varphi(\xi(\tau)))$.

When $f=0$, solutions are called L-harmonic and in this case $h=K\nu$. This illustrates that L-harmonic functions are determined by measures ν which are concentrated on the boundary of E. In fact, the positive solutions to $Lh=0$ in an arbitrary domain $E\subset\mathbf{R}^d$ are in a one-to-one correspondence with finite measures on the minimal part of the Martin boundary of E.

Viewed more generally, if $f(x)\geq 0$ then this represents the solution as $h(x)=P_x(A(0,\infty))$, where A is the additive functional of the process ξ given by

$\mathbf{1}_{[0,\tau)}f(\xi_s)ds + \mathbf{1}_\tau\varphi(\xi_\tau)$. This relationship between additive functionals of a process and solutions to a related differential equation is extended in these works.

These papers deal with the following non-linear analogs to the linear equations above. Describe all solutions to the elliptic PDE in E, $L(u) - u^\alpha = 0$, and the parabolic PDE with measure data, $\partial u/\partial r + Lu - u^\alpha = -\eta$ in $\mathbf{R}_+ \times E$, $u = v$ on $(\mathbf{R}_+ \times E)^e$, $u = 0$ on $\{\infty\} \times E$. (Here the superscript e refers to the extremal part of the Martin boundary for the L-diffusion.) It is assumed in these papers that $1 < \alpha \leq 2$. Note that the non-linear problems need not have a unique solution.

Whereas previously Brownian motion, or more generally an L-diffusion, was known to be the correct probabilistic tool to represent solutions, in the non-linear case the tool used is the superprocess based on L with branching related to α. In a time-inhomogeneous setting, this is a set-indexed, measure-valued process, denoted X_U, where U is a space-time subset of $S = \mathbf{R}_+ \times E$. Let μ be a finite measure on S, $1 < \alpha \leq 2$, and let $(\xi_t, \Pi_{r,x})$ be an L-diffusion in E. Then the superprocess can be defined as a measure-valued Markov process with Laplace functional given by $P_\mu(\exp(-\langle f, X_U\rangle)) = \exp(-\langle u, \mu\rangle)$, where u solves the integral equation $u + \mathscr{E}_U(u) = K_U(u)$. In this setting, $\mathscr{E}_U(u)(r,x) = G_U(u^\alpha)(r,x) = \Pi_{r,x}(\int_r^\infty u(s,\zeta_s)^\alpha ds)$ and $K_U(u)(r,x) = \Pi_{r,x}(u(\tau,\zeta_\tau)$, $\tau < \zeta)$, but this can change depending on context. More generally, the integral equation $u + \mathscr{E}(u) = h$ is termed the $\mathscr{E}$-equation and plays an essential role in these works. As one can see, under some conditions a function u satisfying an $\mathscr{E}$-equation will solve a non-linear equation of the type above. If $U = \{t\} \times E$ then we use the abbreviated notation $X_U = X_t$, and if $U = [0,\infty) \times D$ then we set $X_U = X_D$. The latter are called the exit measures because they capture information about the particles as they first exit the domain D. They are related to the elliptic PDE.

The main results of the three papers may be summarized briefly as follows. An additive functional of the superprocess is used to represent solutions to the above equations. Let $h = Kv$ (or $Kv + G\eta$ in the parabolic case); then conditions are found so that an additive functional A can be constructed with potential $P.(A) = h(\cdot)$, and log-potential $u(\cdot) = -\log P.(\exp(-A))$. The functions h and u are related by the $\mathscr{E}$-equation.

For the parabolic PDE, u is the desired solution. For the elliptic problem let Γ be a closed subset of the boundary, and let v be a Radon measure on $E' \setminus \Gamma$ (E' is the Martin boundary of E) satisfying some conditions. If $h = Kv$, then for nice enough v, $A = A_v$ is defined such that $u(x) = -\log P_x(\mathscr{R} \cap \Gamma = \varnothing, \exp(-A_v))$ solves $Lu = u^\alpha$ in E, where $\mathscr{R}$ denotes the range of the superprocess. This solution is termed the (Γ, v)-solution. It need not solve the $\mathscr{E}$-equation for h.

Two important examples are if $\Gamma = \partial E$ and $v = 0$; then the solution $u_\Gamma(x) = -\log P_x(\mathscr{R} \cap \Gamma = \varnothing)$ is the maximal solution in E. Whereas, if the surface measure σ is defined on the boundary of E and $v = f\sigma$, where f is bounded and continuous and $\Gamma = \varnothing$, then $A_v = \langle X^E, f\rangle$, and the solution $u(x) = -\log P_x(\exp(-\langle X^E, f\rangle))$ has boundary value f.

Before stating in more detail the exact accomplishments of these papers, a brief digression into the relationship between superprocesses and non-linear differential equations is in order. Several authors have used the PDE that occurs in the log-Laplace transform to derive properties of the superprocess. I. Iscoe [Probab. Theory Relat. Fields **71** (1986), no. 1, 85–116; MR 87c:60070] introduced the use of an additive functional of the superprocess in finding solutions to the equation. His functionals are of the form $A_t = \langle X_t, f\rangle + \int_0^t \langle X_s, \varphi\rangle ds$. In particular, he showed that $u(x) = -\log P_x(\exp(-A_t))$ solves $\dot{u} + Lu - u^\alpha = -\varphi$, $u(0) = f$. J.-F. Le Gall [Comm. Pure Appl. Math. **50** (1997), no. 1, 69–103; MR 98c:60144] first characterized all solutions to $\Delta u = u^2$ in the unit circle in $\mathbf{R}^2$ and later for $\mathscr{C}^2$ domains in $\mathbf{R}^2$ in terms of a subset of the boundary and a Radon measure on the boundary. He used the Brownian snake instead of the closely related superprocess. He found that solutions were in a one-to-one correspondence to certain pairs (Γ, ν), as above. The pair (Γ, ν) is called a trace for a solution. The unique characterization of a solution by a trace does not quite carry over to the general case. Dimension 2 is special for the Brownian superprocess as points on the boundary of a smooth enough domain are "hit" by the exit measures. This means that the boundary polar sets for the range of the superprocess are trivial. (Roughly, boundary polar means that the set of points on the boundary are not hit by exiting particles.) He also noted that in higher dimensions the characterization could not be one-to-one. The analysts Gmira, Marcus and Véron have investigated solutions to this type of non-linear differential equations in the last decade [see, e.g., M. M. Marcus and L. Véron, C. R. Acad. Sci. Paris Sér. I Math. **321** (1995), no. 2, 179–184; MR 96f:35045; L. Véron, *Singularities of solutions of second order quasilinear equations*, Longman, Harlow, 1996; MR 98b:35053]. Their work deals with the more general case of $\alpha \geq 1$. Here too, a description of solutions in terms of a trace is given although its definition is no longer probabilistic.

The main results of all three papers build upon many previous papers by the authors. Two classes of sets are involved in most of the theorems. The authors call a subset of E' weakly exceptional if it has Martin capacity 0 (or Green capacity 0 if it is a subset of E), and strongly exceptional if it is polar for the range of the superprocess. (Polar for the range means roughly that the particles do not hit this set of points.) The authors previously have proved that strongly exceptional implies weakly exceptional and that for smooth enough domains the two classes are equivalent. Most of the key theorems of these papers are such that sufficient conditions to ensure that a solution exists are related to the measures involved not charging weakly exceptional sets, and necessity is related to measures not charging strongly exceptional sets. Note that the condition that a measure not charge weakly exceptional sets is a "stronger" (more restrictive) condition than that it not charge strongly exceptional sets.

Previously, the authors [Trans. Amer. Math. Soc. **348** (1996), no. 5, 1959–1987; MR 97d:60135] considered the related problem of finding solutions in a bounded, smooth domain D to the equation $Lv - v^\alpha = -\eta$ in D and $v = \nu$ on ∂D. They showed that if $h = G\eta + K\nu$ and η and ν did not charge weakly exceptional sets (suitably defined) then a solution to the $\mathscr{E}$-equation exists. In contrast, a necessary condition for a solution v, given η and ν, is that these measures not charge strongly exceptional sets. To establish these results, they constructed a class of linear additive functionals, generically denoted by A, whose potential is h $(P_x(A_\infty) = h(x))$ and whose log-potential $(u(x) = -\log P_x(\exp(-A_\infty)))$ is a solution to the $\mathscr{E}$-equation.

It took two papers (the first two reviewed here) to extend these results from smooth domains to arbitrary domains in $\mathbf{R}^d$. To do so required a new construction of their additive functionals. This is done in the first paper. Here, natural (A_{t+} is predictable) linear additive functionals (NLA) are defined and constructed. Again, an NLA has its potential h and its log-potential u. It is shown that h and u are related by the $\mathscr{E}$-

equation and that h is always an exit rule for the underlying L-diffusion. Conversely, h being an exit rule satisfying the condition $h+\mathscr{E}(h)<\infty$ suffices to show that h is the potential for some NLA A. A limit of discrete functionals defined in terms of h gives A.

These results are applied and extended in the second paper to solving the non-linear equations in the time-homogeneous and inhomogeneous cases. Let E be an arbitrary domain in $\mathbf{R}^d$, and η and ν be the measures in the parabolic boundary value problem described previously. Again, set $h=G\eta+K\nu$. Then, if a solution to $u+\mathscr{E}(u)=h$ exists, it can be represented as the log-potential of some NLA A whose potential is h. In this case, the probabilistic representation always applies when a solution exists. Sufficient conditions are given for existence in terms of the spectral measure of h not charging weakly exceptional sets. Necessary conditions for existence are given in terms of these measures not charging strongly exceptional sets. When A exists it is shown that A splits into two pieces, $A_\eta+A_\nu$, the former determined by the events inside the domain, and the latter by events on the boundary. This extends the Iscoe formula. A key result is the proof that three seemingly different classes of functions for h candidates are in fact identical.

In a previous paper [J. Anal. Math. **68** (1996), 15–37; MR 97f:35048] the authors investigated all solutions u to the elliptic equation $Lu=u^\alpha$ in E which are bounded by a harmonic function in E. These solutions are called moderate. They showed for smooth domains that if h is a harmonic function with Martin representation $h=K\nu$ then the $\mathscr{E}$-equation for h has a solution if ν does not charge weakly exceptional sets. In contrast, if u is a moderate solution, and h its minimal harmonic majorant, then the ν appearing in the Martin representation for h does not charge strongly exceptional sets.

However, solutions to the elliptic PDE considered may be more singular than a harmonic function. A typical example is the maximal solution u_Γ defined previously. For E smooth and the process being super Brownian motion, this solution blows up like one over the distance to the boundary squared, whereas any harmonic function blows up at most like one over the distance.

In the third paper under review E is again allowed to be any Riemannian manifold. For each solution u to the elliptic PDE, a pair (Γ,ν) is defined. Again this pair is called the trace of u. Here Γ is a closed subset of E', and ν a Radon measure on $E'\setminus\Gamma$ as described previously. One may think roughly that Γ is where the solution blows up badly and ν corresponds to the "moderate" part of the solution. The main results characterize which pairs are possible traces. The statements are similar to the ones for moderate solutions, but the domains considered are more general. These results hold for domains whose Martin boundary coincides with their Euclidean boundary (actually in slightly more generality). It is shown that for a pair (Γ,ν) to be a trace of some solution it is sufficient that ν has some explosion property and does not charge weakly exceptional sets. Conversely, if (Γ,ν) is the trace of some solution, then ν must have the explosion property with respect to Γ and not charge strongly exceptional sets.

In terms of the solutions the authors show that if (Γ,ν) is a permissible trace, then the (Γ,ν)-solution is indeed a solution. It will be moderate if and only if Γ is empty. Recall that the (Γ,ν)-solution was defined in terms of an additive functional A_ν. The existence of this for the class of ν considered is a consequence of the work in the first paper. The (Γ,ν)-solution will have trace (Γ,ν). If $\Gamma=\varnothing$ then the (Γ,ν)-solution corresponds to a unique solution. However, for an arbitrary solution u, although its trace (Γ, ν) is uniquely defined, the corresponding (Γ, ν)-solution need not be u. It is always the maximal solution with this trace.

This non-uniqueness is addressed in a later paper by the authors [Comm. Pure Appl. Math. **51** (1998), no. 8, 897–936; MR 99f:35046], where a new topology is introduced on the boundary of E, an arbitrary domain in $\mathbf{R}^d$ (or even a Riemannian manifold). This is the fine topology, which is more closely linked to the singularities of solutions. Here, uniqueness of the trace is shown for solutions termed σ-moderate (increasing limits of moderate functions). It is unclear if all solutions are such, but no non-σ-moderate solutions are known. It should be noted that this paper does not use the superprocess but instead a conditioned L-diffusion and is valid for all $\alpha > 1$.

The papers under review are just three of several in the authors' systematic attack on this important problem. Much progress is recorded here in pushing previous results to more general domains, but these are not the final word. For the parabolic problem, the gap between the strong and the weak exceptional sets remains. For the elliptic problem, as indicated, in the authors' more recent paper [op. cit., 1998] a new topology on the boundary of the domain allows one to consider arbitrary domains and get unique probabilistic representations. However, there is still work to do as the gap between σ-moderate solutions and arbitrary solutions remains. Despite this, these papers make interesting reading. They are written in the authors' characteristic style: clearly structured, well motivated and with a generous amount of valuable background material. The authors have worked quite hard to create the proper framework for these questions and in doing so they have created a powerful set of tools for approaching these problems. The tools include a nice mix of diffusion theory, additive functional theory and partial differential equations theory. Readers should be prepared to spend some time looking at references to other papers by the authors when following the proofs. Many theorems appear similar to previous ones as the differences are technical. Typically, only the level of generality has changed. It is a worthwhile endeavor, though. For the interested, I would suggest the elliptic problem first as it seems more self-contained. *John Verzani* (Staten Island, NY)

98e:60135 60J65 60H05 60J60

Donati-Martin, C. (Paris); **Yor, M.** (Paris)

Some Brownian functionals and their laws. (English summary)

Ann. Probab. **25** (1997), *no.* 3, 1011–1058.

In the paper under review the authors develop many themes and re-prove some results appearing in the books [*Some aspects of Brownian motion. Part I,* Birkhäuser, Basel, 1992; MR 93i:60155; *Part II,* Birkhäuser, Basel, 1997; MR 98e:60140] by Yor. In the foreword to *Part II* Yor stated that the aim of *Part I* was to present explicit computations for the laws of a number of interesting functionals of Brownian motion, and that the aim of *Part II* was to look "more into the main technique with which *Part I* was 'working', namely, martingales and stochastic calculus". This article, consisting of ca. 50 pages, can be viewed as a complement and continuation of those two books, although it is written to be self-contained.

In the first section the space-time harmonic functions of Brownian motion are studied. In particular, the well-known representation theorem for positive harmonic functions and the interpretation of the corresponding Doob h-transforms are extended to continuous Gaussian-Markov processes. The case when the space-time harmonic function attains both positive and negative values is also discussed. To such functions we may naturally associate functionals defined by their "principal values". These functionals are generally of the form

$$A_t^f := \lim_{\varepsilon \to 0} \int_0^t f(B_s) \mathbf{1}_{\{|B_s| \geq \varepsilon\}} ds,$$

where B is a Brownian motion and f is an odd function locally integrable in $\mathbf{R} \setminus \{0\}$ and behaving like $1/x$ as $x \to 0$. Note that A^f is well defined by the Hölder continuity of Brownian local time. Section 4 contains a discussion of principal values in general and results for $f(x) = 1/x$ and $f(x) = \coth x$ in particular. From these two particular cases the following identity in law (announced in [M. Yor, op. cit., 1997 (p. 10)]) is seen to hold:

$$\theta_\lambda := \lambda^2 \Big(\Big(\int_0^1 ds \coth(\lambda r_s) \Big)^2 - 1 \Big) \sim \Big(\int_0^1 \frac{ds}{r_s} \Big)^2,$$

where r is the 3-dimensional Bessel bridge from 0 to 0 of length 1 and $\lambda \in \mathbf{R}$. Why it holds is an interesting and challenging problem. Donati-Martin and Yor give a partial explanation in verifying that $\theta_0 \sim \theta_\infty$. It is seen that the functional

$$A_1 := \int_0^{T_1} \frac{ds}{1 - R_s^2},$$

where R is a 2-dimensional Bessel process and T_1 is the first hitting time of 1, is closely connected to θ_λ. In Section 5 it is proved that A_1 has the same law as the first hitting time of $\pi/2$ by a 3-dimensional Bessel process. In analyzing A_1 one is faced with a family of diffusions which is related to processes given by the porous medium equation [S. Benachour et al., Ann. Scuola Norm. Sup. Pisa Cl. Sci. (4) **23** (1996), no. 4, 793–832 (1997); MR 99c:35099].

In Section 2, "Quadratic functionals of Brownian motion", new proofs of some identities in law in [M. Yor, op. cit., 1992] are presented. Especially, it is proved using properties of the Brownian sheet (as announced in [M. Yor, op. cit., 1997 (p. 126)]) that

$$\int_0^1 ds\, (B_s - G)^2 \sim \int_0^1 ds\, b_s^2,$$

where $G = \int_0^1 ds B_s$ and b is a Brownian bridge. It is seen that this technique can also be applied to prove other similar identities, e.g., those arising in mathematical models for polymers.

Section 3 is devoted to extensions of the Ray-Knight theorems. Let l be the local time of B at zero and introduce $\Sigma_\Delta(t) = |B_t| + \Delta^{-1}(2l_t)$, where $\Delta \in C^1$ is strictly increasing with $\Delta(0) = 0$ and $\Delta(\infty) = \infty$. In [M. Yor, op. cit., 1992] the local time process (in the space variable) of Σ_Δ was characterized via a squared Bessel process of generalized dimension using excursion theory (and the martingale approach). Here new proofs are offered based on the analysis of the Sturm-Liouville equation. In particular, taking $\Delta(t) = \delta t$ the authors study the local time process C_a^δ at ∞ as a two-parameter process (in a and δ). It is proved that $(a, \delta) \mapsto C_a^\delta$ is continuous, and the martingale $M_a^\delta = \frac{1}{2}(C_a^\delta - a\delta)$ associated to it is analyzed. The squared Bessel processes of generalized dimension are of some general interest in mathematical finance.

Paavo H. Salminen (Åbo)

97b:60169 60K35 82B20 82B43
Grimmett, Geoffrey (Cambridge)
The stochastic random-cluster process and the uniqueness of random-cluster measures. (English summary)
Ann. Probab. **23** (1995), *no.* 4, 1461–1510.
The random-cluster model is a generalisation of the Ising model, the Potts model and percolation. In this long paper, Grimmett starts with an introduction to the model and a brief description of the main techniques of value such as the FKG inequality and the comparison inequalities. Almost all results are for $q \geq 1$ only, since the FKG inequality is not valid for $q < 1$. I have to warn the reader that what follows is no more than a rough sketch of the main results in this paper. The paper contains many more results and even more beautiful conjectures. The paper is basically self-contained and can serve as an introduction for the non-specialist and at the same time as a very inspiring piece of work for the specialist. Most of the important results are new.

Random-cluster measures are easily constructed on a finite graph, but the definition on an infinite lattice causes some trouble, since there are at least two routes possible. The first is to define the measure first on finite boxes, and then take weak limits. The second is to follow the DLR formalism and study measures which, conditional on the edges outside a finite subgraph, have the correct form. The first measures are called limit random-cluster measures; the second class consists of the random-cluster measures.

Let me summarize the most important issues in this paper sectionwise. In Section 3, the different routes described above are compared, and it is investigated when certain measures are translation-invariant, when one measure stochastically dominates another and when measures are ergodic.

Section 4 is devoted to the uniqueness of random-cluster measures when the two parameters p and q are given. To this end the notion of pressure is introduced. The main result is that for all but (at most) countably many values of the parameter p the random-cluster measure is unique. In two dimensions, uniqueness is obtained for all $p \neq \sqrt{q}/(1+\sqrt{q})$; this strange number is conjectured to be the critical value of p in two dimensions and is the value of a fixed point of a certain mapping involving graphical duality.

In Section 5, Grimmett specialises to the two extremal random-cluster measures with boundary conditions all zeros and ones, respectively. Issues include the possible continuity of the percolation function, and the intuitive fact that if the percolation probabilities in the two extremal measures coincide, then there can be only one random-cluster measure.

The last two sections are devoted to coupling. The "usual" approach from particle systems cannot really be used here, since certain rates are not continuous in the product topology. Grimmett introduces the continuous time Markov process, whose "level sets" have random-cluster measures as their stationary measure. In this way, he obtains an ingenious coupling (for fixed $q \geq 1$) of random-cluster measures for all values of p. All this is done on a finite box first, but in Section 7 he extends this to the whole space. It turns out that there are two different transition semigroups with the same (unique) invariant measure. The first of these semigroups gives rise to a level-set representation of free boundary conditions random-cluster measures, the second to a level-set representation of wired boundary conditions processes.

Ronald Meester (Utrecht)

99j:60157 60K35 60F10 76M35 82C22

Quastel, J. (Davis, CA); **Yau, H.-T.** (New York, NY)

Lattice gases, large deviations, and the incompressible Navier-Stokes equations. (English summary)

Ann. of Math. (2) **148** (1998), *no.* 1, 51–108.

The derivation of different types of continuum equations from mechanical models of particles poses a class of problems important to both mathematics and physics. The classical problem of deriving the Euler or Navier-Stokes equations in their compressible or incompressible versions from the Hamiltonian equations of mechanics is easily the most difficult of these problems. There is not much known in this context unless some noise is present [S. Olla, S. R. S. Varadhan and H.-T. Yau, Comm. Math. Phys. **155** (1993), no. 3, 523–560; MR 94k:60158]. However, if one expands the class of problems to include stochastic dynamics of various kinds, then during the last fifteen years there has been considerable progress in the understanding of the analogous issues [S. R. S. Varadhan, in *Proceedings of the International Congress of Mathematicians, Vol. 1, 2 (Zürich, 1994)*, 196–208, Birkhäuser, Basel, 1995; MR 97d:60163; H.-T. Yau, Doc. Math. **1998**, Extra Vol. III, 193–202 (electronic)].

This article is concerned with the derivation of the incompressible Navier-Stokes (INS) equation from a stochastic lattice gas model in which momenta are conserved. The model roughly speaking involves particles of different types, each moving randomly on a lattice with each type having a different mean velocity. There are two types of interaction. Two or more particles of the same type are not allowed at the same time to be at the same site. Particles of different types that are at the same site can "collide" to become two other types. The model allows only collisions that conserve the total momentum. The basic scale parameter ε is a small positive number that has many roles. Time is speeded up by a factor of ε^{-2}. The density is initially very close to a constant and the difference is of order ε^2 which in addition varies slowly in the scale ε^{-1}. Initially the velocities are of order ε and they also vary slowly in the scale ε^{-1}.

In mathematical terms $\eta(x,v,t)$ equals 1 if at time t there is a particle of type v at x, and 0 otherwise. The particles wait for exponential times with rate 1 and jump from x to $x+y$ with probability $p(y,v)$ that depends on the type. There is of course exclusion that disallows any particle to jump to a site that has already a particle of the same type. We assume that $\sum_y yp(y,v)=v$ so that particles of type v on the average want to move with velocity v. The empirical velocities are $\varepsilon^d\sum_{x,v}\delta_{\varepsilon x}v\eta(x,v,t)$.

It was shown in [R. Esposito, R. Marra and H.-T. Yau, Comm. Math. Phys. **182** (1996), no. 2, 395–456; MR 98h:82042] that the evolution of the rescaled velocities is close to a solution of the INS equation. More specifically, given a smooth solution of the INS equation, the lattice gas that starts from a local equilibrium state that corresponds to the initial value of the equation and the evolution of the stochastic dynamics was shown to stay close to the situation described by the INS equation, as $\varepsilon\to 0$.

In the current work, the results are extended in two very significant ways. The initial value is taken to be arbitrary square integrable functions for the velocities. There may or may not be smooth solutions with this initial value. The suitably scaled empirical velocities, which are random, are here shown to be not far away from the set of weak (Leray) solutions of the INS equation with a probability that goes to 1 as $\varepsilon \to 0$.

This paper derives large deviation bounds as well, i.e. the rate at which the probability of the empirical velocity densities being away from the set of solutions of the INS equation decays. The rate, which is exponentially fast, is of the form $\exp[-C\varepsilon^{-\alpha}]$. For the probability that the velocities do not become divergence free we have a decay rate with the power $\alpha = 2$. On the other hand, for divergence free velocities that fail to satisfy the INS equation the rate is $\alpha = 1$ and the constant that is the rate function for large deviations is explicitly determined as an appropriate negative Sobolev norm. *S. R. S. Varadhan* (New York, NY)

98b:81006 81Pxx 01A75 46N50 47N50 81-01
von Neumann, John
★**Mathematical foundations of quantum mechanics. (English summary)**
Translated from the German and with a preface by Robert T. Beyer.
Twelfth printing.
Princeton Landmarks in Mathematics.
Princeton Paperbacks.
Princeton University Press, Princeton, NJ, 1996. xii+445 pp. ISBN 0-691-02893-1

Many mathematics researchers believe that John von Neumann and David Hilbert were the two greatest mathematicians of the twentieth century. Even those who have other favorite candidates will probably agree that von Neumann is among the top five of such a list. Unlike most mathematics investigators, we can easily gain a glimpse of von Neumann's personality from his writings. Here was a playful man, a man who was bold and fearless. He was not afraid to experiment, to calculate and to reveal his thought processes in public. In the late 1950s and early 1960s when the reviewer and some of his colleagues were in graduate school, von Neumann's book was the only one available (in English) that presented quantum mechanics in a rigorous mathematical fashion. And what a delight it was. It was insightful, motivated, intuitive and understandable physics and it was mathematically consistent. When we took courses in functional analysis, we found that we were seeing the same things over again. Quantum mechanics was Hilbert space analysis and, conversely, much of Hilbert space analysis was quantum mechanics. It was exciting to see that such an elegant, abstract field of mathematics was applicable to revolutionary studies of the natural world. Quantum mechanics was the greatest scientific revolution since Newtonian mechanics. Hilbert space theory was to quantum mechanics as the calculus was to Newtonian mechanics, and von Neumann held the leading beacon of light. Other rigorous books on quantum mechanics and quantum field theory appeared in the mid 1960s such as G. W. Mackey's *The mathematical foundations of quantum mechanics: a lecture-note volume* [Benjamin, New York, 1963; MR **27** #5501] and R. F. Streater and A. S. Wightman's *PCT, spin and statistics and all that* [Benjamin, New York, 1964; MR **28** #4807]. But as good as these books were, they seemed to lack some of the spark of von Neumann's text. We now briefly summarize the contents of this book.

In the preface, von Neumann begins: "The object of this book is to present the new quantum mechanics in a unified representation which, so far as it is possible and useful, is mathematically rigorous." He goes on to explain that his treatment gives a mathematically consistent alternative to that of P. A. M. Dirac's *The principles of quantum mechanics* [Milford, London, 1930; JFM 56.0745.05]. His exposition is a model of succinctness and clarity. For example, the following is taken from his writings in Chapter I. "Quantum theory has made it clear beyond doubt that all elementary processes, that is, all occurrences of an atomic or molecular order of magnitude, obey the 'discontinuous' laws of quanta.... Of great significance is the fact that the general opinion in theoretical physics has accepted the idea that the principle of continuity prevailing in the perceived macroscopic world, is merely simulated by an averaging process in a world which in turn is discontinuous by its very nature. This simulation is such that man generally perceives the sum of many billions of elementary processes simultaneously, so that the leveling law of large numbers completely obscures the real nature of the individual processes."

Chapter I. Introductory considerations. Von Neumann formulates the two theories of quantum mechanics, the Heisenberg matrix mechanics and Schrödinger's wave mechanics, and shows their equivalence. First he considers the Dirac-Jordan method of showing equivalence. This he dismisses as being nonrigorous because of its reliance on delta functions and their derivatives, which he calls "mathematical fictions". He then passes to Schrödinger's method using Hilbert space techniques. He employs the Riesz-Fischer theorem and motivates the need for studying abstract Hilbert spaces.

Chapter II. Abstract Hilbert spaces. Von Neumann defines a (separable) Hilbert space and gives a (now) standard treatment of Hilbert space geometry. He then discusses (linear) operators in Hilbert space, with a special emphasis on projections. He stresses that operators need not be continuous although their domains are always assumed to be dense linear manifolds. Next he states the eigenvalue problem, which has central importance in quantum mechanics. Given a self-adjoint (hypermaximal) operator H, show that H possesses a complete orthonormal set of eigenvectors. Von Neumann points out that although this is possible for finite-dimensional Hilbert spaces, it is not possible in general. He then replaces this problem with one that is equivalent in finite dimensions but is solvable in infinite dimensions. In the finite-dimensional case, the eigenvalue problem is equivalent to diagonalizing H. He shows that this latter problem is equivalent to finding a spectral family $E(\lambda)$ such that H can be written as a Stieltjes integral relative to $E(\lambda)$. This problem can be solved in infinite dimensions and the solution was worked out previously by the author [Math. Ann. **102** (1929), 49–131; JFM 55.0824.02; Math. Ann. **102** (1929), 370–427; JFM 55.0825.02; Ann. of Math. (2) **32** (1931), 191–226; Zbl 002.26703]. The methods and terminology are, for the most part, still used today. Great attention is placed on motivation, and detailed calculations and examples are given to show what holds and what does not hold. Von Neumann then establishes a functional calculus for self-adjoint operators and discusses commuting self-adjoint operators. He next presents his famous theorem that if A and B are commuting self-adjoint operators, then they are both functions of a single self-adjoint operator. This is one of the many cases in which he reveals his thought processes and his willingness to explore various possibilities. In particular, he frequently first gives a heuristic argument without mathematical rigor and then shows how the argument can be made rigorous. This is how von Neumann worked and how he has written this book. One is given a first-hand view of how the master thinks. The road may not be the shortest but we always know where he is headed. In the last section of this chapter, he discusses the trace of operators.

Chapter III. The quantum statistics. Von Neumann begins by postulating that quantum observables correspond to self-adjoint operators on a Hilbert space and (pure) states correspond to unit vectors. Applying the spectral theory developed in Chapter II, von Neumann announces his famous statistical postulate for quantum mechanics. If $A_1, \cdots, A_n$ are commuting self-adjoint operators corresponding to observables and $E_1, \cdots, E_n$ are their respective spectral measures, then for any state corresponding to a unit vector φ, the probability that these observables have simultaneous values in Borel sets $B_1, \cdots, B_n$ is (1) $\|E_1(B_1) \cdots E_n(B_n)\varphi\|^2$. He shows that (1) is equivalent to the fact that if an observable corresponds to the self-adjoint operator A and φ is in the domain of A, then the expectation of the observable in the state φ is $\langle A\varphi, \varphi\rangle$. He then discusses the statistical ramifications of this postulate. Unlike classical mechanics, in which the state of a physical system determines the values of all physical quantities associated with the system, in quantum mechanics the state only determines their probability distributions. Nevertheless, from a given initial state, subsequent states can be calculated causally from Schrödinger's time-dependent wave equation. Why do the operators $A_1, \cdots, A_n$ have to commute? One reason is that the probability (1) would not be additive, otherwise. Von Neumann observes that this deficiency may be caused by an incompleteness in (1) and that there may exist a more general formula, containing (1) as a special case. He then shows that such a generalization of (1) is not possible. In particular, he shows that quantum observables are simultaneously measurable with arbitrarily high accuracy if and only if their corresponding self-adjoint operators commute. He next discusses the uncertainty relations that result when two observables with corresponding non-commuting operators are measured. The interpretation of projections as quantum propositions is developed next. (A later paper with G. Birkhoff [Ann. of Math. (2) **37** (1936), 823–843; Zbl 015.14603] expanded on this topic to create the theory of "quantum logic".) The last section of this chapter is devoted to radiation theory. In this section he shows how transition probabilities between stationary states of an atomic electron can be derived from the statistical postulate. In order to discuss this completely, the theory of electromagnetic radiation is first treated classically and then quantum mechanically. In this treatment, von Neumann elaborates on the earlier work of Dirac. From this development, we witness the beginning phases of quantum electrodynamics. The reasoning is long and involved (forty pages of dense calculations) and is today replaced by the more elegant methods of quantum field theory. But it is quite instructive to see these early calculations and it shows how one of the most difficult paradoxes of quantum theory, namely the dual nature of light as waves and particles, is brilliantly resolved.

Chapter IV. Deductive development of the theory. As shown in Chapter III, the statistical postulate is equivalent to (2) $\operatorname{Exp}(A, \varphi) = \langle A\varphi, \varphi\rangle$, where $\operatorname{Exp}(A, \varphi)$ is the expectation of A in the state φ. Von Neumann now shows that this formula can itself be derived from a few general qualitative assumptions. He argues that the expectation function $\operatorname{Exp}(A)$ must have linearity and positivity properties and, assuming these properties, he shows that there exists a statistical operator (positive, trace-class operator) T such that (3) $\mathrm{Exp}(A) = \mathrm{tr}(TA)$. This gives a generalization of (2) for mixed states, whereas (2) is formulated for pure (extremal) states. (Various investigators have criticized von Neumann's assumption that $\mathrm{Exp}(A)$ should be linear for all observables because this includes observables that may not be simultaneously measurable. However, in 1957, A. M. Gleason [J. Math. Mech. **6** (1957), 885–893; MR **20** #2609] proved that (3) holds under the assumption of linearity only for simultaneously measurable observables.) Von Neumann then shows that there are no dispersion-free (zero variance) expectation functions and that the only extremal expectation functions are those of the form (2). Next he gives an argument against a hidden-variable theory for quantum mechanics. In such a theory, it is assumed that a state φ does not give a complete description of reality. To obtain a complete description, one must augment φ with additional parameters or variables to obtain a complete state. Once this complete state is known dispersions vanish and the theory becomes deterministic. The quantum probabilities given by φ are obtained by averaging over the hidden variables. Von Neumann argues that if a hidden-variable description existed for quantum mechanics, then the present quantum theory must be objectively false and this is contrary to experiment. This constituted the first no-go theorem for hidden variables. (Some of the assumptions of this theorem have been criticized for physical reasons by various researchers and hidden-variable theories have been proposed. Other no-go theorems have been proved under weaker assumptions, including J. S. Bell's famous inequalities [Rev. Modern Phys. **38** (1966), 447–452; MR **34** #8735]. However, the argument over hidden variables in quantum mechanics still rages today [E. G. Beltrametti and G. Cassinelli, *The logic of quantum mechanics*, Addison-Wesley Publishing Co., Reading, Mass., 1981; MR 83d:81008].) The chapter concludes with a discussion of mixed states and how the state of a system can be determined from experiments.

Chapter V. The measuring process. Von Neumann first discusses the peculiar dual nature of the two types of quantum evolution which could not be satisfactorily explained in the previous chapters. For an undisturbed system, the state evolves causally and reversibly according to Schrödinger's time-dependent equation, while for a system that is disturbed by a measurement, the state undergoes a non-causal and irreversible change. In the first type, pure states remain pure, while in the second type, pure states change to mixed states. Von Neumann argues that for this to be consistent, the results of measurements must be independent of the boundary line between the physical system being measured and the observer. To formulate this mathematically, he discusses the structure of composite systems in terms of their components. In modern terminology, this is given by the tensor product of the component Hilbert spaces. Applying this structure of composite systems, the book ends with a discussion of the measuring process.

This book is now in its twelfth printing. The original German text [*Mathematische Grundlagen der Quantenmechanik*, Springer, Berlin, 1932; JFM 58.0929.06] was published sixty-five years ago and was an elaboration of an earlier paper by von Neumann [Nachr. Ges. Wiss. Göttingen Math.-Phys. Kl. **1927**, 1–57; JFM 53.0848.03]. This latter paper appeared only one year after the pioneering work of Heisenberg and Schrödinger. The present English translation was first published forty-two years ago [MR **16**, 654a]. The reviewer first read this book in 1960 and has reread parts of it at least five times. However, it has probably languished, untouched, on his bookshelf for the last twenty-five years. Rereading the text for this review has been a

distinct pleasure and has brought back fond memories of excitement and discovery. How does this classic text now stand in the light of modern developments? One would expect that by now it is hopelessly out-of-date. Of course, more modern treatments have appeared. A representative sample of some of the best includes books by E. Prugovečki [*Quantum mechanics in Hilbert space*, Academic Press, New York, 1971; MR **58** #14457], G. G. Emch [*Algebraic methods in statistical mechanics and quantum field theory*, Wiley-Intersci., New York, 1972; Zbl 235.46085], V. S. Varadarajan [*Geometry of quantum theory*, Second edition, Springer, New York, 1985; MR 87a:81009] and P. Busch, M. Grabowski and P. J. Lahti [*Operational quantum physics*, Springer, Berlin, 1995; MR 96j:81022]. These books are more concise, more direct, and contain simplifications in notation, methods and terminology. They each emphasize different aspects of the subject and they each carry these aspects to more distant frontiers. However, their core material and the germs of most of their ideas are contained in von Neumann's text. Moreover, they do not contain the personality, the motivating thoughts and the seminal calculations of the master himself.
S. Gudder (Denver, CO)

99f:81006 81Pxx 00A30 00A79 81-02

Bub, Jeffrey (College Park, MD)

★**Interpreting the quantum world. (English summary)**

Cambridge University Press, Cambridge, 1997. xiv+298 pp. $49.95.

ISBN 0-521-56082-9

Quantum mechanics is an irreducibly probabilistic theory: due to the noncommutativity of the algebra of observables, or equivalently, the non-Boolean nature of the lattice of propositions, $\mathscr{L}(\mathscr{H})$ (closed subspaces of the Hilbert space), there are no states of a quantum mechanical system in which all dynamical variables have determinate values. This fact raises the problem of the interpretation of the quantum mechanical formalism: the problem of finding a consistent set of rules determining what can be said to be the case with a quantum system in a given state.

In the present book, Bub takes up this question from a realistic point of view. Guided by what he describes as Einstein's realism, he formalizes the problem of interpretation as the problem of specifying a maximal subset of propositions (projections) that correspond to determinate properties of the quantum system. The book consists of three parts: The first part (Chapters 1–3) sets the frame and serves to motivate the constraints that have to be met in specifying a maximal set of determinate observables. Part two is the central Chapter 4 where a uniqueness theorem is formulated and proved. The third part (Chapters 5–8) consists of a fairly comprehensive account of the so-called "no-collapse" interpretations of quantum mechanics, showing that the main interpretations currently on offer are instances of, or can be reconstructed as, "no collapse" interpretations as characterized by the uniqueness theorem.

Chapter 1 summarizes the characteristic non-Boolean structure of the quantum mechanical lattice of propositions as opposed to classical mechanics, and the ensuing difficulties with understanding quantum probabilities in an epistemic sense: in view of the noncommutativity of observables, not all quantities can have simultaneous determinate values. The orthodox interpretation is described as the proposal (ascribed to Dirac and von Neumann) to regard as determinate in a given state (vector) exactly those quantities whose operator has this state as an eigenstate (the so-called eigenvalue-eigenstate link). It is shown how this rule leads to the infamous quantum measurement problem: due to the linearity of the Schrödinger dynamics, a measurement of an indeterminate quantity cannot lead to a definite outcome (a determinate pointer observable). Thus, the orthodox interpretation is not capable of providing an internal account of measurement. Dirac's "quantum jumps" and von Neumann's projection postulate represent the orthodox way of escaping the problem: besides the unitary dynamics of closed systems, there needs to be a description of measurement as a stochastic process effecting the state vector collapse. From the point of view of Einstein's realism, the measurement problem points to the incompleteness of quantum mechanics.

Chapters 2 and 3 consider the incompleteness of quantum mechanics and the possible completions by means of hidden variables. The known no-go theorems due to Bell and to Kochen and Specker are reviewed together with a lucid account of their modern variants. The role of constraints such as non-contextuality, separability and locality is carefully elucidated, thus preparing the ground for an understanding of what will have to be given up in order to maintain the realistic aim: that of providing a probability structure allowing an epistemic interpretation. The essence of the no-go theorems for hidden variables is captured as the problem of constructing the smallest set of observables that generate a Kochen-Specker contradiction: answering this question will contribute to exhibiting those structural features of Hilbert space that are responsible for the crucial difference between quantum and classical mechanics. In the search for a "no collapse" interpretation, the converse quest is pursued of establishing a maximal set of determinate observables without generating a Kochen-Specker contradiction.

The central Chapter 4 gives a precise formalisation of this problem and concludes with a complete solution. A realistic, "no collapse" interpretation sets out to fix a maximal subset of determinate properties, $\mathscr{D}(e,R) \subset \mathscr{L}(\mathscr{H})$, that depends on the given state e (a ray in the Hilbert space $\mathscr{H}$) and a specific chosen "preferred" determinate observable, R. This subset is shown to be uniquely characterized by e and R subject to a set of only four constraints: (1) $\mathscr{D}(e,R)$ is an ortholattice with a sufficient set of $\{0,1\}$-valued homomorphisms, to recover all quantum probabilities as measures over this set of homomorphisms; (2) the eigenspaces r_i belong to $\mathscr{D}(e,R)$; (3) $\mathscr{D}(e,R)$ is invariant under lattice automorphisms that preserve e and R; (4) a weak separability condition. Let e be a ray in an n-dimensional Hilbert space $\mathscr{H}$, R an observable with $m \leq n$ distinct eigenspaces, and $e_{r_i} = (e \vee r_i^{\perp}) \wedge r_i$, $i = 1, \cdots, k \leq m$, the nonzero projections of e onto the eigenspaces r_i. Then the unique maximal "determinate sublattice" $\mathscr{D}(e,R)$ is given by the set $\{p \in \mathscr{L}(\mathscr{H})\colon\ e_{r_i} \leq p \text{ or } e_{r_i} \leq p^{\perp},\ i = 1, \cdots, k\}$. This is the uniqueness theorem. As an illustration, the orthodox interpretation without the addition of the projection postulate is characterized by the choice of the preferred observable $R = I$, the identity operator.

The remaining chapters elaborate on the implications of this fundamental result. In reconstructing the main known interpretations it is made clear what concessions have to be made (nonlocality, contextuality) in order to maintain the structure of a "no collapse" interpretation of the form stipulated by the uniqueness theorem.

Chapters 5 and 6 consider variants of "no collapse" interpretations of "quantum mechanics without observers". It is shown how the measurement problem is avoided. The distinction between dynamical states, represented by rays in Hilbert space evolving according to the Schrödinger equation, and property states, given by truth value assignments in the form of the $\{0,1\}$-valued homomorphisms on $\mathscr{D}(e,R)$, is elucidated and a model for the stochastic evolution of property states given. The role of environmental monitoring is emphasized, deducing the stability of a preferred pointer basis from a "tridecomposition" uniqueness theorem for entangled states of the compound system consisting of object, measuring apparatus and environment. Bohmian mechanics and the different variants of modal interpretations are recovered as observer-free, "no collapse" interpretations with their specific choices of preferred determinate observables. The problem of how a "preferred" observable is (to be) chosen (by nature?) is only briefly addressed, the indication being that this "choice" is a result of contingent features of "our" universe.

Chapters 7 and 8 are devoted to a discussion of the Copenhagen orthodoxy and its modern revivals, the new orthodoxy. Bohr's complementarity interpretation, with its emphasis on the necessity of specifying the full experimental context, allows a reformulation as a "no collapse" interpretation, the corresponding determinate sublattice being exemplified for the Einstein-Podolsky-Rosen and Bell experiment. In the final Chapter 8, Bub argues that environment-induced decoherence and the many-worlds point of view do not, by themselves, avoid or solve the measurement problem. They have to take recourse to the specification of some preferred observables. As a final speculation it is proposed that the quasiclassical domains, defined in the context of the consistent histories approach, may give rise to the choice of a suitable preferred determinate observable.

The book goes a long way in providing rigorous formulations and discussions of problems arising with the interpretation of quantum mechanics. It also indicates areas that require further systematic investigation. Among these are the question as to how to extend the uniqueness theorem to the case of preferred observables that are continuous. The proposed discretization may be regarded at best as a preliminary remedy. Another problem is the quest for Lorentz covariant "no collapse" interpretations. As a final example: the significance of nonideal measurements is explored to some degree, but the extension of the results of this book to the case of general quantum observables represented as positive operator-valued measures seems completely open. References to some basic studies in this direction are given in the bibliography [G. Cassinelli and P. J. Lahti, Found. Phys. Lett. **6** (1993), no. 6, 533–544; Internat. J. Theoret. Phys. **34** (1995), no. 8, 1271–1281; MR 96h:81007].

This book must be seen as a milestone in the search for a coherent, generally acceptable understanding of the quantum world: there seems to be no other account providing a unified reconstruction and systematic assessment (from a realistic point of view) of all major interpretations of quantum mechanics. In making precise the central problems faced by these interpretations, this analysis will undoubtedly provide a source of inspiration to all researchers concerned. The book sets an example of the level of conceptual and mathematical rigor of research into the foundations of modern physics. It is self-contained in the presentation of technical results and proofs. At the same time it is written so as to allow the interested non-expert to follow the main line of philosophical discussion. *Paul Busch* (Hull)

99e:81051a 81Q15 47N50 81Q10 81T17
Bach, Volker (Berlin); **Fröhlich, Jürg** (Zürich); **Sigal, Israel Michael** (Toronto, ON)
Renormalization group analysis of spectral problems in quantum field theory. (English summary)
Adv. Math. **137** (1998), *no.* 2, 205–298.

99e:81051b 81Q15 47N50 81Q10 81V10
Bach, Volker (Berlin); **Fröhlich, Jürg** (Zürich); **Sigal, Israel Michael** (Toronto, ON)
Quantum electrodynamics of confined nonrelativistic particles. (English summary)
Adv. Math. **137** (1998), *no.* 2, 299–395.

Unless prevented by symmetry constraints, atomic or molecular systems prepared in an excited state rapidly return to their ground state by emitting a photon. These spontaneous emissions (or spontaneous absorptions, as the inverse process) in isolated systems are caused by the interaction with the vacuum; each electron is surrounded by a cloud of virtual photons that—in accordance with the uncertainty principle—arise from the vacuum for sufficiently short times. Historically, precisely this problem of spontaneous transitions eventually led to the formulation of quantum electrodynamics. Since then, very efficient computational methods have been developed to treat the atom-photon interaction [see, e.g., C. Cohen-Tannoudji, J. Dupont-Roc and G. Grynberg, *Atom-photon interactions: basic processes and applications*, English translation, Wiley, New York, 1992]. On the other hand, there still exist poorly understood phenomena in the quantum theory of radiation (as, for example, the infrared divergences), and only very few rigorous results are available.

Actually, practically all rigorous studies deal with simplified models constructed in such a way that the systems retain essential physical aspects of the matter-radiation interaction while allowing a more tractable mathematical analysis. In most of the models, the atom or molecule is replaced by a system with explicitly known spectrum. One such substitution is the quantum harmonic oscillator, and the corresponding oscillator-radiation model was studied by A. Arai [J. Math. Phys. **22** (1981), no. 11, 2539–2548, 2549–2552; MR 83b:81032; J. Math. Anal. Appl. **140** (1989), no. 1, 270–288; MR 90j:81035]. A more frequent choice leads to the so-called "spin-boson models", for which the particle Hamiltonian is replaced by a finite-rank operator with N eigenvalues (where $N=2$, usually). Several physical features characterizing the atom-photon interaction have been successfully demonstrated in such spin-boson models. In particular, the existence and the properties of a ground state have attracted some interest [H. Spohn, Comm. Math. Phys. **123** (1989), no. 2, 277–304; MR 90g:82020; A. Amann, Ann. Physics **208** (1991), no. 2, 414–448; MR 92k:82005]. Rather recently, Arai and M. Hirokawa [Sūrikaisekikenkyūsho Kōkyūroku No. 957 (1996), 16–35; MR 99g:81041] proved that for massive photons but without restrictions on the coupling strength there exists a ground state, and, by letting the photon mass tend to zero (while assuming upper bounds on the coupling constant), existence of a ground state is established also without requiring an infrared cutoff. The other principal issue studied within the spin-boson framework is the radiative decay of the excited (spin) states. Here, various methods and techniques have been employed to justify the standard results of formal time-dependent perturbation theory [cf. M. Hübner and H. Spohn, Rev. Math. Phys. **7** (1995), no. 3, 363–387; MR 96i:81292]. Moreover, in a series of papers spin-boson systems with nonzero temperature were investigated [e.g., V. Jakšić and C.-A. Pillet, J. Math. Phys. **38** (1997), no. 4, 1757–1780; MR 98m:82044]. Radiative decay for the realistic one-electron atom but with massive photons has been examined by T. Okamoto and K. Yajima [Ann. Inst. H. Poincaré Phys. Théor. **42** (1985), no. 3, 311–327; MR 87e:81228] using complex scaling methods.

Realistic particle systems are also studied in the papers under review; however, Bach, Fröhlich, and Sigal also make several simplifying assumptions in the model that they consider. Besides the common neglect of the vector character of photons and the ultraviolet cutoff imposed to arrive at well-defined Hamilton operators, the major modification arises from switching off the electron-photon interaction for large separations of electrons from the nuclei. This spatial cutoff amounts to a confining of the electrons, thus allowing an efficient control of the coupling terms. The unperturbed Hamiltonian $H_0 = H_{\rm el} \otimes {\rm id}_{\rm f} + {\rm id}_{\rm el} \otimes H_{\rm f}$ consists of the particle Hamiltonian $H_{\rm el}$ with dynamical electrons but static nuclei, and the standard photon part $H_{\rm f} = \int \omega(k) a^\dagger(k) a(k)\, d^3k$ with dispersion relation $\omega(k) = |k|$. The full Hamiltonian $H(g) = H_0 + W(g)$ includes the electron-photon interaction terms $W(g) = gW_1 + g^2 W_2$ where physically the coupling parameter g is determined by the fine-structure constant. Various assumptions on the coupling functions that are used to derive the respective theorems of this work are collected into three major hypotheses.

The first main result elucidates the ground state problem. For small enough $g \geq 0$ the authors prove that there exists a stable ground state for $H(g)$, i.e., the electrons remain bound to the nuclei although the ground state energy is lowered for $g > 0$. When $g \downarrow 0$, the ground state converges to a ground state of H_0. Furthermore, it is shown that the ground state wave functions decay exponentially for large electronic coordinates; also, sufficient conditions are given that assure uniqueness of the ground state. All these results hold for massless photons; in their proofs, to circumvent the corresponding problems with the continuous spectrum starting at the ground state energy, viz. $\inf \sigma(H) = \inf \sigma_{\rm cont}(H)$, the eigenvalues are isolated from the continuum by introducing an infrared cutoff. Then, estimates uniform in the cutoff parameters are derived so that eventually the limit of vanishing cutoff can be obtained.

The other central results of these papers describe the fate of the excited particle states. For H_0, the excited eigenvalues of the particle Hamiltonian $H_{\rm el}$ are not only embedded into the continuous spectrum $\sigma_{\rm cont}(H_0)$, but each of them also serves as a threshold for a new branch of continuous spectrum. As a consequence, these states are inaccessible to conventional perturbation methods for $H(g)$. The development of a new perturbation scheme for such situations is probably the most important mathematical contribution of this work. Based on renormalization techniques and discussed in detail in the first of the two papers, in contrast to similar schemes in which partition or correlation functions are renormalized, here the renormalization transformations are applied directly on the space of certain Hamilton operators. These operators are obtained from the original (but complex dilated) Hamiltonians by a "Feshbach map", i.e., by an isospectral version of the Feshbach projection method that also is constructed here. By demonstrating that the renormalization transformation is a contractive map on the Banach space of those effective Hamiltonians, it follows that the fixed point of the map leads to an eigenvalue of the original complex dilated Hamiltonian. Via standard arguments these complex eigenvalues can be identified with poles of the resolvent of $H(g)$ on a dense set of states, hence, they are precisely the resonances of $H(g)$. In this way, the authors prove that all excited particle states turn into resonances for $g > 0$; an expansion in g reproduces in lowest order (i.e., $O(g^2)$) the Bethe formula for radiative corrections as the real part of the resonance eigenvalue, whereas the imaginary part of the resonance eigenvalue is identical to the expression from Fermi's Golden Rule generalized to the degenerate case (and physically interpreted as the inverse lifetime of the resonance state). The spectral cuts for $g=0$ open up to cones of width $O(g^2)$ and with cusps at the resonance eigenvalues.

Most of the proofs are lengthy and require heavy technical machinery. The authors have chosen to present the material within two separate and self-contained papers, thus leading to considerable overlap between the two. Readers striving for a quick overview might first consult the more compact articles in which the authors had announced their results [Lett. Math. Phys. **34** (1995), no. 3, 183–201; MR 96i:81291; Found. Phys. **27** (1997), no. 2, 227–237; MR 98i:81048]. *H. Hogreve* (Marseille)

99j:81101 81T05 46L60 47N50 81T40

Wassermann, Antony (Cambridge)

Operator algebras and conformal field theory. III. Fusion of positive energy representations of $\mathrm{LSU}(N)$ **using bounded operators.**

Invent. Math. **133** (1998), *no.* 3, 467–538.

In order to combine the principles of relativistic field theory and quantum physics, quantum field theory (QFT) was developed by W. Heisenberg, P. A. Dirac, etc. about 70 years ago. For a deeper understanding of the underlying principles and for a more concise mathematical formulation, general (or axiomatic) QFT was developed in the 1950s. A first axiomatic approach is given in the Wightman axioms, where a Wightman (quantum) field is an operator-valued distribution satisfying the physically motivated axioms of locality ("Nahwirkungsprinzip"), spectrality, and Poincaré covariance. In the algebraic approach, as it is formulated in the Haag-Araki-Kastler axioms, instead of quantum fields one uses algebras of observables as basic object.

In the latter approach, Doplicher, Haag and Roberts developed a theory of superselection sectors and statistics (DHR theory) in the early seventies. More explicitly, to every bounded space-time region $\mathcal{O}\subset\mathcal{M}$ (Minkowski space) one associates the von Neumann algebra $\mathcal{A}(\mathcal{O})$ of local observables localized in $\mathcal{O}$, and the so-obtained local net of observables has to satisfy further properties (axioms) such as isotony, locality, covariance, Haag duality. Then the C^*-inductive limit $\mathcal{A}=\overline{\bigcup_{\mathcal{O}}\mathcal{A}(\mathcal{O})}$ (algebra of quasilocal observables) is considered, where the overbar denotes the completion in the norm topology. Along these lines, the vacuum representation $\pi_0\colon\mathcal{A}\to\mathcal{B}(\mathcal{H}_0)$ (set of bounded operators acting on the vacuum Hilbert space $\mathcal{H}_0$) is of special interest, and the compositions $\pi_0\circ\rho$ yield a physically relevant class of Hilbert space positive energy representations, where $\rho\colon\mathcal{A}\to\mathcal{A}$ is an endomorphism. The superselection sectors (equivalence classes of irreducible DHR representations of $\mathcal{A}$) are then described by "localized" (DHR) endomorphisms ρ of $\mathcal{A}$, and the statistics (exchange symmetries) are intrinsically encoded in ρ and classified by a statistical parameter λ_ρ. (More precisely, for every superselection sector $[\alpha]$ with finite statistics, a representative DHR endomorphism ρ_α is taken such that ρ_α induces a representation of the observable algebra $\mathcal{A}$ on a copy $\mathcal{H}_\alpha:=(\alpha,\mathcal{H}_0)$ of the vacuum Hilbert space $\mathcal{H}_0$ by $\pi_\alpha(A)\cdot(\alpha,\Psi):=(\alpha,\pi_0(\rho_\alpha(A)\cdot\Psi))$, $A\in\mathcal{A}$, $\Psi\in\mathcal{H}_0$.)

The composition of those endomorphisms gives rise to a "fusion category" of superselection sectors. A further central result in the analysis of superselection sectors is the index-statistics theorem proving that, in any space-time dimension, $\mathrm{Ind}(\rho) = d(\rho)^2$, where $\mathrm{Ind}(\rho)$ is the minimal index of ρ (an extension of the Jones index), and $d(\rho) := |\lambda_\rho|^{-1}$ is the DHR statistical dimension of ρ [see R. Longo, Comm. Math. Phys. **130** (1990), no. 2, 285–309; MR 91j:46083]. Further information on the statistics is contained in the statistics phase $\kappa_\rho := \lambda_\rho/|\lambda_\rho|$ of ρ, and in space-time dimension 4 one has $\kappa_\rho = \pm 1$, a sign labeling the fundamental Fermi-Bose alternative. The discussion of the superselection structure and the statistics is less simple in space-time dimension 2, since Artin braid group statistics enters as a new feature of DHR theory [see K. Fredenhagen, K.-H. Rehren and B. Schroer, Comm. Math. Phys. **125** (1989), no. 2, 201–226; MR 91c:81047].

Using the conformal group as symmetry group (and noticing that it contains the Poincaré group as a subgroup), one is led to a special branch of QFT, called conformal QFT (CFT). In the framework of the Haag-Araki-Kastler axioms, 2-dimensional conformal field theories are described by a net $\mathscr{A}(I)$ of von Neumann algebras acting on a Hilbert space, indexed by a set $\mathscr{I}$ of proper intervals $I \subset S^1$ (unit sphere in $\mathbf{R}^2$), and satisfying the corresponding axioms such as isotony, locality, conformal symmetry and conformal covariance. After CFT was extensively studied between the 1950s and 1970s, the renewed interest in 2-dimensional CFT started with the paper by A. A. Belavin, A. M. Polyakov and A. B. Zamolodchikov [Nuclear Phys. B **241** (1984), no. 2, 333–380; MR 86m:81097], and it is mainly due to the facts that such theories (i) yield a wealth of nontrivial Wightman fields (this is of some interest concerning the existence problem of general QFT since only a very restricted number of such theories are known), (ii) give a better understanding of two-dimensional critical phenomena of statistical mechanics, and (iii) have a rich mathematical structure providing remarkable connections with different areas such as Hopf algebras, low-dimensional topology, knot invariants and subfactors, among many others. Along these lines, using the full strength of the theory of operator algebras, the general spin-statistics relation was derived in [D. Guido and R. Longo, Comm. Math. Phys. **181** (1996), no. 1, 11–35; MR 98c:81121]. The study of CFT from the point of view of operator algebras was also motivated by V. F. R. Jones' suggestion that there might be an operator algebraic explanation of the coincidence between certain unitary representations of the infinite braid group that had turned up independently in the theory of subfactors, exactly solvable models in statistical mechanics and CFT.

The paper under review makes a number of substantial contributions to the study of CFT from the point of view of operator algebras, reconciling the DHR theory, Connes' theory of bimodules and Jones' theory of subfactors with the theory of primary fields of CFT (certain intertwining operators), one of the fundamental concepts in CFT.

A first novel feature is related with the fermionic construction of primary fields. By "compressing" the smeared fermion field $a(f)$, $f \in L^2(S^1, \mathbf{C}^N)$, a vector primary field Φ satisfying the L^2 bound $\|\Phi(f)\| \le \|f\|_2$ is obtained. This is a nonabelian generalization of K.-H. Rehren's construction of bounded Bose fields in two dimensions [Lett. Math. Phys. **40** (1997), no. 4, 299–306; MR 98j:81199]. The most interesting new result is that here all vector primary fields arise by compressing fermions and thus satisfy the L^2 bound above. Furthermore, the fermionic construction yields a proof that the axioms of algebraic QFT are also satisfied in the non-vacuum sectors. This is an extension of the results due to H. J. Borchers, who proved that the axioms are satisfied in the vacuum sector [Comm. Math. Phys. **143** (1992), no. 2, 315–332; MR 92m:46104]. It should be stressed that nontrivial algebraic QFTs in 2 dimensions with finitely many sectors and non-integer statistical dimension are so obtained.

A further highlight of the paper is the treatment of the braiding relations for smeared primary fields. If f and g have disjoint supports in S^1, then $a(f)a(g) = -a(g)a(f)$ and $a(f)a(g)^* = -a(g)^*a(f)$. Similar but more complex "braiding relations" are shown for vector primary fields and their adjoints. The proof of these relations is similar to that of the classical Bargmann-Hall-Wightman theorem given in the textbooks of general QFT.

Since the theory of operator algebras only provides a framework for studying CFT, the author complements it by a detailed and hard analysis of the primary fields and their associated differential equations (transport formulas). Relating positive energy representations of the loop groups $L\mathrm{SU}(N) = C^\infty(S^1, \mathrm{SU}(N))$ to bimodules over von Neumann algebras, and using Connes' tensor product operation on such bimodules ("Connes fusion"), the detailed analysis of the primary fields yields the main thrust of the present paper, namely, the explicit computation of the Connes fusion of positive energy representations. This yields a solution to the "problem of fusion" in CFT (how to "fuse" representations in terms of products of primary fields), since Connes fusion translates directly into a definition of fusion for positive energy representations. Along these lines a computable and manifestly unitary definition of fusion for positive energy representations making them into a tensor category is obtained. The fusion rules so obtained agree with the well-known "Verlinde formulas" [E. Verlinde, Nuclear Phys. B **300** (1988), no. 3, 360–376; MR 89h:81238]. Furthermore, the finiteness of DHR statistical dimension $d(\rho)$ (or Jones index $\mathrm{Ind}(\rho)$) is now a natural consequence. It should also be stressed that the application of Connes fusion to a nontrivial model in QFT is quite new.

{Part I has been reviewed [MR 97e:81143]. Part II is to appear.}

Gerald Hofmann (Leipzig)

99h:81137 81T15 16W30 46L87 81T18

Connes, Alain (Bures-sur-Yvette); **Kreimer, Dirk** (Mainz)

Hopf algebras, renormalization and noncommutative geometry. (English summary)

Comm. Math. Phys. **199** (1998), *no.* 1, 203–242.

In the article under review, the authors bring together two modern research fields in mathematics and physics. In mathematics, foliations provide a large class of examples of noncommutative spaces and lead to an index problem for the transverse hypoelliptic operator [cf. A. Connes and H. Moscovici, Geom. Funct. Anal. **5** (1995), no. 2, 174–243; MR 96e:58149]. The computation of the cocycles in the local index formula turned out to be extremely lengthy even in dimension one. Connes and Moscovici [Comm. Math. Phys. **198** (1998), no. 1, 199–246; MR 99m:58186, see p. D122 in this volume] were looking for an organizing principle for that calculation, which they found in the cyclic cohomology of a Hopf algebra $\mathscr{H}_T$ obtained by the action of vector fields on a crossed product of functions by diffeomorphisms. As for physics, Kreimer [Adv. Theor. Math. Phys. **2** (1998), no. 2, 303–334; MR 99e:81156] discovered that a perturbative quantum field theory carries in a natural way a Hopf algebra structure $\mathscr{H}_R$ given by operations on Feynman graphs. The antipode reproduces precisely the combinatorics of renormalization, i.e. it produces the local counterterms to make the divergent integral corresponding to the Feynman graph finite.

Connes and Kreimer encode the divergence structure of Feynman graphs in terms of rooted trees. This is a new and very convenient concept. A divergent sector in a Feynman graph is represented by a vertex. The root (on top) represents the overall (superficial) divergence. The construction rule for the tree is—in the absence of overlapping subdivergences—to put subdivergences γ_i of a divergence γ into downgoing branches of γ. Disjoint divergences are only indirectly connected via the divergence which contains them as subdivergences. A Feynman graph with overlapping divergences can be expressed as a sum of rooted trees. The coproduct is given by the sum over all possibilities of cutting the tree into pieces, where on the path from any vertex to the root there is at most one cut. The trunk containing the old root forms the right-hand side of the tensor product and the product of cut branches the left-hand side. The same structure (polynomial on the left and linear on the right) appears in the commutative Hopf subalgebra $\mathscr{H}^1$ of $\mathscr{H}_T$ and initiated the joint work of Connes and Kreimer. The precise relation between the two Hopf algebras is given by the operation N of natural growth which maps a tree of n vertices to a sum of n trees each of $n+1$ vertices. These trees are obtained by attaching an additional vertex successively at one of the n previous vertices. The authors prove that the space spanned by 1 and $\{\delta_k\}_{k=1,\cdots,\infty}$, with $\delta_1 = \bullet$ (a single vertex) and $\delta_{k+1} = N^k(\bullet)$, is a Hopf subalgebra of $\mathscr{H}_R$ which is isomorphic to $\mathscr{H}^1$.

The authors provide several results on $\mathscr{H}_R$ derived in analogy to the investigation of $\mathscr{H}^1$ by Connes and Moscovici. First, the Hopf algebra of rooted trees is the solution of a universal problem in Hochschild cohomology: The map $L: \mathscr{H}_R \to \mathscr{H}_R$, which attaches trees as branches to a new root and which is used to prove many properties of $\mathscr{H}_R$, is a 1-cocycle. Next, the extension from $\mathscr{H}^1$ to $\mathscr{H}_T$ by generators of the affine group suggests an analogous extension of $\mathscr{H}_R$. In Section 1 Connes and Kreimer recall that $\mathscr{H}^1$ is the dual of the enveloping algebra of the Lie algebra $\mathscr{A}^1$ of formal vector fields on $\mathbf{R}$ vanishing to order 2 at the origin, and that $\mathscr{H}^1$ itself is isomorphic to the Hopf algebra of coordinates on the group of diffeomorphisms of $\mathbf{R}$ of the form $\psi(x) = x + o(x)$. By analogy, the authors regard $\mathscr{H}_R$ as the Hopf algebra of coordinates on a nilpotent formal group $\mathscr{G}$ whose Lie algebra $\mathscr{L}^1$ they succeed in computing. This group was recently found to be related to the Butcher group in numerical analysis [C. Brouder, "Runge-Kutta methods and renormalization", Preprint, http://xxx.lanl.gov/abs/hep-th/9904014]. It will certainly contain precious information for quantum field theory because the antipode in $\mathscr{H}_R$ governing renormalization is the dual of the inversion operation in $\mathscr{G}$.

In my opinion, this new understanding is a key towards a mathematical foundation of renormalization theory. On the mathematical side, the work of Connes and Kreimer indicates an extension of the Lie algebra of formal vector fields and of the group of diffeomorphisms. Renormalization seems to provide a new mathematical calculus which generalizes differential calculi. This article unifying mathematical and physical concepts gives great impetus to future development.

Raimar Wulkenhaar (Marseille)

97g:81060a 81T30 14J32 32G81 32J17 32S30 83E30
Morrison, David R. (Durham, NC); **Vafa, Cumrun** (Cambridge, MA)
Compactifications of F-theory on Calabi-Yau threefolds. I. (English summary)
Nuclear Phys. B **473** (1996), *no.* 1-2, 74–92.

97g:81060b 81T30 14J32 32G81 32J17 32S30 83E30
Morrison, David R. (Durham, NC); **Vafa, Cumrun** (Cambridge, MA)
Compactifications of F-theory on Calabi-Yau threefolds. II. (English summary)
Nuclear Phys. B **476** (1996), *no.* 3, 437–469.

Originally, superstring theories were believed to exist only in ten dimensions. Recently, however, through non-perturbative advances in the theory, it was realized that there should be an eleven-dimensional theory, which was called M-theory and which has as certain limits some of the known superstring theories upon compactification of one of the dimensions by S^1 or S^1/Z_2. Consequently Vafa [Nuclear Phys. B **469** (1996), no. 3, 403–415; MR 97g:81059] observed that in fact there should be a twelve-dimensional theory, which he called F-theory, which contains both M-theory as well as essentially all known ten-dimensional superstring theories as appropriate limits and compactifications. Each of the superstring theories can of course be compactified also; the most often studied cases are compactifications on T^4 or $K3$ (surfaces), which leave a six-dimensional spacetime, and compactification on Calabi-Yau threefolds down to a four-dimensional space-time.

Part of, and in fact the origin of, the non-perturbative techniques includes the so-called dualities. These are superstring analogues of the strong-weak coupling dualities of $N=2$ super Yang-Mills, which gave rise to the Seiberg-Witten theory. In the context of superstrings, two theories are dual if they give rise to the "same" physics. Although it is in general not possible to actually show that two given theories are dual, there are a series of necessary conditions which can be checked. Paramount among these are the moduli of the theories, which are required to coincide, as well as, in the case in which one of the theories is the heterotic string, the explanation of the gauge group in the corresponding dual theory. This is the context in which this paper was written. In Part I, the authors study compactifications of F-theory on elliptic Calabi-Yau threefolds which admit a $K3$ fibration also. To describe the contents of the paper we must first recall some of the hitherto "known" dualities. One of the first such derived was: a heterotic string compactified on T^4 is dual to a type-IIA string compactified on a $K3$ surface. Later it was discovered that the heterotic string compactified on $K3$ is dual to another heterotic string compactified on $K3$, the so-called heterotic/heterotic duality. Finally, a heterotic string compactified on $K3\times T^2$ (down to four dimensions) turns out to be dual to type II compactified on Calabi-Yau threefolds. Moreover, it was seen from the beginning that the Calabi-Yau threefolds involved should each have an elliptic fibration, or at least a $K3$ one [see P. S. Aspinwall and J. Louis, Phys. Lett. B **369** (1996), no. 3-4, 233–242; MR 97f:81114]. As far as F-theory is concerned, the following is known. F-theory compactified on $K3$ is dual to a heterotic string compactified on T^2; on $K3\times S^1$ this is M-theory compactified on $K3$ which is dual to heterotic strings on T^3; on $K3\times T^2$ this is a type-IIA superstring compactified on $K3$, which is (as mentioned above) dual to heterotic strings on T^4.

Arguing slightly differently, considering F-theory compactified on a Calabi-Yau threefold which is fibered by elliptic $K3$ surfaces, we have a one-dimensional family compactified on $K3$ which is dual to a heterotic string on T^2; considering the entire family then implies that on the heterotic side, this gives an elliptic $K3$ compactification with an appropriate gauge bundle, and this is what the paper studies. Then the most interesting question in this respect is, given a heterotic string compactified on an elliptic $K3$ surface K, find the Calabi-Yau threefold C on which F-theory must be compactified to give the dual of the former. Of particular interest is the gauge group, and in the first paper it is shown how this corresponds to the "bad fibers" of the elliptic fibration on the Calabi-Yau threefold C. Then heterotic compactifications discussed here with gauge group $E_8 \times E_8$ are determined by the instanton bundle with instanton numbers (k_1, k_2), $k_1 + k_2 = 24$. The case $(k_1, k_2) = (24, 0)$ corresponds to breaking one E_8 factor, leaving the other as gauge group, while $(k_1, k_2) = (12, 12)$ corresponds to complete symmetry breaking, i.e., to the trivial gauge group. The authors show that the Calabi-Yau threefold $C = C(k_1, k_2)$ which is dual to the compactification of a heterotic string on $K3$ with instanton numbers (k_1, k_2) is given as a hypersurface in the weighted projective space $\mathbf{WP}^4_{1,1,k_1-12,2k_1-20,3k_1-30}$. Incidentally, the numbers $(16, 8)$ give the manifold the authors briefly discuss in the heterotic string, SO(32) case, and they conjecture that these two theories have indeed the same moduli space. This conjecture was verified by P. S. Aspinwall and M. Gross [Phys. Lett. B **387** (1996), no. 4, 735–742; MR 97j:81234]. A final interesting aspect of Part I concerns the heterotic/heterotic duality in six dimensions. Here an explanation is given. In these cases the Calabi-Yau threefold fibers over a $\mathbf{P}^1 \times \mathbf{P}^1$; the ratio of the two Kähler classes of the base and fiber is essentially the heterotic string coupling constant. The geometric symmetry given by switching the base and fiber (which does not work on the general Hirzebruch surface $\mathbf{F}_n$ but only on $\mathbf{F}_0 = \mathbf{P}^1 \times \mathbf{P}^1$) inverts the heterotic string coupling constant, i.e., is a strong-weak coupling duality.

Part II extends the analysis of Part I in several directions. First of all, the basic duality considered above, F-theory compactified on an elliptic Calabi-Yau threefold and heterotic string compactified on an elliptic $K3$, is rederived from a different point of view. This time the authors begin with a $K3$ which is elliptic (with constant modulus) and two fibers of type II* (corresponding to the gauge group $E_8 \times E_8$ at $z = 0, \infty$ on the base $\mathbf{P}^1$). This is a local description of a single fiber of the Calabi-Yau threefold (which also has a $K3$ fibration), and this local picture is glued to a global one with some elementary arguments. This leads to more insight into the simple correspondence between instanton numbers and the degree n of the rational surface $\mathbf{F}_n$, which is the base of the elliptic fibration. This leads to a recipe for how to build up F-theory duals to heterotic vacua in arbitrary dimensions compactified on elliptic manifolds with given bundle data. This also leads to a reinterpretation of F-theory/heterotic duality in eight dimensions as the statement of exchange symmetry between the fiber and the base of F-theory.

Part II then goes on to study compactifications which were not dealt with in Part I. In addition to obtaining a wide class of heterotic vacua, the authors find theories without any heterotic dual and in fact theories without any tensor multiplet at all. Finally, transitions between the vacua found are discussed. The mathematical structure of such transitions is relatively well understood: degenerations and extremal transitions. The latter are discussed in more detail for elliptic Calabi-Yau's. In particular it is argued that for an elliptic Calabi-Yau over $\mathbf{F}_2$, the degeneration obtained by blowing down the infinite section of self-intersection -2 and smoothing gives rise to an elliptic Calabi-Yau over $\mathbf{P}^1 \times \mathbf{P}^1$, while the same for base $\mathbf{F}_4$ deforms to an elliptic Calabi-Yau over $\mathbf{P}^2$. Finally, some of the less transparent physics of this situation is also discussed. *Bruce Hunt* (Offenbach)

97h:81163 81T30 11F22 11F55 17B67 17B81 32G81 81R10
Harvey, Jeffrey A. (Chicago, IL);
Moore, Gregory [Moore, Gregory W.] (New Haven, CT)
Algebras, BPS states, and strings. (English summary)
Nuclear Phys. B **463** (1996), *no.* 2-3, 315–368.

This physics paper brings together various mathematical fields that have been heavily influenced by string theory: modular and automorphic forms, generalized Kac-Moody algebras, elliptic genera, and rational curves on Calabi-Yau spaces, and proceeds to derive some very interesting relations among these fields.

One of the most striking applications of string theory and conformal field theory to complex algebraic geometry is the phenomenon of mirror symmetry. Mirror symmetry relates two different Calabi-Yau threefolds with the property that the rational curves on one threefold can be computed by considering the variation of Hodge structure on the mirror manifold. This result is best formulated using a generating function for the numbers of rational curves of various degrees, the so-called prepotential $\mathcal{F}$. Although hard to compute from first principles, $\mathcal{F}$ has a classical interpretation on the mirror manifold as a section of a line bundle over the moduli space of complex structures. As such it also carries natural automorphic properties.

In string theory mirror symmetry is a result of the identification of two a priori different string compactifications, the so-called type IIA and IIB strings. In the past two years it has become clear that there are many more of these identifications, collectively known as string dualities. This opens up different ways to explore string theory in mathematics and, in particular, to compute rational curves on threefolds in alternative ways. In this paper the duality with the heterotic string is explored. The dual object is here a holomorphic $E_8 \times E_8$ vector bundle over the product of a $K3$ surface and an elliptic curve. These string compactifications are believed to be dual to type IIA strings on Calabi-Yau threefolds that can be written as $K3$ fibrations. The heterotic string thus gives a third representation of the counting function $\mathcal{F}$.

Let us sketch the important ingredients in this heterotic string theory computation: $\mathcal{F}$ is represented as an integral over the fundamental domain of the moduli space of the elliptic curve E. The integrand is directly related to the elliptic genus of the restriction of the $E_8 \times E_8$ vector bundle to E. Such elliptic genera are by definition modular forms. The authors compute this integral, and prove that it can be written as the logarithm of a function Φ that has an infinite product representation. Some further manipulations relate Φ to the prepotential $\mathcal{F}$. This allows one to express the number of rational curves of given degree on the dual Calabi-Yau threefold directly in terms of the expansion coefficients of the elliptic genus, and thereby relate them to modular forms.

Central in this paper are the automorphic properties of the functions Φ and $\mathcal{F}$ that follow from the modular integral representation. These are necessarily automorphic forms of the discrete group $\mathrm{O}(n,2;\mathbf{Z})$, which is the automorphism group of the Lorentzian lattice of the same signature that appears naturally in the heterotic string. Automorphic forms of these groups having product representations have been studied in detail by R. E. Borcherds [Invent. Math. **120** (1995), no. 1, 161–213; MR 96j:11067] in work related to his studies of the Monster group and related generalized Kac-Moody (GKM) algebras. The authors have essentially discovered a beautiful and natural integral representation of these automorphic forms, from which many of their properties follow directly. Indeed, many of the results of Borcherds are elegantly rederived in the present paper.

These results suggest that behind all this is a natural GKM algebra. Indeed, the automorphic function Φ can be considered as the denominator of this algebra. The relation between GKM algebras and string theory was also crucial in Borcherds' work. The precise form of this algebra, and the interpretation in terms of correspondences of moduli spaces of cycles on the dual Calabi-Yau threefold, are discussed in detail in the follow-up paper by the authors [Comm. Math. Phys. **197** (1998), no. 3, 489–519]. *Robbert Dijkgraaf* (Amsterdam)

98c:81167 81T30 83E30
Hořava, Petr (Princeton, NJ); **Witten, Edward** (Princeton, NJ)
Heterotic and type I string dynamics from eleven dimensions. (English summary)
Nuclear Phys. B **460** (1996), *no.* 3, 506–524.

Prior to the "second superstring revolution", it was known that there were five consistent superstring theories, propagating in ten uncompactified dimensions: the type IIA and IIB theories, the SO(32) type I theory, and the SO(32) and $E_8\times E_8$ heterotic theories. The revolution was in the understanding of the strong coupling behaviour of string theory, and the new non-perturbative physics was organized in terms of powerful "duality" symmetries.

Soon after the revolution, for a time the situation was as follows: The type IIB theory was conjectured to be self-dual, while the type IIA theory at strong coupling was conjectured to be dual to *eleven*-dimensional supergravity. No less dramatic was the conjectured duality between the two SO(32) theories showing that the distinction between open (type I) and closed (heterotic) string theories was merely an artefact of hitherto being limited to perturbation theory. Notably, there was no new knowledge about the $E_8\times E_8$ theory, in contrast to the case for the other four theories.

In the case of the type IIA theory in particular, compactifying the eleven-dimensional supergravity on a circle of radius R_{10} and then sending $R_{10}\to 0$ yields the type IIA string at weak coupling. Conversely, the strong coupling limit of the type IIA string is eleven-dimensional supergravity. The relationship between the string coupling, $g_{\rm IIA}$, and the radius is (in eleven-dimensional length units) simply $g_{\rm IIA}\sim R_{10}$. This relation of type IIA string theory to the eleven-dimensional supergravity began the realization that all of the string theories, and the dualities between them, could possibly be understood in terms of one unifying framework.

The first step in that direction was as follows: After compactification on a circle of radius R_9, to nine dimensions, it has been known for some time that the type IIA and type IIB string are perturbatively T-dual to one another, the physics of one theory at radius R_9 being identical to the physics of the other compactified on a circle of radius α'/R_9, where α' is the inverse string tension.

Combining this T-duality result with the new result about the compactification of eleven-dimensional supergravity on a circle R_{10}, it was realized that both the type IIA and IIB strings, and their dualities, could be summarized in terms of simply compactifying the eleven-dimensional supergravity on a torus, whose geometric data are summarised in terms of R_{10} and R_9. Supplementing this prescription with the knowledge about the T-duality between the strings, relating large and small sizes of compactification, one had a new perspective on the origin of two of the string theories.

If one wishes to be able to *derive* the complete set of stringy properties, including the T-duality facts, instead of putting them in by hand from previous string theory knowledge, then we have to understand more about the short-distance (high-energy) behaviour of the complete theory, going beyond the eleven-dimensional supergravity. Eleven-dimensional supergravity had long been known to have no sensible short-distance physics, which is one of the reasons it lay unused for a long time. Therefore, new short-distance physics is expected to take over to augment the theory into a consistent theory conjectured to exist, called "M-theory".

M-theory would have to possess remarkable properties, giving rise to all of the string theories in various geometrical and dynamical limits, and the dualities between the strings would become geometrical relationships between these limits. The full dynamics of the type IIA/IIB system would simply be captured by the various geometrical limits of M-theory compactified on a torus, their coupling being related to the geometry of the torus. (Indeed, the consequences of this for the type IIB theory were worked out by Aspinwall, and also independently by J. H. Schwarz [Phys. Lett. B **360** (1995), no. 1-2, 13–18; MR 97a:81181a; erratum; MR 97a:81181b; Nuclear Phys. B Proc. Suppl. **49** (1996), 183–190; see MR 97e:81003; Phys. Lett. B **367** (1996), no. 1-4, 97–103; MR 97c:81158], showing that the self-duality of the theory was encoded in the geometry of the torus.) This M-theory would have eleven-dimensional supergravity as its simplest low-energy limit, and the IIA/IIB unified relationship above would then follow.

Even in the absence of complete knowledge of the structure (or even the basic definition) of M-theory, such a point of view is progress, and it was realized that it would be useful to extend this point of view to include the other string theories.

In a seemingly unrelated test of duality, J. Polchinski and Witten [Nuclear Phys. B **460** (1996), no. 3, 525–540; MR 97e:81122] made considerable progress in the duality program by exhibiting more precise evidence for the conjectured duality between the SO(32) type I and SO(32) heterotic strings to be true. However, they did this using purely stringy techniques—the "D-brane technology"—and needed to make no reference to eleven-dimensional physics.

There were therefore two outstanding problems: (1) the urgent need to gain insights about the strong coupling nature of the $E_8\times E_8$ heterotic string, hopefully in terms of a duality symmetry; (2) the desire to understand the $E_8\times E_8$ heterotic string theory and the SO(32) type I/heterotic system in terms of the M-theory conjecture, if possible. This would give a common origin to all string theories.

This groundbreaking paper of Hořava and Witten solves both problems at once. Most simply put, they conjecture that the $E_8\times E_8$ heterotic string at strong coupling is related to M-theory compactified on an "orbifolded" circle—essentially a line interval—of length L_{10}. The coupling of the heterotic string is proportional to L_{10}, and so as the interval shrinks to zero size we obtain again the weakly coupled string, in analogy with the type IIA case.

The philosophy of the proposal is that, operationally, although we do not know the details of M-theory, we know that at long distances (low energy) the theory is simply eleven-dimensional supergravity. Therefore, away from the edges of the interval, the theory should be acceptably described in those terms. At the (ten-dimensional) edges of the interval (i.e., the orbifold fixed points) we expect that new short-distance physics arises to supply the M-theory "orbifold" physics we do not understand.

In order to test whether this conjecture is consistent with known physics the paper proceeds in two major steps. (1) The first step is to constrain the form of the unknown high-energy physics arising at the boundaries by indirect means. This is accomplished by a powerful argument about the structure of the anomalies in eleven-dimensional supergravity theories with boundary. Of course, regardless of the details of the physics giving rise to the massless degrees of freedom of a quantum theory, those states need to satisfy certain relationships amongst themselves in order for the theory to be anomaly-free, and this fact is enough to constrain the properties of the M-theory data. In particular, it is shown that exactly 248 massless states—transforming as vectors under the ten-dimensional Lorentz groups of the boundaries—arise at each boundary. This is enough to explain the appearance of the $E_8 \times E_8$ gauge symmetry: those massless vectors are the gauge generators, each boundary supplying an E_8. (2) The second step, an analysis of how the fundamental heterotic string arises as a macroscopic string solution in ten dimensions, is presented as further motivation for the conjectured duality: It has its origins in the long-known membrane solution of eleven-dimensional supergravity, which wraps one of its two dimensions of extent on the interval, leaving a string in ten dimensions. (A similar mechanism has been shown some time ago to give rise to the type IIA fundamental string, there, the membrane wrapping on the shrunken circle.)

Again in analogy with the type II system, further geometrical insight can be gained by recalling known facts about the properties of the heterotic string compactified on a circle of radius R_9: In nine dimensions, the $E_8 \times E_8$ heterotic string is T-dual to the SO(32) heterotic string compactified on a circle of radius α'/R_9, where α' is the inverse string tension.

Recalling that there is already a strong/weak coupling duality statement about the SO(32) heterotic string, the authors deduce a relationship between M-theory and the SO(32) type I/heterotic string system as well. They explain the implied relationship between the M-theory-on-an-interval geometry and the string theory picture that Polchinski and Witten used to understand SO(32) type I/heterotic duality, showing how that picture arises in a limit of the new proposal.

This paper therefore completes the program of finding a unifying geometrical statement about the origins of the five string theories: the type IIA/IIB system arises from placing M-theory on a torus while the type I/heterotic system (both SO(32) and $E_8 \times E_8$ heterotic) arises from M-theory compactified on a cylinder. The duality program was thus strongly motivated to find a complete definition of M-theory, a program which continues in earnest at the time of writing.

The paper is extremely well written and well organized. It is one of the papers crucial to the "second superstring revolution", and despite the astonishingly rapid pace of development in the field, it is still a useful working reference.

Clifford V. Johnson (Lexington, KY)

98h:81096 81T30 81T13 81T60

Hanany, Amihay (Princeton, NJ); **Witten, Edward** (Princeton, NJ)

Type IIB superstrings, BPS monopoles, and three-dimensional gauge dynamics. (English summary)

Nuclear Phys. B **492** (1997), *no.* 1-2, 152–190.

Since the beginning of the "Second Superstring Revolution", when we achieved understanding of strong coupling properties of supersymmetric string theory (succinctly stated in terms of "strong/weak coupling duality" and "M-theory"), it has been recognised that there may be new things to learn from it about the dynamics of field theories. The chain of reasoning which leads to this expectation is quite standard: field theories naturally arise in string theory in many ways, often as the result of very controllable low-energy limits. In principle, if one can find a way of embedding the field theory of choice into a given string theory configuration, then a study of the properties of the stringy configuration should yield information about the field theory, at both weak and strong couplings, depending upon the nature of the embedding.

Ideally, for detailed study by such methods, one should have a means of controlling as many aspects of the field theory as possible, ultimately developing a complete dictionary which translates between the physics of the "parent" stringy configuration and that of the "daughter" field theory dynamics.

Such a highly desirable scenario has become a reality in recent times for quite complicated situations. At the time of writing, there are a large number of testable non-trivial statements concerning the strong coupling properties of a considerable variety of field theories (especially gauge field theories), which have been made possible via an explicit set of techniques for constructing the required parent string theory configurations. The paper under review here is one of the crucial landmark papers which have assembled the existing knowledge into a sharp working tool, and further, showed us how to use it to discover and organise many new and important phenomena.

The particular means of embedding the field theory into string theory relevant here is rooted in the fact that the low-energy physics of extended objects propagating in the ten space-time dimensions of superstring theory is often well described by a supersymmetric quantum theory defined on an auxiliary space. In particular, an object with p extended dimensions (e.g., a membrane when $p=2$) sweeps out a $(p+1)$-dimensional path in space-time as it propagates. Aspects of the physics of the "p-brane" as it follows this "world-volume" trajectory are captured succinctly in terms of a quantum theory defined on the $(p+1)$-dimensional world-volume itself.

This language is already quite familiar from the old perturbative string theory: The superstring itself, as it sweeps out a $(1+1)$-dimensional world-volume (often called the "world-sheet"), has much of its low-energy (and weak-coupling) properties described in terms of $(1+1)$-dimensional quantum field theories called "sigma-models", containing (among others) ten bosonic fields. The expectation values of the fields in the theory correspond to the position of the string in ten-dimensional space-time, while the couplings between them correspond to expectation values of space-time closed string excitations like the graviton, dilaton and axion, encoding details of the background geometry. Furthermore these sigma models flow to "conformal field theories" in the infrared. In two dimensions, the conformal symmetry group of one of these theories is infinitely large, and the resulting calculability of properties of these theories led to great progress in the understanding of the physics of weakly coupled string theory in the years during and immediately following the "First Superstring Revolution".

In general, the details of the quantum theory depend heavily upon the type of brane in question. Often, we know only a restricted amount of even basic information about such theories, and so progress in determining the physics of the relevant extended objects is somewhat limited. In the case in hand, however, some of the branes studied are "D-branes", the heroes of the Second Superstring Revolution, and the quantum theories governing their low-energy dynamics are supersymmetric gauge theories. This is a consequence of the fact that D-branes are most easily described in perturbative type-II string theory as Dirichlet boundary conditions, specifying that open string sectors are present, but restricted to have the string endpoints lie at specific values of $9-p$ coordinates. This is equivalent to asking that the endpoints be free to move in a $(p+1)$-dimensional subspace of space-time, thus defining the world-volume of the Dp-brane (e.g., a membrane-like D-brane is a "D2-brane").

The dynamics of the extended object thus implicitly defined turns out to be encoded in the physics of the open string sector. For example, the collective excitations of the Dp-brane, describing its motion as a solitonic object propagating in ten dimensions, are described by the massless modes of the open-string sector, which appear as fields in the $(p+1)$-dimensional theory on the world-volume: Expectation values of bosonic scalar fields ($9-p$ of them) arising from oscillations transverse to the brane represent the position of the brane in the $9-p$ transverse coordinates. Meanwhile, the $p+1$ modes arising from fluctuations along the world-volume form components of a vector field in the $(p+1)$-dimensional theory. This vector potential is associated with a U(1) gauge symmetry representing the redundant motion of the center of mass.

Additional complications such as the coincidence of many identical branes are described without departure from the field theory language: n branes oriented in the same way would supply n copies of the fields described above. Succinctly, one replaces the U(1) gauge symmetry with $\mathrm{U}(1)^n$. These fields simply arise from the open strings beginning and ending on the branes, and there is such a family for each brane. There are open strings beginning on one brane and ending on a different one, as well, but these give modes with a lowest mass proportional to the distance the open string has to stretch. When the branes become coincident, therefore, there are new massless fields in the theory and the situation is more complicated, but not unfamiliar in the language of field theory: the $\mathrm{U}(1)^n$ gauge theory is enhanced to a $\mathrm{U}(n)$ gauge theory. Another way to think about this result is to realise that the open strings have a redundancy of n branes upon which to end at any given world-volume point, giving a local internal (gauge) symmetry. The process of separating branes from each other and reducing the amount of gauge symmetry is like the Higgs mechanism: Some of the massless scalar fields representing the position of the branes get expectation values, and give masses to some of the components of the gauge fields, thereby reducing some of the gauge symmetry.

The above recollection of facts serves to show how aspects of the geometry and simple dynamics of certain extended objects in string theory are isomorphic to properties of gauge field theory. These aspects of D-brane technology have been known in the literature for some time, and have been used to study many aspects of the physics of D-branes and the string theories in which they live. Unfortunately, this way of embedding field theory into string theory alone does not work for very complicated or "realistic" field theories. The field theories have half the supersymmetry of the parent type-II string theories, because, as solitons, D-branes in a given orientation provide a linear realisation of only half of the supercharges on their world-volume. There are therefore 16 supercharges in the theory, equivalent to $N = 4$ supersymmetry in four dimensions, which is four times as much supersymmetry as one expects for more "realistic" field theories, which allow richer physics for fewer supersymmetries.

It would be interesting to learn how to embed gauge theories with families of matter multiplets charged under the gauge symmetry (i.e., quark flavours) and with fewer supersymmetries, like $N = 1$ or $N = 2$ in four dimensions. In order to do so, Hanany and Witten show how to combine the features described above with other knowledge. For example, it has been known in the literature for some time that certain p-branes, instead of remaining as extended objects of infinite extent in p directions, can be made to terminate upon another type of extended object known as an NS-fivebrane, forming a junction, making them infinite in only $p-1$ directions. (An NS-fivebrane is another type of soliton of string theory, with five extended dimensions. It is not described perturbatively in terms of the addition of open string sectors, like a D-brane, but instead has a partial description in terms of perturbative closed string theory. This description breaks down near the strongly coupled core of the soliton, as the perturbative string techniques lose their predictive power. This means that some care will need to be exercised when using such branes, as the paper makes very clear.)

That such junctions of branes can be formed can be deduced from properties of string theory, including the use of strong/weak coupling duality. It is possible to deduce in this way that many types of D-branes can end on NS-fivebranes, and further that D-branes can end on certain other types of D-branes. It is a crucial fact that the endpoints of the branes resemble BPS saturated monopoles of the gauge theory living on the world-volume of the branes upon which they end.

Using this knowledge, it is possible to form quite complicated configurations of branes in this manner, and to make sense of much of the dynamics of the configuration and the resulting field theory at low energy. As an example, consider the following type-IIB string construction as presented by Hanany and Witten: A pair of parallel NS-fivebranes of infinite extent are placed a certain distance apart. A family of n parallel and coincident D3-branes is suspended between these two original branes, bridging the gap between them. This arrangement produces a gauge symmetry $\mathrm{SU}(n)$ on the D3-branes. (The $\mathrm{U}(1)$ center of $\mathrm{U}(n)$ is redundant here, referring to the centre of mass of the system.) The fact that there are two separate orientations of branes in the problem further reduces the number of supercharges to eight. The low-energy physics of the configuration as viewed from the D3-branes is described by a field theory which arises on the infinite parts of their world-volume. We therefore have $(2+1)$-dimensional $\mathrm{SU}(n)$ gauge theory with $N = 4$ supersymmetry.

The next important step is to establish further whether one can correctly recover the space of allowed supersymmetric vacua (the "moduli space") of the field theory. This should be obtained here by studying the allowed geometrical deformations of the brane configuration which preserve supersymmetry. In the example above, the "Coulomb branch" of the moduli space (with gauge symmetry $\mathrm{U}(1)^n$) is parameterised classically by the positions of the n D3-branes within the three spatial dimensions of the NS-fivebranes transverse to their two infinite dimensions. In order to make a precise statement about the Coulomb branch, Hanany and Witten notice that the physics of the configuration can be viewed in a different way.

From the point of view of the physics of the world-volume of the NS-fivebranes (restricting to those transverse 3+1 dimensions) the theory is an $N=4$ supersymmetric SU(2) gauge theory (as would have been the case had the NS-fivebranes been coincident) Higgsed to U(1) (by their separation). Furthermore, the endpoints of the n D3-branes form n BPS monopoles of that gauge theory. It turns out that the moduli space of n BPS monopoles of SU(2) gauge theory is already well known. It is parameterised by precisely the same number of parameters as the D3-brane positions, and furthermore the metric on that moduli space is of a known hyper-Kähler manifold. (In the simplest case of $n=2$, the manifold is the Atiyah-Hitchin manifold.)

In this way, Hanany and Witten have—with surprising ease—reproduced dramatically a result known to field theorists: The Coulomb branch of the moduli space of $N=4$ supersymmetric $(2+1)$-dimensional pure SU(n) gauge theory is isomorphic to the moduli space of a collection of n BPS monopoles of SU(2) gauge theory in 3+1 dimensions.

It is possible to make a more complicated theory by adding another NS-fivebrane to the chain and then stretching m D3-branes between it and the previous NS-fivebrane. In this way, one gets a U(m) gauge symmetry from the second family of branes. Hanany and Witten show that there is a family of matter fields in the gauge theory arising from open strings which can stretch from the first family of D3-branes in the chain to the second family. As the open strings see m coincident D3-branes on one side and n on the other, the fields they produce in the spectrum are charged in the fundamental representation of each gauge group. We therefore have gauge group U$(n)\times$U(m) with "bifundamental" matter in the (n,m) representations for our $N=4$ supersymmetric $(2+1)$-dimensional field theory.

In order to establish this latter fact, it has to be realised that the open strings have to cross through the core of the middle NS-fivebrane to do so. It must be stressed that it has to be shown that there is no new physics in the field theory arising from the core of the NS-fivebranes, which is known to be a strongly coupled region in the theory. Hanany and Witten argue convincingly that the simplest result (described above) is indeed the correct one, using properties of string theory duality and field theory as guides. Field theories with product groups with more factors, and with bifundamental matter charged under the neighbouring factors, can be constructed in this way by extending the chain of NS-fivebranes in the obvious way. Using techniques and observations analogous to those in the example above, many powerful statements are made concerning moduli space of vacua of these theories.

Other types of field theories may be constructed by adding m D5-branes to the problem. They can be used either to replace the NS-fivebranes, or in addition to them. (They are chosen to be of infinite extent, and oriented in such a way as to preserve the $N=4$ supersymmetry, while still allowing D3-branes to end on them.) Adding them to the example above between the NS-fivebranes yields m new families ("flavours") of fields to the problem, charged in the fundamental of the SU(n) gauge group. This therefore constructs a sort of supersymmetric QCD in $2+1$ dimensions. The properties of the moduli spaces of such theories are also studied extensively using the geometrical properties of the means of embedding them into type-IIB string theory.

After successfully embedding many field theories into the string theory and establishing a geometrical dictionary of all of its vacua and deformations, Hanany and Witten show how to deduce an important class of new observations about certain allowed string theory deformations, using known properties of the field theory as a guide. One such phenomenon is the creation of an additional D3-brane when a D5-brane and an NS-fivebrane cross one another. The new brane appears stretched between the D5-brane and the NS-fivebrane. The deduction is made using a continuity argument about the BPS spectrum of the field theory, forcing one to conclude that a D3-brane appears to support part of the BPS spectrum which would have otherwise disappeared after the crossing. Further supporting constructions based on string theory arguments are presented later in the paper to support the field theory deduction.

Having established a very powerful set of techniques for embedding the field theories into string theory, the next step the paper takes is to discover whether there are any properties of the strong coupling physics of the field theory which might be established as a consequence of the strong/weak coupling duality properties of the parent configuration. It is shown that the techniques can be used to derive a certain strong/weak coupling duality between families of field theories and classes of certain other field theories, often referred to as a three-dimensional "mirror symmetry" in the literature. (This duality was conjectured in the field theory literature using a number of non-trivial clues from intricate field theory calculations over many years.) To derive this duality, the key feature used—in addition to those already discussed—is the strong/weak coupling duality of the type-IIB string theory, which exchanges this configuration with another in which the roles of the NS-fivebranes and D5-branes are reversed, while the D3-branes remain invariant. This work constituted an important string theory proof of the conjecture, demonstrating the power of the technique, and establishing a new set of tools for use in other contexts.

The paper presents a very general set of techniques for embedding complicated gauge field theories into string theory and studying their properties in that larger context. Following on from this work, using other types of D-branes, and including more types of extended objects, workers in the field now routinely establish results for field theories of various dimensionalities (such as four), with various types of matter content and often with fewer supersymmetries. Many complex field theory dualities and other strong coupling results have been reproduced and extended, by using type-IIA and type-IIB string theory duality as a probe of the strong coupling field theory physics.

This is an extremely important paper. It pulls together many known results from D-brane technology, string duality and the physics of NS-fivebranes to establish a powerful dictionary and an extensive toolbox of techniques for exploiting the interface between string theory and field theory. It is a very readable paper, and will remain a valuable resource in the field. *Clifford V. Johnson* (Lexington, KY)

98j:81248 81T30 81V17 83E30

Banks, T. (Piscataway, NJ); **Fischler, W.** (Austin, TX); **Shenker, S. H.** (Piscataway, NJ); **Susskind, L.** (Stanford, CA)

M theory as a matrix model: a conjecture. (English summary)

Phys. Rev. D (3) **55** (1997), *no.* 8, 5112–5128.

String theory is the leading candidate for a unified theory of all forces of nature, including quantum gravity. The essential idea of string theory is that point particles are generalized to one-dimensional objects—strings. In that way one is led to study the quantum mechanics of the loop space $\mathscr{L}M$ of the underlying space-time manifold M. Because of internal consistency reasons this space-time M must be a 10-dimensional Einstein manifold. The fact that the theory of general relativity emerges naturally from the quantization of loop spaces is one of the fundamental properties of string theory and establishes its claim as a model of quantum gravity.

A string that propagates in space-time sweeps out a two-dimensional Riemann surface or complex curve. Quantum mechanical corrections lead to a sum over topologies of these surfaces. This relation with the theory of algebraic curves has led to many mathematical applications such as quantum cohomology, mirror symmetry, and Gromov-Witten invariants. Also, the quantization of loop spaces has had a profound impact on the representation theory of infinite-dimensional algebras such as the Virasoro algebra and Kac-Moody algebras.

It is, however, good to realize that all these beautiful mathematical applications only concern a perturbative formulation of string theory. One typically constructs series of the form $\sum_g \lambda^{2g-2} A_g$, where λ is the so-called string coupling constant, and where A_g denotes a relevant quantity computed at genus g. The objects A_g are obtained by integrating volume forms on the moduli space of curves $\mathscr{M}_g$ and can be shown to grow roughly as $(2g)!$. Therefore the perturbation series is at best asymptotic to the "real" non-perturbative answer, which is a hypothetical function $A(\lambda)$ with the above Taylor expansion.

For a long time the issue of the non-perturbative formulation frustrated string theorists, since no definition from first principles was available. Indeed it is important to stress that string theory is thought to be more than a theory of strings, in the same way that quantum field theory is more than a theory of particles. Only at weak coupling, $\lambda \sim 0$, do we expect that loop spaces and Riemann surfaces appear as an appropriate framework to encode the asymptotic expansion of the theory.

Ideally one would like to have a well-defined Hilbert space of quantum states of string theory and a Hamiltonian that describes the time-evolution of these states. This formulation should then reduce at weak coupling to the sum over surfaces formulation. (It must be said that this assumption, that quantum gravity is described by a unitary quantum mechanical system, is not universally shared. Especially the work of Hawking on quantum black holes is sometimes seen as a counter-argument to this assumption.)

Matrix theory, as formulated in the paper under review, is the first attempt at a non-perturbative definition of string theory that provides us with exactly such a Hilbert space and Hamiltonian. Although it suffers from some limitations—most notably a lack of a covariant formulation—it has withstood various nontrivial tests during the last year. We will also briefly indicate a few of these more recent developments. (For a review and references see Banks [Nuclear Phys. B Proc. Suppl. **67** (1998), 180–224; MR 99k:81218].) However, to understand the nature of the matrix theory conjecture we first have to discuss the two main ingredients that have generated the current progress in string theory: M-theory and D-branes.

M-theory is the realization, first presented by E. Witten [Nuclear Phys. B **443** (1995), no. 1-2, 85–126; MR 96f:81104], that ten-dimensional string theory has a hidden eleven-dimensional origin. More precisely, string theory on M^{10} with coupling λ has an eleven-dimensional formulation as M-theory on $M^{10} \times S^1$, where the circle has radius λ. This gives a geometric interpretation to the string coupling, putting it on equal footing with the moduli of M^{10}. For a long time this was known to be true for the supergravity theory that emerges as the low-energy approximation to string theory. Indeed, eleven-dimensional supergravity has been a candidate for a unified theory of quantum gravity and M-theory has in some sense reestablished this point of view. That string theory has a similar higher-dimensional origin came as a great surprise, although there has been earlier work hinting at a theory of membranes in eleven dimensions. (These membranes partly explain the "M" in M-theory, as do "magic" and "mystery".)

One of the indications of the eleven-dimensional origin of string theory is the appearance of D-particles. (Remarkably, string theory also contains pointlike objects.) Through the well-known Kaluza-Klein mechanism, momentum in the compactified eleventh direction manifests itself as a charge in ten dimensions. Particles that carry this electric charge are called D-particles. These particles can be considered to be black hole-like solitons with a mass proportional to $1/\lambda$. Their role in string theory has always been enigmatic. This puzzle was solved brilliantly by J. Polchinski [Phys. Rev. Lett. **75** (1995), no. 26, 4724–4727; MR 96m:81185], who showed that a D-particle (and more generally, a D-brane) is simply a point in space where strings can begin or end. (The string therefore has Dirichlet boundary conditions, explaining the "D" in D-brane.)

If we have a collection of N D-particles this gives rise to N^2 different types of open strings, beginning at one and ending at another D-particle. If their excitations are ignored, these open strings give rise to nine Hermitian $N \times N$ matrix coordinates X^i_{IJ} ($i = 1, \cdots, 9$, $I, J = 1, \cdots, N$), one for each space-like direction. Therefore the quantum mechanics of N D-particles is described by a matrix version of a point particle moving in $\mathbf{R}^9$. D-particles can be viewed as noncommutative versions of ordinary particles (that are recovered for $N = 1$). Indeed D-branes have introduced the subject of noncommutative geometry, as developed by Alain Connes, into the heart of string theory. On an even more fundamental level, D-branes and matrix theory point to a deep equivalence between gauge theories and quantum gravity.

If we suppress the fermionic terms necessary for supersymmetry, the matrix theory Lagrangian takes the simple form

$$\int dt \operatorname{Tr}\Big(\sum_i (\dot{X}^i)^2 + \sum_{i<j} [X^i, X^j]^2\Big).$$

This system has a natural $\mathrm{U}(N)$ gauge symmetry, with the X^i in the adjoint representation. It furthermore carries an obvious SO(9) symmetry, that can be extended to the Galilean group ISO(9) by including constant shifts of $\dot{x}^i$.

Matrix quantum mechanics can be obtained by starting with a supersymmetric Yang-Mills theory in ten dimensions. If we dimensionally reduce to $10 - d$ dimensions, for example by choosing the space-time to be $T^d \times \mathbf{R}^{10-d}$ and sending the volume of the torus T^d to zero, we obtain a gauge theory with d adjoint-valued scalar fields—the components of the connection in the compact directions. For the case $d = 9$ we are left with the above nine matrix-valued coordinates. (In that case the remaining one-dimensional connection just serves to project on $U(N)$ invariant states.)

Matrix theory takes these two ingredients, eleven-dimensional M-theory and the matrix formulation of D-particles, and conjectures that in the limit $N \to \infty$ the matrix quantum mechanics gives the exact description of eleven-dimensional M-theory. This seems at first sight very unlikely, since D-particles are only part of the physics of string theory in ten dimensions. How can they account for everything in eleven dimensions?

Before we discuss the origin of this conjecture let us first explain how the correspondence is supposed to work. Let us start with the matrix quantum mechanics. This model is straightforwardly defined—but very difficult to analyze—and just as in any other quantum mechanical system it will lead to a Hilbert space and a Hermitian Hamiltonian. If we consider the commutative limit, $[X^i, X^j] = 0$, the coordinates are labeled by the simultaneous eigenvalues that take values in the symmetric product $S^N\mathbf{R}^9 = \mathbf{R}^{9N}/S_N$. (One can think of this space as $\mathfrak{t}^9/W$, with the $\mathfrak{t}$ the Cartan Lie algebra of $U(N)$ and W the Weyl group.) Although generically the stabilizer of a point in $S^N\mathbf{R}^9$ is just $U(1)^N$, there are strata in this symmetric product where eigenvalues coincide and the stabilizer will be given by $U(n_1) \times \cdots \times U(n_k)$, with $n_1 + \cdots + n_k = N$ a partition of N. The claim is now that the matrix quantum mechanics for each $U(n)$ factor, labeled by a common eigenvalue x^i, has a unique ground state—or more precisely a unique multiplet of the super Galilean group. This ground state can be considered as a bound state of n D-particles. Of course the Hilbert space will contain many more excited states.

The M-theory interpretation of the matrix quantum mechanics is now as follows. We consider a space-time manifold of the topology $S^1 \times \mathbf{R}^{10}$, but with a compact *null* direction. For instance, we can choose the flat metric $ds^2 = dx^+dx^- + (dx^i)^2$ and periodically identify $x^- \equiv x^- + 2\pi$. This gives a discrete momentum $p_- = -i\partial/\partial x^-$ with spectrum $\mathbf{Z}$. Note that picking the null direction breaks the eleven-dimensional (super) Poincaré group to ISO(9). The dual coordinate x^+ can be picked as a "time" direction with conjugate momentum p_+.

Now fix the momentum $p_- = N$. One thing we expect to find in M-theory is a Fock space of supergraviton states. So, for each partition $N = \sum n_i$, we have asymptotic states corresponding to supergraviton multiplets with momentum $p_- = n_i$. We have seen that these particular states are indeed present in the matrix quantum mechanics. One now identifies the M-theory generator p_+ with the matrix Hamiltonian. In the large N limit we expect—but this is certainly the most nontrivial part of the matrix theory conjecture—that the null compactification effectively disappears and that a description of M-theory on flat $\mathbf{R}^{11}$ emerges. In particular one expects to find a (nonlinear) realization of the eleven-dimensional super Poincaré group.

Why do we expect the matrix theory conjecture to hold? A very naive argument observes that in the limit $\lambda \to \infty$, where the eleventh dimension decompactifies, the D-particles become massless and might therefore constitute the natural fundamental degrees of freedom out of which everything else can be constructed. A more precise argument has later been sketched by N. Seiberg [Phys. Lett. B **408** (1997), no. 1-4, 98–104; MR 98j:81299]. Start with M-theory with a *space-like* circle and with a definite momentum N along this compact direction. This means we have a system consisting of N D-particles plus all the open and closed strings. Now consider the limit in which the compactified direction becomes null. A careful scaling analysis then shows that in this limit everything decouples except the open string ground states, leading to the matrix model. However, one still has to take the large N limit to obtain M-theory on $\mathbf{R}^{11}$.

A particularly striking calculation performed in the paper under review is the scattering of two gravitons. At the classical supergravity level this is a standard computation resulting in a velocity-dependent force that scales as v^4/r^7 with v the relative velocity and r the transverse distance. Amazingly, the same result follows directly from the matrix quantum mechanics through a one-loop quantum computation, including the precise numerical coefficient. Although it is now understood that this equivalence is in some sense coincidental, it still shows the remarkable way in which gauge theory and gravity interact.

One aspect of M-theory or string theory that is particularly well understood within the matrix model is the compactification on tori and the appearance of duality symmetries. After compactification on $\mathbf{R}^{11-d} \times T^d$ M-theory is expected to carry a discrete automorphism group $E_d(\mathbf{Z})$, known as the U-duality group. (By this notation one denotes a discrete version of the maximally non-compact version of the Lie algebra labeled by the Dynkin diagram E_d. For example, $E_3(\mathbf{Z}) = \mathrm{SL}_2(\mathbf{Z}) \times \mathrm{SL}_3(\mathbf{Z})$, $E_4(\mathbf{Z}) = \mathrm{SL}_5(\mathbf{Z})$, $E_5(\mathbf{Z}) = \mathrm{O}_{5,5}(\mathbf{Z})$, etc.) These big automorphism groups are one of the most mysterious aspects of M-theory. Ideally one would like to see these groups be realized manifestly. Straightforward ideas from D-branes tell us that within matrix theory these compactifications lead to $(d+1)$-dimensional super Yang-Mills theory on the dual torus $(T^d)^* \times \mathbf{R}$. These gauge theories can then be seen to carry the expected U-duality groups, although often only on the quantum level. For example, in the case $d = 3$, the gauge theory has a manifest "classical" $\mathrm{SL}_3(\mathbf{Z})$ acting on the 3-torus and a "quantum" $\mathrm{SL}_2(\mathbf{Z})$ strong-weak coupling "S-duality". The mathematical consequences of the S-duality where discussed by C. Vafa and Witten [Nuclear Phys. B **431** (1994), no. 1-2, 3–77; MR 95k:81138]: it predicts that the generating function of Euler numbers of moduli spaces of instantons on a four-manifold is a modular form for $\mathrm{SL}_2(\mathbf{Z})$.

Although matrix theory has managed to put some of the pieces of the puzzle of string theory in place, it still leaves much to be desired. Since it makes use of light-cone quantization, various dynamical issues that are often considered the heart of quantum gravity, such as topology change, cosmology, the nature of the vacuum, and the cosmological constant, become inaccessible. (Although it has been possible to use matrix theory to analyze some aspects of black hole physics.) But undoubtably, matrix theory, as proposed in this paper, is a very important first step in the direction of a definition of string theory. *Robbert Dijkgraaf* (Amsterdam)

99e:81204a 81T30 81T60 83E30
Maldacena, Juan (Cambridge, MA)
The large N limit of superconformal field theories and supergravity. (English summary)
Adv. Theor. Math. Phys. **2** (1998), *no.* 2, 231–252.

99e:81204b 81T30 81T60 83E30
Gubser, S. S. (Princeton, NJ); **Klebanov, I. R.** (Princeton, NJ);
Polyakov, A. M. (Princeton, NJ)
Gauge theory correlators from non-critical string theory. (English summary)
Phys. Lett. B **428** (1998), *no.* 1-2, 105–114.

99e:81204c 81T30 81T60 83E30
Witten, Edward (Princeton, NJ)
Anti de Sitter space and holography. (English summary)
Adv. Theor. Math. Phys. **2** (1998), *no.* 2, 253–291.

Recently there has been much progress in understanding non-perturbative aspects of string theory. This has been closely related to many new insights into non-perturbative features of supersymmetric gauge theories. One of the most important tools has been the use of Dp-branes. These non-perturbative objects have a $(p+1)$-dimensional worldvolume and their dynamics is governed, at low energies, by maximally supersymmetric Yang-Mills (SYM) theory with gauge group U(1). A crucial observation by Witten [Nuclear Phys. B **460** (1996), no. 2, 335–350; MR 97c:81162] was that with N parallel branes the gauge group is enhanced from the naively expected $U(1)^N$ to $U(N)$. There have been conjectures that the large N limits of some gauge theories are in fact equivalent to string theory, such as the recent conjectures for matrix models based on the large N limits of 0- and 1-dimensional SYM theories which are the low energy theories on the worldvolume of N D-instantons [N. Ishibashi et al., Nuclear Phys. B **498** (1997), no. 1-2, 467–491; MR 98k:81221] and D0-branes [T. I. Banks et al., Phys. Rev. D (3) **55** (1997), no. 8, 5112–5128; MR 98j:81248, see p. F40 in this volume]. Maldacena has provided yet another conjectured equivalence between string theory and gauge theories, and this was later quantified by Gubser, Klebanov and Polyakov and independently by Witten, who gave a prescription for the calculation of correlation functions in the field theory in terms of supergravity (or string) theory. These three important papers which are reviewed here have been the basis for much further work where Maldacena's conjecture has been (so far successfully) checked and new predictions have been made. This conjecture is closely related to previous studies of large N QCD [G. 't Hooft, Nuclear Phys. B **72** (1974), no. 3, 461–473] and the relation to string theory [Yu. M. Makeenko and A. A. Migdal, Phys. Lett. B **88** (1979), no. 1-2, 135–137; A. M. Polyakov, Nuclear Phys. B **164** (1980), no. 1, 171–188; MR 81c:81060].

Maldacena has made the remarkable conjecture that supergravity in dimensions $D=10$ or 11 on spaces $\mathrm{AdS}_{p+1}\times S^{D-p-1}$ is equivalent to p-dimensional superconformal field theory. The main discussion centres on the example of type IIB supergravity on $\mathrm{AdS}_5\times S^5$. This can be obtained as the near horizon limit of N coincident D3-branes in 10-dimensional Minkowski space. The supergravity solution can be trusted when all curvatures are small and this is the limit of large gN, where g is the string coupling. However, it is now well known that the low energy limit of N coincident D3-branes is the $\mathcal{N}=4$ SYM theory with gauge group $U(N)$.

The connection arises because this low energy limit is equivalent to the near horizon limit and Maldacena's conjecture is that these two descriptions of the same physical system must be equivalent. Since the gauge coupling is given by $g_{\rm YM}^2 \sim g$ this is in fact a duality relation: 4-dimensional large N $\mathcal{N}=4$ $\mathrm{U}(N)$ SYM at large 't Hooft coupling $g_{\rm YM}^2 N$ is described by type IIB supergravity on $(\mathrm{AdS}_5 \times S^5)_N$ with small curvature, where the N indicates that the 5-form field strength is proportional to N.

In the first paper the complex coupling constant of the 10-dimensional theory is identified with the gauge coupling and theta angle. Both theories have an $\mathrm{SL}(2,\mathbf{Z})$ duality and these are identified. The amount of supersymmetry is the same since $\mathcal{N}=2$ supersymmetry in 10 dimensions is the same as $\mathcal{N}=4$ superconformal symmetry in 4 dimensions.

The weakest form of this conjecture would be that the duality is only correct in the $N \to \infty$ limit, but it is natural to expect that finite N corrections can be included. Then quantum corrections in supergravity would correspond (since gN is fixed) to $1/N$ corrections in the large N field theory. The equivalence must then be between gauge theory and string theory.

Maldacena then discusses further examples, including the M-theory 2- and 5-branes as well as cases with reduced supersymmetry where the space is given by the near horizon geometry of intersecting branes.

Gubser, Klebanov and Polyakov give a prescription for calculating the generating functional for connected Green's functions in the SYM theory. They identify this with the minimum of the supergravity action subject to appropriate boundary conditions. This can be generalised to finite N with $1/N$ corrections being due to loop effects in supergravity and corrections of inverse powers of the 't Hooft coupling $g_{\rm YM}^2 N$ corresponding to α' corrections due to string effects. This is of course an important refinement of Maldacena's conjecture since it allows explicit calculations to be performed and nontrivial consistency checks on the validity of the conjecture.

The connection between asymptotic conformal symmetry and AdS spaces is reviewed along the lines of J. D. Brown and M. Henneaux [Comm. Math. Phys. **104** (1986), no. 2, 207–226; MR 87d:83034] for AdS_3. Boundary conditions are chosen to preserve the asymptotic form of AdS_{d+1} and it is shown that diffeomorphisms preserving these boundary conditions are specified by a d-vector ζ^μ defined on the boundary which must obey the conformal Killing equation $\zeta_{\mu,\nu} + \zeta_{\nu,\mu} = (2/d)\zeta^\kappa_\kappa$.

Examples are given of calculations of 2-point functions in 4-dimensional SYM. The first example is of a minimally coupled massless scalar φ in AdS_5. This couples to an operator $\mathcal{O}$ on the D3-brane worldvolume via an interaction $\int d^4x\, \varphi\mathcal{O}$. The 2-point function for $\mathcal{O}$ is calculated by solving the equations of motion to determine φ in AdS_5 for given boundary conditions and evaluating the action with this solution. The result

$$\langle \mathcal{O}(x)\mathcal{O}(y)\rangle \sim N^2|x-y|^{-8}$$

is in agreement with the free field result for small $g_{\rm YM}^2 N$. So the prediction from supergravity is that this result also holds at large $g_{\rm YM}^2 N$. This result can easily be extended to calculate the 2-point function for the stress-energy tensor $T_{\mu\nu}$ which couples to perturbations of the metric $g_{\mu\nu} = \eta_{\mu\nu} + h_{\mu\nu}$ via the worldvolume interaction $\frac{1}{2}\int h^{\mu\nu} T_{\mu\nu}$. The calculation is the same as for the scalar since at quadratic order in the supergravity action, a graviton $h_{xy}(k)$ polarised along the brane with momentum k orthogonal to the xy-plane is described by the minimal massless scalar action. This gives the correct 2-point function of the stress-energy tensor and the normalisation gives the expected result for the central charge (conformal anomaly) $c = N^2/4$.

Calculations using this method can be extended to higher point functions, but this is left to future works. Instead Gubser, Klebanov and Polyakov proceed to analyse the supergravity predictions for the dimensions of operators coupling to higher partial waves of φ in S^5 and then operators coupling to massive string states.

It is shown that operators which couple to the lth partial wave must have dimension $4+l$. Operators with such SO(6) quantum numbers in SYM are of the form

$$\int d^4 x e^{ik\cdot x} \mathrm{tr}[X^{(i_1} \cdots X^{i_l)} F_{\mu\nu} F^{\mu\nu}(x)]$$

and the supergravity prediction is that their bare dimensions are not corrected. The analysis is similar for operators coupling to massive string states at excitation level n and mass given by $m^2 = 4n/\alpha'$. However, now the result is that such operators have, at large $g_{\mathrm{YM}}^2 N$, dimensions of the order $2\sqrt{n g_{\mathrm{YM}} \sqrt{2N}}$, although effects from higher derivative terms in the action have not been taken into account in deriving this. So the conclusion is that it is expected that operators coupling to massless string states have dimensions which are protected by supersymmetry whereas operators coupling to massive string states do not have dimensions protected by supersymmetry and in fact will have unbounded anomalous dimensions at large $g_{\mathrm{YM}}^2 N$.

Witten derives the same results but from a slightly different perspective. Whereas Gubser, Klebanov and Polyakov consider the supergravity background for N branes, Witten simply considers an AdS space with some field theory living on the boundary. He shows the equivalence of the boundary theory and the bulk theory, at least in the case of Euclidean theories. The advantage is that it is then natural to extend the conjecture, as Witten does, to an equivalence between a field theory on the boundary and the theory in the bulk of any suitable space, whether or not the space has any known interpretation in terms of branes.

Witten first considers in detail the relation between the theory on the boundary of a Euclidean AdS space and the theory in the bulk. He derives the result that given the value of a field on the boundary the field is uniquely defined in the bulk. For example, Euclidean AdS_{d+1} can be described as a $(d+1)$-dimensional ball, B_{d+1}, with a particular metric which diverges at the boundary, S^d. If we consider, as the simplest example, a massless scalar field φ with equation of motion $D_i D^i \varphi = 0$ then the fact that there are no normalisable zero modes and the requirement that φ should be regular lead to the result that φ is completely determined by its value on the boundary, φ_0. A metric can be defined on the boundary by multiplying the bulk metric by the square of a function f with a simple zero on the boundary to cancel the second order pole. This boundary metric is then only defined up to the choice of f, which can be redefined by multiplying by any nonzero function on the boundary $f \to e^w f$. Since this transforms the boundary metric by $ds^2 \to e^{2w} ds^2$ it is clear that the boundary does not have a unique metric but only a conformal structure.

Next Witten gives a prescription which relates correlation functions of the boundary conformal field theory to the bulk supergravity theory. Assuming that there is a coupling between φ_0 and a conformal field $\mathcal{O}$ of conformal dimension d, $\int_{S^d} \varphi_0 \mathcal{O}$, the conjecture is that the generating functional of the CFT is equivalent to the partition function of supergravity: $\langle \exp(\int_{S^d} \varphi_0 \mathcal{O}) \rangle_{\mathrm{CFT}} = Z_S(\varphi_0)$. $Z_S(\varphi_0)$ is defined in terms of φ_0 by $Z_S(\varphi_0) = \exp(-I_S(\varphi))$, where I_S is the supergravity action evaluated with φ being the unique extension of φ_0 into the bulk.

Some explicit examples are given. It is shown that the 2-point function $\langle \mathscr{O}\mathscr{O} \rangle$ has the correct form: $\langle \mathscr{O}(x)\mathscr{O}(y) \rangle \sim |x-y|^{-2d}$. The analogous computation is also performed for a free U(1) gauge theory where the 2-point function of the conserved current again agrees with expectations.

A Chern-Simons term in the bulk appears to be problematic since the bulk action is only gauge invariant up to a total derivative which produces a boundary term. However, this boundary term is precisely of the form of the chiral anomaly of the boundary theory.

Witten then considers the case of massive scalars. Now there are no regular solutions approaching a nonzero constant at the boundary. However, by multiplying φ by a suitable mass-dependent power of the function f used to define a metric on the boundary, the asymptotic form of φ can be related to a well-defined scalar φ_0 on the boundary. Now, conformal scaling of the metric will scale φ_0 and so φ_0 is a conformal density with conformal dimension defined in terms of the mass, m, of φ. Again, if φ_0 couples to some operator $\mathscr{O}$ via $\int_{S^d} \varphi_0 \mathscr{O}$, the generating functional for $\mathscr{O}$ is identified with the supergravity partition function evaluated using the unique extension of φ_0 to φ. From the relation between m and the conformal dimension of φ_0, $\mathscr{O}$ must have conformal dimension $\Delta = (d + \sqrt{d^2 + 4m^2})/2$.

Operators in a CFT can be classified according to their dimensions; those with $\Delta < d$ are relevant, $\Delta = d$ are marginal and $\Delta > d$ are irrelevant. This structure is clear in the bulk theory where these limits correspond to $m^2 < 0$, $m^2 = 0$ and $m^2 > 0$ respectively. States with negative m^2 are allowed in AdS spaces and are not tachyonic provided $m^2 \geq -d^2/4$. This corresponds to the lower bound on the dimensions of operators in CFT, $\Delta \geq d/2$.

Now the relation between masses of states in AdS and conformal dimensions of operators in the boundary CFT is used to check that there is indeed a one-to-one correspondence between known Kaluza-Klein states in supergravity on $\mathrm{AdS}_5 \times S^5$ and operators in $\mathscr{N} = 4$ SYM. In fact this can only be checked for operators with dimensions which are protected by some symmetries from receiving corrections in the strong coupling limit. These are expected to correspond to Kaluza-Klein modes in supergravity since they are correspondingly protected from quantum and string corrections to their masses, being in small representations of supersymmetry. This protection is important since string excitations would have a large mass $\sim g_{\mathrm{YM}}^2 N$ which would spoil the predictions. Witten gives three classes of operators where the conformal dimensions are believed to be exact and states in suitable representations are indeed known to exist with precisely the correct masses predicted by the relation between Δ and m^2.

Witten proposes a very natural generalisation of Maldacena's conjecture: a field theory on some compact d-dimensional manifold M should be equivalent to supergravity (or string theory) on a D-dimensional manifold Y where $D = 10$ or 11 corresponding to type IIB or M-theory. Y should have the form $X \times W$ (at least asymptotically), where W is a closed manifold and X has boundary M with the metric on X having a double pole at the boundary. Then the same relationships should follow as between CFT on $M = S^d$ and supergravity on Euclidean $X \times W = \mathrm{AdS}_{d+1} \times S^{D-d-1}$. One important point is that there may be no unique choice of X and in this case the supergravity partition function should be evaluated by summing over the different choices X_i: $Z = \sum_i \exp(-I_S(X_i))$. This creates the possibility of having a phase transition at large N. This can happen because the supergravity action at large N is of the form $I_S(X_i, N) = N^\gamma F(X_i)$ for some fixed $\gamma > 0$. So the partition function at large N will be dominated by the contribution from the manifold X_i that gives the smallest $F(X_i)$. If varying some parameter changes which manifold dominates the partition function then there will be a large N phase transition.

An example of this is field theory on $S^1 \times S^{d-1}$, which is finite temperature field theory, where the S^1 is interpreted as the Euclidean time direction. There are two possible choices of manifold X. One, X_1, is AdS space with some periodic identifications which is topologically $S^1 \times \mathbf{R}^d$, and the second, X_2, is a Schwarzschild black hole in AdS space which is topologically $\mathbf{R}^2 \times S^{d-1}$. The space X_1 allows two different spin structures corresponding to the choice of periodic or anti-periodic conditions on spinors when moving once round the S^1. The space X_2 has a unique spin structure corresponding to the anti-periodic case. So for periodic fermions only X_1 can contribute and there cannot be a phase transition for the partition function $\operatorname{Tr} e^{-\beta H}(-1)^F$. However, the partition function $\operatorname{Tr} e^{-\beta H}$ can have a large N phase transition and Witten demonstrates that this does in fact occur, at least for the case $d = 3$. He shows in this case, using results of S. W. Hawking and D. N. Page [Comm. Math. Phys. **87** (1982/83), no. 4, 577–588; MR 84d:83024], that X_2 only contributes for small β and that for small enough $\beta < \beta_c$, $I(X_2) < I(X_1)$ so that there is a phase transition at some critical temperature $T_c \sim \beta_c^{-1}$. It is expected that this behaviour holds for general dimensions d.

Witten assumes the $d = 4$ case is qualitatively similar to the $d = 3$ case and compares the predictions for the free energy $F = -\ln Z$ via the conjectured correspondence to the expected behaviour of 4-dimensional large N QCD. At low T the vacuum energy is of order N^2 but the particle masses and excitations are of order 1 so in the large N limit $F/N^2 \sim \beta E/N^2$, which agrees with the predictions for the manifold X_1. However, at high T we expect the quarks and gluons to be effectively free so $F \sim N^2$. The temperature dependence is determined by $(3+1)$-dimensional conformal invariance and so at large N, $F/N^2 \sim T^4 \sim \beta^{-4}$, which agrees with the prediction from X_2.

Since the appearance of these three papers there has been much interest in this topic. The correspondence between supergravity and field theory has been used to calculate properties of non-perturbative (non-supersymmetric) QCD in the large N limit. Some examples are the calculation of the linear potential between a quark and an anti-quark characteristic of confinement and also the calculation of the glueball mass spectrum, although this is unfortunately of limited practical use since the calculation is only valid at large 't Hooft coupling. As an introduction to this exciting topic, I would highly recommend the original paper by Maldacena. It is very readable and outlines the main ideas behind the conjecture. To do actual calculations, a more precise formulation is required, and this is exactly what is provided by the important papers of Gubser, Klebanov and Polyakov, and Witten. Both these papers propose the method for calculating correlation functions in terms of the supergravity partition function and provide support for this relation with similar examples. For this I would recommend either excellent paper. For the reader interested in pursuing further generalisations of Maldacena's original conjecture,

Witten's paper is very useful and the discussion of finite temperature field theory and large N phase transitions is particularly interesting. *Douglas J. Smith* (Durham)

99h:81183 81T30 81-01 81-02 83E30
Polchinski, Joseph (Santa Barbara, CA)
★String theory. Vol. I. (English summary)
An introduction to the bosonic string.
Cambridge Monographs on Mathematical Physics.
Cambridge University Press, Cambridge, 1998. xx+401 pp. $49.95.
ISBN 0-521-63303-6

String theory is a remarkable candidate to describe in a unified framework all the fundamental forces observed in nature. It incorporates gravity and gauge symmetry in a natural fashion, leading to consistent theories of quantum gravity. It has been the object of intense study since 1984, when certain supersymmetric versions of the theory, containing chiral fermions, were shown to be consistent by a novel anomaly cancellation mechanism [M. B. Green and J. H. Schwarz, Phys. Lett. B **149** (1984), no. 1-3, 117–122; MR 85k:81173]. Other early results include the description of compactifications to four dimensions, and the appearance of grand unified gauge groups with net number of fermion generations. The two volumes by Green, Schwarz and Witten provide an excellent introduction to string theory at the time [*Superstring theory. Vol. 1,* Second edition, Cambridge Univ. Press, Cambridge, 1988; MR 89f:81001a; *Vol. 2,* 1988; MR 89f:81001b]. Since then, many new properties of string theory have been understood. Some already arise in the perturbative context (for instance, T-duality and mirror symmetry), while others, which have emerged in the last five years, involve non-perturbative string dynamics, and point toward the existence of a unique theory underlying all the different string theories.

This book is the first of two volumes where the author gives a comprehensive introduction to our present understanding of string theory. The first book is centered on the bosonic string, but many of the techniques introduced are useful in the study of the superstrings as well.

The introduction motivates the study of fundamental strings as a way to smooth out the ultraviolet divergences of quantum theories. It also presents the Polyakov action and its quantization in the light-cone gauge, providing a quick computation of the string spectrum and the critical dimension. Chapter 2 discusses the application of conformal field theory techniques to describe the two-dimensional string world-sheet. The introduction of vertex operators to relate space-time and world-sheet physics is explained in a particularly clear way.

The following chapter develops the covariant quantization methods, taking the Polyakov path integral as starting point. The author discusses the gauge fixing procedure, the treatment of ghosts and the computation of the Weyl anomaly and critical dimension. The physical spectrum is computed in Chapter 4, with emphasis on the powerful BRST methods, and with a brief mention of the old covariant quantization.

Chapters 5, 6 and 7 are concerned with the evaluation of string amplitudes. They include a general discussion of the integration measure over the moduli space of punctured Riemann surfaces. Tree level amplitudes are computed in Chapter 6, exploiting the conformal field theory techniques of Chapter 2. Several specific topics are also discussed here, including the Veneziano amplitude, and the appearance of gauge symmetry through Chan-Paton degrees of freedom at the ends of open strings. Chapter 7 discusses one-loop amplitudes. The consistency condition of modular invariance of the torus amplitude is studied in detail. The interpretation of the infrared divergences of open string diagrams as tadpoles for closed string modes is also developed here.

The subject of Chapter 8 is toroidal compactification and T-duality. The discussion for closed strings includes the appearance of enhanced gauge symmetry at self-dual points, and several selected topics, namely the Narain lattice and orbifold constructions. T-duality for open strings motivates the introduction of new extended objects, D-branes, in string theory. Given the essential role of D-branes in string dynamics (especially in the supersymmetric versions of Volume II), their main physical properties are treated at this point. In particular, their tension and world-volume action are computed. The final chapter touches on a series of advanced topics involving higher-genus amplitudes: finiteness and unitarity of string theory, large order behaviour of the perturbative expansion, string field theory, high energy/temperature regime, and noncritical strings.

We would like to stress the pedagogical value of the present book. The approach taken is modern and pleasantly systematic, and it covers a broad class of results in a unified language. A set of exercises at the end of each chapter complements the discussions in the main text. On the other hand, the introduction of techniques and concepts essential in the context of superstrings makes it a useful reference for researchers in the field.

{For Vol. II see the following review [MR 99h:81184, see p. F50 in this volume].}

Angel Maria Uranga (Geneva)

99h:81184 81T30 81-01 81-02 83E30

Polchinski, Joseph (Santa Barbara, CA)

★**String theory. Vol. II. (English summary)**

Superstring theory and beyond.

Cambridge Monographs on Mathematical Physics.

Cambridge University Press, Cambridge, 1998. xx+531 pp. $54.95.

ISBN 0-521-63304-4

This is the second of two volumes where the author provides an introduction to our current understanding of string theory. It is devoted to the study of supersymmetric string theories, with an emphasis on recent results on their non-perturbative dynamics (originated mainly by E. Witten's analysis [Nuclear Phys. B **443** (1995), no. 1-2, 85–126; MR 96f:81104]). The developments in the last five years have modified substantially our viewpoint regarding string theory, uncovering the role of other extended objects, and pointing toward the existence of a unique quantum theory underlying the different superstrings. This book also covers many important perturbative results and phenomenological considerations of string theory that are not contained in other introductory books on the subject.

This volume starts by motivating the introduction of world-sheet supersymmetry as the natural framework to construct string theories with spacetime fermions. The construction of type IIA/B superstrings in ten dimensions is carried out using superconformal field theory techniques, followed by the construction of type I superstrings by orientifolding the type IIB theory. Heterotic strings are considered in the following chapter. The SO(32) and $E_8 \times E_8$ supersymmetric heterotic strings, as well as the seven known non-supersymmetric heterotic strings, are constructed in the fermionic and bosonic formulations. The consistency conditions of all the ten-dimensional theories, modular invariance for closed strings and tadpole cancellation for type I, are explicitly checked. Superstring interactions, comprehensibly described from several points of view, are the subject of Chapter 12. Ten-dimensional low energy effective actions are discussed, with an emphasis on classical symmetries of the supergravity Lagrangians. The cancellation of anomalies in the chiral theories is described in detail. Finally, the basic techniques to compute string amplitudes are introduced, with sample calculations of several tree level and one-loop diagrams.

Chapter 13 centers on toroidal compactifications and T-duality. Its main results are the exchange of type IIA and type IIB theories under T-duality, and the appearance of D-branes. Because of their relevance in non-perturbative string dynamics (uncovered by Polchinski [Phys. Rev. Lett. **75** (1995), no. 26, 4724–4727; MR 96m:81185]), their main properties are discussed at this point, specifically, their tension, RR charge, world-volume effective actions, interactions, and bound states.

In Chapter 14 several ingredients already introduced are exploited to study the strong coupling behaviour of ten-dimensional superstrings. It covers the $\mathrm{SL}(2,\mathbf{Z})$ duality of the type IIB theory, type I/SO(32) heterotic duality, and the appearance of eleven-dimensional M-theory as the strong coupling limit of the type IIA and $E_8 \times E_8$ heterotic superstrings. Many related topics, including the M(atrix) theory definition of M-theory, and the computation of black hole entropy, complete the analysis.

Chapter 15 expands on several advanced techniques in conformal field theory. The subjects covered include minimal models, current algebras, coset constructions, WZW models and rational conformal field theories.

The following three chapters study extensively the phenomenological applications of string theory in four dimensions. Chapter 16 introduces $N = 1$ supersymmetric vacua constructed as toroidal orbifolds, with a description of their low energy supergravity Lagrangian. The alternative approach of compactification on smooth Calabi-Yau manifolds is studied in Chapter 17. After a concise yet complete introduction of the relevant geometrical tools, they are applied to the computation of the massless spectrum and of their low energy interactions from geometrical data. Chapter 18 discusses symmetries in four-dimensional string vacua, illustrating the striking relations between world-sheet and spacetime phenomena. The topics include the gauging of continuous global symmetries in string theory, the realization of C-, P- and T-symmetries in string theory, and gauge coupling unification. Finally, several considerations on supersymmetry breaking, and supersymmetry beyond perturbation theory, are presented.

Chapter 19 covers several advanced topics related to $N = 2$ superconformal field theories on the world-sheet. It describes Gepner models as exact conformal field theories for Calabi-Yau compactifications, and introduces mirror symmetry for type II strings, and its application to the study of topology change in string theory (flop and conifold transition). It also discusses duality of string theories in lower dimensions, most notably heterotic/type IIA duality in six dimensions, and S-duality of toroidal compactification of heterotic string to four dimensions.

The present volume succeeds in giving a detailed yet comprehensive account of our current knowledge of superstring dynamics. The topics covered range from the basic construction of the theories to the most recent discoveries on their non-perturbative behaviour. The discussion is remarkably self-contained (the volume even contains a useful appendix on spinors and supersymmetry in several dimensions), and thus may serve as an introduction to the subject, and as an excellent reference for researchers in the field.

{For Vol. I see the preceding review [MR 99h:81183, see p. F49 in this volume].}

Angel Maria Uranga (Geneva)

97g:81062 81T40 81-02

Di Francesco, Philippe (Gif-sur-Yvette);
Mathieu, Pierre [Mathieu, Pierre[1]] (Quebec, QC); **Sénéchal, David** (Sherbrooke, QC)

★**Conformal field theory. (English summary)**

Graduate Texts in Contemporary Physics.

Springer-Verlag, New York, 1997. xxii+890 pp. $89.00. ISBN 0-387-94785-X

Conformal quantum field theory in two dimensions, and related theories like Chern-Simons and other topological field theories, have largely contributed to the present interaction between mathematics and physics. Although several review articles on conformal field theory are available by now, a comprehensive monograph is still missing. The book under review aims at filling this gap.

There is no generally accepted definition of conformal field theory; in fact, there are several approaches to conformal field theory, and the relation amongst these approaches has not yet been completely clarified. Nonetheless, the comparison of different approaches has been a remarkably powerful source of heuristic ideas, both in mathematics and in physics.

Let us briefly list some approaches to conformal field theory: There is an approach in the framework of Lagrangian field theory which allows one to make rather direct contact with geometrical constructions. A second approach uses the representation theory of infinite-dimensional algebras such as the Virasoro algebra or untwisted affine Lie algebras. A first step towards a mathematical formalization of this approach is the theory of vertex operator algebras. There are attempts to describe conformal field theory in the framework of the algebraic theory of superselection sectors, and, finally, there is an axiomatic approach which tries to describe conformal field theory using modular functors.

The authors of the book under review have chosen the algebraic approach, relying on infinite-dimensional algebras and their representation theory; indeed, many interesting results were first obtained using this approach.

The book is written for physicists and requires only a modest mathematical background; two chapters on finite-dimensional simple Lie algebras and affine Lie algebras make the book almost self-contained from the mathematical point of view. Background from physics is provided in two chapters about quantum field theory and statistical mechanics; a general background in physics is nonetheless helpful for the reader.

Apart from the chapters covering background material, the book contains two types of chapters: some of them introduce the general theory; this is then illustrated in other chapters with concrete examples. The chapters about general aspects of conformal field theories first introduce conformal covariance in arbitrary dimensions and explain the special features that arise in two dimensions. Then the so-called operator formalism is developed and several general aspects of modular invariance are explained. Finally, conformal field theories on manifolds with boundaries are also introduced.

The really strong point of this book, however, is the chapters in which these concepts are illustrated in concrete models; in these parts of the book, the reader can find many fully worked out calculations and explicit details. These chapters also turn the book into a valuable reference tool. The concrete models are first of all the (Virasoro-)minimal models in various guises, including the Coulomb-gas approach and the formulation as coset conformal field theories. There is a separate chapter on the Ising model, which is, in many respects, the simplest example of a conformal field theory. The other class of examples consists of conformal field theories based on (untwisted) affine Lie algebras and includes WZW models and coset theories.

However, despite the impressive length of this book, many important examples of conformal field theories are not covered, such as theories based on $\mathscr{W}$-algebras and superconformal field theories, i.e. theories which are also supersymmetric. As a consequence, no contact can be made in this book with (super-)string theory. The authors plan to address these topics in a second volume.

It is fair to say that the computational part of the book dealing with explicit examples is far superior to the conceptual part. In the latter chapters the presentation is rather traditional and does not provide any novel point of view on the subject. Unfortunately, the other approaches to conformal field theory mentioned above are not explained. This is actually a serious shortcoming, in particular for a reader with a background in mathematics, since a considerable part of the impact of conformal field theory in physics and mathematics comes from the comparison of different approaches.

This criticism can be exemplified in the treatment of the Knizhnik-Zamolodchikov equation. It is derived in a purely representation-theoretic setting, using null vectors; however, its generalization to higher genus, the KZB equation, is not introduced. Unfortunately, the KZ equation is also not presented in its conceptual context, i.e. as the equation that characterizes a projectively flat connection on the sheaf of conformal blocks over the moduli space of complex projective curves. The authors' decision not to provide this background in the main text can be justified; however, in the notes at the end of the corresponding chapter the relevant references are not mentioned either.

Nonetheless, the book under review definitely provides a comprehensive reference on several very important examples of conformal field theories. It is carefully written (one of the few mistakes I found is in figure 17.4; unfortunately this is also reproduced on the cover), and the chapters about concrete examples of conformal field theories will be helpful for anybody working in this field. *Christoph Schweigert* (Zürich)

99d:81172 81U10 35P25 47A40 47N50

Dereziński, Jan (Warsaw); **Gérard, Christian** (Palaiseau)

★**Scattering theory of classical and quantum N-particle systems. (English summary)**

Texts and Monographs in Physics.

Springer-Verlag, Berlin, 1997. xii+444 pp. $89.95. ISBN 3-540-62066-4

Let us imagine several quantum mechanical particles, such as atoms or molecules, interacting with each other through pair potentials. Except for some countable set of energies, these particles gradually change their states due to the interactions, and in the long run they will disintegrate into several clusters and these clusters will move independently of each other. This physically natural phenomenon is the very subject of scattering theory, which mathematicians call asymptotic completeness.

Although the physical intuition is easy to understand, the mathematically rigorous treatment of asymptotic completeness is far from obvious. The first problem is the choice of the Hamiltonian which describes the system in question. This is in general a very difficult problem, in particular in quantum field theory. In the case of non-relativistic quantum mechanics, one has a Hamiltonian in classical mechanics, and the quantum mechanical counterpart is obtained by Schrödinger's method of quantization. One is thus led to consider the operator $\widetilde{H} = -\sum_{j=1}^{N} \Delta_j/(2m_j) + \sum_{1\leq i<j\leq N} V_{ij}(x^i - x^j)$, where $x^j = (x_1^j, \cdots, x_\nu^j)$, $\Delta_j = \sum_{i=1}^{\nu} (\partial/\partial x_i^j)^2$, $m_j > 0$, is the mass of the jth particle, and $V_{ij}(y)$ is a real function decreasing at infinity. Actually one studies the reduced Hamiltonian H by restricting $\widetilde{H}$ to the subspace $\{\sum_{j=1}^{N} m_j x^j = 0\}$, on which the center of mass is fixed. The problem is then formulated as follows: Investigate the asymptotic behavior of $e^{-itH}u$ as $t \to \infty$ for u in the absolutely continuous subspace for H.

Mathematically this is the perturbation problem of the continuous spectrum of self-adjoint operators and has a long history, at the beginning of which one should mention F. Rellich and T. Kato. This study has been active since the 1950s, and T. Ikebe [Arch. Rational Mech. Anal. **5** (1960), 1–34 (1960); MR **23** #B1398] proved the asymptotic completeness of 2-particle systems by showing the existence of a complete family of generalized eigenfunctions spanning the absolutely continuous subspace of H. This fact is itself significant both in mathematics and physics, and in 1963 the monograph of L. D. Faddeev appeared where he used this method to prove the asymptotic completeness of generic 3-particle systems. This method is usually called the stationary theory, and the main idea is to study the boundary value of the resolvent $(H - \lambda \mp i0)^{-1}$ in suitable Banach spaces rigging the original Hilbert space $L^2(\mathbf{R}^\nu)$. The stationary approach was widely used in various problems of scattering theory, but the asymptotic completeness for Schrödinger operators with particle number ≥ 4 remained open until the 1980s.

It was V. Enss [Comm. Math. Phys. **61** (1978), no. 3, 285–291; MR **58** #25583] who invented the time-dependent method which directly studies the unitary group e^{-itH}. His idea agrees with the physical intuition of scattering particles in that he localized the support of wave functions for Schrödinger operators around the orbits of classical particles (propagation properties) and used the analysis on phase space, which has a deep connection to microlocal calculus or the pseudodifferential calculus commonly used in the study of linear partial differential operators. He also noticed the important role of commutation relations in scattering theory and proved the asymptotic completeness for 3-particle systems with general short-range interactions and also long-range interactions including Coulomb potentials. He further found many ideas effective in the study of many-particle systems. His thoughts are seen in [V. Enss, Rev. Math. Phys. **1992**, Special Issue, 85–112; MR 94i:81140].

It was at almost the same time that E. Mourre [Comm. Math. Phys. **78** (1980/81), no. 3, 391–408; MR 82c:47030] presented the general method of commutators for studying the continuous spectrum of self-adjoint operators. His main concern was the analysis of the resolvent, but he also presented an idea of localizing the phase space by functions of operators. While the localization by pseudodifferential operators was effective only in 2-particle systems, this method turned out to be also efficient in the many-particle cases. These ideas were very fundamental, and I. M. Sigal and A. Soffer [Ann. of Math. (2) **126** (1987), no. 1, 35–108; MR 88m:81137] proved the asymptotic completeness for general N-particle systems with short-range perturbations, accommodating these ideas and also Kato's classical work on smooth perturbations. They also derived propagation properties of N-particle systems, which stimulated the study of N-body systems from the viewpoint of Heisenberg's picture. Some years after the paper of Sigal-Soffer, G. M. Graf [Comm. Math. Phys. **132** (1990), no. 1, 73–101; MR 91i:81100] presented a new proof, the main ingredient of which is his clever vector field, which makes it possible to visualize the orbits of quantum mechanical particles in a purely time-dependent manner. Then came the time of harvest. Based on all these ideas, Dereziński [Ann. of Math. (2) **138** (1993), no. 2, 427–476; MR 94g:81209] proved the asymptotic completeness for N-particle systems with long-range perturbations.

Quantum mechanics is the fruition of the commutation relation. For $p = -i\nabla_x$, $q = x$, one has $i[p,q] = 1$. Using this relation, one also has a structure of Lie algebra for $H_0 = p^2, x^2$ and $A = (xp + px)/2$. Namely, $i[H_0, x^2] = 4A$, $i[H_0, A] = 2H_0$, $i[A, x^2] = 2x^2$. These formulas were the starting point of Enss and Mourre. The latter used them to study the resolvent of the Schrödinger operator, while the former used them in such a way that $e^{itH_0}(x/2t - p)^2 e^{-itH_0} = (x/2t)^2$. However, to estimate the unitary group e^{-itH_0}, the formula

$$(d/dt)e^{itH_0}(A/t - x^2/4t^2)e^{-itH_0} = e^{itH_0}(2/t)(x/2t - p)^2 e^{-itH_0}$$

is effective, and it is no exaggeration to say that the central problem of the recent study of N-particle scattering by the time-dependent method is the pursuit of variants of this formula.

The reason I gave the above historical sketch is that this book, written by two active authors, Dereziński and Gérard, both making fundamental contributions in this field, intends to summarize complete achievements, novel results and basic techniques related to the multi-particle systems from the viewpoint of the time-dependent method, and its significance would be properly understood in terms of recognition of the history of scattering theory. The time-dependent method itself is a relatively new idea; however, its mathematical roots are deeply related to the classical idea of quantum mechanics and traditional ideas of estimating solutions of partial differential equations.

In this book, it is stressed that the asymptotic velocity $\lim_{t\to\infty} e^{itH}(x/t)e^{-itH} \equiv p^\infty$ plays a key role in the study of N-particle scattering. In fact, the fundamental plan for the proof of asymptotic completeness is to show that the intercluster coordinates behave like those of free particles and that the size of each cluster has a very low speed of expansion. This problem is closely related with the behavior of classical particles;

hence almost one-third of the book is devoted to studying the scattering of classical particles. To give a general idea of the time-dependent method, this book begins with the classical scattering problem as well as the quantum one of the 2-body case. Chapter 1 deals with the classical scattering with time-decaying potentials. The wave operators in classical mechanics are defined here. In Chapter 2 the time-independent classical 2-body problem is studied, in which the trajectories of the classical particles are studied in detail and the Hamilton-Jacobi equation is solved, which is later used in the quantum mechanical long-range problem. Chapters 3 and 4 are the quantum versions of Chapters 1 and 2, and the quantum 2-body Hamiltonians are studied in detail. The idea of the time-dependent proof of the asymptotic completeness is examined here. By analyzing the classical orbits, one can construct parametrices of Schrödinger operators in the form of Fourier integral operators. These parametrices, due to H. Isozaki and H. Kitada [J. Fac. Sci. Univ. Tokyo Sect. IA Math. **32** (1985), no. 1, 77–104; MR 86j:35125], are reconstructed by using the estimates of classical particles proved in previous sections. Counterexamples of asymptotic completeness are also given. Chapter 5 is devoted to the classical N-body problem. Graf's vector field is constructed here as a gradient of some convex function, the construction of which depends largely on geometrical properties of Jacobi coordinates of the N-body problem. Chapter 6 is devoted to the quantum N-body problem. Those who want to grasp quickly the overview of the N-body scattering theory should begin with the introduction of Chapter 6. The Mourre estimate and the propagation properties are proved in this section. The main concern here is the long-range N-body problem, to which all the results prepared in the book are applied. In particular, the asymptotic Hamiltonian defined by $(p^\infty)^2$ plays a key role here to give a link between asymptotic states and the absolutely continuous subspace.

Asymptotic velocities and asymptotic Hamiltonians are noteworthy notions. The germs of these ideas have already appeared in the works of Enss, Sigal-Soffer and Graf. The authors of this book try to identify them as keystones of the scattering theory. In fact, these notions make the theory transparent and give a new bird's-eye view of the scattering problem.

The explanation of pseudodifferential and Fourier integral operators in the appendix will be useful as a brief review on this subject. As explained earlier, in the N-body problem, from the viewpoint of mathematical technique, pseudodifferential calculus is replaced by some algebra of functions of a small number of operators equipped with the commutator expansion lemma. This is also described in the appendix, and it is very plausible that this idea will be used in many other problems related with the spectral theory.

The N-body Schrödinger operator is a bulky problem whose study requires much preparation. The authors' endeavor of gathering all the necessary notions, established results and related mathematical techniques should be highly rated, which makes this book a first-class guide to the many-body problem for researchers in this field.

Asymptotic completeness is not the ultimate goal of spectral and scattering theory. It is a milestone (of course, a very important one because it sits in the heart of the problems for the Hamiltonians governing the scattering phenomena), and history shows that every time a new idea of the proof of asymptotic completeness is discovered it gives rise to new developments and by-products in the study of the spectral theory. It has already been more than 70 years since the Schrödinger operator was introduced into quantum physics. Although it is physically fully accepted as fundamental machinery, it still has many mathematically mysterious and fascinating properties. The study of the N-particle scattering problem has now been set on a firm basis and many hard problems remain open. The techniques developed thus far are waiting for new developments and applications. The abundance of references will serve as a good guide to related papers. *Hiroshi Isozaki* (Toyonaka)

98k:82099 82B44 60E15 82C32
Talagrand, Michel (Paris)
Rigorous results for the Hopfield model with many patterns. (English summary)
Probab. Theory Related Fields **110** (1998), *no.* 2, 177–276.

Two models of disordered systems have been the focus of attention for theoretical and mathematical physics over the last two decades: the Sherrington-Kirkpatrick model and the Hopfield model. They were proposed in the 1970s as the simplest "solvable" mean field models for spin glasses (the Hopfield model having attracted much attention in the field of neural networks as the simplest model of an associative memory). Both models have been solved on the level of heuristic theoretical physics by use of the "replica trick" and the intricate "replica symmetry breaking" scheme devised by Parisi and coworkers. These solutions reveal a richness of structure that would be hard to suspect a priori in such simple models, and this has not failed to attract attempts to obtain a mathematically rigorous understanding of these models. In the Hopfield model in particular, in various regimes of the parameters of the model (in particular the temperature β^{-1} and the number of stored patterns—or more precisely the ratio α, of the number of stored patterns $M(N)$ over the volume N of the system), simplifications occur that have allowed progress in the rigorous understanding of this system.

The article under review describes a fresh approach by the author to attacking this model on a broad front. Indeed, while most of the major previous results on the model are recovered and often improved, a major breakthrough is achieved towards a rigorous justification of the results of the replica approach. The paper is organized along the different regimes of the model and we will follow this order in this review. Note that we cannot describe all of the results, nor state them in a precise form. A concise summary is given in the first section of the paper itself.

Section 2 deals with the high-temperature regime, more precisely the regime $\beta < 1/(1+\sqrt{\alpha})$. Here a simple and elegant proof of the equality of the quenched and annealed free energy is given, along with sharp bounds on the fluctuations of the free energy.

Section 3 deals with the low-temperature regime for small α. The main result shown here is that the Gibbs measures are concentrated on small balls around the stored patterns in the space of the "overlap parameters". These results are the basic input for what is to follow in the subsequent sections. They are very similar to estimates obtained by the reviewer and V. Gayrard [Probab. Theory Related Fields **107** (1997), no. 1, 61–98; MR 98b:82041] and sharpen the earliest results of that nature obtained by the reviewer, Gayrard and P. Picco [Probab. Theory Related Fields **100** (1994), no. 3, 329–363; MR 96c:60115; J. Statist. Phys. **79** (1995), no. 1-2, 395–414; MR 96c:82033].

Sections 4–8 form the core of the paper and contain the main new results. Roughly speaking, they establish a nontrivial regime of parameters (in particular, $\alpha > 0$, and

$\beta > 1$!) in which it is proven that the formal solution of the model obtained via the replica method by D. J. Amit, H. Gutfreund and H. Sompolinsky [Ann. Phys. **173** (1987), no. 1, 30–67] is correct. This includes in particular an exact expression for the free energy and the expected value of the overlap parameters. The method of proof here is induction over the volume assisted by the a priori estimates from Section 3. This idea as such is not new. Under the name of "cavity method" it was used in the spin glass theory as an alternative approach to the replica method [see M. Mézard, G. Parisi and M. A. Virasoro, *Spin glass theory and beyond*, World Sci. Publishing, Teaneck, NJ, 1987; MR 91k:82066] in a formal way (i.e. without control of error terms). A first attempt to put this method on a rigorous basis in the context of the Hopfield model was undertaken by L. A. Pastur, M. V. Shcherbina and B. Tirozzi [J. Statist. Phys. **74** (1994), no. 5-6, 1161–1183; MR 96a:82027]. The basic idea is simple: take any function of the system with N sites, and try to obtain a recursion expressing the same function for a system of size $N+1$ in terms of functions of the system with N sites. The problem is of course that many such functions may be needed to get a closed system. The nice feature in the so-called "replica-symmetric regime" is that only a few "big" functions are necessary, while all other terms that are produced tend to zero as N tends to infinity. This latter fact, however, has to be proven in turn by induction, and mastering the host of terms that are produced in the processes is a truly admirable achievement. As a true breakthrough result, since its appearance it has already stimulated follow-ups: the reviewer and Gayrard [in *Mathematical aspects of spin glasses and neural networks*, 3–89, Birkhäuser Boston, Boston, MA, 1998; MR 99c:82035] found a simpler proof that replaces the inductive treatment of the "small terms" by a priori estimates, while the author in a recent preprint ["Exponential inequalities and convergence of moments in the replica symmetric regime of the Hopfield model", 1998; per revr.] has extended the range of validity of his results to a region close to the expected optimal one and at the same time proven a corresponding central limit theorem. Also he has applied the same techniques successfully to other disordered mean field models. For a good overview of those results see his contribution to the Proceedings of the ICM '98 [Doc. Math. **1998**, Extra Vol. I, 507–536 (electronic)].

The final section concerns the zero-temperature results, or what is commonly called the "storage capacity" problem. Here the issue is for which values of α the stored patterns are local minima of the Hamiltonian function. Numerical results and replica results indicate a critical value a_c of order 0.14. Here rigorous bounds of C. M. Newman [Neural Networks **1** (1988), no. 3, 223–238] and their recent improvements by D. Loukianova [C. R. Acad. Sci. Paris Sér. I Math. **318** (1994), no. 2, 157–160; MR 95a:92005] are numerically improved.

The paper is written in a rather self-contained way as a purely mathematical paper and requires no a priori knowledge of statistical mechanics or of any of the physics behind the model. Any reader interested in the physical intuition underlying this work is strongly urged to study the paper by Amit et al. cited above.

Anton Bovier (Berlin)

98j:83007 83C05 53C80 83C30
Friedrich, Helmut (Potsdam)
Gravitational fields near space-like and null infinity. (English summary)
J. Geom. Phys. **24** (1998), *no.* 2, 83–163.
For the purposes of observing gravitational radiation given off by isolated systems at large distances, such systems are conventionally modeled in terms of asymptotically simple spacetimes. A main feature of asymptotically simple spacetimes is the attachment of a boundary consisting of the end points of null geodesics of the spacetime. Such a boundary is referred to as null infinity.

Asymptotic simplicity is, essentially, a purely kinematical concept, strongly characterized by the peeling behavior of curvature fields in terms of inverse powers of a radial distance along null directions. In other words, the curvature fields of all asymptotically simple spacetimes have precise fall-off behaviors at null infinity. Whether or not such spacetimes satisfy the Einstein equations with sufficiently generic matter content remains an outstanding question. As a matter of fact, the Einstein equations with vanishing cosmological constant are known to admit a vast number of solutions with asymptotic logarithmic behavior arising from proper peeling data [D. Christodoulou and S. Klainerman, *The global nonlinear stability of the Minkowski space*, Princeton Univ. Press, Princeton, NJ, 1993; MR 95k:83006]. The challenge, thus, resides in demonstrating that there are a good number of solutions with the peeling property (other than trivial ones with vanishing ADM mass), which would make asymptotic simplicity a natural concept for solutions to the Einstein equations.

The formulation of the Einstein equations as an initial value problem requires the prescription of constrained data on a Cauchy surface. All Cauchy surfaces approach asymptotically a set consisting of the end points of the spacelike geodesics of the spacetime, denoted i^0 and referred to as spacelike infinity, also attached to the spacetime. The set i^0 is conventionally viewed as containing a single point. In addition, the Cauchy surface has a concept of infinity associated with it, irrespective of its embedding into the four-dimensional spacetime. In conventional compactification, the infinity of the Cauchy surface is represented as a single point added to the surface.

Breaking with tradition, in the work under review the author formulates a novel description of the concept of infinity of a Cauchy surface, in which such an infinity is represented as a 2-sphere, namely, as a set consisting of a point for every direction in which a geodesic on the surface can run off to infinity. This allows him to reinterpret the spacelike infinity of an asymptotically simple spacetime as a cylinder, as opposed to its usual description as a point i^0.

The benefit of this blow-up of the point i^0 into a cylinder is that the behavior of gravitational fields at null infinity can be seen to relate to properties of their initial values at infinity. In particular, it is seen that asymptotic logarithmic singularities, reported by a number of authors, do not take place if the initial data satisfies certain, apparently mild, regularity conditions at infinity. The conditions can be interpreted as fall-off requirements on the conformal structure of the initial surface, stronger than the ordinary requirements for spatial asymptotic flatness. Since the initial constraints may be solved via the standard York splitting procedure, the conditions apply directly to the free data of the initial value problem and do not affect the solution of the constraint equations in any compact subset of the physical initial surface. The existence of such mild fall-off conditions guaranteeing non-logarithmic asymptotic behavior had been generally unsuspected. The author tests the proposed setting by applying it to the Schwarzschild spacetime. The results are, so far, restricted to time-symmetric initial data, namely, vanishing extrinsic curvature on the Cauchy surface.

The techniques necessary for these results include the use of the conformal Einstein equations, introduced by the author in earlier works [see, e.g., J. Geom. Phys. **17** (1995), no. 2, 125–184; MR 96j:83008]; the introduction of coordinates based on conformal geodesics (as opposed to metric geodesics, which had been in use); the inclusion of conformal connections in the initial value problem, a feature entirely new to the conformal approach; and the use of the spin-frame bundle over the neighborhood of the infinity of the initial surface, which takes full advantage of the blow-up of i^0 to a cylinder.

With these elements, the author formulates a regular conformal initial value problem where both the data and the equations are regular, in the sense that they do not diverge anywhere. This is a critical improvement with respect to the standard conformal initial value problem, where only the equations are regular, but the initial data for the Weyl tensor grows unboundedly at infinity. Furthermore, the equations of the regular conformal initial value problem constitute a symmetric hyperbolic system, everywhere except at the points where null infinity touches spacelike infinity, where they degenerate.

The work under review provides a foothold for the attack of the problem of the existence of good numbers of asymptotically simple solutions to the Einstein equations, as well as a flavor for the degree of complexity involved in such a problem.

Simonetta Frittelli (Pittsburgh, PA)

G. SELECTED REVIEWS OF CLASSIC BOOKS AND PAPERS PUBLISHED 1940–1969

2, 66c 02.0X
Gödel, Kurt
The consistency of the continuum hypothesis.
Notes by George W. Brown on lectures delivered at the Institute for Advanced Study during the autumn term 1938–1939.
Annals of Mathematics Studies, no. 3.
Princeton University Press, Princeton, N. J., 1940. 66 pp.

Gödel shows that, if a certain axiomatic system Σ for set theory is consistent, the system is consistent which results from Σ by introducing as additional axioms the axiom of choice (in a strong form, which provides for the simultaneous choice, by a single relation, of an element from each non-vacuous set of the universe of sets) and Cantor's generalized continuum hypothesis ($2^{\aleph_\alpha} = \aleph_{\alpha+1}$ for any α). His method consists in constructing within Σ a model Δ of Σ, in which model the axiom of choice and the generalized continuum hypothesis hold.

In more detail, the method is this. In an axiomatic system, certain notions are treated as primitive and not defined, and every proposition about those notions must follow from only those properties of them which are expressed by the axioms. For Σ, there are three undefined notions, $\mathfrak{Cls}$ (class), $\mathfrak{M}$ (set) and $\in$ (the membership relation). From the undefined notions, with the aid of the axioms and logical principles, other notions can be defined. In particular, three are defined designated as $\mathfrak{Cls}_l$, $\mathfrak{M}_l$ and ε_l. From the axioms of Σ a set of propositions are deduced which have exactly the forms of the several axioms of Σ, the axiom of choice, and the generalized continuum hypothesis, respectively, except that $\mathfrak{Cls}_l$, $\mathfrak{M}_l$ and ε_l occur throughout in place of $\mathfrak{Cls}$, $\mathfrak{M}$ and ε, respectively. This reduces the consistency of the augmented system to that of Σ. For if a contradition were a consequence of the enlarged axiom list, that is, as axioms for $\mathfrak{Cls}$, $\mathfrak{M}$ and ε, then a contradiction would follow likewise in terms of $\mathfrak{Cls}_l$, $\mathfrak{M}_l$ and ε_l from the aforesaid set of propositions. Since those propositions are theorems of the system Σ, this contradiction would be a consequence of the axioms of Σ alone. In Zermelo's axiomatization of set theory, one of the axioms (the Aussonderungsaxiom) employed as an intuitive notion the notion of a "definite property" of sets. Those elements of a given set which have a given definite property were constituted a subset of the given set. The notion of "definite property" is lacking in clarity, and subsequent axiomatizations of set theory have eliminated this notion as an intuitive presupposition by stating explicitly how "definite properties" are to be constructed. The Skolem "paradox," and now these results of Gödel, depend on this.

Gödel's system Σ is essentially due to Bernays [J. Symbolic Logic **2**, 65–77 (1937)], and is equivalent to von Neumann's system $S^* + \text{VI}$ with Ax. III 3* replaced by Ax. III 3 [J. Reine Angew. Math. **160**, 227–241 (1929)]. A property of a set is treated in extension, as the class of sets which have the property; similarly a property of n sets, that is, a relation among n sets, is treated as the class of the ordered n-tuples of sets which stand in the relation. This makes possible a very succinct delimitation of "definite property." The operations for constructing "definite properties" enter through axioms concerning the existence of classes (Axs. B1-B8). In the notation of propositional functions, they give R under the following eight circumstances, where A and B represent propositional functions already obtained: $R(x,y) \equiv x \in y$; $R(x) \equiv A(x)$-and-$B(x)$; $R(x) \equiv$ not-$A(x)$; $R(x) \equiv$ exists-y-such-that-$A(y,x)$; $R(y,x) \equiv A(x)$; $R(x,y) \equiv A(y,x)$; $R(x,y,z) \equiv A(y,z,x)$; $R(x,y,z) \equiv A(x,z,y)$. Sets are classes (Ax. A1). Only sets can occur as members of classes (Ax. A2). Classes with the same extension are identical (Ax. A3). Classes occur which are not sets, for example, the universal class V; and this is provable using an axiom which has the effect of excluding the existence of infinite sequences of sets such that each set of the sequence has the next as member (Ax. D). The definition of the ordered n-tuples used in treating relations is based on an axiom by which an unordered pair of sets is a set (Ax. A4). The other axioms of Σ are the axiom of infinity (Ax. C1), axioms asserting the existence of a set including the sum of the elements of a given set (Ax. C2) and a set including the set of subsets of a given set (Ax. C3), and the axiom of substitution which takes the place of the Aussonderungsaxiom (Ax. C4).

Some space is devoted to a preliminary development of abstract set theory, partly from these axioms, and partly from these axioms and the axiom of choice. (*8.3 should read: $\overline{\overline{\alpha+1}} \leq \overline{\overline{\alpha}}^2$ for $\alpha > 1$.) The deductions are informal, but could be formalized in the calculus of propositional functions of first order. The operations for constructing sets and classes given by Axs. A4, B1-B8, Gödel writes, after certain modifications, as eight fundamental binary operations which, applied to a pair of sets, yield a set. The idea for the model Δ is to take as sets only those which can be generated by these operations and the operation, applied at certain stages of the generating process, of forming the set of all sets already obtained. This process of generation is expressed in the definition of a function F of an ordinal α by transfinite induction. To give scope to the nine operations, the ordinals are separated into congruence classes $J_0 - J_8$ modulo 9; and the binary character of the fundamental operations necessitates a mapping of the pairs of ordinals on each of the congruence classes $J_1 - J_8$. The values of F taken for α in J_i $(i > 0)$ are sets given by the fundamental operations from pairs of sets obtained as earlier values of F; and the values of F taken for α in J_0 are the sets of all values previously obtained. The sets taken as values of F are called "constructible sets" and constitute a class L. The "constructible classes" are the classes of which both the elements and the intersections with constructible sets are constructible sets. $\mathfrak{Cls}_l$ is the notion "constructible class"; $\mathfrak{M}_l$ is the notion "constructible set"; and $\in_l$ is the $\in$-relation under restriction of the domains of the relation to the constructible classes and sets. It now turns out that all the axioms of Σ are satisfied for the model, that is, they are theorems of Σ when read with $\mathfrak{Cls}_l$, $\mathfrak{M}_l$ and $\in_l$ in place of $\mathfrak{Cls}$, $\mathfrak{M}$ and $\in$. Furthermore the proposition that every set is constructible holds in the model, that is, $V_l = L_l$ is a theorem of Σ, where V_l and L_l are defined in terms of $\mathfrak{Cls}_l$, $\mathfrak{M}_l$ and $\in_l$ in the same way as V and L in terms of $\mathfrak{Cls}$, $\mathfrak{M}$ and $\in$. This shows that the proposition $V = L$ is consistent with the axioms of Σ, if Σ itself is consistent.

The prospectus now calls for showing that the axiom of choice and the generalized continuum hypothesis, relativized for the model, that is, read with $\mathfrak{Cls}_l$, $\mathfrak{M}_l$ and $\in_l$ in place of $\mathfrak{Cls}$, $\mathfrak{M}$ and $\in$, are theorems of Σ. In fact, these propositions can be deduced from the relativized axioms of Σ and the proposition $V_l = L_l$, already proved as theorems of Σ. In other words, dropping the subscript l throughout, the (unrelativized) axiom of choice and generalized continuum hypothesis become theorems of Σ when the proposition $V = L$, that is, every set is constructible, is added to Σ as a new axiom. For the axiom of choice, this is immediately evident, since each constructible set can be assigned as its "order" the least ordinal α for which it is taken as value of F. Then the element to be chosen from any non-vacuous set can be specified as the one having the least order.

To treat the generalized continuum hypothesis $2^{\aleph_\alpha} = \aleph_{\alpha+1}$ (the axiom of choice now being available), Gödel considers the set (written $F``\omega_\alpha$) of all constructible sets of orders less than ω_α (where ω_α is the first number of Cantor's $(2+\alpha)$th number class). $\aleph_\alpha$ is the cardinal of the set of all ordinals less than ω_α. The intersection of this set with the congruence class J_0 has the same cardinal, and on this subset all values taken by F are distinct. Hence $\overline{\overline{F``\omega_\alpha}} = \aleph_\alpha$ (where $\overline{\overline{x}}$ denotes the cardinal number of $x$). Therefore $\overline{\overline{\mathfrak{P}(F``\omega_\alpha)}} = 2^{\aleph_\alpha}$ (where $\mathfrak{P}(x)$ denotes the set of subsets of x). By Cantor's theorem and the comparability theorem (based on the axiom of choice), $2^{\aleph_\alpha} \geq \aleph_{\alpha+1}$. Now consider any member of $\mathfrak{P}(F``\omega_\alpha)$, that is, any subset of $F``\omega_\alpha$. By $V = L$, this subset must be taken as value of F for some ordinal δ. By a detailed examination of the definition of F, Gödel is able to show that the subset must then necessarily be taken as value of F for some ordinal less than $\omega_{\alpha+1}$. Hence $\mathfrak{P}(F``\omega_\alpha)$ is a subset of $F``\omega_{\alpha+1}$. Therefore $\overline{\overline{\mathfrak{P}(F``\omega_\alpha)}} \leq \overline{\overline{F``\omega_{\alpha+1}}} = \aleph_{\alpha+1}$. Therefore $2^{\aleph_\alpha} = \aleph_{\alpha+1}$.

S. C. Kleene (Madison, Wis.)

28 #1118 02.65
Cohen, Paul
The independence of the continuum hypothesis.
Proc. Nat. Acad. Sci. U.S.A. **50** (1963), 1143–1148.
The author announces the following result: There exist families of sets M_1-M_4 such that the ith axioms of the following list and all the axioms of Zermelo-Fraenkel set theory (with the axiom of regularity but without the axiom of choice) are true in M_i $(i = 1, 2, 3, 4)$: (1) There exist non-constructible sets $a \subset \omega$; the axiom of choice; the generalized continuum hypothesis. (2) There is no well-ordering of the set $\mathscr{P}(\omega)$ of all subsets of ω. (3) $\aleph_1 \neq 2^{\aleph_0}$; the axiom of choice. (4) There is a countable set of pairs of elements of $\mathscr{P}(\mathscr{P}(\omega))$ for which no choice-set exists.

These results present the long-awaited solutions of the most outstanding open problems of axiomatic set theory and should be rated as the most important advance in the study of axiomatic set theory since the publication of Gödel's 1940 monograph [*The consistency of the continuum hypothesis*, Princeton Univ. Press, Princeton, N.J., 1940; MR **2**, 66c, see p. G3 in this volume]. It is perhaps premature to say it with certainty but to this reviewer it seems more than probable that the influence of Cohen's discovery will be at least as deep in metamathematics as in the general philosophy of mathematics (and perhaps not only of mathematics).

The present paper contains a series of lemmas needed for the proof that there exists an M_3 satisfying the conditions enumerated above.

The general idea is to start with a denumerable transitive family of sets $\mathfrak{W}$ which is a model for the Zermelo-Fraenkel axioms, including the axiom of regularity and Gödel's axiom $V = L$. Let $\aleph_2$ be the third infinite cardinal of $\mathfrak{W}$ and let a_δ for $\delta < \aleph_2$ be different subsets of ω. Let $\mathfrak{N}$ be the smallest model of Zermelo-Fraenkel axioms containing $\mathfrak{W}$ and all the a_δ's. If the property of $\aleph_2$ to be the third infinite cardinal is preserved in $\mathfrak{N}$, then the continuum hypothesis is false in $\mathfrak{N}$. In order to show that a choice of the a_δ's is possible such that no new properties of the elements of $\mathfrak{W}$ will appear in $\mathfrak{N}$ the author introduces a new and powerful notion called "forcing".

Let the a_δ's be treated first as formal symbols and let us introduce terms F_α whose values will form $\mathfrak{N}$ after the values of the a_δ's have been determined. These terms are obtained in a similar way as Gödel's sets $F`\alpha$ with the difference resulting from the fact that the a_δ's are to be treated along with the void set as the initial elements.

Sentences involving the terms F_α in which all quantifiers are limited to the set $\{F_\beta : \beta < \alpha\}$ are called limited statements; other sentences are called unlimited statements. Let P be a finite set of statements of the form $n \in a_\delta$ or $n \notin a_\delta$ and let $\mathfrak{A}$ be a limited statement. Then the author defines by induction the relation "P forces $\mathfrak{A}$", a typical inductive step, for instance, being: P forces $(x)_\alpha \mathfrak{A}(x)$ if for all $\beta < \alpha$ and all $P' \supset P$ it is not true that P' forces $\neg\mathfrak{A}(F_\beta)$. If $\mathfrak{A}$ has no quantifiers, e.g., if $\mathfrak{A}$ is the formula $F_\alpha \in F_\beta$ $(\beta < \alpha)$, then P forces $\mathfrak{A}$ if and only if P forces a statement ψ obtained by expressing the definition of F_β (e.g., if $F_\beta = \{F_\gamma, F_\delta\}$, then ψ is the statement $F_\alpha = F_\gamma \vee F_\alpha = F_\delta$ which in turn can be replaced by a limited statement in which the symbol "$=$" does not occur). The notion of forcing is further defined for unlimited statements by induction on the number of quantifiers.

These inductive definitions reduce (by a transfinite process) the relation "P forces $\mathfrak{A}$" to relations of the form "P forces $n \in a_\delta$" (or "P forces $n \notin a_\delta$"); by definition, these relations hold if and only if the statement $n \in a_\delta$ (or $n \notin a_\delta$) belongs to P.

The important Lemma 4 proved by the author states that for every P and every (limited or unlimited) statement $\mathfrak{A}$ there is a $P' \supset P$ such that either P' and all its extensions force $\mathfrak{A}$ or P' and all its extensions force $\neg\mathfrak{A}$. From this it easily follows that there is an infinite set P_0 such that for each $\mathfrak{A}$ either $\mathfrak{A}$ or $\neg\mathfrak{A}$ (and not both $\mathfrak{A}$ and $\neg\mathfrak{A}$) is forced by a finite subset P of P_0 and by all extensions of P.

Now the author takes for a_δ the set of integers n such that the formal statement $n \in a_\delta$ belongs to P_0. With this choice of the a_δ's he proves that formulas valid in $\mathfrak{N}$ are precisely those for which the corresponding formal statements are forced by sufficiently big subsets of P_0.

The applications of this result are to appear in a later paper [Proc. Nat. Acad. Sci. U.S.A. **51** (1964), 105–110; MR **28** #2962]. *A. Mostowski* (Warsaw)

17, 1040e 09.0X
Cartan, Henri; Eilenberg, Samuel
Homological algebra.
Princeton University Press, Princeton, N. J., 1956. xv+390 pp.

The title "Homological algebra" is intended to designate a part of pure algebra which is the result of making algebraic homology theory independent of its original habitat in topology and building it up to a general theory of modules over associative rings. The particular formal aspect of this theory stemming from algebraic topology is that of a preoccupation with endomorphisms of square 0 in graded modules. The conceptual flavor of homological algebra derives less specifically from topology than from the general 'naturalistic' trend of mathematics as a whole to supplement the study of the anatomy of any mathematical entity with an analysis of its behavior under the maps belonging to the larger mathematical system with which it is associated. In particular, homological algebra is concerned not so much with the intrinsic structure of modules but primarily with the pattern of compositions of homomorphisms between modules and their interplay with the various constructions by which new modules may be obtained from given ones.

In the recent requisite mathematical terminology, this means that the principal objects of study in homological algebra are functors from categories of modules to other categories of modules. If R and S are two rings (with identity elements that act as the identity on all modules considered) a covariant functor (of one variable) from the category of R-modules to the category of S-modules is a function T attaching to each R-module A an S-module $T(A)$, and to each R-homomorphism $h: A' \to A$ an S-homomorphism $T(h): T(A') \to T(A)$ such that, if k is a second homomorphism $A \to A''$, $T(k \circ h) = T(k) \circ T(h)$ is the identity map $T(A) \to T(A)$ whenever h is the identity map $A \to A$. For a contravariant functor, one has instead $T(h): T(A) \to T(A')$, and $T(k \circ h) = T(h) \circ T(k)$. The definition of a functor in several variables in different categories, and some covariant, some contravariant, is the evident extension of the above, with an additional commutativity requirement for maps acting on arguments in different positions in the functor. The functor T is said to be additive if, for any two homomorphisms h_1 and h_2 between the same modules, one has $T(h_1 + h_2) = T(h_1) + T(h_2)$. As a linear theory, homological algebra deals only with additive functors.

A functor is called exact if (for each variable separately) it maps every exact sequence into an exact sequence. It is the essence of homological algebra to channel any deviation of a functor from exactness into the construction of new functors. For these constructions, the following notions are basic. A module M is called projective if every homomorphism of M into a factor module A/B can be lifted to A; M is called injective if every homomorphism of B into M can be extended to A. Every module is a homomorphic image of a projective module and can be imbedded in an injective module. Hence every module A has a "projective resolution", i.e., an exact sequence $\cdots \to X_n \to \cdots \to X_0 \to A \to 0$ in which the X_i are projective, and an "injective resolution", i.e., an exact sequence $0 \to A \to X^0 \to \cdots \to X^n \to \cdots$ in which the X^i are injective. If projective resolutions are substituted for all the contravariant variables in a functor and injective resolutions are substituted for all the covariant variables a complex is obtained whose homology groups are, up to natural isomorphisms, independent of the particular choice of the resolutions. The nth homology group of this complex is the module value of the nth right derived functor R^nT of the given functor T. Similarly, one obtains the left derived functors L_nT by substituting injective resolutions for the contravariant variables and projective resolutions for the covariant variables. These derived functors have the same variance as T, and vanish for $n < 0$.

The derived functors are related by the so-called connecting homomorphisms which increase the degree by 1 in the case of the right derived functors and lower the degree by 1 in the case of the left derived functors. These result from any exact sequence $0 \to A' \to A \to A'' \to 0$ when A', A, A'' are substituted in turn for any one of the variables. If the functor is covariant in this variable the connecting homomorphism sends $R^nT(A'')$ into $R^{n+1}T(A')$ and, with the usual functor homomorphisms, forms an exact sequence

$$\cdots \to R^nT(A') \to R^nT(A) \to R^nT(A'') \to R^{n+1}T(A') \to \cdots.$$

The case of contravariance, and the case of the left derived functors, are obtained by making the evident appropriate changes.

A covariant functor T of one variable is said to be left exact if, for every exact sequence $0 \to A' \to A \to A'' \to 0$, the sequence $0 \to T(A') \to T(A) \to T(A'')$ is exact. In case T is contravariant, interchange A'' and A' in the last sequence. For functors of several variables, one demands that this property hold for each variable. One says that T is right exact if, instead of the above, the sequence $T(A') \to T(A) \to T(A'') \to 0$ is exact, etc. If T is left exact then the left derived functors L_nT are 0 (except for $n = 0$) and R^0T may be identified with T. If T is right exact the right derived functors R^nT are 0 (except for $n = 0$) and L_0T may be identified with T.

A functor T is said to be right balanced if substitution of any projective module for a contravariant variable, or substitution of any injective module for a covariant variable, yields an exact functor in the remaining variables. The definition of a left balanced functor is obtained by interchanging the words projective and injective. Much of the power of the mechanism of derived functors results from the following facts: if T is right balanced then its full right derived functors R^nT may be identified with the partial right derived functors obtained by using resolutions for any non-empty subset of variables and treating all the other variables as constants in the construction; if T is left balanced then the same result holds for the left derived functors L_nT.

The above basic principles and results are established in chapters I-V. Chapter VI concentrates on the two basic functors of homological algebra and their derived functors. The first basic functor is the twice covariant functor $\otimes_R$ that attaches to each pair (A, B), where A is a right R-module and B a left R-module, the tensor product $A \otimes_R B$, regarded as a module over the ring Z of the integers, in general. This functor is right exact and left balanced. The second basic functor is the contravariant-covariant functor Hom_R that attaches to each pair (A, B) of left R-modules the Z-module $\mathrm{Hom}_R(A, B)$ of all R-homomorphisms of A into B. This functor is left exact and right balanced. The left derived functors of $\otimes_R$ are denoted $\mathrm{Tor}_n{}^R$, and the right derived functors of Hom_R are denoted $\mathrm{Ext}_R{}^n$.

Some of the simpler interpretations and applications of these functors are given in Chapters VI and VII. An R-module A is said to be of projective dimension $\leq n$ if it has a projective resolution $\cdots \to X_i \to \cdots \to X_0 \to A \to 0$ such that $X_i = 0$, for all $i > n$. There is a similar, but less frequently used, notion of injective dimension. One says that a ring R has left global dimension $\leq n$ if every left R-module has projective dimension $\leq n$ or—equivalently—if $\mathrm{Ext}_R{}^{n+1} = 0$. Another equivalent condition is that every left R-module be of injective dimension $\leq n$. The significance of this notion for ring theory is already indicated by the following facts: the rings of global dimension 0 are precisely the semisimple rings, i.e., the rings R such that every R-module is a direct sum of simple submodules; a ring has left global dimension $\leq$ 1 if and only if every left ideal is a projective module; an integral domain has this property if and only if it is a Dedekind ring, i.e., if and only if the classical ideal theory holds for the ring.

The functor $\mathrm{Tor}_1{}^R$ plays an important role in the Künneth relations between the homology groups of a tensor product of complexes and those of the factors. In fact, it arose for the first time exactly in this context. Here, the Künneth relations are obtained in full group theoretic form and under far more general conditions than those obtaining in the original topological situation of the integral homology groups of product spaces.

Chapter VIII introduces the homology theory of augmented rings from which the specific homology (and cohomology) theories of associative algebras, groups, and Lie algebras are obtained by specialization. The structure of an augmented ring consists of a ring R and an R-epimorphism of R onto an R-module Q. The nth homology group of the augmented ring R in a right R-module A is defined as $\mathrm{Tor}_n{}^R(A, Q)$. The nth cohomology group of R in a left R-module A is defined as $\mathrm{Ext}_R{}^n(Q, A)$. These notions apply rather directly to a dimension theory for local rings and graded rings yielding, for instance, various generalizations and extensions of Hilbert's theorem on chains of syzygies of homogeneous polynomial ideals.

Chapter IX discusses the homology and cohomology theory of associative algebras L (with 1) over commutative rings K (with 1). Define L^e as the tensor product algebra $L \otimes_K L'$, where L' denotes the usual anti-isomorph of L. Together with the L^e-epimorphism $L^e \to L$ that sends $a \otimes b'$ onto ab. this is an augmented ring structure. By using a suitable L^e-projective resolution of L, it is shown that the usual cohomology groups $H^n(L, A)$ of L in an L^e-module A are nothing but the groups $\mathrm{Ext}_L{}^n e(L, A)$. This makes the general theory available for the study of the cohomology of algebras and thus gives a high degree of control over it. Of particular interest is the cohomological dimension of L which is defined to be $\leq n$ if and only if $H^{n+1}(L, A) = 0$ for all L^e-modules A.

The homology theories for groups and Lie algebras have certain special features in common, and Chapter X introduces the requisite pre-specialization of the augmented ring theory. The basic structure is that of a supplemented algebra. It consists of a K-algebra L, together with a unitary algebra epimorphism of L onto K. Close relations exist between the homology of L as a supplemented algebra and the homology of L^e, augmented as above. In many situations, this leads to a strengthening of each theory. The homology and cohomology groups of a group G are obtained from the supplemented algebra structure consisting of the group algebra $Z(G)$ and the coefficient sum epimorphism $Z(G) \to Z$.

Chapter XI gives the general foundation for the multiplicative theory. Four "external products" are defined, involving two K-algebras L and M, their tensor product $L \otimes_K M$, and the functors Tor and Ext for these three algebras. "Internal products", involving only one algebra, can be derived from these with the aid of algebra homomorphisms $L \otimes_K L \to L$ or $L \to L \otimes_K L$. In particular, in the case of groups or Lie algebras, one has natural homomorphisms $L \to L \otimes_K L$ by means of which the cup and cap products are derived from two of the external products.

The most complete and most powerful cohomology theory is that of a finite group G. Chapter XII develops the special formal machinery that governs this case. The main point (discovered by J. T. Tate in connection with the application to class field theory) is that the norm map in G-modules $(N(a) = \sum_{x \in G} x \cdot a)$ leads to a tie-up between homology and cohomology by which the homology groups play the role of cohomology groups of negative degree and link up with the usual cohomology groups to form the so-called complete derived sequence of G. The connecting homomorphisms go through this complete sequence, and a product can be defined which extends simultaneously the cup and the cap product and applies without restriction to arbitrary members of the complete derived sequence.

Chapter XIII brings the homology and cohomology theory of Lie algebras into the framework of supplemented algebras. The critical device for doing this is to pass to the universal enveloping algebra; with every Lie algebra L one can uniquely associate an associative algebra L^e in such a way that there is a natural 1–1 correspondence between the representations of L and those of L^e. Assuming that L is free over the ground ring K, one has $L \subset L^e$, and the commutation in L is given by the multiplication in L^e: $[x, y] = xy - yx$. There is a natural algebra epimorphism of L^e onto K whose kernel coincides with the subalgebra of L^e that is generated by L. This is the supplemented algebra structure whose homology and cohomology is that of L. It is, a priori, not at all evident that this actually yields the theory as originally formulated under the inspiration of the differential geometry on Lie groups. The identification requires the use of a certain L^e-projective resolution of K involving the exterior algebra built over L. Interesting by-products are obtained concerning L^e. The cohomological dimension of L^e coincides with the projective dimension of K as an L^e-module. If K is semisimple the cohomological dimension of L^e coincides with its right and left global dimensions. In particular, if L has rank n over K, these dimensions are all equal to n. This generalizes earlier results on polynomial algebras, because if L is abelian then L^e is simply $K[x_1, \cdots, x_n]$.

Chapter XIV takes up the applications of cohomology theory to various extension theories. The simplest case is that of extensions of R-modules. To each such extension $0 \to C \to X \to A \to 0$ one can associate an element of $\mathrm{Ext}_R{}^1(A, C)$ which is 0 if and only if the extension is split. This correspondence yields an isomorphism of the group of equivalence classes of the extensions of C by A (with the Baer composition) onto $\mathrm{Ext}_R{}^1(A, C)$. This is actually the origin of the notation Ext. The other three cases are extensions of K-projective K-algebras with kernels of square 0, group extensions with abelian kernels, and Lie algebra extensions with abelian kernels. In each of these cases one obtains an interpretation of the second cohomology group as a group of equivalence classes of extensions.

Chapter XV gives a complete development of the formalism of spectral sequences. In particular, two spectral sequences are associated with every double complex. The terms E_∞ of these spectral sequences are the graded modules associated with the homology module by the two natural filtrations of the double complex. The terms E_2 of the spectral sequences are the compound homology modules obtained from the double complex by using the two differential operators in succession. In Chapter XVI it is shown how these spectral sequences can be used for establishing connections between partial derived functors and full derived functors, and for investigating the behavior of Tor and Ext under homomorphisms of operator rings. It is also shown how the general method yields the spectral sequences that have been used for examining the relations between the homology and cohomology of a group or a Lie algebra and that of a normal subgroup or an ideal. The rest of Chapter XVI sketches some topological applications concerning spaces with groups of operators.

Chapter XVII reaches a very high degree of complexity. Resolutions of complexes take the place of resolutions of modules, and spectral sequences are obtained which connect the results of first applying a functor and then passing to homology with the results of first passing to homology and then applying the functor. In particular, this leads to a fuller and more general treatment of the Künneth relations.

The appendix by D. A. Buchsbaum proposes an abstract framework of "exact categories" that is capable of accommodating the functor theory of this book as well as additional structural elements that one may wish to introduce. The proposed theory includes an abstract notion of duality which makes it unnecessary, at least in principle, to give separate treatments for covariance and contravariance and for projectivity and injectivity.

The appearance of this book must mean that the experimental phase of homological algebra is now surpassed. The diverse original homological constructions in various algebraic systems which were frequently of an ad hoc and artificial nature have been absorbed in a general theory whose significance goes far beyond its sources. The basic principles of homological algebra, and in particular the full functorial control over the manipulation of tensor products and modules of operator homomorphisms, will undoubtedly become standard algebraic technique already on the elementary level. It is probably with such expectations that the authors have put so much missionary zeal into the systematization of their approach and the cataloguing of the basic results. A probably unavoidable effect of this is that the book cannot be consulted by the uninitiated in a local fashion. The reader is definitely forced to go through it starting at the beginning. Each chapter (with the exception of the last) is followed by a collection of exercises which are designed not so much to strengthen the reader by easy gymnastics (for they are generally not particularly easy) as to point out various ramifications and applications of the general theory.

A few misprints might lead to confusion: p. 80, in the diagram above Prop. 2.3, the top X should have a double prime and the X's in the bottom row should have single primes; p. 101, first line below Lemma 10.1, the middle term of the sequence should have a double prime; p. 152, lines 4 and 7, replace M/J_k by M/MJ_k and M/J_{k-1} by M/MJ_{k-1}; p. 233, diagram (1), reverse the arrow labelled g; p. 319, line

5 from bottom, replace d_r by d_k; p. 322, third line above Theorem 3.2, replace dsx by sdx.
G. Hochschild (Berkeley, Calif.)

10, 595b 10.0X
Selberg, Atle
An elementary proof of the prime-number theorem.
Ann. of Math. (2) **50** (1949), 305–313.

10, 595c 10.0X
Erdös, P.
On a new method in elementary number theory which leads to an elementary proof of the prime number theorem.
Proc. Nat. Acad. Sci. U. S. A. **35** (1949), 374–384.

The prime number theorem (PNT) asserts that

$$\pi(x) \sim x/\log x, \quad x \to \infty,$$

where $\pi(x)$ is the number of primes $p \leq x$. All previous proofs have been by "transcendental" arguments involving some appeal to the theory of functions of a complex variable. Successive proofs have moderated the demands on this theory, or invoked alternative analytical theories (e.g., Fourier transforms), but there remained a nucleus of complex variable theory namely the proposition that the Riemann zeta-function $\zeta(s) = \zeta(\sigma + it)$ has no zeros on the line $\sigma = 1$; and this could hardly be avoided, except by a radically new approach, since the PNT is in a clearly definable sense "equivalent" to this property of $\zeta(s)$. It has long been recognised that an "elementary" proof of the PNT, not depending on analytical ideas remote from the problem itself, would (if indeed possible) constitute a discovery of the first importance for the logical structure of the theory of the distribution of primes. An elementary (though not easy) proof is given, in various forms, in these two papers. Since the papers are in some ways complementary, it seems clearest to review them together.

The starting point is A. Selberg's formula

$$\sum_{p \leq x} \log^2 p + \sum_{pq \leq x} \log p \log q \sim 2x \log x, \tag{1}$$

where p, q run over primes. The formula might be deduced from the PNT or its equivalent $\vartheta(x) \equiv \sum_{p \leq x} \log p \sim x$, but this, on the classical theory of the distribution of primes, would give it a transcendental basis. The significant events to record are: (a) the discovery by Selberg of an elementary proof (indeed with "$\sim 2x \log x$" replaced by "$= 2x \log x + O(x)$," though this refinement is not required for the application); and (b) the recognition that the formula nevertheless embodied facts previously accessible only to the analytical theory of primes. The hope that (1) might therefore be made the basis of an elementary proof of the PNT itself was justified, but the deduction was by no means obvious. The first proof (by Selberg) used, besides (1) (and parts of the elementary theory of primes), the following provisional result (deduced from (1) by Erdös): (2) For any $\lambda > 1$ the number of primes in $(x, \lambda x)$ is at least $Kx/\log x$ for $x > x_0$ $(K = K(\lambda) > 0, x_0 = x_0(\lambda))$. Later it was found possible to argue directly from (1) without the intervention of (2).

Selberg's paper contains his proof of (1), a statement of (2), his deduction of the PNT from (1) and (2), and his final direct deduction of the PNT from (1). Erdös's paper contains a statement of (1), an allusion to Selberg's final proof of the PNT (not published at the time), and an account of other proofs in chronological order; the account includes Erdös's own proof of (2), Selberg's deduction of the PNT from (1) and (2), a sketch of Selberg's simplified proof of (2), and the joint simplified direct deduction of the PNT from (1). It is stated (in both papers) that the method can be adapted to the prime number theorem for arithmetical progressions.

Selberg's proof of (1) is, in accordance with the general spirit of the investigation, expressed in arithmetical form. But in a brief summary it is quicker to treat the formula as the result of equating coefficient-sums in the formal identity $(\zeta'/\zeta)' + (\zeta'/\zeta)^2 = \zeta''/\zeta$, where $\zeta = \zeta(s)$, $\zeta' = \zeta'(s), \cdots$. The left hand side gives substantially the left hand side of (1). As to the right hand side we can write

$$\zeta'' = 2\zeta^3 + b\zeta^2 + c\zeta + \delta,$$

where b, c are certain constants and δ is a Dirichlet's series $\sum d_n n^{-s}$ with coefficient-sum $D(x) \equiv \sum_{n \le x} d_n = O(x^\alpha)$ $(0 < \alpha < 1)$, by elementary theorems on divisor functions. The coefficient-sum of $\zeta''/\zeta = 2\zeta^2 + b\zeta + c + \zeta^{-1}\delta$ can now be estimated with an error $O(x)$, and (1) follows. [This is the general idea; but Selberg, besides avoiding even the formal use of Dirichlet's series, arranges the details rather differently so that the coefficients $d_3(n)$ in ζ^3 do not enter.]

Of the various deductions from (1) the joint simplified proof of the PNT is perhaps the easiest to summarise. In this, (1) is used in the equivalent form

$$\text{(1')} \qquad \vartheta(x)\log x + \sum_{p \le x} \vartheta(x/p)\log p \sim 2x\log x,$$

which may be rewritten as

$$\text{(1'')} \qquad s(x) + \frac{1}{\log x}\sum_{p \le x} \varepsilon_p s(x/p) \to 2,$$

where $s(x) = \vartheta(x)/x$ and $\varepsilon_p = (\log p)/p$; and we wish to deduce that $\vartheta(x) \sim x$, i.e., $s(x) \to 1$. We have, by elementary theory: (3) $\sum_{p \le x} \varepsilon_p \sim \log x$; so the second term in $(1)''$ is essentially a weighted average of $s(x)$. [The situation now recalls Mercer's theorem; a certain linear combination of a function and its average tends to a (finite) limit, and it is required to deduce that the function itself tends to a limit. The proof (by "compensation") has some features in common with Knopp's proof of Mercer's theorem, but it uses special properties of $s(x)$ that have no counterpart in Mercer's theorem, for example the "Tauberian" condition "$xs(x)$ positive and increasing" and the connection of $s(x)$ with the weights ε_p.] Suppose that

$$a = \liminf s(x) < \limsup s(x) = A.$$

We have $a > 0$ by elementary theory; and it is an easy deduction from $(1)''$, (3) that: (4) $A + a = 2$. Consider a (large) x for which $s(x)$ is large (i.e., near A). From $(1)''$, (3) we deduce that (to compensate): (α) $s(x/p)$ is small (i.e., near a) for a set X of primes comprising almost all $p \le x$ (where "almost all" means "except for a subset over which $\sum \varepsilon_p$ is small compared with the same sum taken over the whole set"). Take a particular (small) p_1 in X. Then similarly: (β) $s(x/p_1 p)$ is large for a set Y of primes comprising almost all $p \le x/p_1$. Next, it follows easily from $(1)'$, (4) and the fact that $\vartheta(\cdot)$ is an increasing function that: (γ) points x', x $(x' > x)$ where $s(\cdot)$ is large and small (or vice versa) must be so far apart that $x'/x > (A/a) - \varepsilon$ $(= \lambda$, say) for large x. [This is the sense of lemma 5 of Erdös's paper, but the inequality signs in the conclusion and in most of the proof have become reversed. A list of further misprints (supplied by Erdös) includes: (Lemmas 1, 2) for r read p; (p. 377, line 1) for X read C; (p. 377, line 14 from below) for $\vartheta(x)$ read $\vartheta(x/p)$; (p. 378 (14)) move last square bracket two places to the right; (p. 376, line 9 from below; p. 383, line 3) symbols following $c' > C-$ and $I_i =$ badly printed (but meaning clear).] Now $(\alpha), (\beta), (\gamma)$ involve a contradiction. The proof of this, the most difficult step, is presented in alternative forms. The simpler version is based on a study of the sum $S = \sum \varepsilon_p \varepsilon_q$ taken over all pairs of primes (p, q) for which p is in X, $\lambda^{-1}p < p_1 q \leq \lambda p$. Writing $S = \sum_p \varepsilon_p \sum_q \varepsilon_q$, we have $\sum_q \geq (\lambda p/p_1)^{-1}\{\vartheta(\lambda p/p_1) - \vartheta(\lambda^{-1}p/p_1)\} > c > 0$ by the definition of a, A, since $a\lambda > A\lambda^{-1}$; whence $S > (c-\varepsilon)\log x$, since, by ($\alpha$), $\sum_p$ extends over almost all $p \leq x$. On the other hand, writing $S = \sum_q \varepsilon_q \sum_p \varepsilon_p$, we have

$$\sum_p \leq (\lambda^{-1}p_1 q)^{-1}\vartheta(\lambda p_1 q) < C;$$

whence $S < \varepsilon \log x$, since, by (γ), q can never belong to Y (because x/p_1q is too near to x/p for some p of X), so that, by (β), $\sum_q$ extends over "almost no" primes. This contradiction shows that $a = A$ $(= 1$ since $A + a = 2)$. Selberg, in his own final proof, proceeds on somewhat different lines, by "successive approximation." He first derives from (1) an inequality involving (on both sides) the function $R(x) = \vartheta(x) - x$, and then infers from it (and elementary properties of $R(x)$) that, if $|R(x)/x| < \alpha$ (< 8) $(x > x_0)$, then $|R(x)/x| < \alpha_1$ $(x > x_1)$, where $\alpha_1 = \alpha(1 - k\alpha^2)$ and k is a positive absolute constant; whence by repeated application $|R(x)/x| < \alpha_n$ $(x > x_n)$ where $\alpha_n \to 0$ as $n \to \infty$. [The logical arrangement on p. 311 is confusing, since the proof that (3.2) still holds if $R(n)$ changes sign does not seem to be valid until the order of x'/x has been suitably restricted. Also the argument of the text seems to use a sharpened version of (1), though the full form with error $O(x)$ is not required. The claim in the "final remark" that (1) itself suffices would seem to presuppose the use of the alternative form of inequality referred to in footnote 4 on p. 311.]

It may be useful to view the elementary proof against its analytical background, in order to see how the argument, while omitting all reference to analytical facts known to be inseparable from the truth of the PNT, nevertheless avoids the danger of coming into conflict with these facts (and thereby proving too much). Let f be a function of the complex variable $s = \sigma + it$, to be taken as $-\zeta'/\zeta$ in the application but subject at present only to the following restrictions: (i) $f = \sum a_n n^{-s}$ $(\sigma > 1$; a_n real); (ii) f is regular in $\sigma \geq 1$ except possibly for simple poles on $\sigma = 1$; (iii) $g \equiv -f' + f^2 = \sum b_n n^{-s}$ $(\sigma > 1)$, where $B(x) \equiv \sum_{n \leq x} b_n \sim 2x \log x$. (When $f = -\zeta'/\zeta$, (iii) is equivalent to Selberg's formula (1).) Then, by (iii), g cannot have a double pole at a point $s \neq 1$ of $\sigma = 1$; whence any possible pole of f at such a point must have residue R satisfying $R + R^2 = 0$, i.e., $R = -1$; and the example

$$f(s) = \zeta(s) - \zeta(s - i\alpha) - \zeta(s + i\alpha)$$

$(\alpha > 0)$ shows that this is actually possible. Thus our present assumptions are consistent with the existence of poles of f at points $s \neq 1$ on $\sigma = 1$; and this may be taken as a "reason" why it is possible to give an elementary proof of (1) without becoming involved in the question of the existence of zeros of ζ on $\sigma = 1$. Now introduce the further assumption: (iv) $a_n \geq 0$. Then it can be proved, as in Hadamard's classical argument, that a pole of f at $s = 1 + it$ with $R = -1$ implies a pole at $s = 1 + 2it$ with $R = +1$, and this is impossible; so f can have no pole on $\sigma = 1$ except at $s = 1$. Thus the properties (i)-(iv) of $f = -\zeta'/\zeta$ do embody the essential analytical fact on which previous proofs of the PNT have been based. What Selberg and Erdös do is to deduce the PNT directly from the arithmetical counterparts of (i), (iii), (iv), without the explicit intervention of the analytical fact. In principle this opens up the possibility of a new approach, in which the old logical arrangement is reversed and analytical properties of $\zeta(s)$ are deduced from arithmetical properties of the sequence of primes. How far the practical effects of this revolution of ideas will penetrate into the structure of the subject, and how much of the theory will ultimately have to be rewritten, it is too early to say. *A. E. Ingham* (Cambridge, England)

17, 131d 10.2X
Šafarevič, I. R.
Construction of fields of algebraic numbers with given solvable Galois group. (Russian)
Izv. Akad. Nauk SSSR. Ser. Mat. **18** (1954), 525–578.

L'auteur montre que, pour tout corps (de degré fini) k de nombres algébriques et pour tout groupe résoluble G, il existe des extensions K/k, dont le groupe de Galois soit (à l'isomorphie près) G. Il le fait en donnant une méthode inductive de construction de telles extensions, la démonstration de la possibilité de poursuivre cette construction aussi loin que cela est nécessaire se faisant par les méthodes généralisant celles de ses travaux dans le même journal **18** (1954), 261–296, 389–418 [MR **16**, 571c, 1001a].

Les considérations de la théorie des groupes permettent de simplifier comme suit le problème: si G est une image homomorphe d'un groupe G', la solubilité du problème pour G' entraîne sa solubilité pour G; en combinant ceci avec le théorème d'Ore sur l'existence, dans un groupe résoluble G, d'un sous-groupe invariant nilpotent N et d'un sous-groupe propre G^* tels que $G = G^*/N$, on montre qu'il suffit de savoir résoudre le problème suivant: k/k_0 étant une extension galoisienne de groupe résoluble Γ, scholzienne $(\text{mod}\, M)$, où M est divisible par une puissance suffisamment grande de p, et g étant un p-groupe opéré par Γ, construire une surextension K/k_0 de k/k_0, dont le groupe de Galois soit le produit semi-direct $\Gamma \cdot g$ des Γ et g et qui soit aussi scholzienne $(\text{mod}\, M)$. D'autre part, tout groupe d'exposant divisant p^g opéré par Γ est une image homomorphe d'un groupe de la forme $G_{\Gamma,d,q}{}^{(c)}$, défini comme suit dans le second travail cité de l'auteur: m étant l'ordre de Γ, considérons le groupe libre engendré par dm générateurs $s_{\xi,i}$ $(\xi \in \Gamma; i = 1, 2, \cdots, d)$. Posons, en plus, $s_{\xi,i}{}^{\gamma} = s_{\gamma\xi,i}$. Le groupe opéré par Γ ainsi défini sera noté $G_{\Gamma,d}$, et on notera $G_{\Gamma,d,q}$ le quotient de $G_{\Gamma,d}$ par son sous-groupe caractéristique (donc stable pour Γ) engendré par les puissances p^g-ièmes de ses éléments. $G_{\Gamma,d,q}{}^{(c)}$ est le quotient de $G_{\Gamma,d,q}$ par le c-ième terme de sa suite centrale descendente, considéré comme opéré par Γ.

k/k_0 étant une extension galoisienne, dont le groupe de Galois soit $\Gamma = G_{k/k_0}$, elle est dite scholzienne $(\text{mod}\, M)$ si: 1) tous les diviseurs premiers de M dans k_0 et tous les idéaux premiers à l'infini de k_0 se décomposent complètement dans k/k_0; 2) tous les facteurs premiers dans k du discriminant de k/k_0 sont de degré 1 dans k/k_0

et leurs normes absolues sont toutes $\equiv 1 \pmod{M}$. Un $\alpha \in k$ est dit p-invariant dans k/k_0 si, pour tout $\sigma \in \Gamma$, on a $\alpha^\sigma \equiv \alpha \pmod{k^{(p)}}$, où $k^{(p)}$ est le groupe des puissances p-ièmes des éléments du groupe multiplicatif k^* de k, et on appelle classe de p-invariance toute classe $(\operatorname{mod} k^* k^{(p)})$, dont les éléments sont p-invariants. G étant une extension (au sens de Schreier) donnée de Γ par un groupe g, une surextension galoisienne K/k_0 de k/k_0 sera dite une G-extension de k/k_0 si G_{K/k_0} peut être identifié avec G de manière que $G_{K/k}$ le soit avec g et que l'homomorphisme canonique de G_{K/k_0} sur G_{k/k_0} le soit avec celui de G sur Γ. Une telle extension est dite une p-surextension si l'ordre de g est une puissance de p. Elle est dite une extension centrale (de k/k_0) si g est contenu dans le centre de G, et elle est dite centrale simple si, en plus, g est un groupe cyclique d'ordre premier. A. Scholz avait montré [Math. Z. **30** (1929), 332–356] que toute p-extension centrale simple de k/k_0 est réalisable si k/k_0 est scholzienne $(\operatorname{mod} p^h)$ pour un h suffisamment grand. L'auteur, dans son premier travail cité, a montré que, k/k_0 étant une extension galoisienne scholzienne $(\operatorname{mod} p^h)$ et X étant une classe de p-invariance dans k/k_0, la condition nécessaire et suffisante pour que l'on puisse choisir un $\mu \in X$ de manière que $k(\mu^{1/p})/k$ soit scholzienne $(\operatorname{mod} p^h)$ est que certains invariants (définis par l'auteur) (χ, X) (où χ parcourt certains caractères de degré 1 et d'ordre p de Γ) et $(X)_h$ soient $=1$. Soient Π un p-groupe de Sylow de $\Gamma, \bar{k}$ le souscorps de k appartenant à Π et X une classe de p-invariance dans $k/\bar{k}$. Supposant que k contient les racines p-ièmes de l'unité et que k/k_0 est scholzienne $(\operatorname{mod} M)$, où M se divise par une puissance suffisamment grande de p, l'auteur définit, dans le § 1 du présent travail, certains invariants $[\chi, X]$ (comprenant les (χ, X) comme cas particulier), $(X)_M$ (égal à $(X)_h$ si $M = p^h$) et $[X]_\psi$, et il montre, dans le § 2, que l'égalité à 1 de tous ces invariants est la condition nécessaire et suffisante pour qu'il existe un $\mu \in X$ tel que: 1) $\bar{k}$ soit le plus petit corps intermédiaire de k/k_0 par rapport auquel $K = k(u^{1/p})$ soit galoisien; 2) tous les corps conjugués distincts de K/k_0 soient linéairement disjoints (dans leur ensemble) sur k; 3) le composé de tous ces corps (autrement dit le corps de Galois K' de K/k_0) soit, également, scholzien $(\operatorname{mod} M)$ sur k_0.

Dans le § 3, l'auteur détermine la condition nécessaire et suffisante pour que les p-extensions centrales arbitraires de k/k_0 puissent se réaliser de manière que (en désignant par K/k_0 l'extension centrale considérée de k/k_0) les conditions précédentes 1), 2) et 3) soient satisfaites. Il est, évidemment, nécessaire pour cela que les invariants $[\chi, X]$, $(X)_M$ et $[X]_\psi$ soient identiquement égaux à 1 sur tout leur domaine de définition; mais, en plus, certains autres invariants $(X, Y)^{(i)}{}_{\varphi_1,\varphi_2,\cdots,\varphi_{s-1}}$, définis par l'auteur, doivent aussi être $=1$ identiquement, et l'ensemble de ces conditions est suffisant. Les définitions des invariants $[\chi, X]$, $(X)_M$, $[X]_\psi$, $(X, Y)^{(i)}{}_{\varphi_1,\varphi_2,\cdots,\varphi_{s-1}}$ et les démonstrations des résultats indiqués des §§ 1, 2 et 3 constituent la partie la plus essentielle du travail, mais sont trop compliqués pour qu'on en puisse donner ici même une idée.

Dans le § 4, l'auteur applique le résultat du § 3 pour démontrer l'existence (pour un p fixé) des $\Gamma \cdot G_{\Gamma,d,q}{}^{(c)}$-extensions K/k_0 de k/k_0. Soient $G_{\Gamma,d,q}{}^{(c,r)}$ le quotient de $G_{\Gamma,d,q}{}^{(c)}$ par le groupe des éléments σ de son centre tels que $\sigma^{pq-r} = 1$, et Z_r le noyau de l'homomorphisme naturel de $G_{\Gamma,d,q}{}^{(c,r)}$ sur $G_{\Gamma,d,q}{}^{(c,r-1)}$. Z_r est un groupe abélien opéré par Γ. Π étant un p-groupe de Sylow de Γ, soit $Z_{r,0} = Z_r$, $Z_{r,1} = [Z_{r,0}, \Pi], \cdots, Z_{r,i} = [Z_{r,i-1}, \Pi], \cdots$ la suite des commutateurs successifs de Z_r avec Π, et soit $G_{\Gamma,d,q}{}^{(c,r,i)} = G_{\Gamma,d,q}{}^{(c,r)}/Z_{r,i}$. $Z_r/Z_{r,i}$ est un groupe abélien opéré par Π, et on notera $\mathfrak{A}$ ce groupe avec Π comme groupe d'opérateurs (même si d'autres éléments de Γ préservent $Z_{r,i}$ et opérent, ainsi, dans $Z_r/Z_{r,i}$). L'auteur définit, à l'aide de ${G_{\Gamma,d,q}}^{(c,r,i)}$ en tant qu'extension de $\mathfrak{A}$, un certain groupe ${G_{\Gamma,d,q}}^{(c,r,i)^*}$ opéré par Γ tel que, en vertu d'un théorème de Gaschütz [J. Reine Angew. Math. **190** (1952), 93–107; MR **14**, 445e], $\Gamma \cdot {G_{\Gamma,d,q}}^{(c,r,i)^*}$ possède, comme image homomorphe, un groupe de la forme $\Gamma \cdot {G_{\Gamma,d,q}}^{(c,r,i)}$. D'autre part, il existe un homomorphisme naturel de ${G_{\Gamma,d,q}}^{(c,r,i)^*}$ sur ${G_{\Gamma,d,q}}^{(c,r,i-1)^*}$, et le noyau N^* de cet homomorphisme est un groupe, contenu dans le centre de ${G_{\Gamma,d,q}}^{(c,r,i)^*}$ (donc abélien) et engendré, en tant que sous-groupe stable de ce groupe, par un de ses sous-groupes abstraits N, où tout opérateur appartenant à Π induit l'identité.

L'auteur prouve, en employant certains homomorphismes de ${G_{\Gamma,d,q}}^{(c,r,i)^*}$ dans ${G_{\Gamma,\delta,q}}^{(c,r,i)^*}$ $(d > \delta)$ analogues à ceux de son second travail cité, que si d est suffisamment grand par rapport à δ et si Q/k_0 est une $\Gamma \cdot {G_{\Gamma,d,q}}^{(c,r,i)^*}$-extension de k/k_0, qui est scholzienne $(\bmod M)$, il existe une sous-extension K/k_0 de Q/k_0, qui est une $\Gamma \cdot {G_{\Gamma,\delta,q}}^{(c,r,i)^*}$-extension de k/k_0 telle que tous les invariants $[\chi, X]$, $(X)_M$, $[X]_\psi$ et ${(X,Y)_{\varphi_1,\varphi_2,\cdots,\varphi_{s-1}}}^{(j)}$ de K/k_0 soient égaux à 1 chaque fois qu'ils ont un sens. De lors, la possibilité de construire une $\Gamma \cdot {G_{\Gamma,d,q,}}^{(c)}$-extension de k/k_0 où k/k_0 est scholzienne modulo un entier M_c divisible par une puissance suffisamment grande (et dépendant de c) de p, se démontre par l'induction suivante: on suppose, pour un $x < c$, pour un $r < q$ et pour un i, que, pour tout d, il existe des $\Gamma \cdot {G_{\Gamma,d,q}}^{(x,r,i)^*}$-extensions scholziennes $(\bmod M_c)$ de k/k_0. De lors, il en existe de telles extensions $\overline{K}/k$, où, en plus, tous les invariants indiqués sont $=1$ identiquement. Mais alors, si $\overline{k}$ est le corps intermédiaire de k/k_0 appartenant à Π, on peut, en vertu du § 3, construire, pour tout d, une extension centrale $Q/\overline{k}$ de $\overline{K}/\overline{k}$ telle que, si K est le corps de Galois de Q/k_0, K/k_0 soit une $\Gamma \cdot {G_{\Gamma,d,q}}^{(x,r,i+1)^*}$-extension scholzienne $(\bmod M_c)$ de k/k_0. On a ${G_{\Gamma,d,q}}^{(x,r,0)^*} = {G_{\Gamma,d,q}}^{(x,r)}$ et, pour quelque m, ${G_{\Gamma,d,q}}^{(x,r,m)^*} = {G_{\Gamma,d,q}}^{(x,r+1)^*}$. Ainsi, comme $\Gamma \cdot {G_{\Gamma,d,q}}^{(x,r,m)^*} = \Gamma \cdot {G_{\Gamma,d,q}}^{(x,r+1)^*}$ admet $\Gamma \cdot {G_{\Gamma,d,q}}^{(x,r,m)} = \Gamma \cdot {G_{\Gamma,d,q}}^{(x,r+1)}$ comme image homomorphe, l'existence, pour tout $x < d$, d'une $\Gamma \cdot {G_{\Gamma,d,q}}^{(x,r)}$-extension scholzienne $(\bmod M_c)$ de k/k_0 entraîne celle d'une $\Gamma \cdot {G_{\Gamma,d,q}}^{(x,r+1)}$-extension de k/k_0 de même forme. Comme ${G_{\Gamma,d,q}}^{(x,0)} = {G_{\Gamma,d,q}}^{(x)}$ et ${G_{\Gamma,d,q}}^{(x,q)} = {G_{\Gamma,d,q}}^{(x+1)}$, l'existence, pour tout d, d'une $\Gamma \cdot {G_{\Gamma,d,q}}^{(x)}$-extension scholzienne $(\bmod M_c)$ de k/k_0 entraîne celle d'une $\Gamma \cdot {G_{\Gamma,d,q}}^{(x+1)}$-extension de k/k_0 de même forme, d'où, en vertu de ${G_{\Gamma,d,q}}^{(0,0,0)} = 1$, résulte l'existence des $\Gamma \cdot {G_{\Gamma,d,q}}^{(c)}$-extensions de k/k_0 scholziennes $(\bmod M_c)$.

Le travail se termine par un complément, où est démontrée une généralisation au cas non-galoisien du résultat du travail dans Izv. Akad. Nauk SSSR. Ser. Mat. **18** (1954), 327–334 [MR **16**, 572c] de l'auteur, dont il a besoin dans une démonstration du § 2.

M. Krasner

33 #5590 10.64

Bombieri, E.

On the large sieve.

Mathematika **12** (1965), 201–225.

The grand, or generalized, Riemann hypothesis asserts that the real parts of all zeros of all Dirichlet L-functions $L(s,\chi)$ are at most $\frac{1}{2}$. If this were true, it would follow that, for $(a,q)=1$,

$$E(x;q,a)=\sum_{\substack{n\le x\\ n\equiv a(\operatorname{mod} q)}}\Lambda(n)-\frac{x}{\varphi(q)}=O(x^{1/2}(\log x)^2) \tag{1}$$

uniformly in q; but the best result known at the present time (uniform in q) is the Siegel-Walfisz theorem according to which, given any positive number N, there exists a constant $c=c(N)>0$ such that, if $q\le(\log x)^N$, then $E(x;q,a)=O(x\exp(-c(\log x)^{1/2}))$ uniformly in q. The Siegel-Walfisz theorem has proved to be adequate for some important applications, notably for Vinogradov's three primes theorem; but there are many problems in analytic number theory which depend in a natural way on $E(x;q,a)$ being small uniformly in q for q belonging to longer ranges of type $q\le x^\alpha$ $(0<\alpha<1)$. With such problems little progress was possible, except where the dependence on $E(x;q,a)$ was circumvented by the invention of new methods—Linnik's "dispersion method" is an outstanding example that springs to mind. One important area where the lack was felt especially keenly was the application of sieve methods to binary additive problems of Goldbach's type; and here it came to be realised [see, e.g., Wang Yuan, Sci. Sinica **11** (1962), 1033–1054, Appendix; MR **27** #1424] that substantial progress was possible on the basis of weaker mean value results of the type

$$\sum_{q\le x^{\alpha-\varepsilon}}\max_{(a,q)=1}|E(x;q,a)|=O(x(\log x)^{-A}), \tag{2}$$

with $A>0$ free to be chosen as large as required. For example, Buhštab [Dokl. Akad. Nauk SSSR **162** (1965), 735–738; MR **31** #2226] has shown recently that if (2) holds with $\alpha=\frac{3}{8}$, then every sufficiently large even number is the sum of a prime and an integer having at most three prime factors (this paper appears to contain at least one serious misprint, and many important steps in the proof are omitted). A result (2) of this quality (with $\alpha=\frac{3}{8}$) has actually been claimed, by Barban [Mat. Sb. (N.S.) **61** (**103**) (1963), 418–425; MR **30** #1992] and P'an Ch'eng-tung [Acta Math. Sinica **13** (1963), 262–268; translated as Chinese Math.—Acta **4** (1963), 233–290; MR **29** #82; ibid. **14** (1964), 597–606; translated as Chinese Math.—Acta **5** (1964), 642–652; MR **30** #3871] but an account of this work that is even partially satisfactory does not appear (to this reviewer) to be available in print—a truly remarkable circumstance in view of its importance. Be that as it may, the author of this paper has made a notable contribution to analytic number theory by proving, in convincing detail, that (2) holds with $\alpha=\frac{1}{2}$. Actually, in his Theorem 4 the author proves the even better result that, for any given $A>0$, there exists a constant $B(B=3A+23$ will do) such that, if $X\le x^{1/2}(\log x)^{-B}$,

$$\sum_{q\le X}\max_{y\le x}\max_{(a,q)=1}|E(y;q,a)|=O(x(\log x)^{-A}). \tag{3}$$

Starting out from K. F. Roth's [Mathematika **12** (1965), 1–9; MR **33** #5589] recent far-reaching improvement of the Linnik-Rényi large sieve, the author first reproves Roth's result (in slightly sharper form, but the extent of this improvement is not relevant to the author's main purpose), and, in fact, generalises it to a form which leads to this remarkable inequality for averages of character sums: For any complex numbers a_n, and any finite set Q of positive integers,

$$(4)\quad \sum_{q\in Q}\frac{1}{\varphi(q)}\sum_{\chi}|\tau(\chi)|^2\left|\sum_{Y<n\le Z}\chi(n)a_n\right|^2 \le 7D\max(Z-Y,M^2)\sum_{Y<n\le Z}d(n)|a_n|^2,$$

where, in the inner sum, χ runs through all characters $\bmod q$, $\tau(\chi)=\sum_{a=1}^{q}\chi(a)\exp(2\pi ia/q)$, $M=\max_{q\in Q}q$, and $D=\max_{q\in Q}d(q)$, with $d(n)$ denoting, as usual, the divisor function.

This inequality, of considerable interest in itself, is the author's principal technical innovation in a method which, otherwise, uses with great skill and power the classical analytic techniques described in Prachar's *Primzahlverteilung* [Springer, Berlin, 1957; MR **19**, 393b], especially those of Chapter IX. The main argument is directed towards what the author justly calls a new type of density theorem for zeros of L-functions (and the importance of (4) may be judged from the fact that (4) is used several times, in different contexts and with different coefficient sets a_n, in the course of the argument): If $N(\alpha,T;\chi)$ denotes the number of zeros of $L(s,\chi)$ in the rectangle $\alpha\le\sigma\le 1$, $|t|\le T$ $(\frac{1}{2}\le\alpha\le 1, T\ge 2)$, then

$$\sum_{q\in Q}\frac{1}{\varphi(q)}\sum_{\chi}|\tau(\chi)|^2N(\alpha,T;\chi)=O(DT(M^2+MT)^{4(1-\alpha)/(3-2\alpha)}\log^{10}(M+T))$$

uniformly with respect to Q. (The author conjectures that even $\sum_{q\le X}\sum_{\chi}^{*}N(\alpha,T;\chi)=O(X^{4(1-\alpha)+\varepsilon}T^{1+\varepsilon})$, uniformly in α, $\frac{1}{2}\le\alpha\le 1$, where the asterisk indicates summation over primitive characters $\bmod q$.)

The step from the density theorem to (3) is carried out in Section 3, using the Siegel-Walfisz theorem (to deal with small q) and the classical explicit formula for $\sum_{n\le x}\chi(n)\Lambda(n)$ in terms of the zeros of $L(s,\chi)$. It should be noted that applying (1) to the left-hand side of (3) (with $X=x^{1/2}(\log x)^{-B}$) leads to the estimate $x(\log x)^{-B+2}$, a result, for all practical purposes, no better than (3). Thus Bombieri's theorem leads to a result as good on average as (1). Without a doubt, many important results will follow from his theorem; moreover, his density theorem and extension of the Linnik-Rényi-Roth large sieve should both lead to interesting developments.

H. Halberstam (Nottingham)

36 #3732 10.32

Baker, A. [Baker, Alan]

Linear forms in the logarithms of algebraic numbers. I, II, III.

Mathematika **13** (1966), 204–216; ibid. **14** (1967), 102–107; ibid. **14** (1967), 220–228.

This is a very interesting sequence of papers. We quote from the introduction: "In 1934 Gel′fond and Schneider proved, independently, that the logarithm of an algebraic number to an algebraic base, other than 0 or 1, is either rational or transcendental and thereby solved the famous seventh problem of Hilbert. Among the many subsequent developments, Gel′fond obtained ... a positive lower bound for the absolute value of $\beta_1\log\alpha_1+\beta_2\log\alpha_2$, where β_1, β_2 denote algebraic numbers, not both 0, and α_1,α_2 denote algebraic numbers not 0 or 1, with $\log\alpha_1/\log\alpha_2$ irrational. Of particular interest is the special case in which β_1, β_2 denote integers. In this case it is easy to obtain a trivial positive lower bound, and the existence of a non-trivial bound follows from the Thue-Siegel-Roth theorem. But Gel′fond's result improves substantially on the former, and, unlike the latter, it is derived by an effective method of proof. Gel′fond [*Transcendental and algebraic numbers* (Russian), GITTL, Moscow, 1952; MR **15**, 292e; English translation, p. 177, Dover, New York, 1960; MR **22** #2598] remarked that an analogous theorem for linear forms in arbitrarily many logarithms of algebraic numbers would be of great value for the solution of some apparently very difficult problems of number theory. It is the object of this paper to establish such a result."

Define the height of an algebraic number to be the maximum of the absolute values of the relatively prime integer coefficients in its minimal defining polynomial.

The main result of Part I is Theorem 1.1: Let $\alpha_1, \cdots, \alpha_n$ $(n \geq 2)$ denote algebraic numbers, not 0 or 1, such that (for any fixed branch of the logarithm) $\log \alpha_1, \cdots, \log \alpha_n$ and $2\pi i$ are linearly independent over the rationals Q; suppose $k > n+1$ and let d be any positive integer; then there is an effectively computable number $C = C(n, \alpha_1, \cdots, \alpha_n, k, d) > 0$ such that for all algebraic numbers $\beta_1, \cdots, \beta_n$, not all 0, with degree at most d, we have $(*)$ $|\beta_1 \log \alpha_1 + \cdots + \beta_n \log \alpha_n| > Ce^{-(\log H)^k}$, where H denotes the maximum of the heights of $\beta_1, \cdots, \beta_n$. An immediate consequence is Corollary 1.1: If $\alpha_1, \cdots, \alpha_n$ denote non-zero algebraic numbers and $\log \alpha_1, \cdots, \log \alpha_n, 2\pi i$ are linearly independent over Q, then $\log \alpha_1, \cdots, \log \alpha_n$ are linearly independent over the field A of all algebraic numbers. Starting with Corollary 1.1 for $n = 1$, complete induction gives Corollary 1.2: If $\alpha_1, \cdots, \alpha_n$ denote positive real algebraic numbers other than 1 and $\beta_1, \cdots, \beta_n$ denote real algebraic numbers with $1, \beta_1, \cdots, \beta_n$ linearly independent over Q, then ${\alpha_1}^{\beta_1} \cdots {\alpha_n}^{\beta_n}$ is transcendental. It is remarked that, according to Gel′fond and Linnik, Theorem 1.1 suffices, at least in principle, to settle a conjecture by C. F. Gauss [*Disquisitiones arithmeticae*, § 303, G. Fleischer, Leipzig, 1801; German translation, Springer, Berlin, 1889; reprinting of German translation, pp. 351–352, § 303, Chelsea, New York, 1965; MR **32** #5488; English translation, Yale Univ. Press, New Haven, Conn., 1966; MR **33** #5545] that there are only nine imaginary quadratic fields with class number 1; H. Stark's different approach is mentioned. In order to prove Theorem 1.1, it is first shown that the following modified form of the theorem is sufficient. Theorem 1.1′: Under the hypothesis of Theorem 1.1 there is an effectively computable number C such that for all algebraic numbers $\beta_1, \cdots, \beta_{n-1}$ with degrees at most d, we have $|\beta_1 \log \alpha_1 + \cdots + \beta_{n-1} \log \alpha_{n-1} - \log \alpha_n| \geq e^{-(\log H)^k}$, where H denotes any number not less than C and the height of $\beta_1, \cdots, \beta_{n-1}$. The proof itself then depends on the construction of a function φ of $n-1$ variables; φ is a direct generalization of a function of one variable used in Gel′fond's work (cf. $R(x)$ in C. L. Siegel's *Transcendental numbers* [p. 81, Ann. of Math. Studies No. 16, Princeton Univ. Press, Princeton, N.J., 1949; MR **11**, 330c; German translation, Bibliograph. Inst. Mannheim, 1967; MR **35** #133]) and has the form $\varphi(z_1, \cdots, z_{n-1}) = \sum_{\lambda_1=0}^{L} \cdots \sum_{\lambda_n=0}^{L} p(\lambda_1, \cdots, \lambda_n) {\alpha_1}^{\gamma_1 z_1} \cdots \alpha_{n-1}^{\gamma_{n-1} z_{n-1}}$ with integer coefficients and $\gamma_r = \lambda_r + \lambda_n \beta_r$ $(1 \leq r < n)$.

In Part II it is shown that the number $2\pi i$ can be omitted in Corollary 1.1; this gives Corollary 2.1: Let $\alpha_1, \cdots, \alpha_n$ denote non-zero algebraic numbers; then $\log \alpha_1, \cdots, \log \alpha_n$ are linearly independent over Q if and only if they are linearly independent over A. Also Corollary 1.2 is sharpened to Corollary 2.2: If $\alpha_1, \cdots, \alpha_n$ denote algebraic numbers other than 0 or 1 and if $\beta_1, \cdots, \beta_n$ denote algebraic numbers with $1, \beta_1, \cdots, \beta_n$ linearly independent over Q, then ${\alpha_1}^{\beta_1} \cdots {\alpha_n}^{\beta_n}$ is transcendental. Both corollaries follow from Theorem 2.2: Let again $\alpha_1, \cdots, \alpha_n$ $(n \geq 2)$ denote nonzero algebraic numbers such that $\log\alpha_1, \cdots, \log\alpha_n$ are linearly independent over Q; suppose $k > 2n+1$ and let d be any positive integer; then there is an effectively computable number $C > 0$ such that for all algebraic numbers $\beta_1, \cdots, \beta_n$, not all 0, with degree at most d, we have $(*)$.

In Part III an inhomogeneous analogue is considered. Theorem 3.1: Let $\alpha_1, \cdots, \alpha_n$, $\beta_0, \beta_1, \cdots, \beta_n$ denote non-zero algebraic numbers; suppose $k > n+1$, and let d $[H]$ denote the maximum of the degrees [heights] of $\beta_0, \cdots, \beta_n$; then $|\beta_0 + \beta_1 \log\alpha_1 + \cdots + \beta_n \log\alpha_n| > Ce^{-(\log H)^k}$ for some effectively computable number $C = C(n, \alpha_1, \cdots, \alpha_n, k, d) > 0$. The method of proof can be adapted to $\beta_0 = 0$ and yields Theorem 3.2: Let $\alpha_1, \cdots, \alpha_n$, $\beta_1, \cdots, \beta_n$ denote non-zero algebraic numbers; suppose that either $\log\alpha_1, \cdots, \log\alpha_n$ or $\beta_1, \cdots, \beta_n$ are linearly independent over Q; suppose that $k > n$ and let d $[H]$ denote the maximum of the degrees [heights] of $\beta_1, \cdots, \beta_n$; then $(*)$ holds for some effectively computable number $C = C(n, \alpha_1, \cdots, \alpha_n, k, d) > 0$. Theorem 3.1 implies that $e^{\beta_0}\alpha_1{}^{\beta_1} \cdots \alpha_n{}^{\beta_n}$ is transcendental for any non-zero algebraic numbers $\alpha_1, \cdots, \alpha_n$, $\beta_0, \beta_1, \cdots, \beta_n$. Furthermore, $\pi + \log\alpha$ is transcendental for any algebraic number $\alpha \neq 0$ and also $e^{\alpha\pi+\beta}$ is transcendental for all algebraic numbers α, β with $\beta = 0$. It is conjectured that Theorems 3.1 and 3.2 essentially hold with $k = 1$. Further applications to binary forms and to Liouville's theorem of 1844 concerning the approximation of algebraic numbers by rational numbers are announced.
G. J. Rieger (Munich)

9, 303c 14.0X
Weil, André
Foundations of algebraic geometry.
American Mathematical Society Colloquium Publications, vol. 29.
American Mathematical Society, New York, 1946. xix+289 pp.

Advances in the more arithmetic branches of modern algebra and their application to number theory naturally lead, as we may venture to say today, to problems which to the well-informed mathematician either appeared familiar as part of the heritage of classical algebraic geometry or seemed to be intrinsically adapted to a solution by more conceptual geometric methods. Furthermore, since major parts of the theory of algebraic functions of one variable had been fitted into the system of algebra it was sensible that similar interpretations and attempts at solutions were (and had to be) tried for higher dimensional problems. In order to understand and appreciate the ultimate significance of this book the reader may well keep in mind the preceding twofold motivation for the interest in algebraic geometry. Classical algebraic geometry made free use of a type and mode of reasoning with which the modern mathematician often feels uncomfortable, though the experience based on a rich and intricate source of examples made the founders of this discipline avoid serious mistakes in final results which lesser men might have been prone to make. The main purpose of this treatise is to formulate the broad principles of the intersection theory for algebraic varieties. We find those fundamental facts without which, for example, a good treatment of the theory of linear series would be difficult. The doctrine of this book is that an unassailable foundation (and thereby justification) of the basic concepts and results of algebraic geometry can be furnished by certain elementary methods of algebra. Thus, the reader will agree after some time that he is finding a delicate tool which can serve him to remove the traces of insecurity which occasionally accompany geometric reasoning. Incidentally, the term "elementary" used here and by the author is to be understood in a restricted technical sense, in the sense that general ideal theory and the theory of power series rings are not brought into play too often. The proofs require the general plan of using the "principles of specialization," as formulated algebraically by van der Waerden; and they are by no means elementary in the customary connotation. To some readers the adherence to a definite type of approach, where another author may have deemed it more instructive or appropriate to use slightly different methods, may tend to cloud occasionally immediate understanding by the less adept. However, once the reader has grasped the real geometric meaning of a definition or theorem (he then has to forget occasionally the fine points resulting from the facts that the author imposes no restriction on the characteristic of the underlying field of quantities) he will recognize how skilfully the language and methods of algebra are used to overcome certain limitations of spatial intuition.

The author begins his work with judiciously selected results from the theory of algebraic and transcendental extensions of fields [chapter I, Algebraic preliminaries]. Special emphasis has to be placed on inseparable extensions, which incidentally means a more complete account than is found in books on algebra. The further plan of the book is perhaps best appreciated if one starts to ponder over a more or less heuristic definition of "algebraic variety," and then asks one's self informally how one should define "intersections with multiplicities" of "subvarieties." Then, in view of the principle of local linearization in classical analysis, the author's arrangements of topics is more or less dictated by the ultimate subject under discussion, provided one does not place the interpretation of geometrical concepts by ideal theory at the head of the discussion. Therefore the technical definitions of point, variety, generic point and point set attached to a variety [chapter IV, The geometric language] must be preceded by suitable algebraic preparations [essentially in chapter II, Algebraic theory of specializations] and more arithmetic studies [chapter III, Analytic theory of specializations]. Crucial results in this connection, based on arithmetical considerations, are found in proposition 7 on page 60 and theorem 4 on page 62, where the existence of a well-defined multiplicity is proved for specializations. For further work, the author next introduces the concept of simple point of a variety in affine space by means of the linear variety attached to the point. [See the significant propositions 19 to 21 on pages 97–99.] Next, the intersection theory of varieties in affine space is presented through the following stages of increasing complexity: (i) intersection with a linear subspace of complementary dimension, the 0-dimensional case, with the important criterion for multiplicity 1 in proposition 7 on page 122, and ultimately the criterion for simple points in theorem 6 on page 136; (ii) intersection with a linear subspace of arbitrary dimension, with theorem 4 on page 129 which justifies the invariant meaning of the term "intersection multiplicity of a variety with a linear variety along a variety" [chapter V, Intersection multiplicities, special case]. In chapter VI, entitled General intersection theory, the results for the linear case are extended so as to culminate in the important theorem 2 on page 146 concerning the proper components of the intersection of two subvarieties in a given variety. Furthermore, all important properties of intersection multiplicities are established. Later, in appendix III, it is shown that the properties established for a certain symbol are characteristic for intersection multiplicities and uniquely define that concept. It may be mentioned that the topological definition of the chain intersections on manifolds coincides with the algebraically defined concept of this book. Of course, the underlying coefficient field has to be the field of all complex numbers and further simplifying assumptions on the variety have to be made. However, this comparison cannot be made at the level of chapters V and VI, since there one deals with affine varieties to which the ordinary topological considerations are not directly applicable.

The subsequent chapter VII, Abstract varieties, provides the necessary background for the aforementioned connections and also contains complete proofs of those results which one might have formulated first had one deliberately adopted ideal-theoretic intentions at an early stage. The abstract varieties of this chapter are obtained by piecing together varieties in affine spaces by means of suitably restricted birational transformations. This definition of the author has turned out to be very fruitful for the work on the Riemann hypothesis for function fields and the study of Abelian varieties in general. In the course of the work, the results of the preceding chapters are extended so as to lead up to the important theorem 8 on page 193 related to Hopf's "inverse homomorphism." The chapter ends with a theory of cycles of dimension s, that is, formal integral combinations of simple abstract subvarieties of dimension s. The notion of the intersection product of cycles is also introduced here [page 202], by means of which the investigation of equivalence theories can be initiated. This is done more explicitly in chapter IX, Comments and discussion; apparently the Riemann-Roch theorem for surfaces should now be accessible to a careful re-examination. As a further result, the theory of quasi-divisibility of Artin and van der Waerden is developed in theorems 3 and 4 on pages 224–225 and theorem 6 on page 230. These theorems exhibit the relations between the theory of cycles of highest dimension and the theory of quasi-divisibility, where naturally some of the results in appendix II, Normalization of varieties, are to be added for the necessary integral closure of the required rings of functions. In this appendix the author relates his results on the normalization of algebraic varieties to those of Zariski. At this point the individual reader may well compare the elementary and the ideal-theoretic approach to a group of theorems. In appendix I, Projective spaces, often used properties and facts concerning projective spaces are quickly developed on the basis of the preceding work. This brief discussion not only deals with results which are generally useful in algebraic geometry, but also contains one of the theorems on linear series of divisors which was frequently used in the classical work [see page 266]. Because of the wealth of material and the excellent "advice to the reader" prefacing this rich and important book the reviewer feels that he should mention some of the highlights and not delve into a discussion of technical details. In short, the only way to appreciate this treatise is actually to read it. *O. F. G. Schilling* (Chicago, Ill.)

16, 953c 14.0X
Serre, Jean-Pierre
Faisceaux algébriques cohérents.
Ann. of Math. (2) **61** (1955), 197–278.

This paper contains the basic material for the application of cohomology theory to algebraic geometry; the specific applications themselves are to appear in a forthcoming paper.

Chapter I is concerned with the general theory of sheaves. A sheaf on a space X is a space F with a mapping p (the projection) of F into X which is a local homeomorphism. A continuous mapping s of an open set U of X into F such that $p(s(x)) = x$ for $x \in U$ is called a section of F above U. In all algebraic applications, the sets $F_x = p^{-1}(x)$ have algebraic structures whose laws of composition are assumed to "vary continuously" with x, in a sense easily made precise; the most interesting case is that of a sheaf of modules, where each F_x is a module over a ring A_x, the A_x's forming themselves a sheaf of rings. The notions introduced in sheaf theory are either of an algebraic nature (induced sheaf on a subset of X, prolongation of a sheaf on a subspace, subsheaves, sheaf homomorphisms, factor sheaves, exact sequences of sheaves, tensor product of sheaves) or of a topological nature (cohomology of a space X in a sheaf F of additive groups on X). In defining this last notion, attention must be paid to the fact that X may fail to be a Hausdorff space (this will always happen in algebraic geometry). This fact entails certain difficulties; in particular, the exactness of the cohomology sequence associated to an exact sequence cannot be established in general. This fundamental theorem holds nevertheless for the cohomology of an algebraic variety V, provided only "algebraic coherent" sheaves are considered. Let F be a sheaf of modules over a sheaf A of rings. Then F is said to be of finite type if, for every $x \in X$, there exist a finite number of sections s_i $(1 \leq i \leq p)$ of F over a suitable neighbourhood U of x such that, for all $y \in U$, F_y is generated as a module by the elements $s_i(y)$. Let $s_1, \cdots, s_p$ be sections of F over an open set U. For each $y \in U$, let R_y be the set of elements $(a_1, \cdots, a_p)$ in ${A_y}^p$ such that $\sum_{i=1}^{p} a_i s_i(y) = 0$. Then the sets R_y form a sheaf R over U. If F is of finite type and R of finite type (for all choices of $s_1, \cdots, s_p$, U), then F is called coherent. Coherent sheaves are discussed in relation to the general algebraic operations on sheaves.

Chapter II contains an alternative definition of the notion of an (abstract) variety, due to A. Weil. An algebraically closed field K is given; each affine K-space K^r is endowed with the Zariski topology on this space. For any $x \in K^r$, let O_x be the local ring of x (the set of rational functions on K^r which are defined at x); the rings O_x form a sheaf of rings on K^r. A variety V is defined to be a space V together with a subsheaf O_V of the sheaf of germs of K-valued functions on V, with two axioms of which the first says that X has an open finite covering (U_i) such that each U_i (with its induced sheaf) is isomorphic to a locally closed set in some K^r, with its sheaf of local rings, while the second expresses the condition of biregularity of the birational transformations between the U_i's which are defined by the fact that, if $x \in U_i \cap U_j$, its local ring on U_i is the same as its local ring on U_j (both being the ring $(O_V)_x$). The notion introduced in this manner is slightly more general than the notion defined by Weil, inasmuch as it includes reducible varieties (unions of a finite number of Weil varieties). If O_V is the sheaf which defines the structure of a variety V, then, for $x \in V$, $(O_V)_x$ is called the local ring at x, and is denoted by O_x. By an algebraic sheaf on V is meant a sheaf of modules over O_V. Every sheaf on a closed subvariety W of V may be extended to a sheaf on V which coincides with 0 outside W. This implies that, inasmuch as sheaf theory is concerned, the theory of a subvariety is not essentially more difficult than that of the ambient variety; in particular, sheaf theory on a projective variety may be reduced to sheaf theory on a projective space, in which specific homogeneous coordinates are used to the full. The author studies the theory of sheaves on affine varieties; there, it turns out that, for coherent sheaves, the cohomology is trivial in all dimensions > 0; this seems to indicate that the natural domain of sheaf theory is the study of algebraico-geometric notions in which the completeness of the variety plays a crucial role. By using a finite affine covering of an abstract variety by affine varieties, it is shown that the cohomology sequence attached to an exact sequence $A \to B \to C$ of sheaves is exact provided A is algebraic and coherent.

Chapter III is concerned with the theory of sheaves on a projective variety. The projective space X of dimension r is covered in a natural manner by $r+1$ affine spaces; it is shown that this covering is already sufficient to compute the cohomology groups of a coherent algebraic sheaf; in particular, the cohomology in dimension $> r$ is trivial. More precisely, if V is a closed projective variety of dimension n, then the cohomology of V (in any coherent algebraic sheaf) is trivial in any dimension $> n$. Let S be the ring of polynomials in $r+1$ variables. In order to study the cohomology of X in an algebraic sheaf F, the author associates to F a graded module M over S. Let x be a point of X. Introduce the ring $O_n(x)$ generated by the elements P/Q where P and Q are homogeneous in S, degree of $P = N+$degree of Q and Q and $Q(x) \neq 0$ (n being any integer). The rings $O_n(x)$ form a sheaf $O(n)$; set $F(n) = O(n) \otimes F$. Let $(\Gamma_F)_n$ be the set of sections of $F(n)$ over the whole of X, and Γ_F the direct sum of the groups $(\Gamma_F)_n$, for all integers n. Then Γ_F has an easily defined structure of graded module over the ring S. If F is coherent, then Γ_F satisfies the following condition (TF): there exists a p such that $\sum_{n>p}(\Gamma_F)_n$ is finitely generated. Moreover, in that case, the sheaf F may be recovered from the module Γ_F by a general procedure which associates a sheaf on X to every graded S-module.

To every graded S-module M, the author attaches certain cohomology groups $H^q(M)$ [J. L. Koszul had introduced homology (not cohomology) groups for M [Colloque de Topologie (espaces fibrés), Bruxelles, 1950, Thone, Liège, 1951, pp. 73–81; MR **13**, 109a]; taking into account the self duality of an exterior algebra on a free module, Serre's groups are related in the classical manner to a generalization of Koszul's homology groups]. If F is a coherent sheaf on X, then the cohomology groups of F are the same as those of the module Γ_F. This allows one to obtain a certain number of properties of the cohomology groups $H^q(X;F)$; in particular, these groups are finite-dimensional vector spaces over K, and, for $q>0$, $H^q(X,F(n)) = \{0\}$ for n large.

A new step is accomplished by comparing the cohomology groups $H^q(M)$ of a graded S-module M to certain other groups, which are Ext-groups. Let $M(n)$ be the module obtained from M by lowering the degree by n. The module $H^q(M(n))$ is a vector space over K; denote by $(T^q(M))_n$ the dual of the vector space $H^q(M(-n))$ and by $T^q(M)$ the direct sum of the groups $(T^q(M))_n$ for all integers n; this is again a graded module over S. Let Ω be a free module of rank 1 over S, graded by the condition that the basic element be of degree $r+1$. Then it is proved that, for any $q \neq r$, $T^{r-q}(M)$ is isomorphic to $\operatorname{Ext}^q(M,\Omega)$, while, for $q=r$, there is an exact sequence

$$0 \to \operatorname{Ext}^r(M,\Omega) \to T^0(M) \to M^* \to \operatorname{Ext}^{r+1}(M,\Omega) \to 0,$$

where M^* is the graded module whose elements of degree n are the linear forms on the vector space M_n over K of elements of degree n in M. The passing from $-n$ to n which is involved in the definition of $T^q(M)$, together with the fact that the dual of a vector space $\neq \{0\}$ is $\neq \{0\}$, allows one to infer from this results on the groups $H^q(M(-n))$ for n large, and therefore on the groups $H^q(X,F(-n))$, if F is a coherent sheaf on X. Let V be an irreducible variety of dimension p in X; let F be a coherent sheaf on V such that, for every $x \in V$, F_x is a free module over the local ring of x. Then, if V is free from singularity, $H^q(V, F(-n)) = \{0\}$ for large n and $0 \le q < p$; if V is normal and $p \ge 2$, then $H^1(V; F(-n)) = \{0\}$ for large n.

If F is a coherent sheaf on a projective variety V, it is possible to define the Euler-Poincaré characteristic $\chi(V, F) = \sum_q (-1)^q h^q(V, F)$, where $h^q(V, F)$ is the dimension of $H^q(V, F)$ over K. If V is irreducible of dimension p, and F the sheaf of local rings, then $\chi(V, F) = 1 + (-1)^q p_a(V)$, where $p_a(F)$ is the arithmetic genus of V. If $V = X$, then, for any F, $\chi(X, F(n))$ is a polynomial in n whose degree is the dimension of the carrier of F. *C. Chevalley* (Paris)

18, 509b 14.0X

Hirzebruch, F.

Neue topologische Methoden in der algebraischen Geometrie. (German)

Ergebnisse der Mathematik und ihrer Grenzgebiete (N.F.), Heft 9.

Springer-Verlag, Berlin-Göttingen-Heidelberg, 1956. viii+165 pp.

This monograph is an exposition of the main ideas and concepts centering around the fundamental Riemann-Roch theorem for algebraic manifolds of arbitrary dimension, a theorem whose proof has recently been found by the author [Proc. Nat. Acad. Sci. U.S.A. **40** (1954), 110–114; MR **17**, 535a] and which is presented here in complete form for the first time. An algebraic manifold is understood to mean a compact complex manifold which can be imbedded complex-analytically without singularities in a complex projective space of suitably chosen dimension.

The first chapter is an extremely clear, well-written exposition of those aspects of the theory of characteristic classes and sheaves which are required in the sequel. In particular, the author's systematic treatment of characteristic classes (Pontrjagin and Chern classes) is in itself an important contribution to the general theory. The notation and formalism of multiplicative sequences of polynomials, introduced at the beginning of the first chapter, form the basis of the author's treatment and may be described as follows. Let B denote a commutative ring with unit 1, let $p_1, p_2, \cdots$ be indeterminates and set $p_0 = 1$. The ring $\mathfrak{B} = B[p_1, p_2, \cdots]$ of polynomials in the p_i with coefficients in B is graded by assigning to p_j the weight j: $\mathfrak{B} = \sum \mathfrak{B}_k$, where $\mathfrak{B}_k$ is the module of polynomials of weight k. A sequence $\{K_j\}$ of polynomials in the unknowns p_i with $K_0 = 1$, $K_j \in \mathfrak{B}_j$ $(j = 0, 1, 2, \cdots)$, is a multiplicative sequence if the formal identity $\sum p_j'' z^j = (\sum p_j z^j) \cdot (\sum p_j' z^j)$ where z, p_j, p_j' are indeterminates, implies $\sum K_j(p'') z^j - (\sum K_i(p) z^i) \cdot (\sum K_i(p') z^i)$, where $K_i(p) = K_l(p_1, \cdots, p_l)$ etc. To each formal power series $Q(z) = \sum_{i=0}^{\infty} b_i z^i$, $b_0 = 1$, $b_i \in B$, there corresponds a multiplicative sequence $\{K_j\}$ of polynomials and conversely. This correspondence is established by splitting (formally)

$$1 + p_1 z + p_2 z^2 + \cdots + p_m z^m$$

into a product $\prod_{i=1}^m (1 + \beta_i z)$ of linear factors and using the product $\prod_{i=1}^m Q(\beta_i z)$ to define the K_j for $j \le m$. It is also convenient to replace p_i by c_i, z by x, β_i by γ_i (with $c_0 = p_0 = 1$), and to introduce the relations $z = x^2$, $\beta_i = \gamma_i{}^2$. The most important multiplicative sequences are: (1) $\{L_j\}$ defined by $Q(z) = z^{\frac{1}{2}}/\operatorname{tgh} z^{\frac{1}{2}}$; (2) $\{A_j\}$ defined by $2z^{\frac{1}{2}}/\sinh 2z^{\frac{1}{2}}$; (3) $\{T_j(y; c_1, \cdots, c_j)\}$ defined by

$$Q(y; x) = x(y+1)/[1 - \exp(-x(y+1)] - yx.$$

The second chapter contains an outline of the deep theory of "cobordisme" of R. Thom which is applied by the author to prove that the index $\tau(M)$ of an oriented compact differentiable manifold M of dimension $4k$ is a linear combination of Pontrjagin numbers, namely: $\tau(M) = L_k(p_1, \cdots, p_k)\ [M]$, where $\{L_j\}$ is the sequence (1) and $p_i \in H^{4i}(M, \mathbf{Z})$ is the ith Pontrjagin class of M. The proof of the Riemann-Roch theorem is reduced to this basic result which is called the "index theorem".

In the third chapter, properties of the virtual Todd genus are derived and a method for splitting vector bundles is developed (splitting method). The splitting method together with the calculus of virtual genera provide the main tool for the reduction of the Riemann-Roch theorem to the index theorem. Let M_n denote an almost-complex manifold of complex dimension n with the natural orientation, and let c_i be the ith Chern class of M_n (ith Chern class of the tangent $U(n)$-bundle of M_n). The generalized Todd genus $T_y(M_n)$ of M_n (shortly T_y-genus) is the polynomial of degree n in y with rational coefficients which is equal to $T_n(y; c_1, \cdots, c_n)\ [M_n]$, where $\{T_j(y; c_1, \cdots, c_j)\}$ is the multiplicative sequence (3). Let $T_y(M_n) = \sum_{p=0}^{n} T^p(M_n) y^p$. The Todd genus $T(M_n)$ of M_n is the number $T_0(M_n) = T^0(M_n)$, while $T_{-1}(M_n) = \sum_{p=0}^{n}(-1)^p T^p(M_n) = c_n[M^n]$ is the Euler-Poincaré characteristic of M_n, and

$$T_1(M_n) = \sum_{p=0}^{n} T^p(M_n) = \tau(M_n)$$

is the index of M_n (assumed to vanish for odd n)—a fact which is a consequence of the index theorem. If we set $Q(y; x) = x/R(y; x)$ we obtain from (3):

$$R(y; x) = [\exp(x(y+1)) - 1]/[\exp(x(y+1)) + y].$$

In particular $R(1; x) = \operatorname{tgh} x$, $R(-1; x) = x(1+x)^{-1}$, $R(0, x) = 1 - e^{-x}$. Given $v_1, \cdots, v_r \in H^2(M_n, \mathbf{Z})$, the virtual T_y-genus of the r-tuple $(v_1, \cdots, v_r)$ is defined to be

$$T_y(v_1, \cdots, v_r)_M = \kappa_n[R(y; v_1) \cdots R(y; v_r) \sum_{j=0}^{\infty} T_j(y; c_1(M_n), \cdots)],$$

where κ_n has the following meaning: let $u \in H^*(M_n)$ and let $u^{(n)}$ be the $2n$-dimensional component of u; then $\kappa_n[u] = u^{(n)}[M_n]$. The virtual T_y-genus is a polynomial in y of degree $n - r$ with rational coefficients,

$$T_y(v_1, \cdots, v_r)_M = \sum_{p=0}^{n-r} T^p(v_1, \cdots, v_r)_M y^p,$$

which satisfies the functional equation

$$\begin{aligned} T_y(v_1, \cdots, v_r, u+v) &= T_y(v_1, \cdots, v_r, u) + T_y(v_1, \cdots, v_r, v) \\ &\quad + (y-1) T_y(v_1, \cdots, v_r, u, v) - y T_y(v_1, \cdots, v_r, u, v, u+v) \\ &\qquad\qquad (v_1, \cdots, v_r, u, v \in H^2(M_n, \mathbf{Z})). \end{aligned}$$

For $y = 1, 0, -1$ we obtain functional equations for the virtual index, the virtual Todd genus, and the virtual Euler-Poincaré characteristic respectively. The T_y-characteristic of a continuous $\mathrm{GL}(q, \mathbf{C})$-bundle ξ over M_n is defined as follows. Let $d_0, d_1, \cdots, d_q$ be the Chern classes of ξ, $c_0 = 1, c_1, \cdots, c_n$ the Chern classes of M_n, consider the formal splittings $\sum_{i=0}^{n} c_i x^i = \prod_{i=1}^{n}(1+\gamma_i x)$, $\sum_{i=1}^{q} d_i x^i = \prod_{i=1}^{q}(1+\delta_i x)$, set

$$t_y(\xi) = \exp((1+y)\delta_1) + \exp((1+y)\delta_2) + \cdots + \exp((1+y)\delta_q)$$
$$(t_y(\xi) \in H^*(M_n, \mathbf{Z}) \otimes \mathbf{Q}[y]),$$

and define the virtual T_y-characteristic of ξ with respect to the virtual manifold $(v_1, \cdots, v_r)$, $v_i \in H^2(M_n, \mathbf{Z})$ $(i = 1, \cdots, r)$

$$T_y(v_1, \cdots, v_r|, \xi)_M = \kappa_n[t_y(\xi) \prod_{i=1}^{r} R(y; v_i) \sum_{j=0}^{\infty} T_j(y; c_1(M_n), \cdots)].$$

If ξ is the trivial $\mathrm{GL}(q, \mathbf{C})$-bundle, then

$$T_y(v_1, \cdots, v_r|, \xi) = qT_y(v_1, \cdots, v_r).$$

The T_y-characteristic of ξ satisfies the same functional equation as the T_y-genus.

Now let $\Delta(q, \mathbf{C})$ denote the subgroup of those matrices of $\mathrm{GL}(q, \mathbf{C})$ which leave fixed all r-dimensional linear subspaces L_r of $\mathbf{C}_q$ defined by $z_{r+1} = z_{r+2} = \cdots = z_q = 0$ $(z_1, z_2, \cdots, z_q$ the coordinates of $\mathbf{C}_q)$; $\Delta(q, \mathbf{C})$ is the group of triangular matrices of $\mathrm{GL}(q, \mathbf{C})$ (all coefficients beneath the diagonal vanish). Let X be a complex manifold. An element $\xi \in H^1(X, G_\omega)$, where G_ω denotes the sheaf of germs of holomorphic maps of X into the complex Lie group G, will be called a complex-analytic G-bundle, and a fibre bundle belonging to the equivalence class ξ will be said to be associated to ξ. Let ξ be a $\mathrm{GL}(q, \mathbf{C})$-bundle over X, L a complex-analytic principal bundle over X which is associated to ξ, $E = L \times_G G/\Delta(q, \mathbf{C})$ the bundle associated to L with the flag manifold

$$F(q) = \mathrm{GL}(q, \mathbf{C})/\Delta(q, \mathbf{C})$$

as fibre, and let $\varphi: E \to X$ be the projection. Then E is a complex manifold, φ is an analytic mapping of E onto X, and the structure group of L lifted up to E (bundle induced by φ) can be reduced complex-analytically to $\Delta(q, \mathbf{C})$ in a natural way giving rise to q diagonal complex-analytic $\mathbf{C}^*$-bundles which will be denoted by $\xi_1, \xi_2, \cdots, \xi_q$ in their natural order. Denote by e_0 the point in $F(q)$ defined by the coset $\Delta(q, \mathbf{C})$. The group $\Delta(q, \mathbf{C})$ operates by left translation on $\mathrm{GL}(q, \mathbf{C})/\Delta(q, \mathbf{C})$ but keeps e_0 fixed; hence $\Delta(q, \mathbf{C})$ operates on the tangent space $\mathbf{C}_m(e_0)$ of $e_0 \in F(q)$ and this gives an analytic homomorphism $h: \Delta(q, \mathbf{C}) \to \mathrm{GL}(m, \mathbf{C})$. Let $T(q)$ be the tangential principal fibre bundle of the complex flag manifold $F(q)$; the group $G = \mathrm{GL}(q, \mathbf{C})$ operates by left translation on $F(q)$ and therefore on $T(q)$. Hence we have the fibre bundle $L \times_G T(q)$ over X associated to L which, lifted up to E by φ, is a bundle over E with fibre $\mathrm{GL}(m, \mathbf{C})$; the corresponding $\mathrm{GL}(m, \mathbf{C})$-bundle will be denoted by ξ^Δ. The structure group of ξ^Δ can be reduced complex-analytically to $\Delta(m, \mathbf{C})$ (for example, by using a theorem of Lie since $\Delta(q, \mathbf{C})$ is soluble and the map $\mathbf{h}$ is holomorphic—a remark of A. Borel). This reduction gives rise to m diagonal complex-analytic $\mathbf{C}^*$-bundles equal to $\xi_i \otimes \xi_j^{-1}$ $(i > j)$ with the following ordering: $\xi_i \otimes \xi_j^{-1}$ precedes $\xi_{i'} \otimes \xi_{j'}^{-1}$ if $j > j'$ or $j = j'$ and $i < i'$. Let $\gamma_i = c_1(\xi_i) \in H^2(E, \mathbf{Z})$; then it follows that the Chern class $c(E)$ of E is given by the formula $c(E) = \varphi^* c(X) \cdot \prod_{q \geq i > j \geq 1}(1 + \gamma_i - \gamma_j)$. In particular, if L is the tangential $\mathrm{GL}(n, \mathbf{C})$-bundle of the complex manifold X (of complex dimension n), we set $E = X^\Delta$ and in this case $c(X^\Delta) = \prod_{i=1}^{n}(1 + \gamma_i) \cdot \prod_{n \geq i > j \geq 1}(1 + \gamma_i - \gamma_j)$. Finally a complex manifold

X will be called a complex-analytic split manifold if the structure group $\mathrm{GL}(n,\mathbf{C})$ of its tangent bundle can be reduced complex-analytically to $\Delta(n,\mathbf{C})$ (but the tangent bundle is then not in general the complex-analytic Whitney sum of $\mathbf{C}^*$-bundles). The Chern class of a complex-analytic split manifold X is given by $c(X)=\prod_{i=1}^{n}(1+a_i)$, where $a_i=c_1(\xi_i),\xi_1,\cdots,\xi_n$ the $\mathbf{C}^*$-bundles defined by the $\Delta(n,\mathbf{C})$-bundle. For a complex-analytic split manifold X the Todd genus $T(X)$ may be expressed in terms of virtual genera by the formula

$$\text{(I)}\qquad (1+y)^n T(X)=\sum_{l=0}^{n} y^l \sum_{1\leq i_1<\cdots<i_l\leq n} T_y(a_{i_1},\cdots,a_{i_l})_X.$$

This formula, whose proof is an almost immediate consequence of the definition of T_y-genus, is used in reducing the Riemann-Roch theorem to the index theorem.

Similar considerations must also be developed for the Euler-Poincaré characteristic $\chi(X,W)$ of the cohomology of X with coefficients in the sheaf of germs of holomorphic sections of a complex-analytic vector bundle W over X. In the fourth and final chapter of the book, a virtual χ_y-characteristic is defined which is the analogue in sheaf cohomology of the T_y-characteristic. Let E be a ring which is an extension of the integers $\mathbf{Z}$ and has the integer 1 as unit, and let $\mathbf{Z}\{y\},E\{y\}$ be the rings of formal power series with coefficients in $\mathbf{Z},E$ respectively. The rings $E\{y\}$ and $\mathbf{Z}\{y\}$ are regarded as $\mathbf{Z}\{y\}$-modules, and a d-homomorphism $h\colon E\{y\}\to\mathbf{Z}\{y\}$ is by definition a $\mathbf{Z}\{y\}$-homomorphism. Given an additive homomorphism $h_0\colon E\to\mathbf{Z}\{y\}$, there is one and only one d-homomorphism $h\colon E\{y\}\to\mathbf{Z}\{y\}$ which coincides with h_0 on E. Moreover given a d-homomorphism $h\colon E\{y\}\to\mathbf{Z}\{y\}$ and a fixed element $t\in E\{y\}$, a d-homomorphism h_t is defined by $h_t(u)=h(tu)$, $u\in E\{y\}$. If, for d-homomorphisms h and h' of $E\{y\}$ into $\mathbf{Z}\{y\}$, there exists an element $t\in E\{y\}$ with $h'(u)=h(tu)$ for all $u\in E$,then $h'=h_t$. Now let $f_1,\cdots,f_r,w$ be unknowns, and consider over $\mathbf{Z}$ the ring E generated by these unknowns and by $f_1{}^{-1},f_2{}^{-1},\cdots,f_r{}^{-1}$. A d-homomorphism of $E\{y\}$ into $\mathbf{Z}\{y\}$ is determined by its values on the products $w^\mu f_1{}^{\lambda_1}\cdots f_r{}^{\lambda_r}$, where $\mu,\lambda_1,\cdots,\lambda_r$ are integers, $\mu\geq 0$. Denote by V a compact complex manifold, by $F_1,\cdots,F_r$ complex-analytic line bundles over V, by W a complex-analytic vector bundle over V, and let 1 denote the trivial vector bundle, $\mathbf{T}^{(p)}$ the complex-analytic vector bundle of tangential covariant p-vectors of X. Define two d-homomorphisms h and $\hat{h}$ of $E\{y\}$ into $\mathbf{Z}\{y\}$ by the formulas:

$$h(w^\mu f_1{}^{\lambda_1}\cdots f_r{}^{\lambda_r})=\chi(V,W^\mu\otimes F_2{}^{\lambda_2}\otimes\cdots\otimes F_r{}^{\lambda_r}),h(1)=\chi(V);$$

$$\hat{h}(w^\mu f_1{}^{\lambda_1}\cdots f_r{}^{\lambda_r})=\chi_y(V,W^\mu\otimes F_1{}^{\lambda_1}\otimes F_1{}^{\lambda_1}\otimes\cdots\otimes F_r{}^{\lambda_r}),\hat{h}(1)=\chi_y(V).$$

Here $\chi_y(V,W)=\sum_{p=0}^{n}\chi^p(V,W)y^p$, $\chi_y(V)=\sum_{p=0}^{n}\chi^p(V)y^p$, where $\chi^p(V,W)=\chi(V,W\otimes\mathbf{T}^{(p)})$, $\chi^p(V)=\chi^p(V,1)$. The virtual χ_y-characteristic of the vector bundle W restricted to the virtual manifold $(F_1,\cdots,F_r)$ is defined by the formulas

$$\chi_y(F_1,\cdots,F_r|,W)=\hat{h}_V\left(w\prod_{i=1}^{r}\frac{1-f_i{}^{-1}}{1+yf_i{}^{-1}}\right).$$

If S is a non-singular divisor of V, $F_1=\{S\}$ the corresponding line bundle, it is shown that

$$\chi_y(F_1,\cdots,F_r|,W)_V=\chi_y((F_2)_S,\cdots,(F_r)_S|,W_S)_S.$$

Let

$$\chi_y(F_1,\cdots,F_r|,W)=\sum_{p=0}^{\infty}\chi^p(F_1,\cdots,F_r|,W)_V y^p,$$

$$\chi(F_1,\cdots,F_r|,W)_V=\chi^0(F_1,\cdots,F_r|,W)_V=h_V(w\prod_{i=1}^{r}(1-f_i{}^{-1})).$$

Finally, suppose that V is a complex-analytic split manifold of complex dimension n with the diagonal complex-analytic line bundles $A_1,\cdots,A_n$, and let W be a complex-analytic vector bundle over V; then

$$\text{(II)} \qquad (1+y)^n\chi(V,W)=\sum_{l=0}^{n}y^l\sum_{i_1<\cdots<i_l}\chi_y(A_{i_1},\cdots,A_{i_l}|,W).$$

We remark that the χ_y-characteristic satisfies a functional equation analogous to that for the T_y-characteristic.

Now let $f_1,\cdots,f_r$ be the cohomology classes of the complex-analytic line bundles $F_1,\cdots,F_r$ over V, let the vector bundle W be regarded as a continuous $\mathrm{GL}(q,\mathbf{C})$-bundle ξ, and write $T_y(F_1,\cdots,F_r|,W)_V=T_y(f_1,\cdots,f_r|,\xi)_V$, $T_y(V,W)=T_y(V,\xi)$. The most general formulation of the Riemann-Roch theorem is as follows: Suppose that V is an algebraic manifold, and let $F_1,\cdots,F_r$ be complex line bundles over V, W a complex-analytic vector bundle over V; then (i) $\chi_y(F_1,\cdots,F_r|,W)_V=T_y(F_1,\cdots,F_r|,W)_V$. Taking $r=0$, $y=0$, we obtain (ii) $\chi(V,W)=T(V,W)$. We may still further specialize by taking W to be a complex line bundle F, in which case (iii) $\chi(V,F)=T(V,F)$. Finally, taking $F=1$ (trivial line bundle), we obtain (iv) $\chi(V)=T(V)$. There is little real difference in depth between these various statements; it is almost as difficult to prove (iv) as to prove (i). The statement (ii) is the one which is usually called the Riemann-Roch theorem.

The proof, which occupies the last pages of the author's book, follows fairly quickly on the basis of the machinery described above. For example, formula (iv) may be proved in the following steps. (a) Reduction to algebraic split manifolds. (b) By a theorem of Hodge, true for all compact Kähler manifolds, $\chi_1(V)=\tau(V)$ ($\tau=$ index); by the induction principle of algebraic geometry, χ_1 and τ are shown to coincide in the virtual case. (c) As already pointed out, $T_1(V)=\tau(V)$, and T_1 and τ coincide in the virtual case also. (d) Take $y=1$, $W=1$ (trivial bundle); then formulas (I) and (II) imply that $\chi(V)=T(V)$ for split manifolds and hence, by (a), in general.

Thus formulas (I) and (II) with $y=1$ are the means of reducing the Riemann-Roch theorem to the index theorem which in turn depends on Thom's theory of "cobordisme".

Two obvious questions arise. First, is it possible to give an algebraic proof of the Riemann-Roch theorem which does not make use of Thom's deep theory? Second, is the Riemann-Roch theorem true for a class of manifolds more general than algebraic, for example compact Kähler? There is evidence that the Riemann-Roch theorem is true more generally, but the fact that an induction principle is lacking for non-algebraic manifolds is a serious obstacle in the way of finding a proof.

D. C. Spencer

19, 1195h 14.0X
Borel, Armand
Groupes linéaires algébriques.
Ann. of Math. (2) **64** (1956), 20–82.

This paper contains a comprehensive theory of linear algebraic groups G over an algebraically closed field of arbitrary characteristic. The author first studies diagonalizable groups, i.e. groups whose elements may be simultaneously reduced to diagonal form. Such a group may be defined by monomial relations between the diagonal coefficients of its matrices (these matrices being assumed to be diagonal). A diagonalizable group which is connected is called a torus; torus groups have the important property that their rational representations are all semi-simple.

If x is any invertible matrix, the smallest algebraic group containing x is called the adherent group of x; the author shows that this group always contains the semi-simple and the unipotent component x_s and x_u of x (x_s is semi-simple, x_u is unipotent, which means that $1-x_u$ is nilpotent, and $x=x_sx_u=x_ux_s$). As an application, it is shown that every commutative linear algebraic group is the product of a diagonalizable group by a group of unipotent elements. Moreover, it is also established that rational representations of a linear algebraic group map semi-simple (resp. unipotent) elements upon elements of the same kind.

The author then studies the case where G is solvable. He gives a new proof of the theorem of Lie-Kolchin to the effect that a connected solvable linear algebraic group G may be reduced to triangular form. Moreover, the unipotent elements of G form a normal subgroup G_u and, if T is a maximal torus of G, G is the semi-direct product of G_u and T (i.e. the mapping $(s,t)\rightarrow st$ is an isomorphism of the variety $G_u\times T$ with the variety G). The maximal tori of G are all conjugate to each other in G; if G is nilpotent, it has only one maximal torus T which is the set of semi-simple elements of G and which is in the center of G.

Consider now the general case where G is an arbitrary connected algebraic linear group. The main idea of the author to study this case is to make use of those homogeneous spaces G/H of G which are complete varieties. It is proved that, if a connected solvable group S operates on a complete variety V, then V must contain at least one fixed point for S. On the other hand, a direct construction shows that there always exists a solvable connected subgroup B of G such that G/B is complete (and even projective); B is then maximal in the class of connected solvable subgroups of G. The reviewer has proposed to call Borel subgroups of G the maximal solvable connected subgroups of G; using this terminology, it is established that all Borel subgroups of G are conjugate to each other in G; this theorem had already been stated for Lie groups by Morosov [C. R. Dokl. Acad. Sci. URSS (N.S.) **36** (1942), 83–86; MR **4**, 187b], but Morosov's proof, based on computations in the Lie algebra, was not entirely correct. Making use of this theorem and of previously established results on solvable groups, the author shows that all maximal tori of G are conjugate to each other in G; if T is any one of them, then the centralizer of T is a nilpotent group which is contained in every Borel group which contains T; every semi-simple element is contained in some maximal torus. A group whose elements are all unipotent is nilpotent. The author gives new proofs for Kolchin's results on quasi-compact and anticompact groups.

The last section of the paper is concerned with the theory of Cartan subgroups of a connected algebraic group G (i.e. of maximal nilpotent subgroups C of G with the property that every subgroup of finite index of C is of finite index in its normalizer in G). It is proved that the Cartan subgroups are the centralizers of the maximal tori. For any x in G, denote by x_s the semi-simple component of x; if the centralizer Z of x_s is of smallest possible dimension, then x is called regular; Z is then a Cartan subgroup and is the only Cartan subgroup containing x; conversely, any element which belongs to only one Cartan subgroup is regular. The set of regular elements contains a non-empty open subset of G (in the Zariski topology).

Let f be a rational representation of G; then f maps the Borel subgroups (resp. the maximal tori, the Cartan subgroups, the regular elements) of G upon Borel subgroups (resp. maximal tori, Cartan subgroups, regular elements) of $f(G)$. The reciprocal image of a Borel subgroup (resp. a maximal torus, a Cartan subgroup) of $f(G)$ contains a Borel subgroup (resp. a maximal torus, a Cartan subgroup) of G.

{Since this paper has appeared, the reviewer has established (but not yet published) the following results, which complete those of the author: any Borel subgroup of G is its own normalizer in G; if C is a Cartan subgroup of G, the unipotent elements of C lie in the radical of G (i.e. in the largest connected solvable normal subgroup of G).}

C. Chevalley (Paris)

28 #1199 14.40 10.10

Manin, Ju. I.

Rational points on algebraic curves over function fields. (Russian)

Izv. Akad. Nauk SSSR Ser. Mat. **27** (1963), 1395–1440.

A classical conjecture of Mordell asserts that a curve of genus ≥ 2 defined over a number field has only a finite number of rational points in that field. The author proves the analogue of this conjecture for curves over function fields, and his result can be stated as follows. Let K be a function field, i.e., a finitely generated regular extension of a field k of characteristic 0, and let C be a curve (algebraic, irreducible) of genus ≥ 2, defined over K. If C has infinitely many rational points in K, then there exists a curve C_0 defined over k, a birational transformation $F\colon C_0 \to C$ defined over K, such that all but a finite number of these points are images under F of points of C_0 rational over k.

It is an easy technical matter to reduce the assertion to the case where k is the field of complex numbers, and where K is of transcendence degree 1 over k. In that case, one may view K as the function field of a parameter curve T, and one may view C as the generic member of a pencil of curves on an algebraic surface. The author's theorem is then seen to be equivalent to the following statement. Let V be a projective algebraic surface, and let C_t $(t \in T)$ be a pencil of curves on the surface such that the genus of the generic curve is ≥ 2. If the pencil has infinitely many rational sections, then V is birationally equivalent to a product of a fixed curve and the parameter curve T, and all but a finite number of the sections arise from constant points of the fixed curve.

Finally, one can express the author's theorem in a third equivalent form. Let K be a function field over k again, and assume for simplicity that the transcendence degree is 1. Let P be a point in projective space $\mathbf{P}^n$, rational over K. Then P is the generic point over k of a variety W_P contained in $\mathbf{P}^n$. By deg W_P we shall mean the projective degree. Then the theorem is equivalent to the following assertion. Let C be a non-singular curve defined over K. Then the degrees of the varieties W_P, as

P ranges over the rational points of C in K, are bounded, in a suitable projective embedding. It is actually this final statement which is proved. Its equivalence with the first statement was known [cf. the reviewer, *Diophantine geometry*, Proposition 2 and Corollary of Chapter VII, § 2, Interscience, New York, 1962; MR **26** #119; also, the reviewer and A. Néron, Amer. J. Math. **81** (1959), 95–118; MR **21** #1311].

The proof giving a bound for the above degree proceeds in two steps. The first step consists in obtaining a new theorem on abelian varieties, and the second consists in applying it to the curve, embedded in its Jacobian. It should be pointed out that they represent a major new diophantine method. Siegel's theorem concerning the finiteness of integral points (in the function field case for curves of genus 1) also follows from it. Up to now, the only known proof was Siegel's proof or variants of it, which made use of the Thue-Siegel-Roth theorem.

The theorem on abelian varieties arises from an algebraisation of the theory of Picard-Fuchs differential equations. Let L be a function field over K, assume for simplicity that the dimension of K over k is 1, and let $\partial\colon K \to K$ be a derivation whose constant field is k. Let $\Omega^1 = \Omega^1(L/K)$ be the K-space of differential 1-forms of L over K. If $(x) = (x_1, \cdots, x_n)$ is a transcendence base of L over K, and $\omega \in \Omega^1$, then we can write $\omega = \sum u_i\, dx_i$ with $u_i \in L$. We denote by ∂_x the unique extension of ∂ to L such that $\partial_x x_i = 0$ for all i. Then we can also define ∂_x on Ω^1 by letting $\partial_x\omega = \sum(\partial_x u_i)\, dx_i$. If $\mathscr{Z}$ denotes the group of forms ω such that $d\omega = 0$ and $\mathscr{B}$ the group of forms $d\varphi$ with $\varphi \in L$, then ∂_x is easily seen to operate on $\mathscr{Z}/\mathscr{B}$, and this operation is independent of the choice of transcendence base (x).

Let now V be a (projective) model of L over K, let V_K denote the set of rational points of V in K, and let $P \in V_K$ be a simple point. Let $\mathfrak{o}_P$ and $\mathfrak{m}_P$ be its local ring and maximal ideal in L. If $y \in \mathfrak{o}_P$, we denote by y_P or $y(P)$ the value of y at P, i.e., the residue class in K of $y \bmod \mathfrak{m}_P$. A transcendence base (x) as above will be called a system of quasi-parameters at P if $x_i \in \mathfrak{o}_P$ for all i, and $x_i - x_{iP}$ is a system of parameters. For such (x), we have $(\partial_x y)_P = \partial y_P$.

Let $K[\partial]$ be the ring of polynomial differential operators with coefficients in K. Then by what we said above, $K[\partial]$ operates on $\mathscr{Z}/\mathscr{B}$. If $\omega \in \mathscr{Z}$, we denote by $\overline{\omega}$ the residue class of $\omega \bmod \mathscr{B}$. Let $\omega_1, \cdots, \omega_g$ be a basis for the differential forms of first kind. A Picard-Fuchs equation is a relation of type $\mathscr{L}\colon \sum_{j=1}^{g} \mathscr{L}_j\overline{\omega}_j = 0$ where $\mathscr{L}_j \in K[\partial]$, $\overline{\omega}_j$ is the class of a differential form of first kind $\omega_j \in \mathscr{Z}$. For each choice of transcendence base (x), such a relation has a representative $\mathscr{L}_x\colon \sum \mathscr{L}_{jx}\omega_j = dz_x$ where z_x lies in L.

Let P be a simple point in V_K. A system of quasi-parameters (x) will be called admissible at P if $\partial x_{iP} = 0$ for all i. Given a finite number of simple points, it is easy to show that a system of quasi-parameters admissible for all of them exists. Let P, Q be two simple points in V_K. Let $\mathscr{L}$ be a Picard-Fuchs equation. If (x) is a system of quasi-parameters at P, Q, admissible at both P, Q, and if $\mathscr{L}_x$ is a representative for $\mathscr{L}$ as above, then we let $\mu_x(P, Q) = z_{xQ} - z_{xP}$. This is an element of K, which is easily proved to be independent of the choice of (x), subject to the above conditions. Hence we can denote it by $\mu_{\mathscr{L}}(P, Q)$. The symbol $\mu_{\mathscr{L}}(P, Q)$ is the algebraisation of an integral between P and Q, and in transcendental notation, we would write $\mu_{\mathscr{L}}(P, Q) = \sum_{j=1}^{g} \mathscr{L}_j \int_P^Q \omega_j$ where $\mathscr{L}_j$ is a differential operator with respect to the parameters under the integral sign. Clearly, we have $\mu(P, Q) + \mu(Q, R) = \mu(P, R)$. The symbol also satisfies some obvious functorial properties.

Let $V = A$ be an abelian variety defined over K. Let 0 be the origin. For each Picard-Fuchs relation $\mathscr{L}$, and rational point P in A_K, we define $\mu_{\mathscr{L}}(P) = \mu_{\mathscr{L}}(0, P)$. From the preceding additivity and the functorial properties, one concludes that the map $P \to \mu_{\mathscr{L}}(P)$ is a homomorphism of A_K into the additive group of K (denoted by K^+). Let (B, τ) be Chow's K/k-trace of A. It is easily seen that τB_k is contained in the kernel of this homomorphism, and so is the torsion group of A_K. The main theorem asserts: If a point P of A_K is contained in the kernel of every Picard-Fuchs homomorphism $\mu_{\mathscr{L}}$, then it lies in the group generated by τB_k and the torsion group of A_K. This is a deep statement, which is proved by showing that some integral multiple of P is infinitely divisible in A_K, and hence must lie in τB_k. The proof uses transcendental means, assuming that k is the field of complex numbers, considering algebraic families of abelian varieties parametrized by a curve T (a model of K over k), and using the theory of abelian integrals. Since the module of Picard-Fuchs relations is finitely generated over $K[\partial]$, one may actually choose a finite number of relations and corresponding homomorphisms $\mu_1, \cdots, \mu_s$. The intersection of their kernels is then the group generated as described above, and one obtains an injection of the factor group $A_K/(\tau B_k$+torsion) into the direct product $K^+ \times \cdots \times K^+$ taken s times. The main point of this injection is that it transforms the rather intractable relations of the Mordell-Weil group A_K into essentially linear differential relations in the additive group of K. The preceding theorem is then applied to a curve C of genus ≥ 2 defined over K, embedded in its Jacobian J, and the problem amounts to finding a finite-dimensional k-linear subspace E of L, a function field over K, having the following property. Given a point P in C_K, there exists a non-zero function f in E such that f is defined at P and $f(P) = 0$. Essentially, this shows that such points are all contained in the intersection of C and a hyperplane section of a suitable projective embedding, determined by the k-space E. The proof of this final statement is again very ingenious, using repeatedly the unramified coverings of C induced by isogenies of its Jacobian, in order to eliminate the differential operators from the Picard-Fuchs relations, and end up with functional relations relative to a suitable finitely generated space E. *S. Lang* (White Plains, N.Y.)

33 #7333 14.18
Hironaka, Heisuke
Resolution of singularities of an algebraic variety over a field of characteristic zero. I, II.
Ann. of Math. (2) **79** (1964), 109–203; ibid. (2) **79** (1964), 205–326.

Outre la solution d'un problème célèbre, ce travail donne un des outils les plus puissants de la géométrie algébrique. Ce résultat est atteint avec une surprenante économie de moyens (notion de platitude, structure des anneaux locaux complets, fermeture du lieu singulier), compensée par une étonnante habileté. L'auteur utilise le langage des schémas, et c'est là un point essentiel, mais seulement les notions les plus simples et les plus naturelles: produit fibré, schéma projectif associé à une algèbre graduée. Quant aux résultats, énonçons le plus simple. Soit X un préschéma noethérien réduit, irréductible et "excellent" (p. 161) dont les anneaux locaux sont de caractéristique résiduelle nulle. Soit $U = X - S$, où S est le lieu singulier de X (ensemble des points où l'anneau local n'est pas régulier). Il existe un morphisme surjectif $r: R \to X$ tel que R soit régulier, $r^{-1}(S)$ étant un diviseur à croisements normaux et r induisant un isomorphisme $r^{-1}(U) \xrightarrow{\sim} U$. On pourra prendre pour X une variété algébrique sur un corps de caractéristique nulle car ses anneaux locaux sont "excellents". Un anneau local complet l'est aussi; on sait donc résoudre les singularités des "variétés algébroïdes". Il est d'ailleurs notable que le principe même de la démonstration suppose que l'on étudie simultanément les variétés algébriques et les variétés algébroïdes, ce que permet le langage des schémas. L'auteur obtient également la résolution "locale" des singularités d'un sous-espace analytique complexe de $S \times P$, où S est une variété de Stein et P un espace projectif complexe.

La démonstration se fait par une récurrence subtile qui exige que l'on prouve à chaque cran un résultat plus précis que celui annoncé plus haut; en particulier, on impose que le morphisme résolvant $r\colon R \to X$ soit obtenu en composant un nombre fini d'éclatements "permis". Un éclatement permis est un morphisme $X' \to X$ obtenu en faisant éclater dans X un préschéma régulier D contenu dans le lieu singulier de X et tel que le cône normal de X le long de D soit plat sur D. Pour un tel D, la singularité de X est la même en tous les points de D, en un sens très fort (polynôme de Hilbert). Si X est une hypersurface d'un schéma régulier Y, on voit que D est permis si D est régulier et si X a même multiplicité en tous les points de D (Chapitre II). On prouve ensuite qu'un éclatement permis ne peut que rendre X "moins singulier", le degré de singularité de X en un point $x \in X$ étant mesuré par deux suites d'entiers définies en termes du cône tangent (Chapitre III). Le reste du Chapitre III est consacré au cas où X est le spectre d'un anneau local complet, donné comme quotient d'un anneau local complet régulier R. Il s'agit de trouver des paramètres de R et des équations de X ayant des propriétés très fortes, stables par éclatement permis. Ce résultat de démonstration délicate est fondamental. On l'utilise comme suit: ayant construit un éclatement permis $p\colon X' \to X$ destiné à désingulariser certains points de X, les paramètres et équations évoqués plus haut servent à contrôler les anneaux locaux des points $x' \in X'$ qui sont aussi singuliers que $p(x')$.

La clef du problème consiste alors en une formulation plus précise: les théorèmes A, B, C et D du Chapitre I. Il reste à tirer le feu d'artifice en les démontrant simultanément par une récurrence savante (Chapitre IV). *J. Giraud* (New York)

37 #5215 14.05
Griffiths, Phillip A.
Periods of integrals on algebraic manifolds. I. Construction and properties of the modular varieties.
Amer. J. Math. **90** (1968), 568–626.

This is the first of a sequence of papers whose object is to study the period matrices of an algebraic family V_t of algebraic varieties, parametrized by a generic point t of a variety B. Thus the period matrices are to be thought of as functions on B. This leads to consideration of a mapping $\Phi\colon B \to M$, where M is the modular variety associated with the family V_t, that is, a variety representing the set of inequivalent period matrices associated with the V_t. Part I here is devoted to the study of M (cf. the author [Proc. Nat. Acad. Sci. U.S.A. **55** (1966), 1303–1309; ibid. **55** (1966), 1392–1395; MR **36** #5133; ibid. **56** (1966), 413–416; MR **35** #1812] for statements of some of the results).

The first step is to set up the period matrix space for a variety V, or, more generally, for a Kähler manifold. A period matrix is represented as a point of a Grassmann manifold modulo the action of a discrete group, corresponding to integral changes of homology bases; the period matrix is taken to be for a basis of forms for $H^{p,q}(V)$. Now there is a Lefschetz decomposition $H^q(V)=\sum L^r H_0{}^{q-2r}(V)$, where the H_0 are groups of primitive forms ("primitive" here corresponds to Hodge's "effective") and L is the operator of multiplication by the fundamental form ω of the Kähler structure. The integral $\int \omega^{n-q}\xi\eta$ induces a quadratic form Q on $H_0{}^q(V)$ which can be used to set up generalizations of the Riemann bilinear relations. The period matrix space for primitive cohomology $H_0{}^{q-r,r}(V)$ can now be made into a complex manifold D, defined as a subset of a Grassmannian by these generalized Riemann relations. More specifically, let $\mathbf{F}$ be the flag manifold of nested linear spaces $S^0\subset S^1\subset\cdots\subset S^t\subset W=H_0{}^q(V)$ with $\dim S^r=\dim H_0{}^{q,0}+\dim H_0{}^{q-1,1}+\cdots+\dim H_0{}^{q-r,r}$. The Hodge decomposition $H_0(V)=\sum H_0{}^{s,t}(V)$ determines a point $\Omega=\Omega(V)$ of $\mathbf{F}$, and this point satisfies the generalized Riemann relations. So let D be the set of all flags satisfying these relations, and call D the period matrix space. As an example, the period matrix space for a compact Riemann surface of genus q is analytically isomorphic to the Siegel upper half plane. In general, for a compact Kähler manifold V, with $W=H^q(V,\mathbf{C})$, the group G of real linear automorphisms preserving the quadratic form Q is a real simple Lie group operating transitively on D. D can be represented as an open homogeneous complex manifold $H\backslash G$.

Now assume that V is polarized, that is to say, there is a complex analytic line bundle L over V with characteristic class ω. Primitive cohomology is now defined over the rationals and the quadratic form Q is rational. There is an algebraic automorphism group $\mathbf{\Lambda}_*$, namely, the group of automorphisms of $H^*(V,Z)$ commuting with the algebra of cohomology operations, and also a geometric automorphism group Λ_*, namely, the subgroup of $\mathbf{\Lambda}_*$ induced by homeomorphisms of V on itself preserving ω. Λ_* and $\mathbf{\Lambda}_*$ induce subgroups Λ and $\mathbf{\Lambda}$ of G (operating on D as above). These are discrete subgroups of G operating properly discontinuously on D. Thus D/Λ and $D/\mathbf{\Lambda}$ are analytic spaces and there is a mapping $D/\Lambda\to D/\mathbf{\Lambda}$.

It turns out that $\mathbf{\Lambda}$ is contained in an arithmetic subgroup Γ of G, and the modular variety M is defined as D/Γ. It is shown that M is an analytic space with finite invariant volume, and that a polarized algebraic structure on $(V,\mathbf{L})$ defines a unique point $\Phi(V)$ in M. (Note that $\Omega(V)$ in D is not unique since there are polarization preserving homeomorphisms of V inducing non-trivial action on H^q.) Let $f\colon V\to V'$ be a polarization preserving homeomorphism. The graph of f defines an element $F\in H_{2n}(V\times V',\mathbf{Z})$ and it is shown that $\Phi(V)=\Phi(V')$ if and only if F is of type (n,n).

The next section is devoted to the study of complex torus structures $T_m(V)$ on $H^{2m+1}(V,\mathbf{R})/H^{2m+1}(V,\mathbf{Z})$ (cf. A. Weil [Amer. J. Math. **74** (1952), 865–894; MR **14**, 314e] for another approach to this question). The structures here are defined with the aid of the above described flag manifolds. In particular, if $\dim V\geq 3$, $T_3(V)$ is constructed so that it has a q-convex polarization with $q=h^{2,1}$ (such a polarization is given by a holomorphic line bundle whose characteristic class induces a hermitian metric of signature $(n-q,q)$) and $T_3(V)$ varies holomorphically with V. The more general result for T_m is stated without details of proof.

The final section of the paper is a study of the properties of period matrix domains. The general results are illustrated by working out the details for period matrix spaces of holomorphic 2- and 3-forms. D turns out to be a subset of an algebraic variety X operated on by a complex simple Lie group $\mathbf{G}$. The group G of real transformations preserving the quadratic form Q corresponding to D is a real form of G and D is an open G-orbit with compact isotropy group.

The paper closes with a study of homogeneous line bundles over D and computations of the appropriate p-convex polarizations. In this connection the classical concept of automorphic forms becomes replaced by automorphic cohomology.

A. H. Wallace (Philadelphia, Pa.)

1, 42c 20.0X
Weyl, Hermann
The classical groups. Their invariants and representations.
Princeton University Press, Princeton, N.J., 1939. xii+302 pp.

Les groupes classiques dont parle Monsieur Weyl sont les groupes linéaires général $(GL(n))$ et spécial $(SL(n))$, le groupe orthogonal $O(n)$ et le groupe symplectique $(Sp(n))$ (groupe d'un complexe linéaire). Le chapitre I, en forme d'introduction, pose les notions algébriques de base: corps, anneaux, polynômes, espaces linéaires. Le lecteur y trouvera également une explication détaillée du double aspect de la notion de groupe en mathématiques et même en physique mathématique: d'une part, groupe des automorphismes d'un ensemble préservant certaines relations (telles celles par exemple qu'introduit la géométrie euclidienne); d'autre part, groupe des transitions permettant de passer d'un système de référence à un autre. Cette étude consiste en somme en une mise au point du fameux programme d'Erlangen, complété de manière à fournir la théorie des quantités de divers types (vecteurs, tenseurs, ...), dont chacune se transforme suivant sa loi propre. La notion de ces lois de transformation diverses relatives à un même groupe introduit la notion de représentation d'un groupe, celle-ci à son tour l'idée d'invariants de divers types.

Le chapitre II contient la détermination explicite par des moyens algébriques des invariants de type vectoriel (fonctions d'un certain nombre d'arguments vectoriels) des groupes $GL(n)$, $O(n)$. La méthode consiste à exprimer, au moyen d'identités dues à Capelli, tous les invariants au moyen de ceux qui ne dépendent que de $n-1$ arguments, puis à construire directement ces derniers. Dans chacun des cas envisagés, un "premier théorème principal" fournit les invariants tandis qu'un "second théorème principal" fournit les relations entre eux. L'étude des invariants du groupe $O(n)$ avec un corps de base quelconque est possible en vertu de l'existence de la paramétrisation de Cayley des substitutions orthogonales. Celle-ci à son tour permet de donner dès ce chapitre tout algébrique un aperçu de l'aide que fournissent les considérations infinitésimales de la théorie des groupes de Lie.

Le chapitre III contient essentiellement la théorie des algèbres associatives semi-simples, qui est traitée du point de vue de la théorie des représentations, suivant la méthode inventée par l'auteur. Le lecteur ne perd jamais de vue le lien avec la théorie des invariants et covariants: la théorie de l'algèbre des commutants (commutatoralgebra), dès qu'introduite, est mise en relation avec la notion de covariant; les anneaux de groupe sont introduits à partir d'une première esquisse des relations entre la réduction de l'espace des tenseurs et les représentations du groupe symétrique. De ce point de vue, on s'intéresse surtout à la réduction de l'algèbre des commutants d'une représentation d'un groupe fini: l'auteur procède à cette réduction directement, en montrant qu'elle se poursuit parallèlement à celle de l'anneau de groupe, sans passer par l'algèbre enveloppante de la représentation elle-même.

Les moyens algébriques dont on dispose à ce stage permettent de tenir les promesses sur la foi desquelles les constructions antérieures ont été entreprises: les chapitres IV, V, VI contiennent l'exposé de la réduction complète des représentations tensorielles des groupes $GL(n)$, $O(n)$, $Sp(n)$. Dans le cas de $GL(n)$ on montre que les matrices de la représentation tensorielle de degré f sont celles de l'algèbre des commutants de la représentation correspondante du groupe symétrique; dans le cas de $O(n)$, on peut introduire une algèbre qui joue le même rôle que l'anneau du groupe symétrique; plutôt que d'en passer par l'étude de cette algèbre, on préfère opérer d'abord certaines réductions qui correspondent à l'opération de "rajeunissement" des tenseurs, puis s'aider des opérations induites par les éléments du groupe symétrique dans les sous-espaces qu'on obtient ainsi. Une méthode analogue conduit au but dans le cas de $Sp(n)$. Ici comme au chapitre II, la paramétrisation de Cayley permet de lever toute hypothèse restrictive sur le corps de base. On examine aussi le cas des représentations du groupe orthogonal propre. Par ailleurs ces chapitres contiennent une caractérisation complète des relations algébriques qui doivent exister entre les éléments d'une matrice pour qu'elle soit orthogonale ou symplectique.

À partir du chapitre VII, l'atmosphère générale du livre change quelque peu: l'analyse se substitue progressivement à l'algèbre, amenant dans ses bagages la topologie. On part du problème qui consiste à calculer les caractères des représentations irréductibles qu'on a déjà déterminées. Cette détermination repose sur le célèbre "artifice de l'unitarité" de H. Weyl, qui permet de se ramener à des sous-groupes du groupe unitaire, donc compacts; dans ces groupes compacts, on peut développer des considérations intégrales grâce à l'existence d'un élément de volume invariant. De plus, on prouve de cette manière qu'il n'existe pas d'autres représentations irréductibles du groupe unitaire que celles déjà rencontrées, données par les tenseurs. L'auteur ne manque pas de recouper les résultats qu'il obtient pour $GL(n)$ en les déduisant à nouveau d'une manière purement algébrique, ce qui se trouve possible en raison de la connection avec le groupe symétrique; mais il n'a pas de peine à montrer combien ici la méthode analytique est plus attrayante. Le chapitre se termine sur la détermination des polynômes de Poincaré des groupes en question, dont les coefficients sont définis au moyen des représentations par les tenseurs anti-symétriques: ce n'est qu'au chapitre suivant qu'apparaîtra la signification topologique de ces polynômes.

Le chapitre VIII est consacré à la théorie générale des invariants (Rappelons que les invariants considérés au chapitre II étaient des invariants de type vectoriel, fonctions de un ou plusieurs vecteurs). La théorie classique ne considérait guère que les invariants des formes homogènes, c'est-à-dire les invariants dépendant de tenseurs symétriques. Le cadre de la recherche est élargi par l'auteur, qui se propose la détermination des invariants qui dépendent de "quantités" quelconques, définies par des représentations irréductibles. Ce problème peut être approché soit par des méthodes algébriques, soit par des méthodes analytiques. Côté algèbre, on trouve bien entendu la célèbre méthode symbolique, qui permet de ramener la recherche des invariants de degrés donnés à celle d'invariants de type vectoriel; ce n'est qu'aidée du dehors qu'elle peut quelquefois fournir une base finie pour tous les invariants. Pour générale que soit la méthode symbolique, elle n'est cependant pas la seule méthode algébrique qui puisse donner des résultats, comme il est montré par deux exemples importants. Pour obtenir une base finie du système des invariants de $GL(n)$, on utilise le théorème de Hilbert sur la base d'un idéal; il permet d'exprimer tout invariant comme combinaison de certains d'entre eux avec pour coefficients des polynômes qui ne sont pas nécessairement des invariants. Un artifice algébrique permet d'en déduire une expression à coefficients invariants. On peut déduire de là l'existence d'un nombre fini d'invariants de base pour certains autres groupes, tels notamment le groupe affine, par un procédé que l'auteur appelle "adjunction argument," qui consiste à introduire un ou plusieurs covariants qui expriment algébriquement la notion de l'"absolu" des géométries correspondantes. Côté analyse, on est tout naturellement amené à envisager l'aspect "groupes de Lie" des groupes en question, qui conduit à des équations différentielles pour les invariants. Par ailleurs, si on a affaire à un groupe compact, l'intégration sur le groupe fournit un procédé simple pour remplacer par des invariants les coefficients polynômes qui interviennent après emploi du théorème de Hilbert. D'autre part, l'artifice d'unitarité permet de déterminer toutes les représentations des groupes unimodulaires réel et complexe. Cette question conduit tout naturellement à une étude topologique des groupes classiques; dans le cas du groupe orthogonal, sa connexion particulière conduit à la notion de spineur.

Le chapitre IX et dernier reprend en la résumant et l'étendant l'étude des algèbres associatives qui faisait l'objet du chapitre III. *C. Chevalley*

29 #3538 20.40 20.25
Feit, Walter; Thompson, John G.
Solvability of groups of odd order.
Pacific J. Math. **13** (1963), 775–1029.

In this impressive paper, which occupies the entire issue, the authors present a complete proof of the solvability of finite groups of odd order; thus they give an affirmative answer to the famous Burnside conjecture that all finite non-abelian simple groups are of even order. Not only the result itself, which has many consequences of importance, but the method employed, is significant. It is difficult to state an indication of their long involved proof and explain the reason why the oddness of the group order implies the solvability. The oddness of the group order predominates the proof and is used in all sorts of conceivable ways. The reviewer makes no attempt here to indicate the places in the proof where the oddness of the order is essential.

If there were a non-solvable group of odd order, then a non-solvable group of the smallest possible odd order would be a non-abelian simple group, all of whose proper subgroups are solvable. Such a simple group is called a minimal simple group. The proof of this paper consists in deriving a contradiction from the assumed existence of a minimal simple group of odd order. The proof is an expansion of ideas contained in papers by the reviewer [Proc. Amer. Math. Soc. **8** (1957), 686–695; MR **19**, 248f] and Feit, Hall and Thompson [Math. Z. **74** (1960), 1–17; MR **22** #9539], where special cases of Burnside's conjecture are solved. In these papers special assumptions are placed on the structure of the centralizers, and the minimal simple group to be disposed of satisfies the property that every maximal subgroup is a Frobenius group and is the normalizer of a nilpotent Hall subgroup. From this property one can apply the theory of group characters, the theory of so-called exceptional characters, which provides a certain limitation on the index of a nilpotent

Hall subgroup in a maximal subgroup. One is then able to derive a contradiction if the order of the group involved is odd. Chapter V of this paper is an application of group characters. In the general case of this paper, maximal subgroups may not be Frobenius groups, so that the theory of exceptional characters has to be generalized to take care of situations which may occur in the present case. This generalization is given in Chapter III of this paper. The details are too technical to be presented here, but roughly speaking, given a subgroup H which is "nicely" imbedded in G and given a set S of generalized characters of H which satisfies "nice" properties, one can define a particular isometry τ of S into the ring of generalized characters of G. This makes it possible to study the ring of generalized characters of G by means of $\tau(S)$; thus one obtains a certain information on the structure of G from the structure of H. In Chapter III the authors discuss a tamely imbedded (i.e., "nicely" imbedded) subset, of which H is the normalizer, and a coherent set of characters of H, which is "nice". A particular mapping τ is defined for a coherent set. The main discussion is to find a sufficient condition for a union of several coherent sets to be coherent. Theorem 10.1 is a main result in this direction and is a generalization of a result of Brauer and Feit on exceptional characters [cf. Feit, Illinois J. Math. **4** (1960), 170–186; MR **22** #4784]. The coherency of a certain set often comes from the structure of H. Chapter III contains several results of this nature. All results in this chapter are of a general character and do not refer to a minimal simple group of odd order. In Chapter V, the ring of generalized characters of G is studied by using $\tau(S)$ for certain coherent sets S of generalized characters of various maximal subgroups. A delicate discussion about a not necessarily coherent set comes in also. After about seventy pages of computation, the outcome is not quite a contradiction. But one obtains a contradiction unless there are two maximal subgroups of a special kind, which are interrelated in a particular fashion. These two groups generate G. Their structure and the way they intersect give rise to relations among the set of generators of G. The final chapter, Chapter VI, is a study of these generators and relations (about twenty pages). The authors are then able to show that the particular relations they obtain are not possible in finite groups. One of the distinguished maximal subgroups is isomorphic to an extension of the linear group $y = ax + b$, where a, b are elements taken from a finite field with a restriction that the norm of a over the prime field is 1. From this fact the authors are able to translate the relations among generators into equations in a finite field and then derive a desired contradiction.

Chapter IV is a study of maximal subgroups by means of group-theoretical methods, and the main object is to show that maximal subgroups satisfy nice properties so as to make possible an application of group characters. Here the method the authors employ is quite interesting. Many parts of their argument can be and have been generalized to handle minimal simple groups in general or simple groups in which every proper subgroup involves only a particular class of simple groups [see the work of D. Gorenstein and J. H. Walter in J. Algebra **1** (1964), 168–213; MR **30** #3132].

Let p be a prime divisor of the order of G and S a Sylow p-group of G. Let A be a maximal abelian normal subgroup of S. Assume that the minimal number of generators of A is at least three. The study of subgroups U such that U is normalized by A and $U \cap A = \{1\}$, is important. The first major result is the following transitivity theorem (Theorem 17.1). Let q be a prime different from p. If Q_1 and Q_2 are two q-subgroups, normalized by A and maximal under this restriction, then there is an element x of $C_G(A)$ such that $Q_2 = x^{-1}Q_1x$. This transitivity theorem is true if G is a p-solvable group. Even in the solvable case, this is not quite trivial and is one of the many results the authors have to establish in their course of proof. The fact that this result holds for minimal simple groups is amazing. In this theorem, not only the conjugacy of Q_1 and Q_2, but also the location of the element x which conjugates Q_1 and Q_2, is important. The second step is to prove the existence of a Hall subgroup with respect to a particular set of primes. Let a set of primes π satisfy the property that if p and q are two primes in π, then G contains a proper (hence solvable) subgroup generated by elementary abelian subgroups of order p^3 and q^3, respectively. The existence theorem asserts that if π satisfies the above property, then G contains a Hall π-subgroup. The passage from the transitivity theorem to this existence theorem is not short (about fifty pages) and represents one of the highest points ever achieved in the theory of finite groups. The proof is quite technical and involved. The prime p such that a Sylow p-group does not normalize any p'-subgroup has to be treated separately, but at the end G has a Hall π-subgroup H such that $N_G(H)$ is a maximal subgroup and the intersection of H and its conjugate H^x is not too large.

If a Sylow p-group S does not contain an elementary abelian normal subgroup of order p^3, the structure of S is restricted (and in fact for $p > 2$ a complete classification is available [Blackburn, Acta Math. **100** (1958), 45–92; MR **21** #1349; Proc. London Math. Soc. (3) **11** (1961), 1–22; MR **23** #A208]). This and the above existence theorem are used to study maximal subgroups of G, which are proved to have a structure closely related to Frobenius groups. Another result is that two elements of a nilpotent Hall subgroup H of G are conjugate in G if and only if they are conjugate in the normalizer of H. Also, one has to know that certain sets are tamely imbedded before one is able to apply character theory. The required proof occupies the long last section of Chapter IV (about forty pages).

Chapters I and II are preliminaries, and contain many new properties of finite groups, the importance of which is recognized by the success of this work.

M. Suzuki (Urbana, Ill.)

28 #1252 22.60 20.80
Kostant, Bertram
Lie group representations on polynomial rings.
Amer. J. Math. **85** (1963), 327–404.

Ce mémoire donne un ensemble de résultats très remarquables sur les trajectoires et les invariants polynômiaux de certains groupes linéaires algébriques. Bien que certains théorèmes soient démontrés dans un contexte plus général, l'essentiel concerne le cas du groupe adjoint G d'une algèbre de Lie réductive complexe $\mathfrak{g}$. Soient S l'algèbre des polynômes sur $\mathfrak{g}$ et J la sous-algèbre des éléments de S invariants par G. L'auteur démontre que S est un J-module libre. Plus précisément, S est isomorphe à $J \otimes H$ où H est le sous-espace des polynômes annulés par tous les opérateurs différentiels à coefficients constants sur $\mathfrak{g}$ qui sont homogènes de degré > 0 et invariants par G. Si J^+ est l'idéal de J engendré par les polynômes invariants de degré > 0, alors J^+S est un idéal premier de S et la variété des zéros de J^+S est le cône P des éléments nilpotents de $\mathfrak{g}$. La notion d'élément nilpotent principal et les propriétés de ces éléments [l'auteur, même J. **81** (1959), 973–1032; MR **22** #5693] sont utilisées pour établir une série de propriétés concernant les trajectoires de dimension maximale, c'est-à-dire les trajectoires de dimension $n-l$ où $n = \dim G$ et $l =$ rang de G: (1) Les éléments nilpotents principaux constituent une trajectoire de dimension maximale dense dans P. (2) Pour que la trajectoire O_x d'un point $x \in \mathfrak{g}$ soit de dimension maximale, il faut et il suffit que $P \subset \overline{\mathbf{C}^* O_x}$, où $\mathbf{C}^* = \mathbf{C} - \{0\}$. (3) Les éléments x de $\mathfrak{g}$ dont la trajectoire est de dimension maximale sont les éléments dont la composante nilpotente est un élément nilpotent principal de la sous-algèbre constituée par le noyau de $\mathrm{ad}(x_s)$, où x_s est la composante semi-simple de x. (4) Si $u_1, u_2, \cdots, u_l$ est une base homogène de l'idéal J^+, en associant à chaque trajectoire de dimension maximale les valeurs des u_i sur cette trajectoire, on obtient une bijection de l'ensemble des trajectoires de dimension maximale sur $\mathbf{C}^l$. (5) Chaque trajectoire de dimension maximale O a pour adhérence une variété normale, intersection complète, réunion d'un nombre fini de trajectoires, et O est l'ensemble des points réguliers de $\overline{O}$. (6) Si O est une trajectoire de dimension maximale, l'algèbre des fonctions régulières sur $\overline{O}$ (considéré comme sous-ensemble algébrique de $\mathfrak{g}$) coïncide avec l'algèbre des restrictions à O des fonctions polynômes sur $\mathfrak{g}$. (7) Les idéaux premiers de S engendrés par des éléments de J sont les idéaux premiers de S dont la variété des zéros est stable par G et contient une trajectoire de dimension maximale. L'auteur donne également des résultats sur les facteurs irréductibles de la représentation de G dans S ainsi qu'une généralisation du théorème de Coxeter et Coleman [A. J. Coleman, Canad. J. Math. **10** (1958), 349–357; MR **21** #4990] sur les valeurs propres de la transformation de Coxeter et Killing. *J. L. Koszul* (Grenoble)

16, 1097e 30.0X
Weyl, Hermann
Die Idee der Riemannschen Fläche. (German)
3te Aufl.
B. G. Teubner Verlagsgesellschaft, Stuttgart, 1955. viii+162 pp.

It does not occur frequently that a book which has been a classic for four decades appears in a completely rewritten edition. The event is the more significant as it concerns a work which has undoubtedly had a greater influence on the development of geometric function theory than any other publication since Riemann's dissertation.

The book originated in Weyl's lectures at Göttingen in the winter semester 1911–12. This course grew into the first edition of "Die Idee der Riemannschen Fläche" [Teubner, Leipzig-Berlin, 1913]. An appendix "Strenge Begründung der Charakteristikentheorie auf zweiseitigen Flächen" was added in the second edition [ibid., 1923], which together with its appendix also was reprinted in the United States [Chelsea, New York, 1947]. It was the author's desire to fuse the supplement with the main part of the book that led him to make a complete revision.

It seems appropriate to summarize here the historical background of the new book, as the earlier editions did not fall within the life span of Mathematical Reviews. The reader's interest is, of course, focused on the question: how did the subject of the book, the idea of abstract Riemann surface, come into being? The author's reference to Klein as the originator suggests a closer scrutiny, which turns out to be rewarding and not devoid of surprises.

It is well known that Riemann introduced the surfaces now bearing his name as covering surfaces of the complex plane. Although various passages of his other writings can be interpreted as suggesting a more abstract approach, there is nowhere an explicit mention of it. Klein's first thought in this direction was inspired by Prym. In the preface to the former's book, "Über Riemann's Theorie der algebraischen Functionen und ihrer Integrale" [Teubner, Leipzig, 1882] we find the following significant note: "Er (Prym) erzählte mir, dass die Riemannschen Flächen ursprünglich durchaus nicht notwendig mehrblättrige Flächen über der Ebene sind, dass man vielmehr auf beliebig gegebenen krummen Flächen ganz ebenso komplexe Funktionen des Ortes studieren kann, wie auf den Flächen über der Ebene". In 1923, Klein supplements this acknowledgement with an interesting remark [Ges. Math. Abh., vol. III, Springer, Berlin, p. 479], which seems to have remained quite unnoticed. He says that he had sent his book to Prym and the latter had, in a letter of April 8, 1882, answered "ausdrücklich" that, although he (Prym) did not exactly recall his conversation with Klein, he had certainly not referred to ordinary surfaces as carriers of complex functions. Thus we are faced with the unique situation that the less restrictive idea of Riemann surface is due to some unknown comment made by Prym and misunderstood by Klein.

The formal priority naturally remained with Klein, as he goes on to state: "Die Idee, geschlossene Flächen im Raume der funktionentheoretischer Betrachtung zugrunde zu legen... scheint vor meiner Schrift nicht hervortreten zu sein. Sie hat auch wenig Nachahmung gefunden. Einzig H. Poincaré... und H. Weyl (1913, in seinem Buche, Die Idee der Riemannschen Fläche) scheinen eine Ausnahme zu machen."

But there still was a long way to go from Klein's idea to that of Weyl. First of all, the former only considered closed surfaces, which he covered "dachziegelartig" by overlapping planar regions. Furthermore, his reasoning was purely intuitive, based on physical illustrations, without any attempt at mathematical rigour. Weyl achieved perfection in both respects: he was the first to consider a Riemann surface in its full generality, and he gave the first axiomatic treatment of the concept.

The main advantage of the new notion was that it offered the natural carrier for analytic functions. Analyticity being locally defined, the globally imposed complex plane could only allow an artificially restricted theory of complex functions. Weyl's locally defined surface removed this obstacle and also permitted a full exploitation of Riemann's revolutionary idea of reversing the conventional process by giving the surface a priori and studying functions on it a posteriori.

Oddly enough, it took 35 years to find out whether the introduction of abstract surfaces had added any new specimens to conformal equivalence classes of covering surfaces of the plane. Only in 1948 Behnke and Stein [Math. Ann. **120**, 430–461 (1949); MR **10**, 696c] succeeded in proving that this was not the case: there exists on every open abstract Riemann surface a nonconstant single-valued analytic function. Even more was then clear: the corresponding exponential function gives a conformally equivalent covering surface of the plane punctured at the origin. This result in turn is the best possible, the Picard theorem prohibiting two or more punctures.

For constructing functions on abstract surfaces Weyl used Dirichlet's principle. It is known that Riemann not only failed to give a rigorous proof of this principle but that he did not even make an effort to establish it beyond its intuitive physical meaning; Weierstrass's criticism made no impression on him [Klein, loc. cit., p. 493]. It is believed that Weierstrass left it to his student Schwarz to develop the linear method which then furnished the first strict proof of Riemann's results. Then at the turn of the century Hilbert succeeded in putting Dirichlet's principle on a sound basis, but his proof was still unnecessarily involved. Weyl introduced an essential simplification by his step-function with continuous derivatives. That this contribution, accepted in every text, is due to Weyl, does not seem to have been generally noticed. Dirichlet's principle has then firmly established itself into mathematical thinking, perhaps to the extent that one could regret the oblivion which for some seven decades was the fate of Schwarz's ingenious idea, capable of extensions far beyond its original scope.

In the theory of differentials, the fruitfulness of Weyl's approach is exemplified, as he now remarks, by the fact that the real harmonic rather than complex holomorphic linear differential forms have served as the starting point in recent generalizations to higher dimensions. In particular, the presentation in "die grosse Arbeit von Kunihiko Kodaira" [Ann. of Math. (2) **50**, 587–665 (1949); MR **11**, 108e] reflects the pattern of the 1913 book.

Although Weyl restricted his attention to the theory of Riemann surfaces, the influence of his book has extended a great deal further. The work was written shortly after Brouwer's topological discoveries, and it was the author's ambition to give Riemann's theory the first presentation in which topological concepts and theorems received up-to-date rigorous treatment. The resulting axiomatic way of reasoning has inspired thinking in many doctrines quite remote from complex analysis.

How, then, does the new edition differ from the old one? Roughly speaking, the book is new in treatment while the subject matter has essentially remained unchanged. First of all, the abstract Riemann surface has undergone further development. It was originally defined as a two-dimensional manifold which has a conformal structure and can be triangulated. We owe to Radó [Acta Litt. Sci. Szeged **2**, 101–121 (1925)] the fact that the triangulability is equivalent to the existence of a countable basis for open sets and that the latter in turn is a consequence of the conformal structure. The triangulability axiom can thus be omitted. In the new book the Riemann surface appears as a neighborhood space with conformal structure.

Weyl, however, has come to this definition along another line of reasoning. He had used the combinatorial approach even more extensively than in the first edition, in his paper "Über kombinatorische und kontinuumsmässige Definition der Überschneidungszahl zweier geschlossener Kurven auf einer Fläche" [Z. Angew. Math. Phys. **4**, 471–492 (1953); MR **15**, 460a]. But we now learn that he had subsequently preferred to define the intersection number by "topologizing" his method of constructing Abelian integrals (1913). This led him to drop the axiom of triangulation.

As a consequence of the new starting point, the proofs based on triangulation have been replaced by others using overlapping neighborhoods. This is the essential difference between the two editions. In the function-theoretic part the new treatment of the intersection number is conspicuous. Several notations have been changed. An example is the former "Überlagerungsfläche der Integralfunktionen" which now appears as "Klassenfläche".

The author has purposely avoided any extension of the book beyond its original scope of analyzing the concept of Riemann surface and developing function theory mainly on closed surfaces. Apart from the change of standpoint he discusses none of the new problems, methods, or results that have been developed in the theory of open Riemann surfaces since the appearance of the first edition. This is, in particular, true of Weyl's own method of orthogonal projection [Duke Math. J. **7**, 411–444 (1940); MR **2**, 202a].

It remains to list the contents of the book, the first eleven sections of which are devoted to the concept and topology of Riemann surfaces, the remaining eleven, to functions on Riemann surfaces. (1) Weierstrass's concept of analytic function. (2) Concept of "analytisches Gebilde". (3) Mutual relation of the concepts of analytic function and "analytisches Gebilde". (4) Concept of two-dimensional manifold. (5) Examples of surfaces. (6) Specialization, in particular differentiable and Riemann surfaces. (7) Orientation. (8) Covering surfaces. (9) Differentials and line integrals. Homology. (10) Density and surface integrals. The residue theorem. (11) Intersection number. (12) Dirichlet's integral and harmonic differentials. (13) Construction of a potential flow with a double source. (14) The proof. (15) Elementary differentials. (16) Laws of symmetry. (17) The single-valued functions on F (Riemann surface) as a subclass of additive and multiplicative functions on $\hat{F}$ ("Klassenfläche"). Riemann-Roch's theorem. (18) Abel's theorem. (19) The algebraic function field. (20) Uniformization. (21) Riemann surface and noneuclidian geometry. Fundamental regions. Poincaré's θ-series. (22) Conformal self-mapping of a Riemann surface.

The presentation throughout flows smoothly and beautifully. The absolute contrast with "assembly-line technique" is pleasantly accentuated by an interspersion of a humanistic element. A fascinating example is the motivation of the notation "verstreut" for an infinite point set without points of accumulation; the author tells us he is thinking of a poem by Hermann Hesse: "Sommernacht hat ihre dünnen Sterne verstreut, Jugendgedächtnis duftet im mondhellen Laub".

L. Sario (Los Angeles, Calif.)

19, 951g 31.0X
Hunt, G. A.
Markoff processes and potentials. I, II.
Illinois J. Math. **1** (1957), 44–93, 316–369.

Cet article très important et original établit des connexions étroites entre la théorie générale du potentiel et les processus de Markov homogènes. On sait depuis longtemps déjà les relations entre la théorie newtonienne et le mouvement brownien [voir notamment J. L. Doob, Trans. Amer. Math. Soc. **77** (1954), 86–121; MR **16**, 269a]; le cas traité ici est bien plus général et plus délicat. L'idée essentielle est la suivante: moyennant certaines hypothèses de régularité sur lesquelles nous allons donner quelques détails, si les P_t sont les probabilités de passage d'un processus de Markov homogène, le "noyau" $U = \int_0^\infty P_t\,dt$ satisfait aux principes fondamentaux de la théorie du potentiel, tels que les principes du maximum; inversement à un noyau satisfaisant à ces principes correspond un semi-groupe de probabilités de passage. Cela conduira à des interprétations probabilistes simples des énoncés de la théorie générale du potentiel, et fournira un puissant outil de démonstration, permettant même d'atteindre la théorie "fine".

Soit $\mathscr{H}$ un espace localement compact séparable. L'auteur appelle noyau sur $\mathscr{H}$ toute famille $\{K_r\}_{r\in\mathscr{H}}$ de mesures de Radon positives sur $\mathscr{H}$, telle que la charge $K(r,A)$ de K_r sur le borélien $A \subset \mathscr{H}$ soit une fonction de r mesurable pour toute mesure de Radon. Un tel noyau définit une transformation $f \to Kf$

des fonctions positives mesurables par la formule $Kf(r) = \int K(r, ds) f(s)$ et, sous certaines conditions de finitude, une transformation $\mu \to \mu K$ des mesures positives par la formule $\mu K(A) = \int \mu(dr) K(r, A)$. On définit de manière évidente le produit HK de deux noyaux K et H.

On part d'un semi-groupe de noyaux $P_t(r, A)$ $(0 \leq t < \infty)$ sur $\mathscr{H}$, les mesures définissant chacun d'eux étant des mesures de probabilité $(P_t(r, \mathscr{H}) = 1)$. On suppose que pour toute mesure de probabilité μ sur $\mathscr{H}$ il existe un processus de Markov homogène $X(\omega, t) \colon \Omega \times [0, \infty) \to \mathscr{H}$ qui admet μ pour loi initiale, les P_t pour probabilités de passage, et qui est régulier au sens suivant: 1°) les chemins aléatoires $X(\omega, \cdot)$ sont continus à droite et ont des limites à gauche; 2°) la propriété de Markov "étendue" a lieu [voir R. Blumenthal, ibid. **85** (1957), 52–72; MR **19**, 468g]; 3°) si $\{T_n\}$ est une suite croissante de "temps de Markov" [cf. Blumenthal, loc. cit.], $\lim X(\omega, T_n(\omega)) = X(\omega, \lim T_n(\omega))$ pour presque tout ω pour lequel les $T_n(\omega)$ sont bornés. Il résulte des travaux de Blumenthal [loc. cit] et de Kinney [ibid. **74** (1953), 280–302; MR **14**, 772h] que ces trois conditions sont réalisées dans le cas important où les opérateurs $P_t \colon f \to P_t f$ laissent invariants l'espace des fonctions continues sur $\mathscr{H}$ tendant vers 0 à l'infini, et constituent un semi-groupe fortement continu.

On considère le noyau $U(r, A) = \int_0^\infty P_t(r, A)\, dt$, l'intégrale étant supposée convergente pour tout borélien borné A. On appelle Uf potentiel de la fonction f, μU potentiel de la mesure μ. Lorsque $\mathscr{H} = R^3$ et que les P_t sont les opérateurs de convolution par les distributions de Gauss, U est le noyau newtonien $|r|^{-1}$ (plus exactement $U(r, A) = \int_A |r - s|^{-1}\, ds$); Uf est le potentiel newtonien engendré par f, μU est la mesure ayant pour densité le potentiel newtonien engendré par μ.

Si $X_r(\omega, t)$ est un processus régulier "partant" de r $(X_r(\omega, 0) = r)$ et admettant les P_t pour probabilités de passage, on a la formule fondamentale

$$Uf(r) = \int_\Omega d\omega \int_0^\infty f(X_r(\omega, t))\, dt$$

d'où l'on déduit aisément les énoncés fondamentaux de la théorie du potentiel: principes du maximum pour les potentiels Uf, principe du balayage pour les potentiels μU. A signaler à ce propos que la dualité entre ces principes a été établie indépendamment, d'une façon entièrement différente, par G. Choquet et le reviewer [C. R. Acad. Sci. Paris **243** (1956), 764–767; MR **19**, 848f].

Il y a deux sortes de généralisations de la notion de fonction surharmonique positive: les fonctions et les mesures "excessives". Une fonction $\varphi \geq 0$ est dite excessive si elle est mesurable pour toute mesure de Radon et si $P_t\varphi$ tend partout vers φ en croissant lorsque t décroit vers 0. Exemples: les constantes positives, les potentiels Uf, la probabilité $\Phi_E(r)$ pour que X_r rencontre l'ensemble analytique E (généralisation de la notion de potentiel d'équilibre). On étend à ces fonctions les propriétés connues concernant les enveloppes de familles de fonctions surharmoniques; d'autre part si φ est excessive et X un processus régulier associé aux $P_t, \varphi(X(\omega, t))$ est une semimartingale au sens de Doob [loc. cit.] pourvu que l'espérance mathématique de $\varphi(X(\omega, 0))$ soit finie. Une mesure $\zeta \geq 0$ est excessive si $\zeta P_t \leq \zeta$ pour tout $t \geq 0$. Moyennant une hypothèse de régularité supplémentaire, on démontre un théorème de décomposition de Riesz: toute mesure excessive est de la forme $\xi + \mu U$, avec $\mu \geq 0$ et ξ excessive, analogue aux fonctions harmoniques usuelles en ce sens que $\xi = L\xi$, L étant l'opérateur de "balayage" sur le complémentaire d'un ouvert relativement compact quelconque.

La variable aléatoire $T_r(\omega) = \inf\{t > 0; X_r(\omega, t) \in E\}$, où E est analytique, s'annule avec probabilité 0 ou 1, d'après la propriété de Markov étendue. Selon le cas, le point r sera dit régulier ou irrégulier pour E. Posons $H_E(r, A) = \text{prob}\{T_r(\omega) < \infty; X_r(\omega, T_r(\omega)) \in A\}$; si φ est excessive $H_E\varphi$ est excessive et on a $H_E\varphi \leq \varphi$ partout, $H_E\varphi = \varphi$ en tout point régulier de E. L'opération $\varphi \to H_E\varphi$ généralise l'extrémisation des fonctions surharmoniques selon Brelot [J. Math. Pures Appl. (9) **24** (1945), 1–32; MR **7**, 521]. Le noyau H_E peut être interprété comme la généralisation de la notion classique de masses de Green; il est utile dans la définition précise du balayage. Tout ceci est le point de départ de la théorie fine du potentiel.

En réalité, les P_t étant donnés, on considère non seulement le noyau U, mais toute une famille de noyaux, associés à divers systèmes de "temps d'arrêt" (terminal times). Par exemple, si R_r est le temps où le processus X_r rencontre le borélien $B \subset \mathscr{H}$, on pose $K_t(r, A) = \text{prob}\{X_r(\omega, t) \in A; R_r(\omega) > t\}$; le noyau $V(r, A) = \int_0^\infty K_t(r, A)\, dt$ est l'analogue de la fonction de Green du complémentaire de B. La formule fondamentale pour les potentiels est alors

$$Vf(r) = \int_\Omega d\omega \int_0^{R_r(\omega)} f(X_r(\omega, t))\, dt.$$

Un autre exemple est constitué par les noyaux $U^\lambda(r, A) = \int_0^\infty e^{-\lambda t} P_t(r, A)\, dt$ avec $\lambda > 0$. D'autres exemples sont nouveaux, même dans le cas newtonien; on démontre pour ces noyaux des théorèmes analogues à ceux qui ont été énoncés pour U: c'est la "théorie relative".

Le point de vue inverse est étudié vers la fin de la seconde partie; on part d'un noyau positif V sur un espace localement compact $\mathscr{K}$, appliquant l'ensemble $\mathscr{B}$ des fonctions continues à support compact dans l'ensemble $\mathscr{C}$ des fonctions continues tendant vers 0 à l'infini. On suppose en outre: (i) l'image de $\mathscr{B}$ par V est dense dans $\mathscr{C}$; (ii) si a est une constante ≥ 0, et si f et g sont deux fonctions positives de $\mathscr{B}$, la relation $Vf \leq a + Vg$ a lieu partout si elle a lieu sur le support de f (principe complet du maximum).

Alors on prouve l'existence d'un semi-groupe fortement continu d'opérateurs "sous-stochastiques" K_t tels que $V = \int_0^\infty K_t\, dt$. En ajoutant au besoin un point w à l'espace $\mathscr{K}$, on peut prolonger les K_t à $\mathscr{H} = \mathscr{K} \cup \{w\}$ de façon à obtenir un semi-groupe d'opérateurs P_t satisfaisant à $P_t(r, \mathscr{H}) = 1$. Le noyau V se présente donc comme le noyau de Green relatif au sous-espace $\mathscr{K}$ de $\mathscr{H}$ dans la théorie construite à partir des P_t. Ceci résout l'un des problèmes fondamentaux de la théorie du potentiel.

Ce travail constitue donc un progrès essentiel dans l'étude des aspects "linéaires" de la théorie du potentiel; par contre les questions d'énergie (lorsque le noyau présente certains caractères de symétrie) et les problèmes aux limites y sont délibérément laissés de côté. La lecture n'en est pas facile. Cela tient peut-être en partie à ce que les définitions et hypothèses ne sont pas toujours clairement mises en évidence, et que, malgré la longueur du texte, certaines démonstrations sont seulement esquissées; cela tient surtout à ce que les préliminaires sont longs et font appel à beaucoup de connaissances et à des résultats récents et difficiles tels que les travaux déjà cités de Blumenthal et Kinney, et la théorie des fonctions d'ensembles alternées d'ordre infini, dûe à Choquet [Ann. Inst. Fourier, Grenoble **5** (1953–1954), 131–295; MR **18**, 295g]; les questions de mesurabilité, notamment, posent des problèmes délicats. Le lecteur qui aura fait l'effort de comprendre ces préliminaires pourra apprécier l'élégance avec laquelle la théorie générale s'en déduit, et la fécondité des méthodes utilisées.

J. Deny (Strasbourg)

26 #5255 34.65 57.48

Moser, J.

On invariant curves of area-preserving mappings of an annulus.

Nachr. Akad. Wiss. Göttingen Math.-Phys. Kl. II **1962,** 1–20.

Given the nonlinear mapping

$$\Theta_1 = \Theta + \alpha(r) + F(r,\Theta), \quad r_1 = r + G(r,\Theta), \tag{1}$$

where $d\alpha/dr > 0$ and F, G are of period 2π in Θ. Mappings of this kind, r, Θ being polar coordinates, $0 < a \le r \le b$, arise from the study of area-preserving mappings of an annulus into itself, for example, in connection with the restricted three-body problem studied by Poincaré and G. D. Birkhoff.

The author presents an elegant proof of the existence of closed invariant curves of (1) for small F, G under certain assumptions. Using the definition

$$|h|_s = \sup\left|\left(\frac{\partial}{\partial r}\right)^{\sigma_1}\left(\frac{\partial}{\partial\Theta}\right)^{\sigma_2} h(r,\Theta)\right|, \quad \sigma_1+\sigma_2 \le s,$$

for the sth derivative norm, the following main theorem is proved. For a given $\varepsilon > 0$ and a given integer $s \ge 1$, the mapping (1) has a closed invariant curve (2) $\Theta = \Theta' + p(\Theta')$, $r = r_0 + q(\Theta')$, where p, q are of period 2π with s continuous derivatives satisfying $|p|_s + |q|_s < \varepsilon$ under the following hypotheses: Assume for the mapping (1) that every closed curve $r = f(\Theta) = f(\Theta + 2\pi)$ intersect near a circle and its image curve. Assume further that $b - a \ge 1$ and $c_0^{-1} \le d\alpha(r)/dr \le c_0$ for some constant $c_0 > 1$. Finally, we construct a positive number $\delta_0 = \delta_0(\varepsilon, s, c_0)$ and an integer $l = l(s)$ such that F, G have continuous derivatives up to order l and satisfy the inequalities $|F|_0 + |G|_0 < \delta_0$, $|\alpha|_l + |F|_l + |G|_l < c_0$. Moreover, the mapping induced on the curve (2) is given by $\Theta_1{}' = \Theta' + \alpha(r_0)$.

It is also shown that given any ω in $\alpha(a) + \varepsilon < \omega < \alpha(b) - \varepsilon$ for which $\omega/2\pi$ cannot be closely approximated by rationals: $|n\omega - m2\pi| \ge \varepsilon n^{-3/2}$ for all integers n, m with $n > 0$, there exists a curve (2) with this rotation number $\omega = \alpha(r_0)$.

An analogous theorem was stated by Kolmogorov [Dokl. Akad. Nauk SSSR **98** (1954), 527–530; MR **16**, 924c] under the assumption that the differential equations are analytic. Here only the existence of finitely many derivatives is assumed. For the proof an iterative process is used which converges very rapidly (related paper: J. Moser [Proc. Nat. Acad. Sci. U.S.A. **47** (1961), 1824–1831; MR **24** #A2695]).

The theorem also is of importance for the study of stability of periodic solutions.

H. Ehrmann (Clausthal-Zellerfeld)

16, 139a 35.0X
Leray, Jean
Hyperbolic differential equations.
The Institute for Advanced Study, Princeton, N. J., 1953. 240 pp.

Ce livre est consacré au problème de Cauchy pour les équations aux dérivées partielles. Il constitue une contribution des plus marquantes dans ce domaine par l'originalité de l'exposé et l'intérêt des résultats nouveaux qu'il contient.

La première partie (pp. 1–103) concerne les équations aux dérivées partielles linéaires et à coefficients constants. L'auteur édifie d'abord un calcul symbolique à plusieurs variables. Il commence par définir la transformée de Laplace $\mathscr{L}\mathscr{C}$ des distributions $\mathscr{C} \in \mathscr{D}(R^n)$ qui sont des sommes finies de dérivées de fonctions $f(x), x \equiv x_1, \cdots, x_n$, telles que $\exp(-\sum_1^n x_i\xi_i)f \in L^2(R^n)$ pour $\xi \in \Omega$, domaine convexe donné. Les transformées de Laplace $\mathscr{L}\mathscr{C}$ sont des fonctions $a(\zeta)$, $\zeta = \xi + i\eta$, sommes finies de produits de polynômes en ζ par des fonctions $b(\zeta)$ analytiques, régulières dans le tube de base Ω et telles que $\sup_\eta |b(\xi + i\eta)|$ soit uniformément borné dans Ω. Cela étant, l'auteur nomme produit symbolique de $\mathscr{C}$ par $a(\zeta)$ la distribution $a(p)\mathscr{C} = \mathscr{L}^{-1}a(\zeta)\mathscr{L}\mathscr{C}$, qui s'avère être $\mathscr{L}^{-1}a(\zeta) * \mathscr{C}$; il étudie les propriétés de ce produit symbolique et, en particulier, détermine explicitement pour $a(\zeta)$ donné un fermé convexe C tel que la connaissance de $\mathscr{C}$ dans l'ouvert $\mathscr{O} - \mathscr{C}$ détermine $a(p)\mathscr{C}$ dans $\mathscr{O}$, ouvert quelconque de R^n.

Toutes ces considérations sont basées sur la théorie de la transformation de Fourier et de Laplace dans $L^2(R^n)$. Les résultats sont parallèles à ceux obtenus par L. Schwartz [Comm. Sém. Math. Univ. Lund [Medd. Lunds Univ. Mat. Sem.] Tome Supplémentaire, 196–206 (1952); ces Rev. **14**, 639a] et J. L. Lions [J. Analyse Math. **2**, 369–380 (1953); ces Rev. **15**, 307g] dans le cadre de la théorie des distributions.

La distribution $\mathscr{L}^{-1}a(\zeta)$ est une solution élémentaire au sens de L. Schwartz de l'opérateur linéaire à coefficients constants $a(\partial/\partial x), \partial/\partial x \equiv \partial/\partial x_1, \cdots, \partial/\partial x_n$. Elle permet de résoudre, par composition, le problème de Cauchy pour cet opérateur, formulé dans la théorie des distributions. L'étude de cette solution élémentaire exige l'introduction de l'ouvert Δ, intérieur de l'ensemble des points ξ pour lesquels $a(\xi + i\eta)$ diffère de zéro quel que soit η. La solution élémentaire étudiée possède une forme différente, indépendante de ξ, lorsque ξ appartient à chaque composante connexe (convexe) de Δ; son support est alors le cône réciproque du cône directeur de la composante connexe en question. La considération de l'ouvert Δ et le théorème relatif au support de la solution élémentaire permettent d'obtenir d'une manière très directe (et même de compléter) les résultats de L. Gårding [Acta Math. **85**, 1–62 (1951); ces Rev. **12**, 831g].

L'auteur développe enfin les résultats annoncés dans une note antérieure [C. R. Acad. Sci. Paris **234**, 1112–1115 (1953); ces Rev. **14**, 477c]. Lorsque $a(p)$ est homogène, on sait que la solution élémentaire s'exprime au moyen de périodes d'intégrales abéliennes [cf. Herglotz, Ber. Verh. Sächs. Akad. Wiss. Leipzig. Math.-Phys. Kl. **78**, 93–126, 287–318 (1926); **80**, 69–114 (1928); I. Petrowsky, Mat. Sbornik N.S. **17** (**59**), 289–370 (1945); ces Rev. **8**, 79a; et articles cités par F. Bureau dans Colloque de Géométrie algébrique, Liège, 1949, Thone, Liège, 1950, pp. 155–176; ces Rev. **12**, 338d]. L'auteur donne à ces calculs une forme achevée et rigoureuse; il ne suppose plus que le cône $a(\xi) = 0$ n'a pas de points singuliers; il met les formules finales sous une forme invariante, valable quel que soit l'ordre de l'équation; il précise également un critère de I. Petrowsky [loc. cit.] concernant l'existence de lacunes stables.

Cette premiére partie de l'ouvrage contient de nombreux exemples; en particulier, l'auteur explicite tous ses résultats dans le cas de l'opérateur des ondes.

Les équations aux dérivées partielles hyperboliques, linéaires et à coefficients variables font l'objet de la deuxième partie du livre (pp. 104–219). L'auteur établit d'abord des inégalités applicables à la solution du problème de Cauchy pour l'opérateur $a(x, \partial/\partial x)$ (polynôme en $\partial/\partial x_1, \cdots, \partial/\partial x_n$ dont les coefficients sont fonctions de $x_1, \cdots, x_n$) dans le voisinage de la surface qui porte les données initiales. La méthode utilisée prend comme point de départ une égalité analogue à celle de l'énergie, obtenue dans le cas général après d'intéressantes considérations algébriques; l'opérateur $a(x, \partial/\partial x)$ est supposé hyperbolique dans le domaine considéré, c'est-à-dire qu'en chaque point x de ce domaine, le cône $a_m(x,p) = 0$ ($a_m(x,p)$, termes de degré maximum de $a(x;p)$) contient des points p par lesquels toute sécante réelle coupe le cône en des points réels et distincts. Utilisant les inégalités obtenues, l'auteur étend la solution locale de proche en proche et obtient ainsi une preuve complète de l'affirmation de I. Petrowsky [Mat. Sbornik N.S. **2** (**44**), 815–870 (1937)] disant que le problème de Cauchy admet toujours une solution globale lorsque l'équation est hyperbolique au sens signalé (avec quelques restrictions trop longues à indiquer ici). A cette occasion, il soument le mémoire de Petrowsky à une critique sévère.

Une fois démontrée l'existence de la solution globale, l'auteur donne de nouvelles inégalités plus simples, relatives à cette solution. Il les obtient sous des hypothèses plus générales sur les coefficients de l'équation, qu'il suppose bornés et lipschitziens au lieu d'en postuler toutes les dérivées bornées. Grâce à ces nouvelles inégalités, l'auteur introduit les opérateurs inverses de $a(x, \partial/\partial x)$ qui permettent de résoudre le problème de Cauchy dont les données initiales sont nulles à l'infini. L'auteur fait également une étude complète du domaine de dépendance de la solution du problème de Cauchy. A cette occasion, il reprend et complète des travaux de A. Marchaud [Compositio Math. **3**, 89–127 (1936); Bull. Sci. Math. (2) **62**, 229–240 (1938)] et S. C. Zaremba [ibid. **60**, 139–160 (1936)] sur les fronts d'émission par un ensemble donné. Il montre que les fronts d'émission qui limitent les domaines de dépendance sont le lieu de trajectoires définies par les équations de Jacobi, de Lagrange ou de Hamilton issues d'un principe variationel.

Toutes les considérations précédentes sont enfin étendues aux opérateurs hyperboliques sur une variété ou aux systèmes linéaires hyperboliques à coefficients variables.

La troisième partie du livre (pp. 220–238) est relative aux équations et systèmes d'équations hyperboliques non linéaires. On y trouve seulement un théorème local d'existence et un théorème global d'unicité. Contrairement à I. Petrowsky [Mat. Sbornik N.S. **2** (**44**), 815–870 (1937)] l'auteur utilise une méthode d'approximations successives inspirée des travaux de J. Schauder pour les équations du second ordre [Fund. Math. **24**, 213–246 (1935)]. *H. G. Garnir* (Li)

17, 853d 35.0X
Hörmander, Lars
On the theory of general partial differential operators.
Acta Math. **94** (1955), 161–248.

Ce travail contient de nombreux résultats importants et entièrement nouveaux.

Notations: sur Ω ouvert de R^ν $(x=(x^1,\cdots,x^\nu)\in\Omega)$ on donne un opérateur différentiel

$$\mathscr{P}=\sum a^\alpha(x)D_\alpha,\ a^\alpha=a^{\alpha_1\cdots\alpha_k},\ D_\alpha=D_{\alpha_1}\cdots D_{\alpha_k},$$

$D_{\alpha_k}=\partial/i\partial x^{\alpha_k}$. Soit $P(x,\xi)$ le polynôme en $\xi=(\xi_1,\cdots,\xi_\nu)$ correspondant; si $P^\alpha(x,\xi)$, x fixé, est une dérivée en ξ du polynôme $P(x,\xi)$, on désigne par $\mathscr{P}$ l'opérateur différentiel correspondant.

On désigne par $\mathscr{D}_P$ l'espace des (classes de) fonctions $u\in L^2(\Omega)$ telles que $\mathscr{P}u$ (pris au sens des distributions sur Ω) soit dans $L^2(\Omega)$, espace muni de sa structure hilbertienne naturelle. Soit $\mathscr{D}_{P_0}$ l'adhérence dans $\mathscr{D}_P$ de l'espace $C_0{}^\infty$ des fonctions indéfiniment différentiables à support compact dans Ω. L'A. désigne par P (resp. P_0) l'opérateur $\mathscr{P}$ défini sur $\mathscr{D}_P$ (resp. $\mathscr{D}_{P_0}$); P (resp. P_0) est dit maximal (resp. minimal). Soit $\mathscr{P}$ et $\mathscr{Q}$ deux opérateurs différentiels. On dit que $\mathscr{P}$ est plus fort que $\mathscr{Q}$ sur Ω si $\mathscr{D}_{P_0}\subset\mathscr{D}_{Q_0}$ (espaces construits sur Ω).

Théorème 1. Une condition nécessaire et suffisante pour que $\mathscr{P}$ soit plus fort que $\mathscr{Q}$ sur Ω borné (cas des coefficients constants) est qu'il existe une constante C telle que

$$\widetilde{Q}(\xi)\le C\widetilde{P}(\xi),$$

où $\widetilde{P}(\xi)^2=\sum|P^\alpha(\xi)|^2$ (id. pour $\widetilde{Q}(\xi)$). Cette condition est indépendante de Ω. La démonstration repose essentiellement sur l'important Théorème 2. Soit Ω un ouvert borné, $\mathscr{P}$ à coefficients constants (l'A. écrit alors $\mathscr{P}=P(D)$. Quel que soit α, il existe une constante k telle que, pour tout $u\in C_0{}^\infty(\Omega)$,

$$\|\mathscr{P}u\|\ge k\|\mathscr{P}^\alpha u\| \tag{1}$$

($\|f\|$ est la norme dans $L^2(\Omega)$).

La démonstration du théorème 2 repose sur des intégrales d'énergie (méthode de Friedrichs, Lewy et Leray dans les cas hyperboliques). Dans le cas particulier $\alpha=0$, l'A. donne une autre démonstration, utilisant la transformation de Laplace et un lemme voisin d'un lemme de Malgrange [C.R. Acad. Sci. Paris **237** (1953), 1620–1622; MR **15**, 626]. La méthode des intégrales d'énergie permet d'attaquer le cas des opérateurs à coefficients variables. Soit $p(x,\xi)$ la partie principale du polynôme $P(x,\xi)$; on dit que $\mathscr{P}$ (ou p) est de type principal si les $\partial p(x,\xi)/\partial\xi_i$ ne peuvent s'annuler simultanément pour $x\in\Omega$, $\xi\ne0$.

Théorème 3. Soit $\mathscr{P}$ à coefficients variables continus, de type principal, d'ordre m, $p(x,\xi)$ réel à coefficients une fois continûment différentiables. Alors tout point de Ω a un voisinage ω dans lequel

$$\|\mathscr{P}u\|^2\ge k\sum_{|\alpha|<m}\|D_\alpha u\|^2,\ u\in C_0{}^\infty(\omega). \tag{2}$$

Des théorèmes 2 et 3 résulte que $\mathscr{P}$ applique $\mathscr{D}_P$ sur L^2 d'où facilement l'existence de problèmes aux limites bien posés. Le théorème 2 entraîne aussi l'existence d'une solution élémentaire (théorème de Malgrange, loc. cit.). Le théorème 3 entraîne l'existence locale d'un noyau élémentaire. Par utilisation des inégalités (1), l'A.

résout ensuite (entre autres) les problèmes suivants: condition pour que Qu soit dans $L^P(\Omega)$ (resp. continue après modification éventuelle sur un ensemble de mesure nulle, etc.) pour tout $u \in \mathscr{D}_{P_0}$; condition nécessaire et suffisante pour que l'injection de $\mathscr{D}_{P_0}$ dans $\mathscr{D}_{Q_0}$ soit complètement continue.

Un opérateur $\mathscr{P}$ (toujours à coefficients constants) est dit complet (Gårding) si "η réel et $P(\xi + t\eta) = P(\xi)$ pour tout ξ et t" entraînent $\eta = 0$. Alors si u est dans le domaine minimal de $\mathscr{P}^n$ quel que soit n, $\mathscr{P}$ étant complet, u est indéfiniment différentiable dans Ω. Un problème naturel est maintenant de comparer $\mathscr{D}_P$ et $\mathscr{D}_Q$. On a: Théorème 4. Une condition nécessaire et suffisante pour que $\mathscr{D}_P \subset \mathscr{D}_Q$ est que ou bien $\mathscr{Q} = a\mathscr{P} + b$, $a, b =$ constantes, ou bien $\mathscr{P}$ et $\mathscr{Q}$ sont deux opérateurs différentiels ordinaires, avec ordre $\mathscr{Q} \leq$ ordre $\mathscr{P}$. La démonstration de ce théorème fait appel à des résultats non triviaux de géométrie algébrique et utilise entre autres une idée de B. L. van der Waerden.

Il résulte du Théorème 2 que: $u \in \mathscr{D}_{P_0}$ et $\varphi \in C_0{}^\infty$ entraîne $\varphi u \in \mathscr{D}_{P_0}$. On dit que $\mathscr{P}$ est de type local si la même propriété a lieu pour $u \in \mathscr{D}_P$. L'A. donne cinq conditions équivalentes (et non trivialement équivalentes) nécessaires et suffisantes pour que $\mathscr{P}$ soit de type local.

La terminologie suivante a été adopteé au Colloque International sur les équations aux dérivées partielles, Nancy, Avril 1956: un opérateur différentiel D est dit hypo-elliptique si $Du = S$ entraîne que u est indéfiniment différentiable en même temps que S. Alors: Théorème 5. Une condition nécessaire et suffisante pour que $\mathscr{P}$ soit hypo-elliptique est que $\mathscr{P}$ soit complet et de type local. La démonstration repose entre autres sur la construction d'une solution élémentaire d'un opérateur complet et vérifiant l'une des cinq conditions équivalentes dont on parle çi dessus. A l'aide des propriétés de la solution élémentaire ainsi construite, l'A. donne des estimations pour les majorations des dérivées successives des solutions de $Pu = 0$. Il généralise ainsi considérablement les résultats de Holmgren pour l'équation de la chaleur.

Signalons encore que l'A. démontre l'existence d'une "solution nulle" dans toute direction caractéristique, $\mathscr{P}$ étant quelconque à coefficients constants, et qu'il étudie brièvement la théorie spectrale des opérateurs self-adjoints hypo-elliptiques.

J. L. Lions (Nancy)

19, 280a 35.0X

Malgrange, Bernard

Existence et approximation des solutions des équations aux dérivées partielles et des équations de convolution.

Ann. Inst. Fourier, Grenoble **6** (1955–1956), 271–355.

Since 1953, the author has been announcing interesting results in a series of notes [C. R. Acad. Sci. Paris **237** (1953), 1620–1622; **238** (1954), 196–198, 2219–2221; **240** (1955), 1958–1960; MR **15**, 626f; **16**, 127a; **17**, 404g]. In the present monograph, he presents the detailed treatment of these, together with extensions and generalizations. It is indeed a worthy addition to the growing literature in modern differential equations in the spirit of Leray, Hörmander, Gårding, Schwartz and others. Certain of his results (especially those of Chapter I) are closely parallel to those obtained independently by Hörmander [Acta Math. **94** (1955), 161–248; MR **17**, 853d, see p. G51 in this volume] and L. Ehrenpreis [Amer. J. Math. **76** (1954), 883–903; **77** (1955), 286–292, 293–328; **78** (1956), 685–715; Proc. Nat. Acad. Sci. U.S.A. **41**

(1955), 945–946; Ann. of Math. (2) **63** (1956), 129–159; MR **16**, 834a, 1123a, 1122d; **18**, 746g; **17**, 877b, 876g].

The keystone of this common method is the use of the Fourier transform to convert a differential equation into a division problem for a function algebra. [If $\mathfrak{A}$ is an algebra and F and G are in $\mathfrak{A}$, then under what conditions is there $H \in \mathfrak{A}$ with $F = HG$?] The solution of this leads in turn to an inverse for the differential operator, but which may be defined only on a proper subspace of the relevant function space. If this operator can be represented there by means of a linear functional defined on the subspace, then the Hahn-Banach theorem may be applied to this functional and one may obtain an extension of the inverse operator to the whole space.

In a preliminary section, the author introduces the multitude of function spaces in which he works, and proves a number of basic lemmas. Among these, one has special importance. Let μ be a distribution with compact support ($\mu \in \mathscr{E}'$) and let D be the operator on C^∞ functions defined by $D(f) = \mu * f$. Let $V(\mu)$ be the subspace generated by the exponential polynomials

$$Q(x) = P(x_1, x_2, \cdots, x_m) \exp\left(\sum_1^m \alpha_j x_j\right)$$

which lie in Null(D). Let $\mathscr{F}\mu$ be the entire function which is the Fourier transform of μ. Let ν be another distribution in $\mathscr{E}'$. Then, the meromorphic function $\mathscr{F}\nu/\mathscr{F}\mu$ is entire if and only if $V(\mu) \subseteq V(\nu)$. This result (also obtained by Ehrenpreis) is extended to the case in which ν and μ are replaced by matrices with distribution entries; one then has $\mathscr{F}\nu = A \cdot \mathscr{F}\mu$, where A is a matrix with entire entries.

In Chapter I, the author deals with a general finite order partial differential operator D having constant coefficients. If $\mathbf{F}$ is a fixed function space, one seeks to characterize Null(D) and $D(\mathbf{F})$ as subspaces of $\mathbf{F}$. He first shows that, for several choices of $\mathbf{F}$, D has an "elementary solution". This is a distribution E such that $DE = \delta$. This is then applied to yield other results. Theorem 2: If $\mathbf{F}$ is $C^\infty(\Omega)$ where Ω is an open convex set, then Null(D) is the closure of the linear span of the exponential polynomials Q which it contains. He points out that the word "exponential" cannot be dropped, since Null(D) may contain no polynomials, but obtains a necessary and sufficient criterion for this in terms of an algebraic variety associated with the Fourier transform of $D\delta$. Theorem 3: With the same choice of $\mathbf{F}$, $D(\mathbf{F}) = \mathbf{F}$. In particular, for any $g \in C^\infty$, the equation $D(f) = g$ has a solution in C^∞. The same results are also obtained for other function spaces. The convexity of Ω enters in because of the use of a theorem of Lions connecting the support of two distributions with that of their convolution [J. Analyse Math. **2** (1953), 369–380; MR **15**, 307g]. A generalization of Theorem 3, with Ω merely open, is obtained provided that D is a local operator on $\mathscr{E}'$.

Chapter II deals with the more general operators defined by $D(f) = \mu * f$, where μ is a fixed distribution in $\mathscr{E}'$. In this, his results do not go as far as those of Ehrenpreis. As before, the null space of D is generated by the exponential polynomials which it contains. However, D is not in general an onto mapping; $D(f) = g$ does not always have a solution in the same function class as g. This, however, is true if $\mathbf{F}$ is chosen as $\mathfrak{E}$, the space of entire functions, and $\mu \in \mathfrak{E}'$. The corresponding division problem is Theorem 1: if F and G are entire functions of exponential type (m-variables) and F/G is entire, then it is also of exponential type.

In the last Chapter, the author takes up the study of general differential operators whose coefficients need not be constant. The results are stated in terms of general differentiable varieties, fibre spaces, sheaves, and currents; the results are interesting but difficult to describe briefly, without introducing the complex notation of the author. The general scope can be indicated. [See also Ehrenpreis, Proc. Amer. Math. Soc. **7** (1956), 713–718, 1131–1138; MR **18**, 584d; **19**, 36g] Say that an operator D is elliptic if Df is of class C^∞ only when f is; say that D is analiptic (=analytic-elliptic) if Df is analytic only when f is. Define the envelope of an open set $\mathcal{O} \subset \Omega$ to be the union of $\mathcal{O}$ with the relatively compact components of its complement in Ω. Say that $\mathcal{O}$ is replete if it coincides with its envelope. Say that an operator D has the Runge property on Ω if, for every replete open precompact subset $\mathcal{O}$ of Ω, the solutions of $Df=0$ on $\mathcal{O}$ which can be extended as solutions of $Df=0$ on Ω are dense in the space of all solutions of $Df=0$ on $\mathcal{O}$. The author formulates general conditions under which an operator D has the Runge property, and connects it with the general existence theorem for solutions of $D(f)=g$. Specialization: Let D be an operator with constant coefficients. D is elliptic [analiptic] if and only if D has an "elementary solution" E which is of class C^∞ [which is analytic] at the origin. If D is analiptic, then D has the Runge property, D is elliptic, and the equation $Df=g$ always has distribution solutions. The author also shows that a certain class of operators with analytic coefficients (called by him Petrowsky operators) are analiptic. There are a number of applications, using special choices of D; one obtains, for example, theorems of Behnke and Stein on holomorphic functions on analytic varieties, and theorems of de Rham dealing with harmonic functions on real varieties. The chapter concludes with a discussion of the existence of elementary kernels, in the sense of Schwartz, for general operators, proving in particular that this is the case for Petrowsky operators. *R. C. Buck*

14, 637f 42.4X
Calderon, A. P.; Zygmund, A.
On the existence of certain singular integrals.
Acta Math. **88** (1952), 85–139.

In his paper on conjugate functions [Math. Z. **27**, 218–244 (1927)] M. Riesz deals with Hilbert's operator

$$\tilde{f}(x)=\frac{1}{\pi}\mathrm{PV}\int_{-\infty}^{\infty}\frac{f(t)}{x-t}\,dt=\lim_{0<\lambda\to\infty}\tilde{f}_\lambda(x)\quad(-\infty<x<\infty)$$

where

$$\tilde{f}_\lambda(x)=\left(\int_{-\infty}^{x-1/\lambda}+\int_{x+1/\lambda}^{\infty}\right)K(x-t)f(t)\,dt,$$

the kernel $K(y)=(\pi y)^{-1}$. He uses complex variable methods. These are his basic results: Let $1<p<\infty$, $f(t)\in L^p(-\infty,\infty)$; then (1) $\int_{-\infty}^{\infty}|\tilde{f}_\lambda(x)|^p\,dx\le A_p{}^p\int_{-\infty}^{\infty}|f(t)|^p\,dt$ (that $A_2=1$ was known before; the best possible value of A_p for $p\neq 2$ is still unknown), (2) $\tilde{f}(x)=\lim\tilde{f}_\lambda(x)$ exists for almost all x, (3) $\tilde{f}_\lambda(x)\to\tilde{f}(x)$ also in the mean of order p; in (1), $\tilde{f}_\lambda(x)$ can equivalently be replaced by $\tilde{f}(x)$. The problem was subsequently solved by real variable methods, which are harder; again the difficult case $p=1$ and related ones were treated by various writers.

The authors, dedicating their work to Riesz on his 65th birthday, give a complete theory of the corresponding problem in the n-dimensional space E^n, yet in a remarkably generalised form. They had to use real variable methods. This required auxiliary results which are of great interest in themselves and for which the reader must be referred to the paper. In three chapters, they discuss the mean convergence of the singular integral, its pointwise convergence, and the application of the results to some potentials, respectively. In an addendum results on periodic functions and on linear and quadratic forms are stated, and reference is given to related recent work. Let P and Q be points in E^n, $P-Q$ be the point whose coordinates are equal to the components of the vector QP, $|P|$ the distance of P from the origin. Let $\Omega(P)$ be defined for the points P of Σ, the surface of the unit sphere, and be such that $\int_\Sigma \Omega(P)\,d\sigma = 0$, $|\Omega(P)-\Omega(Q)| \le \omega(|P-Q|)$ (P and Q on Σ), where $\omega(t)$ is increasing $(0 \le t < \infty)$, $\omega(t) \ge t$, $\int_0^1 \omega(t)t^{-1}\,dt < \infty$. Finally, for any P, let $K(P) = |P|^{-n}\Omega(P/|P|)$; $K_\lambda(P) = K(P)$ for $|P| \ge 1/\lambda$, $=0$ otherwise; and

$$\widetilde{f}_\lambda(P) = \int_{E^n} K_\lambda(P-Q) f(Q)\,dQ.$$

In chapters 1 and 2 the authors, starting with the case $p=2$ which, in analogy to a familiar procedure, is dealt with by Fourier transforms in E^n, prove all the results corresponding to the above Riesz theorems on $\widetilde{f}_\lambda(P)$ and $\widetilde{f}(P)$ for $f(P) \in L^p(E^n)(1<p<\infty)$; and also results on the cases $p=1$ and ∞, e.g.: if $p=1$, then $\widetilde{f}(P) = \lim \widetilde{f}_\lambda(P)$ exists almost everywhere in E^n; also $\widetilde{f}_\lambda(P) \to \widetilde{f}(P)$ in the mean of order $1-\varepsilon$ $(\varepsilon > 0)$ over any set S of finite measure, and

$$\int_S |\widetilde{f}_\lambda(P)|^{1-\varepsilon}\,dP \le \varepsilon^{-1} c |S|^\varepsilon \left[\int_{E^n} |f(P)|\,dP\right]^{1-\varepsilon} \quad (c = c(n,\Omega)),$$

which corresponds to the Kolmogoroff-Littlewood result in E^1. Other theorems concern cases like $|f(P)| \le M$; $\int_{E^n} |f(P)|(\log^+ |f(P)|+1)\,dP < \infty$ and similar ones; and the singular integral $\widetilde{f}(P) = \int_{E^n} K(P-Q)\,d\mu(Q)$, where $\mu(P)$ is a completely additive function of Borel sets and is of bounded total variation over E^n.

The results are now applied to

$$u(x,y) = \int_{E^2} R^{-1} f(Q)\,dQ$$

$$(P=(x,y), Q=(s,t), R=|P-Q|=\{(x-s)^2+(y-t)^2\}^{1/2}),$$

that is the Newtonian potential of masses with density $f(s,t)$ in E^2. Theorem 1. If $f \in L(E^2)$ and $|f|\log^+|f|$ is integrable over any finite circle, then (a) over almost every line $y = $ constant the integral converges and is an absolutely continuous function of x; and $\partial u/\partial x = \iint_{E^2} R^{-3}(s-x)f(s,t)\,dsdt$ almost everywhere in E^2; (b) if $f \in L(E^2)$ and $|f|^q$ is integrable $(q>2)$ over any finite circle, then $u(x,y)$ has a complete differential almost everywhere. In the proof of (a) it is first shown that, for $z \to 0$ $(z>0)$, $\partial u(x,y,z)/\partial x - \widetilde{f}_z(x,y)$ converges to 0 and $\widetilde{f}_z(x,y)$ to a function $\widetilde{f}(x,y)$ in the mean of order 1 over any set of finite measure; here

$$u(x,y,z) = \iint_{E^2} (R^2+z^2)^{-1/2} f(s,t)\,dsdt,$$

$$\widetilde{f}_z(x,y) = \iint_{R\geq z} R^{-3}(s-x) f(s,t)\,dsdt$$

(cf. Riesz's second method of proof). The theorem is shown not to admit of much improvement. Theorem 2 concerns its analogue in E^n, while Theorem 3 concerns the logarithmic potential $u(P) = \int_{E^2} \log|P-Q|^{-1} f(Q)\,dQ$ in E^2, and Theorem 4 the potential $\int_{E^n} |P-Q|^{-n+2} f(Q)\,dQ$ in E^n. *H. Kober*

33 #7774 42.11
Carleson, Lennart
On convergence and growth of partial sumas of Fourier series.
Acta Math. **116** (1966), 135–157.

La découverte spectaculaire que contient cet article est la validité de l'hypothèse de Lusin, à savoir que la série de Fourier d'une fonction de carré sommable est convergente presque partout. En désignant par $s_n(x)$ la somme partielle d'ordre n de la fonction $f \in L^1(-\pi,\pi)$, les meilleurs résultats précédemment connus étaient les suivants: (1) si $f \in L^1$, $s_n(x) = o(\log n)$ p.p. (Hardy); (2) si $f \in L^2$, $s_n(x) = o(\sqrt{\log n})$ p.p. (Kolmogorov-Seliverstov 1925, Plessner 1926); (3) si $f \in L^p$, $1 < p < 2$, $s_n(x) = o((\log n)^{1/p})$ p.p. (Littlewood-Paley 1931). La cohérence de ces résultats, et la grande difficulté du dernier (voir A. Zygmund [*Trigonometric series*, second edition, Vols. I et II, Cambridge Univ. Press, New York, 1959; MR **21** #6498]), avaient fait penser aux spécialistes qu'ils étaient vraisemblablement les meilleurs possibles; d'où un scepticisme, justifié jusqu'il y a quelques mois, à l'égard des démonstrations proposées de l'hypothèse de Lusin.

L'auteur énonce les résultats suivants: (1) si $f(\log^+|f|)^{1+\delta} \in L^1$ pour un $\delta > 0$, $s_n(x) = o(\log\log n)$ p.p.; (2) si $f \in L^p$ $(1 < p < 2)$, $s_n(x) = o(\log\log\log n)$; (3) si $f \in L^2$, $s_n(x)$ est convergente p.p. vers $f(x)$. Il démontre (1) et (3), et donne un simple aperçu de la démonstration de (2). Les résultats (1) et (2) ne sont sans doute pas les meilleurs possibles; on peut conjecturer que $s_n(x) = \sigma(\log\log n)$ p.p. dès que $f \in L^1$ (et l'exemple classique de Kolmogorov montrerait alors que c'est le meilleur résultat possible) et que $s_n(x)$ est convergente p.p. dès que $f \in L^p$, $p > 1$.

Les démonstrations sont très délicates, et forcent l'admiration. L'idée de départ est la suivante. Au lieu d'étudier $s_n(x) = \sum_{-n}^{n} \hat{f}_m e^{imx}$, on peut étudier $\sum_n^\infty \hat{f}_m e^{imx}$, ou aussi bien la transformée de Hilbert de $f(x)e^{-inx}$, ou la transformée approchée $s_n{}^*(x) = \int_{-4\pi}^{4\pi} f(t)e^{-int}(x-t)^{-1} (-\pi < x < \pi)$; si $s_n{}^*(x) = O(w_n)$, on a aussi $s_n(x) = O(w_n)$. Généralement pour un sous-intervalle ω^* de $[-2\pi, 2\pi]$, on pose $s^*(x,n,\omega^*) = \int_{\omega^*} f(t)e^{-int}(x-t)^{-1}\,dt$; convenons de dire que n est adapté à ω^* si $n|\omega^*|$ est multiple de 2π, et que ω^* entoure x si x est plus près du milieu de ω^* que de ses extrémités. Alors $s^*(x,n,\omega^*)$ est majorée par la transformée de Hilbert maximale de $f(t)e^{-int}$, convenablement définie; cette remarque est utilisée seulement pour $n = 0$, et les efforts de l'auteur consistent à se ramener, par étapes, à ce cas.

On commence par se donner un entier N, et on se restreint aux n entre 2^{N-1} et 2^N. On définit (c'est très laborieux) un ensemble exceptionnel $X \subset [-\pi,\pi]$. Pour $x \notin X$ on définit une suite finie de couples $(n_j,\omega_j{}^*)$ de sorte que (1) $n_0 = n$, $\omega_0{}^* = [-4\pi,4\pi]$; (2) $\omega_j{}^* \subset \omega_{j-1}^*$, et $\omega_j{}^*$ entoure x; (3) n_j soit adapté à $\omega_j{}^*$; (4) le dernier des n_j soit nul, soit $n_\nu = 0$. Pour majorer $s_n{}^*(x)$, on écrit $s(x,n_j,\omega_j{}^*) = s(x,n_j,\omega_{j+s}^*) + R_j(x)$,

$$e^{in_jx}s^*(x,n_j,\omega_j{}^*) = e^{in_{j+1}x}s^*(x,n_{j+1},\omega_{j+1}^*) + S_j(x),$$

et par conséquent

$$(*)\qquad |s_n{}^*(x)| = |s^*(x,n,\omega_0{}^*)| \le \sum_{j=0}^{\nu-1}(|R_j(x)| + |S_j(x)|) + |s^*(x,0,\omega_\nu{}^*)|.$$

On s'efforce simultanément de rendre la mesure de X petite et le second membre de $(*)$ pas trop grand pour $x \notin X$.

Ce schéma général vaut pour les trois théorèmes de l'auteur. C'est la définition de X qui change d'un cas à l'autre. Les techniques utilisées pour les majorations sont raffinées mais classiques (transformées de Hilbert maximales, fonctions harmoniques dans un demi-plan, convolutions, inégalités de Young sur les transformées de Fourier). Un outil nouveau, qui peut trouver d'autres applications, est une série associée en chaque point x à une $f \in L^2(-\pi,\pi)$ et à un $b > 0$; le premier terme de la série est la somme des termes de la série de Fourier (complexe) de f dont le module dépasse b, soit $p_0(x)$; le second terme, soit $p_1(x)$, est obtenu en restreignant $f - p_0$ à celui des intervalles $[-\pi,0[$, $[0,\pi[$ qui contient x, en développant $f - p_0$ en série de Fourier à fréquences paires sur cet intervalle, et en supprimant tous les termes de module $\le b_j$ et ainsi de suite... La série $\sum p_j(x)$ est ce qu'on obtient en regardant f, au voisinage de x, avec des microscopes de plus en plus puissants, mais avec des analyseurs harmoniques incapables de mettre en évidence des harmoniques d'amplitudes supérieures à b.

L'article de l'auteur est très difficile à lire. Il est souhaitable qu'on trouve, ou bien une démonstration plus rapide par une autre méthode, ou bien quelques théorèmes généraux, suggérés par la méthode de l'auteur, d'où les théorèmes sur la convergence des séries de Fourier découleraient sans trop de peine. *J.-P. Kahane* (Orsay)

21 #1507 45.00

Kreĭn, M. G.

Integral equations on the half-line with a kernel depending on the difference of the arguments. (Russian)

Uspehi Mat. Nauk **13** (1958), *no.* 5 (83), 3–120.

The primary aim of this paper is a detailed study of the integral equation

$$(A)\qquad \chi(t) - \int_0^\infty k(t-s)\chi(s)ds = f(t) \quad (0 \le t < \infty);$$

the homogeneous equation

$$(B)\qquad \chi(t) - \int_0^\infty k(t-s)\chi(s)ds = 0 \quad (0 \le t < \infty);$$

and the associated homogeneous equation

$$\chi(t)-\int_0^\infty k(s-t)\chi(s)ds=0. \tag{C}$$

In addition, there is a fairly complete study of the discrete analog of (A), namely, the infinite system of equations

$$\sum_{k=0}^{\infty}\alpha_{j-k}\xi_k=\eta_j \ (j=0,1,2,\cdots), \tag{1}$$

where the coefficients satisfy $\sum_{j=-\infty}^{\infty}|\alpha_j|<\infty$. Under special restrictions, consideration is also given to the integral equation of the first kind

$$\int_0^\infty k(t-s)\chi(s)ds=f(t)\quad (0\le t<\infty). \tag{2}$$

Fully eighteen pages at the close of the paper are devoted to examples illustrating methods and techniques for carrying out the detailed steps in solving specific equations of types (A) and (2).

In the consideration of (A), (B), (C) it is assumed that the kernel $k(t)\in L_1(-\infty,\infty)$. The complex-valued functions $f(t)$ and $\chi(t)$ are considered as belonging to the Banach space E, which, in the present paper, may be any one of the spaces $L_+{}^{(p)}$ $(p\ge 1)$, $C_+{}^0, C_+, M_+{}^u, M_+{}^c, M_+$. The space $L_+{}^{(p)}$ is the space of all measureable functions $f(t)$ $(0\le t<\infty)$ with $\int_0^\infty |f(t)|^p dt<\infty$, and norm $\|f\|_p=(\int_0^\infty |f(t)|^p dt)^{1/p}$; M_+ is the space of all bounded functions $f(t)$ $(0\le t<\infty)$, with norm $\|f\|_M=\sup_{0<t<\infty}|f(t)|$. The spaces $M_+{}^c, M_+{}^u$ are subspaces of M_+, the first consisting of all continuous functions, the second of all uniformly continuous functions. Finally C_+ is the space of all continuous functions $f(t)$ $(0\le t<\infty)$ for which $f(\infty)=\lim_{t\to\infty}f(t)$ exists, and $C_+{}^0$ is its subspace for which $f(\infty)=0$. Thus, $C_+{}^0\subset C_+\subset M_+{}^u\subset M_+{}^c\subset M_+$.

The author arrives at three principal theorems, which we state below.

Theorem 1: Let $k(t)\in L_1(-\infty,\infty)$. Then for arbitrary $f\in E$, equation (A) will have one and only one solution $\chi\in E$ if and only if

$$1-K(\lambda)\ne 0\quad(-\infty<\lambda<\infty),\qquad \nu=-\mathrm{ind}\,(1-K)=0, \tag{$*$}$$

where $K(\lambda)$ is the Fourier transform of $k(t)$ and $\nu=-(2\pi)^{-1}\int_{-\infty}^{\infty}d_\lambda \arg(1-K(\lambda))$ is the index of (A). The solution will be given by

$$\chi(t)=f(t)+\int_0^\infty \gamma(t,s)f(s)ds, \tag{3}$$

wherein the resolvent kernel has the formulation

$$\gamma(t,s)=\gamma(t-s,0)+\gamma(0,s-t)+\int_0^\infty \gamma(t-r,0)\gamma(0,s-r)dr \tag{4}$$

$$(\gamma(t,0)=\gamma(0,t)=0\quad\text{for}\quad t<0).$$

The functions $\gamma(t,0)$, $\gamma(0,t)$ belong to $L_1(0,\infty)$ and are in turn determined uniquely from the relations

$$(1-K(\lambda))^{-1}=G_+(\lambda)G_-(\lambda)\quad(-\infty<\lambda<\infty), \tag{5}$$

$$(6)\qquad G_+(\lambda) = 1 + \int_0^\infty \gamma(t,0)\exp(i\lambda t)\,dt,$$

$$G_-(\lambda) = 1 + \int_0^\infty \gamma(0,t)\exp(-i\lambda t)\,dt,$$

$$(7)\qquad G_+(\lambda) \neq 0 \quad (\operatorname{Im}\lambda \geq 0), \quad G_-(\lambda) \neq 0 \quad (\operatorname{Im}\lambda \leq 0).$$

The factorization (5) with the factors normalized so that $G_\pm(\infty) = 1$ is unique whenever $(*)$ is satisfied and conversely.

Theorem 2: If $k(t) \in L_1(-\infty,\infty)$ and $1 - K(\lambda) \neq 0$ $(-\infty < \lambda < \infty)$, then (B) will have non-trivial solutions in E if and only if $\nu > 0$. (The same statement applies to (C) but with $\nu < 0$.) These solutions will be the same for all the component spaces of E and, in fact, will have a basis consisting of functions $\varphi_j(t)$ $(j = 0, 1, \cdots, \nu - 1)$, absolutely continuous, with derivatives belonging to $L_1(0,\infty)$, tending to zero as $t \to \infty$, and related, in sequence, by

$$(8)\qquad \varphi_{k+1}(t) = \frac{d\varphi_k}{dt}, \quad \varphi_k(0) = 0 \quad (k = 0, 1, 2, \cdots, \nu - 2), \ \varphi_{\nu-1}(0) \neq 0.$$

As a by-product of the proof of Theorem 2, the author furnishes an analytical procedure for determining the $\varphi_j(t)$.

The applicable theorem for (A) when the index $\nu \neq 0$ is Theorem 3: If $k(t) \in L_1(-\infty,\infty)$ and $1 - K(\lambda) \neq 0$, then for $\nu > 0$ and $f \in E$, equation (A) will possess an infinite set of solutions $\chi \in E$. However, if $\nu < 0$, then (A) will have either no solution or a single unique solution, the latter obtaining when $\int_0^\infty f(t)\psi_j(t)\,dt = 0$ $(j = 0, 1, 2, \cdots, |\nu| - 1)$, where the $\psi_j(t)$ represent a basis for the solutions of (C).

The paper contains a long introductory section, followed by four principal chapters, a page or two of supplementary remarks and a very complete bibliography. The introduction traces the history of equations (A), (B), (2) and the particular restrictions employed in their treatment, citing in detail the work of N. Wiener, E. Hopf, E. Reissner, V. Fok and I. Rapoport. The problem of factorization of $(1 - K(\lambda))^{-1}$, as represented by (5) of Theorem 1 and due originally to Wiener and Hopf, is treated in Chapter I, consideration being given successively to factorization on the real line (needed in Theorem 1), on a strip, and on a circle (used in solving (1)); in addition, there is an independent section on factoring entire functions of the form $1 + \int_{a_-}^{a_+} g(t)\exp(i\lambda t)\,dt$ on the real line, where $a_\pm$ are finite and $g \in L_1(a_-, a_+)$. The proofs of Theorems 1, 2, 3 and the solution of systems of type (1) are discussed in Chapters II, III, a by-product of the analysis being a study of the spectrum of the operator K in E where $K\varphi = \int_0^\infty k(t-s)\varphi(s)\,ds$. Chapter IV studies (A), (B) for the Wiener-Hopf type case where $k(t)$ is assumed to satisfy a condition of the form $k(t)\exp(h(t)) \in L_1(-\infty,\infty)$ $(h > 0)$. New material here is reflected in a series of theorems concerned with obtaining asymptotic representations of the solutions of (B) in terms of exponentials with polynomial coefficients. This is followed by the aforementioned section on detailed examples for special cases.

J. F. Heyda (Cincinnati, Ohio)

12, 31d 46.3X
Schwartz, L.
Théorie des distributions. Tome I. (French)
Actualités Sci. Ind., no. 1091 = Publ. Inst. Math. Univ. Strasbourg 9.
Hermann & Cie., Paris, 1950. 148 pp.

This is a generally clear, carefully organized, and detailed account of the basic aspects of the theory of "distributions" due to the author, and described by him in earlier publications [see especially Ann. Univ. Grenoble. Sect. Sci. Math. Phys. (N.S.) **21** (1945), 57–74 (1946); these Rev. **8**, 264e]. This theory provides a convenient formalism for many common situations in theoretical and applied analysis, but its greatest significance may be in connection with partial differential equations, particularly those of hyperbolic type, where its adaptability to local problems gives it an advantage over Hilbert space (and other primarily global) techniques. On the other hand, the latter, when applicable, are considerably more powerful than any techniques as yet provided by distributions. A second volume in press is apparently to deal with real harmonic analysis and other applications of the general theory, and presumably will contain the bibliography needed for the identification of the references indicated in the present volume.

A "distribution" is a continuous linear functional on the space $\mathscr{D}$ of functions on Euclidean space which are of class C^∞ and vanish outside compact sets, the continuity being relative to the topology of uniform convergence of the functions and all their derivatives, all the functions simultaneously vanishing outside the same fixed compact set. For example, any Lebesgue-Stieltjes measure defines a distribution via integration; and conversely every distribution which assigns nonnegative values to nonnegative functions arises in this way. Any function f integrable on compact sets likewise defines a distribution F by the equation $F(g) = \int fg$, and it is convenient to identify this distribution with the function. Every distribution T has "partial derivatives" defined by the equation $(\partial T/\partial x_k)(g) = -T(\partial g/\partial x_k)$; when T is a function this definition agrees with the usual one. All distributions are therefore indefinitely differentiable, and in particular, if the Dirac "function" at a point a is defined as the distribution arising from a measure of one concentrated at a, an explicit meaning has been attached to the derivatives of all orders of the Dirac function. The "support" of a distribution is the complement of the unique maximal open set such that the distribution vanishes on all functions vanishing outside it.

The set $\mathscr{D}'$ of all distributions is a topological linear space, the dual of the space $\mathscr{D}$. The (dual) topology in $\mathscr{D}'$ reduces to the following: A sequence T_n of distributions converges to a distribution T if $T_n(g) \to T(g)$ uniformly on bounded sets of elements g of $\mathscr{D}$, such a set consisting of functions vanishing outside a fixed compact set which, together with their derivatives, are uniformly bounded ("bounded" as applied to subsets of $\mathscr{D}$ has the usual meaning assigned in the theory of topological linear spaces). Then $\mathscr{D}'$ is complete, $T(g)$ is jointly continuous on $\mathscr{D}'$ and $\mathscr{D}$, and most important of all, differentiation is a continuous linear operation on $\mathscr{D}'$, so that convergent series in $\mathscr{D}'$ can be differentiated term by term.

Every distribution coincides locally, i.e., inside a conditionally compact open set, with a derivative (of some finite order) of (the distribution arising from) a continuous function. A distribution whose support consists of a single point is a finite linear combination of the Dirac function and its derivatives. If S and T are distributions in spaces X and Y respectively, then there exists a unique "direct product" distribution $U = S \times T$ on $X \times Y$ such that $U(k) = S(g)T(h)$, if $k(x,y) = g(x)h(y)$; for measures, the result is the usual product measure. The map $(S,T) \to S \times T$ is jointly continuous on the spaces of distributions on X and Y to the space of distributions on $X \times Y$. This concept of direct product is to be used in the later treatment of convolutions of distributions. In general, distributions cannot be multiplied in the usual way, i.e., so as to generalize the multiplication of functions (formal generalizations of this type have actually been employed in physics) but an arbitrary distribution can always be multiplied by a function of class C^∞. A significant problem here is that of division of distributions by functions. This problem is not yet fully resolved, but it is always possible to divide by arbitrary monomials in the coordinates (in Euclidean space).

The book closes with a sketch of applications to partial differential equations. These are treated mainly in terms of their "elementary solutions," the elementary solution at a point a of a partial differential operator Γ being defined as a distribution T such that $\Gamma(T)$ is the Dirac function at a (this differs somewhat from the classical definition, but is formally the same as a definition employed in theoretical physics).

I. E. Segal (Chicago, Ill.)

12, 833d 46.3X
Schwartz, Laurent
Théorie des distributions. Tome II. (French)
Actualités Sci. Ind., no. 1122 = Publ. Inst. Math. Univ. Strasbourg **10**.
Hermann & Cie., Paris, 1951. 169 pp.

This volume is a detailed treatment of the harmonic analysis of distributions on Euclidean spaces or toruses, and like volume I [Théorie des distributions. Tome I. Actualités Sci. Ind., no. 1091 = Publ. Inst. Math. Univ. Strasbourg 9, Hermann et Cie., Paris, 1950; these Rev. **12**, 31d, see p. G60 in this volume] is unusually clearly written and well organized. A summary of the major results of the volume appeared in Ann. Univ. Grenoble. Sect. Sci. Math. Phys. (N.S.) **23**, 7–24 (1948); these Rev. **10**, 36a. The results are qualitatively similar on the whole to the well known ones for functions, but serve to extend these results to maximal (though not maximum) domains of applicability. The techniques are partly standard, and partly based on a novel use of the theory of topological linear spaces, which use is appropriate in the present connection in view of the introduction of at least 14 distinct types of spaces of distributions. While the present approach is more systematic and algebraic than the common ones in harmonic analysis on the reals, its analytical force does not seem very substantially greater, and for the most part it would seem to amount to a much greater improvement in form than in essential content. On the other hand, the treatment of partial differential equations with constant coefficients given in the present volume is significantly simpler and more intelligible than the conventional ones.

The first of the two chapters concerns convolutions of distributions, which are first defined for distributions with compact supports by the formula $(S * T)(f) = (S_a \times T_b) f(a+b)$, where $\times$ indicates the direct product, and are shown to have the expected properties. Later the convolution is defined for distributions with supports A and B such that $A \cap (C - B)$ is compact for all compact sets C, an extension required for the treatment of differential equations. An intensive study is made of the regularization of a distribution T by an infinitely differentiable function f, this being defined as the convolution of T with f and being itself an infinitely differentiable function. Convergence of distributions and boundedness of a set of distributions are shown to be equivalent to corresponding properties of the regularizations of the distributions, and by these and related results the investigation of harmonic properties of distributions is reduced in considerable part to the corresponding investigation for functions. The limitations imposed on a distribution by restrictions of various types on its derivatives are studied, and classes of distributions analogous to the classical L_p spaces and to the Stepanoff almost periodic functions are treated. The chapter ends with an interesting treatment of equations such as $A * T = B$, where A and B are given distributions, based on the use of elementary solutions, this being in the case of the equation cited a distribution E such that $A * E$ is the Dirac "function". Applications are made to the analyticity of solutions of homogeneous elliptic equations and to a generalized notion of superharmonic function.

The second chapter treats the Fourier transform for distributions. A periodic distribution always has a Fourier series which converges to the distribution, and any sequence of coefficients which grow sufficiently slowly is the Fourier coefficient sequence of some distribution. An arbitrary distribution on Euclidean space will not have a Fourier transform, but a "temperate" (originally, "spherical") distribution has a Fourier transform, and the class of all such is mapped bicontinuously and univalently onto itself by the Fourier transform. Among other results, the Parseval formula holds, essentially as a matter of definition of the Fourier transform. Fourier transforms of distributions with compact supports and "positive definite" distributions are characterized, in extension of well-known results of Paley and Wiener and of Bochner. The chapter closes with a number of interesting applications to partial differential equations of all types with constant coefficients, and to other equations of the convolution type described in the preceding paragraph.

The choices of the linear topological spaces involved in the foregoing are limited in various ways, but in many respects formally similar results could be obtained with rather different spaces. For example, for a number of purposes in harmonic analysis it would seem possible and even convenient to replace infinite (local) differentiability by infinite differentiability in L_2.

The book has excellent summaries preceding each chapter, very good indexes of nomenclature and notations, and includes a bibliography pertinent to the first as well as to the present volume. *I. E. Segal* (Princeton, N. J.)

37 #778 47.00

Sz.-Nagy, Béla; Foiaş, Ciprian

Analyse harmonique des opérateurs de l'espace de Hilbert. (French)

Masson et Cie, Paris; Akadémiai Kiadó, Budapest, 1967. xi+373 pp.

The theory of self-adjoint operators on Hilbert space reached a definitive form during the first third of the century through the work of Hilbert, von Neumann, Stone and others. A comparable completeness of the theory for unitary operators awaited the development of the seminal ideas of Beurling by Lax, Helson, Lowdenslager, and Halmos to mention a few. About twenty years ago M. S. Livšic initiated a study of "almost unitary" operators and introduced the notion of characteristic operator function to study such operators. In the subsequent work of many mathematicians including M. S. Brodskiĭ and M. G. Kreĭn, this theory has enjoyed rather striking accomplishments. About fifteen years ago the first author showed how to associate a unique unitary operator, the minimal unitary dilation, with each contraction operator and proceeded to develop a theory for contractions based on this correspondence. He was joined in this work by the second author and the theory was developed in more than a dozen papers. This book presents a detailed exposition of their theory for contractions along with its relation to the aforementioned theories stemming from Beurling and Livšic.

In Chapter I several constructions are given of the minimal (strong) unitary dilation for a contraction. The similar result for one-parameter semigroups of contractions is proved along with consideration of the question of dilating more general abelian semigroups of contractions. Whether a unitary dilation exists in the general case is still an open question.

In Chapter II various geometrical and spectral properties of the unitary dilation of a contraction are studied. In particular, the result due to the first author that the minimal unitary dilation of a completely non-unitary contraction is absolutely continuous is proved. A classification of contractions is introduced and it is established that the operators in one of the classes have many proper invariant subspaces.

A functional calculus for contractions with bounded analytic functions is studied in Chapters III and IV. Contractions T satisfying $\varphi(T) = 0$ for some φ in H^∞ are studied and results paralleling those obtained in the finite-dimensional case using the minimum polynomial are proved. Dissipative and accretive operators are next studied using the Cayley transform.

The representation of invariant subspaces using operator valued functions is derived in Chapter V along with the relation between the inclusion of invariant subspaces and the problem of factorization. A rather complete treatment is given of what is known about the latter problem including new results on the class of operator valued functions that admit a scalar factor. This is the natural class of operator functions to which techniques involving the trace and determinant can be applied.

In Chapter VI the characteristic operator function $\Theta_T(\lambda)$ for a contraction T is introduced and the (canonical) functional model for the contraction defined in terms of it. The relation of the spectrum of T to properties of $\Theta_T(\lambda)$ is derived.

The relation between invariant subspaces for T and the factorization of $\Theta_T(\lambda)$ is studied in Chapter VII in considerable detail. (The determination of usable conditions that a factorization of $\Theta_T(\lambda)$ yield a proper invariant subspace is one of the basic unsolved problems in this theory.) The authors obtain again in this context the results of Chapter II on the existence of invariant subspaces.

In Chapter VIII the authors study the class of contractions T for which $I - T^*T$ is in the trace class. They show that $\Theta_T(\lambda)$ admits a scalar factor in this case and obtain good results.

In the last chapter the authors answer several questions using their theory. In particular, they determine which contractions are similar to a unitary operator and which contractions in a certain class are unicellular.

As we indicated at the beginning of this review there is at present a considerable amount of activity in the study of non-self-adjoint operators both by the authors of the book at hand and others. The authors have performed a valuable service in organizing and presenting so much of it in this book. Particularly valuable is the bibliography and the notes and comments at the end of each chapter. The authors have asked the reviewer to point out that the following were omitted from the bibliography: D. B. Lowdenslager, Ann. of Math. (2) **78** (1963), 450–454 [MR **27** #5094] and H. Nakano, Acta Sci. Math. (Szeged) **22** (1961), 286–288 [MR **25** #459].

There are several books on related topics that have appeared in the last few years: see H. Helson, *Lectures on invariant subspaces* [Academic Press, New York, 1964; MR **30** #1409]; I. C. Gohberg and M. G. Kreĭn, *Introduction to the theory of non-self-adjoint linear operators in Hilbert space* [(Russian), Izdat. "Nauka", Moscow, 1965; MR **36** #3137]; *Theory of Volterra operators on Hilbert space and its applications* [(Russian), Izdat. "Nauka", Moscow, 1967; MR **36** #2007] and P. D. Lax and R. S. Phillips, *Scattering theory* [Academic Press, New York, 1967; MR **36** #530].

We conclude by stating that we feel this to be an extremely valuable and a necessary book for the analyst working in operator theory and for many mathematicians in a diverse number of related areas. *R. G. Douglas* (Ann Arbor, Mich.)

2, 296d 48.0X

Hodge, W. V. D.

The theory and applications of harmonic integrals.

Cambridge University Press, Cambridge, England; *Macmillan Company, New York*, 1941. ix+281 pp.

The theory of harmonic integrals on a Riemannian manifold was developed by Hodge in the course of his investigations on the integrals attached to an algebraic variety. It is possible indeed to put a finger on the exact spot of the transcendental theory of algebraic varieties which, in last analysis, is responsible for the appearance of this stimulating book: it is the question of whether or not there exist double integrals on an algebraic surface (or, more generally, m-fold integrals on an m-dimensional algebraic variety) which are everywhere finite and which have all their periods equal to zero. This question remained unsolved for a long time until Hodge proved that such integrals do not exist (subsequently it has been recognized that a similar theorem holds, more generally, for integrals of closed forms on a topological manifold, as a corollary of results obtained previously by de Rham). However, the new theory, although an outgrowth of specific problems in algebraic geometry, goes far beyond algebraic geometry proper. Indeed a glance at the table of contents shows that of the five chapters which make up the book, only one deals with algebraic varieties. The rest is an ingenious blend of analysis pure and simple, topology, tensor calculus, differential geometry and continuous groups. The book therefore should also prove of considerable interest to specialists in the fields just mentioned. Since the original investigations of Hodge are scattered in many notes in which the topic is at times presented in a somewhat condensed fashion, the present book will be welcome by those who wish to gain a better understanding of the central points of the theory of harmonic integrals. The book is exceedingly well written, there is great emphasis on rigor, and the original material is presented anew in a systematic and unified form.

The first chapter, "Riemannian manifolds," is introductory and presents the material from differential geometry and topology which is needed later and used extensively throughout the book. The presentation of this auxiliary material is so well done that the reader has no need of consulting the textbooks given in the references, unless he wants details of some proofs. From the topological material, the intersection theory of Lefschetz and the properties of product manifolds play a specially important role later on.

The second chapter, "Integrals and their periods," deals with the theory of integrals on a topological manifold, due to de Rham. Here we find the main properties of closed and null differential forms and the following two theorems of de Rham: (1) The periods of an integral of a closed p-form on the R_p independent p-cycles of the manifold can be assigned arbitrarily. (2) If the periods are all zero then the closed form is necessarily a null form; in other words, the periods determine the closed form uniquely, to within an arbitrary null form. In the case of algebraic integrals on an algebraic variety, this second theorem is the original theorem of Hodge.

From the standpoint of the general theory, the most important chapter is the next one, entitled "Harmonic integrals." As to degree of generality, the notion of an harmonic integral lies somewhere between the general notion of an integral on a topological manifold, developed in the previous chapter, and the notion of an algebraic integral on an algebraic variety discussed in the next chapter. To define an harmonic integral we need not only a topology but also a Riemannian metric on the manifold. The harmonic integrals are then special integrals satisfying certain conditions relative to the given Riemannian metric. On the other hand, the real and imaginary parts of an algebraic integral on an algebraic variety are harmonic integrals with respect to suitable metrics which can be introduced on the variety. This is not the place to go into the details of the definition of an harmonic integral. Suffice it to say that a closed p-form is defined to be harmonic if a certain dual $(n-p)$-form (n is the dimension of the manifold) is also closed. For example, in the case $n=2$, the linear element being given by $ds^2=\lambda(dx^2+dy^2)$, the dual of the form $Pdx-Qdy$ is $Qdx+Pdy$, and the forms are harmonic if they are both closed, that is, if $\partial P/\partial y+\partial Q/\partial x=\partial P/\partial x-\partial Q/\partial y=0$. In this case the forms are the real and imaginary part of an analytic integral. The main theorem proved in this chapter is the existence theorem for harmonic integrals with preassigned periods. The proof requires a deep analysis and makes use of the theory of Fredholm integral equations. In the course of the proof it is shown directly that if the periods of an harmonic integral are all zero then the differential form is identically zero. From this the second theorem of de Rham follows readily.

The fourth chapter, "Applications to algebraic varieties," contains some of the most interesting results of the author. We have already mentioned that the real and imaginary part of an algebraic integral are harmonic integrals. However, not every harmonic integral arises from an algebraic integral. Moreover, an harmonic integral on an algebraic variety is not a birationally invariant concept, since there are infinitely many ways of defining a metric on the variety. Nevertheless, from the properties of harmonic integrals it is possible to extract properties which are birationally invariant, either in the absolute sense, or in the more restricted, relative sense. It is this birational aspect of the theory of harmonic integrals that is emphasized in this chapter. If the variety V_m lies in a projective space S_r, there is a metric on V_m induced by the representation of S_r as a real analytic locus of $2r$ dimensions, due to Mannoury. It is necessary to assume that V_m is free from singularities in order to have a topological manifold. The fact that the $2m$ real coordinates of V_m are the real and imaginary parts of m complex uniformizing parameters implies certain special properties for the metric which play a role in the proofs. For instance, if $T^{j_1,\cdots,j_q}_{i_1,\cdots,i_p}$ is a tensor, then also

$T^{j_1,\cdots,j_h+m,\cdots,j_q}_{i_1,\cdots,i_p}$ is a tensor. The harmonic integrals, or the corresponding differential forms, are classified as effective and ineffective. A p-fold harmonic form is effective if a certain associated $(p-2)$-form (obtained by covariant differentiation) is a null form. It is shown that there are exactly $R_p - R_{p-2}$ linearly independent effective p-fold harmonic forms. The ineffective harmonic forms are expressible in terms of the effective forms of lower order $p-2, p-4, \cdots$, and of certain differential forms which are defined projectively in terms of the ambient space S_r. This classification of the harmonic integrals is in close connection with Lefschetz's classification of the p-cycles on V_m as effective and ineffective: the effective p-fold integrals have non-vanishing periods only on the $R_p - R_{p-2}$ effective p-cycles. The properties of the period matrix of the effective integrals are studied and it is shown that this matrix is invariantively related not only to the system of hyperplane sections of the original projective model but also to the entire continuous system containing the hyperplane sections. We shall mention one more application. A p-cycle Γ may be homologous to a cycle lying on some $(p-k)$-dimensional subvariety of V_m $(k \leq [\frac{1}{2}p])$. It is shown that in that case the periods of certain harmonic integrals relative to Γ must be euqal to zero. If $p = 2q$ and $k = q$ we have thus necessary conditions that Γ be an algebraic cycle. Unfortunately it is not yet known whether these conditions are also sufficient. It is an important question for possible applications to algebraic geometry.

The last chapter deals with applications to continuous groups. It is shown how a metric can be defined on any manifold which admits a continuous group of transformations into itself, provided the group satisfy certain general conditions (semi-simple, closed, transitive Lie group). The invariant integrals of the group are studied and it is shown that the harmonic integrals lead to group invariants which have been discussed by Cartan. Several sections deal with special groups: the unimodular group, the orthogonal group and the symmetric group.

O. Zariski (Baltimore, Md.)

18, 511a 53.3X
Serre, Jean-Pierre
Géométrie algébrique et géométrie analytique.
Ann. Inst. Fourier, Grenoble **6** (1955–1956), 1–42.

Soit X une variété algébrique définie sur le corps complexe C. On peut étudier X de deux points de vue: 1° algébrique (on s'intéresse aux anneaux locaux de fonctions rationnelles, aux applications rationnelles, ou aux applications "régulières" de X); 2° analytique, ou "transcendant" (on s'intéresse aux fonctions holomorphes et aux applications analytiques de X considéré comme espace analytique). Ce mémoire a pour but de comparer les deux points de vue; le classique théorème de Chow, qui dit que tout sous-espace analytique fermé d'une variété algébrique projective est une variété algébrique, est un exemple de l'équivalence essentielle des deux théories. D'autres résultats de ce genre sont établis ici. Voici un résumé incomplet des divers paragraphes.

§ 1: rappel de généralités concernant les "espaces analytiques".

§ 2: toute variété algébrique X (au sens général de A. Weil) définit un espace analytique X^h et une bijection $X \to X^h$ qui est continue quand on munit X de la topologie de Zariski et X^h de la topologie usuelle. Pour que X soit complète, il faut et il suffit que X^h soit compacte. On a un homomorphisme $\mathcal{O}_x \to \mathcal{H}_x$ des anneaux locaux (algébrique, resp. holomorphe) en un point $x \in X$, et l'homomorphisme $\hat{\mathscr{O}}_x \to \hat{\mathscr{H}}_x$ des anneaux complétés est un isomorphisme. Ceci permet d'appliquer au couple d'anneaux $(\mathscr{O}_x, \mathscr{H}_x)$ la théorie des "couples plats" traitée en Annexe, et de comparer "dimension algébrique" et "dimension analytique". Si Y est une sous-variété algébrique (localement fermée) de X, toute fonction holomorphe en $x \in Y$ qui s'annule sur Y appartient à l'idéal de $\mathscr{H}_x$ engendré par les fonctions rationnelles, définies en x, qui s'annulent sur Y. Signalons la prop. 9 (démontrée par Samuel): si une application régulière $f: X \to Y$ induit un isomorphisme $X^h \to Y^h$ des espaces analytiques associés, f est un isomorphisme birégulier.

§ 3: tout faisceau algébrique $\mathscr{F}$ sur X définit un faisceau analytique $\mathscr{F}^h$ sur X^h et une application $\mathscr{F} \to \mathscr{F}^h$; $\mathscr{F}^h$ est un foncteur exact de $\mathscr{F}$, et l'application $\mathscr{F} \to \mathscr{F}^h$ est injective; si $\mathscr{F}$ est algébrique-cohérent, $\mathscr{F}^h$ est analytique-cohérent. Pour tout entier $q \geq 0$, on a un homomorphisme naturel des groupes de cohomologie $\varepsilon: H^q(X, \mathscr{F}) \to H^q(X^h, \mathscr{F}^h)$. Les trois théorèmes fondamentaux concernent le cas où X est une variété algébrique projective (i.e. plongeable dans un espace projectif, avec éventuellement des singularités): (1) si $\mathscr{F}$ est algébrique-cohérent sur X, ε est un isomorphisme; (2) si $\mathscr{F}$ et $\mathscr{G}$ sont algébriques-cohérents sur X, tout homomorphisme analytique $\mathscr{F}^h \to \mathscr{G}^h$ provient d'un homomorphisme algébrique $\mathscr{F} \to \mathscr{G}$; (3) tout faisceau analytique-cohérent sur X^h est isomorphe à un $\mathscr{F}^h$, où $\mathscr{F}$ est algébrique-cohérent sur X, et $\mathscr{F}$ est unique (à un isomorphisme algébrique près).

§ 4: utilisant un lemme de Chow, l'auteur prouve que tout sous-ensemble analytique compact d'une variété algébrique (quelconque, non nécessairement projective) est algébrique, et que toute application holomorphe d'une variété algébrique compacte dans une variété algébrique est algébrique-régulière (corollaire: unicité de la structure de variété algébrique donnant naissance à une structure d'espace analytique compact). Soit G un groupe algébrique; soit $\mathscr{G}$ (resp. $\mathscr{G}^h$) le faisceau des germes d'applications régulières (resp. holomorphes) de X (resp. X^h) dans G; on a une application $\varepsilon: H^1(X, \mathscr{G}) \to H^1(X^h, \mathscr{G}^h)$ des classes de fibrés principaux algébriques dans les classes de fibrés principaux analytiques (de groupe G). Si X^h est compacte, ε est injective, et même tout isomorphisme analytique de fibrés algébriques de base X^h est un isomorphisme algébrique. Si X est algébrique projective, ε n'est pas nécessairement une bijection; toutefois ε est une bijection dans les cas suivants: (α) G est le groupe additif C; (β) $G = \mathrm{GL}_n(C)$, ce qui, pour $n = 1$, permet d'étendre aux variétés normales un théorème prouvé par Kodaira-Spencer quand X est non-singulière: tout fibré analytique de groupe C^* provient d'un diviseur; (γ) G est un sous-groupe de $\mathrm{GL}_n(C)$ tel qu'il existe une section rationnelle $\mathrm{GL}_n(C)/G \to \mathrm{GL}_n(C)$ (il en est ainsi quand G est résoluble, ou est l'un des groupes $\mathrm{Sp}_n(C)$ et $\mathrm{SL}_n(C)$).

L'Annexe est consacrée aux modules plats et aux couples plats d'anneaux, avec application à la théorie des anneaux locaux. *H. Cartan* (Paris)

18, 590f 55.0X
Papakyriakopoulos, C. D.
On Dehn's lemma and the asphericity of knots.
Proc. Nat. Acad. Sci. U.S.A. **43** (1957), 169–172.
In 1910 Max Dehn published what purported to be a proof of the proposition that if D is a combinatorial singular 2-cell in a triangulated 3-manifold M, and D has no

singularities on its boundary, then the boundary of D is unknotted [Math. Ann. **69** (1910), 137–168, p. 147]. This was known as the Dehn lemma. It can be stated more precisely as follows:

Lemma. Let σ^2 be a 2-simplex with boundary β. Let f be a piecewise linear mapping of σ^2 into the triangulated 3-manifold M. Suppose that there is a neighborhood U of β in the space σ^2, such that (1) $f|U$ is a homeomorphism and (2) $f(U)\cap f(\sigma^2-U)=0$. Then the polygon $f(\beta)$ is the boundary of a polyhedral disk in M.

Nineteen years after the appearance of Dehn's paper it was found by Kneser [Jber. Deutsch. Math. Verein. **38** (1929), Abt. 1, 248–260, p. 260] that Dehn's proof of the lemma was erroneous; and since then the lemma has represented a celebrated problem. In the paper under review, the author announces an affirmative solution of this problem, and gives indications of a proof which will later be published in full. The proof makes use of a result due to Johansson [Math. Ann. **115** (1938), 658–669] to the effect that the lemma reduces to the case in which M is orientable.

As Dehn correctly proved, it is a consequence of the lemma that the knot-type of the circle in the 3-sphere is characterized by the knot-group. That is to say, if K is a (polygonal) knot in S^3, and the fundamental group $\pi_1(S^3-K)$ is infinite cyclic, then K is the boundary of a polyhedral disk.

Let F be a non-empty, closed, proper subset of S^3. If there is a 2-sphere S^2 in S^3-F such that each of the components of S^3-S^2 intersects F, then we say that F is geometrically splittable. If $\pi_1(S^3-F)$ is the free product of two non-trivial groups, then we say that F is algebraically splittable. If U is an open subset of S^3, and the second homotopy group $\pi_2(U)$ is $=0$, then we say that U is aspherical.

Theorem. Let U be a non-empty, open, proper connected subset of S^3. Then U is aspherical if and only if S^3-U is not geometrically splittable.

Corollary. If F is a non-empty, connected, closed proper subset of S^3, then every component of S^3-F is aspherical.

This is the solution of a problem of Eilenberg [Fund. Math. **28** (1936), 233–242, p. 241].

Corollary. If F is a connected graph, or a knot, then S^3-F is aspherical.

By a link we mean a finite union of disjoint knots. The following result is the solution of a problem of Higman [Quart. J. Math. Oxford Ser. **19** (1948), 117–122; MR **9**, 606a].

Theorem. Let K be a link in S^3. The following three statements are equivalent: (i) S^3-K is not aspherical. (ii) K is geometrically splittable. (iii) K is algebraically splittable.

The proofs of the above results are based in part on the following theorem. The Sphere Theorem. Let M be an orientable triangulated 3-manifold with boundary. (The boundary of M may be empty, and M is not necessarily compact.) Suppose that $\pi_2(M)\neq 0$; and suppose that M is combinatorially imbeddable in a 3-manifold N which has the property that the first homology group of every nontrivial subgroup of $\pi_1(N)$ has an element of infinite order. Then there is a polyhedral 2-sphere S in M such that S is not homotopic to zero in M.

The following result verifies a conjecture of Hopf. Theorem. If U is an open connected subset of S^3, then $\pi_1(U)$ has no element of finite order.

Finally, it is shown that if K is a knot in S^3, then $\pi_1(S^3-K)$ has either one end or two ends, according as K is knotted or unknotted. *E. Moise* (Princeton, N.J.)

25 #4530 55.40
Adams, J. F.
On the non-existence of elements of Hopf invariant one.
Ann. of Math. (2) **72** (1960), 20–104.

In this paper the author solves in the negative the famous Hopf problem of deciding whether or not there exist maps $S^{2n-1} \to S^n$ of Hopf invariant 1 whenever $n \neq 2, 4, 8$ [H. Hopf, Fund. Math. **25** (1935), 427–440]; such maps are constructed for $n = 2, 4$, and 8. Chapter 1 (Introduction) gives a sketch of the history of the problem. In particular, it is recalled that the non-existence of a map $S^{2n-1} \to S^n$ of Hopf invariant 1 is equivalent to the following statement: If $K = S^m \cup e^{n+m}$ is obtained by attaching an $(n+m)$-cell e^{n+m} to S^m by a map $S^{n+m-1} \to S^m$, then the Steenrod square $Sq^n\colon H^m(K; Z_2) \to H^{n+m}(K; Z_2)$ is zero. Now, for $n \neq 2^r$, J. Adem has proved that Sq^n is a sum of (non-trivial) iterated Steenrod squares [Proc. Nat. Acad. Sci. U.S.A. **38** (1952), 720–726; MR **14**, 306e; *Algebraic geometry and topology,* pp. 191–238, Princeton Univ. Press, Princeton, N.J., 1957; MR **19**, 50c]. Thus, since $H^q(K; Z_2) = 0$ for $m < q < n+m$, Sq^n must be zero on $H^m(K; Z_2)$ for $n \neq 2^r$. It was also known that Sq^{2^r} is not expressible as a sum of iterated lower squares. What the author shows is that for $k \geq 3$, $Sq^{2^{k+1}}$ is decomposable in terms of secondary cohomology operations: $Sq^{2^{k+1}} = \sum_{i,j} a_{i,j,k} \Phi_{i,j}$, where the $a_{i,j,k}$ are primary, and the $\Phi_{i,j}$ are secondary cohomology operations. The formula is valid when applied to a cohomology class in the kernel of all the squares Sq^r with $0 < r < 2^{k+1}$, and is to be interpreted as a congruence modulo a suitable subgroup which, in the case $K = S^m \cup e^{n+m}$, turns out to be 0. Although it was perhaps not unlikely that the Hopf problem would have a negative answer, it is certainly quite surprising that secondary operations suffice to make Sq^{2^r} decomposable.

After some preliminaries in homological algebra (Chapter 2), the paper starts with a general treatment of stable secondary cohomology operations (Chapter 3). The vague notion that secondary operations correspond to relations between primary operations is given an axiomatic and precise formulation. If A denotes the Steenrod algebra (over Z_2, say), and C is a graded free A-module on free generators $\{c_\alpha\}$, then a relation is a (homogeneous) cycle for some A-map $d\colon C \to A$ of degree 0. For instance $Sq^2 Sq^2 + Sq^3 Sq^1 = 0$ corresponds to $C =$ free A-module on c_1, c_2 with $\deg(c_i) = i$, and $d\colon C \to A$ given by $c_1 \to Sq^1$, $c_2 \to Sq^2$. The cycle z is $z = Sq^3 c_1 + Sq^2 c_2$.

It is shown [Theorems 3.6.1 and 3.6.2], using Eilenberg-MacLane spaces, that a secondary cohomology operation Φ is indeed uniquely determined by such a pair (d, z), modulo primary operations. Φ is defined on those $u \in H^*(X; Z_2)$ such that the composition

$$C \xrightarrow{d} A \xrightarrow{u} H^*(X)$$

is zero. [$u(1) = u$, u is an A-map.] It raises degrees by $\deg(z) - 1$. The indeterminacy is the subgroup $Q(z, X)$ generated by the images of z under all possible A-maps $C \to H^*(X)$ of degree $\deg(u) - 1$. Now, if $z = \sum a_k z_k$, where z, z_k are cycles for $d\colon C \to A$, then

$$\Phi = \sum a_k \Phi_k, \text{ modulo primary operations,}$$

where Φ, Φ_k are secondary cohomology operations corresponding to (d, z) and (d, z_k) respectively. This fact [Theorem 3.7.1 of the paper] is exploited to produce relations between explicitly constructed secondary operations $\Phi_{i,j}$. The definition of $\Phi_{i,j}$ is as follows: Let C be free on $c_0, c_1, \cdots$, where c_i is of degree 2^i. The map d sends c_i into Sq^{2^i}. The Adem relations [loc. cit.] imply that for $i \le j$, $i \ne j-1$, there is a relation of the form

$$Sq^{2^i} Sq^{2^j} = \sum_{0 \le s < j} b_s Sq^{2^s},$$

which means that $z_{i,j} = Sq^{2^i} c_j + \sum_{0 \le s < j} b_s c_s$ is a cycle for d. The operation $\Phi_{i,j}$ is associated with the pair $(d, z_{i,j})$.

The author shows that for each $k \ge 3$, there is a relation $\sum a_{i,j,k} z_{i,j} = 0$, where $\deg(a_{i,j,k} z_{i,j}) = 2^{k+1} + 1$. Hence, there is a relation $\sum a_{i,j,k} \Phi_{i,j} = 0$, modulo primary operations of degree 2^{k+1}.

Explicit calculations of the values of $\Phi_{i,j}$ in infinite-dimensional complex projective space then yield the formula

$$\sum a_{i,j,k} \Phi_{i,j} = Sq^{2^{k+1}}.$$

The calculations are rather complicated and involve some more secondary operations and relations between them, but the exposition is very clear and self-contained.

M. A. Kervaire (Paris)

3, 316e 56.0X
Hopf, Heinz
Fundamentalgruppe und zweite Bettische Gruppe.
Comment. Math. Helv. **14** (1942), 257–309.

This paper is, in the reviewer's mind, one of the most important contributions to combinatorial topology in recent years. It gives far reaching results concerning the relations between the fundamental group, the first and second homology and cohomology groups, and the products between these groups, with beautiful and simple methods. The work is based on some new constructions in groups which are undoubtedly of real significance by themselves. The paper is in three main parts: the group theory; determination of the second homology group $\mathfrak{B}^2$ (all groups are with integer coefficients) modulo the group $\mathfrak{S}$ of "spherical homology classes" [see below]; and the determination of the products for dimensions not greater than 2 (omitting considerations of torsion). In each case, the formulas are in terms of the fundamental group $\mathfrak{G}$, and pure group-theoretic constructions, but with geometric meanings.

Let $\mathfrak{G}$ be a group with finite sets of generators and relations. Let Γ be the set of all $\alpha = (X_1, Y_1, \cdots, X_n, Y_n)$, X_n, $Y_n \in \mathfrak{G}$. Two elements are added by placing them in juxtaposition, and α^{-1} is α with the X_i and Y_i interchanged. Let Δ be the subset of Γ with elements α with the following properties: For each X_i and Y_i fixed representations as products of elements of $\mathfrak{G}$ may be chosen so that $C(\alpha) = X_1 Y_1 X_1{}^{-1} Y_1{}^{-1} \cdots Y_n{}^{-1}$, when written out, is also of the form $U_1 V_1 U_1{}^{-1} V_1{}^{-1} \cdots V_m{}^{-1}$, with fixed representations of the U_i and V_i, and with each $V_i = 1$. (Hence $C(\alpha) = 1$.) Then Γ falls into cosets modulo Δ; these form a group $\mathfrak{G}^*$. If F is a homomorphism of a free group $\mathfrak{F}$ onto $\mathfrak{G}$, with kernel $\mathfrak{R}$, and $\mathfrak{C}(\mathfrak{R})$ is the subgroup generated by all $frf^{-1}r^{-1} (f \in \mathfrak{F}, r \in \mathfrak{R})$, then Δ consists of all α such that a_i and b_i exist with $F(a_i) = X_i$, $F(b_i) = Y_i$, and $C(a_1 \cdots b_n) \in \mathfrak{C}(\mathfrak{R})$. Hence, if $\mathfrak{C} = \mathfrak{C}(\mathfrak{F})$, $\mathfrak{G}^* \cong \mathfrak{C}/\mathfrak{C}(\mathfrak{R})$. Let $\mathfrak{G}_1{}^*$ be the subgroup of $\mathfrak{G}^*$, using only those $\alpha \in \Gamma$ with $C(\alpha) = 1$. Then $\mathfrak{G}_1{}^* \cong (\mathfrak{C} \cap \mathfrak{R})/\mathfrak{C}(\mathfrak{R})$. Examples: If $\mathfrak{G}$ is free, $\mathfrak{G}_1{}^* = 0$. If $\mathfrak{G}$ is a free Abelian group with p generators, $\mathfrak{G}_1{}^*$ is a free Abelian group with $\frac{1}{2}p(p-1)$ generators.

It is easily seen that any 2-chain Y in the complex K may be determined as the image of an oriented 2-cell E; the boundary curve ρ of E is mapped into a closed path r, called a "homotopy boundary" (h.b.) of Y. Let $\mathfrak{G}$ and $\mathfrak{F}$ be the fundamental groups of K and K^1 (= 1-dimensional part of K); let F be the natural homomorphism of $\mathfrak{F}$ onto $\mathfrak{G}$, with the kernel $\mathfrak{R}$. Then r is a h. b. if and only if it determines an element of $\mathfrak{R}$. The different h. b.'s of Y differ by elements of $\mathfrak{C}(\mathfrak{R})$; Y is a cycle if and only if $r \in \mathfrak{C}$. If Y is a "spherical cycle," obtainable as the image of a 2-sphere, this is equivalent to mapping the above ρ into a point; thus $r \in \mathfrak{C}(\mathfrak{R})$. The mappings are homomorphic; it follows that

$$\mathfrak{B}^2/\mathfrak{S}^2 \cong (\mathfrak{C} \cap \mathfrak{R})/\mathfrak{C}(\mathfrak{R}) \cong \mathfrak{G}_1{}^*.$$

As a corollary, if $\mathfrak{G}_1{}^* \neq 0$, $\mathfrak{B}^2 \neq 0$. The structure of $\mathfrak{G}_1{}^*$ is determined in various examples, with the help of specific complexes with $\mathfrak{S}^2 = 0$.

The third part of the paper is carried out for intersections in a manifold, but holds for products in a complex. The basic formula is the following: Let U be a 1-cocycle, and Z a 2-cycle. If r is a h. b. of Z, then $r \in \mathfrak{C}$, and corresponds to an element of $\mathfrak{G}$ of the form $C(X_1, \cdots, Y_h)$. Each X_i and Y_i defines 1-cycles X_i' and Y_i'. Then

$$U \cap Z \sim \sum_{i=1}^{h} [(U \cdot Y_i') \cdot X_i' - (U \cdot X_i') \cdot Y_i'].$$

A similar formula holds for the $\cup$ product. The proof runs as follows. Since r is a h. b., the path $X_1 Y_1 X_1{}^{-1} Y_1{}^{-1} \cdots Y_h{}^{-1}$ is the image of the boundary ρ of a simplex E under a mapping f. If we identify the parts of E going into X_1 and $X_1{}^{-1}$, etc., a closed oriented surface ξ^2 is formed; $Z \sim f(\xi^2)$. Since $U \cap f(\xi^2) \sim f(f' U \cap \xi^2)$ (f' = dual of f), it is sufficient to prove the formula for surfaces ξ^2; in ξ^2 it reduces to well-known facts.

When the chains are expressed as above in terms of $\mathfrak{G}$ (the products act trivially on elements of $\mathfrak{S}$), a product is defined, as follows. Let $\mathfrak{G}' = \mathfrak{G}/\mathfrak{C}$ (=first reduced homology group), with character group (into the integers) $\mathrm{Ch}\,\mathfrak{G}'$ (=first reduced cohomology group). Then the product is between $\mathrm{Ch}\,\mathfrak{G}'$ and $\mathfrak{G}_1{}^*$, into $\mathfrak{G}'$, or between $\mathrm{Ch}\,\mathfrak{G}'$ and $\mathrm{Ch}\,\mathfrak{G}'$ into $\mathfrak{G}_1{}^*$. The product may be defined algebraically, with the help of results of E. Witt [J. Reine Angew. Math. **177**, 152–160 (1937)]. Various corollaries and examples are given, in particular, for 3-dimensional manifolds.

H. Whitney (Cambridge, Mass.)

22 #987 57.00

Bott, Raoul

The stable homotopy of the classical groups.

Ann. of Math. (2) **70** (1959), 313–337.

This is a detailed account of the proof of the important results of the author on the stable homotopy groups of the classical Lie groups as announced in an earlier note [Proc. Nat. Acad. Sci. U.S.A. **43** (1957), 933–935; MR **21** #1588]. The main steps in the treatment of the various groups have been unified in a single general proposition (Theorem I) which contains the periodicity theorems as special cases. This theorem and its proof are a striking illustration of the fact that geometrical methods in homotopy theory still have some interest.

Theorem I is as follows. Let M be a closed compact connected Riemannian manifold of class C^∞. Let $v=(P,Q,h)$ consist of a pair of points P,Q on M and a homotopy class h of paths from P to Q. Let $\Omega_v M$ be the set of piecewise differentiable curves from P to Q on M, in the homotopy class h, parametrized from 0 to 1 proportionally to arc length. $\Omega_v M$ is given the topology induced by the metric

$$\rho(c,c')=\max\rho_0(c(t),c'(t))+|J(c)-J(c')|,$$

where ρ_0 is the distance on M and $J(c)$ is the length of c. If $s\in\Omega_v M$ is a geodesic segment and $t\in I$, denote by $\lambda(s(t))$ the dimension of the vector space of Jacobi fields along s vanishing at $P=s(0)$ and $s(t)$. (The definition of a Jacobi field is recalled on page 319.) Define the index of a geodesic segment $s\in\Omega_v M$ to be the finite number $\sum_{0<t<1}\lambda(s(t))$. Let $|v|$ denote the minimum of the positive integers which occur as the index of a geodesic segment in $\Omega_v M$, and let M^v be the subspace of $\Omega_v M$ consisting of the geodesic segments of minimal length. Theorem I: If M is a symmetric space (in the sense of E. Cartan), then M^v is also a symmetric space (in particular a Riemannian manifold), and the inclusion $i\colon M^v\to\Omega_v M$ induces isomorphisms $i_*\colon\pi_k(M^v)\to\pi_k(\Omega_v M)\approx\pi_{k+1}(M)$ for $0<k<|v|-1$. In dimension $k=|v|-1$, i_* is an epimorphism.

A procedure to compute $|v|$ in practice is the subject of § 6.

The proof of Theorem I is essentially in two steps. Step I: In § 3, a homotopy formulation of Morse theory is given following an idea inaugurated by R. Thom [C. R. Acad. Sci. Paris **228** (1949), 973–975; MR **10**, 558b]. Thom's argument is made more specific using J. H. C. Whitehead's notion of a CW-complex and is extended so as to include a more general type of critical sets, i.e., non-degenerate critical manifolds, a notion due to the author [Ann. of Math. (2) **60** (1954), 248–261; MR **16**, 276f]. Step II: Previous work of the author, and joint work with H. Samelson is then used to show that the Morse theory in its new formulation applies to the symmetric spaces and yields the result.

The paper ends with explicit computations of $|v|$ for the symmetric spaces arising from the classical Lie groups (§ 7 and § 8).

Although results from Morse theory and previous work of the author arc used, the proofs can be followed if these results are taken for granted. Moreover, precise references are given whenever the proofs are not self-contained and the reconstitution of the full details is a routine matter. *M. A. Kervaire* (New York)

22 #990 57.00
Milnor, John
Differentiable structures on spheres.
Amer. J. Math. **81** (1959), 962–972.

In a previous paper [Ann. of Math. (2) **64** (1956), 399–405; MR **18**, 498d] the author produced differentiable manifolds which are homeomorphic but not diffeomorphic to the 7-sphere S^7. Use was made of the fact that S^3 is parallelizable. Thus it was not clear a priori that similar examples could be produced in dimensions $\neq 7,15$.

In this paper a general method of constructing differentiable manifolds which are topological spheres is exhibited. The method consists in matching $D^{m+1} \times S^n$ with $S^m \times D^{n+1}$ by a suitable diffeomorphism $f: S^m \times S^n \to S^m \times S^n$ of the boundaries. (D^q is the q-disk in euclidean q-space.) An important special case of such diffeomorphisms f is obtained from differentiable maps $f_1: S^m \to \mathrm{SO}_{n+1}$, $f_2: S^n \to \mathrm{SO}_{m+1}$ of spheres into rotation groups by setting $y' = f_1(x) \cdot y$ and $x = f_2(y') \cdot x'$, where $(x', y') = f(x, y)$. The resulting differentiable $(m+n+1)$-manifold is denoted by $M(f_1, f_2)$, and a simple sufficient condition on f_1, f_2 is given for $M(f_1, f_2)$ to be homeomorphic to S^{m+n+1}.

An invariant λ (of the J-equivalence class) of $M(f_1, f_2)$ is constructed and turns out to be of interest for dimensions of the form $m = 4r - 1$, $n = 4(k-r) - 1$. (Hence, $m + n + 1 = 4k - 1$.) The value $\lambda(M(f_1, f_2))$ is computed explicitly in terms of the Pontryagin classes of the $\mathrm{SO}_{4(k-r)}$-, resp. SO_{4r}-, bundles over S^{4r}, resp. $S^{4(k-r)}$, defined by f_1, resp. f_2, as characteristic maps. These Pontryagin classes in turn are known by a theorem of R. Bott [Bull. Amer. Math. Soc. **64** (1958), 87–89; MR **21** #1590]. If $M(f_1, f_2)$ is diffeomorphic to S^{4k-1}, then $\lambda(M(f_1, f_2)) = 0$. Thus every $(4k-1)$-dimensional $M(f_1, f_2)$ homeomorphic to S^{4k-1} with a λ different from zero yields a non-standard differentiable structure on S^{4k-1}.

The problem of evaluating the possible values of λ (for a given k) is shown to amount to solving the following number-theoretic problem. Let $s_k = 2^{2k}(2^{2k-1} - 1)B_k/(2k)!$, where B_k is the kth Bernoulli number. Does there exist an integer r satisfying

$$k/3 < r \le k/2,$$

such that the greatest odd factor $d_{k,r}$ in the denominator of

$$(2r-1)!(2(k-r)-1)!s_r s_{k-r}/s_k$$

is > 1? If such an r exists, the number $d_{k,r}$ is shown by the author to be a lower bound for the number of distinct differentiable structures on S^{4k-1}. The author has checked that the answer to the above question is affirmative for $k = 2$ and $4 \le k \le 14$. For instance, $d_{5,2} = 73$, showing that S^{19} has at least 73 distinct differentiable structures. Similarly, S^{31} is shown to possess at least 16,931,177 distinct differentiable structures.

As usual the author's exposition is excellent.

{Reviewer's note: The above number-theoretic formulation is obtained by supplementing Lemma 5 of the paper under review with the statement on lines 9 and 8 from the bottom of page 348 of the paper by A. Borel and F. Hirzebruch [Amer. J. Math. **81** (1959), 315–382; MR **22** #988].} *M. A. Kervaire* (New York)

24 #A3662 57.34

Adams, J. F.

Vector fields on spheres.

Bull. Amer. Math. Soc. **68** (1962), 39–41.

This paper solves the celebrated vector-field problem for spheres. Precisely, the problem is to determine the maximum integer $k(n)$ for which there exist k continuous vector fields on S^{n-1} (the unit sphere in R^n), linearly independent everywhere. One method of constructing such vector fields is to use linear functions. The maximum number $l(n)$ which can be constructed in this way was determined many years ago by Hurwitz, and is a question of linear algebra. The result of the author may be put as follows: $k(n) = l(n)$. This result includes, as a special case, the determination of those spheres which are parallelizable.

Stated in the language of fibre bundles, the problem is to determine when the fibration

$$O_{n,k} = O(n)/O(n-k) \to O(n)/O(n-1) = S^{n-1}$$

admits a cross-section ($O(n)$ denoting the orthogonal group). Now the real projective space P_{n-1} sits naturally in $O(n)$ as the set of all reflections, and hence the "stunted projective space" $P_{n,k} = P_{n-1}/P_{n-k-1}$ (obtained by pinching P_{n-k-1} to a point) sits inside $O_{n,k}$. Thus we get a diagram

$$\begin{array}{ccc} P_{n,k} & \xrightarrow{i} & O_{n,k} \\ & \searrow f \quad \swarrow g & \\ & S^{n-1} & \end{array}$$

This diagram was introduced by N. E. Steenrod and J. H. C. Whitehead in their early work on this problem [Proc. Nat. Acad. Sci. U.S.A. **37** (1951), 58–63; MR **12**, 847c], and it has remained the starting point of all subsequent attacks on the vector-field problem. Steenrod and Whitehead took the mod 2 cohomology of the diagram and showed that (i) i^* was an isomorphism in a suitable range of dimensions and (ii) for certain values of (n,k), the homomorphism f^* had no left inverse compatible with the Steenrod operations Sq^i. From this it follows at once that g^* has no left inverse and hence the map g has no right inverse (for the appropriate values of (k,n)). The point of the method is that the space $O_{n,k}$, which is rather complicated, is replaced by the much simpler space $P_{n,k}$.

Further work was done on the problem by James [Proc. London Math. Soc. (3) **8** (1958), 507–535; MR **20** #7267; ibid. (3) **8** (1958), 536–547; MR **20** #7268; ibid. (3) **9** (1959), 115–140; MR **21** #1596]. In particular, James improved on (i) above by showing that $P_{n,k}$ was an S-retract of $O_{n,k}$ (i.e., a retract after suitable suspension). Thus, if g has a right inverse, f has a right S-inverse and the problem is therefore reduced to a study of the S-type of $P_{n,k}$. The problem can now be dualized by the Spanier-Whitehead duality. It was observed by the reviewer [ibid. (3) **11** (1961), 291–310; MR **24** #A1727] that $P_{n,k}$ is a Thom complex over a manifold and that its S-dual is again a stunted projective space.

It is at this stage that the author takes up the problem, and what he in fact proves is that the map

$$h: P_m/P_{m-1} \to P_{m+l(m)+1}/P_{m-1}$$

has no left inverse. Since h is the S-dual of an appropriate f, the problem is then solved. The proof is analogous to the Steenrod-Whitehead method, but instead of mod 2 cohomology, the author uses the Grothendieck ring K_R of real vector bundles. This is the "extraordinary cohomology theory" introduced by the reviewer and Hirzebruch [Proc. Sympos. Pure Math., Vol. III, pp. 7–38, Amer. Math. Soc., Providence, R.I., 1961; MR **25** #2617]. If h had a left inverse then the induced map h^* in the K_R groups would have a right inverse, and this would be compatible with any natural operations in the theory. Now using representations of the orthogonal group, one can introduce various operations in K_R. This was done for the "exterior powers" λ^i by Grothendieck. The λ^i, however, have complicated formal properties, and the author introduces new operations ψ^i which are certain combinations of λ^i and have much simpler properties. The relation between the λ^i and ψ^i is the same as that between elementary symmetric functions and power sums (or between Chern classes and the Chern character).

The author calculates the various K_R groups involved and the way the ψ^i operate. These calculations show that h^* has no inverse compatible with the ψ^i.

The much easier case of the complex $P_{n,k}$ (related to the fibration $U(n)/U(n-k) \to S^{2n-1}$) was treated by rather similar methods by the reviewer and Todd [Proc. Cambridge Philos. Soc. **56** (1960), 342–353; MR **24** #A2392].

M. F. Atiyah (Oxford)

36 #7157 57.36 28.00

Anosov, D. V.

Geodesic flows on closed Riemannian manifolds of negative curvature. (Russian)

Trudy Mat. Inst. Steklov. **90** (1967), 209 pp.

The author gives a detailed exposition and complete proofs of his results concerning U-systems announced previously [Dokl. Akad. Nauk SSSR **145** (1962), 707–709; MR **26** #716; ibid. **151** (1963), 1250–1252; MR **29** #561]. These yield the structural stability (roughness, in the sense of Andronov and Pontrjagin) and ergodicity of the flows named in the title. A dynamical system (a flow for continuous time, a cascade for discrete time) of class C^1 on a compact connected smooth manifold W is called a U-system ($= C$-system) if at each point w the tangent space splits as a direct sum $X_w{}^k + Y_w{}^l + Z_w{}^1$ of invariant subspaces, where $X_w{}^k$ and $Y_w{}^l$ are nonzero subspaces on which the induced system is exponentially contracting and expanding, and $Z_w{}^1$ (omitted for a cascade) is a one-dimensional subspace tangent to the flow. This splitting depends continuously upon w, but in general not smoothly even for an analytic cascade. The argument of V. I. Arnol′d and Ja. G. Sinaĭ [ibid. **144** (1962), 695–698; MR **26** #6535] for a particular example shows that the property of being a U-system is invariant under C^1-small perturbations.

That the geodesic flow on a closed Riemannian manifold of negative curvature is a U-flow was proved essentially by Hadamard and É. Cartan. Smale's conjecture that such a flow is structurally stable follows from Therem 1: Any U-system is structurally stable. The proof for U-flows involves constructing local cross-sections and an analog of the Hadamard-Perron theory of invariant manifolds of perturbed systems. The discrete case is reduced to the continuous one by the standard construction of the suspension of a cascade, namely, a flow on a manifold of one dimension higher admitting a cross-section on which the flow induces a cascade conjugate to the given one; the suspension of a U-cascade is a U-flow. {A direct proof of Theorem 1 for cascades was appended to S. Smale's article in Bull. Amer. Math. Soc. **73** (1967), 747–817; MR **37** #3598, see p. G76 in this volume.} The same methods give Theorem 3: The periodic orbits of a U-system having an invariant (Lebesgue) measure form a dense set.

G. A. Hedlund [ibid. **45** (1939), 241–260] and E. Hopf [Ber. Verh. Sächs. Akad. Wiss. Leipzig **91** (1939), 261–304; MR **1**, 243; Russian translation, Uspehi Mat. Nauk **4** (1949), 129–170; MR **10**, 718; Math. Ann. **117** (1940), 590–608; MR **2**, 106] proved the ergodicity of geodesic flows for surfaces of negative curvature and manifolds of constant negative curvature. The general case follows from Theorem 4: If a U-system of class C^2 has an invariant measure, then it is ergodic. For his ergodic analysis, the author constructs foliations $\mathfrak{S}^k, \mathfrak{S}^l$ of W tangent to the fields X^k, Y^l, and for a U-flow foliations $\mathfrak{S}^{k+1}, \mathfrak{S}^{l+1}$ tangent to X^k+Z^1, Y^l+Z^1. These foliations are invariant (sheets are mapped to sheets), and integral manifolds of the tangent fields generating them are open subsets of single sheets. Individual sheets of $\mathfrak{S}^k, \mathfrak{S}^l$ for a U-cascade, and of $\mathfrak{S}^{k+1}, \mathfrak{S}^{l+1}$ for a U-flow, are dense in W if an invariant measure exists; in general, some finite union of sheets is dense. Theorem 2: A U-system has countably many periodic orbits. The crucial technical fact permitting the extension of Hedlund and Hopf's results is Theorem 10: For a U-system of class C^2, the foliations $\mathfrak{S}^k$ and $\mathfrak{S}^l$ are absolutely continuous (in the sense that local maps from one sheet to another have well behaved Jacobians). This says that a certain coordinate transformation sends null sets to null sets, so that Hopf's method can be applied.

From this point, only systems of class C^2 and having an invariant measure are considered. Theorem 11: For a U-cascade, $\mathfrak{S}^k$ and $\mathfrak{S}^l$ are metrically transitive, that is, each sheet or its complement has measure zero. Theorem 14: For a U-flow, either (i) the flow is the suspension of a U-cascade on a submanifold of W of codimension 1, or else (ii) $\mathfrak{S}^k$ and $\mathfrak{S}^l$ are metrically transitive (and then each sheet is dense in W). Alternative (i) holds precisely when the flow fails to have continuous spectrum, and then any nonconstant eigenfunction is continuous. From Ja. G. Sinaĭ's theory of dynamical systems with transversal foliations [Izv. Akad. Nauk SSSR Ser. Mat. **30** (1966), 15–68; MR **33** #5847] it then follows that an arbitrary U-cascade and a U-flow with continuous spectrum is a K-system, hence has countable Lebesgue spectrum, is mixing, and has positive entropy. By a theorem of V. I. Arnol′d [Sibirsk. Mat. Ž. **2** (1961), 807–813; MR **31** #1432], reproved here, the geodesic flows of the title have continuous spectra, hence they are K-systems, a result obtained earlier for particular cases by Sinaĭ.

The author gives an example of an ergodic automorphism of the 2-torus having small analytic perturbations leaving the euclidean measure invariant but altering the entropy.

The required facts from ergodic theory, including Sinaĭ's theory, are summarized in an appendix. *M. Eisenberg* (Amherst, Mass.)

37 #3598 57.47 34.00

Smale, S.

Differentiable dynamical systems.

With an appendix, "Anosov diffeomorphisms", by John Mather.

Bull. Amer. Math. Soc. **73** (1967), 747–817.

In the author's Mexican paper of 1960 [Bol. Soc. Mat. Mexicana (2) **5** (1960), 195–198; MR **25** #5252] he formulated for the first time what should be the fundamental problem of the qualitative theory of ordinary differential equations (= flows) on a compact differentiable manifold $M=M^n$, $n \geq 1$. As revised in the light of further developments, this can be stated now as follows. One starts by considering the space $\mathscr{X}$ of all flows with the C^r-topology, $r \geq 1$. The fundamental problem then consists of (1) defining an equivalence relation E between flows which as much as possible preserves the orbit structure, (2) finding a Baire set $\Sigma \subset \mathscr{X}$, and (3) classifying the equivalence classes of Σ modulo E by means of numerical and algebraic invariants. The present paper contains as main original contribution an attack to this problem in dimension $n>2$ and among other things gives a survey of the extensive literature that sprung from this approach to differential equations. Previous work of the reviewer [Ann. of Math. (2) **69** (1959), 199–222; MR **21** #753] on S^2, later extended to M^2, showed that for $n = 2$ the above program was very successful taking for E a very stringent equivalence relation, namely, a trajectory-preserving homeomorphism of the manifold onto itself. In this case the set Σ is open and dense and coincides with the set of all E (structurally)-stable flows, i.e., Σ is the set of all flows X such that X has a neighborhood in $\mathscr{X}$ such that all flows on it are E-equivalent to X. The concept of structural stability was introduced in 1937 by Andronov Pontryagin, its importance unsuspected for two decades.

For $n > 2$ the original contributions of this paper—to be explained in detail later on—concerning the fundamental problem are: (a) the choice of a certain E—the Ω-equivalence—to replace the old E; (b) the formulation of the conjecture that the set of all flows satisfying Axiom A is a Baire set in $\mathscr{X}$; (c) the spectral decomposition theorem; (d) the Ω-stability theorem. Items (c) and (d) are in the line of item (3) of the fundamental problem.

In what follows we give a brief mention of the four parts into which this paper is divided, but otherwise our aim is only to explain what (a)–(d) are.

Definition: A differentiable dynamical system on M is a differentiable action of a Lie group G on M, i.e., a homeomorphism $G \to \mathrm{Diff}(M)$ such that the induced map $G \times M \to M$ is differentiable. Here $\mathrm{Diff}(M)$ stands for the group of all diffeomorphisms of M onto itself with the topology of uniform C^r-convergence. Flows correspond to the case $G = R$. When $G = Z$ then the consideration of the suspended flow of a diffeomorphism, defined on a manifold N of dimension $\dim M + 1$, shows that this case is essentially a special situation of the case $G = R$ on N. This case $G = Z$ is the one that is studied at Part I of the paper and covers seventy percent of it. In Part II the results of Part I are translated for flows, this being essentially what was abstracted above. Part III of the paper deals with flows of special type like Hamiltonian flows, a brief mention being made to the nowadays classical work of Kolmogoroff, Arnold and Moser. Part IV covers only seven percent of the paper and considers the case when G is an abelian Lie group more complicated than Z or R. A relevant feature is then a result of R. S. Palais [Bull. Amer. Math. Soc. **67** (1961), 362–364; MR **24** #A185] showing a strong form of structural stability for all actions with G compact.

We now proceed to explain what (a)–(d) are and we do this in the context of Part I, i.e., when $G = Z$, so that we deal with diffeomorphisms rather than with flows. We need several definitions. First of all remark that instead of the space $\mathscr{X}$ of all flows we have the set $\mathrm{Diff}(M)$ of all diffeomorphisms of M onto itself with the C^r-topology. The orbit of $x \in M$ relative to $f \in \mathrm{Diff}(M)$ is the subset $\{f^m(x) | m \in Z\}$. The finite orbits are called periodic orbits and their points periodic points. A point $x \in M$ is called non-wandering when given any neighborhood U of x there are infinitely many values $m \in Z$ such that $f^m(U) \cap U \neq \varnothing$. The set of all non-wandering points $\Omega = \Omega(f)$ is easily seen to constitute a closed invariant subset of M. By invariant we mean that $f(\Omega) = \Omega$. We now can define the Ω-equivalence mentioned in (a). Definition: Two diffeomorphisms f, $f \in \mathrm{Diff}(M)$ are said to be Ω-equivalent or Ω-conjugate if there is a homeomorphism $h: \Omega(f) \to \Omega(f')$ such that, on Ω, $hf = f'h$. We proceed now to introduce the crucial concept of hyperbolic structure. Definition: If $\Lambda \subset M$ is closed and invariant and M has a Riemannian structure, then f is said to be hyperbolic on Λ if $Df: T_\Lambda(M) \to T_\Lambda(M)$ splits continuously, i.e., $T_\Lambda(M) = E^s + E^u$ with E^s, E^u subbundles of locally constant dimension, invariant under Df, and in addition, Df contracts E^s and expands E^u. This means that there are constants $0 < c$, $0 < \lambda < 1$ for which $\|Df^m(x)(v)\| < c\lambda^m\|v\|$, $v \in E^s$, $m \in Z$ and $\|Df^{-m}(x)(v)\| < c\lambda^m\|v\|$, $v \in E^u$, $m \in Z$, where the norm comes from the Riemannian structure. When $\Lambda = M$, then f is said to be an Anosov diffeomorphism. Axiom A: A diffeomorphism $f \in \mathrm{Diff}(M)$ is said to satisfy Axiom A if $\Omega(f)$ has a hyperbolic structure and the periodic points of f are dense in Ω.

Definition: An indecomposable set for $f \in \mathrm{Diff}(M)$ is a closed invariant set which is not the union of two other closed and invariant sets of f.

Item (c) is explained now by the statement of the spectral decomposition theorem: If $f \in \mathrm{Diff}(M)$ satisfies Axiom A, then there is a unique way of writing $\Omega = \Omega(f)$ as the finite union of disjoint, closed, invariant, indecomposable subsets ("pieces") $\Omega = \Omega_1 \cup \cdots \cup \Omega_K$ in such a way that on each piece f is transitive, i.e., has a dense orbit. Furthermore, M can be written as a finite disjoint union of invariant subsets $M = \bigcup_{i=1}^K W^s(\Omega_i)$, where $W^s(\Omega_i) = \{x \in M \mid f^m(x) \to \Omega, m \to \infty\}$.

The paper gives a good sketch of the proof of this important theorem modulo the generalized stable manifold theorem which is not proved. This is a generalization of the usual stable manifold theorem relative to a hyperbolic periodic point p to the case where p is replaced by Ω. The situation concerning the spectral decomposition theorem is quite analogous to the decomposition of a finite-dimensional vector space into the eigenspaces of a linear map. To state the very important Ω stability theorem mentioned in (d), we need two more definitions. Axiom B: We say that $f \in \mathrm{Diff}(M)$ satisfies Axiom B if it satisfies Axiom A and $W^s(\Omega_i) \cap W^u(\Omega_j) \neq \varnothing$ implies that there exist periodic points $p \in \Omega_i$, $q \in \Omega_j$ such that $W^s(p)$ and $W^u(q)$ have a point of transversal intersection, $\Omega^s(\Omega_i)$ is as above and $W^u(\Omega_j) = \{x \in M \mid f^m(x) \to \Omega_j, m \to -\infty\}$. Definition: $f \in \mathrm{Diff}(M)$ is said to be Ω-stable if there is a neighborhood $N(f)$ of f such that whenever $f' \in N(f)$, then f and f' are Ω-conjugate.

We now state the Ω-stability theorem. The set of $f \in \mathrm{Diff}(M)$ which satisfy Axioms A and B is open and such f's are Ω-stable. The paper gives only a very sketchy indication of the proof of the theorem. This theorem is related to the theorem of Anosov to the effect that every Anosov diffeomorphism is structurally stable.

After the present paper was published, Abraham and the author showed by examples that conjecture (b) is false, and this has the implication that Ω-stability is not a generic property, i.e., a property that is possessed by a Baire set of diffeomorphisms in $\mathrm{Diff}(M)$. As the author himself puts it, this means that the basic framework of the present paper has to be modified and one has to find a weaker equivalence to put in the place of Ω-equivalence and this in a way consistent with the new breed of examples. *M. M. Peixoto* (Rio de Janeiro)

38 #5243 57.50

Atiyah, M. F.; Singer, I. M.

The index of elliptic operators. I.

Ann. of Math. (2) **87** (1968), 484–530.

The earlier proof of the Atiyah-Singer index theorem as given in the book by R. Palais [*Seminar on the Atiyah-Singer index theorem*, Ann. of Math. Studies, No.

57, Princeton Univ. Press, Princeton, N.J., 1965; MR **33** #6649] used cobordism theory and was in this respect modelled on the proof of the Riemann-Roch theorem due to the reviewer [*Neue topologische Methoden in der algebraischen Geometrie*, Ergeb. Math. Grenzgeb., Heft 9, Springer, Berlin, 1956; MR **18**, 509b, see p. G26 in this volume; second edition, 1962; MR **25** #1155; third edition in English, Springer, New York, 1966; MR **34** #2573]. This proof did not lend itself to certain generalizations (for example, in the equivariant case) because the corresponding cobordism theories are not known. The index theorem of the paper under review includes the equivariant case: Suppose a compact Lie group G operates differentiably on a compact differentiable manifold X such that the action is compatible with a linear elliptic problem on X. Then the index of this elliptic problem is an element of the representation ring $R(G)$. In the simplest case the elliptic problem is given by an elliptic linear differential operator $D: C^\infty(E) \to C^\infty(F)$, where E, F are complex vector bundles over X. The group G operates on the kernel and on the cokernel of D. Thus we have two finite-dimensional representations of G. The difference of the two representations as elements of $R(G)$ is the index of the elliptic problem. The symbol of such an elliptic problem is an element of $K(BX, SX)$ if one forgets about the G-action. (For the notation, see the review of the book of Palais [loc. cit.].) If a G-action is given, then equivariant K-theory must be used and the symbol is an element of $K_G(BX, SX)$. The authors define K and K_G for a locally compact space as the corresponding reduced groups of the one-point-compactification of the space. With this understanding we have $K_G(BX, SX) = K_G(TX)$, where TX is the total space of the covariant tangent bundle. Using "enough operators" (pseudo-differential operators) and homotopy properties which ensure that the index depends only on the symbol, the authors define the analytical index a-ind: $K_G(TX) \to R(G)$. This is a homomorphism of $R(G)$-modules and has by its very definition the property that the index of an elliptic problem equals a-ind of its symbol. The definition of an $R(G)$-homomorphism t-ind: $K_G(TX) \to R(G)$ in topological terms is given in § 3. The "index theorem" is the main theorem 6.7 of the paper and asserts that a-ind and the topological index t-ind coincide. For the definition of t-ind and for the investigation of properties of a-ind and t-ind, the following construction is fundamental. Suppose X, Y are G-manifolds with $X \subset Y$ and X compact. Then $TX \subset TY$ and the normal bundle W of $TX \subset TY$ equals $N \oplus N$ lifted to TX, where N is the normal bundle of X in Y. Since $N \oplus N$ carries a complex structure, so also does W. The de Rham complex of the exterior powers of W can be tensored with any complex of vector bundles over TX lifted to W which has compact support in TX. The result is a complex of vector bundles over W with compact support, because the de Rham complex is canonically trivialized outside a zero-section of W. In this way we get a homomorphism $K_G(TX) \to K_G(W)$. Since W may be regarded as a tubular neighborhood of TX in TY and since the one-point-compactification of TY maps onto the one-point-compactification of W, we have a homomorphism $K_G(W) \to K_G(TY)$. The composition is an $R(G)$-homomorphism $i_!: K_G(TX) \to K_G(TY)$, where $i: X \to Y$ denotes the embedding. (Of course, many details are omitted in this review. Constructions using the alternating sum of the exterior powers occur in several earlier papers of Atiyah and others.)

To define t-ind for a compact differentiable G-manifold X we embed X in a real representation space E of G. This is always possible [R. Palais, J. Math. Mech. **6** (1957), 673–678; MR **19**, 1181e]. Let i be the embedding. We have $i_!: K_G(TX) \to$

$K_G(TE)$. Under the embedding j of the origin P in E we have $j_!: K_G(TP) = K_G(P) = R(G) \to K_G(TE)$. $j_!$ is an isomorphism. This is a special case of the equivariant form of the Bott periodicity theorem [the first author, *K-theory*, Benjamin, New York, 1967; MR **36** #7130; Russian translation, Izdat. "Mir", Moscow, 1967; MR **36** #7131; Quart. J. Math. Oxford Ser. (2) **19** (1968), 113–140; MR **37** #3584].

t-ind is defined by $j_! \circ$ (t-ind) $= i_!$. The authors show that the definition is independent of the choice of the embedding. t-ind is the identity of $R(G)$ if (A1) X is a point. The diagram

(A2)
$$\begin{array}{ccc} K^G(TX) & \xrightarrow{i_!} & K_G(TY) \\ \text{t-ind} \searrow & & \swarrow \text{t-ind} \\ & R(G) & \end{array}$$

commutes for any inclusion $i: X \to Y$ with X, Y compact G-manifolds. An index function ind is given if we have for every compact differentiable manifold X an $R(G)$-homomorphism $K_G(TX) \overset{\text{ind}}{\to} R(G)$. If such an index function ind satisfies (A1) and (A2), then ind = t-ind (Proposition 4.1). For the analytical index, (A1) is trivial. To prove the main theorem, axiom (A2) has to be checked for the analytical index. This is not easy because for an operator D on X with symbol $\gamma(D)$ we have to construct an operator on Y with symbol $i_!\gamma(D)$ and show that the index of this new operator equals the index of D. This construction is the essential analytical part of the proof of the main theorem. "Once this has been done, we can take Y to be a sphere, and the general index theorem is reduced to the case of operators on the sphere. For these the problem is easily solved." At this point one recognizes that the whole proof has "in spirit, at least" much in common with Grothendieck's proof of the Riemann-Roch theorem. Since it is difficult to verify (A2) directly, it is shown that certain axioms, (B1), (B2″) and (B3), imply (A2) for any index function. In § 8 the axioms (B1) and (B2″) are proved for the analytical index; in § 9 the axiom (B3) is proved. A special case of (B3) is the behaviour of the index function if one takes the Cartesian product of two G-manifolds with elliptic problems. The analytical index behaves multiplicatively in this case. (B3) generalizes this multiplicative property to differentiable fibre bundles. In this case the bundle of "indices along the fibres" may not be trivial over the base and enters essentially in the formulation (B1) is an excision axiom: Let U be a (non-compact) G-manifold and $j: U \to X$, $j': U \to X'$ two open G-embeddings into compact G-manifolds X, X'. Then the following diagram commutes.

$$\begin{array}{ccc} & K_G(TX) & \\ j^* \nearrow & & \searrow \text{ind} \\ K^G(TU) & & R(G) \\ j'^* \searrow & & \nearrow \text{ind} \\ & K_G(TX') & \end{array}$$

Observe that the one-point-compactifications of TX and TX' map onto the one-point compactification of TU. j^* and j'^* are induced by these maps.

The excision property of the analytical index was observed by R. T. Seeley [Trans. Amer. Math. Soc. **117** (1965), 167–204; MR **30** #3387]. The axiom (B2″) is a normalisation axiom for certain operators on S^1 and S^2. Information on operators on other spheres follows by using excision and the multiplicative property. The idea of the proof for the essential property (A2) of the analytical index is sketched by the authors as follows (§ 1): let $i: X \to Y$ be an inclusion of compact manifolds (we forget the G-action). Let U be a tubular neighborhood of X and Z its double. Then the excision property (B1) shows that ind $i_! A = \operatorname{ind} k_! A$, where $A \in K(TX)$ and $k: X \to Z$ is the inclusion in the double. Z is fibred over X by spheres. The multiplicative property and the information on spheres gives the desired result ind $i_! A = \operatorname{ind} A$.

This paper contains an impressive amount of analysis. The theory of pseudo-differential operators (Hörmander, Kohn-Nirenberg, Seeley) is essential to have "enough operators" to realize all elements of $K_G(TX)$ as symbols and to carry through all constructions.

The analytical index was "calculated" in this paper by topological terms (topological index defined by K-theory). In the following Parts II [M. F. Atiyah and G. B. Segal, Ann. of Math. (2) **87** (1968), 531–545; MR **38** #5244] and III [M. F. Atiyah and I. M. Singer, Ann. of Math. (2) **87** (1968), 546–604; MR **38** #5245] this topological index will be interpreted in two steps. In Part II the topological index is expressed in terms of fixed-point sets of G. This is done in K-theory. It leads to a general "Lefschetz fixed-point theorem" where the fixed point set of an element of G is a disjoint union of submanifolds. For isolated fixed points this is contained in the fixed point theorem of the first author and R. Bott [Ann. of Math. (2) **86** (1967), 374–407; MR **35** #3701], a theorem on differentiable maps g with isolated simple fixed points, where g need not be invertible. In Part III the result is reformulated in cohomological terms. If G is the identity, this gives the index theorem in its well-known cohomological form [R. Palais, loc. cit.]. In general, it gives the cohomological form of the fixed-point theorem. *F. Hirzebruch* (Bonn)

15, 445b 60.0X
Doob, J. L.
Stochastic processes.
John Wiley & Sons, Inc., New York; Chapman & Hall, Limited, London, 1953.
viii+654 pp.

In this valuable book the author defines a stochastic process as "any process running along in time and controlled by probabilistic laws", or, more precisely, as "any family of random variables $\{x_t : t \in T\}$" where "a random variable is... simply a measurable function". He observes that "probability theory is simply a branch of measure theory, with its own special emphasis and field of application", and adheres uncompromisingly to this point of view throughout the book. It follows that one really is working all the time with genuinely probabilistic arguments, instead of performing analytical tricks with distribution functions, but a thorough familiarity with measure theory has to be assumed. A Supplement of 23 pages summarises what is needed, but it is rather an examination which must be passed by a candidate wishing to be admitted to Chapter I, than a course in itself. Two of the topics discussed in the Supplement deserve special mention.

(1) Construction of measures on product spaces. Doob's treatment of this is based on a paper by C. T. Ionescu Tulcea [Atti Accad. Naz. Lincei. Rend. Cl. Sci. Fis. Mat. Nat. (8) **7**, 208–211 (1950); these Rev. **12**, 85e] which constructs a measure on the cartesian product of countably many replicates of an abstract space X in terms of functions q_n on $X^n \times \mathscr{F}_X$ to the real line ($\mathscr{F}_X$ is a Borel field of subsets of X) which determine generalised Baire functions on X^n and probability-measures on $(X, \mathscr{F}_X)$.

When $q_n(x_1, \cdots, x_n; A)$ is independent of $x_1, \cdots, x_{n-1}$ we have a "Markov-chain" measure; when it depends on none of the x's we have "direct product" measure. The theorem is used to derive (i) the existence of the direct product measure on X^T (where T is arbitrary) and (ii) Kolmogorov's theorem concerning the existence of measures on R^T (where R is the real line) (the last requires the author's theory of conditional distributions discussed below, and here the reader is left to fill in the gaps of a highly condensed argument). This construction of Markov-chain measure would run into difficulties when X is abstract and T is not the set of positive integers. If T is "all integers" then "backwards" transition-probabilities must be given at the outset, and if T is R then "two-sided" transition probabilities are required.

(2) Representation theory. (A thorough grasp of this is essential to an understanding of the author's point of view.) $\mathbf{T}$ is called a single-valued measure-preserving point transformation of one probability-space $(\Omega, \mathscr{F}, q)$ into another $(\widetilde{\Omega}, \widetilde{\mathscr{F}}, \widetilde{q})$ when $\mathbf{T}$ is a single valued map of Ω into $\widetilde{\Omega}$, $\mathbf{T}^{-1}$ maps $\widetilde{\mathscr{F}}$ onto $\mathscr{F}$ and $q(\mathbf{T}^{-1}\widetilde{\Lambda}) = \widetilde{q}(\widetilde{\Lambda})$ for all $\widetilde{\Lambda} \in \widetilde{\mathscr{F}}$. Such transformations are discussed in detail, and an important theorem is proved, the essence of which is that the corresponding mapping from random variables on $(\widetilde{\Omega}, \widetilde{\mathscr{F}})$ to random variables on $(\Omega, \mathscr{F})$ is "onto". In the constantly recurring application, $(\Omega, \mathscr{A}, q)$ is the basic probability-space, $\{x_t : t \in T\}$ is a stochastic process (S.P.) thereon, $\mathscr{F}$ is $\mathscr{B}(x_t : t \in T)$ (the smallest Borel field with respect to which all the x's are measurable), $\widetilde{\Omega}$ is the function-space $R^T, \widetilde{x}_t$ is the tth coordinate function thereon, $\widetilde{\mathscr{F}}$ is the usual Borel field of subsets of R^T, $(\mathbf{T}\omega)(t) \equiv x_t(\omega)$ and $\widetilde{q}$ is the probability-measure induced by $\mathbf{T}$. This provides the link between measure-theory and distribution-theory throughout the book. Everywhere in the book the basic probability space is supposed to be complete. A note on completeness in the $\widetilde{\Omega}$ space would have been helpful here.

Chapter I ("Introduction and probability background") follows naturally on the Supplement. Some topics (e.g., convergence concepts) are given little space, but conditional expectations, probabilities and distributions receive 22 pages of great importance. If $(\Omega, \mathscr{A}, \mathbf{P})$ is the basic probability space, then $\mathbf{P}(M|\mathscr{F})$, the conditional probability of $M \in \mathscr{A}$ with regard to the Borel field $\mathscr{F} \subseteq \mathscr{A}$, is defined as the class of random variables y satisfying: (i) y differs at most on an $\mathscr{A}$-set of $\mathbf{P}$-measure zero from an $\mathscr{F}$-measurable function, (ii) y is $\mathbf{P}$-integrable and (iii) $\int_A y d\mathbf{P} = \mathbf{P}(A \cap M)$ for all $A \in \mathscr{F}$. We have a CD($\mathscr{F}$) (conditional distribution with regard to $\mathscr{F}$) when there exists a real-valued function P on $\mathscr{A} \times \Omega$ such that (i) for each fixed $\omega \in \Omega$ it determines a probability-measure on $(\Omega, \mathscr{A})$, (ii) for each fixed $M \in \mathscr{A}$ it determines a function on Ω differing from an $\mathscr{F}$-measurable function at most on an $\mathscr{A}$-set of $\mathbf{P}$-measure zero and (iii) $P(M, \omega)$ is a random variable in the class $\mathbf{P}(M|\mathscr{F})$ for each $M \in \mathscr{A}$. Notes on the non-existence (in general) of CD($\mathscr{F}$) are contained in the Appendix. In the text the author proceeds at once to introduce two weaker concepts. Let y_i $(i = 1, \cdots, n)$ be random variables and let $\mathscr{B}(y_1, \cdots, y_n)$ be defined as above. A CD($\mathscr{F}$) "of the y's" differs from a CD($\mathscr{F}$) only in that M is now restricted to be an element of $\mathscr{B}(y_1, \cdots, y_n)$.

A CD($\mathscr{F}$) "of the y's in the wide sense" is the weakest of the three. It (p) is a function on $\mathscr{B}_n \times \Omega$ ($\mathscr{B}_n$ being the class of Borel sets in R^n) and has exactly similar properties except that in place of (iii) we require that $p(Y, \omega)$ is to be in the class

$$\mathbf{P}(\{[y_1(\omega'), \cdots, y_n(\omega')] \in Y\}|\mathscr{F})$$

for every $Y \in \mathscr{B}_n$. The author proves two important existence theorems. (1) A CD($\mathscr{F}$) "of the y's in the wide sense" always exists, and one can arrange that $p(Y, \cdot)$ is $\mathscr{F}$-measurable for all $Y \in \mathscr{B}_n$. (2) Under a condition $(*)$ a CD($\mathscr{F}$) "of the y's" always exists, and one can arrange that $P(M, \cdot)$ is $\mathscr{F}$-measurable for all $M \in \mathscr{B}(y_1, \cdots, y_n)$.

The condition $(*)$ in (2) is that $\{[y_1(\omega), \cdots, y_n(\omega)] : \omega \in \Omega\}$ is to be a Borel set in R^n. As the author remarks, this condition $(*)$ is a useful one (especially if we are working in the representation space, when it is trivially verified), but it is a shade mysterious. To avoid overburdening the text, remarks about extensions of the theorem (2) are relegated to the Appendix. The reviewer heartily wishes the author had been positively verbose on this topic. If, as is often the case, $\mathscr{F} \equiv \mathscr{B}(x_1, \cdots, x_m)$ where the x's are further random variables, then representation theory plus theorem (2) appears to give a stronger result than theorem (1) and leads to the natural counterpart to the Tulcea theorem mentioned above. In a future edition it would be nice to have this topic treated in extenso.

Chapter I finishes with an excellent survey of characteristic functions, and here there are important new inequalities suggested by earlier work by Wintner.

In Chapter II ("Definition of a S.P.-principal classes") the author gives a solution to the problem raised by the non-measurability of $\sup x_t$ when t ranges through a continuum. A S.P. is "separable" when there exists (i) a (dense) denumerable subset S of the parameter set T and (ii) an $\mathscr{A}$-set Z of **P**-measure zero such that, outside Z, and for every open interval I which meets T, the sets

$$\{x_t(\omega) : t \in I \cap T\} \quad \text{and} \quad \{x_t(\omega) : t \in I \cap S\}$$

have the same supremum and infimum. A stronger concept, "separability with regard to the closed sets", is also defined but is only rarely used later in the book.

If a S.P. $\{x_t : t \in T\}$ can be replaced by a separable one $\{\tilde{x}_t : t \in T\}$ on the same probability space and with the same parameter set, such that

$$\mathbf{P}(\{x_t(\omega) \neq \tilde{x}_t(\omega)\}) = 0, \quad t \in T,$$

and such that each $\tilde{x}_t$ is measurable with regard to the completion of $\mathscr{B}(x_t : t \in T)$ under **P**, then the original S.P. can be replaced by this separable "standard modification" for all practical purposes. That this is always possible is the content of Theorem 2.4 of Chapter II; as the author says, "this result is the best that could be hoped for". In constructing such a separable "standard modification" of a S.P. it will in general be necessary to allow the new random variables to assume infinite values, even if the original random variables were finite-valued.

If it is desired to integrate the sample functions of a S.P. then separability is not enough. Accordingly, Doob calls a S.P. "measurable" when (i) T is Lebesgue measurable and (ii) $x_t(\omega)$ is measurable in (t, ω) with regard to $\mathscr{L} \otimes \mathscr{A}$ (presumably after the latter has been completed under the product of Lebesgue measure and **P**-measure) (here $\mathscr{L}$ is the Borel field of Lebesgue measurable subsets of R). In the important Theorem 2.6 the author shows that under a condition $(**)$ there is a separable measurable "standard modification" of a S.P. $\{x_t : t \in T\}$ whenever T is Lebesgue measurable. (In general the new S.P. will not be finite-valued.) A sufficient condition $(**)$ is that x_t be continuous-in-probability at almost all $t \in T$. If $T \in \mathscr{L}$ and is of positive Lebesgue measure, it is shown by an example that some such condition as $(**)$ is essential; thus, unlike separability, measurability imposes a restriction on the finite-dimensional distributions (though one which will usually be satisfied in practice).

There follow some theorems on the integrals of sample functions, and a summary without proofs of the earlier solution to the above problems by Doob and Ambrose. (This second method, by extension of the basic probability space, is not used in the book.) The remainder of Chapter II is devoted to the definitions and simplest properties of the eight classes of S.P.'s discussed in the rest of the book. The reader will find his comprehension of the theory so far developed severely and excitingly tested when he works through Doob's proof that a Markov process when reversed in time is still a Markov process.

An adequate review of the remaining ten chapters (500 pages) would be impossible within present space- and time-limitations, and only an indication of their contents can be given. Chapter III ("Processes with mutually independent random variables") deals in 50 pages very thoroughly with the zero-one law, series of independent random variables and their convergence, the law of large numbers, infinitely divisible distributions (an improved treatment utilising Doob's inequalities for characteristic functions), the central limit theorem and a theorem on the futility of gambling systems. "Stable laws" and the law of the iterated logarithm are about the only topics not discussed in this field.

Chapter IV ("Processes with mutually uncorrelated or orthogonal random variables") is a probabilistic version of the theory of orthogonal functions. For the proof of the Riesz-Fischer theorem and an important theorem on power-series required later for the factoring of a spectral density the reader is referred to classical texts on orthogonal functions and trigonometrical series.

Chapter V ("Markov processes—discrete parameter") is concerned with stationary Markov processes for which the phase-space X (in which the random variables assume their values) is abstract and T is $\{n:n=1,2,3,\cdots\}$. Strictly speaking the author here departs from his definition of a Markov process in Chapter II (in which X was the real line) but there is never any doubt about the existence of the conditional probabilities, etc., for these are supposed given a priori and a probability-measure is built up from them as in Tulcea's theorem. Specifically it is supposed that we are given a function p on $X\times\mathscr{F}_X$ ($\mathscr{F}_X$ being a Borel field of subsets of X) which determines for each $\xi\in X$ a probability-measure on $(X,\mathscr{F}_X)$ and for each $A\in\mathscr{F}_X$ a function measurable with regard to $\mathscr{F}_X$. $p^n(\xi,A)$ is the corresponding n-step transition probability. The author commences with the classical case in which $X=\{1,2,\cdots,N\}$. After some interesting remarks on card-shuffling he then proceeds straight to the general case, giving a treatment based on but more general than that of W. Doeblin [Bull. Math. Soc. Roumaine Sci. **39**, no. 1, 57–115, no. 2, 3–61 (1937)], and depending on the following "Hypothesis D": There is a non-null finite-valued measure φ on $(X,\mathscr{F}_X)$, an integer $\nu\geq 1$ and a positive ε such that $p^\nu(\xi,A)\leq 1-\varepsilon$ if $\varphi(A)\leq\varepsilon$.

This is a limitation even in the important special case when X is countable, and as no separate discussion is given of the latter, the beginner (who will, however, not read so far as this) might conclude that Hypothesis D is necessary there. But this hardly matters, as an excellent discussion of the countable case is already available in another book [W. Feller, Probability theory and its applications, v. I, Wiley, New York, 1950; these Rev. **12**, 424a]. Doeblin's treatment of the general case is further amplified by an analysis of the possible classes of D-triples $(\varphi,\nu,\varepsilon)$ associated with such a process, and the chapter concludes with very general forms of the law of large numbers and the central limit theorem for Markov processes.

In Chapter VI ("Markov processes—continuous parameter") T is the half-line $\{t : t \geq 0\}$, the processes are (to begin with) of the "jumping" type, and once again the chapter begins with a careful account of the properties of such systems when X is finite, and the emphasis is again on the stationary case. After dealing with a finite phase-space the author then allows X to be any Borel set in k-dimensional Euclidean space. Hypothesis D again appears, but only at one point not affecting the main argument. The author claims that the results of this section are valid "with moderate changes even if X is an abstract space (with no topology defined on it)" and the restriction to a Euclidean X is presumably to make available probabilistic methods. (As noted above, Tulcea's theorem will not suffice for the construction of a probability measure if T is a half-line and X is abstract and if only "forwards" transition-functions are given. When X is Euclidean, the required measure can be constructed by an application of Kolmogorov's theorem.) The author discusses the character of the discontinuities of the sample functions and extends his (1945) method of transforming a "dishonest" into an "honest" Markov process. Some topics receiving much attention at the moment (semi-group methods, Lévy's analysis of the possible pathologies, and Feller's necessary and sufficient condition for the occurrence of an infinity of transitions in a finite time) are mentioned only in the Appendix. Just after the book went to press, the author's theorems on the limiting behaviour of the transition-functions near $t = 0$ were sharpened by A. N. Kolmogorov [Moskov. Gos. Univ. Učenye Zapiski **148**, Matematika **4**, 53–59 (1951); these Rev. **14**, 295b]. The chapter concludes with an extension of important work by K. Itô [e.g., Mem. Amer. Math. Soc. no. **4** (1951); these Rev. **12**, 724a] on Markov processes of the diffusion type and associated stochastic differential equations.

The hundred pages on "Martingales" (Chapter VII) form the core of the book. A S.P. $\{x_t : t \in T\}$ of integrable real-valued random variables is a (semi-)martingale when $\mathbf{E}(x_{t_{n+1}} | x_{t_1}, \cdots, x_{t_n}) = (\geq) x_{t_n}$ (with probability one) whenever $t_1 < t_2 < \cdots < t_{n+1}$. When the parameter-set T is $\{m : m = 1, 2, \cdots\}$, a (semi-)martingale can be viewed as a rationalisation of one's intuitive notion of (an advantageous) a fair game (x_m being the gambler's fortune after the mth play). This latter notion motivates the analysis throughout the chapter and leads to theorems of astonishing scope. Thus, (semi-)martingales should preserve their nature under transformations suggestively called "optional stopping" and "optional sampling", and this is proved (under suitable conditions). There follow fundamental inequalities for martingales and semi-martingales and some very general consequences in the form of convergence theorems for both types of S.P. The Appendix contains a detailed account of the links between these results and earlier work (particularly that of E. S. Andersen, B. Jessen, P. Lévy and J. Ville). There are applications to the theory of sums of independent random variables, the law of large numbers, integration theory, the theory of derivatives, the likelihood ratio in statistics, sequential analysis and (after a long discussion of continuous-parameter martingales, in which separability naturally plays a role) to sample-function continuity.

Chapter VIII ("Processes with independent increments") is the continuous-parameter equivalent of chapter III and contains an account of the Brownian-motion process and the Poisson process (with an interesting invariance theorem for the latter) before proceeding to detailed proofs of theorems due to Lévy concerning the "centering" of the general process with independent increments, the character of the distribution functions and the continuity of the sample functions.

Chapter IX ("Processes with orthogonal increments") is what the author would call the "wide sense" version of Chapter VIII. Throughout this and the remaining chapters of the book a consistent and elegant use is made of the fact that the random variables of finite (absolute) second moment on an arbitrary fixed basic probability space can be grouped into equivalence classes which constitute the elements of a not necessarily separable Hilbert space. Stochastic integrals of the Wiener type ($\int \Phi(t)dy_t$) with regard to a S.P. $\{y_t : t \in T\}$ of orthogonal increments are defined in the spirit of this idea, and a generalisation is then given of an extension by K. Itô [Proc. Imp. Acad. Tokyo **20**, 519–524 (1944); these Rev. **7**, 313c] to integrals of the type $\int \Phi(t,\omega)dy_t$. In Doob's integral $\{y_t : t \in T\}$ is a martingale and for each t the random variable $\Phi(t,\omega)$ is (very crudely speaking) determined by the y's carrying a parameter-suffix $\leq t$. The construction is such that Doob is then able to define $\int_a^t \Phi(s,\omega)dy_s$ so as to make this "indefinite integral" itself a martingale. Finally the author characterises those martingales which can be represented in this way, using a y-process of Brownian-motion type.

Chapters X and XI ("Stationary processes—discrete/continuous parameter") contain parallel results in the relation indicated. The discussion commences with a valuable account of 1–1 measure-preserving point transformations, measure-preserving set transformations, invariant sets and metric transitivity, and the equivalence of these concepts with their images in the theory of strictly stationary S.P.'s. There is also an account of isometric and unitary transformations in relation to "wide sense" stationary processes (S.P.'s stationary "to the second order"). The Birkhoff ergodic theorem is proved as a strong law of large numbers for strictly stationary processes, and the L^2-ergodic theorem as a LLN for stationary (wide sense) processes. The auto-covariance function of a stationary (wide sense) process is represented as a Fourier transform and the Cramér-Loève harmonic analysis of a (wide sense) stationary process is obtained following a method of H. Cramér. All the analogous theorems for a continuous parameter are proved in Chapter XI, and in the Appendix the connexion with the von Neumann-Wintner and Stone representation theorems for unitary operators in Hilbert space is explained in detail. Each chapter contains a section on the estimation of the covariance and the spectrum (although of course the very recent work of U. Grenander and M. Rosenblatt [Ann. Math. Statistics **24** (1953), 537–558; MR **15**, 448d] is not covered) and an analysis of linear operations and rational spectral densities. Chapter XI concludes with an account of the representation theorem for S.P.'s with stationary (wide sense) increments.

The final Chapter (XII) is concerned with "Linear least squares prediction—Stationary (wide sense) processes". This rather specialised topic is, as the author remarks, somewhat out of place in the book but it follows very naturally on the immediately preceding chapters. General solutions of the prediction problem are given in both the discrete- and continuous-parameter case, and the chapter concludes with the approximation of an arbitrary random variable (with finite second absolute moment) by linear combinations of the random variables constituting a stationary (wide sense) S.P., and with a short account of multidimensional prediction.

There is a substantial historical Appendix and a bibliography of 160 items. The index is largely one of definitions, and this review will (it is hoped) be found more informative than the list of contents.

Very few readers will work steadily through the whole of this difficult book, but it cannot fail to exercise a decisive influence on the development of its subject.

D. G. Kendall (Oxford)

30 #1521 60.66
Spitzer, Frank
Principles of random walk.
The University Series in Higher Mathematics
D. Van Nostrand Co., Inc., Princeton, N.J.-Toronto-London, 1964. xi+406 pp.

This book presents a complete, interesting and quite self-contained treatment of random walk. "Random walk" is a very special class of stochastic process; more specifically, it is (1) a class of sums of independent random variables (vectors), as well as (2) a class of Markov chains (with stationary transition probabilities). The author explains why this high degree of specialization is worthwhile, as follows: "The theory of such random walks is far more complete than that of any larger class of Markov chains. Random walk occupies such a privileged position primarily because of a delicate interplay between methods from harmonic analysis on one hand, and from potential theory on the other hand." Although "harmonic analysis", the relevance of which to random walk stems from the above described feature (1) of random walk, is an indispensable tool, the emphasis is on "potential theory" in connection with the feature (2). Probabilistically speaking, "potential theory is basically concerned with the probability laws governing the time and position of a Markov process when it first visits a specified subset of its state space." In the last decade since Doob [Trans. Amer. Math. Soc. **77** (1954), 86–121; MR **16**, 269a] showed that the classical (Newtonian and logarithmic) potential theory corresponds to Brownian motion in a natural way, the potential theory associated with Markov processes has been greatly developed by many authors in various settings. {For example, see Hunt [Illinois J. Math. **1** (1957), 44–93; MR **19**, 951g, see p. G45 in this volume; ibid. **1** (1957), 316–369; MR **19**, 951g, see p. G45 in this volume; ibid. **2** (1958), 151–213; MR **21** #5824] for transient continuous-parameter processes, Doob [J. Math. Mech. **8** (1959), 433–458; erratum, ibid. **8** (1959), 993; MR **21** #5825] for transient discrete-parameter processes, and Kemeny and Snell [J. Math. Anal. Appl. **3** (1961), 196–260; MR **25** #3563] for recurrent discrete parameter processes.} Naturally, the specialization to random walk enables us to make further analysis within the framework of the associated potential theory. This was first illustrated by Itô and McKean [Illinois J. Math. **4** (1960), 119–132; MR **22** #12317] in showing that the so-called Wiener test in the classical potential theory has a counterpart in the theory of (simple) random walk of dimension ≥ 3. Independently of Kemeny and Snell (cited above), the author [Proc. Fourth Berkeley Sympos. Math. Statist. and Prob., Vol. II, pp. 515–534, Univ. California Press, Berkeley, Calif., 1961; MR **25** #5552] studied two-dimensional recurrent random walk to obtain a theory similar to the logarithmic potential theory. Finally, the author [J. Math. Mech. **11** (1962), 593–614; MR **25** #2655] established the potential theory on one-dimensional recurrent random walk which, the reviewer believes, is one of the deepest studies in modern probability theory. Of particular interest are the ratio limit theorems for recurrent random walk, mainly due to Kesten and the author [J. Analyse Math. **11** (1963), 285–322; MR **28** #5478]. These topics form the core of the book.

The table of contents is as follows. Chapter I. The classification of random walk. Chapter II. Harmonic analysis. Chapter III. Two-dimensional recurrent random walk. Chapter IV. Random walk on a half line. Chapter V. Random walk on an interval. Chapter VI. Transient random walk. Chapter VII. Recurrent random walk.

Chapter I is mainly concerned with the definition of random walk and some basic concepts like recurrence and aperiodicity. Since the word "random walk" has been used in various different contexts in references, let us here state the author's definition. Let R be the space of all lattice points of d-dimensional Euclidean space. Analytically, a random walk of d dimensions is a (transition) function $P(x,y)$, defined over $R\times R$, such that $0\le P(x,y)=P(0,y-x)$ and $\sum_{x\in R}P(0,x)=1$. Three subclasses of random walk, which are called "simple" ($d\ge 1$), "Bernoulli" ($d=1$) and "left (or right) continuous" ($d=1$), serve often as examples throughout the book. The first two sections give a direct analytic treatment of random walk. Some elementary results on recurrence are obtained. The useful concept of "aperiodicity" is introduced; its definition differs from the conventional one in the theory of Markov chains, as in Feller's book [*An introduction to probability theory and its applications*, Vol. I, second edition, Wiley, New York, 1957; MR **19**, 466a]. It is (in Section 3) after the author points out that such a direct analytic treatment is insufficient even for a simple recurrence problem (which is an immediate consequence of the strong law of large numbers) that he introduces the measure-theoretical formulation. This may be an instructive manner to the novice. The rest of the chapter deals with a theorem of Dvoretzky and Erdős [Proc. Second Berkeley Sympos. Math. Statist. and Prob., 1950, pp. 353–367, Univ. California Press, Berkeley, Calif., 1951; MR **13**, 852b] and the strong (or individual) ratio theorem on recurrent random walk (Chung and Erdős [Mem. Amer. Math. Soc. No. 6 (1951); MR **12**, 722g]); the latter is crucial for the proof of the deeper ratio theorems in Chapter VII.

Chapter II contains the elements of harmonic analysis, with applications to periodicity and recurrence. The local central limit theorem and its variants are proved. The recurrence criterion by Chung and Fuchs [ibid. No. 6 (1951); MR **12**, 722e] is applied, for example, to show that every genuinely (i.e., non-degenerate) three- or higher-dimensional random walk is transient. The renewal theorem is also given in this chapter for the convenience of later reference.

Chapter III deals with the potential theory on two-dimensional recurrent random walk. The harmonic measure $H_A(x,y)$, the transition probabilities of the imbedded process $\Pi_A(x,y)$, the (exterior) Green's function $g_A(x,y)$ for a specified set A and the general relations between them are introduced. Then the author proves the existence of the potential kernel

$$A(x,y)=\lim_{n\to\infty}\sum_{k=0}^{n}[P_k(0,0)-P_k(x,y)]$$

for aperiodic two-dimensional recurrent random walk and studies the limiting behavior of $A(x,y)$. $H_A(x,y)$, $\Pi_A(x,y)$ and $g_A(x,y)$ for a finite set A can be represented by means of $A(x,y)$, so that their limiting properties are derived easily from those of $A(x,y)$. This gives a sample of the potential theory on random walk. {Similar results are obtained for the transient case in Chapter VI and for the one-dimensional recurrent case in Chapter VII.} Section 13 is very instructive; there one finds the counterparts in the theory of random walk ($d=2$) of three problems in logarithmic potential theory (the Liouville theorem, the exterior Dirichlet problem and the integration of Poisson's equation). Section 15 is reserved for some special problems concerning simple random walk ($d = 2$). Finally, the author discusses a ratio limit theorem in two dimensions due to Kesten and the author (cited previously).

Chapter IV gives what has been known as the fluctuation theory (E. Sparre Andersen [Math. Scand. **1** (1953), 263–285; MR **15**, 444f; ibid. **2** (1954), 195–223; MR **16**, 839e], the author [Trans. Amer. Math. Soc. **82** (1956), 323–339; MR **18**, 156e], Baxter [J. Analyse Math. **9** (1961/62), 31–70; MR **24** #A3676] and others). The subject is concerned with the time and position of the first visit to the half-line, the maximum of successive partial sums, and so on. Thus here again one encounters some potential theory (or rather its germ), but in a different context from the preceding chapter. Actually such a theory, involving some analytic function theory, is not only itself interesting, but also it is a necessary step in the full course of the one-dimensional (either transient or recurrent) potential theory (Chapters VI and VII). This chapter ends with the proof of the arcsine law.

In Chapter V the author gives a survey of the two-sided absorption problems. The main result (Theorem 1 of Section 22) is concerned with the asymptotic behavior of the absorption probabilities [the author and C. Stone, Illinois J. Math. **4** (1960), 253–277; MR **22** #8547], and it is applied to the problem of finding the asymptotic property of the (interior) Green's function for a finite interval and the occupation-time problem.

Chapter VI deals with the potential theory of transient random walk, except for the last section (where one finds some elements of a more general potential theory and boundary theory on transient Markov chains by Doob—cited previously—and others). In the transient case, the potential kernel is the (interior) Green's function $G(x, y)$ for the whole space. It is easy to express $\Pi_A(x, y)$ and $H_A(x, y)$ in terms of $G(x, y)$ when A is a finite set. Therefore, the central problem turns out to be to investigate the behavior of $G(0, x)$ for large x; this is carried out in Section 24. There occurs a new problem in the transient case: "When is the total hitting probability $H_B(x)$ of the set B identically one?" The study of such sets (called "recurrent") leads one to the notion of capacity. In Section 26 the author extends the characterization of recurrent sets of simple random walk ($d = 3$) by Itô and McKean (cited previously) to random walk with mean zero and finite second moments. The proof depends heavily on a detailed estimate of $G(0, x)$ for large x ($G(0, x)$ is proportional to $|x|^{-1}$).

Chapter VII is the one-dimensional version of Chapter III. An aperiodic one-dimensional recurrent random walk is called "type I" or "II" according as $\sigma^2 = \sum x^2 P(0, x) = +\infty$ or $< \infty$. (Type II random walk must have zero mean, so that σ^2 is its variance.) In both cases the potential kernel $a(x) = \lim_{n\to\infty} \sum_{k=0}^{n} [P_k(0, 0) - P_k(x, 0)] < \infty$ exists for every x, but different methods of proof are required. Although the results on the kernel $a(x)$ in type I random walk are the same as in the two-dimensional case (Chapter III) with one single exception (Proposition 2 of Section 30), one needs more delicate arguments based on results from Chapter IV to reach these results in the one-dimensional case. For type II random walk one meets with some new, interesting phenomena on the asymptotic behavior of $a(x)$. The book finishes with the ratio limit theorems in one dimension (also due to Kesten and the author) and its related conjecture by Kesten.

This book certainly covers almost all major topics in the theory of random walk. It will be invaluable to both pure and applied probabilists, as well as to many people in analysis. References for the methods and results involved are very good. A useful interdependence guide is given. Excellent choice is made of examples, which are mostly concerned with very concrete calculations. Each chapter contains complementary material in the form of remarks, examples and problems which are often themselves interesting theorems.

{The arguments on page 34 and pages 302–303 require minor changes. All other misprints remain quite minor and will be easily corrected by the reader.}

T. Watanabe (Princeton, N.J.)

24 #A3025 62.30
James, W.; Stein, Charles
Estimation with quadratic loss. 1961
Proc. 4th Berkeley Sympos. Math. Statist. and Prob., Vol. I, pp. 361–379. *Univ. California Press, Berkeley, Calif.*

This paper contains, in Sections 3–4, another chapter in the remarkable series of results by Stein on admissibility of invariant translation parameter estimators. {The authors state: "The results of Section 2 were obtained by the two authors working together. The remainder of the paper is the work of C. Stein."} Previous work on admissibility in this and related contexts is due to Blyth [Ann. Math. Statist. **22** (1951), 22–42; MR **12**, 622f], Blackwell [ibid. **22** (1951), 393–399; MR **13**, 260b], Hodges and Lehmann [Proc. 2nd Berkeley Sympos. Math. Statist. and Prob. (1950), pp. 13–22; Univ. California Press, Berkeley, Calif., 1951; MR **13**, 479e], Karlin [Ann. Math. Statist. **29** (1958), 406–436; MR **23** #A1418], and Stein [ibid. **30** (1959), 970–979; MR **22** #278], the last using in the general one-parameter case (quadratic loss) the method Blyth had used in the normal case. This method proceeded by computing the Bayes risk r_σ relative to an a priori law ξ_σ with suitably large tails (the Cauchy, for Stein) depending on a scale parameter σ, and showing under a slight integrability assumption that (1) $\bar{r} - r_\sigma = O(\sigma^{-2})$ as $\sigma \to \infty$, where $\bar{r}$ is the minimax risk; if r' is the risk function of a procedure better than the invariant minimax procedure t^* on a set A of positive Lebesgue measure, one has (2) $\bar{r} - \int r' d\xi_\sigma > c/\sigma$ with $c > 0$, and this contradiction to (1) yielded almost admissibility of t^* (and hence admissibility in the absolutely continuous case). The analogous argument in the case of a k-dimensional translation parameter yields c/σ^k in (2) and hence fails, and indeed Stein showed for $k \geq 3$ that t^* is inadmissible in the normal case [Proc. 3rd Berkeley Sympos. Math. Statist. and Prob. (1954/55), Vol. I, pp. 197–206, Univ. California Press, Berkeley, Calif., 1956; MR **18**, 948c], and in fact more generally (as stated in the present paper). When $k = 2$, the last cited paper also proved admissibility in the normal case when $k = 2$ by employing the differential inequality method of Hodges and Lehmann. In the present paper Stein obtains the bivariate admissibility result in general by choosing ξ_σ so that, roughly speaking, ξ_1 has a logarithmic singularity in the neighborhood of a density one point of A, so as to obtain $g(\sigma)/\sigma^2$ with $g(\sigma) \to \infty$ as $\sigma \to \infty$ in (2), and (because also of the behavior at ∞ of ξ_1) $o(g(\sigma))/\sigma^2$ in (1). These again yield the desired contradiction. The details of this last calculation, which are not included in the paper, are much longer and more delicate than when $k = 1$. The (incompletely solved) question of admissibility for quadratic loss relative to a matrix of rank $< k$ is discussed briefly.

The main result of Section 2 is the computation, in the multivariate normal case with identity covariance matrix and loss $\|\varphi - \theta\|^2$, of the best estimator of the form $(1 - b/\|x\|^2)x$ when σ^2 is known, and of an analogous form when the covariance matrix is unknown. Certain bounded loss functions are also considered, to illustrate inadmissibility of x when $k \geq 3$.

In Section 5 Stein considers the problem of estimating the covariance matrix of a normal k-vector ($k \geq 2$) with known expectation. The best estimator invariant under the full linear group is shown to be not minimax for a variety of loss functions, and an explicit minimax estimator is computed for one choice of loss function. The method is to study the estimator which is best invariant under the group of lower triangular matrices; this group is transitive on the parameter space and, since it is solvable, the invariance-minimax theorem holds [the reviewer, Ann. Math. Statist. **28** (1957), 573–601; MR **19**, 1097c]. *J. Kiefer* (Ithaca, N.Y.)

30 #943 85.57 57.48
Arnol′d, V. I.
Small denominators and problems of stability of motion in classical and celestial mechanics. (Russian)
Uspehi Mat. Nauk **18** (1963), *no.* 6 (114), 91–192.

This paper is not only a survey article on the recent work on problems in celestial mechanics but also contains the proofs of a number of new results. In particular, the xistence of a set of almost periodic solutions for the n-body problem is established, a result which has been announced by V. I. Arnol′d at the International Congress for Mathematicians, Stockholm, 1962 [Proc. Internat. Congr. Math., 1962, Inst. Mittag-Leffler, Djursholm, 1963; MR **28** #1]. This is the first result which establishes a set of solutions which for all times remains bounded and collision free and which, moreover, forms a set of positive measure in phase space. Another topic of great interest in this paper is the existence of an adiabatic invariant for all time for a charged particle in a magnetic field (a problem which usually is treated by asymptotic methods, yielding estimates for finite time intervals only).

After an instructive and interesting introduction the author discusses (Chapter I) the classical machinery of perturbation theory, the difficulties of small divisors and its significance. In a heuristic manner the statements of Kolmogorov and of the author are discussed and the implications of these results for the stability of periodic solutions of Hamiltonian systems are explained.

In Chapter II questions of adiabatic invariants as they are relevant for the motion of a charged particle in a magnetic field are discussed. The problem is to study the solutions of differential equations which are described by a Hamiltonian $H(p, q, \mu t)$ which varies slowly in time: H is a smooth function of the phase-space variables p, q, and $\mu t = \lambda$, and μ is a small parameter. Adiabatic invariants are functions $J(p, q, \lambda)$ which have the property that they change little over long time intervals, e.g., $0 < t < \mu^{-1}$. Problems of this sort were basic in connection with the "old" quantum theory of N. Bohr (see P. Ehrenfest [Ann. Physik (4) **51** (1916), 327–352], or M. Born [Z. Physik **40** (1926), 167–192]). After having been abandoned, the question appeared again in the study of charged particles in a magnetic field, "magnetic traps" (see J. Berkowitz, C. Gardner [Comm. Pure Appl. Math. **12** (1959), 501–512; MR **21** #6923]; C. Gardner [Phys. Rev. (2) **115** (1959), 791–794; MR **21** #7715]; M. Kruskal

[*La Théorie des Gaz Neutres et Ionisés* (Grenoble, 1959), pp. 275–284, Hermann, Paris, 1960; MR **22** #8177]). The new feature of Arnol′d's result is that it guarantees adiabatic invariance for all times, at least for some problems of one degree of freedom in which the parameter λ enters periodically.

The most remarkable part of the paper is Chapter III, which contains a discussion of the n-body problem and a precise formulation of the new result concerning almost periodic solutions of the n-body problem. It can be stated as follows: For the majority of initial conditions for which the osculating ellipses are of small eccentricity and inclination the mutual perturbation of the mass points result only in small changes for all times, provided that $n-1$ masses are small compared to one (sun). Actually the majority of these solutions are proven to be almost periodic by use of the theorems explained in Chapter II. However, the application involves many difficulties since this problem is highly degenerate. For the successful treatment of these difficulties the expansion of the perturbation function up to fourth order in the eccentricities is needed, which was available in Le Verrier's calculations of the last century.

Chapters IV and V contain the details of the technique of the proof and will not be discussed here since much of it is parallel to the author's earlier paper [Uspehi Mat. Nauk **18** (1963), no. 5 (113), 13–40; MR **29** #328]. The last chapter contains a number of unsolved problems, as well as a discussion of some current research of workers in this area.

It is to be hoped that this remarkable paper and exceptional work helps to arouse the interest of more mathematicians in this subject. *J. Moser* (New Rochelle, N.Y.)

6, 235k 90.0X

von Neumann, John; Morgenstern, Oskar

Theory of games and economic behavior.

Princeton University Press, Princeton, New Jersey, 1944. xviii+625 pp.

The book presents a new approach to the problem of rational behavior. The view taken is that the problem of rational behavior cannot be regarded as a simple maximum problem, but is of a type that arises in the theory of games of strategy. The main reasons for this attitude may be briefly summarized as follows. An essential characteristic of a social-economic organization is that the participants enter into relations with each other and the result for each participant depends not only on his own actions but also on the actions of others. For example, in a social exchange economy where several persons exchange goods, the increase of utility a participant can achieve will depend on the behavior of all the other participants. In general, the situation that arises may be characterized as follows. Each participant wishes to maximize a certain function of several variables. He controls, however, only a partial set of these variables, while the values of the remaining variables depend on the actions of the other participants, who may have different or even opposite interests. The problem of a rational choice of the variables for each participant cannot be regarded as a simple maximum problem. Instead, we have a peculiar mixture of several conflicting maximum problems. Problems of the same type arise in the theory of games of strategy.

After a discussion of the nature of the problem of economic behavior, the authors turn to a development of the theory of games. The main features of the rigorous definition of the concept of a game are as follows. A game consists of a sequence of moves, $M_1, \cdots, M_\nu$. Each move M_k consists of a number α_k of alternatives, denoted by $A_k(1), \cdots, A_k(\alpha_k)$. There are two kinds of moves, personal and chance moves. The move M_k is a personal move if any of $A_k(1), \cdots, A_k(\alpha_k)$ can be freely chosen by one of the players specified by the rules of the game. The move M_k is a chance move if the choice of a particular alternative is made by a chance mechanism constructed in such a way that the probability of choosing any particular alternative is equal to a value specified by the rules of the game. Denote by σ_k the alternative chosen in the move M_k. The alternatives $A_k(1), \cdots, A_k(\alpha_k)$ and the number α_k may depend on the choices $\sigma_1, \cdots, \sigma_{k-1}$ in the preceding moves. Similarly, whether M_k is a personal or a chance move, the player who makes the choice if M_k is a personal move, and the probabilities assigned if M_k is a chance move, may depend on $\sigma_1, \cdots, \sigma_{k-1}$. Furthermore, the total number ν of moves may itself depend on the choices σ_i, etc. The rules of the game also specify the outcome of the play (the amount each player gets when the play is completed) as a function of $\sigma_1, \cdots, \sigma_\nu$. A game is called an n-person game if the number of players is n. Denote by $F_i(\sigma_1, \cdots, \sigma_\nu)$ the outcome of the play for player i $(i = 1, \cdots, n)$. A game is called a zero-sum game if $\sum_{i=1}^n F_i(\sigma_1, \cdots, \sigma_\nu) = 0$ identically in $\sigma_1, \cdots, \sigma_\nu$. It is shown later that the general game can be reduced to a simple normalized form which is equivalent to the original formulation of the game. This reduction is made possible by considering the strategies of the players instead of the individual moves. By a strategy of a player is meant a complete plan which specifies what choices he should make in every possible situation. Since the number ν of moves is assumed to be bounded, the number of possible strategies for player i is finite, say β_i. Then the choice of a strategy by player i is characterized by the choice of a value of a variable τ_i which can take the integral values $1, \cdots, \beta_i$. The expected value $K_i(\tau_1, \cdots, \tau_n)$ of the outcome of the game for each player can be shown to be a single-valued function of the strategies chosen by the n players. The normalized form of a general n-person game is then as follows: n functions $K_1(\tau_1, \cdots, \tau_n), \cdots, K_n(\tau_1, \cdots, \tau_n)$ and n positive integers $\beta_1, \cdots, \beta_n$ are given. The variable τ_i can take the integral values $1, \cdots, \beta_i$ $(i = 1, \cdots, n)$. Player i chooses a particular value of τ_i and this choice is made in complete ignorance of the choices of the other $n-1$ players. After the values $\tau_1, \cdots, \tau_n$ have been chosen, player i receives the amount $K_i(\tau_1, \cdots, \tau_n)$. All further discussions are based on this normalized form of a game. The zero-sum restriction now means that $\sum_{i=1}^n K_i(\tau_1, \cdots, \tau_n) = 0$ identically in $\tau_1, \cdots, \tau_n$.

The discussion of zero-sum two-person games provides a foundation for the general theory of n-person games. Denote by $K(\tau_1, \tau_2)$ the outcome of the game for player 1. Then the outcome for player 2 is given by $-K(\tau_1, \tau_2)$. The fundamental problem is that of a suitable choice of the value of τ_i by player i $(i = 1, 2)$. Player 1 wishes to maximize $K(\tau_1, \tau_2)$ and player 2 wishes to minimize $K(\tau_1, \tau_2)$. For any particular choice of τ_1, player 1 can be sure that he will get at least $\min_{\tau_2} K(\tau_1, \tau_2)$, but he cannot be sure that he will get more than this amount. Thus, if player 1 chooses a value τ_1' of τ_1 for which $\min_{\tau_2} K(\tau_1, \tau_2)$ takes its maximum value, he is certain to receive an amount at least $v_1 = \max_{\tau_1} \min_{\tau_2} K(\tau_1, \tau_2)$. We shall refer to such a choice τ_1' as a minimax strategy of player 1; similarly, a choice of a value τ_2' of τ_2 for which $\max_{\tau_1} K(\tau_1, \tau_2)$ takes its minimum value is a minimax strategy of player 2. If player 2 follows a minimax strategy, he can be sure to receive an amount at least $-v_2 = -\min_{\tau_2} \max_{\tau_1} K(\tau_1, \tau_2)$. If player 1 does not follow a minimax strategy, his gain may be less than v_1. It is shown that $v_1 \leq v_2$ always holds. A game for which $v_1 = v_2$ is called strictly determined. This case is of decisive importance, since the common value v of v_1 and v_2 may be interpreted as the value of the game (for player 1) and the minimax strategies may be regarded as good strategies. If a game is strictly determined and if each of the players follows a minimax strategy, player 1 will get the amount v and player 2 the amount $-v$. For strictly determined games the choice of minimax strategies by the two players creates a perfectly stable situation in the sense that even if one of the players were able to discover his opponent's strategy, he still would have no reason to deviate from his own minimax strategy. No such stable situation exists for games which are not strictly determined.

Thus a satisfactory solution can be given only for strictly determined games; for other games there are difficulties even in defining the meaning of a good strategy. Therefore the problem is reformulated as follows. Instead of choosing a particular value of τ_1, player 1 considers all possible values $1, \cdots, \beta_1$ and chooses merely the probabilities $\xi_1, \cdots, \xi_{\beta_1}$ with which he is going to use them. Similarly, player 2 chooses the probabilities $\eta_1, \cdots, \eta_{\beta_2}$ with which he is going to use the values $1, \cdots, \beta_2$ of τ_2. Thus, the two choices are characterized by vectors $\xi = (\xi_1, \cdots, \xi_{\beta_1})$ and $\eta = (\eta_1, \cdots, \eta_{\beta_2})$. The expected value of the outcome $K(\tau_1, \tau_2)$ is then $K^*(\xi, \eta) = \sum_{j=1}^{\beta_2} \sum_{i=1}^{\beta_1} K(i, j)\xi_i \eta_j$. The main theorem is that the game corresponding to the outcome function $K^*(\xi, \eta)$ is always strictly determined. Player i is said to have a pure strategy if he chooses a particular value of τ_i, and a mixed strategy if he chooses merely the probability distribution of the possible values of τ_i. Thus if both players are permitted to use mixed strategies, the game is always strictly determined. A number of special cases and examples are discussed.

The authors next turn to the discussion of zero-sum n-person games with $n > 2$, for which the problem has essentially new features. In fact, when $n = 2$ there is no possibility of coalitions, while when $n > 2$ some players may find it advantageous to form a coalition against the others. The possible formation of coalitions and the compensations that may be necessary between coalition partners play a decisive role. In the general theory the formation of coalitions and the payment of compensations are governed entirely by the characteristic function of the game, defined as follows. Let S be any subset of the set I of all players, $-S$ the complement of S. Suppose that all players in S cooperate with each other, and all players in $-S$ cooperate with each other. Then each of the groups can be regarded as one player and the game reduces to a zero-sum two-person game. Thus the game will have a definite value $v(S)$ for S. The function $v(S)$ is called the characteristic function. A game is called inessential if for any subset S we have $v(S) = \sum_{i \in S} v(i)$, where i denotes the subset consisting of the single player i. In such a game nothing can be gained by forming coalitions and, therefore, the value of the game for player i is given by $v(i)$. A distribution or imputation is a set of n numbers $\alpha_1, \cdots, \alpha_n$, where α_i is the amount allotted to player i. For any inessential game the solution consists of the single imputation $v(1), \cdots, v(n)$.

A game which is not inessential is called essential. For essential games coalitions and compensations play a decisive role. A solution of an essential game will be a set of imputations rather than a single imputation; furthermore, there will be, in general, several solutions. The definition of a solution is based on the notions of effective sets and domination. A subset S of the set I of all players is said to be effective for the imputation $\alpha = (\alpha_1, \cdots, \alpha_n)$ if $\sum_{i \in S} \alpha_i \leq v(S)$. An imputation

α dominates another imputation β if there exists a nonempty set S such that S is effective for α and $\alpha_i > \beta_i$ for all i in S. A solution is defined as a set V of imputations satisfying the following two conditions. (1) No imputation in V is dominated by another imputation in V. (2) Any imputation not in V is dominated by at least one imputation in V. The reason that a solution V will consist of more than one imputation is that domination is not a transitive relation. A solution V may be regarded as a possible standard of rational behavior, possessing a certain inner stability which would be upset by omission of any of the two restrictions imposed on a solution. Condition (1) is needed to eliminate inner inconsistencies in V. The desirability of condition (2) can be seen as follows. For any set of imputations which satisfies (1) there will be an imputation β outside V such that β dominates an imputation α in V. The players can hardly be expected to maintain faith in V as a standard of rational behavior unless there is another imputation γ in V which dominates β. If a solution V is accepted as a standard of behavior, no imputation conforming with the accepted standard can be overruled by another. On the other hand, any imputation not conforming with the accepted standard can be overruled by an imputation conforming with this standard. There is nothing objectionable in the fact that there will usually be no unique solution. This simply means that different "established orders of society" or "accepted standards of behavior" may be built on the same physical background (the same set of rules describing a game). Each of these standards of behavior will have the characteristics of inner stability, but the different standards may well be in contradiction with each other.

The mathematical difficulties in finding solutions for games on the basis of their characteristic functions are considerable even for small n. The complexity of the problem increases rapidly with increasing n. Although there is hardly any doubt that every game has at least one solution, no general existence theorem has yet been established. A complete and exhaustive discussion is given for the essential 3-person game, for which all solutions are known. There is also a detailed discussion of zero-sum 4-person games. No systematic and exhaustive theory is available for $n > 4$. However, a number of special cases are treated, in particular, the so-called decomposable games and simple games. It is shown that all solutions of decomposable games can be determined if all solutions of the constituents are known. Furthermore, a simple game is discussed for which all solutions are determined for any value of n.

In dealing with problems of behavior in a social economy, the zero-sum restriction is not very realistic. An extension of the theory to general non-zero-sum games is achieved by re-interpreting an n-person general game Γ as an $(n+1)$-person zero-sum game in the following way. Let $K_i(\tau_1, \cdots, \tau_n)$ $(i = 1, \cdots, n)$ be the outcome functions of the n-person general game Γ. A fictitious player $(n+1)$ is introduced by defining $K_{n+1}(\tau_1, \cdots, \tau_n) = -\sum_{i=1}^{n} K_i(\tau_1, \cdots, \tau_n)$. In this way we obtain an $(n+1)$-person zero-sum game $\overline{\Gamma}$, which is called the zero-sum extension of Γ. It is shown that the only solutions of $\overline{\Gamma}$ which can be interpreted as solutions of Γ are those which are in accordance with the fact that the fictitious player cannot have even an indirect influence on the course of the game. It turns out that a solution $\overline{V}$ of $\overline{\Gamma}$ can be regarded as a solution of Γ if for any imputation $\alpha = (\alpha_1, \cdots, \alpha_{n+1})$ in $\overline{V}$ we have $\alpha_{n+1} = v(n+1)$. The solutions of all general games with $n \leq 3$ are derived and the results are applied to a social exchange economy with 2 and 3 participants. These economic applications deserve special attention, since even in these simple cases the new approach deviates from the ordinary viewpoint. Consider first the case of two participants: 1, the seller; 2, the buyer. Suppose that the transaction is the sale of one unit A. Let u denote the value of A to 1 and $v (v > u)$ to 2. Then the commonsense result is that the price p may be anywhere between the two limits u and v. The theory of games leads to exactly the same result. However, if the conditions of the problem are changed by allowing transactions involving any or all of s units of a commodity $(s > 1)$, the result implied by the theory of games will differ from the commonly accepted viewpoint as expressed, for example, in the Boehm-Bawerk theory. While the number of units transferred according to the theory of games will be the same in both theories, the Boehm-Bawerk theory restricts the price to a narrower interval. The reason is that the ordinary viewpoint assumes the existence of a unique price valid for all transfers, while the theory of games allows for premiums and rebates (compensations) which obliterate the unique price. Another qualitative discrepancy occurs in the case where one seller, 1, sells a unit A to either buyer 2 or buyer 3. Let u be the value of A for 1, v for 2 and w for 3, where $u < v \leq w$. The solution V given by the theory of games is the sum of the following two sets of imputations: the set of all imputations $(\alpha_1, \alpha_2, \alpha_3)$ which satisfy the condition $(*)$ $v \leq \alpha_1 \leq w$; $\alpha_2 = 0$; $\alpha_3 = w - \alpha_1$, and the set of imputations for which $(**)$ $u \leq \alpha_1 \leq v$; α_2, α_3 are certain monotonic decreasing functions of α_1. In the ordinary treatment only imputations satisfying $(*)$ are considered. The imputations satisfying $(**)$ can be interpreted as arising from a coalition of the two buyers against the seller. Also, the case of divisible goods is discussed for the 3-person market. Some remarks are made about the general n-person market.

The book will be of interest to sociologists and economists who are interested in a rigorous foundation of the theory of rational behavior, in particular, the theory of exchange in a general market. [It may be of interest to mention that the theory of games has applications to statistics as well, since the general problem of statistical inference may be treated as a zero-sum two-person game.] The authors make an endeavor to present the material in a form which will make it understandable to readers without a special knowledge of any part of advanced mathematics. In spite of this, the book is not really elementary, because of the intricate nature of many of the mathematical deductions. Nonmathematical readers will find the excellent verbal expositions given parallel with every major mathematical deduction very helpful.

A. Wald (New York, N. Y.)

AUTHOR INDEX

REVIEWER INDEX